Laser Processing:
Surface Treatment and Film Deposition

NATO ASI Series

Advanced Science Institutes Series

A Series presenting the results of activities sponsored by the NATO Science Committee, which aims at the dissemination of advanced scientific and technological knowledge, with a view to strengthening links between scientific communities.

The Series is published by an international board of publishers in conjunction with the NATO Scientific Affairs Division

A Life Sciences	Plenum Publishing Corporation
B Physics	London and New York
C Mathematical and Physical Sciences	Kluwer Academic Publishers
D Behavioural and Social Sciences	Dordrecht, Boston and London
E Applied Sciences	
F Computer and Systems Sciences	Springer-Verlag
G Ecological Sciences	Berlin, Heidelberg, New York, London,
H Cell Biology	Paris and Tokyo
I Global Environmental Change	

PARTNERSHIP SUB-SERIES

1. Disarmament Technologies	Kluwer Academic Publishers
2. Environment	Springer-Verlag / Kluwer Academic Publishers
3. High Technology	Kluwer Academic Publishers
4. Science and Technology Policy	Kluwer Academic Publishers
5. Computer Networking	Kluwer Academic Publishers

The Partnership Sub-Series incorporates activities undertaken in collaboration with NATO's Cooperation Partners, the countries of the CIS and Central and Eastern Europe, in Priority Areas of concern to those countries.

NATO-PCO-DATA BASE

The electronic index to the NATO ASI Series provides full bibliographical references (with keywords and/or abstracts) to more than 50000 contributions from international scientists published in all sections of the NATO ASI Series.
Access to the NATO-PCO-DATA BASE is possible in two ways:

– via online FILE 128 (NATO-PCO-DATA BASE) hosted by ESRIN,
Via Galileo Galilei, I-00044 Frascati, Italy.

– via CD-ROM "NATO-PCO-DATA BASE" with user-friendly retrieval software in English, French and German (© WTV GmbH and DATAWARE Technologies Inc. 1989).

The CD-ROM can be ordered through any member of the Board of Publishers or through NATO-PCO, Overijse, Belgium.

Series E: Applied Sciences - Vol. 307

Laser Processing: Surface Treatment and Film Deposition

edited by

J. Mazumder
Department of Mechanical and Industrial Engineering,
University of Illinois,
Urbana-Champaign, IL, U.S.A.

O. Conde
Department of Physics,
University of Lisbon,
Lisbon, Portugal

R. Villar
Department of Materials Engineering,
Instituto Superior Técnico,
Lisbon, Portugal

and

W. Steen
Department of Mechanical Engineering,
University of Liverpool,
Liverpool, U.K.

Kluwer Academic Publishers

Dordrecht / Boston / London

Published in cooperation with NATO Scientific Affairs Division

Proceedings of the NATO Advanced Study Institute on
Laser Processing: Surface Treatment and Film Deposition
Sesimbra, Portugal
July 3–16, 1994

Library of Congress Cataloging-in-Publication Data

```
NATO Advanced Study Institute on "Laser Processing: Surface Treatment
  and Film Deposition" (1994 : Sesimbra, Portugal)
    Laser processing : surface treatment and film deposition / edited
  by J. Mazumder ... [et al.].
        p.    cm. -- (NATO ASI series. Series E, Applied sciences ; vol.
  307)
    "Proceedings of the NATO Advanced Study Institute on 'Laser
  Processing: Surface Treatment and Film Deposition', Sesimbra,
  Portugal, July 3-16, 1994."
    "Published in cooperation with NATO Scientific Affairs Division."
    ISBN 0-7923-3901-0
    1. Lasers--Industrial applications--Congresses.  2. Metals-
  -Surfaces--Congresses.  3. Thin films--Congresses.  4. Laser
  ablation--Congresses.    I. Mazumder, J.   II. North Atlantic Treaty
  Organization.  Scientific Affairs Division.  III. Title.
  IV. Series: NATO ASI series.  Series E, Applied sciences ; no. 307.
  TA1677.N42  1994
  620'.44--dc20                                                    95-26539
```

ISBN-13: 978-94-010-6572-6   e-ISBN-13: 978-94-009-0197-1
DOI: 10.1007/978-94-009-0197-1

Published by Kluwer Academic Publishers,
P.O. Box 17, 3300 AA Dordrecht, The Netherlands.

Kluwer Academic Publishers incorporates the publishing programmes of
D. Reidel, Martinus Nijhoff, Dr W. Junk and MTP Press.

Sold and distributed in the U.S.A. and Canada
by Kluwer Academic Publishers,
101 Philip Drive, Norwell, MA 02061, U.S.A.

In all other countries, sold and distributed
by Kluwer Academic Publishers Group,
P.O. Box 322, 3300 AH Dordrecht, The Netherlands.

Table of Contents

Chapter 3. Laser Alloying

Chapter 4. Laser Cladding

Chapter 5. Surface Treatment

Chapter 6. Laser Thin Film Deposition

Chapter 7. Laser Ablation

Chapter 8. Application to Industry

INTRODUCTION

The NATO Advanced Study Institute entitled "Laser Processing: Surface Treatment and Film Deposition" was held July 3 through July 16, 1994 at Sesimbra, Portugal. This NATO ASI is a follow up of the previous on the similar topic held September 2 through September 13, 1985 at San Miniato, Italy. One significant difference in this program is the topic on ablation and film deposition. Experts in the field and scientists and engineers newly entering the field reviewed the recent progress and innovations related to the science and technology of the laser surface treatment and film deposition. The papers presented in this ASI are contained in this proceedings.

This ASI proceedings is a comprehensive book dealing with the topic of laser surface treatment and film deposition. First, it provides an overview of the various processes. Second, fundamental understandings of the processes are built with the discussion on the coupling mechanism of laser materials interaction followed by theories on non-equilibrium synthesis, microstructure and phase selection, interface response for rapid solidification, modeling of transport phenomena (heat, mass and momentum), and hydrodynamics of the process. Third, theoretical discussion was complemented by discussion on various experimental measurement techniques for temperature, plasma electron density and microstructure for theory validation and on-line process control. Fourth, properties of many a laser processed materials are also discussed. Finally, specific applications of laser surface modification techniques such as alloying, cladding, laser chemical vapor deposition, ablation, cleaning and heat treatment are described.

One distinct difference in this ASI compared to the last one is how much of the work described related to the on-line diagnostics and model validation. This indicates an advancement of the field and a transition toward implementation of the processes in the real world. New findings and future research directions including suggestions for improvements of future NATO ASI's are summarized at the beginning of this book.

We thank all the chairmen, lecturers and participants for organizing a comprehensive program, illuminating presentations and stimulating discussions. Many scientific acquaintances have become friends and future collaborators due to this intense interaction during the two-week period. In spite of the inviting sandy beach and blue Atlantic waters of Sesimbra, scientific curiosity seemed to have prevailed most of the time.

We would especially like to thank our major sponsor NATO Scientific Affairs Division who funded the lion's share. We would also like to acknowledge the financial and moral support from: Laser Institute of America, Luso-American Foundation, JNICT (Junta Nacional de Investigacao Cientifica e Tecnologica), INVOTAN Commission, French Embassy (Lisbon) and British Council (Lisbon), and Setubal Tourist Office. We are also very grateful to Dr. J. Choi, Ms. Tammy Smith and Ms. Dianna Barnett of the University of Illinois at Urbana-Champaign for their help in the organization and preparation of this study Institute and the book.

J. Mazumder, University of Illinois at Urbana-Champaign
O. Conde, University of Lisbon
R. Vilar, Instituto Superior Tecnico
W.M. Steen, University of Liverpool

GRADUATE STUDENT REPORT
Xiangli Chen

SUMMARY OF DISCUSSIONS AT THE NATO ASI
SESIMBRA, PORTUGAL, JULY 15, 1994

1. What have you heard here first?

* Developments of laser surface treatments in other countries (e.g. Ukraine) and laboratories.

* Different specific research topics:
  - binary alloy manufacturing with surface deposition
  - possibilities of designing properties
  - metallic particles in polymers
  - nanoscale microstructures
  - art cleaning by laser
  - possibilities of beam shaping

* Problems in understanding the energy coupling mechanisms in the classic methods of laser surface treatment (hardening, remelting, alloying, and cladding).

* Dominance of high power CO_2 lasers in classic fields of laser surface treatment.

* Laser safety research projects.

* High quality thin films deposited by Excimer or Nd:YAG laser ablation processes.

* Learning about basics and recent developments in other fields of laser surface treatment and film deposition.

2. What are the most important conclusions you can make regarding the status of the field from the tutorial lectures you have attended?

* A clear need for process control system to provide QA control which industry requires to turn laboratory processes into commercially viable projects. Simple examples are the need to measure surface temperature and beam quality. Alternatively, it might be to electronically record the quality of the process to provide long-term QA record for the manufacturer.

* It was felt that laser surface treatment was still a technology driven science rather than by the practical needs of industry.

* The use of laser surface processing was much more widely accepted in the electronic industry than in mechanical industries.

* Laser pulse shapes need to be better understood as their characteristics do affect the quality of the process even in apparently "crude" processes such as ablation.

* Need to build a stronger link between the science of surface processing and the industrial engineering community.

* Need to determine the macroscopic physical properties a customer (an industrialist) requires and then try and design the structure from the microscopic properties upwards to yield the desired results.

* Enormous progress in the field of laser surface processing since the laser NATO ASI in this area. Last time, the meeting was primarily about the electronic industry. This time, there was a wider audience with interests in metals, ceramics, concrete, and a wide range of applications.

* Need to get more industrialists at the meeting.

* There is a need to retain and expand slightly the overview lectures since they give new students and industrial delegates a greater breadth of knowledge in their new subject.

* The large number of groups interested in improving machine tools with laser coating, cladding, or hardening.

* The very large number of applications of laser chemical vapor deposition (LCVD) in micro-electronics and micro-mechanics.

* The importance of mathematical modeling
    - simple models are needed especially to assist in on-line process control of industrial applications.
    - complex models are needed to help the scientific specialists to understand the underlying science. However, the complex models must never loose sight of the underlying physics they try to simulate.
    - Both types of models are essential.

* There is a need for multi-disciplinary teams to solve many of the problems in laser materials processing.

3 Where do you think the future direction for laser processing research should be?
   a) Scientific
   b) Technical

* Transformation hardening, welding, melting and cladding will find more applications.

* Scientific understanding of the interaction mechanisms, modeling for applications, simple models.

*   Higher power lasers will allow larger area surface treatment that avoids overlapping problems.

*   Process control, diagnostics, on-line monitoring, and in-situ analysis.

*   Laser machining, micromachining, CAD/CAM interfacing with laser tools, expert systems.

*   A future process development should start from the desired materials properties from industry, find and define a microstructure, and perform process design.

*   UV laser processing.

*   Modification of tool steels.

*   Multi-disciplinary team approach.

*   LCVD will find increasing applications in microelectronics and micromechanics.

*   Applications of magnetic force in cladding.

*   A balance between simple and sophisticated models.

*   Ablation process needs further study for better understanding.

*   Development of inexpensive beam shaping device, especially for smaller wavelength.

*   Development of optical fiber for CO_2 laser.

*   Design of new alloys and substitution of expensive alloys by laser processing.

4. Any other questions and comments?

*   Contributing authors should be informed in advance on how much time they have for presentation.

*   Lecturers from industry should be invited because:
    - industry should be better informed of the new developments made in university and research centers. (e.g., wear and corrosion resistant coatings, in-situ processing and analysis)
    - scientists should be aware of the current industrial needs.
    - a two way dialogue will enhance technology transfer.

*   It is recommended that the Institute director, Prof. J. Mazumder, write a review article to be published in one of the trade magazines that have a wide circulation in industry.

*   We have heard a lot on metal and ceramic processing. It is suggested that other topics such as plastics and biomedical applications be included as well.

* We'd like to see an increased NATO budget for the ASI in the future. The ASI on the topic of laser processing should be held more often due to the rapid growth of the field.

* Some lectures lacked information on up-to-date research.

* Remarks on organization:
    - more social events are suggested to enhance interaction between participants.
    - program seems a little too heavy. Even though the contributed papers are a good element in providing different aspects of a field covered by the lectures, it may be helpful to be more selective.
    - could have had more coverage in LCVD and industrial applications.
    - improved poster session time schedule.

LASER SURFACE TREATMENT

An Overview

W.M. STEEN

Mechanical Engineering Department
University of Liverpool, UK

The laser is an ideal tool for surface treatment. Radiant energy from a laser is absorbed in the top few atomic layers of an opaque material, where it can either heat the surface or excite the surface atoms, leading to pyrolytic or photolytic processes. It is also chemically clean. The principle industrial advantages of laser surface treatment are:

* The thermally affected region is easily controlled in depth, extent and time above temperature, due to the ease with which this form of energy can be shaped and switched.

* The process is chemically clean.

* There is no need to touch the workpiece.

* Automation is usually possible due to the lack of environmental disturbance while the radiant energy is delivered to the process. This allows many forms of in-process sensing and hence control.

With these many and significant advantages it is surprising to find that the industrial use of the laser for surface treatment is slow to take off. Possibly this is simply due to the lack of engineers sufficiently fluent in the application of lasers that a critical mass of them has not yet been generated to relate the potential to the problems. This is something to look forward to; but come it will; the advantages are too prominent.

1

J. Mazumder et al. (eds.), Laser Processing: Surface Treatment and Film Deposition, 1–19.
© 1996 *Kluwer Academic Publishers.*

The laser was invented in 1960 by Maiman and has brought to industry a new form of energy. In surface treatment its action can be by heating (pyrolytic processing), which is the most common route, or by direct interaction of the photons in the process (photolytic processing). The intensity can be varied from the most intense industrial energy source available today when focussed to any required value when defocussed. When pulsed the power can be exceedingly high, for example 100MW, but the energy may be small due to the short duration of the pulses. With such pulses instantaneous ablation may occur leading to stress effects, but with no significant heating of the surrounding material (shock hardening, see Section 4.4.). The power from the laser can be altered in picoseconds allowing the power to be shaped in time as well as in spatial extent. All this adds up to one of the most flexible forms of energy; it can be used to generate any required thermal experience in a workpiece surface. The depth of the treatment is decided by conduction for nonmelting processes, conduction and convection for melting processes, evaporation depth for vaporisation processes and opacity for photolytic processes.

1. Range of Processes [1]

It is not surprising to find with this capability, that there is a wide range of possible surface treatments with the laser. They depend on the time and intensity of the treatment, with some treatments depending also on the laser wavelength. Figure 1 illustrates the range of processes. Currently the range of laser surface treatments include:

Non melting processes:
* Surface heating for transformation hardening and annealing.
* Surface stress formation for non contact bending.
* Surface domain refinement for control of magnetic properties.
* Annealing of ion implanted surfaces.

Melting processes
* Surface melting for homogenisation, microstructure refinement, enhanced solid solubility, metallic glass formation, and surface sealing of porous materials.
* Surface Alloying for improved corrosion, wear and/or cosmetic properties.
* Surface cladding for similar reasons.

* Surface particle injection - another form of cladding.

* Surface machining by melt blowing - lasercaving, a form of engraving.

<u>Vaporisation processes</u>

* Engraving , marquetry and microlithography

* Marking.

* Shock hardening.

* Surface texturing for improved paint flow.

* Paint stripping, cleaning.

* Laser Physical vapour deposition (LPVD)

<u>Other processes - photolytic processes</u>

* Laser Chemical Vapour Deposition (LCVD)

* Laser activated polymerisation - Stereolithography

* Enhanced electroplating and cementation

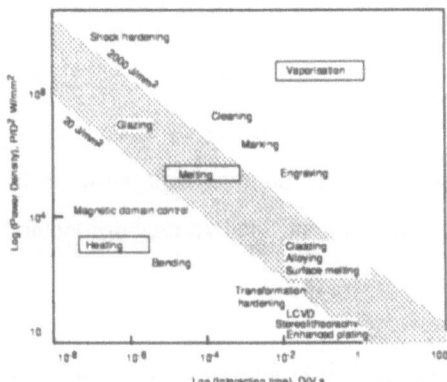

Fig. 1 Range of Laser Surface Treatment

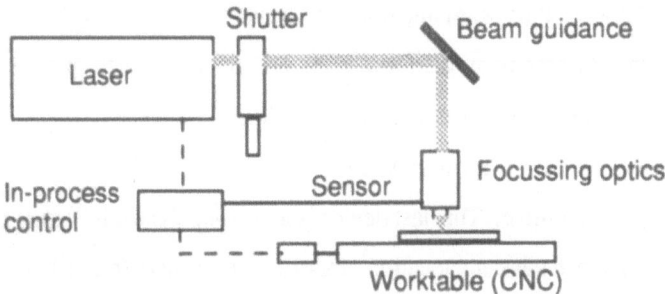

Fig. 2 General arrangement for laser surface treatment

2. Non Melting Processes

(Typical energy loading- Specific energy <20 J/mm²)

2.1 LASER HEAT TREATMENT [2]

The objective in laser surface heating is to heat the surface uniformly over the area to be treated to a temperature exceeding the transformation temperature (Ae3 temperature) but not exceeding the melting point and over a period of time sufficient for the heat to be conducted into the material so that the transformation temperature is exceeded at the depth required. The optical arrangement for most laser processes is illustrated in Figure 2. The beam delivery may be by mirrors, as illustrated, or by optical fibre. Optical fibres are available for Nd:YAG lasers, whose wavelength is 1.06μm; but power transmission fibres for the longer wavelength 10.6μm wavelength from a CO_2 laser are not available at present; hollow waveguide sapphire tubes or shiny tube or square section waveguides are an alternative.

There are various subtleties to the process; ensuring uniform heating by beam shaping; ensuring power absorption by absorbing coatings or using the Brewster angle; ensuring the depth of treatment is correct; allowing for interrun annealing and finally arranging a method for automation.

Beam shaping (Fig. 3): For spots larger than 3mm or so it is not possible to get uniform heating with a defocussed beam due to the transverse power distribution in the beam - known as the beam "mode"; thus beam integration techniques are required. They include rastering or scanning the near focussed beam at high enough speed to avoid melting; Passing the beam down a reflective tube of the required shape - known as a "kaleidoscope" - and imaging the beam at the exit from the kaleidoscope with lenses; reflecting or focussing the beam off an integrating mirror. This last device is a mirror polished in segments each segment forming an image in the same location - for example a square from a barrel shape mirror segment. Another method is to use specially shaped lenses or even holographically etched mirrors - "kinoforms".

Enhanced absorption: Components are usually painted with an absorbing material such as black paint, graphite and other patented mixtures; they may also be etched or sand blasted. The reflectivity of the surface must be reduced to around 40% or less for the process to be stable. Normal machined steel will have a reflectivity of around 90%. Thus if the remaining 10% is sufficient for the heating process then as the temperature rises there may well be changes in the reflectivity due to oxidation, recrystallisation or simply heating. If the reflectivity changes to say 80% then the power absorbed has doubled and there is a great risk of thermal runaway resulting in surface melting. The other reason for applying some technique for enhanced absorption is the loss of efficiency and the safety implications of trapping the reflected power. It is possible to avoid the need for coatings if the beam strikes the surface at the Brewster angle - that angle at which the reflected and refracted beams are at right angles. At this angle light polarised in the plane of incidence has a greatly reduced reflectivity; typical reflectivities may be as low as 10-20%. For example, this is particularly useful for hardening inside small holes such as valve guides.

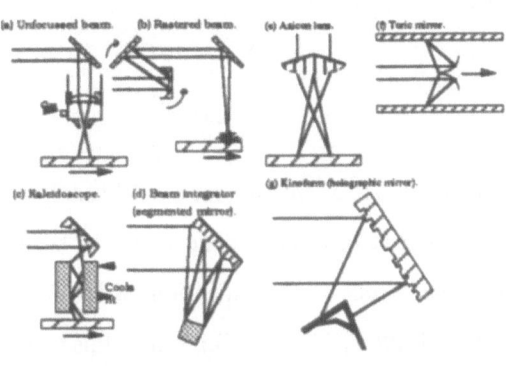

Fig. 3 Some methods of beam shaping

The depth of treatment, d (mm), is described by the normal laws of heat conduction. As a rule of thumb the depth has been found to be almost proportional to the parameter $P/(DV)^{1/2}$, where P = absorbed power, Watts, D = incident spot size diameter, mm and V = traverse speed, mm/s; the constants have to be found for each material. For example En8 steel fits the formula: $d = -0.10975 + 3.02 \ P/(DV)^{1/2}$.

When covering an area wider than a single track, it is necessary to overlap successive tracks. The heat at the edges of this second track is sufficient to cause tempering of the Martensite in the first track and hence some softening occurs. This gives a surface with hard and softer regions often found to be ideal for wear surfaces in which the softer regions act as traps for the oil and wear debris.

The process of transformation hardening relies on the heating of the material into the transformed region and then quenching fast. For steel the material is heated into the Austenite region and then quenched to form Martensite. Since the process with the laser is fast there is little thermal penetration and hence a fast quench without the need for auxiliary water or oil is usually possible. Since the quench is faster than alternative processes (induction, flames, arcs) the Martensite formed is finer and the residual stress is more compressive. Both these features mean that the laser treated surface may have improved fatigue and wear properties with a reduced friction coefficient.

Applications are numerous and are usually justified on grounds of reduced distortion, improved process speed or removal of the need for a quenchant.

2.2.LASER BENDING [3],[4]

When a near focussed laser beam is passed at speed (e.g. 4m/min. for a 2kW beam; fast enough to avoid melting) over the surface of a piece of metal, the metal may be heated locally to the point where the compressive stress due to the expansion on heating is sufficient to cause plastic flow of the heated and softened region. On cooling this region will not plastically flow, due to heat flow into surrounding areas and the strain being higher when the material is colder and thus stronger. Hence a controlled distortion occurs. With each pass the bend may be around 1-5°. The extent of bending has been calculated by Vollertsen(3). He assumed all the intitial thermal compression is absorbed by plastic flow and that the material can be likened to a two layer bend beam - the top layer creating the bending force and the bottom layer bending. The resulting equation was:

$$\alpha_B = 3 \frac{\alpha_{th}}{\rho c_p} \frac{PA}{v} \frac{1}{s^2}$$

Where: α_b = bend angle per pass, °/pass; α_{th} = thermal conductivity, mm²/s; ρ = density, g/mm³; c_p = specific heat, J/g°C; P = incident power, W; A = absorptivity; v = traverse speed, mm/s; s = thickness of sheet being bent, mm.

It is interesting to note that this approximate relationship does not include the beam diameter since the effect depends on the volume of heated material and hence simply the delivered power. Secondary effects occur on bending, such as work hardening and thickening of the bent region which affects the linearity of this relationship. Fig. 4 shows the angle as a function of the number of passes illustrating this non linearity.

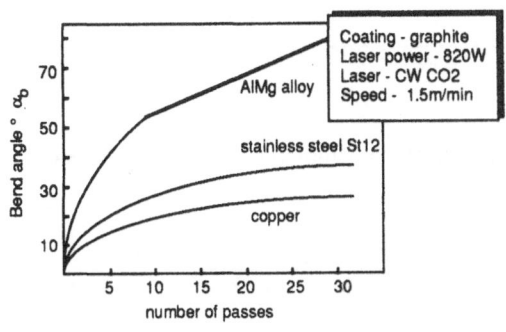

Fig. 4 Relationship between bend angle and number of passes [3]

Precise corners or curves may be created by this non contact bending method, depending on whether the traverse is in the same or adjacent locations. The application of the process to bending sheets to allow fit up during welding, particularly for ship plates and car bodies, or for shaping beams in space where heavy bending equipment is not available or for rapid prototyping of small sheet components is being considered. An interesting feature of the process is that if the whole structure is heated through (i.e. for thin materials or small shafts) the bending direction may be reversed due to reduced plastic flow under compression and some plastic flow under tension on cooling. This means that bending away from the beam might occur. This feature is used for the straightening of long rod shaped components such as knitting needles and thin shafts.

2.3.SURFACEDOMAIN REFINEMENT [5]

When a transformer steel (Si containing steel) is traversed by a fast moving (10-100m/s) focussed laser beam the stresses generated will divide the magnetic domains in such steel into smaller domains. The thermal shock is believed to cause slip planes to form in the microstructure, thereby producing these new magnetic domains. By adjusting the spacing of the laser traverses the size of the domains can be controlled. This finer structure will have less hysteresis loss in an alternating magnetic field, as in a transformer. The power loss from transformers using this steel can be reduced by approximately 10%. This represents a possible saving of 0.01-0.1% of the power transmitted through the transformer.

2.4. ANNEALING OF ION IMPLANTED SURFACES [6],[7].

Integrated circuits may lose their epitaxial form after ion implantation. This can be reformed by laser annealing.

3. Melting Processes
(typical energy loading - Specific energy <2000 J/mm²)

3.1. LASER SURFACE MELTING

When the incident laser energy can not be conducted away fast enough the surface melts; firstly to form a conduction limited melt pool and subsequently a keyhole due to the onset of evaporation and boiling. The pool dimensions and quench rate vary with the traverse speed and the laser power density. At very high speeds a melt pool, may cool at rates of approximately 10^{6}°C/s; this is sufficient for some alloys with a deep eutectic to form metallic glasses. More normal cooling rates are typically 10^{3}°C/s. These high quench rates result in a refined microstructure. Solidification theory predicts that the scale of the microstructure is proportional to the reciprocal of the square root of the cooling rate (interdendritic arm spacing μ 1/√[cooling rate]). The high quench rate may also affect the phases formed, often allowing increased solid solubility. The solidification mechanism is almost always dendritic or cellular dendritic. The style of solidification is determined by the ratio of the temperature gradient,G °C/m to the solidification rate, R, m/s. The surface of the melt pool has steep thermal gradients which cause strong stirring forces due to out of balance surface tension forces - Marangoni forces. This causes homogenisation of the melt pool, which is of particular relevance when melting non homogeneous materials, such as cast irons. The flow of material decides the shape of the weld bead and hence the surface finish. A surface roughness of 10-25µm is typical. Applications of surface melting are in hardening cast irons to form ledeburite[9], sealing thermal barrier coatings of Yttria stabilised zirconia and other ceramics and hardening of tool steels[10]. In the case of surface melting ceramics the use of sharp pulses (e.g. 20ns, 20MW) from an excimer laser (249nm, in the ultra violet) have been shown to both melt a 10μm layer, and hence seal against corrosion, and also to smooth the surface reducing the heat transfer coefficient. Metal surfaces can gain strength and durability and withstand friction with less damage when treated with short pulses of ultraviolet light. In stainless steel the thin melt skin on resolidifying drives the chromium to the surface and thus improves the material for medical prostheses, such as artificial hip joints. In applications for space where lubrication is not possible the improvement in friction properties is useful and may be due to the ultra fine microstructure of the resolidified layer[11]

3.1.1. *Selective laser sintering* (SLS)[12]

SLS is a rapid prototyping process developed by DTM Corp (Austin Texas) that creates 3D models by melting one layer at a time from computerised data. The material may be a bed of powdered nylon or polycarbonate plastic for artificial limbs and engineering models; it may also be metal powder, though such models tend to be very porous. There are approximately 30 different approaches to rapid prototyping which is fast becoming a major design tool or even manufacturing method for small batches. Many of these techniques use lasers (see Section 5.3. - stereolithography and Section 3.3 - laser cladding).

3.2. SURFACE ALLOYING

Surface alloying is a variation on surface melting which offers the potential of vast improvements in the surface properties. If another material is added to the surface melt pool then the resulting "weld bead" will be an alloy. The alloy will be homogeneous due to the Marangoni stirring, except at high traverse speeds when incomplete mixing may occur. The new material can be added by electroplating, vacuum evaporation, preplaced powder coating, thin foil application, ion implantation, diffusion - as in boronising or carburising -, powder blowing, wire feed or a reactive gas shroud - such as acetylene, nitrogen or oxygen. However the exact composition of the alloy depends on the depth of melting into the substrate. This is not so easily controlled due to variations in the beam coupling with the surface. Some applications are the formation of TiN by melting Ti in a nitrogen atmosphere, the formation of high silicon or chromium cast irons on cheaper cast iron substrates, the formation of stainless steel on steel or the hardening of aluminium by alloying with Ni or Si. A new material is metal coated ceramic which when treated this way forms an alloy between the metal and ceramic, giving a tougher surface than the ceramic alone, thus reducing crack tendency.

Although few of these processes are in production at present this process and that of laser cladding has put engineers in the position that they can choose the material they need for the surface and the material they need for the bulk separately. The choice is exhausting to contemplate, but the challenge to our knowledge of materials is stimulating!

3.3. SURFACE CLADDING

The laser is unique in being able to clad small areas with a fusion bond and low dilution. This can be achieved by melting a preplaced bed of powder, feeding wire or by blowing powder into a laser generated melt pool.

In the case of preplaced powder cladding it is relatively difficult to get a fusion bond with low dilution since the powder bed must be melted down to the substrate at which point the melt will refreeze due to the additional thermal load of the relatively high thermal conductivity of the substrate; this resolidified layer must be remelted before the substrate melts to form a strong fusion bond. However the powder bed does have a reduced reflectivity and hence beam absorption is not a problem with this method, also powder utilisation is high. In the case of the blown powder process the clad will not form unless the powder strikes a molten surface, hence a fusion bond is assured. However variations in the reflectivity of the substrate can be a problem and hence the clad zone is often surrounded by a reflective dome which acts as an optical feed back system and gas shroud, Fig. 5. The process competes with other hardfacing techniques, such as plasma spraying, TIG and D-Gun, by allowing precise placement of the clad, low thermal spillage and low dilution. The low dilution permits thinner coatings of the expensive clad layer without loss of the desired properties. The thinner coatings in turn cause reduced distortion.

Applications are becoming more frequent as the process is better understood. Some examples are: cladding shroud interlocks on turbine blades, valve seats and poppet valves on car engines, internals of pipes, and valve seats for oil or water applications. More recently the process has become one of the many alternatives for rapid prototyping, by building up 3D shapes by layer cladding[13]. This aspect forms the basis of an industry in laser repair of undersized or worn components (See also Sections 3.1.2 and 5.3).

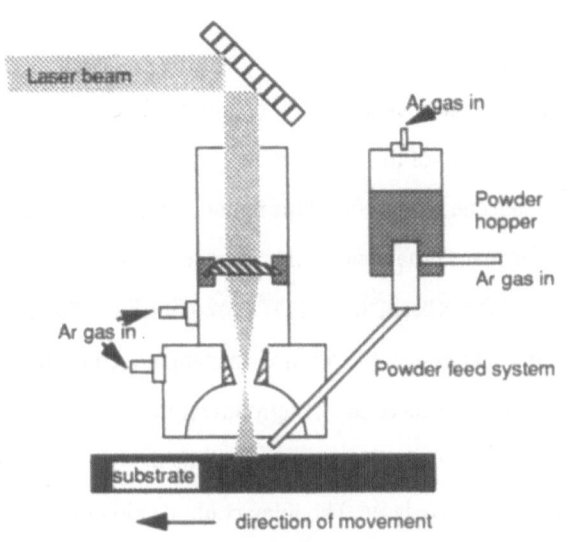

Fig. 5 The arrangements for cladding using blown powder

3.4. PARTICLE INJECTION

An alternative to cladding and alloying is to blow high melting point particles into low melting point substrates - such as TiC, SiC, WC or Al_2O_3 particles into Al - to form a type of "Macadam" surface. The usual application is to make a hard, lightweight material.

3.5. SURFACE MACHINING

By laser melting the surface of a metal with a pulsed or continuous beam and blowing hard with a side or axial jet, the melt can be removed and thus leave a small groove or pit. This has been exploited in such processes as "lasercaving". The process is slow and has limitations on the side wall angle, but is capable of engraving metal with remarkable detail.

4. Vaporisation Processes

(Typical Energy loading - Specific Energy > 2000 J/mm²)

4.1. ENGRAVING

In addition to melt blowing,(described in Section 3.5) engraving can be done by vaporisation or by photolytic decomposition.

4.1.1. Marquetry

Materials such as wood do not melt but evaporate or volatilise. By scanning a focussed laser beam ($\approx$1-500W) over a copper mask the image of the mask can be carved into the wood to a depth of several millimetres. The detail can be very fine. One remarkable aspect is that this can be done without charring and so in many ornamental boxes, plaques etc. the carving may be subsequently traversed by an out of focus beam to create a charred finish to enhance the contrast. An alternative to using a mask is to scan the beam via a computer controlled mirror system (as in Section 4.2. Marking) and carve the pattern directly rather than by rastering over a mask. The carved surface can be inlaid by marquetry using the same mask or programme to cut out the fitting piece, also by laser. The inlay could be either metal, wood, ceramic or plastic.

4.1.2. Microlithography

The photons from short wavelength radiation such as ultraviolet light or X-rays have energies similar to the bond energy between carbon and hydrogen or carbon and carbon in organic materials ($\approx$ 4eV) and hence these photons can break these bonds. If sufficient photons arrive simultaneously then the material can be broken into volatile components without heating. Thus it is possible to carve the human hair, or make very small machined details in microelectronic components. The use of synchrotrons to generate X-rays, by a form of laser radiation, allows the machining of metallic parts for micromechanisms and micro electric contacts.

4.2.MARKING

The engraving, melting or photolytic processes have found a huge market in marking. Marking is usually done with a pulsed laser (either Nd:YAG, excimer or CO_2 TEA - transversely excited atmospheric - laser) to limit the thermal spillage and increase the penetration of the mark. The mark may be a melt or alloyed track on metal or a change of colour or engraving on plastic, ceramic or wood. It is applied by imaging a mask on the article to be marked or by rastering the beam in the required pattern with galvanometer driven mirrors. The latter being slower but more flexible. Using strong pulses of very short duration (≈ 10ns) marks can be put onto cable, bottles, or some product "on the fly" without stopping the process. The increase need for quality control and the threat of legislation has meant that many manufactured components are now routinely marked by laser.

4.3. SHOCK HARDENING

Striking a surface with an extremely short pulse (≈ 1ns) and large energy (>5J, e.g. an iodine laser) with laser fluxes of greater than 10^{11} Jmm^{-2}s^{-1} results in the surface atoms being explosively evaporated, with hardly any surface melting. This surface "explosion" causes a shockwave which has similar compressive cycles to shot peening but without the risk of debris being buried in the surface. The temperatures and pressures calculated by Saraday[14] rose to values of 12k°C and 80kbars. The process can be amplified by doing it under water or with a thin film on the surface. For processes which require great cleanliness and yet a surface residual compressive stress to enhance fatigue resistance this may be one of the few alternatives.

4.4. SURFACE TEXTURING [15]

Cold rolled steel strip which is to be pressed for white body applications has to be roughened to enhance the tool grip and to aid in lubrication during pressing. This is done by rolling in a cold roll whose rolls have been sand blasted. If this roll is systematically "roughened" by pattern pitting with a pulsed laser beam, then the resulting strip can not only be gripped in the press but lubrication is enhanced and there is less wastage due to galling with the press tool. The pressed sheet retains some of the pattern. This pattern causes paint to adhere better and to flow more uniformly over the surface without the usual "waviness" - a longer wavelength roughness sometimes called "the orange peel effect". This gives an improved appearance similar to that achieved after many layers of paint. Thus many car manufacturers now use laser textured sheet steel for their body panels. This waviness is also due to a waviness developed during the grinding of the roll and hence great care has to be taken during roll grinding. The process was a development from gravure printing methods which began using lasers in the 1980s. The uniformity of crater formation improves the quality and hence the low order mode beams from a CO2 laser are prefered. Two types of crater have been noted; a crater with a lip and a crater with a central peak. Typical processing conditions with a 2.5kW CO2 laser would be to pulse, by chopping, at up to 30kHz and creating a crater every 200μm in x and y directions - a covering rate of approximately 10cm^2/s.

4.5. PAINT STRIPPING AND CLEANING

By choosing the correct frequency of the laser beam, dirt on stonework, brickwork or art, paint on aircraft or graffiti on metal, ceramic or wood will absorb the beam and be evaporated while the substrate will not do so to the same extent and will be unaffected. Control of the power and pulse shape of the laser beam can result in the removal of the surface layer with little to no thermal penetration of the substrate; typical pulse lengths are of the order of 30μs. This is an important new development in cleaning technology. It means that aircraft can be paint stripped (e.g.CO2 laser with 6J pulses at 1kHz, 30μs, with cleaning rates of up to 1m^2/min) without the risk of highly corrosive paint removers damaging the aircraft, or causing a disposal problem, artwork can be cleaned without damaging the paint layers beneath the dirt with solvents (excimer laser with J/pulse often applied with a feed back spectroscopic analysis to check what is being removed as it is being removed)[16], stonework can be cleaned without affecting the patina and buildings can be cleaned without the mess of sand blasting, water sprays and with reduced scaffolding requirements (Nd:YAG 1mJ/pulse)[17,18]. Tatoos can be removed the same way. In this process the embedded ink - usually carbon, or the like, absorbs the beam and evaporates within the tissue which then dissolves it over a period of a few weeks (Ti-Sapphire laser frequency tuned to the colour to be removed; pulse ms and power).

4.6. LASER PHYSICAL VAPOUR DEPOSITION (LPVD)

The laser can be used to evaporate a difficult material in a clean manner in a vacuum chamber. For example, platinum may be evaporated by laser and will then condense on a target. There is some directionality with the vapour emission. Currently the explosive nature of laser ablation is being used to ablate and condense superconducting alloys. It appears there is no distillation or similar causes affecting the composition. If the target is heated, possibly with a second laser then the deposited film will probably crystallise in the required form to be super conducting[19].

5. Other processes - Photolytic processes

5.1. LASER CHEMICAL VAPOUR DEPOSITION (LCVD)[20]

Metals, metal oxides, carbides or nitrides can be deposited from volatile compounds of metals such as halides, hydrides or chelates by pyrolytic or photolytic decomposition in a laser beam or on a laser generated hot spot. The process is slow, usually measured in μm/ minute; if it is speeded up there is a tendency to form a concentration boundary layer and a consequent change to mass transport control from chemical control of the process. This results in the formation of dust rather than a thin adherent film. The process is used for "writing" printed circuit boards or other components in electronics. The coating of tools with TiC or the corrosion protection of metal with SiO_2 films. If deposition is made in an inert atmosphere then the metal may be deposited; if it is done in a nitrogen atmosphere then a nitride may form and so on. Thus silicon may be deposited from SiH_4 and H_2, or SiO_2 could be deposited from $SiCl_4$ and H_2O.

5.2. LASER ACTIVATED POLYMERISATION - STEREOLITHOGRAPHY [21]

An ultraviolet laser (He-Cd or excimer) can be used to polymerise certain plastic monomers (acrylic or epoxy). Photosetting resins are frequently used for masks in the manufacture of integrated circuits and circuit boards as well as in electrochemical etching. The exposure of the varnish may be achieved by laser in a manner similar to that used for marking (Section 4.2.). In a recent development a laser beam is scanned by galvanometer driven mirrors over a smooth pool of resin in a pattern described from a CAD (Computer Aided Design) package and hence sets the resin to the shape of a layer of a 3D article. The model to be made is sliced in virtual reality within the computer each layer is printed one on top of the other to make the 3D article with great precision. These models are used for prototyping or as lost wax models for casting into metal. The design time is typically reduced from several months to several days by this route. One advantage of rapid prototyping is the high level of flexibility in design, for example, artificial limbs can be made to personalised designs giving much better fits, easy duplication and overall process efficiency.

5.3.ENHANCED ELECTROPLATING AND CEMENTATION

An argon laser directed onto a cathode within an electrolysis cell will cause an increase in deposition rate of several times probably due to the photons exciting the deposition radicals. It has been considered as an alternative route for the printing of circuit boards. It is not always necessary to have an electric potential to drive the process; simple cementation may be sufficient.

Summary

The laser is offering industry a surface treatment tool of unprecedented versatility, from which a number of new processes have developed and from which many more are expected.

Bibliography

Draper.C.W., Mazzoldi.P. editors "Laser Surface Treatment" Proc. NATO Advanced Study Institute, San Miniato, Italy, Sept. 1985 NATO ASI series E Applied Sciences Number 115 publ Nijoff, Dordrecht, Holland 1986.

Steen. W. M. "Laser Material Processing" publ Springer-Verlag, London, Berlin, Heidelberg ISBN 3-540-19670-6 1991.

References

1. Steen. W. M. (1991) *Laser Material Processing*, publ Springer-Verlag, London, Berlin, Heidelberg.

2. Gregson.V.G. (1984) Laser heat treatment, in M. Bass (ed), *Laser Material Processing*, North Holland Publishing Co., London, pp.

3. Vollertsen.F. (1994) An analytical model for laser bending, in *Lasers in Engineering*, Gordon and Breach Science Publishers, USA, Vol 2 pp261-276.

4. Namba.Y. (1986) Laser forming in space in C.P. Wang (ed), *Int. Conference on Lasers*; pp403-407.

5. Gillner.A., Wissenbach.K., Beyer.E., Vitr.G. (1988) Reducing core loss of high grain oriented electrical steel by laser scribing, in H. Hugel (ed) *5th Int conf. on Lasers in Manufacturing (LIM 5)* , IFS Publishers, Stuttgart, Germany, pp137-144.

6. Park.H.G., Rose.K. (1981)Characterisation of small area laser beam annealing, *Journ Electron. Mater.* **10**, No. 5, 823.

7. Narayan.J., Young.R.T., White.C.W. (1978) A comparative study of laser and thermal annealing of boron implanted silicon, *J.App.Phys.*, **49**,7, 3912-17.

8. Flemings.M. (1974) *Solidification Processing*, McGraw Hill Book Co., New York.

9. Steen.W.M., Chen.Z.D., West.D.R.F. (1987) Laser surface melting of cast irons and alloy cast irons, in *Ind. Laser Annual Handbook* , Springer-Verlag Publishers, Berlin. pp87-96.

10. Colcao.R., Vilar.R. (1993) Laser surface melting of bearing steels in, Martelucci S., Chester, A.N., Scheggi, A.M. (eds), *Proc NATO ASI Laser Applications for Mechanical Industry* , Kluwer Academic Publishers, Dordrecht, publ Kluwer Academic Publ., Dordrecht **238, Series E,** 305-314.

11. Anon, (1993) UV strengthens metal surfaces, in *Ind. Laser Review*, Pennwell Publishers, pp5-6.

12. Pera.L., Marinsek.G. (1993) The role of the laser in rapid prototyping, in Martelucci, S., Chester, A.N., Scheggi, A.M. (eds) in *Proc NATO ASI Laser Applications for Mechanical Industry* , Kluwer Academic Publ., Dordrecht **238,** Series E, 293-303.

13. Murphy.M., Steen.W.M. Lee.C. (1994) Rapid manufacture of metal components by laser surface cladding, in Proc 26th Int CIRP seminar LANE'94, Erlangen.

14. Saraday.I., Magnusson.C., Wei.Y., Meijer.J., Wisselink.F. (1993) Microstructural changes in steel irradiated by nanosecond iodine laser pulses in Proc Int conf Lasers and Electron beams (ICALEO '93), Orlando Fl. USA.

15. Hector.L.G., Sheu.S. (1993) Focused energy beam work roll surface texturing science and technology, *Journal of Material Processing and Manufacturing Science*, **2**, 63-117.

16. Asmus.J.F. (1987) Light for art conservation, in *Interdisciplinary Science reviews* , **12**, No.2, 171-179.

17. Watkins.K.G., J.H.Larson, D.C.Emmony, W.M.Steen (1994) Laser cleaning in art restoration: a review, in Proc NATO ASI.

18. Larson.J.H. (1985) The conservation of stone sculture in museums, in *The conservation of building and decorative stone*, Butterworths Publishers, **2**, pp 197.

19. Schultz.L., Roas.B.,Schmitt.P., Endres.G. (1990) High-Te superconductor thin films by laser deposition, in H.Bergmann, R. Kupfer (eds), *Proc 3rd Europ. conf. on laser treatment of materials*, Sprechsaal publ grp., **1**, No.2 pp835-848.

20. Azer.M.N., Kar.A., Mazumder.J. (1990) Theoretical and experimental studies of laser chemical vapour deposition of Ti from TiBr$_4$ in *Proc 6th Int. conf. on Lasers in Manufacturing (LIM 6)* Birmingham, UK pp193-204.

21. Benedict.G.F. (1990) Stereolithography - The new design tool for the 1990s in *Proc 6th Int. conf. on Lasers in Manufacturing (LIM 6)* Birmingham, UK pp249-261.

BASIC COUPLING MECHANISMS IN LASER SURFACE TREATMENT

H. HÜGEL, F. DAUSINGER, W. BLOEHS, B. GRÜNENWALD
Institut für Strahlwerkzeuge
Universität Stuttgart
Pfaffenwaldring 43
70569 Stuttgart, Germany

1. Introduction

The functioning of a large number of workpieces is mainly determined by the properties of their surfaces. This is the case, for example, whenever friction and resulting wear mechanisms are involved or where the workpiece is exposed to high temperature and/or corrosive ambient conditions. Well established thermal processes like induction or flame hardening and plasma techniques for alloying or cladding are industrially used to achieve the required properties. These treatments yield chemical and mechanical alterations of the material in layers close to the surface with hardly affecting the bulk material of the workpiece. This approach allows the design and fabrication of workpieces of improved performance with an economical use of expensive and strategic materials, at the same time.

In this field, laser technology is, in principle, offering many advantages. They are primarily due to the facts that the laser beam, i.e. the heat source, is easily pointed to well-defined areas at the surface and the energy input per unit time and area is well controllable. This results in a low thermal load, and hence thermal distortion of the workpiece and in smaller efforts for finishing work. In addition, the achievable quality of the modified layers is often higher with laser treatment. Nevertheless, because of higher costs compared to conventional methods, laser surface processing technologies have not reached wide spread applications in industry, so far.

To increase the economy of any laser treatment, its costs have to be reduced. This may be obtained, partially by lowering the investment and running costs of the

21

J. Mazumder et al. (eds.), Laser Processing: Surface Treatment and Film Deposition, 21–46.

laser source and of the handling system by technical improvements and further technological developments. A good example in that direction is the simplified beam guiding through flexible glass fibers for 1 μm wavelength. On the other hand, the gain in economy by an increase in efficiency of a well-conducted process may be, by far, the more promising approach. Since the basic phenomenon underlying laser materials processing is the absorption of electromagnetic energy, the problem of energy coupling plays the most decisive role. The knowledge of basic mechanisms, therefore, is a necessary starting point for all the measures to optimize laser surface treatment processes and to make them competitive with other technologies.

This contribution is dealing with investigations that are aimed at a better understanding of phenomena involved in energy coupling during hardening, surface melting, cladding and alloying. All these processes require high laser power and have represented applications typical for CO_2-lasers, so far. The coming-up of Nd:YAG lasers operating at a power level of several kW, however, renders it more and more important to include the consequences of a shorter wavelength with respect to the handling equipment (glass fibre) and energy coupling (increased absorptivity for metals). For this reason, the discussion of wavelength effects has received major concern.

There are some more processes standing for surface treatments, as well. Among them are such that, e.g., involve a depth-restricted removal or deposition of material at the surface (paint stripping or plating as typical examples). The energy coupling phenomena occurring there, however, are quite different from those involved in the class of technologies mentioned before. That is why they cannot be discussed that way and will not be included in this contribution.

2. Absorptivity and energy coupling

Mechanisms determining the energy coupling of laser radiation into a workpiece represent the key for the efficiency of a process and, hence, for a decision whether this treatment will be considered for an industrial application or not. In general, estimates of the absorbed laser power are being made by using available data of the absorptivity which, however, often might not particularly well describe the real situation. This has to do with the fact, that the absorptivity is a direct measure for the in-coupled energy only, if the area under the laser spot is plane and smooth - what is true for hardening and to some extent for remelting, but no longer for cladding and alloyng processes. In addition, modifications at the surface due to oxidation effects and the supply of alloying material may introduce considerable alterations. Apart from these more practical implications it has to be noted that there exists a considerable lack of theoretical data, especially for elevated temperatures - for the processes under discussion values close to the melting temperature are of relevance - low wavelengths and alloys of technical importance. In view of this background attempts had been undertaken to calculate and experimentally verify data of the absorptivity and further to measure the energy coupling rate under conditions as close as possible to the real treatment process.

2.1. ABSORPTIVITY / FRESNEL ABSORPTION

The well-known Fresnel formulas allow the calculation of the absorptivity depending on polarization and angle of incidence as a function of the two optical properties, the index of refraction and extinction coefficient. Their temperature and wavelength dependency was taken into account by an approach that is outlined in detail in [1]. Some main results shall be presented and discussed herein.

Fig. 1 compares experimental data on iron of other authors with the calculated ones for normal incidence. It can be seen an excellent agreement in all the wavelength

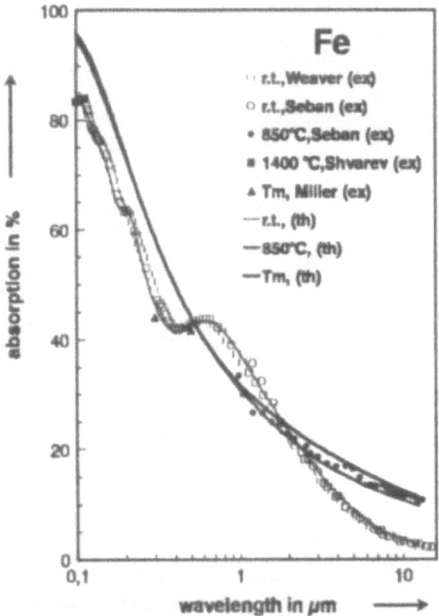

Fig. 1.   Absorption of iron at perpendicular incidence: experimental data from literature and theoretical approximations based on 2-band Drude theory.

24

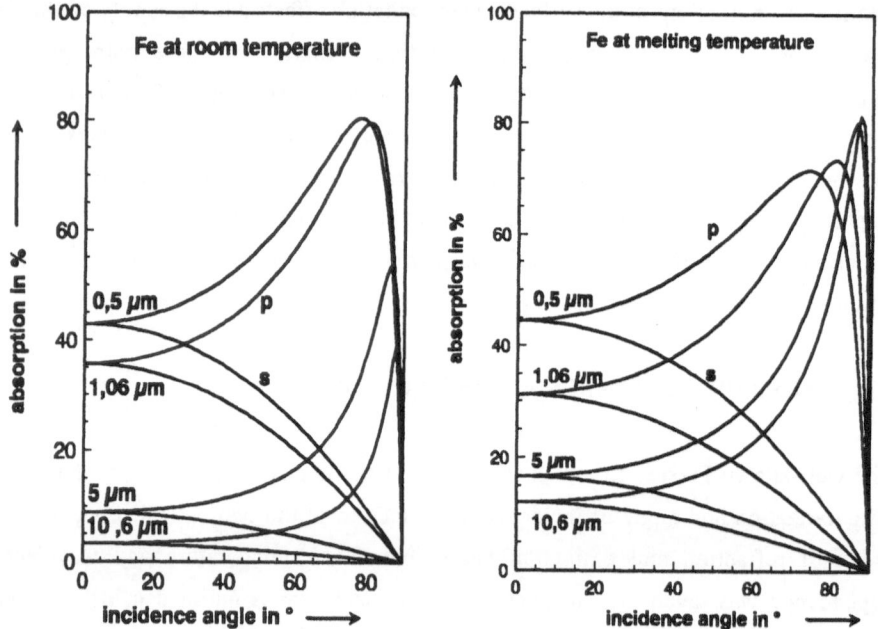

Fig. 2. Absorption of iron depending on angle of incidence, polarization and wavelength. Left: at room temperature, right: at melt temperature.

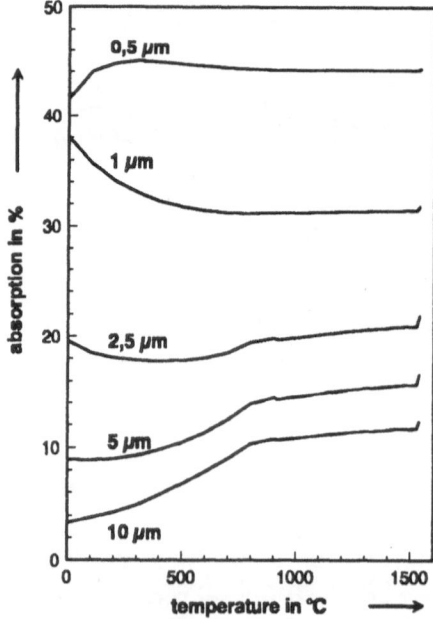

Fig. 3. Temperature dependence of absorption at perpendicular incidence of iron: calculated data from electrical conductivity values.

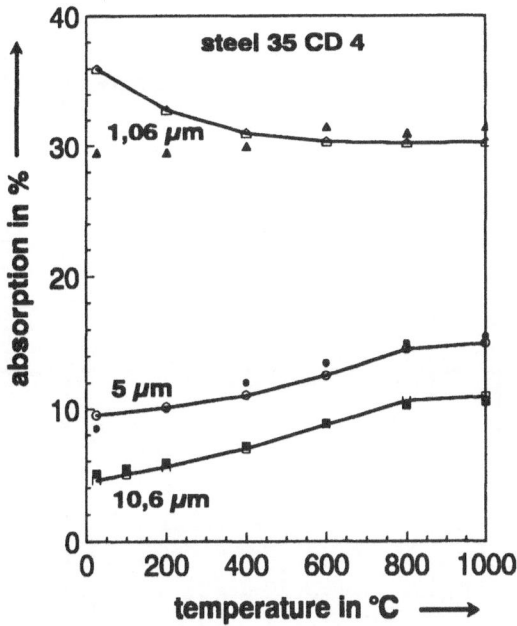

Fig. 4.   Temperature dependence of absorption in steel calculated from electrical conductivity values for different wavelengths, experimental data after Stern.

regions for room temperature, and between 0.5 μm and 15 μm also for high temperatures. The absorptivity at room and melting temperature depending on the angle of incidence and polarization for four wavelengths of industrial interest is presented in Fig. 2. The effects of wavelength and temperature on the absorption of iron are summarized in Fig. 3 and a particular steel in Fig. 4.

One major conclusion to be drawn from these results is the fact that using Nd:YAG-lasers in place of CO_2-lasers (which might be well feasible in the kilowatt regime) the gain in coupling efficiency will not amount a factor of up to eight as could be concluded from judging by the data at room temperature. Instead, a factor of two appears more realistic for the whole range of beam incidence.

2.2.   ENERGY COUPLING

With respect to changes of the coupling rate by oxidation, up to now, no strict and correct prediction is possible since this effect not only depends on the temperature, but also on the dwell time - both parameters being closely related to a specific process. It hence appears particularly useful to calorimetrically measure the coupling rate under conditions as close as possible to each process. As an example, the measured coupling

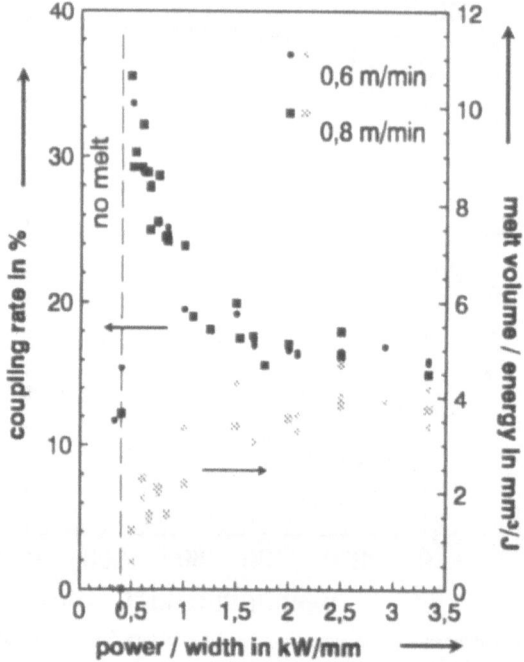

Fig. 5   Coupling rate and melt efficiency at remelting of mild steeel (16 MnCr 5) with CO$_2$-laser in air.

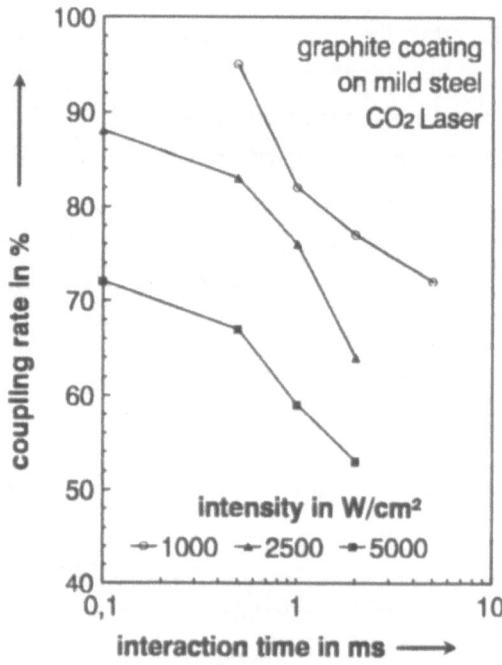

Fig. 6   Dependence of coupling rate of coated surfaces on interacton time and incident intensity.

rate for a remelting process in air is presented in Fig. 5. The decrease of energy coupling with increasing (related) power is due to the fact that an enhanced stirring of the melt pool limits the growth of oxide layers.

An other point of concern is the time-dependent behaviour of coatings. Their use is a widespread means to attain a reasonable coupling when hardening with CO_2-lasers. As can be seen from Fig. 6 they degrade with increasing intensity as well as with interaction time. It becomes evident from these findings, that show a rather strong dependence on the process parameters, how careful one has to use cited values !

Yet an other modification will be introduced in the mechanism of energy coupling when powder is fed into the interaction zone during alloying or cladding. In order to have well-defined conditions for a thorough study on that subject, both theoretical and experimental investigations were performed for the coupling of laser radiation into powder layers [2]. At values far below those to reach surface melting, coupling rates of up to 35 % for pure metallic and about 45 % for an alloy powder were measured. There, as well, a strong effect of oxidation was found. Raytracing calculations revealed that multi- reflections between the powder grain cause these high values. As a consequence, the coupling rate decreases with decreasing porosity as is shown in Fig. 7. A somehow contradictory result is given in Fig. 8 indicating an

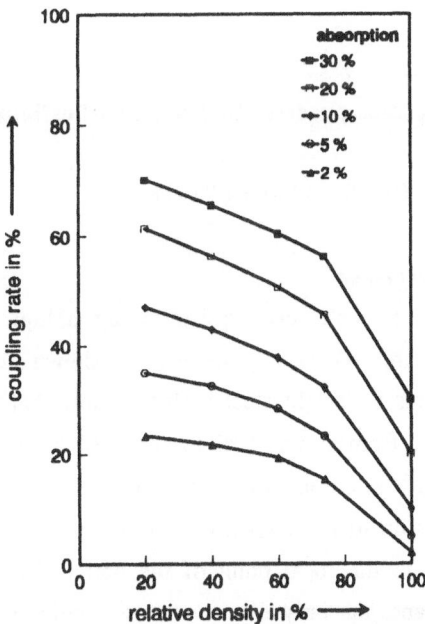

Fig. 7   Calculated coupling rate in powder layers depending on density and absorption rate of powder material.

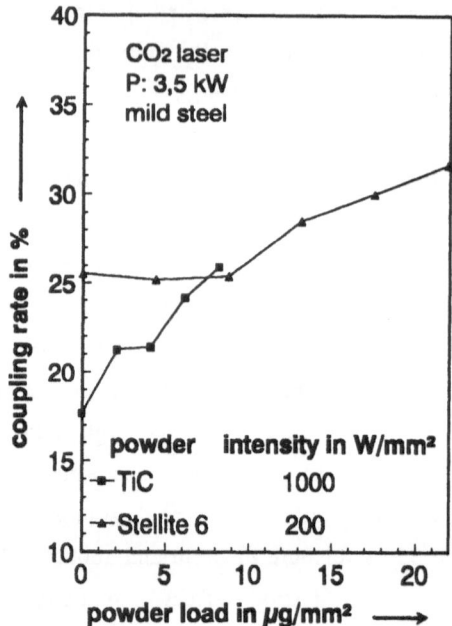

Fig. 8   Contibution of powder supply to coupling rate in alloying and cladding.

increase of coupling rate with density. Judging from this as well as from the rather poor knowledge of the phenomena involved here more investigations are needed on that point.

3.   Measures to increase process efficiency and flexibility

3.1.   MEANS TO INCREASE ABSORBTIVITY

3.1.1.   *Oblique beam incidence*

The possibility to use the enhanced absorption by oblique incidence of parallel polarized laser beam during CO_2-laser hardening was shown a couple of years ago [3]. Absorption measurements on bright steel surfaces show that the absorptivity of the surface can be increased from around 5 % to about 30 % by this effect [4] when an angle of incidence of 75° was used. If a laser beam hits the surface at a certain angle of incidence, the distortion of the intensity distribution is depending on the relative position of the surface to the focal point of the optic. If the surface is positioned directly in the focal plane, the only distortion is a change of aspect ratio (width to height) of the beam profile, whereas outside the focus an intensity maximum can

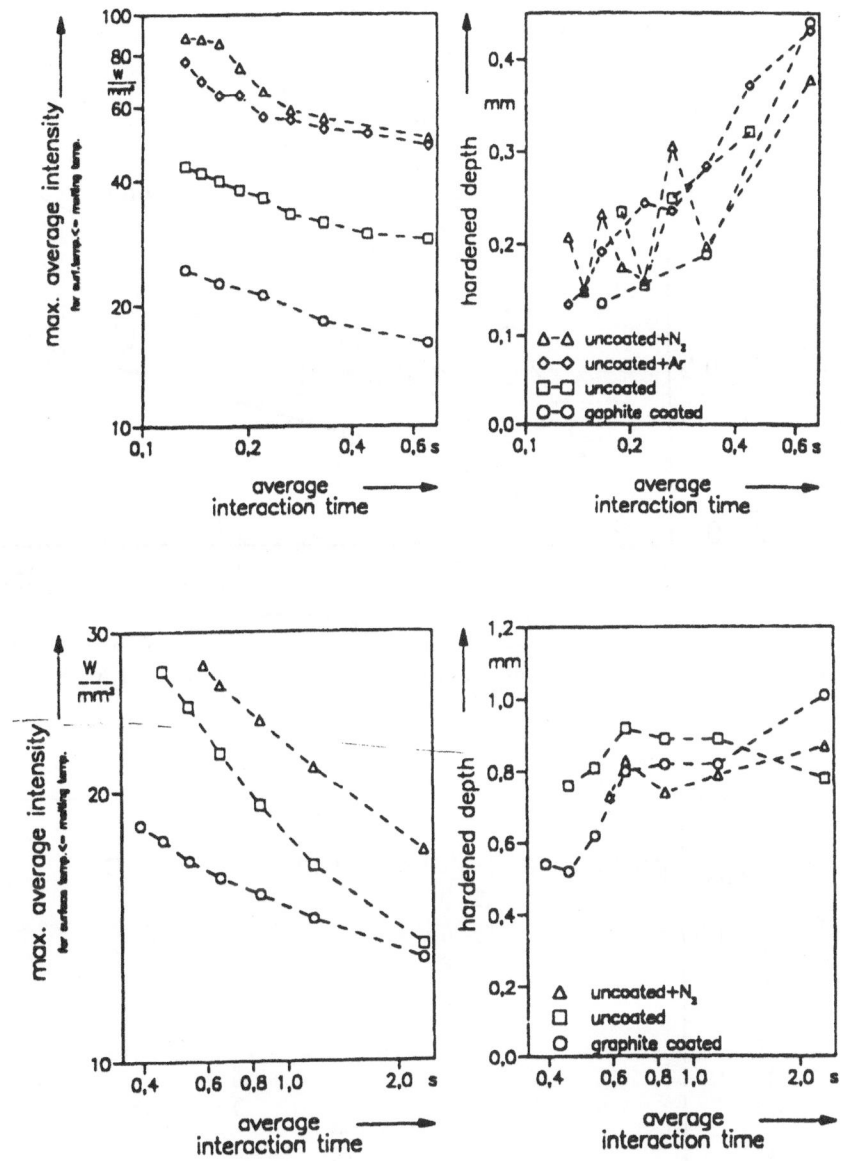

Fig. 9. Maximum intensity (without surface melting) and hardening depth dependent on interaction time. Top: CO_2-laser, angle of incidence 75°, bottom: Nd:YAG-laser, normal incidence.

appear, leading to surface melting when hardening with this setup. Because of this effect, for hardening with inclined incidence a focussing optic with long focal length should be used, forming a large spot and a large focus-depth (Rayleigh-length). To utilize this effect of increased absorption, hardening cannot be done with kaleidoscope

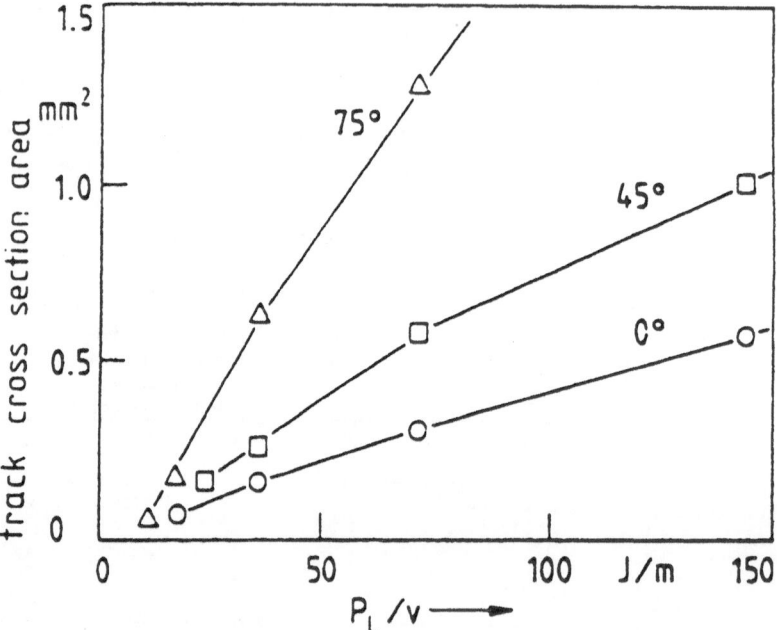

Fig. 10. Dependence of remolten area on angle of incidence and on line energy P_L/v.

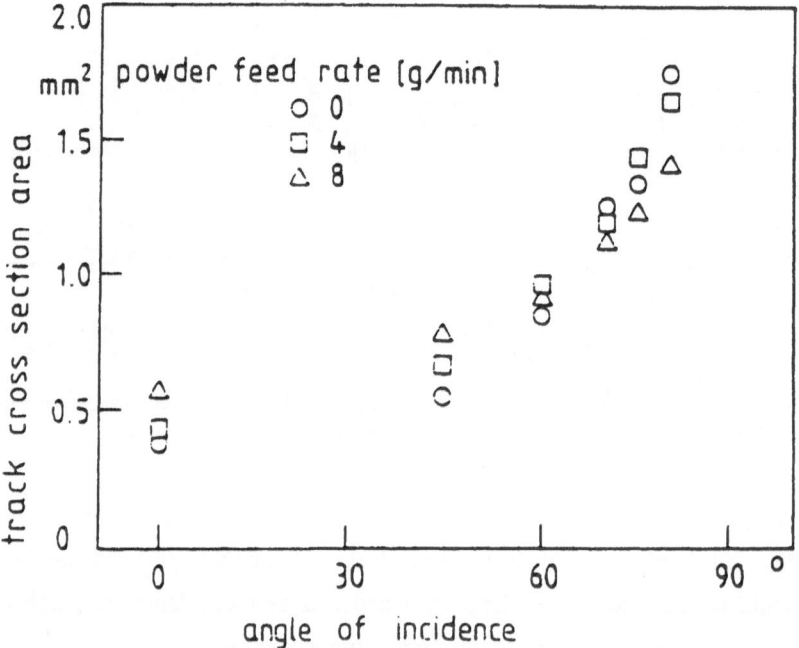

Fig. 11. Effect of coupling rate in dependence on angle of incidence and powder feed rate.

or facet-integrators due to their restriction to a single focal plane.

In hardening experiments, the linear polarized laser-beam of a CO_2-laser with 5 kW maximum output power and a laser-mode near TEM_{01*} was focussed with a mirror optic of 1 m focal-length onto the surface of a perlitic steel (C 45) [5]. The experiments were done with an angle of incidence of 75°. Scanning direction and plane of incidence were chosen perpendicular to each other, with the polarization direction in the plane of incidence. Due to the long focal-length of this optic, the whole irradiated zone was in a region with nearly constant beam diameter of 2.8 mm. The experiments were performed with uncoated, cleaned surfaces and with graphite coated surfaces. With the uncoated surfaces a couple of attempts were made, using a shielding gas, either argon or nitrogen. At each scanning velocity, the maximum laser power with which the surface could be irradiated without surface melting was determined. These samples were examined and the results are presented in Fig. 9 top. It can be seen, that the lowest power is needed with graphite coating. Nearly double that value was required without coating and without shielding gas (i.e., oxidizing atmosphere!). With shielding gas at least 3 to 4 times more power was necessary. The achieved hardening depths at the same velocity but different surface conditions were all at the order of 0.5 mm due to the same limiting condition (i.e., melting temperature at the surface and given spot size).

These investigations have demonstrated that, by applying linearly polarized laser light at large angle of incidence, it is possible to achieve hardening with a CO_2-laser, even without coating. Further examples are given by Shibata [6] and Miyamoto [7]. It has to be noted, however, that more energy is needed.

As in hardening, the energy coupling can be enhanced also in remelting processes by using oblique incidence of parallel polarized CO_2-laser light. This is demonstrated in Fig. 10 showing an increase of the molten area by a factor of approximately five when the angle of incidence is increased from 0° to 75°.

The same positive effect is found in cladding and alloying. As seen from Fig. 11, the coupling rate, and hence, the cross section of the tracks substantially increases with growing angle of incidence. For angles below 45° an improved energy coupling is obvious due to the powder.

3.1.2. *Use of lasers with shorter wavelengths*

As already pointed out in chapter 2, the absorptivity depends highly on the wavelength of the laser. For metals, in general, absorptivity increases with shorter wavelength. Today, Nd:YAG-lasers are available in the power range of more than 2 kW and, therefore, may be considered as candidates for laser surface processing, as well. Absorptivity of a bright steel surface is around 30 % for Nd:YAG-lasers (at melting temperature), about three times more than for CO_2-lasers. If oxidation of the surface is admitted, the absorptivity increases to values of about 60 %.

The results of successful hardening experiments with a cw-Nd:YAG-source are presented in Fig. 9 bottom. A defocused beam of 7.5 mm spot diameter was used at normal incidence. Both surface conditions, coated and uncoated, were investigated. Corresponding to the experiments discussed along Fig. 9 top, the maximum power, respectively intensity was searched at which no melting occurred. The results demonstrate that hardening without coatings at normal incidence is possible with a wavelength of 1.06 μm. On the other hand, the beneficial effects of oxidation and coating are still present. - The fact that the intensity in the case of the graphited surface approaches that of the uncoated one is an indication of the degradation (burning away) of the layer at long interaction times (see also Fig. 6).

The coupling rates achieved in remelting with CO_2- and Nd:YAG-lasers are compared. As expected, the laser with the shorter wavelength yields a far higher energy coupling.

3.2. ADAPTED INTENSITY DISTRIBUTION

3.2.1. *Intensity profile for homogeneous surface temperature*

Today, customary beam shaping systems are utilized that are producing a homogeneous intensity on the surface of the workpiece. Since the intensity distribution influences the hardening result to a large extent, theoretical attempts were made to understand this effect and to improve the hardening process. The principal statement of [8] is that a most advantageous intensity distribution is such one yielding a homogeneous temperature inside the irradiated area. Consequently, not only case depth would be maximal, but also the process efficiency, i.e., the hardened material volume per unit input energy.

First, the effects of different beam shaping shall be demonstrated by calculations out of [8, 9]. Fig. 12 shows an intensity profile for homogeneous temperature distribution in detail. Obviously, higher intensity values are necessary in regions where cooling by heat conduction occurs. The steep intensity rise at the front of the laser spot is required to heat up the material almost instantaneously to the required temperature. The higher elbow-rest regions of this armchair like distribution equalizes the lateral losses by heat conduction. The calculated case depths for different intensity distribution are summarized in Fig. 13 from where the following conclusions can be drawn:

- case depth becomes more shallow at increasing travel speed for all intensity profiles,
- case depth depends on intensity distribution,
- case depth is maximum for homogeneous temperature field at the surface,
- for a given case depth, the travel speed is maximum for the optimized profile, i.e., resulting from a homogeneous temperature field.

An efficiency quantification of the transformation hardening process may be made by the ratio of transformed material volume per input energy. Out of the calculations it can be seen, that for a given intensity profile the energy consumption per hardened volume generally increases with velocity, but is least for an intensity distribution that induces a homogeneous temperature field. When using a modified facet-integrator mirror, for example, a profile close to the ideal one could be created [10]. In this way, significantly deeper tracks were produced than with a homogeneous distribution, see Fig. 14.

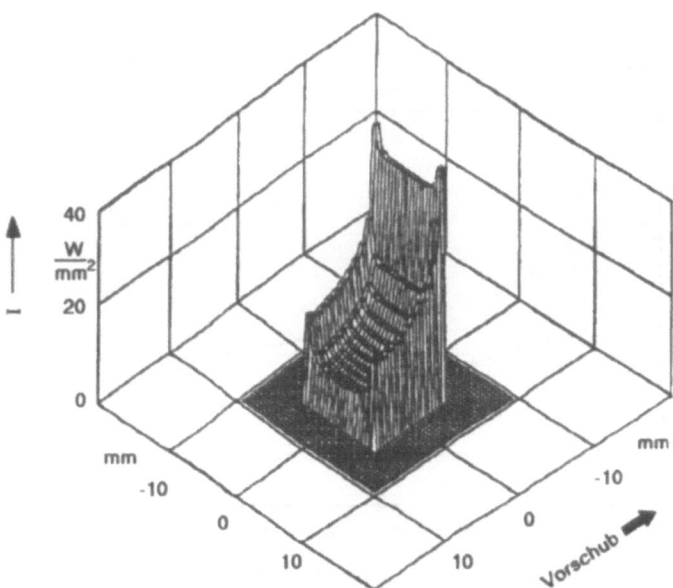

Fig. 12.   Detailed view of calculated intensity distribution for homogeneous temperature inside the illuminated area after Burger. Approach in experiments is by using facette or oscillating mirrors.

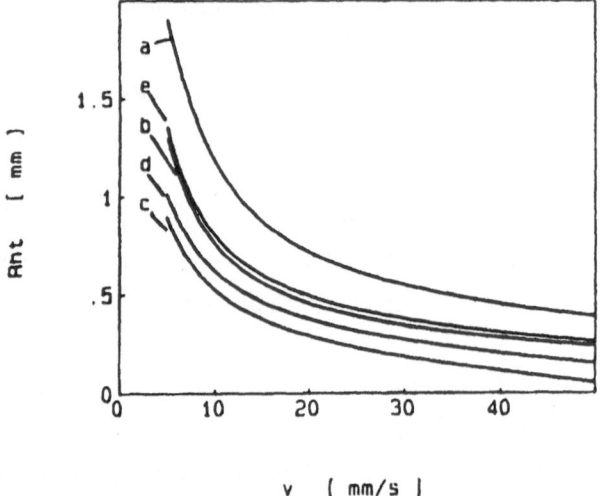

Fig. 13. Hardened case depth, calculated for different intensity profiles (size:10 x 10 mm²) after Burger. Laser power was chosen to reach 1450 °C as maximum temperature.

a →   special profile for homogeneous temperature distribution
b →   homogeneous profile
c →   profile of an oscillator system (beam diameter 1 mm, ampl. ± 4mm).
d →   profile of an oscillator system (beam diameter 2 mm, ampl. ± 4mm).
e →   profile of an oscillator system with sinusoidal-triangular movement.

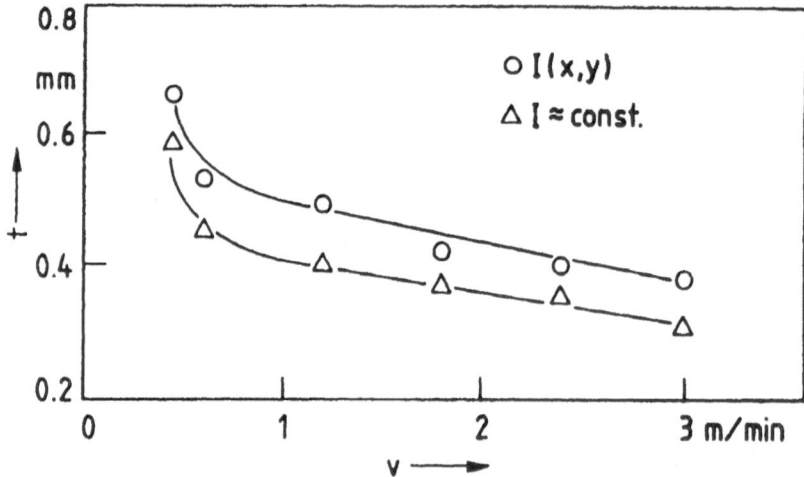

Fig. 14. Experimental results of case depth in dependence on travel speed for homogeneous and optimized intensity distribution (graphite coating, no surface melting).

3.2.2. *Adaption of intensity profile to workpiece geometry*

A flexible beam shaping system has been developed at the IFSW to meet the needs described in chapter 3.2.1. Therein the incoming laser beam is focused through an optic with long focal length (f = 1m). Then it is deflected in two directions by plane mirrors that are mounted on galvanometer scanners. Simultaneously to the scanning of the laser beam, the output power of a RF-exited 5 kW CO_2-laser is changed using a specially designed electronic unit that directly controls the RF-generators. The movements of the galvanometer scanners and the laser power are controlled by a single microcomputer containing a digital to analog conversion board. So the intensity distribution at the surface can be shaped like a TV picture [11, 12].

Intensity shaping for plane surfaces.   To demonstrate the effect of beam shaping on the hardening results, three different shapes were chosen and realized. The size of all three patterns was 10 mm x 10 mm at a normal incidence. The focal spot was positioned at the surface of the workpiece, with a diameter of 2 mm. The first profile was formed to achieve a rectangular uniform distribution, whereas the second and third profile were shaped to obtain distributions with a maximum at the leading edge of the beam spot. During the hardening experiments (graphit coated mild steel type C45) with the profiles 2 and 3 the average laser power was kept constant at 1800 W whereas with profile 1, it was varied in a range between 1500 W and 2400 W, with the higher power at the higher traveling speeds. It appears that profile 3 (with the highest maximum) is best for deep hardening at low speeds whereas profile 1 delivers best results at higher speeds. These findings confirm the theoretical approach addressed in chapter 3.2.1 according to which the intensity maximum at the leading edge of the profiles has to be a function of travel speed.

Intensity shaping for complex geometries.   The particular advantage of this type of beam shaping optic lies in its flexibility for hardening of components with complex geometries [13]. The time to achieve a required hardening result is reduced due to the degrees of freedom that are given by this approach.

One example is hardening along a *90 ° edge*. The requirements for the hardened zone are similar to those of gear and rack teeth, with a hardened layer at both flanks and no melting of the edge. With a conventional integrating optic this is possible only by producing two adjacent tracks on each flank, with the disadvantage that the first track will be annealed when producing the second track. Alternatively, the laser beam can be

36

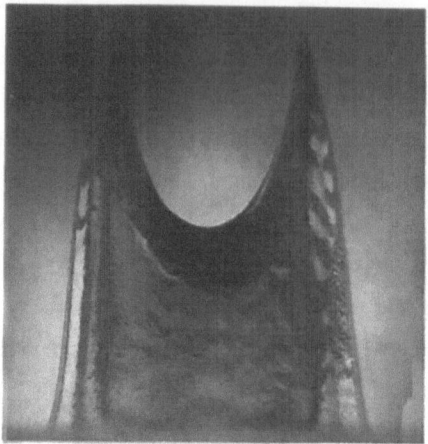

Fig. 15.   Shaped intensity distribution for instantaneous hardening of both sides of teeth or edges.

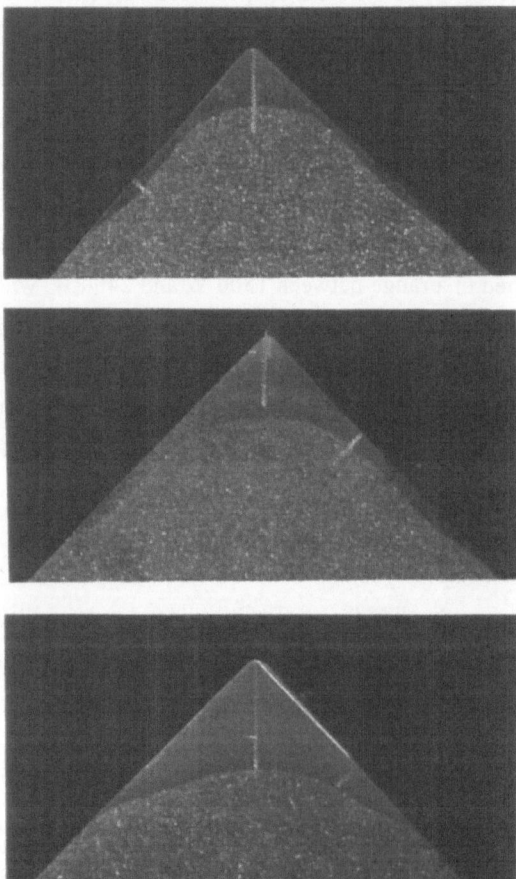

Fig. 16.   Different possibilities to form a hardened zone, when employing an intensity profile similar to that of Fig. 15.

split into two beams to treat both flanks, simultaneously. For this split-beam technique, a specially designed optic is required [14].

With the flexible beam shaping optic an intensity distribution was achieved, where the center can be positioned directly onto the edge with an angle of incidence of 45° on the flanks. The profile, see Fig. 15, was shaped to yield maximum power at the sides of the distribution and less power in the middle to prevent the edge from melting. The optics was handled by a robot that travelled the pattern along the workpiece and generated the necessary orientation of the beam. The hardening was done with a graphite coated C45 mild steel. Typical hardening zones depending on spot size and travel speed are shown in Fig. 16.

If standard ring geometries like *ball seats* are to be hardened, there will be an overlapping zone. To prevent the inevitable annealing effects there, different strategies are possible. It would be ideal if a ring shaped intensity distribution were available with the dimensions of the area that has to be hardened. In this case, the hardening zone could be heated as a whole. When rotating the workpiece very quickly, one comes extremely close to such an ideal distribution even with standard beam shaping. This process strategy, however, cannot be employed for asymmetric workpieces or such with large volume. With the flexible beam shaping, the problem is solved by fixing the workpiece, and letting the laser beam "rotate" fast on the circular path.

As an example, the ball seat of a hydraulic component was hardened by this strategy. With the laser beam a cycle with the outer diameter (7 mm) of the ball seat was formed, see Fig. 17, and the hardening was done by heating the whole seat, at once. This was done in an irradiation time of 0.8 s with 3000 W laser power. The material was already heat treated before, so an oxide layer gave sufficient absorption enhancement for the CO_2-laser. A hardened depth of 0.5 mm was reached (Fig. 18).

3.2.3. *Aspects of intensity profiling after fiber transmission*
When applying Nd:Yag-lasers, the opportunity is given to use transmissive fibers for beam guiding. Due to multi-reflections, the laser beam comes out of a step-index fiber with a round and homogeneous distribution. This "beam shaping" property of fibers can beneficially be used for laser processing, and a variety of simple concepts was examined; Fig. 19 gives an overview.

Hardening can be done without any additional optics when using a glass fiber directly for material processing. Due to its small dimensions, small and complex geometries become accessible for laser treatment. Disadvantageous of this way of

38

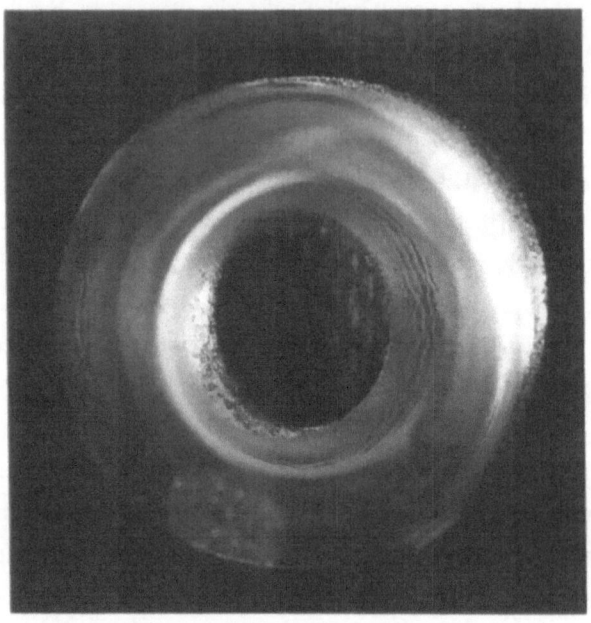

Fig. 17.   Shaped intensity distribution for hardening ring geometries.

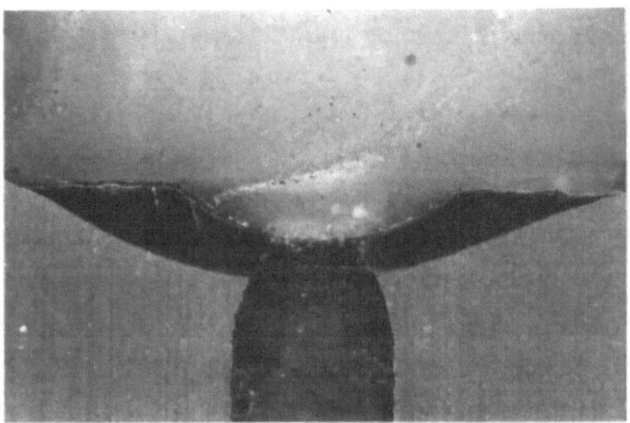

Fig. 18.   Cross section of a ball seat, hardened with the profile of Fig. 17.

processing is the fact that only a very short distance between fiber end face and working area can be admitted. Otherwise, the spot diameters would become too large because of the considerable divergency of the laser radiation behind the fiber (in the range of 70 to 100 mrad).

When using one single focussing lens, the laser beam can be collimated behind the fiber. In this way, the beam diameter can be kept practically constant over a range of several hundreds of millimeters distance. This means that processing conditions do not change within the above-mentioned range, a large "focussing depth" is obtained. For the experiments shown in Fig. 9 bottom, a focussing optics with internal collimation was used. For this optical concept, an image of the fiber endface can be found in the focus. Consequently, the intensity distribution in the focus is the already mentioned "top hat" profile. Since for hardening purposes larger spot diameters are required (larger than fiber core diameter!), the optic has to work out of focus. That is why the spot diameter of 7.5 mm in Fig. 9 bottom becomes possible. In this case, however, the intensity distribution found at the workpiece in a certain distance from the focal, respectively image plane, is Gaussian like.

A further optical concept is a telescopic imaging optic designed at the IFSW. Its purpose is to project a magnified image of the "top hat" distribution onto the workpiece surface. The magnification factor ranges from 1 to 16. Examples of measured distributions are shown in Fig. 20. Here, however, focal plane and image plane do not coincide.

After beam analysis hardening was done with a continuous-wave Nd:YAG-laser on mild steel C45. No inert gas or absorbant coatings were employed and surface melting was avoided. As the Gaussian like intensity distribution of the free propagating beam is identical to that cases of the defocussed beam, the hardening results in both cases are the same. The comparison of focussing optic and telescopic imaging optic, however, shows some differences in hardening results caused by different intensity profiles. As shown in Fig. 21 at the same travel speed and with the same laser power, wider tracks can be hardened with homogeneous profile; the tracks, however, are more shallow.

3.2.4. *Beam Combining*

Using special focussing devices, it is possible to combine two independent laser beams stemming from two separate sources. By such an arrangement it is possible to individually vary the beam properties (power, polarization, focal parameters, mode of operation) yielding a wide variety in possible space and time dependent intensity

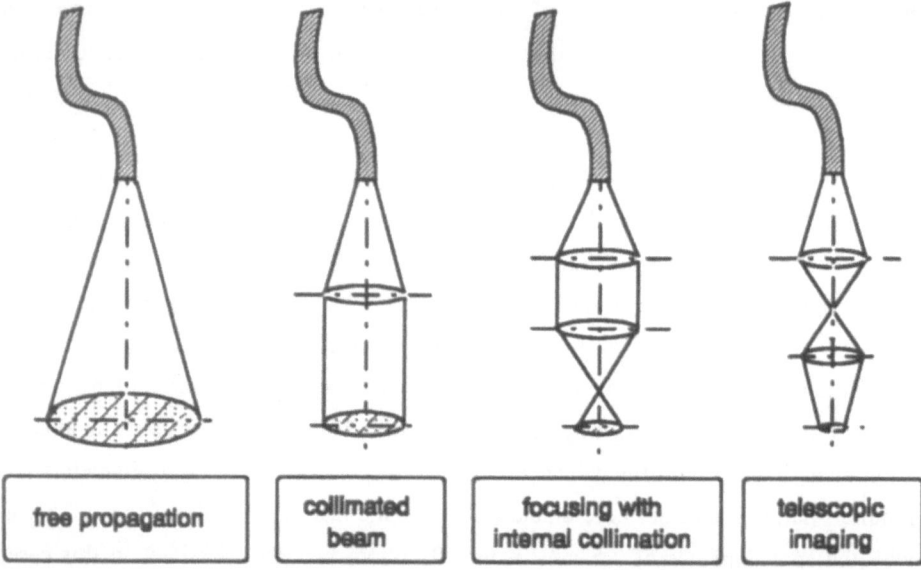

| free propagation | collimated beam | focusing with internal collimation | telescopic imaging |

Fig. 19.   Simple beam shaping concepts after fiber transmission of Nd:YAG-laser.

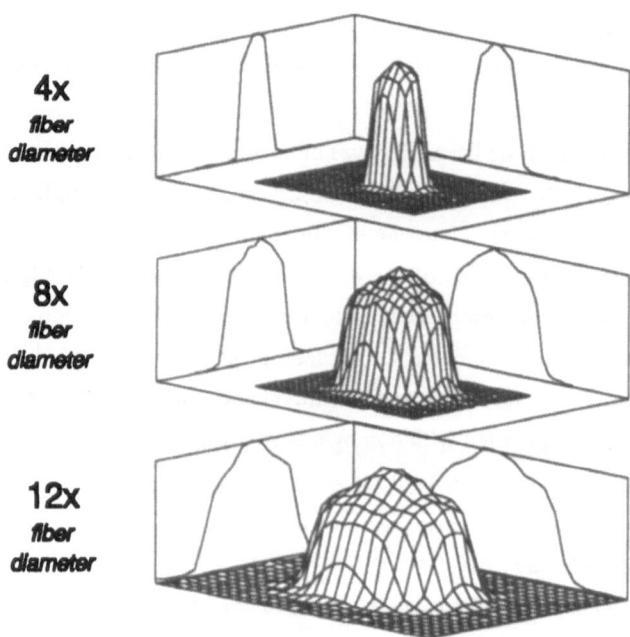

4x
fiber
diameter

8x
fiber
diameter

12x
fiber
diameter

Fig. 20.   Measured intensity profiles obtained by telescopic imaging optic (core diameter 0.4 mm, magnification 4, 8 and 12).

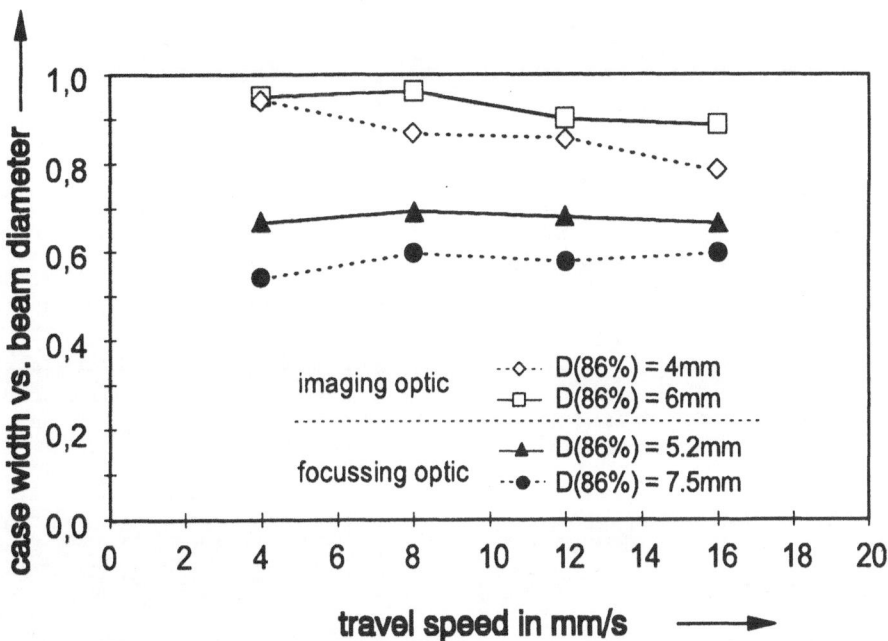

Fig. 21.   Relation of case width and beam diameter as a function of travel speed for focussing optic and imaging optic.

distributions at the workpiece. Hence, this technique is a most useful tool to study the effects of intensity distribution and resulting temperature cycles on process efficiency and quality in laser surface treatments. In [15] such an apparatus has been presented allowing to combine two high-power CO_2-lasers.

Beam combining for high process efficiency.   One possibility is the combination of the two laser beams with same diameters in such a way that a focal area is produced with widths in the range of up to 16 mm, see Fig. 22. For cladding with this intensity distribution, an optimal adjustment of the powder-gas stream to the resulting melt pool is necessary in order to get a good track quality, a high coupling rate and a high cladding efficiency. A rectangular nozzle with inside dimensions of 3 x 10 mm, resulting in a high powder efficiency was used. The cladding tracks show a homogene-ous shape and - at optimized parameters - they are characterized by minimized dilution and high cladding efficiency (> 80 %). In Fig. 23 a micrograph is presented of the cross section of a single track (Stellite 21) produced with an average laser intensity of 6.5 kW/cm², a powder load of 40 g/min and a traverse speed of 0.2 m/min. The track shows a width of 14 mm, a build-up of 1.5 mm and a dilution with the substrate below 3%.

42

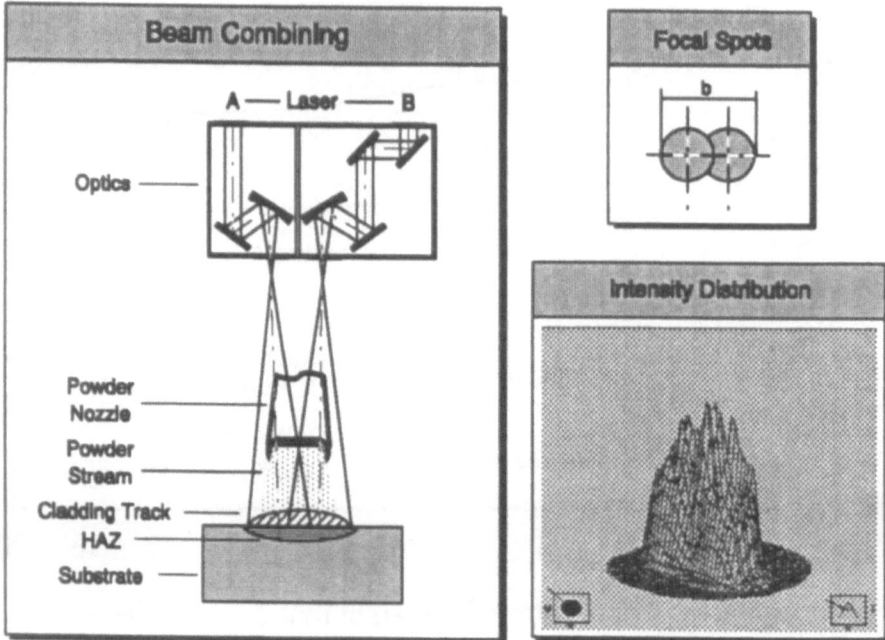

Fig. 22   Beam combining optic for CO_2-lasers and resulting intensity distribution for cladding with high efficiency.

Fig. 23.Cross section of a single track of Stellite 21 produced with a wide focal area (b = 15.1 mm) as shown in Fig. 22 and a broad powder nozzle (3 x 10 mm).

Beam combining for simultaneous postheating.   By cladding with heterogeneous powder mixtures of carbide particles (WC, TiC) and metallic binders (Co- or Ni-based) it is possible to produce ceramic/metal composite layers showing high resistance against abrasive and erosive wear. Due to the reduced ductility of such layers, however, the high local temperature gradients during the cladding process result in high thermal stresses exceeding the rupture strength and leading to the formation of cracks in the layers. An effective reduction of the local temperature gradient and of the crack formation can be achieved using a second laser beam for simultaneous postheating [16].

By variation of the distance between the processing and the postheating beam, as well as of the laser power and the diameter of the second beam, the resulting intensity distribution can be varied. In Fig. 24 some configurations of the resulting intensity distribution are shown. For a powder mixture of Stellite 21 and 50 wt. % WC, the effect of the intensity distribution on the time-temperature profile at the layer surface in traverse direction is exemplarily shown in Fig. 25. Due to the high temperature gradient in the region of solidification for the one-beam cladding process the produced composite layers are characterized by hot cracks. The simultaneous postheating yields a constant temperature level in the critical temperature range. Its height and extension, of course, are strongly dependent on the intensity distribution. With an optimized temperature profile crackfree claddings can be produced. This is

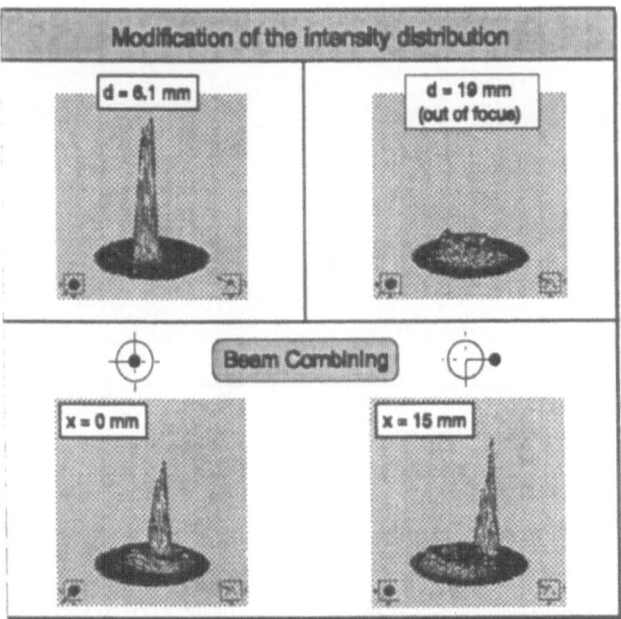

Fig. 24.   Beam combining and resulting intensity distribution for simultaneous postheating.

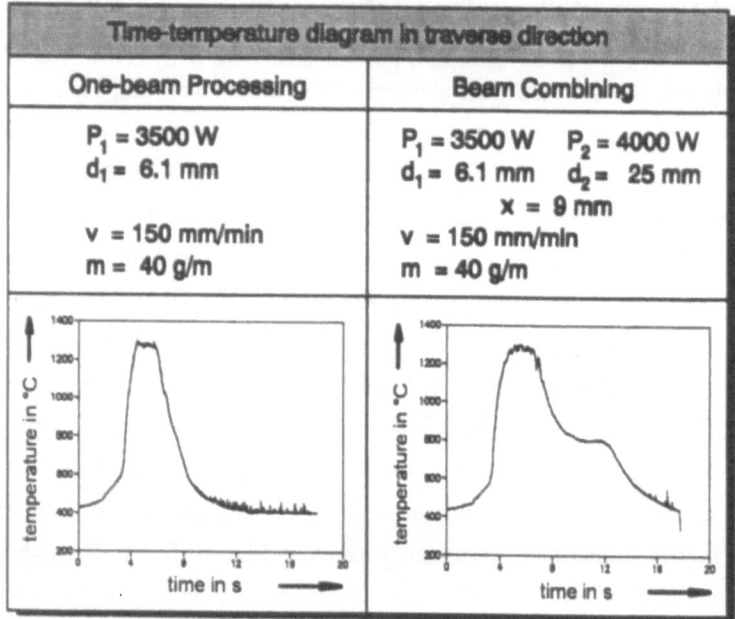

Fig. 25. Comparison of the time-temperature profile in travel direction measured with a pyrometer.

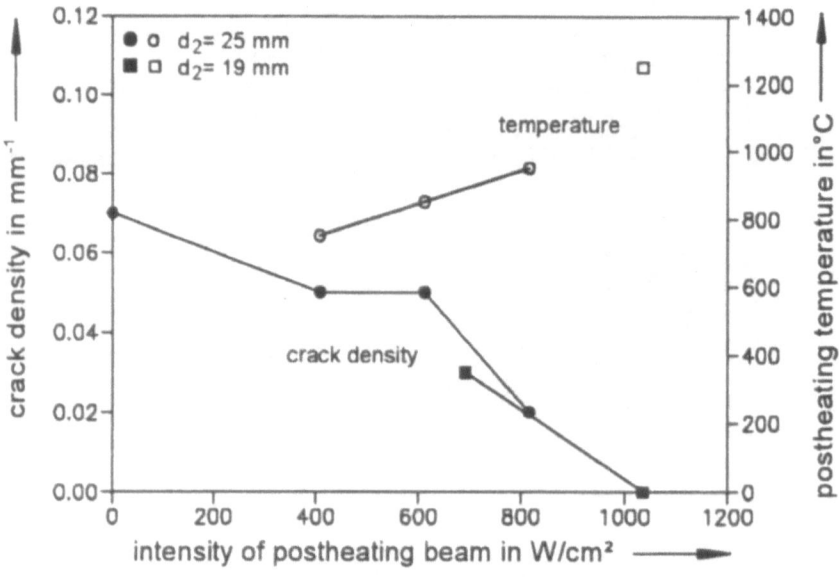

Fig. 26. Temperature in the postheating 'plateau' and crack density (number of cracks per unit length) versus intensity of the postheating beam (d_2).

demonstrated in Fig. 26 showing the influence of the intensity of the postheating beam on the postheating temperature plateau and the crack density. With increasing intensity the postheating temperature rises resulting in a linear decrease of the crack density. For the conditions presented, the cladding layers become crackfree for a postheating temperature of 1200 °C.

4.   References

1.   DAUSINGER, F.; SHEN, J.: *Energy Coupling Efficiency in Laser Surface Treatment*. ISIJ International, 33 (1993), p. 925 -933.

2.   HAAG, M.: CO_2 *Laser Absorption Characteristics of Metal Powders*. Universität Stuttgart, diploma thesis, 1994 (Inst. f. Strahlwerkzeuge: IFSW 94-9).

3.   DAUSINGER, F.; RUDLAFF, TH.:: *Novel Transformation Hardening Exploiting Brewster Absorption*. In: Arata, Y. (Ed): Proc. of Laser Advanced Materials Processing (LAMP), Osaka, 1987. Japan: High Temp. Society, 1987, p. 323-328.

4.   DAUSINGER, F.; BECK, M.; LEE, J.H.; MEINERS, E.; RUDLAFF, T.; SHEN, J.:*Energy Coupling in Surface Treatment Processes*. Journal of Laser Applications 1 (1990) No. 3-4, p. 17 - 21.

5.   RUDLAFF, TH.: *Arbeiten zur Optimierung des Umwandlungshärtens mit Laserstrahlen*. Stuttgart: Teubner-Verlag, 1993, Forschungsberichte des IFSW.

6.   SHIBATA, K.; SAKAMOTO, H.; MATSUJAMA, H.: *Absorptivity of Polarized Beam During Laser Hardening*. In: Proc. of the Intern. Congress of Lasers and Electro-Optics (ICALEO 91), Orlando, Florida (USA), 1991. Laser Institute of America (LIA), 1991, p.409 - 413.

7.   MIYAMOTO, I.; MARUO, H.: *Novel Laser Beam Shaping Optics:LSV Optics Applications to Transformation Hardening and Ceramic Joining*. In: Proc. of the Intern. Congress of Lasers and Elektro-Optics (ICALEO 92), Orlando, Florida (USA), 1992. Laser Institute of America (LIA), 1992, p. 88 - 102.

8.   BURGER, D.: *Beitrag zur Optimierung des Laserhärtens*. Universität Stuttgart, Fakultät Fertigungstechnik, thesis, 1988.

9.   BURGER, D.: *Optimierung der Strahlqualität beim Laserhärten*. In: Waidelich, W. (Hrsg.):Optoelektronik in der Technik. Proc. of the 9th Intern. Congress LASER 89. Berlin: Springer-Verlag, 1989, p. 492 - 495.

10.   HUONKER, M.: *Untersuchungen zum Laserstrahlhärten mit Facettenspiegel und Strahlscanner*. Universität Stuttgart, Studienarbeit, 1991 (Inst. für Strahlwerkzeuge: IFSW 91-16).

11.   RUDLAFF, TH.; DAUSINGER, F.: *Hardening with Variable Intensity Distribution*. In: Proc. of the 3rd European Conference on Laser Treatment of Materials (ECLAT 90), Erlangen, 1990. Coburg: Sprechsaal Publishing, 1990, p. 251 - 265.

46

12. BLOEHS, W.; RUDLAFF, T.: *Laserstrahlhärten mit variabler Intensitätsverteilung*. In: Schriftliche Fassung der Vorträge zum Fertigungstechnischen Kolloquium am 1./2. Oktober 1991 in Stuttgart. Berlin: Springer-Verlag, 1991, p. 140 - 142.

13. RUDLAFF, TH.; BLOEHS, W.; DAUSINGER, F.; HÜGEL, H.: *Hardening with CO_2-Lasers and Flexible Beam Shaping Optic*. In: Proc. of the Intern. Conference on Laser Advanced Materials Processing (LAMP 92), Nagaoka, Japan, 1992. p. 673 - 678.

14. CANTELLO, M.; CRUCIANI, D.: *Optics for Laser Treatment of Complex Geometry Components*. In: Proc. of the Intern. Congress of Lasers and Electro-Optics (ICALEO 86), Arlington, Virginia (USA), 1986. Laser Institute of America (LIA), 1986, p. 11 - 18.

15. BEA, M.; GLUMANN, C.; GRÜNENWALD, B.; RAPP, J.; GIESEN, A.; HÜGEL, H.: *Flexible Beam Delivery for Improved Laser Applications*. In: Laser Materials Processing: Industrial and Micrelectronic Appliccations. SPIE Vol. 2207, 1994, to be published.

16. GRÜNENWALD, B.; SHEN, J.; DAUSINGER, F.; HÜGEL, H.: *Laser Cladding with Composite Powders Using Pyrometric Temperature Control and Beam Combining*. In: Proc. of the 26th Intern. Symposium on Automotive Technology and Automatisation (ISATA'93), Aachen, 1993, p. 287 - 294.

Non-Equilibrium Synthesis by Laser for Tailored Surfaces

J. Mazumder
Department of Mechanical and Industrial Engineering
University of Illinois at Urbana-Champaign
1206 W. Green Street
Urbana, IL   61801

Abstract

Rapid heating and cooling rate associated with laser processing provides an unique opportunity for non-equilibrium synthesis of materials which can provide compositions with extended solid solution. With proper selection of alloying elements and process parameters, materials with tailored properties can be synthesized. This paper discusses the theoretical basis for the mechanisms and the experimental approach for non-equilibrium synthesis of materials tailored for various applications such as wear, high temperature oxidation and corrosion resistance.

Non-equilibrium partitioning due to rapid solidification is explained based on the atom trapping theory. A mathematical model involving heat and mass transfer and non-equilibrium partition co-efficient is used to predict the non-equilibrium phase diagram and establish a relationship between the process parameters and the final alloy composition. Experimental procedure is discussed and the microstructure and properties of non-equilibrium alloys for various applications are described. Microstructure was characterized using various electron optical techniques such as Transmission Electron Microscopy (TEM), Scanning Electron Microscopy (SEM), Auger Electron Spectroscopy (AES) and Energy Dispersive X-Ray Analysis (EDAX). Room temperature wear properties were characterized by a line contact Block on Cylinder method, high temperature wear tests were similar to simulated engine test. High temperature oxidation properties were characterized by using Perkin-Elmer Thermo-Gravimetric Analyzer (TGA) where dynamic weight change were monitored at 1200°C. Corrosion properties were evaluated by a potentio-dynamic method using a computer controlled Potentiostat manufactured by EG&G.

J. Mazumder et al. (eds.), Laser Processing: Surface Treatment and Film Deposition, 47–75.
© *1996 Kluwer Academic Publishers.*

1. Introduction

Synthesis of nonequilibrium metallic phases has been an area of great interest to the materials processing community since early 1960. Inherent rapid cooling rates in laser processing are being used to engineer non-equilibrium microstructures which cannot be rivaled by other processes. This lecture will discuss the phenomena involved and its application in designing materials with tailored properties.

What is non-equilibrium Synthesis?

This is a synthesis method to produce binary or higher order materials where kinetics of the process affects the transport of the constituent elements during phase transformation resulting in a composition or crystallographic configuration which is different from what is observed when the elements arranges themselves with the lowest possible Gibbs Free energy, which is the equilibrium condition.

Figure 1 illustrates the phenomena. Phase diagram under equilibrium condition is illustrated by the solid line whereas the no-equilibrium phase diagram is represented by the dotted line. One can observe the shrinkage of the phase field under non-equilibrium condition. Any alloy composition between the solidus lines of the equilibrium and non-equilibrium phase diagram will be a non-equilibrium alloys with extended solid solution.

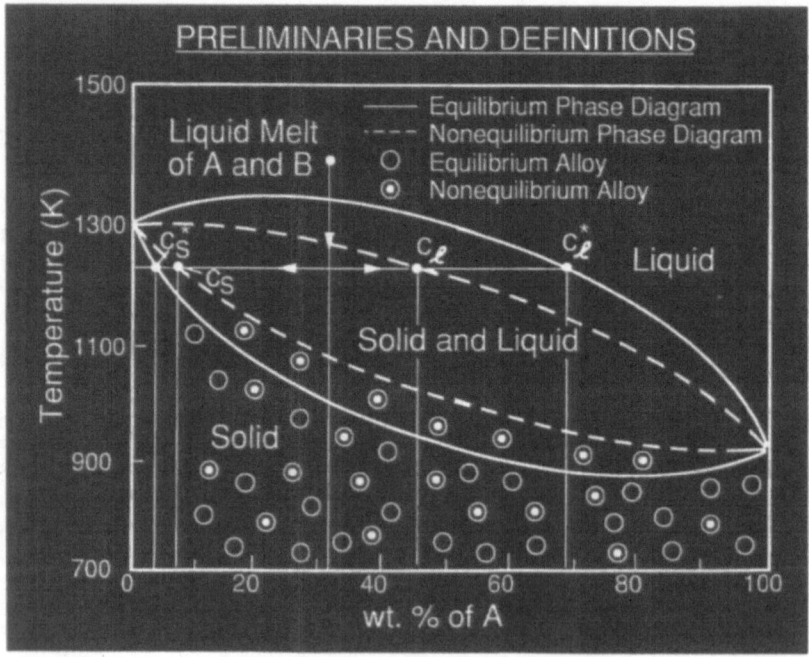

Figure 1. Non-equilibrium phase diagram

Sometimes, besides composition deviation from the normal crystallographic configuration may be a result of non-equilibrium processing. Martensite in hardened steel is an example. Recently, nano-crystalline phases are produced by various techniques, which neither has long range order nor has short range order [1]. Almost 50 Vol.% of these materials are grain boundaries. They are also a product of non-equilibrium synthesis. However, physics of this process needs further investigation before the phenomena can be explained definitively.

Why non-equilibrium Synthesis?

A wide variety of microstructure, including extended solid solution and amorphous phase can be generated. Many surface related failures such as corrosion, oxidation, wear, and fatigue may be minimized or eliminated using these techniques. In this process, compositionally graded structure can be produced to minimize thermal stress due to the difference in coefficient of expansion. Another advantage of this method is that, enormous savings of alloying elements, which are often expensive or strategically important, can be achieved since only the surface is being modified. Figure 2 shows the dependency of USA and European community on foreign imports for commercially important materials [2].

Mineral/Metal	Imports as Percentages of Apparent Consumption						Sources of Imports
	0	20	40	60	80	100	
Manganese				U.S.		95	Brazil, Gabon, South Africa, Australia
			E.C.			100	
Cobalt						97	Zaire, Zambia, Norway, Finland
						100	
Platinum Group					92		South Africa, USSR
						100	
Chromium					89		South Africa, USSR, Turkey, Philipines
						100	
Aluminum (ores and metals)					86		Jamaica, Surinam, Australia, Guinea, Sierra Leone
				82			
Tin					86		Bolivia, Malaysia, Indonesia, Thailand, Zaire, Nigeria
			65				
Nickel				70			Canada, Norway, New Caledonia
						100	
Zinc			55				Canada, Peru, Mexico, Australia
						100	
Tungsten		39					China, Canada, Peru, Thailand, Australia
						99	

Figure 2. Dependence of USA and European Community on Foreign Imports for commercially important elements; numbers represent percent relative to amount consumed [2].

Inherent rapid solidification involved in laser alloying and cladding provides a unique opportunity for non-equilibrium synthesis of novel alloys with extended solid solution. This has been experimentally observed for various binary, ternary and quaternary alloy systems. Compositions of alloys produced by laser aided synthesis often exceeds solid solubility limit far beyond that expected from equilibrium phase diagram. These alloys with non-equilibrium phases offer the possibility of new materials with novel properties. These processes allow the surface properties of a structure to be tailored to the surface requirement of the application without sacrificing the bulk characteristics of the structure.

Moreover, the great control provided by the laser gives laser processing the potential of synthesizing nonequilibrium materials in near net shape. In fact such a goal has recently been realized and the proof of the concept is already available in processes such as Laser Glazing [3]. There are other methods [4] of producing nonequilibrium metallic materials using Rapid Solidification Technology (RST) such as atomization, twin roll quenching, and melt spinning. However, in order to make an engineering component, these materials produced by conventional RST need to be consolidated, but in the consolidation process, temperature and pressure cycles induce phase transition in a metastable material. Thus, laser processing has far reaching implications in synthesis of non-equilibrium materials.

How to make Tailored properties by non-equilibrium synthesis?

The potential of laser beams to produce metastable phases derives from their ability to induce rapid localized heating and melting and almost equally rapid quenching of the melt. Cooling rate up to 10^{14} K/s has been reported [5-7]. Laser processing with cooling rate around 10^5 K/s is quite common [8]. In general, under such high cooling rates, two phases can be produced: glassy phases or metastable crystalline phases with composition and crystallographic configuration different from those predicted by the equilibrium phase diagram. Formation of glassy phases at cooling rate of around 10^6 K/s is only possible for typically eutectic alloys of three or more elements, often including metalloids. Higher cooling rates (10^{10} to 10^{14} K/s) enable glass formation in many binary and ternary alloys. But such versatility comes at the expense of the thickness of the quenched region. In fact, nano-crystalline compounds are being made at such high cooling rate by laser ablation technique [9,10]. However, one of the interesting observation during the formation of nano-crystalline phases by condensation from vapor phase is that the compound are mostly stoichiometric, although atomic lines are in abundance in the ablation plume. Nucleation and formation of stoichiometric compounds with few nano-meter grain size is not clearly understood yet. This lecture will mainly concentrate on the synthesis from liquid phase.

The strategy for non-equilibrium synthesis for tailored properties are as follows:

1) Selection of Phases:
   For example, Face Centered Cubic (F.C.C.) structure with large number of available slip planes will be beneficial for ductility whereas brittle non-cubic phases with limited number of available slip planes will promote hardness and wear resistance. A combination of them with duplex phases are often needed to provide adequate toughness during service with reasonable wear resistance. For desired mechanical properties selection of phases is extremely important.

2) Selection of Elements
   It is important both for promoting certain phases as well as protection against chemical degradation such as corrosion. For example Chromium promotes Body Centered Cubic(B.C.C) phase for ferrous alloys and Cr_2O_3 forms passive layer to inhibit corrosion at temperature up to 800^oC. Reactive elements such as Yttrium and Hafnium are known to stabilize Al_2O_3 at temperature above 800^oC, leading to high temperature oxidation resistance.

3) Selection of Process parameters.
   Although it is well known that Y and Hf can improve high temperature properties, but dissolution of these high atomic number reactive elements are rather difficult. Processing techniques promoting atom trapping and extended solid solution are key to the incorporation of such alloys. As explained above, inherent high cooling rates and strong convection associated with laser melting and solidification are helpful for this purpose. The important consideration is the synthesis of information from phase transformation kinetics and processing transport phenomena so that the chosen process parameters promote the desired phases and chemistry.

What was done before?

Considerable progress has been made in developing tailor made materials by Laser Surface Modification (LSM) in this decade. Laser Surface Alloying (LSA) and Laser cladding techniques are two major LSM techniques for synthesis of unique materials. Review articles by Draper [11,12] and Steen [13] provide an excellent survey of the field. The author and his co-workers at the University of Illinois at Urbana-Champaign in the past few years developed corrosion resistant Fe-Cr-Ni alloys [14], wear resistant Fe-Cr-C-Mn alloys [15] and oxidation resistant Ni-Cr-Al-Hf alloys [16] using LSM. Results of the work carried out by the author and his group on non-equilibrium synthesis for tailored properties are summarized in an earlier paper [17]. This paper examines recent examples for the theoretical and experimental studies for design and synthesis of alloys by laser processing for wear, high temperature oxidation resistance and corrosion .

2. Experimental Procedure

As mentioned earlier high cooling rate can either be achieved by self quenching during rapid solidification of a laser melted pool or condensation from vapor phase. The first is generally associated with laser surface modification and the later is achieved by laser ablation. Synthesis of tailored materials from vapor phase using laser ablation is at its infancy. However, formation of nano-crystalline phases and congruent evaporation of compounds by this technique have generated considerable interest in laser ablation.

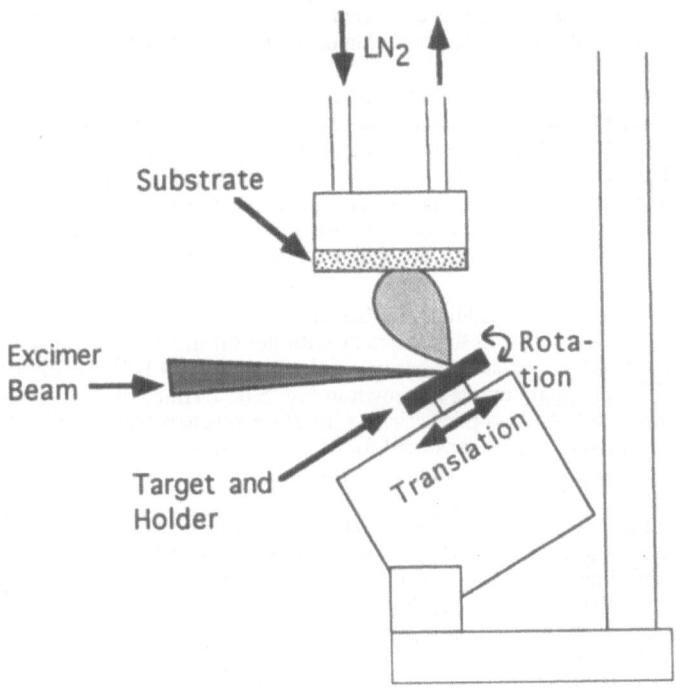

Figure 3. Laser ablation set up with the plume

Recently nano-crystalline NbAl3 powders were synthesized by laser ablation by the authors and co-workers [9]. NbAl3 was chosen as the material for this study because it exhibits a high strength to weight ratio and good oxidation resistance, but is brittle in conventional forms. Nanocrystalline

$NbAl_3$ holds good promise of retaining the former characteristics while rectifying the latter. The concept is the high volume fraction of grain boundary will provide ductility by grain boundary sliding. An excimer laser beam is focused to a small area on a $NbAl_3$ target creating a plasma/vapor plume. The ablated material interacts with the background gas in the processing chamber, condensing into particulates. These are collected on a cooled substrate, which can be extracted from the chamber and scraped to recover the nano-phase powder. The laser ablation experimental set up is shown in Fig. 3. Various values of excimer pulse energy and background (helium) gas pressure were studied. Materials studies were performed to establish the variation in properties of the synthesized materials with change in processing conditions. Microstructural characterization and chemical analysis of nanocrystalline powders were performed using a transmission electron microscope equipped with an energy discursive spectroscometer (EDS). TEM photographs were used to measure the particle diameters. X-ray diffraction patterns and analyzed to reveal the crystal structures. Deposition substrates were weighed before and after a processing run to evaluate the process yield.

During the nano-crystalline synthesis, laser absorption spectroscopy was employed to measure various physical properties of the neutral niobium atoms within the plume, as shown in Fig 4. A probe beam from a cw, tunable wavelength dye laser is directed through the plume. The absorption strength may be used to calculate atomic density. Probing of population densities of different energy levels provides data on the electron temperature. The high resolution of the dye laser employed in these experiments permits accurate resolution of the spectral line shape, which gives data on the velocity distribution of species within the plume via the Doppler effect.

While ablation is relatively recent, laser cladding and alloying processes have been employed for almost two decades for synthesis of novel materials. Elemental powders of various sizes (2 - 44 μm) were mixed in order to obtain a powder mixture of the desired composition of the cladding. The powders were mixed in the required proportion and then mechanically stirred before drying. The drying process is carried out by heating the powders in an open boat in a tube furnace at 200°C with a steady flow of argon for about eight hours followed by slow cooling to room temperature. Prealloyed powders are also used sometimes, depending on their availability.

The experimental apparatus for laser cladding consists of two units working simultaneously. The laser system, first unit, produces a beam that interacts with the substrate and the powder to form the clad. The cladding process was carried out using either an AVCO HPL 10 kW continuous wave CO_2 laser with F7 Cassegrain optics as shown in Fig. 5 or a Trmpf 6KW Rf excited CO_2 laser with parabolic mirror as focusing optics. Preheating stage was necessary for cladding of Fe-Cr-W-C alloys and pressurized chamber was used for Mg-Al cladding. The cladding process and its mechanism is discussed in earlier publications [17,18].

Microstructural characterization of the clad material was performed using optical microscopy, Scanning Electron Microscopy (SEM), (JEOL 35C, HITACHI S-800) Transmission Electron Microscopy (TEM), (Philips 400, 420, and 430T) and STEM (Vacuum Generators, HB-5) equipped with an Energy Dispersive X-ray (EDX) microanalysis system. The EDX spectra were obtained using a stationary probe approximately 10Å in diameter. X-ray intensities were converted into weight fractions and atomic fractions using a standardless thin film analysis software program available at the materials research lab of the University of Illinois. Special precautions were taken to obtain data from very thin regions of the sample. It is, however, possible to improve on the accuracy of the data obtained by the above technique by using thin foils of stoichiometric intermetallic line compounds of elements used in this system and then comparing the EDX intensities [19].

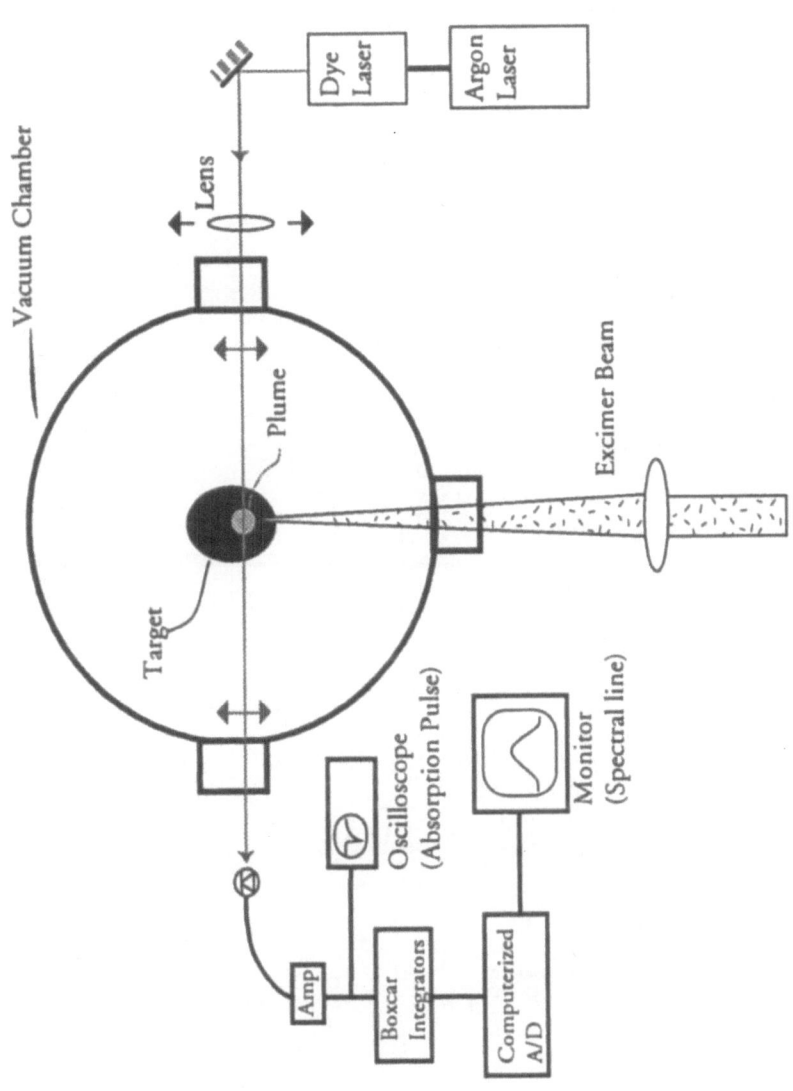

Figure 4. Absorption spectroscopy apparatus for diagnostic of laser ablation deposition of nano-crystalline Nb Al₃

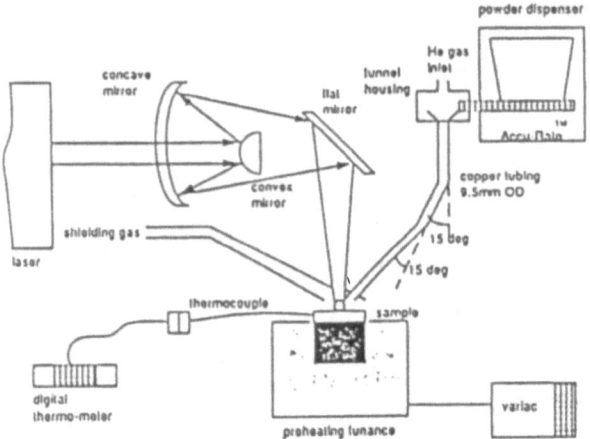

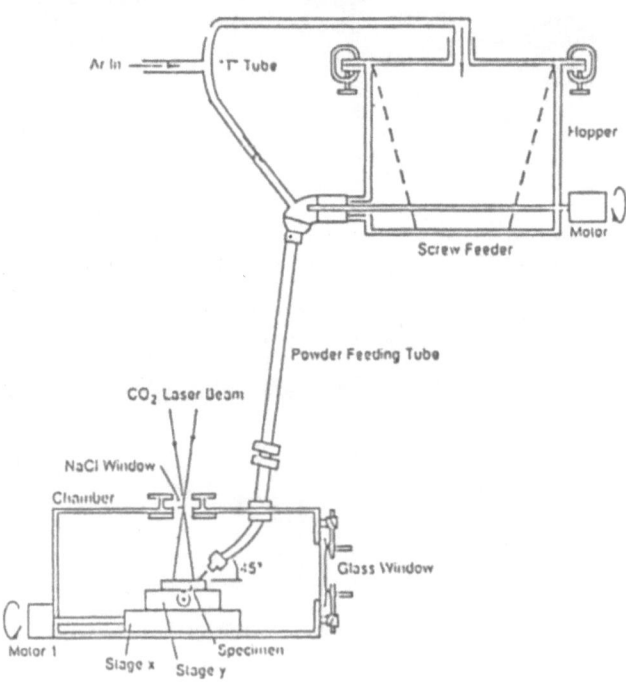

Figure 5. Optics and general layout for laser cladding.

The substrate material was abraded off the cladding and then finally thinned mechanically to a thickness of 0.3 mm. Three millimeter discs were punched out from these strips. A twin jet chemical polishing technique (Fischione) was utilized for TEM sample preparation.

A preliminary friction and wear test for Fe-Cr-W-C alloys were performed with a block-on-cylinder machine. Before the wear test, a calibration was done with a multimeter and a chart recorder by applying the load which was known, and then a chart was made. The test was monitored with measuring the voltage change by the multimeter under the load given. The load was gain-controlled by a load transducer controller. Similarly, the friction force was calculated with measuring the voltage change by a multimeter and a friction force transducer controller. The load applied on the lever was proportional to actual test load at the block-on-ring contact to 10:1, that is, 1 lb applied on the lever is equivalent to 10 lbs actual test load. A disk which was used for the test was set at 24 rpm (0.032 m/s) was performed for 1 hour per sample under the boundary lubrication condition. The lubricant was used a straight paraffinic mineral oil, Sunvice 21. During the operation, temperature and humidity were checked and controlled. Simulated engine tests were carried out to study the wear properties of Cu- alloy and Nickel alloys on Aluminum.

Oxidation resistance of the Ni-Cr-Al-Hf claddings and the substrate was measured using a Perkin-Elmer TGA-7 analyzer with computerized system control and data acquisition capability. The analyzer itself during operation consists of a furnace surrounding a Pt sample pan that is suspended from a microbalance by a Pt wire. Samples are stood up on their width and held between two Pt retaining wires (see Ref. [20]). Three tenths ml/s of dry air flows past the sample and 0.6 ml/s of Ar is used to protect the microbalance during operation. Since the Ar outlet from the balance is concentric to the air inlet to the furnace, some dilution of the air occurs. While the amount of dilution near the sample is not known, the conditions were identical for each run. The sample is heated from ambient at 100°C/min to 1200°C and held at 1200°C for eight hours. In situ cooling is then performed at a rate of 10°C/min to 950°C before cooling to ambient at a rate of 25°C/min. Details of oxidation test are available in Ref. [20].

Thermogravimetric TGA samples are sectioned from the claddings using a low speed saw with a silicon nitride blade. Following sectioning, all of the faces of the sample are abraded progressively through 600 grit using wet SiC paper. The samples were cleaned successively with acetone and trichloroethane prior to testing.

The surfaces and cross sections of the TGA samples were examined using the Physical Electronics model 595 Auger Microprobe. The surface of an as-oxidized broad face is scanned both before and after approximately 10 minutes of sputtering with Ar ions at 3 kV. A cross section was taken perpendicular to the cladding direction at a distance of ~ 1.25 mm from one end of the oxidized sample. (A sputtering time of 30 minutes at 3 kV was required to clean the cross-sectional samples because of the presence of the epoxy mounting material.) The amplitude of the counting peak for a characteristic electron energy is proportional to the concentration of the corresponding element. By dividing the amplitude by the appropriate sensitivity factor, the relative amounts of different elements are found. Photomicrographs of the entire cross section exposed were taken on an optical microscope at 250x to allow measurement of the thickness of unoxidized material.

Differential Thermal Analysis (DTA) of Ni-Cr-Al-Hf alloys were also performed in order to obtain the γ' dissolution temperature of the clad and the substrate material. The DTA used was Perkin-Elmer system 4/7, with a high temperature cell attachment. Small pieces of the clad material weighing between 10.0 and 70.0 mg are put into a sample cup which then is filled with Al_2O_3. The reference cup is filled to approximately the same level with Al_2O_3 before the DTA run. The heating rate is 20°C/min, and a flowrate of argon gas of about 45 ml/min was used as the atmosphere. Similar tests were also carried out for Mg-Al alloys.

The corrosion behavior of laser clad $Mg_{27}Al_{73}$ was evaluated by using a EG&G PARC Model 351 corrosion measurement system. The cleaning procedure and operation sequence could remarkably influence the results of corrosion tests. A fresh surface was obtained by abrading the sample on 320-600 grit silicon carbide paper, rinsing in deionized water and ultrasonic cleaning for five minutes in acetone. If necessary, the samples were cleaned with acetone or methanol again. After the corrosion cell was set up, the 3.5% NaCl electrolyte was added to the cell. The test was started as soon as the whole surface of the sample was immersed. For the sake of comparison, corrosion tests were also performed for Mg and AZ91B samples also. To allow comparisons of different materials, every test was run using the same procedure and the same parameters. The details of the tests are available in earlier publications[17,21,22].

3. Theory and Mechanisms:

The extension of solid solubility during non-equilibrium synthesis by rapid solidification is mainly due to the atom trapping at the solidification front. Convective mass transport in the liquid pool provides almost an homogeneous melt pool to solidify from [23]. Some of the atoms get trapped during rapid solidification resulting in a phase with extended solid solubility. All three transport phenomena play a role in providing non-equilibrium phases. Where classical heat transfer can be applied to estimate temperature gradient, the sharp gradient in laser processing leads to thermocapillary flow with high liquid velocity in the meltpool [24,25], which in turn promotes convective mass transport [23]. The important boundary condition at the solid-liquid interface involving the phenomena of atom trapping is the non-equilibrium partition coefficient.

For process such as laser surface alloying and cladding, mass transport is a very important phenomena. A mathematical model for mass transport during laser surface alloying would help clarify several aspects of the problem. A set of processing conditions could be tested for uniformity of mixing and for the resultant average compositions in liquid state. The mass flux necessary to obtain a desired average composition in the liquid state could be computed. Powder loss during alloying would then be estimable, knowing the actual mass flux during laser processing. The model could be used to simulate the effect of varying the method of supplying alloying elements. Having predicted an average liquid pool composition and measured the actual solid state composition, the effective solute partitioning coefficient at the solid liquid interface could be calculated. This effective could then be compared to the value determined from the equilibrium phase diagram to check if conditions of local equilibrium existed ahead of the solid liquid interface. This would be a useful check as local equilibrium is assumed in predicting possible compositions during rapid solidification from equilibrium phase diagrams.

Two-Dimensional Transient Model for Mass Transport in Laser Surface Alloying

Chande and Mazumder [23] used the momentum transfer model described in Ref. 24 &25; and coupled that with diffusion equation to estimate the mass transfer during laser surface alloying since convection dominates the process.

For mass transfer calculations, the following assumptions are made:

1. The effect of alloying on solute mass diffusivity was neglected because accurate high temperature data were unavailable; mass transport by diffusion was negligible compared to that by convection and it simplified the formulation.
2. Mass flux was constant and uniform across the width of the pool at the surface. This rate and distribution could be altered to allow for any other experimental condition such as wire feed or nonuniform powder addition.

3.   There was no transfer of alloying elements across the solid liquid interface. This was a very good assumption that was verified experimentally using electron probe micro analysis (EPMA) technique.

Melting of powder particles introduced into the melt was found to be practically instantaneous [23]. Thus mass flux could be considered as being added in the liquid state.

The result of the mass transfer calculations indicate the importance of fluid flow in determining solute distribution during laser surface alloying. The small effect of changing solute diffusivity on the solute distribution, the uniform mixing even when mass flux was present over only a part of the pool surface and the nature of the three-dimensional plots all show that the pattern of fluid flow controls the resultant solute distribution. The effect of laser-substrate interaction time on solute distribution during surface alloying is shown in Fig. 6.

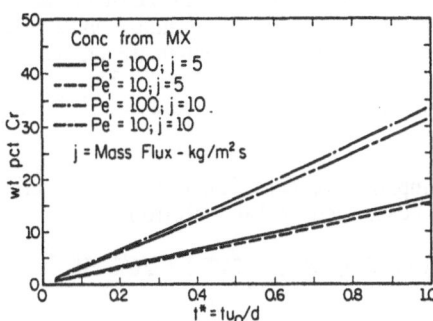

Figure 6. Summary of mass transfer calculations showing growth of average composition with increasing interaction time for uniform mass flux of 5 and 1^ ᵏᵷ/m²-s and Pe' values of 100 and 1000

One-Dimensional Transient Model for Laser Cladding

To study the extension of solid solubility during laser cladding, a diffusion model is developed by Kar and Mazumder [8,26,27] under the following assumptions:

a.   The thermal conductivity and the thermal diffusivity for a mixture is the sum of the volume-averaged value of the respective transport properties of each element of the mixture,
b.   The mass diffusivity of each element in the liquid phase is the average value of self-diffusivity over the room temperature and the initial temperature with modified activation energy for the mixture,
c.   The cladding pool and the substrate are in thermally perfect contact,
d.   There is no mass diffusion in the solid phase,
e.   The solute segregated at the solid-liquid interface moves into the liquid phase by diffusion only. This is because a boundary layer is formed near the interface where diffusion of solute atoms is dominant,
f.   The cladding melt forms a uniform solution of composition equal to that of the cladding powder mixture before its solidification begins,
g.   Only 50% of the laser energy is absorbed by the cladding material. Studies [18] show that the amount of laser energy absorbed by different materials is 37-60%.

The procedure to solve this problem can be found in Ref. [26] and the lecture by Professor Kar in this meeting, where the model has been discussed in detail and applied to the Ni-Hf system. The importance of studying the Ni-Al system [27] compared to that of Ref. [26] can be understood by analyzing the equilibrium phase diagrams of the Ni-Al and the Ni-Hf systems. The equilibrium phase diagrams of various systems can be divided into several regions based on the sign of the slope of the phase diagrams. Each of these regions can be divided further into smaller sections based on the assumption of constant equilibrium partition coefficient (k_e) which is true over a small range of the concentration around the point of interest on the equilibrium phase diagram. In Ref. [26], we have presented the effects of that regime of the equilibrium phase diagram that has negative slope. The study of Ref. [27] examines the effects of that region of the equilibrium phase diagram that has positive slope.

The model predictions have been compared with experimental data. Laser cladding was performed on nickel substrates with a mixture of -Ni-Hf powder of nominal composition by weight, 74% Ni and 26% Hf in one case and 74% Ni and 26% Al in the other case. STEM analysis of these samples shows that the concentration of Hf in the primary phase and Al in the solid solution regions is in excess of that predicted by the equilibrium phase diagram. The present model predicts the composition of the extended solid solution quite well. These results have been presented in Table 3.1. A small fraction of the NiHf alloy was found experimentally contain 9.3 percent (by wt.) Hf in the Ni matrix.

TABLE 3.1. Comparison of theoretical results with experimental results
of extended solid solution.

Nominal Composition of the Cladding, wt.%	74% Ni, 26% Hf	74% Ni, 26% Al
Laser Power, (Kw)	5	5
Laser Beam Diameter, mm	3	3
Workpiece Speed, in/min	50	50
Initial Pool Mean Temperature ($\overline{T_2}$), °K	1862	2420
Composition in the Solid Solution, wt.%		
Theoretical Results	7.15 Hf	30.76 Al
Experimental Results	6.5 Hf	29 Al

Also, the above model was used to obtain the nonequilibrium phase diagrams for Ni-Hf and Ni-Al alloy systems as shown in Figs. 7 and 8. Deviation of nonequilibrium phase diagram from the equilibrium one shows the extension of solid solubility that can be obtained during laser cladding. The nonequilibrium phase diagram has been plotted in the neighborhood of the melting point of the cladding powder. It can be seen from Fig. 7 that the width of the solid-liquid region between the equilibrium solidus and liquidus has reduced considerably. The nonequilibrium solidus line of this figure shows the extension of Hf concentration in Ni that can be obtained due to the rapid cooling in laser cladding. This phenomenon can be understood from the fact that the equilibrium phase diagram of Ni-Hf has a negative slope at the point which corresponds to the nominal composition of the cladding powder. Due to this, Hf is rejected fro the Ni matrix when the solution of Ni-Hf solidifies and this results in increasing the concentration of Hf in the liquid phase. But since the solid phase retains more Hf than its equilibrium composition, the liquid phase will have less Hf than the equilibrium value and hence the extended solid solution (that is, the nonequilibrium) phase diagram will shrink.

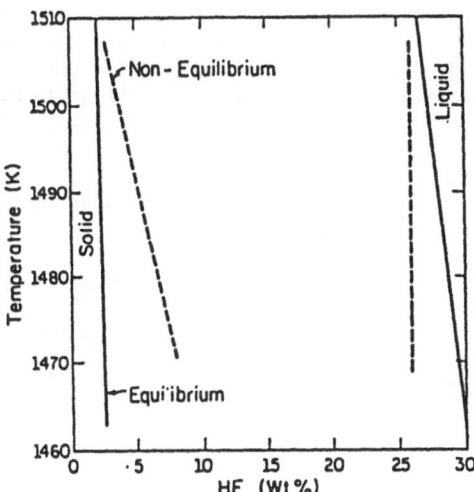

Figure 7 Comparison of the nonequilibrium phase diagram of the extended solid solution of nickel-hafnium with its equilibrium phase diagram.

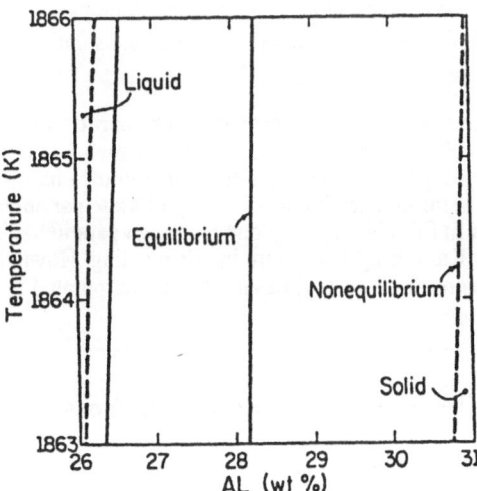

Figure 8 Comparison of the nonequilibrium phase diagram of the extended solid solution of nickel-aluminum with its equilibrium phase diagram.

On the contrary, it can be seen from the Fig. 8 that the width of the solid-liquid region between the equilibrium solidus and liquidus has increased considerably. The nonequilibrium solidus line of this figure shows the extension of Al concentration in Ni that can be obtained due to the rapid cooling in laser cladding. This phenomenon can also be understood from the fact that equilibrium phase diagram of Ni-Al has a positive slope at the point which corresponds to the nominal composition of the cladding powder. Due to this, Al is retained in the Ni matrix whereas Ni is rejected into the liquid phase as solidification proceeds. This causes an extension of Al concentration in the solid phase and lowers its weight fraction in the liquid phase and thus enlarges the solid-liquid region between the solidus and the liquidus lines.

The important boundary condition at the solid-liquid interface involving the phenomena of atom trapping is the non-equilibrium partition coefficient. In the above model non-equilibrium partition co-efficient based on a dilute solution assumption, developed by Aziz [28] was used. Recently Kar and Mazumder have developed the following expression for non-equilibrium partition co-efficient for concentrated solution [29].

$$k = \frac{C_s}{C_{li}} + \frac{k_m^* + \beta}{1 + \beta}\left(1 - \frac{C_l}{C_{li}}\right). \tag{1}$$

where

$$ua^2/[(a+b)D_{ABs}]$$

D_{Abs} = Interdiffusion coefficient of species A with respect to species B in the solid phase
u = Speed of the freezing front
β = Dimensionless solidification rate
  = $ua^2/[(a+b)D_{ABs}]$
C_s = Molar concentration of A in the previous solid phase [see Fig. 1(b)].
C_{li} = Molar concentration of A in the liquid phase at the interface
k = Nonequilibrium partition coefficient for concentrated solutions
k_m^* = Modified equilibrium partition coefficient for concentrated solutions

Aziz has also developed an expression for the non-equilibrium partition co-efficient(k_{ne}) for concentrated solution [30] by solving a transcendental equation which has four variables. They include parameters related to temperature, composition of solute, chemical potential and dimensionless speed of interface at infinite driving force. Many of these parameters are difficult to determine wheras the expression in Eq. (1). has only one unknown parameter. This topic will be discussed in details by Professor Kar during his lecture in this meeting. However, the expression for the non-equilibrium partition co-efficient(k_{ne}) has not been incorporated in the mass transfer model yet.

4. Applications in Tailored surfaces.

Synthesis of Nano Crystalline Niobium Aluminides

Materials synthesized varied from thin films (at low helium pressure) to powders with mean grain sizes in the range 5–10 nm. Powder with the exact stoichiometric composition and crystal structure (DO_{22}) of $NbAl_3$ was obtained at the laser pulse energy 270 mJ and p(He)=10.0 Torr.

The mean particle diameter and variance of the size distribution were found to increase with increasing He pressure. The atomic ratio (Al/Nb) decreased from the stoichiometric composition with increasing He pressure. Changes in the material characteristics with changing laser pulse energy were dramatic for a change of 270 mJ/pulse to 320 mJ/pulse, less significant for a change of 320 to 370 mJ/pulse. Increases in laser pulse energy produced decreases in both mean particle diameter and width of the particle size distribution. The atomic ratio (Al/Nb) also decreased from the congruent composition with increasing energy, but no change in atomic ratio was seen between 320 mJ to 370 mJ. The crystal structure of the deposited material was that of $NbAl_3$, irrespective of changes in pressure or pulse energy.

Synthesis of Fe-Cr-W-C Alloys for Improved Wear Resistance

There is considerable contemporary interest to develop cobalt free ferrous wear resistance material. Fe-Cr-C-X (X = Mn, W, Mo) alloys are excellent candidates. The conventional Fe-W-C tool steels have typical microstructural problems such as inhomogeneous microstructure with a coarse dispersion of carbides. Recent work by the author and co-workers revealed superior wear properties of laser clad Fe-Cr-Mn-C alloys with metastable M_6C carbides [15]. Therefore, application of laser synthesis technique for Fe-Cr-W-C alloys was the logical extension of the ongoing work at the University of Illinois at Urbana-Champaign.

There is little information available in the literature about the quaternary system of Fe-Cr-C-W. In 1950's Goldschmidt presented the first schematic diagram of Fe-Cr-C-W quaternary system (see Ref. [31]). Through their works the ternary M_6C carbide and $Cr_{23}C_6$ carbide were identified and it was shown that $Cr_{23}C_6$ could be changed by Fe and W to make $Fe_{21}W_2C_6$ carbide. Not many types of carbides has been presented in the equilibrium Fe-Cr-C-W system. Recently, Uhrenius et al. [32], reported equilibrium associated with carbides and austenite in Fe-Cr-C-W system. Their study was mainly concerned with equilibrium involving $M_{23}C_6$, however M_6C and cementite were also treated. Meanwhile, Bergstom has studied the region of existence of M_6C carbide [33,34]. The region around M_6C in quaternary system was observed, and the existence of two other carbides, which are $M_{12}C$ and M_4C were reported.

The purpose of the non-equilibrium synthesis of Fe-Ce-W-C alloying system was to get a fine and uniform distribution of a complex type of different carbide precipitates in cubic matrix and to develop a crack-free clad in order to develop a cobalt free alloy with improved tribological properties. The relationships between the process parameters, carbide morphology, and wear properties were studied to evaluate the role of process parameters on tribological properties of these non-equilibrium alloys [35].

Table 4.1 shows the size, purity, weight percent and atomic percent of the powders used for this laser synthesis technique. Table 4.2 shows the laser process parameters used for synthesis of Fe-Cr-W-C alloys. Figure 5a shows the experimental setup.

Table 4.1 Size and Purity of the Element Powders

Element	Size (μm)	Purity (%)	wt (%)	at (%)
Fe	6-9	99.9	62.5	51.9
Cr	2	99.5	25.0	22.3
C	0-300 mesh	99.5	6.25	24.2
W	0.5	99.9	6.25	1.60

Table 4.2 Laser Process Parameters

Process parameter	operating range
Laser power (kw)	3-8
Beam diameter (mm)	2-4, underfocused
Traverse speed (mm/s)	10.58-25.04
Powder feed rate (gm/s)	0.1342-1.3403
Preheating temperature	Room temp. → 800°C

Microstructure. A scanning electron micrograph (SEM) of the laser clad sample shows uniform microstructure throughout the clad region. The corresponding SEM EDX analysis from the precipitates show that the precipitates are rich in Cr with small amounts of Fe and W. This suggests that the precipitates consist of complex carbides. The general results of the electron probe microanalysis of the laser clad layer and substrate are shown in Fig. 9. It shows uniform distribution of W throughout the clad region. It also shows a sharp concentration gradient across the interface. In addition, near the interface, there is about an equal concentration of Fe and Cr. This region of interface is of interest for detailed investigation.

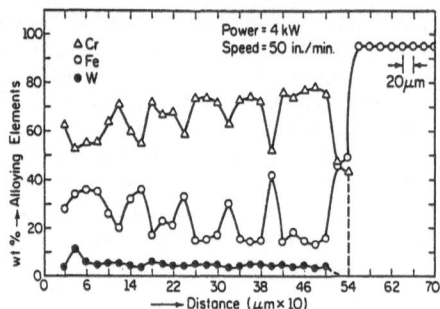

Figure 9. EPMA concentration profile of Fe-Cr-W-C cladment.

The overall structure of the laser melted zone was found to be uniform throughout. However, the interface zone which is adjacent to the clad/substrate interface, is approximately 8 μm. The composition of this zone was found to be rich in iron (the substrate). This part of the laser clad region exhibited columnar epitaxial growth. The rest of the clad zone consists of complex carbide precipitates.

This can be explained on the basis of the heat and mass-transfer model proposed by Li and Mazumder [18]. In the case of powder delivered directly to the laser beam interaction region, the total specific energy of the laser beam would be consumed by simultaneously melting the powder and the substrate. Because of the unique fluid flow mechanism in the laser clad region, the melt pool region would have a homogeneous liquid solution [23]. The rate of cooling in the laser cladding process is approximately the same as in the RSP process (10^5 to 10^6 K/sec). The melted region would rapidly solidify. During solidification, the latent heat energy would evolve and be conducted to the substrate. This latent heat energy would melt the additional layer of the substrate but would not have enough time to remix with the remaining melt. That is why this banded zone is rich in the substrate, i.e. iron phase. Away from the interface (substrate/clad) and towards the top surface of the clad region, because of high mixing, the structure is almost uniform and there is not much dilution of the laser clad alloy region.

Interfacial structure between the laser clad/substrate interface is presented in Fig. 10. It revealed high density of dislocations in the substrate near the interface (Fig. 10a). From the substrate interface, growth of eutectic type structure towards the laser clad region was observed (Fig. 10a). The average thickness or diameter of eutectic rod (dark region) is about 0.08 μm and average inter laminar spacing is about 0.16 μm. In the eutectic structure, high density of dislocations were observed at the interface between the second phase and depleted matrix (Fig. 10b). Interfacial dislocations may be due to mismatch between the second phase and depleted matrix. The selected area diffraction pattern for one of the precipitate was taken. The calculated interplanar spacings of the second phase from the diffraction pattern do not match with the equilibrium phases (such as σ phase, χ-phase or carbide precipitates) of the binary Fe-Cr and ternary Fe-Cr-C system. Therefore, this second phase appeared to be new metastable phase. Because of inherent rapid melting and solidification during laser process, metastable phases were also previously observed for Fe-Cr-Mn-C system [15]. Metastable phases contained distortions and strains. Therefore, identification of phase is complex. In addition, volume fraction of the second phase is very small, thus the identification of the phase is based on STEM and convergent-beam microdiffraction analysis. Convergent-beam patterns (CBPs) were obtained from the precipitate at different zone axis. Zone axis patterns were used to determine the crystal point and space group which are discussed in Ref. 17. The deduced projection group for this CBP is $2_R mm_R$ point group is m3m which belongs to the cubic family.

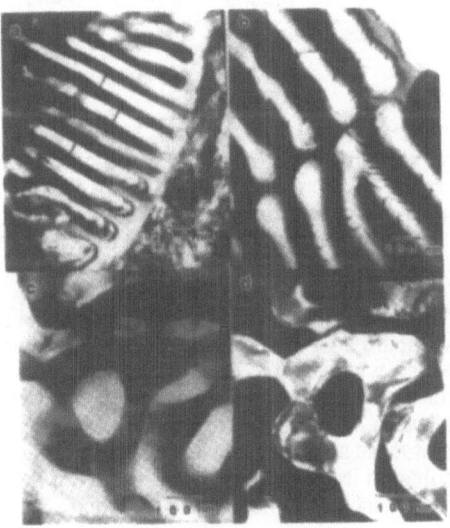

Figure 10. TEM micrographs of clad Fe-Cr-W-C.

At this point, one should note that these laser clad carbide-rich materials tends to be rather brittle. In order to avoid cracking, preheating was introduced during laser cladding. It was found that most samples preheated over 550°C were crack free. Process parameters for preheated laser clad are given in Table 4.3.

Table 4.3 Data Chart for the Synthesis of Fe-Cr-W System (Preheated)

sam#	beam dia. (mm)	laser power (kw)	travese speed (ipm)	powder feedrate (om/s)	preheat temp. (C)	specific energy (kJ/cm^2)	power density (kw/cm^2)	interaction time (sec)
#1-1	4	3	22	0.5368	600-621	8.05	18.75	0.4295
#1-2	4	4	23	0.5368	618-667	10.27	25	0.4108
#2-1	4	5	28	0.5368	612-646	10.546	25	0.3375
#3-1	4	4	47	0.5368	586-591	5.025	25	0.201
#3-2	4	4	51	0.8052	590-613	4.6325	25	0.1853
#3-3	4	4	43	1.0736	590-678	5.4925	25	0.2197
#7-1	4	4	23	0.5368	663-699	10.27	25	0.4108
#7-2	4	4	25	0.5368	698-748	9.4475	25	0.3779
#7-3	4	4	25	0.5368	725-812	9.4475	25	0.3779
#8-1	3	3	26	0.5368	688-702	8.9925	33	0.2726
#8-2	3	4	28	0.5368	700-735	12.376	44.4	0.2531
#13-1	3	3	27	0.5368	426-454	8.6625	33	0.2625
#16-1	4	4	25	0.5368	298-351	9.4475	25	0.3779
#17-1	4	4	25	0.5368	384-457	9.4475	25	0.3779
#19-1	4	4	25	0.5368	484-527	9.4475	25	0.3779
#19-3	4	4	26	0.5368	513-652	9.085	25	0.3634

Figure 11 presents a typical scanning electron micrographs of the preheated laser clad sample. While details convergent beam electron microscopy is underway to identify all the phases, a preliminary micro-chemical analysis using auger electron spectroscopy technique is employed for initial identification of the various carbide phases. Many different carbides such as M_7C_3, M_6C, $M_{23}C_6$ as well as MC_2 and MC were detected.

Microhardness. Microhardness analysis was carried out to understand the hardness distribution in the cross section of the laser clad Fe-Cr-W-C alloys. Figure 12 shows the hardness distribution for samples without preheating whereas Fig. 13 shows the hardness distribution for preheated sample. Note the sharp hardness increase at the interface for samples without preheating compared to that of samples with preheating. This is due to change in cooling rate at the interface due to preheating. Microstructure of these samples indicates that hardness is strongly related to carbide precipitate size and its distribution in the matrix. It was observed that bigger carbide precipitate leads to higher hardness since it increases the volume fraction of carbide in the matrix which is soft.

Wear Analysis. Figure 14 shows the wear scar width scar width for laser clad (without preheating) Fe-Cr-W-C with respect to laser clad Fe-Cr-Mn-C and Stellite 6 as well as mild steel and plasma deposited Stellite 6. A remarkable improvement in wear data can be noted from this figure. Figure 15 shows the wear scar against specific energy for preheated laser clad Fe-Cr-W-C alloys. With the increase of specific energy scar, width increases as expected since it reduces the cooling rate. However, the wear data for preheated samples is comparable to unpreheated sample. Therefore, this laser synthesis technique with preheating provides superior crack free wear resistance material compared to conventional stellite alloys.

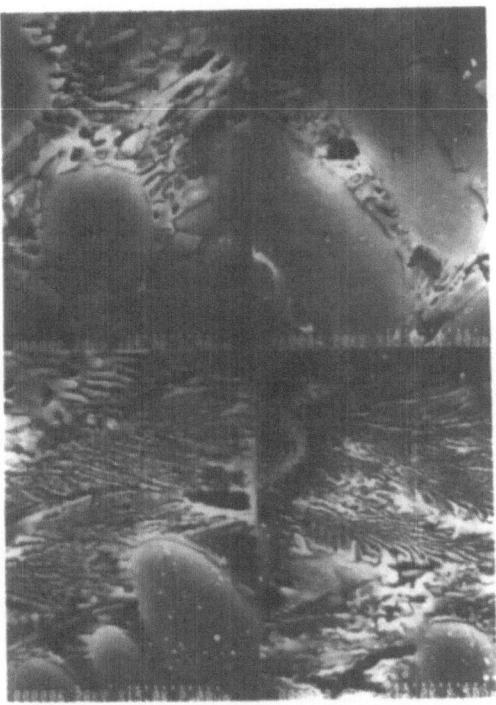

Figure 11. SEM micrographs of clad Fe-Cr-W-C.

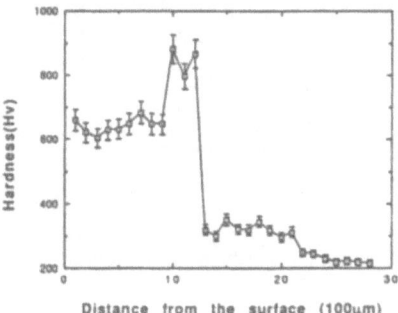

Figure 12. Hardness distribution as a function of distance for the sample without preheating
(laser power = 6 kW; beam diameter = 4 mm, powder feedrate = 0.671 gm/s,
traverse speed = 25.4 mm/s).

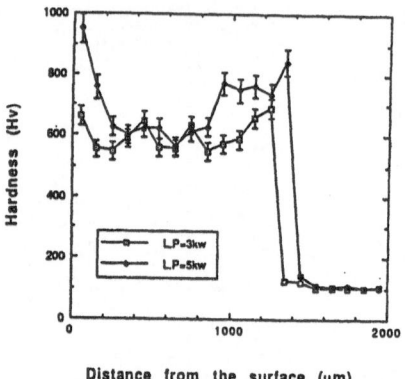

Figure 13. Effect of laser power on hardness distribution (beam diameter = 4 mm, powder feedrate = 0.5368 gm/s, traverse speed = 10.58 mm/s).

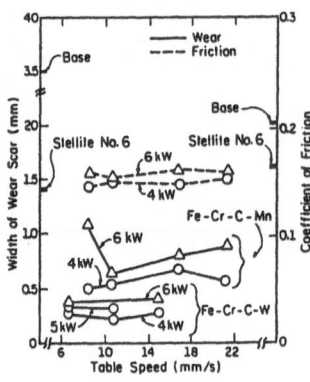

Figure 14. Friction and wear of laser clad surfaces as a function of table speed and laser power.

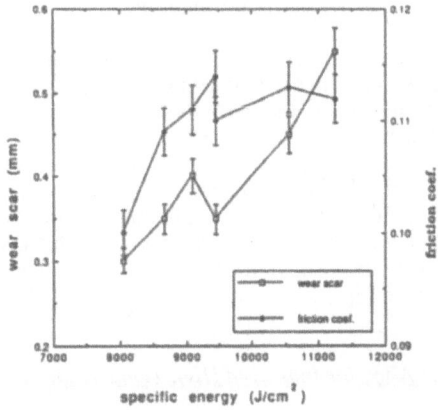

Figure 15. Friction and wear of laser cladded surface as a function of specific energy.

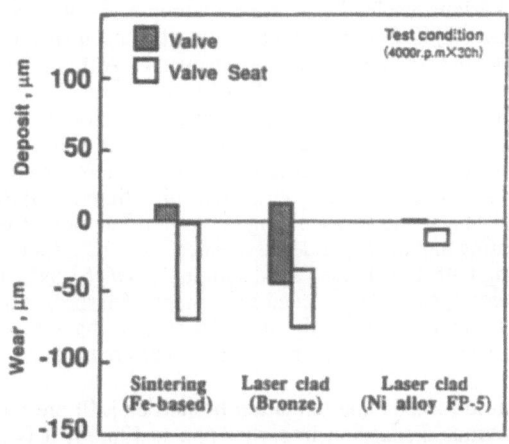

Figure 16. Comparison of wear resistence of Fe-base, Cu-base, and Ni-base materials fabricated by sintering and laser cladding. The Ni-base cladding shows superior wear resistence over Fe-base and Cu-base materials.

Aluminum - Copper interface with extended solid solubility.

Often aluminum components are used due to their light weight but surface properties does not match the service needs. Conventional hardfacing techniques poses some problem due to its low malting point and tendency to form intermetallic compounds. Both of these problem can be solved by Laser surface modification. Non-equilibrium synthesis provides an interface with extended solid solubility and low specific energy input minimizes dilution by excessive melting of substrate. Recently a series of papers are published by the authors and his co workers on the laser cladding of Bronze [36-40]and Nickel alloy FP-5 on Al-333. Careful examination of the microstructure at the interface revealed that solid solubility of Cu in Al was almost doubled resulting in a tough interface capable of providing good wear resistance at 200°C. Relative wear properties derived from simulated engine test are shown in Fig 16.

Synthesis of Ni$_{70}$Al$_{20}$Cr$_7$Hf$_3$ Alloy for Improved High Temperature Oxidation Resistance

Why nonequilibrium structure with extended solid solution of hafnium? Alloys for coatings for turbine blades require superior mechanical properties and oxidation resistance at elevated temperatures (1200°C) and aggressive environments. M-Cr-Al-RE (M = Ni, Co, Fe and RE = reactive elements) systems are widely used for such coatings. Coatings using the right proportion of the elements tend to form Al$_2$O$_3$ rich scales. Al$_2$O$_3$ is the coating of choice because of its limited volatility, sluggish growth kinetics, relatively inert behavior and also because oxygen diffusion through the oxide scale is extremely slow. The adherence of the Al$_2$O$_3$ to the alloy however is very poor. It has been well established that the addition of trace level of reactive elements such as hafnium, yttrium, etc. to the coating alloy composition greatly enhances oxide scale adherence [42-44]. Various mechanisms have been proposed regarding the method by which this beneficial effect is obtained [42-51]. However, the exact mechanisms for different systems are not well understood and this is a realm of extreme controversy. Although, there are controversies regarding the particular mechanism or group of mechanisms responsible for improved high temperature properties, there is a general consensus that the addition of rare earths improves the high temperature service life of the alloy and they perform best when they are in a finely dispersed form. However the solid solubility limit of Hf in these alloy systems is very low, of the order of 1 wt% maximum [52]. Conventional techniques such as directional solidification or investment casting cannot be applied because the slow cooling rates would induce an undesirable amount of segregation. The rapid heating and cooling rates associated with the laser cladding process can be utilized to produce cladding with an extended solid solution of Hf [53,54]. The process variables, such as laser power, traverse speed, powder feed rate, shape of laser beams, initial temperature, and geometry of the substrate and the powder delivery system, have a profound effect on the amount of extension of the solid solubility and hence need to be controlled.

Details of the microstructure [53,54] and oxidation properties [20] are already published else-where. Based on the previous discussion in earlier papers [20,53] it is possible to suggest a possible microstructure evolution process for this system. The possible phase transformation sequence could be presented as follows.

$$L \rightarrow \gamma'_p + L \rightarrow \gamma'_p + (E) \approx \gamma'_p + (\gamma'_E + \delta_E)$$

where L = liquid phase; E = eutectic phase; γ' = primary γ' phase (ordered f.c.c.) Ll$_2$; γ'_E = eutectic γ' phase (ordered f.c.c.); δ_E = other eutectic phase (heavily faulted, f.c.c.); δ_p = primary δ phase. As the microchemistry studies presented earlier will indicate, the γ' phase has a composition consistent with the stoichiometric ration of the type Ni$_3$ (Cr, Al, Hf) and the δ_p phase has (Al, Hf)$_3$Ni$_2$ composition.

Recently, since the earlier publication, a more dramatic contrast was observed under cyclic oxidation condition. Figure 17 shows the weight change per unit surface area as a function of time during thermal cycling in slowly flowing air. Each cycle in Fig. 17 represents one hour. This figure shows that while the substrate (Rene 80) suffers tremendously from spalling the clad samples exhibit negligible change in weight. One can conclude that adherence is significantly improved for Al₂O₃ this laser clad Ni-Cr-Al-Hf alloy.

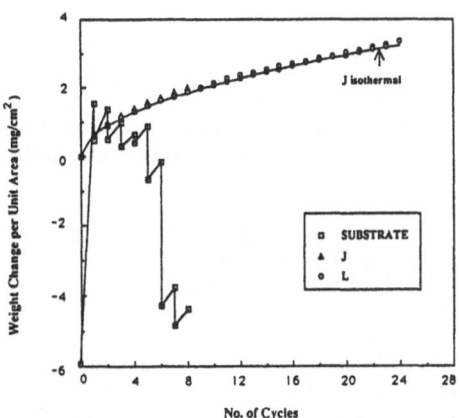

Figure 17. Weight change per unit area as a function of time during thermal cycling in slowly flowing air.

Synthesis of Mg-Al Alloys for Improved Corrosion Resistance

Although magnesium and its alloys have some very attractive properties such as low density and superb machinability, they have not yet found widespread application as an engineering material due to their poor environmental performance and mechanical properties. The extreme position of magnesium in the electrochemical series coupled with the fact that, unlike aluminum, magnesium is unable to form protective self healing passivating surface films in corrosive environments, make it vulnerable to galvanic attack. When coupled with a nobler metal, magnesium undergoes anodic dissolution and hence exhibits poor corrosion resistance.

Rapid solidification processing (RSP) has been identified as the most promising approach to improve corrosion properties of Mg alloys [55-57]. Aluminum is a well known solid solution strengthener in magnesium and is widely used in Mg alloys. Das and Chang reported that a Mg-Al based alloy prepared using RSP exhibited a remarkably reduced corrosion rate [58]. RS Mg-23.4wt%Al had a corrosion rate at least two orders of magnitude lower than the corresponding chill-cast material [59]. From the investigation of binary RS Mg alloy ribbons, G. L. Makar [60] found that Al is the only element to cause a significant decrease in corrosion rate of Mg alloys with increasing Al content up to 28.8wt% and above. Also the rare earth alloying additions increase corrosion resistance in Mg-Al based alloys [60,61].

In the RSP category, laser processing is an important method to improve the corrosion behavior of materials. Earlier work indicated that laser surface melting improves the intergranular corrosion of stainless steel [62]. For Mg alloys, the corrosion rate of Mg-Zn alloy was reduced by laser remelted RSP because of the solution of the finely distributed interdendritic nonequilibrium Mg₇Zn₃ phase [63]. The corrosion resistance of Mg-Li based alloy was substantially increased by laser surface treatment which caused a homogeneous fine-grained microstructure [64].

The laser cladding technique provides a unique means of synthesizing nonequilibrium alloys due to its inherent rapid solidification rate, which promotes the formation of either an amorphous phase or nonequilibrium crystalline phases. Proper selection of alloying elements during laser cladding provides a passivating film system, which greatly reduces the galvanic action, thereby increasing the corrosion resistance of the alloys. Under the laser cladding category, Mg-Zr system and Mg-Al system have been already studied and reported in the earlier publication [21,22]. This paper reports some of the important information,

To reduce oxidation, Mg cladding is carried out in a chamber in which the shielding argon pressure is above ambient pressure. The chamber dimensions are 35 cm x 35 cm x 35 cm. The chamber remains fixed on a stationary stage during cladding. Figure 2b depicts the translation system and the powder delivery system of the chamber. The translation system allows movement in two dimensions using stages driven by stepping motors which are manipulated by a microprocessor controller, a monitor and a keyboard.

Modifications of the powder delivery gas apparatus were made for this application. If the chamber is pressurized to about 3.4 kPa (0.5 psi) above ambient pressure, a pressure gradient inside the powder feeding tube results. This back pressure restricts and can even stop the powder flow. A nozzle was designed to supply a gas jet around the powder flow tube and to eliminate the back pressure in the feeding tube. An additional problem was the pressure gradient across the screw feeder: if the gradient is positive, the powder flow stopped; if negative, the powder flow is uncontrollable. Figure 5b shows a "T" tube joint which is used to eliminate this pressure gradient between the powder dispenser and delivery tube. After a new nozzle and the "T" tube were adopted, the powder flow improved remarkably. It was found that helium is superior to argon in protecting the NaCl window from the Mg vapor during cladding.

Table 4.4 indicates the nominal compositions of the three groups of binary alloys chosen for laser cladding experiments. The powder size and purity are: -325 mesh and 99.8% for Mg, 20 microns and 99% for Al, respectively. Both atomized and ground Mg powders (same nominal particle size) were tried for composition B but it was found that only atomized powder is suitable for cladding. Mg substrate purity is 99.8%. The specimen size is about 1 cm x 4 cm x 6 cm. Before cladding, the substrate was polished with 180 grit emery paper and sand blasted.

Before cladding, the positions of nozzle and substrate are adjusted in order to feed the powder directly into the interface between the laser beam and the substrate. The air remaining in the chamber is purged by flowing argon for five minutes before closing the valve. The powder delivery and laser beam are controlled simultaneously. During laser cladding, the argon gas pressure was 700 kPa (100 psi) at the tank regulator, the overall flow meter reading was 25 (relative units), the "T" tube flow meter reading was 80 cm^3/sec (10 SCFH). The helium gas protecting the NaCl window was at 140 kPa (20 psi). Table 4.5 shows the range of the cladding parameters.

Microstructure and microchemistry are reported in details in the Ref 22. Only corrosion properties are summarized.

Corrosion Test Results for Mg$_{27}$Al$_{73}$. The corrosion test results for laser clad alloy Mg$_{27}$Al$_{73}$, commercially produced alloy AZ91B and pure Mg in 3.5 wt% NaCl electrolyte, are presented in Tables 4.6 & 4.7, and Fig. 18. The corrosion potential data of laser clad Mg-2wt%Zr and Mg-5wt%Zr [21] in 3.5% NaCl are also provided in Table 4.6 for comparison.

These results indicate that laser clad Mg$_{27}$Al$_{73}$ is superior to laser clad Mg-2wt%Zr, Mg-5wt%Zr, cast AZ91B and Mg in corrosion properties. The corrosion rate (C.P.) of Mg$_{27}$Al$_{73}$ is one order of magnitude lower than that of AZ91B and two orders of magnitude lower than that of Mg. In potentiodynamic tests, the clad materials passivated spontaneously during the 3600 seconds initial delay while Mg and AZ91B have no potential region over which the sample surface remained passive. Figure 18a shows that the $Mg_{27}Al_{73}$ sample remained passive over a potential region of 200 mV where the current was quite stable. The passive film began breaking at about -1.0 V of potential. In the case of AZ91B, when the applied potential was above -1.6 V, the passive film was breaking as indicated by the current increasing rapidly with a slightly increase in potential (see Fig. 18b). The passive film was broken completely when the potential was about -1.4 V. Similar results were found for Mg (see Fig. 18c). These effects indicate that the passive film on $Mg_{27}Al_{73}$ is much more stable than the passive film on AZ91B. In Table 4.7, the slope ($\Delta V/\Delta I$) at E_{corr} for $Mg_{27}Al_{73}$ is much larger than the corresponding slopes for AZ91B and Mg, which again indicates that the laser clad alloy $Mg_{27}Al_{73}$ is more corrosion resistant than cast AZ91B and cast pure Mg in 3.5% NaCl solution.

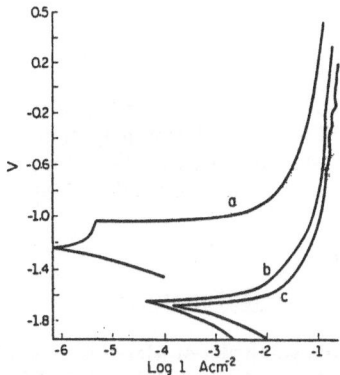

Figure 18 Corrosion results of potentiodynamic polarization for alloys
      (a) laser clad $Mg_{27}Al_{73}$.
      (b) cast AZ91B.
      (c) cast Mg.

5. CONCLUDING REMARKS

It can be concluded from three preceding examples that the inherent rapid solidification and strong convection current in laser processing can be successfully utilized for the synthesis of homogeneous alloys with extended solid solution and metastable phases to provide certain desired properties.

By Laser ablation we can generate nano-crystalline NbAl3 intermetallic compounds which demonstrates unique crystallography.

Fe-Cr-W-C alloys produced by non-equilibrium synthesis with laser exhibited wear properties far superior to Stellite 6. From the microstructure analysis it is postulated that the metastable M_6C carbides, which contains high volume fraction of metals (Cr, W & Fe), are mainly responsible for improved wear properties. Lower specific energy input results in better wear properties.

Ni-Cr-Al-Hf with extended solid solution of Hf in f.c.c. γ'(Ni_3Al) seems to improve the Al_2O_3 adherence remarkably resulting in superior high temperature oxidation resistance compared to arc melted materials of similar composition and conventional superalloys such as Rene 80.

Mg-Al alloys produced by non-equilibrium synthesis with lasers consist of eutectic phases along with primary Mg and aluminum phases for Mg-rich ($Mg_{53}Al_{47}$) and Al-rich ($Mg_{27}Al_{73}$) alloys, respectively. The volume fraction of eutectic phase can be controlled by laser parameters. $Mg_{27}Al_{73}$ alloys exhibited superior aqueous corrosion (in 3.5% NaCl electrolyte) properties compared to commercial alloy AZ91B and pure Mg. Refinement of microstructures seems to play an important role in generating improved corrosion properties.

This laser based technique, thus provides a way to synthesize "tailor made" materials. The predictive capability provided by the mathematical model now makes it a powerful process and takes the guess work out. This process can produce such "tailor made" materials in near net shape by depositing subsequent layers.

Acknowledgment

This work represents several years of group effort at the University of Illinois funded by Office of Naval Research, Air Force Office of Scientific Research, Quantum Laser Corporation, and DAEWOO Heavy Industries. Transmission electron microscopy studies by Dr. J. Singh, Dr. S. Sircar and Dr. Y. Liu are greatly appreciated. Several other group members include: J. Choi, C. Ribaudo, A.Wang and Dr.Kar.

References

1. Birringer, R.and H. Gleiter, Encyclopedia of Materials Science and Engineering, Edited by R.W.Cahn(Pergamon Press, Oxford, U.K.)mVol1.(1988) 339

2. Swager, W. L., Batelle Today, 18 (1980) 3-6.

3. Breinan, E. M., and B. H. Kear, "Rapid Solidification Laser Processing at Higher Power Density," Chapter 5 in Laser Materials Processing, M. Bass (ed.), North Holland Publishing Co. (1983).

4. Savage. S. J., and F. H. Froes, "Production of Rapidly Solidified Metals and Alloys," J. of Met., 36(4) (1984) 20-33.

5. Bloembergen, N. Laser Solid Interactions and Laser Processing, S. D. Ferris, H. J. Leamy and J. M. Poate (eds.), AIP Conference Proc., 50 (1979) 1-10.

6. Von Allmen, M., E. Huber, A. Blatter, and K. Affostler, Int. J. of Rapid Solidification, 7 (1984) 15-28.

7. Affotler, K., and M. Von Allmen, Appl. Phys. A, 33 (1984) 93.

8. Kar, A., and J. Mazumder, J. Appl. Physics, 61(7) (1987) 2645-2655.

9. J.Mazumder, T.Duffey, T.McNeela, T.Yamamoto and A.L.Schawlow; Laser Materials Processing-IV, J.Mazumder, K.Mukherjee and B.L.Mordike(editors), To be published by TMS, Warrendale, Pa. 1994

10. T.McNeela, M.S.Thesis, University of Illinois, 1994

11. Draper, C. W., and J. M. Poate, "Laser Surface Alloying," Inter. Metal. Reviews, 30(2) (1985) 85.

12. Draper, C. W., Lasers in Metallurgy, K. Mukherji and J. Mazumder (eds.), published by The Metallurgical Society of AIME, Warrendale, PA (1981) 67-92.

13. Steen, W. M., "Surface Engineering with a Laser," Metals and Materials (1985) 730.

14. Chande, T. C., A. Ghosh, and J. Mazumder, Surface Engineering, 3(1) (1987) 53-58.

15. Singh, J., and J. Mazumder, Metallurgical Transactions, 18A (1987) 313.

16. Singh, J., K. Nagarathnam, and J. Mazumder, High Temperature Technology, 5(3) (1987) 131-137.

17. Mazumder, J., J.Choi, C.Ribaudo, A Wang and A.Kar: International Conference on Beam Processing of Advanced Materials, J.Singh and S.M.Copley(Editors), (1993) 41-67

18. Li, L. J., and Mazumder, J., "A Study of the Mechanism of Laser Cladding Processes," Laser Processing of Materials, K. Mukherjee, and J. Mazumder (eds.), Proc. The Metal. Soc AIME, Warrendale, PA (1984) 35-50.

19. Twigg, M. K., and H. L. Fraser, Ultramicroscopy (in press).

20. Ribaudo, C., S. Sircar, and J. Mazumder, Met. Trans. A, 20A (1989) 2389-2497.

21. R. Subramanian, S. Sircar, and J. Mazumder,"Laser Cladding of Zr on Mg for Improved Corrosion Properties," Journal of Materials Science, U.K.Vol.26,(1991) 951-956

22. Wang,A., S.Sircar and J.Mazumder, Journal of Materials Science, U.K.Vol.28(1993) 5113-5122

23. Chande, T., and J. Mazumder, J. Appl. Physics, 57 (1985) 2226.

24. Chan, C. L., Mazumder, J., and Chen, M. M., A Two-Dimensional Transient Model for Convection in Laser Melted Pools, Metall. Trans., vol. 15A, .(1984) 2175-2184

25. Chan, C. L., Mazumder, J., and Chen, M. M.,Vol.64. No.11,(1988) 6166-6174

26. Kar, A., and Mazumder, J., One-Dimensional Finite-Medium Diffusion Model for Extended Solid Solution in Laser Cladding of Hf on Nickel, Acta Metall., vol. 36, pp. 701-712, 1988.

27. Kar, A., and Mazumder, J., Extended Solid Solution and Nonequilibrium Phase Diagram for Ni-Al Alloy Formed during Laser Cladding, Metall. Trans. A, vol. 20A, pp. 363-371

28. Aziz, M.J., J.Appl.Physics, Vol 53,(1982) 1158-1168

29. Kar, A., and J.Mazumder; Acta Metall .mater. Vol 40, No.8, (1992) 1873-1992

30. Aziz, M.J. in Science and Technology of Rapidly quenched alloys (edited by M.Tenhover, W.L.Johnson and L.E. Tanner) Mat. Res. Soc. Pittsburgh, Pa. (1987) 25

31. Goldschmidt, H. L., J. Iron Steel Institute, 170 (1952) 189-204.

74

32. Uhrenius, B., and S. Frondell, Metal Science, 11 (1977) 73-81.

33. Bergstöm, M., Material Science and Engineering, 27 (1977) 271-286.

34. Bergstöm, M., Material Science and Engineering, 27 (1977) 257-269.

35. Mazumder, J., and J. Singh, US Patent Appl. Ser. # 133, 346, dated Dec. 15, 1987

36. Transmission Electron Microscopy Study of Martensites in Laser-Clad Ni-Al Bronze on Aluminu Alloy AA333, Metallurgical and Materials Transactions A, Vol. 25A, no number, January 1994, pp. 37-46.

37. TEM Crystal and Defect Structure Study of Martensite in Laser Clad Ni-Al Bronze, Acta metall. mater., Vol. 42, No. 5, 1994 (only date was a year), pp. 1755-1762

38. TEM Study of Precipitates in Laser Clad Ni-Al Bronze, Acta metall. mater., Vol. 42, No. 5, 1994 (only date was a year), pp. 1763-1768

39. Transmission Electron Microscopy Crystal Structure Study of the Cr-Rich Phase in a Laser-Clad Ni Alloy, Metallurgical and Materials Transactions A, Vol. 25A, no number, March 1994, pp. 487-497

40 Laser Cladding of Ni-Al Bronze on Al Alloy AA333, no reprint, accepted for publication in Metallurgical Transaction)

41 Processing, Microstructure, and Properties of Laser-Clad Ni Alloy FP-5 on Al Alloy AA333, Metallurgical and Materials Transactions B, Vol. 25B, no number, June 1994, pp. 425-434

42. Wood, G. C., and F. H. Stott, "High Temperature Corrosion," Rapp R. A. (ed.), Conference held at San Diego, CA, March 1981, Int. Corrosion Conf. Series, NACE-6, published by National Association of Corrosion Engineers, Houston, TX, p. 227 (1983).

3. Whittle, D. P., and D. H. Boone, "Surface and Interface in Ceramic and Ceramic-Metal Systems," J. Pask and A. Evans (eds.), Material Science Research, 14, Plenum Press, New York, 487.

44. Whittle, D. P., and J. Stringer, Philos. Trans. R. Soc. London, A295 (1980) 309.

45. Allam, I. M., D. P. Whittle, and J. Stringer, Corrosion and Erosion of Metals, K. Nateson (ed.), TMS-AIME, Warrendale, PA, p. 103 (1980).

46. Smeggil, J. G., and A. W. Funkenbusch, "A Study of Adherent Oxide Scale," United Technologies Research Center Report No. R85-916564-1 (1985).

47. Tien, J. K., and F. S. Pettit, Metall. Trans., 2 (1972) 1587.

48. Golightly, F. A., F. H. Stoot, and G. L. Wood, Oxidation of Metals, 10(3) (1976) 163-187.

49. Antill, J. E., and K. A. Peakall, J. Iron Steel Inst., 205 (1967) 1136-1142.

50. Pfeiffer, H., Werkst. Korros., 8 (1957) 574.

51. Giamei, A. F., and D. L. Anton, Metall. Trans. A, 16A (1985) 1997-2005.

52. Nash, P., and R. F. West, Metal Science, 15 (1981) 347.

53. Sircar, S., C. Ribaudo and J. Mazumder, Met. Trans. A, 20A (1989) 2267-2277.

54. Mazumder, J. S. Sircar, C. Ribaudo, and A. Kar, J. of Laser Applications, 1 (1989) 27-42.

55. A. Joshi, et al., "Literature Review and Interim Technical Report--Rapid Solidified Magnesium Alloys," AFWAL Contract F33615-85-C-5032, Lockheed Missiles and Space Company, Inc., Report No. LMSC--F083181, January, 1986.

56. R. E. Lewis and A. Joshi, "Rapidly Solidified Magnesium Alloys For High-Performance Structural Applications A Review," Processing of Structural Metals by Rapid Solidification, F. H. Froes and S. J. Savage (eds.), published by ASM, pp. 367-378 (1987).

57. A. Joshi and R. E. Lewis, "Role of SRP on Microstructure and Properties of Magnesium Alloys," Advances in Magnesium Alloys and Composites, H. Paris and W. H. Hunt (eds.), The Minerals, Metals, & Materials Society, pp. 89-103 (1988).

58. S. K. Das and C. F. Chang, "High Strength Magnesium Alloys by Rapidly Solidification Processing," Rapidly Solidified Crystalline Alloys, S. K. Das, B. H. Kear & C. M. Adam (eds.), published by Metal Society, Inc., Warrendale, PA, pp. 137-156 (1985).

59. F. Hehmann, et al., "Effect of Rapid Solidification Processing on Corrodibility of Magnesium Alloys," presented in The Third in a Series of International Conferences on (P/M) Aerospace Materials, organized by Metal Powder Report, in Lucerne, Switzerland, Nov. 2-4, 1987.

60. G. L. Maker and J. Kruger, "The Effect of Alloying Elements on the Corrosion Resistance of Rapidly Solidified Magnesium Alloys," Advances in Magnesium Alloys and Composites, H. Paris and W. H. Hunt (eds.), The Minerals, Metals, & Materials Society, pp. 105-121 (1988).

61. C. F. Chang, S. K. Das and D.Raybould, "Rapidly Solidified Mg-Al-Zn-Rare Earth Alloys," Rapidly Solidified Materials, P. W. Lee and R. S. Carbonara (eds.), American Society for Metals, pp. 129-135 (1986).

62. T. R. Anthony and H. E. Cline, "Laser Surface Melting of Stainless Steel for Corrosion Protection," SPIE, Laser Application in Materials Processing, 198 (1979) 82-91.

63. T. Z. Kattamis, "Solidification Microstructure of Laser Processed Alloys and its Impact on Some Properties," Laser in Metallurgy, K. Mukherjee and L. Mazumder (eds.), published by the Metallurgical Society of AIME, Warrendale, PA, pp. 1-10 (1981).

64. R. Kh. Kalimullin and Yu. Ya. Kozhevnikov, "Structure and Corrosion Resistance of a Mg-Li Base Alloy After Laser Treatment," Metal Science and Heat Treatment, translated from Russian, Plenum Publishing Corporation, NY, 27(3-4) (1985) 272-274.

MICROSTRUCTURE AND PHASE SELECTION IN RAPID LASER PROCESSING

P. GILGIEN and W. KURZ
Swiss Federal Institute of Technology, Lausanne
Centre de Traitement des Matériaux par Laser (CTML)
CH-1015 Lausanne, Switzerland

1. Introduction

Laser processing is a high technology treatment which aims to produce specific properties locally on the surface of materials. These properties are strongly dependent on the microstructure formed during the processing, and since laser treatment usually induces local melting, the microstructure is formed during the resolidification of the liquid pool. Therefore, in order to progress further than an empirical approach to the optimisation of processing conditions and the development of appropriate alloys for laser treatment, it is necessary to understand the mechanisms of the formation of microstructure during solidification.

To this end, thermodynamics does not suffice to explain why a particular microstructure is formed, as is shown by such well-known phenomena as the change from a stable to a metastable phase when the solidification speed increases, or the growth of a plane eutectic front at low speeds in a binary alloy whose composition is far from eutectic. In order to understand this behaviour, the growth kinetics of the various microstructures must be examined on a microscopic scale. The aim of this paper is to present a method by which the mechanism of microstructure selection during laser processing may be understood. In the first instance, some aspects of laser processing which are important for solidification are underlined. Then the mechanism of the competition between microstructures which arises during solidification is presented, followed by a description of the models for growth kinetics which are needed for dealing with the most frequent cases. Finally, the idea of a microstructure selection map is presented and illustrated using the Al-Fe system.

J. Mazumder et al. (eds.), Laser Processing: Surface Treatment and Film Deposition, 77–92.
© *1996 Kluwer Academic Publishers.*

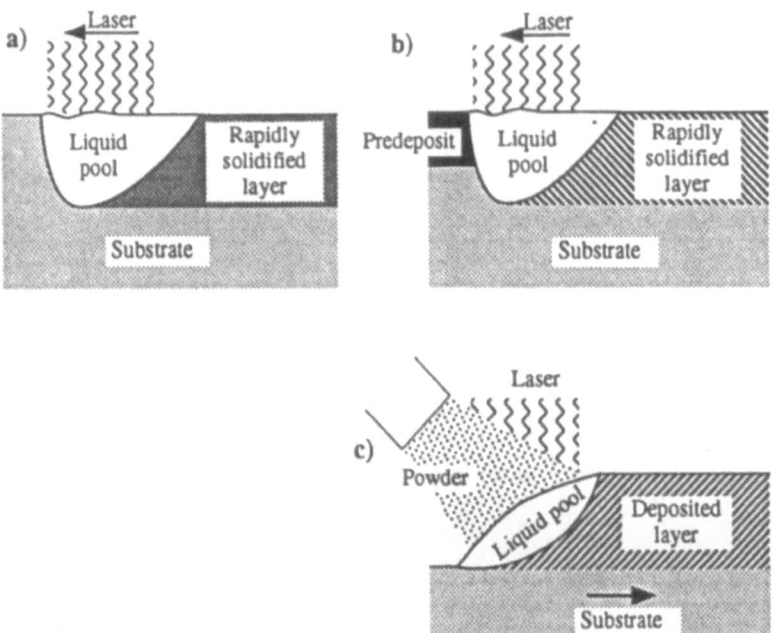

Figure 1: Diagrammatic representations of various laser surface treatments: a) remelting, b) alloying, c) cladding.

2. Solidification during Laser Processing

Most treatments of materials by laser involve a local superficial melting of the sample, followed by resolidification. It is useful to discuss first the general physical and geometrical characteristics of these laser treatments in order to model the solidification process.

Figure 1 shows diagrammatically the main laser treatments which involve a passage into the liquid state: (a) simple remelting which produces a rapidly solidified surface layer; (b) fusion of a previously deposited surface layer resulting in an alloy; (c) cladding a base with a deposit by injection and fusion of powdered material imported into the laser beam. These three cases have certain properties in common:

1. The substrate, being large in comparison to the liquid pool, remains globally cold. The heat is therefore mainly evacuated from the liquid pool into the substrate, whose large mass ensures rapid cooling. Solidification occurs from the coldest to the warmest part, i.e. with positive temperature gradient, which is typical of directional solidification. The temperature gradient being high,

the temperature drops rapidly after the passage of the solidification front so that there are no reheating phenomena and little ripening.

2. Solidification takes place starting from the substrate, without requiring nucleation since the liquid–solid interface exists throughout the process. This situation can be considered as equivalent to an abundant nucleation of all phases. The solidification velocity, V_s, is zero at the bottom of the laser trace (at the interface between the remelted and unmelted zones), and it increases towards the sample surface, where it is less than or equal to the beam scanning velocity, V_b (Figure 2b).

3. In the stationary state of a laser treatment, there appears, under the laser beam, a thermal field which moves across the sample at the same speed as the beam. The solidification front follows the thermal field, so that the growth velocity is the imposed solidification parameter for laser treatments, as in all processes of directional solidification, and in opposition to the case of equiaxed growth in which the parameter which controls solidification is the undercooling of the liquid.

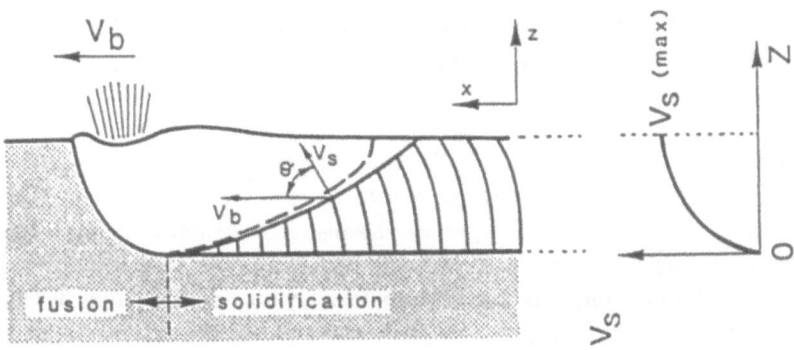

Figure 2: a) Schematic longitudinal section of the centre of a laser trace. The leftward-curving lines show the direction of growth of the microstructure visible on micrographs. The local solidification velocity is given by $V_s = V_b \cos \theta$. b) Evolution of the solidification velocity as a fonction of position in the trace.

Although its geometry is a little more complex, the process of laser welding shows essentially the same solidification parameters. It should also be noted that, on a macroscopic scale, in all the laser treatments mentioned a stationary state occurs very rapidly, so that the solidification models can be restricted to steady state growth.

Furthermore, laser remelting is a very useful process in which to study rapid solidification, since it is possible, using geometrical arguments, to determine the local solidification speed from a metallographic longitudinal section taken in the middle of the resolidified trace which shows the direction of growth (Figure 2a). Figure 2b shows the typical evolution of the growth velocity in a laser trace. It must be noted that although the growth velocity is dependent on the beam scanning velocity, V_s cannot be augmented indefinitely by raising V_b: apart from the difficulty of augmenting the power transmitted to the material by the beam, the thermal conductivity of the material treated effectively limits heat extraction from the liquid pool. For this reason, the form of the thermal field (and hence that of the liquid pool) becomes elongated when V_b increases, so that V_s remains lower than V_b, even on the sample surface. The maximum possible solidification speed for metals is of the order of 2 m/s for continuous laser processing: it can be higher for pulsed laser treatments.

These characteristics of laser processing provide a framework for the solidification which must now be modelled.

3. Selection of Microstructures

Basically, a solidification microstructure is characterised by its phase and growth morphology. This may be a plane front, cells or dendrites of a single phase or of multiphase eutectic.

As we have seen in the preceding Section, the growth velocity of the microstructure is imposed by the beam scanning velocity and the thermal conductivity of the material. Thermodynamically, every physical transformation requires a driving force whose strength at low transformation speeds is proportional to the speed. In the case of solidification, this force is provided by the undercooling of the solid–liquid interface, i.e. the difference between the equilibrium temperature at which the two phases would coexist without solidification and the effective temperature of the interface. Therefore, the solidification velocity being given by the process, there exists a corresponding equilibrium temperature (or standard temperature) and an undercooling, which together determine a growth temperature.

Since there is a stationary temperature field if the coordinate system is fixed to the melt pool, this growth temperature itself corresponds to a position on the sample, which means that at the given speed the interface may occur in that place. This argument is illustrated in Figure 3. Logically, in the case of directional solidification with a positive temperature gradient and if there is no barrier to nucleation, it is the structure whose growth temperature is highest for a given speed which would develop the 'farthest forward in the liquid pool'. In theory, the interface of the other structures with lower solidification temperatures, could develop behind the first

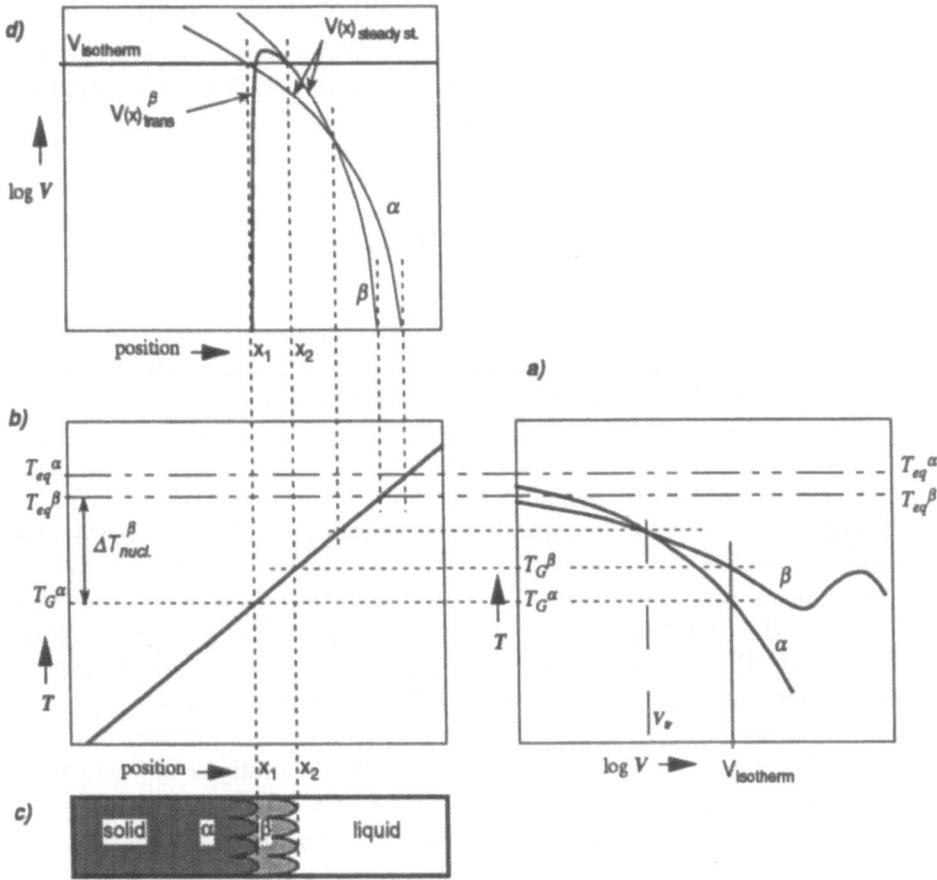

Figure 3: Illustration of the selection of microstructure during directional solidification (V_s given). a) Growth temperature of two competing structures as a function of the velocity. b) Field of stationary temperature in the sample. c) Sample with the possible positions of the two structures. d) Variation of the velocity in the sample during the transition $\alpha \rightarrow \beta$. The structure α grows as far as point X_1 at the speed of the isotherms. At this point structure β nucleates, then accelerates to reach its equilibrium position at X_2. From then on, β continues to grow at the speed of the isotherms. The curves α and β are the same as those in a).

interface, but this situation has little meaning since at that point the sample is already solid. The selection criterion for microstructures during directional solidification and abundant nucleation can thus be written as follows:

At any given speed, the structure which develops is the one whose growth temperature is the highest.

This criterion was already formulated in 1926 in a slightly different way for the case of equiaxed growth [1]. To apply it, the growth temperature must be known as a fonction of speed for all the competing structures. In the following Section models of growth kinetics are used to establish these relations.

4. Modelling Rapid Solidification

Let us consider recent analytical models of growth kinetics of plane front, dendrites and eutectic. Cellular growth may be neglected here since it takes place only over a small range of speed, and it has not been well described by any analytical model to date. For more information on all these models, see [2].

The simple fact that the solidification of a system travels at non-zero speed implies that the system is not in a state of thermodynamic equilibrium. However, when the front progresses slowly, we can consider a local equilibrium to exist at the interface, and hence that the respective compositions of the solid and liquid in contact be given by the equilibrium phase diagram of the system. When the solidification speed is high, this local equilibrium at the interface is lost because the atoms no longer have time to diffuse and to adjust the composition of the two phases in contact. In this case, we refer to solute trapping because the concentration of the solid increases beyond the equilibrium composition. Thus the ratio of the compositions of the solid and liquid in contact at the interface, called the partition coefficient k, varies with the speed. Aziz has developed a relation for dilute solutions [3]:

$$k_v = \frac{k_e + \frac{a_0 V_s}{D_i}}{1 + \frac{a_0 V_s}{D_i}}, \tag{1}$$

where k_e is the equilibrium partition coefficient, a_0 is a characteristic value of the thickness of the interface and D_i is the coefficient of diffusion across the interface, supposed equal to the coefficient of diffusion in the liquid. On the other hand, Boettinger and Coriell [4] established an equation describing the variation of the slope of the liquidus of the phase diagram with the speed under the solute trapping effect, also valid for dilute solutions:

$$m_v = \frac{1 - k_v[1 - \ln(k_v/k_e)]}{1 - k_e} m_e, \tag{2}$$

where m_e is the slope of the liquidus of the equilibrium phase diagram. Using these two relations, for a given speed we can define a 'kinetic' phase diagram which gives the respective compositions of the solid and liquid in contact at an interface moving at that speed. Figure 4 shows a simple eutectic phase diagram and its evolution for various solidification speeds. The movement of the liquidus causes a change of eutectic temperature and composition with the speed.

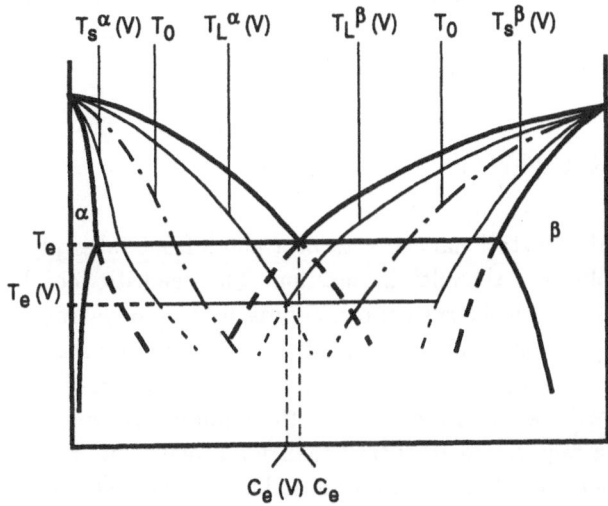

Figure 4: Simple eutectic phase diagram showing the kinetic evolution for a given velocity V. The liquidus and solidus converge to the lines T_0 as the velocity increases. Due to the change of the liquidus, the eutectic composition and temperature vary with the speed.

4.1. PLANE FRONT GROWTH

In the stationary state, a phase which develops with a plane front must have the initial composition of the system, C_0, otherwise solute would be created or destroyed. The growth temperature of a plane front, $T_p(V_s)$, is thus the temperature of the kinetic solidus of the system minus the attachment kinetic undercooling, which becomes large at very high solidification speeds. For dilute solutions, this is expressed mathematically by the equation [3]

$$T_p = T_f + C_l^* m_v - (R_g T_f / \Delta S_f) V / V_0, \qquad (3)$$

where T_f is the melting temperature of the pure phase, $C_l^* = C_0 / k_v$ is the composition of the liquid in contact with the interface, R_g is the gas constant, ΔS_f is the molar melting entropy and V_o is the velocity of sound in the alloy. The last term on the right-hand side of equation (3) (attachment kinetic undercooling) represents the driving force necessary for the atoms to move from their random positions in the liquid to the crystalline positions of the solid. The lower curve in Figure 5 shows the variation of the growth temperature of a plane front with velocity. At low speed, T_p is the same as the equilibrium temperature of the solidus, then because of the solute trapping effect, T_p increases, nearly reaching the temperature T_0 then decreases monotonically under the influence of the attachment kinetic undercooling. We should mention here that the growth of a plane front becomes unstable in time when $dT_p/dV_s > 0$. In this region instabilities occur which produce bands [5, 6, 7] whose study is beyond the scope of this paper.

4.2. DENDRITIC GROWTH

We have seen that the stationary growth of a plane front takes place at the temperature of the solidus of the alloy in question. The dendritic morphology may be seen as a means by which the system can approach thermodynamic equilibrium by letting the solid eject solute laterally and thus grow in steady state with a smaller composition than the nominal composition of the alloy. This is reflected in the phase diagram by a growth temperature closer to the liquidus. However, the creation of a curved interface requires more energy than for a plane interface, and dendritic growth results effectively from a compromise between the efficiency of the solute ejection in allowing growth as near as possible to the temperature of the liquidus, and the minimisation of the curvature of the interface which raises the energy of the system. In the rest of this Section, we present the principal stages of the KGT model [8] for dendritic growth in columns including kinetic effects at high speeds.

Experiment has shown that the tip of a dendrite approaches a paraboloid of revolution. Ivantsov solved the equation of the diffusion field about such a shape, and established the composition of the liquid in contact with the tip of the dendrite, C_l^* [9]:

$$C_l^* = \frac{C_0}{1 - (1 - k_v)\mathrm{Iv(Pe)}},$$ (4)

where $\mathrm{Pe} = RV_s/2D$ is the dimensionless Péclet number with R the radius of the dendrite tip, and Ivantsov's function is given by

$$\mathrm{Iv(Pe)} = \mathrm{Pe} \cdot \exp(\mathrm{Pe}) \cdot E_1(\mathrm{Pe}),$$ (5)

with E_1 the first integral exponential function. Various solutions of this equation are given in [2].

This expression does not give C_l^* as a single function of the growth velocity since the radius of the dendrite remains unknown. For a unique determination, one uses an empirical criterion which states that a dendrite grows with a radius equal to the smallest wavelength of perturbation which can destabilise a plane front (criterion of marginal stability: for details see [10, 11]). This gives

$$-\frac{4\pi^2\Gamma}{R^2} - \frac{\eta_c 2m_v \mathrm{Pe}(1-k_v)C_l^*}{R} - G = 0, \tag{6}$$

with the Gibbs–Thomson parameter $\Gamma = \sigma/\Delta s_f$ where σ is the surface tension at the solid–liquid interface, Δs_f is the entropy of melting by volume, G is the temperature gradient and η_c is a parameter resulting from the stability analysis, defined by

$$\eta_c = 1 - \frac{2k_v}{\sqrt{1 + (2\pi/\mathrm{Pe})^2} - 1 + 2k_v}. \tag{7}$$

The numerical solution of equation (6) by iteration of the Péclet number gives the radius of the dendrite tip as a function of the velocity of growth. The growth temperature of the dendrite, T_d is then given by

$$T_d = T_f + m_v C_l^* - \frac{2\Gamma}{R}. \tag{8}$$

The third term of this equation represents the undercooling required to create the curvature of the interface. The velocity affects the growth temperature through the terms m_v (loss of local equilibrium at the interface) and C_l^* (chemical undercooling ensuring the diffusion of the solute).

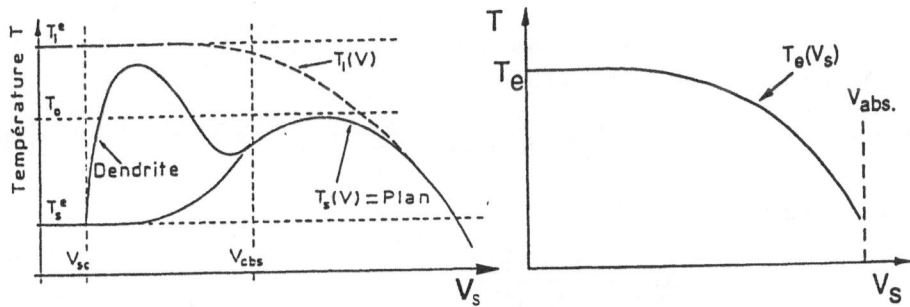

Figure 5: Temperature–velocity curves indicating the variation of the interface temperatures of the dendritic front and the plane front with the growth velocity.

Figure 6: Temperature–velocity curve indicating the variation of the interface temperature of a eutectic front in a stationary state with the growth velocity.

Figure 5 illustrates the behaviour of the dendritic growth temperature as a function of the growth velocity. The dendritic growth model only produces solutions in the range of velocities between V_{sc} and V_{abs}. These bounds are actually the physical limits of dendritic growth: at temperatures below $V_{sc} = GD/\Delta T_0$, where ΔT_0 is the difference between the liquidus and the solidus at C_0, the temperature gradient is high enough to stabilise the plane front by preventing any protuberance from developing ahead of it. At the velocity of absolute stability, V_{abs}, the diffusion distance of the solute ahead of the dendrites is of the same order as the radius of the tip of the dendrites, so that the tips are no longer a better morphology than a plane front for the ejection of solute. As a plane interface is more advantageous in terms of energy, plane front growth again takes place above $V_{abs} = \Delta T_0 D / k_v \Gamma$. It should be noted that the velocity of absolute stability is situated in that part of the plane front velocity which is temporally unstable (see §4.1). The dendritic morphology is then replaced by an oscillating interface which leads to a banded structure rather than to a plane front! Further, the fact that V_{abs} is inferior to the maximum value of the growth temperature of the plane front justifies the neglect of the kinetic undercooling in modelling dendritic growth.

4.3. BINARY EUTECTIC GROWTH

The coupled growth of two adjacent phases constitutes a eutectic structure: the solute rejected by each phase is used by the other. Here we present a summary of the TMK model [12] of eutectic growth at high and low velocities, limited to the case of the growth of lamellar regular eutectics, and presupposing a phase diagram in which the respective partition coefficients of the two phases are equal ($k_\alpha = k_\beta$).

Solving the diffusion equations and computing the diffusion flux in front of the two phases produces a relation between the undercooling, ΔT, the velocity of solidification, V_s and the interlamellar distance, λ:

$$\Delta T = K_1 \lambda V_s + \frac{K_2}{\lambda}, \tag{9}$$

where K_1 and K_2 are given by

$$K_1 = \frac{\bar{m}_\nu \Delta C_0 P}{f_\alpha f_\beta D},$$

$$K_2 = 2\bar{m}_\nu \left(\frac{\Gamma_\alpha \sin \theta_\alpha}{f_\alpha |m_\nu^\alpha|} + \frac{\Gamma_\beta \sin \theta_\beta}{f_\beta m_\nu^\beta} \right),$$

where $\bar{m} = |m_\nu^\alpha| m_\nu^\beta / (|m_\nu^\alpha| + m_\nu^\beta)$ is the mean kinetic slope, ΔC_0 is the difference in concentration of the ends of the eutectic line, f is the volume fraction of the respective phases, θ_α and θ_β are the angles of contact at the junction of the solid and liquid phases caracteristic of the form of the interface, and P is a mathematical series which may be approximated by [13]

$$P \simeq 0.3251 (f_\alpha f_\beta)^{1.63} \cdot [1 - 0.205 \cdot \exp(-24 f_\alpha f_\beta)]. \tag{10}$$

As for dendritic growth, a criterion is needed in order to determine which interlamellar distance λ the system chooses at any speed. Experiment and theory show that for a given speed, regular eutectics grow with an interlamellar distance very close to the value which would give minimum undercooling ΔT. Applying this criterion, the undercooling of a eutectic front is given by

$$\Delta T = \sqrt{K_1 K_2}\,\sqrt{V_s}. \tag{11}$$

Thus the growth temperature can be expressed as a function of the velocity by the simple equation

$$T_e = T_{\mathrm{eut}}(V_s) - \sqrt{K_1 K_2}\,\sqrt{V_s}, \tag{12}$$

where T_{eut} is the kinetic eutectic temperature, which allows for the effects of loss of local equilibrium at the interface (see Figure 4).

Figure 6 shows the shape of the curve $T_e(V_s)$: as the growth velocity tends to 0, the growth temperature tends to the eutectic equilibrium temperature. When the velocity increases, the growth temperature decreases monotonically until it reaches the solidus of one of the two phases so that the plane front becomes stable (V_{abs}).

5. Microstructure Selection Maps

The models presented in the preceding Section can be used simultaneously to calculate maps for the selection of microstructures when the velocity and composition are known. For a fixed composition, the growth kinetics of all the structures in competition must be calculated, giving the respective variation of growth temperature with velocity. The criterion of highest growth temperature is used to decide which structure will effectively grow for a given condition, as seen in Figure 7 for an Al-1.2 at%Fe alloy. By repeating this procedure for a range of compositions, it is possible to elaborate a map showing the lines of transition between microstructures in the composition–growth-velocity plane. Figure 8 shows such a map calculated for the Al-Fe system (broken lines) and superimposed on the transitions determined by experiments (continuous lines) of directional solidification at high and low speeds.

The *same information* can be shown on a composition–temperature diagram, in conjunction with the phase diagram of the system. Figure 9a shows in this form the transitions of the Al-Fe system calculated at low speed and Figure 9b is a magnified view of the zone of transition from the stable Al-Al$_3$Fe eutectic to the metastable Al$_6$Fe eutectic. Lines of isovalues of the interlamellar distance calulated for the two eutectics have been added to give an idea of the size of the microstructure obtained. The representation of a microstructure selection map in the composition–temperature plane may lead to confusion as it produces a hybrid diagram: the higher part of the diagram, above the eutectic line, is an equilibrium diagram, but the lower part is a kinetic diagram. Furthermore, in general it does not provide directly applicable information since the growth temperature is not a known and controlable parameter of the directional solidification procedure. It is, however, frequently used to indicate the form and size of the coupled zone of a system in relation to its phase diagram.

The calculation and usage of these microstructure selection maps is subject to several considerations. First of all, it should be noted that the velocities of transition are highly sensitive to small differences of growth temperature. This phenomenon is easily understood from Figure 7 which has a logarithmic scale for the velocity as compared with a linear temperature scale: for example if the growth temperature curve of the metastable eutectic were uniformly raised by as little as 1 K as a result of an inaccurate value of the metastable eutectic temperature of the system, then the velocity of transition between the two eutectics would be lowered by a factor of 10! This high sensitivity of the transition velocities to the calculated growth temperatures requires that the thermo-physical data for the system under consideration be extremely precise, which is rarely the case. Certain of the data, as for example the coefficient of diffusion in a liquid metal or the metastable extensions of the phase diagrams, are often inadequately known. Despite this, even in these cases, the use of growth models to complement experimental data on velocities of transition between microstructures often enables the missing thermophysical parameters to be determined. This procedure, which was used for the Al-Fe system presented in Figures 8 and 9, provides a useful means of obtaining parameters which would be difficult or impossible to measure directly.

Although in general the calculation of microstructure selection maps has to be completed experimentally, these maps enable us to account for the appearance and disappearance of microstructures and to predict (at least semi-quantitatively) the effect of a change in composition or in one of the parameters of the process on the structure of solidification, provided that there is no nucleation problem for the phases under consideration.

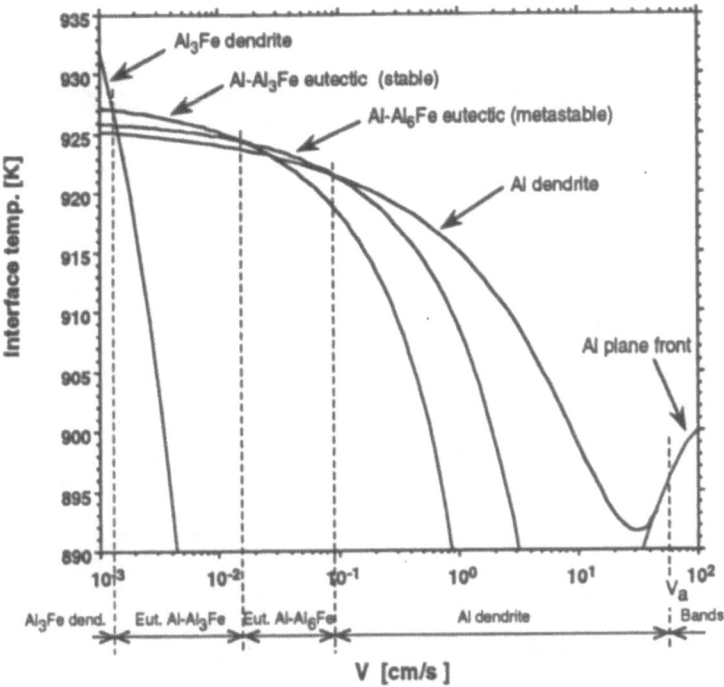

Figure 7: Growth temperature as a function of solidification speed for the competing structures in an Al-1.2 at%Fe alloy. For each structure, the domains of stability, determined by means of the criterion of maximal growth temperature, are indicated at the foot of the diagram.

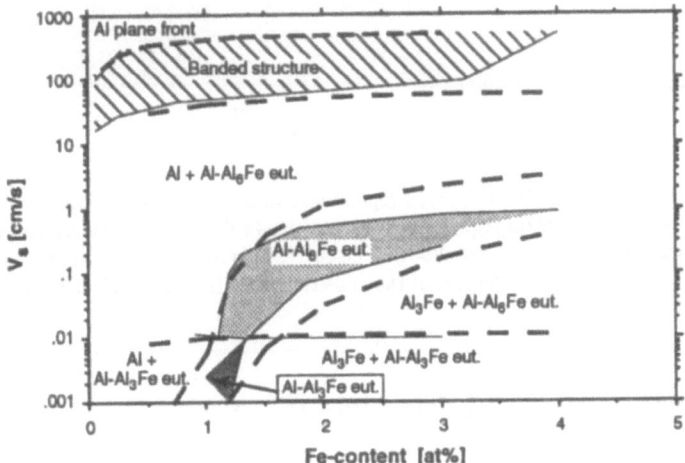

Figure 8: Microstructure selection map for the Al-Fe system. Transitions determined experimentally and by calculation are shown by continuous and broken lines, respectively.

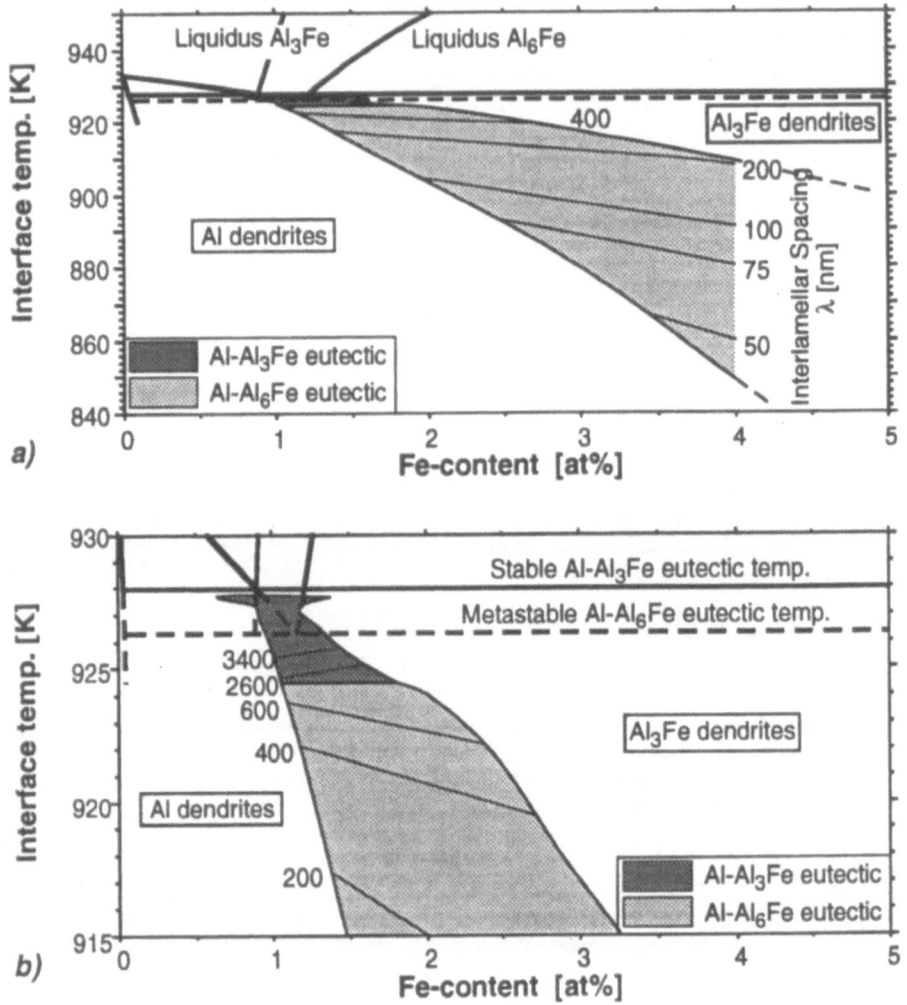

Figure 9: a) The data from Figure 8 represented in the temperature–composition plane. To simplify the diagram, only the transitions calculated at low velocity for the appearance of eutectics are shown. The curves of isovalues of the interlamellar distances give an idea of the size of the eutectic microstructures. b) Detail of a).

6.   Conclusions and Future Prospects

The control of properties obtained by laser treatment is necessarily subordinate to the understanding of the microstructure of solidification upon which these properties depend. Knowledge of the phase diagram of the alloy treated, and of the thermal conditions prevailing during the laser treatment is not sufficient: it is indispensable to consider growth kinetics at a microscopic level since this is where the mechanisms of microstructure selection take place. Using analytical models of solidification of plane, dendritic and eutectic fronts, it is now possible to understand the majority of the microstructures of solidification that can be observed experimentally. Even if the microstructure selection maps thus obtained cannot be used in a totally predictive way because of the imprecise state of what is known about thermophysical properties of alloys, they allow a rational approach to the search for optimal alloys and conditions to supersede empiricism.

It is possible to pair the models presented here with numerical calculations of the temperature field during solidification processing (by finite difference or finite element methods). On the other hand, a realistic treatment of industrial alloys requires the modelling of multi-component alloys. Kinetic growth models already exist, but for such alloys phase diagram information is often missing. To deal with this problem, it is proposed to couple growth kinetics calculations with a commercial programme of phase diagram calculation. The comparison of results from the modelling of solidification structures with experimental observation will perhaps lead to a better understanding of multi-component phase diagrams, and especially of their metastable parts.

7. References

[1] Tammann,G. and Botschwar A.A. (1926) *Z. Anorg. Chem.* **157**, 26.

[2] Kurz, W. and Fisher D.J. (1989) *Fundamentals of solidification*, Trans Tech Publications, Aedermannsdorf (Switzerland).

[3] Aziz, M.J. (1982) *J. Appl. Phys.* **53**, 1158.

[4] Boettinger, W.J. and Coriell, S.R. (1986) in P.R. Sahm, H. Jones, C.M. Adams (eds) *Science and Technology of the Undercooled Melt*, NATO ASI Series E-No114, Martinus Nijhoff, Dordrecht, pp. 81-108.

[5] Carrard, M., Gremaud, M., Zimmermann, M. and Kurz, W. (1992) *Acta metall. mater.* **40**, 983.

[6] Karma, A. and Sarkissian, A. (1993) *Phys. Rev. E* **47**, 513.

[7] Yankov, E.Y., Copley, S.M., Todd, J.A. and Yankova, M.I. (1991) in M.O. Thompson, M.J. Aziz, G.B. Stephenson and D. Cherns (eds) *Kinetics of Phase Transformations*, Materials Research Society Proceedings **205**, 307-312.

[8] Kurz, W., Giovanola, B. and Trivedi, R. (1986) *Acta metall.* **34**, 823.

[9] Ivantsov, G.P. (1947) *Doklady Akademii Nauk SSSR* **58**, 567.

[10] Langer, J.S. and Mller Krumbhaar, H. (1983) *J. Cryst. Growth* **42**, 11.

[11] Kurz, W. and Trivedi, R. (1990) *Acta Metall.* **38**, 1.

[12] Trivedi, R., Magnin, P. and Kurz, W. (1987) *Acta Metall.* **35**, 971.

[13] Magnin, P. and Trivedi, R. (1991) *Acta Metall.* **39**, 453.

SOLIDIFICATION OF AgCu ALLOYS AT HIGH GROWTH RATES PRODUCED BY CONTINUOUS LASER MELT QUENCHING

S.M. COPLEY, J.A. TODD M. YANKOVA AND E.Y. YANKOV
Department of Metallurgical & Materials Engineering
Armour College of Engineering & Science
Illinois Institute of Technology
Chicago, IL 60616

This paper describes research begun in the late 1970s at the University of Southern California and continued since 1990 at the Illinois Institute of Technology with the objective of understanding the process of solidification at high growth rates. Until recently, this research has focussed on AgCu alloys as a model system. High growth rates have been attained by moving AgCu specimens on a rotary stage under a stationary CO_2 continuous wave laser beam.

The interest in AgCu alloys was stimulated by the research of Duwez et al. showing that eutectic growth could be suppressed by a novel high rate quenching technique, "splat cooling," to produce metastable extended solid solutions [1]. Several investigations of the constitution and microstructure of splat cooled AgCu alloys also contributed to this interest [2-4].

The approach of producing high growth rates by continuous laser melt quenching grew out of a long standing interest in applying lasers to materials processing, which also led to the National Science Foundation sponsored workshop entitled "Laser Materials Interactions: Needs and Opportunities for Basic Research," co-organized by one of the authors (Copley) at the University of Southern California in 1977 [5]. In 1978, USC received a major equipment grant for a high power CO_2 laser. This led to the establishment of a high power laser lab, which was moved to the Illinois Institute of Technology in 1990. Our research in high rate solidification, laser machining of ceramic materials and in laser chemical vapor deposition including recent work at IIT was summarized by articles in the proceedings of a recent international symposium [6].

This paper will be organized into eight sections: (1) Structural Studies of Laser Melted and Resolidified Trails; (2) Literature Review; (3) Thermodynamic Considerations; (4) Interface Response Functions; (5) Coupled Growth; (6) Banding Phenomenon; (7) Summary and Recommendations for Future Work; and (8) References.

J. Mazumder et al. (eds.), Laser Processing: Surface Treatment and Film Deposition, 93–119.
© 1996 *Kluwer Academic Publishers.*

1. Structural Studies of Laser Melted and Resolidified Trails

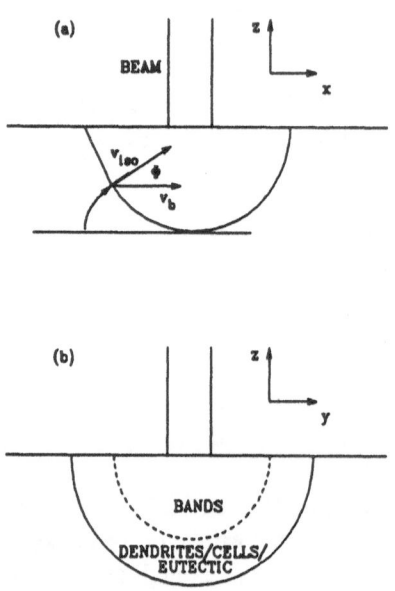

Figure 1. Schematic diagrams showing: (a) the angular relationship, Φ, between the beam velocity, v_b, and the isotherm velocity, v_{iso}, shown here at the midplane of the melt pool. The dark lines show the approximate position and path of the solid/liquid interface; (b) microstructures observed in the melt pool.

Presented here are observations of laser melted and resolidified trails in Ag-Cu alloys obtained at beam velocities ranging from 0.7 to 34 cm s^{-1}. When the laser beam is scanned along the surface (positive x-direction) with a velocity v_b a melt pool with the shape given in Fig. 1a moves with the beam. The isotherms move with the velocity, $v_{iso} = v_b \cos \Phi$. Calculations employing the thermal analysis of Kou, Hsu and Mehrabian [7] for the beam velocities of this investigation and for the thermal properties of AgCu alloys, indicate that, Φ, the angle between the isotherm velocity and beam velocity should be small at the surface where microstructural studies were made. This prediction was supported by observations of trail cross-sections. Based on these observations, and the calculations which include latent heat effects, the isotherm velocity for observations near the surface in our investigation was taken to be 27 cm s^{-1}.

For a Ag-38.3 at. pct Cu (near eutectic) alloy and a beam velocity of 34 cm s^{-1}, a region of dendritic, eutectic and cellular microstructures occurs at the bottom of the laser trail, (z-y cross-section, Fig. 1b) followed by an abrupt change to a banded microstructure in the upper part of the trail (Fig. 2). The bands are approximately 1 micron wide and are composed of a light-etching region, 750 nm wide, and a dark-etching region, 250 nm wide.

A transmission electron micrograph of the boundary between the dark and light etching regions is shown in Fig. 3 [8]. The light etching region is comprised of a single phase, Ag-rich fcc metastable solid solution, while the dark etching region contained a lamellar coupled growth structure, spacing 10 nm, comprised of a Cu-rich phase, 1.5 nm wide, and a Ag-rich phase, 8.5 nm wide.

Beck et al. employed energy dispersive spectrometry to statistically determine the difference, if any, in Ag concentration between the light- and dark-etching regions in a Ag-50 at. pct Cu alloy scanned at 10 cm s^{-1} [9]. They applied the null hypothesis two-tailed test. The null hypothesis tested was that adjoining light- and dark-etching

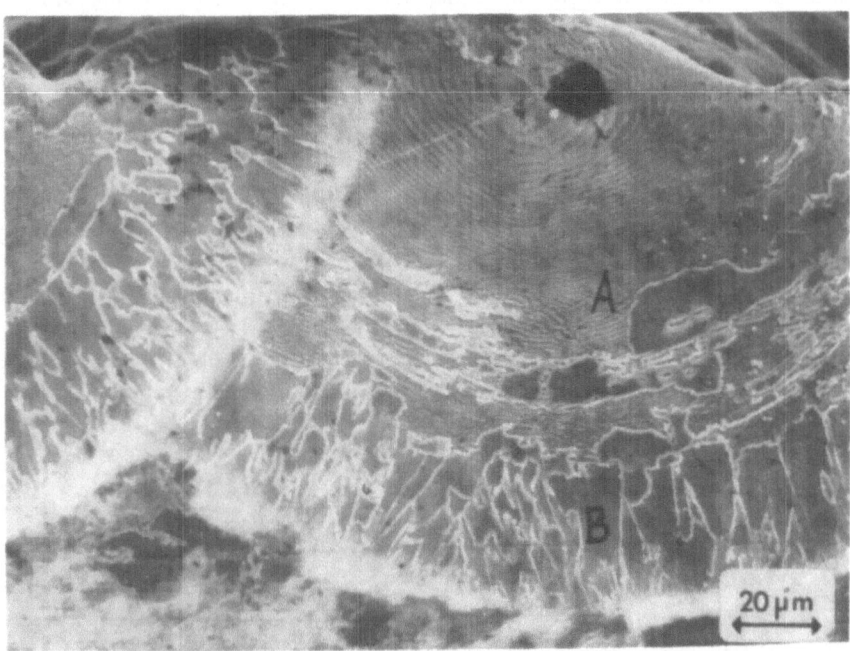

Figure 2. SEM micrograph showing the transition from the region containing dendritic, cellular and eutectic microstructures (B) to the banded region (A).

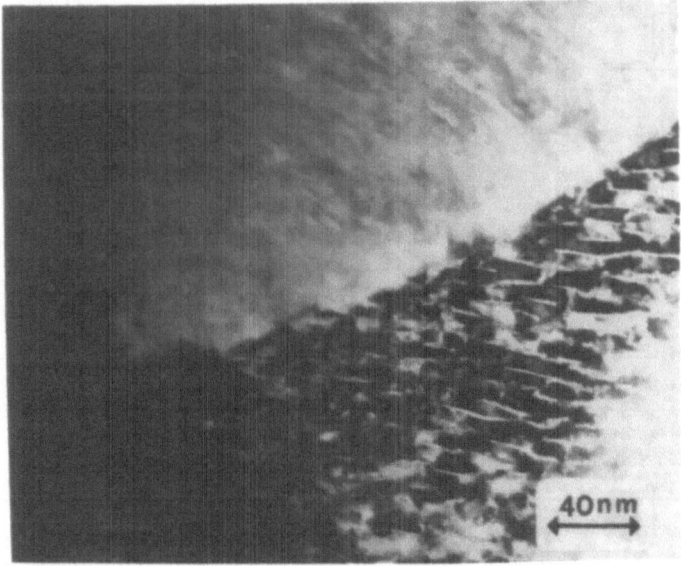

Figure 3. TEM micrograph showing the boundary between the Ag-rich fcc metastable solid solution and the lamellar coupled growth structure [8].

regions had, on the average, the same concentration of Ag. Using 99.9 pct confidence limits, they concluded that the light-etching regions had a significant increase in Ag concentration, i.e., there was less than 1 chance in 1000 that the null hypothesis was actually true. The average difference in Ag concentration was determined to be 1.1 at. pct. Beck et al. noted, however that this value should be treated as a "lower limit" on the concentration difference because the sampling volume of the electron probe was too large to allow an accurate determination of the concentration difference between the light- and dark-etching regions. Subsequently, these results were verified in a Ag-38.3 at. pct Cu alloy scanned at 34 cm s^l [10].

A schematic of a single band is presented in Fig. 4, together with the compositions of the phases determined by X-ray diffraction; selected area diffraction (SAD) and energy dispersive X-ray spectroscopy (EDS) in the transmission electron microscope for the Ag-38.3 at. pct Cu alloy scanned at 34 cm s^l (Table 1).

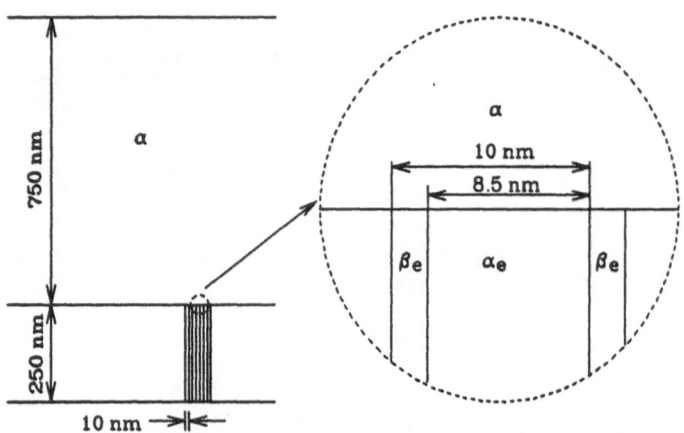

Figure 4. Schematic diagram of a single band.

TABLE 1. Compositions of the phases determined by X-ray diffraction; selected area diffraction (SAD); and energy dispersive X-ray spectroscopy (EDS).

Phase	(X-ray)[10] at. pct Cu	(SAD) at. pct Cu	(EDS) at. pct Cu
α	40.0 ± 1.2	41 ± 5	-
β_e	97.8 ± 0.2	94 ± 5	77.6*
α_e	-	-	32.3 ± 0.5

* Too fine to be resolved

The X-ray diffraction analysis was made by Beck et al., who measured (111) intensities of fcc phases from multiple overlapping trails using a scanning diffractometer [10]. The largest intensity was observed for a peak corresponding to a phase with a lattice spacing of 0.3923 ± 0.0006 nm. The intensity from this phase was found to decrease during annealing indicating its metastable character. Based on Linde's results [2], the composition of this phase is 40.0 $\pm$ 1.2 at. pct Cu. This phase is thought to be the Ag-rich fcc phase, α, comprising the light etching region of the bands. A peak was also observed corresponding to a lattice spacing of 0.3628 $\pm$ 0.0001 nm. The intensity of this peak was observed to grow slightly during annealing. This peak probably includes overlapping intensities from the fcc Cu-rich phase corresponding to the narrow lamellae in the dark etching region, β_o, and the equilibrium fcc Cu-rich phase, which are thought to have about the same lattice spacings and compositions. The composition of the fcc phase(s) corresponding to this peak is 97.8 $\pm$ 0.2 at. pct Cu. A third peak reported by Beck et al. is now thought to be an artifact produced by the complex and possibly partly aged structure [10].

The SAD analysis was unable to isolate the Ag-rich fcc phase in the dark etching region from that in the light etching region due to the aperture size available in the transmission electron microscope, i.e., it was not possible to obtain separate spot patterns that could be analyzed for each phase. The measured lattice parameter was 0.392 $\pm$ 0.002 nm for the Ag-rich fcc phase in the light etching region because this phase comprised a much higher volume fraction of the specimen. The composition of the phase corresponding to this peak is 41 $\pm$ 5 at. pct Cu. The lattice parameter obtained for the fcc Cu rich phase in corresponding to the narrow lamellae in the dark etching region was 0.365 $\pm$ 0.002 nm. The composition of the phase corresponding to this peak is 94 $\pm$ 5 at. pct Cu.

The EDS analysis was performed with scanning transmission electron microscope with a probe diameter of approximately 4 nm, which was larger than the width of the narrow lamellae in the dark-etching region (1.5 nm). Thus the excited volume must have contained the fcc Ag-rich phase as well as the fcc Cu-rich phase accounting for the low average Cu concentration measured. However, the Ag-rich phase was approximately 8.5 nm wide, so that the determination of its composition was possible. The analysis indicated that its composition was 32.1 $\pm$ 0.5 at. pct Cu.

To summarize, results obtained by a variety of techniques for the Ag-38.3 at. pct Cu alloy (near eutectic) scanned at 34 cm s^{-1} indicate that the microstructure of laser melted and resolidified trails varies from dendrites/cells/eutectic at the bottom to bands at the top. This is attributed to the change in growth rate of the solid-melt interface (and the local thermal gradient) as growth occurs from the bottom to the top of the pool (Fig. 1a). The TEM results show that the banded region is comprised of alternating regions of metastable fcc solid solution and coupled growth. Both the X-ray diffraction results and the SAD results indicate that a major constituent in the banded region near the surface is a metastable fcc solid solution. The X-ray results indicate that the average solute content of this constituent (40.0 at. pct Cu) exceeds the average solute content of the alloy (38.3 at. pct Cu) suggesting that some macroscopic segregation may be occurring from the bottom to the top of the trail. The coupled growth region comprises a Ag-rich fcc phase and a Cu-rich fcc phase with average solute contents of 32.3 and 97.8 at. pct Cu, respectively. It can be easily shown by transport considerations that a coupled growth structure with this spacing and with phases differing in composition by this amount must have been formed at a solid-melt interface and not subsequent to solidification in the solid state.

Investigations of Ag-33.8 at. pct Cu and Ag-50 at. pct Cu alloys show that the light-etching bands (metastable fcc solid solution) and the dark-etching bands differ in composition with the dark-etching bands containing slightly more Cu.

2. Literature Review

The primary objective of this section is to recognize the results of others that have helped build our understanding of the solidification of Ag-Cu alloys at high growth rates. A further objective is to place our own results in the context of research going on elsewhere.

The possibility of forming metastable extended solid solutions at highly undercooled solid-melt interfaces was first demonstrated in 1960 by Duwez, Willens and Klement, Jr. in Ag-Cu alloys [1]. They used the "splat cooling" technique to attain large undercoolings in the melt prior to nucleation of the solid phase. In this technique, a small volume of melt is quenched by flattening it against a massive heat sink.

In 1969, Baker and Cahn reported an experiment that unequivocally demonstrated solidification could occur under conditions where local equilibrium at the solid-melt interface was impossible [12]. In the Cd-Zn system, the Zn-rich solid solution has a retrograde solidus, i.e. there is a maximum in the equilibrium/metastable equilibrium solidus composition beyond which solidification at an interface in local equilibrium can not occur. Baker and Cahn splat cooled alloys with Cd concentrations exceeding this maximum to form the metastable extended solid solutions. Not only did their results show that local equilibrium could be violated during local solidification but it gave insight into the type of kinetic theory required to describe solidification at high undercooling (growth rates). Their results showed that solidification could occur even when the chemical potential of the solute (Cd) increased in passing from the melt to the solid phase. They pointed out that such passivity on the part of the Cd could be explained by theories where solute atoms seek the interface and are subsequently buried by the next layer of crystal growth.

The conceptual and theoretical framework that has guided our research to understand the process of solidification at high growth rates was presented at the ASM Seminar on Solidification, October 1969, in a key paper by Baker and Cahn [13]. Briefly, they: (1) employed thermodynamics to elucidate the possible ranges of melt and solid compositions that could result in solidification as a function of interface temperature; (2) elucidated the mathematical analysis of N-component solidification; and (3) offered considerable insight into the nature of the kinetic theory that could be employed to describe solidification at high growth rates.

They pointed out that mathematical analysis of the solidification of an N component alloy requires solving 2N differential equations: 2 for heat flow and 2N-2 for interdiffusion in the solid and melt phases. The solutions are subject to initial conditions and boundary conditions. Of these, 2N+2 boundary conditions describe the solid-melt interface.

The interface boundary conditions specify: (1) temperature continuity; (2) conservation of heat; (3) conservation of mass (N equations); and (4) interface response (N equations called "interface response functions"). In a binary alloy, the interface response functions are of the form $f(T,v,X_S,X_L) = 0$: where T is the interface temperature; v is the interface velocity; X_S is the solute content of the solid at the interface; and X_L is the solute content of the melt at the interface.

The banded microstructure was first reported in 1973 by Elliott, Gagliano and Krauss [11]. They used single laser pulses of a Nd-glass laser to melt quench small volumes of Ag-Cu alloys to form both the banded structure and metastable extended solid solutions. In Elliott et al.'s approach, a Nd-glass laser pulse is used to melt a small volume of alloy at the surface of the sample. In this case, the sample serves as: a heat sink; a seed for crystallization of the equilibrium phases; and, a possible catalyst for nucleation of metastable phases.

The feasibility of forming continuous trails of metastable solidification products by the "continuous laser melt quenching" (CLMQ) technique was first reported in 1980 by Beck, Copley and Bass [14]. A more detailed paper followed in 1981 [9]. In this technique, a small volume of material at the surface of a sample was continuously melted and resolidified by moving the sample at a constant velocity under the beam of a continuous wave (CW) CO_2 laser. The sample serves the same purposes as in the pulsed laser experiment of Elliott et al.; however, in contrast to the pulsed laser and splat cooling experiments, it was hypothesized that in continuous laser melt quenching solidification would occur at a quasi-steady state. In this state, the amount of undercooling and the compositions of solid and melt would be determined at a particular point on the solid-melt interface by the velocity at which the sample was moved under the laser beam and the spatial distribution of intensity of the laser beam. Beck et al. found evidence of the banded structure and metastable extended solid solutions in Ag-25, 50 and 75 at. pct Cu alloys at a beam velocity of 10 cm s^{-1} [9,14]. In an earlier, study Copley et al. had found no evidence of metastable extended solid solutions or bands in samples comprising an Ag-10 at. pct Cu substrate with a 5 micron electroplated Cu coating continuously laser melt quenched at beam velocities ranging from 0.7 to 2.1 cm s^{-1} [15]. The trails contained dendrites surrounded by a fine eutectic structure.

Beck, Copley and Bass extended the range of velocities investigated in detail for continuous laser melt quenching to 34 cm s^{-1} for a wide range of compositions [10]. Their results for an Ag-38.3 at. pct Cu alloy (near eutectic composition) were published in 1982 and have been discussed in Section 1. Following Sundquist [16], they applied the two-parameter solution model of Lumsden [17] to generate the first metastable phase diagram for the Ag-Cu system showing the metastable solidi lines, liquidi lines and the T_o curve (the locus of composition/temperature points where the melt and the solid phase have equal Gibbs free energies). They used this diagram to interpret their observations. They correctly identified the bands as arising from a planar, oscillating interface motion due the diffusional instability first predicted by Baker and Cahn [15].

Jackson, Gilmer and Leamy [34], and Aziz [18] presented models for solute redistribution during rapid solidification in 1980 and 1982, respectively. The models,

100

which give the same equation for the partition coefficient as a function of interface velocity and melt composition, are limited to dilute solutions. They have one unknown parameter that must be evaluated, v_D, the maximum speed of interdiffusion at infinite driving force. Coupled to a second interface response function giving the interface temperature as a function of interface velocity and melt composition and macroscopic heat- and solute-diffusion equations, this equation forms a complete description of one-dimensional growth for dilute solutions.

Coriell and Sekerka used linear perturbation theory to study morphological instability for rapid directional solidification at a constant velocity under conditions where there is significant departure from local equilibrium at an initially planar solid-melt interface. In 1983, they reported that oscillatory instabilities can occur leading to a three-dimensional segregation pattern where periodic solute variations in the transverse directions are modulated by a periodic variation in the growth direction [19].

In 1984, Murray published a metastable phase diagram for Ag-Cu based on a three-parameter solution model [20]. This diagram predicts retrograde solubility for both the Ag-rich and Cu-rich solid solutions, and is the diagram used in this study.

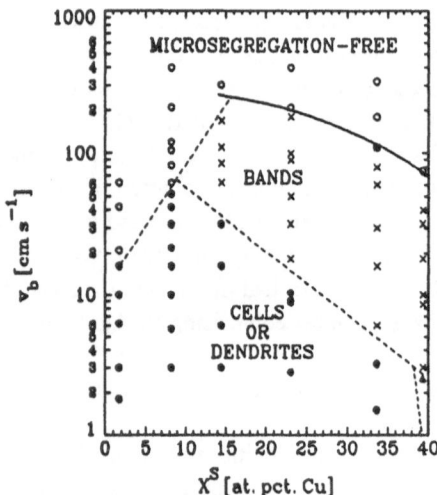

Figure 5. Diagram indicating observed microstructures as a function of beam velocity and atomic percent Cu [21].

Figure 5 summarizes the results of detailed microstructural studies on rapidly solidified hypoeutectic Ag-Cu alloys published in 1984 by Boettinger et al. employing electron beam melting and resolidification [21]. At low velocities, Boettinger et al. observed cellular or dendritic growth for compositions up to 40 at. pct Cu giving way to coupled lamellar growth near the eutectic composition. In dilute alloys, a transition from cellular-dendritic to microsegregation-free growth took place at a critical velocity that increased with increasing Cu content. Beyond 9 at. pct Cu, the banded structure intruded. No composition difference was detected between the light- and dark-etching bands; however-

er, analysis was not carried out on alloys as concentrated as the Ag- 38.3 and 50 at. pct Cu alloys investigated by Beck et al., where such concentration differences were reported [9,10].

Boettinger et al. interpreted their results as follows [21]:
* In dilute alloys (less than the solidus retrograde of 9 wt pct Cu), the transition from cellular to microsegregation free solidification is satisfactorily predicted by the absolute stability branch of the morphological stability theory of the solid-melt interface [22].

* There are maximum growth velocities for cellular and eutectic growth, which decrease with increasing Cu content. These were attributed to the increasing difficulty of solute redistribution due to the decreasing interface temperature and the temperature dependence of the liquid diffusion coefficient. A maximum growth velocity for eutectic growth was calculated; however, the calculation does not take solute trapping into account and thus overestimates the amount of undercooling.

* The occurrence of microsegregation free solidification (> 9 wt. pct Cu) is due to solute trapping.

* Banding occurs when the maximum velocity for cellular (or eutectic) growth is less than the velocity of the isotherms. Consequently, the solid-melt interface falls behind the isotherms. As it falls further behind, a melt zone of the same composition as the average alloy composition forms which is well below the T_o temperature. At some point, solid of composition equal to the liquid forms and rushes forward with a flat interface to catch up with the advancing isotherms.

Both the Baker and Cahn's diffusional instability [13] as proposed by Beck et al. [10], and Coriell and Sekerka's oscillatory morphological instability [19] were rejected by Boettinger et al. as providing a satisfactory explanation for bands.

In 1988, Aziz and Kaplan extended the continuous growth model for solute trapping in dilute solutions by Aziz [18] to include concentrated solutions [23]. It gives general interface response functions, which can be used to calculate simultaneously the velocity of a planar interface, the temperature of the interface and chemical compositions of the solid and melt phases at the interface. As in the dilute solution version, it has one unknown parameter that must be evaluated, v_D, the maximum speed of interdiffusion at infinite driving force.

The results of a related investigation by one of the authors (Todd) of the interphase precipitation reaction in vanadium steels [24], along with the development of general interface response functions applicable to concentrated as well as dilute solutions by Aziz and Kaplan, rejuvenated our interest in the banding phenomenon in Ag-Cu alloys. Todd proposed that the development of banded microstructures in eutectic alloys is a solidification analog to the formation of sheets of precipitates by the interphase precipitation reaction in eutectoid systems [25]. Although the two systems are modeled differently, they have several important similarities. In both cases, solute balance is preserved while solute is redistributed parallel to the growth direction, producing equally spaced sheets containing a second phase. Interface motion is cyclic involving a long period during which growth proceeds with a monotonically decreasing velocity leading to a nucleation event at the interface producing a sheet of precipitates or a band of coupled growth.

This interest led to (1) an effort to determine v_D and to estimate the chemical potentials for Ag-Cu alloys so that the general Aziz-Kaplan interface response functions could be evaluated; (2) an effort to solve the two interface response functions and to display the results in an easily applied form (i.e., solve two simultaneous transcendental equations in four unknowns by specifying two of the unknowns and solving for the other two); (3) a more detailed investigation of the banded microstructure in a Ag-38.3 at. pct Cu alloy scanned at 34 cms[-1]; and (4) applying the results to the banding phenomenon. In this research, which was

102

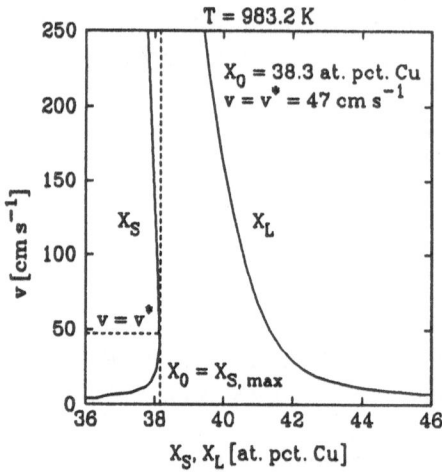

Figure 6. Condition for microsegregation-free solidification for an Ag-Cu alloy with composition $X_0 = 38.3$ at. pct. Cu.

presented and published in the period from 1990 to 1992, Copley and Todd collaborated with E.Y. Yankov and M.I. Yankova.

Yankov, Yankova, Copley and Todd (1991) proposed a method for determining v_D based on the relationship between: (1) the experimentally determined beam velocity, $v_b(X_0)$, for an alloy with composition X_0 to change from banded to microsegregation-free growth (Fig. 5); and (2) the interface velocity, $v^*(X_0)$, at which the solid of of maximum solute content with composition X_0 grows from the melt according to the Aziz-Kaplan interface response functions, as illustrated in Fig. 6 [26]. In this figure, compositions of the solid phase, X_S, and liquid (melt) phase, X_L, co-existing at the solid-melt interface, are plotted on the abscissa and the corresponding interface velocities, v, are plotted on the ordinate. With increasing undercooling, the X_S and X_L curves move to the right. Yankov et al. proposed that the parameter v_D could be determined for a specific composition, X_0, from the condition that $v^*(X_0) = v_b^*(X_0)\cos \Phi$, where Φ is the angle between the beam velocity and the isotherm velocity (see Fig. 1).

For example, in Fig. 6, solutions for the interface response functions are displayed for an undercooling of 68.8 K (interface temperature, T = 983.2 K). The undercooling of 68.8 K was selected by increasing the undercooling to the point that the X_S curve was just tangent to the vertical dashed line corresponding to the alloy composition of interest, $X_0 = 38.3$ at. pct Cu. Following the proposal by Yankov et al., the parameter v_D, which acts as a scaling parameter for the ordinate of the graph, has been adjusted to give $v^* = 47$ cm s^{-1}. This equals the product of $v_b = 80$ cm s^{-1} (see Fig. 5), and cos $\Phi = 0.59$. This process can be repeated at other compositions to evaluate $v_D(T(X_0))$.

For hypoeutectic Ag-Cu alloys, a fit for $v_D(T(X_o))$ is given by the equation $v_D(T(X_0)) = A(T(X_0)/3 - B)$ where $A = A' \cos \Phi$, $A' = 1$ cm s^{-1} K^{-1}, cos Φ is evaluated using data reported by Boettinger et al. [21], and B = 314.5 K. With this parameter and chemical potentials based on Murray's extrapolated free energy functions for Ag-Cu, solutions for the Aziz-Kaplan interface response functions giving the solid and melt compositions as a function of velocity similar to those shown in Fig. 6 can be calculated for various interface temperatures.

Yankov, Todd and Copley (1991) reported results of a detailed microstructural study of the banded structure in a Ag-38.3 at. pct Cu alloy scanned at 34 cms^{-1} [8]. They showed for the first time that the dark-etching bands in this alloy contained a lamellar coupled growth structure. The spacing of the lamellae observed in the dark etching bands was 10 nm, one half the smallest spacing previously reported by Boettinger et al. for continuous eutectic growth at 2.5 cms^{-1}. Boettinger et al. [21] and others [27] have proposed that coupled growth is impossible beyond a certain limiting velocity, which in the case of Ag-Cu was suggested by Boettinger et al. to be in the range 1-10 cms^{-1}.

Yankov, Copley, Yankova and Todd (1991) presented a model for steady state coupled lamellar growth from the melt [28]. Although, neglecting curvature effects, and using a simplified treatment of the diffusion fields similar to that used by Zener [29] and Tiller [30], it used the general interface response functions developed by Aziz and Kaplan to include the effect of solute trapping in an analysis of coupled growth for the first time. Applied to the Ag-Cu system, the model was shown to fit the available experimental data including observations of coupled growth in the dark-etching bands.

Yankov, Copley, Todd and Yankova (1992) presented a model which explains the origin of banding phenomenon and the transition from banding to microsegregation free growth [31]. It begins with a time dependent analysis of microsegregation free growth (Ag-rich solid solution, light etching band) based on: (1) Murray's analytical free energy functions [20]; and (2) the Aziz-Kaplan interface response functions [23] using Murray's results [20] with v_D determined by the method described by Yankov et al. [26]. This growth proceeds at a monotonically decreasing rate with a build up of solute in front of the solid-melt interface, and a decrease in solute in the solid formed at the interface. The melt-solid interface falls behind the advancing isotherms and thus decreases in temperature. At a critical point, determined experimentally in the theory by the width of the light-etching band, the Cu-rich phase nucleates initiating coupled growth. Quasi-steady-state coupled growth, described by Yankov et al.'s model [28], proceeds until the solute build up is exhausted thus assuring solute balance over one cycle of banded growth. At this point, the interface accelerates and catches up with the advancing isotherms and then the cycle is repeated. This model has been further developed in several recent papers. [32,33]

The availability of experimental techniques such as continuous laser melt quenching and electron beam melting and resolidification along with the development of interface response functions have stimulated the interest of a number of researchers in solidification at high growth rates. Their research has focussed on understanding the ranges of experimental parameters -- growth rate, gradient and composition governing the occurrence of planar, dendritic, cellular, coupled and banded growth. Most of this research has employed interface response functions for dilute solutions combining Jackson et al.'s [34] and Aziz's [18] equation for the partition coefficient with an equation for the interface temperature developed by Boettinger et al. [35,36]. These will be referred to here as the dilute solution interface response functions.

Undoubtedly, the most comprehensive research effort has been that of Kurz at the Swiss Federal Institute of Technology at Lausanne and his collaborators. Since Professor Kurz is presenting this research here at the Advanced Studies Institute, I will discuss their results only briefly.

104

Kurz and his collaborators have developed models for dendritic, eutectic (coupled), banded and planar front growth. At a particular growth rate, they have postulated that the type of growth that will be favored will be that which grows with minimum undercooling (maximum interface temperature).

The essentials of their models that might be applied to the Ag-38.3 at. pct Cu alloy are described here, based on the recent paper by Carrard, Gremaud, Zimmermann and Kurz [37], and are illustrated schematically in Fig. 7. In this diagram, undercooling (interface temperature - equilibrium liquidus temperature) is plotted on the ordinate and growth rate on the abscissa for a specified alloy composition (in this case Ag-38.3 at. pct Cu). At low velocities corresponding to growth near the bottom of the melt pool, dendritic or a mixed dendritic/coupled growth may be favored. In the intermediate velocity regime, this may be replaced by coupled growth. As the velocity is increased banded growth intrudes giving way at the highest velocities to planar front growth.

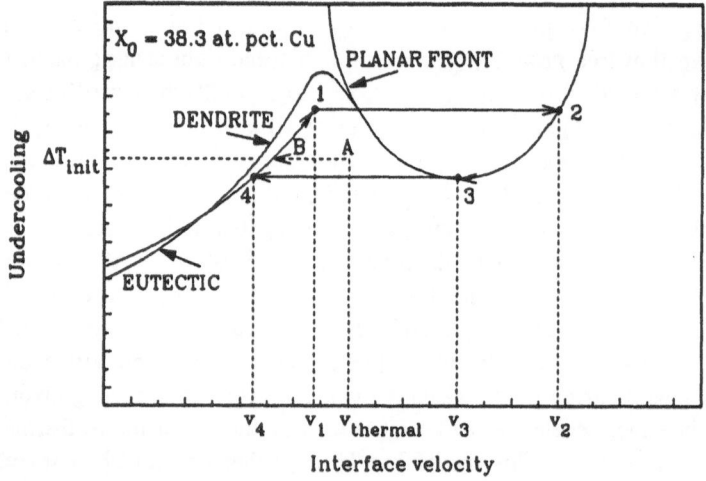

Figure 7. Schematic diagram illustrating undercooling versus interface velocity for planar front, dendritic, coupled and banded growth.

To determine undercooling versus velocity for dendritic growth, Carrard et al. [37] employ the model developed by Kurz, Giovanola and Trivedi [38] with modifications described in their paper. The modified model uses the Aziz-Kaplan interface response functions; however, v_D is determined by a different procedure than that described by Yankov et al. [26].

To determine undercooling versus velocity of coupled growth, Carrard et al. [37] use the model developed by Trivedi, Magnin and Kurz [39], and modified to include solute trapping by Zimmermann, Carrard and Kurz [40]. The modified model uses the dilute solution interface response functions, which may not reliably predict the behavior of the Ag-rich and Cu-rich solid solutions in coupled growth. It predicts a limit on the maximum growth rate for coupled growth that is not predicted by our model as described by Yankov et al. [28].

To predict banded growth, Carrard et al. [37] use the model of banded growth proposed by Gremaud, Carrard and Kurz [41]. This model recognized that bands can be regarded as an example of unstable behavior observed in many mechanical, electrical, biological and hydrodynamic systems where the characteristic function connecting the driving force (in the case of bands, undercooling) and the associated flux (in the case of bands, growth rate) has a region of negative slope (see Fig. 7). This type of instability has been described by Kubin and Estrin [42].

Carrard et al. [37] point out that stable planar front growth is not possible in the region between v_1 and v_3, because the slope of the undercooling versus growth rate curve is negative. They propose that growth occurs instead by alternating periods of steady state growth as illustrated in Fig. 7. Carrard et al.'s model uses the frozen temperature approximation (FTA), which assumes that the thermal properties of the solid and melt phases are the same and neglects latent heat effects -- i.e., it decouples the processes of thermal and mass transport.

According to their model, an interface with undercooling, ΔT_{init}, finding itself in a thermal field moving at a velocity, $v_{thermal}$, in the unstable region (A) propagates either in a coupled growth mode or a planar front mode. If, for example, it chooses initially the coupled growth mode (B), then it will fall increasingly behind the thermal field and its undercooling will increase. Consequently, its velocity will increase until it reaches v_1, the limiting velocity for coupled growth. At this point, it is proposed that it accelerates instantaneously to v_2, and propagates in the planar front mode. As v_2, is greater than the velocity of the thermal field, the interface undercooling now decreases. Consequently, its velocity will decrease until it reaches v_3. At this point, it is proposed that it decelerates instantaneously to v_4 and propagates again in the coupled growth mode. Since v_4 is less than the velocity of the thermal field, its undercooling and hence its velocity will increase causing a repeat of the cycle.

Carrard et al.'s model [37] provides an attractive qualitative explanation for banded growth but does not provide a satisfactory basis for a quantitative analysis of the phenomenon. It does not attempt to attain a true time dependent description of the phenomenon but instead describes it in terms of steady state solutions. It does not account for the observed difference in composition between the light-etching and dark-etching bands. [9,10] Also the validity of the FTA at high growth velocities is doubtful based on recent work by Huntley and Davis [43] and Karma and Sarkissian [44]. However, both Huntley's and Davis' and Karma's and Sarkissian's analysis of banded growth use the dilute solution approximation and thus cannot reliably predict the behavior of the Ag-38.3 at. pct Cu alloy.

In summary, our approach to understanding solidification at high growth rates in near eutectic, Ag-38.3 at. pct Cu alloy, can be distinguished from the work of others by one or more of the following features:
* We have used the Aziz-Kaplan interface response functions in our models and calculations describing planar front single phase, coupled growth and banded growth. It will be shown in Section 4 that serious errors arise from applying the dilute solution interface response functions at concentrations exceeding about 10 at. pct Cu. The value of v_D used in evaluating these functions has been determined experimentally from measurements of the interface velocity at the transition from banded to

microsegregation-free (planar front) growth.

* We have developed a model for banded growth using a time dependent solution of the mass transport and thermal transport equations for the single phase, planar front growth.

* We have developed a model for banded growth that accounts for observed differences in the composition between the light-etching and dark etching band by applying conservation of mass to the materials solidified during each cycle.

Recent results suggest the importance of including the effects of latent heat generation at the solid-melt interface in our models. [43,44]

3. Thermodynamic Considerations

To understand the solidification of Ag-Cu alloys at high growth rates, a knowledge of free energy of the solid and melt phases as a function of interface temperature and composition is required. Since considerable undercooling occurs during solidification at high growth rates, free energies at such temperatures must be extrapolated from higher temperatures where equilibrium measurements of thermodynamic quantities can be made.

Figure 8(a) plots the free energy of the solid and melt phases versus at. pct Cu for Ag-Cu alloys at 1030 K calculated using the three-parameter extrapolation model of Murray. [20] The eutectic temperature is 1052 K and thus the diagram corresponds to an undercooling of 22 K.

Figure 8(b) shows the relationship between the free energy versus composition diagram and the phase diagram for Ag-Cu alloys at 1030 K. It shows the use of the common tangent rule to determine the composition of the solid phase in metastable equilibrium with the melt phase at 1030 K. This is the composition of the Ag-rich solid solution that would co-exist with melt at 1030 K if the Cu-rich solid solution failed to nucleate. Also shown is a point on the T_0 curve determined by the point of intersection of the solid and melt free energy curves. This represents an upper bound on the interface

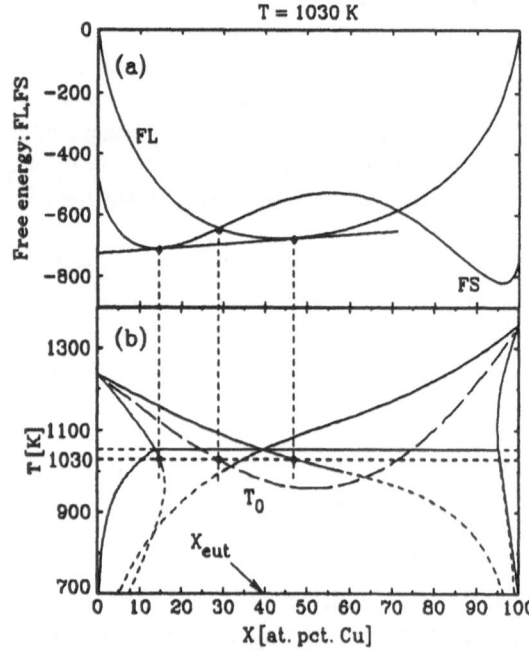

Figure 8. Ag-Cu system: (a) free energy versus composition diagram for solid and melt phases at 1030 K; (b) phase diagram showing T_0 curve and metastable solidi and liquidi.

temperature for a melt to form a solid phase of the same composition. By applying these constructions at a series of temperatures, the T_0 curve and the metastable solidi and liquidi curves for Ag-Cu are determined. Both the Ag-rich and Cu-rich solid solutions exhibit retrograde solubility. Hence for each phase there is an upper limit on solute content beyond which metastable equilibrium with a melt phase is not possible.

Baker and Cahn [13] describe thermodynamics as follows -- "the science of the impossible. It enables you to tell with certainty what cannot happen. Thermodynamics is noncommittal about the things that are possible." This is illustrated in the case of solidification from the melt by the Baker-Cahn diagram for the Ag-rich solid solution at 1030 K shown in Fig. 9(b). It shows the range of compositions (composition lying outside the closed loop) that cannot be formed from a specified melt composition (plotted on the abscissa). Alternatively, it indicates a wide range of compositions that can be formed from the specified melt but is noncommittal regarding which one will actually form. With decreasing interface temperature (increasing undercooling) the closed loop expands. Again shown in Fig. 9(a), is the free-energy versus composition diagram so the construction of the Baker-Cahn diagram can be described.

Points on the loop of the Baker-Cahn diagram are determined using the tangent to curve rule. Referring to Fig 9(a), the free energy change per mole reacted for a reaction in a closed system that transfers small amounts of components of a specified composition (for the example here, Ag-38.3 at. pct Cu) from a melt phase of this composition to an Ag-rich solid solution phase can be obtained graphically by first constructing the tangent to the free energy curve of the melt phase at the composition Ag-38.3 at. pct Cu and then reading the vertical distance from this tangent to the solid free energy curve at the composition of the solid phase of interest. It can be seen in the case of the Ag-38.3 at. pct Cu melt that transfer of a small amount of components with the melt composition to solid phases with compositions lying between points 1 and 2 results in an increase in negative free energy. Thus formation of solid phases with

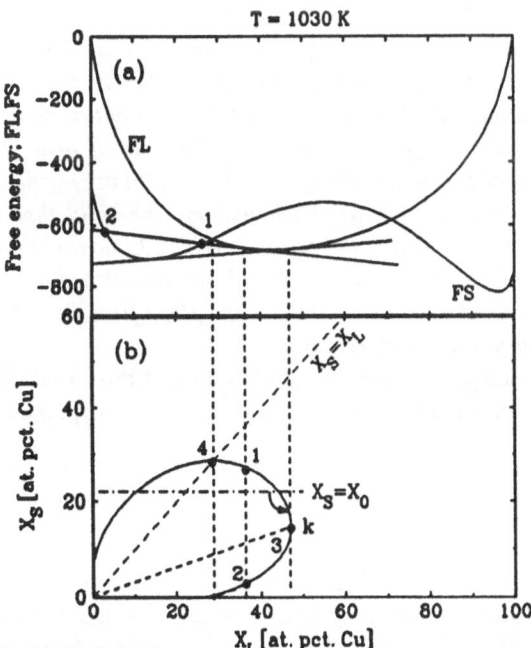

Figure 9. Ag-Cu system: (a) free energy versus composition diagram for solid and melt phases at 1030 K; (b) Baker-Cahn diagram showing the range of solid compositions can be formed from a melt of a specific composition.

compositions in this range from such a melt is thermodynamically possible. For all other solid compositions, it is impossible. Points 1 and 2 on the Baker-Cahn diagram lie on the closed loop and indicate the range of possible solid composition for an Ag-38.3 at. pct Cu melt. By repeating this construction for the entire range of liquid compositions all the points on the loop are determined.

Two points on the closed loop, Fig. 9(b) deserve special attention. The fraction X_S/X_L at point 3 corresponds to the equilibrium partition coefficient. It is determined by applying the common tangent rule as illustrated in Fig. 9(a) to find X_S and X_L. It corresponds to the melt of maximum solute content that can form a solid phase at 1030 K. Point 4 indicates the T_0 composition at 1030 K. It corresponds to the solid of maximum solute content that can form from the melt at 1030 K, and lies on the $X_S = X_L$ line.

Steady state solidification requires that the composition of the solid formed from the melt equal the alloys overall composition X_0. For example, the range of melt compositions that can form a solid composition of $X_0 = 22$ at. pct Cu lie on the dash-dot line within the closed loop in Fig. 9(b). Included in this range of possible compositions is the composition point for diffusionless solidification where the composition of the solid is equal to the composition of the liquid from which it is formed. However, it is very likely that there will be a diffusional layer ahead of the interface, and the composition of the melt at the interface, X_L, will differ from X_0.

Baker and Cahn noted a possible diffusional instability that might interfere with steady state solidification occurring at composition points at or near the loop boundary between points 3 and 4 as indicated by the down arrow in Fig. 9(b). Adapting their words to our figure -- *If momentarily the composition of the solid drops below X_0, conservation of mass requires that the liquid be enriched in the minor component. Thus a downward fluctuation ($X_S < X_0$) from the horizontal line should result in a shift to the right (X_L increasing). If the system finds itself near the boundary, it can not re-establish steady state (climbing to the horizontal line) without decreasing X_L. But, because the system is near the boundary, the solid that is forming is below the average composition X_0 and the excess is rejected into the liquid making a reduction in X_L unlikely. Thus, steady state cannot be re-established at this temperature, and since we impose a velocity on the system, the plane front interface either breaks up or lags behind to a lower temperature.*

Beck et al. identified banded growth to be a manifestation of this instability [10]. In Section 4, we will offer additional arguments in support of this hypothesis.

4. Interface Response Functions

At high interface velocities, equilibrium conditions no longer apply and solute trapping occurs. The solid composition, liquid composition, velocity and temperature now become functionally related. The Aziz-Kaplan [23] interface response functions relate these four variables as follows.

$$f(X_L, X_S, v, T) = 0; \qquad g(X_L, X_S, v, T) = 0$$

The first interface response function describes the partition coefficient:

$$K(v) = \frac{X_S}{X_L} = \frac{\dfrac{v}{v_D} + g_e}{\dfrac{v}{v_D} + 1 - (1 - g_e)X_L} \tag{1}$$

where X_S, X_L = atomic fraction of solute in the solid and liquid, respectively; v_D = velocity of solute-solvent redistribution at infinite driving force; g_e = partitioning parameter:

$$g_e(X_L, X_S, T) = \exp\left[-\frac{\left(\Delta\mu_B' - \Delta\mu_A'\right)}{RT}\right] \tag{2}$$

and μ' = redistribution potential:

$$\mu'(X, T) = \mu(X, T) - RT \ln X \tag{3}$$

The second interface response function gives the velocity:

$$v = v_o \left[1 - \exp\left[\frac{\Delta G_{DF}}{RT}\right]\right] \tag{4}$$

where v_0 = sound wave velocity, 3×10^5 cm s^{-1}, and

$$\Delta G_{DF} = X_S \Delta\mu_B + (1 - X_S)\Delta\mu_A \tag{5}$$

By comparing the velocities for microsegregation-free growth determined experimentally by Boettinger et al. [21] (Fig. 5), with the interface velocity $v^*(T(X_0))$ at which the solid of maximum solute content with composition X_0 grows from the melt the following expression was derived: $v_D(T(X_0)) = A(T(X_0)/3 - B)$ where $A = 0.5$ cm s^{-1} K^{-1}, $T(X_0)$ = interface temperature (K), which is a function of X_0, and $B = 314.5$ K.

It should be noted that in Fig 5, the transition from bands to microsegregation-free growth was reported by Boettinger et al. [21] in terms of the beam velocity, v_b. To obtain the interface velocity at which microsegregation-free growth first occurs, the beam velocity, v_b, must be multiplied by cos Φ (see Fig. 1), which according to Boettinger et al. equals 0.5. [21]

Incorporation of the expression for v_D, Eq. (6), gives the "scaled" Aziz-Kaplan functions which are used throughout this paper. In Fig. 6, the solutions to the

110

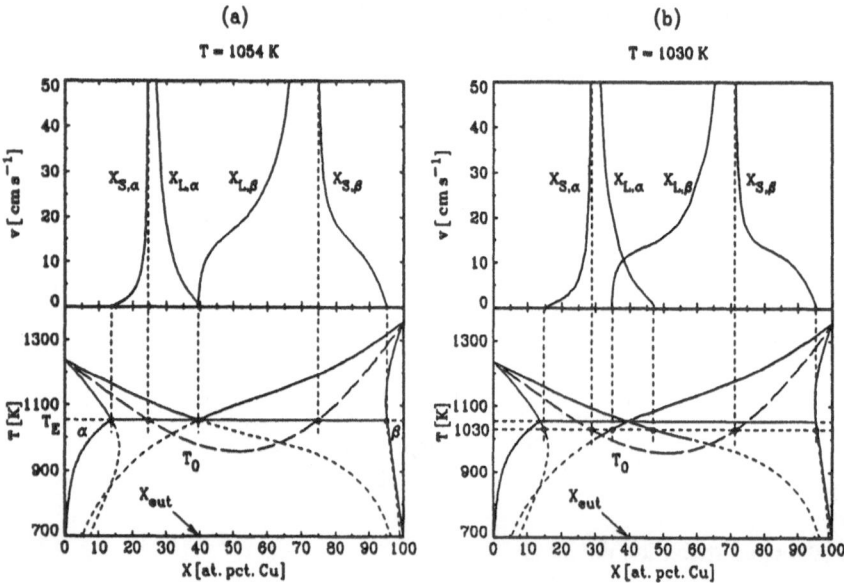

Figure 10. Relationships between solutions of the scaled Aziz-Kaplan interface response functions and the phase diagram at the eutectic temperature, 1054 K, and at 1030 K.

scaled Aziz-Kaplan interface response functions for the Ag-rich solid solution were displayed for a specific interface temperature by plotting velocity on the ordinate, and the composition of the solid and melt phases on the abscissa. In Fig. 10, solutions to the scaled Aziz and Kaplan interface response functions are plotted in a similar manner for both the Ag-rich, (α), and Cu-rich, (β) phases for temperatures of 1054 K (eutectic temperature) and 1030 in Figs. 10(a) and 10(b), respectively. The relationships between these solutions and the phase diagram are also illustrated in Fig. 10. At zero velocity and 1054 K, $X_{L,\alpha} = X_{L,\beta}$ = eutectic composition and $X_{S,\alpha}$, $X_{S,\beta}$ correspond to the eutectic compositions of the α (Ag-rich) and β (Cu-rich) solid solution phases, respectively. At high velocities, the compositions of the α and β phases approach those given by the T_0 curve. When the interface is cooled below the eutectic temperature (Fig. 10(b)), the compositions of the solid and liquid phases at zero velocity are given by the metastable extensions of the solidi and liquidi (dashed lines). As the undercooling increases, the plot for the α phase shifts to the right of the velocity-X_S,X_L diagram, while that for the β phase shifts to the left.

Figure 11 shows the relationship of solutions to the Aziz-Kaplan interface response functions to the Baker-Cahn diagram discussed in Section 3 (see Fig. 9). In

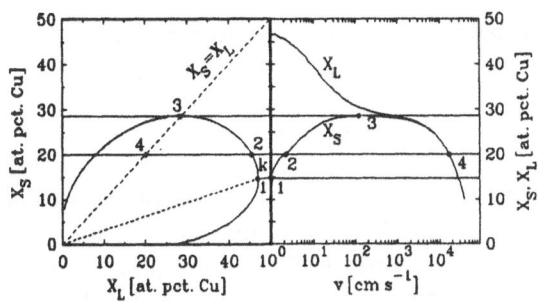

Figure 11. A comparison of solutions to the Aziz-Kaplan interface response functions to the Baker-Cahn diagram.

this figure, which has been drawn for the Ag-rich solid solution at 1030 K, velocity has been plotted on a logarithmic scale so that solutions could be shown at high velocities where the compositions of the solid and melt phases become very nearly equal.

It has already been noted that for a specific undercooling and melt composition, the Baker-Cahn diagram indicates a broad range of solid compositions that *can* form but is noncommittal regarding which one *will* form. Such prediction is in the realm of kinetic theory not thermodynamics. The Aziz-Kaplan interface response functions are derived based on kinetic theory and their solutions specify which solid compositions, within the range of allowed compositions, will actually form as indicated in Fig 11.

The solid composition that grows at zero velocity (metastable equilibrium) is represented by point one. The corresponding melt composition is the melt of highest solute content from which a solid phase can form at 1030 K. Point 3 corresponds to the solid of maximum solute content that can form from a melt phase at 1030 K. Yankov et al. [26] have identified the velocity, $v^*(X_0)$ (point 3, Fig. 11), at which this phase grows as marking the threshold for microsegregation free growth for an alloy of composition X_0. This is the basis of their method for determining v_D.

As we have discussed elsewhere in detail [32], solids forming with a planar interface at $v \geq v^*$ like at point 4 are stable with respect to compositional and thermal perturbations. In contrast, those forming with a planar interface at $v \leq v^*$ like at point 2, are unstable with respect to compositional fluctuations. If, for example, the solute content of such a solid decreases slightly, conservation of mass requires that the melt content increase. However, a melt of higher solute content will solidify at a lower velocity producing a solid of still lower solute content. Thus steady state cannot be re-established at 1030 K and the interface will lag behind to a lower temperature. But, this is precisely the Baker-Cahn instability. Solutions to the Aziz-Kaplan interface response functions lying on either the X_S curve or the Baker-Cahn loop between points 1 and 3 are all subject to this instability, which will be shown in Section 6 to lead to banded growth.

In Section 3, it was pointed out that with increasing undercooling, the closed loop on the Baker-Cahn diagram defining the range of solid compositions that can form from a specified melt composition increases in size, and the X_S and X_L curves describing solutions to the Aziz-Kaplan interface response functions move in the direction of increasing solute content. Thus point 3 can be interpreted as describing

112

either the solid of maximum solute content that can form in a steady state process from a melt with composition X_0 at 1030 K or velocity at which a solid of minimum undercooling (maximum interface temperature) that can grow in a steady state process from a melt with composition X_0 at 1030 K. This observation establishes the relationship between point 3 in Fig. 11 and point 3 in Fig. 7. The conclusions drawn from both diagrams are the same: (1) for velocities greater than or equal to that at point 3, an alloy with composition X_0 undergoes microsegration-free growth; and (2) for velocities less than that at point 3, steady-state growth is not possible and banding is observed. It is apparent that the diffusional instability first noted by Baker and Cahn [13], to which Beck et al [10] correctly attributed the banding phenomenon, is the same instability as that discussed by Gremaud et al. [41], Carrard et al. [37], and is analogous to the instability discussed by Kubin and Estrin [42].

The importance of using the general Aziz-Kaplan interface response functions for alloys with high concentrations of solute is illustrated in Fig. 12. In this graph, k(v), the partition coefficient, is plotted versus interface velocity for various Ag-Cu alloy compositions. The solid line has been calculated using the scaled Aziz-Kaplan general interface response functions (Eqn. (1)). The dashed line has been calculated using the dilute solution interface response functions. While the agreement at 5.7 at. pct Cu is close, for the higher solute content alloys a considerable error results from using the dilute solution interface response functions.

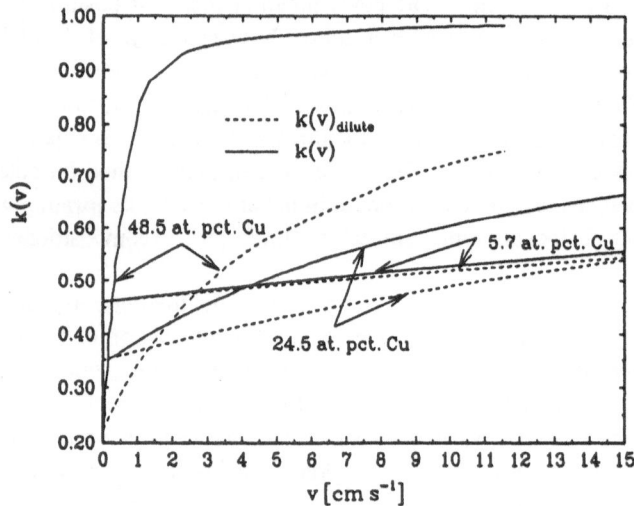

Figure 12. Comparison between the dependence of the partition coefficient on the interface velocity for different compositions calculated according to the dilute solution interface response functions (dashed lines) and the partition coefficient, valid over the entire range of compositions, calculated using the scaled general Aziz-Kaplan interface response functions.

Application of the scaled Aziz-Kaplan interface response functions will now be illustrated in the case of: (1) Coupled Growth at High Velocities; and (2) Banded Growth.

5. Coupled Growth

The coupled growth model developed by Yankov et al. has been described in detail elsewhere [28,32] and will be summarized only briefly here. It is used to describe the coupled growth of two lamellar phases at a planar interface as illustrated in Fig. 13.

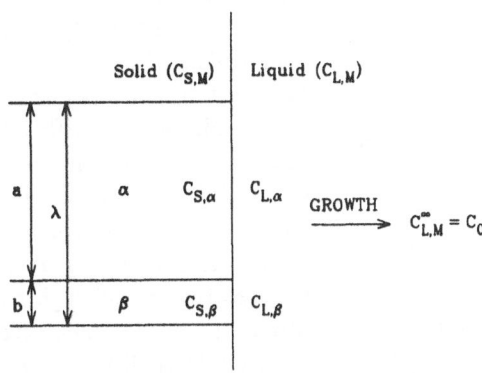

Figure 13. Model for coupled growth.

The major features of the model are:

* Both phases move at the same velocity and interface temperature.

* The mean compositions for each phase at the melt-solid interface are specified by Aziz-Kaplan interface functions.

* An approximate transport treatment is employed that limits diffusion parallel to the interface at high velocities by defining a diffusion layer with a thickness which decreases with increasing velocity.

* The undercooling associated with interphase boundary formation is calculated using the free-energy extrapolations developed by Murray [20].

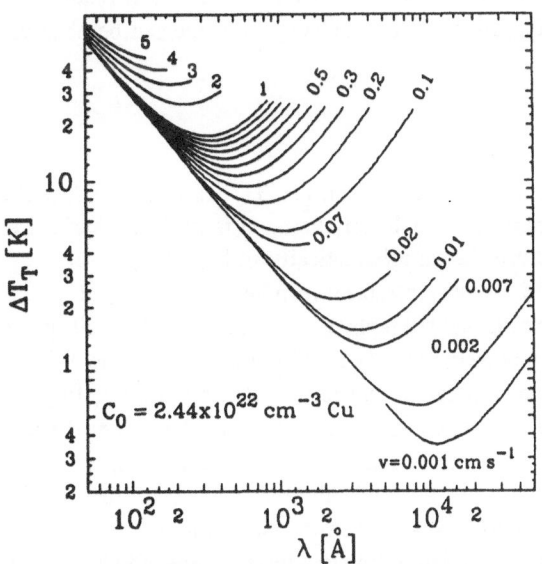

Figure 14. Total undercooling vs lamellar spacing for different growth rates assuming that the spacing for a specific velocity will be that which grows at a minimum total undercooling.

The model predicts the total undercooling for solidification for a eutectic or near eutectic alloy as a function of lamellar spacing and growth rate. Figure 14 shows that the lamellar spacing that grows with minimum undercooling decreases with increasing growth rate. Boettinger et al. found experimentally that the maximum velocity for coupled growth in Ag-Cu alloys was 2.5 cm s^{-1} with a spacing of 20 nm. [21] This model predicts the same velocity for a spacing of 20 nm but does not predict that this growth rate should be a maximum for the Ag-Cu system. Boettinger et al. calculate that

114

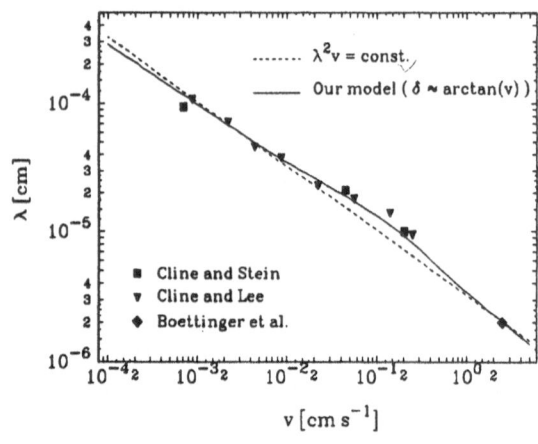

Figure 15. Lamellar spacings at minimum total undercooling vs growth rate. The comparison of our model (solid line) and classical theory (dash-dot line) to experimental data is shown.

for a velocity of 2.5 cm s⁻¹, there is an undercooling of 254 K, which they suggest causes a decrease in the diffusion coefficient leading to a termination of steady state eutectic growth. Our model predicts a total undercooling of 27 K at this velocity. The high undercooling calculated by Boettinger et al. arises from neglecting the effect of solute trapping in their analysis. The observation of a coupled growth structure in Ag-Cu with a spacing of 10 nm, see Fig. 3, although not formed at steady state, suggests the feasibility of attaining lower spacings than 20 nm.

In Fig. 15, the spacings that are realized at the minimum total undercoolings are plotted against growth rate and are compared to the existing experimental data for Ag-Cu from Cline and Stein [45], Cline and Lee [46] and Boettinger et al. [21] It can be seen that the model curve fits the results better than the classical relationship, especially in the region of $v = 0.1$ cm s⁻¹ where a change in slope can be observed.

6. Banding Phenomenon

The banded growth model developed by Yankov et al. [31-33] has also been described in detail elsewhere and will summarized only briefly here. It is used to described the development of the banded structure shown schematically in Fig. 4.

Having scaled the Aziz-Kaplan interface response functions, it is now possible to follow what will happen if one attempts to continuously laser melt quench the Ag - 38.3 at. pct alloy at an isotherm velocity of 27 cm s⁻¹, which is less than $v^* = 47$ cm s⁻¹ (i.e., under conditions similar to that represented by point 2, Fig. 11). We have done this using a finite difference solution to the time dependent diffusion equation in the melt in front of the moving interface. In our calculations, we employ the frozen temperature approximation to decouple thermal and mass transport. A temperature gradient of 1 K per micron was used in the calculations.

The onset of growth requires undercooling to a temperature of 983.2 K, i.e., until the nose of the X_s curve reaches the X_0. If solidification were to begin before this point is reached, then the melt would be flooded with rejected solute and growth would stop. When the interface reaches 983.2 K growth begins at 47 cm s⁻¹. Thus

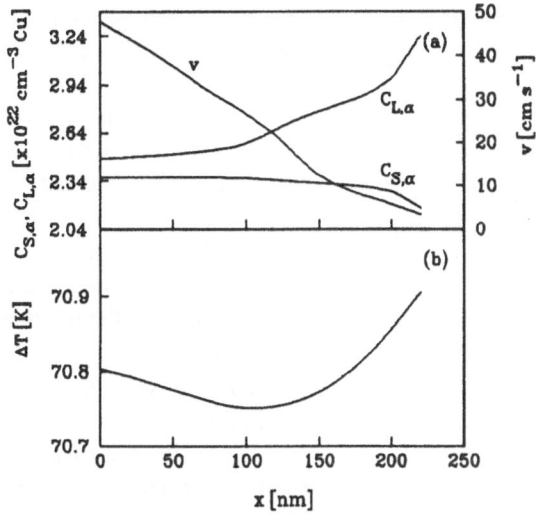

Figure 16. Interface temperature, solute compositions and velocity during the first cycle of single phase growth.

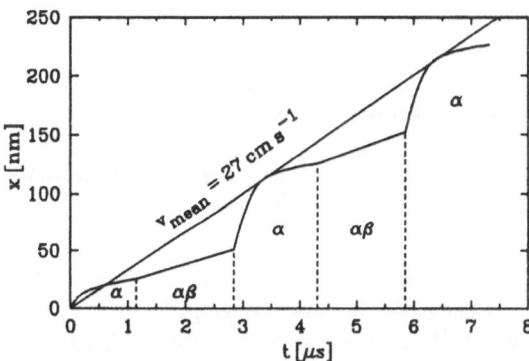

Figure 17. Position of the solid-melt interface shown as a function of time.

the interface will run ahead of the isotherms into regions of higher temperature. This cau ses the X_S and X_L curves to shift in the direction of lower solute content. Accordingly, the concentration of solute in the solid decreases causing the concentration in the liquid to increase and the interface to slow down.

With time, Fig. 16 shows that: (1) the solute concentration in the liquid builds; (2) the corresponding solute concentration in the solid decreases; and (3) the interface velocity decreases monotonically. After growing for about 220 nm: (1) the interface has almost slowed to a stop; (2) the solute content of the liquid has increased into the hypereutectic range; and (3) a large amount of solute has been stored in the solute profile in front of the interface. At this point, there are three possibilities: (1) coupled growth of the Ag-rich and Cu-rich solid solutions, which requires nucleation of the Cu-rich phase at the interface; (2) breakdown of the plane interface resulting in cellular or dendritic growth; or (3) a decrease in interface temperature to the point that growth can begin again, which would result in a modulated solute profile. In the 38.3 at. pct Cu alloy, the first of these possibilities is observed. Coupled growth of the Ag-rich and Cu-rich phases is initiated and continues until the solute stored in front of the interface is exhausted. At this point single phase growth is reestablished.

116

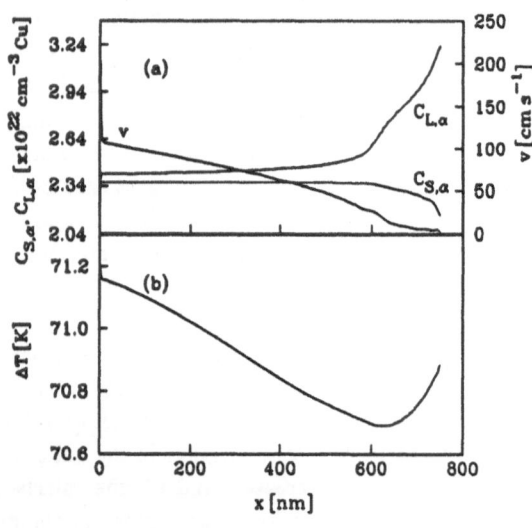

The main features of our previously published analysis of banded growth in a Ag-38.3 at. pct Cu alloy are illustrated in Fig. 17, which plots interface position versus time. After the first cycle, the following sequence of events is repeated. Growth of the Ag-rich solid solution occurs first at a high rate, which then decreases as the interface moves into warmer regions. Solute builds up in the melt at the interface resulting in nucleation of the Cu-rich solid solution and coupled growth. This coupled growth is treated using the model described in Section 5 as occurring at steady state at a velocity of 18 cm s⁻¹, which is less than the isotherm velocity. The variation of interface temperature, solute compositions and velocity during the second and subsequent cycles of growth are shown in Fig. 18.

7. Summary and Recommendations for Future Work

The availability of experimental techniques such as continuous laser melt quenching and electron beam melting and resolidification along with the development of interface response functions have significantly increased our understanding of the process of solidification at high growth rates. A coherent interpretation of planar front, coupled growth, dendritic/cellular growth and banded growth is emerging.

There are several areas where further research is needed. Theories of banded growth should be extended to include latent heat effects, interphase response functions for concentrated solutions, nucleation of second phases and a complete time dependent solution. Theories of coupled growth should be extended to include interphase response functions and curvature effects. Extrapolated free energies and kinetic parameters are needed for a wider range of alloy systems and methods for estimating such parameters in multicomponent alloy systems should be developed.

8. References

1. Duwez, P., Willens, R.H., and Klement Jr., W. (1960) Continuous series of metastable solid solutions in silver-copper alloys, *J. Appl. Phys.* **31**, 1136-37.
2. Linde, R.K. (1966) Lattice parameters of metastable silver-copper alloys, *J. Appl. Phys.* **37**, 934.
3. Stoering, R. and Conrad, H. (1969) Metastable structures in liquid quenched and vapor quenched Ag-Cu alloys, *Acta Met.* **17**, 933-48.
4. Boswell, P.G. and Chadwick, G.A. (1977) The structure of the gamma-prime extended solid solution in a splat-cooled Ag-50 at. % Cu alloy, *J. Mat. Sci.* **12**, 1879-94.
5. Bass, M. and Copley, S.M. (1977) Laser materials interactions: needs and opportunities for basic research, NSF Workshop Report.
6. J. Singh and S.M. Copley, (1993) *International Conference on Beam Processing of Advanced Materials*, The Minerals, Metals & Materials Society, Warrendale, PA.
7. Kou, S., Hsu, S.C., and Mehrabian, R. (1981) Rapid melting and solidification of a surface due to a moving heat flux, *Metall. Trans.* **12B**, 33-45.
8. Yankov, E., Todd, J.A., and Copley, S.M. (1991) New evidence for the microstructural nature of the banded structure produced during rapid solidification of Ag-Cu alloys, in P.K. Liaw, J.R. Weertman, H.L. Marcus, and J.S. Santer (eds), *Morris E. Fine Symposium*, The Minerals, Metals & Materials Society, Warrendale, PA, 29-31.
9. Beck, D.G., Copley, S.M., and Bass, M. (1981) The microstructure of metastable phases in Ag-Cu alloys generated by continuous laser melt quenching, *Metall. Trans.* **12A**, 1687-92.
10. Beck, D.G., Copley, S.M., and Bass, M. (1982) The constitution and phase stability of overlapping melt trails in Ag-Cu alloys produced by continuous laser melt quenching, *Metall. Trans.* **13A**, 1879-89.
11. Elliott, W.A., Gagliano, F.P., and Krauss, G. (1973) Metastable phases produced by laser melt quenching, *Metall. Trans.* **4**, 2031-37.
12. Baker, J.C. and Cahn, J.W. (1969) Solute trapping by rapid solidification, *Acta Met.*, **17**, 575-78.
13. Baker, J.C. and Cahn, J.W. (1971) Thermodynamics of solidification, in T.J. Hughel and G.F. Bolling (eds), *Solidification*, ASM, Metals Park, OH, 23-58.
14. Beck, D.G., Copley, S.M., and Bass, M. (1980) Constitution and microstructure of Ag-Cu alloys produced by continuous laser melt quenching, in C.W. White and P.S. Peercy (eds), *Laser and Electron Beam Processing of Materials*, Academic Press, New York, 734-39.
15. Copley, S.M., Bass, M., Van Stryland, E.W., Beck, D.G., and Esquivel, O. (1978) Microstructures of surface alloyed Ag-Cu films produced by laser melt quenching, in R. Cahn (ed), *Proc. III Int. Conf. Rapidly Quenched Metals*, **Vol. 1**, The Metals Society, London, 147-50.
16. Sundquist, B.E. (1966) The calculation of thermodynamic properties of miscibility-gap systems, *Trans. TMS-AIME*, **236**, 1111-22.

118

17. Lumsden, J. (1952) *Thermodynamics of Metals*, The Institute of Metals, London, 335.

18. Aziz, M.J. (1982) Model for solute redistribution during rapid solidification, *J. Appl. Phys.*, **53**, 1158-68.

19. Coriell, S.R. and Sekerka, R.F. (1983) Oscillatory morphological instabilities due to non-equilibrium segregation, *J. Cryst. Growth*, **61**, 499-508.

20. Murray, J.L. (1984) Calculations of stable and metastable equilibrium digrams of the Ag-Cu and Cd-Zn systems, *Metall. Trans.*, **15A**, 261-68.

21. Boettinger, W.J., Shechtman, D., Schaefer, R.J., and Biancaniello, F.S. (1984) The effect of rapid solidification velocity on the microstructure of Ag-Cu alloys, *Metall. Trans.*, **15A**, 55-66.

22. Mullins, W.W. and Sekerka, R.F. (1964) Stability of a planar interface during solidification of a dilute binary alloy, *J. Appl. Phys.*, **35**, 444-51.

23. Aziz, M.J. and Kaplan, T. (1988) Continuous growth model for interface motion during alloy solidification, *Acta metall.*, **36**, 2335-47.

24. Todd, J.A. and Su, Y. (1989) A mass transport theory for interphase precipitation with application to vanadium steels, *Metall. Trans.*, **15A**, 1647-55.

25. Todd, J.A. (1991) The interphase precipitation reaction in HSLA steels, *J. Metals*, **43**, 45-48.

26. Yankov, E.Y., Yankova, M.I., Copley, S.M., and Todd, J.A. (1991) Method for predicting microsegregation-free solidification with applications to Ag-Cu alloys, *Appl. Phys. Lett.*, **59**, 2106-08.

27. Trivedi, R., Magnin, P., and Kurz, W. (1987) Theory of eutectic growth under rapid solidification conditions, *Acta metall.*, **35**, 971-80.

28. Yankov, E.Y., Copley, S.M., Yankova, M.I., and Todd, J.A. (1991) Coupled growth in near- and far-from-equilibrium modes, in P.K. Liaw, J.R. Weertman, H.L. Marcus, and J.S. Santer (eds), *Morris E. Fine Symposium*, The Minerals, Metals & Materials Society, Warrendale, PA, 33-37.

29. Zener, C. (1946) Kinetics of the decomposition of austenite, *Trans. AIME*, **167**, 550-95.

30. Tiller, W.A. (1958) Eutectic growth, in *Liquid Metals and Solidification*, American Society for Metals, Metals Park, OH, 276-318.

31. Yankov, E.Y., Copley, S.M., Todd, J.A., and Yankova M.I. (1992) New insight into the banded microstructure of rapidly solidified Ag-Cu alloys: experiment and theory, *Mat. Res. Soc. Symp. Proc.*, **205**, 307-12.

32. Copley, S.M., Yankov, E.Y., Todd, J.A. and Yankova, M.I. (1992) Coupled growth in eutectic systems, in E.J. Mittemeijer (ed), *Proceedings of the First ASM Heat Treatment and Surface Engineering Conference in Europe, Mat. Sci. Forum*, **102-04, Pt. 1**, Trans Tech Publications, Switzerland, 417-32.

33. Copley, S.M., Yankov, E.Y., Yankova, M.I., and Todd, J.A. (1993) Interface response in metallic alloys, in J. Singh and S.M. Copley (eds), *International Conference on Beam Processing of Advanced Materials*, The Minerals, Metals & Materials Society, Warrendale, PA, 131-40.

34. Jackson, K.A., Gilmer, G.H., and Leamy, H.J. (1980) Solute trapping, in C.W. White and P.S. Peercy (eds), *Laser and Electron Beam Processing of Materials, Mat. Res. Soc. Symp. Proc.*, Academic Press, New York, 104-110.

35. Boettinger W.J. and Perepezko, J.H. (1985) in S.K. Das, B.H. Kear and C.M. Adams (eds), *Rapidly Solidified Crystalline Alloys, Proc. TMS-AIME Northeast Regional Meeting*, TMS-AIME, Warrendale, PA, 21-59.

36. Boettinger, W.J. and Coriell, S.R. (1986) in P.R. Sahm, H. Jones and C.M. Adam (eds) in *Science and Technology of the Undercooled Melt - Rapid Solidification Materials and Technologies*, NATO ASI Series E, **114**, Nijhoff, Dordrecht, 81-.

37. Carrard, M., Gremaud, M., Zimmermann, M., and Kurz, W. (1992) About the banded structure in rapidly solidified dendritic and eutectic alloys, *Acta metall. mater.* **40**, 983-96.

38. Kurz, W., Giovanola, B., and Trivedi, R. (1986) Morphological stability of a planar interface under rapid solidification conditions, *Acta metall.* **34**, 1663-70.

39. Trivedi, R., Magnin, P. and Kurz, W. (1987) Theory of eutectic growth under rapid solidification conditions, *Acta metall.* **35**, 971-80.

40. Zimmermann, M., Carrard, M. and Kurz, W. (1989) Rapid solidification of Al-Cu eutectic alloy by laser remelting, *Acta metall.*, **37**, 3305-13.

41. Gremaud, M., Carrard, M. and Kurz, W. (1991) Banding Phenomena in Al-Fe alloys subjected to laser surface treatment, *Acta metall. mater.*, **39**, 1431-43.

42. Kubin, L.P. and Estrin, Y. (1985) The Portevin-Le Chatelier effect in deformation with constant stress rate, *Acta metall.*, **33**, 397-407.

43. Huntley, D.A. and Davis, S.H. (1993) Thermal effects in rapid directional solidification: linear theory, *Acta metall. mater.*, **41**, 2025-43.

44. Karma, A and Sarkissian, A. (1993) Interface dynamics and banding in rapid solidification, *Phys. Rev. E*, **47**, 513-33.

45. Cline, H.E. and Stein, D.F. (1969) *Trans. AIME*, **245**, 841-.

46 Cline, H.E. and Lee, D. (1970) Strengthening of lamellar vs. equiaxed Ag-Cu eutectic, *Acta metall.*, **18**, 315-23.

SURFACE RESPONSE FUNCTIONS DUE TO LASER IRRADIATION

W.W. DULEY
GWP²
University of Waterloo
Waterloo, Ontario
Canada   N2L 3G1

ABSTRACT

Many non-metallic materials exhibit both photochemical and photothermal effects when irradiated with intense laser radiation. The relative role of these processes depend primarily on laser wavelength with photochemical processes dominating in the ultraviolet. A quantitative estimate of the importance of photochemical and photothermal processing can be obtained from a simple physical model developed in this paper. It is shown that the adaptation of an irradiated surface leads to radiation resistance and to unique surface morphologies/chemical properties.

1.   INTRODUCTION

Laser sources can initiate both photochemical and photothermal effects in condensed media. The relative importance of these two effects depend on a variety of factors including laser wavelength, pulse duration, intensity and the photochemical/photothermal response of the irradiated material. In addition, exposure to laser radiation can result in radiation conditioning or hardening, such that the response of the medium to subsequent irradiation may be quite different from its initial response.

This paper explores some of the fundamental limitations of materials processing with lasers as they relate to the physical and chemical response of the irradiated medium. Some general constraints on the relative rate of ablation in photochemical and photothermal regimes are also discussed. The question of radiation resistance is shown to exhibit both geometrical as well as physico-chemical characteristics.

2.   PHOTOCHEMICAL VS. PHOTOTHERMAL MATERIAL REMOVAL

The relative rate of photon-driven vs. thermally-driven processes can be estimated from a simple model. Laser radiation with a Gaussian intensity distribution

$$I(r,t) = I_0(t) \exp\left[- r^2/d^2\right] \tag{1}$$

irradiates a planar semi-infinite surface consisting of molecular absorbers. The absorption coefficient α at the laser wavelength λ is

J. Mazumder et al. (eds.), Laser Processing: Surface Treatment and Film Deposition, 121–128.

$$\alpha = \sigma_\lambda N \qquad (2)$$

where N is the density of absorbers in the surface and σ_λ is the cross-section per absorber. It will be assumed that $\alpha \gtrsim 10^4$ cm^{-1} so that the energy deposited in the surface can be considered to represent a surface heat source in any thermal conduction model. For simplicity $I(r,t)$ will be taken to be independent of time after initiation ie. $I(r,t) = I(r)$, for $t > 0$. Then the surface temperature rise ΔT at the centre of the focal spot is (Duley 1976)

$$\Delta T \ (r = 0, \ t) = \frac{\epsilon I_0 d}{K \sqrt{\pi}} \ \tan^{-1} \left[\frac{4 \kappa t}{d^2} \right] \qquad (3)$$

where ϵ is the absorptivity, K is the thermal conductivity and κ is the thermal diffusivity. For a laser pulse of infinite duration

$$\Delta T \ (r = 0, \ \infty) = \frac{\epsilon I_0 d \sqrt{\pi}}{2K} \qquad (4)$$

The rate of a thermally-induced reaction at the focus will be driven by $\Delta T(t)$. Assuming a first order rate constant of the Arrhenius form (Busch et al. 1978)

$$\delta = A \ \exp \ [-\Delta E / kT] \qquad (5)$$

where A (sec^{-1}) is the preexponential frequency factor, ΔE is the activation energy, k is Boltzmann's constant and T is °K,

$$\delta(t) = A \ \exp \ [-\Delta E / k \ (T_0 + \Delta T(t))] \qquad (6)$$

where T_0(°K) is the surface temperature at $t = 0$.

The rate of a photon-driven reaction (eg. photodissociation, photo-rearrangement) is

$$\rho_\lambda^{(n)} (t) = \sigma_\lambda^{(n)} F_\lambda^{(n)} (t) \qquad (7)$$

where $F_\lambda(t)$ is the photon flux, and $\sigma_\lambda^{(n)}$ is the nth-order absorption cross-section. At the centre of the focal spot

$$F_\lambda(t) = \frac{\lambda \epsilon I_0(t)}{hc} \qquad (8)$$

where h is the Planck constant and c is the speed of light. With $I_0(t)$ constant $\rho_\lambda^{(n)}$ is independent of time. Then, the ratio

$$R_\lambda^{(n)}(t) = \frac{\rho_\lambda^{(n)}}{\delta(t)} \tag{9}$$

yields a quantitative estimate of the relative importance of an nth-order photo-process and a thermally driven reaction with an activation energy ΔE. Taking $d = \lambda$, $R_\lambda^{(n)}(t)$ becomes

$$R_\lambda^{(n)}(t) = \frac{\sigma^{(n)}}{A} F_\lambda^n \exp\left[\frac{T^*}{[T_0 + \Delta T(t)]}\right] \tag{10}$$

where $T^* = \Delta E/k$ and

$$\Delta T(t) = \frac{\epsilon I_0 \lambda}{K\sqrt{\pi}} \tan^{-1}\left[\frac{4\kappa t}{\lambda^2}\right]^{1/2} \tag{11}$$

with

$$\beta = \tan^{-1}(\eta)^{1/2} \tag{12}$$

and $\eta = 4\kappa t/\lambda^2$, equation 10 becomes

$$R_\lambda^{(n)}(t) = \gamma^{(n)} \exp\left[\frac{T^*}{(T_0 + \beta I_\lambda')}\right] \tag{13}$$

where

$$\gamma^{(n)} = \frac{\sigma^{(n)}}{A} F_\lambda^n \tag{14}$$

and

$$I_\lambda' = \frac{\epsilon I_0 \lambda}{K\sqrt{\pi}} \tag{15}$$

As $t \to \infty$; $\beta = \pi/2$ and

$$R_\lambda^{(n)}(\infty) = \gamma^{(n)} \exp\left[\frac{T^*}{(T_0 + \epsilon I_0 \lambda \sqrt{\pi}/2K)}\right] \tag{16}$$

Thus, assuming constant photon flux, thermal effects become more dominant as time increases. Conversely, photon-dominated reaction rates are most significant for $t < \lambda^2/4\kappa$.

Figure 1 shows a plot of $R_\lambda^{(n)}(t)/\gamma^{(n)}$ vs. η for various values

124

of I'_λ and with $T^* = 5000$ K. Since the activation energy, ΔE, is typically 20% of the bond energy (Busch et al. 1978) $T^* = 5000$ K would correspond to a bond energy of 2.16 eV and the thermal reaction rate or dissociation rate would involve breaking a bond with this energy. The strong dependence of $R^{(n)}_\lambda/\gamma^{(n)}$ on η at small η clearly shows the dominance of photon effects at short times.

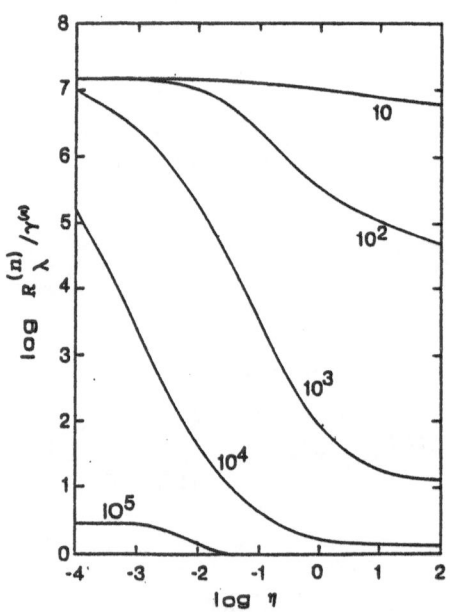

Figure 1.   Plot of $R^{(n)}_\lambda/\gamma^{(n)}$, vs. η for various values of I'_λ and assuming $T^* = 5000$ °K and $T_0 = 293$ °K.

If we consider that the solid is a collection of molecular units that absorb photons, redistribute this energy among internal vibrational modes, and then transfer this energy over a longer timescale to the surrounding medium, then the rate of thermal dissociation becomes

$$\delta_m = \nu_v \exp\left[-D/kT\right] \tag{17}$$

where D is the dissociation energy for a bond with frequency ν_v. In this equation, T is to be interpreted as the intramolecular vibrational temperature following absorption of one or more photons. For a molecule with N atoms,

$$T = \frac{q\,h\nu}{k(3N-6)} \tag{18}$$

where q is the number of photons absorbed by the molecule over a timescale corresponding to the intermolecular energy delocalization time, τ_R. Assuming constant incident photon flux,

$$q = n\sigma^{(n)} F^n \tau_R \tag{19}$$

and

$$\delta_m = \nu_v \exp\left[\frac{-(3N-6)\,D}{h\nu n\sigma^{(n)} F^n \tau_R}\right] \tag{20}$$

Taking $\nu_v = 10^{13}$ Hz, N = 100, D = 3 eV, hν = 5 eV, $\sigma^{(n)} = \sigma^{(1)} = 10^{-21}$ m^2, n = 1, and F = 10^{29}/m^2/sec as typical of parameters for an organic polymer such as polyimide irradiated at 248 nm

$$\delta_m = 10^{13} \exp\left[\frac{-1.76 \times 10^{-6}}{\tau_R}\right] \tag{21}$$

The value to be assigned to τ_R will depend on the surrounding medium and its state of aggregation. For a free molecule in a gas under collisionless conditions, τ_R will be either an IR radiative lifetime (~1-10 msec) or the laser pulse length τ_p whichever is shorter. Then the probability of thermal dissociation per pulse is

$$p^{(n)} = \delta_m \tau_R \tag{22}$$

$$= \nu_v \tau_R \exp\left[\frac{-(3N-6)\,D}{nh\nu\sigma^{(n)} F^n \tau_p}\right] \tag{23}$$

since photons are absorbed only while laser radiation is present, while dissociation persists over the relaxation time, τ_R.

Vibrational energy delocalization times for molecules in condensed media are typically 1-100 psec (Bondeybey 1984). Then the probability of thermal dissociation per n-photon absorption is

$$p_1^{(n)} = \nu_v \tau_R \exp\left[-\frac{(3N-6)\,D}{nh\nu}\right] \tag{24}$$

where τ_R = 1-100 psec. The overall dissociation probability for a laser pulse with $\tau_p > \tau_R$ is approximately

$$p^{(n)} = \sigma^{(n)} F^n \tau_p p_1^{(n)} \tag{25}$$

For $v_v = 10^{13}$ Hz, $\tau_R = 100$ psec, $N = 100$, $D = 3$ eV, $h\upsilon = 5$ eV and $n = 1$, $p_1^{(1)} = 10^3 \exp(-176.4)$ which is vanishing small. However, under the same conditions, but with $N = 10$, $p_1^{(1)} = 5.6 \times 10^{-4}$ per event. Taking $\sigma = 10^{-21}$ m^2, $F = 10^{29}$ /m^2/sec and $\tau_p = 100$ nsec, $p^{(1)} = 5.6 \times 10^{-3}$ per pulse. This simple estimate of dissociation probability ignores any statistical effects in the photon flux or re-distribution of vibrational energy within the excited molecule.

3. RADIATION RESISTANCE

In many applications of lasers for materials processing, the material to be worked is subjected to a number of overlapping laser pulses or to an extended dwell time within the laser beam. It is a general principle in such an instance that the material will attempt to adapt to its radiative environment. This adaptation can take the following forms: -

i)   Development of a surface that minimizes exposure to incident laser radiation, for example by assuming an orientation at oblique incidence with respect to the beam.

ii)   Development of a shielding plasma or gas flow that acts to intercept the incident beam.

iii) Development of a mass flow that provides strong convective cooling of the laser-heated surface.

iv)   By structural and/or chemical modification of the absorbers exposed to laser radiation.

Mechanism (i) is geometrical in nature and is readily observed in the formation of cone-like structures in solids heavily irradiated with excimer laser radiation (Krajnovich and Vazquez 1993). It also occurs in the formation of a keyhole during penetration welding of metals with high power CO_2 laser radiation (Duley 1983).

The second mechanism is observed when processing materials such as metals at laser intensities exceeding $\sim 10^{10}$ watt/m^2 at 10.6 μm and $\sim 10^{11}$ watts/m^2 at excimer laser wavelengths. It arises when rapidly vaporized material is ionized by multiphoton absorption or by inverse bremmstrahlung followed by avalanche ionization.

Rapid heat loss due to convection occurs in the flow of liquid metal around the laser keyhole and in conduction welding (Mazumder and Kar 1993).

The last adaptation mechanism to an intense radiative environment involves physical and chemical changes in irradiated material that allow it to withstand the applied radiative flux without further substantial alteration. A simple example is the graphitization of amorphous carbon films upon exposure to laser radiation (Armeyev et al. 1989). In this case, adaptation involves a change in bonding from the sp^3 hybridized orbitals characteristic of diamond-like carbon to the sp^2 bonding associated with graphite. The effect of radiation is then to increase the proportion of sp^2 bonded material thus decreasing the bandgap energy and increasing the conductivity (Robertson and O'Reilly 1987). This permits heat to be more readily conducted away from the laser focus.

Mechanism (iv) can involve both photochemical and photothermal effects. On a molecular basis, weak bonds are the first to be broken leading to the evolution of molecular fragments. The parent molecule then undergoes a rearrangement, or a reaction with its environment, to yield a product with fewer bonds that are susceptible to dissociation under the ambient radiative conditions. Eventually, the dissociation probability per photochemical or photothermal event is reduced to the point where dissociation does not occur in this ambient radiative field over the timescale of the interaction.

In the special case of organic materials, this process initially involves the loss of volatile species such as H, OH, CO, etc. Generally this can lead to a significant dehydrogenation of the parent material. Since bond energies for ring carbons are typically 11 eV, while peripheral functional groups are bonded to these rings by $\lesssim$ 5 eV (Busch et al. 1978) dehydrogenation and the loss of other peripheral functional groups will facilitate the formation of extended aromatic ring structures. Such compact aromatic ring structures are known to be radiation resistant (Gonzalez-Hernandez et al. 1988).

Compositional and morphological modifications leading to enhanced radiation resistance have also been observed on the surface of $YBa_2Cu_3O_{7-x}$ films exposed to many overlapping 308 nm excimer laser pulses. (Foltyn et al. 1991). After high levels of irradiation, the exposed surface was found to have evolved a columnar cone-like structure oriented in the direction of the incident laser beam. The chemical composition of these cones was observed to differ significantly from that of the un-irradiated

YBCO surface. The overall effect of these compositional and structural changes was to reduce the evaporation rate/pulse to $\simeq$ 10% of its original value.

ACKNOWLEDGEMENTS

This research was supported by grants from NSERCC, OCMR, and MRCO.

REFERENCES

Armeyev, V.Y., Chapliev, N.I., Konov, K.I., Ralchenko, V.G., Strelinitsky, V.E. and Volkov, Y.Y. 1989. Proc. SPIE vol. 1352, p. 200.

Bondeybey, V. 1984. Ann. Rev. Phys. Chem. 35, 591.

Busch, D.H., Shull, H. and Conley, R.T. 1978. "Chemistry" Allyn and Bacon Inc., Boston.

Duley, W.W. 1976. "CO_2 Lasers: Effects and Applications" Academic Press, N.Y.

Duley, W.W. 1983. "Laser Processing and Analysis of Materials" Plenum Press, N.Y.

Foltyn, S.R., Dye, R.C., Ott, K.C., Peterson, E., Hubbard, Hutchinson, W. Muenchausen, R.E., Esther, R.C., and Wu, X.D. 1991. Appl. Phys. Lett. 59, 594.

Gonzalez-Hernandez, J., Asomoza, R., Mena, A.R., Rickards, J., Chao, C.S.S., and Pawlik, D. 1988. J. Vac. Si. Technol. A6, 1798.

Krajnovich, D.J. and Vazquez, J.E. 1993. J. Appl. Phys. 73, 3001.

Mazumder, J. and Kar, A. 1993. Proc. NATO ASI "Laser Applications for Mechanical Industry" eds. Martellucci, S., Chester, A.N. and Scheggi, A.M., Kluwer Academic Publ., Dordrecht, p. 47.

Robertson, J. and O'Reilly, E.P. 1987. Phys Rev. B35, 2946.

MODELING IN LASER MATERIALS PROCESSING: MELTING, ALLOYING, CLADDING

A. Kar
Center for Research and Education in Optics and Lasers (CREOL)
Mechanical and Aerospace engineering Department
University of Central Florida
Orlando, Florida 32826, USA

J. Mazumder
Center for Laser-Aided Materials Processing (CLAMP)
Mechanical and Industrial Engineering Department
University of Illinois at Urbana-Champaign
Urbana, Illinois 61801, USA

With the advent of high power lasers, the laser technology has taken an important place in the areas of manufacturing and materials processing. To utilize this technology in the economical and efficient ways, a proper understanding of the phase changes occurred during laser processing is required. Both theoretical and experimental studies are required to achieve this goal. This paper presents several mathematical models for various types of laser processing. Most of the laser processing involves the melting, vaporization, and solidification of materials. Due to the inherent rapid cooling rate, novel microstructures with metastable phases are produced during laser processing. Laser cladding is a technique to coat a substrate with a thick layer of other materials to improve the surface properties of the substrate. If the cladding powder is a mixture of more than one type of materials, the resulting coating is usually found to have metastable phases with nonequilibrium compositions. The solidification rate, which affects the composition of the nonequilibrium alloys, is obtained by solving the energy conservation equations in the melt, solidified clad, and substrate regions. Also, the mass transfer equation is solved to determine the distribution of solute atoms in the liquid and solid regions. Finally, the model is used to obtain the nonequilibrium phase diagrams for Ni-Al, Ni-Hf, and Nb-Al systems. The partition coefficient is found to be an important parameter for this model, and for this reason, an expression for the nonequilibrium partition coefficient for the concentrated binary systems is also presented in this paper.

1. Introduction

Laser technology has drawn considerable interest in using lasers for materials processing and manufacturing. It not only makes manufacturing processes simpler and more economical but also provides a unique way of improving the surface

J. Mazumder et al. (eds.), Laser Processing: Surface Treatment and Film Deposition, 129–155.
© 1996 *Kluwer Academic Publishers.*

properties of materials. Due to the inherent rapid solidification rate in laser alloying and cladding processes, it is often possible to obtain novel materials without being restricted by the equilibrium phase diagram. The high cooling rate results in extended solid solution to produce metastable alloys. Recently, laser cladding technique has been applied to increase the solid solubility of reactive elements such as Hf in nickel superalloy [1] for improved high temperature properties.

To understand the formation of extended solid solution during laser cladding, we need to understand the energy, momentum, and mass transport of solute atoms. The heating and cooling rates are determined by the heat transfer equation while the extent of mixing and the redistribution of solute atoms in the molten cladding pool are obtained from the momentum and mass transport equations. This paper presents a mathematical model for binary alloys to calculate the composition of solute in the extended solid solution which is usually formed during laser cladding. It allows the construction of a nonequilibrium phase diagram of a given system from its equilibrium phase diagram.

There is very little information in the available literature on mathematical modeling of laser surface cladding and alloying with rapid cooling rate. Three-dimensional heat transfer models for various material processing with CW laser were presented by several authors [2-5]. It was found experimentally that surface tension arising due to very high temperature gradient in laser melted pools causes convection. This aspect was modeled by Chan et al. [6]. They studied the effects of surface tension on the cooling rate, surface velocity, surface temperature, and the pool shape. Chande and Mazumder [7] examined the distribution of solute by diffusion and convection in a laser melted pool after it was delivered to such a pool.

The problem of extended solid solution due to rapid cooling has been studied by various investigators using thermodynamic variables such as free energy and chemical potential. The thermodynamics of nonequilibrium solidification has been examined very well by Baker and Cahn [8]. Boettinger and Perepezko [9] discussed the process of rapid solidification from the thermodynamics point of view. Boettinger et al. [10], also used the response function approach of Baker and Cahn [8] and stability analysis for microsegregation-free solidification. Further details on rapidly solidified materials can be found in Refs [11, 12]. Mathematical modeling of extended solid solution during laser cladding has been discussed in Refs. [13-15]. Yankov et al. [16, 17] have studied the microstructural nature of the banded structure produced during rapid solidification of Ag-Cu alloys.

Response functions are found to be very important for modeling the rapid solidification phenomena. These functions are essentially the constitutive laws governing the dependence of temperature on the solute concentration, and the ratio of the solid-phase solute concentration to the liquid-phase solute concentration, at the solid-liquid interface during solidification. The latter response function is known as the partition or segregation coefficient. Several theoretical models developed to understand nonequilibrium partitioning can be found in the literature. One of the earliest works on this type of modeling is due to Jackson [18]. His model is based on reaction rate theory, and on the assumption that the reaction rate theory can be applied independently to each species and that the rate of crossing the solid liquid interface from one phase to another by the atoms of each species is proportional to the mole fraction of the species in that phase. Borisov [19] developed a model based on a thermodynamic approach under the assumption that the chemical potentials of both species in a binary system decrease during solidification. The inadequacies of these two models have been discussed in details by Baker and Cahn [20]. However, an atomistic theory of Chernov [21] predicts the correct deviation of the solute concentration in the solid phase from local equilibrium. Cahn et al. [22] reported a model which is called Baker's model. In this model the diffusive flux is assumed to be proportional to the driving force and then the steady state mass diffusion equation is solved in a reference frame fixed at the solid-liquid interface. By using rate equations for the transfer of solute and solvent atoms across the interface, Jackson et al. [23] developed a model for determining the nonequilibrium partition coefficient. In a series of papers [24-28], the nonequilibrium thermodynamic processes that occur during laser annealing have been discussed and an expression for nonequilibrium partition coefficient has been obtained. By using chemical rate theory and the mass diffusion equation, Aziz [29] derived an expression for the nonequilibrium partition coefficient. His expression compares well with those of Refs.[22,23] under suitable conditions. Besides the analytical approaches mentioned above Monte Carlo simulation [30-33] has also been carried out for crystal growth, and in particular, Ising model has been used to simulate impurity trapping during rapid solidification. However, these models are applicable for dilute solutions.

Since laser cladding involves concentrated solutions in many cases, an understanding of the nonequilibrium partition coefficient for concentrated solutions is needed. Aziz [34,35] has derived an expression for nonequilibrium partition coefficient for concentrated solutions by correcting the fluxes and driving forces in his [29] continuous growth model. The advantages of the nonequilibrium partition coefficient of this paper are that it has only one unknown parameter, namely, the characteristic velocity, and that one does not need to solve a transcendental equation [34, 35] to determine the nonequilibrium partition coefficient. If the characteristic velocity is treated as an adjustable parameter, the proposed nonequilibrium partition coefficient can be determined solely by using the equilibrium phase diagrams, and for this reason, the results of this study can be implemented easily for studying solute redistribution during rapid solidification.

2. Mathematical Model

In laser cladding process, the substrate material, which is to be clad, is moved at a constant rate while the cladding powder is poured onto it using a pneumatic powder delivery system, and melted simultaneously by a laser beam [see Fig. 1(a)]. The molten pool of the cladding material which forms just below the laser beam solidifies by dissipating heat to the surrounding air, adjacent cladding, and the solid substrate as it moves away from the laser beam. The shape of the cladding melt pool and the solidified cladding on the substrate is influenced by laser power, laser beam diameter, thermo-physical properties of the cladding powder and the substrate, temperature of the substrate, relative motion between the cladding powder delivery system and the substrate, the cladding powder feed rate and the interaction time between the cladding and the laser beam. In the present model the strip of cladding BCDEB [see Fig. 1(a)] is assumed to be semicylindrical in shape [see Fig. 1(b)] and based on this the initial pool mean temperature is determined [see equation (5)]. The liquid cladding pool and the substrate were taken to have finite thickness in the position direction of the x-axis and they are considered infinitely large in the other two directions [see Fig. 1(c)].

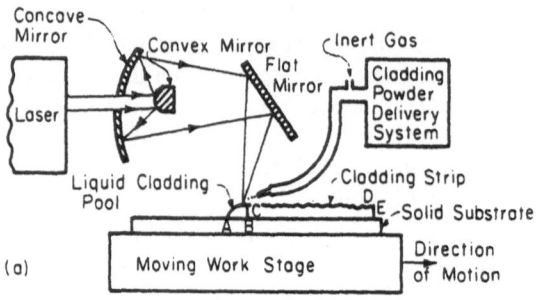

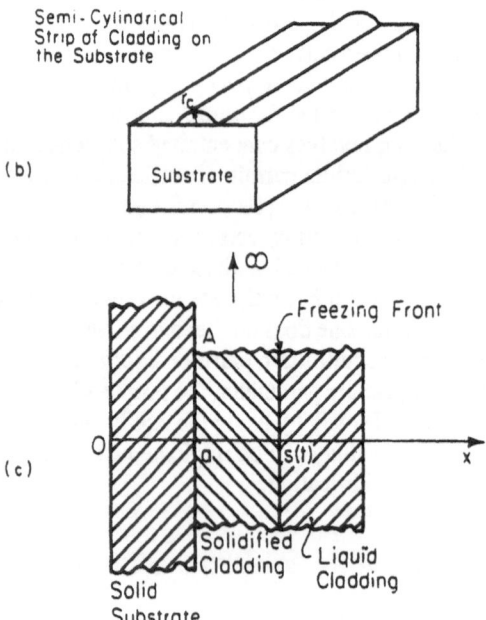

Fig. 1. (a) Schematic diagram of laser surface cladding.
(b) Three-dimensional view of the cladding and the substrate.
(c) Geometric configuration used in the present model. The
model substrate and the solidification of the model cladding
have been shown after rotating the pool ABCA [see Fig. 1(a)]
by 90° clockwise.

This paper is concerned with heat and mass transfer in the cladding melt ABCA to model extended solid solution during laser cladding. This liquid pool solidifies by conducting away heat to the substrate and to the solid cladding across AB and BC, respectively. Also, it loses some energy to the ambient gas across the free surface CA. To carry out one-dimensional heat transfer calculations, we assume the pool extends up to infinity along AB. Due to this assumption the cladding melt solidifies along AB and this freezing front moves upward in the direction of BC. The solid substrate has been considered to have finite width, but it is infinitely large along AB. The geometric configuration of this pool has been shown in Fig. 1(c) by rotating the pool 90 deg in the clockwise direction. In this figure the freezing front has been shown to be planar. This is true when a pure metal solidifies. For a cladding melt of an alloy system, the freezing front develops curvature due to surface tension and also there could be dendrite and cellular growth at the solid-liquid interface. Moreover, rapid cooling may lead to nucleation in the bulk of the liquid phase. The dendrites will affect the diffusion of solute atoms at the freezing front. However, to simplify the mass transfer analysis, the freezing front has been assumed to be planar due to the small size of the dendrites. Besides these considerations, the present model has been developed under the following assumptions:

1. The thermal conductivity and the thermal diffusivity for a mixture is the sum of the volume-averaged value of the respective transport properties of each element of the mixture,
2. The mass diffusivity of each element in the liquid phase is the average value of self-diffusivity over the room temperature and the initial temperature where the activation energy for diffusion in the mixture is taken to be the average value of the self-diffusion activation energy,
3. The cladding pool and the substrate are in thermally perfect contact,
4. There is no mass diffusion in the solid phase,
5. The solute segregated at the solid-liquid interface moves into the liquid phase by diffusion only,
6. The cladding melt forms a uniform solution of composition equal to that of the cladding powder mixture before its solidification begins,
7. Only 50% of the laser energy is absorbed by the cladding material. Studies [12] show that the amount of laser energy absorbed by different materials is 37-60%.

With these assumptions, the one-dimensional heat conduction equations for the substrate, solidified cladding, and the liquid cladding regions are solved. The solutions of these heat transfer equations are used to obtain the velocity of the solid-liquid interface and then the mass transfer equation is solved to determine the distribution of solute atoms in the liquid phase. From this, the concentration of the solute atoms in the solid phase is computed by using an expression for nonequilbrium partition coefficient for dilute solution developed by Aziz [29].

2.1. GOVERNING EQUATIONS AND BOUNDARY CONDITIONS FOR HEAT AND MASS TRANSFER

In this section the governing heat conduction equations for energy transport in the substrate, solidified cladding, liquid cladding regions, and the mass transfer equation for diffusion of solute atoms in the liquid cladding pool are presented along with boundary and initial conditions.

2.1.1. *Heat Transfer Equations and Boundary Conditions*
The governing equations for energy transport are:

(i)   Solid substrate region

$$\frac{\partial^2 T_1'}{\partial x^2} = \alpha_1 \frac{\partial T_1'}{\partial t}; \quad t \geq 0, \ 0 \leq x \leq a; \tag{1}$$

134

(ii)   Solidified cladding region

$$\frac{\partial^2 T_2'}{\partial x^2} = \alpha_2 \frac{\partial T_2'}{\partial t}; \quad t \geq 0, \, a \leq x \leq s(t); \tag{2}$$

(iii)   Liquid cladding region

$$\frac{\partial^2 T_3'}{\partial x^2} = \alpha_3 \frac{\partial T_3'}{\partial t}; \quad t \geq 0, \, s(t) \leq x \leq b; \tag{3}$$

The auxiliary conditions are

$$k_1 \frac{\partial T_1'}{\partial x} = h_1 \left(T_1' - T_r \right) \text{ at } x = 0 \text{ and } t \geq 0, \tag{4a}$$

$$T_1'(a,t) = T_2'(a,t), \tag{4b}$$

$$k_1 \frac{\partial T_1'}{\partial x} = k_2 \frac{\partial T_2'}{\partial x} \text{ at } x = a \text{ and } t \geq 0, \tag{4c}$$

$$T_2'(s(t),t) = T_3'(s(t),t) = T_I\left(C_1, \dot{s}(t)\right), \tag{4d}$$

$$-k_3 \frac{\partial T_3'}{\partial x} = h_3 \left(T_3' - T_r \right) \text{ at } x = b \text{ and } t \geq 0, \tag{4e}$$

$$k_2 \frac{\partial T_2'}{\partial x} - k_3 \frac{\partial T_3'}{\partial x} = \rho L \dot{s}(t), \text{ at } x = s(t) \text{ and } t \geq 0, \tag{4f}$$

and

$$s(0) = a. \tag{4g}$$

Here $\dot{s}(t) = ds / dt$. The initial conditions for the above problems are obtained from the following considerations.

The cladding has been assumed to melt almost instantaneously as soon as it is exposed to the laser beam to reach a uniform temperature $\overline{T}_2$. $\overline{T}_2$ is calculated by taking a lumped-parameter energy balance of the laser energy imparted to the cladding and the energy required to melt the cladding powder and raise its temperature to $\overline{T}_2$. This yields the following expression for $\overline{T}_2$

$$\overline{T}_2 = T_r + \frac{1}{C_p}\left(\frac{2pf}{\pi r_c^2 v\rho} - L\right). \tag{5}$$

This expression is derived by considering the cladding to be formed in the shape of a semi-cylindrical strip [see Fig. 1(b)]. To obtain the initial condition for the solid substrate, a linear temperature profile has been constructed between the temperature $\overline{T}_2$ at $x = a$ and T_r at $x = 0$. Therefore, the initial conditions are

$$T_1'(x,0) = \left(\overline{T}_2 - T_r\right)\frac{x}{a} + T_r, \tag{6a}$$

and

$$T_3'(x,0) = \overline{T}_2. \tag{6b}$$

2.1.2. Mass Transfer Equations and Boundary Conditions

The concentration dependence of T_I couples the energy equations (1-3) with the mass transfer equation and hence they must be solved simultaneously. The diffusion of solute atoms in the cladding melt pool is governed by the following mass transfer equation

$$\frac{\partial^2 C}{\partial x^2} = \frac{1}{\overline{D}}\frac{\partial C}{\partial t}, \tag{7}$$

which is subject to the following initial and boundary conditions

$$C(x,0) = C_0, \tag{8a}$$

$$-\overline{D}\frac{\partial C}{\partial x} = 0 \text{ at } x = b, \tag{8b}$$

and

$$-\overline{D}\frac{\partial C}{\partial x} = \dot{s}(t)\left[k^*(t) - 1\right]C(x,t) \text{ at } x = s(t). \tag{8c}$$

Now the heat and mass transfer problems have to be solved simultaneously to determine the concentration, temperature, and cooling rate at the solidification front.

2.1.3. *Method of Solution*

To solve the above heat and mass transfer problems, we need to know two response functions, which are T_I and $k^*(t)$ that appear in expressions (4d) and (8c), respectively. The model of this paper is based on the following expressions for T_I and $k^*(t)$.

The interface temperature, T_I, which has been derived by Boettinger and Perepezko [9], is taken to be

$$T_I = T_m^* + m_1 C_1^* + \frac{m_1 C_1^*}{1 - k_e}\left[k_e - k^*\left(1 - \ln\frac{k^*}{k_e}\right)\right] + \frac{m_1}{1 - k_e}\frac{\dot{s}(t)}{v_0}.$$

Here v_0 represents the velocity of sound in the liquid phase. The velocity of sound n a medium depends on its temperature and composition. In this study v_0 has been iken to be equal to the velocity of sound in the pure solvent. T_m^* and m_1 are determined from the equilibrium phase diagram by taking a segment of the liquidus around the point of interest. This segment is considered linear whose slope and intercept with the temperature axis are m_1 and T_m^* respectively. C_1^* is obtained by evaluating $C(x,t)$ at $x = s(t)$ and by expressing it in terms of mole fraction.

The partition coefficient, normally defined as the ratio of the concentration of solute in the solid phase to that in the liquid phase at the solid-liquid interface is constant for equilibrium solidification. When solidification progresses rapidly, it is no longer constant. For rapid cooling rate, the partition coefficient $k^*(t)$ has been derived by Aziz [15] which is

$$k^*(t) = \frac{C_s}{C_l} = \frac{\beta + k_e}{1 + \beta}$$

where $\beta = \dot{s}(t)\lambda / D^*$. In the present model the interdiffusivity D^* is taken to be equal to the mass diffusivity $\overline{D}$ and the interatomic distance l is taken to be a 4 Å, which is the average diameter of an atom. To use the expression (31) for determining the concentration of solute in the solid phase at the solid-liquid interface, we have to know $\dot{s}(t)$ and C_l. For this, equation (22) and its time-derivative are solved for $s(t)$ and $\dot{s}(t)$ respectively. Similarly, C_l is obtained from equation (30) by evaluating $C(x,t)$ at $x = s(t)$.

Using these two response functions, the solidification problem can be solved to obtain the temperature fields and solute concentration distribution. The solution techniques and the solutions have been discussed in Ref. [14].

2.2. MODEL FOR NONEQUILIBRIUM PARTITION COEFFICIENT

In this section, we will analyze the solute segregation or partition at the freezing front under equilibrium and nonequilibrium cooling conditions, and derive the corresponding partition coefficient.

2.2.1. *Equilibrium Partition Coefficient*

An expression for equilibrium partition coefficient for concentrated solutions can be obtained by noting that the flux of a species from solid to liquid is equal to that from liquid to solid under equilibrium cooling condition. The flux of the solute A from the solid to liquid phase, and the liquid to solid phase are, respectively, given by the following two expressions.

$$J_{A,sl} = -D_{ABs} \frac{dC_{As}}{dx}, \tag{9}$$

$$J_{A,ls} = -D_{ABl} \frac{dC_{Al}}{dx}, \tag{10}$$

In order to evaluate the concentration gradients in the rate Eqs. (9) and (10), the solid and the liquid phases of interest are shown in Fig. 2a. In this figure, the solidification process is modeled by considering a small region, mushy zone, of width a + b, a > 0, and b > 0, around the interface located at O, i.e., at x = 0 where solid and liquid phases coexist. $x \leq -a$, $x \geq b$, and $-a \leq x \leq b$ represent the solid, liquid, and mushy regions, respectively. The transfer of the species A from solid phase to liquid phase is modeled by considering the transport of A in the region $-a \leq x \leq 0$. Similarly, the transport of A in the region $0 \leq x \leq b$ is considered in order to determine the flux of A from liquid phase to solid phase. Assuming that C_{As} and C_{Al} vary linearly in the regions $-a \leq x \leq 0$ and $0 \leq x \leq b$, respectively, the concentration gradients can be written as

$$\frac{dC_{As}}{dx} = \frac{C_{si} - C_s}{a} \tag{11}$$

and

$$\frac{dC_{Al}}{dx} = \frac{C_l - C_{li}}{b} \tag{12}$$

where C_s and C_l are, respectively, the concentrations of a in the previous solid and liquid phases which are formed at temperature $T_i + \Delta T$ (Fig. 2b). Considering ΔT to be very small, C_s and C_l can be expanded in Taylor series around the temperature, T_i, to obtain

138

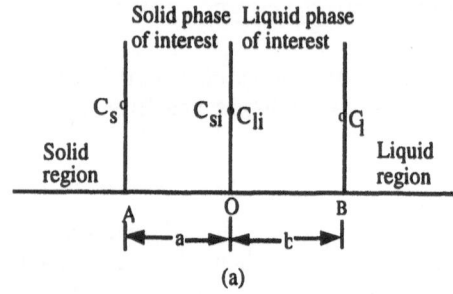

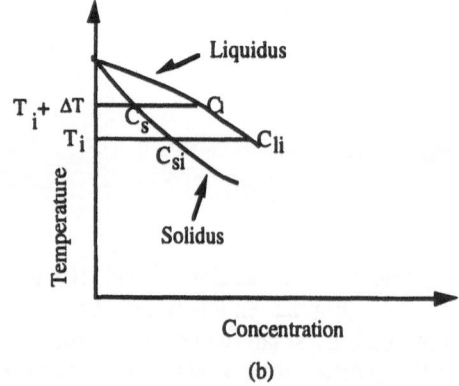

Fig. 2 (a) Geometric model for the interface during solidification.
(b) A representative phase diagram illustrating concentrations
at two different temperatures.

$$C_s = C_{si} + \Delta T \; m_s$$

and

$$C_l = C_{li} + \Delta T \; m_l,$$

which can be used in Eqs. (11) and (12) to write

$$\frac{dC_{As}/dx}{dC_{Al}/dx} = -\frac{m_s b}{m_l a} \tag{13}$$

Under equilibrium cooling condition, Eqs. (9), (10), and (13) yield the following
equation for modified equilibrium partition coefficient for concentrated solutions:

$$k_m^* = \frac{D_{ABl} \; a}{D_{ABs} \; b} \tag{14}$$

where $k_m^* = m_s^*/m_l^*$ which follows from the definition for modified solute distribution coefficient given by Trivedi and Kurz [36].

2.2.2. Nonequilibrium Partition Coefficient

To obtain an expression for nonequilibrium partition coefficient, we examine the conservation of mass at the interface, i.e., at the plane $x = 0$. Under quasi-steady condition, the one-dimensional diffusion equation can be written as

$$u \frac{dC}{dx} = \frac{dJ}{dx} \tag{15}$$

when the origin moves with the interface at a constant speed, u. Since the concentration field is discontinuous at $x = 0$, any finite difference approximation of dC/dx at $x = 0$ would be inappropriate. For this reason, dC/dx at $x = 0$ is taken to be the distance-weighted average of dC_{As}/dx and dC_{Al}/dx, where dC_{As}/dx and dC_{Al}/dx are defined in the regions $-a \leq x \leq 0$ and $0 \leq x \leq b$, respectively. Under this assumption and by noting that the diffusion in the solid region is negligibly small, Eq. (15) can be written as

$$u \left[\frac{a}{a+b} \frac{dC_{As}}{dx} + \frac{b}{a+b} \frac{dC_{Al}}{dx} \right] = \frac{1}{a} [J_{A,sl} - J_{A,ls}] \tag{16}$$

By using Eqs. (9), (10), (13) and (14) in Eq. (16), we obtain

$$k_m = \frac{k_m^* + \beta}{1 + \beta} \tag{17}$$

where $k_m = m_s/m_l$ which is defined as the modified nonequilibrium partition coefficient for concentrated solutions, and $\beta = ua^2/[(a+b) D_{ABs}]$. Equation (16) can also be derived by expanding k_m in Taylor series around $u = 0$ as shown in the next section. From Eq. (17), the expression for nonequilibrium partition coefficient, $k = C_{si}/C_{li}$, for concentrated solutions can be written as

$$k = \frac{C_s}{C_{li}} + \frac{k_m^* + \beta}{1 + \beta} \left(1 - \frac{C_l}{C_{li}} \right) \tag{18}$$

This equation involves the concentrations C_s and C_l which are, respectively, two points on the solidus and the liquidus of the nonequilibrium phase diagram at temperature $T_i + \Delta T$, where T_i is the temperature at which k is to be calculated. However, C_s and C_l are not known a priori when a given volume of liquid solidifies, i.e., when a thin layer of solid is formed for the first time. For this reason, C_s and C_l are estimated for the first instance of solidification by satisfying the following conditions:

$$C_s \rightarrow C_s^* \text{ as } \beta \rightarrow 0,$$

(19a)

$$C_s \rightarrow C_{lo} \text{ as } \beta \rightarrow \infty,$$

(19b)

$$C_l \rightarrow C_l^* \text{ as } \beta \rightarrow 0,$$

(19c)

$$C_l \rightarrow C_{lo} \text{ as } \beta \rightarrow \infty,$$

(19b)

and then, for the subsequent solidification stages, i.e., temperature steps, C_s and C_l are determined by solving the mass balance equation in the liquid region. Although the conditions (19a-d) can be satisfied in many different ways, we choose the following two expressions in this study to satisfy these four conditions:

$$C_s = C_{lo} + e^{-\beta} (C_s^* - C_{lo})$$

(20a)

and

$$C_l = C_{lo} + e^{-\beta} (C_l^* - C_{lo})$$

(20b)

It should be noted that, for dilute solutions, if we assume that $C_l^* = C_{lo}$, and if we consider that Eqs. (20a-b) hold good not only for the first instance of solidification but also during the entire solidification process, then we obtain the following expression for nonequilibrium partition coefficient for dilute solutions:

$$k_d = 1 - (1 - k_e) e^{-\beta}, \tag{21}$$

where $k_d = C_s/C_l$.

2.2.3. *Nonequilibrium Partition Coefficient by Taylor Series Expansion*
In order to derive Eq. (17) by using Taylor series expansion, we define the modified nonequilibrium partition coefficient for concentrated solutions as

$$k_m = \frac{m_s}{m_l} \tag{22}$$

following the definition of modified solute distribution coefficient given by Trivedi and Kurz [36].

Also, we note that, in general, k_m is a function of concentration and the solidification speed, i.e., $k_m = k_m (C, u)$. Under equilibrium cooling condition, $u \rightarrow 0$ and the modified equilibrium partition coefficient for concentrated solutions is given by

$$k_m^* = k_m \, (C, u)\big|_{u \to 0} = \frac{m_s^*}{m_l^*} \tag{23}$$

Expanding $k_m \, (C, u)$ in Taylor series around $u = \varepsilon$, where $\varepsilon \to 0$ and using the expression (23), we obtain

$$k_m \, (C, u) = k_m^* + u \, \frac{\partial k_m}{\partial u}\bigg|_\varepsilon \tag{24}$$

In order to evaluate $\dfrac{\partial k_m}{\partial u}\big|_\varepsilon$, we note that

$$k_m \, (C, u) \to k_m^* \text{ as } u \to 0,$$

and

$$k_m \, (C, u) \to 1 \text{ as } u \to \infty,$$

that is, the function $k_m \, (C, u)$ has two parallel asymptotes represented by the lines $k_m = k_m^*$ and $k_m = 1$. We know that any function will have the same gradient in the asymptotic regions if the corresponding asymptotes are parallel. Therefore, we can write

$$\frac{\partial k_m}{\partial u}\bigg|_\varepsilon = \frac{\partial k_m}{\partial u}\bigg|_{u \to 0} \tag{25}$$

Approximating $\dfrac{\partial k_m}{\partial u}\big|_{u \to 0}$ by $(1 - k_m \, (C, u))/V_c$ in Eqs. (24) and (25), and defining $\beta = u/V_c$, we obtain

$$k_m = \frac{k_m^* + \beta}{1 + \beta} \tag{26}$$

which is identical to Eq. (17).

3. Results and Discussion

We will now discuss the results that have been obtained by using the above-mentioned models.

3.1. EXPERIMENTAL VERIFICATION

The results of the above heat and mass transfer model (section 2.1) have been compared with experimental data. Laser cladding was performed on nickel substrate with a mixture of Ni-Hf powder, and Ni-Al powder. Also, niobium substrate was coated with Nb-Al alloy by using the laser cladding technique. Electron Probe Micro-Analysis (EPMA) and Scanning Transmission Electron Microscope (STEM) analysis of these samples show the concentration of the solute atoms (Hf or Al in this study) in the matrix of the substrate (Ni or Nb in this study) in excess of that predicted by the equilibrium phase diagram. These experimental results are compared with the theoretical predictions in Table 1.

Table 1. Comparison of theoretical and experimental results for extended solid solution.

Experimental Conditions	Ni-Hf System	Ni-Al System	Nb-Al System
Nominal composition of the cladding powder (Wt%)	74% Ni, 26% Hf	74% Ni, 26% Al	50 % Nb, 50 % Al
Laser power (kW)	5	5	6
Laser beam cross-section	Circular with diameter 3 mm	Circular with diameter 3 mm	Rectangular with sides 4 mm and 4.5 mm
Workpiece speed (mm/s)	21	21	9.31
Initial speed pool temperature, $\overline{T}_2$ (K)	1862	2420	1881
Composition in the solid solution, (wt%). Theoretical results: Experimental results:	7.15 % Hf 6.5 % Hf	30.76 % Al 29 % Al	39.48 % Al 40.63 % Al

The theoretical results differ a little from the experimental results, which may be due to some of the assumptions of this model. For example, it has been assumed that all of the solute dissolves in the solvent before the cladding melt starts solidifying. But the time required to melt and freeze the cladding powder may not be sufficient for dissolution of all solute atoms to occur. Also, the present model utilizes an expression for the nonequilibrium partition coefficient which is applicable to dilute solutions. Apart from these conditions, the presence of a two-phase zone between the solidus and the liquidus lines and the surface tension driven flow causing convection in the liquid pool will affect the mixing of solute in the liquid phase and thus alter its composition in the solid phase. Paucity of high temperature liquid metal data also contributes to the numerical error.

3.2. PARAMETRIC RESULTS

The above model was used to study the effect of various important process parameters such as laser power, laser-cladding interaction time, cladding thickness, cladding powder delivery rate on the composition of hafnium in nickel matrix. Results were obtained for ambient temperature T_r = 293K and the heat transfer coefficients h_1 = 1W/cm^2-K and h_3 = 0.01 W/cm^2-K.

Figure 3 shows the concentration of hafnium (in wt%) in nickel matrix for different values of laser power and cladding thickness measured in units of laser beam

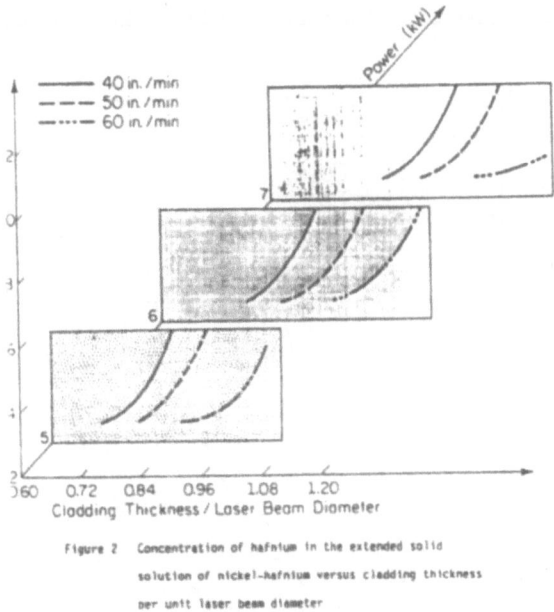

Figure 2   Concentration of hafnium in the extended solid
solution of nickel-hafnium versus cladding thickness
per unit laser beam diameter

Fig. 3. Concentration of hafnium in the extended solid solution of nickel-hafnium versus cladding thickness per unit laser beam diameter.

diameter. The laser beam diameter used in this study is 3 mm. As can be seen from this figure, the concentration of hafnium increases with cladding thickness for a given speed of the workpiece and laser power. This is due to the fact that the initial pool mean temperature, $\overline{T}_2$, varies inversely with the cladding thickness as evident from Eq. (5). The closer is $\overline{T}_2$ to the melting point of the cladding powder the higher would be the cooling rate and this will result in a higher concentration of solute in the solid phase. This was discussed in detail by Kar and Mazumder [13]. However, it would be erroneous to conclude that one can enrich the solid phase with solute to any concentration by increasing the cladding thickness indefinitely for a given laser power and a given speed of the workpiece. There is a critical value of the cladding thickness of which $\overline{T}_2$ becomes equal to the melting point of the cladding powder

144

for a given workpiece speed and laser power. This critical cladding thickness can be obtained from equation (5). If the cladding thickness is greater than this critical value, then there will be some unmelted powder between the substrate and the cladding melt and hence the substrate will not be clad. Thus for a given workpiece speed and laser power, the concentration of solute in the solid phase increases with cladding thickness as long as it is lower than the critical thickness value.

Figure 4 is concerned with the variation of solute concentration in the solid phase as the cladding powder feed rate changes. The cladding powder feed rate is determined by using the expression $\frac{1}{2}\pi r_c^2 v \rho$ and assuming that the cladding takes the shape of a semi-cylindrical strip [see Fig. 1(b)]. Cladding powder feed rate is a very important process parameter since it is related to cladding thickness and velocity of the

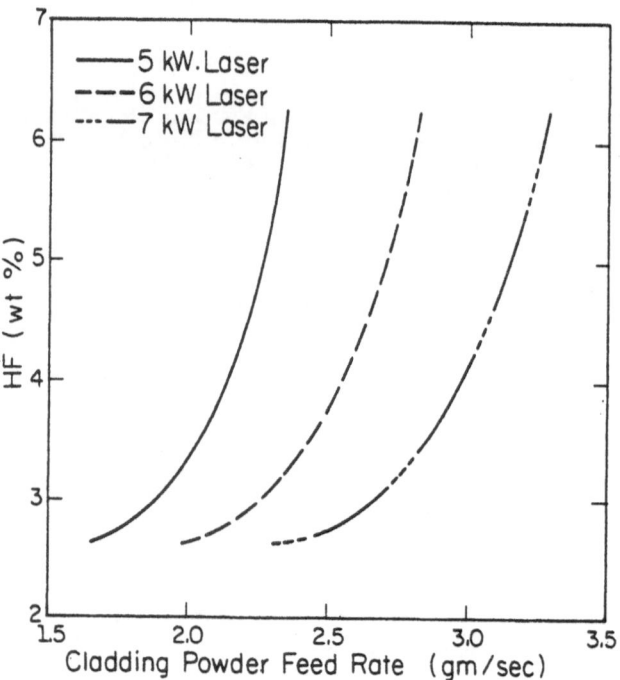

Fig. 4. Concentration of hafnium in the extended solid
solution of nickel-hafnium versus cladding powder feed rate.

workpiece as well as the initial pool mean temperature. Figures 5 and 6 indicate the effect of specific laser energy on solute concentration in the solid phase. Specific laser energy is defined as laser power required to produce a cladding of unit mass per unit time. It is determined by the relation $P / \left(\frac{1}{2}\pi r_c^2 v \rho \right)$. The importance of selecting specific laser energy as a parameter is that it allows the representation of the solute concentration data for any combination of laser power, workpiece speed, and

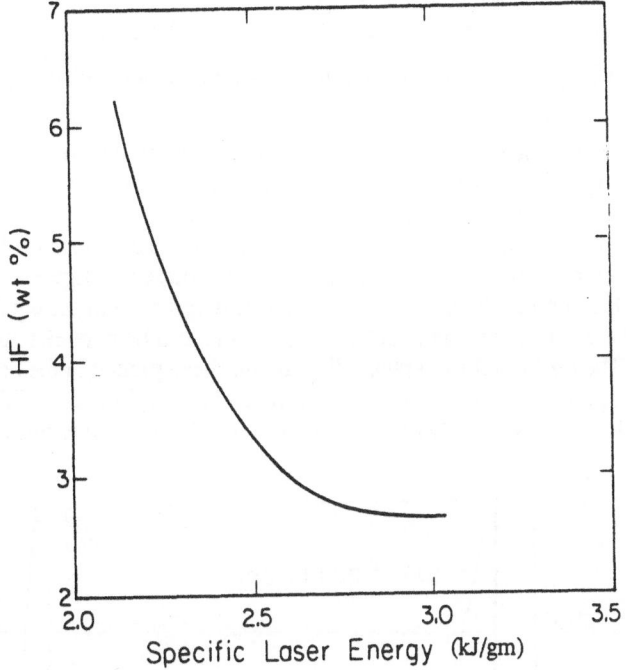

Fig. 5. Concentration of hafnium in the extended solid
solution of nickel-hafnium versus specific laser energy.

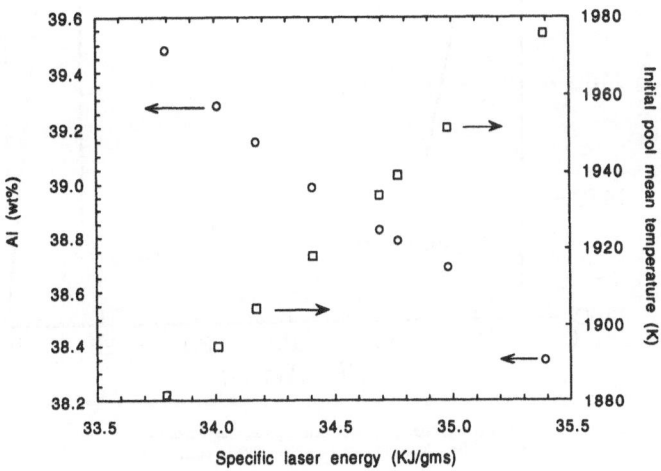

Fig. 6. Concentration of Aluminum in the extended solid solution of Nb-Al
and initial pool mean temperature versus specific laser energy.

cladding thickness parameters on one graph. This is because the initial pool mean temperature is proportional to specific laser energy. Thus the initial pool mean temperature $\left(\overline{T}_2\right)$ and hence the thermal characteristic of the cladding melt pool will not be different as long as the laser specific energy remains the same for any choice of the process parameters. This effect is also observed for Nb-Al system as indicated in Fig. 6, which shows how various process parameters can be combined into one group to represent the experimental data on one graph.

Figures 7 and 8 show the nonequilibrium phase diagrams for Ni-Hf and Ni-Al alloy systems. The characteristic parameters for these two figures are presented in Table 2. Deviation of the nonequilibrium phase diagram from the equilibrium phase diagram shows the extension of solid solubility that can be obtained during laser cladding. The nonequilibrium phase diagram has been plotted in the neighborhood of the melting point of the cladding powder. It can be seen from Fig. 7 that the width of the solid-liquid region between the equilibrium solidus and liquidus has reduced

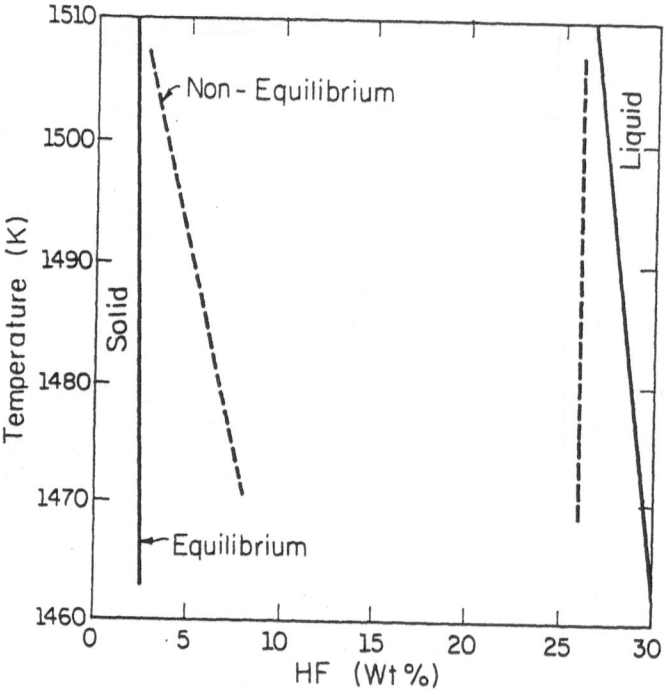

Fig. 7. Comparison of nonequilibrium phase diagram of the extended solid solution of nickel-hafnium with its equilibrium phase diagram.

considerably. The nonequlibrium solidus line of this figure shows the extension of Hf concentration in Ni that can be attained due to rapid cooling in laser cladding. This phenomenon can be understood from the fact that the equilibrium phase diagram of Ni-Hf has a negative slope at the point which corresponds to the nominal composition of the cladding powder. Due to this, Hf is rejected from the Ni matrix when the solution of Ni-Hf solidifies, and this results in incraesing the concentration of Hf in the liquid phase. But since the solid phase retains more Hf than its equilibrium composition, the liquid phase will have less Hf than the equilibrium value, and therefore, the extended solid solution (the nonequilibrium) phase diagram will shrink.

On the contrary, it can be seen fron Fig. 8 that the width of the solid-liquid region between the equilibrium solidus and liquidus has increased considerably. The nonequilibrium solidus line of this figure shows the extension of Al concentration in Ni that can be obtained due to rapid cooling in laser cladding. This phenomenon can

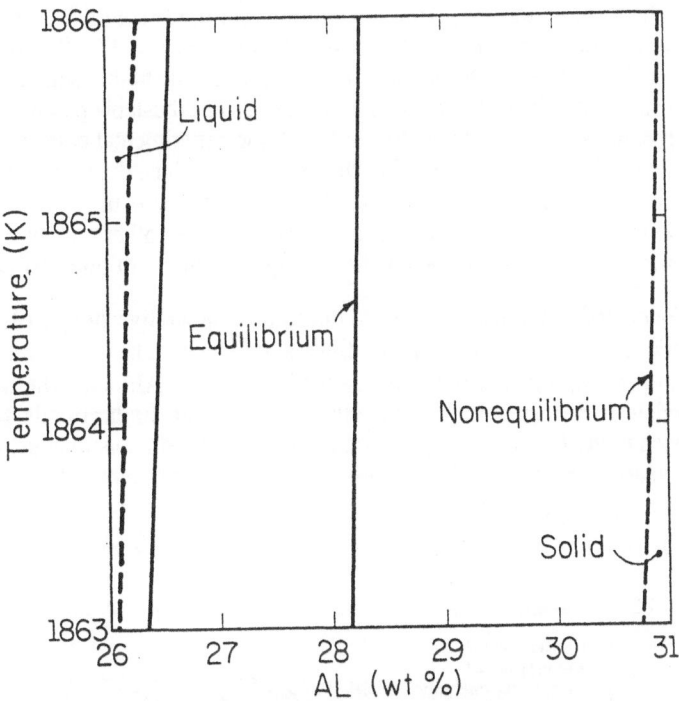

Fig. 8. Comparison of nonequilibrium phase diagram of the extended solid solution of nickel-aluminum with its equilibrium phase diagram.

Table 2. Solidus concentration of the solute, Solidus/Liquidus temperature at the substrate-cladding interface, and the solidification speed at this interface. Nonequilibrium conditions for Figs. 7 and 8.

Solute Elements	Solute Concentrations (wt%)	Solidus/Liquidus Temperature (K)	Interface Speed (cm/s)
Hf	3.19	1505	3.26
	4.83	1499	2.88
	6.23	1489	2.74
	7.15	1480	2.68
	9.58	1453	2.55
Al	30.83	1864	6.1
	30.74	1863	5.6
	30.56	1862	4.5

148

also be understood from the fact that the equilibrium phase diagram of Ni-Al has a positive slope at the point which corresponds to the nominal composition of the cladding powder. Due to this, Al is retained in the Ni matrix, whereas Ni is rejected into the solid phase as solidification proceeds. This effect increases Al concentration in the solid phase and lowers its weight fraction in the liquid phase, and thus enlarges the solid-liquid region between the solidus and liquidus lines.

We will now discuss some results concerning the nonequilibrium partition coefficients obtained by using the models of this paper. There are experimental data [37-41] for the nonequilibrium partition coefficient for dilute solutions, which have been compared with the results of various models in Refs. [24, 27, 29, 40] by assuming suitable values for the characteristic velocity. Due to the lack of adequate data for the characteristic velocity, it can be viewed as an adjustable parameter, and the model predictions can be shown to agree with the experimental data by suitably selecting a value for this parameter. For this reason, and since the theoretical results have been compared with the experimental data in the literature, the results obtained from Eq. (21) will be compared only with the predictions of other models. Figure 9 represents the results of various models for the equilibrium partition coefficient, $k_e =$ 7×10^{-4}. Cohen and Flemings [42] carried out some comparative studies of nonequilibrium partition coefficients for dilute solutions with $k_e = 0.44$ which are presented in Fig. 10, together with the result of this study. Although the results of various models do not agree with one another, they exhibit similar trends, that is, they approach towards unity (complete trapping) as the nondimensional growth velocity increases and towards the equilibrium partition coefficient as β tends to zero. It can be seen from Figs. 9 and 10 that the differences in the results of various models decrease as the value of k_e increases from 7×10^{-4} to 0.44.

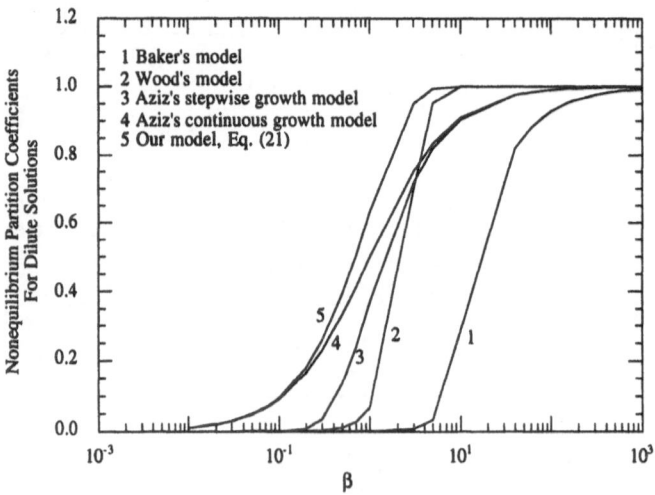

Fig. 9. Bismuth segregation in silicon. Results plotted assuming $k_e = 7 \times 10^{-4}$.

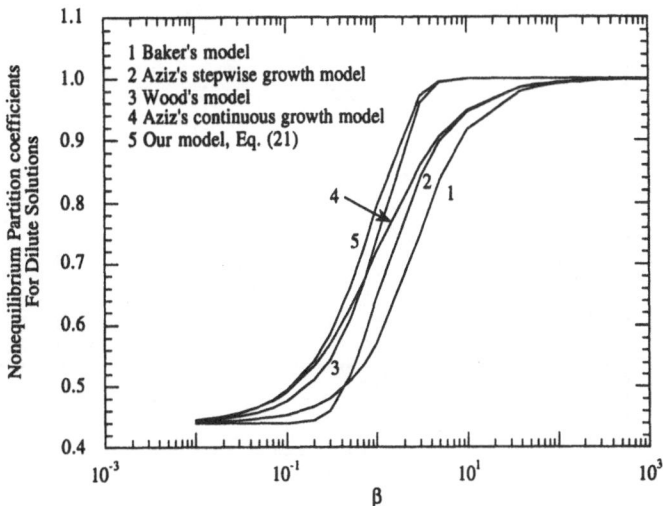

Fig. 10. Comparative studies of various models for $k_e = 0.44$.

The nonequilibrium partition coefficient for concentrated solutions obtained from Eq. (18) of this study has not been compared with the results that can be obtained from Aziz's [34, 35] equation, because k is determined in his model by solving a transcendental equation which has four variables m_l, k, T and β. In his model, β is related to temperature, composition of solute, chemical potential, and β_0 where β_0 is the dimensionless speed of the interface at infinite driving force. By adjusting the parameters of Aziz's model and Eq. (18) of this paper, the results of both models may probably be shown to be equivalent, but nevertheless, no meaningful conclusion can be drawn from such a comparison unless all the relevant parameters of a given system are known.

However, some results obtained from Eq. (18) are plotted in Fig. 11 in order to show the effects of β in solute trapping during the solidification of concentrated solutions. Since C_s/C_{li} and C_l/C_{li} vary with β, we approximate them as

$$\frac{C_s}{C_{li}} = 1 - e^{-\beta}\left(1 - \frac{C_s^*}{C_l^*}\right),$$

and

$$\frac{C_l}{C_{li}} = 1 - e^{-\beta}\left(1 - \frac{C_l^*}{C_{li}^*}\right),$$

for numerical computations in this study. Using these two expressions, Fig. 11 is plotted for $C_s^*/C_l^* = 0.2$ and for different values of k_m^* and C_l^*/C_{li}^* as shown in the figure, and k is found to reach unity when β is about 5.

150

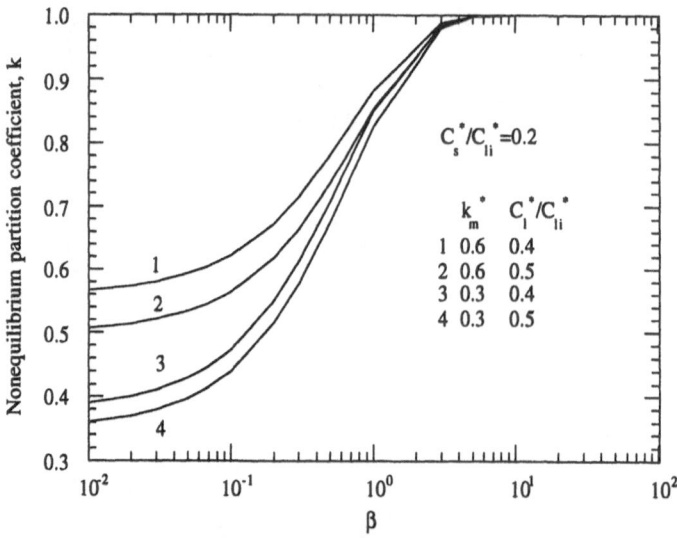

Fig. 11. Parametric studies on nonequilibrium partition coefficients for
concentrated solutions.

4. Conclusions

The present work examines the extension of solid solubility based on the transport of
energy and mass. Solute transport has been considered to take place only in the
liquid phase while the energy transport has been considered in both solid and liquid
phases. The effect of nonequilibrium cooling rate on solute segregation at the
freezing front has been taken into account by considering a nonequilibrium partition
coefficient. Using this, the mass transfer problem has been solved for solute
distribution in the liquid phase and the heat transfer problem has been solved to
obtain the velocity of the solid-liquid interface, its location, and the freezing
temperature of the interface. These mathematical solutions have been utilized to
study the effect of various process parameters on the concentration of solute in the
solid phase.

It is found that the same composition of solute is obtained in the alloy for
different cladding thicknesses by varying the laser power and the speed of the
workpiece. This is because the initial pool mean temperature of the cladding material
and hence the cooling rate are the same even though other parameters are different.
So, it can be concluded that the choice of the initial pool mean temperature
determines the composition of the alloy. The composition of solute in the solid
phase can be presented by a single curve for all possible values of the process
parameters if it is plotted against the specific laser energy or the initial pool mean
temperature. Also, it has been found that if the equilibrium liquidus has a negative
slope at the nominal composition of the cladding powder, then the nonequilibrium
phase diagram shrinks from the equilibrium phase diagram, resulting in a thin solid-
liquid region between the solidus and liquidus. On the other hand, if the

corresponding slope is positive, the nonequilibrium phase diagram enlarges resulting in a thicker solid-liquid region between the solidus and liquidus than the equilibrium phase diagram.

Expressions for nonequilibrium partition coefficient for dilute and concentrated solutions with only one unknown parameter are presented. For dilute solutions, the differences in the results of various models are found to decrease for higher values of the equilibrium partition coefficient, k_e. For concentrated solutions, the nonequilibrium partition coefficient, k is found to reach unity when the nondimensional characteristic velocity, β is about 5.

Acknowledgment

This work was made possible by a grant from the Air force Office of Scientific Research (Grant No: AFOSR 85-0333 and 88-0061). Dr. A. Rosenstein is the Program Manager.

Nomenclature

a	width of the substrate.
b	width of the substrate and the cladding melt.
B	Laser beam diameter.
B_{As}	Mobility of species A in the solid phase.
C	concentration of solute in the liquid phase (in section 2.1).
C	Molar concentration of species A (in section 2.2).
C_{Al}	Concentration of species A in terms of moles per unit volume in the liquid phase.
C_{As}	Concentration of species A in terms of moles per unit volume in the solid phase.
C_{Bs}	Concentration of species B in terms of moles per unit volume in the solid phase.
C_1	concentration of solute in the liquid phase at the solid-liquid interface.
C_1	Molar concentration of A in the previous liquid phase (see Fig. 1b).
C_1^*	concentration (mole fraction) of solute in the liquid phase at the solid-liquid interface (in section 2.1).
C_1^*	Liquidus composition of species A in the equilibrium phase diagram at temperature $T_i + \Delta T$ (in section 2.2).
C_{1i}^*	Liquidus composition of species A in the equilibrium phase diagram at temperature T_i.
C_{1i}	Molar concentration of A in the liquid phase at the interface.
C_{lo}	Initial concentration of species A in the liquid region.
C_p	average specific heat of the cladding material.
C_s	concentration of solute in the solid phase at the solid-liquid interface (in section 2.1).

C_s Molar concentration of A in the previous solid phase, see Fig. 2b (in section 2.2).

C_s^* Solidus composition of species A in the equilibrium phase diagram at temperature $T_i + \Delta T$.

C_{si} Molar concentration of A in the solid phase at the interface.

D average mass diffusivity of solute in the liquid phase.

D^* interdiffusivity of solute at the interface.

D_{ABl} Interdiffusion coefficient of species A with respect to species B in the liquid phase.

D_{ABs} Interdiffusion coefficient of species A with respect to species B in the solid phase.

D_{As} Intrinsic diffusion coefficient of component A in the solid phase.

D_{Bs} Intrinsic diffusion coefficient of component B in the solid phase.

f fraction of laser energy absorbed by the cladding material.

h_1 heat transfer coefficient at the substrate boundary.

h_3 heat transfer coefficient at the cladding surface boundary.

J Flux of species A.

$J_{A,ls}$ Flux of species A from liquid to solid at the interface.

$J_{A,sl}$ Flux of species A from solid to liquid at the interface.

k Nonequilibrium partition coefficient for concentrated solutions.

$k^*(t)$ Nonequilibrium partition coefficient (in section 2.1).

k_d Nonequilibrium partition coefficient for dilute solutions.

k_e Equilibrium partition coefficient

k_i Thermal conductivity of the ith region, i = 1, 2, 3 represents the solid substrate, solidified cladding, and liquid cladding region, respectively.

k_m Modified nonequilibrium partition coefficient for concentrated solutions.

k_m^* Modified equilibrium partition coefficient for concentrated solutions.

L latent heat of fusion of the cladding material.

m_1 liquidus slope (K/mol fraction of solute).

m_1 Reciprocal of the liquidus slope at temperature T_i.

m_1^* Reciprocal of the liquidus slope at the solidification temperature (T_i) under equilibrium cooling condition.

m_s Reciprocal of the solidus slope at temperature T_i.

m_s^* Reciprocal of the solidus slope at the solidification temperature (T_i) under equilibrium cooling condition.

P Laser power.

R Universal gas constant.

r_c Radius of the semi-cylindrical strip of cladding.

s Position of the solid-liquid interface.

T Absolute temperature.

T_I Temperature at the solid-liquid interface (in section 2.1).

T_i Interface temperature (in section 2.2).

T_i	Temperature of the ith region, i = 1, 2, 3 represents the solid substrate, solidified cladding, and liquid cladding region, respectively (in section 2.1).
T_r	Ambient temperature.
$\overline{T}_2$	initial pool mean temperature.
T_m^*	intercept of liquidus line on temperature axis.
u	Speed of the freezing front.
v	Speed of the workpiece.
v_0	Speed of sound in the cladding melt.
V_c	Characteristics velocity of species A.
x	Cartesian coordinate axis pointing from the solid to the liquid phase.
x_{As}	Mole fraction of component A in the solid phase.
x_{Bs}	Mole fraction of component B in the solid phase.
ΔT	A small and positive number such that C_s and C_l refer to the concentrations at $T_i + \Delta T$ (see Fig. 2b).

Greek Symbols

α_i	Average thermal diffusivity in the i-th region i = 1, 2, 3 represents the solid substrate, solidified cladding, and liquid cladding region, respectively.
β	Dimensionless solidification rate.
λ	Characteristic length.
ρ	Average density of the cladding material.

References

1. J. Singh and J. Mazumder, Acta metall. **35**, 1995 (1987).
2. J. Mazumder and W. M. Steen, J. appl. Phys. **51**, 941 (1980).
3. E. Cline and T. R. Anthony, J. appl. Phys. **48**, 3895 (1977).
4. S. Kou, S. C. Hsu and R. Mehrabian, Metall, Trans. **12B**, 33 (1981).
5. M. F. Ashby and K. E. Easterling, Acta metal. **32**, 1935 (1984).
6. C. Chan, J. Mazumder and M. M. Chen, Metall. Trans. **15A**, 2175 (1984).
7. T. Chande and J. Mazumder, J. appl. Phys. **57**, 2226 (1985).
8. J. C. Baker and J. W. Cahn, Solidification, pp. 23-58, Am. Soc. Metals, Metals Park, Ohio (1971).
9. W. J. Boettinger and J. H. Perepezko, Proc., Rapidly Solidified Crystalline Alloys, TMS-AIME, N.J. (1985).
10. W. J. Boettinger, S. R. Coriell and R. F. Sekerka, Mater. Sci. Engng **65**, 27 (1984).
11. B. H. Kear, B. C. Giessen and M. Ghen (editors) Rapidly Solidified Amorphous and Crystalline Alloys, Proc. M.R.S. Vol. 8, Boston, Mass. (1981).
12. L. J. Li and J. Mazumder, Laser Processing of Materials (edited by Mukherjee K. and J. Mazumder), pp. 35-50. Proc. Metal. Soc. AIME, Los Angeles, Calif. (1984).
13. A. Kar and J. Mazumder, J. appl. Phys. **61**, 2645 (1987).

154

14. A. Kar and J. Mazumder, Acta Metall. **36**, 701(1988).
15. A. Kar and J. Mazumder, Met. Trans. **20A**, 363(1989).
16. E. Y. Yankov, J. A. Todd, and S. M. Copley, in Proc. Morris E. Fine Symp., P. K. Liaw, J. R. Weertman, H. L. Marcus, and J. S. Santner, eds., TMS, Warrendale, Pennsylvania, 1991, P. 29.
17. E. Y. Yankov, S. M. Copley, M. I. Yankov, and J. A. Todd, in Proc. Morris E. Fine Symp., P. K. Liaw, J. R. Weertman, H. L. Marcus, and J. S. Santner, eds., TMS, Warrendale, Pennsylvania, 1991, P. 33.
18. K. A. Jackson, *Can J. Phys.*, **36**, 683 (1958).
19. V. T. Borisov, Soviet Phys. Dokl. **7**, 50 (1962).
20. J. C. Baker and J. W. Cahn, Acta Metall., **17**, 575 (1969).
21. A. A. Chernov, Growth of Crystals (Consultants Bureau, New York, 1962), **3**, p. 35.
22. J. W. Cahn, S. R. Coriell, and W. J. Boettinger, in Laser and Electron Beam Processing of Materials, Ref. 7, p. 89.
23. K. A. Jackson, G. H. Silmer, and H. J. Leamy, in Laser and Electron Beam Processing of Materials, C. W. White and P. S. Peercy, eds., Academic Press, New York, 1980, p. 104.
24. R. F. Wood, J. C. Wang, G. E. Giles, and J. R. Kirkpatrick, in Laser and Electron Beam Processing of Materials, C. W. White and P. S. Peercy, eds., Academic Press, New York, 1980, p. 37.
25. R. F. Wood, Appl. Phys. Lett. **37**, 302 (1980).
26. R. F. Wood, and G. E. Giles, Phys. Ref. **B23**, 2923 (1981).
27. R. F. Wood, J. R. Kirkpatrick, and G. E. Giles, Phys. Rev. **B23**, 5555 (1981).
28. R. F. Wood, Phys. Rev., **B25**, 2786 (1982).
29. M. J. Aziz, *J. Appl. Phys.*, **53**, 1158 (1982).
30. G. H. Gilmer and P. Bennema, *J. Appl. Phys.*, **43**, 1347 (1972).
31. G. H. Gilmer and K. A. Jackson, in Crystal Growth and Materials, E. Kaldis and H. J. Scheel, eds., North-Holland, New York, 1977, p. 80.
32. G. H. Gilmer, in Materials Research Society Symposium, Elsevier, New York, 1983, Vol. 13, p. 249.
33. G. H. Gilmer, Mat. Sci. Engrg., **65**, 15 (1984).
34. M. J. Aziz, Appl. Phys. Lett., **43**, 552 (1983).
35. M. J. Aziz, in Science and Technology of Rapidly Quenched Alloys, M. Tenhover, W. L. Johnson, and L. E. Tanner, eds., Materials Research Society, Pittsburgh, Pennsylvania, 1987, p. 25.
36. R. Trivedi and W. Kurz, Metall. Trans. A. **21A**, 1311 (1990).
37. P. Baeri, G. Foti, J. M. Poate, S. U. Campisano, and A. G. Cullis, Appl. Phys. Lett. **38**, 800 (1981).
38. C. W. White, S. R. Wilson, B. R. Appleton, and F. W. Young, Jr., *J. Appl. Phys.* **51**, 738 (1980).
39. C. W. White, B. R. Appleton, B. Stritzker, D. M. Zehner, and S. R. Wilson, in Laser and Electron-Beam Solid Interactions and Materials Processing, J. F. Gibbons, L. D. Hess, and T. W. Sigmon, eds., North Holland, New York, 1981, p. 59.
40. M. J. Aziz, J. Y. Tsao, M. O. Thompson, P. S. Peercy, and C. W. White, Phys. Rev. Lett. **56**, 2489 (1986).

41.   P. Baeri, J. M. Poate, S. U. Campisano, G. Foti, E. Rimini, and A. G. Cullis, Appl. Phys. Lett., **37**, 912 (1980).
42.   M. Cohen and M. C. Flemings, in Rapidly Solidified Crystalline Alloys, S. K. Das, B. H. Kear, and C. M. Adam, eds., Metall. Soc., Warrendale, Pennsylvania, 1985, p. 3.

SCOPING ANALYSES AND LIMITATIONS OF TRANSPORT PHENOMENA MODELING FOR LASER SURFACE MODIFICATION

MICHAEL M. CHEN
Department of Mechanical Engineering & Applied Mechanics
University of Michigan, G. G. Brown Laboratory
Ann Arbor, MI 48109-2125, USA

Abstract: This paper presents a discussion of the basic relationships for commonly encountered transport phenomena in laser surface modification, with the laser serving as the heat source for the processes. In recent years numerical modelings of the transport phenomena have contributed much to our understanding. On the other hand, in quantitative details and some key qualitative details there are still some disagreement among the different computations. It will be demonstrated in this paper that much important qualitative and semi-quantitative information can be obtained more economically, and perhaps with greater insight, by approximate scoping analyses. Such an analysis usually can directly yield scaling information, valuable for process optimization. In addition, a scoping analysis of neglected or poorly known mechanisms can also point out difficulties and limitation in the predictability of certain processes. Understanding these limitations *a priori* will permit a worker to judge the validity and limitations of numerical modeling results and properly evaluate the potential worth of planning modeling efforts.

1. Introduction

In recent years numerical modeling of the fluid flow, heat and mass transfer phenomena in laser materials processing has become increasingly popular[1-16]. This is due to the fact that in most of these processing technologies, the laser primarily serves as an efficient and easily controllable deliverer of high intensity thermal power, which produces the desired heating and melting required for materials processing. Largely due to thermally induced surface tension variations, the melt is known to develop an intense recirculating flow that also plays strong roles in the redistribution of absorbed energy subsequent cooling and solidification, and mixing of alloying components. It is thus recognized that with a greater understanding of the details of the transport phenomena,

157

J. Mazumder et al. (eds.), Laser Processing: Surface Treatment and Film Deposition, 157–175.
© *1996 Kluwer Academic Publishers.*

rational process optimization or fine tuning of process parameters can be more easily accomplished.

Unfortunately, in spite of the many papers already available in the literature, no clear, consistent conclusions have emerged from these studies beyond the basic qualitative fact that significant velocities can be attained in the melt pool due to thermocapillarity. Some interesting phenomena, such as very high velocities near the cold corner [10, 14], were found in some computations and not in others. Even the basic parameters governing the depression and/or elevation of the liquid surface, and their relationship to undercut and humping in the re solidified solid, have not yet been clearly established. The preponderance of strictly numerical approaches, and the lack of any study with sufficient attention to matters of accuracy, resolution, and completeness of parameter ranges have inhibited the formation of physical insight, let alone the transfer of such insight to material scientists and process developers, who could benefit from such insight.

The development of physical insight can best be accomplished by researchers with a good understanding of both components of this interdisciplinary science. This author, as a member of the transport phenomena community, dedicates this paper to the members of the materials and processing community. It is hoped that it can serve two functions:

(a) It will attempt to show that in many instances, good semi-quantitative analysis of the transport phenomena is possible without resorting to full scale numerical analysis or exact analytical solutions. Frequently such a semi-quantitative analysis will yield better physical insight and clearer scaling laws than numerical results. Coupled with actual processing experience, such scaling laws are quite adequate (some may even view them as desirable) for decisions on process improvement. Considering the difficulties in prescribing the exact physical properties and boundary conditions, and in view of the mathematical and physical uncertainties discussed in (b) below, such semi-quantitative results may be as accurate as the detailed numerical computations.

(b) It will attempt to make readers aware of the physical and mathematical pitfalls that are present in modeling the thermo-fluid phenomena associated with laser processing. For example, the Marangoni number, a parameter governing the intensity of thermocapillary flows, is usually on the order of 10^4 for typical laser melt pools. The mathematical property of the governing equations for thermocapillary flows are such that at these high Marangoni numbers, extremely fine mesh, considerably beyond what has been employed for in the literature, is required for valid solutions. This and related difficulty has been pointed out by Chen [14]. This fine structure of the cold wall region has been recently verified by Canwright in a very careful analysis [17]. In addition, the physics of the tri-phase line in the presence of melting and solidification with dendritic micro structure is sufficiently unclear to preclude a meaningful continuum formulation of the so-called contact angle boundary condition. The roles of surfactants and/or other surface contaminants in either liquid or solid form, including the oxides of the alloy constituents, are also unclear. These problems render the results of casual computations suspect, as pointed out by this author [14]. These difficulties are highlighted by the fact that after nearly ten years of fluid flow modeling for the weld pool, no definitive quantitative statements can be made on pool surface elevation or depression. Some authors predicted an elevation of the free surface at the pool center [7] under conditions when others predicted a depression.

2. Length Scale for Diffusive Spread

The basic tool for semi-quantitative analysis is scale analysis, which, like dimensional analysis, can be performed by examining the governing differential equations. This is done *arithmetically*, without the need for actually solving the differential equation. It should be stated that in order for the results to be clear and unequivocal, the governing equations have to be sufficiently simplified that only the barest minimums of terms are present. This simplification can usually be performed intuitively or with heuristic arguments. However, the neglect of any term implies an asymptotic consideration, and occasionally the neglect of certain term or terms may lead to mathematical difficulties or loss of essential physics. Readers interested in such subtleties may consult standard tests [18] or this authors' introductory discussion [19].

2.1. THE BASIC RELATIONSHIP

Partly as an example of the method and partly because of its importance in laser surface modification, the length scale of the diffusive spread of heat, mass and momentum will be considered first. The question we wish to pose is this: Given a quantity of thermal energy or a quantity of a solute species located at a point (or a line, or a surface), what is the extent of spread after an elapsed time interval t? These two phenomena are governed by the physical properties heat diffusivity $\kappa \equiv k/\rho c$ (sometimes denoted by the symbol α) or mass diffusivity D respectively. Another similar process is the diffusive spread of momentum. One may imagine a surface temporarily possessing a different tangential velocity from the bulk of the fluid it is adjacent to. What is the thickness of the fluid layer which after a time interval t would be viscously retarded or accelerated by the surface? The *diffusivity* for this momentum transfer phenomenon is the kinematic viscosity, $\nu \equiv \mu/\rho$.

The governing equations for all three diffusive phenomena are similar, so for simplicity the heat conduction equation will be used as the surrogate. It is also immaterial whether one considers a one-dimensional, two-dimensional, or three dimensional spread, with the following differential equations:

$$\frac{\partial T}{\partial t} = \kappa \frac{\partial^2 T}{\partial x^2} \quad \text{or} \quad \frac{\partial T}{\partial t} = \frac{\kappa}{r}\frac{\partial}{\partial r} r \frac{\partial T}{\partial r} \quad \text{or} \quad \frac{\partial T}{\partial t} = \frac{\kappa}{r^2}\frac{\partial}{\partial r} r^2 \frac{\partial T}{\partial r} \tag{1}$$

If T_S denotes the temperature scale, which can be taken to be difference between the maximum temperature and the room temperature, and t_S denotes the time scale, which can be taken to be time interval t, and δ_T denotes the heat diffusion scale, which is a measure of the extent of the diffusive spread of heat, then the magnitudes of the derivatives in Eq. (1) can be estimated by corresponding ratios of these scales, yielding the semi-quantitative algebraic equation

$$\frac{T_s}{t_s} \sim \kappa \frac{T_s}{\delta_T^2} \tag{2}$$

from which one can easily solve for δ_T in terms of the other quantities

$$\delta_T \sim \sqrt{\kappa t_s} \tag{3}$$

Similar results can be obtained for the diffusive spread of solute and viscous diffusion of momentum:

$$\delta_D \sim \sqrt{D t_s} \tag{4}$$

$$\delta_V \sim \sqrt{v t_s} \tag{5}$$

While the above results have been obtained from strictly macroscopic continuum considerations, it is useful to note that there is a clear and simple molecular interpretation. In gases where all three transport phenomena can be described by kinetic theory for molecules, or in condensed matter where heat transfer can be described by a similar kinetic theory for phonons and/or electrons, it can be shown that the diffusivities are on the order of $f_c \lambda^2$, where f_c is the collision frequency and λ is the mean free path. Accordingly, the extent of diffusive spread is simply the *distance made good in a sequence of random-walk steps*:

$$\delta \sim \lambda \sqrt{f_c t_s} \sim \lambda \sqrt{n_c} \tag{6}$$

where n_c is the total number of collisions or steps.

Coupled with conservation of energy, mass, or momentum, the results (3) - (5) are amazingly versatile, and can be used to obtain scaling laws for maximum temperatures and the size of the heat affected zone. With the additional use of integral methods, quantitative results with accuracies in the 10 - 20% range can be obtained. This is comparable to those which are achievable with analytical and numerical methods, unless extreme care has been used in choosing property values and geometry. The following results for specific applications will also serve as examples for situations not discussed here.

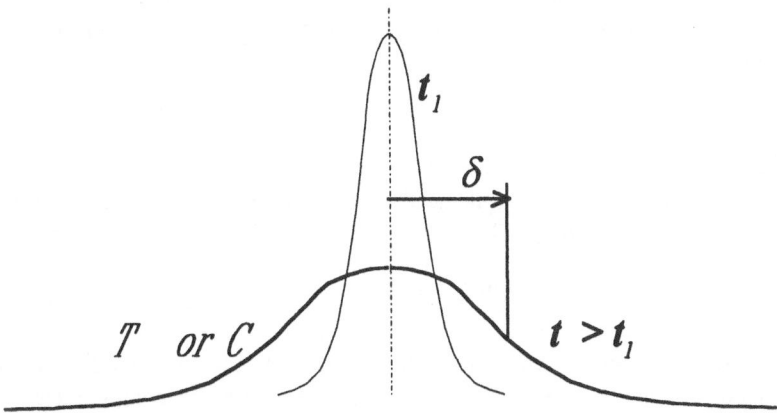

Fig. 1 Temperature (or Concentration) Distribution due to Point-wise Deposition of Energy (or Solute). δ is shown as width at half-height

2.2. EXAMPLE: APPLICATION TO DEPOSITION OF ENERGY OR SOLUTE MASS AT A POINT

The above results describe the basic scaling of the spreading rates for diffusive effects. The results are quite universal and can be applied to many physical situations of interest. The only additional information needed are (1) the conservation laws (2) a simple relationship to relate the speed and distance of a moving system to the time scales in the diffusive spread relationships, Eq. (3)-(5). The following examples will illustrate the use of conservation laws, whereas sub-section 2.3 will briefly discuss steadily moving systems.

2.2.1. *Three-dimensional Spread*

When a fixed amount of energy E (joules) is delivered at a given location, the temperature at that point rises to a high value, then decreases as the heated zone spreads, as shown in Fig. 1. The scaling law for the decay of ΔT, the temperature elevation above the baseline, can be easily obtained by combining Eq. (3) with the conservation of energy:

$$\rho c \Delta T \, \delta_T^{\;3} \;\sim\; E$$

whence

$$\Delta T \;\sim\; \frac{E}{\rho c \sqrt[3]{\kappa t_s}} \tag{7}$$

Clearly some simplification is possible by canceling the specific heat and density contained in the definition of heat diffusivity. However, we chose to leave the equation in the unsimplified form to enhance a more intuitive perception that it is just the conservation of energy.

Approximate Quantitative Estimate: One can obtain a relatively easy approximate, but nevertheless quantitative estimate of the temperature by using the well-known *integral method*. To perform this analysis, one needs to make a reasonable assumption regarding the *shape* of the temperature profile:

$$T(r,t) \;\equiv\; \Delta T_{max}(t)\, f(\eta) \quad \text{where} \quad \eta \equiv \frac{r}{a\sqrt{\kappa t}} \tag{8}$$

Eq. (8) in fact separates the temperature dependence into an amplitude function of time and a shape function in terms of a spatial coordinate that is continuously re-scaled according to the diffusive spread $\sqrt{\kappa t}$. a is a dimensionless constant. For convenience but not mathematical necessity, it is desirable to normalize $f(\eta)$ such that $f(\eta) = 1$. The adjustable constant a further permits the volume integral of $f(\eta)$ to be normalized to unity, or any other convenient number. Clearly, conservation of energy requires the volume integral of the internal energy to be equal to the deposited energy E:

$$\int \rho c \, \Delta T \, dV \;=\; E \tag{9}$$

which leads to the result

$$\Delta T \;=\; \frac{E}{I_f \,\rho c \sqrt[3]{\kappa t_s}} \quad \text{where} \quad I_f \equiv 4\pi \int_0^\infty f \, dr^2 \tag{10}$$

The power of the integral method is such that as long as a reasonable shape function $f(\eta)$ is assumed, relatively good estimates of the temperatures can be obtained. It is of course well known that the exact solution has the shape of the error function. However, any shape resembling the Gaussian would lead to a good approximate estimate.

2.2.2 *Two-dimensional Spread*

We briefly discuss here the two-dimensional counterpart of the above, since it will be used later to obtain the scaling laws and approximate solutions for the well-known Rosenthal solution. Since the derivation is identical to those of Eqs. (7) and (10), we shall write down the results without derivation. For the 2-D problem, the energy deposition is expressed as the specific energy deposition per unit length $\varepsilon \equiv dE / d\ell$. Hence the counterparts of Eqs. (7) and (10) are, respectively

$$\Delta T \;\sim\; \frac{\varepsilon}{\rho c\, \kappa t_s} \tag{11}$$

$$\Delta T \;=\; \frac{\varepsilon}{I_f\, \rho c \kappa t_s} \quad \text{where} \quad I_f \equiv 2\pi \int_0^\infty f\, dr \tag{12}$$

2.3. STEADILY MOVING SYSTEMS

Most of the modalities of laser surface processing involve steadily moving systems. The typical arrangement is for the workpiece to be moving steadily under a focused or defocused laser. In these instances the time available for the diffusion of heat is given by a length scale L and the workpiece traverse velocity U, which is sometimes called the scanning velocity even though scanning by the laser beam is rarely performed.

$$t \;\sim\; L/U \tag{13}$$

Hence

$$\delta_T \sim \sqrt{\kappa L / U} \tag{14}$$

The appropriate length scale to use depends on the problem. These will be illustrated by two examples.

2.3.1. *Heating of a Moving Workpiece with a Steady Point Source*

This is the situation of the well-known "Rosenthal Solution" [20] employed frequently in welding analysis. The physical configuration involves a workpiece moving at a constant velocity U steadily past the laser beam, focused on the surface. The laser thus represents a stationary point heat source Q *(W)*. Immediately under the beam, melting and other complications are usually present (see Section 3 below), so the heat conduction equation on which the present analysis and the Rosenthal solution are based no longer applies. Heat conduction analysis is more meaningful for heat affected zone estimates somewhat downstream of the bean location. If the distance downstream is denoted by L, then the diffusive time scale is L/U, and if the diffusive spread $\sqrt{\kappa t} = \sqrt{\kappa L/U}$ is small relative to L, then conduction in the streamwise direction is small Vs the radial conduction from the track of the laser on the surface, with a the specific energy deposition per unit length ε computed from Q and U.

$$\varepsilon \;=\; Q/U \tag{15}$$

Hence from Eq. (11), we have

$$\Delta T \;\sim\; \frac{Q/U}{\rho c \kappa L/U} \;\sim\; \frac{Q}{\rho c \kappa L} \tag{16}$$

2.3.2. *Heating of a Moving Workpiece by Uniform Heat Flux*

Another situation of interest to laser surface modification is the steady heating of a workpiece by a locally uniform surface heat flux, as the workpiece is moving steadily under laser irradiation. This can be viewed as representing the configuration of transformation hardening, when a high powered laser is defocused to deliver a moderate heat flux over a relatively large area. The length L then represents the beam width in the direction of motion, as shown in Fig. 2.

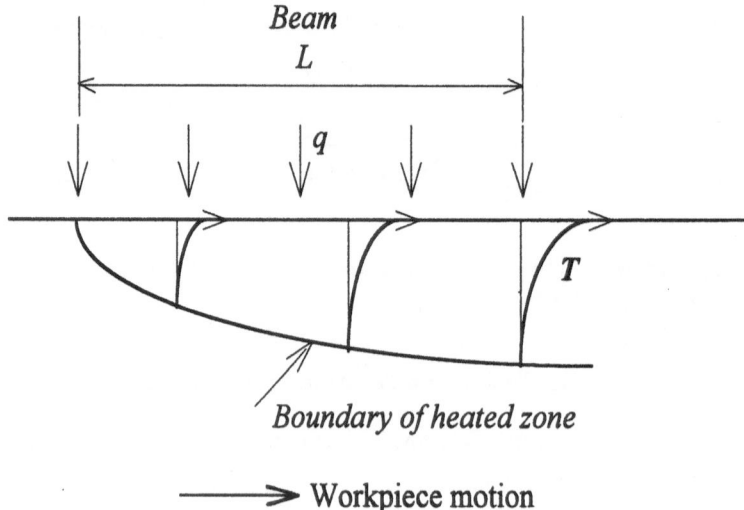

Fig. 2 Heating of a Moving Workpiece by a Broad, Defocused Laser

If the length and the width of the irradiated region is large versus the depth of the heated layer, whose thickness is on the order $\sqrt{\kappa t} \sim \sqrt{\kappa L / U}$, then conservation of energy would lead to

$$\Delta T ~ \sim ~ q \sqrt{\frac{L}{\rho c k U}} \tag{17}$$

In both of these cases, if desired, the integral method can be used to obtain an approximate, but nevertheless quantitative estimate of the temperatures. If should be noted, however, that while all the previously discussed cases had a Gaussien temperature profile, the temperature profile for the present case more resembles an exponential function or a parabola. Both should yield results with about 20% accuracy, which is probably comparable to the so-called exact solutions, because of other certainties and the difficulty of the exact solutions to fit the experimental conditions.

The case considered above is concerned primarily with the heated region. After the workpiece has passed under the beam, it enters a cooling region the temperature would gradually decrease. A similar derivation, using the total energy deposited per unit width (i.e., transverse length measure), would yield the scaling law for cooling.

The two cases considered in Sections 2.3.1. and 2.3.2. obviously represent two asymptotic cases of laser surface treatment. If the beam diameter is small relative to $\sqrt{\kappa t}$ and L, then 2.3.1. applies. If the beam diameter and length are large relative to $\sqrt{\kappa t}$, then 2.3.2. applies. If the beam diameter is comparable to $\sqrt{\kappa t}$, then more exact analysis is necessary, though either 2.3.1. or 2.3.2. might give good semi-quantitative estimates.

3. Melt Pool Formation and Melt Motion

In some laser surface modification schemes, the temperature is so high that local melting occurs. This section deals with the problems of making estimates under these conditions. As a rule, vaporization is not a significant concern in typical laser surface modification conditions. It will not be discussed in this paper.

3.1. EFFECTS OF THE LATENT HEAT OF FUSION

For most alloys of interest the latent heat is small, though not negligible, compared with the energy vested in sensible heat, equal to $c\Delta T$. Therefore while the latent heat is important in exact analysis, it can be neglected in semi-quantitative analysis without seriously affecting carefully considered conclusions [4].

3.2. LIQUID MOTION IN MELT POOL -- ITS IMPORTANCE AND ORIGIN

Early attempts at modeling welding processes often overlooked liquid motion in the weld pool [e.g., 20]. It is now known that melt pool flows play important roles in the process and should not be neglected. Among the important processes affected by liquid motion include the determination of width-to-depth ration of the pool, the formation of undercut and humping, and the mixing of added material with the original alloy.

There are several basic causes of melt pool motion. The first that might come to mind is thermal buoyancy. For the small dimensions involved, this has been demonstrated [2] to be unimportant. Electromagnetic forces, important for high current arc welding, are not present for laser processes. Mechanical vibrations due to unsteady workpiece support can potentially be important in determining the finish of the re-solidified surface. This will be briefly discussed in Section 4. Recoil forces due to vaporization can be important for welding, but is not likely to be important for the conditions of laser surface modification. The pressure fields and shear stresses due to the motion of the overlaying gas may also play a role.

The most important class of forces responsible for melt motion is probably surface tension variations associated with temperature and composition non-uniformities. These phenomena are quite complex and the following Subsection will be devoted to the discussion of thermocapillary flow, or flow due to thermally induced surface tension variations. At present very little is known about the compositional counterpart, soluto-capillary flow in laser melt pools, even thought its importance is strongly suspected.

3.3. THERMOCAPILLARY FLOW IN MELT POOLS

Most of our knowledge on thermocapillary flow in laser melt pools has come from studies on weld pools. Much of the following discussion has been taken from Chen [9].

For most materials, surface tension decreases with increasing temperature. Hence in a small melt pool, surface tension at the rim is greater than that at the center, thus fluid is pulled along the surface toward the rim, as shown in Fig. 3. It should be mention at this point that it is by no means clear that the fluid will return to the center near the bottom. Depending on the precise shape of the solid-liquid interface, the fluid may separate before reaching the bottom and return to the center at an intermediate depth, thus creating a multi-cell flow pattern, as schematically shown in Fig. 3. It is for this reason that great precision is needed in determining the shape of the solid-liquid interface. More on this will be discussed later.

When the surface tension of the molten material has a positive temperature coefficient, the flow direction is reversed, as expected.

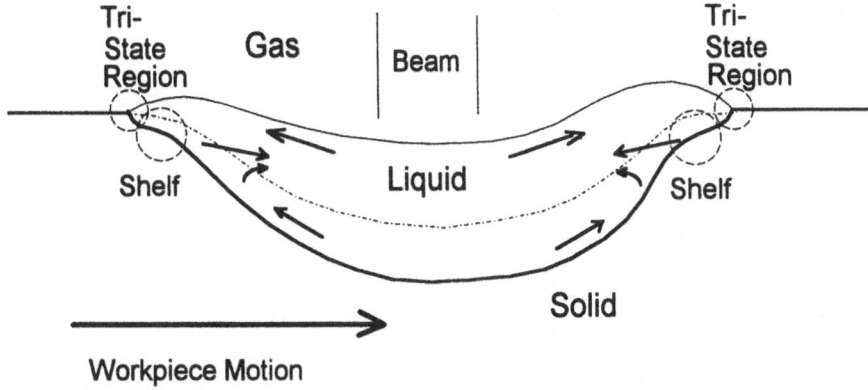

Fig. 3. Schematic representation of the flow field in a melt pool caused by laser heating.

A modest depression of the free surface is illustrated in Fig. 3. This is in agreement with general expectations and the results of most computations, though the issue is not yet unequivocally established, as some researcher showed an elevation in the center [11].

The type of scaling analysis discussed in previous sections can be used here to estimate the order-of-magnitude of the amplitude of velocities and other quantities of interest. The starting point of such an analysis is the thermocapillary boundary condition:

$$\mu \frac{\partial u}{\partial y} = -\gamma \frac{\partial T}{\partial x} \tag{18}$$

where γ is the temperature coefficient of surface tension.

For a scale analysis, Eq. (18) implies

$$\mu \frac{U_s}{\delta_v} \sim \gamma \frac{\Delta T}{W} \qquad \text{where} \quad \Delta T \quad \text{is} \quad T_{max} - T_{melt} \tag{19}$$

Here W denotes the melt pool width. For a given temperature difference, this result still has two unknowns, the velocity scale U_S and the viscous boundary layer thickness δ_v. The additional equation required is an equation for the viscous boundary layer:

$$\delta_v \sim \sqrt{vW/U_s} \tag{20}$$

Although the solution is straight forward, it is clearer to display the results in terms of dimensionless parameters Marangoni number and Prandtl number.

$$\text{Ma} \equiv \frac{\gamma \Delta T W}{\mu \kappa} \tag{21}$$

$$\text{Pr} \equiv \frac{\nu}{\kappa} \tag{22}$$

$$\frac{\mu U_s}{\gamma \Delta T} \sim \frac{\delta_v}{W} \sim \left(\text{Ma}/\text{Pr}\right)^{-1/3} \tag{23}$$

The derivation of Eq. (23) has been discussed in [6, 14] For typical melt pool conditions, Ma is on the order of $(10)^4$.

Eq. (23) can be used for semi-quantitative estimates for the magnitudes of thermocapillary velocities in the melt pool. For the past decade, numerous attempts have also been made to numerically predict the flow and the velocities. While most such studies have been able to produce weld pools with shapes roughly similar to Fig. 3, and with velocities of the order of m/s, same as predicted from Eq. (23), detailed examinations showed that there were considerable inconsistencies in both qualitative and quantitative details of the flow, as well as in the details of the shape of the melt pool. These difficulties were pointed out by Chen [14]. Some of the physical difficulties associated with formulation will be discussed later. We shall first deal with the mathematical difficulties, which is most responsible for the inconsistencies observed. It was pointed out by Chen [6, 14] that the flow structure approaches a singularity in the cold corner in the limit of large Marangoni numbers. Recent studies by Canright [17] confirmed this and further showed that the local length scale of the cold corner region is W/Ma. For an accurate computation, the mesh size should always be smaller than the length scale of the region. Given the fact that Ma is of the order 10^4, this means that practically all of the published models of weld pools have not had a sufficiently fine spatial resolution to accurately compute the velocities!

It may be wishfully argued that the details of the flow in such a small corner region cannot be expected to influence the large scale motion in the pool. Unfortunately this is a very shaky argument! Indeed all careful computations [14] show that the highest velocities in the pool do occur in the corner region. Furthermore, it has been demonstrated [12, 14] that the local heat flux at the solid-liquid interface also increases rapidly near the corner region, due to the locally high velocities. Such a locally high heat flux is likely to have a strong influence on the shape of the liquid-solid interface near the corner. For example, one possible consequence of the high corner heat flux is the local recession of the solid surface, leading to the formation of a small shelf, as shown speculatively in Fig. 3. Whether the shelf is actually present or as prominent as shown is immaterial, it is certainly true that these small details of the pool shape may have a strong influence on flow separation and hence the overall recirculating pattern of the flow. These are qualitative features and go beyond just a small quantitative inaccuracy. Accurate resolution of these features is important before meaningful conclusions can be drawn.

Besides mathematical difficulties, there are also a number of unsettled questions regarding the physics of the problem, including

Mushy Zone: It is well known that the solidification of alloys usually involves a porous dendritic structure consisting of both solid and liquid phases. This is frequently called the mushy zone. On the other hand most of the modeling studies of melt pools in the literature have assumed a simple liquid-solid phase change at a melting point temperature. Current state of understanding for the science of solidification has not reached the point of providing a simple method of describing the mushy phase transition with the accuracy required for flow separation and flow pattern prediction. The are also potential numerical difficulties in resolving the thin mushy zones.

Contact Angles: One of the boundary conditions which is needed for the computation of the free surface is the contact angle, which specifies the geometrical relationships of the surfaces at the tri-state line (see Fig. 3). While some data are available for selected combinations of liquid, gas and solid, few data exist for liquids that are the molten phase of the adjacent solid. Indeed, how the three included angles accommodate the requirement that the solid-liquid interface must also lie on an isothermal surface given by the heat conduction equation, especially in the presence of dynamic cooling-solidification and heating-melting phenomenon, is far from understood. In other words the physical understanding of the process is really inadequate to permit a complete mathematical formulation of the problem at this time, even for a pure substance. In the presence of a dendritic mushy interface region, even the basic definition of contact angle is now in question.

Soluto-capillarity and Surface Contamination: Many investigators suspect that surfactants and other contaminants, including the metal's own oxides, play major roles in determining the flow through their influence on the surface tension. This can be especially important since the outward thermocapillary flow tends to concentrate these surface contaminants near the cold corner. The possible existence of a thin sold layer floating on top of the liquid has been postulated [12]. A different implication of these problems is that the cold-corner singularity of pure thermocapillary flow may also have to be modified by soluto-capillary considerations. However, physical understanding for these effects does not exist at this moment.

4. Surface Depression, Humping, Undercut, and Other Surface Features

4.1. SURFACE DEPRESSION PREDICTIONS

Another uncertainty in thermocapillary flow calculations is the prediction of pool surface depression and/or elevation. In principle, the specification of the contact angle completes the mathematical description of the problem. However, to by-pass the lack of contact angle data, many investigators have adopted the assumption that the mass of the molten material in the pool is equal to the mass of the original solid. While this assumption is clearly valid for a stationary workpiece in time-dependent melting/solidification in the absence of vaporization, it is not valid for a steady-state system, whose formulation is not history dependent.

This fact can be clearly understood by referring to Fig. 3 and assuming that it represents a two dimensional pool. Here an additional hypothetical free surface is indicated by the dot-dash curve. Clearly this pool, with a thin liquid layer, does not have the same mass as the original solid. However, as long as the height of the solid on the exit side (right hand side) is the same as the height of the solid on the inlet side, mass is conserved for a steady state process. Indeed, even if a pool surface is so severely depressed so that only an infinitesimally thin layer of liquid is coating the solid, it is still possible for the mass to be conserved in a steady state sense. Similarly, a bulging liquid pool with an instantaneous mass greatly in excess of the original solid mass would also satisfy the conservation of mass.

In conclusion, there does not appear an easy way to by-pass the difficulty associated with the need to specify the contact angle at the tri-state line. This conclusion applies to both two-dimensional and three dimensional problems.

4.2. HEIGHT PREDICTION FOR THE RE-SOLIDIFIED METAL

It is natural for the casual investigator to associate pool surface depression or elevation to features of the re-solidified surface, such as undercut and humping. The two-dimensional discussion of Section 4.1, however, shows that no such simple association exists. In that example, either a depressed melt pool or a bulging melt pool can result in a re-solidified surface identical in height to the original solid surface, since the height is governed by the conservation of mass. Therefore, to use a depressed pool surface to explain undercut, or a bulging surface to explain humping, as has been done by many modeling studies, appears to be of questionable reliability.

Careful examination of the governing equations as they are known now indicates that there is no physical condition that would permit one to predict the height of the re-solidified solid in three dimensional flow. In two dimensional flow the heights on the inlet and exit side must be equal by virtue of conservation of mass. In 3-D, however, conservation of mass can be satisfied as long as both undercut and humping, of comparable volumes, exist. This is also true of ripples or chevrons, as long as the average height obeys the conservation of mass.

Since humping, undercut, ripples and chevrons are known to exist in solid surfaces, clearly the governing equations as they are known now must be modified to account for surface height variations of the solid. How this should be accomplished is not known at present.

4.3. RIPPLES AND CHEVRONS

It is not difficult to see that ripples or chevrons in the re-solidified solid are consequences of pool oscillations. These oscillations may be due to mechanical vibrations of the workpiece support, or due to oscillations of the shield gas, amplified by resonance or other similar dynamic phenomena. Another important source for the oscillations is hydrodynamic instability arising out of the coupling between the flow and heat transfer. Precise prediction of these oscillations, and the relationship between the oscillations and the surface features of the re-solidified solid are not within current capabilities.

4.4. SURFACE HEIGHT PREDICTION IN LASER CLADDING AND ALLOYING

We shall end this section in a positive note. In laser cladding and alloying, significant amount of material is added to the surface. The added material may be of sufficient quantity as to overwhelm the theoretical uncertainties discussed above. However, care should be exercised in the choice of assumptions and in interpreting the results of modeling efforts.

5. Mixing in Melt Pools

For stationary materials, mixing is accomplished by diffusion. The time required for mixing is given by re-arranging Eq. (4):

$$t \sim D d^2 \tag{24}$$

Here d denotes the length scale representing the separation of the two species to be mixed. For materials with low coefficients of diffusion, mixing is usually accomplished as a combination of diffusion and stirring. The latter serves the function of breaking up each material into smaller packets, so that the effect distance to be bridged by diffusion processes is shortened.

It should be mentioned that many large-amplitude motions, such as waves or oscillations, may not contribute to mixing no matter how chaotic it may appear to the eye. On the other hand, stirring does not need to be chaotic. A very effective stirring for mixing is illustrated in Fig. 4. Here the two materials to be mixed are inside the annular region between two concentric cylinders. After the outer cylinder has been rotated by four revolutions, the distribution of the material, in the absence of diffusion, is shown in (b). It is seen that the two components have been folded and stretched into thin films adjacent to each other. The distance that must now be bridged by the diffusion process is only d/4, or d/n, where n is the number of folds. One can then estimate that the mixing time is now

$$t \sim D d^2 / n^2 \tag{25}$$

This mechanism is in fact the same mixing mechanism in dough kneading. The same mechanism is present in the laser-melted pool. Many pool modeling calculations [4] show that a material particle executes a toroidal spiral motion as shown in Fig. 5. The number of turns of such spiral is given roughly by the ratio of thermocapillary velocity and the scanning velocity

$$n \sim U_{TC} / U_{Sc} \tag{26}$$

This is at least one of the explanations for the efficient mixing process in the laser melt pool. It is seen that this mechanism does not depend on the presence of any visible chaotic condition or turbulence.

6. Concluding Remarks

It is hoped that the discussions in this paper can convince non-thermo-fluid specialists of the utility of semi-quantitative analyses in understanding the essential physics of laser-material interactions relevant to surface modification, and of the need for careful examination of the physical principles and mathematical requirements for a meaningful numerical analysis of the melt pool.

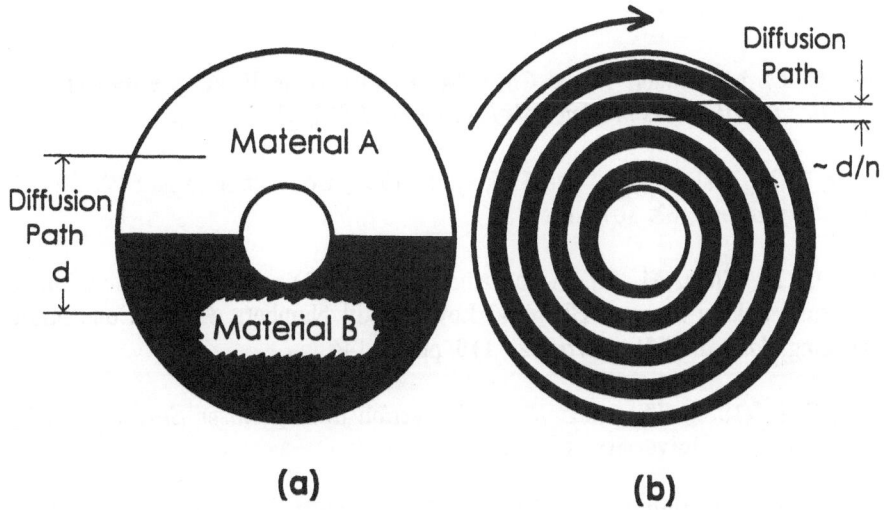

Fig. 4 Illustration of Enhanced Mixing Process

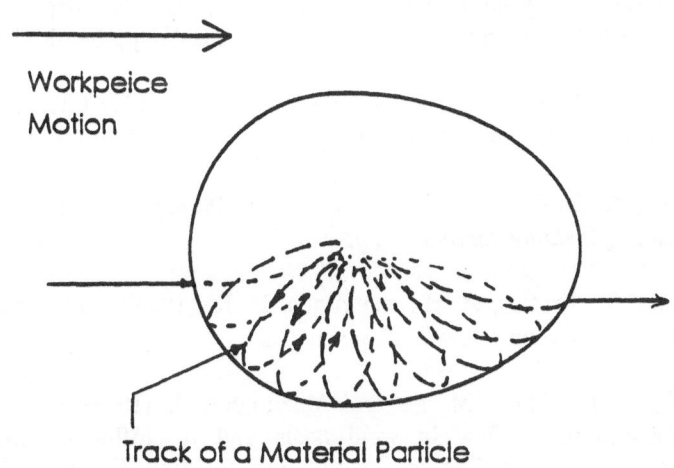

Track of a Material Particle

Fig. 5 Toroidal Spiral Motion in the Melt Pool as a Mechanism for
Enhanced Mixing

7. References

1. Chan, C. L., Mazumder, J. and Chen, M. M. (1983), in E. A. Metzbower (ed.), *Application of Lasers in Materials Processing*, ASM, p150.

2. Oreper, G. M., Eagar, T. W. and Szekely, J. (1983) Convection in Arc Weld Pools, *Welding Journal*, **63**, 307-312.

3. Chan, C. L., Chen, M. M. and Mazumder, J. (1983) Asymptotic Solution for Thermocapillary Flow at High and Low Prandtl Numbers due to Concentrated Heating, *Journal of Heat Transfer*, **110**, pp140-146

4. Chan, C. L. (1986) *Thermocapillary Convection during Laser Surface Heating*, Ph.D. Thesis, University of Illinois.

5. Chan, C. L., Mazumder, J. and Chen, M. M. (1988), Effect of surface tension driven convection in a laser melted pool: 3-dimensional perturbation model, *J. App. Phys.*, **64**, pp6166-6174.

6. Chen, M. M. (1987) Thermcapillary Convection in Materials Processin", in S. K. Samanta, R. Komanduri, M. M. Chen, (Eds), *Interdisciplinary Issues in Materials Processing and Manufacturing*, ASME, pp 541-558.

7. Paul, A. and DebRoy, T. (1988) Free Surface Flow and Heat Transfer in Conduction Mode Laser Welding, *Metallurgical Transactions*, **19B**, pp. 851-858.

8. Zacharia, T., Eraslan, A. H., and Aidun, D. K. (1988a) Modeling of Non-Autogenous Welding, *Welding Journal*, **67**, pp19s-27s.

9. Zacharia, T., Eraslan, A. H., and Aidun, D. K. (1988b) Modeling of Autogenous Welding, *Welding Journal*, **67**, pp53s-62s.

10. Zehr, R. L., Chen, M. M. and Mazumder, J. (1988) Generic structure of thermocapillary flow in weld pools and its influence on computational strategy, in *Modeling of Casting and Welding Processes IV*, Engineering Foundation, New York.

11. Cipriani, F. D., Morvan, P. B., Garino, A. (1990) Thermocapillary Effects during Laser Surface Melting, presented at Advance Computational Methods in Heat Transfer, pp 15-25

12. Zehr, R. L. (1991) *Thermocapillary Convection in Laser Melted Pools during Materials Processing*, Ph.D. Thesis, University of Illinois.

13. Zacharia, T., David, S. A., and Vitek, J. M. (1991) Effect of Evaporation and Temperature-dependent Material Properties on Weld Pool Development, *Metallurgical Transaction*, **22B**, pp. 233-241.

14. Chen, M. M., (1992) Generic Structure of Flow and Temperature Fields in Welding and High Energy Beam Processing, in I. Tanasawa and N. Lior, (Eds.), *Heat and Mass Transfer in Materials Processing*, Hemisphere Publishing Co., New York.

15. Mazumder, J. and Kar, A. (1992) Transport Phenomena in Laser Materials Processing, in I. Tanasawa and N. Lior, (Eds.), *Heat and Mass Transfer in Materials Processing*, Hemisphere Publishing Co., New York, pp. 81-106.

16. Pirch, N., Kreutz, E. W., Olier, B. and He, X. (1994) The Modelling of Heat, Mass and Solute Transport in Surface Processing with Laser Radiation, *This Volume*.

17. Canright, D., 1994, ``Thermocapillary flow near a cold wall", Phys. Fluids, Vol. 6, 4, pp. 1415-1424.

18. Kervorkian, J. and Cole, J. D. (1981) *Perturbation Methods in Applied Mathematics*, Springer-Verlag.

19. Chen, M. M. (1990) Scales, Similitude, and Asymptotic Considerations in Convective Heat Transfer, Annual Review of Heat Transfer, **3**, pp 233-91, Hemisphere Publishing Corp., New York.

20: Rosenthal, D. (1946) The theory of Moving Sources of Heat and Its Application to Metal Treatment, Trans. ASME, **68**, pp 849-866.

The modelling of heat, mass and solute transport in surface processing with laser radiation

N.Pirch, E.W. Kreutz, B. Ollier

Lehrstuhl für Lasertechnik

Rheinisch-Westfälische Technische Hochschule Aachen

Steinbachstr. 15, 52074 Aachen, Fed. Rep. Germany

X. He

Institut für Werkstoffkunde und -technik

TU Clausthal

Agricolastr. 6, 38678 Clausthal-Zellerfeld, Fed. Rep. Germany

Abstract-The basic physical processes of surface processing with laser radiation are shortly outlined. The state-of-the art concerning modelling of thermal material processing with laser radiation is reviewed. Some recent results due the Stefan-Problem in surface melting, mass transfer mechanisms in laser surface alloying and 3D free surface deformation calculations because of density variations due to phase transformation are presented.

1. Introduction

Thermal materials processing with laser radiation [1] involves the photon-matter interaction with subsequent absorption of radiation and rapid thermalization of the absorbed optical energy. Depending on the processing variables as power density distribution of the laser beam, interaction time and matter properties different physical processes [2] as optical absorption, heat conduction , phase transition, fluid dynamics, evaporation kinetics and plasma dynamics are involved determining the processing results. In surface melting the surface temperature distribution originates in surface tension gradients on the melt pool surface which initiates the so-called thermocapillary Marangoni flow [3]. At the solidification front the convection causes an compaction of the isotherms near the surface and increases thereby the quenching rate for the subsequent solidification and transformation processes. The geometry of the phase transformation interface especially the inclination angle of the interface against the surface which represents the local solidification rate is dominantly influenced by the resulting heat source along the interface due to the heat of transformation and the temperature gradient in the direction normal to the interface. The temperature gradient determines the modification of the geometry of the phase transformation interface for the additional heat source by the heat of transforma-

177

J. Mazumder et al. (eds.), Laser Processing: Surface Treatment and Film Deposition, 177–193.

tion. This effect is represented by the Stefan number [4]. The Stefan number results from a dimensional analysis as the ratio of the heat flows given by the product of heat of phase transformation with processing velocity and the temperature difference between the temperature of phase transformation and operating temperature on a typical length. The intermixing of solute in the melt pool and the resulting concentration distribution along the phase transition interface in laser surface alloying results essentially from the momentum boundary layer at the solidification front and the partition coefficient [5]. The local cooling and solidification rate and the concentration distribution along the solidification front govern the possible formation of metastable phases, their composition and the scale of metallurgical microstructure. During thermal surface processing with laser radiation several transport processes occur simultaneously. It is impossible to directly observe these complicated physical processes due to the small treated zone and the short laser-material interaction time. Therefore, for an optimum performance, mathematically modelling the processes for better understanding the processes and metallurgical mechanisms is necessary.

2. Influence of Heat of Transformation on the Interface of Phase Transition

The analysis of heat conduction processes involving phase change is essentially nonlinear due to the discontinuity of the enthalpy at the melting temperature for pure substances, which implicates a discontinuity for the heat flux across the moving interface of phase transition

$$-\lambda \, \partial_n T \,|_l = -\lambda \, \partial_n T \,|_s + \rho L v_n \qquad (1)$$

Due to the nonlinearity of the so called Stefan problem [6] very few analytical solutions may obtained [7,8]. Due to the importance of the problem for many practical applications of engineering interest a large number of numerical methods have been developed employing Finite Differences or Finite Element Methods whereat the different numerical approaches may be divided in front tracking and fixed domain methods [9]. In spite of the effort in numerical modelling of heat conduction processes the most commonly used solution for the heat conduction problem for a moving heat source is the so called Rosenthal solution [10] because of its mathematically and numerical simplicity. The Rosenthal

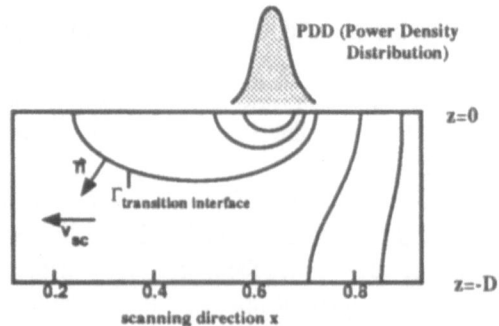

Fig.1 Schematic representation of laser surface melting

solution assumes temperature independent thermophysical material properties, neglects the heat of transformation but incorporates the additional adiabatic boundary conditons at the inferior side of a finite sheed by the method of fictitious sources. Recently the method of Rosenthal has been modified in order to solve the Stefan problem arising in surface melting with laser radiation [11].

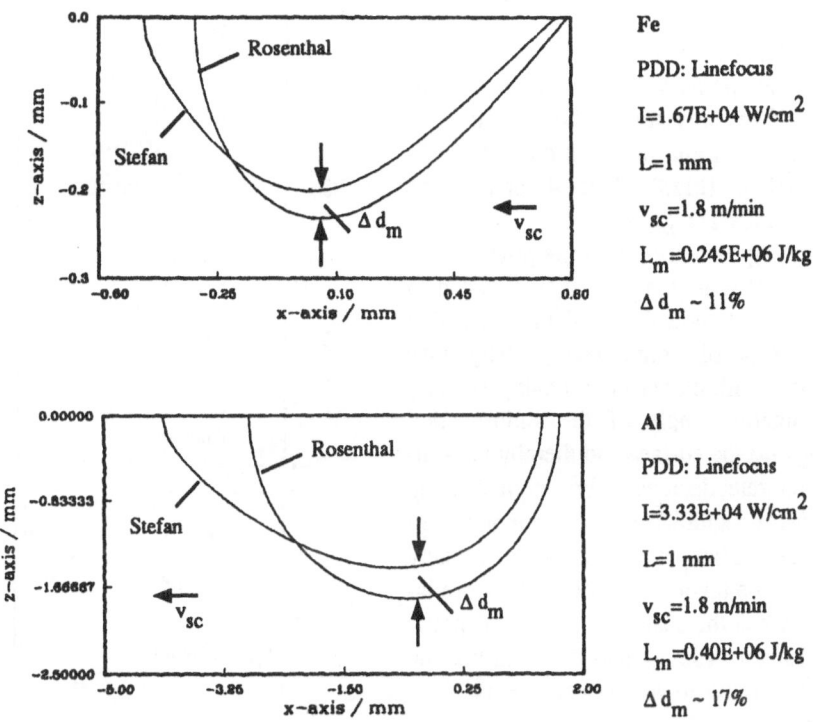

Fe

PDD: Linefocus

$I=1.67E+04 \ W/cm^2$

$L=1 \ mm$

$v_{sc}=1.8 \ m/min$

$L_m=0.245E+06 \ J/kg$

$\Delta d_m \sim 11\%$

Al

PDD: Linefocus

$I=3.33E+04 \ W/cm^2$

$L=1 \ mm$

$v_{sc}=1.8 \ m/min$

$L_m=0.40E+06 \ J/kg$

$\Delta d_m \sim 17\%$

Fig.2 Geometry of phase transition interface for pure heat conduction (Rosenthal solution) and taking heat of transformation (Stefan) into account.

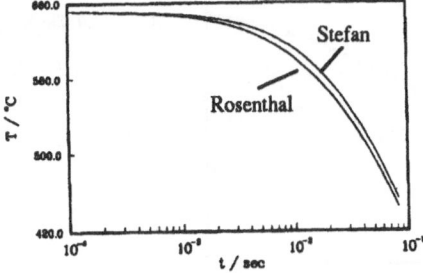

Fig.3 Time temperature variation z=0, (variables see Fig.2 Al)

The process of phase transition consumes at the melting front the latent heat and deliveres the stored energy at the solidification front. This means that the phase change takes effect at the melting front as a heat sink and at the solidification front as a heat source. For this reason the melting front, as figured out by pure heat conduction, moves in the

180

feeding direction in the high temperature region (Fig.2). The solidification front is likewise the melting front but here for the reason of delivered latent heat displaced in the feeding direction in the low temperature region . The inclination angle of solidification front against the surface and thereby the solidification rate ist lowered (Fig.2). The melting depth decreases for Fe by 11% and for Al by 17% for the listed process variables. The time-temperature variation after solidification differes only slightly between the Rosenthal solution and the solution of the Stefan Problem (Fig3). Fig. 4 shows the transition interface geometry as a function of the scanning velocity. The laser power is for the different scanning velocities adjusted to get a constant melt depth of the Rosenthal solution d_{mR} of 1mm. The investigations shows that with increasing scanning velocity the inclination angle of the solidification front against the surface and thereby the solidification rate decreases. For high sanning velocities the solidification rate amounts approximately 50% of the scanning velocity and has to be taken into account for correlations between the scale and phases of metallurgical microstructure and the solidification rate in rapid solidification experiments.

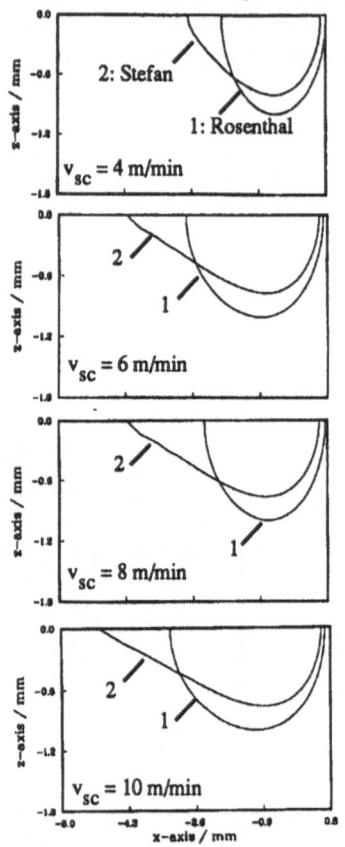

Fig.4 Transition interface geometry and scanning velocity, Al, Linefocus, L=400 μm, d_{mR}=1 mm.

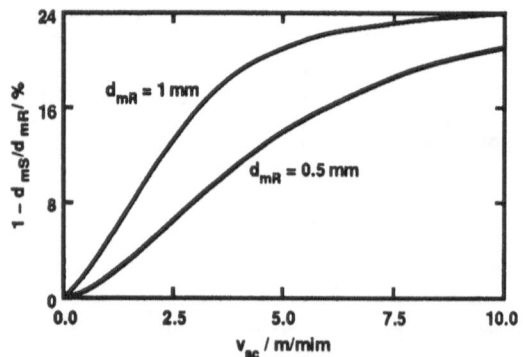

d_{mR} Melt Depth of Rosenthal Solution

d_{mS} Melt Depth of Stefan Problem

Fig.5 Difference in melt depth between Rosenthal solution and Stefan Problem, Al, Line Mode, L= 400 μm as a function of scanning velocity.

The difference concerning the melt depth between the Rosenthal solution and the Stefan Problem amounts up to 24% in dependence of sanning velocity and the melt depth of the Rosenthal solution (Fig.5).

3. Melt Dynamics and Surface Deformation

3.1 MATHEMATICAL MODELLING

Fig.6 shows schematically a cross section of the melt pool and the surface deformation during the laser melting process. After liquefaction the surface first bulges beneath the laser beam because of the volume expansion by the solid liquid phase transition. Assuming a negaive surface tension coefficient (as for metals) thermocapillary effects initiate

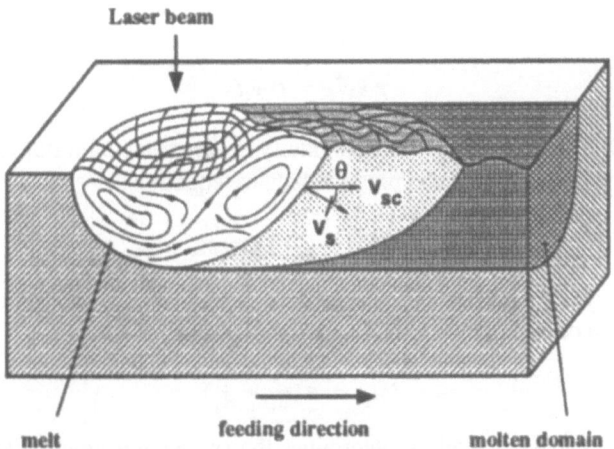

Fig.6 Schematic representation of the laser melting process, illustrating flow dynamics, surface

deformation and solidification.

a surface flow directed radially outward from the center of the laser beam to the outer boundary of the melt and, from there, it propagates because of the shear forces into deeper regions of the melt. At the edge the flow is directed down recirculates along the solid liquid interface and moves upwards to the melt pool surface. Previous theoretical and numerical investigations [12-17] have shown the importance of the fluid flow within the liquid region (due to buoancy, thermocapillary or electromagnetic forces) and its effect on the heat transfer mechanisms.

The velocity and the pressure field in the melt pool satisfies the incompressible Navier Stokes equations with the Boussinesq approximation

$$\partial_t(\rho\vec{v}) + div\ (\vec{v} \otimes \vec{v}) - \mu(\nabla \otimes \vec{v} + (\nabla \otimes \vec{v})^T) = -\rho\vec{g}b_l(T - T_{ph}) \qquad (2)$$

$$div\ \vec{v} = 0 \qquad (3)$$

with the gravitational accelaration g and the liquid thermal expansion coefficient b_l. The density discontinuity because of the phase transformation at the solid liquid interface gives rise to lokal (excess) negative pressure at the (melting) solidification front leading to micro convection and thereby to microsegregation as pointed out by Langbein [18]. The excess pressure gradient is modelled by a generalized δ-function localized on the solid liquid interface. A more suitable description is derived from the local mass conservation at the phase transition interface leading to a modified no-slip condition

$$v_n = \rho_s / \rho_l \cdot (\vec{v}_{sc}, \vec{t}) \tag{4}$$

$$v_t = (\vec{v}_{sc}, \vec{t}) \tag{5}$$

At the free selfadjusting surface the boundary conditions for the fluid dynamics are derived from the continuity of the momentum flux vector. Taking into account capillarity effects the projection onto the tangential vector at the surface leads to the balance of the shear forces

$$\mu \, \partial_n \vec{v}_t = \sigma' \cdot \partial_t T \tag{6}$$

inducing a flow at the surface. The continuity of the momentum flux in the direction normal to the surface results in the pressure balance equation

$$P_{gas} = (P_{fl} - C_o) - 2\mu \partial_n v_n + P_{kap} \tag{7}$$

where the capillary pressure p_{cap} is given at a point of the melt pool surface with the principal radii of curvature R_1 and R_2 by the equation

$$P_{kap} = \sigma' \cdot (\frac{1}{R_1} + \frac{1}{R_2}). \tag{8}$$

The pressure balance equation determines the surface deformation as a function of the pressure variation on the melt pool surface, molten volume, density discontinuity at phase transformation, surface temperature distribution, surface tension coefficient σ' and the outer boundary of the melt pool surface the so called triple point line. The equation is derived from the physical principle of minimal surface energy.

The parabolic heat diffusion equation

$$\partial_t(\rho cT) + div(\rho cT \, \vec{v} - \lambda \, \nabla \, T) = 0 \tag{9}$$

requires to have appropriate initial and boundary conditions. Because the absorption length of laser radiation is small compared to the thermal penetration depth the absorption of the laser radiation can be modelled as a von Neumann boundary condition on the curved top surface

$$- \lambda \, \partial_n T = Q_{Laser}. \tag{10}$$

The latent heat release which is associated with the phase transformation results in a discontinuity of the heat flux at the solid liquid interface (equation 1).

This hydrodynamic description of the laser melting process is valid for pure materials. The processsing of multiconstituent systems is characterized by the existence of a multiphase region the so called mushy region which seperates the complete resolidified region from the pure liquid region. Improved models are developed for the complex solid formation within the mushy region [19,20]. Because of the high solidification rates in laser surface remelting the mushy region is reduced on a boundary layer of a few microns along the phase transformation interface.

A dimensional analysis show that the bulk equations are governed by the Peclet number Pe and the Reynols number Re. The Peclet number controlls the influence of the thermocapillary flow on the heat transfer and thereby the temperature field and the melt pool geometry. The influence of the Peclet number is well documented by computation results for aluminium and iron [21]. For the comparison the laser intensity is adjusted to get identical melt depths keeping the remaining processing parameters constant. Despite of the comparable flow velocities the computation exhibit for iron a strong influence of the thermocapillary flow on temperature field and melt pool geometry which is not observed for aluminium as expected because of the high thermal conductivity. The Reynols number can be physically interpreted as ratio of the diffusive to the convective momentum transfer. This characteristic number marks the extension of the momentum boundary layers at the solid liquid interface due to the no-slip condition and at the free surface the so called Marangoni boundary layer. The surface tension number S enters the governing equations by the balance equation of the shear forces controlling the coupling between the surface tension gradients and the gradient of the tangential velocity in the direction normal to the melt pool surface (equation 6).

Fig. 7 shows the isotherms and the velocity field for pure heat conduction ($\sigma'=0$) and a negative and positive surface tension coefficient σ' which arise by the addition of surface active elements. These elements can significantly influence the temperature dependence of the surface tension up to the case that the surface tension coefficient will be completely reversed as experimentally demonstrated by Heiple et al. [22]. For the negative surface tension coefficient the flow field in the longitudinal cross section near the surface is directed outwards from the center of the melt pool to the low temperature region at the edge of melt pool. The flow field in the surface layer enhances significantly the heat flux in flow direction causing thereby a redistribution of the absorbed optical power. This leads to a depression of the maximum surface temperature given by the Rosenthal solution for pure heat conduction in the range of 30%. The velocity field in the surface layer reduces by one or two orders of magnitude to the scanning velocity at the solid liquid interface changing rapidly the heat transfer from a convective dominated to a diffusively controlled one. The temperature gradients at the solidification front are enhanced as demonstrated by the compaction of the isotherms near the edge of the melt pool. For the case of the positive surface tension coefficient the fluid is driven in a surface layer from the edge of the melt pool to the high temperature region beneath the center of the laser beam, is directed down and moves along the solid liquid interface to the edge of the melt pool. Two counter-rotating vortices develope. The asymmetry in their shape is a consequence of the scanning velocity. The enhanced heat flux in the beam direction shortens the extension of the melt in scanning direction and increases the melt depth. Both flow pattern result in a temperature homogenization in the melt but with different consequences for the solidification process. For the case of a negative surface tension coefficient the maximum surface temperature is lowered and the quenching rate increased. In contrast for the reversed surface tension coefficient the maximum surface tempe-

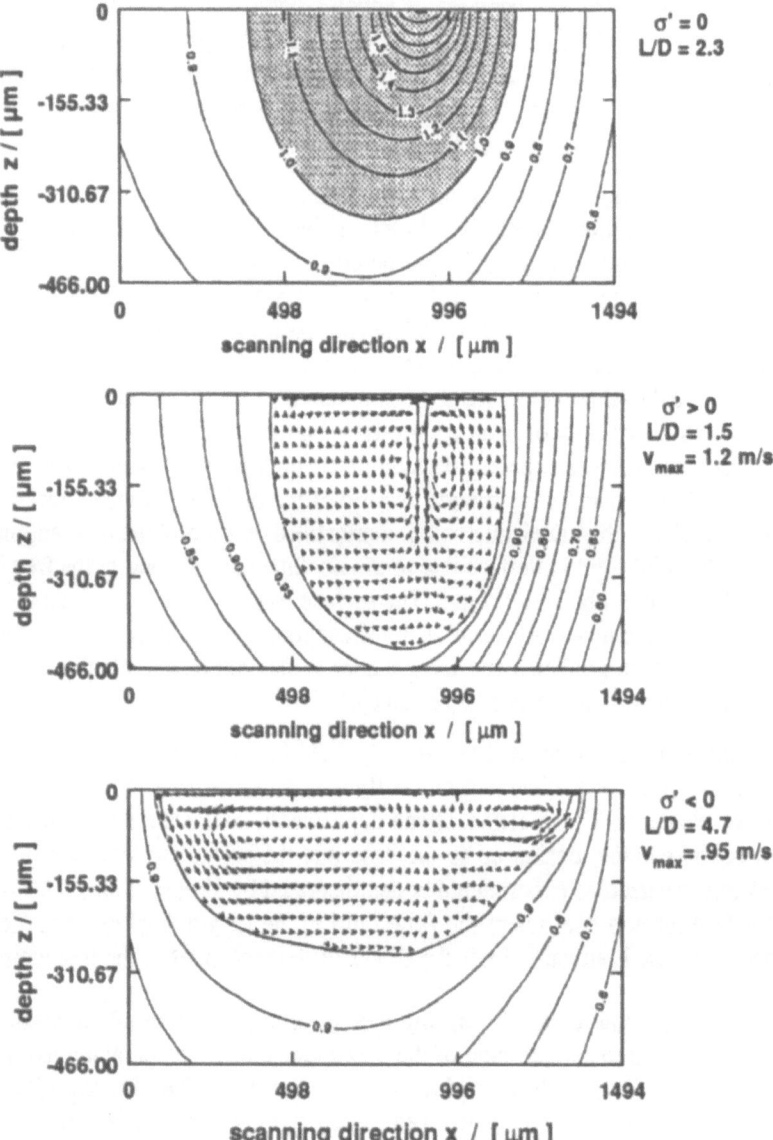

Fig.7 Isotherms (T/T_m), melt pool geometry and thermocapillary flow field for S=0 (Rosenthal solution, S=396 ($\sigma'>0$) and S=-264 ($\sigma'<0$), in the plane y=0 (GGG50: Linefocus, a=200 μm b=800 μm, Pe=3.1, Re=25.5.

rature is lowered, approximately equal the temperature depression for the negative surface tension number, but the quenching rate at the solidification front is roughly not affected by the convection. Due to the enhanced heat flux in the direction normal to the surface the temperature average in deeper regions increases.

3.2 FREE SURFACE CALCULATIONS

The assumption of a flat surface in modelling of fluid dynamics in laser surface melting is in general done for reasons of numerical expense but implies that the maximum solidification rate near the surface has to be equal to the scanning velocity of the laser beam. Experiments performed for eutectic alloys [23] however show that the maximum solidification rate becomes significantly reduced depending on the scanning velocity. Considering the melt pool surface as deformable the theoretical investigations show that the influence of surface deformation on the inclination angle of the solidification front against the workpiece surface may explain this disagreement. Because the local quenching rate as well as the local solidification rate governs the development of metallurgicall microstructures the melt flow, heat of transformation and the free-selfadjusting melt pool surface have to be taken into account in a micro-macroscopic simulation of laser surface melting [24].

First investigations on surface deformation due to fluid dynamics was done by Anthony and Cline [25] within the frame of a one dimensional model, predicting the spacing and the hights of the surface ripples for the surface tension driven melt dynamics. The model assumes steady state conditions and is for this reason not to be able to explain the kinetics of ripple formation. The pressure differences between two adjoining points at the melt pool surface equals the potential difference in the gravity field. In addition, the non-linear convective momentum flux is assumed to be small compared to the pressure gradient in the Navier Stokes equations. Numerical results for the pressure profile at the melt pool surface and the velocity field of the thermocapillary flow field show that these assumptions don't hold [26]. Mazumder [27] showed the surface geometry of a deformed laser weld pool. However, nothing was mentioned about how the result was obtained. Indeed in the same investigation a flat non-deformable melt pool surface is assumed using rectangular co-ordinates for integration of the Navier Stokes equations. As pointed out by Tsai and Kou [28] the boundary conditions at the melt pool surface and hence its deformation cannot be treated properly in the numerical frame of rectangular and cylindrical co-ordinates. Picasso [29] used the Finite Element Method for free surface modeling in a two-dimensional model for the thermocapillary flow in surface melting with laser radiation including free surface deformation but neglects volume expansion due to the density discontinuity in phase transition and further thermal expansion in the heated melt. Pirch and Kreutz [15] showed that the surface deformations due to pressure variations from thermocapillary flow amounts approximately a few microns and can be neglected in process simulation. The volume expansion of the molten volume results in a surface deformation of about 10% of the melt depth [15]. The melt depth is lowered due to the surface deformation and equals approximately the surface deformation [15,30]. The free surface calculations [15,29,30] are based on two-dimensional models or three-dimensional axissymmetric models. Zacharia [31,32] used in their investigations the linear approximation of the pressure balance equation for free surface calculations. Fig. 8 shows the surface deformation in laser surface melting in a three-dimensional model [33]. The surface deformation calculation bases on integration of the pressure balance equation (equation 7) using Finite Element Method. The resulting non-linear algebraic equations are solved by Picard Iteration. As in the case for a two-dimensional model the surface deformation amounts approximately 10% of melt depth. The surface geometry exhibits in the low temperature region due to the temperature dependence of surface ten-

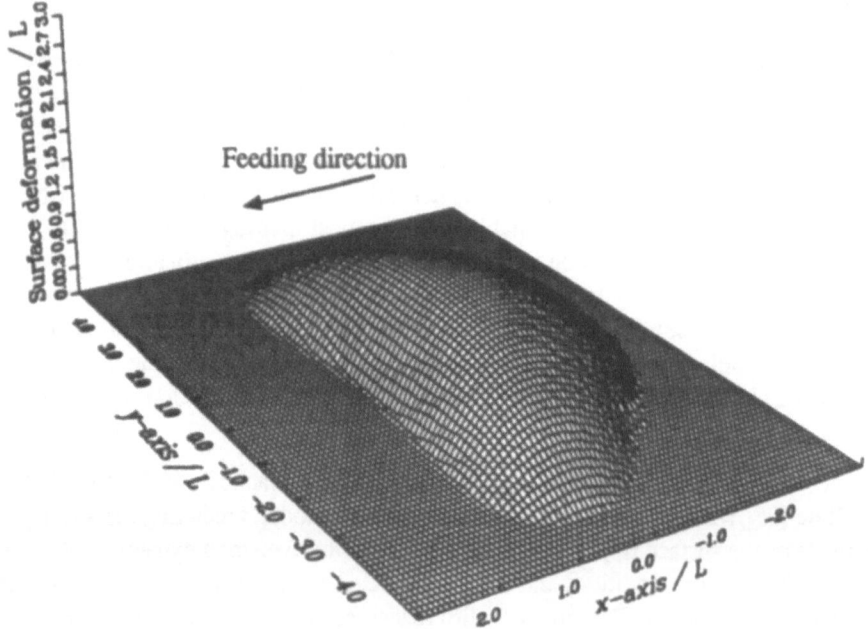

Fig.8 Three-dimensional surface plot of surface deformation in laser surface melting, Linefocus,

a=6mm, b=2.2mm, P=5kW, linear polarized, $\alpha = 70°$.

sion the least curvature. The local inclination angle of the laser beam exhibits on the deformed melt pool surface a variation up to 10°. In case linear polarized laser radiation strikes the surface at approximately the Brewster angle the absorbed heat flux is significantly influenced by the surface deformation due the marked absorption maximum of linear polarized laser radiation as a function of the inclination angle according to the Fresnel formulas.

4. Mass transfer mechanisms in laser surface alloying (LSA)

In the one-step process of LSA the alloying element is supplied to the melt in powder form, using a carrier gas (Fig.9). Absorption and scattering of the laser radiation on the powder particles may lead to a substantial variation in the transmitted power density distribution, q_{trans}, on the surface in comparison to the original power density distribution of the laser radiation [34]. Parameters influencing this interaction are the grain size and grain geometry, the refractive and absorption index of the alloying element, the density of the powder particles and the interactive volume of the powder particles in the beam path. The absorption of radiation by the powder particles results in an increase of their temperature prior to contact with melt surface. Depending on the wetting properties which apply between the melt and the alloying element, the thermal contact results in fusion of the powder particles in case the surface temperature of the melt pool is sufficiently high. The typical times required for this fusion process can be calculated by means of an analytical solution for the heat conduction problem of a spherical particle in

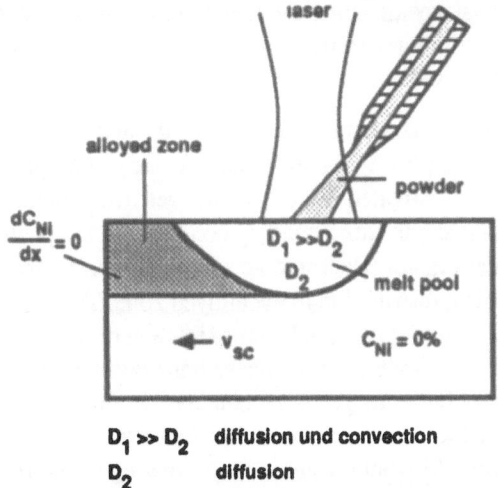

D₁ >> D₂   diffusion und convection
D₂   diffusion

Fig.9 Schematic representation of one-step laser surface alloying with powder spraying.

a heating bath at a specified bath temperature [6]. These times are in the milliseconds range and are dependent on the bath temperature, the grain size and the thermophysical properties of the alloying element. A typical time scale for LSA is provided by the quotient of the melt pool extension in feeding direction and the scanning velocity. This is a measure of the duration of existence of a volume element in the liquid phase. In the context of this scale, the sub-process of adsorption, fusion anf admixture of the powder particles takes place almost instantaneously. The admixture of the allying element can thus be represented in model form via a Neuman boundary condition on the melt pool surface $\Gamma_{l/g}$

$$- D \frac{\partial C}{\partial z} = q(x,t),$$ (11)

whereby D is the diffusion coefficient of the alloying element in the base material. The particle flow into the melt, q(x,t), is the result of the powder particle density on the melt pool surface and the particle speed. Admixture of the alloying element in the melt is effected via diffusive and convective transport mechanisms, whereby convective transport is the dominant transport mechanism when a thermosolutal or Marangoni flow applies, on account of the low diffusion coefficient constant ($D \cong 10^{-9} m^2/s$). On solidification, the concentration of the solute in the melt adjacent to phase transition interface, C_{int}, reduces to the value kC_{int}, which is determined by the distribution coefficient k, due to the lower level of solubility in the solid matter. Neglecting the diffusion in the solid phase, a balance of the solute flux at the solidification front results in the boundary condition,

$$(1- k) v_n C = - D \frac{\partial C}{\partial z}$$ (12)

or on account of the condition that the solid does not contain the alloying element along the fusion front

$$v_n C = - D \frac{\partial C}{\partial z}$$ (13)

188

for the diffusion equation which provides the basis for the particle transport. The local solidification and fusion rate, v_n, is obtained by projecting the scanning velocity onto the outward-oriented normal vector along the phase transition interface. The extent of the mushy region adjacent to the phase transition interface amounts a few microns and may be neglected for the purpose of macroscopic calculation of the concentraion distribution.

This investigation concerning solute transport in the melt via convection differs from that performed by Chande [34] by the different boundary conditions at the phase transition interface. Chande investigated the temporally resolved solute transport in the melt but due to the vanishing solute flux boundary condition at the phase transition interface the solidification process is not taken into account and thereby the investigation admits no prediction of the solute distribution in the alloyed zone. The simulation of LSA in this investigation [35] suffers under the rough approximation of the phase transition interface by the so called step approximation of the phase transition interface as the closest approximation of the interface in the frame of rectangular co-ordinates. The numerical approximation of the solute flux boundary condition (equation 12,13) in the numerical approach used by Chande [35] would result in an oscillating solute flux along the step like interface. For this reason the Finite Element Method was used in this investigation to treat the boundary conditions at the phase transition interface (equation 12,13) properly by an interface adapted meshing.

The simulation is performed for an negative surface tension coefficient with Al as substrate and Cu as the alloying addition using the thermophysical material properties from [36]. The resulting flow pattern exhibits two counter rotating vortices. The maximum ve-

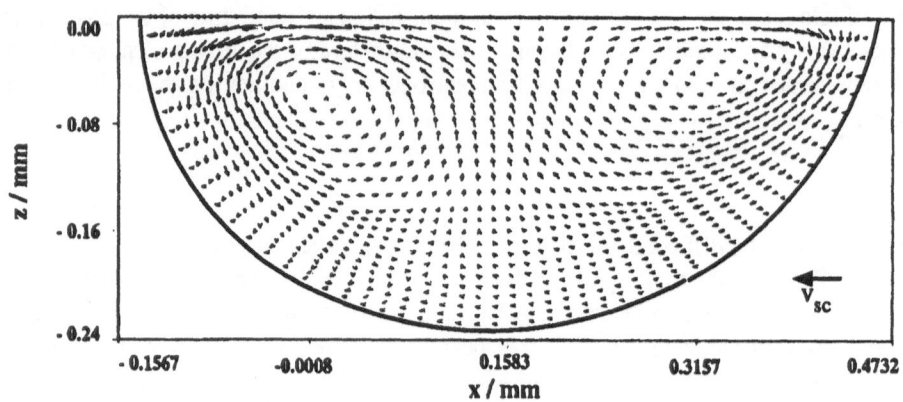

Fig.10 Melt pool geometry and thermocapillary flow field for Al: $\sigma'=-0.35E-3$ N/m/K, $P_L=33E+4W/cm^2$,

Linefocus, $a=400 \mu m$, $v_{sc}=1.8m/min$.

locity amounts .65 m/s in the Marangoni boundary layer. The streamline field (Fig.11) shows two areas of closed streamlines, the extension of which is asymmetric, due to the feed movement. The trajectories show that, in spite of the complex vortex structure of the Marangoni convection a molten volume element crystallizes at the end of its path at a point of the solidification front, the depth of which is virtually identical to the starting point of its path on the fusion front. If the volume elements were to be numbered in sequence along the fusion front, this sequence would remain unchanged after crystallisation at the solidification front, despite the convection. This result also remains valid for

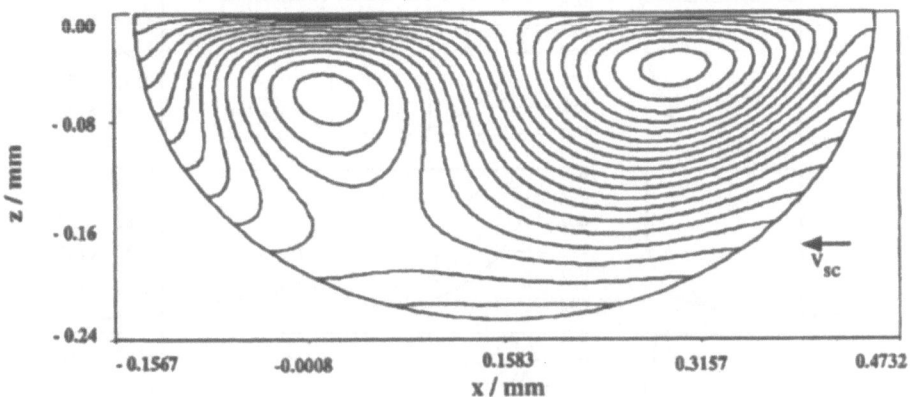

Fig.11 Melt pool geometry and streamlines, variables see Fig. 10.

cases in which the surface flow is oriented from the edge of the molten bath towards the centre, as applies in the case of a positive surface tension coefficient σ' (Figure 13). No real intermixing of the molten material takes place. The resultant temperature-time cycle for the various strata of the material in the melt corresponds to the form of the trajectory. A volume element close to the surface, for example, is drawn under the vortex located in front of the fusion front after fusion (Figure 2), and is thus initially kept away from the area of maximum temperatures. Only when the volume element is caught by the upward flow of the following vortex does it pass through the high-temperature area on the surface. The diffusion constant, D, is in the range of approximately 10^{-9} m^2/s [2]. The resolution of the FEM network which would thus be required for integration of the mass transport equation would exceed the available computer capacity. For this reason, the model calculations were carried out for maximum diffusion coefficients of $5*10^{-7}$ m^2/s. As a supplementary measure, the model calculations were carried out for greater diffusion coefficients, in order to determine a trend from the results and to carry out extrapolations on the basis of the results for smaller diffusion coefficients.

The results (Fig.12) reveal the concentration distribution for smaller diffusion coefficients to possess an increasing level of solute distribution at the phase transition interface and the surface of the melt pool, and a constant level of concentration in the melt. The concentration profile reveals a maximum over the depth of the alloyed-on case in the area of the fusion depth, the characteristic of which becomes more pronounced as the diffusion coefficient diminishes (Fig.12). This phenomenon of a pronounced maximum level of concentration distribution above the fusion boundary was observed in experiments on aluminium, onto the surface layer of which nickel was alloyed in a one-stage process [37]. The results show that the convection in the melt pool has a dominant influence on the form of the concentration profile over the depth of the alloyed region, and that it may be possible to utilize this phenomenon in industrial applications, in order to produce required gradient layers or double layers.

190

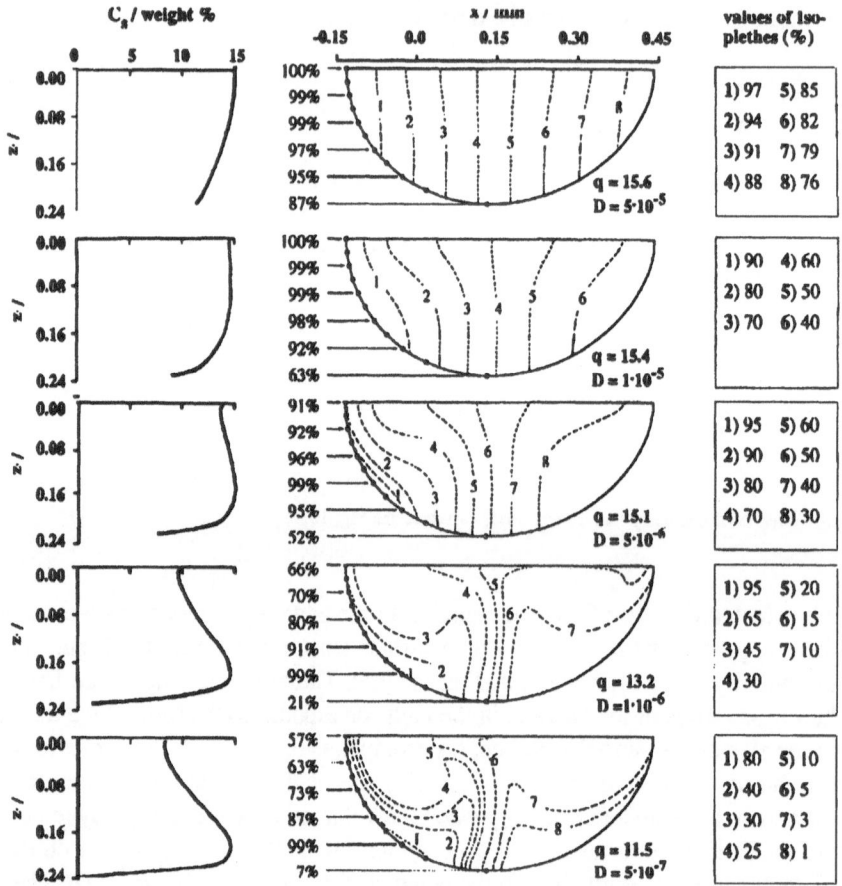

Fig.12 Isoplethes of Cu Concentration distribution in the melt pool and resulting Cu distribution over the surface layer depth, variables see Fig.10.

5. Discussion & Summary

The investigation concerning the influence of heat of transformation on the geometry of the solidification front shows that in dependence of the process parameters and thermophysical material properties the solidification rate is decreased up to 50% of the scanning velocity. The quenching rate calculated by pure heat conduction and taking heat of transformation into account coincide approximately.

The surface deformation is dominated by the volume shrinkage. The ratio of the surface deformation to the melt depth in laser surface melting equals roughly the ratio of the density discontinuity at phase transition to the density in the solid phase. The local inclination angle of the laser beam exhibits on the deformed melt pool surface a variation up to $10°$. In case linear polarized laser radiation strikes the surface at approximately the Brewster angle the absorbed heat flux is significantly influenced by the surface deformation due the marked absorption maximum of linear polarized laser radiation as a function of the inclination angle according to the Fresnel formulas.

The solute transport in the melt and thereby the resulting concentration profile in laser surface alloying is because of the small diffusion cefficients dominantly influenced by melt pool convection. It may be possible to utilize this effect in industrial applications, in order to produce required gradient layers or double layers.

6. References

1. M. Bass, Ed., "Laser Materials Processing", North Holland Publ. Co. Amsterdam, New York, Oxford (1983).

2. A. Gasser, E.W. Kreutz, K. Wissenbach, "Physical Aspects of Surface Processing with Laser Radiation", SPIE Proc. 1020 (1989), 70-83 and references herein .

3. C.G.M. Marangoni, "Über die Ausbreitung der Tropfen einer Flüssigkeit auf der Oberfläche einer anderen", Ann. Phys. 143, (1871) 337-354.

4. N.Pirch, E.W. Kreutz, "Analytical Solution of Stefan Problem for Surface Remelting with CO_2-Laser Radiation", DPG Frühjahrstagung, Greifswald (1993), Kuzzeitphysik: Fachvortrag K10.6.

5. Xuhui He, "Simulation und experimenteller Nachweis von Legierungsprozessen beim Laseroberflächenlegieren", Thesis, Tech. Univ. Clausthal, (1994).

6. J. Stefan, "Über einige Probleme der Theorie der Wärmeleitung", Sitzumgsber. Wien, Akad. Mat. Natur. 98 (1989), 473-484, 614-634, 965-983, 1418-1442.

7. H.S. Carslaw, J.C. Jaeger, "Conduction of Heat in Solids", Oxford, Oxford University Press, (1959).

8. H. Budhia, F. Krieth, "Heat Transfer with Melting or Freezing in a Wedge", Int. J. Heat Mass Trans. 16, (1973), 195-221.

9. M. Rappaz, "Modelling of Microstructure Formation in Solidification Processes", Int. Mat. Rev. 34, (1989) 93-123.

10. D. Rosenthal, "The Theory of Moving Sources of Heat and its Application to Metal Treatments", Transactions of the ASME (1949), 849-866.

11. N. Pirch, E.W. Kreutz, "Integration of the Stefan Problem arising in Surface Remelting with Laser Radiation, Rapport, Lehrstuhl für Lasertechnik, RWTH Aachen, Fed. Rep. Germany (1994).

12. W.D. Bennon, F.P. Incropera, "The evolution of Macrosegregation in Statically Cast Binary Ingots", Met. Trans. 18B, (1991) 482-507.

13. O. Besson et al., "Numerical Modelling of Electomagnetic Casting Processes", J. Comp. Phys. 92, (1991) 482-507.

14. C. Chan et al., "Threedimensional axissymmetric model for convection in laser melted pools", Mat. Science and Techn. 3, (1987) 306-311.

15. E.W. Kreutz, N. Pirch, "Melt Dynamics and Surface Deformation in Processing with Laser Radiation", SPIE Proc. 1502, (1991) 160-176.

16. S. Kou, Y.H. Wang, 3D Convection in Laser Melted Pools, Met. Trans. 17A, (1986) 2265-2277.

17. N. Ramanan, S.A. Korpella, "Fluid Dynamics of Stationary Weld Pools, Met. Trans. 21A (1990) 45-57.

18. D. Langbein, "Mikrokonvektion an Erstarrungsfronten", Proc. Workshop zur Erstarrungsdynamik", Aachen, Fed. Rep. Germany, (1981) 6-23, P.R. Sahm (ed.).

19. J. Ni, C. Beckermann, "A Volume-Averaged Two-Phase Model for Transport Phenomena during Solidification", Met. Trans. 22B, (1991) 349-360.

20. W.D. Bennon, F.D. Incropera, "Numerical Analysis of Binary Solid-Liquid Phase Change using a Continuum Model", Num. Heat Trans. 13, (1988) 277-296.

21. E.W. Kreutz, N. Pirch, "Melt Dynamics in Surface Processing with Laser Radiation: Calculations and Applications", SPIE Proc. 1276, (1990) 343-360.

22. C.R. Heiple et al., "Surface Active Element Effects on the Shape of GTA, Laser and Electron Beam Welds", Welding Journal, 62, (1983) 72-77.

23. M. Rappaz et al., "Solidification Front and Growth Rate during Laser Remelting", ECLAT Proc. (1986) 43-54, (publ. by DGM, "Laser Treatment of Materials", B.L. Mordike, ed.).

24. M. Rappaz, "Modelling of Microstructure Formation in Solidification Processes", Int. Mat. Rev. 34, (1989) 93-123.

25. T.R. Anthony, H.E. Cline, "Surface Rippling Induced by Surface Tension Gradients during Surface Melting and Alloying", J. Appl. Phys. 48, (1977) 3888-3894.

26. D. Morvan, F.D. Cipriani, A. Roux, "Thermocapillary Effects in a Melted Pool", Workshop on Mathematical Modelling" Aachen, Fed. Rep. Germany (1990).

27. J. Mazumder, "Overview of Melt Dynamics in Laser Processing", Optical Engineering, 30, (1991) 1208-1219.

28. M.C. Tsai, S. Kou, "The Advantages of orthogonal curvilinear coordinates in simulating Marangoni Convection in Defromed Weld Pools", (publ. "Modeling of Casting and Welding Processes IV, The Metallurgical Society, Warrendale, PA, (1988) 409-420.

29. M. Picasso, "Numerical Simulation of Laser Remelting Including Convection Effects and Free Surface Calculation", Proc. Engineering Foundation Conference on Modelling of Casting, Welding and Advanced Solidification Processes, Davos, Switzerland, M. Rappaz, M.R. Özgü, K.W. Mahin, TMS-AIME, Warrendale (1990) 153-158.

30. M.C. Tsai, S. Kou, "Weld Pool Convection and Expansion due to Density Variations", Num. Heat Trans. 17A, (1990) 73-89.

31. T. Zacharia, H. Eraslan, D.K. Aidun, " Modeling of Non-Autogenous Welding", Weld. J. (1988) 70s-75s.

32. T. Zacharia, H. Eraslan, D.K. Aidun, " Modeling of Autogenous Welding", Weld. J. (1988) 53s-62s.

33. N. Pirch, E.W. Kreutz, "Free Surface Calculations", Rapport, Lehrstuhl für Lasertechnik, RWTH Aachen, Fed. Rep. Germany (1994).

34. H. Frerichs, "Untersuchungen zur Energieeinkopplung beim einstufigen Beschichten von Metallen mit CO2 -Laserstrahlung, Diploma Thesis, Lehrstuhl für Lasertechnik, RWTH Aachen, Fed. Rep. Germany (1991).

35. T.S. Chande, "Laser Surface Alloying Low Carbon Steel Using Chromium and Nickel Powder Feed: Mechanisms, Microstructures, Properties and Models", Thesis, University of Illinois at Urbana-Champaign, USA, (1984) 41-43.

36. Smithells Metals Reference Book, E.A. Brandes, (Ed), ASM, Ohio (1958).

37. E.W. Kreutz, N. Pirch, M. Rozsnocki, "Solidification in Laser Surface Alloying of Al and AlSi10Mg with Ni and Cr", in Laser Treatment of Materials, B.L. Mordike (ed.), ECLAT Proc., DGM (1992) 269-279.

ONE-DIMENSIONAL THERMAL MODEL INCLUDING THE DEPENDENCE OF ABSORPTIVITY ON TEMPERATURE USING HAGEN-RUBENS EQUATION

A. M. DEUS, R. VILAR

Materials Engineering Department
Instituto Superior Técnico, Technical University of Lisbon
Av. Rovisco Pais 1, 1096 Lisboa, Portugal

ABSTRACT. In laser surface treatment of materials it is a matter of practical importance to determine relations combining the various processing parameters. For a particular laser setup and for a certain practical requirement, such a relation is quite useful for the purpose of process optimisation. For instance, in pulsed laser irradiation, when the requirement is that a definite value of temperature is attained at the surface (e.g. the melting temperature), a relation between power density and pulse duration should be looked for. If heat conduction is one-dimensional and the material properties are considered as temperature independent, a first estimate can be done by using a simple analytical model, which gives a temperature rise at the surface proportional to $t^{1/2}$. Considering the effect of the various properties on the thermal field, the absorptivity plays a special role, because it controls directly the heat input to the sample, while being itself a function of surface temperature. In order to take this dependence into account, the one-dimensional heat conduction equation was solved for a semi-infinite geometry using variable absorptivity given by Hagen-Rubens equation and assuming constant thermal conductivity and thermal diffusivity. It is shown that the surface temperature satisfies an Abel integral equation with an approximate analytical solution that gives a temperature rise at the surface linear with t. Comparison is made between analytical approximate expressions, numerical solutions, previous models and experimental results.

1.   Introduction

During the last three decades, lasers have been widely used for materials processing. Since many applications rely on the thermal effects of laser-material interaction, it becomes very important to obtain information about the temperature field as a function of processing parameters and materials properties. For that purpose, several models have been proposed, both analytical and numerical. Despite the fact that extremely sophisticated numerical models have become available, analytical expressions continue to be used. This is often due to simplicity and also because they allow to obtain simple relations between the process parameters that give a better insight on the phenomena involved.

In some cases, analytical models can take into account the fact that parameters affecting the thermal field depend themselves on temperature. For instance, the Kirchoff Transform [1] is performed to account for the dependence of thermal conductivity on temperature. Nevertheless, for many materials, the variable which more strongly affects the final result is the absorptivity, because it controls directly the amount of energy that is transferred into the material. Considering that the absorptivity

J. Mazumder et al. (eds.), Laser Processing: Surface Treatment and Film Deposition, 195–201.
© 1996 *Kluwer Academic Publishers.*

varies linearly with temperature, a one-dimensional model was proposed by Sparks and Loh [2].

However, the dependence of absorptivity on temperature can be expressed in a different way by means of Hagen-Rubens equation [3, 4], in which A is expressed as a function of the wavelength of radiation and the electrical resistivity of the material. The main goal of the present paper is to present an analytical expression for the surface temperature of a laser irradiated material in which the variation of absorptivity with temperature follows Hagen-Rubens equation. The heat conduction regime will be considered as one-dimensional, an approximation that is valid when the characteristic heat conduction distance is negligible as compared to the diameter of the laser beam. Moreover, we will apply the present analysis to the estimation of pulse duration leading to melting on the surface of a laser irradiated material, for constant power density.

2. Assumptions

We will develop our model based on the following assumptions:

a) $r >> 2\sqrt{\alpha t} \Rightarrow$ one-dimensional heat conduction.

b) $L >> 2\sqrt{\alpha t} \Rightarrow$ semi-infinite solid.

c) $\delta << \sqrt{\alpha t_p}, \Rightarrow$ surface heating.

d) The surface is polished and clean.

e) The thermal conductivity and thermal diffusivity are taken as constants.

f) The electrical resistivity varies linearly with temperature.

g) The absorptivity varies with temperature according to Hagen-Rubens equation.

h) The laser pulse shape is square.

3. Starting equations

For one-dimensional heat conduction with constant thermal properties and semi-infinite geometry we write:

$$\frac{\partial T}{\partial t} = \alpha \frac{\partial^2 T}{\partial x^2}, \quad 0 \le x < \infty \tag{1}$$

The initial and boundary conditions are:

$$T(x, t = 0) = 0 = T(x = \infty, t) \tag{2}$$

$$-K \frac{\partial T}{\partial x}\bigg|_{x=0} = A\left(T(x = 0, t)\right) I \tag{3}$$

We consider the Hagen-Rubens law [3] (which gives good agreement with experimental values for the absorptivity of various metals when $\lambda > 5$ μm [4]):

$$A = 2 \sqrt{\frac{4 \pi \varepsilon_o c \rho}{\lambda}} = 0.365 \sqrt{\frac{\rho}{\lambda}} \tag{4}$$

which is the first term of Bramson's equation [4]. We assume that the electrical resistivity varies linearly with temperature:

$$\rho(T) = \rho_o \left(1 + \frac{T}{\chi} \right) \tag{5}$$

4.   Solutions

The solution for the surface temperature can be obtained by applying the Duhamel theorem [1]. The result is:

$$T_o(t) = \frac{\eta E}{2} \int_0^t \sqrt{\frac{T_o(s) + \chi}{t - s}} \, ds \tag{6}$$

with:

$$E \equiv \frac{2I}{K} \sqrt{\frac{\alpha}{\pi}} \tag{7}$$

and:

$$\eta \equiv 2 \sqrt{\frac{4 \pi \varepsilon_o c \rho_o}{\chi \lambda}} \tag{8}$$

which is an integral Abel equation. An exact solution for the case $\chi = 0$ and finite η can be found:

$$T_o(t) = \mu t \tag{9}$$

with:

$$\mu \equiv \left(\frac{\pi \eta E}{4} \right)^2 \tag{10}$$

which we will refer to as the linear solution.

4.1   ADIMENSIONAL VARIABLES

We introduce now a characteristic time:

$$t^* = \frac{\chi}{\mu} \tag{11}$$

Adimensional temperature and time can be defined, respectively, as:

$$\varphi_o \equiv \frac{T_o}{\chi} \tag{12}$$

$$\tau \equiv \frac{t}{t^*} \tag{13}$$

By doing so, we obtain an Abel equation independent on materials properties:

$$\varphi_o(t) = \frac{2}{\pi} \int_0^\tau \sqrt{\frac{\varphi_o(s)+1}{\tau-s}} \; ds \tag{14}$$

The linear approximate solution can now be written as:

$$\varphi_o(\tau) = \tau \tag{15}$$

By introducing this expression into the RHS of Abel equation, a more accurate analytical solution is obtained, still in a closed form:

$$\varphi_o(\tau) = \frac{2}{\pi} \left[\sqrt{\tau} + (1+\tau) \; arctg\left(\sqrt{\tau}\right) \right] \tag{16}$$

which will be referred to as the approximate analytical solution. Note that this expression tends to the linear solution (15) when $\tau \gg 1$. Also, for $\tau \ll 1$, equation 16 can be written as:

$$\varphi_o(\tau) = \frac{4}{\pi} \sqrt{\tau} \tag{17}$$

which is the well known Carslaw and Jaeger solution [1] for constant materials properties, with absorptivity taken at room temperature.

5.   Results and discussion

Calculations were made to obtain the adimensional surface temperature as a function of adimensional time for laser irradiation of copper (figure 1) and aluminum (figure 2). In both figures we display the results for:
     a) equation 15 (linear solution);

b) equation 16 (approximate analytical solution);

c) numerical solution of Abel equation (14);

d) numerical solution of the heat conduction equation taking K, α and A as temperature dependent;

e) Carslaw and Jaeger solution (equation 17);

f) Sparks and Loh equation [2];

g) experimental value of melting threshold energy density for a 100 ns pulse of 10.6 μm laser radiation [2, 5]. An adimensional pulse duration was calculated as:

$$\tau_p = \frac{\pi \rho_o \, \alpha}{t_p \, \lambda} \left(\frac{0.365 \, J_m}{2 \, \chi \, K} \right)^2 \tag{18}$$

h) adimensional melting temperature.

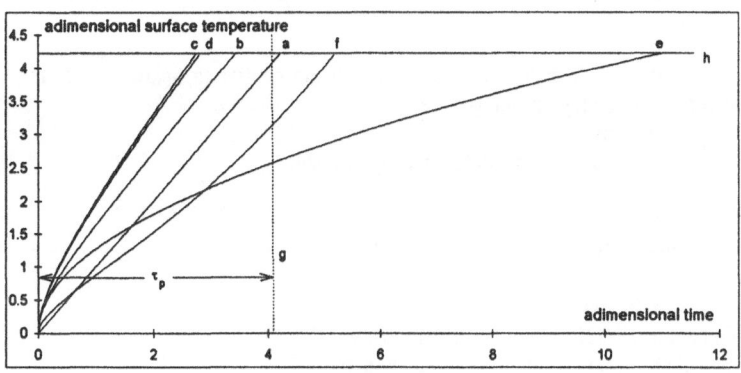

Figure 1. Results for copper. Data from references [2, 6, 7]. Curves a-g: see text.

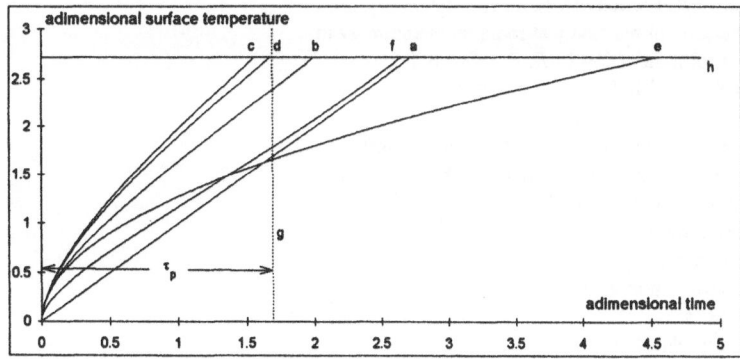

Figure 2. Results for aluminum. Data from references [2, 6, 7]. Curves a-g: see text.

The experimental values for adimensional melting temperature and adimensional pulse duration are represented by horizontal and vertical lines, respectively, for better comparison with the results obtained by the models a-f.

We see that the approximate (a) and the linear (b) solutions match well the experimental results (g), in contrast to Carslaw and Jaeger expression (e). Solutions (a) and (b) predict values for the pulse duration similar to those obtained by numerical calculations (c) and (d) and also to the results of Sparks and Loh (f), yet showing greater simplicity.

We see in both cases that temperature dependence of the thermal properties also influence the results, but not as much as the absorptivity. So, the linear solution gives a particularly simple way of estimating melting thresholds, avoiding large errors that usually arise using the constant absorptivity approximation.

6.   Conclusions

1. An integral Abel-type equation independent on material properties was derived. Its numerical solution can be used to obtain the surface temperature of a laser irradiated material with temperature dependent absorptivity following Hagen-Rubens equation, in good agreement with experimental results.

2. Simple analytical solutions were obtained, giving also a reasonable estimation of melting thresholds.

3. The analytical formulae give results which are similar to Sparks and Loh's, yet showing the following advantages:
    a) simplicity
    b) electrical resistivity data is widely available.

7.   Nomenclature

α: thermal diffusivity, $m^2 \ s^{-1}$.
χ: temperature coefficient of electrical resistivity, K.
δ: optical absorption depth of laser radiation, m.
ε_o: permittivity of vacuum, $C^2 \ N^{-1} \ m^{-2}$.
λ: wavelength of laser radiation, m.
ρ: electrical resistivity, Ω m.
ρ_o: electrical resistivity at room temperature, Ω m.
A: absorptivity ($A = 1 - R$, R: reflectivity).
c: velocity of light in vacuum, $m \ s^{-1}$.
I: power density, $W \ m^{-2}$.
J_m: energy density (melting threshold), $J \ m^{-2}$.
K: thermal conductivity, $W \ m^{-1} \ K^{-1}$.
L: thickness of sample, m.
r: beam radius, m.
T: temperature, K.
T_o: surface temperature, K.
t: time, s.
t_p: pulse duration, s.
x: distance to surface, m.

8.   References

1. H. S. Carslaw and J. C. Jaeger, 1959, *Conduction of Heat in Solids*, 2nd edition, p.10, 30, 75-6, Oxford University Press, Oxford.
2. M. Sparks and E. Loh, Jr., (1979), *Temperature Dependence of Absorptance in Laser Damage of Metallic Mirrors: I. Melting*, J. Opt. Soc. Am. 69(6), 847.

3. E. Hagen and H. Rubens, (1903), *Über Beziehungen des Reflexions- und Emissionsvermögens der Metalle zu ihrem elektrischen Leitvermögen*, Annln. Phys., **11**, 873.

4. M. Bramson, 1968, *Infrared Radiation*, p. 127, Plenum Press, New York.

5. J. O. Porteus, M. J. Soileau and C. W. Fountain, (1976), *Slip Banding in Al Crystals Produced by 10.6-μm Laser Pulses*, Appl. Phys. Lett. **29**, 156.

6. Y. S. Touloukian *et al.*, *Thermophysical properties of matter*, vol. 1 (1970); vol. 4 (1970); vol. 10 (1973), IFI/Plenum, New York.

7. J. R. Davis *et al.*, *"Metals Handbook"*, 1990, vol. 2, 10th edition, ASM International, USA.

MODELING OF LASER CHEMICAL VAPOR DEPOSITION OF THIN FILMS

A. Kar
Center for Research and Education in Optics and Lasers (CREOL)
Mechanical and Aerospace engineering Department
University of Central Florida
Orlando, Florida 32826, USA

J. Mazumder
Center for Laser-Aided Materials Processing (CLAMP)
Mechanical and Industrial Engineering Department
University of Illinois at Urbana-Champaign
Urbana, Illinois 61801, USA

Laser Chemical Vapor Deposition (LCVD) provides an excellent way of modifying surfaces. The deposition rate is found to be very high in the LCVD technique compared to the conventional Chemical Vapor Deposition (CVD) technique. The usefulness of the deposited film depends on its shape and morphology which are affected by the laser process parameters. This paper presents a mathematical model to study the effects of the laser parameters and processing conditions on the shape and size of the film. In many cases, the film is found to have volcano-like morphology which is also explained by this model.

The LCVD process involves several physical phenomena such as the laser-matter interactions, transport processes, chemical reaction kinetics, and adhesion of the film material to the substrate surface. Depending on the laser-matter interactions, the LCVD technique is divided into two types that are referred to as the photolytic and pyrolytic LCVD techniques. In the photolytic process, the reactant molecules absorb the laser to undergo dissociation, whereas the substrate absorbs the laser energy in the pyrolytic process to create a hot spot at the substrate surface where the reactant molecules undergo thermal decomposition. The model of this paper is concerned with the pyrolytic LCVD technique.

To model the LCVD process, the diffusion of various species is considered to be the dominant transport mechanism inside the LCVD chamber, and the three-dimensional transient mass diffusion equation is solved. The thermal decomposition of the reactant molecules at the hot substrate surface is considered to follow Arrhenius' law of the chemical reactions. This model allows to select the appropriate process parameters to obtain a good quality film, and provides a means of determining the activation energy and pre-exponential factor from the film thickness and film deposition temperature data. The Damkohler number is shown to affect the morphology of the film. Also, an optimum condition is found to exist for depositing thin films by using the LCVD technique.

J. Mazumder et al. (eds.), Laser Processing: Surface Treatment and Film Deposition, 203–235.
© 1996 *Kluwer Academic Publishers.*

1. Introduction

A mathematical model of a given LCVD process is a theoretical representation of the process, which allows us to analyze various aspects of the process. It can be used to evaluate the relative importance of various process parameters, such as the laser power, wavelength and beam diameter, the pressure of the gas inside the deposition chamber, etc. Also, the models provide basic understanding of the chemical kinetics and the film deposition mechanisms. The results of the mathematical model can be used to minimize the number of experiments required to obtain the maximum amount of information about the LCVD process, to select a proper set of the process parameters for experiments, and to design an LCVD system optimally.

There are several physical phenomena, such as the laser-matter interactions, transport processes, chemical reaction kinetics, and adhesion of the film material to the substrate surface, that occur during a typical LCVD process. Depending on how the laser beam interacts with the substrate or reactant molecules, the LCVD process is divided into two categories that are referred to as the pyrolytic and photolytic LCVD techniques. In the photolytic process, the reactant molecules absorb the laser to undergo dissociation, whereas the substrate absorbs the laser energy in the pyrolytic process to create a hot spot at the substrate surface where the reactant molecules undergo thermal decomposition. This paper is concerned with the pyrolytic LCVD technique.

In pyrolytic LCVD processes, the internal energy of the reactant molecules increases as they collide with the substrate surface at the hot spot. This process eventually leads to the thermal dissociation of the reactant at the gas-substrate interface and deposition of a thin film on the substrate surface. In some cases, the reactant is so chosen that it absorbs the laser beam to rise in temperature, and dissociates in the gas phase due to the thermal decomposition reactions, and eventually a thin film is deposited on the substrate. In both cases, the temperature of the reactant molecules is raised to induce the chemical reactions, and therefore, the heat and mass transfer processes are coupled to each other. In the case where the substrate is heated, the absorptivity of the surface changes as the film is deposited, which alters the temperature of the localized hot spot, and consequently, the chemical reaction and the film deposition rate are affected. This phenomenon also couples the heat and mass transfer processes during the pyrolytic LCVD. For this reason, the heat and mass conservation equations have to be solved simultaneously to model the LCVD process. However, if the chemical reaction rate is very slow, the heat and mass transfer equations can be solved separately by assuming that these two transfer processes are uncoupled. Even in other cases when the chemical reaction rate is not very slow, the solutions obtained by solving the decoupled heat and mass transfer equations can be utilized sequentially to set up an iterative scheme for modeling the LCVD process.

In order to model the pyrolytic LCVD process, we have to solve the mass, energy, and momentum conservation equations in a particular coordinate system depending on the configuration of the system, such as the shape and direction of propagation of the laser beam, the orientation of the substrate with respect to the laser beam, etc., as well as the process conditions, such as the reactant flow rate, the pressure inside the reaction chamber, etc. The laser beam and substrate are usually perpendicular or parallel to each other. Also, the LCVD reactors can be operated in the batch or continuous mode. In the batch mode, the deposition chamber is filled with the reactant molecules, the inlet and outlet valves of the chamber are closed, the film is deposited for a certain period of time, the reaction product and other gases are removed from the reactor, and then the chamber is charged with fresh reactant molecules to repeat the deposition process. Due to the absence of the gas flow into and out of the LCVD reactors during the deposition process, the forced convection does not exist in batch reactors. In the continuous operation mode, however, the reactant gas flows continuously into and out of the LCVD reactors during the deposition process, and the effect of forced convection can be important. At low pressures inside the LCVD reactors, the natural convection is usually not important, and diffusion is the most dominant transport mechanism in the absence of the forced convection. In this paper, the LCVD chamber is assumed to operate in the batch mode, and the laser beam is considered to be perpendicular to the substrate surface.

To model the pyrolytic LCVD process, it is convenient to divide the deposition chamber into at least two regions, such that the concentration and temperature change more rapidly in one region than in the other region. This type of multi-region analysis reduces the number of computational grids for numerical modeling, and also allows us to obtain analytical solutions by selecting a suitable geometry for the inner region that surrounds the laser beam. There can be several species, that is, different types of molecules, in the LCVD chamber, whose concentration, velocity, and temperature distributions have to be determined in order to model the deposition process. Usually, the mass conservation equation is solved for each species, and by assuming that all of the species are under the local thermodynamic equilibrium, the momentum and energy conservation equations are solved to determine the velocity and temperature fields, respectively, at various points in the gas phase. The derivations of the mass, momentum, and energy conservation equations for a single component system as well as for a mixture of gases can be found in Bird, et al., (1960). However, it should be noted that the mass, momentum, and energy conservation equations are difficult to solve in practice, because they are coupled nonlinear partial differential equations. For this reason, the conservation equations and boundary conditions are simplified in this study by using appropriate approximations in order to model the pyrolytic LCVD process.

2. Literature Review

Availability of high-power lasers and the economical advantages of using laser as a tool for manufacturing and materials processing have led to many interesting applications of laser technology (Mazumder, 1987). Considerable efforts are being made in various industries to implement laser technology for manufacturing and materials processing, and in various research facilities to understand the physics of such processes. Layer glazing (Breinan and Kear, 1983) and laser cladding (Singh and Mazumder, 1987; Kar and Mazumder, 1987, 1988) have already demonstrated the feasibility of synthesizing novel materials in near net shape using laser. These processes can generate relatively thick coatings of metastable materials, whereas LCVD can produce thin (Å to several hundreds of μm) layers of metals, and ceramics (Allen, 1981; Chou, et al., 1989; and Mazumder and Allen, 1980). Metals and ceramics (such as titanium and titanium nitride) can be co-deposited using this technique to produce composites. The heat transfer in the substrate plays an important role in laser processing of materials. Mathematical models of such heat transfer processes allow us to examine the effect of various parameters on the performance of the process and control such manufacturing processes.

In the literature, one can find that a lot of work has been done to determine the distribution of laser energy in the substrate materials. These studies consider either constant thermophysical properties for the substrate material or infinite or semi-infinite geometry. One of the earliest works on the energy distribution in the substrate during laser heating is due to Cline and Anthony (1977). They considered infinite geometry and constant thermophysical properties for the substrate material and solved the heat conduction equation by using Green's function for a Gaussian beam moving at a constant velocity. Lax (1977; 1978; and 1979) studied the temperature rise under steady-state condition due to a stationary Gaussian beam in a semi-infinite cylindrical medium. He considered the linear (constant thermophysical properties) case in Lax (1977) and the nonlinear (temperature dependent thermal conductivity) case by the method of Kirchoff transformation in Lax (1978) and Lax (1979). Hess, et al. (1980) presented a quasi-steady state solution for the temperature distribution in a radially infinite cylindrical medium with temperature dependent thermal conductivity by neglecting the effect of scan velocity of the laser beam. Bell (1979) developed a one-dimensional thermal model for laser annealing over a wide range of laser pulse durations and absorption coefficients. Nissim, et al. (1980), studied the effects of scanning elliptical or circular CW laser beam on the temperature distributions in semiconductors. For an infinite medium, they modeled the linear case following the formulation of Cline and Anthony (1977) and the nonlinear case by the method of Kirchoff transformation.

Kokorowski, et al. (1955) studied the temperature distributions during CW laser annealing with a stationary beam beyond the melt temperature. They solved the steady-state heat conduction equation in a radially infinite cylindrical medium with temperature dependent thermal conductivity. Moody and Hendel (1982) presented a numerical algorithm for temperature distribution in an infinite medium during laser heating to generalize the models of Lax (1977; 1978; and 1979), Cline and Anthony (1977), Hess, et al. (1980), Nissim, et al. (1980), and Kokorowski, et al. (1955). Recently, Kant (1988) modeled the heating of a multi-layered cylindrical medium resting on a half-space with a stationary laser beam. He considered the medium to be radially infinite and the materials of the medium to have constant thermophysical properties. Besides these works, various studies concerning the laser heating of different types of materials, such as metals, semiconductors, thermally insulating materials, thin films, etc., can be found in Gibbons and Sigmon (1982), Abraham and Halley (1987), Sanders (1984), Ferrieu and Auvert (1983), Kwon and Kim (1982), Kim, et al. (1981), Calder and Sue (1982), Burgener and Reedy (1982), Meyer, et al. (1980a), Meyer, et al. (1980b), El-Adawi and Elshehawey (1986), and Liarokapis and Raptis (1985). It can be seen that a lot of work has been done to model the energy distribution in the substrate due to laser heating by considering the substrate infinite or semi-infinite. Although the results of such models can be applied to substrates of dimensions much larger than the laser spot size, the infinite or semi-infinite medium approximation will not hold good for small substrates. A transient and three-dimensional thermal analysis for laser heating of uniformly moving finite slabs can be found in Kar and Mazumder (1989), which is discussed in the next section.

In LCVD processes, the laser heating aspect of laser-materials interactions will have to be considered in conjunction with the chemical reactions among various species and their distributions inside the LCVD chamber. In conventional chemical vapor deposition (CVD) processes, the substrate material is heated to a high temperature in the environment of reactant molecules which are usually in the vapor or gaseous state. The reactant molecules undergo thermal decomposition to produce the depositing materials which subsequently adhere to the surface of the substrate to form a film. Although a similar principle is used in LCVD, it differs from the CVD processes because only a very small area of the substrate is heated during LCVD. LCVD can be achieved by relying on the thermal decomposition reaction (pyrolysis) or on the photo-chemical dissociation reaction (photolysis) of the reactant molecules. A combination of these two processes can also be used in LCVD. Several studies have been carried out to model the CVD process. Sitrl. et al. (1974) performed thermodynamic equilibrium calculations for CVD processes. Bloem and Giling (1978) studied CVD by using diffusion and surface kinetic models. Coltrin, et al. (1984; 1986) developed a mathematical model for a CVD process by considering gas phase chemical kinetics coupled with fluid mechanics. Jensen (1987) presented a summary of various studies on different types of CVD processes. Choi, et al. (1987) analyzed a heat transfer problem related to the Modified Chemical vapor Deposition (MCVD) processes. Yarmoff and McFeely (1988) have studied experimentally the process of CVD of tungsten on silicon from tungsten hexafluoride. Holstein (1992) has reviewed the design and modeling aspects of CVD reactors.

The process of LCVD has also been studied theoretically by several investigators. Piglmayer, et al. (1984) studied temperature distributions in laser induced pyrolytic deposition. Zeiger and Ehrlich (1989) have used a three-dimensional Green function to model a thermally driven localized surface reaction. Herman, et al. (1983) developed a model for laser induced pyrolytic deposition processes. Allen, et al. (1986) used a finite difference technique to study the laser heating and film formation by considering temperature dependent thermophysical and optical properties of the substrate material. Skouby and Jensen (1988) used the Galerkin finite element method to examine the effects of mass transfer and chemical kinetics on the shape of the deposit on a semi-infinite substrate in laser-assisted chemical vapor deposition with a Gaussian and stationary laser beam. Copley (1988) has presented a mass transport model to analyze the effects of diffusion and convection during laser chemical vapor deposition. Allen, et al. (1986) have pointed out that convection during pyrolytic LCVD is negligible for low deposition rates and/or small spot sizes. Also, Allen, et al. (1984) have suggested that the diffusion process dominates if $\xi r_0 \leq 1.0$, where

$$\xi = \frac{RTj_f S'}{DP}, \quad S' = -S'_{q+1} + \sum_{i=1}^{q+p} S'_i,$$

and $S'_i = \pm S_i$. The plus and minus signs in $\pm S_i$ correspond to the products and reactants, respectively. However, Copley (1988) has shown that the reactant pressure obtained by considering the transport of the species due to both diffusion and convection differs from that predicted by the diffusion analysis by 10% of the total pressure if $\xi r_0 = 0.1$.

McWilliams, et al. (1984) have used the LCVD technique to deposit films for the interconnection of very large scale integrated (VLSI) gate arrays. The topology of the deposited spot in LCVD is of interest. Allen (1981) found experimentally that the shape of the nickel film formed on quartz substrates in LCVD from nickel carbonyl is volcanic. Moylan, et al. (1986) observed that the copper film deposited on silicon substrate in LCVD has volcanic shape. Similarly, Chou, et al. (1989) observed volcano-like titanium deposits in LCVD of titanium on S.S. 304 from titanium tetrabromide. The mathematical model of Skouby and Jensen (1988) also yields nickel deposits with volcanic morphology for certain operating parameters.

Following the work of Kar, et al., (1991), we will discuss a three-dimensional transient model for mass diffusion with chemical reactions during the LCVD of pure titanium on SS 304 substrate from titanium tetrabromide ($TiBr_4$) in order to carry out parametric studies for predicting the thickness of the deposited film, and to understand the effects of various physico-chemical parameters on the thickness and the morphology of the film.

Laser-induced chemical processing (LCP) of materials involves unique processing characteristics which are of fundamental importance in many areas of technology (Bäuerle, 1986). Among the applications of LCP, laser-induced chemical vapor deposition (LCVD) has been one of the most extensively studied, both experimentally and theoretically. The extremely high deposition rates (Allen, et al., 1983) achieved in LCVD together with the strongly localized heat and chemical treatments, make LCVD an attractive technique for the deposition of a wide variety of materials useful in micromechanics, microelectronics, etc. The morphology of the deposited material plays an important role for many applications, and usually, the flat-topped deposits with sharp edges are desired. However, when the energy distribution in the laser beam used to induce film deposition is Gaussian, volcano-like shapes (Allen, 1981; Chou, et al., 1989; Moylan, et al., 1986; Conde, et al., 1990) have very often been observed. The explanations given to account for these volcanoes have been based on high chemical reaction rates, convection from the surface, low sticking coefficients and melting, and even evaporation, at the center of the heated spots. By studying the effects of convection on the deposition rate of Ni from $Ni(CO)_4$, Allen et al. (1984) explained the occurrence of volcano-shaped profiles. The three-dimensional transient mass transfer model for LCVD of titanium shows volcano-like Ti deposits for high Damkohler numbers, that is, when the chemical reaction rate is much faster than the diffusion rate of various species inside the LCVD reactor. Similarly, Zeiri et al. (1991) obtained volcano-shaped patterns when the rate of decomposition is large compared to the rates at which the precursor molecules diffused into the irradiated region by using a Monte Carlo method to simulate the photolytic decomposition of Ni(CO)$_4$. The role played by the sticking coefficient, which depends on the surface coverage and temperature, has been taken into account in the model presented in Konstantinov (1990) and considered by Moylan, et al. (1986) as an important parameter in determining deposit shapes.

we will discuss a model, which has been presented in Conde, et al. (1992b); and by considering a given chemical reaction kinetics for the deposition of TiN and a temperature dependent sticking coefficient, we determine the profiles of TiN films formed during LCVD and compare them with experimental data (Conde, et al., 1992a). Titanium nitride possesses many interesting properties such as high thermal and electrical conductivities, high melting point, good thermodynamic stability, and low mass diffusivity. Titanium nitride (TiN) thin films have several applications such as gold reflective coatings (in jewelry industry) or as a diffusion barrier or in metallization (in integrated circuit technology) and for wear or corrosion resistant coatings (in tool industry). In micro-electronics, TiN thin films have frequently been used as a diffusion barrier between Si and Al for very large scale integrated circuits (VLSI). This broad range of applications has resulted in the development and understanding of the LCVD technique to deposit TiN films.

3. Mathematical Model

We will now analyze the governing equations and boundary conditions concerning the determination of the temperature field, and deposition of Ti and TiN films. The techniques to solve these equations are also discussed in this section.

3.1. GOVERNING EQUATIONS AND BOUNDARY CONDITIONS FOR TEMPERATURE DISTRIBUTION

In LCVD processes, the temperature field has to be obtained to determine the chemical reaction zone on the surface of the substrate. The chemical reaction zone is defined as that region on the surface of the substrate in which the temperature is higher than or equal to the chemical reaction temperature. Also, the temperature field can be used to determine the thermal stress of the deposited film. The temperature field is obtained by solving the heat conduction equation with temperature-dependent thermophysical properties of the substrate. Figure 1a shows the finite-slab geometry and the Cartesian coordinate system under consideration. In this study, the substrate is considered to move at a constant velocity in the negative x-direction and the laser beam is stationary. For the purpose of mathematical analysis, the frame of reference is so chosen that the substrate stays stationary and the laser beam moves with respect to the substrate at a constant velocity in the positive x-direction. The energy transfer equation in the slab is given by

$$\rho(T) \, C_p(T) \frac{\partial T}{\partial t} (x,y,z,t) = \nabla[k(T) \, \nabla T \, (x,y,z,t)] \qquad (3.1.1a)$$

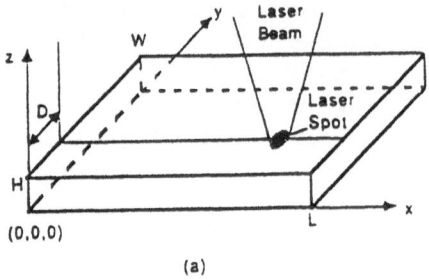

(a)

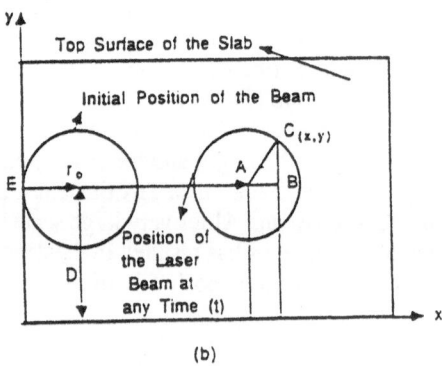

(b)

FIG. 1. (a) Geometric configuration of the substrate and the relative position of the laser spot. (b) Top-view of the laser-substrate interaction zone at the initial time and at a later time.

Equation (3.1.1a) is solved to satisfy the following initial and boundary conditions at the surfaces of the slab. Both convective and radiative heat losses through the boundary surfaces of the substrate are considered in this analysis.

$$T(x,y,z,0) = T_i \qquad (3.1.1b)$$

$$k(T)\frac{\partial T}{\partial x} = h_{xo}(T - T_a) + \sigma\varepsilon(T^4 - T_a^4) \text{ at } x = 0 \qquad (3.1.1c)$$

$$-k(T)\frac{\partial T}{\partial x} = h_{xl}(T - T_a) + \sigma\varepsilon(T^4 - T_a^4) \text{ at } x = L \qquad (3.1.1d)$$

$$k(T)\frac{\partial T}{\partial y} = h_{yo}(T - T_a) + \sigma\varepsilon(T^4 - T_a^4) \text{ at } y = 0 \qquad (3.1.1e)$$

$$-k(T)\frac{\partial T}{\partial y} = h_{yw}(T - T_a) + \sigma\varepsilon(T^4 - T_a^4) \text{ at } y = W \qquad (3.1.1f)$$

$$k(T)\frac{\partial T}{\partial z} = h_{zo}(T - T_a) + \sigma\varepsilon(T^4 - T_a^4) \text{ at } z = 0 \qquad (3.1.1g)$$

$$k(T)\frac{\partial T}{\partial z} = \frac{2P(1-R)}{\pi r_0^2} e^{-2r^2/r_0^2} - h_{zh}(T - T_a) - \sigma\varepsilon(T^4 - T_a^4) \text{ at } z = H \quad (3.1.1h)$$

R represents the reflectivity of the composite medium which is made up of stainless steel 304, titanium film, and TiBr4 vapor. Such a medium is encountered in LCVD of pure titanium from titanium tetrabromide on stainless steel (SS 304). The high temperature data for the absorptivity A, which is 1-R, of the composite medium is obtained from the low temperature data (Kar and Mazumder, 1989). A is considered constant and equal to the absorptivity of pure titanium in this analysis. While setting up the boundary condition (3.1.1h), the loss of the absorbed laser energy in the thin metallic film is considered negligibly small. Also, the effect of optical absorption depth is ignored in this model because it is usually very small for metallic materials. It should be noted that the location of the laser beam, r varies with time. Figure 1b shows the top-view of the location of the laser beam on the top surface of the substrate at time zero and at a later time t. The substrate moves to the left with a uniform velocity U, that is the laser beam moves in the positive x-direction with a constant velocity U with respect to the substrate. So r is given by

$$r^2 = (x - r_0 - Ut)^2 + (y - D)^2 \quad (3.1.2)$$

where D is the y-coordinate of the center of the cross-section of the laser beam on the top surface of the substrate.

3.1.1. Method of Solution for Temperature Distribution

The above conduction problem is nonlinear because the thermophysical properties of the substrate are considered to vary with temperature and radiation is taken to be one of the modes of heat loss from the substrate. To obtain an analytic solution of this problem, the governing Eq. (3.1.1a) is linearized by using the Kirchoff transformation (Carslaw and Jaeger, 1986) and the boundary conditions are linearized by expanding the nonlinear terms in Taylor series. The linearization process is carried out after reformulating the problem in terms of the following non-dimensional variables:

$$x^* = x/L; \ y^* = y/W; \ z^* = z/H; \ t^* = t/\tau \text{ where } \tau = L/U .$$

$$T^* = T/T_d ; \ \bar{T}_a^* = \bar{T}_a/T_d;$$

In this model, $\bar{T}_a$ is taken to be equal to T_d. Also, we consider that the thermal conductivity of the substrate depends on temperature in the following way:

$$k(T) = pT^q$$

where q and log p are respectively the slope and the intercept of log k(T) versus T graph. The non-dimensional thermal conductivity is defined as

$$k^* \left(T^* \right) = \frac{k(T)}{k(T_d)} = \frac{T^q}{T_d^q} = T^{*q} \tag{3.1.3}$$

To simplify the governing Eq. (3.1.1a) using the Kirchoff transformation, we define a transformed temperature T'* such that

$$\frac{\partial T'^*}{\partial z^*} = k^*(T^*) \frac{\partial T^*}{\partial z^*} \tag{3.1.4}$$

Substituting Eq. (3.1.3) into Eq. (3.1.4) and integrating the latter with respect to temperature from 1 to T*, we obtain

$$T^{*q+1} = 1 + (q+1)T'^* \tag{3.1.5}$$

The nonlinear term on the left hand side of Eq. (3.1.5) is expanded around $\overline{T}_a^*$ by Taylor series to obtain the following linear relation between the original temperature T^* and the transformed temperature T'*.

$$T^* = \frac{1}{(q+1)\overline{T}_a^{*q}} \left\{ 1 - \overline{T}_a^{*q+1} + (q+1)T'^* \right\} + \overline{T}_a^* \tag{3.1.6}$$

Eq. (3.1.6) is used to simplify the boundary conditions. For this purpose, we first non-dimensionalize the boundary conditions (3.1.1c) through (3.1.1h) and then linearize the radiative heat loss terms by expanding the nonlinear terms around $\overline{T}_a^*$ by Taylor series. Thus, we obtain the following effective heat transfer coefficients H_i at the boundary surfaces of the substrate.

$$H_i = h_i + \sigma \varepsilon \, T_d^3 \, (\overline{T}_a^{*2} + T_a^{*2}) (\overline{T}_a^* + T_a^*)$$

where i = 1, 2, 3, 4, 5 and 6 which denote the x = 0, x = L, y = 0, y = W, z = 0 and z = H planes, respectively. The resulting boundary conditions can now be expressed in terms of the transformed temperature T'*, and are linear.

The governing Eq. (3.1.1a) is linearized by noting that $\alpha(T)$ is almost constant for many materials because it varies slowly with temperature. In this study, $\alpha(T)$ is considered constant which is taken to be the temperature-weighted average value of the thermal diffusivity over the temperature ranging from the ambient temperature to the melting point of the substrate. This heat transfer problem can be solved by successively applying Fourier transform (Sneddon, 1972) in the x, y, and z directions to obtain an expression for the temperature profile which is given in Kar and Mazumder (1989).

3.2. GOVERNING EQUATIONS AND BOUNDARY CONDITIONS FOR TITANIUM FILM DEPOSITION

Heat and mass transfer are two important phenomena that have to be considered for modeling the LCVD processes. Besides this, the mechanisms of the chemical reactions occurring during LCVD and the associated chemical kinetics should also be known along with other parameters such as the sticking probability, diffusion coefficient, absorptivity. The following reactions have been established (Funaki, et al., 1961) to occur during thermal decomposition of $TiBr_4$. The LCVD of Ti films on S.S. 304 under the following assumptions.

1. Reactions take place only on the substrate surface within a region to be called chemical reaction zone where the temperature is greater than or equal to the Ti film forming chemical reaction temperature (1173 K).
3. Reactions that may occur on the substrate surface outside the chemical reaction zone are not considered in this model because the work by Chou et al. (1989) shows that the chemical reactions occurring outside the chemical reaction zone do not have significant effects on the deposition of Ti film on SS 304 substrate.

Using these assumptions, the following irreversible reactions (Funaki, et al., 1961) are studied.

$$(E) \quad TiBr_4(g) \xrightarrow{k_1} TiBr_2(s) + Br_2(g)$$

$$(F) \quad 3TiBr_2(s) \xrightarrow{k_2} 2TiBr(s) + TiBr_4(g)$$

$$(G) \quad 4TiBr(s) \xrightarrow{k_3} 3Ti(s) + TiBr_4(g)$$

The rate of production of various species is given by

$$\dot{r}_1 = m_1 \left[\frac{1}{4} k_3 \left(\frac{C_3}{m_3} \right)^{n_3} + \frac{1}{3} k_2 \left(\frac{C_2}{m_2} \right)^{n_2} - k_1 \left(\frac{C_1}{m_1} \right)^{n_1} \right] \quad (3.2.1)$$

$$\dot{r}_2 = m_2 \left[k_1 \left(\frac{C_1}{m_1} \right)^{n_1} - k_2 \left(\frac{C_2}{m_2} \right)^{n_2} \right] \quad (3.2.2)$$

$$\dot{r}_3 = m_3 \left[\frac{2}{3} k_2 \left(\frac{C_2}{m_2} \right)^{n_2} - k_3 \left(\frac{C_3}{m_3} \right)^{n_3} \right] \quad (3.2.3)$$

$$\dot{r}_4 = m_4\left[\frac{3}{4}k_3\left(\frac{C_3}{m_3}\right)^{n_3}\right] \tag{3.2.4}$$

$$\dot{r}_5 = m_5\left[k_1\left(\frac{C_1}{m_1}\right)^{n_1}\right] \tag{3.2.5}$$

In the above expressions (3.2.1) through (3.2.5), as well as in the following expression, the subscripts i = 1, 2, 3, 4, and 5 refer to the species $TiBr_4$, $TiBr_2$, TiBr, Ti, and Br_2, respectively. Thus $\dot{r}_i$ is the mass of the i-th species produced per unit area of the substrate surface per unit time.

Also, the following Arrhenius equation for the reaction rate constant is used,

$$k_j = k_{0j}\exp[-E_j/(RT(x,y,0,t))] \qquad \text{for } j = 1, 2, \text{ and } 3 \tag{3.2.6}$$

The concentration of various species such as $TiBr_4$, $TiBr_2$, TiBr, Ti, and Br_2 inside the LCVD chamber and the thickness of the deposited Ti film are determined by solving five three-dimensional transient, and nonlinear mass diffusion equations. Figure 2 shows the LCVD chamber, the substrate, and the Cartesian coordinate system under consideration.

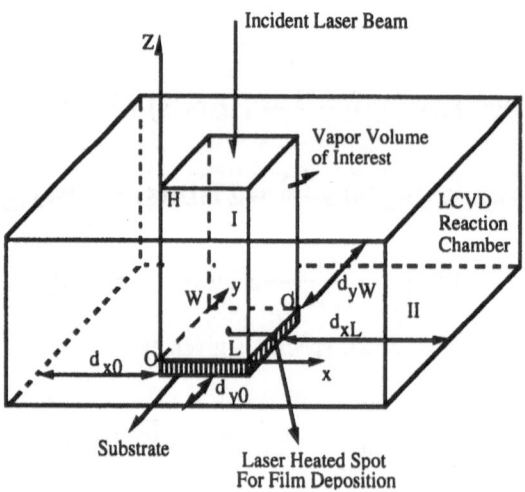

Figure 2. Idealized LCVD chamber and the mass transfer region of interest.

OLO'W represents the top surface of the substrate, and the point O the origin of the chosen coordinate system. In Fig. 2, the LCVD chamber is divided into two regions. The top surface of the substrate forms the base of region I and this region extends up to the top inner surface of the LCVD chamber. Region II constitutes the rest of the chamber; thus, the length, width, and height of the region I are L, W, and H, respectively.

The governing equation for mass transfer of the i-th species in the region I is given by

$$\frac{\partial C_i}{\partial t} = D_i \left(\frac{\partial^2 C_i}{\partial x^2} + \frac{\partial^2 C_i}{\partial y^2} + \frac{\partial^2 C_i}{\partial z^2} \right) \quad \text{for } i = 1, 2, 3, 4, \text{ and } 5, \tag{4.27a}$$

where C_i represents $C_i(x, y, z, t)$, that is the concentration of the i-th species at any point, (x, y, z) in Cartesian coordinates at any time t.

Equation (3.2.7a) is solved to satisfy the following boundary and initial conditions.

$$D_i \frac{\partial C_i}{\partial x} = \beta_{ix0} (C_i - C_{i\infty}) \quad \text{at } x = 0 \tag{3.2.7b}$$

$$-D_i \frac{\partial C_i}{\partial x} = \beta_{ixL} (C_i - C_{i\infty}) \quad \text{at } x = L \tag{3.2.7c}$$

$$D_i \frac{\partial C_i}{\partial y} = \beta_{iy0} (C_i - C_{i\infty}) \quad \text{at } y = 0 \tag{3.2.7d}$$

$$-D_i \frac{\partial C_i}{\partial y} = \beta_{iyW} (C_i - C_{i\infty}) \quad \text{at } y = W \tag{3.2.7e}$$

$$-D_i \frac{\partial C_i}{\partial z} = \dot{r}_i (1-\gamma_i) \quad \text{at } z = 0 \tag{3.2.7f}$$

$$\frac{\partial C_i}{\partial z} = 0 \quad \text{at } z = H \tag{3.2.7g}$$

$$C_i(x,y,z,t_0) = f_i(x,y,z) \tag{3.2.7h}$$

The mass transfer coefficients β_{ix0}, β_{ixL}, β_{iy0}, and β_{iyW} are obtained from the following considerations. We refer to Fig. 2 where the diffusion of various species is considered in the region I in this model, and the region II represents the rest of the vapor volume in the LCVD chamber. At the interfaces of these two regions, the mass flux and the concentration of each species are continuous. Considering these two conditions at $x = 0$, we can write

$$D_i \frac{\partial C_i}{\partial x} = D_{iII} \frac{\partial C_{iII}}{\partial x}$$

and

$$C_i = C_{iII}$$

Assuming that the concentration varies linearly with x in the region II, it can be shown from the above two interface conditions that

$$\beta_{ix0} = D_{iII}/d_{x0}$$

and similarly, the interface conditions at $x = L$, $y = 0$, and $y = W$ yield

$$\beta_{ixL} = D_{iII}/d_{xL}, \beta_{iy0} = D_{iII}/d_{y0}, \text{ and}$$
$$\beta_{iyW} = D_{iII}/d_{yW}, \text{ respectively,}$$

where the distances d_{x0}, d_{xL}, d_{y0}, and d_{yW} are as shown in Fig. 2.

3.2.1. *Method of Solution for Titanium Film Deposition*

Equation (3.2.7a) and the boundary and initial conditions (3.2.7b-h) represent a nonlinear and coupled mass transfer problem because of the mass generation term in the boundary condition (3.2.7f). The mass transfer problem is also coupled with the heat transfer problem because of the Arrhenius equation (3.2.6). Since the pyrolytic decomposition of TiBr$_4$ is considered in this model, the heat transfer problem mainly involves the transfer of laser energy into the substrate and its redistribution in the substrate. The deposits in LCVD can affect the transfer of laser energy into the substrate by altering the thermophysical and optical properties of the substrate surface. In particular, the thermal conductivity, density, specific heat, and absorptivity of the substrate are often changed when films are deposited. For thin and metallic deposits, the loss of the absorbed laser energy in the film can be considered negligibly small. Although the absorptivity depends on the film thickness at the substrate surface, it is taken to be constant in this model because the exact variation of the absorptivity as the deposition continues is not known. This approximation results in one-way coupling between the heat and the mass transfer problem, and it is possible to solve the heat transfer problem before solving the mass transfer problem.

The heat transfer problem is solved by adopting the mathematical technique of section 3.1 for a stationary laser beam irradiating the center of the top surface of the substrate. From the solution of the heat transfer problem, the temperature distribution at the substrate surface is obtained and it is used in Eq. (3.2.6) in order to solve the mass transfer Eq. (3.2.7a) by using the Fourier transform (Sneddon, 1972). The Fourier transform is first appliedin the x and y directions, then it is applied in the z direction under the assumption that the concentration of various species remains unchanged at the substrate surface over a small time interval. The expression for the distribution of various species inside the LCVD chamber is given in Kar, et al. (1991).

3.3. GOVERNING EQUATIONS AND BOUNDARY CONDITIONS FOR TITANIUM NITRIDE FILM DEPOSITION

Although a considerable amount of work has been reported on the deposition of titanium nitride films by various chemical vapor deposition (CVD) systems (Kato and Tamari, 1975; Sadahiro, et al., 1977; Kim and Chun, 1983; Motojima and Kohno, 1986; Itoh, et al., 1986; Motojima and Mizutani, 1988; Rong, et al., 1989; and Michalski and Wierzchon, 1989), the dominant mechanisms and rate limiting processes are still not fully known. In a conventional CVD reactor, TiN is produced according to the following overall chemical reaction:

$$TiCl_4(g) + 2H_2(g) + \frac{1}{2}N_2(g) \rightarrow TiN(s) + 4HCl(g) \qquad (3.3.1)$$

The CVD of TiN films has been investigated as a function of deposition parameters such as the substrate temperature and the composition and flow rates of the reactants. For instance, Sadahiro et al. (1977) found that at constant temperature, the amount of TiN deposited increases linearly both with TiCl4 content and with the product p_{H_2} x $p_{N_2}^{1/2}$. But, the order of the chemical reaction and the reaction pathways yielding the final TiN solid product still remain controversial (Rong, et al., 1989). However, it is generally accepted (Kato and Tamari, 1975; Sadahiro, et al., 1977; Rong, et al., 1989; and Michalski and Wierzchon, 1989), that above a threshold temperature of about 900°C, TiCl4 is reduced by H2 to TiCl2 and that the adsorption of the reactant species TiCl2 controls the process. This same kinetic mechanism was assumed by Jang and Paik (1983) while studying the CVD of Ti from a TiCl4 and H2 gas mixture. They also found that the amount of deposited Ti varies linearly with the partial pressure of TiCl4.

In the present model, we consider that TiN is produced in pyrolytic LCVD through the overall chemical reaction given by the expression (3.3.1), and by assuming Sadahiro's kinetics (1977), the rate of production of TiN can be written as

$$R_{TiN} = k(T) \, C_{H_2} \, C_{N_2}^{1/2} \, C_{TiCl_4} \,, \qquad (3.3.2)$$

where rate constant, $k(T)$ follows the Arrhenius equation

$$k(T) = k_0' \exp\{-E/RT(x, y, 0, t)\}$$

Here k_0' and E are, respectively, the Arrhenius constant and activation energy of the chemical reaction (3.3.1).

Since the mole fractions of H2 and N2 inside the LCVD reactor were at least one order of magnitude higher than the TiCl4 mole fraction in the experimental study (Conde, et al., 1990), we can neglect any variation of C_{H_2} and C_{N_2} during the deposition process and replace them in Eq. (3.3.2) by their initial values $^0C_{H_2}$ and $^0C_{N_2}$, respectively. So, Eq.(3.3.2) can be written as

$$R_{TiN} = A \, k_0' \, \exp\{-E/RT(x, y, 0, t)\} \, C \qquad (3.3.3)$$

where $A = {}^0C_{H_2}\ {}^0C_{N_2}^{1/2}$ = constant, and we have discarded the index TiCl4, that is, $C \equiv C_{TiCl_4}$.

Eq.(3.3.3) shows that the rate of production of TiN is determined by the concentration of TiCl4 and therefore, the thickness of the deposited TiN film may be obtained by solving the following mass diffusion equation:

$$\frac{\partial C}{\partial t} = D\left(\frac{\partial^2 C}{\partial x^2} + \frac{\partial^2 C}{\partial y^2} + \frac{\partial^2 C}{\partial z^2}\right) \tag{3.3.4}$$

In order to define the boundary conditions which have to be satisfied for solving the mass transfer equation (3.3.4), we consider a geometrical configuration as shown in Fig. 2 where the boundary conditions can be written in the same way as in Section 3.2.

3.3.1. Method of Solution for Titanium Nitride Film Deposition
As in section 3.2, Eq. (3.3.4) is solved by applying the Fourier transform (Sneddon, 1972) in the x and y directions. However, the mass diffusion equation of section 3.2 has been solved (Kar, et al., 1991) by applying the Fourier transform in the z direction under the assumption that the surface source term remains unchanged over a small time interval. This assumption holds good for slow chemical reactions in which the reaction rate varies slowly with time. However, one will need an extremely small time interval and, therefore, a large computational time to obtain results for fast chemical reactions if one uses the solution technique of Section 3.2.1. For this reason, after applying the Fourier transform in the x and y directions, Eq. (3.3.4) is solved by using the Laplace transform technique in stead of applying the Fourier transform in the z direction. The expression for the distribution of TiCl4 concentration during the LCVD of TiN is given in Conde, et al. (1992b).

4. Results and Discussion

We will now discuss the results, which are obtained from the models of section 3, concerning the temperature distribution, and deposition of Ti and TiN films.

4.1. RESULTS FROM THERMAL ANALYSIS

The thermal analysis of section 3.1 can be used to carry out parametric studies of laser heating of finite slabs. Various parameters such as the wavelength of the laser beam, laser power, shape of the laser beam, diameter of the laser beam, speed of the laser beam relative to the substrate, dimension of the substrate, the thermophysical and optical properties of the substrate, and the conditions of the medium surrounding the substrate can affect the temperature distribution in the substrate. Some typical results are presented in this section. Also, two simple expressions for the variation of the peak temperature with the laser power, and with the substrate velocity are obtained.

For all of the results in this section, the value of r_0 is taken to be 2 mm. The laser beam is located at $y^* = 0.5$ and moves in the positive x-direction with respect to the substrate. The values of the heat transfer coefficients in the boundary conditions (4.2c) through (4.2h) are determined from the following considerations. The slab considered in this study is very thin in the z-direction. Its length, width, and height are 1 cm, 1 cm and 0.1587 cm, respectively. Since the slab is exposed to the laser beam at the $z = H$ plane, the surfaces at $z = 0$ and $z = H$ are expected to heat up more than the other four surfaces for the same ambient conditions at all the six sides of the slab. So the convective activities will be more at the planes $z = 0$ and $z = H$ than at the other surfaces. In this study, the convective heat loss from the surfaces at $x = 0$, $x = L$, $y = 0$, and $y = W$ is considered to be due to free convection and the heat transfer coefficients are taken to be 5 W/m^2-K (Incropera and Dewitt, 1985) at these four surfaces. The heat transfer coefficients at $z = 0$ and $z = H$ planes are determined by assuming that the Biot (Bi) numbers are equal at all the surfaces, where the characteristic lengths in the Bi numbers are taken to be the length, width and height of the slab for the surfaces perpendicular to the x, y and z axes, respectively.

4.1.1. Temperature Distribution

Figure 3 represents the temperature distribution on the top surface of the substrate which is plotted by letting x^*, y^*, and the temperature (T^*) increase along Ox^*, Oy^*, and OT^* directions, respectively,. Here, the point O represents one of the corners of the square containing the lines $O'V$ and $O''V$ as two of its sides in the plane of $O'VO''$. For improved clarity in representing the surface temperature field in three-dimensional plots, views from the two planes, one from the plane $T'O'V'$ located at $x^* = 1$ and the other from $T''O''V$ located at $y^* = 1$ are shown. Figure 2 shows the temperature fields for laser power, $P = 600$ W, laser scanning speed relative to the substrate, $u = 0.25$ cm/s, and for the nondimensional time, $t^* = 0.5$.

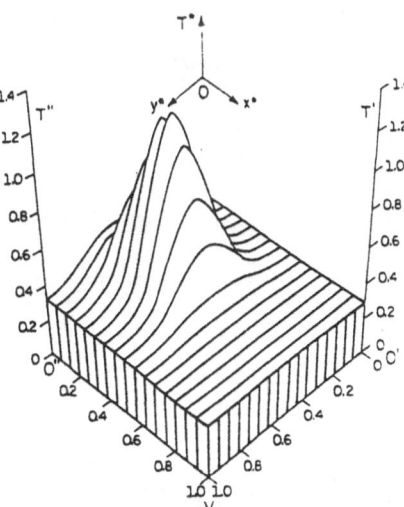

Figure 3. Surface temperature distribution along $y^* = 0.5$ at $t^* = 0.5$ for laser beam of power 600 W and moving in the x direction at a constant velocity 0.25 cm/s relative to the substrate.

It can be seen from this figure that the shape of the surface temperature field has Gaussian-like structure because of the consideration of Gaussian laser beam as the source of heat in this study. The temperature and length of the heated zone ahead of the laser beam in the x* direction are found to increase as t* increases. This is so because at a low scanning speed the Fourier number (Fo) is large, that is, the conduction rate is higher than the heat storage rate. Consequently, the substrate material which is in front of the laser beam is heated up due to the heat conducted away from the laser heated spot. Hence, the laser energy is progressively imparted to points on the substrate which are at higher temperatures than the preceding points. For the very same reason, the laser heated zone in the y* direction increases as t* increases for low scanning speed.

4.1.2. *Film Width*
The knowledge of the width of the laser heated zone is very important in LCVD. The chemical reaction that generates the film forming material will take place wherever the temperature is more than or equal to the chemical reaction temperature. As explained above, the width of the heated zone will increase as the laser beam scans the substrate at a low scanning speed. This means that the width of the zone, to be referred to as the chemically reactive zone, over which the temperature can be larger than or equal to the film forming chemical reaction temperature will increase progressively along the scanning direction for low scanning speed of the laser beam. Thus, the film width will not be uniform for low scanning speed of the laser beam as deposition progresses. On the other hand, if the scanning speed is high the Fourier number will be small, that is, the conduction rate will be lower than the heat storage rate. This reduces the area of the chemically reactive zone due to less conduction of heat from the laser heated spot. Because of this, a narrow film of constant width can be deposited on the substrate by increasing the scanning speed of the laser beam. This concept is reflected in the results presented in Figure 4, which shows the width

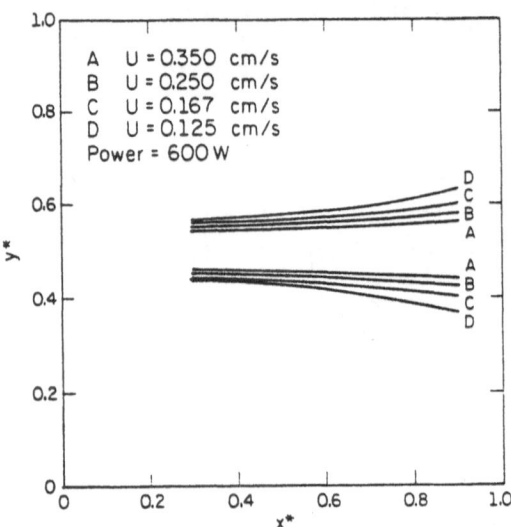

Figure 4. Variation of the width of the heated zone at the substrate surface along the laser beam scanning direction for various scanning speeds relative to the substrate and for laser power 600 W.

of the chemically reactive zone and its variation in the scanning direction for various scanning speeds and laser powers. The chemically reactive zones are symmetrical around the line, y* = 0.5 because the laser beam is located at y* = 0.5 in this study and are bounded by the curves, A, B, C, and D for the scanning speeds 0.35, 0.25, 0.167, and 0.125 cm/s, respectively. It can be seen that the chemically reactive zone becomes wider and less uniform in width as the scanning speed decreases.

4.1.3. *Processing Conditions*

It is established from Figure 4 that the chemically reactive zone becomes narrower as the laser scanning speed increases. However, there will be a critical scanning speed at which the two boundary curves of the chemically reactive zone will collapse into one, giving rise to the narrowest possible film deposition region. With any other scanning speed higher than this critical speed, the width of the chemically reactive zone cannot be reduced any further. So the film deposition process has to be operated at a scanning speed lower than the critical speed. The critical scanning speed is

defined as the one at which the nondimensional peak temperature T_p^* is unity, where

the peak temperature refers to the temperature at the center of the laser beam on the

top surface of the substrate. The line $T_p^* = 1$ is referred to as the line of the

narrowest chemical reaction zone in Figure 4. This figure is plotted on the

logarithmic scales which show the linear variation of the peak temperature (T_p^*)

with the laser scanning speed (U) for different powers of the laser beam at various locations on the top surface of the substrate. The points of intersections of the line of the narrowest chemical reaction zone with the curves of Figure 5 give the critical

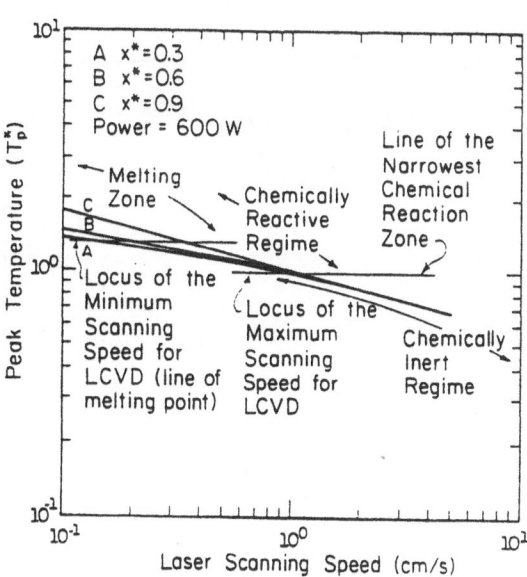

Figure 5. Variation of the peak temperature with laser scanning speed relative to the substrate at various locations on the top substrate surface for laser power 600 W.

scanning speed of the laser beam. The region which is to the right of the critical speed is referred to as the chemically inert regime because the operating conditions of this region do not raise the surface temperature of the substrate to the film-forming chemical reaction temperature. The region to the left of the critical speed is referred to as the chemically reactive regime where the operating conditions are such that films of finite width can be deposited. However, the surface temperature of the substrate can reach its melting temperature at a low scanning speed for a given operating condition as shown in Figure 5.

4.1.4. *Relationships among Control Parameters*

Figure 6 represents the variation of the peak temperature with laser power at various locations on the surface of the substrate for different scanning speeds of the laser beam. This figure shows that the peak temperature increases as the scanning speed decreases for a given operating condition. It should be observed that the peak temperature (T_p^*) varies linearly with laser power (P) on linear scales and with U on logarithmic scales in the chemically reactive regime. These relations can be expressed as

$$T_p^* = \gamma \, U^\delta$$

at different locations on the top surface of the substrate for various powers of the laser beam, and

$$T_p^* = \eta \, P + \beta$$

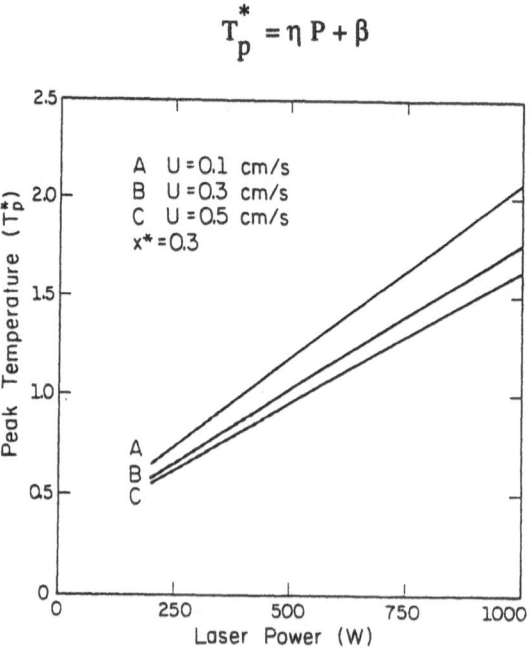

Figure 6. Variation of the peak temperature with laser power for various scanning speeds relative to the substrate at x* = 0.3 on the top surface of the substrate.

at different locations on the top surface of the substrate for various scanning speeds of the laser beam.

The values of γ, δ, η, and β are given in Tables 1 and 2 for various operating conditions.

Table 1. Values of γ and δ for the expression $T_p^* = \gamma U^\delta$

Power (W)	x*	γ	δ
	0.3	0.985	-0.141
600	0.6	1.006	-0.163
	0.9	1.006	-0.244
	0.3	1.101	-0.146
700	0.6	1.125	-0.169
	0.9	1.127	-0.253
	0.3	1.217	-0.150
800	0.6	1.244	-0.174
	0.9	1.247	-0.259

It should be noted that the expression, $T_p^* = \gamma U^\delta$ is applicable in the chemically reactive regime. Since the substrate temperature decreases as the scanning speed increases, there will be a critical scanning speed, say U^*, at and above which the substrate temperature will remain at its initial temperature. So the slopes of the curves of Fig. 5 will be zero for scanning speeds higher than U^*. This physical aspect is not reflected by the equation $T_p^* = \gamma U^\delta$ because this expression is obtained from the results of the chemically reactive regime.

Table 2. Values of η and β for the expression $T_p^* = \eta P + \beta$

x*	U(cm/s)	η	β
	0.1	0.1783(-2)*	0.2925
0.9	0.2	0.1480(-2)	0.2924
	0.5	0.1342(-2)	0.2925
	0.1	0.1955(-2)	0.2925
0.6	0.3	0.1574(-2)	0.2924
	0.5	0.1413(-2)	0.2925
	0.1	0.2461(-2)	0.2924
0.9	0.3	0.1747(-2)	0.2925
	0.4	0.1486(-2)	0.2924

*The numbers inside parentheses are order of magnitude exponents.

It should be noted that when P = 0, the peak temperature must be equal to the initial temperature of the substrate. This is indeed the case as evident by the values of β in Table 2 because the initial temperature (T_i) of the substrate and the thermal decomposition temperature (T_d) of the film-forming chemical reaction are taken to be 343 K and 1173 K, respectively, in this study for which the non-dimensional initial temperature of the substrate is 0.2925.

4.2. RESULTS FOR TITANIUM FILM DEPOSITION

The mathematical model of section 3.2 is used for carrying out the mass transfer analysis during LCVD. Various parameters such as the wavelength of the laser, laser power, shape of the laser beam, thermophysical and optical properties of the substrate and the depositing film, partial pressure of the species inside the LCVD chamber can affect the deposition process. For the results concerning the Ti film deposition, the laser beam radius is taken to be 1.537 mm at the top surface of the substrate and L, W, and H are, respectively, 1 cm, 1 cm, and 14.605 cm. d_{x0}, d_{xL}, d_{y0}, and d_{yW} are taken to be equal to 10.93 cm. The center of the laser beam is located at x = 0.5 cm, y = 0.5 cm, and z = 0, that is at the center of the top surface of the substrate and results are obtained by varying x and keeping y and z fixed at 0.5 cm and 0, respectively. The Arrhenius constants k_{0j}, j = 1, 2, 3 for reactions E, F, and G are taken to be 6.5 x 10^3 cm/s, 8.7 x 10^3 cm/s, and 1.1 x 10^4 cm/s, respectively, which are calculated by using an equation from Johnston (1966). Although the solution of the mass transfer problem has been obtained for any order of the chemical reactions, E, F, and G; results are presented for $n_1 = n_2 = n_3 = 1$. The Damkohler numbers De, Df, and Dg for the chemical reaction E, F, and G, respectively, have been calculated at the thermal decomposition temperature (1173K) of the reaction G. Also, when $t_0 = 0$, the initial distribution of various species is taken to be $f_1(x,y,z) = C_{01}$; $f_i(x,y,z) = 0$ for i = 2, 3, 4, and 5; and $C_{i\infty}$ is taken to be C_{01} for i = 1, and zero for all other values of i.

4.2.1. *Volcano-Like Ti Film*
Figure 7 represents the variation of the deposited Ti film thickness with x for various

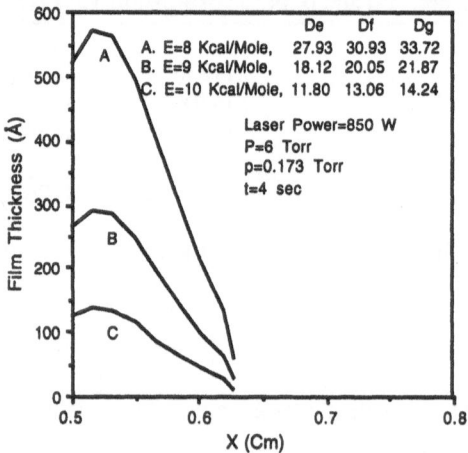

Figure 7. Spatial variations of the thickness profile and the morphology of titanium deposits with activation energies.

parameters as shown in the figure. The complete morphology of the deposits can be obtainedby rotating the film thickness profile around the vertical axis. The effects of the activation energy E are examined this figure. It can be seen from this figure that the film thickness decreases as the activation energy increases. For a given value of the total pressure and the partial pressure of TiBr4, the Damkohler number, which is defined as the ratio of the chemical reaction rate to the diffusion rate, decreases as the activation energy increases. We can, therefore, say that the film thickness decreases as the Damkohler number decreases. The thickness profiles are found to have depressions at x = 0.5 cm, that is, at the center of the deposits. This kind of volcanic morphology of the deposits has been reported in many studies as mentioned earlier. Here again, we find that the depression decreases as the Damkohler number decreases. For large Damkohler numbers, the reaction rate is much faster than the diffusion rate and hence there will be depletion of $TiBr_4$ near the center of the deposits as the diffusion-limited chemical reaction progresses which will cause volcano like deposition. More results on Ti film deposition can be found in Kar, et al. (1991).

4.3. RESULTS FOR TITANIUM NITRIDE FILM DEPOSITION

As pointed out earlier, LCVD often yields films with volcano-like shapes. Among the various factors which can cause this type of profile, the low sticking coefficient at the center of the laser heated spots plays a major role. We therefore consider a temperature dependent sticking coefficient for TiN, $\gamma_{TiN}(T)$, defined in such a way that $\gamma_{TiN}(T) = 1$ below a certain temperature T_m, $\gamma_{TiN}(T) = 0$ above a temperature T_M ($T_m < T_M$), and between these two temperatures, the sticking coefficient decreases linearly as the temperature increases, that is

$$\gamma_{TiN}(T) = \begin{cases} 1 \text{ for } T < T_m, \\ 1 + (T_m - T)/(T_M - T_m) \text{ for } T_m \leq T \leq T_M, \\ 0 \text{ for } T > T_M. \end{cases}$$

In this study, we investigate the effect of the sticking coefficient of TiN on the thickness and morphology of the TiN deposits using the above-mentioned temperature dependent sticking coefficient. The results presented in Figures 6 and 7 have been obtained for T_m =1473 K and T_m = 1640 K. Tm is chosen to be 1473 K because it is the one usually quoted as the maximum substrate temperature for which TiN films with good properties can be deposited in CVD reactors. T_M represents a temperature slightly under the substrate melting temperature (1658 K).

4.3.1. *Morphology of TiN Film and Experimental Verification*
Figures 8 and 9 represent the variation of the deposited TiN film thickness with x for various parameters. It should be noted that the titanium nitride, which is produced due to chemical reaction in the central region of the volcano shown in Figure 8, but

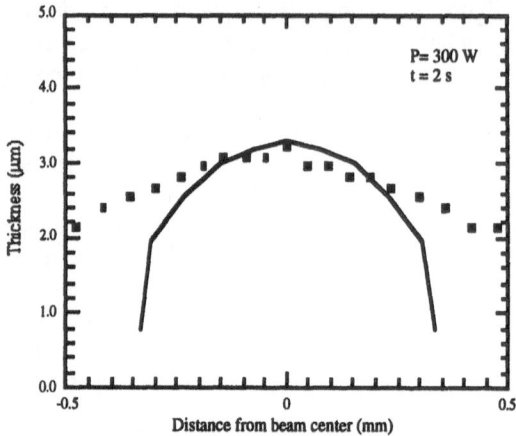

Figure 8. Spatial variation of the thickness of TiN deposits.
■ experimental; ——calculated: r_0 =1.0 mm, T_m =1473 K, T_M =1640 K,
E=12.2 Kcal/mole, k_0 =7.4x10^2 cm/s.

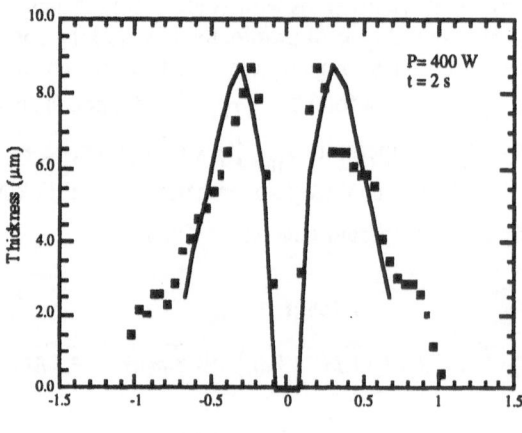

Distance from the beam center (mm)

Figure 9. Spatial variation of the thickness of TiN deposits.
■ experimental; ——calculated: r_0 =1.0 mm, T_m =1473 K, T_M =1640 K,
E=11.0 Kcal/mole, k_0 =8.4x10^2 cm/s.

does not deposit there due to fully nonsticking condition [$\gamma_{TiN}(T)$=0], is considered to deposit entirely on the top of the inner wall of the volcano. This assumption is due to the fact that deposition is not experimentally observed outside the heated spots. In general, an excellent qualitative agreement is found between the theoretical and the experimental results in Figs. 8 and 9. More results on TiN film deposition can be found in Conde, et al. (1992b).

4.3.2. *Optimum Processing Conditions*

Figure 10 shows the TiN film growth rate as a function of the substrate (SiO_2) temperature for various laser powers. Each of the three curves in Figure 10 shows the conditions under which maximum deposition rate can be attained. At the beginning of the LCVD process, the film deposition rate is controlled by the chemical reaction at the laser-heated spot on the substrate surface, which is represented by the AB portion of the curves in Figure 10. The deposition rate attains its maximum value at the point B, and then decreases as shown by the BC portion of the curves where the process is diffusion-controlled.

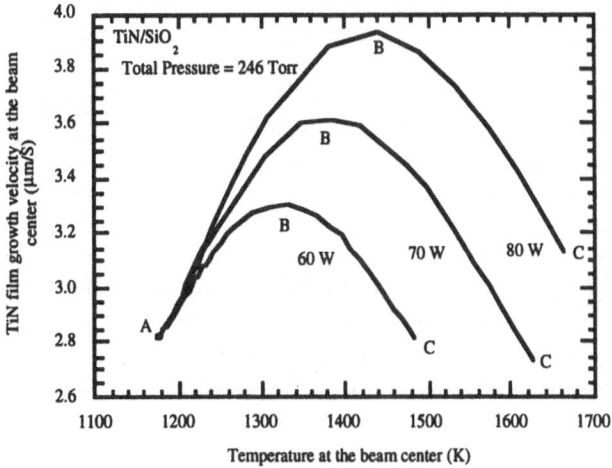

Figure 10. Variation of the TiN dot deposition rate with temperature. AB and BC are, respectively, the chemical- and diffusion-controlled regimes. Laser beam radius is 3 mm. p_{TiCl_4} = 7 Torr, p_{N_2} = 101 Torr, and the rest is H_2.

5. Summary

A three-dimensional and transient heat conduction model is developed in Cartesian coordinates for slabs having finite dimensions, and moving at a constant velocity. The temperature-dependent thermophysical properties of the material of the slab are considered and both convective and radiative losses of energy from the slab to the surrounding medium are taken into account. The laser beam is considered to be Gaussian in shape. Based on these considerations, an analytic expression for the three-dimensional and transient temperature field is obtained.

The surface temperature field has Gaussian-like structure because of the Gaussian laser beam considered in this study. The width of the chemically reactive zone is found to depend on the scanning speed of the laser beam. The width of this zone becomes more uniform and decreases as the scanning speed increases. The critical scanning speed for the narrowest film deposition is determined. Also, the lower and the upper limits of the scanning speeds for LCVD processes are obtained. The peak temperature is found to vary linearly with the lower power for given scanning speeds and the logarithm of the peak temperature is shown to vary linearly with the logarithm of the scanning speed in the chemically active regime for given laser power.

The three-dimensional, transient, and nonlinear mass diffusion is considered in order to model the deposition of Ti film in Cartesian coordinates. The film is produced due to the pyrolytic decomposition of $TiBr_4$ at the surface of SS 304 which is heated with a 10.6 mm continuous wave CO_2 laser beam. The deposited films are found to have volcanic shape for higher Damkohler numbers, that is, when the chemical reaction rate is much faster than the diffusion rate of various species inside the LCVD chamber. This implies that volcanic deposits will be formed for diffusion-limited chemical reactions in which depletion of the reactant occurs due to the rapid chemical reaction rate and the slow diffusion rate.

The deposition of TiN dots by pyrolytic LCVD with a cw TEM_{00} CO_2 laser beam has been modelled by solving the three-dimensional, transient mass diffusion problem using an integral transform technique. In general, excellent agreement has been found between the model predictions and the experimental results. The volcanic shape observed under some conditions has been reproduced by assuming a temperature dependent sticking coefficient for TiN which decreases linearly from one to zero when the temperature increases from 1473 K to 1640 K. Best results are obtained with an activation energy, $E = 11.0 \pm 1.2$ Kcal/mole and an Arrhenius constant, $k_0 = (8.4 \pm 1.6) \times 10^2$ cm/s. The results indicate that the mass transfer model can be used to determine important chemical parameters with a relatively high accuracy.

Acknowledgments

This study is based on a research work supported by the National Science Foundation (Grant Nos. NSF MSM84-12118 and NSF MSS88-15417). O. Conde of the Department of Physics at the University of Lisbon was involved in the work on TiN film deposition.

Nomenclature

k_0' Arrhenius constant.

h_{ij} Heat transfer coefficient at the i=j plane of the substrate. For example, h_{xL} represents the heat transfer coefficient at the x=L plane of the substrate.

$^0C_{H_2}$ Initial concentration of hydrogen.

$^0C_{N_2}$ Initial concentration of nitrogen.

$\overline{T}_a$ Temperature around which the Taylor series expansion is carried out to linearize the boundary conditions.

∇ The Del operator, $\nabla \equiv \hat{i} \dfrac{\partial}{\partial x} + \hat{j} \dfrac{\partial}{\partial y} + \hat{k} \dfrac{\partial}{\partial z}$ in Cartesian coordinates.

$\hat{i}$	Unit vector along the x axis.
$\hat{j}$	Unit vector along the y axis.
$\hat{k}$	Unit vector along the z axis.

α	Thermal diffusivity.
β_{ijk}	Mass transfer coefficient of the i-th species at the j=k plane. For example, b_{ixL} represents the mass transfer coefficient of the i-th species at the plane x=L.
C	Concentration of $TiCl_4$ in region I (see Fig. 2).
C_0	Initial concentration of $TiCl_4$.
C_{01}	Distribution of $TiBr_4$ concentration at time $t_0 = 0$.
C_i	Concentration of the i-th species in units of mass per unit volume. $i = 1, 2, 3, 4,$ and 5 represent the species $TiBr_4$, $TiBr_2$, $TiBr$, Ti, and Br_2, respectively (in Section 4.4).
C_i	Concentration of the i-th species in units of mass per unit volume. $i = H_2$, N_2, and $TiCl_4$ represent the species H_2, N_2, and $TiCl_4$, respectively (in Section 4.5).
C_{iII}	Concentration of the i-th species in region II (see Fig. 2).
$C_{i\infty}$	Concentration of the i-th species at a location far away from region I (see Fig. 2).
$C_p(T)$	Temperature-dependent heat capacity of the substrate.
C_∞	Concentration of $TiCl_4$ at a location far away from region I (see Fig. 2).
D	Diffusion coefficient of $TiCl_4$ in region I (in section 3.3).
D	Y-coordinate of the center of the laser beam on the top surface of the substrate (in section 3.1).
De	Damkohler number for the reaction E, $De = C_{01}^{n_1-1} k_{01} H/D_1$.
Df	Damkohler number for the reaction F, $Df = C_{01}^{n_2-1} k_{02} H/D_2$.
Dg	Damkohler number for the reaction G, $Dg = C_{01}^{n_3-1} k_{03} H/D_3$.
D_i	Diffusion coefficient of the i-th species in region I (in Section 3.2).
D_{iII}	Diffusion coefficient of the i-th species in region II (in section 3.2).
E	Activation energy.
ε	Thermal emissivity of the substrate.
E_j	Activation energy of the j-th reaction. $j = 1, 2,$ and 3 refer to the reactions E, F, and G, respectivly (in Section 3.2).
γ	Sticking coefficient of $TiCl_4$ at the substrate surface.
γ_{TiN}	Sticking coefficient of TiN at the substrate surface.
H	Thickness of the substrate.
k(T)	Reaction rate constant (in Section 3.3).
k(T)	Temperature-dependent thermal conductivity of the substrate.
k_0	Modified Arrhenius constant, $k_0 = k_0' {}^0C_{H_2} {}^0C_{N_2}^{1/2}$

k_{0j}	Arrhenius constant for the j-th reaction. $j = 1, 2,$ and 3 refer to the reactions E, F, and G, respectivly (in section 3.2).
k_i	Surface reaction rate constant with respect to the concentration of the i-th species. $i = 1, 2,$ and 3 represent the species $TiBr_4$, $TiBr_2$, and $TiBr$, respectively.
L	Length of the substrate.
m_i	Molecular weight of the i-th species $i = 1, 2, 3, 4,$ and 5 represent the species $TiBr_4$, $TiBr_2$, $TiBr$, Ti, and Br_2, respectively .
m_i	Molecular weight of the i-th species. $i = 1,$ and 2 represent $TiCl_4$ and TiN, respectively (in Section 3.3).
n_j	Order of the j-th reaction with respect to the j-th species. $j = 1, 2,$ and 3 refer to the reactions E, F, and G, respectively (see Section 3.2). Also $j = 1, 2,$ and 3 represent the species $TiBr_4$, $TiBr_2$, and $TiBr$, respectively .
P	Power of the incident laser beam.
P	Total pressure inside the LCVD chamber (in section 4.2.1).
p	partial pressure of $TiBr_4$ inside the LCVD chamber (in section 4.2.1).
P_{H_2}	Partial pressure of hydrogen inside the deposition chamber.
P_{N_2}	Partial pressure of nitrogen inside the deposition chamber.
R	Reflectivity
r	Radial distance from the center of the laser beam on the top surface of the substrate.
$\rho(T)$	Temperature-dependent density of the substrate.
r_0	Laser spot radius at the substrate surface at $1/e^2$ point.
γ_i	Sticking probability of the i-th species at the substrate surface.
R_{TiN}	Mass of TiN produced per unit area of the substrate surface per unit time.
σ	Stefan-Boltzmann constant.
S_i	Reaction coefficient of the i-th species (Mole number for the i-th species in the chemical reaction). $1 \le i \le q$, $q + 1 \le i \le q+p$, and $q+p+1 \le i \le q+p+n$ represent the reactant, product, and inert species, respectively, that is, $\sum_{i=1}^{q} S_i X_i = \sum_{i=1}^{p} S_{q+i} X_{q+i}$.
T	Temperature.
t	Time variable.
t_0	Initial time.
$T(x,y,0,t)$	Temperature at the top surface of the substrate at any point (x,y) at time t.
T_a	Ambient temperature inside the LCVD chamber.
T_D	Threshold temperature for the chemical reaction (3.3.1).
T_d	Dissociation temperature for pyrolytic LCVD (K).
T_i	Initial temperature of the substrate.
T_M	Upper limit of temperature for the linear variation of $\gamma_{TiN}(T)$.

T_m	Lower limit of temperature for the linear variation of $\gamma_{TiN}(T)$.
U	Velocity of the laser beam with respect to the substrate.
W	Width of the substrate.
x	Distance along the x axis in Cartesian coordinates.
y	Distance along the y axis in Cartesian coordinates.
z	Distance along the z axis in Cartesian coordinates.
$f_i(x,y,z)$	Concentration distribution of the i-th species in region I (see Fig. 1) at the initial time t_0.

References

Abraham, E., and Halley, J. M., 1987, "Some Calculations of Temperature Profiles in Thin Films with Laser Heating," *Appl. Phys.*, Vol. A42, p. 279.

Allen, S. D., 1981, *J. Appl. Phys.*, Vol. 52, p. 6501.

Allen, S. D., Goldstone, J. A., Stope, J. P., and Jan, R. Y., 1986, *J. Appl. Phys.* Vol. 59, p. 1653.

Allen, S. D., Jan, R. Y., Edwards, R. H., Mazuk, S. M., and Vernon, S. D., 1984, "Optical and Thermal Effects in Laser Chemical Vapor Deposition," *Proc., SPIE on Laser Assisted Deposition, Etching, and Doping*, S. D. Allen, ed., SPIE-The Int. Soc. for Optical Engineering, Washington, Vol. 459, pp. 42-48.

Allen, S. D., Trigubo, A. B., and Jan, R. Y., 1983, *Mater. Res. Soc. Symp. Proc.*, Vol. 17, p. 207.

Bäuerle, D., 1986, *Chemical Processing with Lasers*, Springer Ser. in Mater. Sci., Vol. 1, Springer, Berlin.

Bell, A. E., 1979, *RCA Review*, Vol. 40, p. 295.

Bird, R. B., Stewart, W. E., and Lightfoot, E. N., 1960, *Transport Phenomena*, John Wiley and Sons, New York.

Bloom, J., and Giling, L. J., 1978, in *Current Topics in Materials Science*, E. Kaldis, ed., North-Holland, Amsterdam, Chapter 4.

Boyd, R. D., and Vest, C. M., 1975, "Onset of Convection Due to Horizontal Laser Beams," *Appl. Phys. Lett.*, Vol. 26, pp. 287-288.

Breinan, E. M., and Kear, B. H., 1983, in *Proc., Laser Mat. Proc.*, M. Bass, ed., North-Holland, Amsterdam, pp. 235-296.

Burgener, M. L., and Reedy, R. E., 1982, "Temperature Distributions Produced in a Two-Layer Structure by a Scanning CW Laser or Electron Beam," *J. Appl. Phys.*, Vol. 53, pp. 4357-4363.

Calder, I. D., and Sue, R., 1982, "Modeling of CW Laser Annealing of Multilayer Structures," *J. Appl. Phys.*, Vol. 53, pp. 7545-7550.

Carslaw, H. S., and Jaeger, J. C., 1986, *Conduction of Heat in Solids*, 2nd ed., Clarendon, Oxford, pp. 10-11.

Choi, M., Baum, H. R., and Greif, R., 1987, *J. Heat Transfer*, Vol. 109, p. 642.

232

Chou, W. B., Azer, M. N., and Mazumder, J., 1989, "Laser Chemical Vapor Deposition of Ti from TiBr4," *J. Appl. Phys.*, Vol. 66, pp. 191-195.

Cline, H. E., and Anthony, T. R., 1977, *J. Appl. Phys.*, Vol. 48, p. 3895.

Coltrin, M. E., Kee, R. J., and Miller, J. A., 1984, *J. Electrochem. Soc.* Vol. 131, p. 425.

Coltrin, M. E., Kee, R. J., and Miller, J. A., 1986, *J. Electrochem. Soc.*, Vol. 133, p. 1206.

Conde, O., Ferreira, M. L. G., Hochholdinger, P., Silvestre, A. J., and Vilar, R., 1992a, "CO_2 Laser Induced CVD of TiN," *Appl. Surf. Sci.*, Vol. 54, pp. 130-134.

Conde, O., Kar, A., and Mazumder, J., 1992b, "Laser Chemical Vapor Deposition of TiN Dots: A Comparison of Theoretical and Experimental Results," *J. Appl. Phys.*, Vol. 72, pp. 754-761.

Conde, O., Mariano, J., Silvestre, A. J., and Vilar, R., 1990, *Proc., 3rd European Conf. on Laser Treatment of Materials*, H. W. Bergmann and R. Kupfer, eds., Sprechsal Publ. Group, Coburg, p. 145.

Copley, S. M., 1988, "Mass Transport During Laser Chemical Vapor Deposition," *J. Appl. Phys.*, Vol. 64, pp. 2064-2068.

de Heer, J., 1986, *Phenomenological Thermodynamics with Applications to Chemistry*, Prentice-Hall, Inc., Englewood Cliffs, New Jersey, p. 60.

El-Adawi, M. K., and Elshehawey, E. F., 1986, "Heating a Slab Induced by a Time-Dependent Laser Irradiance - An Exact Solution," *J. Appl. Phys.*, Vol. 60, pp. 2250-2255.

Ferrieu, F., and G., Auvert, 1983, "Temperature Evolutions in Silicon Induced by a Scanned CW Laser, Pulsed Laser, or an Electron Beam," *J. Appl. Phys.*, Vol. 54, pp. 2646-2649.

Funaki, K., Uchimura, K., and Kuniya, Y., 1961, Kogyo Kagaku Zasshi Vol. 64, p. 1914.

Gibbons, J. F., and Sigmon, T. W., 1982, "Solid Phase Regrowth," in *Laser Annealing of Semiconductors*, J. M. Poate and J. W. Mayer, eds., Academic Press, New York, Ch. 10, pp. 325-381.

Gordon, J. P., Leite, R. C. C., Moore, R. S., Porto, S. P. S., and Whinnery, J. R., 1965, "Long-Transient Effects in Lasers with Inserted Liquid Samples," *J. Appl. Phys.*, Vol. 36, pp. 3-8.

Hauf, W., and Grigull, U., 1970, in *Advances in Heat Transfer*, J. P. Hartnett and T. F. Irvine, Jr., eds., Academic Press, New York, Vol. 6, pp. 134-366.

Herman, I. P., Hyde, R. A., McWilliams, B. M., Weisberg, A. H., and Wood, L. L., 1983, in *Mat. Res. Soc. Symp.*, R. M. Osgood, S. R. J. Brueck, and H. R. Schlossberg, eds., North-Holland, New York, Vol. 17, pp. 9-18.

Hess, L. D., Forber, R. A., Kokorowski, S. A., and Olson, G. L., 1980, in *Proc., Soc. of Photo-Optical Instrumentation Engineers*, J. F. Ready, ed., Soc. of Photo-Optical Instrumentation Engineers, Washington, Vol. 198, pp. 31-34.

Hirschfelder, J. O., Curtiss, C. F., and Bird, R. B., 1954, *The Molecular Theory of Gases and Liquids*, Wiley, New York.

Holstein, W. L., 1992, "Design and Modeling of Chemical Vapor Deposition Reactors," *Prog. Crystal Growth and Charact.*, Vol. 24, pp. 111-211.

INCO Alloys International, Huntington, WV 25720, data supplied.

Incropera, F. P., and Dewitt, D. P., 1985, *Fundamentals of Heat and Mass Transfer*, 2nd ed., John Wiley and Sons, 1985, p. 8.

Itoh, H., Kato, K., and Sugiyama, K., 1986, *J. Mater. Sci.*, Vol. 21, p. 751.

Jang, H., and Paik, Y. H., 1983, *J. Kor. Inst. Met.*, Vol. 21, p. 38.

Jensen, K. F., 1987, *Chem. Eng. Sci.*, Vol. 42, p. 923.

Johnston, H. S., 1966, *Gas Phase Reaction Rate* Theory, Ronald Press, New York, p. 138.

Kant, R., 1988, *J. Appl. Mech.*, Vol. 55, p. 93.

Kar, A., and Mazumder, J., 1987, *J. Appl. Phys.*, Vol. 61, p. 2645.

Kar, A., and Mazumder, J., 1988, *Acta Metal.*, Vol. 36, p. 702.

Kar, A., and Mazumder, J., 1989, "Three-dimensional Transient Thermal Analysis for Laser Chemical Vapor Deposition on Uniformly Moving Finite Slabs," *J. Appl. Phys.*, Vol. 65, pp. 2923-2934.

Kar, A., Azer, M. N., and Mazumder, J., 1991, "Three-dimensional Transient Mass Transfer Model for Laser Chemical Vapor Deposition of Titanium on Stationary Finite Slabs," *J. Appl. Phys.*, Vol. 69, pp. 757-766.

Kato, A., and Tamari, N., 1975, *J. Cryst. Growth*, Vol. 29, p. 55.

Kim, D. M., Kwong, D. L., Shah, R. R., and Crosthwait, D. L., 1981, "Laser Heating of Semiconductors – Effect of Carrier Diffusion in Nonlinear Dynamic Heat Transport Process," *J. Appl. Phys.*, Vol. 52, pp. 4995-5006.

Kim, M. S., and Chun, J. S., 1983, *Thin Solid Films*, Vol. 107, p. 129.

Kokorowski, S. A., Olson, G. L., and Hess, L. D., 1955, in *Proc. of Laser and Electron-beam Solid Interactions and Materials Processing*, J. F. Gibbons, L. D. Hess, and T. W. Sigmon, eds., Elsevier North-Holland, New York, pp. 139-146.

Konstantinov, L., Nowak, R., and Hess, P., 1990, *Appl. Surf. Sci*, Vol. 46, p. 102.

Kwong, D. L., and Kim, D. M., 1982, "Pulsed Laser Heating of Silicon: The Coupling of Optical Absorption and Thermal Conduction During Irradiation," *J. Appl. Phys.*, Vol. 53, pp. 366-373.

Lax, M., 1977, *J. Appl. Phys.*, Vol. 48, p. 3919.

Lax, M., 1978, *J. Appl. Phys. Lett.*, Vol. 33, p. 786.

Lax, M., 1979, in *Proc., Laser-Solid Interactions and Laser Processing 1978*, S. D. Ferris, H. J. Leamy, and J. M. Poate, eds., American Inst. of Physics, New York, pp. 149-154.

Leidler, K. J., 1987, *Chemical Kinetics*, Harper & Row, New York, pp. 262-265.

234

Liarokapis, E., and Raptis, Y. S., 1985, "Temperature Rise Induced by a CW Laser Beam Revisited," *J. Appl. Phys.*, Vol. 57, pp. 5123-5126.

Mazumder, J., 1987, in *Proc., Interdisciplinary Issues in Mat. Proc. and Manuf.*, S. K. Samanta, R. Komanduri, R. McMeeking, M. M. Chen, and A. Tseng, eds., ASME, New York, pp. 599-630.

Mazumder, J., and Allen, S. D., 1980, in *Proc., Society of Photo-Optical Instrumentation Engineers*, J. F. Ready, ed., (Soc. of Photo-Optical Instr. Eng., Washington), Vol. 198, pp. 73-80.

McWilliams, B. M., Chin, H. W., Herman, I. P., Hyde, R. A., Mitlitsky, F. ,Whitehead, J. C., and Wood, L. L., 1984, "Wafer-Scale Laser Pantography: VI. Direct Write Interconnection of VLSI gate arrays, *Proc., SPIE on Laser Assisted Deposition, Etching, and Doping*, S. D. Allen, ed., SPIE-The Int. Soc. for Optical Engineering, Washington, Vol. 459, pp. 49-54.

Meyer, J. R., Bartoli, F. J., and Kruer, M. R., 1980a, "Optical Heating in Semiconductors," *Phys. Rev. B*, Vol. 21, pp. 1559-1568.

Meyer, J. R., Kruer, M. R., and Bartoli, F. J., 1980b, "Optical Heating in Semiconductors: Laser Damage in Ge, Si, InSb, and GaAs," *J. Appl. Phys.*, Vol. 51, pp. 5513-5522.

Michalski, J., and Wierzchon, T., 1989, *J. Mater. Sci.*, Lett., Vol. 8, p. 779.

Moody, J. E., and Hendel, R. H., 1982, *J. Appl. Phys.*, Vol. 53. p. 4364.

Motojima, S., and Kohno, M., 1986, *Thin Solid Films*, Vol. 137, p. 59.

Motojima, S., and Mizutani, H., 1988, *J. Mater. Sci.*, Vol. 23, p. 3435.

Moylan, C. R., Baum, T. H., and Jones, C. R., 1986, *Appl. Phys.*, Vol. A40, p. 1.

Nissim, Y. I., A. Lietoila, R. B. Gold, and J. F. Gibbons, 1980, *J. Appl. Phys.*, Vol. 51, p. 274.

Piglmayer, K., Doppelbauer, J., and Bäuerle, D., 1984, *Mat. Res. Soc. Symp.*, by A. W. Johnson, D. J. Ehrlich, and H. R. Schlossberg, eds., Elsevier, New York, Vol. 29, pp. 47-54.

Rong, C. Z., Sheng, Du Y., and Fang., M. H., 1989, *Surf. Eng.*, Vol. 5, p. 315.

Sadahiro, T., Cho, T., and Yamaya, S., 1977, *J. Japan Inst. Met*, Vol. 41, p. 542.

Sanders, D. J., 1984, "Temperature Distributions Produced by Scanning Gaussian Laser Beam," *Appl. Opt.*, Vol. 23, pp. 30-35.

Singh, J. and Mazumder, J., 1987, *Acta Metal.*, Vol. 35, p. 1995.

Sitrl, E., Hunt, L. P., and Sawyer, D. H., 1974, *J. Electrochem. Soc.*, Vol. 121, p. 919.

Skouby, D. C. and Jensen, K. F., 1988, *J. Appl. Phys.*, Vol. 63, p. 198.

Sneddon, I. N., 1972, *The Use of Integral Transforms*, McGraw-Hill Book Co., New York.

Yarmoff, J. A., and McFeely, F. R., 1988, *J. Appl. Phys.*, Vol. 63, p. 5213.

Zeiger, H. J., and Ehrlich, D. J., 1989, "Lateral Confinement of Microchemical Surface Reactions: Effects on Mass Diffusion and Kinetics," *J. Vac. Sci. Technol.*, Vol. B7, pp. 466-480.

Zeiri, Y., Atzmony, U., Bloch, J., and Lucchese, R. R., 1991, *J. Appl. Phys.*, Vol. 69, p. 4110.

FUNCTIONALLY GRADIENT COATING LAYERS PRODUCED BY LASER ALLOYING / CLADDING

J.H.ABBOUD, D.R.F.WEST, AND R.D.RAWLINGS

Department of Materials, Imperial College

London SW7 2BP, United Kingdom

ABSTRACT

Investigations have been made of functionally gradient materials (FGMs) produced on Ti-based and Ni-based substrates by depositing a series of superimposed alloyed or clad layers using a CO_2 laser with a continuous powder feed facility. With titanium substrates the objective has been to enhance the resistance to high temperature oxidation and to erosion through the formation of FGM coating layers containing aluminium. In the alloying process, aluminium powder is fed into the laser generated melt pool, while in cladding the feed mixture contains aluminium and titanium powder. Layers with Al contents up to 44at% covering the range of Ti_3Al and TiAl have been produced. Also composite layers have been produced incorporating partially dissolved ceramic particles (eg TiB_2) in a titanium aluminide matrix. Observations are reported of microstructural and compositional features of the FGMs together with data on hardness, oxidation, and erosion resistance.

J. Mazumder et al. (eds.), Laser Processing: Surface Treatment and Film Deposition, 237–254.
© 1996 *Kluwer Academic Publishers.*

1. INTRODUCTION

The term functionally gradient materials (FGMs) is applied to components whose composition and structure vary progressively as a function of position. Recently high power CO_2 lasers in conjunction with a powder feed technique, have been used to produce FGMs by the successive deposition of several alloyed / clad layers superimposed in a direction normal to the substrate [1-4]. The present paper reviews results recently reported on producing functionally gradient materials based on the Ti-Al and Ni-Al systems using alloying or cladding technique by superimposition of two or three alloyed / clad layers; a powder feed technique was used with a change in process parameters, normally powder flow rate or powder composition, for successive layers.

2. EXPERIMENTAL PROCEDURE

A 2kW continuous wave CO_2 laser operated between 1.7 to 1.85kW with 3mm beam diameter has been used. In alloying, commercial pure aluminium powder of average particle size of ~ 75µm was blown into the laser melted zone of a near α (IMI-685 Ti) alloy while in cladding a mixture of titanium ~ (100µm) and aluminium powder was blown into the melted zone of commercial purity (CP) Ti substrate. The process was carried out with an effective argon shrouding device to minimize oxygen contamination.

After laser processing, transverse sections were prepared and standard methods of metallography applied to obtain compositional analysis of each alloyed / clad layer using energy dispersive X ray spectrometry (EDS) with ZAF correction. Optical microscopy, scanning electron microscopy (SEM) and transmission electron microscopy (TEM) observations were made. Oxidation tests were carried out in air for a range of times and temperatures and the oxide scale thickness was measured on transverse sections. Erosion tests were carried out using quartz sand as the erodant.

3. RESULTS AND DISCUSSION

3.1 ALLOYING OF NEAR α Ti 685 ALLOY WITH ALUMINIUM

Fig.1 shows the composition profile of the transverse section of two alloyed layers containing ~ 25 and 46at%Al, the latter resulting from a higher Al powder flow rate than the former; the data shows reasonable compositional uniformity in the layer.

Fig.2 a and b and Table 1 show data for two superimposed alloyed layers. The interface between the heat affected zone and the first layer was planar, while it was epitaxial between the first and the second layer.

Table 1: Data for FGM produced by two superimposed alloy layers.

Layer No.	Dimensions, mm		Composition, wt%			Hv,500g	Structure
	Width	Thickness	Ti	Al	Zr	kg/mm^2	
1	2.11	0.15	Bal.	30	3.8	525$\pm$25	$\alpha_2+\gamma$
2	1.8	0.25	Bal.	36	3.6	410$\pm$10	predominantly γ
HAZ	–	0.1	Bal.	6	5	380$\pm$10	α'
Substrate	–	–	Bal.	6	5.0	350$\pm$10	$\alpha+\beta$

3.2 CLADDING OF (Ti-6Al-4V+Al) ALLOY USING PREMIXED POWDERS

Several Ti-Al-V clad layers were produced by blowing a mixture of powders of Ti-6Al-4V alloy and Al at a certain ratio into the laser melted zone of the CP Ti substrate (Fig.3). Some dilution of the initial powder mixture with the molten titanium substrate (typically ~10%) occurred and this can be controlled by appropriate selection of powder flow rate and traverse speed. EDS analysis of the clad layers showed a reasonably homogeneous composition with fluctuations of ~ $\pm$1wt%. Light microscopy showed coarse prior β grains of width about 0.5-1mm; martensitic features were observed within these grains. The use of polarized light showed a cellular / dendritic structure within the prior β grains. The cell / dendrite secondary arm spacings were between 5 and 10μm which suggests a cooling rate of about 10^3 °C/s.

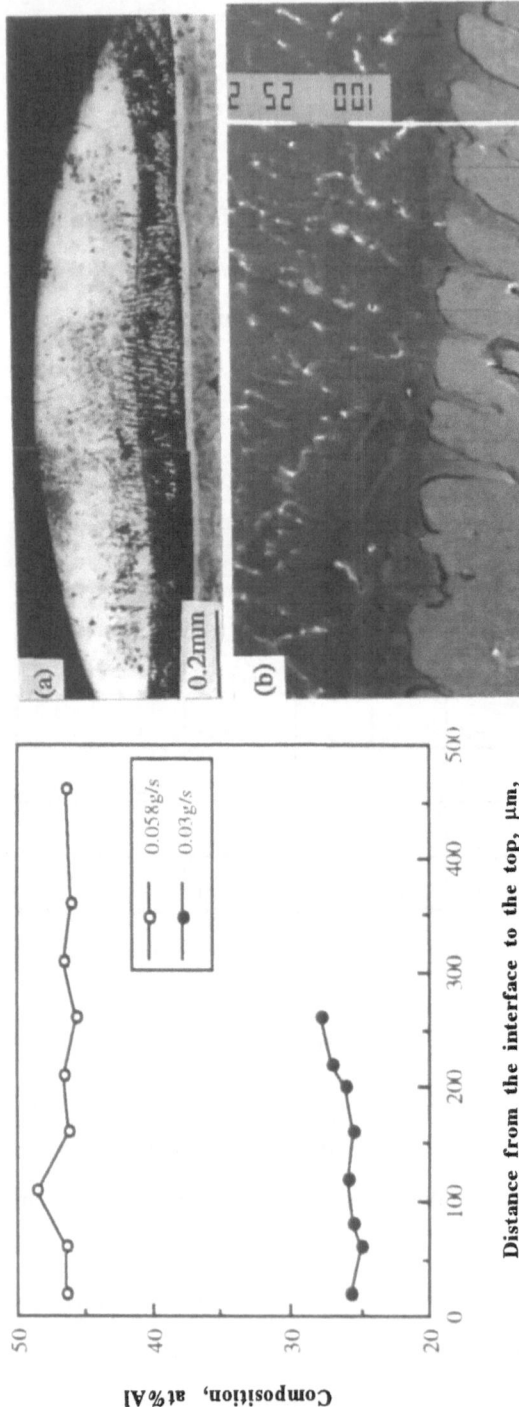

Fig.1: EDS analysis across the thickness of two alloyed layers produced at different flow rate (1.8kW,3mm, 7mm/s).

Fig.2: Two vertically overlapping layers produced by laser alloying, (a) light micrograph,(b) SEM micrograph (processing condition for layer 1:1.8kW,3mm, 0.033g/s, 5mm/s followed by remelting at 3.2 mm/s and for layer 2:1.8kW, 3mm, 0.033g/s, and 7mm/s).

Fig.4a shows a transverse section of two clad layers deposited on each other using the ᵌme laser processing parameters. The clad powder composition was Ti-15wt%Al-3.6wt%V. EDS analysis showed that the Al content was 11wt% and 18wt% in the first layer and the second layer (Fig.4b and Table 2) respectively, showing more dilution in the first layer compared with the second layer. Both layers consisted of coarse prior β grains of 0.2-0.5mm width showing a martensitic structure resulting from the β to α' transformation during cooling.

Table 2: Data for FGM produced by vertical deposition of two clad layers (1.8kW, 3mm, 7mm/s, 0.1g/s).

Layer	Composition (wt%)			Dimension, mm		Hv, 500g
	Ti	Al	V	Width	Thickness	kg/mm^2
1	Bal	11.0	2.8	2.5	0.43	485
2	Bal	18.0	3.5	2.5	0.43	535

The microstructures and the compositions of three deposited clad layers using various powder mixtures and the same laser processing parameters are shown in Fig.5a and Table 3.

Table 3: Data for FGM (wt%) produced by deposition of three superimposed layers, 1.8kW, 3mm, 7mm/s, and 0.1g/s

Layer	Structure	Ti	Al	V	Hv, 500g
1	α'	Bal.	14.5	3.7	500±25
2	α'$_2$	Bal.	17	3.7	550±50
3	α$_2$+(α$_2$+γ)	Bal.	30	2.8	525±25

The martensite in the first layer was fine and in the second layer was coarse. The third layer exhibits a cellular / dendritic structure with cell / dendrite arm spacings between 5 and 10μm. Epitaxial growth was observed at the interfaces between the heat affected zone and the first layer, and between the first and second layer; planar growth was observed between the second and the third layer. TEM showed that the martensite in the first layer was acicular disordered (α') (Fig.5b) while in the second layer it was lath

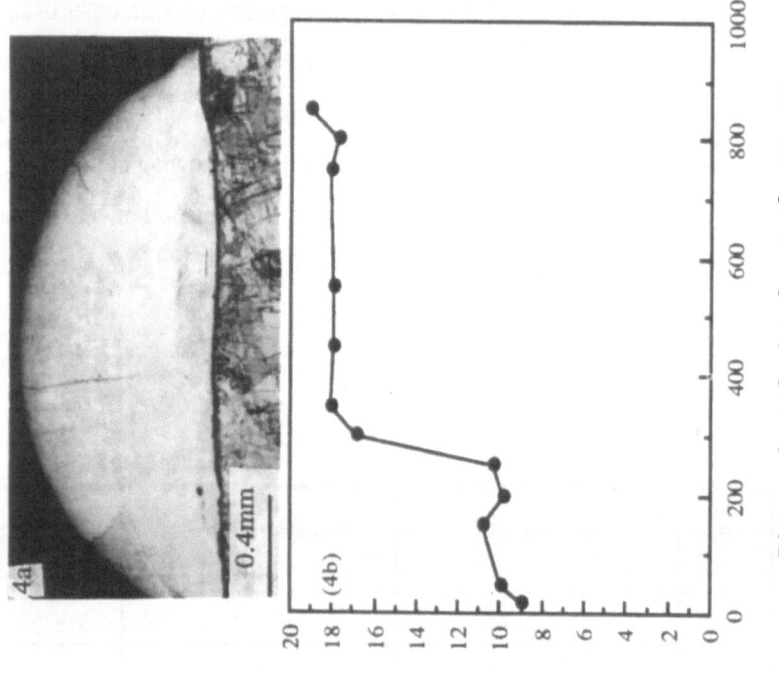

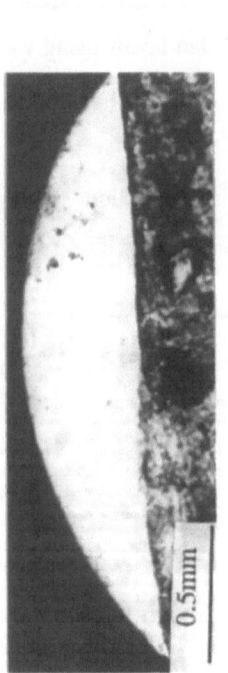

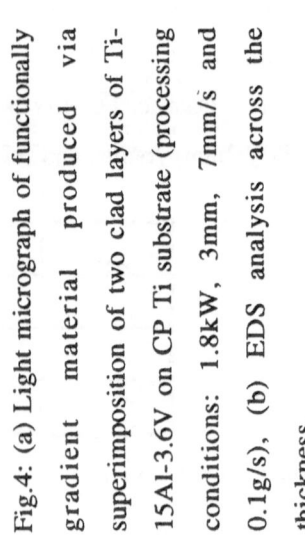

Fig.3: Light micrograph of transverse section of clad layer deposited on commercial purity Ti (processing conditions: 1.8kW, 3mm/s, 7mm/s, and 0.11g/s).

Fig.4: (a) Light micrograph of functionally gradient material produced via superimposition of two clad layers of Ti-15Al-3.6V on CP Ti substrate (processing conditions: 1.8kW, 3mm, 7mm/s and 0.1g/s), (b) EDS analysis across the thickness.

type ordered (α'_2) showing antiphase boundaries APB's (Fig.5c). The third layer showed cells / dendrites of α_2 and intercellular / dendritic regions consisting of a mixture of lamellar $\alpha_2+\gamma$ with spacings of ~ 100-200nm (Fig.5d).

3.3 CLADDING WITH PARTICLE INJECTION

Single track composite surface layers have been produced by injecting either ceramic particles or a mixture of metal powder and ceramic at a certain ratio, into the laser melted zone. In both cases the ceramic particles were partially dissolved leading to the enrichment of the matrix with the components of the ceramic. For example, the injection of SiC or TiB_2 into the melt zone of CP Ti substrate leads to the enrichment of the matrix with Si and C, or B respectively and consequently the formation of Ti_5Si_3, TiC, and TiB, respectively [5,6]. A titanium aluminide composite was produced by blowing a mixture of Ti+10wt%Al+25wt%TiB_2 into the laser melted zone of CP Ti. Porosity was observed in the transverse sections; cracks were also seen on the top surface of the clad. SEM examination showed needle-like particles identified as TiB nucleated at the TiB_2 / matrix interface and within the matrix.

Fig.6a shows transverse section of two superimposed clad layers produced using the same laser processing parameters and two different powder mixtures. The first mixture was Ti-20wt%Al-25wt%TiB_2, and the second mixture Ti-30wt%Al-25wt%TiB_2. There was an increase in average Al content from ~ 25wt% in the first layer to ~35wt% in the second layer as determined by EDS. Partial dissolution of the TiB_2 and the formation of TiB at the interface between TiB_2 and the matrix and within the matrix was observed in both layers (Fig.6b). TEM analysis (Figs.6c and d) of the first layer showed that TiB crystals had a faulted structure. Selected area diffraction from the matrix showed a hexagonal pattern with superlattice reflections. Dark field imaging using superlattice reflections showed antiphase domain boundaries (APB's) which indicates the transformation of α to $\alpha_2+\gamma$ during cooling. The microstructure of the second layer as observed by SEM was interpreted as a mixture of α'_2, γ, and TiB.

Fig.7 illustrates the compositional change during the production of the two layer sample. The feed mixture compositions, designated as X_1 and X_2 (wt%) are shown in the Ti-Al-B compositional triangle, assuming no preferential loss of elements during

244

Fig.5: (a) Light micrograph of functionally gradient material produced via superimposition of three clad layers (processing conditions: 1.8kW, 3mm, 7mm/s, 0.1g/s), (b-d) TEM micrographs of layers 1, 2 and 3, respectively.

processing. The production of layer 1 involves a dilution of ~15% from the substrate and the average layer composition is shown as point A_1 on the line joining Ti and X_1. A_1 lies within the composition triangle corresponding the three constituent TiB_2, TiB and Ti-Al(ss) (solid solution). Microstructural observations showed ~ 15% of undissolved TiB_2; a lever rule calculation (extrapolating the line from TiB_2 through A_1) defines the position of the TiB/Ti-Alss side of the triangle. The TiB+Ti-Al(ss) area of the layer contains ~20wt%Al (M_1) in reasonable agreement with the analysed value of 25wt% Al. A similar procedure can be applied to layer 2, with the average layer composition, A_2 lying on the line joining A_1 and X_2 determined by the dilution, which is ~ 20%. There is ~20wt% of the undissolved TiB_2 leading to an Al content of ~ 34w% (M_2) from the TiB+Ti-Al(ss) mixture, in good agreement with the analysed value of 35wt%Al. This graphical procedure can be used as a guide to the design of FGMs assuming that the appropriate processing conditions can be experimentally determined. Other considerations to be taken into account to refine this procedure are the influence of track geometry and the need to produce partially overlapping, side by side, layers if a large area of FGM is to be build up on a substrate.

3.4 FUNCTIONALLY GRADIENT TITANIUM-ALUMINIUM CLAD PRODUCED USING TWO SEPARATE FEEDERS

As an alternative to using a premixed powder a technique employing two separate feeders has been developed. Fig.8 a and b and Table 4 refer to three superimposed clad layers deposited on CP Ti using constant Ti feed rate and different Al feed rates. EDS analysis showed reasonable homogeneity in the clad layers. The first layer consisted of a dendritic structure interpreted as predominantly γ (TiAl) while the second and the third layer consisted of $TiAl_3$ and Al(ss). The proportion of $TiAl_3$ was much higher in the second layer than in the third layer (Fig.9b).

Table 4: Data for three FGM Ti-Al clad layers deposited on CP Ti using two separate feeders, 1.8kW, 3mm, 5mm/s.

246

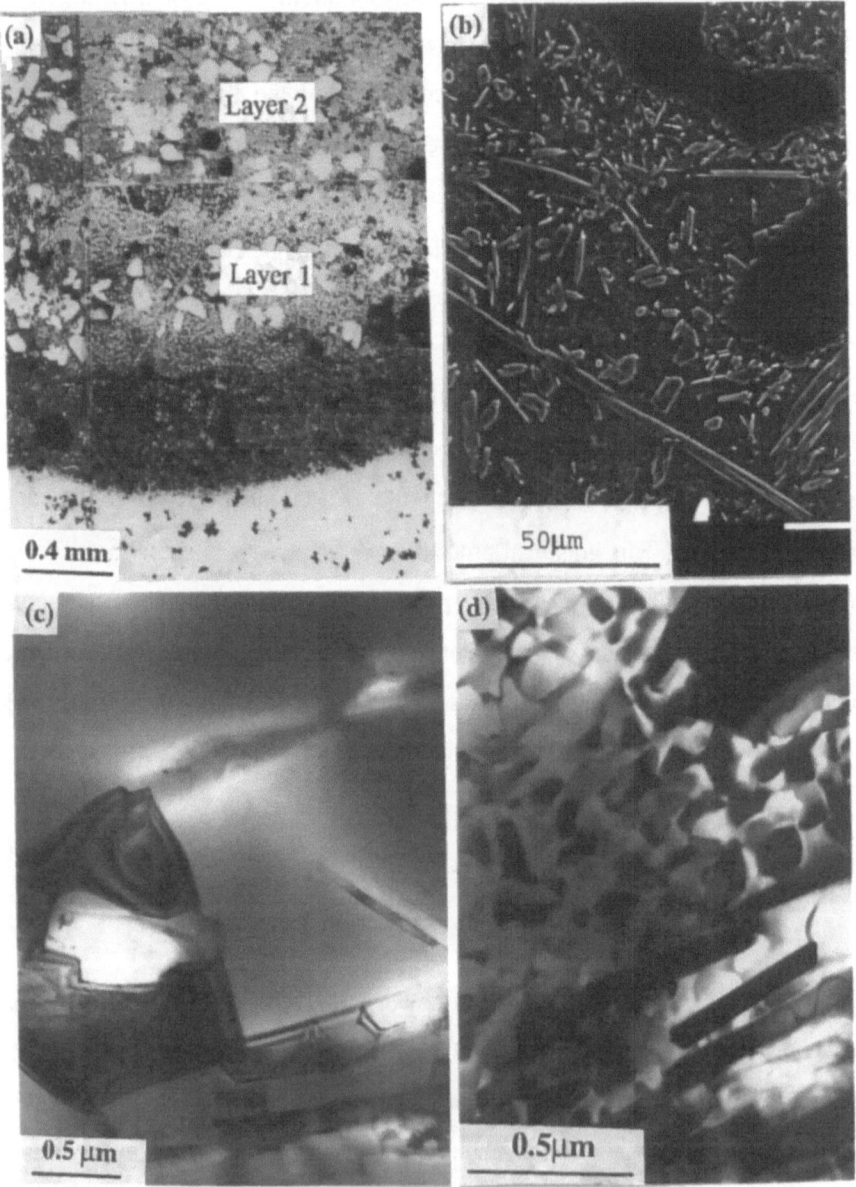

Fig.6 (a) Light micrograph showing transverse section of FGM produced 1 superimposition of two Ti-Al-TiB$_2$ clad layers on CP Ti substrate (processing conditions:1.7kW, 3mm, 5.3mm/s and 0.07 g/s.) , (b) SEM micrograph (first layer), (c-d) TEM micrographs (first layer).

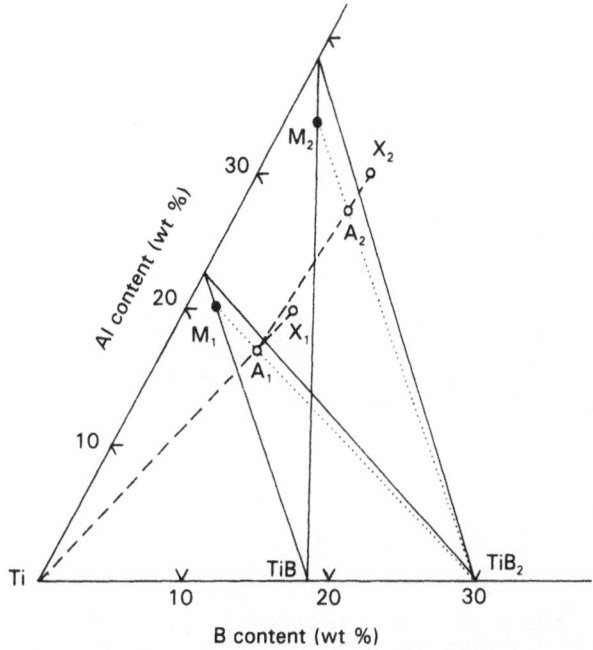

Fig.7 Ti–Al–B composition triangle showing an approximate quantitative representation of a double-clad layer on a Ti substrate: X_1 and X_2 are the compositions of the powder feed mixtures of Ti + Al + TiB$_2$; A_1 and A_2 are the approximate average compositions of clad layers based on estimated dilutions; M_1 and M_2 are the approximate compositions of TiB + Ti–Al solid-solution matrices based on estimated proportions of undissolved TiB$_2$. The subscripts 1 and 2 refer to the two layers.

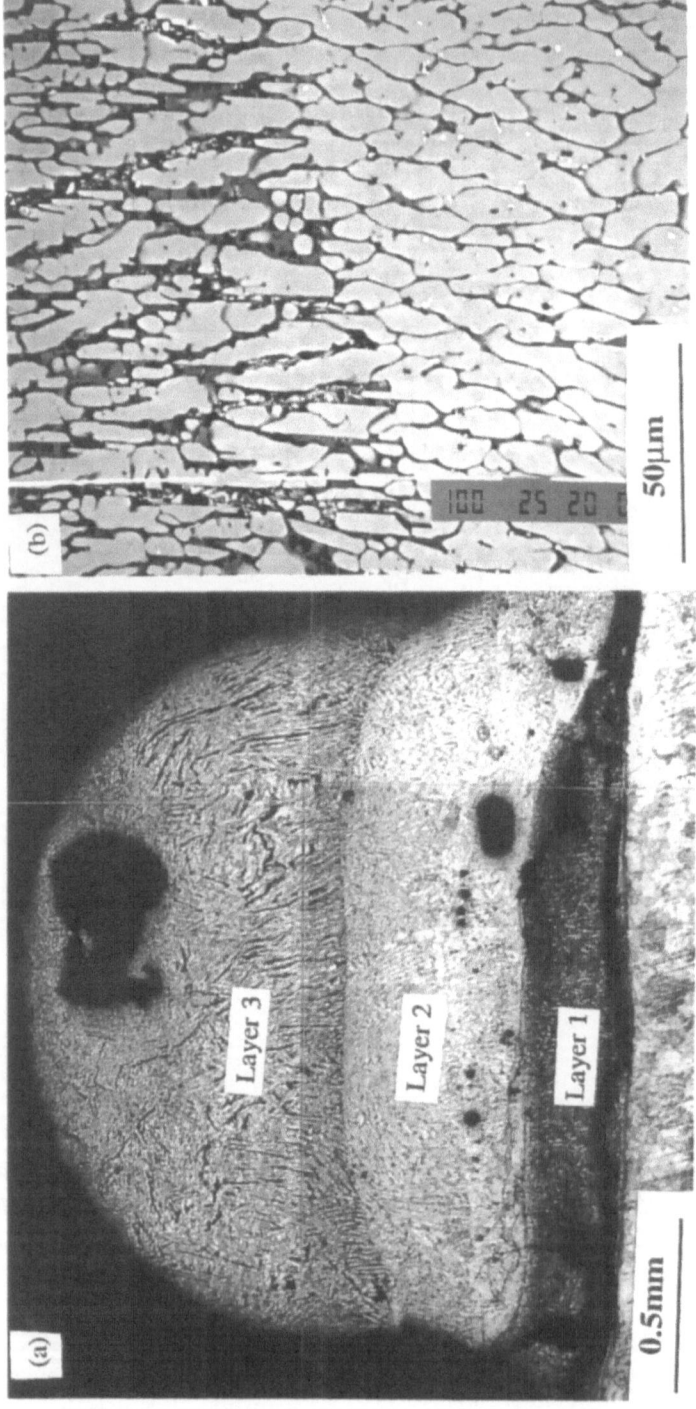

Fig.8: (a) Light micrograph of functionally gradient material produced via superimposition of three Ti-Al clad layers using two separate feeders (processing condition:1.8kW, 3mm, 5mm/s), (b) SEM showing interface between layers 2 and 3.

Layer no.	Ti flow rate	Al flow rate	Composition	
	g/s	g/s	At%Ti	At%Al
1	0.05	0.042	Bal.	60
2	0.05	0.067	Bal.	77.5
3	0.05	0.075	Bal.	92

3.5 CLADDING OF NICKEL SUBSTRATE USING PREMIXED POWDERS

Laser cladding of nickel-aluminide on a Ni-Cr based alloy substrate (IN625) has been carried out by blowing a mixture of nickel and aluminium powder onto the laser melted zone. Fig. 9a shows a single clad layer produced using a powder mixture of Ni-15wt%Al. The clad thickness was nearly 1mm and the width 2mm. Cracks or porosity have not been observed in transverse sections but there were some transverse surface cracks. The structure as observed by SEM consisted of primary dendrites surrounded by an apparently single phase region and also interdendritic regions of eutectic structure (Fig.9b). The dendrites were rich in Al (~17wt%) while the interdendritic regions were rich in Cr (Table 5). The structure is interpreted as consisting of primary β phase with a rim of γ' and interdendritic $\gamma+\gamma'$; TEM observations of the eutectic are illustrated in Fig.9c in which the γ regions have precipitated γ' in the solid state. EDS analysis data of laser produced Ni-Al clad layer is shown in in Table 5.

Table 5: EDS analysis (wt%) of laser produced Ni-Al clad layers

Position	Ni	Al	Cr	Co	Mo	Si	Fe
Average	Bal.	10.65	8.2	3.2	3	2.8	1.2
Dendrites	Bal.	17	4.7	2	-	1.8	1
Interdendritic	Bal.	7	10	4	1.4	4	1.8

Functionally gradient Ni-Al-SiC on a Ni alloy (IN 625) substrate has been produced [1] in a similar way to that described for producing Ti-Al-TiB$_2$ in section 3.3. Three vertically overlapped tracks were produced. The first was deposited from a premixed powder of Al+10wt%SiC; the second using Al+30wt%SiC; and the third layer using Al+50wt%SiC. The average Al content increased from ~ 27.6 near the substrate to 52 wt% in the top layer. Virtually complete solution of SiC particles ocurred in all but the surface region leading to enrichment of the matrix with Si and C. The microstructure varied in a complex way from the first layer to the last layer .

3.6 PROPERTIES ASSESMENT OF TITANIUM BASED MATERIALS

Significant improvement in the oxidation resistance of Ti can be achieved by the alloying of titanium with aluminium [7]. Laser produced Ti-Al alloyed / clad layers with various aluminium contents can enhance the oxidation resistance of CP Ti and near α–Ti alloys [8]. For example, the results of oxidation tests of laser surface Ti-Al alloyed layers produced on IMI 685 substrate oxidised in air at 900°C for 100hr showed a decrease of about five times in the oxide thickness on Ti-50at%Al as compared with untreated substrate; any oxide formed on the surface of Ti-65%Al alloyed layer (Table 6) was too thin to be observed with the technique employed. Composition analysis by EDS analysis showed that the oxide scale was mainly $(Ti, Zr)O_2$ on the Ti-30at%Al layer while it was a mixture of $(Ti,Zr)O_2$ and Al_2O_3 on the Ti-40 and on the Ti-50%Al layer.

Table 6:Thickness (μm) of external and intermediate oxide layers formed on substrate and on laser surface Ti-Al alloyed layers (at%) exposed in air at 900°C

Oxide scale	substrate	Ti-30%	Ti-40%	Ti-50%	Ti-65%
External	50	30	20	10	-
Intermediate	350	40	-	-	-

This result is similar to that previously reported by Hirose et al [9]. More recently oxidation tests have been carried out on laser produced functionally gradient Ti-Al clad layers (three layers) at different times (up to 310hr) and temperatures (800, 900 and 1000°C) [4]. The results showed that at 300hr and 800°C, the oxide thickness was

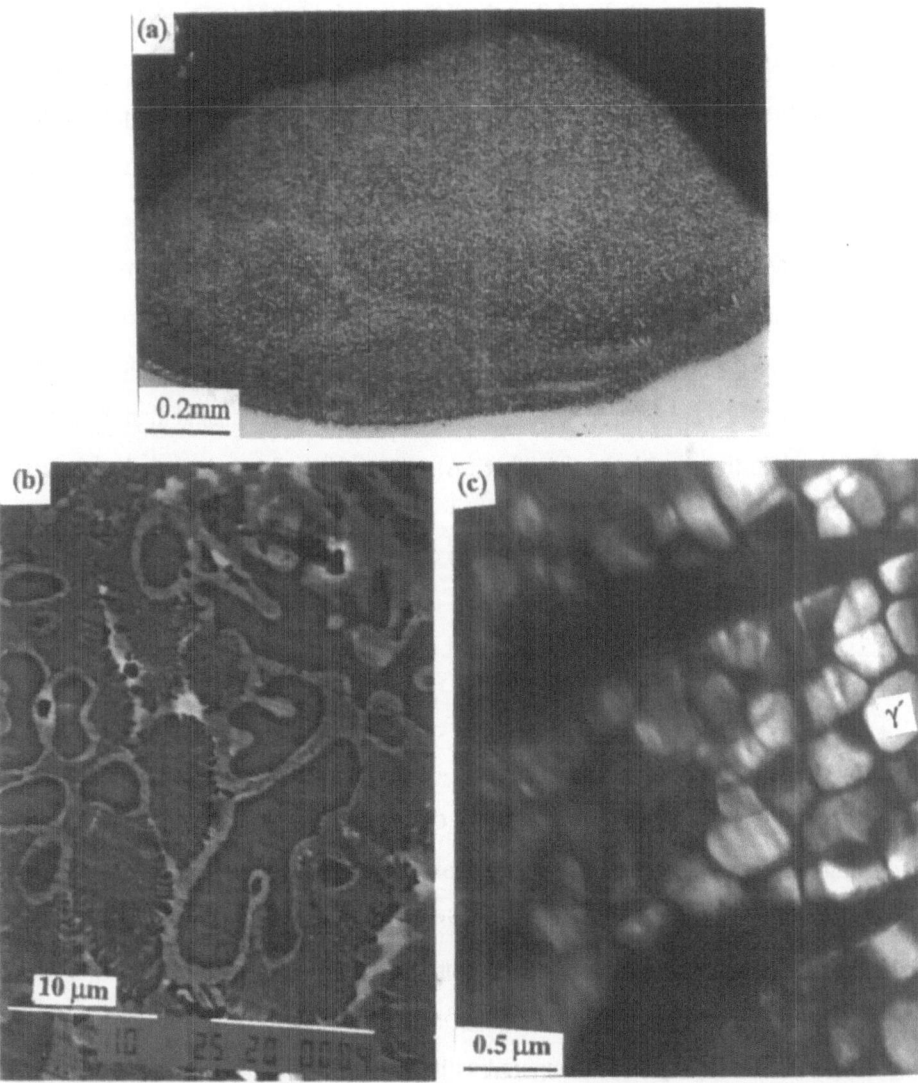

Fig.9: (a) Light micrograph showing transverse section of clad layer deposited into Ni based alloy (1.8kW, 3mm/s, 5mm/s, and 0.3g/s) (b) SEM micrograph and (c) TEM micrograph.

reduced by a factor of ~2, 3 and 9 as the Al content increased from 14.5% to 17% and to 30 wt%Al, in the first, second, and third layers, respectively (Fig.10) .

Laser surface alloying of titanium with aluminium or SiC particle injection led to some beneficial effect on the erosion resistance [7]. A single track from both the alloyed layer and the injected area was subjected to severe erosion testing and the weight loss compared with that of untreated titanium. More than 25% improvement in the erosion resistance was obtained. Also erosion testing on laser produced functionally gradient Ti-Al clad layers (three clad layers superimposed on each other) showed that the second layer (Ti-17wt%Al-3.7wt%V) had a better erosion resistance than the substrate (CP Ti) while the third layer (Ti-30wt%Al-2.8wt%V) had inferior erosion resistance (Fig.11).

4. CONCLUSIONS

1. Surface alloying or cladding utilising a continuous feed of premixed component powders into a laser generated melted pool can be used to produce functionally gradient coating layers on a substrate. The technique involves deposition of superimposed layers with different feed rates. The material consists of essentially discrete layers rather than a smooth gradient of composition and structure.

2. Cladding involving minimum dilution by the substrate is preferable for the production of FGM layers as compared with alloying where dilution is substantial [2].

3. Metal-ceramic composite layers can be produced by incorporating particles of ceramic such as TiB_2 or SiC with the feed mixture and employing laser processing conditions that lead to partial solution of ceramic.

4. Improvements in oxidation and / or erosion resistance can be obtained, but further work is needed to established appropriate conditions for producing FGM layers covering substantial areas and thicknesses.

5. Phase diagram data can be used to provide a guide to designing FGMs.

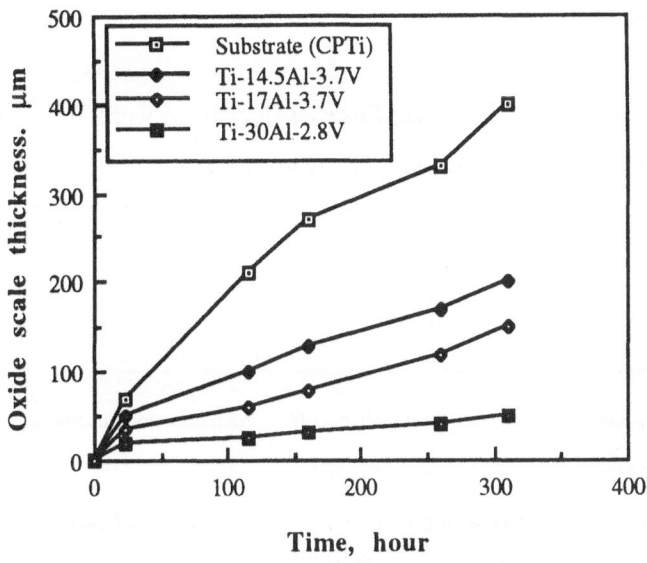

Fig.10: Oxide thickness vs time for CP Ti substrate and three layers of functionally gradient cladding exposed at 800°C for various times

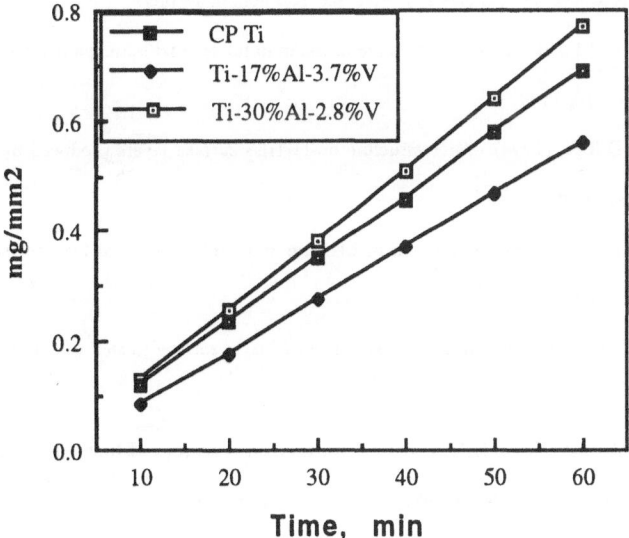

Fig.11: Weight loss vs time for CP Ti substrate and two laser clad layers eroded by SiO_2 with particle size 150-212µm. 140l/min, air flow and 45° impact angle

254

5. ACKNOWLEDGEMENTS

Acknowledgements are made to SERC for financial support the work, IMI Titanium Ltd for supplying Ti alloys, and to INCO for providing Ni alloy.

REFERENCES

1. Mohammed Jasim,K., Rawlings, R.D., and West, D.R.F, (1993), Metal-ceramic functionally gradient material produced by laser processing, J.Mater. Sci. , 28 , 2820.

2. Abboud, J.H., D.R.F.West, D.R.F., Rawlings, R.D., (1994) , Functionally gradient layers of Ti-Al-based alloys produced by laser alloying and cladding, Mat. Sci. Technol., in the press.

3. Abboud,J.H., West, D.R.F, and R.D.Rawlings, R.D., (1994), Functionally gradient Ti-Al composite produced by laser cladding, J.Mater.Sci., in the press.

4. Abboud, J.H.,West, D.R.F., and R.D.Rawlings, R.D., (1994), Microstructure and properties of functionally gradient Ti-Al produced by laser cladding, Mat. Sci. Technol., in the press.

5 Abboud, J.H. and West, D.R.F., (1991), Microstructure of titanium injected with SiC particles by laser processing, J. Mat. Sci. Lett., 10, 1149.

6. Abboud J.H. and West, D.R.F., (1994), Microstructure of Ti-TiB$_2$ surface layers produced by laser particle injection, J.Mater. Sci. Lett., 13, 457.

7.Abboud, J.H. and West, D.R.F., (1994), Oxidation of titanium-aluminides produced by laser surface alloying, J.Matter.Sci.Lett., 11, 1478.

8.Abboud, J.H. and West, D.R.F., (1993), Properties assessment of laser surface treated titanium alloys, Surface Eng, 9, 221.

9.Hirose,A.,Ueda,T.,and Kobayashi, K.F., (1993), Wear and oxidation properties of titanium aluminides formed on titanium surface by laser laser alloying,Mater. Sci. Eng., A160, 143.

CARBIDE AND BORIDE LASER ALLOYING
OF HIGH SPEED STEELS

R. EBNER[1], K. RABITSCH[1], B. KRISZT[1], B. MAJOR[2]
[1] Christian Doppler Laboratory for „Laser Application in Materials
Research" at the Institute for Metals Technology and Materials
Testing, Montanuniversität Leoben, Austria
[2] Institut for Metals Research, Polish Academy of Science,
Krakow, Poland

Abstract

Laser alloying can be used to create extremely high alloyed materials with novel
structures and properties on high speed steel (HSS) substrates. The development
of these materials occurs predominantly on an empirical basis today, because there
exists only little experience with rapidly solidified laser alloyed structures. To
study the influence of the chemical composition on the solidification structure a
high speed tool steel was alloyed with vanadium carbide (VC) and vanadium
boride (VB_2) powder. The solidification structures were characterized by means of
scanning electron microscopy, electron probe microanalysis and X-ray diffraction.
Thermodynamic calculations were made to predict the changes of the constitution
due to laser alloying. A comparison of the solidification structure and the
constitution of the materials revealed a good correspondence for carbide alloyed
HSS. Boride alloying leads to very complex microstructures consisting of various
metallic and intermetallic phases. Thermodynamic calculations are not possible
because of lack of the thermodynamic functions of the various phases.

1. Introduction

High speed steels (HSS) belong to the group of ledeburitic tool steels. They are
mainly used as materials for metal cutting tools. The fabrication route of
conventionally produced HSS consists of melting, casting into small ingots
followed by hot working to form semi finished products like rods, plates etc. Tools
are usually manufactured from these semi-finished products by application of
various machining techniques, for instance turning, milling, grinding etc.

The microstructure of HSS changes several times during fabrication of HSS
tools. The solidification structure consists of metallic dendrites which are
surrounded by one or more carbide eutectics. During hot working the carbide

255

J. Mazumder et al. (eds.), Laser Processing: Surface Treatment and Film Deposition, 255–266.
© 1996 Kluwer Academic Publishers.

networks of the eutectics break up into isolated spherical carbide particles which usually have a diameter of some micron. In the fully heat treated condition HSS exhibit a secondary hardened martensitic matrix with various types of embedded primary carbides. Common HSS contain about 5 to 10 vol.% of primary carbides [1-4] mainly to improve the wear resistance.

One major advantage of HSS is their good machinability in the annealed condition which makes it relatively easy to fabricate even complex tool shapes. Another benefit is the high toughness of HSS in the heat treated condition which favours them as tool materials in case of intermittent cutting. HSS are therefore widely used for metal cutting tools like mills, drills etc. but there are also some other important applications e.g. cutting of paper or wood.

Unfortunately HSS are less wear and tempering resistant than some other cutting tool materials like cemented carbides or ceramics which restricts the application of HSS to comparably low cutting speeds. HSS are heat resistant up to about 560 °C. Higher temperatures lead to a coarsening of secondary hardening carbides in the martensitic structure accompanied by a softening of the material. The wear resistance of HSS depends on the stability of the structure and the amount of wear resistant particles which are embedded in the metallic matrix [5]. A higher wear resistance can therefore be expected from HSS with an improved matrix stability (e.g. more stable and higher amount of secondary carbides) on the one hand and from a higher amount of wear resistant coarse and hard particles on the other hand. This requires an increase of the alloying element contents.

But due to the limited melt temperatures and the low cooling rates in conventional metallurgy upper limits in the chemical composition can hardly be exceeded. These limits can be extended to higher concentrations by using advanced processes like powder metallurgy or laser surface treatment. All these processes take advantage of high cooling rates which cause fine microstructures consisting of supersaturated crystals.

Laser alloying belongs to the group of metallurgical laser surface treatment processes and is usually used to increase the alloying element content at the surface of a substrate by simultaneous surface melting and addition of alloying elements. The laser beam acts as a high power density heat source to create a small superheated melt pool, in which the additional material is injected to form the surface alloy. The chemical composition of the surface alloy therefore depends on the composition of the materials involved and on the degree of mixture. This means, that the substrate and the additions must be adapted to each other to achieve useful structures and properties.

Laser alloying is an outstanding metallurgical process because of the special combination of process features. The high melt pool temperature, the strong convection and the rapid solidification can be used to create new materials with novel microstructures and properties [3,6-8]. The high melt pool temperature favours the dissolution of stable phases like intermetallic compounds which enables the production of alloys with very high alloying element contents. The convection in the melt pool causes a good mixing of the molten substrate with the materials added, and the rapid solidification leads to the formation of fine solidification structures [3,6-8].

The development of new laser alloyed materials occurs predominantly on an empirical basis because of the lack of experience with the new high alloyed and rapidly solidified materials. The first and most important step concerning the properties of laser alloyed materials is the solidification process. The solidification microstructure depends on the chemical composition, but also on the solidification parameters like crystal growth rate and temperature gradient [9]. Solidification is usually discussed on the basis of equilibrium phase diagrams taking into account, that some deviations from the equilibrium conditions occur with increasing solidification rate.

Typical laser alloying conditions lead to cooling rates of about 10^3 to 10^4 K/s and to crystal growth rates lower than 20 mm/s. In the light of solidification theory this means that the situation at the solid/liquid interface of the growing crystals can be described with the equilibrium partition coefficient taking the solute pile-up in the melt ahead of the growing crystals into account. Especially dendritic crystal growth is accompanied by a small undercooling compared to the equilibrium [9]. It can be therefore expected that equilibrium diagrams might be used to get a rough estimation of the solidification microstructure of new laser alloyed materials. This is confirmed by previous results which indicate a good correlation between equilibrium phase diagrams and the solidification microstructure [6,8,10].

Contrary to most binary systems, the constitution of multicomponent systems is usually unknown which is one of the missing links between the chemical composition and the solidification microstructure. In the latest years, thermodynamic concepts and software packages were developed to predict phase diagrams of multicomponent systems [11,12]. As shown previously thermodynamic calculations can be successfully applied to predict the constitutional changes on laser alloying of a plain carbon steel and a high speed steel substrate [10].

But it has to be noticed that the necessary thermodynamic functions are available only for a limited number of elements and phases. For instance most of the data for boron and borides are still missing today.

2. Experimental

In the alloying experiments a common AISI M2 (DIN S 6-5-2) HSS in an annealed condition was used as substrate material. The chemical composition of the substrate is summarised in Table 1. The additions chosen were vanadium carbide (VC) and vanadium boride (VB_2).

TABLE 1: Typical chemical compositions of the substrate material in wt.%

Substrate	%C	%Si	%Mn	%Cr	%W	%Mo	%V
AISI M2 (DIN S 6-5-2)	0.89	<0.45	<0.40	4.10	6.30	5.00	1.90

Laser alloying was done using a CNC controlled 6 axes machine and a 2.5 kW CO_2 laser with a TEM_{20} beam mode. All experiments were performed with a defocussed beam with power densities of about 10^5 W/cm^2. The additional material was injected in form of powder, using inert gas as carrier. The resulting cross sections of the melt tracks were nearly semicircular with a width of about 2 mm and a depth of about 1 mm.

The microstructure of the melt tracks was studied with a scanning electron microscope (SEM) using secondary as well as backscattered electrons for image generation. The chemical composition and the homogeneity of the tracks were measured by means of electron beam microanalysis using an energy dispersive X-ray spectrometer, which is attached to the scanning electron microscope. The determination of the chemical composition of the rapidly solidified phases is often impossible because of their small size.

X-ray diffraction (XRD) was applied for a crystallographic characterization of the phases present. But XRD pattern of laser alloyed tracks usually contain a large number of reflections from various unknown phases, which leads to some uncertainties.

Thermodynamic calculations were performed to calculate the constitution of the alloy systems. These calculations were done with the Thermo-Calc software package [11].

The hardness of the melt tracks was determined by means of Vickers hardness testing with a load of 29.43N (HV3).

3. Results

The present study focusses on the effects of laser alloying on the solidification microstructure. But it has to be noticed that the as-solidified microstructures are usually not suitable for applications in cutting tools. Reasons are the high amount of retained austenite which lowers the hardness and the networks of brittle eutectics which impair the toughness of the material. A post alloying heat treatment is therefore necessary to modify the microstructure and the properties of the laser alloyed material to achieve appropriate material properties for cutting applications.

Metallographic sections of the melt tracks revealed, that the added carbides and borides dissolve completely up to relatively high alloying element contents. It can be therefore assumed, that the solidification starts from a nearly homogeneous melt. Electron beam microanalyses revealed a uniform alloying element distribution within the melt tracks. Usually, the differences in the chemical compositions are smaller than the accuracy of the used energy dispersive X-ray analysis system. This indicates, that laser alloying produces homogeneous melt tracks.

Carbon as well as boron are known to form intermetallic compounds with most metallic elements, but the solubility of boron is usually significantly lower than that of carbon in most metals. The structure of boron and carbon containing alloys is complex, because not only borides and carbides from the binary systems but also ternary carbo-borides may precipitate from the melt. Especially boron is known to form numerous stable and metastable compounds with almost all metallic elements. The resulting mixture of different phases in the alloyed tracks make a complete metallographic and crystallographic characterization of the phases difficult. Usually a large number of partially overlapping peaks appear in the X-ray diffraction pattern of laser alloyed melt tracks.

For the discussion of the effect of alloying element additions the microstructure of the laser remelted M2 HSS is taken as a reference:

3.1 LASER REMELTING OF M2 HSS:

Laser remelting of the M2 HSS causes a fine solidification microstructure consisting of metallic dendrites which are surrounded by a carbide eutectic, fig.1. The dendrite arm spacing is about 3 to 5 μm in the laser remelted HSS compared to about 80 to 120 μm in conventional HSS. As a result of the supersaturation of the metallic dendrites a high amount of retained austenite up to 50 vol.% is present after cooling to room temperature, which is responsible for the relatively low hardness of about 650 to 750 HV3 [6].

The volume fraction of the eutectic is about 5 % which is about one half of the eutectic in conventional M2 HSS. XRD studies of the laser remelted M2 HSS revealed that the carbides of the eutectic are mainly M_6C, MC and M_2C carbides are present in small amounts.

The constitution of the M2 HSS was reported by Horn and Brandis [13]. These results are in accordance with the calculated constitution of the M2 HSS [14]. Based on the constitution and the solidification microstructure the following conclusion can be drawn concerning the solidification of the laser melted M2:

The solidification starts with the precipitation of δ-ferrite which subsequently transforms completely to austenite. During further cooling a eutectic decomposition of the melt into M_6C and austenite occurs. The solidification ends with the eutectic reaction of the retaining melt into MC and austenite. M_2C carbides are not present in equilibrium diagrams. But as shown by Golczewsky and Fischmeister [15] the presence of M_2C carbides can be predicted by taking microsegregation into account.

More details about the structure of the laser remelted M2 HSS are given in [6,16].

3.2 LASER ALLOYING OF M2 WITH VANADIUM CARBIDE:

Vanadium carbide is a thermodynamic stable intermetallic compound. The vanadium content in common HSS is therefore limited to about 3 wt.%, because a higher vanadium content leads to the formation of carbides in the carbon containing melt prior to solidification followed by a separation due to density differences. On adding vanadium carbide in the laser alloying process, melt tracks with vanadium contents up to ≈50 wt.% can be achieved [3,6-8]. But it has to be recognized, that in melt tracks with the highest vanadium contents some undissolved particles might be present.

The microstructure of the laser alloyed M2 changes with increasing vanadium carbide additions as follows: Vanadium carbide additions cause an increasing amount of eutectic in the interdendritic space. There also occurs a change in the morphology of the eutectic from a federy to a spherical shape. At more than about 4 wt.% vanadium a new type of eutectic is created which forms eutectic cells besides the metallic dendrites. Reaching about 8 wt.% vanadium, the structure gets fully eutectic. Exceeding the vanadium content above 8 wt.%, the solidification starts with the precipitation of monocarbides. Figs. 2 to 4 show the structure of melt tracks with about 10, 15 and 27 wt.% vanadium indicating a significant increase of the amount of monocarbides combined with a change of their morphology.

X-ray diffraction studies reveal, that increasing vanadium carbide additions are accompanied by a reduction of the M_6C carbides whereas the amount of MC rises significantly [17].

For comparison, an isopleth along the alloy path M2+VC was calculated, fig. 5. This equilibrium phase diagram shows, that the solidification starts with the precipitation of metallic dendrites below about 8 wt.% vanadium and with the precipitation of monocarbides at higher vanadium contents. Fig.5 indicates that the metallic crystals are δ-ferrite dendrites for vanadium contents up to about 4 wt.%. These δ-ferrite dendrites subsequently transform in a peritectic reaction to austenite. Between about 4 and 8 wt.% the solidification starts with the precipitation of austenite dendrites. It has to be mentioned, that the liquidus temperature rises very rapidly on exceeding about 8 wt.% vanadium, which is responsible for the undissolved powder particles in the highest alloyed melt tracks.

A comparison of the calculated isopleth and the microstructure of the laser alloyed melt tracks shows a good correspondence. This is true for the primary phases in the melt tracks but also for the chemical composition of the melt tracks with a fully eutectic struture.

Vanadium carbide additions can be used to increase the amount of carbides and to improve the hardness. Fig.6 shows, that the hardness increases from about 700 HV3 in the remelted M2 HSS to about 1040 HV3 in the highest alloyed melt tracks.

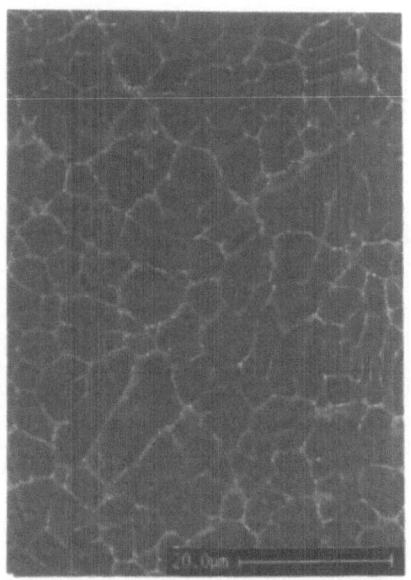

Fig. 1: Solidification structure of laser remelted M2 HSS; ≈1,8 wt.% V

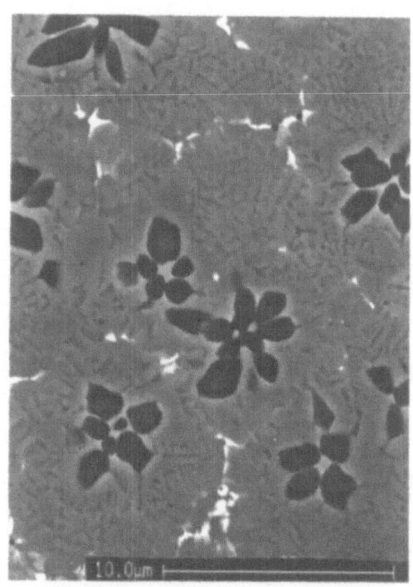

Fig. 2: Solidification structure of M2 HSS laser alloyed with VC; ≈10 wt. % V

Fig. 3: Solidification structure of M2 HSS laser alloyed with VC; ≈17 wt. % V

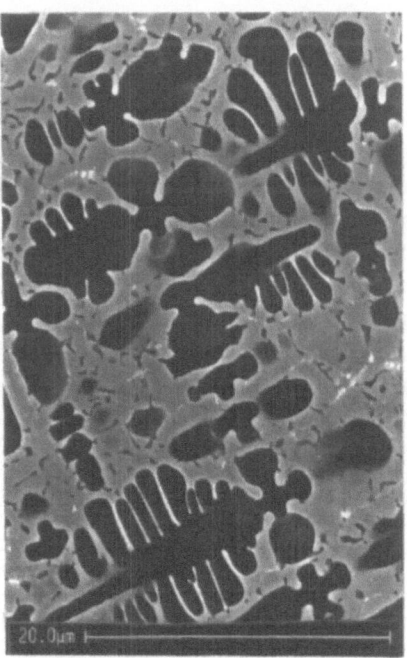

Fig. 4: Solidification structure of M2 HSS laser alloyed with VC; ≈27 wt. % V

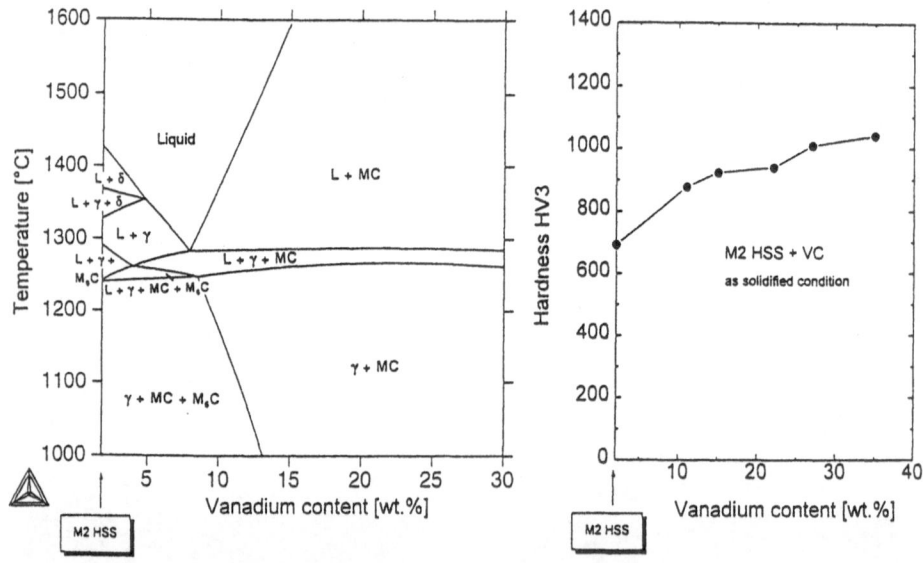

Fig.5:   Isopleth through the multicomponent
         system Fe-C-Cr-W-Mo-V along the alloy
         path M2+VC

Fig.6:   Hardness of the M2
         HSS laser alloyed
         with VC

3.3 LASER ALLOYING OF M2 WITH VANADIUM BORIDE:

Boron is not used as alloying element in common HSS although boron forms stable and hard intermetallic phases. One of the reasons is that even very low boron additions lead to a eutectic with iron at about 1150 °C which is lower than the conventional hardening temperatures of HSS. Moreover boron additions impair the ductility and toughness. In spite of these disadvantages boron seems to be an interesting alloying element especially for an increase of the amount of hard and wear resistant particles.

Boron is known as element that forms a variety of intermetallic phases with metals which makes the alloy development extremely difficult [18,19]. Support of the alloy development by means of thermodynamic calculations is almost impossible in case of boride alloying because the thermodynamic functions of the various phases are more or less unknown. It is therefore necessary to start with the thermodynamic modeling of simple systems. First results on this topic were recently published for the ternary system Fe-B-C. Experimental investigations within this system also reveal a good correlation between the microstructure of laser alloyed tracks and the calculated constitution [20].

Vanadium boride additions have a marked effect on the solidification structure of the laser alloyed M2 HSS as indicated by the figs. 7 to 10. Increasing VB_2 additions cause a decrease of the volume fraction of dendrites whereas the amount of eutectic phases rises. The morphology of the eutectic changes

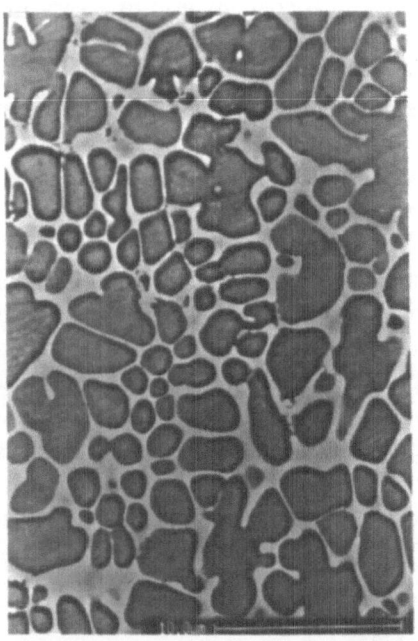

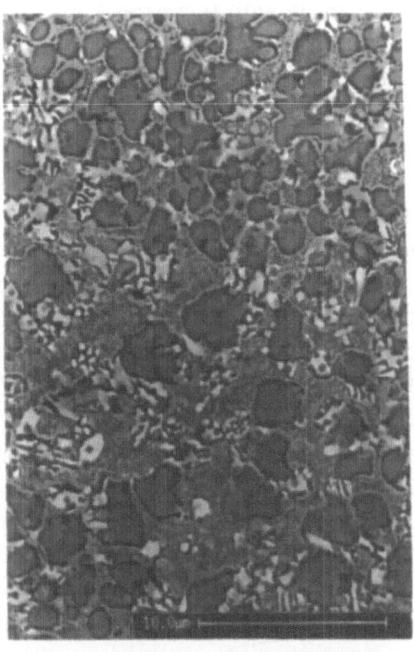

Fig. 7: Solidification structure of M2 HSS laser alloyed with VB₂; ≈3,8 wt. % V

Fig. 8: Solidification structure of M2 HSS laser alloyed with VB₂; ≈6,3 wt. % V

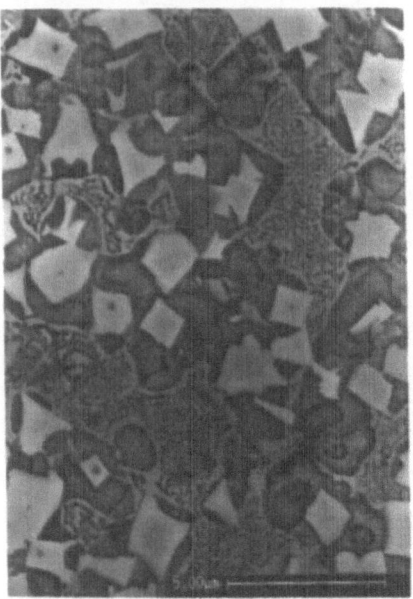

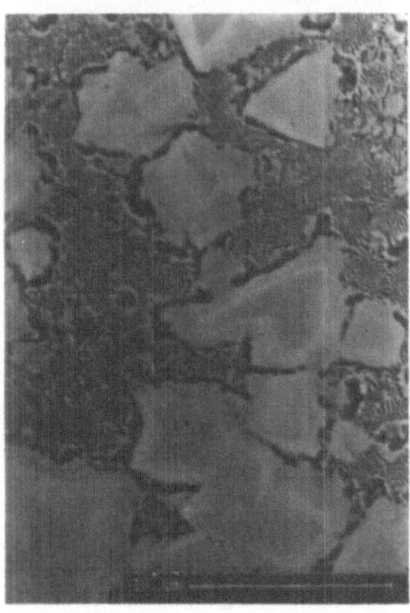

Fig. 9: Solidification structure of M2 HSS laser alloyed with VB₂; ≈9,4 wt. % V

Fig.10: Solidification structure of M2 HSS lase alloyed with VB₂; ≈11,2 wt. % V

264

continuously with incresing VB_2 additions. Low VB_2 contents lead to a very fine eutectic as shown in fig. 7. A further increase of the VB_2 additions rises the content of the eutectic phases . Fig. 8 shows that the eutectic consists of various phases. Also the dendrites are not homogeneous as indicated by the etching behaviour. A detailed investigation of the dendrites revealed that some fraction of δ-ferrite is still present in the as-solidified condition. Vanadium is a strong ferrite stabilizing element whereas boron has nearly no influence on the ferrite to austenite transition because of ist low solubility in iron. Medium and higher VB_2 additions lead therefore to higher amounts of δ-ferrite in the solidification structure. At high VB_2 additions the solidification starts with the precipitation of borides which is followed by a precipitation of metallic crystals, fig. 9. These metallic crystals are mainly formed around primary borides and in some cases also directly from the melt. After some amount of precipitation of metallic crystals, the remaining melt solidifies in various eutectic reactions. The solidification starts at the highest VB_2 additions with the precipitation of primary borides and ends with some eutectic reactions, fig. 10. Metallic dendrites are no longer present in the microstructure.

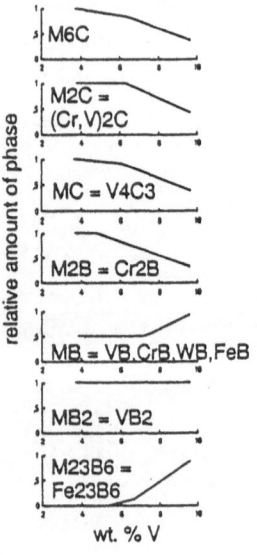

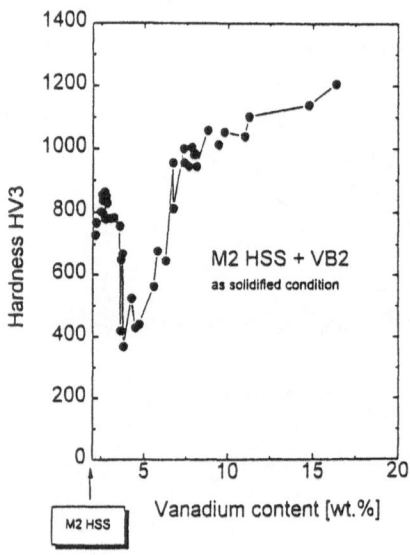

Fig.11: Relative changes of the amount of the various intermetallic phases on laser alloying of M2 HSS with VB_2

Fig.12: Hardness of the M2 HSS laser alloyed with VB_2

Fig. 7 to 10 indicate a very complex structure of the VB_2 alloyed M2 HSS that consists of various metallic and intermetallic phases. XRD studies reveal at least seven types of intermetallic carbide and boride phases. The amounts of the various phases are changing with increasing VB_2 additions as shown in fig. 11. Increasing VB_2 additions lead to a reduced amount of the carbide phases whereas the amount of boride phases increases.

The structural changes on laser alloying lead to hardness variations. Fig. 12 shows the dependence of the hardness as a function of the vanadium content of the laser alloyed melt track. The hardness increases at low VB_2 additions until a significant hardness drop occurs on exceeding about 4 wt.% vanadium. This hardness loss corresponds to the presence of δ-ferrite in the dendrites. A further increase of the vanadium content leads to a rise of the hardness up to about 1150 HV.

4. Conclusion

The following conclusions can be drawn from the presented results :

- Laser alloying enables the production of extremely high alloyed materials in surface near regions of high speed steel substrates.
- Powder additions can be used to achieve homogeneous melt tracks. Powder with a low melting temperature can be completely melted in the melt pool up to very high contents, powder of stable chemical compounds may lead to some undissolved particles in the melt track at high additions.
- The amount of hard intermetallic phases in the laser alloyed tracks on high speed steel substrates can be increased significantly by additions of carbides or borides.
- A good correspondence between equilibrium phase diagrams and the rapidly solidified structures was found.
- Thermodynamic calculations seem to be a proper tool to evaluate the constitution of multicomponent alloys. Some uncertainties due to microsegregation during solidification might occur. This can lead to the precipitation of phases, which are not expected from equilibrium diagrams.
- The microstructures of boride alloyed HSS are extremely complex and contain a variety of different metallic crystals and intermetallic compounds.

Acknowledgement

The authors would like to thank the Austrian Industries, the „Forschungsförderungsfonds für die Gewerbliche Wirtschaft" and the „Jubiläumsfonds" of the Austrian National Bank for their financial support.

266

5. References

1. Karagöz, S.: Thesis, Montanuniversität Leoben (1982)
2. Riedl, R.: Thesis, Montanuniversität Leoben (1984)
3. Hackl, G., R.Ebner, F.Jeglitsch: Z. Metallkde., Bd.83, Heft 6 (1992), 368
4. Hackl, G.: Thesis, Montanuniversität Leoben, (1991)
5. Karagöz, S., H. F. Fischmeister: Proc. of the First International High Speed Steel Conference, Leoben (1990), 41
6. Ebner, R.: Proc. of the 2nd Conference on Advanced Materials and Processes, Vol.1 "Advanced Processing", Cambridge (1991), Ed. T.W.Clyne and P.J.Withers, 115
7. Ebner, R., B. Kriszt: Proc. International European Conference on Tooling Materials, Interlaken (1992), 289
8. Ebner, R., B. Kriszt: Proc. 4th European Conference on Laser Treatment of Materials ECLAT'92, Göttingen (1992), 187
9. Kurz, W., D. J. Fisher: Fundamentals of Solidification, Trans Tech Publications, (1989)
10. Ebner, R.: Proc. ICALEO'93 (1994), Orlando, Florida, p.857
11. Sundmann, B., B. Jansson, J. O. Anderson: CALPHAD 9, (1985), 153
12. Hillert, M.: Proc. AIME "Calculation of Phase Diagrams and Thermochemistry of Alloy Phases", Milwaukee, (1979), 1
13. Horn, E., H. Brandis: DEW-Technische Berichte 11, 1971, 147
14. Golczewski, J.A., H.F. Fischmeister: Z. Metalkd. 84 (1993), 557
15. Golczewski, J.A., H.F. Fischmeister: Z. Metalkd. 84 (1993), 860
16. Ebner, R., G. Hackl, E. Brandstätter, F. Jeglitsch: Proc. of the First International High Speed Steel Conference, Leoben (1990), 81
17. Kriszt, B., R. Ebner, B. Major, R. Ciach: Proc. of the FEE, Czestochowa, (1993), in press
18. Ebner, R., K. Rabitsch, B. Major, R. Ciach: Proc. of the FEE, Czestochowa, (1993), in press
19. Rabitsch, K., R. Ebner, B. Major: Scripta Metallurgica et Materialia 30 (1994), p.253
20. K. Rabitsch, R. Ebner, B. Major: Sonderbände der Praktischen Metallographie 25 (1994), in press

INNOVATIVE INTERMETALLIC COMPOUNDS BY LASER ALLOYING

I. SMUROV
Ecole Nationale d'Ingénieurs de Saint-Etienne,
58, rue Jean Parot, 42023 Saint-Etienne Cedex 2,
France

M. IGNATIEV
Institut de Science et de Génie des Matériaux et Procédés,
Centre National de la Recherche Scientifique,
B.P.5 Odeillo, 66125 Font-Romeu Cedex,
France

ABSTRACT. Mathematical modeling of heat and mass transfer in pulsed laser alloying is used to simulate the dynamics of interphase boundaries during complete thermal cycle, convective mass transfer, concentration fields of admixture and so on. Several original solutions are proposed for laser alloying of systems with large differences in physicochemical properties such as : Al-Sn(In), Fe-Sn(In, Pb), Al-Fe. Among them: (a) combined continuous wave laser and pulse-periodic laser action; (b) using of shielding covers to avoid the removal of alloying elements from the zone of treatment. The obtained results indicate the formation of nonequilibrium phases and/or submicron sized precipitations. The developed method of laser surface alloying with Sn is applied for the problem of wear resistance improvement for bearing steels working at severe operating conditions.

1. Introduction

Concentrated energy fluxes, such as laser, electron beam, plasma, and so on, have unique possibility for fast melting and intensive remixing of alloying elements (for example, [1,2]), these processes are characterized by extremely high cooling rates that are able to quench the melt [3]. As a result nonequilibrium and metastable phases could be obtained [4-19]. As the classical example, composition of Ni-Hf alloys produced by laser cladding process exceeds the equilibrium solid solubility limit [4,5].

Present laser alloying processes may be divided into two classes. The first one concerns with action of short energy pulses (width range from some nanoseconds up to a few microseconds) on system "thin film-based alloy" (for example, [20-24]). These processes of "laser implantation" are characterized by: (a) thin modified layers (thickness less than 1 μm); (b) diffusion mechanism of mass transfer, and therefore; (c) strong decrease of alloying elements concentration with depth of modified layer.

267

J. Mazumder et al. (eds.), Laser Processing: Surface Treatment and Film Deposition, 267–326.
© 1996 *Kluwer Academic Publishers.*

The second class uses duration of energy pulses from 1 up to 20 ms or continuous wave radiation. The main features of these processes are the following: (a) thickness of modified layer varies from 0.1 up to a few millimeters; (b) mainly convective mechanism of mass transfer; (c) opportunity to regulate the features of mass transfer and, therefore, concentration of alloying elements in a wide range : from quite nonuniform up to rather uniform distributions [3,25]. In the present article only the second class of laser alloying processes is considered.

2. Main Directions of the Research

The objective of the present research is the elaboration of protective coatings and surface layers with unique performances on metallic materials by laser alloying.

To reach the goal, the work is oriented toward the following directions :

(1) Process Development, (2) Process Monitoring, (3) Heat-and Mass Transfer Phenomena, (4) Materials Elaboration and Characterization, (5) Materials Properties.

In the frames of these directions the following problems are considered :

(I) PROCESS DEVELOPMENT :

Modification of traditional alloying processes for the given materials system to obtain the required composition and properties of elaborated surface layers.

Usually, the following ways are used to insert alloying elements into the zone of laser beam interaction with base material : by powders flow; from pre-deposited coatings; from gas (plasma) phase; from the layer of transparent fluids (for example, [1,2,26-29]). As a rule, these methods are not efficient for laser alloying when components with big differences in physical properties need to be remixed.

In the present article several original solutions are proposed to solve this problem, among them: (a) combined action of the different types of lasers (for example, pulse-periodic and continuous wave lasers); (b) using of shielding covers to avoid the destruction and evaporation of alloying element from the surface of base metal; (c) initial preparation of thick (more than 100 μm) multilayers coatings ("puff-pastry") of alloying elements.

Combined Continuous + Pulse Periodic Laser Alloying. Laser alloying was provided in the conditions of combined continuos + pulse-periodic laser-laser action: both - 2.5 kW CO_2 and 400 W Nd:YAG lasers were focused in the same zone [30]. The irradiation of solid state laser was delivered throw an optical fiber of about 10 m length. CO_2 laser beam was oriented normal to the surface, Nd:YAG - at the angle of 45°. The combined action of two lasers provides intensive remixing of alloying elements with material of the substrate (Figure 1) even in case of big difference of their physical properties (Figure 2). For example, the concentrations of doping element are equal at the middle part of the modified zone and at its bottom after combined laser alloying with Sn of Ti substrate. (Figure 2, points A and B).

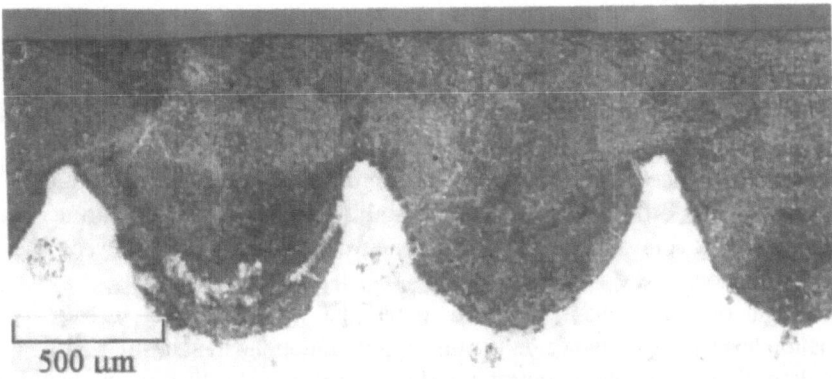

Figure 1. Cross section photo of Ti alloying from predeposited graphite coating by combined continuous CO_2 + pulse-periodic Nd: YAG laser action. The average power of Nd:YAG laser composed 8% from the one of CO_2 laser.

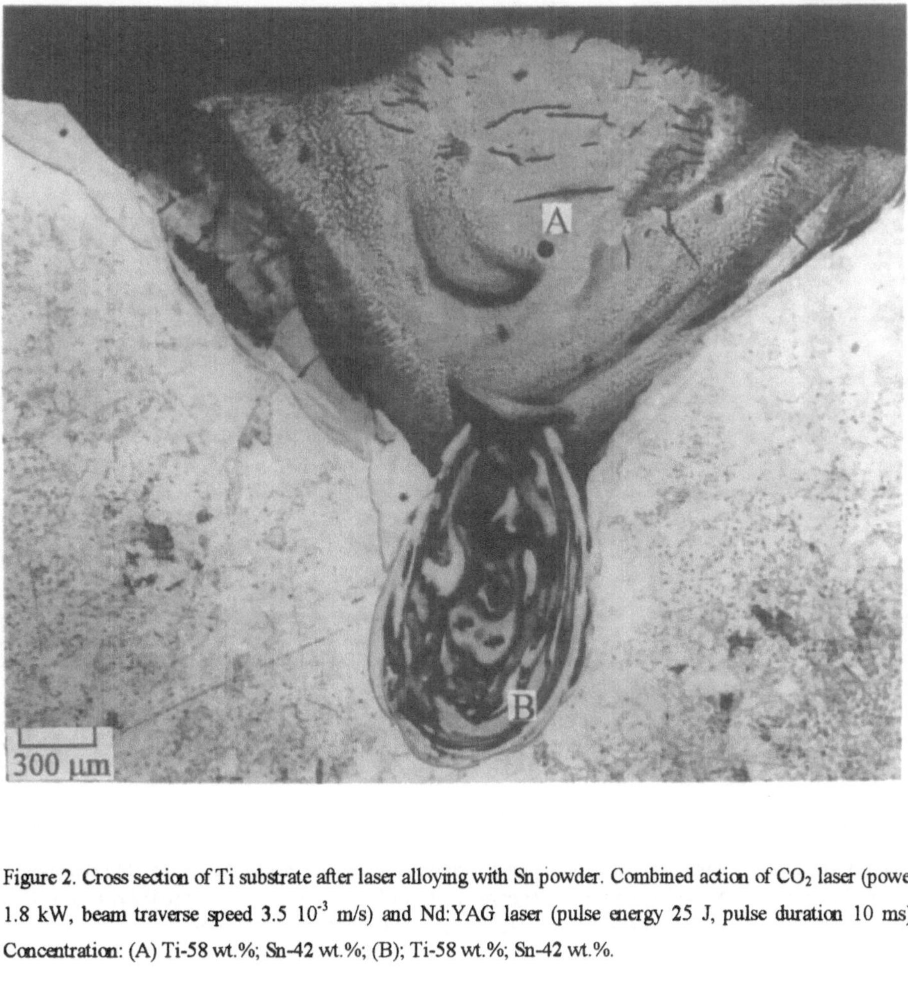

Figure 2. Cross section of Ti substrate after laser alloying with Sn powder. Combined action of CO_2 laser (power 1.8 kW, beam traverse speed 3.5 10^{-3} m/s) and Nd:YAG laser (pulse energy 25 J, pulse duration 10 ms). Concentration: (A) Ti-58 wt.%; Sn-42 wt.%; (B); Ti-58 wt.%; Sn-42 wt.%.

(II) PROCESS MONITORING :

the complex real time process monitoring based on high-speed diagnostics : photo recording, pyrometry, spectroscopy, opto-acoustic signals, and so on.

The utilization of real-time pyrometry in laser alloying is discussed in [31]. The example of photo recording data is presented below.

The high speed photocamera is used for recording the laser plume dynamics. It allows to study the influence of plume instability on mass transfer during laser alloying process (by comparison with metallographycal examination of alloyed zones). Plume rotation along the irradiated surface can be clearly observed (Figure 3a) when slit diaphragm of photocamera is placed in parallel with the target surface. This phenomenon has strong influence on the uniformity of doping elements remixing. For the best definition of the plume pulsations (Figure 3b), the slit diaphragm should be placed perpendicular to the target surface. It could be shown that the destruction of the treated surface is connected with the frequency and intensity of the plume pulsations.

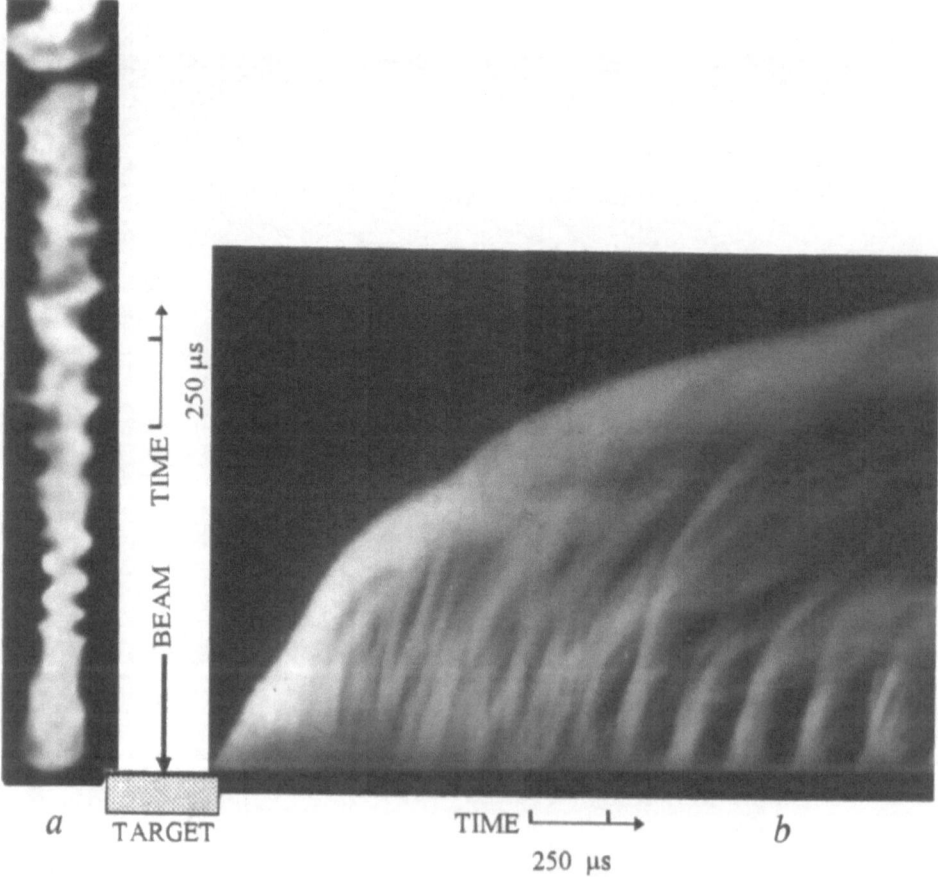

Figure 3. High speed slit photographs of laser plume near the irradiated surface of the system : Mo coating-iron substrate (thickness of Mo coating 30 μm, energy per pulse 14.5 J, duration 4 ms, Nd/YAG laser). (a) slit is placed in parallel with the target surface; (b) slit is placed perpendicular to the target surface.

(III) HEAT AND MASS TRANSFER PHENOMENA :

3.1. *Determination of the correlations between* : (a) conditions of treatment, (b) macroscopic coupled heat-and mass transfer, (c) resulting composition and microstructure.

3.2. Evaluation of the advantages of the *combined energy beams* treatment to produce innovative protective coatings.

Numerical simulation includes: transient models of melting-solidification, surface evaporation, melt hydrodynamics, convective mass transfer of alloying elements, near-surface plasma formation, interaction of energy flow with plume, and so on.

(IV) MATERIALS ELABORATION AND CHARACTERIZATION :

Surface alloying, remelting, remixing by single and combined energy beams treatment.

Obtained materials characterization by the following methods : SEM, TEM, Auger spectroscopy, etc.

(V) MATERIALS PROPERTIES :

Evaluation in the aims of potential industrial applications of the structural and functional properties of the surface layers and more especially their fatigue, wear, corrosion and thermal shock resistance.

3. Heat and Mass Transfer in Pulsed Laser Alloying

By means of a typical process monitoring, based on the detection of electromagnetic radiation from the zone of laser action, it is possible to obtain information about various surface phenomena and about the material removal (vapor and plasma flows, droplets of melt and so on), but no information from the inside of irradiated metal is available. Analysis of the treatment results after the end of energy action (usually on the base of different cross sections) can help to reconstruct mainly integral parameters, but not the dynamics of the process. That is why numerical simulation of heat and mass transfer in laser machining is the unique method to obtain the required information, to "look inside" the laser action zone.

The calculation of temperature fields on the base of liner mathematical models is a comparatively simple task. The main difficulties appear during the simulation of phase transformations of melting (solidification) and evaporation. The corresponding mathematical models are essentially nonlinear.

3.1. HEAT TRANSFER PHENOMENA : DYNAMICS OF PHASE BOUNDARIES IN PULSED LASER ACTION

3.1.1. The Two Phase Fronts Model

The main peculiarity of the model [3,32,33] is the consideration of the movement of two phase fronts : melting and evaporation. Including the evaporation phenomenon into the model prevent the rise of temperature up to nonrealistic values and makes it possible to consider comparatively high values of energy density flows. Together with heating and melting also cooling and solidification are taken into account, that permit to analyze the complete thermal cycle. It makes it possible to determine the dynamics of melt thickness in a wide range of values of energy density flow and duration of action. This information (and corresponding temperature distribution) is necessary to determine the conditions to obtain materials with required properties. One of the advantages of the present approach is an opportunity to analyze the dynamics of phase fronts velocities on the base of the exact determination of the phase boundaries positions.

It is assumed that the energy flow is absorbed on the irradiated surface; convection and radiation mechanisms of heat losses from both the sides of the slab are considered; melting (solidification) is determined by the classical Stefan boundary condition.

From our point of view the present model can be considered, on the one hand, as one of the simplest to take into account both melting and evaporation phenomena; on the other hand, it can be used in a comparatively wide range of energy density flow and pulse duration values [33]. For a nanosecond pulse length it is better to simulate the evaporation phenomena based on the Knudsen layer approach [34,35].

The used mathematical model can be written in the following form:

$$\frac{\partial^2 T_1}{\partial x^2} = \frac{1}{a_1}\frac{\partial T_1}{\partial t} \qquad S_1(t) < x < S_2(t) \quad , \qquad t_m < t < 0$$

$$q_0(t) - \alpha_g[T_1(x,t) - T_g] - \sigma[\varepsilon_1 T_1^4(x,t) - \varepsilon_g T_g^4] = -\lambda_1 \frac{\partial T_1}{\partial x} + \rho_1 L_v \frac{dS_1}{dt}$$

$$\frac{dS_1}{dt} = \frac{V_*}{T_1^{1/2}(S_1(t),t)}\exp[-\frac{T_*}{T_1(S_1(t),t)}] \quad , \qquad x = S_1(t)$$

$$\lambda_1 \frac{\partial T_1}{\partial x} = \lambda_2 \frac{\partial T_2}{\partial x} - \rho_2 L_m \frac{dS_2}{dt} \quad , \quad T_1 = T_2 = T_m \, , \; x = S_2(t), \, S_2(T_m) = 0$$

$$\frac{\partial^2 T_2}{\partial x^2} = \frac{1}{a_2}\frac{\partial T_2}{\partial t} \, , \qquad S_2(t) < x < L \tag{1}$$

$$-\lambda_2 \frac{\partial T_2}{\partial x} = \alpha_f(T_2 - T_f) + \sigma(\varepsilon_2 T_2^4 - \varepsilon_f T_f^4) \, , \qquad x = L$$

$$T(x, t = 0) = T_0$$

where : $T_1(x,t)$ temperature of liquid phase; $T_2(x,t)$ temperature of solid phase; x, t distance and time, respectively; $S_1(t)$, $S_2(t)$ positions of evaporation and melting phase boundaries, respectively; $q_0(t)$ absorbed energy density flux; $a_{1,2}$ thermal diffusivity of liquid and solid phase, respectively; $\lambda_{1,2}$ thermal conductivity of liquid and solid phase, respectively; $\rho_{1,2}$ densities of liquid and solid phase, respectively; L_v specific heat of evaporation; L_m latent heat of melting; α_{gf} convection heat losses coefficient of irradiated and rear surfaces of the slab; $\varepsilon_{1,2}$ emissivities of irradiated and rear surfaces of the slab; ε_{gf} emissivities of the environment near the irradiated and rear surfaces of the slab; T_m temperature of melting; t_m starting time for melting; L thickness of the slab; T_0 initial temperature.

Constants V_*, T_* are determined by the Herz-Knudsen law of evaporation:

$$V_* = \frac{P_v}{2\rho_1(2\pi k / m)^{1/2}} \exp[\frac{L_v}{T_v(k / m)}] , \qquad T_* = \frac{L_v}{(k / m)}$$

Where : $k = 1.38 \ 10^{-23}$ J/K, Boltzmann constant; m atomic mass of slab material; T_v boiling temperature corresponding to the pressure P_v [3].

System (1) is solved numerically, the phase change fronts are tracked continuously and the latent heat release is treated as a moving boundary condition. In both regions of liquid and solid phases the moving uniform grids consists of a fixed number of points. Each grid point moves with a different velocity. The Crank-Nicolson's technique of various derivatives is used [36-38].

3.1.2. Results and Discussion

The main physical results obtained by the model (1) are the following [33]: The simulation of action of constant energy density flux shows that it is necessary to distinguish three different transient periods: the smallest one - for the surface temperature, the largest one - for the melt thickness (and melting front velocity), and between them - for the evaporation front velocity. The values of surface temperature calculated taking into account evaporation and neglecting it may differ in several times. The interaction of melting and evaporation fronts may result in sharp nonmonomone behavior of melting velocity even if laser pulse shape is smooth.

Several results of heat processes dynamics for pulsed laser action on steel slab 1 mm thickness with the following heat transfer constants: $a = 5.5 \ 10^{-6}$ m^2/s; $\lambda = 29$ W/(mK); $L_m = 2.7 \ 10^5$ J/Kg; $L_v = 7.1 \ 10^6$ J/Kg; $T_m = 1808$ K; $T_v = 2750$ K; $T_0 = 300$ K ; $\varepsilon = 0.42$ are presented in Figures 4-11.

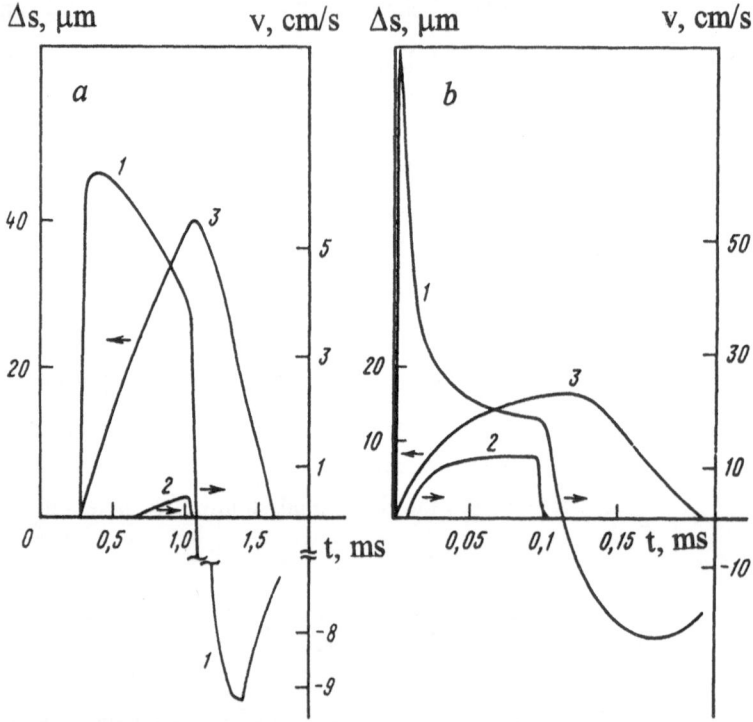

Figure 4. Time dependence of : velocities of melting (1) and evaporation (2) fronts, melt thickness (3) for the action of rectangular laser pulse on steel : (a) $q_0=10^5$ W/cm^2, pulse duration $\Delta\tau=1$ ms; (b) $q_0=10^6$ W/cm^2, pulse duration $\Delta\tau=0.1$ ms. In both cases absorbed energy per pulse equals 10 J.

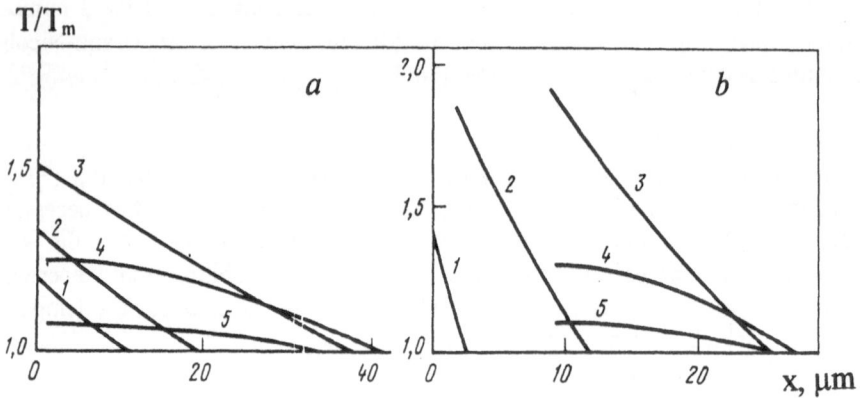

Figure 5. Temperature distributions in the melt:
(a) $q_0=10^5$ W/cm^2, pulse duration $\Delta\tau=1$ ms; curve 1 - t = 4.5 10^{-4} s, 2 - 6 10^{-4} s, 3 - 9.6 10^{-4} s, 4 - 1.05 10^{-3} s, 5 - 1.2 10^{-3} s. (b) $q_0=10^6$ W/cm^2, pulse duration $\Delta\tau=0.1$ ms; curve 1 - t = 6.6 10^{-6} s, 2 - 3.2 10^{-5} s, 3 - 9.75 10^{-5} s, 4 - 1.15 10^{-4} s, 5 - 1.4 10^{-4} s. Left border of the curves corresponds to the current position of evaporation front.

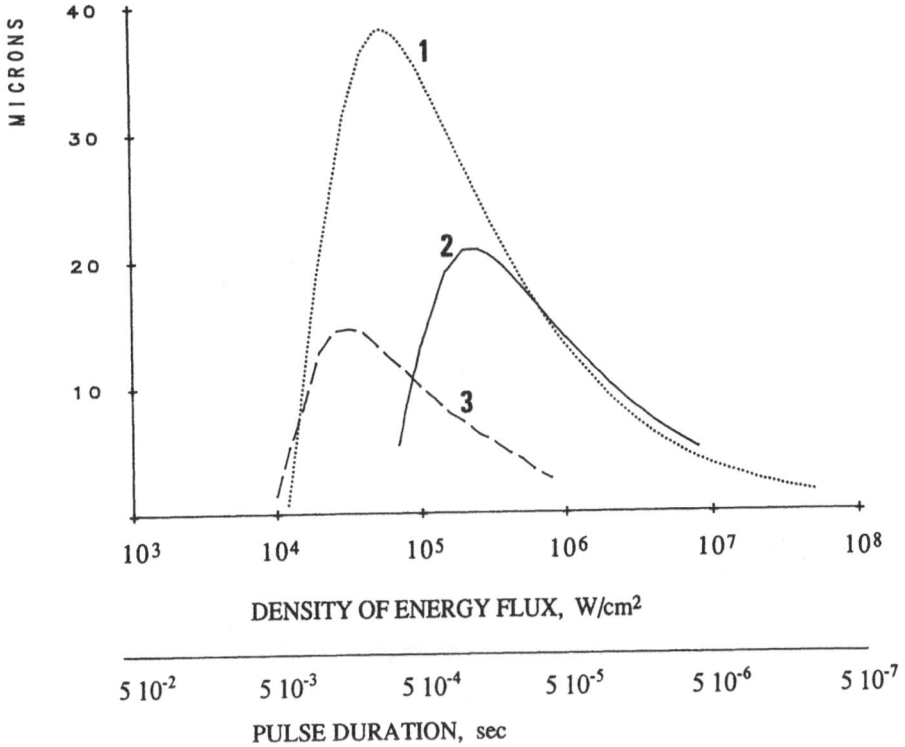

Figure 6. Maximum values of melt thickness produced by laser pulse versus density of energy flux and pulse duration (fixed energy density per pulse 50 J/cm²). Curve 1 - Ti; 2 - steel; 3 - ZrO₂.

Let us analyze the behavior of maximum melt thickness for a given (constant) energy input into the material, but for different energy densities fluxes and corresponding values of pulse duration. Two asymptotic cases are easy to be considered: the first, corresponding to low values of energy flux densities and large pulse duration, and the second one, exactly opposite, for high densities of energy flux and short pulse duration. In the first case, a large part of energy will be spent on material's heating $(t_m \sim \Delta t)$ up to the disappearing of melting phenomenon. In the second, reverse case, strong evaporation will be the most important under the point of view of energy losses and as $L_v >> L_m$ only a comparatively small part of energy input will go on melting. Therefore, if in the both asymptotic cases (on the scale of q_0) the melt thickness tends to zero, than in between, there must exist a maximum (Fig. 6).

For laser alloying applications, for example, alloying from gas or liquid phase, by powders flow, etc. it is important to know the life-time of the melt on the irradiated surface. This time is equal to $(t_{sol} - t_m)$, where t_{sol} is the time of the end of solidification (disappearing of liquid phase). The calculations show that together with the melt thickness maximum there exists a maximum value of the melt lifetime (Fig. 7).

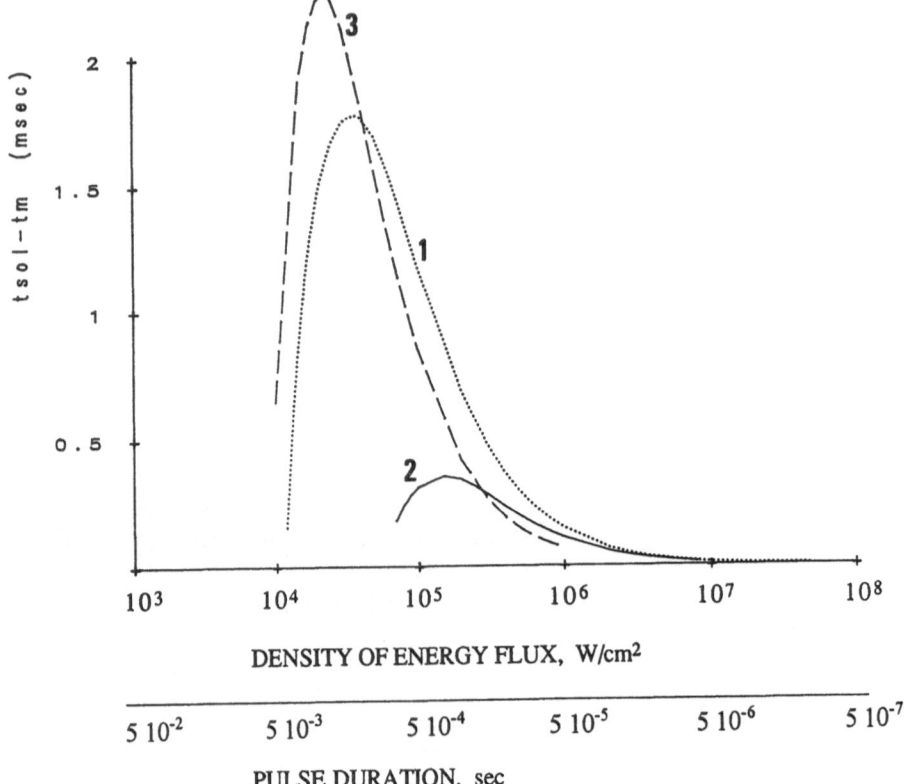

Figure 7. Melt lifetime versus density of energy flux and pulse duration (fixed energy density per pulse 50 J/cm²). Curve 1 - Ti; 2 - steel; 3 - ZrO₂.

In the case of pulse-periodic action, it is easier to find optimum parameters of duty cycle, when the characteristic time is normalized on pulse duration. The extreme value of melt lifetime normalized on the pulse duration $(t_{sol} - t_m)/\Delta\tau$ is about 2 for steel, about 3 for titanium. With q_0 increasing, the curves are tend to unity because $t_m \rightarrow 0$ $(t_m \sim q_0^{-2})$ and $t_{sol} \sim \Delta\tau$.

As a rule the real shape of Nd:YAG laser pulse corresponding to the case of chaotic (free) generation is approximated by a smooth curve. But in practice a radiation pulse with duration about 10^{-3} s consists of a sequence of spikes with duration about 10^{-6} s and the duty ratio (pulse period to pulse duration ratio) about 2-5. In free generation pulse, the spike's amplitudes are different; in, so-called, ordered generation - the amplitude of the individual spikes and the time interval between successive spikes remain constant during the considerable part of the pulse (Fig.9). In practice it is important to understand in which case it is necessary to consider this real spike structure and when it appears possible to approximate its shape by a smooth curve. On the other hand this is the question of the result's reproducibility of laser pulse action with chaotic generation.

The produced analysis have shown that the spike structure of laser pulse may induce strong oscillations of surface temperature and evaporation intensity (Figs.10,11), on the other hand its influence on melting dynamics often is weak.

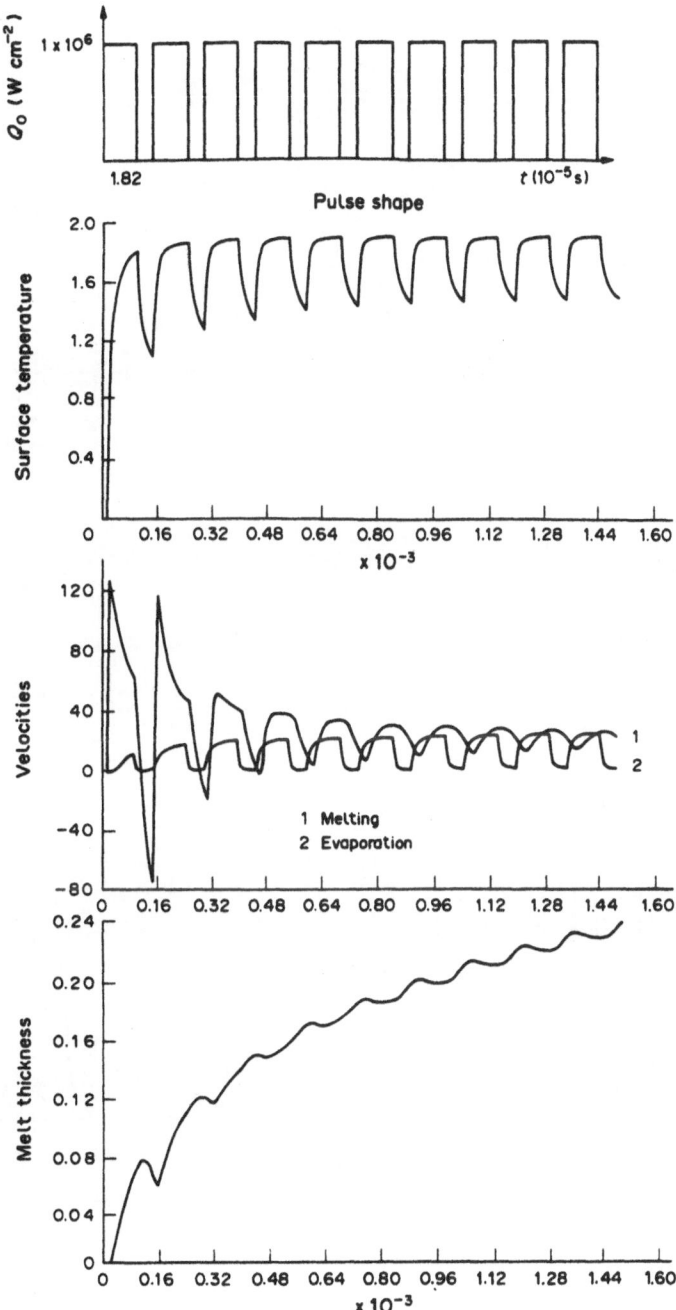

Figure 8. Heat processes corresponding to the pulse periodic laser action on steel slab. Shape of energy pulse: rectangular, pulse duration: 1.82×10^{-5} s , duty cycle: 1.5; maximum value of energy density flow: 10^6 W/cm^2.

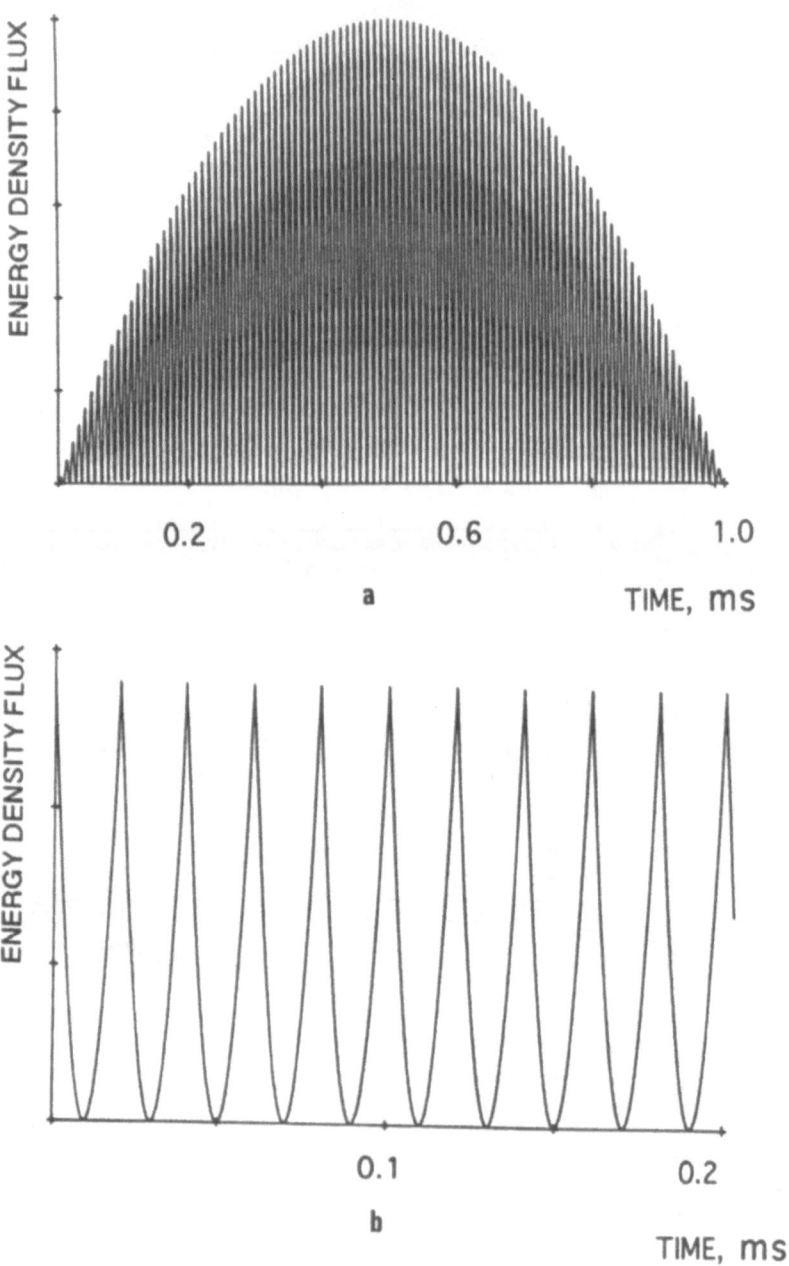

Figure 9. Model spike structures of laser pulses used in simulation: (a) - free generation, (b) - ordered generation.

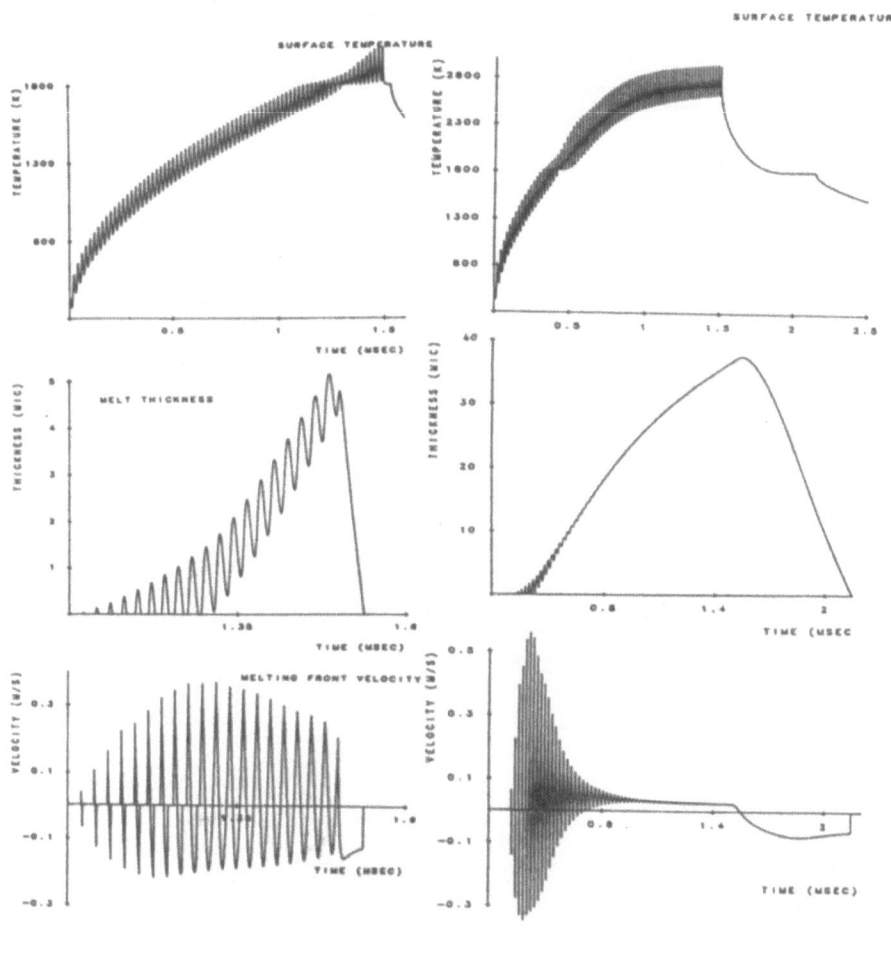

a   b

Figure 10. The action of laser pulse (duration 1.5 ms) with regular (ordered) generation (from Fig.9b); (a) q_{max} = 4.6 10⁴ ; (b) q_{max} = 1.2 10⁵ W/cm². Respectively, from up to down: surface temperature (K); melt thickness (μm); melting front velocity (m/s). Material - steel.

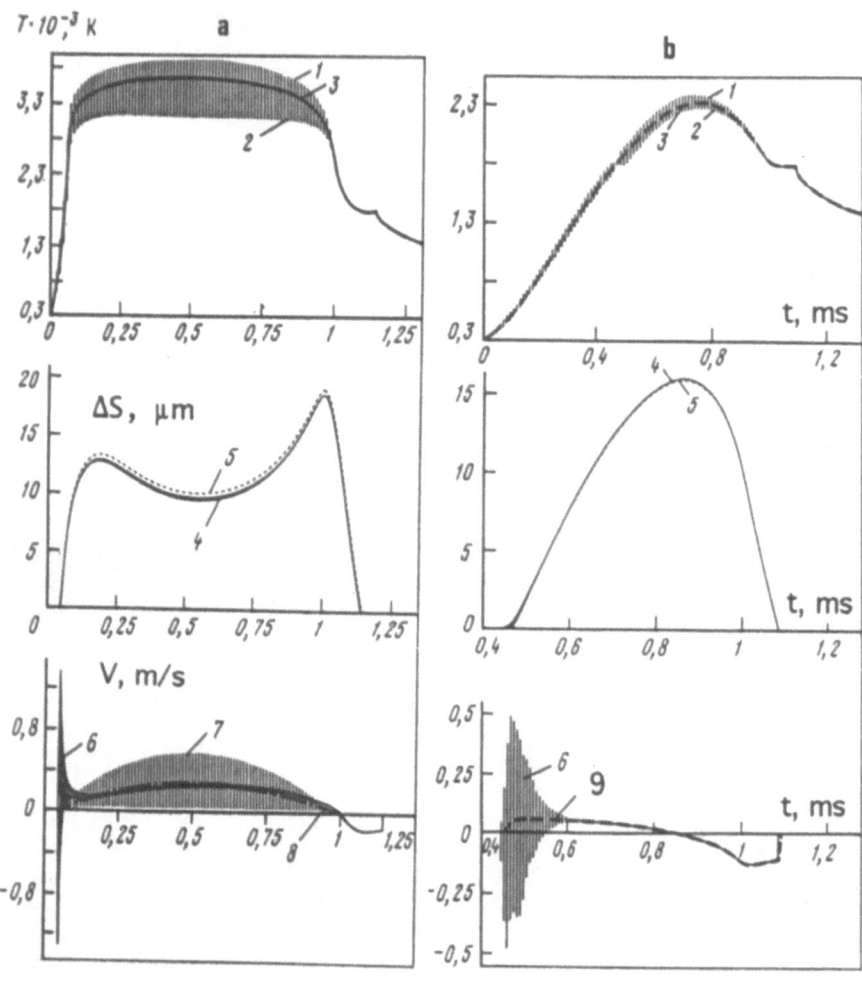

a   b

Figure 11. The action of laser pulse (duration 1 ms) with chaotic generation (from Fig.9a); (a) - q_{max} = 4·10⁶,

(b) - q_{max} =2·10⁵ W cm⁻²; curves: 1,2 - upper and low boundaries of the temperature's oscillations, 4 - melt
thickness, 6 - melting front velocity, 7 - evaporation front velocity. Dashed curves: 3 - temperature, 5 - melt
thickness, 8 - evaporation, 9 - melt front velocity correspond to the action of the similar parabolic pulse with the
smooth shape and equal energy input.

3.1.3. Two - dimensional melting-solidification problem.

The proposed 2D model is an extension of the above mentioned 1D model but with some simplifications. The main of them is that evaporation is assumed to be weak enough to neglect the movement of evaporation phase boundary, but enough strong to be taken into account as a heat loss [39]. Also the free surface of the melt is supposed to be flat. From the physical point of view it corresponds to the action of comparatively low energy density flux, when surface evaporation is weak, the amount of evaporated mass is small, and the reactive pressure of evaporation does not curve too much the free surface of the melt. In technological applications it corresponds, for example, to pulsed laser alloying (by solid state Nd:YAG - laser sources, $\lambda = 1.06$ μm) with energy density flow $q_0 = 5 \ 10^4 - 10^5$ W cm^{-2} and corresponding pulse duration $\tau = 1$-10 ms.

The used mathematical model (cylindrical symmetry is assumed) can be written in the following form:

EQUATION OF HEAT TRANSFER

$$\rho(t)(C_p(T) + \lambda_m \delta(T - T_m))\frac{\partial T}{\partial t} = \text{div}(\lambda(T)\,\text{grad}\,T) \qquad (2)$$

$$T = T(r,z,t) \ ; \quad t > 0 \ , \quad 0 < z < L \ , \quad r > 0$$

BOUNDARY CONDITIONS

$$-\lambda\frac{\partial T}{\partial z}\bigg|_{z=0} = q_0 f(r) - q_1$$

$$q_1 = \frac{V_* \exp(-T_* / T(r,z=0,t))}{T^{1/2}(r,z=0,t)}\rho L_v + \sigma\varepsilon T^4(r,z=0,t)$$

$$-\lambda\frac{\partial T}{\partial z}\bigg|_{z=L} = \sigma\varepsilon T^4(r,z=L,t) \ ; \quad \frac{\partial T}{\partial r}\bigg|_{r=0} = 0 \ ;$$

$$T(r=0,z,t) = T(r,z=0,t) = T_0$$

INITIAL CONDITION

$$T(r,z,t=0) = T_0$$

Here: r - radial coordinate, z - directed inside the material; $\delta(T\text{-}T_m)$ - delta function which is used for the treatment of latent heat evolution; $f(r)$ - radial distribution of absorbed energy flux. The rest nomenclature are the same as for 1D model.

The time discretization in heat conduction equation is based on the two level backward Euler (fully implicit) scheme. For spatial discretization was employed the control-volume finite-difference method [40]. The general iterative source based method on the fixed grid was used for the treatment of latent heat evolution [41]. To solve the linear algebraic equations, the tridiagonal matrix algorithm was used. The convergence at a given time step is declared when the residuals of enthalpy become less then 10^{-8}.

The results of heat process simulation under the action of laser beam with Gaussian spatial distribution $q=q_0 \exp(-kr^2)$ on Titanium slab 270 µm thickness are presented in Figure 12. The duration of laser pulse equals 7 ms, $q_0 = 1.5 \ 10^5$ W/cm^2, $k= 500$ cm^{-2}. Material properties and parameters of laser action remain the same in Figures 12-15.

The energy flux q_0- q_1 which is spent on the processes of heating and melting of metal decreases in time due to heat losses on evaporation.

The surface temperature in the middle of the heating zone $T(r=0, z=0, t)$ reaches the saturation value of about 4000 K.

During the initial period of laser action (curve 0, Fig. 12), both: heat flux (a) and surface temperature (b) copy the spatial distribution of the beam. Later on, because of sufficient nonlinearity of the task, they start to differ rather from the Gaussian shape. The real spatial distribution of surface temperature is important for a number of applications, for example, it determines thermocapillary convection of the melt. That is why to simulate the convective mass transfer in pulse laser alloying, one need to take into account surface evaporation.

After the end of laser action the solidification starts on the edges of the pool, while melting continuous on the pool's bottom.

On the final stage of solidification, the heat losses from the melt free surface induce the solidification of the pool from its peripheral parts towards the center and from the surface down into the bulk. The solidification front forms a circle and the last volume of the melt solidifies on some depth inside the plate.

The influence of evaporation phenomenon and slab thickness on melting/solidification dynamics is presented in Figures 14,15.

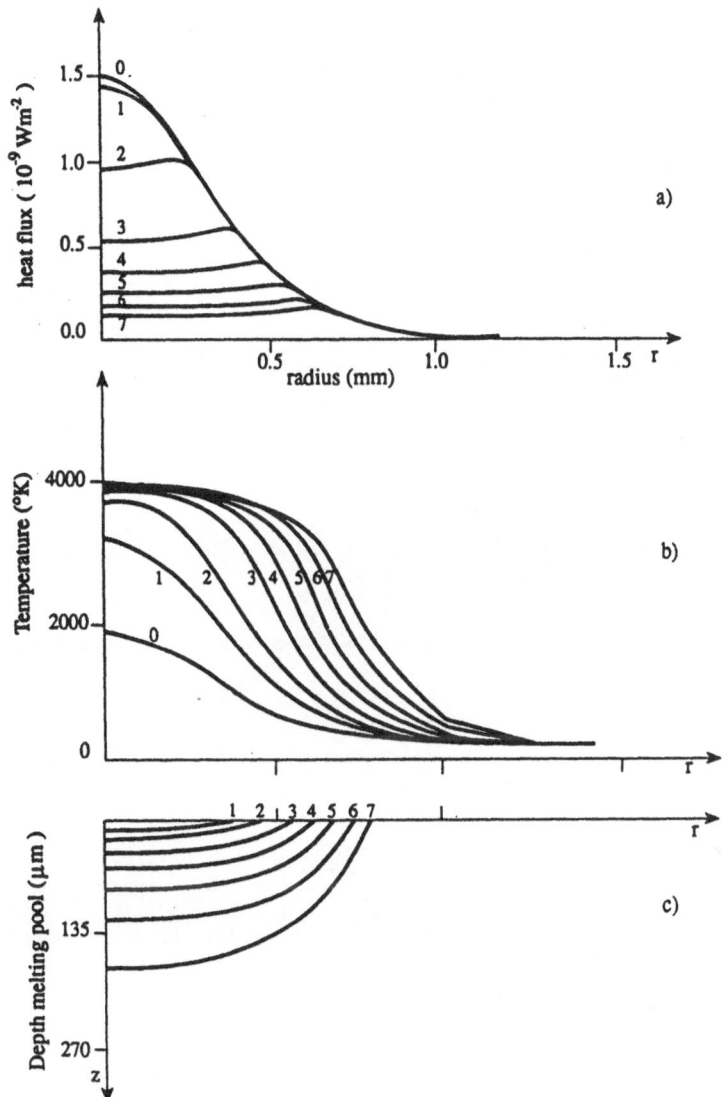

Figure 12. Distributions of (a) heat flux (q_0- q_1), (b) temperature along the irradiated surface;
(c) shape of melted pool. The curves correspond to the following moments of time:
0 - 3 10^{-2}; 1 - 0.1; 2 - 0.2; 3 - 0.5; 4 - 1; 5 - 2; 6 - 4; 7 - 7 ms.

284

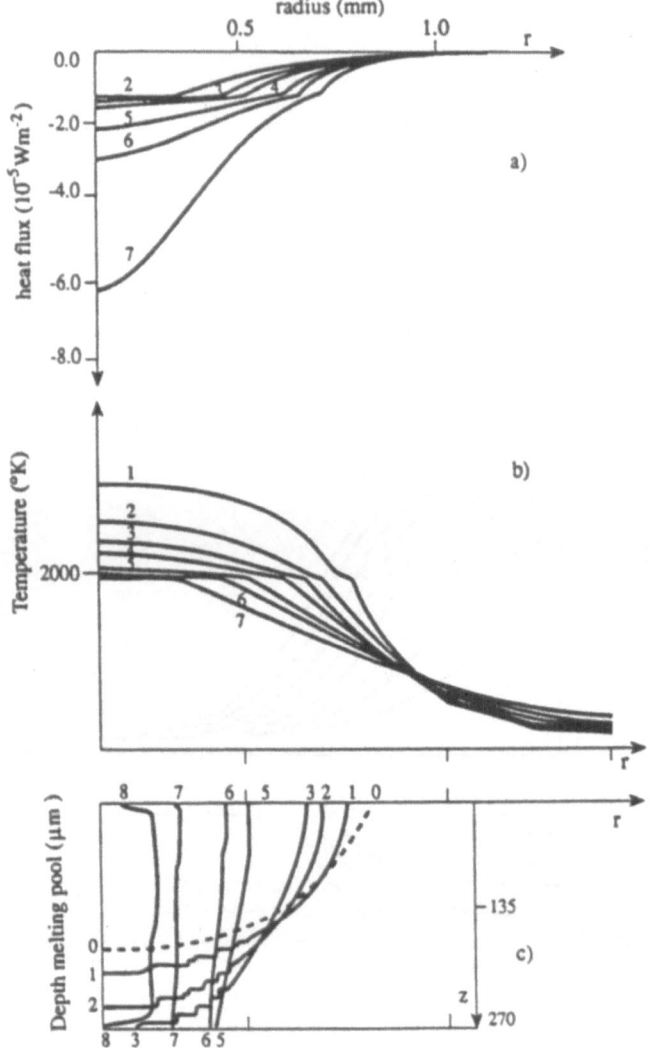

Figure 13. Distributions of (a) heat flux (q_0-q_1), (b) temperature along the irradiated surface;
(c) shape of melted pool. The curves correspond to the following moments of time:
0 - 7; 1 - 8; 2 - 10; 3 - 12; 4 - 14; 5 - 20; 6 - 25; 7- 35; 8 - 40 ms.

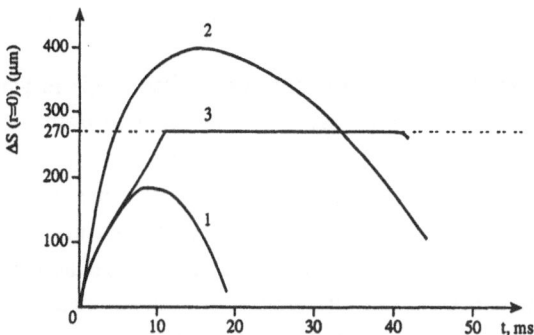

Figure 14. Melt thickness in the center of irradiated zone versus time : curves 1,3 - surface evaporation is taken into account; 2 - evaporation is neglected; 1,2 - thickness of the slab equals 1 mm; 3 - thickness equals 270 μm.

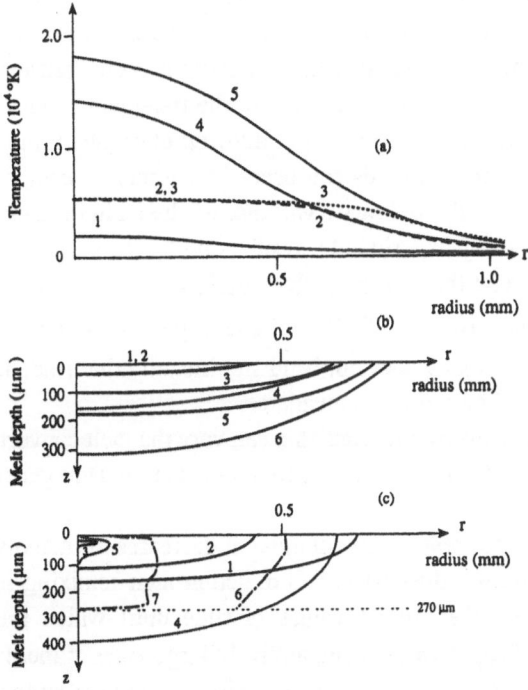

Figure 15. Surface temperature distribution (a) and shape of the molten pool (b) during pulsed laser action, (c)- after the end of laser action.

(a) curve 1 - t=310^{-2} ms; curves 2,4 - 2 ms; curves 3,5 - 7 ms; curves 2,3 evaporation is taken into account; curves 1,4,5 evaporation is neglected.

(b) curves 1,2 - t=0.2 ms; curves 3,4 - 2 ms; curves 5,6 - 7 ms; curves 1,3,5 evaporation is taken into account; curves 2,4,6 evaporation is neglected.

(c) curve 1 - t=8 ms; curve 2 - 14 ms; curve 3 - 19 ms; curve 4 - 14 ms; curve 5 - 44 ms; curve 6 - 20 ms; curve 7 - 40 ms; curves 1,2,3,6,7 evaporation is taken into account; curves 4,5 evaporation is neglected; curves 1-5 - thickness of the slab equals 1 mm; 6,7 - thickness equals 270 μm.

3.2. CONVECTIVE MASS TRANSFER IN PULSED LASER ALLOYING

Laser alloying has a great potential to improve surface properties of materials, as hardness, high temperature strength, wear and corrosion resistance (including high temperature). Up to the present time laser alloying is not yet widely used in industrial applications because of a number of reasons, among them are the following : (a) a great number of influencing factors (partially unknown) that makes difficult a good understanding and optimizing of the process; (b) poor repeatability of the results, including way of melt remixing and distribution of alloying elements; (c) sharp and unknown variations of surface properties, in particularly, surface tension; (d) intensive evaporation and thermal decomposition of alloying elements, and so on.

3.2.1. Main influencing factors for convective mass transfer in pulsed laser alloying
The main influencing factors for convective mass transfer in pulsed laser alloying generally do not differ deeply from the ones for pulsed laser welding, or surface melting; it is possible to say that pulsed laser alloying is more "sensitive" to them. As it will be shown later, even a slight variation of the treatment conditions can change the flow pattern and therefore distribution of alloying elements in the zone of action. So that, different concentration fields can occur in the molten pool, even if showing the same geometry practically. Below we will discuss the main influencing factors and will illustrate them both by experimental results and by theoretical estimations. It is necessary to underline that the present analysis deals with comparatively low values of energy density flux, $q_0 = 10^4 - 5 \cdot 10^5$ W/cm^2 and large pulse duration $\tau = 1 - 10$ ms. So, surface evaporation usually does not bend significantly the irradiated surface, and the keyhole or drilling effects are not considered.

It is possible to list the following reasons that can vary the melt convection:
- *energy input per pulse* (it increases the activity of remixing) or generally, intensity-time profile of laser pulse;
- *diameter of the zone of action*; when enlarged, it decreases the activity of remixing;
- *duration of action*: during the first period of action melt remixing is weak, so it is possible to speak about the critical values of time until which there will be no remixing practically. That is why a comparatively large zone of melting without any remixing could be obtained. Later on, the velocity of remixing strongly increases in time and the final result will be a quasi homogeneous distribution of doping elements;
- *spatial distribution of the beam*: sharp variations of energy flux induce sharp variations of surface temperature and therefore of surface tension; spatial distribution of the beam can determine the number of cells in the molten pool;
- *influence of surface evaporation* on the way of remixing: it can smooth out the surface temperature gradients; induce forced convection under the reactive pressure of evaporation: also, it can vary the chemical composition of the thin upper layer which is responsible for surface tension phenomena;

- *surface thermo-chemistry*: in our experiments was especially illustrated by the flow of different gases CO_2, N_2, Ar, He, air, etc.[62];

- the *properties of the material*: mainly the dependence of surface tension $\sigma(T,C)$ on temperature (generally non-linear) and on concentration of surface active elements (also generally non-linear) ; the temperature dependencies of absorbtivity, density and viscosity of the melt;

- the *angle of incidence of laser beam*: for high values of energy density flux, when surface evaporation is intensive, it can induce the propagation of the melt in the direction of the beam; in the general case the absorptivity depends on the angle of incidence.

Generally these factors can be separated on several groups:

- the factors that modify the surface temperature distribution: energy input, diameter of the zone of action, pulse duration (these three determine the energy density flux), spatial distribution of the beam, the angle of incidence, and the total surface heat losses (including evaporation). This group determines the features of, so called, thermocapillary convection $\sigma(T,C) \sim \sigma(T)$; the flow pattern is determined by the surface temperature and by the function $\sigma(T)$;

- the factors that determine the dependence of surface tension on the concentration of surface active element (SAE) : ambient gas atmosphere (SAE in the gas phase), the presence of surface coatings (initial deposition of SAE on the irradiated surface), initial presence of SAE in the base metal (for example, volume sulfur concentration in steel). This group determines the features of, so called, concentration-capillary convection $\sigma(T, C) \sim \sigma(C)$;

- when concentration of SAE and surface temperature vary sharply and simultaneously, the flow pattern is determined by both of terms $\sigma = \sigma(T, C)$;

- the reactive pressure of evaporation can bend the free surface, remove the melt from the zone of action, and induce the transition of melt convection from the surface tension driven to the another type of forced convection driven by evaporation induced phenomena [42].

3.2.2. Simulation of Convective Mass Transfer in Pulsed Laser Alloying

In pulsed periodic laser alloying solid state Nd:YAG - laser sources ($\lambda = 1.06$ μm), are usually used; whose main variables during processing are: pulse-on time 1 - 10 ms, energy density flow 10^4 -10^6 W cm^{-2} in the focused spot.

Depending on the above chosen range of values, both convection and surface deformation induced by reactive pressure are usually negligible. So the surface tension convection (Marangony effect) plays the main role in melt hydrodynamic, induced by surface temperature gradient and corresponding surface tension gradient [43-53].

3.2.2.1. Mathematical Model.

In a number of publications (for example, [43, 54-57]) it is pointed out that even a low content of solute element in the alloyed region has a strong influence on the dependence of surface tension σ versus temperature T and versus concentration C in comparison with the behavior of pure material. That is why, for the precise modeling of fluid dynamic caused by surface tension, it is necessary to take into account both thermocapillary and concentration capillary convection [43]. Nevertheless, for some metals, Fe as an example, when alloyed with Mo, V, Cr up to 20 at. %, the change of surface tension of the liquid phase (in the temperature range near the melting point) is of the order of 10% [56,57]. Where this is the case, the thermocapillary convection i.e., $\sigma(T,C) \sim \sigma(T)$ plays the main role.

Under the conditions of unsteady melting, the increase in the size of the molten metal pool with time, deeply effects the hydrodynamic processes, since the rate of propagation of the fusion front into the metal is comparable with the thermocapillary velocity; moreover, at the initial stage of melting, it considerably exceeded that. So that, to determine the velocity field in the pool, while taking into account the propagation of its boundaries with time, it is necessary to solve, both the hydrodynamic problem and the unsteady problem of heat conduction (with a Stefan type boundary condition at the melting front), keeping in mind the convective heat transfer too.

For practical applications, the parameters of pulsed laser treatment are optimized to obtain shallow pools (which diameter D is much larger than depth S). As an example let us consider: $q=10^5$ W cm^{-2}, $\tau=1$ ms, when $D = 1$ mm $>> S = 0.1$ mm. For these typical melting regimes, the thermal problem can be considered separately from the hydrodynamic one; the latter will be used as a basis for calculating the thermocapillary velocity field.

This opportunity occurs when heat conduction in the shallow pool predominates over convective heat transfer, i.e., when the melt velocities are fairly low (or the pool is shallow enough). The thermal problem determines the hydrodynamic problem that, in its turn, as a first approximation, does not react to the previous one. The present chapter is concerned with an analysis of the melt motion in this particular case.

Let us consider the two-dimensional (because of the cylindrical symmetry) melting of the massive metallic body by laser irradiation (surface heat source). We assume here that the thermal conductivity λ_i and thermal diffusivity a_i of the liquid phase ($i=1$) and solid phase ($i=2$), the dynamic viscosity η and kinematic viscosity ν of the melt, and the density ρ are constants; that the surface tension depends linearly on the temperature: $d\sigma/dT = -\alpha = $ const, and that the free surface of the liquid is not curved.

In the cylindrical coordinate system r, z, ϕ, where the z axis is directed into the target material and the laser beam is absorbed on the surface $z=0$, the phase transition front is determined by the equation $z=S(r,t)$, where t is time.

The melting and convection are considered to take place in a thin layer of melt, which maximum depth S_{max} is small if compared to the radius of the pool $R(t)$ determined from the equation $S(R,t)=0$.

Due to this, it will be possible to assume, in the heat - conduction equation, that the terms corresponding to the conductive heat transfer in the radial direction, are small enough if compared to heat conduction values along the z axis.

The hydrodynamic problem presents two characteristic linear dimensions, one of them exceeding the other. So the condition that the inertia terms appear small when compared to the viscosity terms, has the form $Re^* \ll 1$, where $Re^*=(S/r_0)^2 (v_r r_0 /v)$ is the reduced Reynolds number, r_0 is the radius of laser beam (for example, for Gaussian spatial distribution: $q(r)=q_0 \exp(-kr^2)$, $r_0 \sim (k)^{-1/2}$), v_r and v_z are the melt velocity components along the corresponding axes, respectively.

For liquid metals the Prandtl number $Pr=v/a_1 < 1$; so, in this particular case, we obtain the product $Re^* Pr \ll 1$. It means that in the melt the convective heat transfer is small if compared to conduction (Pe number is much less than unity). Because of this, the heat - conduction equation for the liquid phase does not contain the velocity components, i.e., the thermal and hydrodynamic problems are not coupled. The solution of the thermal problem determines the coordinates of the phase - transition surface and the temperature distribution along the free surface of the melt, which are used in the boundary conditions for the Stoke's equations.

As a consequence of the above-mentioned reasons, for the heat task, the appropriate governing equations are:

$$\frac{\partial T_1}{\partial t} = a_1 \Delta T_1 \quad , \qquad\qquad \frac{\partial T_2}{\partial t} = a_2 \Delta T_2 \qquad\qquad (3)$$

$$-\lambda_1 \frac{\partial T_1}{\partial z}\bigg|_{z=S(r,t)} = -\lambda_2 \frac{\partial T_2}{\partial z}\bigg|_{z=S(r,t)} + \rho_2 L_m \frac{\partial S / \partial t}{1+(\partial S / \partial r)^2}$$

$$-\lambda_1 \frac{\partial T_1}{\partial z}\bigg|_{z=0} = q_0 f(r)$$

$$T_1(r,z = S,t) = T_2(r,z = S,t) = T_m$$

$$T_2(r = 0, z, t) = T_2(r, z = 0, t) = T_2(r, z, t = 0) = T_0$$

Here: Δ - Laplace operator; T_i - temperature of the liquid and solid phase, respectively; T_m - melting temperature; T_0 - initial temperature; L_m - latent heat of melting; $f(r)$ - spatial distribution of absorbed energy flux q_0. To obtain the boundary condition on the melting front (Stefan type boundary condition), the temperature is assumed to be equal to T_m.

The solution of the task (1) for the Gaussian spatial distribution of energy flow, in case of neglecting the radial heat transfer (i.e., $\partial T / \partial r \ll \partial T / \partial z$), was obtained in [3, 58, 59].

The main idea of the solution method is based on the following assumptions: (a) the molten pool is shallow: $S \ll r_0 \sim (k)^{-1/2}$; (b) the time $t \ll r_0^2/a$ (a - thermal diffusivity coefficient, r_0 – laser beam radius); so that $\partial S / \partial r \ll 1$; $\partial T / \partial r \ll \partial T/\partial z$; $\partial^2 T /\partial r^2 \sim (1/r) \partial T/\partial r \ll \partial^2 T /\partial z^2$. The heat conduction equation and the boundary condition on the melting front can now be written in the form:

$$\frac{\partial T_i}{\partial t} = a_i \frac{\partial^2 T}{\partial z^2} \tag{4}$$

$$-\lambda_1 \frac{\partial T_1}{\partial z}\bigg|_{z=S(t)} = -\lambda_2 \frac{\partial T_2}{\partial z}\bigg|_{z=S(t)} + \rho_2 L_m \frac{dS}{dt}$$

In [60] the radial heat transfer was considered to be a small correction (order of magnitude of the value S/R) to the heat transfer inside the bulk of material. The approximate solution of the task (3) was found in the form

$$T_i^{(1)}(r, z, t) = T_i^{(0)}(r, z, t) + \delta T_i(r, z, t), \quad \delta T_i \ll T_i^{(0)}, \quad \delta = S / R \ll 1$$

where the phase boundary position $S^{(1)}(r, t)$ is satisfied by the correlation

$$(S^{(1)}(r, t) - S^{(0)}(r, t)) \ll S^{(1)}(r, t), S^{(0)}(r, t)$$

The values indexed by (0), are the solution of the task (3) without radial heat transfer.

Another limitation of the present model is neglecting of surface evaporation that could lead to the increase of surface temperature up to unreal values, exceeding boiling temperature. The parameters of laser action respecting the limitations of the model are in the range of: energy density flow $q=10^4$ - 10^5 W cm^{-2}, pulse duration - a few milliseconds.

The solution of the one-dimensional problem (3) can be obtained both analytically [3] and numerically [3, 32, 33].

Another consequence of the shallow molten pool assumption is the following: Re*<<1, $v_z << v_r$, $\partial v_j /\partial r << \partial v_j /\partial z$, $j=r,z$; so that the hydrodynamic task can be presented in the following form :

$$\frac{\partial v_r}{\partial t} = -\frac{1}{\rho}\frac{\partial p}{\partial r} + v\frac{\partial^2 v_r}{\partial z^2}$$

$$\frac{\partial v_z}{\partial t} = -\frac{1}{\rho}\frac{\partial p}{\partial r} + v\frac{\partial^2 v_z}{\partial z^2} \tag{5}$$

$$\frac{1}{r}\frac{\partial}{\partial r}(rv_r) + \frac{\partial v_z}{\partial z} = 0$$

$$\frac{\partial v_r}{\partial z}\bigg|_{z=0} = \frac{\alpha}{\eta}\frac{\partial T(r,0,T)}{\partial r}$$

$$v_r[z = S_1(r,t)] = v_z[z = S_1(r,t)] = v_z[z = 0]$$

Here p - is a pressure.

The approximate analytical solution of the system (5) is based on the method discussed in details in [3, 58-60].

One of the main assumptions of the shallow molten pool approach is that both Pe*=PrRe* and Re* are much less than unity. The latter is more critical because for liquid metals Pr<1. For Gaussian spatial distribution of energy flux: $q(r)=q_0\exp(kr^2)$, it is possible to show that in the frame of the above mentioned assumptions Re*~ $(\alpha/4v^2\rho\lambda)q_0kS^4$, where S is the pool depth. As an example, for $q_0=510^4$ W cm^{-2}, $k=25$ cm^{-2}, $t=1.5$ ms - Re*~0.1. To obtain a simple and at the same time accurate estimation for $S(t)$ is a difficult task. If we will use the very rough estimation $S \sim (at)^{1/2}$, that gives essentially larger values of melt depth, the criteria will have the following form Re*~$(\alpha a^2/4v^2\rho\lambda)q_0kt^2$ <<1. As a general comment it is possible to note, that the smaller are the considering periods and corresponding melt thickness, the better is the accuracy of the obtained solution. Concerning the below reported results is necessary to underline, that starting nearly from the middle of the considered periods, they are mainly qualitative.

   3.2.2.2. Gaussian spatial distribution of laser beam. It is necessary to underline some peculiarities of the initial stage of the process. After the beginning of melting when $t > t_m$ the derivations $\partial v_r/\partial t$ and $\partial v_z/\partial t$ are close to zero, but the melting front velocity increases nearly proportionally with time [3, 59]. It means that on the initial stage of melting (and alloying) it is necessary to take into account the moving of the phase boundary, because $\partial S/\partial t >> v_r, v_z$. The special feature of the phase front evolution resides from the fact that during the initial stage the bath rapidly increases in a radial direction and after having reached the value, close to the radius of the heating spot, continue to grow mainly deep inside of the metal [3,60].

   The flow of the melt has a vortex structure. In the region near the center of the heating spot the liquid moves to the surface, whereas on the free surface, $z=0$, it moves from the center to the pool edges (Fig. 16), since with $\alpha>0$ the thermocapillary force is directed to the side of decreasing temperature. By solving the equation $v_r=0$ the coordinate of the flow turning can be found. The analysis of this formula [59] shows that this coordinate depends weakly on r and t and is equal approximately to $z\sim S(r,t)/3$ (see the dashed line in Fig. 16). The position of the vortex center is determined by the approximate relation $r_c\sim R(t)/2$, $z_c\sim S(r_c,t)/3$.

   Before going over to the analysis of the mass transfer of alloying elements from the surface $z=0$ into the melt volume, note that the admixture can reach the bottom of the pool only in the case when the maximum positive value v_z^{max} (attainable at the point $r_m\sim 3R(t)/4$, $z_m\sim S(r_m,t)/3$) exceeds the velocity of the melting front dS/dt (see Fig. 17). It showed be emphasized that this estimate affords only the necessary (but not sufficient) condition for deep mass transfer; a more detailed analysis requires the study of the trajectories of liquid particles.

   Because of the transient character of the melting process (see Figs. 18, 19), the shape and the dimensions of the vortex vary with time. Consequently, the shape and dimensions of trajectories of liquid particles change (Fig. 20).

   3.2.2.3. Concentration Fields. The simple estimations show that the typical diffusion propagation depth for the laser action with pulse duration of the order of a ms is much less than the convection propagation length : $(D \cdot \tau)^{1/2} \sim 5\ \mu m << v_z \cdot$ $\tau \sim 100\ \mu m$. So it is possible to neglect the diffusion mass transfer of alloying elements in comparison with the convective one (on the distance comparable to the molten pool depth).

   The concentration fields in the zone of laser alloying are determined mainly by: the characteristic features of the hydrodynamic of the melt, by the motion of the melting front and by the method how the alloying elements are introduced into the melt

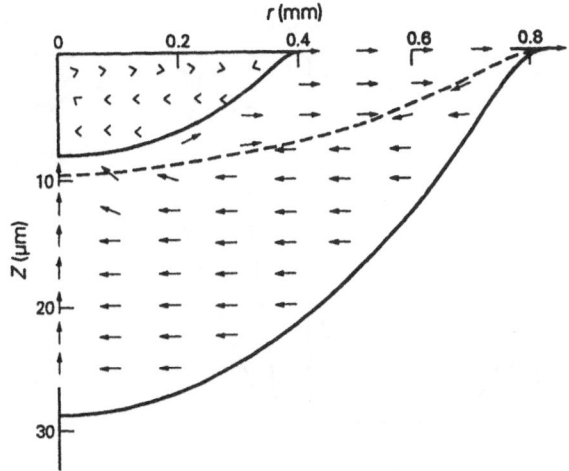

Figure 16. Molten pool shape and field of directions for velocity vector in surface tension melt flow of titanium (q_0=5.10^4 W cm^{-2}, k=100 cm^{-2}) t=0.47 ms (>) and 1 ms (−>).

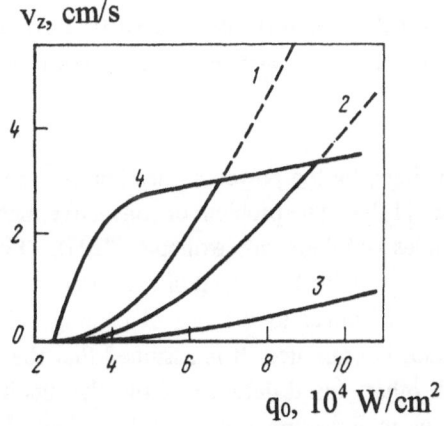

Figure 17. The Vz component of melt velocity in the spot $r=3R(t)/4$, $z=S(r,t)/3$, for t=1 ms versus absorbed energy density flux q_0. Material Ti; curve 1 corresponds to k=200 cm^{-2}; curve 2 - k=100 cm^{-2}; curve 3 - k=20 cm^{-2}; curve 4 - velocity of melting front $S(r,t)/t$.

294

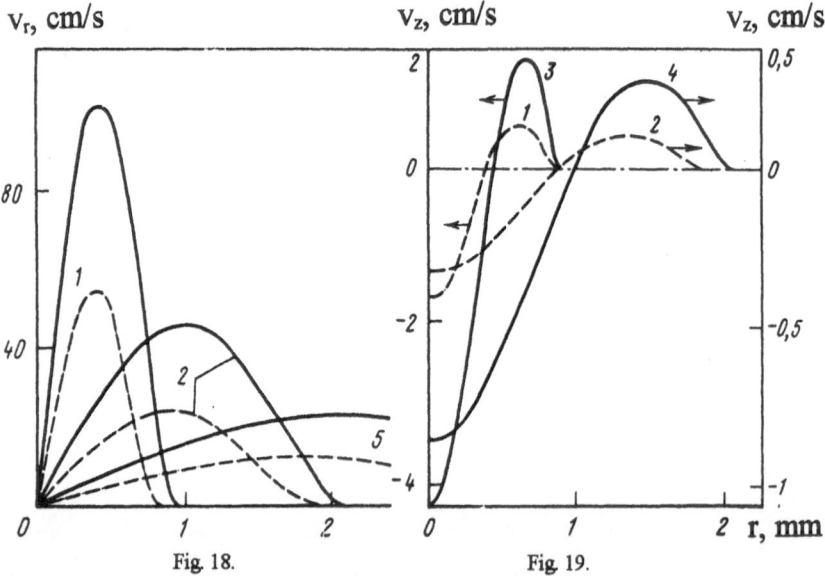

Fig. 18.   Fig. 19.

Figure 18. The Vr component(for $z=0$) versus radius for $t=1$ ms (dashed curves) and $t=1,4$ ms (solid line curves). Material Ti, $q_0=5 \cdot 10^4$ W/cm^2; curve 1 correspond to $k=100$ cm^{-2}; curve 2 to $k=20$ cm^{-2}; curve 3 to $k=5$ cm^{-2}.

Figure 19. The Vz component of melt velocity (on the depth $z=S(r,t)/3$) versus radius for $t=1$ ms (dashed curves) and $t=1,4$ ms (solid line curves). Material Ti, $q_0=5 \cdot 10^4$ W/cm^2; curves 1 and 3 correspond to $k=20$ cm^{-2}; curves 2 and 4 to $k=100$ cm^{-2}.

(from predeposited coatings, by the powders injection on the melt free surface, from the gas or liquid phase [1,2]). The problem of convective mass transfer was solved by the method of particles (markers, for example [61]). For the alloying from predeposited coatings, towards the beginning of the laser melting of the surface layer, which has the thickness identical to the one of coating being modeled, a constant concentration of particles is presented. It is assumed that the markers do not interact with the melt or each other; the difference of the thermophysical properties of the coating and the substrate is disregarded in the calculations. The latter corresponds to the coating's thickness much smaller than the typical depth of the melt. In this case the average heat transfer properties and the viscosity of the mixture "coating+ substrate" will not differ rather from the corresponding properties of the basic metal. If on the one hand the considering period is much larger than the typical time of coating's material to left the surface of the melt, if also on the other hand the values of $d\sigma/dT$ for the material of the coating and the substrate are not far one from the other, the influence of the admixture concentration on the surface tension driven convection is then not so critical. In this case the existence of the admixture on the melt free surface can cause the quantitative, but not qualitative difference in the dynamic of convective mass transfer. An alternative approach is to consider the melting and remixing of the two layers system.

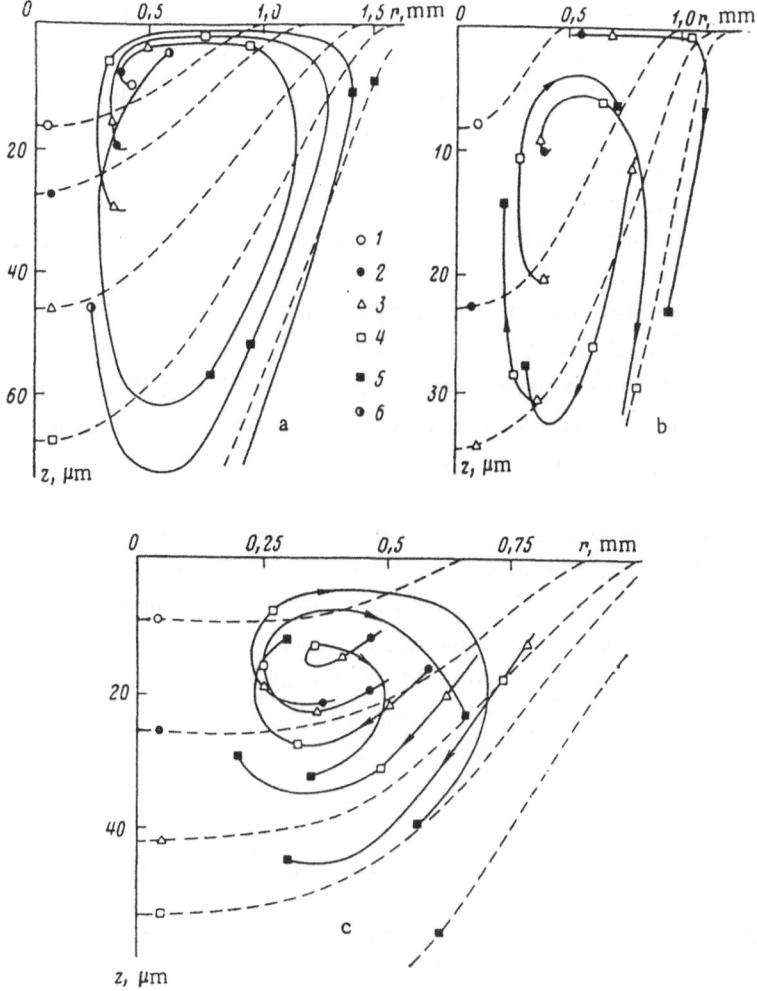

Figure 20. Trajectories of particles of the melt : material - Ti; (a) $q_0=5.10^4$ W/cm^2, $k=50$ cm^{-2}; (b) $q_0=10^5$ W/cm^2, $k=100$ cm^{-2}; (c) $q_0=10^5$ W/cm^2, $k=150$ cm^{-2}. The time is indicated by the positions of symbols 1-6 that correspond to the following moments : symbol (1) - $t=0.2$ ms, 2- 0.4 ms, 3- 0.6 ms, 4 - 1 ms, 5 - 1.6 ms, 6 - 2.3 ms.

In the present approach only the period of laser action is considered, but in reality it is followed by the periods of cooling and solidification. It is possible to analyze some of their typical features on the base of heat model (1), taking into account the processes of evaporation and heat losses from the irradiated surface (but without any kind of hydrodynamics). These results can be used to estimate the relaxation of surface temperature, and therefore - thermocapillary force. At the end of laser pulse (absorbed energy density flow $q_0 =10^5$ W cm^{-2} ; pulse duration $\tau=1$ms; material - steel) the irradiated surface temperature reaches 1.5 T_m and the melt thickness - nearly 40 μm; 0.05 ms later the surface temperature decreases up to 1.25 T_m

(mainly because of the cooling from the surface), but melt thickness increases on 3 μm because of the melt overheating; 0.2 ms later the surface temperature is 1.07 T_m and melt thickness - 34 μm (see Figure 5). The complete solidification of the melted layer takes place 0.6 ms after the end of laser pulse. The typical feature of the melt solidification in the processes of pulsed laser action is that it takes place at practically zero temperature gradient in the liquid phase [33]. It is necessary to underline that this result is obtained on the basis of one-dimensional heat task without any hydrodynamics. The absence of the temperature gradient inside the pool means its absence also on the melt free surface (in any case the temperature on the molten pool boundary equals T_m). Also it is possible to show that surface heat losses will reduce the temperature gradient in the radial direction (see Chapter 3.1.3). Without the driving force viscous melt flow stops, and it is not necessary to consider the transformation of the concentration fields during period of solidification. The period of the surface temperature decreasing (because of heat transfer into the bulk of metal and because of surface heat losses) up to practically melting point value is much smaller than both melting and solidification periods (also during this period the temperature gradient and therefore the driving force is smaller than before). It seems possible to neglect the movement of the melt during this short period and therefore - its influence on the concentration field transformation. This rough qualitative analysis corresponds to the assumptions of the shallow molten pool, i.e., neglecting of the inertia terms in Navier-Stokes equations in comparison with the viscous ones. It is possible to expect that sufficiently viscous melt flow will spot in absence of the driving force. As a matter of fact, however, the condition $Re^* \ll 1$ is often not satisfied on the final stage of remixing, when surface tension gradient induces a strong flow pattern, i.e., fluid flow will continue the laser power is turned off.

The influence of solidification induced phenomena on the melt convection (and elements redistribution) is the subject of independent investigations.

In the frames of the above mentioned assumptions the concentration fields of alloying elements for the pulse laser action with Gaussian spatial distribution were obtained in [3, 59]. The remixing of the predeposited coatings (Fig. 21) and the prescribed alloying element flux on the surface of the melt (Fig. 22) were considered.

When alloying material is absorbed by melt free surface (alloying by powders flow, from gas phase, etc.) not only the redistribution of the admixture over the molten pool takes place, but also an increase in the quantity of the doping substance in the laser remelted zone.

To obtain the universal form of presentation of concentration fields of alloying elements the following procedure is used. The maximum value of concentration in the pool for a fixed moment is equalized to unity. Three ranges of concentration and corresponding zones are distinguished : the first one for the concentration

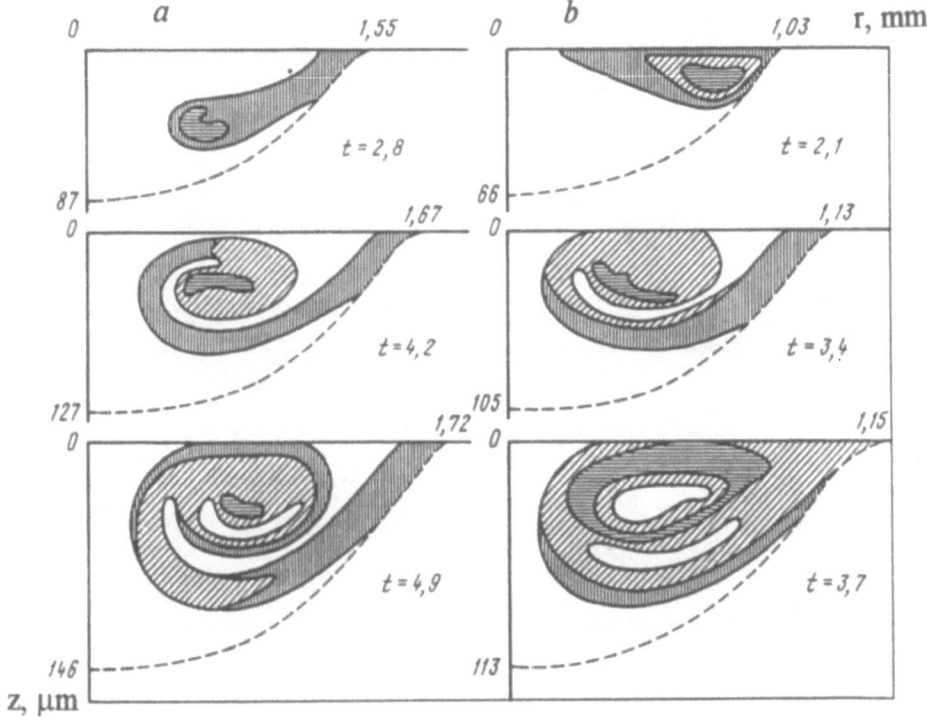

Figure 21. Concentration fields of the alloying element at indicated instants of time (ms). Alloying from predeposited coatings (thickness 15 μm); base metal Ti; $q_0 = 5 \cdot 10^4$ W/cm^2; (a) k=50 cm^{-2}; (b) k=100 cm^{-2}.

range 0.5-1 is shown by the horizontal hatching; for 0.1-0.5 - by inclined hatching; the zone with the concentration less than 0.1 is shown by vertical hatching. The actual concentration values in the above indicated zones in different figures may differ rather depending on time and/or process parameters.

In the majority of publications, as well as in the previous chapters, the convective mass transfer is analyzed in particular for the liner decrease of surface tension on temperature $(d\sigma/dt = C_1 < 0)$. The concentration fields for the linear increase of surface tension on temperature $d\sigma/dt = C_2 > 0$ previously were obtained in [62] (Fig. 23).

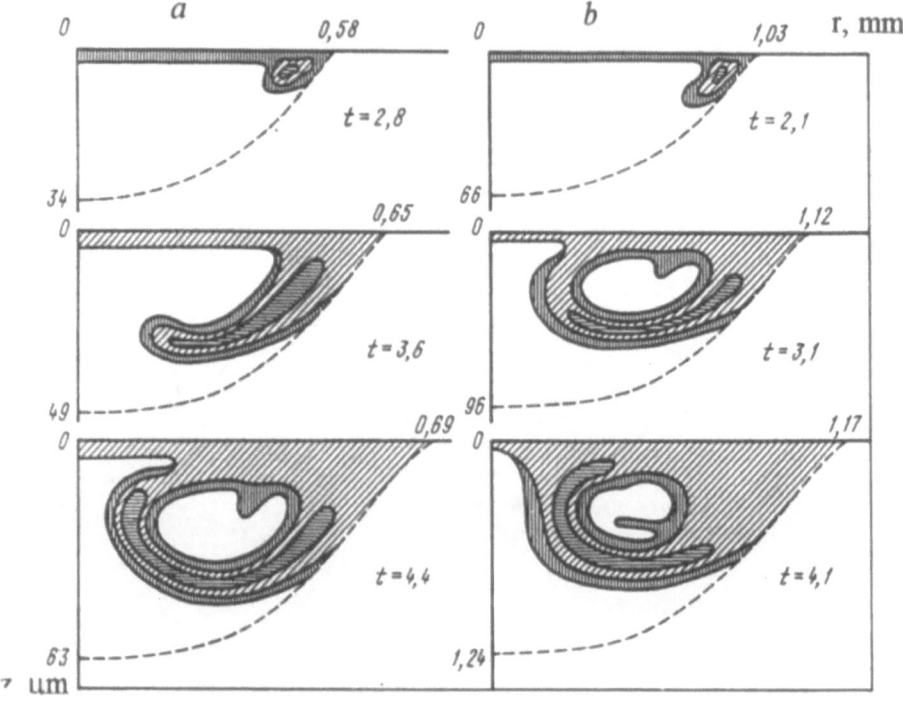

Figure 22. Concentration fields of the alloying element at indicated instants of time (ms). Absorbed constant flux of alloying material on melt free surface; base metal Ti; (a) $q_0=2.5\cdot10^4$ W/cm^2; (a) k=150 cm^{-2}; (b) $q_0=5\cdot10^4$ W/cm^2; k=100 cm^{-2}.

3.2.2.4. Multi-Cells Flow Pattern. It is necessary to underline that the single cell model (3,5) can predict only some qualitative (not quantitative) features of convective mass transfer in pulse alloying, that is why agreement with experimental results (Fig. 23, for example) was obtained only on qualitative level. Moreover, in case of $d\sigma/dt=C_2>0$ the single cell model can explain some features of convective mass transfer, that are observed in our experiments, mainly in the central part of the pool. In the peripheral part of the pool often can be observed another cell (vortex) rotating in the opposite direction (see Fig. 24). On the contrary, in case of $d\sigma/dt=C_1<0$, the single cell model can explain the penetration of alloying elements mainly in the peripheral part of the pool [59,60,62].

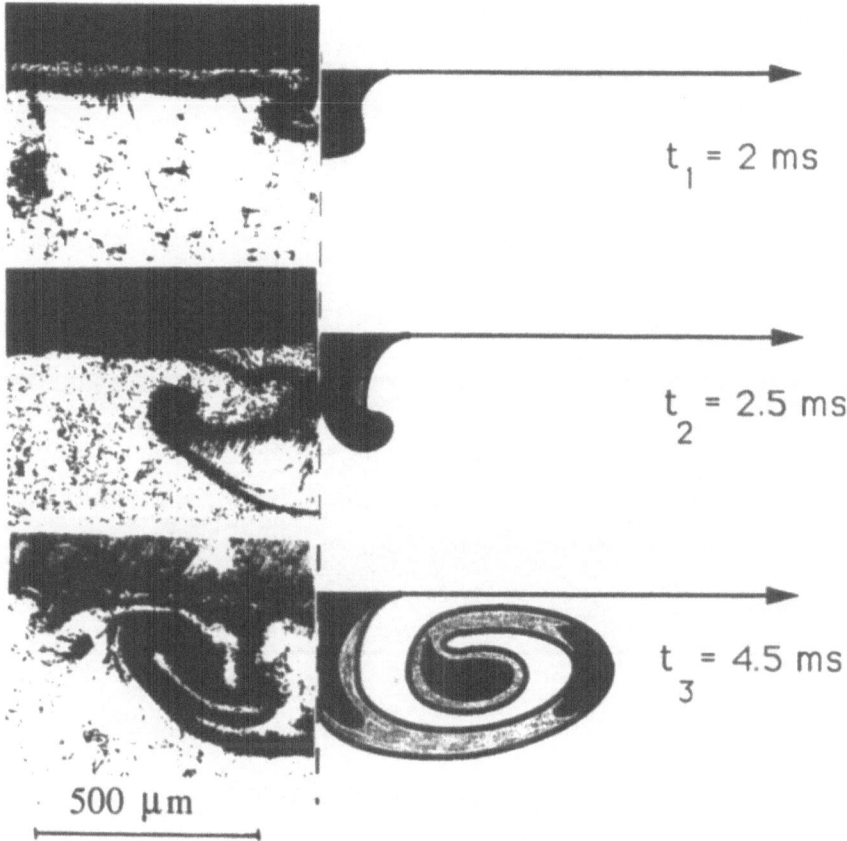

Figure 23. The propagation of admixture in pulsed laser alloying from pre-deposited coatings for positive value of surface tension coefficient. Numerical simulation: $q_0 = 5 \ 10^4 \ W/cm^2$, $k = 50 \ cm^{-2}$, the corresponding instants of time: 2, 2.5, 4.5 ms. Experiment: graphite coating 10 μm thickness on Ti, pulsed Nd:YAG laser.

More realistic explanation of convective mass transfer can be done on the basis of multi-cells flow pattern. The physical reasons to increase the number of cells are mainly the following: (a) non monotone temperature distribution along the free surface (can be induced by non monotone spatial distribution of the beam); (b) non monotone temperature dependence of surface tension on temperature, and on concentration of surface active elements; (c) the non stability of a single cell structure when one dimension sufficiently exceeds another; (d) the interaction of surface tension driven and free convection; (e) the influence of surface evaporation and its non stability. The latter varies the surface tension gradient (by modifying the chemical composition of the thin upper layer, which is responsible for surface tension phenomena), and at the same time induces another type of forced convection in the zones of maximum temperatures, on the other hand.

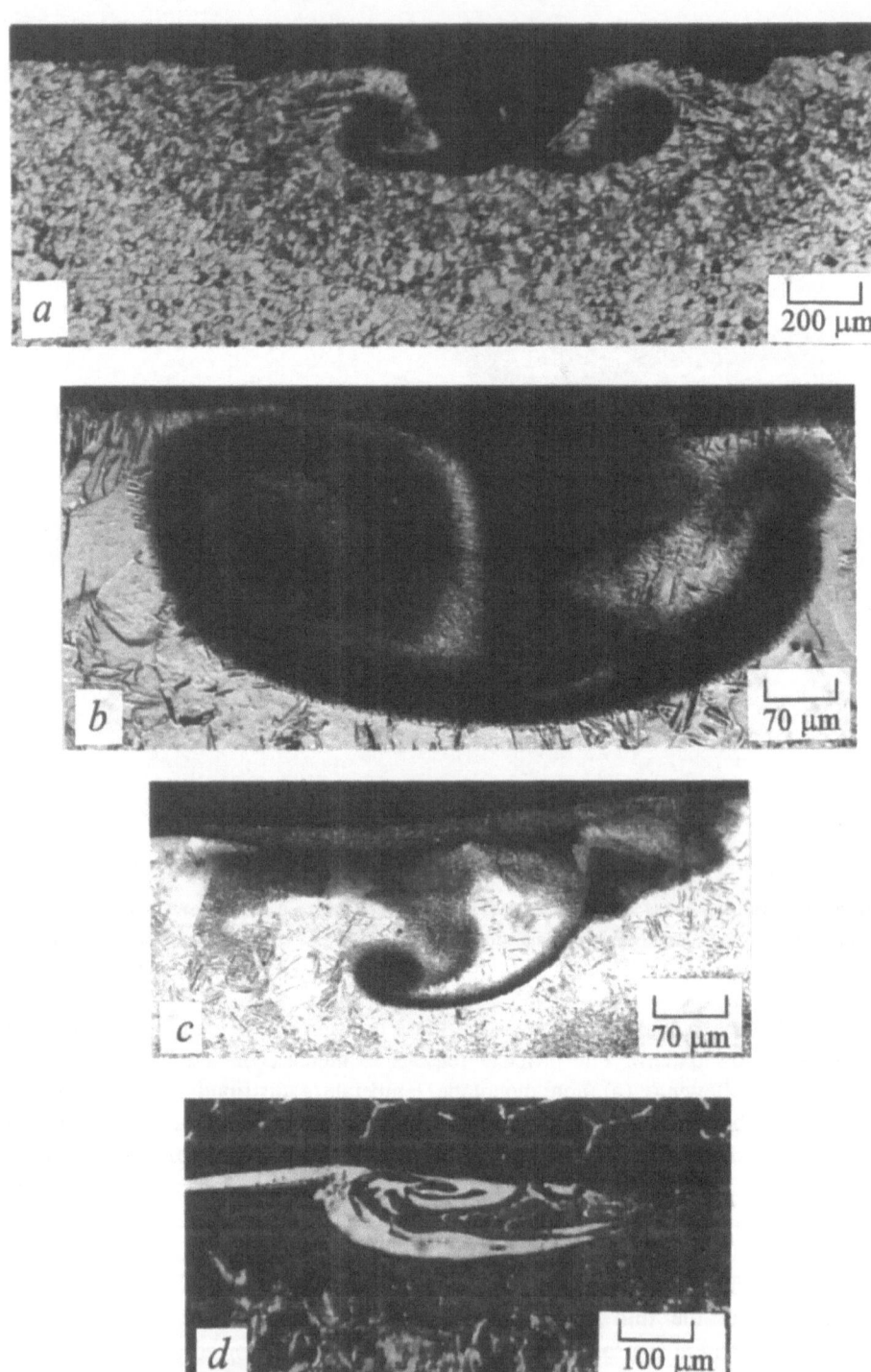

Figure 24. Pulsed laser alloying of Ti from graphite coating (a, b, c) and Fe from Mo coating (d). Cross-sectional photos : (a) all the zone of action; (b) central part; (c) right part; (d) left part.

3.2.2.4.1. Influence of Spatial Distribution of the Beam. The present chapter deals with the analysis of the relationships between the typical laser beam intensity profiles, the corresponding melt flow pattern, and concentration fields of alloying elements.

Flow Pattern. When the spatial distribution of the laser beam has a number of local extremes, the flow in the pool has a multi vortex structure [60]. In the region near the maximum of energy density flux the melt moves towards the surface and at the free surface it moves from the regions of temperature maximum towards the regions of temperature minimum. Since the surface tension force acts in the direction of increasing surface tension, i.e., for the typical situation $d\sigma/dT<0$ (for example Cr, V, Mo, W - on the Fe substrates) in the direction of temperature decreasing. For the energy spatial distribution with a local minimum in the spot center and a local maximum on some distance from it, the flow structure consists of two toroidal vortexes, separated by the position of energy flow maximum (Fig. 25a). For the energy spatial distribution with two local maximums: one in the middle of the zone of action and the second - on some distance from it, the flow structure consists of three toroidal vortexes, which are separated by the positions of local extremes of energy flow (Fig. 25b). In the position of local minimum of surface temperature the flow moves deep inside the molten pool; on the contrary in the position of local maximum of surface temperature the melt rise to the free surface. So the neighboring vortexes always rotate in opposite directions. The number of vortexes and that of local extremes of spatial distribution of energy flow is just the same. The vortexes are heavily flattened in the direction of z axis.

The dependence of radial component of melt velocity $V_r(r, z=0, t)$ on the radius is presented in Fig. 26. Plain lines 1,2 and dashed line 2 (Fig. 26a) do not intersect in one point on x -axis because of a small shift of surface temperature extreme. The parameters of Gaussian spatial distributions were chosen to be close to the nonmonotonous ones in the central regions of the zones of action.

Let us make the rough estimation of the influence of nonmonotonous spatial distribution of laser beam on the absolute values of v_z , which are responsible for the alloying elements transport inside the material. There are mainly two kinds of parameters: the characteristic change of energy density flow $q = =q_0 f(r)$ between two neighboring local extremes - Δq and the distance between them - Δr. Let us compare two cases of laser action: the first - nonmonotonous spatial distribution of laser beam, with corresponding mean value q_0 and spot action radius r_0 ; the second one - for the Gaussian spatial distribution $q=q_0\exp[-(r/r_0)^2]$ with the same values of q_0 and r_0 . The melt thickness will have the same order of magnitude, because of the approximate equal energy input. In this case the estimation of the ratio of axial velocities (1 - for the nonmonotonous, 2 - for Gaussian spatial

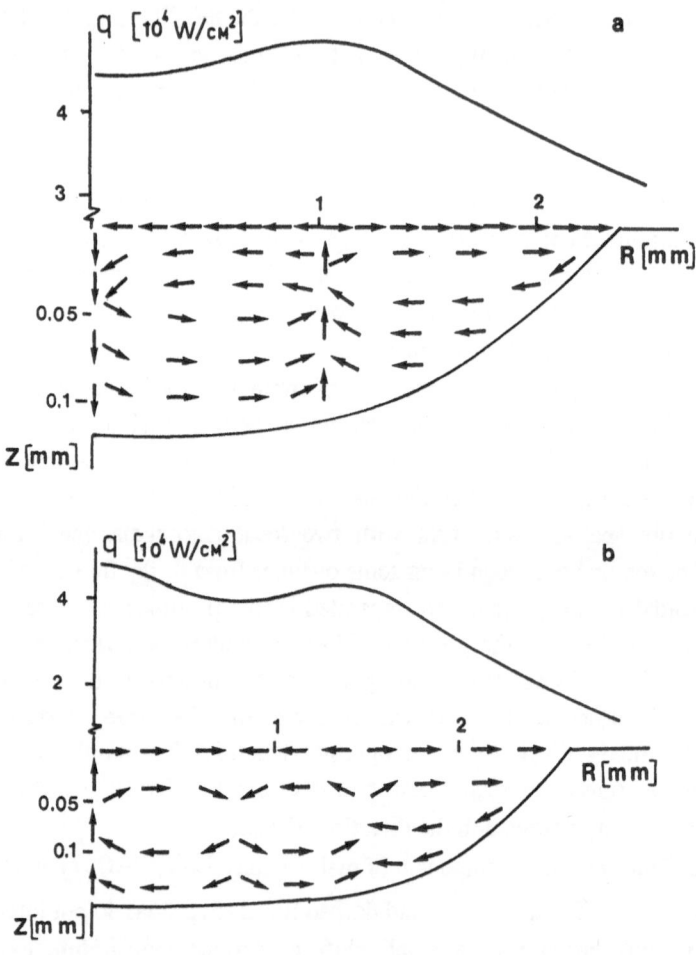

Figure 25. Field of directions for velocity vector of surface tension melt flow of titanium at time moment t=3.5 ms (a), t=4.3 ms (b) for the absorbed energy flux radial distribution, shown on upper curves.

distribution) can be presented in the form $v_{z_1}/v_{z_2} = (\Delta q/q_0)(r_0/\Delta r)^2$. From this estimation it is possible to see that for the constant average values of q_0 and r_0 and nonvariable Δq, by increasing the number of local extremes (i.e., decreasing Δr) an intensification of the surface tension remixing is possible. On the other hand it is possible to decrease the liquid velocity in the corresponding volume of the melt by decreasing the Δq value.

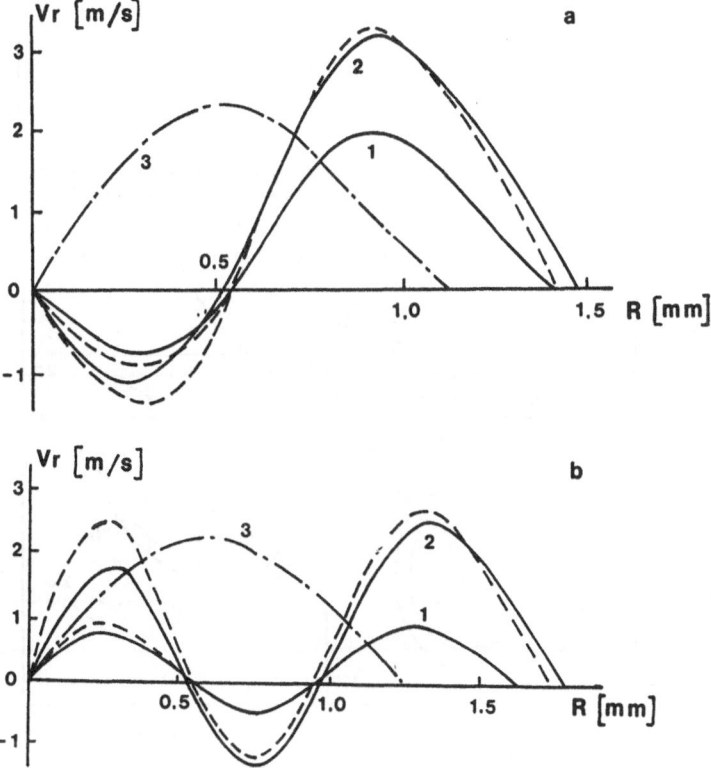

Figure 26. Radial component of melt velocity V_r (r, $z=0$, t) versus radius (material Ti); plain line curves - taking into account radial heat transfer; dashed curves - neglecting it; 3 - for Gaussian spatial distribution $k=90$ cm^{-2} (a), $k= 68$ cm^{-2} (b).

a - spatial distribution of laser beam is similar to the one in Fig. 25a, curve 1 - t= 2.5 ms; curves 2,3 - t=3 ms;
b- spatial distribution of laser beam is similar to the one in Fig. 25b, curve 1 - t= 2.0 ms, curves 2,3 - t=3.5 ms.

Concentration Fields: (a) Continuous coatings. The results of numerical simulation of convective mass transfer during pulsed laser action on Fe and Ti base materials are depicted in Figs. 27-30. The concentration field of alloying element is normalized on the value of its initial concentration in the coating (taken as unity). The distribution of alloying element in the molten pool is presented by the curves corresponding to the constant values of concentration. In all these cases the coating is continuous, it means that the coating radius is larger than the one of the heating zone. The corresponding shapes of molten pool are presented in the frames (windows) in reduced scales, because the areas of spreading of alloying element and the dimensions of molten pool do not always coincide. The transformation of the molten pool shape in time makes it smoother because of the radial heat transfer, less corresponding to the non-monotone spatial distribution of laser beam.

304

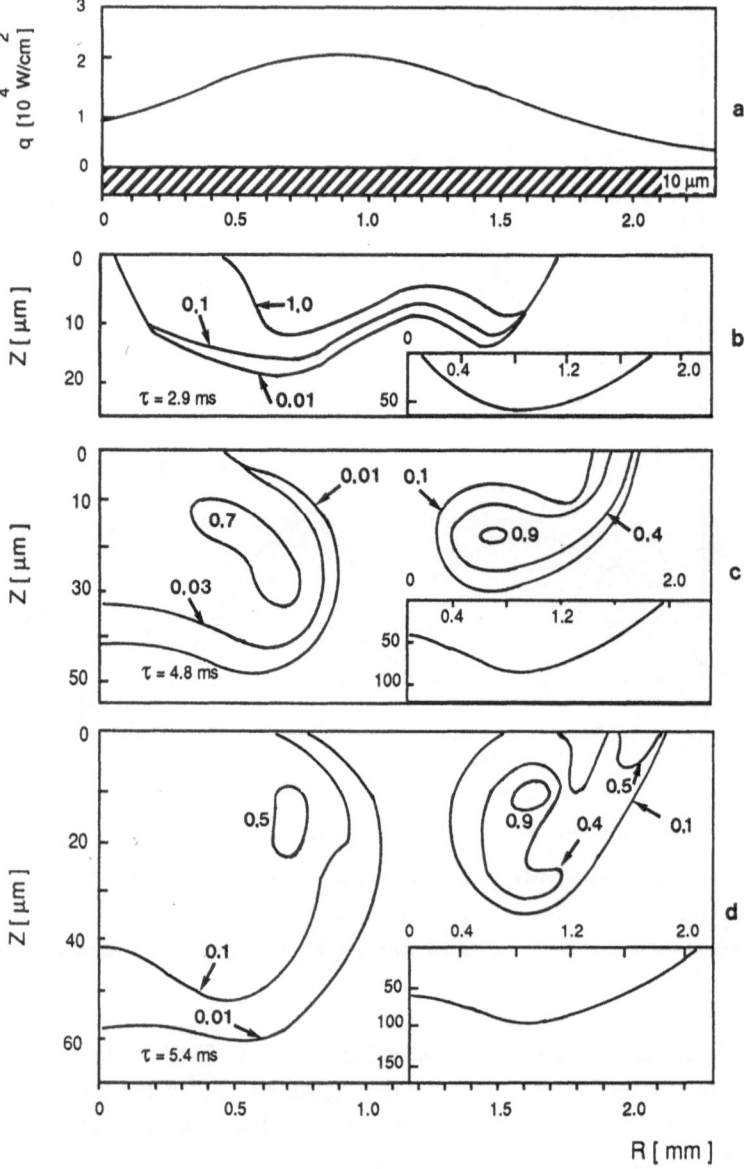

Figure 27. Concentration fields of the alloying element at corresponding time instants t=2.9 (b), 4.8 (c), 5.4 (d) ms. Base metal Ti; coating thickness 10 μm; continuous coating; (a) - spatial distribution of energy density flow and of the coating initial position. The corresponding shapes of molten pool are shown in the frames in the reduced scale.

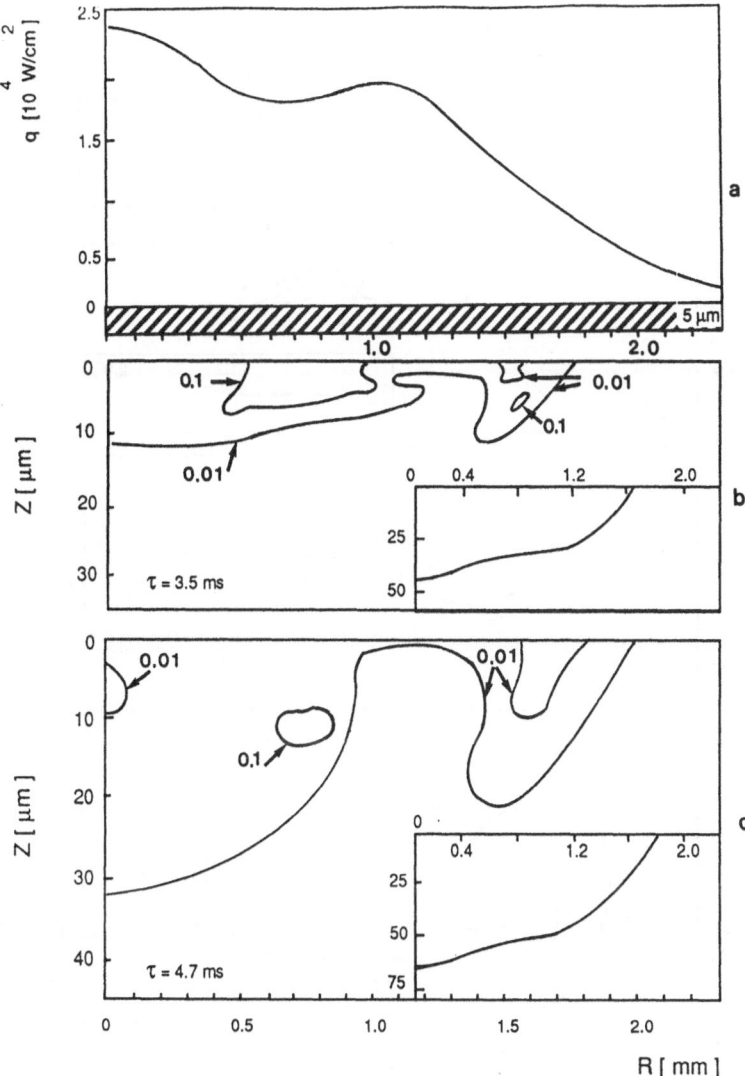

Figure 28. Concentration fields of the alloying element at corresponding time instants t=3.5 (b), 4.7 (c) ms. Base metal Ti; coating thickness 5 μm; continuous coating; (a) - spatial distribution of energy density flow and of the coating initial position. The corresponding shapes of molten pool are shown in the frames in the reduced scale.

Let us first consider the common (for all the figures) regularities. Because of the vortex structure of the melt convection, the admixture initially is carried along the free surface from the zones, corresponding to the local maxima of energy flow to its local minima. In the zones of surface temperature maximum the

306

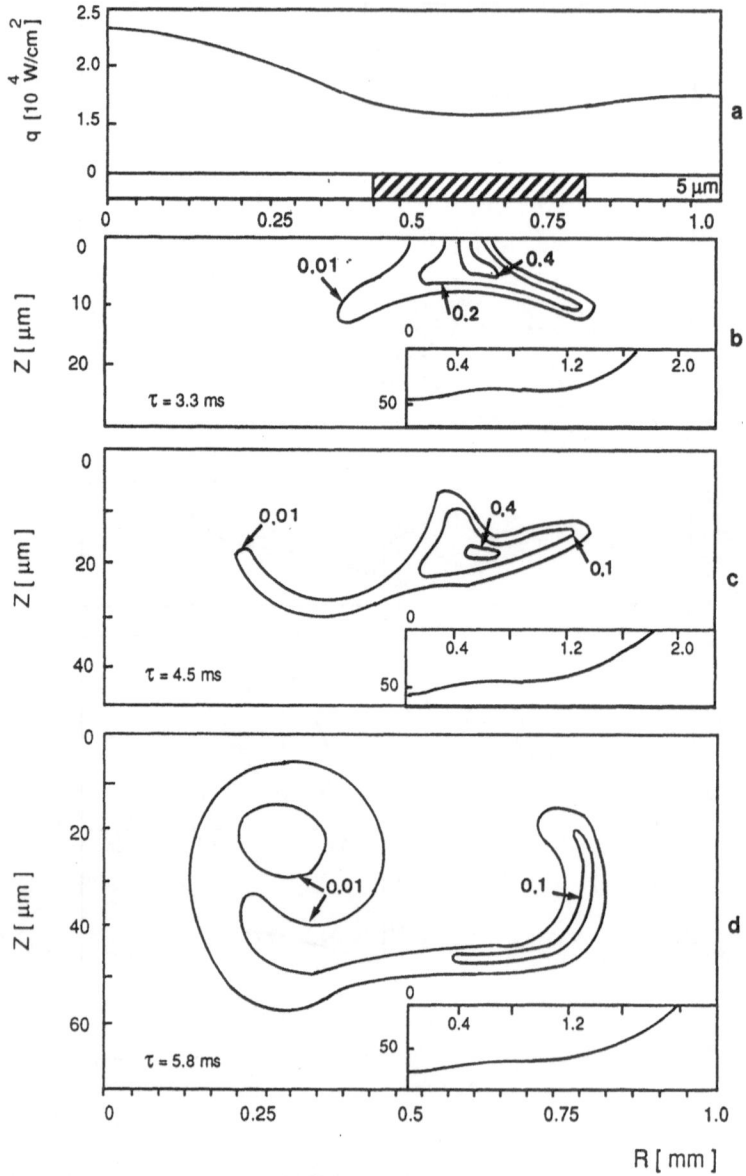

Figure 29. Concentration fields of the alloying element at corresponding time instants: t=3.3 (b), 4.5(c), 5.8(d) ms. Base metal Ti; coating thickness 5 μm; ring shape coating; a - spatial distribution of energy density flow and of the coating initial position. The corresponding shapes of molten pool are shown in the frames in the reduced scale.

thickness of the coating first decreases and then it practically disappears. On the contrary, in the zones of minimum values of surface temperature, the alloying material is collected and enters deeply into the bulk of the melt.

For the cases of Gaussian spatial distribution or with the local minimum in the center, the admixture enters the bulk of the melt along the melting front boundary. For the spatial distribution with two local maximum values - it can enter also the zone far from the edges of the pool. The existence of the zones with local temperature minima is not enough for a good remixing of the materials; it is necessary to produce also a large temperature gradient, which causes a large value of melt velocity. So the regions of the largest energy flow gradients are the most promising for an intensive remixing of the materials.

It is necessary to underline that the entering of the new amounts of the alloying element into the molten pool takes place from the surface regions, where the melt increases its size and remixes the coating. For example, for the spatial energy distribution with a local minimum in the center (Fig. 27), the melted zone initially has the shape of the ring, which width increases in both directions - to the spot center and in the opposite. During this stage the new amounts of the alloying element are introduced into the melt from both the sides of the ring. Furthermore, when the ring transforms into the circle, new amounts of alloying element are introduced only from the peripheral part of the melted zone.

The exact positions of local extremes of energy flow spatial distribution can change both the values of concentration in a chosen point and its average values in the whole zone of individual vortex. The latter is because the entering of the alloying elements from the zone of one vortex to the zone of the other is rather small (in the present model is not possible at all). The positions of local extremes divide the coating on the separate remixing zones, presenting different depths and, as a consequence, different volumes.

   (b) Non-continuous coatings. The peculiarities of convective mass transfer can be more distinctly revealed, if one will consider not continuous coating, but in the shape of a ring, placed in one or another position of the irradiated zone (Figs. 29,30). The coating is placed in the position of local minimum of energy flow, where the melt moves deep inside the molten pool (Fig. 29). It is possible to see that the volume occupied by the alloying element first decreases, because of its concentration in the zone of surface temperature minimum; and then the melt surface becomes free from the alloying element at all. In a period corresponding to Fig. 29b the alloyed zone appears in a form of nonregular ring width of about 0.5 mm situated on the depth of about 20 μm. Later on, the alloyed zone is transformed into a spiral shape, but the melt surface is still free from it. This kind of alloying can be promising for the preparing alloyed layers inside the treated materials (multilayer structures), not on its surface.

   Let us consider (Fig. 30) the "ring shape" coating to be placed in the position of energy flow maximum, where the melt moves from inside the molten pool to its surface and then parallel to the surface towards the center of the irradiated zone and radially outward. On the initial stage of the process the alloying element

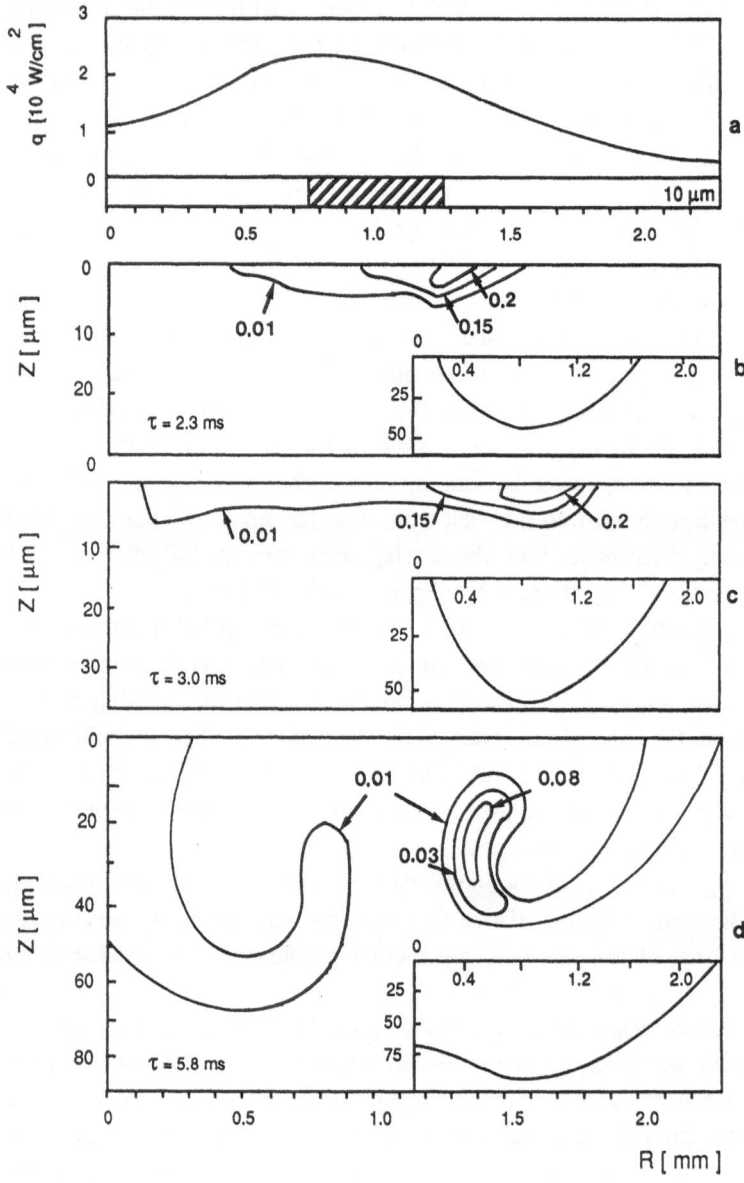

Figure 30. Concentration fields of the alloying element at corresponding time instants t=2.3 (b), 3.0 (c), 5.8 (d) ms. Base metal Ti; coating thickness 10 μm; ring shape coating; a - spatial distribution of energy density flow and of the coating initial position. The corresponding shapes of molten pool are shown in the frames in the reduced scale.

covers all the melt surface (Fig. 30c). Later on, it learns the surface and penetrates inside the molten pool near its boundaries. Because of the vortex structure of the melt convection the alloyed zones have a spiral shape. On this stage the melt surface is free from the alloying element except a small circle in the center and a ring near the boundary (Fig. 30d).

   Conclusion. In the processes of pulsed laser alloying of metallic materials, the spatial distribution of laser beam intensity plays an important role (together with the pulse duration and energy density flux values). Namely spatial distribution of laser energy determines the structure of melt convection. In the cylindrically symmetric case the number of toroidal vortexes in the molten pool equals to the number of local extremes of energy flux. In each vortex the melt moves near the surface from the region of local maximum of temperature towards the neighboring local minimum, then deeply into the melt; in the regions of surface temperature maximum the melt raises to the surface.

The structure of spatial distribution of energy flow determines the alloying element distribution in the molten pool. Furthermore, the alloying elements are carried inside the melt in all the regions of local minimal values of energy density flow (even if far away from the edges of molten pool), and the most intensive remixing takes place in the zones with the largest values of energy and corresponding temperature gradients.

By varying the positions of local extremes of energy flux (for all the same average values of treatment parameters) it is possible to change sufficiently the distribution of alloying elements in the zone of action. The existence of a number of local extremes of the spatial distribution of energy flow leads to a more homogeneous distribution of alloying element in the zone of action than for the Gaussian one (with the same energy input); in addition this often occurs with a smaller penetration depth.

3.2.3. Flow Pattern for Nonlinear Dependence of Surface Tension

In the above discussion it was assumed that the dependence of surface tension versus temperature is a linear function. It corresponds to the positive or negative but always constant values of surface tension coefficient $d\sigma/dT$. Generally this assumption is not correct and surface tension coefficient depends both on temperature and on amount of SAE. As a first approximation $d\sigma/dT$ could be considered as a linear function of temperature: positive in the lower temperature range and negative in the higher temperature range. The increase of SAE concentration shifts the curve in the direction of positive values.

The temperature dependence of $d\sigma/dT$ generally results in multi-cells flow pattern instead of a single sell. So, both non monotone distribution of surface temperature, and non monotone dependence of surface tension versus temperature (and SAE

310

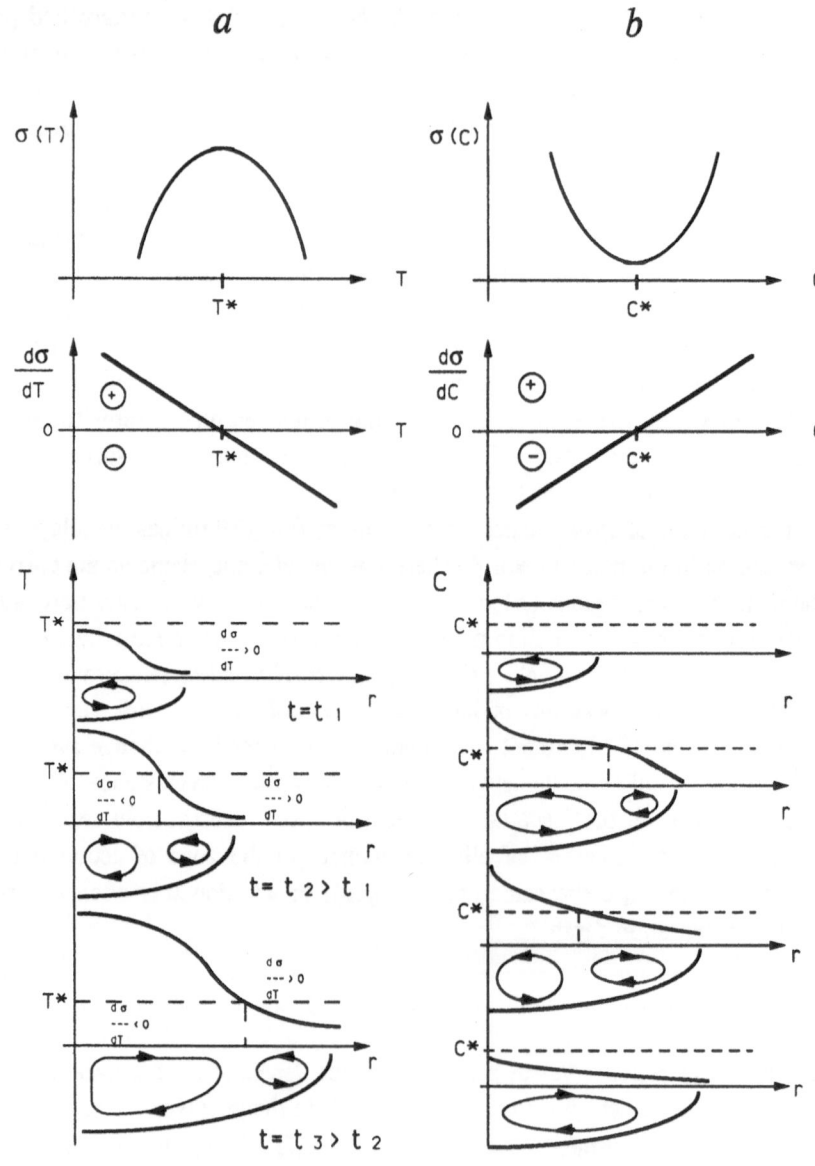

Figure 31. The flow pattern for the parabolic dependence of surface tension on temperature (a) and on concentration of surface active elements (b).

concentration) lead to the multi-cells melt convection. The later corresponds better to the experimentally result in pulse laser alloying, where multi-cells could be often observed.

In dependence on the surface temperature distribution and the value of temperature T^* where $d\sigma/dT$ changes the sign, different directions of melt rotations in a fixed point could be obtained (Fig. 31).

4. Materials Elaboration and Characterization

4.1. EXPERIMENTAL PROCEDURE

2.5 kW carbon dioxide (λ=10.6 μm) and Nd:YAG pulse-periodic (λ=1.06 μm, pulse duration 0.1-15 ms, pulse energy 1-30 J) lasers were used in the experiments. The specimens without additional absorption coatings were mounted on numerical control (CNC) X-Y table and were used in laser alloying. During laser treatment Ar gas was used to protect metal surface from oxidation. Simultaneously with continuous wave laser also the pulse-periodic laser was involved in combined action processes [30] (see Chapter 2). Irradiation of solid state laser was delivered through an optical fiber. CO_2 laser beam was oriented normally to the surface, Nd:YAG to the angle of 45°. The samples have been specially prepared before treatment by the methods which were mentioned above in Chapter 2. The pure metals including less than 0.1% of impurities were used. Cross-sections of the treated samples were metallographically polished and etched to allow microstructural examination and microhardness measurements. The distribution of elements in the laser treated regions was examined using CAMEBAX X-ray microprobe.

The following system "base metal-alloying element" are considered: Al-Sn(In), Fe-Sn(In,Pb), Al-Fe. Here base metal is indicated on the first position, alloying element - on the second, alternative alloying element - in the brackets.

4.2. SYSTEM Al-Sn(In)

Al-Sn system. According to equilibrium phase diagram for Al-Sn [63,64] these elements are not miscible in solid state. The attempt to mix Al with Sn is interesting from practical point of view to make the alloy with high performance properties.

The typical modified zone, obtained in our experiments, is sufficiently uniform. There are small (0.7-1.5 μm) Sn-rich precipitations ("white" regions) in matrix of melted pool. Most of these inclusions have practically spherical shape at the top and middle part of molten pool. The elongate needles of Sn-rich regions (directed along the temperature gradient) are observed at the bottom of alloyed zone. The presence of Sn-rich precipitations is explained by full miscibility in liquid state in total range of concentrations (at temperature higher 930K) and full immiscibility in solid state. The combined action of CO_2 laser and Nd:YAG pulse-periodic laser sharply intensify the convective mass transfer. It leads to uniform distribution of alloying element in the molten pool. The fast cooling of the melt leads to the formation of Sn-rich precipitations.

The average Sn concentration was measured on the square 10x10 μm by X-ray microprobe analysis. It varies in the range 19-23 wt.%. The presence of Sn (2-3 wt.%) was also found in the matrix itself. SEM examinations did not reveal any nonhomogenity of matrix at magnification up to 8000.

Sn was not found inside the treated zone when only CO_2 laser was used. In case of action of pulse-periodic laser only, the distribution of Sn was strongly nonuniform. In some regions the average Sn concentration have reached 31 wt.%. and, as a rule, the size of Sn-rich precipitations was smaller (about 0.7 μm).

Al- In system. Al and In are also not miscible in solid state, but in comparison with Al-Sn system, there exist the region of immiscibility in liquid state (between 930K and 1180K) [63,64]. The structure of alloyed zone is different from the structure of Al-Sn alloys discussed before. It consists of two strongly contrasted regions that for convenience could be named as "white" and "black" areas.

The size range of the first ones is 0.8-4 μm, the second ones are smaller: 0.4-1.5 μm. Black areas are strongly chemically etched. The average concentration of In varies from 1 up to 7 wt.%. Maximum concentration is observed at the top part of melted pool, minimum - at the bottom. There are a lot of spherical pores, probably as a result of gas bubbles solidification inside the molten pool. At the pool's bottom In was found (about 1 wt.%) only near these pores.

Using only CO_2 laser leads to the decrease of total amount of In: only 3 wt.% are observed near the surface of treated layer.

Big difference between the results of laser alloying for systems Al-Sn(In) may be explained by the presence, in one case, of immiscibility domain in liquid phase (Al-In) and full miscibility in liquid phase in the other case (Al-Sn). Note, that the observed maximum of In concentration in the alloyed zone is not far from the lower limit of mutual (Al-In) immiscibility in the liquid phase.

4.3. SYSTEM Fe-Sn(In,Pb)

Fe-Sn system. The proposed method concerns with using Cr as a shielding cover to avoid the destruction and evaporation of soft metals coatings. Two systems have been chosen for investigations: (1) 50 μm Cr layer on 70 μm Sn layer; (2) 50 μm Cr layer on 50 μm Sn layer. Pure iron was used as a base material. The energy and pulse duration of pulsed-periodic laser were varied from 5 to 8 J and from 6 to 14 ms, respectively.

Analysis of cross sections of treated zones have shown the sharp influence of: (a) pulse energy, (b) pulse duration, and (c) thickness of Cr and Sn layers, on the depth of melting pool and concentration of alloying elements. Laser alloying is practically

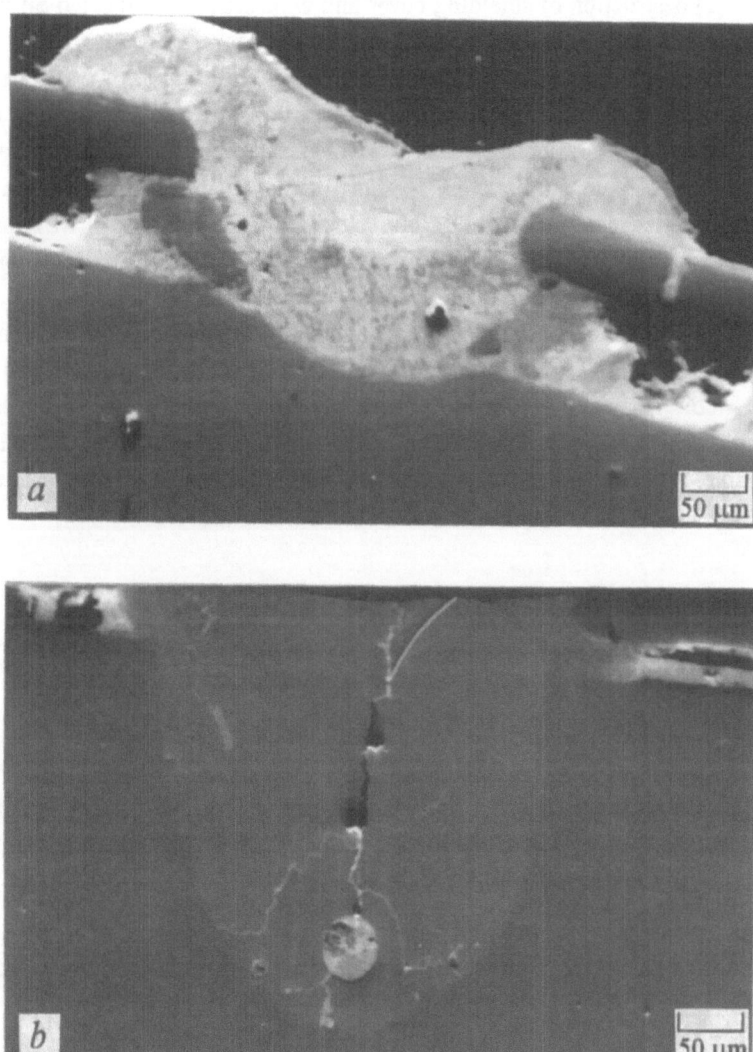

Figure 32. Cross section of laser alloyed zone: base material Fe. Thickness of Sn coating: (a)-70 μm; (b)-50 μm. Thickness of Cr shielding cover 50 μm. Nd:YAG laser parameters: pulse duration 14 ms, pulse energy 8 J.

absent at 8 J pulse energy and 14 ms pulse duration for system 50 μm Cr+70 μm Sn. In this case, Cr shielding cover was partially destroyed and overheated Sn melt have flowed out (Fig. 32a).

314

Only thin layer (about 30 μm) of Fe was melted. Laser alloying of required quality could not be reached at any variations of treatment parameters (inside the above mentioned range) for given thickness of the coatings. Two typical situations were observed: (1) destruction of shielding cover and expulsion of overheated Sn melt; (2) formation of alloyed zone with strongly destroyed top part.

A small decrease of Sn coating thickness (from 70 to 50 μm) allows to optimize the alloying process (Fig. 32b). The distribution of Sn and Cr becomes enough uniform (fluctuations of concentrations less than 0.5 wt.%) except for individual comparatively large inclusions of pure Sn. Figs. 33, 34 show the dependencies of maximum Sn and Cr concentrations (maximum equilibrium solubility of Sn in Fe is 18 wt.% at 900°C and 7 wt.% at 600°C) and dimension of melted pool on pulse energy and pulse duration. Note, that the fluctuations of alloying elements concentration sharply increase up to 5 wt.% at lower pulse energy and shorter pulse duration; microhardness in remelted zone varies from 330 to 450 Hv. One of the advantages of pulsed and pulsed-periodic laser alloying is the opportunity to vary easily the concentration of alloying elements and the dimensions of melted pool by the parameters of laser pulses. For example, it is possible to obtain maximum Sn concentration up to 30 wt.% at 150 μm depth. The corresponding microhardness can reach 600 Hv.

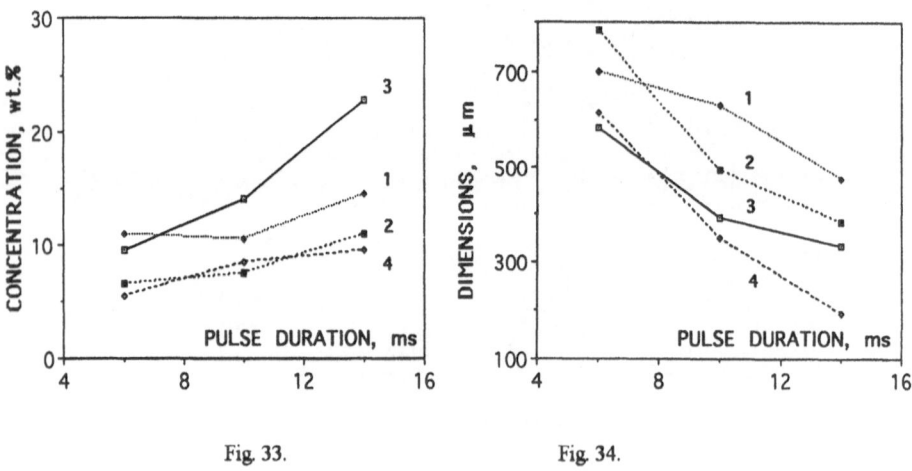

Fig. 33.   Fig. 34.

Figure 33. Maximum Sn and Cr concentrations inside alloyed zones versus laser pulse duration: curves (1), (2) -Cr, curves (3), (4) - Sn. Pulse energy: curves (1), (3) - 5 J; curves (2), (4) - 8 J. Base material Fe. Thickness of coatings: Sn - 50 μm, Cr - 50 μm.

Figure 34. Diameter (1), (3) and maximum depth (2), (4) of laser alloyed zone versus pulse duration. Pulse energy: curves (1),(2) - 8 J; curves (3),(4) - 5 J. Base material Fe. Thickness of coatings: Sn- 50 μm, Cr- 50 μm.

The other feature of laser alloying from double layer coatings is a large depth of melted pool which reaches the value of 783 μm for E=8 J, t=6 ms (Fig. 34). The corresponding depth of the molten pool (for exactly the same parameters of laser action) for pure Fe is about 150 μm, i.e. 5 times smaller. It indicates on significant influence of forced convection of overheating Sn melt on heat transfer processes.

The results of CO_2 laser alloying (power 1.6 kW and beam traverse speed 6.7 10^{-3} m/s) of Fe from the layer of Sn powders differ from the above mentioned. The modified zone consists of intermetallic crystals distributed in a-Fe solid solution of Sn (8 wt.%). The compound of intermetallic crystals (Fe 59 wt.%-Sn 41 wt.%) corresponds to Fe_3Sn phase. This phase is stable between 620°C and 830°C at equilibrium conditions [65]. The chosen method of laser alloying allows probably to fix Fe_3Sn as a metastable phase. Small (about 1-2 μm) inclusions of pure Sn are also observed inside the melted pool.

Fe-In system. In case of CO_2 laser alloying of Fe with In (Fig. 38), the molten pool consists of two alternating strongly contrasted small regions (size 1-2 μm).

The average In concentration varies from 3 up to 7 wt.%. The equilibrium miscibility of In in Fe is less than 1 wt.% in solid state [65]. Intensive remixing of Fe and In in the molten pool, that is followed by fast cooling, leads probably to the exceeded solid solution of In in Fe. The main feature of equilibrium phase diagram for these system is the wide range of immiscibility in liquid state (from 6.5 wt.% up to 95 wt.% in temperature range 1770-2050K). The maximum In concentration in alloyed zone is close to its lower limit of immiscibility with Fe in liquid state. The same situation was also observed for system Al-In (see previous Chapter) which zones of alloying have the similar structure.

Fe-Pb system. The Fe and Pb are not miscible both in solid and in liquid phase. The conventional methods (including traditional laser alloying) are inefficient to remix Fe and Pb. The forces of chemical nature on the atomic level prevent the remixing of these elements in the melt. Special method of laser alloying was developed to solve this problem. As a result, cross section of the alloyed zone consists of an elliptical shape region of pure Pb surrounded by pure Fe. Small alloyed Fe-Pb regions were found in the different parts on their boundary. It is, as a rule, thin (20-30 μm) layers with 4-7 wt.% of Pb concentration. For example, this layer is located near the bottom of treated zone (Fig. 39) and Pb concentration fluently varies from 100 wt.% (top region) to 7 wt.% (near the border with base metal, i.e. Fe).

SEM investigations at 8000 magnification did not reveal any nonhomogenity of the above mentioned layer. Probably, remixing of Fe and Pb was reached inside the overheating melt (in spite of thermodynamic barrier of mutual immiscibility of Fe and Pb in liquid phase at equilibrium conditions). Later the composition was fixed by extremely high cooling rates (that were realized inside the bulk of base metal) that lead to submicron sized Pb precipitations in Fe.

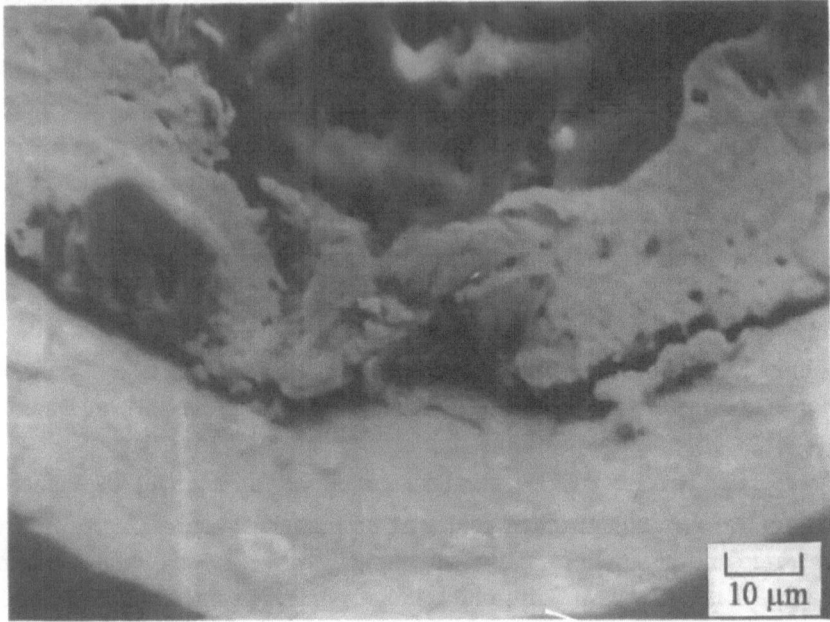

Figure 35. SEM-micrograph of cross section of laser alloyed zone. The layer alloyed with Pb (curved down light homogeneous layer) is located between the pure Pb (upper partially destroyed part) and pure Fe (down part of dark color.

4.4. SYSTEM Al-Fe

The influence of high cooling rate on the microstructure of Al-Fe binary alloys have been investigated in order to develop of light alloys having high temperature strength [7,14-17,66]. For example, the Al-8 wt.% Fe alloy was made by splash-quenching technique [14]. The metastable intermetallic phases and completely supersaturated (up to 8.3 wt.% at equilibrium miscibility about 0.05 wt.%) small α-Al grains (0.5 mm in diameter) were observed after solidification of splats [14]. Vapor quenching by co-sputtering onto a substrate at 77K yields supersaturated Al solid solution up to about 20 wt.% Fe [64]. The microstructure of rapidly solidified Al-Fe alloys subjected to fast surface CO_2 laser remelting was investigated in [6]. Rapid solidified Al-Fe alloys have the microstructure of precipitation free solid solution at Al concentration less than 6 wt.% at high speed of laser beam traverse [6].

In our experiments typical modified zone consists of two regions: (a) upper part of melted pool which has the usual shape; (b) down part - narrow and deep as if in case of deep penetration welding. Al distribution is practically uniform (13-14 wt.%) in the region (a). Its structure corresponds to Al-Fe solid solution plus $AlFe_3$; microhardness varies in the range 400-450 Hv. Fe distribution in the region (b) is nonuniform and varies from 4 to 12 wt.%. There are a lot of Fe-rich inclusions inside of zone (b) : small "needles" at the top part and large "drops" near the bottom.

The microhardness of the region (b) is also quite nonuniform and varies in the range 700-1400 Hv (it depends on the process parameters). This result differ from the one obtained by splat-quenched Al-8 wt.%. Fe alloy which has the maximum microhardness about 260 Hv [67]. Maximum value of microhardness (1400 Hv) correspond to the maximum value of Fe concentration (12 wt.%) and was found in the strongly contracted with the surrounding material areas of "white" color : or in the above mentioned "drops", or near melted pool boundaries in zone (b). The dimensions of these areas can reach up to 70x100 μm. The behavior of Al-Fe alloy from the above mentioned areas, during annealing at temperature 730K associated with small (about 10%) increase of microhardness and did not reveal any other changes. Probably, the main reason of a little microhardness increase was the formation and the slow growth of the particles of metastable $FeAl_6$ phase like it was observed in [5]. Mentioned should be paid to the fact that this region could not be etched metallographically by Keller's reagents. The results of these preliminary investigations have shown that it is, probably, a supersaturated solid solution of Fe in α-Al formed in a macro-volume (up to 70x100 μm).

5. Materials Properties

The precise analysis of the above mentioned intermetallic compositions is not yet completed. The work is in the progress. One of the main questions is ether the overequilibrium solubility, or submicron sized precipitations are obtained. Nevertheless, the evaluation of the functional properties of the elaborated surface layers in the aims of potential industrial applications already have been done for several intermetallic systems. As an example, the application of Fe-Sn system to obtain solid lubricant coatings is considered below.

5.1. IMPROVEMENT OF WEAR RESISTANCE OF BEARING STEELS BY LASER ALLOYING WITH Sn

Recently, many methods are used to overcome wear problem   [68]. However, a number of applied tasks concerning with reliability and durability are not solved up to now. For example, there are no satisfactory methods to improve wear resistance of friction joints working at severe conditions : high loads and contact pressures, wash-out of lubricant from the contact area, corrosive medium, etc. In this case, the nature of a friction surface interaction is closed to dry friction. It leads to high linear wear and fast destruction of mechanical parts. The wear behavior of friction joints of an oil-drilling equipment (bearings, seals, gates, etc.) are a typical illustration of the mentioned problem. For example, the durability of turbo-drill multirows bearings is only 4-6 hours [69]. Tbibotechnical joints working at severe conditions have, as a rule, high wear tolerance : from some hundreds microns up to several millimeters.

318

Consequently, the thickness of modified layer must be comparable with the wear tolerance.

Sn is a well-known component of anti-friction alloys. However, the low melting point and high vapor pressure make difficulties for laser alloying with Sn. The new developed methods of laser alloying allows to overcome this problem and to produce the solid lubricant composition in surface layers of bearing steels.

5.1.1. Experimental Procedure

A 2.5 kW CO_2 laser and Nd:YAG pulsed-periodic (pulse duration τ=1-10 ms, pulse energy E=1-30 J) were used. Argon with flow rate 50 l/min was directed coaxial with laser beam to protect metals from oxidation during laser treatment. The steel used for samples and bearing rings has the following composition : 0.5% C, 0.8-1.0% Is, 0.5-0.6% Mo, 0.1% V.

Sn coatings were covered by galvanic Cr shielding coatings and this double-layers' system on steel base was remelted by laser action at optimum treatment parameters (see paragraph 4.3).

The cross-sections of the treated samples were metallographically polished and etched to allow microstructure examination and microhardness measurements. The distribution of Sn in the laser treated regions was examined by CAMEBAX X-ray microprobe.

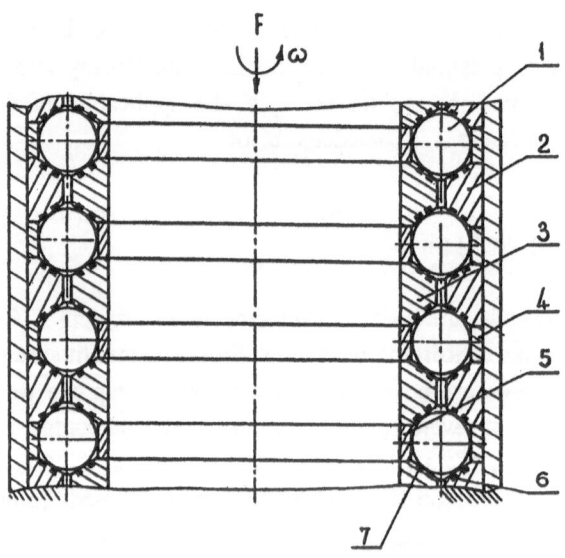

Figure 36. Design of multirows turbo-drill bearing: (1)-balls; (2)-external rings; (3)-internal rings; (4, 5)-distance rings; (6, 7)-alloyed regions.

The scheme "pin-on-disk" was used for wear tests at sliding friction (dry and oil lubricant) [70]. Disk flat surfaces were alloyed with Sn by pulse-periodic Nd:YAG laser. The test conditions were the following: (1) pin material - high alloyed steel with 60 HRC hardness and 6 mm diameter; (2) load - 2 MPa for dry friction, and 4 MPa for friction with oil lubricant; (3) sliding velocity - 0.2 m/s for dry friction, and 0.8 m/s for friction with oil lubricant; (4) test duration - 1 hour; (5) lubricant - paraffin oil. The friction coefficient was measured as the main parameter of the tests.

The wear velocity and service life time were investigated for multirows turbo-drill bearings (Fig. 36) after CO_2 laser alloying of their races. All the experiments were carried out at conditions closed to the real exploitation parameters : axis load - F=15 kN; startup contact pressure - σ=2500 MPa; frequency of rotation ω=700 min^{-1}; ambient medium - water at temperature T=300 K.

5.1.2. Results of Wear Tests

The main task of "pin-on-disk" wear test is to study a friction behavior of alloyed layers at no-lubricant (dry friction). In this case, the steady state is observed from the beginning of the test (Fig. 37a).

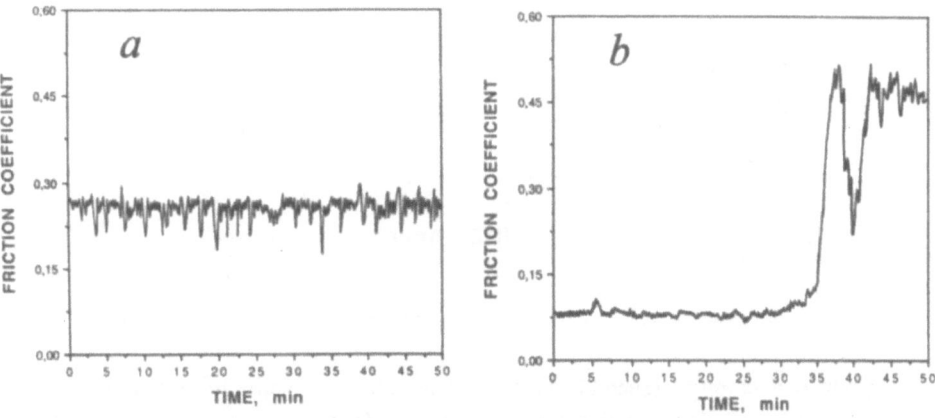

Figure 37. Friction coefficient versus time for "pin-on-disk" wear test. Bearing steel alloyed with Sn by pulse-periodic Nd:YAG laser. (a)-dry sliding friction; (b)-oil lubricant sliding friction.

The average value of friction coefficient (0.232) is low enough to avoid galling and adhesive wear during one hour. The mechanism of friction behavior is the following : Sn emerges at contact area as a result of precipitations from intermetallic compounds or solid solution, and of small inclusion's outlet on the disk surface. The first process is caused by high temperature at contact area. The second one concerns with mechanical destruction of thin solid layers. The steady state for friction coefficient behavior is explained by intensive Sn transfer from one point of the disk surface to another. It keeps the friction coefficient at a low level due to Sn solid lubricant properties.

The wear test with oil lubricant, under the conditions of increased load and sliding velocity, leads to the total wear of modified layer after 30 minutes. As a result, the friction coefficient sharply increases from very low value (0.08) up to 0.45 (Fig. 37b).

Based on the results of above mentioned tests, optimum parameters of laser alloying were chosen for treatment of turbo-drill external and internal races. The first (short) stage of bearing wear test (Fig. 38, curve 2) corresponds to the alignment of mating parts.

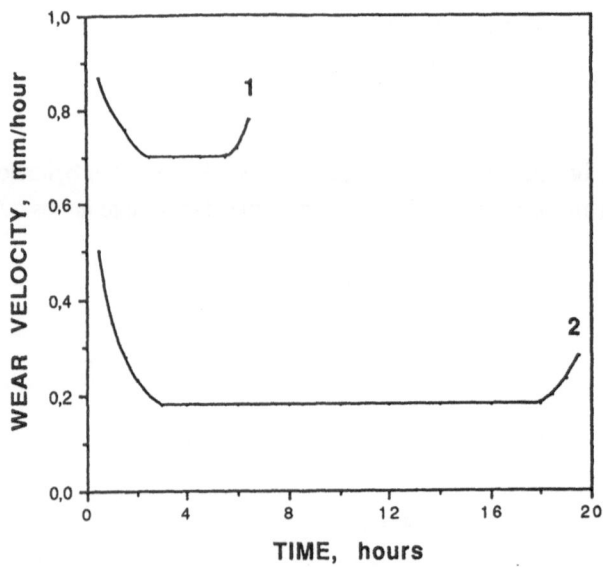

Figure 38. Wear velocity versus time for turbo-drill race. (1)-induction hardening; (2)-laser alloying with Sn.

The wear rate stability (second stage) is guarantied by solid lubricant feed from alloyed layers and Sn intensive transfer between friction parts. Geometry of laser traces (width, thickness and position on bearing races) was chosen to provide a good lubrication within the required wear tolerance. It keeps the friction surface from intensive wear and destruction (Fig. 39).

The sharp increase of wear velocity is observed only after total wear of Sn-containing zones. The comparison with the conventional method of induction hardening (Fig. 38, curve 1) have shown 2.5 times increase of service life time and 3 times decrease of wear velocity. Full destruction of the bearing is observed after 6 hours of the wear test in the case of induction hardening of external and internal rings (Fig. 40b). At the same time the bearing with laser alloyed layer keep the operation ability (Fig. 40a).

Figure 39. SEM-micrograph of friction surface (top view) after 4 hours of wear test. (a)- laser alloying with Sn;
(b)- induction hardening.

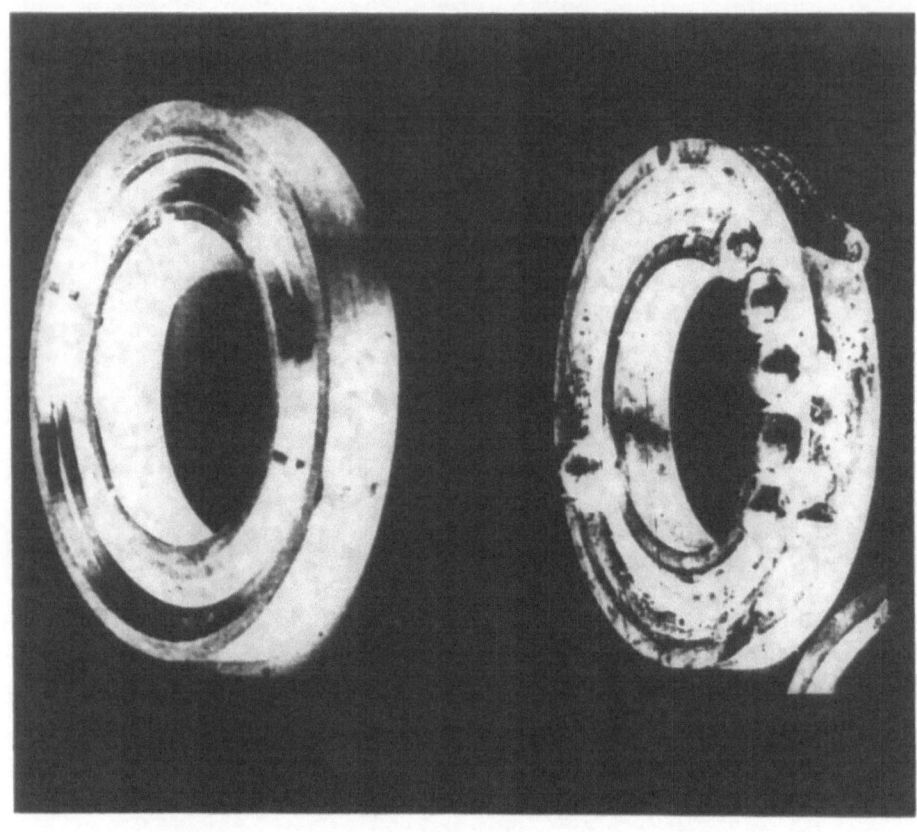

a   b

Figure 40. Parts of turbo-drill bearings after 6 hours of wear test. (a)- laser alloying with Sn; (b)- induction hardening.

6. Conclusion

Heat and mass transfer in pulsed laser alloying was analyzed by mathematical modeling. The complete thermal cycle is simulated taking into account melting/solidification and evaporation phenomena. The concentration fields of admixture resulting from convective mass transfer are obtained.

1. The following methods were used to remix the components with large differences of physicochemical properties: (a) combined simultaneous action of pulsed-periodic and continuous wave lasers; (b) using of a shielding cover to avoid the

removal of alloying elements from the alloying zone; (c) using of thick combined multilayers coatings.

2. As a result, several innovative intermetallic compounds were obtained. The main results are presented in the Table.

System	Equilibrium solubility or miscibility	Maximum concentration in matrix after laser alloying	Additional Information
Al-Sn	immisible in liquid phase	3 wt.% of Sn	Sn rich precipitations (0.7-1.5 μm) are observed. Average Sn concentration in 10x10 μm region is in the range 19-23 wt.%.
Al-Fe	equilibrium solubility of Fe in Al is less than 0.1 wt.%	12 wt.% of Fe	Microhardness up to 1400 H_v, very slight response to etching.
Fe-Sn-Cr (Sn under Cr shielding cover)	equilibrium solubility of Sn in Fe is about 7 wt.% at 600°C	30 wt.% of Sn	Maximum Sn concentration (about 30 wt% at 150 μm depth) corresponds to its nonuniform distribution in the alloyed zone. Also uniform Sn concentration about 10 wt.% (fluctuations 0.5 %) plus Sn rich precipitations can be obtained. Microhardness can reach 600 H_v.
Fe-In	equilibrium miscibility of In in Fe is less than 1 wt.%	7 wt.% of In	Alternating strongly contrasted regions (1-2 μm) with In concentration of 7 and 3 wt.% respectively.
Fe-Pb	absolute immisibility	7 wt.% of Pb	Alloyed layer of about 30 μm thickness on the boundary between pure Fe and pure Sn.

3. The above mentioned results indicate that by means of laser alloying it is possible to remix a wide range of metals with sufficiently different physicochemical properties. For the systems with a region of immiscibility in liquid phase (for example, Al-In, Fe-In), maximum concentration of alloying element (In) in the alloyed zone is close to its lower limit of immiscibility. Zones of alloying have the similar structure for both systems. Alloying element rich precipitations are typical for the system with full miscibility in liquid state and full immiscibility in solid state (Al-Sn).

As a general comment, it is possible to note that composition of alloyed zone is strongly influenced by mutual miscibility (or immiscibility) of metals in the liquid phase.

4. Laser alloying of bearing steel with Sn allows to produce Sn-containing layers which demonstrate a good performance at conditions of sliding friction: low friction coefficient, elimination of galling and fatigue wear. Durability of oil turbo-drill after

324

laser alloying with Sn is increased in 2.5 times in comparison with induction hardening.

Acknowledgments

The authors would like to thank Professor G. Flamant, Ing. R. Flamand from Institut de Science et de Génie des Matériaux et Procédés, C.N.R.S., France and Dr.Ing. L. Covelli, Dr. A. Lashin from Istituto di Tecnologie Industriali e Automazione, C.N.R., Italy for their useful participation in the present work.

References

1. G.Herziger and E.W.Kreutz (1986) *Phys. Scripta* **T 13** 139.

2. C.W.Draper and J.M.Poate (1985) Laser surface alloying, *Int. Met. Rev.* **30**, 85.

3. A.A.Uglov, I.Yu.Smurov, A.M.Lashin, A.G.Guskov (1991) *Modeling of Thermal Processes under Pulsed Laser Action on Metals*, "Nauka" Publishers, Moscow.

4. S. Sarkar, J. Singh and J. Mazumder (1989) *Acta Met.* **37**, 1167.

5. J. Mazumder, J. Choi, C. Ribaudo, A. Wang and A. Kar (1993) in J. Singh and S.M. Copley (eds.) *Intern. Conference on Beam Processing of Advanced Materials*, 41-67.

6. M.Gremaud, M.Carrard and W.Kurz (1990) *Acta Metall.Mater.* 4381, 2587.

7. J.A.Michael and J.D.Budal (1986) *J.Mater.Res.* **411,** 401.

8. J.A.Knap and D.M.Follstaedt (1985) *Phys.Rev.Lett.* 4551, 1591.

9. J.Eridon and G.S.Was (1988) *J.Mater.Res.* **431**, 626.

10. M.Libera, P.Pedro, R.E.Spjut and J.B.Vander Sande (1988) *J.Mater.Res.* **431**, 441.

11. A.S.Ubhi, T.N.Baker, P.Holdway and A.W.Bowen (1988) in P.G.Lacombe, R.Tricot and Beranjer (eds.), *Proc. Sixth Int. Conf. on the Processing of Titanium*, Cannes, June 1988, Pt **III**, Societe Francaise de Metallurgie, 1613.

12. J.H.Abboud and D.R.F.West (1992) *J.Mater.Sci.Lett.* 4111, 1322.

13. P.Matteazzi, F.Miani, I.Yu.Smurov, A.M.Lashin. (1990) in *Proc. of the IIIrd European Conference of Laser Treatment of Materials* (ECLAT-90), Germany, 927.

14. M.H.Jacobs, A.G.Dogget and M.J.Stowell (1974) *J.Mater.Sci.* 491, 1631.

15. M.De Sanctis, A.P.Woodfield and M.H.Loretto (1988) *Int.J.Rapid Solidif.* 441, 53.

16. A.Toneje and A.Bonefacic (1969) *J.Appl.Phys.* 4401, 419.

17. W.J.Boettinger, L.A.Bendersky and J.G.Early (1986) *Metall.Trans.* **17A**, 781.

18. M.Ignatiev, E.Kovalev, V.Titov, A.Uglov, I.Smurov and S.Sturlese (1992) in *Proc. of the IVth European Conference on Laser Treatment of Materials* (ECLAT-92), Göttingen, DGM, Oberursel, 241.

19. Smurov I., Covelli L., Flamant G., Balat M., Ignatiev M. (1993) in *Proc. of the International Conference on Surface Engineering*, March 9-11 1993, Bremen, Germany.

20. C.W.Draper, E.N.Kaufmann, L.Buene (1982) *Surface and Interface Anal.* **4**, 8.

21.C.W.Draper, L.S.Meyer, Buene L. (1981) *Applic. of Surface Science* **4**, 276.

22.C.W.Draper, F.J.Broeder, D.C.Jacobson (1982) in B.R.Appleton, G.R.Celler (eds.) *Laser and Electron Beam Interaction with Solids*, Amsterdam: North-Holland, 419.

23.D.V.Folstaedt, ibid, 377.

24.Y.BikovskiyI, V.Nevolin and V.Fominskiy (1991) *Ion and Laser Implantation of Metallic Materials*, Energoizdat, Moscow.

25.A.A.Uglov, I.Yu.Smurov, K.I.Tagirov, A.G.Guskov (1992) *Int. J. Heat Mass Transfer* **35**, 783.

26.A.I.Manohin, A.A.Uglov, A.F.Gorbach, I.Yu.Smurov, L.I.Mirkin (May 1986) *Sov. Phys. Dokl.* **31**, 449.

27.A.A.Uglov, M.B.Ignatiev, I.Yu.Smurov (1987) *Phys. and Chem. of Mater. Treatment*, No. 2, 88.

28.A.A.Uglov, M.B.Ignatiev, A.G.Gnedovets, I.Yu.Smurov (1989) *Journal de Physique*, Colloque C5, supplement au n5, **50**, C5-727.

29.I.Smurov, A. Uglov, P. Matteazzi, F. Miani, S. Tosto. (1991) in *Proc. of the 10th International Symposium on Plasma Chemistry*, Bochum, Germany, 1.4-28.

30.I.Smurov, L.Covelli in *Proc. of the IVth European Conference on Laser Treatment of Materials* (ECLAT-92), Göttingen, 1992 (DGM, Oberursel, 1992), p. 251.

31. M.Ignatiev, I.Smurov, G.Flamant, (1994) *Meas. Sci. Technol.* **5**, 563-573.

32. I.Yu.Smurov, A.A.Uglov, A.M.Lashin, P.Matteazzi, L.Covelli, and V.Tagliaferri. (1991) *Int. J. Heat Mass Transfer* **34**, 961-971.

33.I.Smurov (1993) Heat processes in pulsed laser action, in S.Martellucci et al (eds.) *Laser Applications for Mechanical Industry*, NATO ASI Series E: Applied Sciences, Vol. **238**, Kluwer Academic Publishers, Dordrecht, pp. 165-206.

34.V.I.Mazhukin, I.Smurov, G.Flamant, in *Proc. of ICALEO'93*, LIA **77**, Florida, USA, 184-192.

35.V.I.Mazhukin, I.Smurov, G.Flamant, C.Dupuy (1994) *Thin Solid Films* **241**, 109-113.

36.Hastaoglu M.A. (1986), *Int .J. Heat Mass Transfer* **29**, 495-499.

37.Benard C., Gobin D., Zanoli A. (1986) *Int. J. Heat Mass Transfer* **29**, 1669-1675.

38.Blom J.G., Sanz-Serna J.M., Verwer J.G. (1988) *Journal of Computational Physics* **74**, 191.

39. Smurov I., Lashin A., Poli M., in *Proc. of ICALEO'92*, LIA **75**, Florida, USA, 121-129.

40. S.V.Patankar (1980) *Numerical Heat Transfer and Fluid Flow*, Hemispher Publ. Corp., New York.

41. A.W. Date (1992) *Numer. Heat Transfer* **21B**, 231-251.

42. I.Smurov I., L.Aksenov, G.Flamant, in *Proc. of ICALEO'93*, LIA **77**, Florida, USA, 242-249.

43. V.G.Levich. (1962) *Physico-Chemical Hydrodynamics*, Prentice-Hall, Cliffs, N.J.

44. T.R.Anthony and H.E.Cline (1977) *J.Appl. Phys.* **48**, 3895.

45. T.Chande and J.Mazumder (1982) *J. Appl. Phys. Lett.* **41**, 42.

46. C.L.Chan, J.Mazumder, M.M.Chen (1987) A Three -Dimensional Axisymmetric Model for Convection in Laser Melted Pool, *Mat.Scien. Engineering* **3**, 306-311.

47. C.L. Chan, J. Mazumder, and M.M Chen (1988) *J. Appl. Phys.* **64**, 6166.

48. J.Mazumder (1991) Overview of Melt Dynamics in Laser Processing, *Optical Engineering* 30, 1208-1219.

49. G.M.Oreper and J.Szekely (1984) *J. Fluid Mech.* **147**, 53.

50. J.Srinivasan and B.Basu (1986) *Int. J. Heat Mass Transfer* **29**, 563.

51. B.Basu and A.W. Date (1990) *Int. J. Heat Mass Transfer* **33**, 1149.

52. V.Babu, S.A. Korpela, and N.Ramanan (1990) *J. Appl. Phys.* **67**, 3990.

53. N.Ramanan and S.A. Korpela (1990) *Metall.Trans. A* **21 A**, 45.

54. C.R. Heiple and J.R.Roper (1982) *Weld. J.* **61**, 97s.

55. C.R. Heiple and P. Burgarrdt (1985) *Weld. J.* **64**, 159s.

56. G.S.Ershov and Yu.B.Bychkov (1983) *Properties of Metal Melts and their Interaction in Steelmaking Processes* , Metallurgia, Moscow.

57. T. Zacharia, S.A.David, J.M. Vitek, T.DebRoy in S.A.David and J.M.Vitek (eds.) *Recent Trends in Welding Science and Technology TWR'89.* Proceedings of the 2nd International Conference on Trends in Welding Research, Gatlinburg, Tennessee, USA 14-18 May 1989, 25-30.

58. I. Smurov, A. Guskov (1989) in *Physico-Chemical Processes of Materials Machining by Concentrated Energy Flows* , Nauka, Moscow, 25-36.

59. A.A.Uglov, I.Yu.Smurov, K.I.Tagirov, and A.Guskov (1992) *Int. J. Heat Mass Transfer* **35**, 783-793.

60. I.Smurov, L.Covelli, K.Tagirov, L.Aksenov (1992) *J. Appl. Phys.* **71**, 3147-3158.

61. D.Potter (1973) *Computational Physics*, A Wiley-Interscience Publications, Imperial College, London.

62. I. Smurov, L Covelli in *Proc. of ICALEO'92*, LIA **75**, Florida, USA, 265-277.

63. *Equlibrium Diagrams of Aluminium Alloy Systems* (1961) The Aluminium Development Association, London.

64. Landolt-Bornstein (1992) *Numerical Data and Functional Relationship in Science and Technology*, editor in chief: O.Matter, Group **IY**, Macroscopic and Technical Properties of Materials, Volume **5**: *Phase Equilibria, Crystallographyc and Thermodynamic Data of Binary Alloys*, Subvolume **A**.

65. O.Kubaschevski (1982) *Iron Binary Phase Diagrams*, Springer-Verlag, Berlin.

66. J.D.Cotton, M.J.Kaufman (1991) *Metall. Trans.* **22A**, 927.

67. G.Thursfield and M.J.Stowell (1974) *J. Mater. Sci.* **491**, 1644.

68. *Industrial Tribology* (1983) M.Jones and D.Scott (eds), Elsevier, Amsterdam.

69. Ignatiev, M., Kovalev, E., Ionesyants, Yu., Kuzin, B., Uglov, A., Voloshin, V. and Titov,V. (1989) *Westnik Mashinostroeniya*, No. **9**, 15-20.

70. Ignatiev, M., Kovalev, E., Melekhin, I., Smurov, I. and Sturlese, S. (1993) *Wear* **166**, 233-236.

GAS SURFACE ALLOYING OF Ti6Al4V ALLOY BY LASER

A. ZAMBON*, E. RAMOUS*, M. MAGRINI*, M. BIANCO** and
C. RIVELA**
* DIMEG - Università di Padova - Italy
** Istituto RTM - Vico Canavese (Torino) - Italy

Abstract

The optimization of surface alloying treatments of Ti6Al4V under CO_2 and N_2 atmospheres with a pulsed Nd-YAG laser has been performed, trying to maximize alloyed layer hardness and minimize surface roughness.

The samples were examined by light microscopy, microhardness profiles, roughness measurements and XRD. Samples obtained under a CO_2 atmosphere exhibited higher hardness and lower surface roughness than those obtained under a N_2 one. Pin-on-disk wear tests were also performed to compare the wear behaviour of carburized and nitrided samples. Corrosion tests in different environments have been carried out on the nitrided surfaces in order to evaluate their behaviour in comparison to untreated surfaces.

1. Introduction

The utilization of titanium and its alloys, though being favourably influenced by good mechanical properties and low density, is sometimes negatively affected by their high friction coefficient and poor wear resistance.

Surface modification processes, such as CVD (chemical vapour deposition) or PVD (Physical Vapour Deposition) besides affording small nitrided thickness to be obtained, need long times to be carried out. Plasma or Ion nitriding permit to reduce the process times. In general all the mentioned processes produce layer thicknesses as low as few microns, while a layer thickness of hundreds of microns is needed in the case of wear resistant applications under high contact pressures.

Nitrogen absorption from the environment could be improved if carried out in the liquid state: the formation of Nitrides would take place upon cooling of the melted layer, helped by the fact that their melting temperatures are quite higher than that of the alloy being processed.

Liquid phase alloying of a surface layer can be achieved if a high power density is delivered onto a specimen, causing a sudden surface temperature increase, up to values close to the boiling temperature of the material, while the bulk temperature remains relatively low, so that its microstructure is unaffected. Laser nitriding of titanium alloys shielded by a N_2 atmosphere (or even in air) has already been described [1], [2],

327

J. Mazumder et al. (eds.), Laser Processing: Surface Treatment and Film Deposition, 327–335.
© 1996 Kluwer Academic Publishers.

[3]. Another means of improving the above mentioned surface properties is to promote carbon absorption from a surrounding atmosphere again by means of liquid phase alloying. Titanium carbides formation during laser treatments has already been studied by Walker et al. [1].

Till now expensive cw CO_2 lasers have mainly been used, but Nd-YAG or excimer lasers have become a cheap means to perform such treatments.

Aim of the present work was the evaluation of the feasibility of surface modification of a titanium alloy by means of nitrogen and carbon absorption promoted by Nd-YAG laser surface melting, and the optimization of the operative parameters evaluated on the basis of the obtained microstructures, layer hardness and hardened depth, surface roughness, wear resistance in pin-on-disk tests and of the corrosion behaviour in different environments, compared with that of untreated samples.

2. Experimental

In all the tests was used the well-known alloy Ti6Al4V. A Nd-YAG Lumonics JK series 700 laser unit was used (rated mean power 0-250W), whose main characteristics are listed below:

Pulse Energy   0.1 - 35   J
Pulse duration   0.5 - 20   ms
Frequency   0.2 - 500 Hz

The focal length of the lens was 190.5 mm, and its diameter 75 mm.

Another parameter controlled by the operator is, besides the above mentioned, the traverse speed of the worktable.

The training of the 1.06 μm laser beam onto the target was checked by means of a He-Ne laser fitted in the main laser unit, while its resonating cavity permitted to deliver rectangular pulses. The pulsed laser beam impinged perpendicularly on the specimen surface, which was traversing at constant speed. As preliminary tests in which a chamber was used to keep a controlled atmosphere above the surface being treated did not show significant improvements with respect to a nozzle fitted in close proximity to the lased track, this last configuration was adopted, delivering the gas under an angle of 45°, in the same direction of the traverse speed.

3. Results and Discussion

3.1. N_2 GAS SHIELDING

Preliminary tests were carried out in order to determine the proper parameters combination aimed at obtaining a surface as smooth as possible, without cracks.

The influence of the pulse frequency and of the traverse speed on the surface roughness was soon clear. Both parameters define, with the spot diameter, the density d_i (which represents, in other words, the number of laser pulses that hit one definite point of the surface during the treatment) whose optimum value as concerns the surface roughness was determined to be 200.

The ON/OFF ratio, that is the fraction of the pulse period during which the laser beam is actually striking the specimen surface, was determined to be 0.06. Different combinations of pulse energy, duration and frequency together with worktable traverse speed were tested and their influence evaluated by means of microstructure examinations and microhardness and roughness profiles.

The results of the afore said tests can be summarized as follows:

- the frequency seems to be the most important parameter when attempting to avoid surface cracks formation (good values are over 100 Hz), whereas surface smoothness seems to be related to the density, as already pointed out. In specimen treated with the same density, frequency and mean power, surface roughness seems to decrease with pulse duration, that is increasing single pulse power (P_i);

- comparing different combinations of the operative parameters, and of correlated ones, it was found that alloyed layer thickness, the ON/OFF ratio being constant, is related to the product $P_i * d_i$: values of 6000W were related to the layer thickness of say 10μm, 30000W to about 15 μm, 480000W to 160μm;

- surface hardness is related to specific power P_s (P_i/beam section).

The above mentioned characteristics are obviously related to the microstructure and to the composition of the samples.

Columnar dendrites

Random oriented dendrites layer

50 μm

Figure 1. Typical microstructure in nitrided sample.

As the thermal cycle in laser processing takes fractions of a second to be completed (and this is particularly true for pulsed laser processing), the resulting microstructures are typically non-equilibrium ones.

In the micrography of figure 1, three zones are distinguishable:

— a shallow layer of columnar dendrites, oriented perpendicularly to the specimen surface is noticeable, which on the basis of the N-Ti phase diagram, consists of TiN dendrites with Nitrogen amount over 30 at%, whose solidification temperature ranges between 2950 and 3290 °C [4]. On the basis of the surface dendrites orientation, it can be reasonably supposed that the solidification begins at the surface and not at the liquid-solid interface.

— a second random oriented dendrites layer immediately under the columnar layer, involving a peritectic reaction (at about 2350 °C) between α-Ti and δ-TiN. According to [2] this dendritic layer consists of non stoichiometric TiN_x, with x ranging between 0.65 and 0.85.

— a third zone, where Nitrogen amount is relatively low, in which takes place a peritectic transformation between α-Ti and β-Ti, which during cooling transforms in martensite.

The resulting microhardness profiles showed surface hardness values ($DPH_{0.3}$) ranging between 700-1000 kgf/mm^2. The 550 kgf/mm^2 boundary generally ranged between 100-160 μm (figure 2a). On the nitrided tracks surface roughness was determined as well (figure 2b).

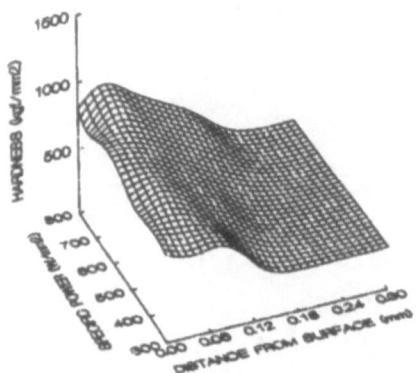

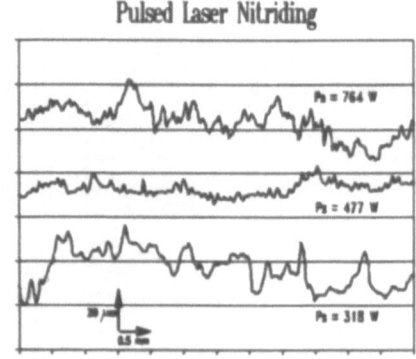

Figure 2a. Fit of microhardness profiles in nitrided specimen

Figure 2b. Corresponding surface roughness profiles

3.2. CO$_2$ GAS SHIELDING

The microhardness values of the specimens obtained with a CO$_2$ atmosphere were quite higher than those of the nitrided ones obtained with similar working parameters (figure 3a). Moreover the surface roughness generally resulted lower (figure 3b).

with the same working conditions this result not only concerned the value of the surface microhardness, but also the hardened layer thickness was higher (figure 4a). Moreover surface roughness resulted lower (figure 4b).

High surface hardness accompanied by low roughness usually result in good tribological behaviour.

The association of high hardness values and relatively low roughness may be caused by the additional energy coming from the concomitant exothermic reactions of carbides formation and, in particular, of titanium oxidation. In effects CO_2 dissociates at elevated temperatures in C and O_2 to a smaller extent and, to a far greater extent in CO and O_2, so that oxygen is available to combine with the titanium of the substrate.

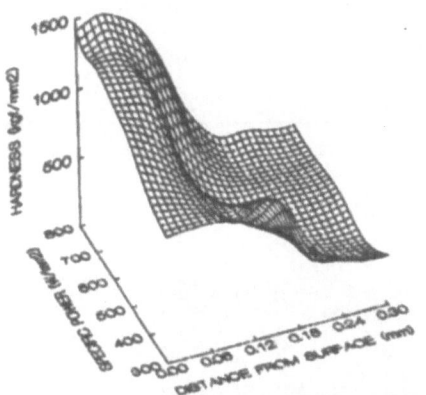

Figure 3a. Fit of microhardness profiles in carburized specimens

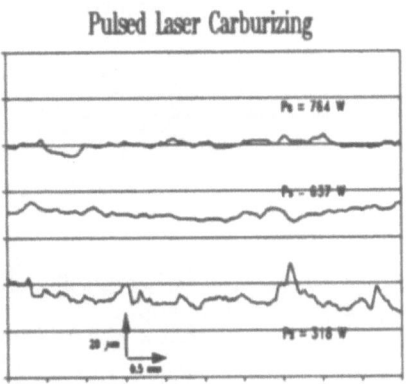

Figure 3b. Corresponding surface roughness profiles

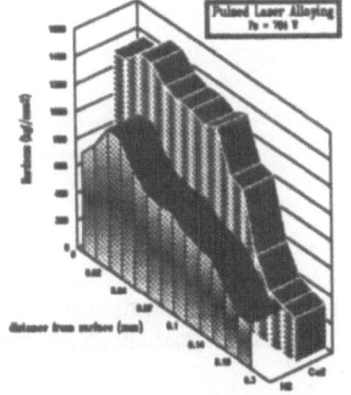

Figure 4a. Comparison of microhardness profiles of nitrided and carburized layers

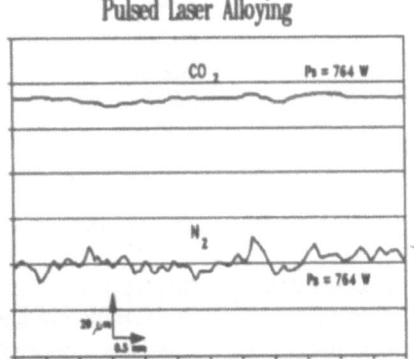

Figure 4b. Corresponding surface roughness profiles

Indeed the concomitant presence of γ-TiC and δ-TiO was verified by means of X Ray diffraction analysis performed on carburized specimens [5].

For microstructure examination a tint etching procedure was adopted after usual metallographic polishing. The reagent consisted of Sodium Molibdate 3g, HCl 5ml, Ammonium Difluoride 2g, in 100ml H_2O, which colors in brownish yellow the TiC phase and in greenish blue the titanium alloy matrix.

The alloyed layer showed a dendritic morphology consisting of TiC, and a heavily corroded interdendritic one, which, on the basis of former XRD results performed on the lased surface, reasonably consisted of the relatively low-melting TiO (figure 5).

Surface ⟶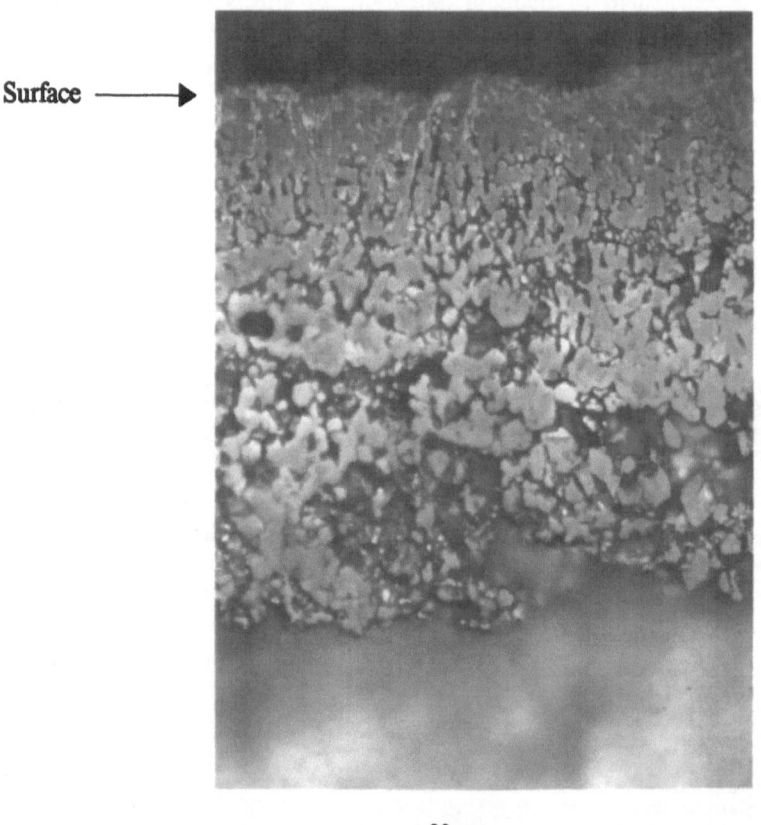

20 µm

Figure 5. Carburized layer, with dendritic TiC and former interdendritic TiO.

It can be reasonably supposed that the different brown tones proceeding from the laser treated surface towards the bulk, are due to a carbon concentration gradient so that non stoichiometric TiC phases are produced.

4. Wear Tests.

A pin-on-disk tribometer was used to perform wear tests without lubricant on annular lased tracks produced with the N_2 and CO_2 atmospheres. For both the samples series the counter consisted of an AISI 52100 10 mm ball. The normal loads were 10 N and 20 N, the speed 0.15 m/s and the sliding distance up to 6 km.
The specimens produced under a CO_2 atmosphere proved to be more wear resisting than the nitrided ones which underwent a dramatic failure (figure 6a, b): in the 20 N

load test the overall tribological behaviour of the nitrided layer was even worse than that of the untreated Ti6Al4V alloy sample, whose overall performance was roughly the same in the two test conditions, probably due to surface delamination and consequent production of the "third body", which acted as an abrasive on the two original counterparts.

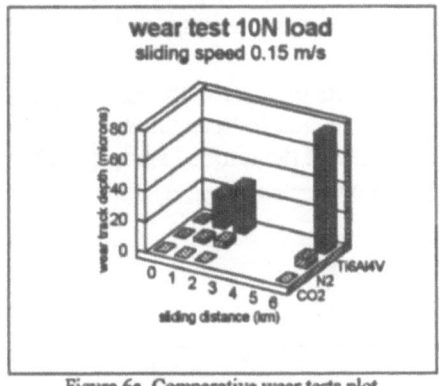

Figure 6a. Comparative wear tests plot.

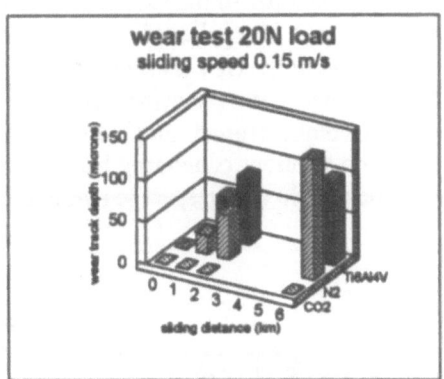

Figure 6b. Comparative wear tests plot.

5. Corrosion Tests

The results of the corrosion tests, performed on nitrided samples, are summarized in figures 7 and 8, as a function of pulse power (P_i), and compared to those of the untreated alloy. The corrosion tests refer to 1 cm^2 surfaces.

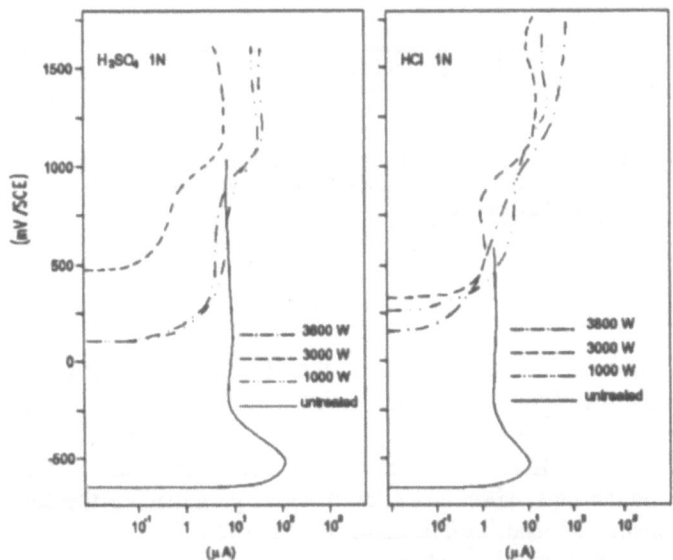

Figure 7. Potentiostatic curves for nitrided samples in H_2SO_4 and HCl 1N.

Figure 7 shows the anodic polarization curves of the different samples in H_2SO_4 1N and HCl 1N. An increase of the corrosion potential is evident for the nitrided surfaces in general, compared to that of the untreated one. At the same time the corrosion rate at equilibrium is lower in the case of the nitrided samples: in the H_2SO_4 1N solution the equilibrium current (i_0) decreases of two orders of magnitude, from 10 μA to 0.1 and even 0.05 μA in the case of the sample obtained with 3000 W pulse power whose behaviour can be explained on the basis of an observed more uniform and smoother surface and in the HCl 1N solution from 1 to 0.1 μA.

Such behaviour is more evident in figure 8, which reports the concentrations of Ti ions vs time, coming from a sample treated with P_i = 1000 W and from an untreated one in sulphuric solutions containing NaCl. The corrosion rate of the nitrided surface appears to be negligible with respect to that of the untreated one.

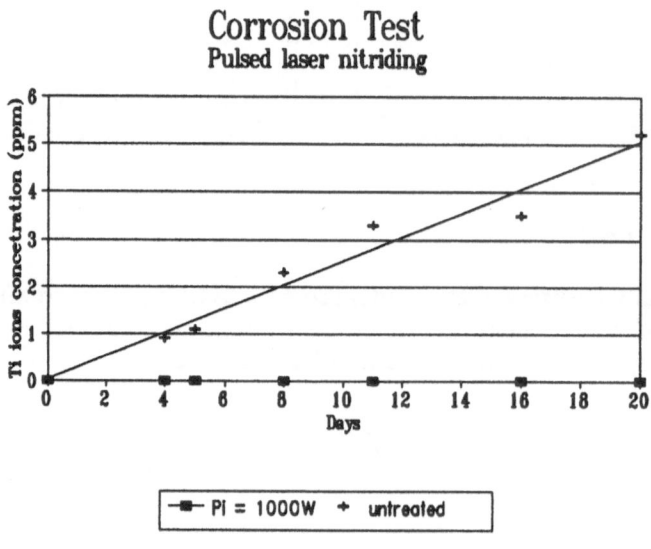

Figure 8. Ti ions concentration in 1N H_2SO_4 + 1 wt% NaCl solutions from corrosion tests of a nitrided surface (P_i = 1000 W) compared to that of an untreated one.

6. Conclusions

Laser surface nitriding and carburizing of the Ti6Al4V alloy has been performed by means of a pulsed Nd-YAG laser, using Nitrogen and Carbon Dioxide gas flows.

The carburized specimen exhibited higher hardness and lower surface roughness compared to that of nitrided specimen. Diffraction analysis as well microscopy observations of the latter specimen alloyed layers permitted to identify a compound structure consisting of dendritic TiC and interdendritic TiO.

Wear tests results showed a better behaviour of the carburized samples compared to the nitrided ones. Corrosion tests confirmed improved properties of the treated surface.

7. References

1) Walker, A., Folkes J., Steen W.M. and West, D.R.F. (1985), Laser Surface Alloying of Ti Substrates, Surface Eng. 1985, 1, 23-29

2) Bell, T., Bergmann, H.W., Lanogan, J., Morton,. P.H. and Staines A.M. (1986), Surface engineering of Titanium with Nitrogen, Surface Eng. 1986, 2, 133-143

3) Cantello, M., Bianco, M., Lorenzi, M., Giordano, L. and Ramous, E. (1990) Nitrurazione Superficiale della Lega Ti6Al4V Assistita da Laser in Associazione Italiana di Metallurgia (ed.), Proc. XXIII AIM Nat. Cong. (Italy) sept. 1990 Associazione Italiana di Metallurgia (Milano, Italy), pp.427-433

4) ASM Handbook, Alloy Phase Diagrams, (1992) Bakes, H. (ed.), ASM International, Materials Park, Ohio, vol. 3, p. 2.229

5) Bianco, M., Rivela, C., Zambon, A. and Costantini, G. (1993) Carburazione della Lega Ti6Al4V Mediante Laser Associazione Italiana di Metallurgia (ed.), Proc. XIV National Cong. Heat Treatments (Italy) May 1993, Associazione Italiana di Metallurgia (Milano, Italy), pp.73-83

LASER SURFACE ALLOYING OF ALUMINIUM 2014 ALLOY WITH Mo FOR ENHANCED CORROSION RESISTANCE

M A McMahon[†], K. G. Watkins[††], W. M. Steen[††], R. Vilar[†††],
M. G. S. Ferreira[††††]

[†] *Department of Materials Science and Engineering,*
University of Liverpool, UK

[††] *Department of Mechanical Engineering, University of Liverpool, UK*

[†††] *Dipartamento de Engenharia Materias,*
Instituto Superior Tecnico, Lisbon, Portugal

[††††] *Dipartamento de Engenharia Quimica,*
Instituto Superior Tecnico, Lisbon, Portugal

Abstract

The incorporation of Mo in commercial aluminium 2014 alloy has been investigated via laser surface alloying (LSA) of predeposited plasma spayed layers using a CO_2 laser.

The microstructure and composition of the LSA materials produced was investigated via SEM, TEM, SAD, STEM, WDS, XPS and XRD and the electrochemical properties were investigated via anodic potentiodynamic polarisation in chloride electrolyte. It was found that there was a significant increase in the critical pitting potential following LSA treatment. A complex microstructure was displayed consisting of Mo precipitates in an aluminium matrix showing enhanced solid solubility of Mo in aluminium.

1 Introduction

The highly localised interaction between lasers and materials makes laser processing an attractive alternative in the realisation of coating strategy, particularly for the production of corrosion resistant surfaces. The following advantages compared with more conventional techniques can be identified: (i) a dense surface layer with no porosity can be produced, (ii) integrity of bonding between the surface layer and the substrate is complete, (iii) there is no deleterious effect on bulk mechanical properties, (iv) high vacuum conditions are not required since oxidation effects can be simply avoided by the use of an inert gas shroud, (v) there is no fundamental restriction on component shape, (vi) surface microstructures which are unobtainable by any of the main coating or surface treatment technologies can be produced, (vii) limited areas of a component surface can be

J. Mazumder et al. (eds.), Laser Processing: Surface Treatment and Film Deposition, 337–358.
© 1996 *Kluwer Academic Publishers.*

processed without affecting surrounding areas and hence these processes are applicable to the limited area processing of critical regions of a component.

Limitations in the use of aluminium alloys include wear, temperature and corrosion resistance considerations. Corrosion problems are influenced by the effect of second phase precipitates produced as a result of alloying for strength with pitting corrosion, crevice corrosion, intergranular corrosion, exfoliation corrosion and stress corrosion cracking amongst the principal forms of attack.

Under equilibrium processing conditions elements such as Cr, Mo, W or Ti with a known passivating effect on the progress of corrosion in aluminium are either insoluble or nearly insoluble in the metal and hence techniques for their incorporation in non-equilibrium concentrations are required. Since during laser surface alloying (LSA) the cooling rate during resolidification of melted surface layers is high (ca 10^4 -10^6 K/s, compared with ca 10^2 K/s by conventional alloying techniques) [1] the technique can be employed to produce such non-equilibrium microstructures via increased cooling rates from the molten state.

Formation of metastable solid solution of Cr, Mo, W, Ti and other elements in aluminium has been studied by a number of non-laser deposition techniques and this has formed a theoretical basis for the selection of suitable alloying element additions to promote enhanced resistance to pitting in aluminium [4]. Non-equilibrium solid solutions of Cr, Mo, W, Zr, Nb, Ta or Ti in Al have been produced by a variety of techniques which include ion implantation [5-6], RF magnetron sputtering [7 -10] and DC magnetron sputtering [11,12]. Enhancement of the critical pitting potential [13] of aluminium in aqueous chloride electrolyte has been observed by these means and, indeed, these investigations have led to new propositions on the mechanism of corrosion in aluminium alloys, particularly in the case of pitting corrosion [4]. Correlation has been observed between the pH of zero charge of the hydrated oxide of the added element, pH_{zch}, and the critical pitting potential, E_{crit}, and this has been taken as being linked with the establishment of conditions for the adsorption of chloride ions on the aluminium oxide surface as a key stage in the corrosion process. Correlation of E_{crit} with the solubility in chloride electrolyte of the hydrated oxide of the added element was also observed. With particular reference to Mo, Moshier et al [7,10] observed that the addition of 6 atomic per cent Mo to Al by RF-magnetron sputtering increased the critical pitting potential (as determined by anodic potentiodynamic polarisation in deaerated 0.1 N KCl solution at pH 8) from -650 mV (SCE) for pure Al to -210 mV (SCE) for the alloy. Using a different strategy, Shaw et al [14] produced uptake of Mo of 12 atomic per cent in the passive film on Al by potentiostatically polarising the metal in deaerated 0.1M Na_2MoO_4 for 2 hours. The critical pitting potential in 1000 ppm chloride solution was determined as being as high as 33 - 67 mV(SCE).

In Foley's model [15] for pit initiation and propagation in aluminium and its alloys in chloride electrolytes there are four stages, as follows:
Stage 1 : The adsorption of the reactive anion on the oxide-covered aluminium,

Stage 2: The chemical reaction of the adsorbed anion with aluminium ion in the aluminium oxide lattice or the precipitated aluminium hydroxide,

Stage 3: The thinning of the oxide film by dissolution,

Stage 4: The direct attack of the exposed metal by the anion, possible assisted by an anodic potential. (This stage is sometimes referred to as pitting propagation).

Hence, if conditions need to be created for chloride ions to be adsorbed on the oxide surface for pitting to take place, the local pH in the electrolyte must become lower than pH$_{zch}$ since only then will the surface have a net positive character favourable for the adsorption of the anion. The potential will need to be increased to a higher value to achieve these conditions for added elements with oxides with a low pH$_{zch}$ compared with those with a higher pH$_{zch}$. Hence, addition of Mo with pH$_{zch}$ of 1.95 [5] should result in a higher E$_{crit}$ than for pure aluminium where the pH$_{zch}$ is greater than 8, provided the character of the oxide is dominated by that of the added element. Low pH$_{zch}$ elements such as Mo, Cr, and W also display low solubility of their oxides in chloride electrolytes and hence, if oxide thinning plays a role in pit initiation, the beneficial effect of elements such as Mo in raising E$_{crit}$ may also be related to the lowering of the solubility of the oxide produced by additions of these elements. Alternative theories of the influence of Mo concern the formation of oxidised molybdenum species within the passive layer which may have an inhibitive effect on the progress of corrosion [10, 14].

2. Experimental

2.1. LASER SURFACE ALLOYING (LSA)

AISI 2014 aluminium alloy of composition: 4.65 % Cu, 0.75 % Si, 0.54 % Mg, 0.12 % Fe, 0.11 % Mn, 0.11 % Zn, 0.01 % Ti and Al balance (where compositions are in weight per cent) was used as the substrate material. The heat treatment condition was T-651. For LSA with Mo an Electrox 2kW CO_2 laser was used. Mo was provided in the form of a predeposited plasma sprayed layer of thickness 100 μm containing 25wt % Mo and 75wt % Al. In preliminary tests, this predeposited layer composition was found to give the most consistent alloying results. LSA was carried out at a laser output power of 1.4 kW with a spot size of 1 mm and with the rate of movement of the coated substrate beneath the stationary beam of 20 - 70 mm/s. Area coverage was produced by overlapping adjacent tracks by 50%.

2.2. METALLOGRAPHIC AND STRUCTURAL INVESTIGATION

Surface samples and cross-sections were examined in a Philips 501 SEM. Further detailed microstructural and compositional examinations were carried out on a Philips EM400 TEM, a JEOL 2000FX TEM with Scanning mode attachment, a VG Scientific Scanning Auger microscope with x-ray photoelectron spectroscopy (XPS), JEOL JXA-50A electron probe microanalyser, and a computer controlled McLean diffractometer.

For surface sectioned TEM samples, 3 mm discs were trepanned from thin surface sections of both as received and surface treated samples by spark erosion,. Initial thinning was achieved by grinding with 1200 grit SiC paper. The final thin area was achieved by either electropolishing using a Struers Tenupol or by ion beam thinning using a Gatan ion mill. The electrolytes used was: 70% Methyl alcohol, 30% Conc. Nitric acid, cooled with Liquid nitrogen to -20 °C, operation at 40 V. The structure of the phases present in the laser treated materials was studied using Philips EM400 and JEOL 2000FX TEMs to obtain transmission electron micrographs and selected area electron diffraction (SAD) patterns from the materials. The composition was measured by energy dispersive X-ray (EDX) microanalysis in the STEM mode on the JEOL 2000FX using Link System software. X-ray element mapping was also performed by setting energy windows corresponding to the peak energies of the elements of interest.

For samples for X-ray photoelectron spectroscopy (XPS), a square approximately 8 mm x 8 mm and < 1 mm thick was cut from the surface of each of the specimens using a diamond cut-off wheel. In most cases the surfaces were polished to 1200 grit or better except when the as-laser processed surface was to be analysed. In all cases the samples were cleaned, firstly in acetone and then rinsed in distilled water and dried by hot air. The surfaces to be examined were either allowed to oxidise naturally in air or were polarised in 1M NaCl. After polarisation the specimens were rinsed in distilled water and dried by hot air. Typically the time elapsed between polarising the specimens and insertion into the microscope was 10 minutes.

A JEOL JXA-50A electron microprobe was used to carry out quantitative compositional analysis by both EDX and wavelength dispersive spectroscopy (WDS). In the case of the EDX a Co standard was used for calibration and Link Analytical QX200 data analysis software was used to calculate the composition. For a greater level of accuracy WDS measurements were carried out. Pure standards were used for each of the elements concerned. The counts were collected for 100 seconds at the maximum pulse height position and at +0.5° and -0.5° background positions for each of the elements concerned and their standards. The X-ray intensity ratios or K values were obtained from the count data for each of the elements and used to calculate the respective weight percentages. The ZAF correction programme used was a version of the Frame 3 programme described in NBS Technical Note 796. X-ray element maps were also obtained on the JEOL JXA-50A electron microprobe by WDS.

Phase analysis was carried out from X-ray spectra obtained using a McLean diffractometer. The incident beam was focussed using a solar slit collimator and 2° front slit to give an area of ~1 cm^2 in the plane of the sample surface. A Ni foil was used to filter the diffracted beam to eliminate Kβ radiation. Computer software running on a Philips P 3302 PC was used to drive the diffractometer and to collect and plot the X-ray diffraction data from the scans.

A VG Scientific scanning Auger electron microscope was used to carry out surface analysis. The incident beam consisted of Al X-rays (λ = 8.339 A) operating at 15 kV and

20 mA. A Pass Energy (PE) of 100 eV was used for general scans between Binding Energies (BE) of 1300 eV and 0 eV with a step size of 1 eV. For scans of the individual elements PE = 10 - 20 eV, the step size was 0.05 - 0.1 eV and a minimum of 30 scans were collected for each analysis; in some cases 60 scans were collected to enable clear resolution of the peak position. In all cases the dwell time of each step was 200 ms.

2. 3. ELECTROCHEMICAL INVESTIGATION

A wire was attached to the back of sections of each of the samples and these specimens were then mounted in *Scaniplast* cold setting resin. Two hours before testing the specimens were first abraded to 1200 grit SiC paper, washed with soap, rinsed with distilled water and dried with hot air. The areas chosen for testing, between 0.4 and 1.5 cm², were masked using *Alcomit* lacquer.

The chloride electrolytes used for testing were made from distilled water and AnalaR grade NaCl. The concentration of the electrolytes used were 0.1, 0.5 and 1 molar solutions. During testing the electrolyte was maintained at a temperature of 30 ˚C by means of a thermostatically controlled water bath. Some of the tests involved using deaerated solutions, which were obtained by passing pure nitrogen through the electrolytes for at least two hours prior to testing. During the tests the cells were sealed to prevent the ingress of air.

Resistance to pitting was measured in both the as received and laser processed conditions using a conventional three electrode cell arrangement. Potentiodynamic and potentiostatic polarisation measurements were carried out in accordance with ASTM G5-69. A Sycopel Scientific Ltd. Autostat was used under computer control. Potentiodynamic anodic pitting scans were made from 0 mV to 1000 mV at a scan rate of 10 mV/min. Prior to starting the tests the free corrosion potential (Ecorr) was monitored in the test electrolyte for half an hour. Potentials were measured against a saturated calomel electrode (SCE) and a platinised titanium rod was used as the auxiliary electrode.

Measurement of the free corrosion potential versus time in accordance with ASTM G-69 was made using a CIL A to D converter, multiple channel data logging card fitted to an Amstrad 3086 PC. Logging was carried out for time periods of 150 hours and in excess of 300 hours (2 weeks). A 1 molar NaCl solution was used for each of the immersion tests and the temperature was kept at 30 ˚C by using a thermostatically controlled water bath.

3. Results and Discussion

3.1. MICROSTRUCTURE OF THE SURFACE ALLOY

The microstructure of the laser surface alloyed material was dendritic. Figure 1 shows the typical microstructure of the surface alloy in cross section. At the limit of the melt depth

(~400-500 μm) growth started with a region of planar front solidification. However, in comparison with laser surface melted samples of the same material [16], the planar front region was very small and in some cases very difficult to distinguish from the adjacent partially melted grains in the heat affected zone (HAZ) of the substrate; this is shown in Figure 2. The columnar grains that started to grow at the fusion boundary consisted of cellular dendrites orientated in the general direction of heat flow. Cooling rates were estimated to be of the order of 10^6 K/s during solidification adjacent to the fusion boundary but increased to be of the order of 10^7 K/s just below the surface. The columnar growth was epitaxial, with some grain boundaries being continuous, from the HAZ in the substrate, through the melt zone to the free surface.

Intermetallic particles were dragged into the melt pool from the top edge of the melted region as shown in Figure 3. In addition Figure 4 shows that convection in the melt then helped to distribute the micron sized intermetallic particles within the grains. A banding structure was also found in the alloyed layer. The bands were marked out by intermetallic particles lying in arcs parallel to the fusion line as illustrated in Figure 5.

A more detailed inspection of the microstructure was carried out by TEM Figure 6 shows the basic cellular dendritic structure. Sub-micron sized intermetallic particles were also nucleated and frozen inside the primary α-Al dendrites; they were not just confined to the interdendritic phase. These intermetallics, shown in Figure 7, were found to consist mainly of faceted crystals seemingly randomly orientated within the α-Al, as indicated by the accompanying SAD pattern in Figure 7 (b). Examination of the calculated plane spacings from the SAD patterns and the data from the X-ray diffraction scans shown in Figure 8 indicated that the faceted particles were similar in structure to the intermetallic $Al_{12}Mo$ and Al_5Mo phases; these results are summarised in Table 1. Two other types of intermetallic were identified; a spherical particle shown in Figure 9 and a star shaped particle shown in Figure 10. No match was made with existing data for the spherical particles. From the SAD pattern shown in Figure 10 (b) it was found that the dendritic star crystals were $MgAl_2O_4$.

3.2. COMPOSITION OF ALLOYED LAYER

It was found that the melt tracks were quite homogeneous and a good degree of mixing was achieved after the LSA treatment. Composition profiles, measured by WDS for Cu and Mo, are shown in Figure 11. Each point on the composition profiles represents an area averaged value because the minimum probe size for analysis was about 2 μm. Even so it can be clearly seen that complete mixing has been achieved to the limit of the melt depth. It is apparent that the concentration of Mo in Al has been significantly increased as a result of laser surface alloying. The composition profile across an α-Al dendrite obtained by STEM is shown in Figure 12. This shows that up to 2.5 wt.% Mo was retained in solid solution, and that the interdendritic phase contained only a slightly higher concentration of Mo. From EDX analysis the star shaped particles shown in Figure 10 were found to be oxide particles that were essentially $MgAl_2O_4$ in type but

50 µm

Figure 1. SEM micrograph showing the microstructure of the surface alloyed layer.

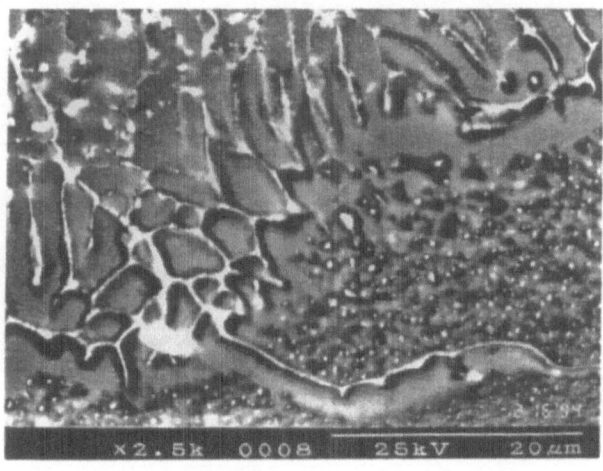

Figure 2 Back-scattered electron (BSE) microgaph showing the fusion boundary region of the LSA 2014/Mo alloy.

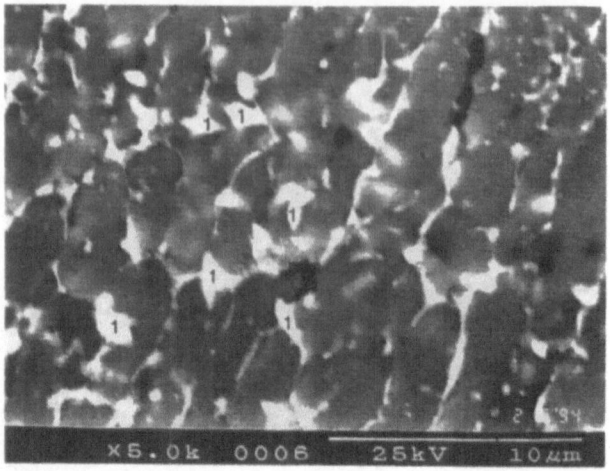

Figure 3. BSE micrograph of LSA 2014/Mo taken just below the surface and at the edge of the melt zone.

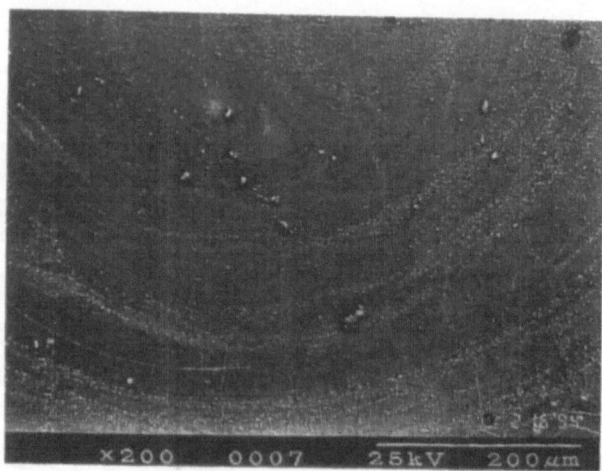

Figure 4. BSE micrograph showing the convection contours which are highlighted by the clusters of intermetallic particles.

Figure 5. BSE micrograph of the banded structure observed close to the limit of the melt depth in LSA 2014/Mo alloy.

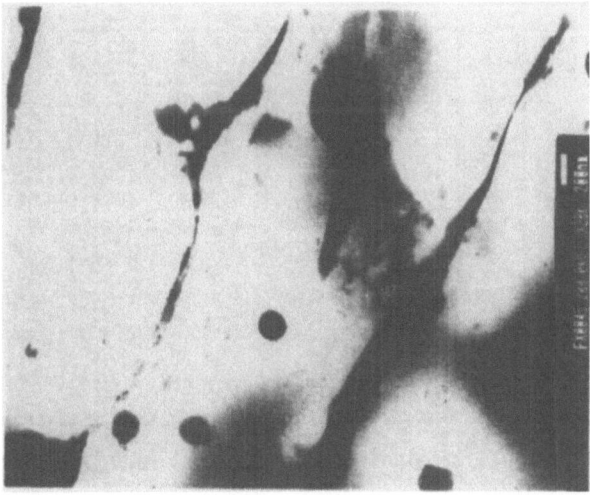

Figure 6. TEM micrograph of cellular dendrites in LSA layer.

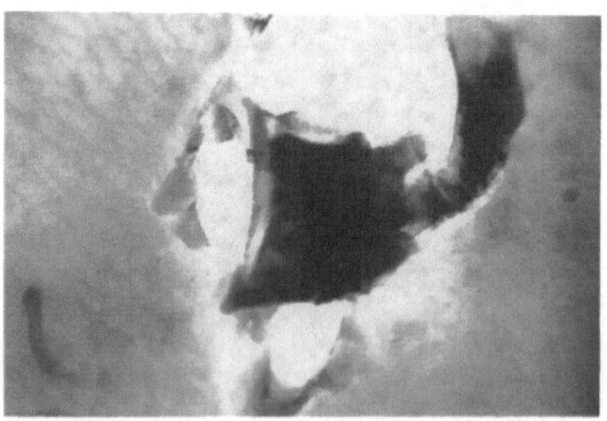

Figure 7. TEM bright field image of angular intermetallic particle.

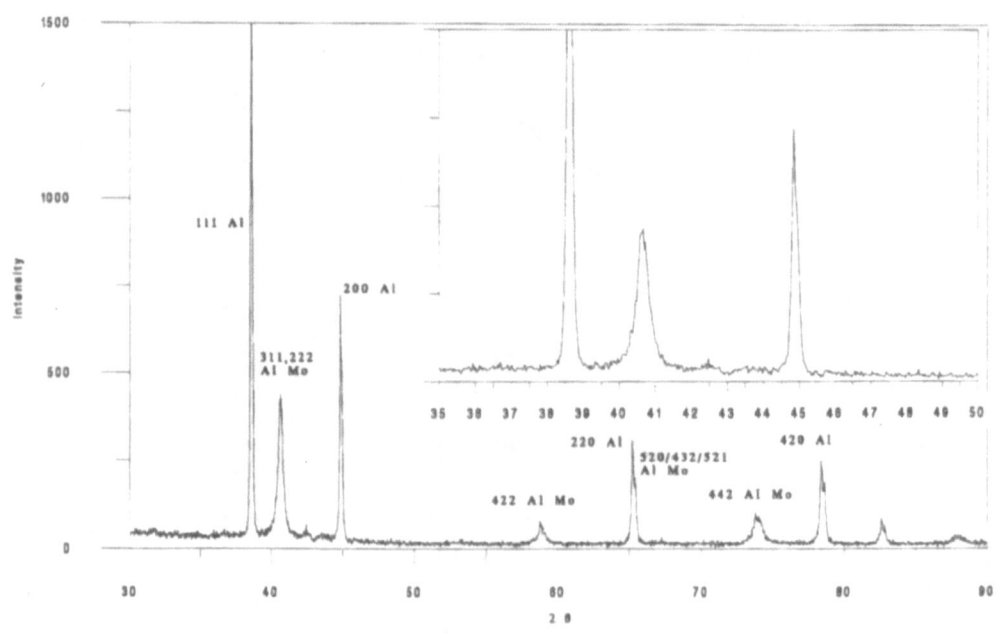

Figure 8. XRD profile for Mo alloyed layer

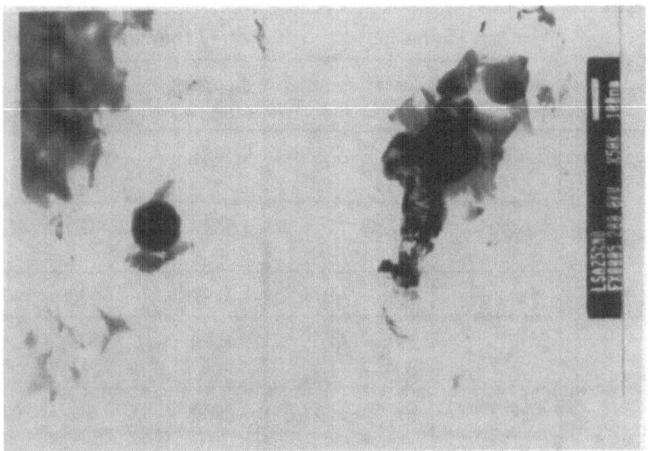

Figure 9. TEM bright field image of spherical intermetallic particle.

210 nm

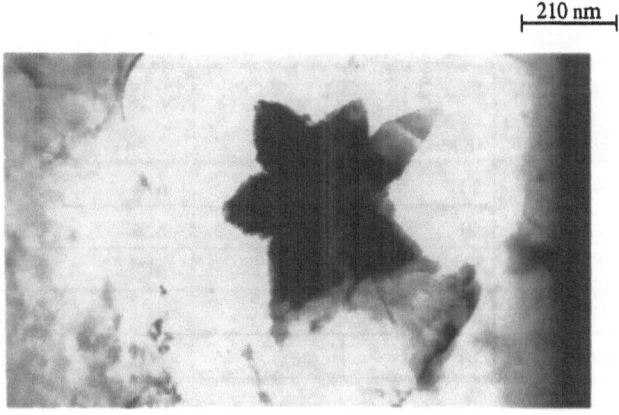

Figure 10. TEM micrograph of Al$_2$MgO$_4$ type particle

SAD Pattern	Electron Diffraction		X-ray Diffraction	
	d-spacing (ang.)	(hkl)	d-spacing (ang.)	(hkl)
1. Al [011]	7.4074	100 ($Al_{12}Mo$)	2.3061	311, 222 ($Al_{12}Mo$)
	2.4591	221, 300 ($Al_{12}Mo$)	1.5697	422 ($Al_{12}Mo$)
	2.344	111 (Al)	1.4287	220 (Al)
	2.027	200 (Al)	1.4199	520, 432, 521 ($Al_{12}Mo$)
2.	4.3478	100 (Al_5Mo)	1.2809	442 ($Al_{12}Mo$)
	2.3585	111 (Al_5Mo)	1.2178	420 (Al)

Table 1. Electron diffraction data from the two types of angular precipitates observed in the LSA material compared with diffraction data

Test solution	Laser scan velocity (mm/s)	E_{corr} (SCE)	E_{pit} (SCE)	E_{PD}
1 M NaCl	20	- 844	- 614	$230_{(2)}$
	50	- 794	- 624	$170_{(4)}$
	70	- 730	- 612	$118_{(4)}$
0.1 M NaCl	20	- 834	- 519	$315_{(3)}$
0.1 M KCl	20	- 890	- 512	$378_{(5)}$

Table 2. Results of potentiodynamic anodic polarisation in various deaerated electrolytes.

with a significant amount of additional Mo. The faceted precipitate shown in Figure 8 was found to have a more complex composition to that of $Al_{12}Mo$ because it also contained Cu and Mg in approximately the same atomic concentrations as Mo.

3.3 POTENTIODYNAMIC POLARISATION

A typical polarisation curve for the LSA material in 1M NaCl is shown in Figure 13. Table 2 shows the average values for E_{corr}, E_{pit}, and E_{PD} in each of the test electrolytes (where E_{corr} is the free corrosion potential, E_{pit} is the critical pitting potential and E_{PD} = (E_{pit} - E_{corr})). The figure in parenthesis in each row shows the number of tests performed. The passive current was typically of the order of 0.5 $\mu A/cm^2$ for all of the tests. It is shown that the effect of LSA with Mo was to decrease E_{corr} and increase E_{pit} and hence increase the relative potential difference, E_{PD}, required to cause pitting. For corrosion in 1M NaCl it can be seen that as the scanning velocity was increased the pitting resistance decreased. The value of E_{pit} was -617 ± 8 mV (SCE) for all the samples tested but E_{corr} increased with increasing scan velocity, hence E_{PD} decreased. The corrosion resistance of the surface alloy in the 0.1 M chloride solutions was better than in the 1M NaCl solution and E_{pit} increased to -516 ± 10 mV (SCE).

For comparison the as-received plasma sprayed coating and pure Mo were corrosion tested in 0.1M and 1M NaCl. The plasma sprayed coating exhibited no corrosion resistance, with the current rapidly increasing even at low applied potentials. Pure Mo, however, suffered from general corrosion in the 1M NaCl solution. The current increased with potential but very slowly indicating low corrosion rates. In the 0.1M NaCl solution the pure Mo seemed to remain passive over a very large range. At low potentials the surface darkened slightly but the current did not increase until the applied potential was nearly 1V.

The free corrosion potential, E_{corr}, was monitored for up to 1 week in 1M NaCl and the results are shown in Figure 14 along with the measured pitting potential. It can be seen that E_{corr} stayed below the pitting potential for the majority of time of the test but never attained a steady state value. Inspection of the samples after immersion showed a tendency for preferential attack along the track-track interfaces.

During anodic polarisation pits were observed to form where intermetallic particles had grouped together; this can be seen in Figure 15. The pits grew by the dissolution of the aluminium between the intermetallic particles. Pits did not, however, form preferentially at existing surface defects such as the exposed gas pores shown in Figure 16.

3.4. X-RAY PHOTOELECTRON SPECTROSCOPY

The oxide that formed on a freshly abraded surface of the LSA alloy under normal atmospheric conditions was found to contain Mo. Studies using XPS showed that the Mo in the air formed oxide was in both the metallic state and in one of its oxidation states, probably the Mo^{6+} state. After polarisation in chloride electrolytes the XPS data

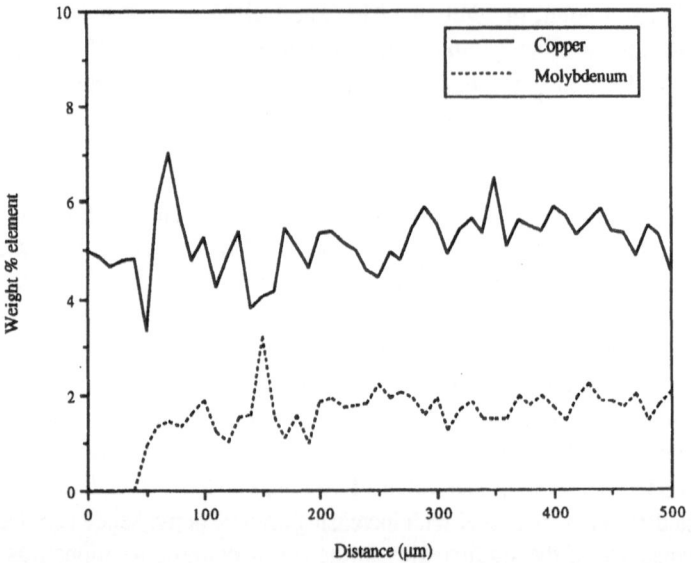

Figure 11. WDS depth profile from substrate (0 μm) to surface (500 μm) for LSA surface.

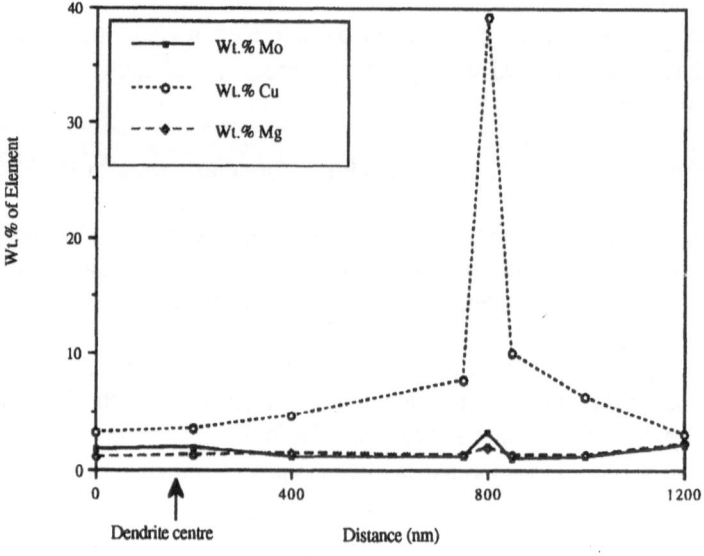

Figure 12. EDX composition profile across a dendrite arm.

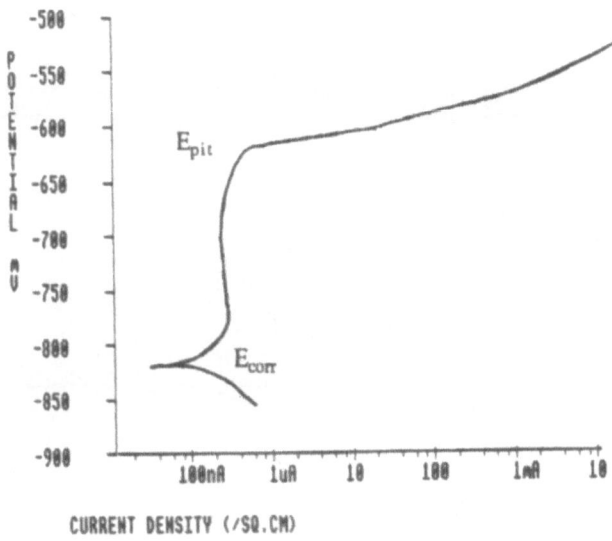

Figure 13. Potentiodynamic anodic polarisation curve for LSA 2014/Mo alloy.
Tested in 1 M NaCl, 30 °C, deaerated for 2 hrs with N_2, 10 mV/min.

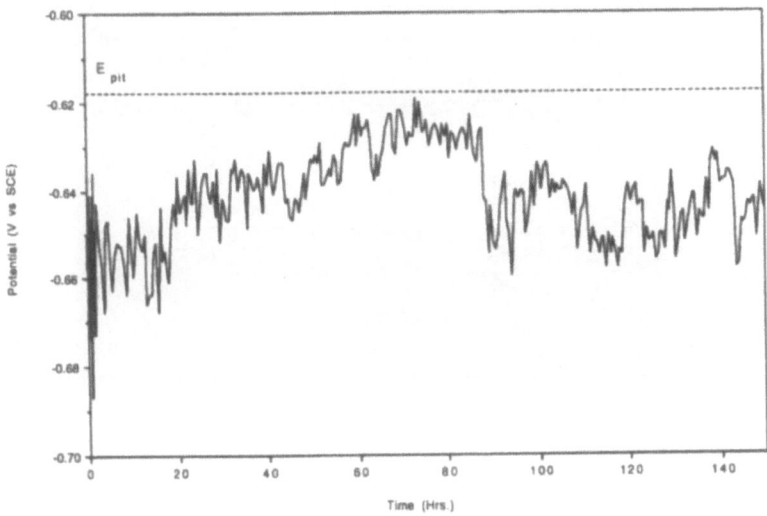

Figure 14. Free corrosion potential versus time for LSA material

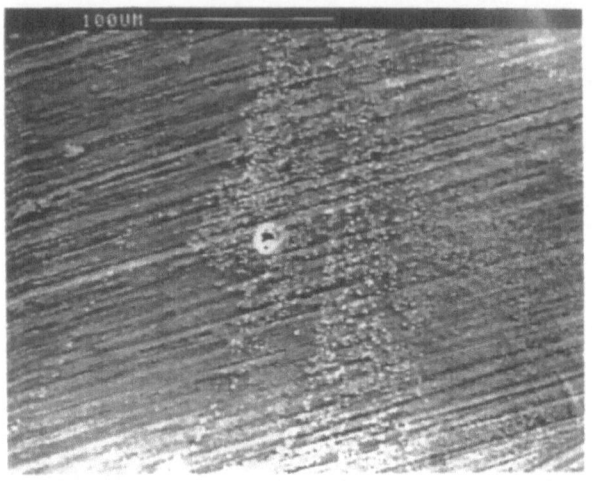

Figure 15. SEM micrograph of corrosion pit formed amidst a cluster of intermetallic particles

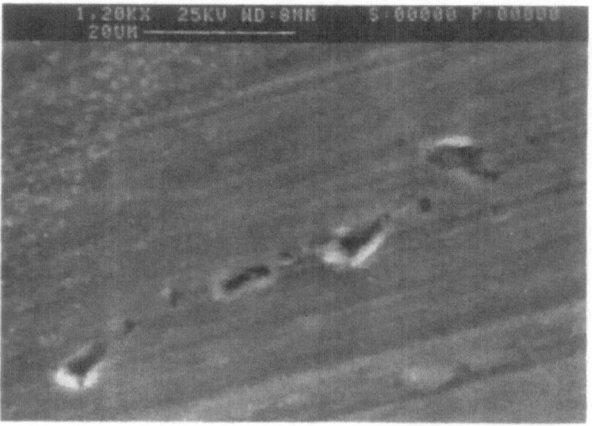

Figure 16. SEM micrograph of the surface of the LSA material after exposure to 1 N NaCl solution

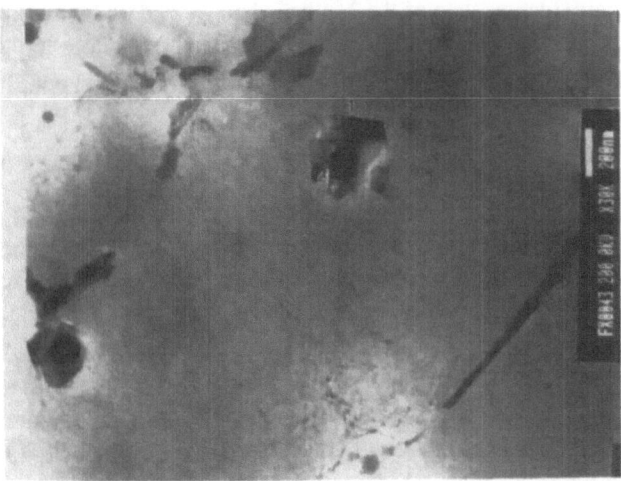

Figure 17. TEM micrograph showing how intermetallic particles were frozen into the cellular dendrites in LSA 2014/Mo alloy. It is possible that the larger particle (1) was nucleated by the spherical particle (2).

indicated that the oxidation state of the Mo had changed and that the Mo was present in more than one form, as a combination of the Mo^{4+} and Mo^{6+} states forming a complex molybdate or oxide. Mo was not found in the metallic state after polarisation.

3.5 CELLULAR DENDRITIC GROWTH

The equilibrium solubility of Mo in Al is very low (~0.08 wt.%) so that on solidification it is rejected from the solidifying liquid very quickly producing large constitutional supercooling and the dendritic structure nucleates. The solute concentration was very even in the melt, which is apparent by the presence of intermetallic particles at the fusion boundary and which is also confirmed by the composition profile in the results (see Figure 11). However, Figure 17 also seems to show that there was not sufficient time in the melt for the Mo to completely segregate to the interdendritic boundaries. This meant intermetallic particles solidifying at the solid/liquid interface were frozen into the α-Al. Figure 17 shows two such cases of this where the nucleation of an intermetallic particle on the cell boundary has interrupted the growth of the interdendritic phase and another is seen to have formed in the centre of the dendrite arm. Both of the intermetallic particles seem to have been nucleated by the presence of the smaller spherical precipitate. In fact all of the angular precipitates were formed on or adjacent to one of the spherical precipitates.

The true identity of the spherical precipitates is not known. Clear electron diffraction could not be obtained in order to index its structure because, as Figure 17 shows, in many cases more than one angular precipitate would be nucleated on the spheres. Electron diffraction from the angular precipitates, however, showed similarities with X-ray powder diffraction data which indicated that the intermetallic $Al_{12}Mo$ phase was present in the new surface alloy.

A third form of precipitate was also identified in the results and these were the $MgAl_2O_4$ star shaped particles. Ar gas was used to shroud the surface alloying operation but as already explained for the surface melting if the gas pressure is not sufficiently high then oxygen can be adsorbed and very quickly diffuse into the liquid. However, oxygen may already have been present because the plasma sprayed coatings were very porous. Therefore this makes it very difficult to remove all of the oxygen and explains why the $MgAl_2O_4$ particles were found in all of the surface alloyed samples prepared with predeposited plasma sprayed coatings.

3.6. CORROSION OF AS-RECEIVED ALUMINIUM ALLOY

The 2000 series alloys are known to be very susceptible to pitting corrosion because the principal alloying addition is copper which has a recognised detrimental effect on the corrosion resistance of aluminium. The corrosion properties of this group of aluminium alloys has been the subject of extensive research and the results presented in this study show good agreement with the literature.

Urushino and Sugimoto [17] measured the pitting potential of the 2024 alloy to be in the range of -750 mV to -600 mV (SCE) for different aging times between 0 and 10 days. E_{pit} for the alloy aged for 2 days was the same as that of the 2014 alloy used in this study, which was aged under the same conditions. They also showed however, that the 2024 and 2017 alloys had two pitting potentials. A lower Epit corresponding to grain boundary corrosion and a second, higher, Epit which marked the onset of grain matrix pitting. This type of behaviour was not observed for the 2014 alloy either by potentiodynamic or potentiostatic conditions in this study. Ambat and Dwarakadasa [18] have measured E_{pit} for the 2014 alloy to be -705 mV (SCE) in 3.5% NaCl pH 8. They later showed [19] that the effect of increasing the pH was to decrease Epit. They measured E_{pit} for the 2014 alloy in 3.5% NaCl, pH 10 to be -762 mV (SCE) and with a passive current density of 3.5 $\mu A/cm^2$ which is approximately 50 mV lower than the results of this study where the pH is close to a neutral solution.

3.7 CORROSION OF LSA MATERIAL

In this section the results for the corrosion of the LSA 2014 alloy with the predeposited 25 wt.% Mo- 75 wt.% Al plasma sprayed coating (LSA 2014/Mo) will be discussed with reference to results obtained by workers using other non-equilibrium techniques.

The anodic pitting potential was increased after the LSA treatment. From the results it has been assumed that the Mo present in both the α-Al and the oxide film of the surface alloyed layer was responsible for the increase. In the 0.1 NaCl solution the pitting potential was found to be ~-520 mV (SCE) which is comparable to the work of Natishan et al [5] who reported a pitting potential of -560 mV for an alloy containing 4 at.% Mo ion implanted into pure Al, though no data was presented for Ecorr. However, Moshier et al [7] did observe similar values of Ecorr (-930 mV (SCE) in deaerated 0.1 N KCl) to those in this study (-890 mV (SCE) in 0.1 M KCl) for an Al-6 a/o Mo ion implanted alloy but reported a pitting potential of -210 mV (SCE).

The Mo in solid solution has the effect of raising the free corrosion potential of the Al, which in turn should decrease the rate of galvanic dissolution at cathodic second phase particles such as $CuAl_2$. The presence of Mo rich intermetallic particles, however, caused pitting to occur and thus prevented the true pitting potential of the Al-Mo solid solution from being measured. Similar behaviour has been reported by Davis et al [20] who found that pits spontaneously formed at intermetallic particles in heat treated Al-Mo but is in contrast to the work of Shaw et al [21] who showed that the presence of intermetallic particles in Al-W alloys prepared by magnetron sputter deposition had only a minor effect on the corrosion resistance of the alloys.

It is evident from the results that the presence of the Mo has increased the corrosion resistance of the aluminium alloy. At the present time, however, it is not possible to determine whether the Mo has been incorporated in the oxide as part of the Al_2O_3 structure or as an intermediate layer between the metal surface and the outer oxide layer. It is possible that the Mo could exist in the oxide either in the metallic state or in one of its oxidation states. Its effect on the corrosion resistance could then be one of strengthening the oxide structure as observed by Davis et al [20] so that anodic dissolution is retarded, or if Mo were present in the oxide film as MoO_4^{-2} it could possibly behave as an effective barrier against the diffusion of chloride ions to the alloy surface. In this study it has been shown that the oxide that forms in air contains the Mo^{6+} ion and further consideration of the Al 2p peaks indicates that the oxide may be $Al_2(MoO_4)_3$ when compared to the work of Moshier et al [8]. It was also shown that after polarisation in chloride electrolytes that the Mo in the film had been oxidised to more than just one state, possibly containing the Mo^{4+} ion. This then seems to support the findings of Moshier et al who said that as the potential was increased the amount of Mo^{6+} decreased and the amount of hydrated Mo compound increased. They proposed that breakdown occurs when the hydration of the film reaches such an extent that the molybdate ion is so severely depleted as to allow the ingress of the chloride ion to the metal surface.

Urquidi and MacDonald [22] have used a solute-vacancy interaction model to explain the increase in breakdown potential when Mo^{6+} is present in passive films on stainless steels. It may also be possible to apply the model to the present work where Mo^{6+} was found in the oxide film. The model assumes that Mo^{6+} can form complexes with the negatively charged cation vacancies in the oxide film. This means there would be a decrease in the free cation vacancy concentration by reducing the flux of these vacancies from the film/solution interface to the metal/film interface. This phenomenon then leads to an increase in the breakdown potential. Urquidi and MacDonald [22] also showed that the model predicts that the induction time for the onset of pitting at any given applied potential was increased due to the presence of Mo^{6+}. This can then be used to explain why the potential in the immersion tests was quite stable indicating good corrosion resistance.

Since pitting is believed to be initiated by the adsorption of an anion at the oxide/solution interface the presence of Mo may also affect the pitting resistance by decreasing the pHzch of the oxide film on the alloy [6]. Cl- ions will only adsorb, however, when the surface has a net positive charge (i.e. when the local surface pH < pHzch). By decreasing the pHzch of the oxide it means that the potential must be increased to a higher value before the local pH at the surface falls below the pHzch, and Cl- can be adsorbed. This can be further coupled with the findings of Davis et al [20] that the oxide dissolution is affected by the presence of Mo because the value of pHzch for metal oxides is related to their solubility [4]. Therefore, if the characteristics of the oxide film were still the same as aluminium oxide but with a lower pHzch due to the Mo being present, then the oxide solubility would also be decreased for the same reason. Hence, if oxide thinning plays a role in pit initiation, the beneficial effect of Mo in raising Epit may be related to the decreased dissolution rate of the oxide.

4. Conclusions

1. After LSA of Mo in 2014 alloy, the solid solubility of Mo in the α-Al dendrites was increased to nearly 2 wt.% Mo compared to the equilibrium concentration of only 0.08 wt.% Mo.

2. $Al_{12}Mo$ intermetallic particles were present in the alloyed layer and clustered on the convection contours, whilst much smaller precipitates (mostly Al_5Mo) were able to form within the cellular dendritic structure. The intermetallic phase accounted for the majority of the Mo in the new surface alloy. Oxide particles similar to Al_2MgO_4, but containing Mo, were also found in all the surface alloy samples, and it is thought that they were due to the presence of oxygen in the voids of the predeposited plasma sprayed coatings.

3. The corrosion resistance of the 2014 alloy was improved after LSA resulting in an increase in E_{pit} of over 100 mV and up to 400 mV. Pits were observed to form in the proximity of the intermetallic particle clusters, which is believed to be the reason why the increase in E_{pit} was hindered.

4. The shift in E_{pit} to more noble values is believed to be due to the presence of Mo in the surface oxide film. XPS analysis showed that Mo was present in the air formed oxide, most probably in the Mo^{6+} state, and that after polarisation in chloride electrolytes a more complex oxide formed, that possibly also contained Mo^{4+}.

Acknowledgements

This work is funded by the EC BRITE-EuRam programme (Project BE 4305) under contract BREU-0494.

References

1. Draper, C. W. and Poate, J. M. (1985) *Int Metals Reviews* **30** 85.
2. Bonora, P. L. et al, (1980) *Electrochim. Acta* **25** 1497.
3. McCafferty, E., Moore, P. G. and Pearce, P. G. (1982) *J. Electrochem. Soc* **129** 9
4. Szklarska-Smialowska, Z. (1992) *Corrosion Science* **33** 1193
5. Natishan, P. M., McCafferty, E. and Hubler, G. K. (1986) *J Electrochem Soc* **133** 1061
6. Natishan, P. M., McCafferty, E. and Hubler, G. K. (1988) *J Electrochem Soc* **135** 321
7. Moshier, W. C., Davies, G. D., Ahearn, J. S. and Hough, H. F. (1986) *J Electrochem Soc* **133** 1063
8. Moshier, W. C., Davies, G. D. and Cote, G. O. (1989) *J Electrochem Soc* **136** 356
9. Moshier, W. C., Davies, G. D. Fritz, T. L. and Cote, G. O. (1990) *J Electrochem Soc* **137** 422
10. Moshier, W. C., Davies, G. D., Ahearn, J. S., and Hough, H. F. (1987) *J Electrochem Soc* **134** 2677
11. Frankel, G., Russak, M., Jahnes, C., Mirzamaani, M. and Brusic, V. [1989] *J Electrochem Soc* **136** 1243
12. Yoshioka, H., Yan, Q., Asami, K. and Hashimoto, K. (1991) *Materials Science and Engineering* **A134** 1054-1057
13 Hoar, T. P. and Mears, D.C. (1966) *Proc Roy Soc Series A* **294** 486
14. Shaw, B.A, Davies, G. D. Fritz, T. L. and Oliver, K. A. (1990) *J .Electrochem Soc.* **137** 359
15. Foley, R. T. (1986) *Corrosion* **42** 277
16. McMahon, M. (1994) *PhD Thesis: The Microstructure and Corrosion Properties of Laser Processed Aluminium Alloys* University of Liverpool
17. Urushino, K. and Sugimoto, K. (1979) *Corrosion Science* **19** 225
18. Ambat, R. and Dwarakadasa, E.S. (1992) *Corrosion Science* **33** 681

358

19. Ambat, R. and Dwarakadasa, E.S. (1993) *Brit. Corrosion Journal* **28** 142
20. Davies,G. D., Moshier, W. C., Fritz, T. L. and Cote, G. O. (1990) *J .Electrochem Soc.* **137** 422
21. Shaw, B.A, Davies, G. D. Fritz, T. L. Rees, B. J. and Moshier, W. C. (1991) *J . Electrochem Soc.* **138** 3288

22. Urquidi, M. and MacDonald, D. (1985) *J Electrochem Soc.* **132** 555

LASER ALLOYING OF COPPER AND ITS ALLOYS

S.P. GADAG , R. GALUN, A. WEISHEIT and B.L. MORDIKE
Institut für Werkstoffkunde und Werkstofftechnik,
Technische Universität Clausthal
D-38678 Clausthal, Germany

Abstract:

Laser surface alloying of copper, yellow brass and Al Bronze were carried out by duplex process involving powder alloying followed by remelting using an inclined plasma jet spraying nozzle. The aim of this investigation was to study the rapidly solidified non-equilibrium phases of copper alloys and their properties. The copper and brass substrates were surface alloyed with the powders viz., a) Ni to form silver white colour cupro-nickel and nickel silver surface layers respectively; b) Al to form the typical bronze surface with copper and c) mixture of Al:Si (2:1 wt.%) on Cu to form either a cladding of Al-Si binary alloy or a ternary Al-Si-Cu alloy. For comparison, laser alloying of pure Si on Al-Bronze was also carried out. The surface alloy had a two phase $\alpha+\gamma$ field with high microhardness of 405-460 $HV_{0.1}$, compared to the substrate two phase $\alpha+\beta$ field of 215 $HV_{0.1}$ hardness. d) Mixture of Pb and Sn (1:1 wt.%) was tried on brass to get a typical gunmetal surface with an additional top lubricating Pb layer suitable for sliding/bearing tribological applications. Preliminary studies of microhardness, EDX, SEM, and XRD results have revealed suitability of the alloyed coats (Si in Al-Bronze; Al-Si on Cu and Pb-Sn in brass) for further work on abrasion, erosion and sliding wear studies.

Key Words:
Copper, Laser alloying, Nickel-Silver, Cupro-Nickel, Al-Bronze, Special Bearing-Bronze, Lubricating layer, Lead globule, Basket Weave,Silicon needle, Rossette, Banded structure.

J. Mazumder et al. (eds.), Laser Processing: Surface Treatment and Film Deposition, 359–377.
© 1996 *Kluwer Academic Publishers.*

1. Introduction:

Pure copper although has an excellent electrical conductivity and corrosion resistance, it can not provide sufficient strength, hardness, and hence wear resistance especially at the elevated temperature. Laser surface alloying /cladding of copper is a most promising innovative technique among the various surface treatment processes to overcome these problems. It can sucessfully improve its elevated temperature properties, retaining core ductility and toughness as well as bulk thermal and electrical properties. Hence, laser alloying of copper and its alloys is suitable for wear, erosion and corrosion resistant marine applications.

Attempts on laser surface alloying of copper or its alloys are very limited in literature. However laser remelt hardening response of copper to surface residual stress using KrF excimers [1] and those of brass or bronze to improved microhardness using pulsed or continuous lasers have been made [2,3]. Previous investigations on laser melting of copper alloys [4] and Cr alloying of copper [5] seem to indicate that a maximum of 200 μm depth from the surface can be achieved. Only laser quenched aluminium bronzes have been investigated for their corrosion [6] and erosion resistance [7]. It seems, there is need for most of the mechanical properties of laser treated copper alloys to be explored thoroughly.

Often, copper with its high reflectivity (over 98%) and thermal conductivity, can pose an obstacle for CO_2 laser alloying unless its reflectivity is reduced at least by 50% for a 3 kW output power of the laser. This was overcome in this investigation by graphite coating of the substrate and using a plasma jet spray for powder injection through inclined nozzle into the molten pool of the copper or using the substrate precoated with the powder bed and spraying graphite on its top. The feasibility of laser alloying of copper and its alloys and process optimisation for sound, crack-free homogeneous copper surface alloy/clad formation have been investigated.

2. Experimental:

Commercial 99% purity copper rectangular specimens of 100x50x15 mm were sand blasted using alumina powder and washed in a steam of running tap water, cleaned with alcohol and dried under an air blower. A similar procedure was also adopted for 33% Zn containing brass specimens of identical dimension. A thin graphite coat was sprayed on these specimens to improve surface absorptivity to CO_2 laser radiation.

The specimens of copper and its alloys (brass and bronze) were surface alloyed by a high power, 5 kW CO_2 CW laser by duplex process involving powder alloying in the first step with a plasma spraying 30° inclined nozzle followed by remelting in the second step. Commercial high purity powders: Ni (-45 +22), Al (-100, +63), Si (-94, +45), Pb,Sn were used for the surface alloying. Laser alloying of Zn or brass can pose some practical difficulties due to substantial increase of vapour pressure of zinc (B.P. of Zn ≈ 906 °C) on melting. It might lead to explosive vapourisation of Zn followed by immediate oxidation of its vapour while laser alloying. Therefore, a thin layer of 150-200 μm, powder/binder mixture was precoated for alloying of brass specimens to avoid the explosive action of zinc in contact with air by former method.

3. Results and Discussion:

The structure and properties of rapidly quenched following binary and ternary copper alloys formed by laser surface alloying as mentioned in previous section-2, are taken up for detailed investigated in this paper: 1) Cupro-nickel and Nickel-silver, 2) Al-Bronze and Al-Si Bronze, 3) Al-Si Clad or Al-Si-Cu alloy and 4) Pb-Sn-Cu alloy.

3.1 CUPRO-NICKEL AND NICKEL-SILVER

The continuous solid solubility of Ni in copper helps laser alloying to easily form a single phase FCC-α solid solution of cupro-nickel and nickel silver alloys in copper and brass respectively (Fig.1).

Copper was surface alloyed with Ni using- Power: 2.5 and 3 kW; Speed: 300 - 500 mm/min., (270 mm focal distance) at 0.6 RPM and 4-8 LPM powder feed rate to form a continuous α solid solution of FCC structure. Brass was laser alloyed with Power: 3 and 3.5 kW, Speed: 300 and 400 mm/min., at 1-2 RPM and 4-6 LPM powder feed to form similar FCC α solid solution. Laser alloying of Ni was intended to form corrosion resistant silver white with geenish tinge coloured surface alloys of cupro-nickel on copper and nickel-silver on brass (Fig. 2). Structure was, by and large, columnar with dispersed cellular `honey comb' pattern here and there for both the surface alloys.

Effect of laser processing parameters on the structure and properties viz depth/width of alloying, concentration profile, micro-hardness were investigated. The depth of alloying increased from 70 to 240 μm with increasing heat input from 3.5 to 8 kJ /cm . By and large, it linearly increased with the laser heat input but seemed to decrease with gas flow rate from 6 LPM to 4 LPM at a constant powder feed rate of 0.6 RPM and focal distance (270 mm).

The micro-hardness range of cupro-nickel was 115-125 $HV_{0.1}$ compared to 70 $HV_{0.1}$ of it's copper substrate. The micro-hardness of nickel silver alloy was slightly higher and ranged from 130-140 $HV_{0.1}$ as compared to 90-95 $HV_{0.1}$ of its brass substrate. The concentration of cupro-nickel alloy seemed to decrease (51 to 9 wt%Ni) with increasing depth (70 to 240 μm). There was a negative concentration gradient of 0.25 wt% per μm, across the depth alloyed zone (Fig. 3). The chemical composition of Ni-Silver surface alloys was Cu- 27-28%Ni -15%Zn.

The most interesting feature of this alloy is the absence of heat affected zone between the alloy and substrate. Hence the alloy-coat is insensible to heat treatment and can be an excellent material for cladding application in corrosive environments. For example, Cu-25% Ni alloys are used for cladding on both sides of *ETP* (electrolytic tough pitch) copper in the manufacture of United States dime. Cu-10%Ni has good resistance to marine biofouling and is used to minimise biofouling on intake screens, sea water pipework, marine craft and offshore structures. The cupro-nickel claddings have high resistance to corrosive fatigue and corrosive-erosion of rapidly moving parts in sea water. A small addition of iron provides extra strength and improves corrosion resistance to velocity effects in sea water [8]. Hence, it can be extended for selective surface alloying of condensor, distiller and heat exchanger tubes in Naval vessels and coastal power plants.

3.2 ALLOYING OF AL AND AL-SI IN COPPER AND BRASS

3.2.1 *Cu-Al Binary System*
Laser alloying of Al in Cu formed an α-solid solution and under equilibrium conditions solubility of Al at room temperature is about 9% [9]. Laser alloying with 3.5 kW power, 200 mm/min., by duplex process involving alloying followed remelting, an alloy depth of 200 μm was obtained. For this lasing condition, there were some undissolved Al particles at both the corners at the top of alloy pool. This was followed by 60 μm depth of martensite in the HAZ in Fig. 4. and a thin layer of 14 μm remelted and solidified pure Cu superficial layer were noticed. The overall composition of Al was 15 wt% for this sample. i) On the otherhand, laser alloying of Al with same power and alloying speed but with lower melting speed of 100 mm/min., resulted in a homogeneous solid solution of Al in Cu which had partially transformed to two phase, $\alpha+\beta$ at top and bottom of the alloy pool due to favourable cooling rates required for the transformation as shown in Fig. 4. The chemical composition for this sample was Cu-3%Al. ii) On further increasing the power to 4 kW for both alloying and remelting at the same speed, a convex shaped clad of 400 μm depth and 1.9 mm width was formed. iii) On increasing the power of remelting during the second step to 4 kW while retaining the power required for alloying in the first step at 3.5 kW and scan speed at 200 mm/min., a complete dissolution of Al was possible with nearly a homogeneous solid solution and a crack free alloy zone of 185 μm depth and 50 μm depth of martensitic β phase. The chemical composition of this was Cu-14.5%Al. Its matrix micro-hardness of $(\alpha+\beta)$ phase ranged from 649-681 $HV_{0.1}$; $\approx 579\ HV_{0.1}$ in the transition zone ; 210-300 $HV_{0.1}$ in the martensitic HAZ ; and ≈ 115 $HV_{0.1}$ in the transition zone between HAZ and substrate (Fig. 5).

3.2.2 *Cu-Al-Si Ternary System*
Laser alloying of Al-Si in Cu was possible at 3 kW power and 250 mm/min., for 7.2 kJ/cm heat input (1 RPM / 4 LPM powder feed rate) to form a solid solution of α-Cu containing 2.3%Al-2.1%Si with 245 μm depth of alloy. There was small circular cavity at the top for this lasing condition. On increasing the power to 3.5 kW and scanning at

slightly lower speed 200 mm/min., (0.7 RPM / 4 LPM feed rate) resulted in nearly homogeneous solid solution of crack free alloy zone depth of 208 μm. The composition from EDX analysis was Cu- 2.3%Al-1.9%Si for this sample. The matrix microhardness of this composition was only 150-160 $HV_{0.1}$ compared to 75 $HV_{0.1}$ of Cu substrate. The structure for this composition, with suitable etching revealed cellular/columnar bands at the centre and feathery and fine dendritic structure at the two edges of the alloyed coat. A thin white superficial layer of α-Cu was visible at the bottom of alloyed pool. But a higher power of 4 kW and slightly higher scan speed at 300 mm/min., resulting in heat input of 8 kJ/cm gave maximum depth of 321μm alloy of α-Cu- 7.9%Al-7.8%Si, a typical of hypereutectic structure of Al-Si alloy with primary silicon needles at the top and silicon rosette in the bottom of melt pool (Fig. 6). The width of alloy coat was 1.23 mm. The micro-hardness of this type of Al-Si cladded Cu alloy ranged from 700-960 $HV_{0.1}$ (Fig. 7), which is an order of magnitude higher than that of the copper substrate (68-70 $HV_{0.1}$). This type of surface structure is most ideally suited for abrasion resistant applications.

An overlapped area scanning at 3.5 kW, 410 mm/min,. by duplex alloying/melting showed alternate structure of cellular/columnar bands in longitudinal to laser scanning direction as shown in Fig. 8. Columnar grains seemed to have a 45° preferred orientation to the direction of powder feeding. The composition in the cellular region of α-Cu-1.5%Al-2.7%Si while that in columnar region was Cu-1.8%Al-2%Si alloy. The dendritic feature of columnar cells were discernible. The micro-hardness of cellular region ranged from 140-170 $HV_{0.1}$; that of columnar zone was 115 $HV_{0.1}$.

3.2.3 *Al in Brass*
Typical microstructure of laser alloyed Cu-8%Al-20%Zn is shown in Fig. 9. Processing with a power of 3 kW, 400 mm/min., while alloying and with the same power but at 200 mm/min., for remelting. (1 RPM/5 LPM powder feed rate) a ternary eutectic Cu-8%Al-20%Zn, α-solid solution formed cored dendrites in the alloyed zone. The heat affected zone (HAZ) had a fine granular structure. Sometimes β'-phase had mixed feathery and β' spikes in the transition zone or a fine network of β'-spikes formed a basket weave pattern in the melt zone.

3.3 LASER REMELTING AND SI ALLOYING OF CU-10% AL-5% NI -4% FE

CO_2 laser power of 3 kW, 300-400 mm/min., with 4 LPM shielding gas was used for remelting of 10%Al-Bronze containing 5%Ni-4%Fe.This material had a very good response to laser melting and hardening like many ferrous materials. Melt depth as large as 1 mm to 1.2 mm could be easily achieved with moderate powers (Fig. 10). The structure in the remelted zone consisted of feathery structure of retained β-phase of 290-310 $HV_{0.1}$ as compared to 200 $HV_{0.1}$ of the substrate (α+β) phase field. On alloying of 3%Si into Al-Bronze using 3.5 kW power at 200 mm/min., the hardness could be increased to as high as 460 $HV_{0.1}$ microhardness (Fig.11). The surface finish of as lased

Al-bronze was quite good free from surface ripples and cracks. These are most suitable for wear, corrosion and erosion resistant marine applications like ship propellors and submergible parts of vessels.

3.4 ALLOYING OF LEAD AND TIN

Pre-coated powder of 50% Pb -50% Sn on brass using an organic binder, was alloyed in to 33% Zn containing brass with 3 kW power and 500 mm/min., scan rate for both alloying as well as remelting. The alloyed structure consisted of 15-20 μm thick top lead layer followed by Cu-21%Sn-22%Zn ternary alloy of bearing bronze surface. There was a third layer consisting of fine cored dendrites with Pb globules interspersed in the matrix as shown in Fig. 12. This type of bronze with an additional top lead layer is most ideally suited for bearing and sliding applications.

4. Conclusions:

1] Laser alloying of copper and it's alloys was sucessfully achieved by duplex laser processing involving laser alloying followed by remelting with suitable surface preparation.

2] The continuous solid solubility of nickel makes laser alloying easier to form silver white Cupro-nickel and silver white greenish tinge colour Nickel-silver alloys. The micro-hardness of nickel-silver was slightly higher (130-140 $HV_{0.1}$) than that of cupro-nickel (115-125 $HV_{0.1}$). The concentration of nickel decreased with the depth of alloying.

3] Laser alloying of 3%Al in Cu by duplex alloying/melting with 3.5 kW power at 100 mm/min. speed, had a homogeneous solid solution of Al to form partially transformed two phase ($\alpha+\beta$) field, which resembled the typical structure of commercially popular aluminium bronze (DIN:17 665). The micro-hardness of this partially transformed alloy was nearly twice that of the commercial aluminium bronze.

4] A higher matrix micro-hardness of 649-680 $HV_{0.1}$ was obtained by laser alloying of 14.5%Al to form a complete homogeneos solid solution of the alloy. This was higher than even that of laser alloyed aluminium bronze with 3% silicon (450 $HV_{0.1}$).

5] Laser alloying of Al up to 8% by wt. in yellow brass had a complex structure: cored dendrites in alloyed melt zone and granular structure in the HAZ. The transition zone had mixed feathery and spike structures which formed a *basket weave* pattern.

6] Laser alloying of 7.9%Al-7.8%Si in copper formed a cladding of hypereutectic Al-Si alloy with primary silicon needles at the top and rosettes at the bottom of the alloy pool. This matrix had a very high micro-hardness 700-950 $HV_{0.1}$ without any cracks, among the copper alloys. This is most suitable for abrasion resistant applications.

7] The laser alloying of Pb-Sn in brass formed a *special bearing bronze* surface with top lubricating lead layer, second layer of 21%Sn-22%Zn-Cu ternary bronze and a third layer of fine cored dendrites with dispersed Pb globules in the alloyed melt zone. This type of special bronze with the top lubricating layer is ideally suited for bearing and sliding applications.

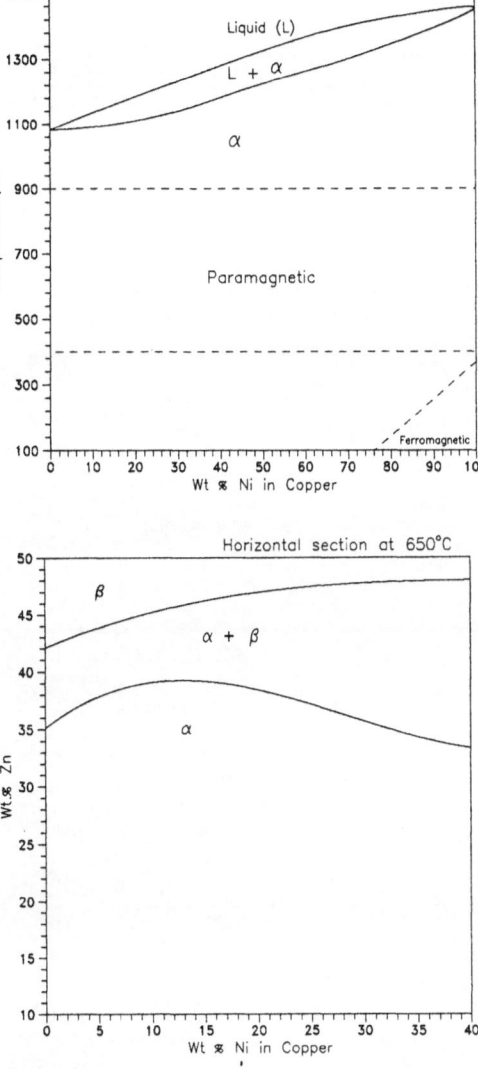

Figure 1.
Equilibrium diagrams: a) Cu—Ni binary; b) Cu—Zn—Ni ternary systems.

366

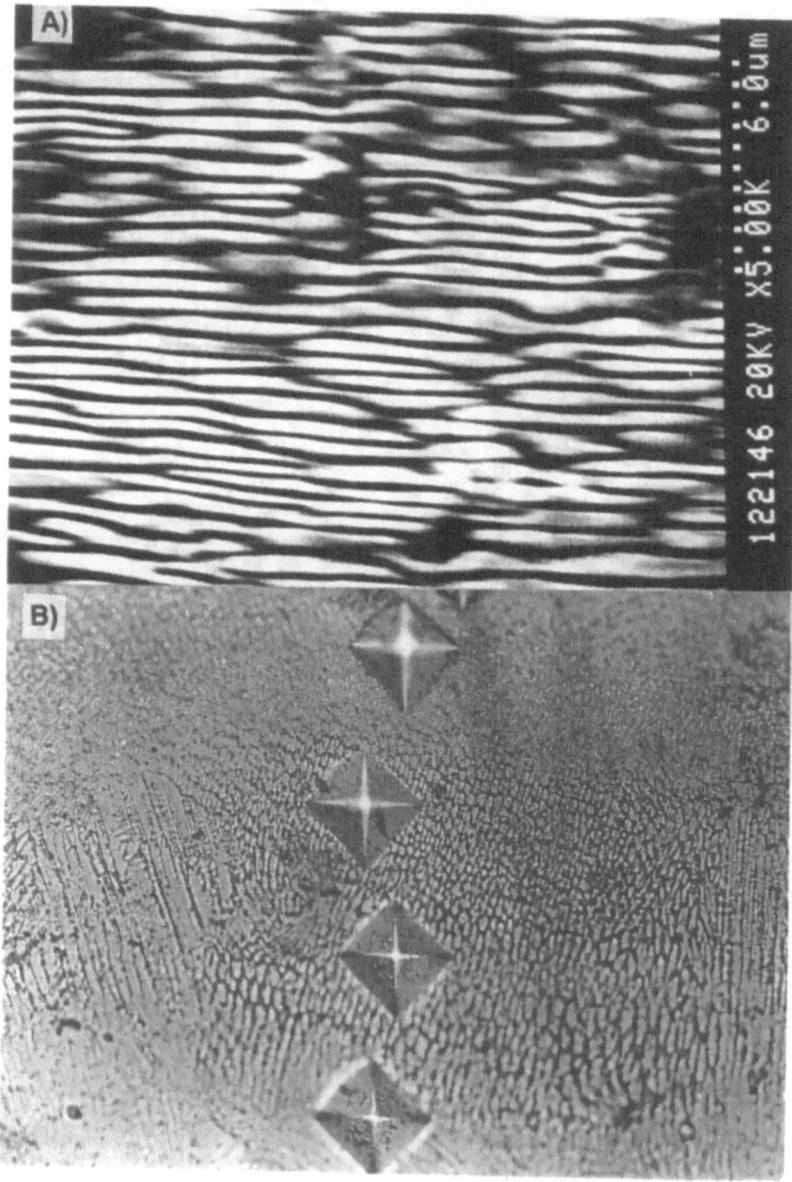

Figure 2. Laser alloyed structure: A) Cupro-nickel ; B) Nickel Silver (500x) surface alloys.

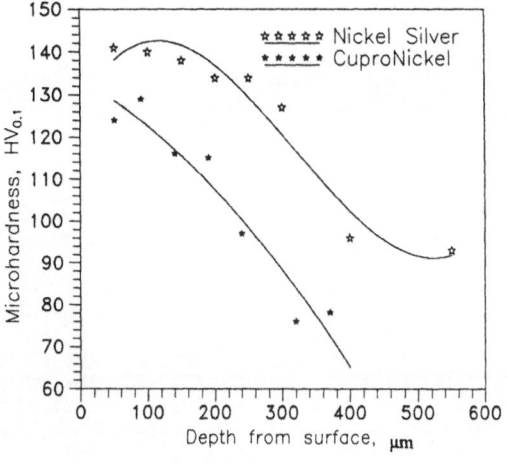

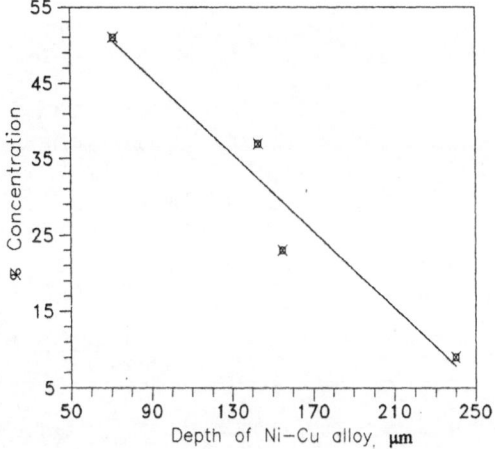

Fig. 3.
Microhardness traverse and concentration profiles: laser alloying of Ni.

368

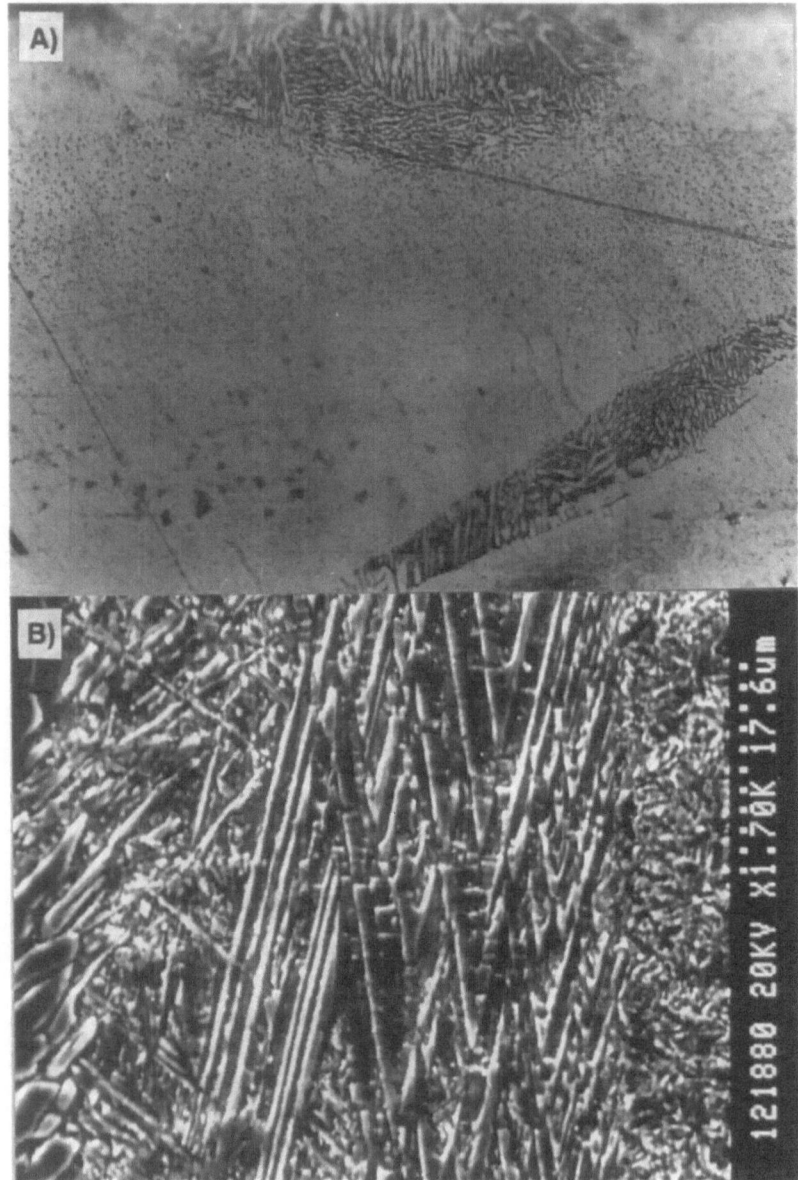

Figure 4. Laser alloyed structure, Al-Cu :
A) Melt zone: partially transformed $(\alpha+\beta)$ phase, 500x ; B) HAZ: Martensite.

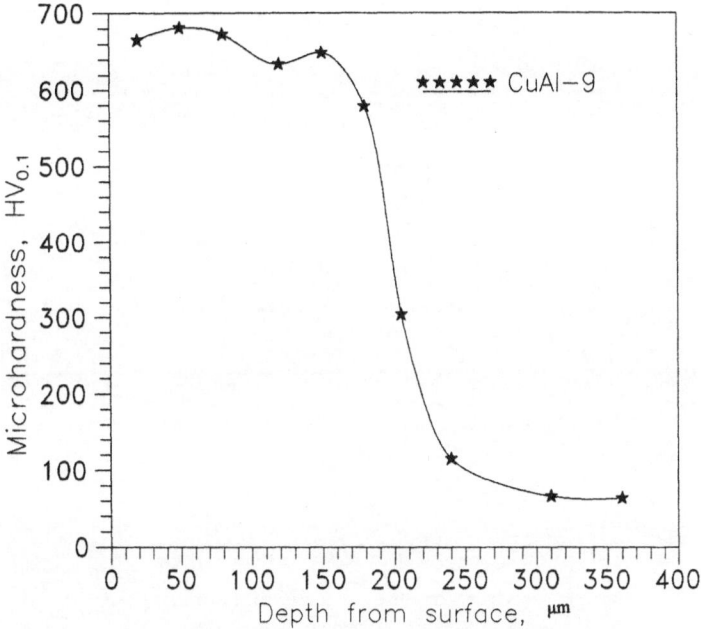

Fig. 5.
Microhardness traverse on laser alloying Al in Cu.

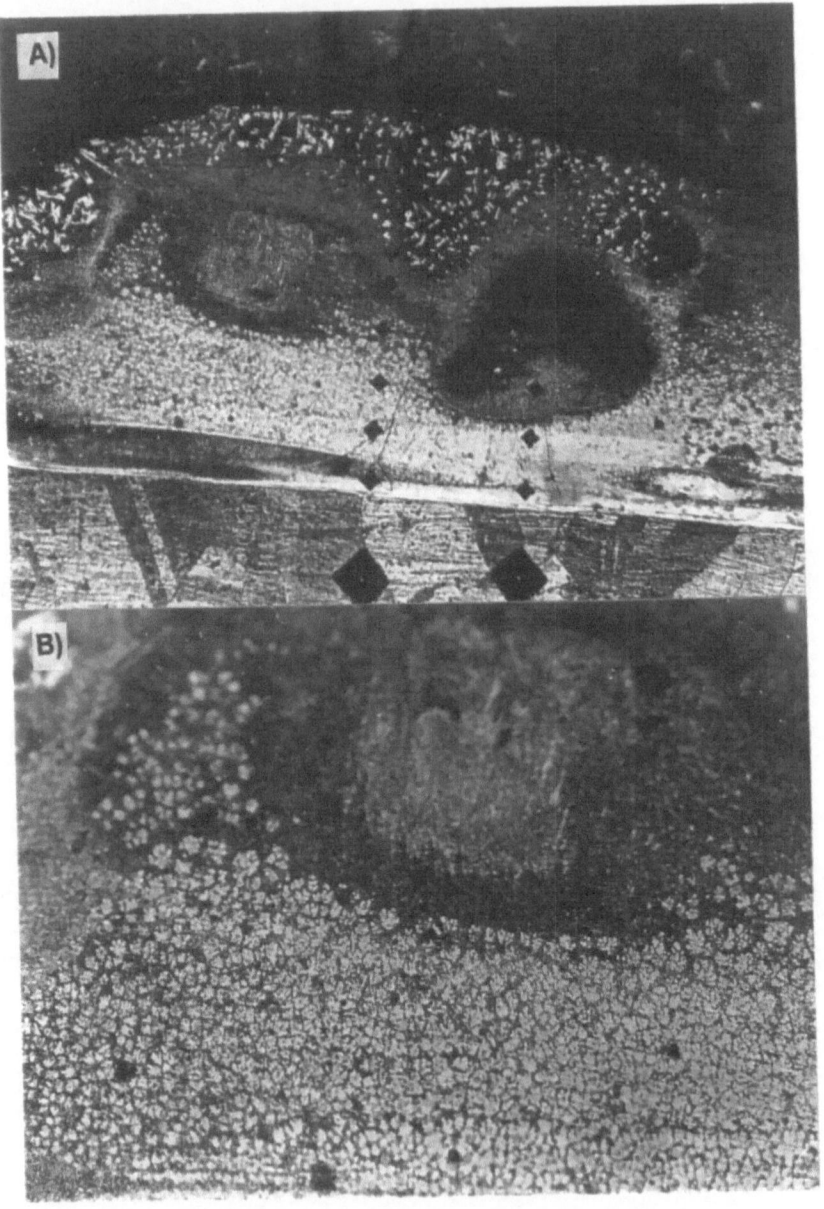

Figure 6. Laser cladding of copper with hypereutectic Al-Si alloy (Cu- 7.9% Al-7.8% Si):
A) Overview at 200x ; B) Silicon rosettes at 500x .

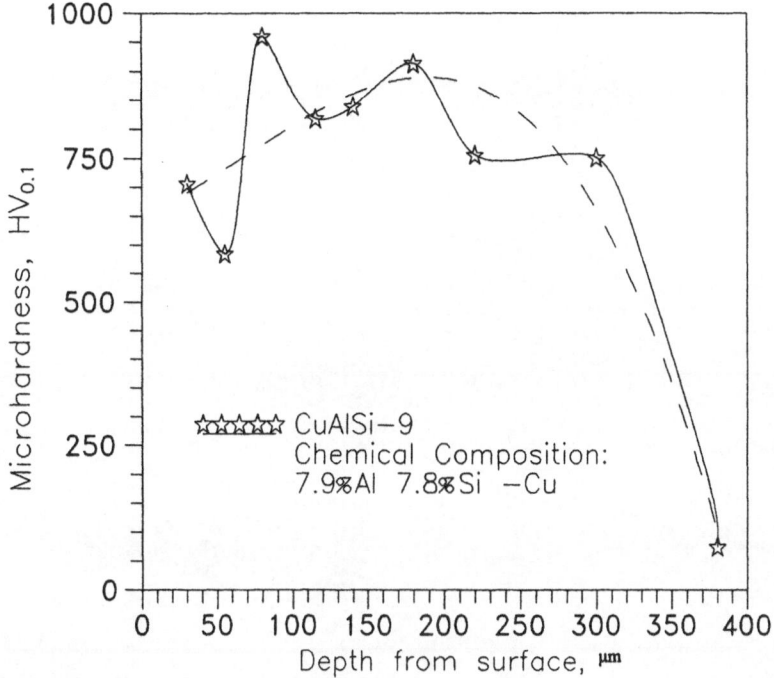

Fig. 7.
Microhardness traverse on laser alloying Al, Si in Copper.

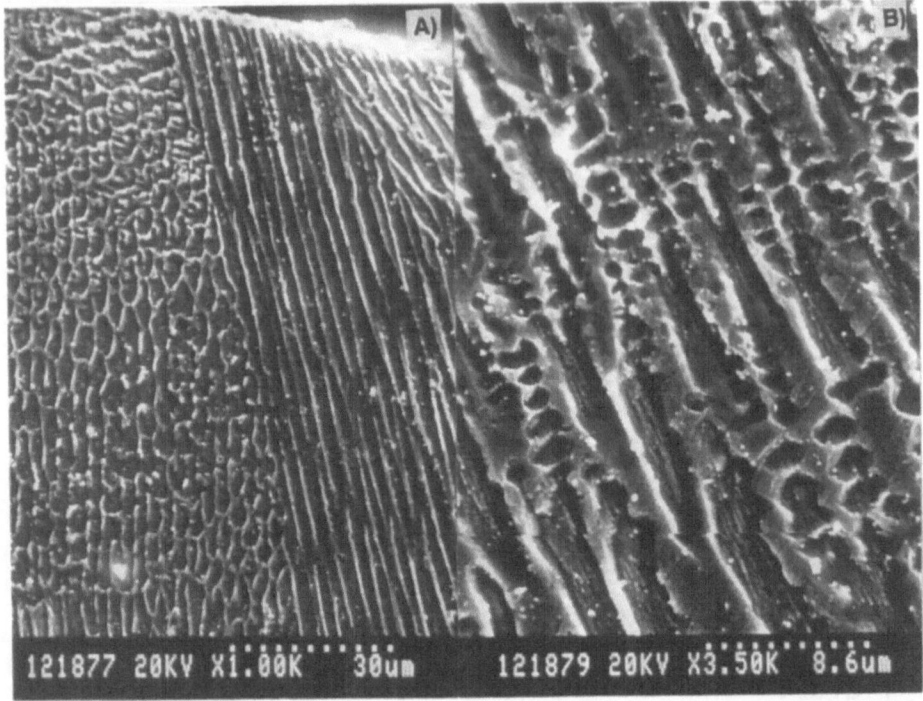

Figure 8. Laser alloying of copper with Al-Si (Cu- 1.5%Al-2.7%Si solid solution): Structures in longitudinal section: A) Banded cellular/columnar ; B) Dendritic feature.

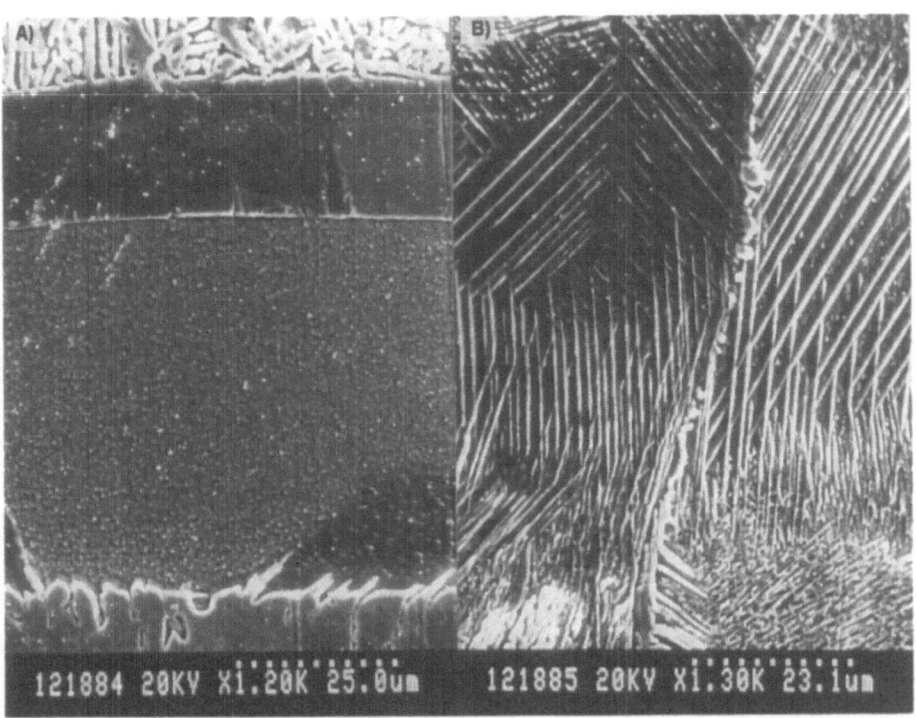

Figure 9. Laser alloyed structure Cu- 8%Al-20%Zn ternary system:
A) Alloy zone/ HAZ ; B) Transition zone.

374

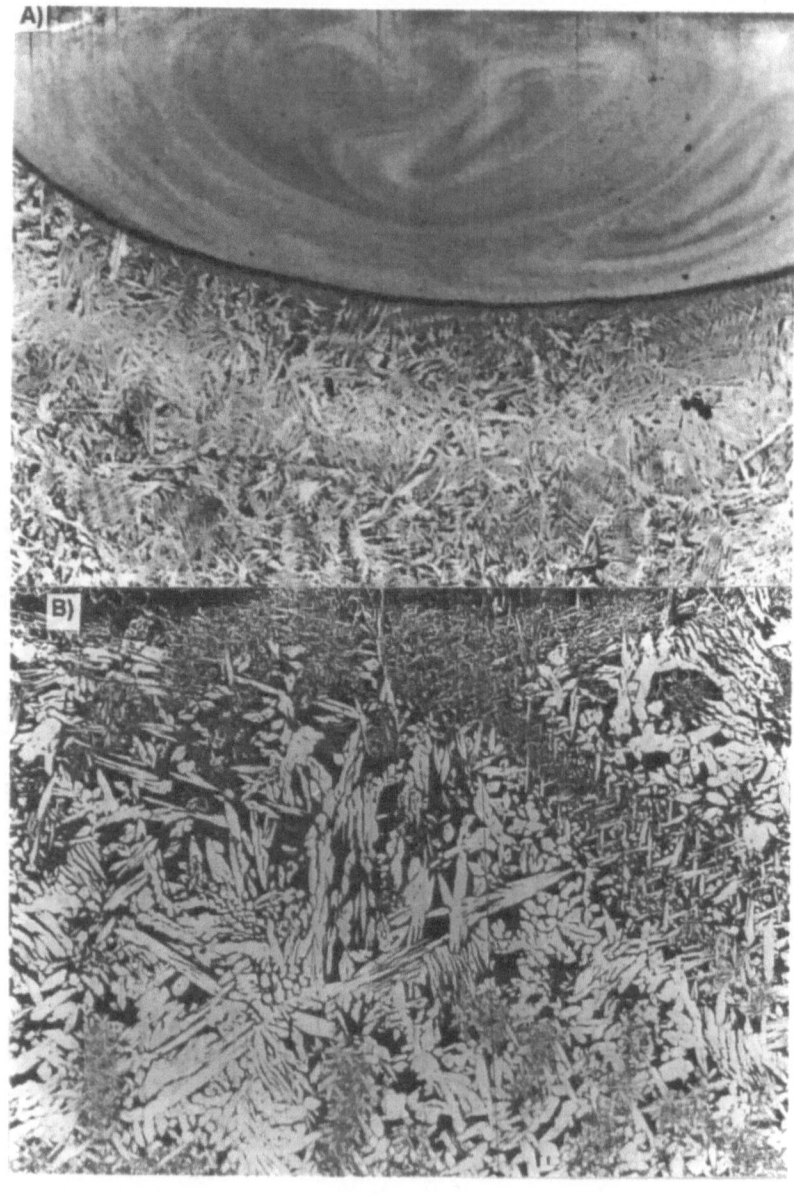

Figure 10. Laser melted structure, 10%Al-Bronze at 500x : A) Melt zone ; B) HAZ, Substrate.

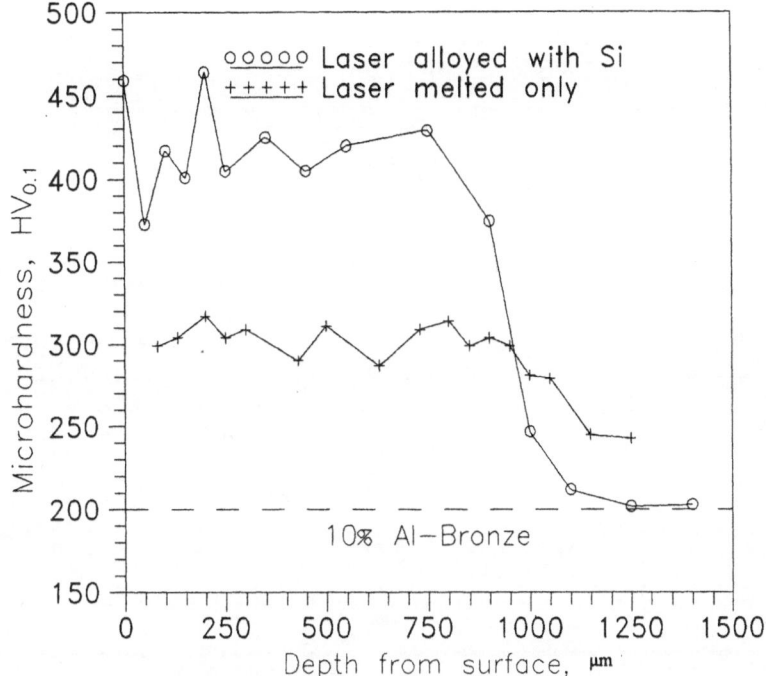

Fig. 11.
Microhardness traverse after laser treatment.

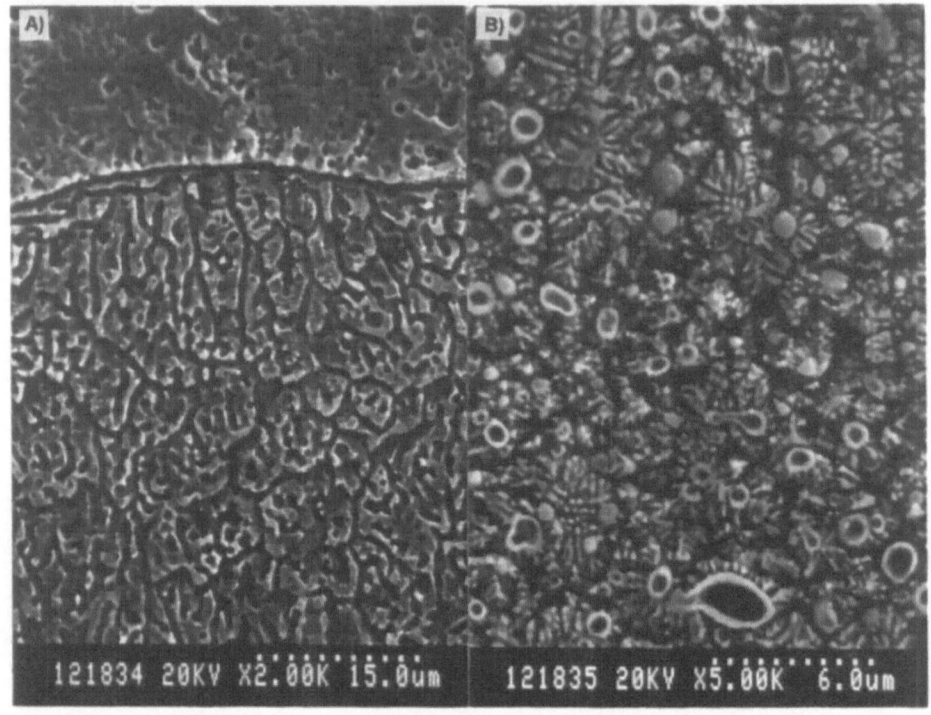

Figure 12. Laser alloyed structure, special Bronze surface:
A) Alloy zone with lubricating lead layer ; B) Alloy zone with fine dendrite/lead globule.

Acknowledgements:

♦ Dr S.P. Gadag, gratefully acknowledges the financial support of **Internationales Büro, KFA Jülich** for his stay under *Indo-German bilateral research co-operation between KFA/DoEd* for this project. He also acknowledges *'General Consulting, KEHL and the German Industry'* for their sponsorship and encouragement to attend the NATO Advanced Study Institute on Laser Processing held on July 3-16, 1994 in Sesimbra, Portugal. The authors wish to thank Herr **Rainer Bosse** for his technical assistance in metallography/Photography.

References:

1. Panagopoulos, C. and Michaelides, A. (1992) Laser surface treatment of copper, *J. Materials Science*, **27**, 1280-1284.

2. Anchev, V. (1990) Laser hardening of copper-based alloys, *Key Engineering Materials*, **46/47**, 103-110.

3. Benkissen, G., Horn,G.,Semjonov,S and Kaps, R. (1992) Randschichthärten von heteogenen Mehrstoffaluminiumbronzen durh Laserstrahlumschmelzen, *Metall* No.4, **46**, 324-328 (in German).

4. Barten, G. Fritsch, H.U and Bergmann, H.W. (1983) Microstructure of laser melted copper alloys, *SPIE,* **455**, 120-127.

5. Hirose, A. and Kobayashi, Kojiro F (1994) Surface alloying of copper with chromium by CO_2 laser, *Materials Science &. Engineering -A* **174**, 199-206.

6. Draper, C.W, Woods, R.E., and Meyer, L.S. (1980) Enhanced corrosion resistance of laser surface melted aluminium bronze D (CDA-614), *Corrosion*, No.8, **36**, 405-408.

7. Draper, C.W, Vandenbee, J.M, Preece, C.M. and Clayton, C.R. (1982) Characterisation and properties of laser quenched aluminium bronzes, in B.H. Kear, B.C. Giessen and M. Cohen (eds) Proc. M.R.S Annual meeting on *Rapidly solidified Amorphous and Crystalline alloys*, Elsevier Publshers, 529-533.

8. Powell, C.A. (1994) Preventing Biofouling with copper-nickel alloys, *Materials World*, No.4, **2**, 181-183.

9. West, E.G(1982) *Copper and its alloys*, Ellis Horwood Ltd. series in Industrial Metals, Chichester, England.

Laser Cladding of Low Melting Point Alloys

M.ELLIS[+], D.C.XIAO[+], W.M.STEEN[+], C.LEE[-], K.G.WATKINS[+] and W.P.BROWN[*]

[+]*Dept. of Mechanical Engineering* [-]*Dept. of Industrial Studies University of Liverpool, Liverpool L69 3BX, UK.*
[*]*Glacier Vandervell Ltd., Riccarton, Kilmarnock, Strathclyde, Scotland KA1 3NA.*

1. Introduction

Laser cladding has been used for some years as a commercial hardfacing method [1]. The two main laser cladding (and alloying) techniques are:

i) Preplaced powder [2]
Here powder is preplaced on the substrate with a binding agent to "stick" the powder into place until it is melted by the laser. The whole process is normally shrouded in an inert gas such as argon. Unfortunately there is only a small operating region of process parameters where low dilution and a good fusion bond are possible.

ii) Blown powder [3]
Blown powder has a much larger operating region than preplaced powder. Powder is blown directly into a laser generated meltpool as it scans across the surface of the substrate. The powder melts and a clad track is built up. Complete coverage is achieved by overlapping the tracks.

Most work on blown powder laser cladding has been done on cladding hard, wear resistant materials onto a cheap, bulk substrate such as mild steel. Previous studies of blown laser cladding have concentrated on conventional hardfacing alloy systems (for example Co alloys such as stellite [4], Ni-Cr alloys[5] and stainless steels [6]) onto mild steel. Very little research has been done on cladding softer materials (with a correspondingly lower melting point) onto steel, though some work has been done on laser cladding Cu alloys [7,8]. Work has been done on laser cladding and alloying aluminium alloys onto non-ferrous substrates. This includes the laser surface alloying of titanium with aluminium [9] and cladding an aluminium substrate with a SiC/Al metal matrix composite [10]. As far as the authors are aware, there have been no previous attempts to clad aluminium alloys onto a ferrous substrate.

J. Mazumder et al. (eds.), Laser Processing: Surface Treatment and Film Deposition, 379–394.
© 1996 *Kluwer Academic Publishers.*

Work has been done at Liverpool to show how it is possible to clad copper and aluminium alloys, which have a relatively low melting point, onto a mild steel substrate. The main problem in laser cladding these alloys is their inability to wet the substrate effectively when molten. A technique to clad these alloys to mild steel has been developed - partial width cladding. Partial width cladding involves cladding at lower power densities, with a bigger beam size than would normally used when cladding conventional hardfacing alloys. This enables the molten clad alloy to wet the substrate more effectively.

2. Experimental

A powder feeder system has been previously built at Liverpool [11] and is shown in as a schematic in fig. 1. Powder was removed from the bottom of a hopper by a screw feed into a stream of argon gas. The flow of powder from the hopper was determined by the the speed of the motor driving the screw feed. The motor speed was controlled by a microcomputer. The powder stream then flows into a powder feeder tube where it was directed onto the substrate, into a laser generated meltpool. The two alloy powders used in this investigation were an Al-Sn alloy (Al-Sn10Si4Cu1) and a Pb-Bronze alloy (Cu-Pb25Sn1.5). The two alloy powders had melting points of approximately 650 and 1000 degrees Celsius respectively.

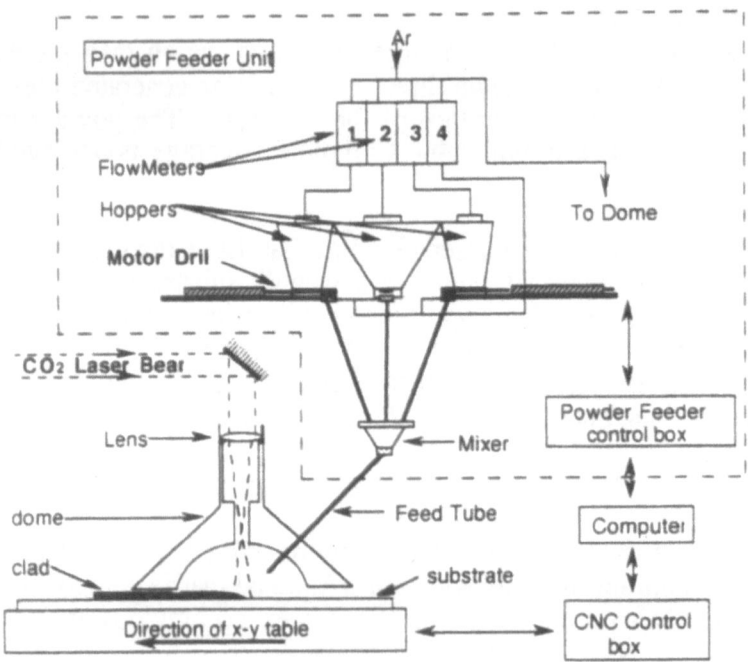

Fig.1 - Schematic of Powder Feeder Unit and Laser Cladding Operation

The laser meltpool was created by a partially focused laser beam from either an ECOSSE AF5 5kW CO_2 laser or an ELECTROX 2kW CO_2 laser. An optical dome (fig. 2) was placed over the meltpool in order to position accurately the powder feeder tube with relation to the meltpool. The dome also provided a shroud of argon gas over the meltpool. The substrate used was 10mm thick EN3 mild steel plate, the surface of which had been ground, shot-blasted and degreased. The range of processing parameters used in the investigation is shown in Table 1.

TABLE 1. Processing Parameters used in this Investigation

	Copper Alloy	Aluminium Alloy
Power (P)	600-3400 W	900-2700W
Beam Diameter (D)	3-10 mm	2.5-10mm
Traverse Speed (v)	3-24 mm/s	2-16 mm/s
Powder Feed Rate (R)	0.1-1.4 g/s	0.02-0.07 g/s
Powder Size	90-300 microns	90-212 microns

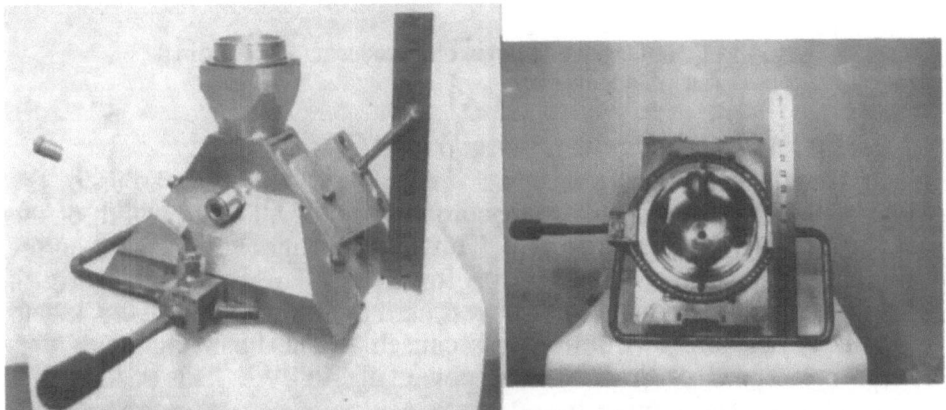

Fig.2 - Photographs of the Cladding Dome

3. Results and Discussion

3.1. COPPER ALLOY

3.1.1. *Single Track Cladding*
It has already been shown that it is possible to produce defect free single track clads for a range of processing conditions (see [8] and fig. 3). In order to produce total coverage of a substrate, these tracks need to be overlapped.

382

The three axes represent:
  i) Beam Power /Beam Size, P/D [W/mm]
  ii) Powder Feed Rate [g/s]
  iii) Cladding Speed, V [mm/s]

The alloy used was Cu-Pb25Sn1.5

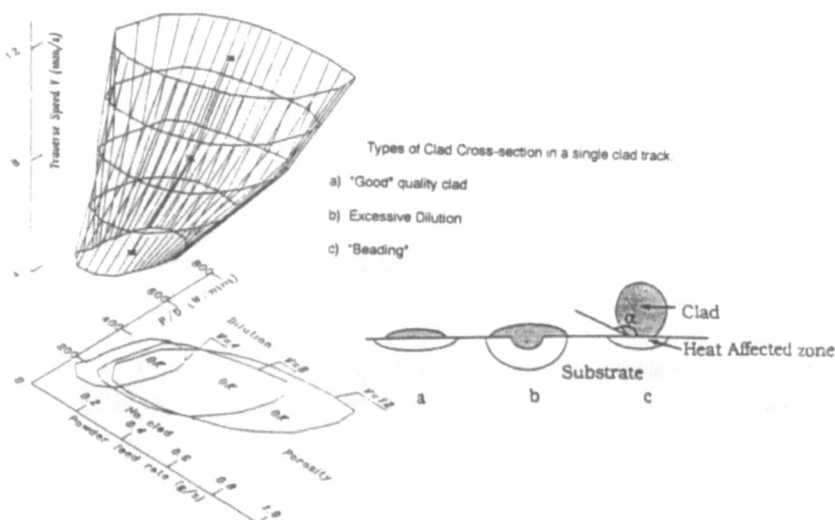

Types of Clad Cross-section in a single clad track.

a) "Good" quality clad

b) Excessive Dilution

c) "Beading"

Fig.3 - Plot of Operating Region for Producing "Good" Single Track Cladding [8].

3.1.2. *Full Width Cladding and Partial Width Cladding*

When a small beam size (and hence high power density) was used, the width of the single track clad was approximately equal to the width of the laser beam (**full width cladding**). However, during overlap cladding the individual tracks were unable to fuse together. This is shown in fig. 4 where a very large amount of power was used to produce a single track clad and a large amount of dilution was caused. (The beam diameter was 2.5mm at a speed of 8mm/s and a power of 1460W. This results in a power density, $P/d^2 = 234$ W/mm^2 and a specific energy, $P/vd=73$ J/mm^2.) A single track clad of width 2.3mm was produced. However, on overlapping the tracks, the small beam was not able to remelt the previous tracks due to the high thermal conductivity of the previously clad copper tracks. Mild steel has a thermal conductivity of 63 W/m/K whilst using the rule of mixtures it is estimated that the Pb-Bronze alloy has a thermal conductivity of 299 W/m/K. This means that as the clad layer is built up a large heat sink is created, which prevents the previous track from becoming sufficiently hot enough to remelt.

Fig 4 - Full Width Cladding of Cu-Pb25Sn1.5

P=1460W, v=8.0mm/s, d=2.5mm, powder flow rate = 0.25 g/s and %overlap=50%

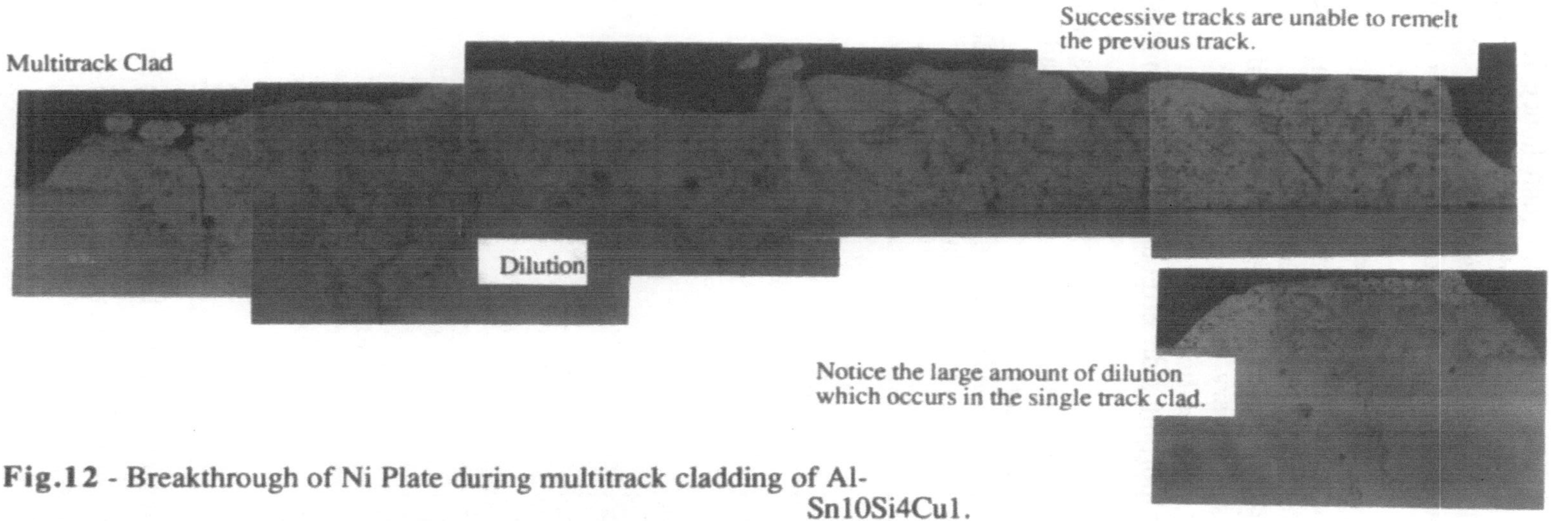

0.5 mm

Successive tracks are unable to remelt the previous track.

Multitrack Clad

Dilution

Notice the large amount of dilution which occurs in the single track clad.

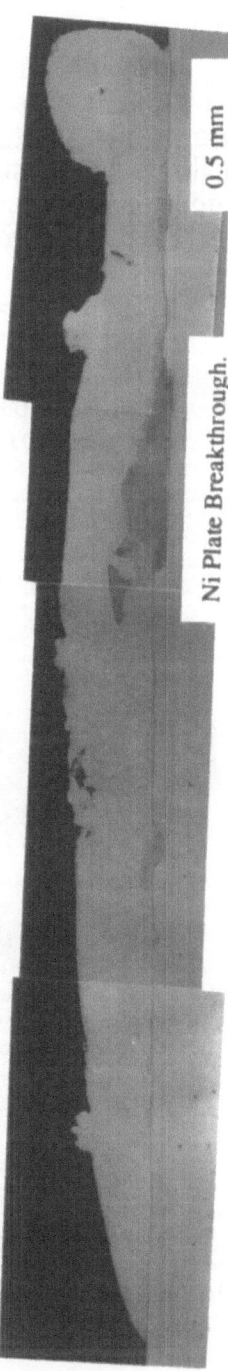

Fig.12 - Breakthrough of Ni Plate during multitrack cladding of Al-Sn10Si4Cu1.

This is due to the large meltpool size during *partial width* multitrack cladding.

0.5 mm

Ni Plate Breakthrough.

P=1400W, v=6.0mm/s, d=5.5mm and %overlap=50%

It was necessary to use a large beam (and hence lower power density) to reheat and melt the previously clad tracks and enable the tracks in the overlap clad to key in together (**partial width cladding**). However this reduced power density produced a single track clad width of 1.4mm (Beam diameter = 6 mm, cladding speed = 3.0 mm/s, power = 1420 W results in a power density, $P/d^2 = 40$ W/mm^2 and a specific energy, $P/vd = 79$ J/mm^2). This method of cladding with a large beam size and low power density to enable tracks to key in together is known as **partial width cladding**. It should be noted that the specific energy in both the given examples was very similar. This effect is therefore a function of power density, not specific energy.

There are two important side effects caused by partial width cladding:

i) The clad becomes thicker than expected (typically a single track clad height of 0.4-0.5 mm produced an overlapped clad height 1.5 mm for a nominal % overlap of 50%).

ii) Dross is built up at the side of the clad which causes problems when overlap cladding. This effect is illustrated in fig. 5.

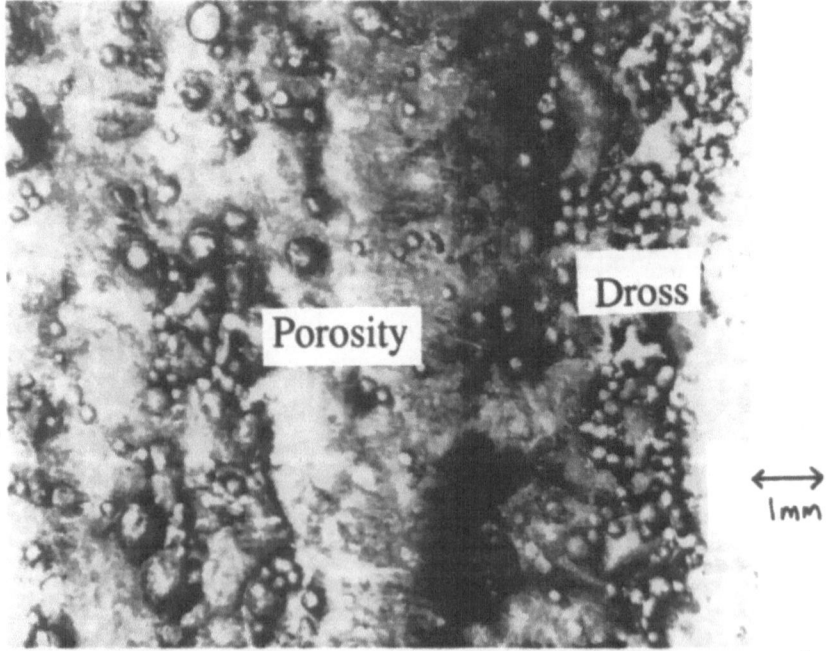

Fig.5 - Macrograph of Surface of an Overlapped clad of Cu-Pb25Sn1.5

The clad was formed by *partial width cladding*. The build up of dross at the side of the clad caused porosity to occur in the main body of the clad.

Unfortunately as the substrate heats up, more and more dross is produced from partially melted powder at the side of the track. Eventually enough dross can be built up so that it is unable to be effectively melted by the laser beam as the clad is built up. This results in the clad becoming porous as the molten clad is unable to wet the substrate properly. It is necessary then to decrease the powder flow rate and decrease the advance distance for each new track. This increases the heat input per unit area whilst keeping the clad dimensions constant.

3.2. ALUMINIUM ALLOY

3.2.1. *Single Track Cladding*
When cladding molten aluminium onto mild steel, it was found that there was a reaction between the molten Al alloy and the steel substrate. A brittle intermetallic layer was formed at the interface (See fig. 6). Ni plated mild steel substrate has been successful in eliminating the problem in single track clads (see fig. 7). The Ni plate acts as a barrier keeping the molten Al alloy and the steel substrate apart.

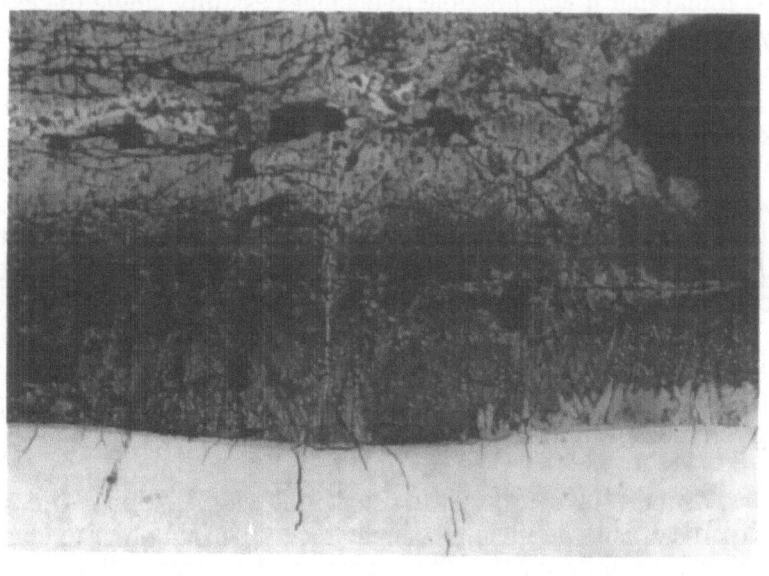

0.1 mm

Fig. 6 - Micrograph of the interface between the mild steel substrate and the Al-Sn10Si4Cu1 clad.

A brittle intermetaalic layer is formed at the interface, which readily cracks.

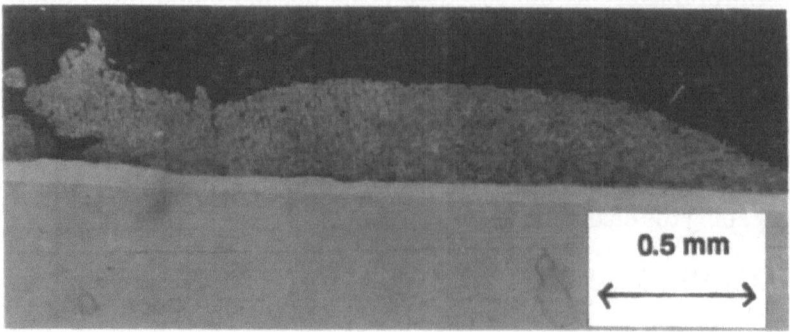

Fig.7 - Micrograph of a cross-section of a single track of Al-Sn10Si4Cu1
clad onto Ni plate

The Ni plate acts as a barrier to prevent the brittle intermetallic layer from forming.

3.2.2. *Full Width Cladding and Partial Width Cladding*
It was found that it was possible to clad successfully single tracks of the Al
alloy onto Ni plate by both partial width and full width cladding . However
it has proved impossible to produce defect free overlapped clads with full
width cladding (i.e. the beam diameter is approximately equal to the single
track clad width). With all other process conditions constant, the clad broke
through the Ni plate at 1000W, but the clad was unable to wet the Ni plate
properly at 900W. This was unacceptable as complete surface coverage
was not achieved at 900W and the Ni plate was partially destroyed at
1000W (see fig. 8). The lack of wetting at 900W was a surface tension
effect. If borax (a flux) was placed on the surface during cladding at 900W,
then the surface tension of the molten Al alloy was reduced and the molten
Al alloy was able to wet the Ni plate. However, the clad then burnt through
the Ni plate and porosity was caused due to the water of crystallisation
being released from the borax (fig. 9).

It is therefore necessary to use partial width cladding (i.e. where the
beam diameter is larger than the single track width) in order to clad Al alloy
onto Ni plated mild steel. This heats up the surrounding Ni plate and
enables the molten Al alloy to wet the substrate. The large spot size also
remelts the previously clad Al tracks. This means that the meltpool is bigger
than the width of a single track clad and able to absorb more powder from
the powder stream[1] . The clad height therefore becomes much bigger than

_____

[1] It should be noted that the powder stream was very wide and covered any
size of meltpool.

Fig 8a Al-Sn10Si4Cu1 Full Width Cladding at 1000W

Overlapped Clad. P=1000W,
v=13mm/s, d= 2.5 mm, powder flow rate = 0.03 g/s & %Overlap= 50%.

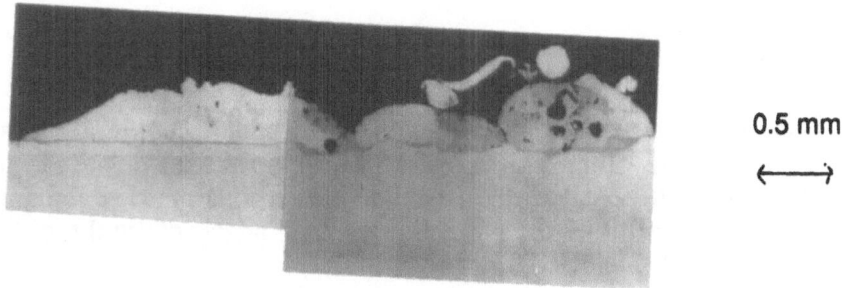

0.5 mm

←——→

Fig 8b - Al-Sn10Si4Cu1 Full Width Cladding at 900W

Overlapped Clad. P=900W,
v=13mm/s, d= 2.5 mm, powder flow rate = 0.03 g/s & %Overlap= 50%

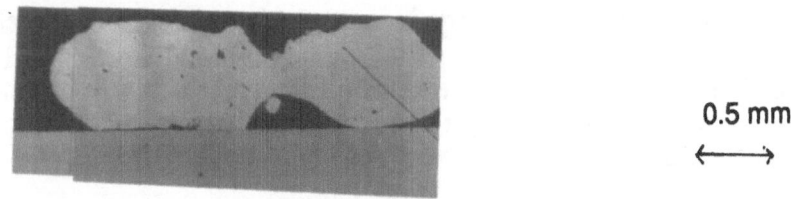

0.5 mm

←——→

Fig 9 Al-Sn10Si4Cu1 Full Width Cladding at 900W with Borax.

Overlapped Clad of the Al alloy onto Borax coated, Ni plated, mild steel.
P=900W, v=13mm/s, d= 2.5 mm, powder flow rate = 0.03 g/s &
%overlap=50%

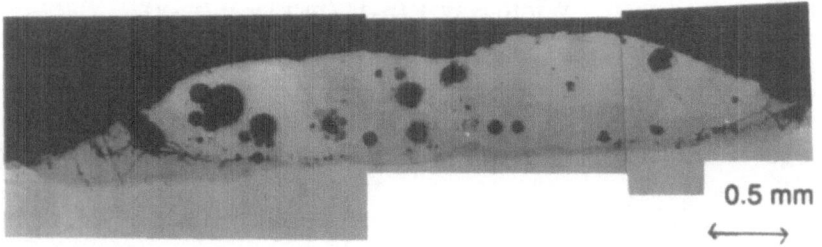

0.5 mm

←——→

expected - this is illustrated in fig. 10 and section 4.1 . A model has been developed to predict the size of the meltpool from the dimensions of the clad (see section 4). Unfortunately, the large meltpool size means that the Ni plate is "attacked" several times by molten aluminium (see fig. 11). This eventually leads to the breakthrough of the Ni plate (fig. 12).

4. Model

It has been noted in overlap cladding that the meltpool becomes larger than the single track clad width. This model was developed to calculate the size of the meltpool can be calculated from the dimensions of the overlapped clad. The number of times the Ni plate has been "attacked" by the molten Al alloy can then be calculated.

If the meltpool absorbs powder from the powder stream at a constant rate/unit area of meltpool, then the size of the meltpool can be calculated from the multitrack clad thickness and this is done below.

4.1. EXPECTED THICKNESS OF CLAD

In order to demonstrate the large increase in the thickness of the clad it is first necessary to define the "expected" height of a multitrack clad (see fig. 10 and below). How the expected clad height deviates from experimentally derived values of clad height is also shown in fig. 10, along with a graphical definition of the variables used in the model.

Assume $A_1 = A_2 = ... = A_n = A$ where A is the cross-sectional area of a single track clad.

Also the cross-section of a single track clad can be approximated to a semi-ellipse, hence:

$$A = \pi hw/4 \qquad\qquad (1)$$

For n tracks $\qquad$ Area = nA $\qquad$ (i.e. a multitrack clad)

If n is large: $\qquad$ Width = w + (n-1) (w-x) $\approx$ n (w-x)

$$\text{Average ht of clad, } h_m = \frac{\text{Area}}{\text{Width}} = \frac{nA}{n\,(w-x)} = \frac{A}{(w-x)}$$

$$(2)$$

Fig. 10

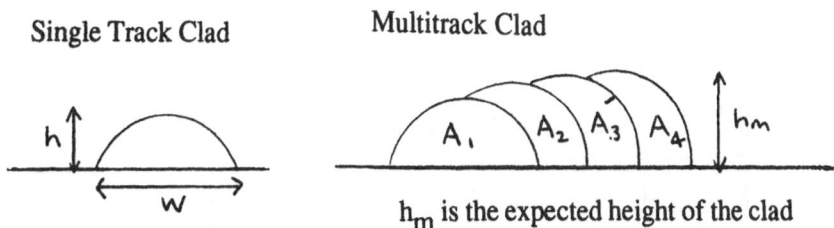

Single Track Clad

Multitrack Clad

h_m is the expected height of the clad

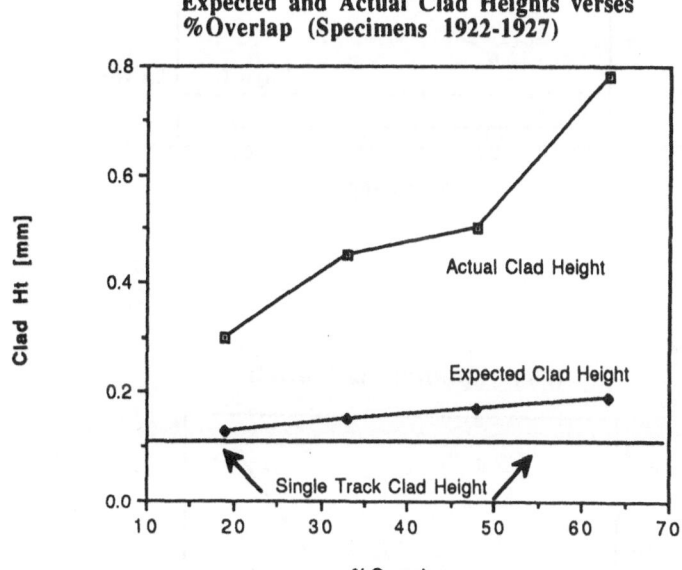

Expected and Actual Clad Heights verses %Overlap (Specimens 1922-1927)

Clad Ht [mm]

Actual Clad Height

Expected Clad Height

Single Track Clad Height

%Overlap

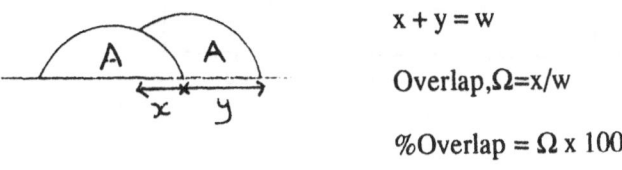

$x + y = w$

Overlap, $\Omega = x/w$

%Overlap $= \Omega \times 100$

Fig. 11 - Plots of calculated meltpool size and the number of times the Ni plate has been "attacked" against %Overlap.

The graphs below are produced from experimental data of clad heights (from fig.9) using the model outlined in section 4.

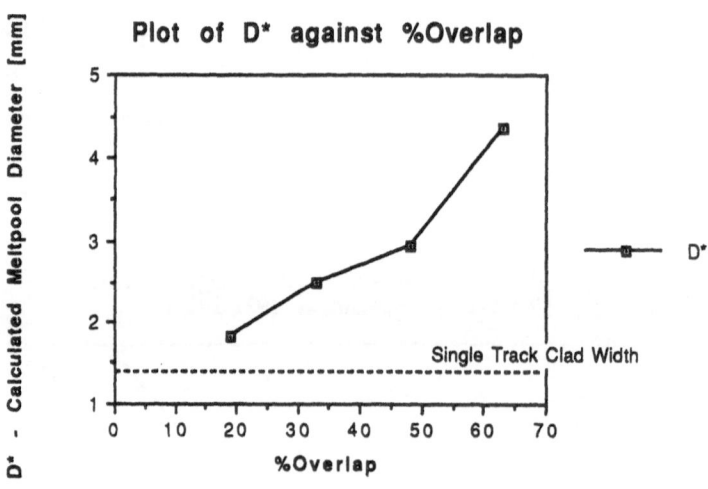

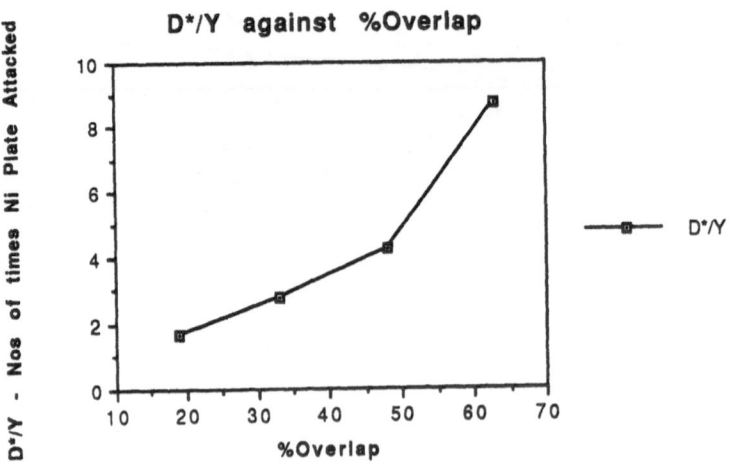

But as $A = \pi h w / 4$:

$$h_m = \frac{\pi h w}{4(w-x)}$$

(3)

also $\Omega = x/w$ therefore:

$$\text{Average height of clad, } h_m = \frac{\pi h}{4(1-\Omega)}$$

(4)

h_m is the expected height of an overlapped clad - where the meltpool size is equal to the single track clad width.

For partial width cladding (i.e. a meltpool size bigger than the single track width) the meltpool becomes large and hence absorbs more powder from the powder stream. The clad height therefore becomes larger than the expected value, h_m, above (see fig. 10).

4.2 CALCULATION OF MELTPOOL SIZE

If it is assumed that the meltpool is circular, then:

Meltpool area $= \pi d^2/4$, where d = meltpool diameter.

The amount of powder arriving per unit area per unit time is defined as q [kg m^{-2} s^{-1}]. An assumption of the calculation is that this is constant across the meltpool. If all the powder entering the meltpool is melted:

$$\text{Mass of powder melted per second} = \frac{\pi q d^2}{4}$$

(5)

In t seconds the beam traverses distance, l

$$\text{Mass of melted powder} = \frac{\pi q t d^2}{4} = A^* l \rho$$

(6)

where A^* = cross-sectional area of melted powder

ρ = density

Therefore:

$$A^* = \frac{\pi q d^2}{4 \rho v} \quad \text{where } v = \frac{1}{t} \text{ is the cladding speed}$$

(7)

If y is the advance distance (i.e. the distance moved forward for a new track) then the experimentally found height of the clad, $h_e = A^*/y$ and

$$h_e = \frac{\pi q d^2}{4 \rho v y}$$

(8)

d, the meltpool diameter, can be calculated from this equation if q is known. All the other parameters are known experimental values.

4.3. CALCULATION OF q FROM SINGLE TRACK DIMENSIONS

For a single track clad $A = \pi h w/4$. (9)

It has been shown previously (7) that if q is constant then:

$$A = \frac{\pi q d^2}{4 \rho v}$$

(10)

Combining (9) and (10) gives:

$$q = \frac{h \rho v}{w}$$

(11)

4.4. FINAL CALCULATION OF MELTPOOL DIAMETER, d, IN TERMS OF THE EXPERIMENTAL MULTITRACK HEIGHT, h_e

Combining (8) and (11) gives:

$$d = \sqrt{\frac{4 \, h_e \, w \, y}{\pi \, h}} \qquad (12)$$

or

$$d = \sqrt{\beta \, h_e \, y \, \frac{w}{h}} \qquad (13)$$

where $\qquad \beta$ = shape factor of the multitrack meltpool

(circular meltpool, $\beta = 4/\pi$)

h_e = height of multitrack clad

w/h = aspect ratio of single track clad

y = advance distance during multitrack cladding

A plot of calculated meltpool size and the number of times the Ni plate is attached by the meltpool is given in fig. 11.

4.5 THE MODEL'S ASSUMPTIONS

It should be noted that this simple model makes several assumptions, the main two being that q is constant across the meltpool and that the meltpool is circular.

4.5.1. Constant q

The assumption here is that powder flow density is constant across the meltpool area. However, the authors have found very little published data on the distribution of powder from a screw fed, powder flow system. Video work has shown the powder area to be at least twice the diameter of the beam - the largest possible meltpool size. If the area over which the powder is spread is large compared to the size of the meltpool then the variation in powder flow density across the meltpool will be cut a minimum.

4.5.2. Circular Meltpool

The assumption that the meltpool is circular is somewhat of a simplification. Studying Rosenthal's moving point heat source equations, it becomes clear that the meltpool (melting point isotherm) will have a ovoid shape. Even this is a simplification as it does not take into account such factors as latent heat of fusion and convection currents in the meltpool, both factors which will alter the shape of the meltpool.

One of the model's possible uses is to use a vision analysis system to analyse the shape and size of the meltpool and use the information to control the cladding conditions. A clad of a given height can then be produced. This would mean that the shape factor of the meltpool could be continually monitored and altered in any control algorithm.

5. Conclusions

It can be seen that cladding low melting point alloys onto a high melting substrate presents unusual problems depending on the composition of the cladding alloy. Partial width cladding overcomes these problems but this results in a much thicker than expected clad caused by a large meltpool. Controlling the size of the meltpool is the key to controlling the dimensions of the resulting clad. A simple model is given which presents a relationship between the size of the meltpool and the thickness of the clad.

6. References

1. McIntyre, M. (1983) Proc. 2nd International Conference on Applications of Lasers in Materials Processing, Los Angeles, USA.
2. Rocca, A.V. et al., (1977) Development Activities for the Industrial Applications of High Power Lasers to Metal Working, Int. Seminar on Laser Processing of Engineering Materials, June 1977.
3. Clark, J., Vilaneuva, L.T. and Steen, W.M. (1979) Report to BSC Scottish Laboratories, Imperial College London.
4. Liu, C.A., Humphries, M.J. and Mason D.W. (1983) *Thin Solid Films*, **107**, 251
5. Lugscheider, E., Oberlander, B.C. and Meinhardt, H. (1990) ECLAT '90, Erlangen, Germany.
6. Takeda, T., Steen,W.M. and West D.R.F. (1985), Laser '85 Opto-Electronik, Munich, Germany.
7. Takeda, T. (1987) Proc. LAMP'87, Osaka, Japan.
8. Xiao, D.C., Ellis, M., Steen, W.M., Lee, C., Watkins, K.G. and Brown, W.P. (1993) *Laser Cladding of Pb-Bronze*, Proc. ICALEO '93, Orlando, Florida, USA.
9. Aboud, J.H. and West, D.R.F. (1991), *Materials Science and Technology*, **7** ,353-356.
10. Hegge, H.J., Boetje, J. and De Hosson (1990) *J. of Materials Science*, **20**, 2267.
11. Jeng, J.Y. (1992) *PhD Thesis*, Laser Group, Dept. of Mech. Eng., University of Liverpool, UK.

Catchment Efficiency For Novel Nozzle Designs Used In Laser Cladding And Alloying

S. Carty[1], I. Owen[1] & W.M. Steen[1]·
[1]Laser Laboratory, Department of Mechanical Engineering, University of Liverpool, PO Box 147, Liverpool. L69 3BX.
B. Bastow[2]·
[2]Building B125, British Nuclear Fuels Limited, Sellafield, Cumbria,
J.T. Spencer[3]·
[3]British Nuclear Fuels Limited, Company Research Laboratories, Springfield Works, Preston.

1. Introduction

This paper describes the laser alloying and cladding nozzles that are being investigated, with the objective of alloying and cladding in any direction and at any angle of inclination with the highest possible efficiency. Measurement of the powder delivery from the end of the nozzle to the substrate has formed the basis of this work, because without it, laser alloying and laser cladding would not be possible. The objective is to produce a nozzle that can be used in an industrial situation.

The catchment efficiency is defined as:

$$\frac{\text{Mass of clad layer/s}}{\text{Mass of powder ejected from the nozzle/s}} \times 100 \qquad (1)$$

The designs of three distinct types of nozzle are described; an annular nozzle, a multiple injector nozzle and a venturi type nozzle. A high catchment efficiency can lead to an improvement in the cost effectiveness of the process.

The cladding efficiency is influenced by the powder flow from the end of the nozzle to the substrate, and can be described in terms of the divergence and the velocity of the powder when it impinges upon the surface. The nozzles are being tested at a number of orientation from pointing vertically downwards to vertically upwards, since not all industrial applications of laser cladding and laser alloying can be undertaken vertically downwards; for example, repair by cladding the inner surfaces of a large cylindrical vessel.

J. Mazumder et al. (eds.), Laser Processing: Surface Treatment and Film Deposition, 395–410.
© 1996 *Kluwer Academic Publishers.*

Testing of these nozzles has been undertaken using 304L stainless steel, 310S stainless steel, nickel and chromium powders. Results obtained so far have shown that cladding efficiency is affected by varying the supply rate of the powder, the velocity of the powder, the divergence of the powder and the traverse speed of the nozzle[1]. Cladding efficiency ratings of between 75-95 per cent have been achieved using the nozzles described.

2. Discussion

2.1 THEORETICAL MODEL

All the arrangements investigated to date have the powder introduced from the side into the beam. Whether a powder particle is caught in the clad or not could be decided by:

- Does it strike the melt pool
- Does it stick to the melt pool
- Does it melt in the laser beam prior to arrival at the melt pool

The temperature of a powder particle on the leading edge of the beam will be cold and therefore solid. A solid/solid interaction would be expected to result in a ricochet provided the kinetic energy of the particle is sufficient. Thus, on the whole, the powder is unlikely to melt before arrival in the region of the clad (item 3 above).

It is considered, as a first approximation, that catchment will occur only when there is a liquid/solid collision, this means the particles, which are expected to be solid, must strike the melt pool to be trapped. Thus the problem is divided into two parts:

- How large is the melt pool?
- What percentage of the powder strikes it?

The relationship between these two properties can be seen in figure 1.

a) The size of the melt pool can be approximated by Rosenthals' equation for a moving point source.

$$Tm - To = (q/4\pi k) \, e^{-(vx/2\alpha)} \, e^{-(vr/2\alpha)} \qquad (2)$$

Where Tm = Melting point (k)
To = Ambient temperature (k)
α = Thermal diffusivity
v = Traverse speed (mm/s)
x = Distance along the traverse direction
r = Radius of the melt pool

---

[1] In this investigation the nozzle was stationary and the substrate was moved, but in the industrial application the substrate will be stationary and the nozzle will move.

$$q = \text{absorbed power}$$
$$k = \text{thermal conductivity}$$

If x = 0 then:

$$Tm - To = (q/4\pi k)\, e^{-(vr/2\alpha)} \tag{3}$$

If $e^{-(vr/2\alpha)}$ is replaced by a Taylor series expansion then neglecting terms of higher order:

$$(4\pi k(Tm - To)r)/q \approx 1 \tag{4}$$

Therefore the radius of the melt pool can be approximated by the formula:

$$r = q/(4\pi k(Tm - To)) \tag{5}$$

b) Assuming the top hat modes for both powder and power the area which the powder strikes on the surface can be equated as:

$$Apow = B\, C\, (\pi\,/\,4) \tag{6}$$
$$\text{Where } Apow = \text{Area of powder striking the surface}$$
$$B = Dp$$
$$C = Dp\,/\, \text{Cos}\,\theta$$

Area of the laser beam is

$$Abeam = \pi\, W^2 \tag{7}$$
$$\text{Where } Abeam = \text{Area of beam striking the surface}$$
$$W = \text{Beam radius}$$

W is assumed to be equal to the radius r from equation 5.

c) It can be deduced that catchment efficiency can be predicted using the following:

$$CE = Abeam\,/\,Apowder \times 100 \tag{8}$$

$$CE = \frac{\text{Cos}\,\theta\, q^2}{Dp^2\, \pi^2\, k^2\, (Tm - To)^2} \times 100 \tag{9}$$
$$\text{Where } CE = \text{Catchment Efficiency}$$

2.2. EXPERIMENTAL WORK

This work only concerns a small section of the area that is being researched at the current time. The area of catchment efficiency is of great importance to blown powder laser cladding, as this controls the economic viability of the process. The catchment efficiency determines how much of the powder is wasted during the process and so how much money is lost due to this excess powder and also the time and money that is spent recovering this wasted powder.

and so how much money is lost due to this excess powder and also the time and money that is spent recovering this wasted powder.

For the catchment efficiency to be optimised the following parameters have to be controlled as far as possible: the divergence of the powder flow once it has left the nozzle and the velocity the powder possesses when it hits the substrate. If these parameters are not controlled then the powder may miss the melt pool altogether or rebound off the substrate.

This section covers the characteristics of catchment efficiency that have been investigated and are considered to be important to the overall optimisation of blown powder laser cladding. The first covers the different types of nozzle that have been tested, the amount of powder ejected from the nozzles in different orientations and with different process parameters and the catchment efficiency of the different nozzles when the process parameters are varied.

2.2.1. *The Nozzles Under Investigation*

The Annular Nozzle.   The annular nozzle (fig 2) consists of three pipes, which are arranged concentricity. The innermost pipe allows the laser beam to pass through and supplies a small amount of gas that is used to stop powder falling back onto the optics during alloying and cladding operations that are vertically upwards.

The middle pipe supplies the stainless steel powder that is used during cladding, it is delivered by up to four separate inlets. The powder is delivered by a carrier gas, in this case argon. The outermost pipe delivers argon, which is used to contain the powder within the influence of the laser beam prior to cladding.

It is feasible to exchange the middle and outer pipes for other versions that allow the powder and shrouding gas to be aimed at different heights below the nozzle. The powder can also be delivered into the outer pipe, which is similar to the system used by the Marchione et al. [1].

The Multiple Injector Nozzle.   The multiple injector nozzle (fig 3) consists of a large diameter cylinder with three openings that allow three narrow diameter delivery pipes to supply powder for alloying and/or cladding. The delivery pipes are independently aimed at the melt pool by means of a ball and socket joint. The pipes can also be extended towards the melt pool or retracted away from the melt pool, whichever is required.

The Venturi Nozzle.   The venturi nozzle (fig 4) consists of a thick walled tube with four sets of three delivery pipes drilled through the wall. The laser beam passes through the centre of the tube. A large gas flow passes through this tube that acts in two ways. The first is to protect the optic systems particularly when cladding vertically upwards, second is to deliver the powder to the melt pool. The powder is brought to the nozzle via the delivery holes by a relatively small amount of carrier gas and then it is drawn into the main stream of gas in the centre of the tube. The

The Venturi Nozzle.　　The venturi nozzle (fig 4) consists of a thick walled tube with four sets of three delivery pipes drilled through the wall. The laser beam passes through the centre of the tube. A large gas flow passes through this tube that acts in two ways. The first is to protect the optic systems particularly when cladding vertically upwards, second is to deliver the powder to the melt pool. The powder is brought to the nozzle via the delivery holes by a relatively small amount of carrier gas and then it is drawn into the main stream of gas in the centre of the tube. The delivery holes are arranged so that powder can be delivered either below, above or at the focal point of the laser.

The Beam Bender.　　To allow these nozzles to be used at different angles, the beam bender (fig 5) was developed. This consists of a flat mirror set at 45 degrees to the horizontal that delivers the beam to a rotatable off-axis parabolic mirror that is used for aiming the beam at the substrate. The nozzles are attached to the end of the beam bender.

The powder was continuously supplied to the nozzle under investigation by a screw feeder with pneumatic delivery using argon[2]. The powder feeder output (g/s) can be varied by adjusting the argon flow rate into the system and the screw feeder rpm.

2.3. THE MASS FLOW OF POWDER EJECTED FROM THE END OF A NOZZLE

The mass of powder that is ejected from a nozzle is controlled by the amount of carrier gas (l/min) flowing into the powder feeder, the speed of rotation of the screw feeder (RPM) in the hopper and in special circumstances the shrouding gas used by the nozzle. These mass measurements are easily calibrated at different angles of inclination for a nozzle.

The initial mass flow involved testing 304L and 310S stainless steels, chromium and nickel powders, at vertically downwards and horizontally (the chemical composition of the 304L stainless steel can be found in figure 18). These powders were either dried or not dried (i.e. assumed to be damp). The mass flow rate of powder (g/s) for nickel, 304L stainless steel and 310S stainless steel shows no major differences when the powder is dried compared to when it is not. The chrome powder mass flow however doubled when this powder was dry. The undried powder was also split into two separate parts, one of which was sieved to $>45\mu m$ and $<90\mu m$, but this sieving made no difference to the mass flow. However, by sieving the powder to the smaller diameter, the flow characteristics of the particles from the end of the nozzle were improved, because more particles reached the substrate, since the larger particles that previously were too heavy to be carried to the substrate were no longer present. Previous work by Jeng[2] has also shown that particles that are greater than $75\mu m$ diameter are also less likely to melt and therefore produce lower quality clads, so smaller particles therefore have two advantages. The data obtained showed that for all

---

[2]Argon was used because it was readily available, but any gas that is inert to the process could be used.

of the powders that the powder flow rate did not vary within the time frame of the investigation.

The variance in the flow of powder (g/s) for all of the powder types increases as the powder feeder RPM increases, but this could simply be due to the increased vibration of the powder feeder as the RPM increased from 100-1000 (fig 6). The mass flow rate for all of the powders drops as the carrier gas flow rate is increased, which is possibly due to the powder being compacted in the hopper by the gas and thus not flowing as well into the screw threads of the powder feeder (fig 7).

The mass flow from the powder feeder can be increased by simply increasing the shrouding gas flow above 50 l/min, if the shrouding gas completely surrounds the carrier gas delivery pipe. In this case the powder is "sucked" out of the hopper using the venturi effect, and this works if there is no carrier gas present and the screw feeder turned off.

2.4. THE CATCHMENT EFFICIENCY OF THE NOZZLES

Work on the catchment efficiency has been carried out using the multiple injector nozzle directed vertically downwards and horizontally to this. A comparison of results has been produced which compares the efficiency of this nozzle when the injectors are moved closer to the melt pool. The annular nozzle has also been tested vertically downwards using a single powder input.

The data from the mass flow experiments is used as the basis for the catchment efficiency that is: evaluated as:

$$\frac{\text{Mass of clad layer/s}}{\text{Mass of powder ejected from the nozzle/s}} \times 100 \qquad (10)$$

The mass of powder ejected from the nozzle is supplied by the work described previously and the mass of the clad layer is determined by weighing the sample before and after cladding.

The catchment efficiency is affected by several process parameters; the mass flow rate, the divergence of the powder jet, the traverse speed of the nozzle and the velocity of the particles.

2.4.1. *Multiple Injector with Three Delivery Pipes when Cladding Vertically Downwards*

The mass flow determines how much powder is available for the cladding process, when cladding vertically downwards with a constant carrier gas velocity and traverse speed, the efficiency rating increases as the mass flow increases, this is due to the powder travelling with a lower velocity when it hits the substrate and so it does not bounce off (fig 8). It is expected that as the mass flow rate increases past 500RPM then the efficiency will drop as too much powder will be being used and so it will not all be able to melt and form into a clad layer.

The speeds of the particles in this investigation have been previously measured at between 7mm/s-84mm/s using high speed photography. It was suggested by Weersinghe[3] and Steen[4] that particles with velocities up to 1600mm/s can be used for cladding.

The divergence of the particles is important in the cladding process because as the particles leave the delivery pipe they begin to diverge. As this divergence happens, it means that fewer particles will hit the melt pool and so the efficiency rating will drop. A new type of nozzle is being developed which is expected to allow partial control over the divergence of the powder flow, a schematic of which can be seen in fig 9. Figure 10 shows how this new type of nozzle will be combined with the existing multiple injector nozzle, to facilitate accurate powder delivery to the melt pool.

If all the process parameters are kept constant and the velocity of the substrate (traverse speed) is increased, the efficiency of the process will fall off very quickly. This is due to the smaller melt pool that directly affects catchment efficiency, at higher traverse speed.

Another important characteristic of traditional blown powder laser cladding is that the single jet of powder is directed at the melt pool, but show variations in clad shape and catchment efficiency with direction. The multiple injector jets are also aimed directly at the melt pool, and show a reduced effect of variations in direction. However, when cladding in different directions, the efficiency ratings were found to vary slightly; since one or more of the delivery pipes may have been slightly mis-aligned or the powder flow may not have been exactly balanced between the three delivery pipes (fig 11).

A further variable that is involved with the cladding efficiency is the distance between the substrate and the end of the delivery pipe. As the powder flows from the delivery pipe it diverges at a set rate, so the closer the delivery pipe is placed to the melt pool, the less divergence will be able to happen (fig 12). The problem of the delivery pipes clipping the beam and being heated up by back reflection becomes important at this stage. A new type of nozzle (fig 10) is being developed which produces a focused powder flow from a delivery pipe and which still allows for a large stand off distance between the substrate and the nozzle.

2.4.2. *Multiple Injector with One Delivery Pipe when Cladding Horizontally*
The minimum flow of carrier gas required to produce a powder flow that can be used for laser cladding horizontally (fig 13), is considerably more than that which is required to clad vertically downwards (downwards 0 l/min, horizontally 2.5 l/min).

To produce a powder flow from three delivery pipes with only one powder hopper has not been possible (using our equipment), because the combined carrier gas flow rate required is too much for the hopper to withstand, so only one delivery pipe can be used at any one time, these were labelled A, B and C (fig 14). Pipe A is always positioned above the melt pool, pipe B is below and behind the melt pool and pipe C is below and in front of the melt pool. Depending upon which direction cladding is taking place pipes B and C are interchanged.

The process variables that are involved in cladding horizontally are the same as those involved when cladding vertically downwards (which were described in the previous section).

As the nozzle is moved closer to the substrate the efficiency rating increases, in the same way as when the nozzle was cladding vertically downwards. These results were obtained using delivery pipe A (fig 15).

Apart from the result that cladding is possible horizontally, the most important result was that the powder can be fed into the melt pool from below and behind, i.e. using pipe B and a clad track can be formed. The catchment efficiency rating when using pipe B is not as high as pipe A. Pipe C was not investigated since it also feeds from below the melt pool, but feeds from in front. Since this is a better position to delivery the powder from when compared to B, it is expected to produce better results than the B position.

Different mass flow rates have been tested using this nozzle and again the efficiency level increased as the flow rate increased, this can be seen in fig 16. As the traverse speed increases it also adversely affects the catchment efficiency when cladding horizontally, as it did when cladding vertically downwards, which is shown in figs 15-16.

2.4.3. *Annular Nozzle with One Delivery Pipe when Cladding Vertically Downwards*
The annular nozzle differs from the multiple injector nozzle in that powder is being supplied from an infinite number of delivery pipes, which means that the same amount of powder should be delivered to the melt pool from all directions, and clad tracks of equal quality can be formed in any direction.

The nozzle has only been tested with one powder delivery input to the annular powder chamber, and so it was found that only one third of the chamber became occupied by this powder, which in turn meant that powder was only being delivered from one third of the total possible area. The results in fig 17 show how the efficiency drops when the powder is only delivered from behind the melt pool. To overcome this problem two more delivery pipes have been introduced at 120° to the original and the efficiency ratings have immediately improved in the initial trials with three inputs.

When one powder input was being used the carrier gas and shrouding gas could not direct all of the powder flow into the melt pool, whereas using the same process variables, but increasing the number of delivery pipes has improved the aiming of the powder and so the efficiency has increased.

When the traverse speed is increased the catchment efficiency drops, as it did for the multiple injector nozzle.

3. Conclusions

304L and 310S produce similar cladding efficiencies as long as the size distribution and density of the particles for each are similar.

The shape of the powder flow from the nozzle is affected by a combination of both the carrier gas and shrouding gas velocities, furthermore and to a certain extent this shape can be predicted.

Cladding can be successfully undertaken horizontally, as demonstrated in fig 13. It is hoped that by building upon this success cladding can be undertaken vertically upwards with one or more of the nozzles.

The main conclusions are summarised in the following table:

	Annular Nozzle With 1 Delivery Pipe Vertically Downwards	Multiple Injector Nozzle With 3 Delivery Pipes Vertically Downwards	Multiple Injector Nozzle With 1 Delivery Pipe Horizontally
Increase Mass Flow Rate (100-500RPM)	Increase Catchment Efficiency	Increase Catchment Efficiency	Increase Catchment Efficiency
Increase Mass Flow Rate (past 500RPM)	Decreases Catchment Efficiency	Decreases Catchment Efficiency	Decreases Catchment Efficiency
Increase Traverse Speed	Decreases Catchment Efficiency	Decreases Catchment Efficiency	Decreases Catchment Efficiency
Decrease Stand-off Distance	Increases Catchment Efficiency	Increases Catchment Efficiency	Increases Catchment Efficiency
Different Delivery Pipe Position	Catchment Efficiency Maximised When Placed In Front of Melt Pool	N/A	Catchment Efficiency Maximised When Placed Above Melt Pool

The most important characteristic that improves catchment efficiency is the aiming of the powder into or just in front of the melt pool.

4. References

1. Marchione, T., Lagrange, P.A. and Vetter, P.A. (1992) Developpement d'une buse coaxiale multidirectionelle de revetements et d'alliages de surface par laser avec apport de poudre, *MAT-Tec'92*, 203-210.

2. Jeng, J.Y. (1992) Computer controlled laser surface treatments of stainless steels for corrosion resistance, PhD Thesis, Liverpool University.

3. Weersinghe, V. M. and Steen W. M. (1984) Computer Simulation Model For Laser Cladding, Imperial College, University of London.

4. Steen, W. M. (1985) Laser Surface Cladding, Liverpool University.

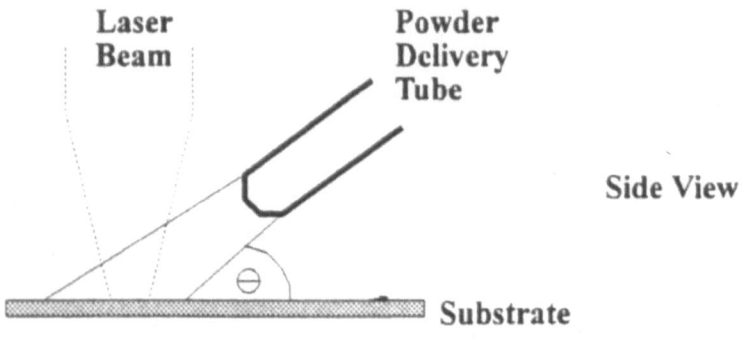

Side View

Substrate

Top View

Figure 1: Basis for Theoretical Model

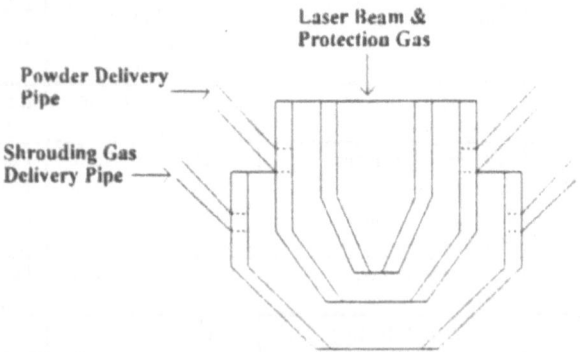

Figure 2: The Annular Nozzle

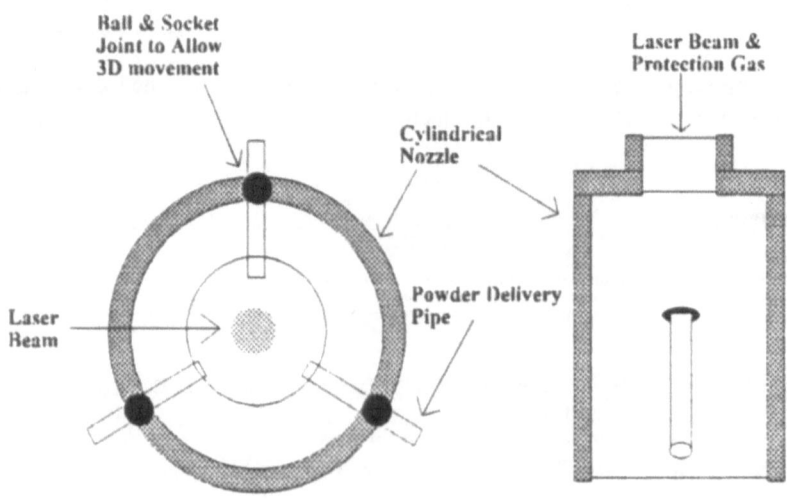

Bottom View

Side View

Figure 3: The Multiple Injector Nozzle

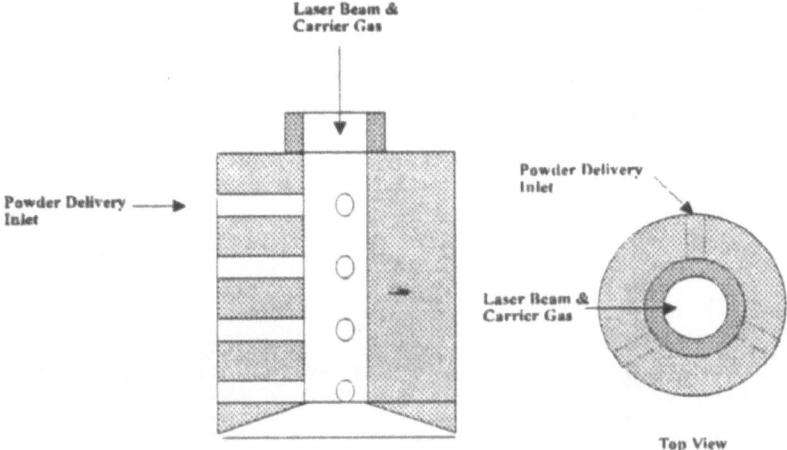

Figure 4: The Venturi Nozzle

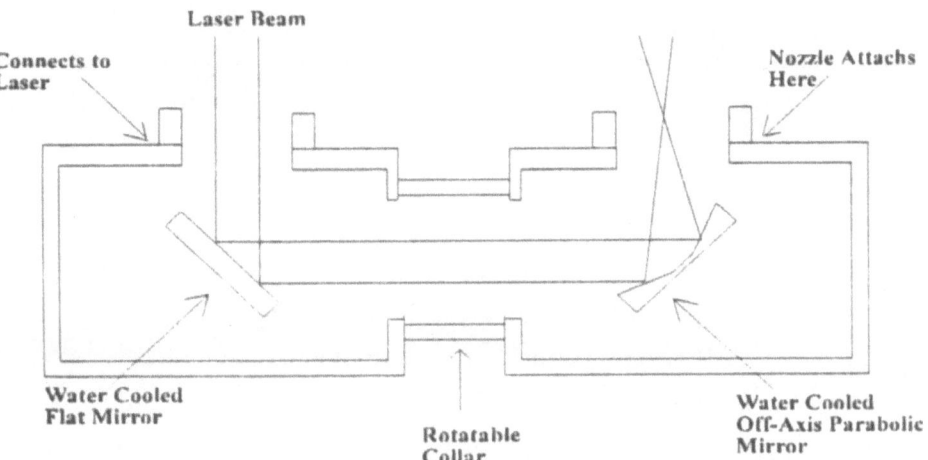

Figure 5: The Beam Bender

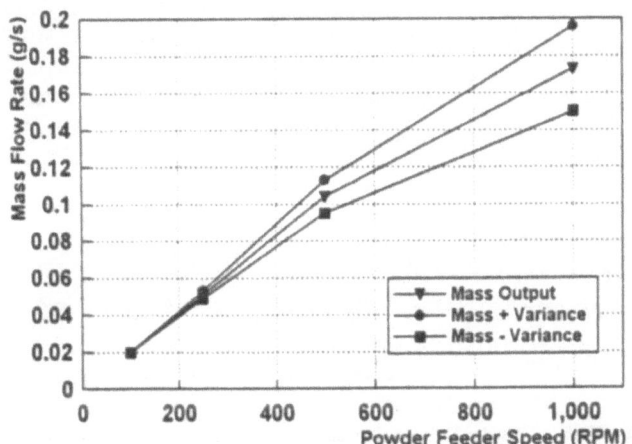

Figure 6: Powder Feeder Speed vs. Mass Flow Rate with a Constant Carrier Gas Supply

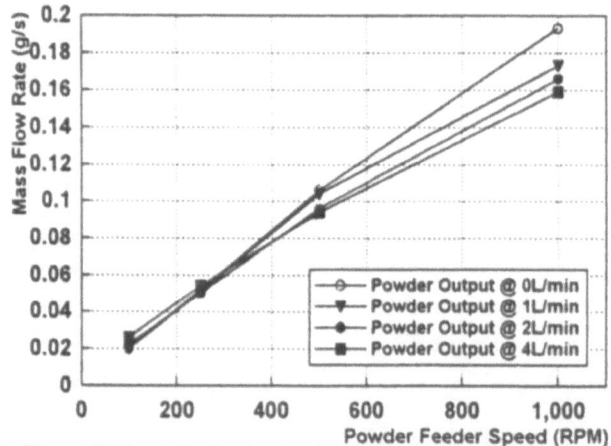

Figure 7: Mass Flow Rate at Different Carrier Gas Flow Rates Vertically Downwards

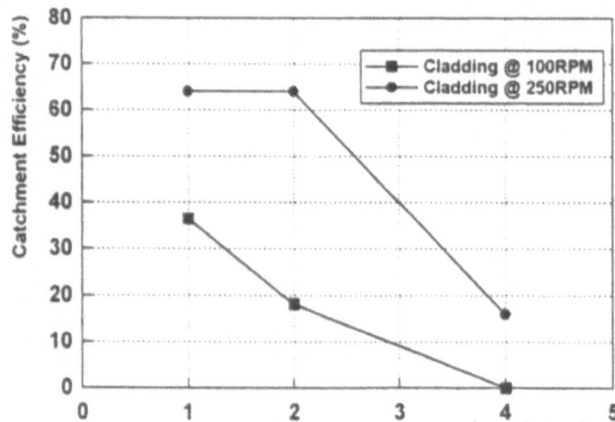

Figure 8: Mass Flow Rate vs. Catchment Efficiency when Cladding Vertically Downwards

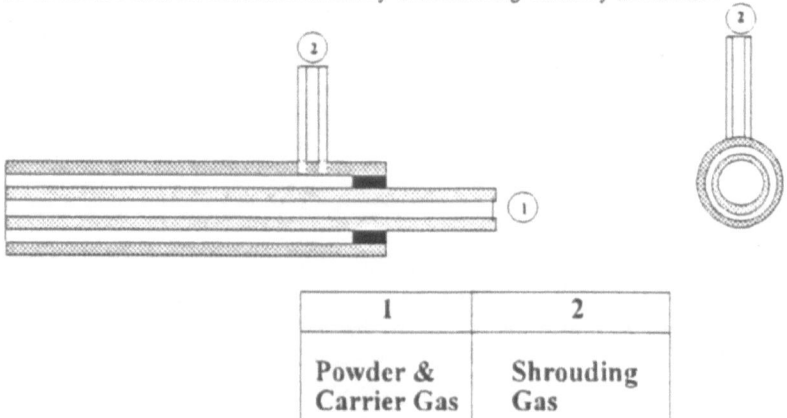

1	2
Powder & Carrier Gas	Shrouding Gas

Figure 9: The New Type of Powder Delivery Nozzle

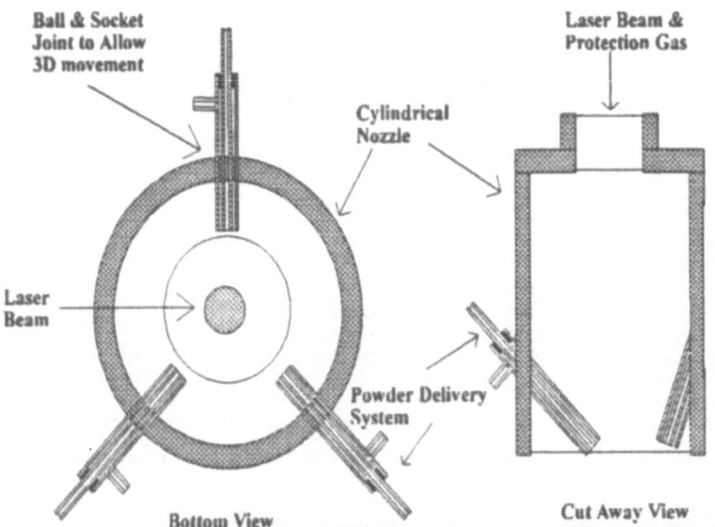

Figure 10: The New Multiple Injector Nozzle Incorporating the New Powder Delivery Nozzle

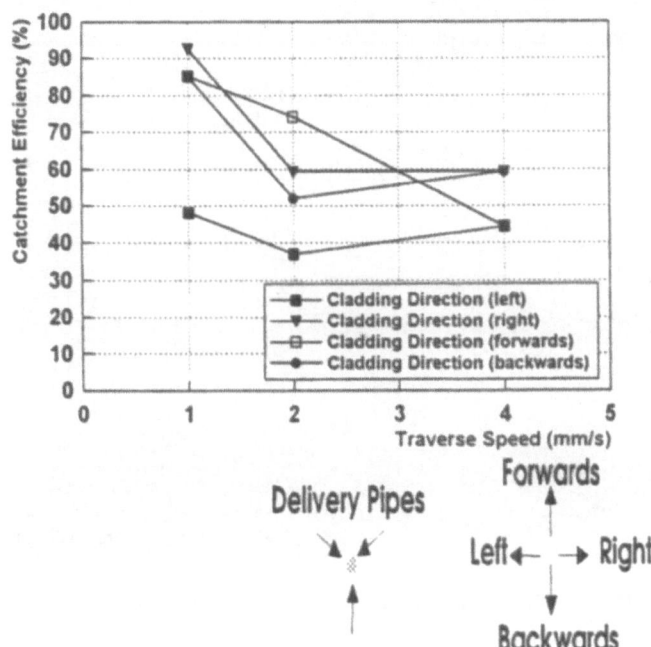

Figure 11: Traverse Speed vs. Catchment Efficiency for Different Cladding Directions Vertically Downwards

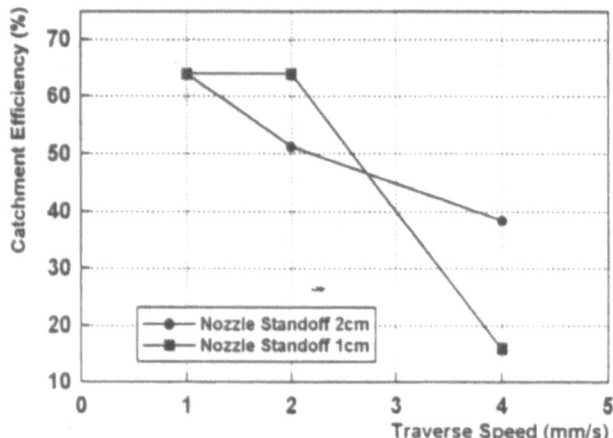

Figure 12: Stand-off Distance vs. Traverse Speed when Cladding Vertically Downwards

Figure 13: Cladding Horizontally Using the Multiple Injector Nozzle (powder delivery position A)

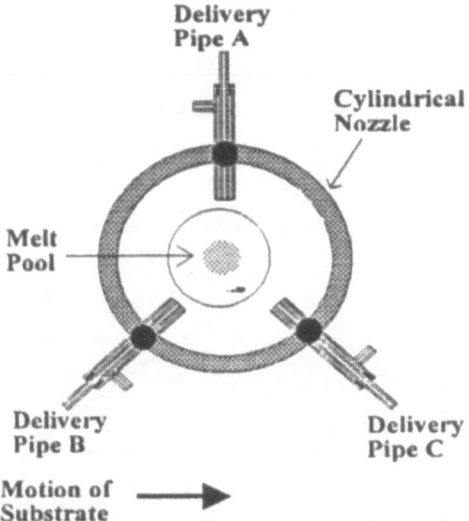

Figure 14: Schematic to Show the Arrangement of Delivery Pipes when Cladding Horizontally

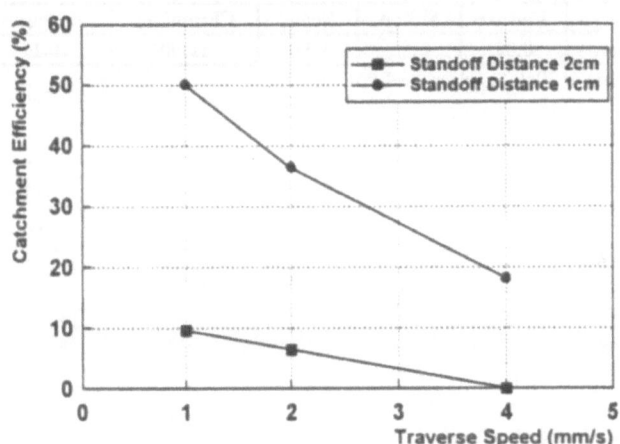

Figure 15: Stand-off Distance vs. Catchment Efficiency when Cladding Horizontally Using Pipe A

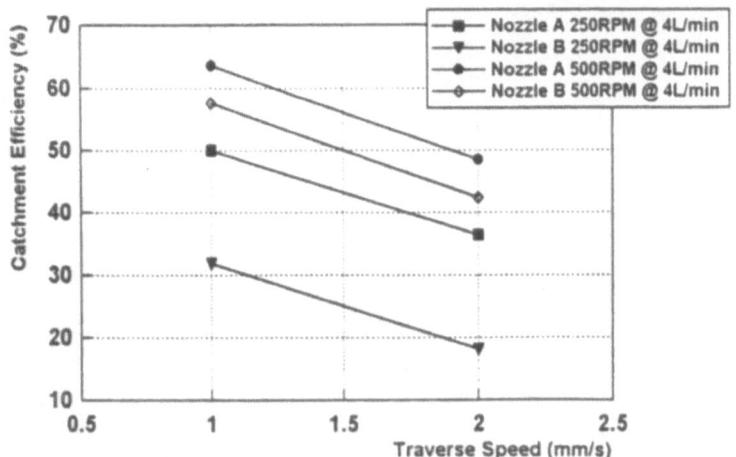

410

Figure 16: Nozzle Position vs. Catchment Efficiency when Cladding Horizontally

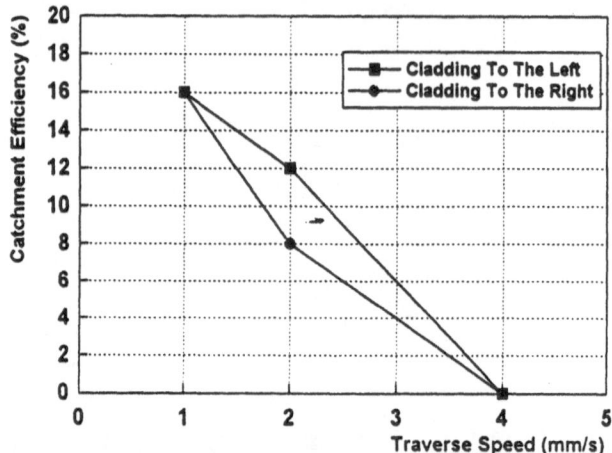

Figure 17: Catchment Efficiency vs. Traverse Speed for Annular Nozzle Cladding Vertically Downwards

Element	Carbon	Silicon	Nickel	Chromium	Manganese	Iron
Percentage	0.02	0.83	9.90	19.00	0.17	BAL

Figure 18: Analysis of 304L Stainless Steel Powder

LASER PROCESS ADAPTED POWDER DELIVERY SYSTEM

B. GRÜNENWALD, W. HENNIG, *St. NOWOTNY, F. DAUSINGER,
H. HÜGEL

University of Stuttgart　　　　　　　　*Fraunhofer-Institut für Werkstoff-*
Institut für Strahlwerkzeuge　　　　　*physik und Schichttechnologie*
Pfaffenwaldring 43　　　　　　　　　　*Helmholtzstraße 20*
70569 Stuttgart, Germany　　　　　　　*01069 Dresden, Germany*

ABSTRACT. The controlled and continuous powder supply is one of the main precon-
ditions for an efficient process of laser cladding, alloying and particle injection. Powder
feeders available from thermal spraying are in general used for laser processing. They
fulfill the typical requirements of laser surface treatments only incompletely. For this
reason, a new powder feeding device was developed and tested for several applications.

1.　　Introduction

In laser cladding, alloying and particle injection the wear and corrosion properties of
metallic surfaces are modified by the addition of hardfacing alloys. For these laser
surface treatment methods several processing techniques have been developed. The
adding material can either be supplied to the substrate during the laser irradiation (one-
step process) or preplaced on the substrate surface in different form before laser
processing (two-step process). For both processing techniques different possibilities of
material supply have been reported. However, the one-step process offers a higher
degree of flexibility than the two-step process.

The most widespread one-step process involves the injection of powder into the melt
pool through a nozzle by means of an inert gas stream. This is shown in Fig. 1 for the
example of laser cladding. The blown powder method of material supply is charac-

411

J. Mazumder et al. (eds.), Laser Processing: Surface Treatment and Film Deposition, 411–420.
© 1996 *Kluwer Academic Publishers.*

412

terized by high process quality concerning track performance and surface finish as well
as high process flexibility concerning workpiece geometry and availability of alloy
materials. An appropriate choice and adjustment of the surface layer properties is
possible by using different kinds of metallic and carbidic hardfacing alloys as well as
composite powders. The resulting surface quality of this processing technique is
strongly dependent upon the exact dosage of the powder and the defined supply to the
laser generated melt pool at the workpiece surface. This will be shown for the laser
cladding process.

For the powder supply, hence, several basic requirements arise:
- Infinitely variable powder mass flow rate in the range between 1 and 100
  g/min with small ramp-up and ramp-down times.
- High long- and short-time constancy of the feeding rate with deviations
  smaller than 5 %.
- Online measurement and control of the mass flow rate.
- Possibility of steering of the powder feeding rate according to the temperature
  and geometry of the workpiece.

These requirements have to be fulfilled in order to make use of the typical advantages
of the laser for surface treatment processes like precise and localized processing, high
flexibilty as well as easy automatisation and integration into production lines. Therefore, a new powder delivery system has been developed being characterized by high
flexibilty and accuracy. By this delivery system an online control and a modulation of
the powder feed rate is possible.

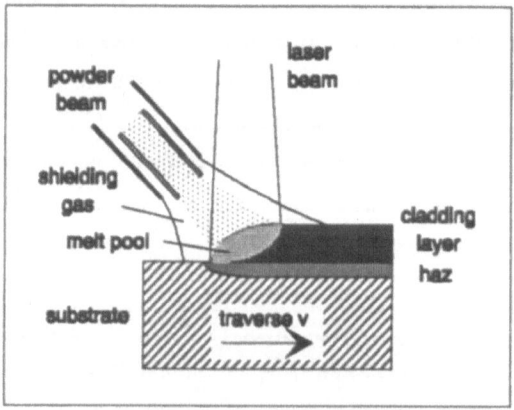

Fig. 1   Schematic of the one-step laser cladding process.

2.   Role of Powder Feed Rate in Laser Cladding

The quality of laser cladded layers is dependent upon the processing parameters and the powder feed rate. Small deviations of the mass flow rate result in large variations of the geometry and the microstructure of the produced layers [1].

2.1.   TRACK GEOMETRY

Cladding experiments were performed on a low alloyed steel using a 5 kW CO_2-laser with a constant power of 3500 W at the workpiece and Stellite 21 with a mean particle size between 45 and 90 µm as cladding powder. In order to see the influence of traverse speed and mass flow rate on track performance and microstructure different beam diameters resulting in different intensity values were investigated.

In Fig. 2 the characteristic decrease of the dilution of the cladding tracks with the substrate material with increasing powder feed rate is shown. This behaviour is independent of the laser intensity and the traverse speed and can be explained by the energy balance of the cladding process [2]. Cross sections of the resulting cladding tracks are presented in Fig. 2(a) and (b). Here the strong influence of the mass flow rate on the layer thickness and the dilution can be seen. High quality tracks with low dilution as shown in Fig.2(a) are characterized by a small melting depth and a high build-up. Small feeding rates, however, result in cladding tracks with a high dilution (Fig.2(b)) and a small layer thickness.

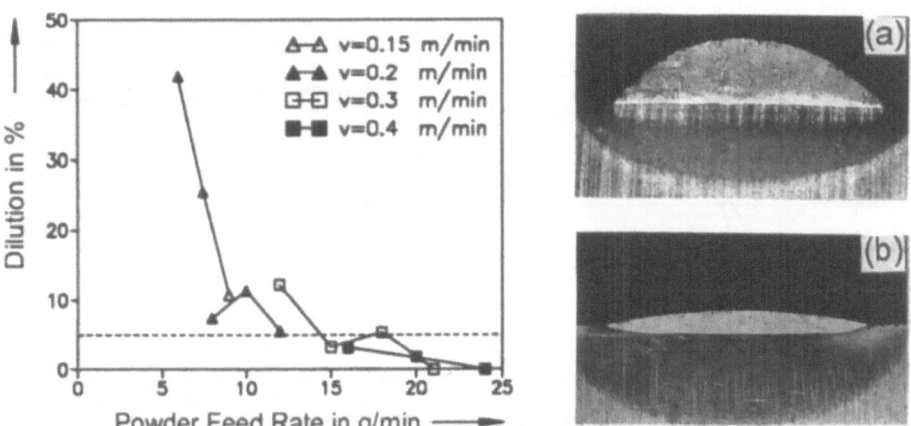

Fig. 2   Dilution with substrate material versus powder feed rate for laser cladding with Stellite 21 (I = 15 kW/cm²) and single tracks for (a) high and (b) low feed rate.

414

2.2.   MICROSTRUCTURE AND HARDNESS

Stellite 21 which was used as cladding material is characterized by high hot-hardness, high ductility as well as high resistance against thermal and mechanical load. These good mechanical properties are based on the dispersion of hard Cr-carbides (Cr_7C_3) in the ductile Co-based metal matrix.

Therefore, several investigations were done characterizing the microstructure by measuring the content of precipitated, interdendritic Cr-carbide particles in the cladding layers as function of powder mass flow rate. In Fig. 3(a) the amount of Cr-carbides in the Stellite layers versus the powder feed rate is shown. According to the decreasing dilution with increasing feeding rate the carbide content increases. As a result the microstructure and the mechanical properties of the cladding layers change. This can be seen in Fig. 3(b) where the microhardness of the cladding layers is plotted versus the measured carbide content. The hardness of each layer was characterized by an average value of a hardness linescan from the layer surface down to the substrate. Only for high carbide contents, that means minimized dilution with the substrate material, the maximum hardness of Stellit 21 (450 HV0.1) is reached. Low feed rates result in a high dilution, low carbide contents as well as low hardness values.

This means that not only the track geometry but also the resulting microstructure and the mechanical properties are highly influenced by a proper choice and a constant value of the powder mass flow rate.

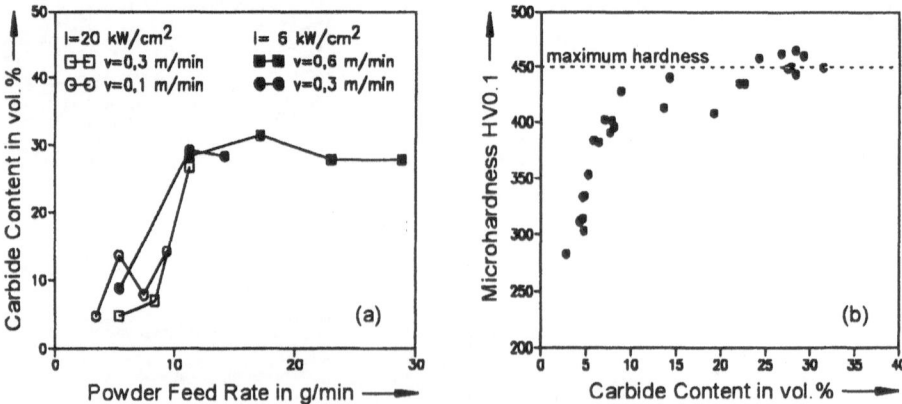

Fig. 3   (a) Carbide content versus powder feeding rate. and (b) microhardness as function of carbide content for cladding with Stellite 21.

3.   Powder Delivery System

3.1.   DOSAGE PRINCIPLE AND MECHANICAL SET-UP

In Fig. 4(a) the mechanical set-up together with the powder dosing principle and the integration of the feeding device into a laser machine tool is shown. As can be seen the powder flow is controlled independently of the gas flow by a specially shaped bar (scraper) pushing the powder from a rotating disc with an annular slot which is coupled with a DC motor. The homogeneous distribution of the powder on the dosing disc is guaranteed by the static pressure of the remaining powder in the powder storage hopper, the width of the ring slot and the shape and position of the scraper. Consequently, this leads to a dependence of the powder mass flow rate merely on the number of revolutions of the motor respectively the dosing disc. The intermixing of the powder with the carrier gas is provided by an injector outside the dosage chamber. The carrier gas flow is totally independent of the powder mass flow rate and can be kept at a minimum necessary for shaping the powder beam. For special cases a gas-free powder delivery is possible, as well.

The powder feeder permits the dosage of powders with grain sizes > 20 μm. The mass flow rate can be variied between 1 and 150 g/min. Due to the dosage principle all kinds of laser cladding powders including edged ceramic powders can be fed. The measured limits of tolerance are in all cases less than 5 %. Due to its low weight and compact construction the powder feeding device can be fixed to the z-axis of the laser work station (Fig. 4(b)). Since the necessary ramp-up and ramp-down times of the powder feeder are very short the time lag of the entire system is below 1 s. The technical details are summarized in Fig. 5.

3.2   ONLINE MEASUREMENT AND CONTROL OF MASS FLOW RATE

In the literature different ways for an on-line measurement of the powder flow rate using acoustic, optical or mechanical sensors have been described [3,4,5,6,7]. The important feature of the feeding device presented herein, however, is the integrated online feeding rate measurement and control system. The measuring principle is based on the continuous recording of the weight of the powder feeder during the feeding process using a special balance integrated into the feeding device as shown in Fig. 4(a). Therefore, the actual powder mass flow rate can be calculated as weight loss per time

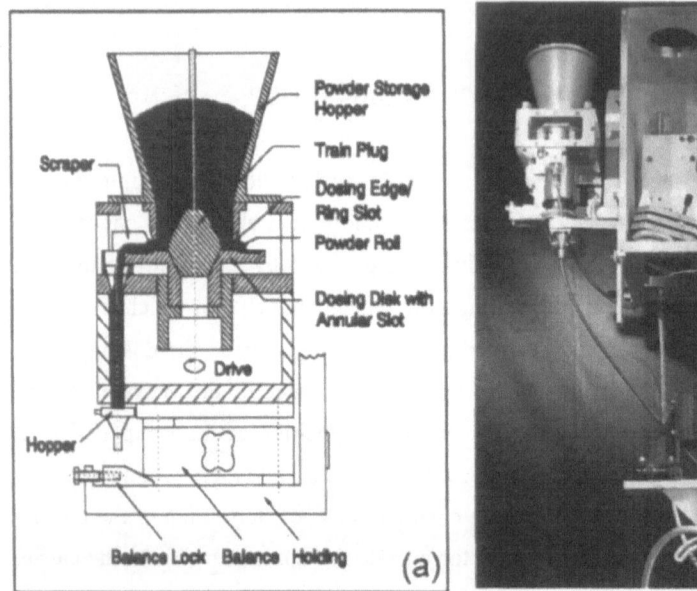

Fig. 4 (a) Mechanical set-up showing the powder dosing principle and (b) the feeding device attached to the z-axis of a laser work station.

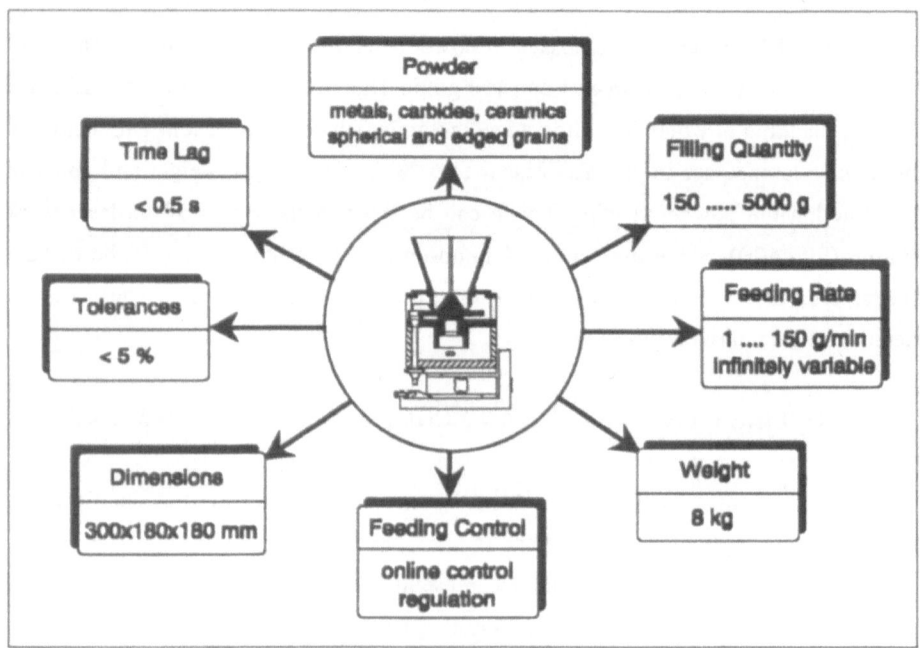

Fig. 5 Technical data of the powder feeding system.

unit and can be controlled online. Due to the set-up of the balance and the low weight of the entire feeding device a resolution of the strain gauge signal between 20 and 100 mg is achieved. The electric signal of the balance is processed by a strain gauge amplifier which is connected via an A/D card to a personal computer. For the online measurement and control of the powder flow rate a special computer program has been developed with the operational details given in [8,9]. The regression times for calculating the powder feeding rate out of the weight loss of the entire system are usually 5 to 20 s. For deviations of the actual powder feeding rate from the set value smaller than the limits of tolerance the transmitted motor speed is not changed. For deviations larger than the set limits the installed PID-controller calculates a new motor speed value. The command variable of the powder feeding control loop is the nominal feed rate given in g/min. The controller output is the motor respectively dosing disc speed whereas the regulating variable is presented by the actual feeding rate value.

The controlled powder mass flow rate as function of time for a feed rate of 50 and 10 g/min is shown in Fig. 6. The high feeding rate is characterized by a high constancy with the standard deviation of the actual to the nominal value being in the range of 2 %. The low feeding rate of 10 g/min also shows a constant mass flow rate yet with a higher standard deviation of 4 %. Fig. 7 further illustrates the effect of the feed rate control loop. The actual mass flow rate is calculated and compared with the nominal one.

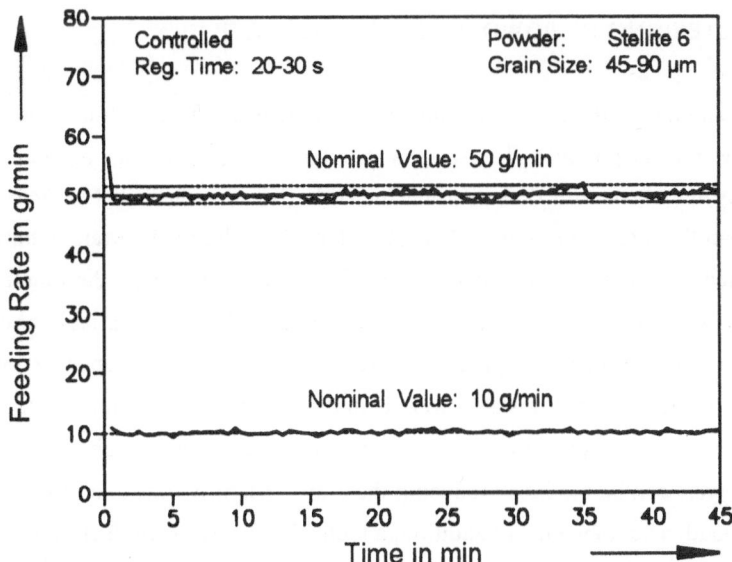

Fig. 6   Controlled feeding rate as function of time for a powder mass flow of 50 and 10 g/min.

418

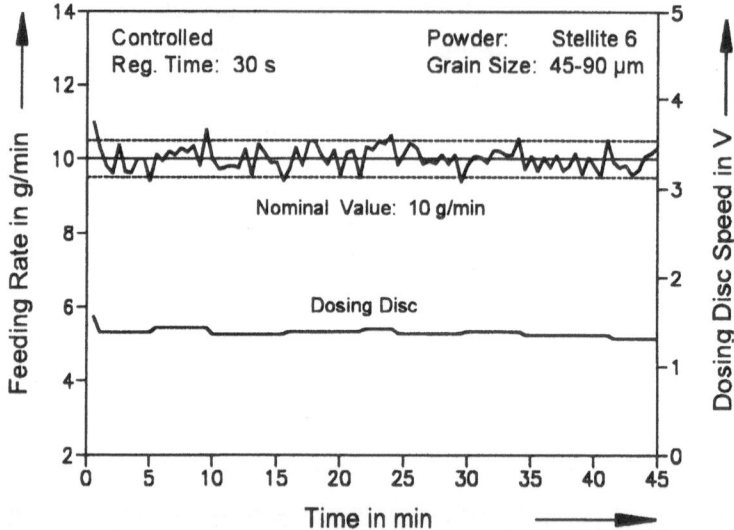

Fig. 7   Controlled powder feed rate and dosing disc speed for 10 g/min.

As soon as the limit of tolerance is exceeded the motor speed is changed according to the output of the PID-controller. Therefore, the control system guarantees the long-time constancy also for low mass flow rates during the whole treatment process. The short-time constancy, on the other hand, is guaranteed by the mechanical set-up of the feeding device and the revolution control of the motor.

3.3.   STEERING OF POWDER MASS FLOW RATE

In combination with the powder feeding control system a time-controlled steering of the mass flow rate is possible. This means that the feeding rate can be increased and decreased linearly within defined time intervals (linear ramping). For time intervals longer than the regression time and constant feed rates the feeding rate control system is activated and controls the actual powder feed rate according to the nominal value. This steering option can be applied for generating a well-defined variable powder feed rate adapted to the cladding process, as required e.g. for odd-shaped or cylindrical workpieces (Fig. 8), repair tasks as well as in the presence of substrate heating by heat conduction during the process. By linear ramping of the powder feeding rate geometry-adapted layers can be produced reducing the after-machning time and production costs of laser cladded components. In addition, a highly automated manufacturing is possible due to the integration of the powder feeding device into the CNC system of the laser machine tool.

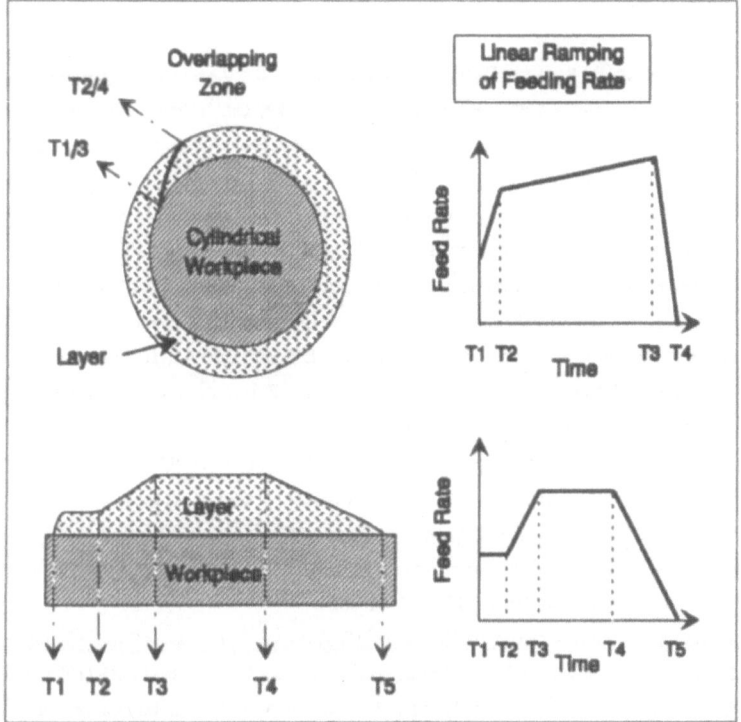

Fig. 8   Schematic diagram of examplerary cladding processes with steering of the powder feeding rate.

4.   Conclusions

For the example of laser cladding the important implications of the powder feed rate on the process quality has been shown. As a consequence, a novel powder feeding system designed for the requirements of laser surface treatment has been developed and its beneficial properties have been demonstrated. It permits the feeding of metallic and ceramic powders with grain sizes > 20 μm and is characterized by feed rates of between 1 and 150 g/min. The accuracy of the powder feeder is better than 5 %. The online measurement of the powder mass flow rate makes it possible to achieve fast and precise setting, long-time constancy as well as a control of the feed rate during the entire treatment process. In addition, the steering option of the feed rate can be applied for generating well-defined cladding layer geometries.

Due to its compact construction and low weight the entire powder feeding device can easily be attached to a laser working head. By an integration of the feeding device into the control system of the laser machine tool an automated manufacturing is possible.

420

5. References

[1]   GRÜNENWALD, B.; SHEN, J.; DAUSINGER, F.; NOWOTNY, ST.: *Laser Cladding with a Heterogeneous Powder Mixture of WC/Co and NiCrBSi*. In: Mordike, B.L. (Hrsg.): Proc. of the 4th Europen Conference on Laser Treatment of Materials (ECLAT'92), Göttingen, 1992. Oberursel: DGM Informationsgesellschaft, 1992, S. 411-416.

[2]   SHEN, J.: *Optimierung von Verfahren der Laseroberflächenbehandlung bei gleichzeitiger Pulverzufuhr*. In: Forschungsberichte des IFSW. Stuttgart: Teubner-Verlag, 1993.

[3]   Powder Flow-Rate Measuring and Controlling Apparatus, Patent, EP 0 216 572, A1, 1.9.86.

[4]   Pulverdosiervorrichtung für einen Brenner zum thermischen Beschichten eines Grundwerkstoffs mit einem pulverförmigen Zusatzwerkstoff, Patent, DE 4 030 539, A1, 16.4.92.

[5]   Verfahren und Vorrichtung zur Überwachung eines Pulverstromes in einer Pulverapplikationsanlage, Patent, EP 0 438 976, A2, 28.12.90.

[6]   Powder Weighing Mixer and Method Thereof, Patent, EP 0 290 999, A2, 9.5.88.

[7]   JENG, J.-Y.; QUAYLE, B.; MODERN, P.J.; STEEN, W.M.: *Computer Control of Laser Multi-Powder Feeder Cladding System for Optimal Alloy Scan of Corrosion and Wear Resistance*. In: Proc. of the Intern. Conference on Laser Advanced Materials Processing (LAMP'92) - Science and Applications, Nagaoka (Japan), 1992, S. 819-824.

[8]   NOWOTNY, ST.; GRÜNENWALD, B.; HENNIG, W.; DAUSINGER, F.: *Regulated Powder Supply for the Laser Surface Treatment with Additional Materials*. Laser und Optoelektronik 25 (1993) 6, S. 71-77.

[9]   GRÜNENWALD, B.; NOWOTNY, ST.; HENNIG, W.; DAUSINGER, F.; HÜGEL, H.: *New Technological Developments in Laser Cladding*. In: Proc. of the Intern. Congress of Lasers and Electro-optics (ICALEO'93), Orlando, Florida (USA). Laser Institute of America, 1994.

A SIMPLE CORRELATION BETWEEN THE GEOMETRY OF LASER CLADDING TRACKS AND THE PROCESS PARAMETERS

R. COLAÇO, L. COSTA, R. GUERRA, R. VILAR

Instituto Superior Tecnico

Departamento de Engenharia de Materiais

Av. Rovisco Pais, 1, 1096 Lisboa Codex, Portugal

Laser cladding is a complex process and full featured cladding models require large comp. ing times. However, from a practical standpoint, it is useful to establish simple correlation between the laser cladding parameters and the characteristics of the clad track, that can be used to select the processing parameters while avoiding the expensive and time consuming experiments. In this paper we present simple relationships that correlate the height and width of tracks obtained by the blown powder technique with the scanning speed, powder feed rate and catchement efficiency of the powder, for a given spot diameter and laser power. These relationships can be used to optimize the values of the parameters in order to obtain tracks with a shape of the cross section that minimizes defects due to overlapping or excessive dilution.

I. Introduction

Laser cladding is a very promising technique to coat structural materials with a wear and/or corrosion resistant material without altering the mechanical properties of the bulk. This process is best suited to coat medium to small areas with a moderate thickness coating. It has significant advantages over alternative methods such as plasma spraying or arc welding. The advantages include easy control of the coating thickness, minimal heat load to the substrate, low roughness, excellent adhesion between the coating and the substrate, minimal dilution of the coating's material by the substrate's material, and fine grained microstructures with superior hardness, wear and corrosion properties which result from the high cooling rates involved [1]. Laser cladding can be performed either by melting a powder preplaced on the substrate or by blowing the powder into the molten pool produced by the laser beam. The powder which impacts the liquid surface is

421

J. Mazumder et al. (eds.), Laser Processing: Surface Treatment and Film Deposition, 421–429.
© 1996 *Kluwer Academic Publishers.*

retained, whereas particles falling on the solid substrate bounce away and are lost [3]. The blown powder method is fundamentally more flexible and superior to the preplaced powder technique in that localized areas of components of complex geometry can be clad with better control of dilution and clad thickness [2]. Cladding of large areas is possible by overlapping single-clad tracks.

In order to achieve outstanding wear, oxidation and corrosion resistance, cladding alloys present complex and carefully balanced compositions and contain often expensive alloying elements. Hence, dilution of the cladding material by the substrate's material must be avoided if the desired properties are to be preserved. However some dilution (2% to 5%) is necessary in order to ensure a good fusion bonding between the substrate and the coating [4, 5]. Also, the contact angle α between the cladding track and the substrate (Fig. 1) should be obtuse in order to avoid interrun porosity when a surface is coated by overlapping successive tracks [6].

Weerasinghe and Steen [7] studied the influence of the process parameters on the cladding quality and concluded that dilution depends on the specific energy (power/(beam diameter x traverse speed)) and on the injected powder mass flow (q). The second relevant parameter (the angle α) depends on the dilution, the spot diameter, the scanning speed (v), the powder feed rate, the catchment efficiency and the powder's material density (ρ).The same authors [8] proposed a numerical model which can predict the clad track dimensions given the cladding parameters. In their model, a circular shape for the track's cross section is assumed based on experimental observations. Hoadley and Rappaz [9] determined the clad liquid surface shape using a force balance equation and calculated the track height on the basis of a mass balance.

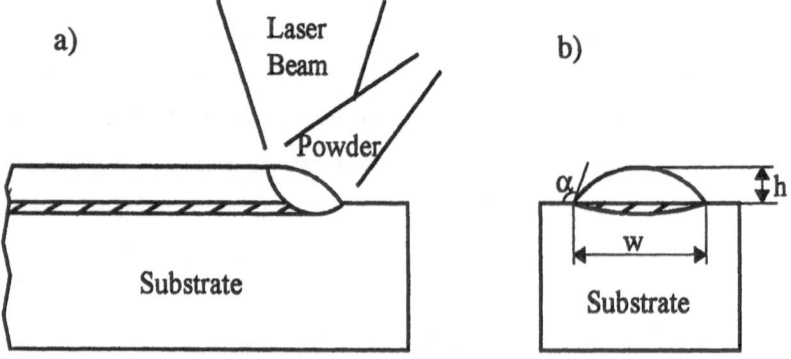

Figure 1. Laser cladding track. a) longitudinal section; b) cross section.

In the present paper, we describe a simple theoretical correlation between the cladding parameters and the dimensions and shape of the clad track that is valid for low dilution. This correlation can be used to systematize the choice of values for the operating parameters in order to ensure good quality tracks.

II. The Model

This model is based on the assumption that the catchment efficiency (η) does not depend on the processing parameters. Sometimes this hypothesis can be quite limiting, but the model is still applicable if we know the dependence of η on the processing parameters. We also neglect the dilution in the model calculations and the predictions may fail for experimental conditions for which dilution is high.

Considering that the cladding process is stationary, the area (A) of the cross section of the clad track is given by the following mass balance:

$$A = \frac{q}{\rho\,v}\eta \tag{1}$$

where q is the powder feed rate, ρ the density of the powder's material, v the scanning speed and η the catchment efficiency of the powder.

It was experimentally shown [7] that the variation of the track width (w) with the scanning speed can be expressed by an equation of the type:

$$w = a - b.v \tag{2}$$

where a and b are constants that depend on the laser power, the beam diameter, the powder mass flow rate and the powder particle velocity. However the clad track width is approximately equal to the width of the molten pool produced on the substrate when there is no powder injection, the other parameters being similar [8].

Since the process is stationary, one can neglect time and the scanning direction in the determination of the geometry of the cross section of the clad track. We will also assume that the track is formed in the liquid state and that solidification does not alter significantly its shape. So, let us consider that the profile of the transverse cross section of the liquid track (S) is given by:

$$S: y = f(x) \tag{3}$$

in the two dimensional reference frame illustrated on Fig. 2

The position of the contact line between the three different single-phase regions, which in our case corresponds to the points (w/2) and (-w/2), is previously known, since as we have seen before, we can use the width values obtained with no powder injection. The shape of the interface at x=±w/2 determines the contact angle between the solid and the liquid $(\tan(\alpha)=(df/dx)_{x=w/2})$. This shape is determined by the normal component of the force balance over the interface which is given by [10]:

$$\Delta p(x) + \Delta(\mu\overline{\overline{S}} \cdot \overline{n} \cdot \overline{n}) = \gamma \left[1 + \left(\frac{df}{dx} \right)^2 \right]^{-\frac{3}{2}} \frac{d^2f}{dx^2} \tag{4}$$

where ΔQ represents the jump at the interface of a property Q, p is the pressure, m the viscosity, $\mu\overline{\overline{S}}$ the viscous stress tensor and $\overline{n}$ the unit normal pointing out of the liquid. The right-hand side of equation (4) represents the surface tension (g) times the mean curvature of S. If we neglect the gravitational forces ($\Delta p(x)=\Delta p$) and the variation of γ with temperature (T) and concentration (C) - which lead to the sometimes called Marangoni effect - and if $\overline{\overline{S}} = 0$ (the static case), the balance given by equation (4) states that the interface S has a constant curvature radius. Davis [10] pointed out that this static balance is a good approximation even in dynamical conditions if γ is large enough, a condition which is generally observed for liquid metal/gas interfaces [11]. This allows to understand the observations made by several researchers [4, 8] that the profile of the cross section of a clad track is always an arc of a circle.

Therefore, we can write the expression for f(x) that describes the interface shape as a circumference of radius R, centered at the point (0,h-R) (see Fig. 2)

$$S: f(x) = \sqrt{R^2 - x^2} + (h - R) \tag{5}$$

The integration of eq. 5 between -w/2,w/2 gives the area of the cross section of the cladding track, which can be related with the operating parameters by equation (1). We have:

$$\int_{-w/2}^{w/2} f(x)dx = A \quad \Rightarrow \quad \frac{w}{2}\sqrt{R^2 - \frac{w^2}{4}} + R^2 \arcsin\left(\frac{w}{2R}\right) + w(h - R) = A \qquad (6)$$

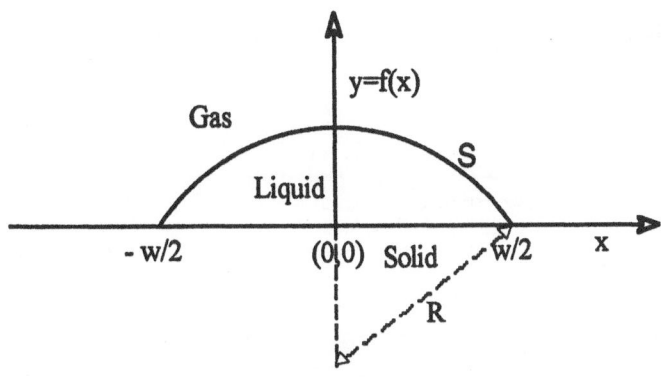

Figure 2. Liquid/gas interface.

By geometrical considerations we take:

$$h - R = -\sqrt{R^2 - \left(\frac{w}{2}\right)^2} \implies h = R - \sqrt{R^2 - \left(\frac{w}{2}\right)^2} \qquad (7)$$

Combining equations (6) and (7), we can calculate R by finding the root of the equation:

$$2R\sin\left[\frac{A + \frac{w}{2}\sqrt{R^2 - \left(\frac{w}{2}\right)^2}}{R^2}\right] - w = 0 \qquad (8)$$

Equations (1), (2), (7) and (8) fully describe the dependence of the shape and the dimensions of the track on the processing parameters.

To find the operating conditions that avoid interrun porosity when a surface is coated by overlapping successive tracks, we introduce the condition h≤w/2 in equation (8), which gives

$$A \leq \frac{\pi}{8}w^2 \qquad (9)$$

Combining equations (1) and (2) with (9), we find the condition

$$q < \frac{\pi \rho}{8\eta} v(a - bv)^2 \qquad (10)$$

which limits the operating region for avoidance of interrun porosity. This region is represented grafically in Fig. 3.

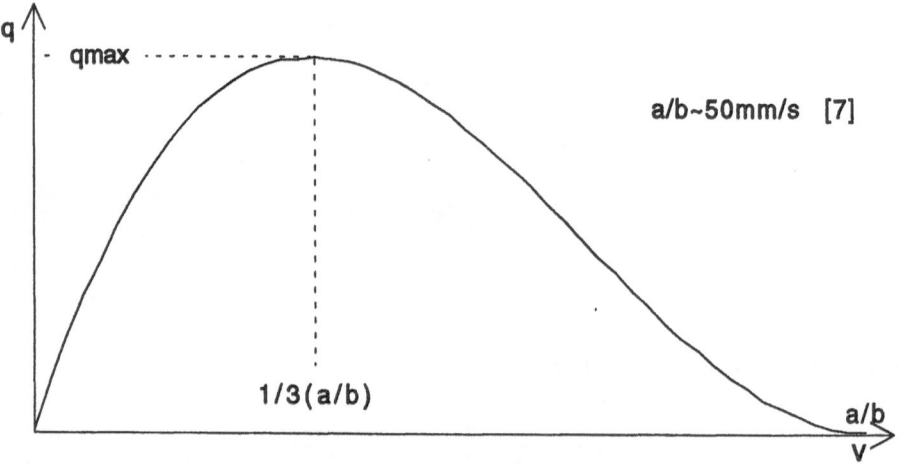

Figure 3. Operating region for interrun porosity free cladding.

III. Comparison with experimental results

To test the validity of the model, its predictions were compared with the results of laser cladding experiments performed using a 3 kW CO_2 fast axial flow laser with a TEM_{01}^* mode. The beam was focused by a water-cooled copper mirror. A steady flow of stellite 6 powder was blown into the melt-pool by a flux of argon. A second gas jet, coaxial with the laser beam, was used to protect the copper mirror from fumes and spattered metal. The substrate, a plain carbon steel, was placed under the stationary beam on a CNC-XY table.

Six scanning speeds (4, 6, 8, 10, 12 and 14 mm/s) and two powder feed rates (0.2 and 0.3 g/s) were found appropriate to produce the clad tracks on the basis of previous experiments. A laser power of 2000 W and a beam diameter of 2.5 mm were selected to ensure a good fusion bond without significant dilution for the all values of the cladding parameters used. Before cladding, the substrate was blasted with glass microspheres to remove surface oxides and to improve the coupling of energy to the material.

Preliminary experiments showed that:

$$w = 2.56 - 0.05*v \quad (r = 98\%) \tag{16}$$

for the variation of the width of the molten pool (w in mm) - no powder injction - as a function of the scanning speed (v in mm/s), and,

$$\begin{cases} \eta = 0.34 - 0.016*v & (q = 0.2 \text{ g} / \text{s}, \ r = 99\%) \\ \eta = 0.333 - 0.0165*v & (q = 0.3 \text{ g} / \text{s}, \ r = 99\%) \end{cases} \tag{17}$$

for the variation of the powder particles catchement efficiency (h), as a function of the scanning speed (v in mm/s).

The predicted values for the tracks width can be calculated by equation (16). To calculate the tracks height we will insert the results of equation (17) in equation (1) and use the w and A values to calculate R by equation (8). Finally h can be calculated by equation (7).

Fig. 4.a and b compare the values of w and h for each track (for a powder feed rate of 0.2 g/s) calculated on the bases of this method and measured experimentally.

IV. Conclusions

1. In this work it was presented a physical explanation for the circular shape of the cross sections of the laser clad tracks that has been reported by several researchers. Based on that it was possible to correlate width and height of the tracks with the operating parameters such as the powder feed rate, the scanning speed and the catchement efficiency of the powder. This relations allow to choose some parameter values in order to obtain clad tracks with the characteristics that are required for a given application. Once that is done the power of the laser beam that ensures an optimal fusion bond can be chosen if the thermal characteristics of the injected powder and substrate are known (see, for instance, [5]).

2. The results of this simple model are in good agreement with the experimental ones. However, one shortcoming should be kept in mind and it emerges from the fact that an accurate knowledge of the surface melt width and catchement efficiency as functions of the other cladding parameters is required.

a)

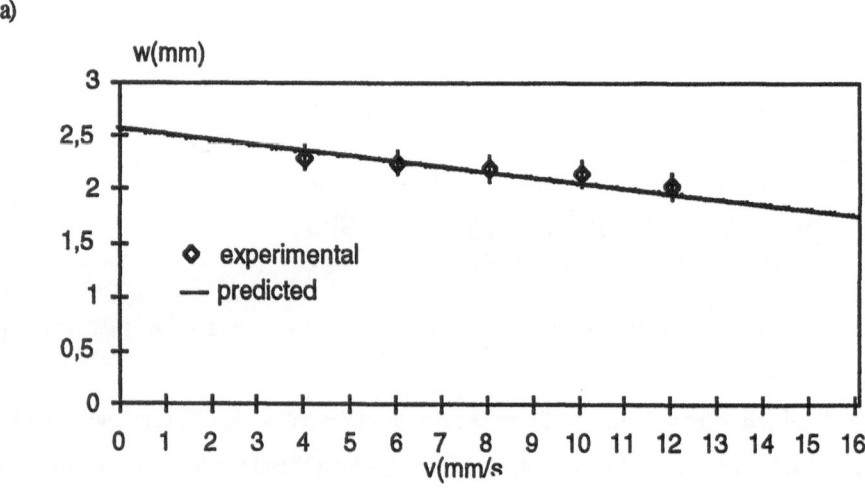

b)

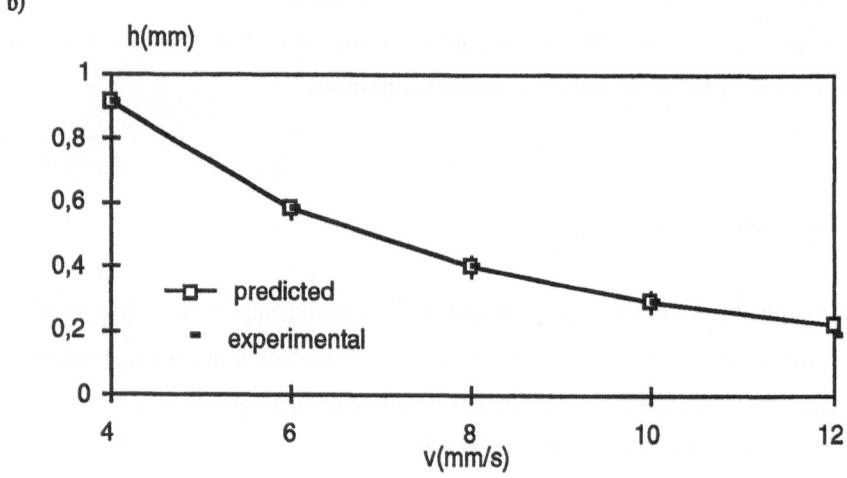

Figure 4. Predicted and experimental values for w (a) and h (b) at a powder feed rate of 0.2.

References

1. E. Lugsheider and B. C. Oberländer (1992) *A Comparison of the Properties of Coatings Produced by Laser Cladding and Conventional Methods*, Surface Modification Technologies V, Ed. T.S.Sudarshan and J.F.Braza, The Institute of Materials, pp. 383-400.

2. W. M. Steen (1986) *Laser Surface Cladding*, Laser Surface Treatment of Metals, Ed. C.D. Draper, P. Mazzoli,P.; eds., Martinus Nijhoff, Dordrecht, pp. 369-87.

3. A. F. A. Hoadley, C. F. Marsden and M. Rappaz (1991) *A Computer Study of the Laser Cladding Process*, Proceedings of the 5th Engineering Foundation Conference on 'Modeling of Casting, Welding and Advanced Solidification Processes', Davos, Switzerland, TMS, Warrendale - Penn., pp. 123-30.

4. R. Colaço, T. Carvalho, R. Vilar (1994) *Laser Cladding of Stellite 6 on Steel Substrates*, High Temp. Chemical Processes 3, pp. 21-29.

5. E. Lugscheider, B. C. Oberländer and S. E. Leising (1992) *Influence of Laser Cladding Parameters on the Microstructure and Properties of Claddings*, Surface Modification Technologies V, Ed. T.S.Sudarshan and J.F.Braza, The Institute of Materials, pp. 881-88.

6. W. M. Steen and J.Powell (1981) *Theoretical Modeling of the Laser Cladding Process* , Inter. J. of Materials Eng. **2**, pp. 157-62.

7. V. M. Weerasinghe and W. M. Steen (1983) *Laser Cladding With Pneumatic Powder Delivery*, 2nd Conference on 'Applications of Lasers in Materials Processing', ASM, pp. 166-74.

8. V. M. Weerasinghe and W. M. Steen (1983) *Computer Simulation Model for Laser Cladding"*, Transport Phenomena in Materials Processing, ASME, New York, NY, pp. 15-23.

9. A. F. A.Hoadley and M. Rappaz (1992) *A Thermal Model of Laser Cladding by Powder Injection* , Metall. Trans. B **23**, pp. 631-642.

10. S. H. Davis (1983) *Contact-Line Problems in Fluid Mechanics*, J. App. Mech **50**, pp. 977-82.

11. B. J. Keene (1993) *Review of Data for the Surface Tension of Pure Metals*, Inter. Materials Review **38**, No.4, pp. 157-92.

Surface treatment by laser processing and tribological application

A.Bernard VANNES *, J.M.PELLETIER **
*MMP-CALFETMAT-URA 447
Ecole Centrale de LYON
69193 ECULLY Cedex (FRANCE)
**GEMPPM-CALFETMAT-URA 341
INSA de LYON
69621 VILLEURBANNE Cedex (FRANCE)

I. Introduction

On a technical point of view, coating is made to reach some objectives termed "working properties". More precisely, it is possible to deal with durability, resistance of damage, improvement of the behaviour of fatigue failure, friction, wear and corrosion. In this case, the antagonists and the operating conditions must be defined.

The process of coating induces a change in the properties on the surface of the material. The thickness of the affected layer depends on the work conditions of the piece.

Among all the processes that fit, the laser technology looks like a plausible vector of energy.

Firstly, we will describe what we look for and how we want to proceed (thanks to accurate examples, we can make sure that the objectives will be reached). Then, we will study different processes of coating, realized under a laser beam.

II. Characterization of the treatments

There are many kinds of properties induced by coating.

II.1 Mechanical properties

Through a couple of examples, we will define the principle of the approach :

   a. *friction, and more precisely fretting (1)(2)(3)*

This concerns all the detachable structures subject to vibrations : this appeals to notions of "dry friction". This concept is often tackled with a simulator that reproduces the operating conditions.

In fretting conditions, two samples slide against each other. Generally, one is flat, the other is cylindrical or spherical. This permits us to know the contact area with a good precision. Normal load N acts on the contact. Value of the displacement (defined by +/- x μm) is carefully controlled. The tangential force F is continuously measured during the test.

In these conditions, in a plane representation (tangential force F, imposed displacement D), a cycle is obtained, which describes the mechanical behaviour of the tribological system, for a given normal load.

J. Mazumder et al. (eds.), Laser Processing: Surface Treatment and Film Deposition, 431–452.
© 1996 Kluwer Academic Publishers.

Three kinds of cycles have been observed (figure 1):
- a closed cycle, indicating an elastic response of the system, without actual displacement between the surfaces,
- an open cycle corresponding to sliding between surfaces,
- an intermediate cycle, elliptic, where slip is partial.

For a given test, the steady state regime is always obtained after a run-in period of 10 to 100 cycles. The 3D representation (F, D, N)(figure 2), where n is the cycle number, is called the friction log. This representation constitutes the mechanical response of the system and allows to determine the evolution over the time.

When the test is repeated for various normal loads, a synthetic representation can be obtained, which is called the running conditions friction map (RCFM)(figure 3), usually defined for a limited number of cycles (<5000). RCFM evidences three areas :
- sticking,
- partial slip,
- gross slip.

This map is establish for a given couple of materials. In our case, the antagonist part is a cylindrical sample (AISI 52100 steel), treated to about 850 Hv. It will be considered as the reference material for the present study.

In the two first areas of the map, the main material response is cracking, while in the third one, wear fragments are generally observed, leading to a wear phenomenon.

In order to facilitate interpretation, a 2D representation (figure 4) is performed : F_{max}/N is plotted as a function of the effective width of the cycle. Three different zones can be defined :
- a linear increase, for $d<d_{V1}$, corresponding to sticking,
- a non-linear increase $(d_{V1}<d<d_{V2})$; in some cases, a maximum is finally achieved, which is related to partial slip.
- a plateau $(d>d_{V2})$; gross slip is obtained and then the friction coefficient μ can be defined, according to Coulomb, by :

$$\mu = F/N$$

with   F : tangential load
           N : normal load

This simple curve make it possible to get the threshold values. By performing at least three times this experience, the fretting map can be plotted. Let us mention that it has been shown, elsewhere, that the boundaries between the different zones are straight lines. From these intermediate curves, two parameters can be deduced :
- size of the zone (Δ) in which the material will be submitted to a fatigue phenomenon,
- friction coefficient μ.

b. *abrasive wear* (4)(5)

Abrasive wear and derived forms, voluntary or otherwise, affect many areas (aeronautics, hydrolic and pneumatic industries, etc...). The impact of particles on a surface can cause severe erosion damage. The erosion of a surface by abrasive particles in an inert fluid depends on the number of particles sticking the surface, their velocity and their direction relative to the surface. A test bench (figure 5) has been built with some possibility of behaviour.

The samples are subject to controlled erosion (figure 6), which can be described by the depth of the central crater, the worn volume (rectified from the secondary effects of the interaction) or an equivalent area.

II.2 Physico-chemical corrosion

In this case, we will take into account some determinations of electro-chemical properties which refer to natural potentials of corrosion after stabilization, thanks to normalized systems or current densities of corrosion obtained by voltamperometric curves.

II.3 General data

The data of chapter II.1/II.2 correspond to particular properties which are no intrinsic properties of materials. Actually, they strongly depend on the environment and the using conditions. To know well the properties of such a treatment, a group of intrinsic properties must be associated. J.F. CARTON (6) proposes a methodology, a proceeding and a presentation of the main properties.

Various authors (6)(7)(8) have noted that two different kinds of coatings may be considered from a mechanical point of view i.e. "thick" ones and "thin" ones. A coating will be said to be thick if the thickness of the layer is great compared with the corresponding contact size (noted 2a and defined in the conditions of the chapter II.1). Mid-width "a" is given by Hertz's theory as

$$a^2 = (\frac{4PR}{\pi E^*})$$

With the previous values, "a'" is about 115 μm. Consequently, beyond a thickness of 230 μm, the coating will be said to be "thick". This means that the contact induced stress field is almost entirely located in the coating. The use of Hertz's theory remains justified. The base material has little influence.

If the coating thickness is smaller or of the order of magnitude of contact width 2a, the coating will be considered to be thin. Hertz's theory can no longer be applied. A calculation may be made if the assumtion is made that there is strain continuity at the substrate-coating interface and also the rigidity of the cylinder is infinite. Mid-width contact then is given by [9] as :

$$a = \left(3P\,Re \times \frac{1-2v_R}{(1-v_R)^2} \times \frac{1-v_R^2}{E_R} \right)^{\frac{1}{3}}$$

with E_R, n_R are respectively the elastic modulus and the Poisson's coefficient of the coating.

If v_R value is also assumed to be about 0,3 and to the previous values (corresponding to T=1) are used for the other parameters, "a" is given by a simplified formula as :

$$a = 14\left(\frac{e}{E_R} \right)^{\frac{1}{3}}$$

It can be noted that the elastic modulus and layer thickness condition the "a" value. Figure.7 presents a qualitative approach. Strain continuity at the interface is assumed again.

For a given strain ε_i at the substrate-coating interface :

- when the coating is thin, it has little influence on the stress field and only transmits it. Strains are imposed by the substrate. To minimise stresses in the layer, the elastic modulus of the layer must be smaller than that of the substrate.

- when the coating is thick, the substrate has little influence. The strains at the interface are imposed by the coating. The coating deformation is low if the coating modulus is higher than the modulus of the substrate and he coating deformation is high if the coating modulus is lower.Consequently, the layer modulus must be higher in order to make stresses lower.

The previous remarks underline the important role played by coating-substrate binding forces. The transmission of efforts will be greatly influenced by those forces. Defects or degradation at the interface will have a strong influence on the mechanical behaviour of a coating. Different kinds of degradation exist : wear, cracking in the volume, etc...

The tests described in the chapter II.1 allow to illustrate them.

There are numerous parameters to quantify wear (volume loss, weight loss, wear rate...). In our case, we chose a maximum wear depth, noted Z_U and defined for a given number of cycles.

To characterise cracking, two different parameters are necessary :
- the first one is connected with the crack initiation phasis. This parameter, noted n_A, is a minimum crack initiation time and it is given in number of cycles.
- the second one concerns the propagation. It is a maximum crack depth which is defined for a given number of cycles. This value is measured by observation on metallographic cross sections of samples.

To quantify cracking at the interface, it is necessary to introduce another parameter which is defined as a maximum crack length at the interface, for a given number of cycles. This parameter is also characteristic of cracking. It can therefore be joined to the parameter defined for volume cracking. Then there is only one parameter defined as a maximum crack length either at the interface or in the volume (noted Z_F).

All these parameters are defined for a given number of cycles. As coating life in contact is limited, coating degradation may induce a change in running conditions.

Consequently, the parameter value must be considered before and after coating degradation because of the change in running conditions. Another parameter is then defined which corresponds to the minimum time for unchanging running condition (noted n_S in number of cycles). Previous parameters will be defined for the number of cycles associated with n_S. When the running condition does not change this parameter is equal to the maximum number of cycles for the tests (i.e. one million cycles).

The parameters we have just seen can be considered as *output parameters*.

To reach our goal, another step is necessary i.e. the identification of the parameters conditioning fretting behaviour. The values of these parameters, which can be considered as the tribo-system *input parameters*, condition the ones of the output parameters. To achieve such an identification a damage criterion is associated with each type of degradation.

Before fretting wear takes place, there is a formation of a tribologically transformed structure. This structure is induced by shear plastic strains accumulated in materials. The Tresca criterion is useful to predict plastification in this case.

In a partial slip regime, crack initiation is generally related to the variations of the lateral stress σ_X. The maximum value of this stress is $2\mu P_0$. Using a fatigue criterion enables the taking into account of cyclic loading. The Dang Van criterion (10) is used here. For a given material, a two-dimensionnal boundary between high cracking probability and low cracking probability is defined. This criterion is written :

$$\left| \tau' - a + bp \right| \leq 0$$

with τ' shear stress amplitude
       a   fatigue limit in alternative torsion

       p   hydrostatic pressure : $p = \dfrac{1}{3}(\sigma_1 + \sigma_2 + \sigma_3)$

           where s_1, s_2, s_3 are principal stresses for x,y,z axes.

       $b = \dfrac{3}{f}(a - \dfrac{f}{\sqrt{3}})$ with f as the fatigue limit in rotative bending.

A coating binding characterisation must be made. Interfacial cracks are generally related to shear stresses. A damage criterion is then possible to determine and is similar to wear criterion :

$$\tau \leq \sigma_{adh}$$

σ_{adh} can be defined as an interfacial fracture stress.

A damage boundary is associated with each criterion. In plasticity, damage boundary is K, shear yield stress. K is proportional to the σ_Y traction yield stress which is more commonly used. The yield stress is difficult to measure in the case of coatings. If this value is not available, the coating hardness may be used because its influence is similar. At the worst, the hardness should be weighted with a factor of influence, which is sometimes difficult to know.

The values of a and b in fatigue criterion are influenced by various parameters. The variations in local yield stress (or hardness) induce changes in the "a" value, as Deperrois (11) explained it. The rotative bending fatigue limit (noted σ_D) for a material either coated or not is useful in order to quantify the changes on the damage boundary as defined by a and b.

Residual stresses have an influence on the p value. The base material is considered without residual stresses. In this case, these stresses can be introduced in the Dang Van criterion, using Deperrois' work. This gives :

$$p = \frac{1}{3}\left(\sigma_1 + \sigma_2 + \sigma_3\right) + \frac{1}{3}\left(\sigma_{r1} + \sigma_{r2} + \sigma_{r3}\right)$$

where σ_{r1}, σ_{r2}, σ_{r3} are principal residual stresses for the x,y,z axes. They may be considered as equivalent to an hydrostatic pressure noted p_r. Each component of the residual stress term may also be written introducing superficial yield strength as these stresses can not be higher than yield strength. Using ultimate yield strength, it gives :

$$p_r = \frac{1}{3}\left(\sigma_{r1} + \sigma_{r2} + \sigma_{r3}\right) = \frac{1}{3}\sigma_y\left(\alpha + \beta + \gamma\right)$$

with $\sigma_{r1} = \alpha.\sigma_Y$; $\sigma_{r2} = \beta.\sigma_Y$; $\sigma_{r3} = \gamma.\sigma_Y$ and α, β, γ are algebric values. They are negative for compressive stresses and positive for tractive ones.

Another parameter related to cracking phenomena is Δ, the width of the stick domain plus the partial slip domain on the RCFM. This parameter is not included in fatigue criterion but is of great interest in fretting. This makes possible the evaluation of the relative importance of cracking phenomena compared to wear, as Δ is defined for the domains where fatigue phenomena prevail This parameter is defined for a particular load, for instance the one leading to a plastification (in static state) or the one corresponding to T=1 i.e. the maximum load associated with elastic behaviour of steel.

In our case, Δ was measured for a normal effort of 900 N. This allows to determine the importance of cracking phenomena compared with wear.

Binding is also difficult to define and to quantify. The binding boundary stress (noted σ_{adh}) defined as a rupture shear strength was introduced in criterion as a damage boundary. The difficulty is to achieve the measurement of that parameter. There are numerous tests but the results are often dispersive and the comparisons are difficult.

Polar representation (6)

A synthesis becomes necessary because the parameters previously defined are numerous. It can be achieved by using a polar representation. Some conventions must be defined in order to make this clear :

- Each parameter is written as a ratio of the coating value and the base material value. The ratio is defined so that an improvement in fretting behaviour will be associated with a ratio value lower than 1.

- A reference circle of radius 1 will be drawn so that the plan will be divided into two parts; an inner disk will be associated with an improvement and an outer part associated with a worsening of the reference parameter.

The diagramm (figure 8) is divided into four sectors :

- sector 1 concerns the intrinsic properties of the treated layed : elastic modulus, type of coating (thin or thick), yield strength, fatigue strength.
Depending on whether the coating is thin or thick, the elastic modulus ratio will be defined in a different way, as justified in § II.3.

$$E^{*}_{thick} = \frac{E_{ref}}{E} \text{ or } E^{*}_{thin} = \frac{E}{E_{ref}}$$

Using the previous analysis, the ratios associated with yield strength and fatigue strength are defined as :

$$\sigma^{*}_{Y} = \frac{(\sigma_{Y})_{ref}}{\sigma_{Y}}; \sigma^{*}_{D} = \frac{(\sigma_{D})_{ref}}{\sigma_{D}}$$

Usually the yield strength of layers is unknown. This ratio may be estimated by the ratio of hardness values. Fatigue strength values are provided using rotative bending fatigue tests.
- sector 2 concerns the coating-substrate interaction; residual stresses and binding strength.

σ_r^{*} is defined using p_r formula results. For this parameter, the reference value can no longer be the value associated with steel which is supposed to be 0. The reference value must be the yield strength of the coated surface. Thus a σ_r ratio can be defined as :

$$\sigma_r^* = \frac{\frac{1}{3}\sigma_Y(\alpha+\beta+\gamma)}{\sigma_Y} = \frac{1}{3}(\alpha+\beta+\gamma)$$

This parameter varies from -1 to +1. If the residual stresses are compressive, this ratio varies from -1 to 0. If they are tractive, the ratio varies from 0 to +1. The reference value is 0. In order to have a reference value equal to 1, a corrected ratio σ_{rc}^* is defined as :

$$\sigma_{rc}^* = \sigma_r + 1 = \frac{1}{3}(\alpha+\beta+\gamma)+1$$

The binding stress was previously discussed, the associated ratio is defined as :

$$\sigma_{adh}^* = \frac{(\sigma_{adh})_{ref}}{\sigma_{adh}}$$

- sector 3 concerns the running conditions : load rate, friction coefficient and the Δ parameter.

The load rate is defined as : $T^* = \frac{0,6P_0}{\sigma_Y}$.

σ_Y is the coating yield strength and P_0 the maximum contact pressure. The calculation of this pressure depends on the layer thickness. If the yield strength value is not available, it is estimated by using the hardness value.

The ratios concerning μ and Δ are respectively :

$$\mu^* = \frac{\mu}{(\mu)_{ref}} \; ; \Delta^* = \frac{\Delta}{(\Delta)_{ref}}$$

- sector 4 gathers the tribo-system input parameters associated with the response of the material : running condition steady state, crack initiation time, crack length, worn depth. The corresponding ratios are respectively :

$$n_S^* = \frac{(n_S)_{ref}}{n_S} \; ; n_A^* = n_S^* \frac{n_S - n_A}{(n_S)_{ref} - (n_A)_{ref}} \; ;$$

$$Z_F^* = n_S^* \frac{Z_F}{(Z_F)_{ref}} \; ; Z_U^* = n_S^* \frac{Z_U}{(Z_U)_{ref}}$$

The values of the degradation ratios are corrected by n_S^* ratio in order to possible a comparison between degradations which have different time durations.

This will be applied for various treatments.

III. Laser treatments

III.1. Proceedings and source

The following systems were used for laser cladding, surface alloying or melting :

• *CW CO₂ laser*

- a CW CO_2 laser, supplied by CILAS (CI 4000), operating here between 2.4 kW and 3.2 kW;
- a 10" ZnSe focusing lens; the focal point is located in the sample and the laser beam diameter on the specimen surface is about 4 mm. Parallel tracks are achieved with a shift δ=1.5 mm between two adjacent treatments.
- a powder injection system (Plasma Technik, Twin 10C); powders have been injected, through a coaxial nozzle, with argon gas acting as support.

Specimens are mounted on a numerically controlled X-Y table. Scanning speed under the laser beam is in the range 10-20 mm/s.

• *CW Nd -YAG laser* (12)

- a CW YAG-Nd laser supplied by NEC, operating between 800 and 1100 W .
- The other parameters remain the same as the CO_2 laser ones, and the re-covering rate is 30%.

• *pulse Nd-YAG laser*

In this case, the approach is quiet different. A source with "Q-switch" is used; it works between 1 and 99kHz, and the pulse durations are about 100 ns. The delivered average power is about 30 W.

This process is made to show that it is possible to realize steady coatings not deeper than a few micrometers.

III.2. Treated materials

III.2.1 Metal Matrix Composites (MMC) for fretting approach (13)

Different materials are used as substrate (steels, brass, bronze) and as powder :

	C	Fe	Cr	Ni	Mo	Mn	Si	Cu	Zn	Pb	Sn	WC/W₂C	S	P	B
Cast Iron	3,2		0,27		trace	0,76	2,4						<0,1	<0,6	
Steel A	0,38	97,1	1,15		0,3	0,7	0,3						0,035	0,035	
Steel B	0,18	98,7	0,2	0,1		0,7							0,035	0,035	
brass								61	37	2					
bronze								92			8				
powder A												100%			
powder B	0,7	4,2	14	73,9			4,2								3

Panel 1 : chemical analysis of used alloys

The particles of the powder A are embedded in a 5 µm-depth nickel layer.

Typical granulometry of the powders is 40-100 µm. Powder B contains tungsten carbides (HV=2500), spherical in shape. Each particle is embedded in an electroless nickel layer (thickness = 5 µm). Except in the case of a coating with powder A on a steel, a mixture of both powders A+B (50%-50%) is always injected into the laser beam.

After a selection of suitable processing conditions, good MMC coatings are achieved, i.e.: sound layers (without pores or cracks), homogeneous layers (uniform distribution of carbide particles), good bonding between MMC coating and substrate, without deleterious dilution for the addition element.

Let us mention that experimental parameters depends only slightly on the composition of the substrate : suitable parameters for cladding on a steel are also suitable for manufacturing a MMC coating on a bronze or on a brass. Typical values are as follows :

- laser power : $2400 < P < 3000$ W
- scanning speed : $10 < v < 20$ mm/s
- powder feedind rate : $8 < F < 16$ g/mn
- diameter of the laser beam on the substrate : $\Phi = 4$ mm.

An example of MMC coating on a cast iron is given in figures 9 and 10. The heterogeneous microstructure of the substrate is in favour of a good bonding between the two materials. A partial melting occured in the interfacial zone. X-ray diffraction experiments confirm this result: patterns of the cladding and of the initial powder are very similar and MMC on different substrates exhibit identical patterns.

Compared to more conventional manufactoring routes of MMC, laser cladding has an attractive advantage : interaction time between reinforced particles and liquid matrix is very limited (typically = 100 µm). Therefore, the interfacial reaction between carbides and metallic pool can have only a very slight deleterious effect. Chemical protection of the hard particles is not requiered.

In addition, thermal spraying has an advantage over more classical solidification techniques : its dynamic feature. Indeed, particles are introduced into the melt pool, in which large convection movements take place : these movements cause efficient mixing and, consequently, induce a uniform distribution of undissolved reinforced carbides.

III.2.2. Metal matrix composite for erosion test (4)

The studied substrate is the the steel A of the panel 1 and the mixed powders are the same as before (A and B).

- A first serie of samples was realized with the CW CO_2 laser described in the previous chapter. Two kinds of pieces were manufactured :

    • a set with 20 % in weight of carbides (i.e. 10 % in volume),

    • a set with 40 % in weight of carbides (i.e. 20 % in volume).

Each of these sets have a coating's thickness between 1000 to 1200 µm.

- A second serie identical with the set with 40 % in weight was realized thanks to the CW Nd-YAG laser (NEC), which delivers a power of 900 W. The caracteristics of the coating are identical with the ones of the set obtained under the CO_2 beam. It can be noticed that the power of the Nd-Yag is obviously lower than the CO_2 one.

III.2.3. Surface reforming (14)

In this part, the Cast Iron (Panel 1) was chosen. The Nd-YAG laser in "Q-switch" mode was used in order to realize two sets :

- First set : the refusion on an even 5-μm thick layer is obtained by scanning a beam which delivers an energy of 0,1 J/mm2 during 0,3 μs, and whose frequency is 10 kHz. The speed scanning is 40 mm/s, which makes a repetition rate of about 15 by melted volume unit. (figure 11). On a top of view, made by scanning electron microscopy, we can observe the baring of graphite lamellaes (figure 12), while on a cross section the absence of a had affected zone is clearly evidenced (figure 13).

This means that the life of the melt pool is about 20 μs. The mean value of the hardness of the zone is 700 $Hv_{(0,25)}$. The area is mainly composed of austenite with a high rate of carbon and of trace of martensite (determination with the method of the grazing X-ray).

- Second set : the refusion is obtained on a 35-μm thick layer.(figure 14) In this case, the mean value of the hardness is 760/800 $Hv_{(0,25)}$. As the micrography of a transversal view shows it, the layer is irregular, the craters correspond to the emergence of graphite's lamellae. In comparison with the previous case, the only parameter that changes is the scanning speed (10 mm/s). This means that each element of the volume undergoes 70 shocks. The duration of life of the melt pool can be estimated at 150 μs.

It can be noticed that observations with an electron microscope did not allow to display the heat affected zone (HAZ) which ensures the transition between the melt zone and the substrate.

III.3. Analysis of the results

These polar diagrams (figures 15 and 16) summarizes the main results.

The calculation of the modulus and the yield strength (yield point) was made with an original method based on the principle of Hertz' hardness.(15) This method gives local measurements which are representative of the properties of the matrix, in the case of the composites (MMC).

Thanks to these values and the complementary tribological data, it is possible to draw the polar representation. To make it, a reference must be chosen. To take the substrate as a reference allows to underline the advantages of the treatment, for the representations concerning the sector 1 (intrinsic properties). This can be interesting. In return, concerning the tribological problems (fretting), the contributions of the treatments have a sens only if they are compared with the fixed objectives. That is why the chosen reference (especially for sector 3 and 4) is a mid-carbon steel qualified in a thesis' work .

In these conditions, the figures present the associated diagrams; they underline well the contribution of the treatments. It can be noticed that the treatments are associated with thick coatings, whereas the melting treatments on cast iron correspond to thin coating. In this case let us remarck the evolution of fretting log for given processing conditions. Figures 17 and 18) illustrate the plots for melting 1 (thin layer = 5 μm) and 2 (thin layer = 35 μm), respectively; they can be compared to figure 2, which corresponds to an untreated cast iron, in the same sollicitation conditions. Friction coefficient is low and its stability cn be noticed

The figure 19 presents the erosion behaviour by particles in the coatings. The improvement is significant, there is no difference between a CO_2 and a Nd-YAG coating. The achieved response of the structures has something in common with the behaviour of fragile materials.

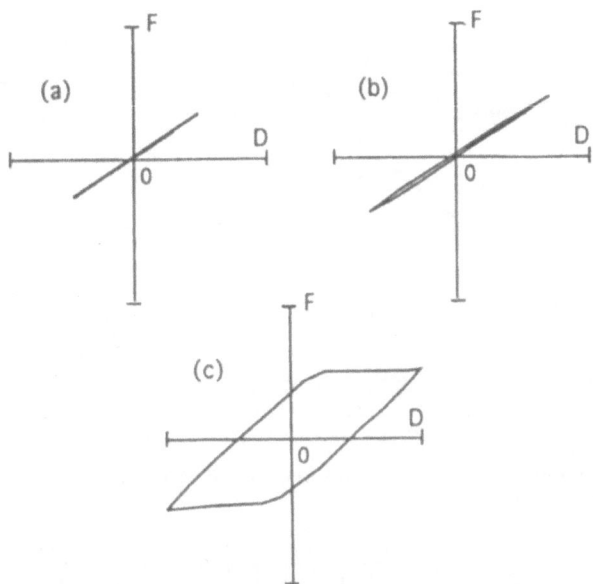

figure 1 : Variation of the tangentiel force as a fonction of the displacement : (a) stick · regime, (b) partial slip regime, (c) gross slip regime.

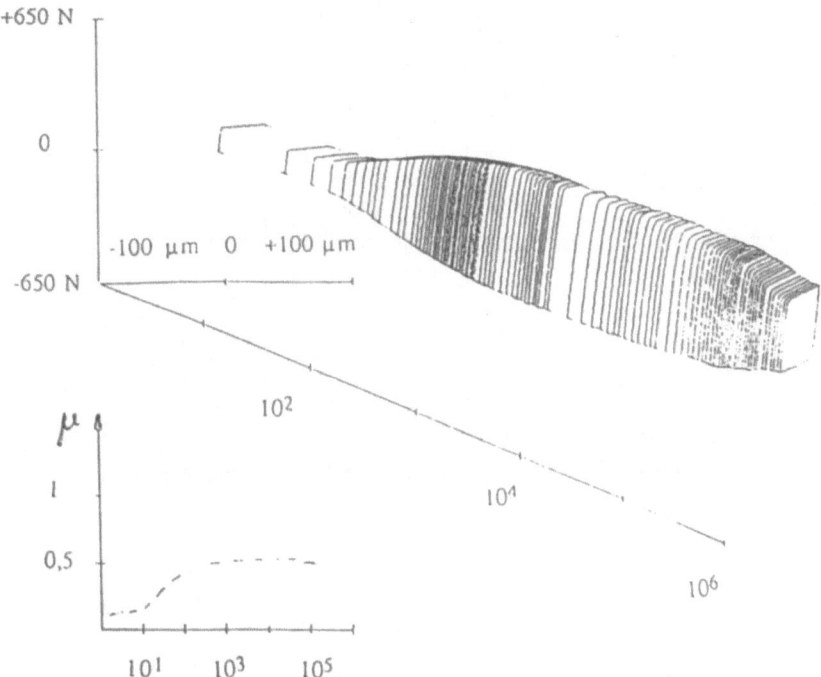

figure 2 : Friction log (untreated cast iron against AISI 52100, N= 650 N, D = ± 18 µm, frequency = 5 Hz, n = 250 000 cycles)

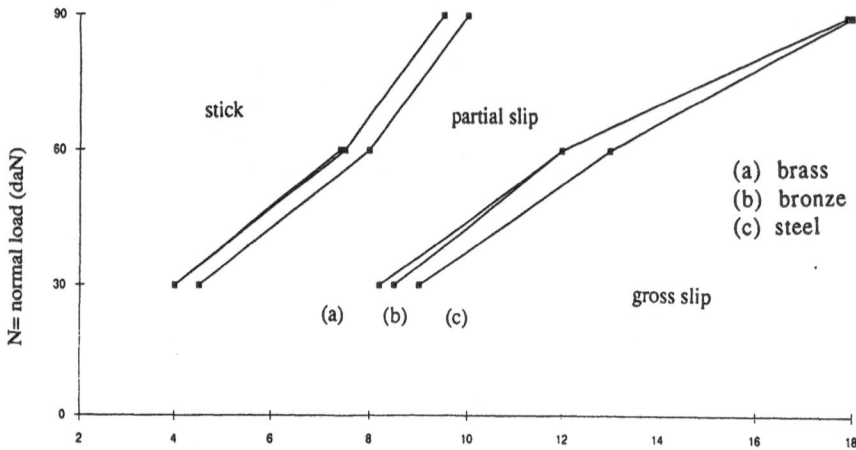

figure 3 : Running conditions friction map (MMC coating on brass, bronze or steel against AISI 52100)

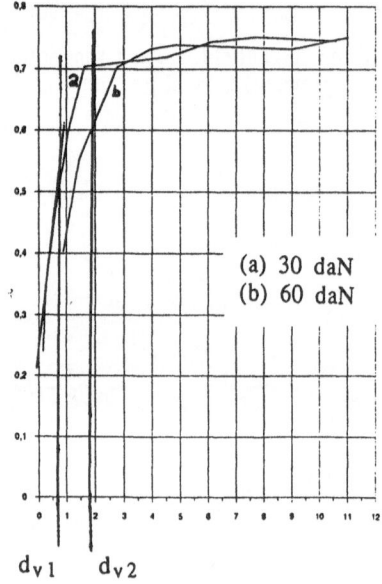

figure 4 : Evolution of the ratio Fmax/N versus effective displacement (MMC coating on brass)

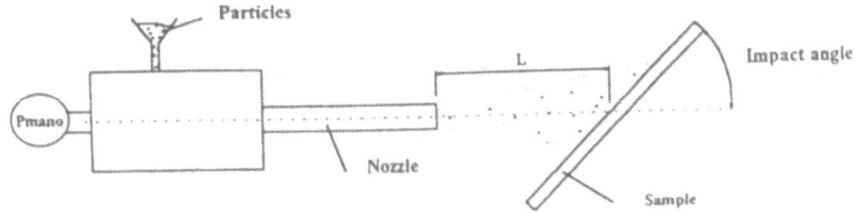

figure 5 : Diagram of principle for erosion test

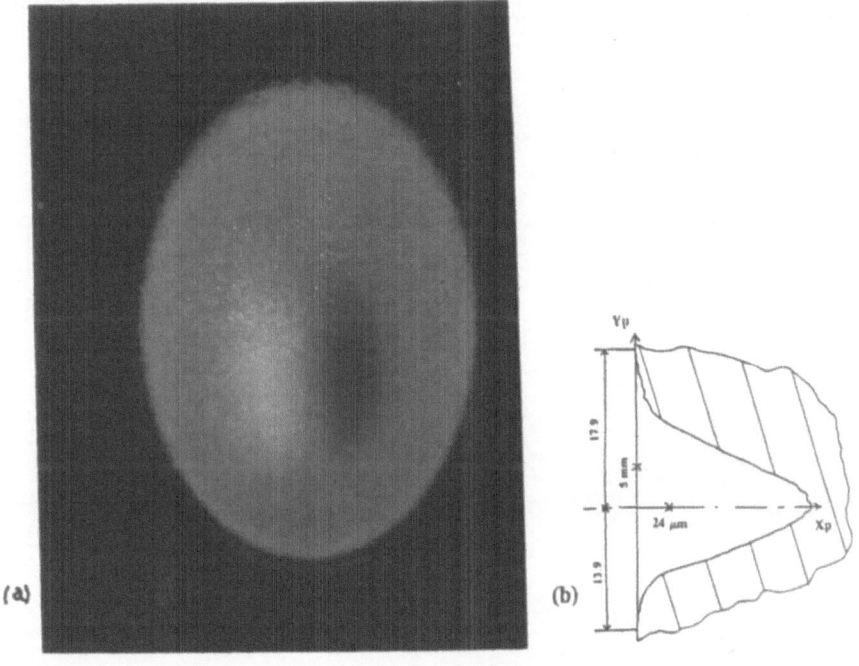

figure 6 : Impact of a sand jet : (a) top of view, (b) profil of crater

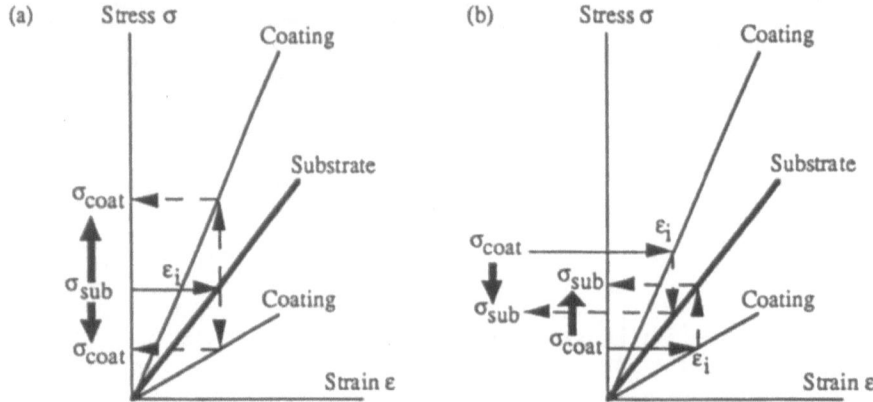

figure 7 : Role of the elastic modulus on the stress transmission : (a) thin coating, (b) thick coating

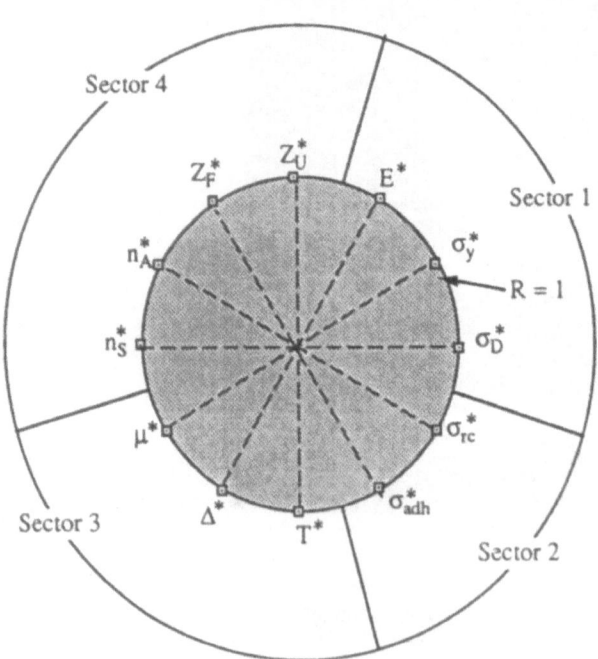

figure 8 : Principle of polar diagram

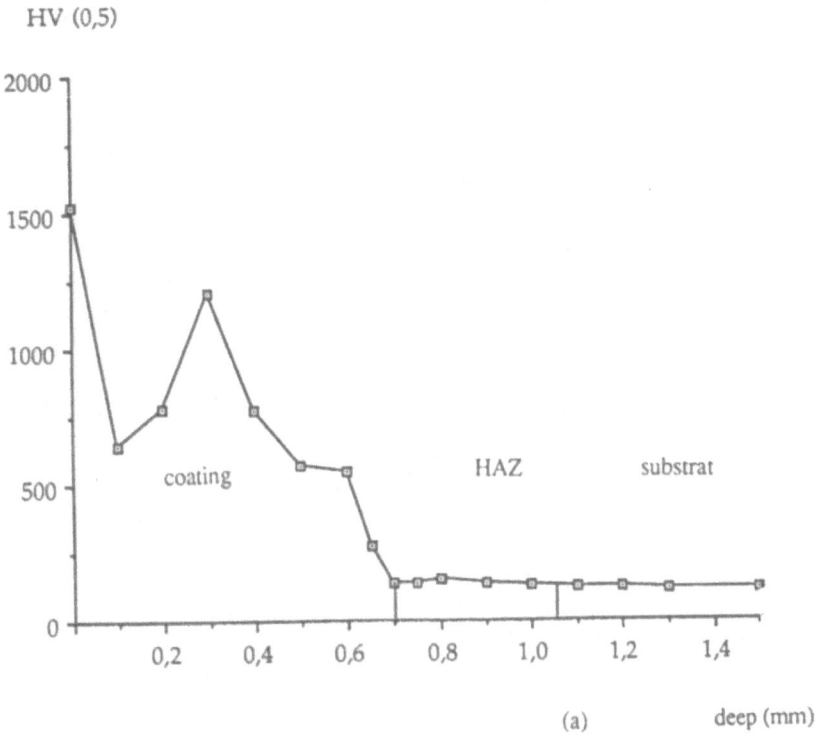

(a)

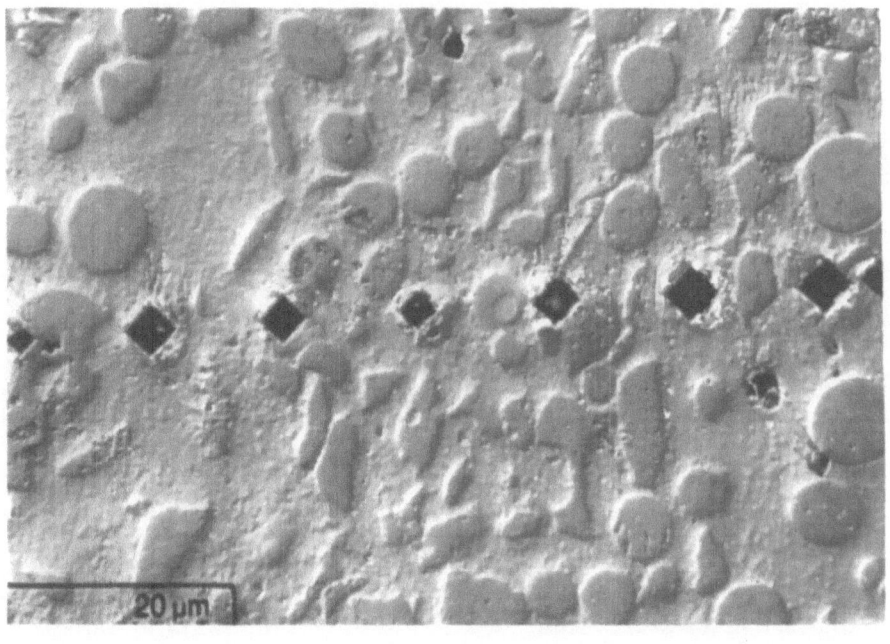

(b)

figure 9 : MMC coating on brass : (a) microstructure, (b) profil of hardness

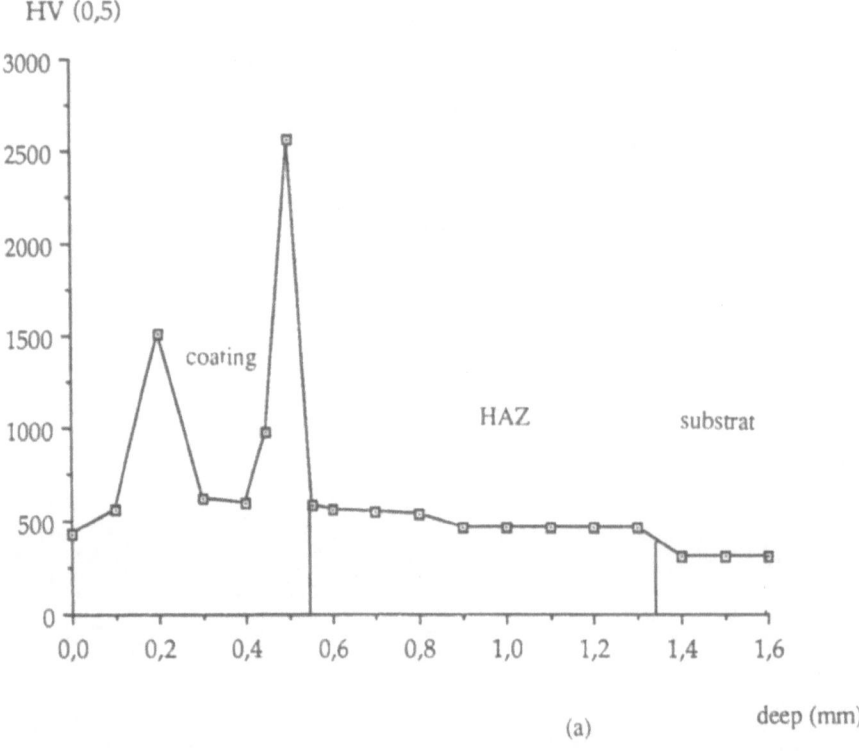

(a)

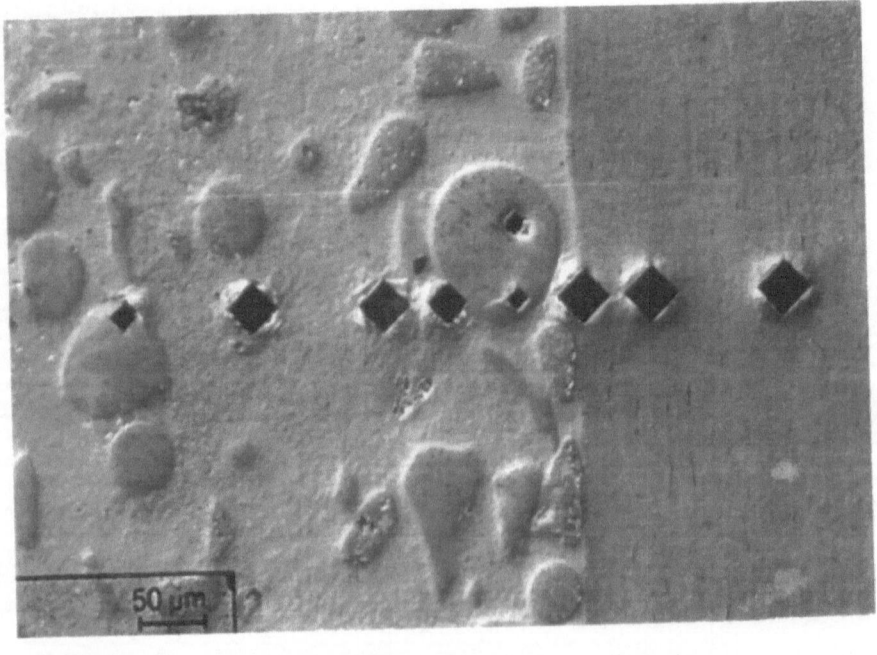

(b)

figure 10 : MMC coating on steel : (a) microstructure, (b) profil of hardness

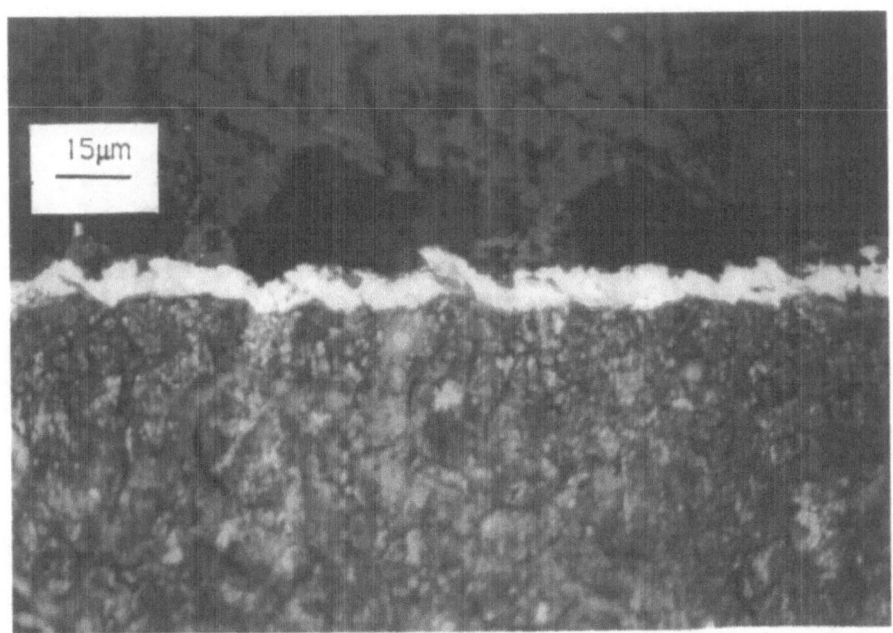

figure 11 : Cross section of melting cast iron treated by Nd-YAG laser
(speed scanning = 40 mm/s)

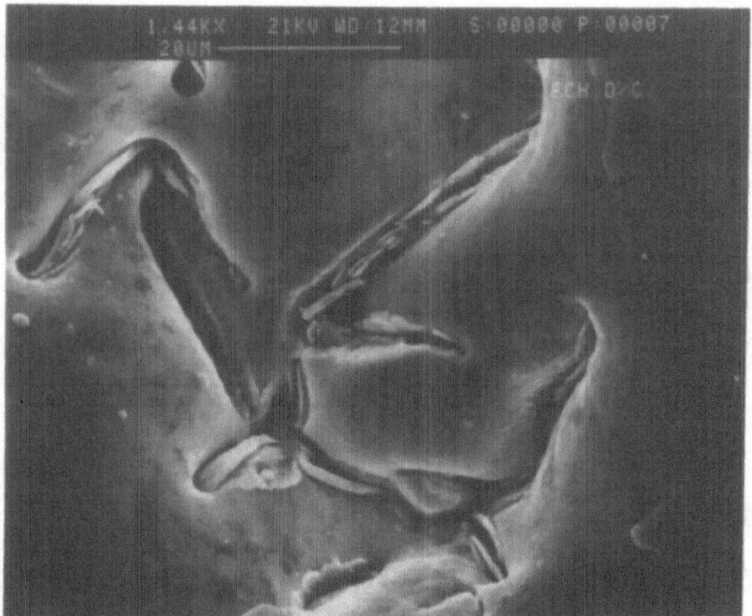

figure 12 : Top of view by SEM of melting cast iron treated by Nd-YAG laser
(speed scanning = 40 mm/s)

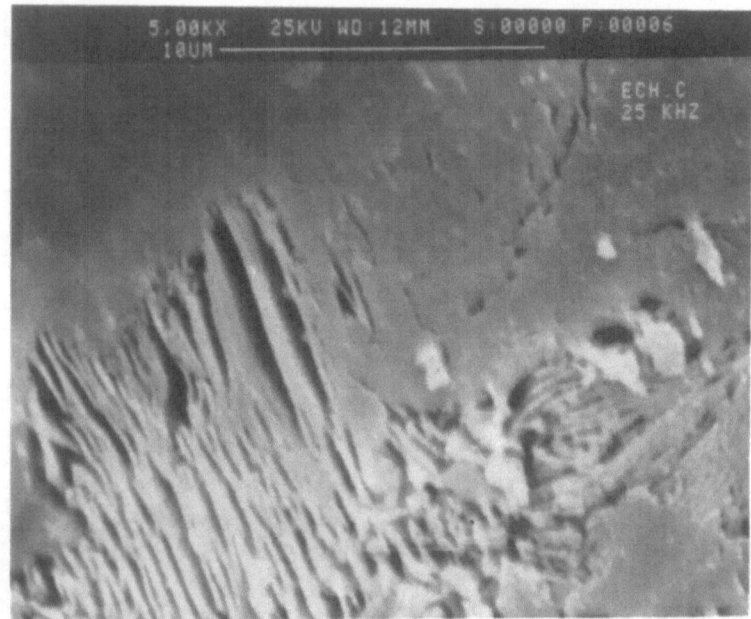

figure 13 : Cross section by SEM of melting cast iron treated by Nd-YAG laser
(speed scanning = 40 mm/s)

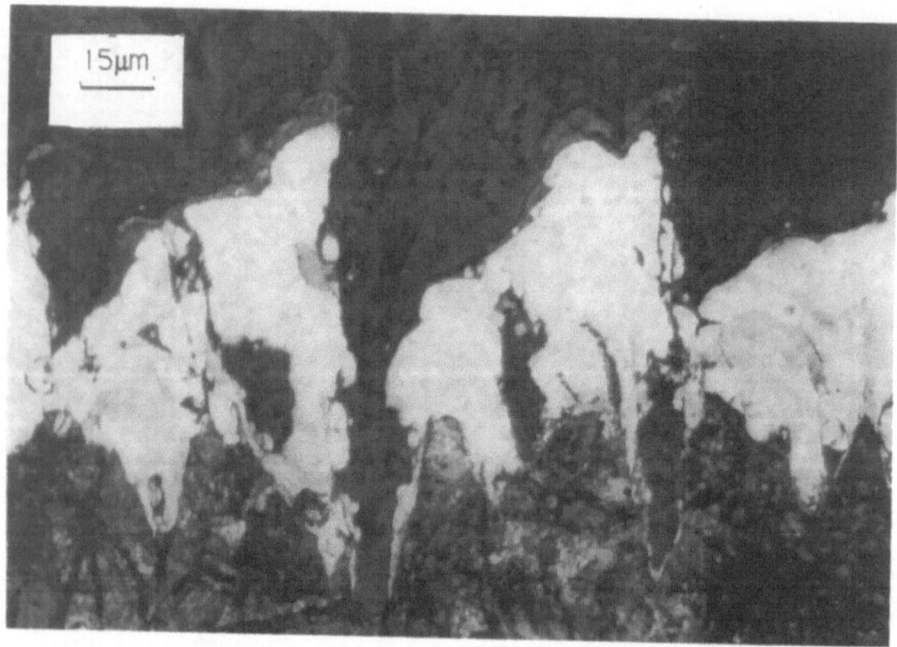

figure 14 : Cross section of melting cast iron treated by Nd-YAG laser
(speed scanning = 10 mm/s)

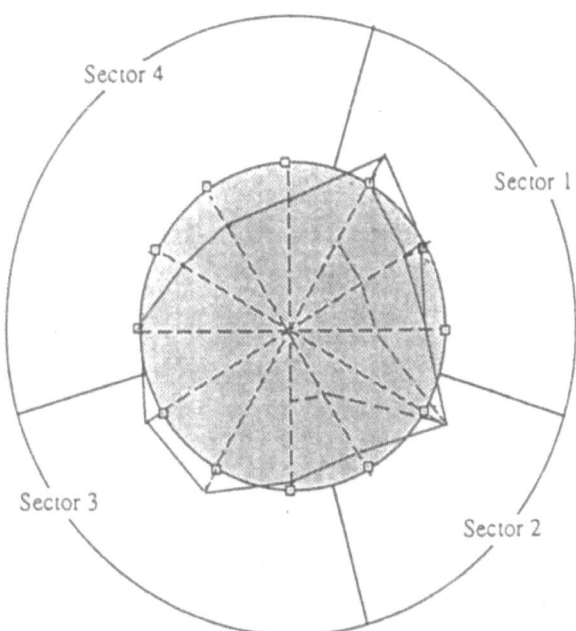

figure 15 : Polar diagram of MMC coating (thick layer) (on brass, bronze or steel substrat) (—steel comparison, --- substrat comparison) (n= 250 00 cycles)

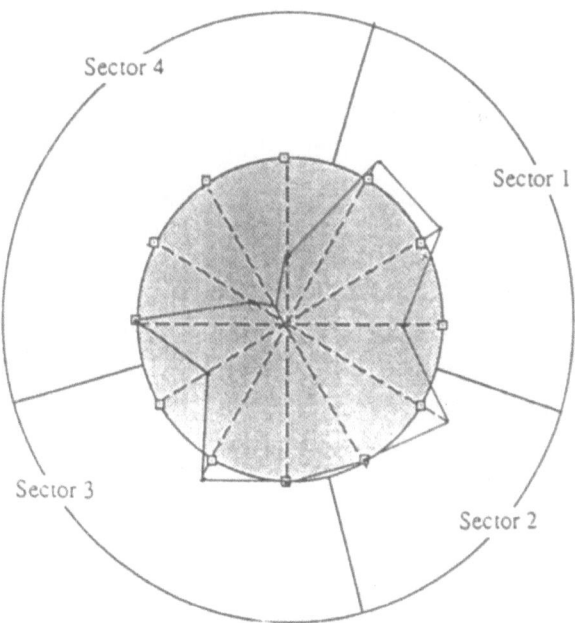

figure 16 : Polar diagram of melting cast iron (thin layer = 5 µm) (untreated cast iron comparison)(n = 250 000 cycles)

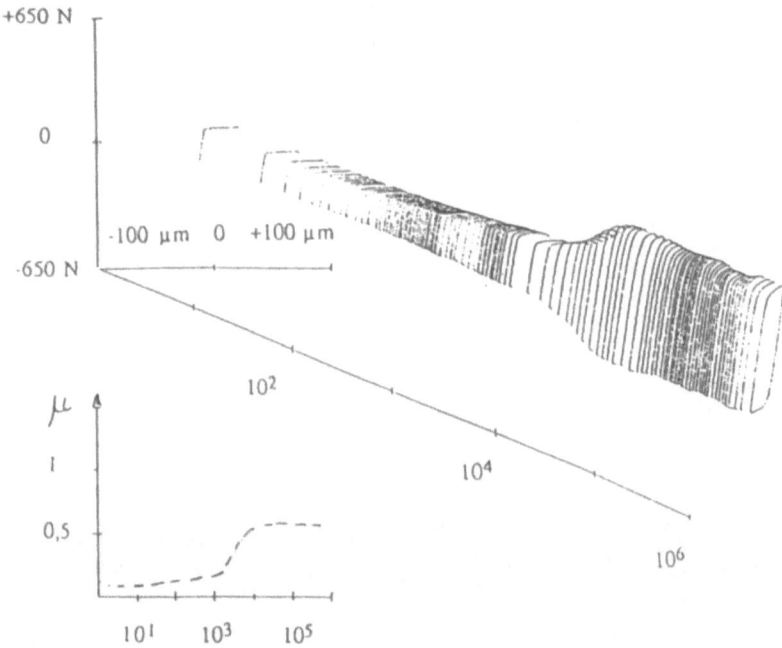

figure 17 : Friction log of treated cast iron (thin layer = 5 μm) (N= 650N, D = ± 18 μm, frequency = 5 Hz, n = 250 000 cycles)

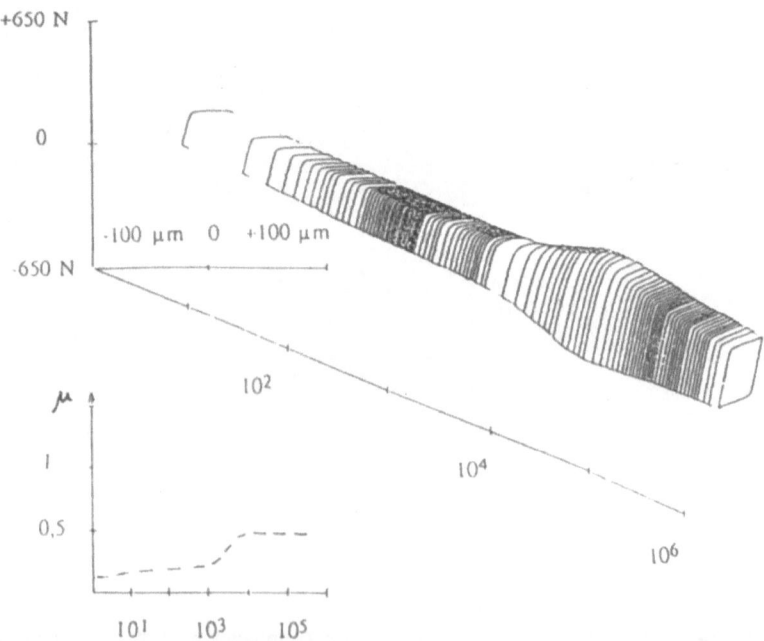

figure 18 : Friction log of treated cast iron (thin layer = 35 μm) (N= 650N, D = ± 18 μm, frequency = 5 Hz, n = 250 000 cycles)

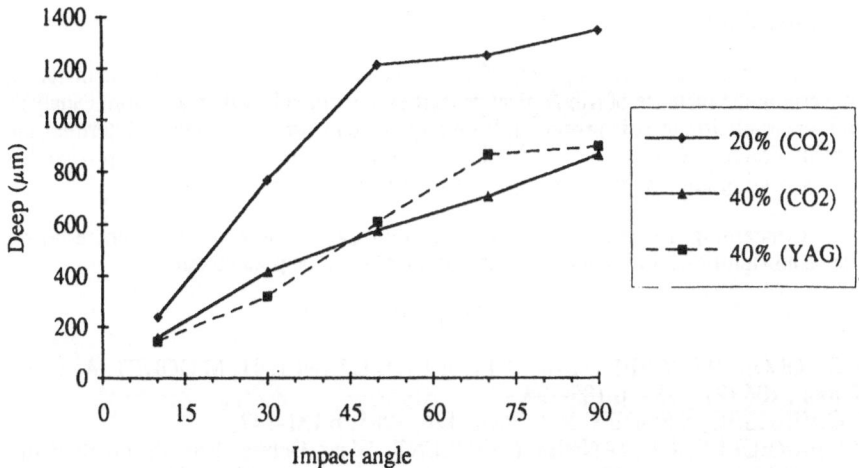

figure 19 : Maximum deep of crater versus impact angle (project weigt of sand = 2 kg)

452

IV. Conclusion

Coatings are interesting only if the objectives are well defined. For instance, concerning the increase of tribological properties, a method based on a sectorial analysis and on simulation tests is proposed. It gives precise informations and it underlines the contribution of the treatments concerning the response of the materials. It is proved that a thick coating imposes its own responses for all the kinds of substrate.

Concerning the "thin" coatings realized on Cast Irons with low average power lasers, an significant improvement of the behaviour can be pointed out.

Bibliography :

1-C.COLOMBIE, Y.BERTHIER, A.FLOQUET, L.VINCENT, M.GODET, Asme J. Tribol., 106 (2), 1984, p.185-194.
2-O.VINGSBO, S.SODEBERG, Wear, 126, 1988, p.131-147.
3-S.FAYEULLE, A.B.VANNES, L.VINCENT, Wear, Debris : from the cradle to the grave, Tribology series, Elsevier editor, Amsterdam, vo 11, 1992
4-P.CHEVALLIER, Etude de l'érosion par des particules solides : conception et réalisation d'un simulateur, recherche des paramètres pertinents, application à des structures homogènes ou complexes, thèse de doctorat ECL n°94-24, 5 juillet 1994.
5-P.CHEVALLIER, A.B.VANNES, Effect on a sheet surface of an erosive particle jet upon impact, accepted in WEAR.
6-J.F.CARTON, Traitements superficiels et tenue en service des assemblages démontables : base d'une méthodologie de choix en tribologie, thèse de doctorat ECL n° 93-26, 23 septembre 1993.
7-J.M.LEROY, Modélisation thermoélastique des revêtements de surface utilisés dans les contacts non lubrifiés, thèse de doctorat, INSA de LYON, 1989, n°89ISAL 0090.
8-M.GODET, Y.BERTHIER, J.M.LEROY, L.FLAMAND, L.VINCENT, Mechanical Aspect of coatings, Tribology Series, Ed. D.DOWSON and al., Elsivier, 1990.
9-K.L.JOHNSON, Contact Mechanics, Cambridge Press, 1985.
10-K.DANG VAN, Sciences et Techniques de l'Armement, 47, Vol. 3, 1973.
11-A.DEPERROIS, Sur le calcul des limites d'endurance des aciers, thèse de doctorat, Ecole Polytechnique Paris, 1991.
12-F.LEMOINE, Réalisation de revêtemets en alliage réfractaire à l'aide de lasers YAG-Nd continu ou impulsionnel : modélisation statistique et analytique de la section des cordons obtenus, thèse de doctorat ECL n° 94-14, mars 1994.
13-L;QUIBLIER, J.M.PELLETIER, P.SALLAMAND, Elaboration de revêtements de surface à fort coefficient de frottement sous faisceau laser pour des applications tribologiques, Rapport IFTS-CALFETMAT, INSA de LYON, septembre 1993.
14-S.KNECHT, P.KNU, Etude du fonctionnement d'un laser YAG-Nd à courte durée d'implulsion et haute fréquence, Application aux traitements de fonte et de titane, Studienarbeit, Ecole Centrale de LYON-Université de DARMSTHAD, juin 1994.
15-V.MALAU, A.B.VANNES to be published.

LASER SURFACE TREATMENT OF TOOL STEELS

R. VILAR, R. COLAÇO, A. ALMEIDA
Departamento de Engenharia de Materiais
Instituto Superior Técnico
Av. Rovisco Pais, 1096, Lisboa, Portugal.

Abstract

Laser surface treatment (LST) is a promising technique to improve the wear and corrosion resistance of materials. In the case of tool steels, laser surface treatment is carried out preferably in the liquid state to allow for complete dissolution of alloy carbides. In this paper, the main requirements for materials used in different types of tools and the advantages of using surface engineered materials for these applications are presented. The application of laser melting to the treatment of tool steels is exemplified for AISI 420 and 440C Cr steels and sintered AISI T15 HSS. Usually, the laser melted layers contain martensite, retained austenite and carbides. In steels containing large proportions of ferrite-forming alloying elements δ-ferrite may also be observed. The laser treatment of sintered steel leads to the elimination of residual porosity. The proportion of retained austenite in laser surface melted steels is much higher than in conventionally treated steels. However, the hardness of the steel is high because the austenite is strengthened by solid solution, dislocations and the small grain size. The high volume fraction of retained austenite usually prohibits the application of tool steels in the laser treated condition. Austenite may be eliminated by double or triple tempering treatments at temperatures in the range 550 to 650 °C. During tempering, carbides precipitate within austenite and martensite, and austenite transforms to martensite. Strong secondary hardening is often observed and the temperature of the secondary hardening peak of laser surface melted (LSM) steels is higher than after conventional heat treatment.

1. Introduction

Tools used in modern industry are usually complex and costly and they are submitted to extremely severe service conditions. To compensate for the high capital investment tools must be reliable and have a long service life, qualities that are primarily determined by the appropriateness of tool design and the ability of the tool material to withstand imposed service loads without deformation, fracture and excessive wear. To accomplish their task satisfactorily, tool materials need to be simultaneously hard, tough and wear resistant throughout their operating temperature range. Hardness is an important property, since it is a measure of the material's resistance to plastic deformation and, in general, wear resistance increases with hardness. However, tool materials must also present high machinability to reduce production costs, and high toughness to reduce the risk of premature failure under shock loading; these

453

J. Mazumder et al. (eds.), Laser Processing: Surface Treatment and Film Deposition, 453–478.
© 1996 *Kluwer Academic Publishers.*

requirements are irreconcilable with high hardness. As a consequence, either a compromise is imposed which restricts the performance of the tool or, preferably, the material's properties are modified during the fabrication cycle. Since in most tool applications conditions of use are more demanding at the surface than in the interior, the behavior of tools may be greatly enhanced by adequate surface treatment.

Surface treatment techniques aim at designing a composite with enhanced surface properties. This composite may be created by modifying the surface region of the material or by coating it with a layer of another material [1]. The diversity of requirements for tool materials and the economic importance of tools lead to the development of a plethora of surface modification processes for tool manufacturing. Induction hardening consists of austenitizing a surface layer of the steel and transforming the austenite to martensite by water quenching. It may be used for localized hardening of large tools, but it is of limited interest because surface oxidation is difficult to avoid and hardening is moderate. Carburizing and nitriding involve the addition of carbon and nitrogen, respectively. Both treatments are lengthy and require complex and costly equipment, special steels and additional heat treatment. Chromium electroplating is used to protect tools from mild corrosion and wear, but requires immersion of components in water-based solutions which may cause corrosion. More recently, thin coatings of TiN, TiC and other materials were applied by CVD and PVD techniques. The main limitations of these coatings are their very low thickness and moderate adherence, but owing to their high hardness and low friction coefficient they improve tool life considerably. Good results are also obtained with ion implantation.

Laser surface treatments present characteristics that are particularly well adapted to tool manufacturing, since they allow a remarkable improvement in the surface hardness and the wear and corrosion resistance of materials, without degradation of their bulk properties [2, 3]. Also, different areas of the same tool may be treated with the same equipment but using slightly different processes to tailor the surface properties to local requirements. Some laser assisted processes involve chemical reactions [4, 5], but, more frequently, the surface treatment relies only on the thermal effects of laser radiation.

One of the most interesting characteristics of laser surface treatment is its inherent versatility [6]. Tool steels may be hardened by martensitic transformation, without melting. Laser hardening has been extensively used to improve the fatigue [7] and the wear resistance [8-10] of steels, with negligible surface roughness and distortion. Nevertheless, processes involving melting seem more promising, since they allow the production of a large variety of novel structures. Laser surface melting can be carried out without modifying the chemical composition [11-13] or alloying elements can be added to the molten pool [14, 15]. The surface of the workpiece may be coated with a layer of another material by laser cladding [16-18] or LCVD [4, 5, 19] or a surface layer of a metallic matrix composite may be synthesized by particle injection [20-22]. In any case, due to the high solidification rates involved in LSM processes, the resulting microstructures are fine and frequently contain non-equilibrium phases and supersaturated solid solutions with interesting properties. In particular, laser treated materials are very often hard and present good resistance to wear, corrosion and oxidation. The coating grows epitaxially on the substrate, leading to excellent adherence. Since the laser energy is applied locally, laser surface treatment is particularly well suited to the treatment of small areas, a possibility that does not exist with most other surface engineering methods. Extended area coverage may be achieved by overlapping treated areas, but this procedure may lead to difficulties [23, 24]. In some cases (e.g. for wear resistance) overlapping is not necessary and the full area may be covered with a pattern of non-overlapping tracks [25].

Laser surface melting of tool steels leads to controlled dissolution of carbides and elimination of inclusions and carbide stringers [11, 26-32]. Usually, the hardness of the material will be improved, but often large amounts of austenite are retained [33, 34], resulting in low hardness, unpredictable wear behavior and risk of grinding cracking and dimensional changes in service. However, retained austenite may be eliminated by appropriate treatments [16, 33, 35].

2. Requirements for Tool Steels

A tool is a device used to produce components by cutting, forming or molding materials. The properties required for tool materials vary according to the application involved. For our purposes, it is helpful to distinguish machining tools (lathe cutting tools, milling tools, drills, saw blades, etc.), cold working tools (such as gages, clamps, jigs, blanking, piercing and drawing dies, etc.) and hot working tools, which include stamping and forging dies, die casting, plastic and glass molds and extrusion dies.

Tool material requirements are related either to performance in service and long lifetime or to the manufacturing process [36] (table I). An obvious requirement for all tool materials is that applied stresses should not exceed the yield strength of the material. Also, the material must keep suitable properties over long-term exposure to the operating temperature. In this respect, the steel must present high secondary hardening temperatures and negligible softening at the operating temperature. Toughness is critical in applications such as shearing, punching and interrupted machining, where shock loading conditions are prevalent. Usually toughness decreases with increasing hardness, since hard materials present low ductility. Mechanical fatigue strength is not usually a problem in tool materials, since they have a high yield strength, but thermal fatigue resistance may be a major concern when tools are submitted to rapidly alternating heating and cooling during service. Other important properties are oxidation and corrosion resistance.

Properties related to manufacturing are also important. High hardness materials are difficult to machine and tool steels must be processed in the annealed condition and then hardened by heat treatment. The demand for extremely smooth polishing and high quality surface textures produced by chemical etching is an important requirement in plastic molds. Both are surface-related properties, which will depend on the surface treatments applied to the material.

To perform their task, tools must come into contact with the workpiece and exert on it the load needed to deform or to cut it. During processing the tool and the workpiece surfaces slide over each other under strong pressure. As a consequence, tools are submitted to drastic wear conditions and wear resistance is the single main requirement for tool steels. Wear is a system property and the exact wear mechanisms involved in a particular situation will depend on factors as diverse as the roughness of the surfaces in contact, the properties of the bulk materials, their chemical composition and microstructure, the presence of surface layers, the contact pressure, the type and speed of relative motion, the environment, etc. [37]. Since these parameters may vary widely, the study of wear in practical situations and the design of wear resistant materials suited to those situations is a challenging task. Wear maps [38, 39] successfully summarize the mechanisms of wear encountered in materials in relation to operating conditions and material properties.

TABLE 1. Types of tools: working conditions and main requirements (adapted from [36]).

Type of tool	Working conditions and requirements
Cold forming tools	High stresses (high yield strength) High pressure sliding contact (sliding wear resistance) Shock loading (toughness)
Hot forming tools	The same as above, at high temperature Oxidation (oxidation resistance) Thermal fatigue (resistance to thermal fatigue)
Shearing tools, cutting blades	High stresses (high yield strength) Sliding wear (sliding wear resistance) Abrasion (abrasion wear resistance) Shock loading (toughness)
Die-casting, plastic and glass molds	High stresses (high temperature strength) Cyclic stresses (fatigue, surface fatigue) Abrasion (abrasion wear resistance) Erosion (erosion wear resistance) Oxidation (oxidation resistance) Thermal fatigue (resistance to thermal fatigue) Machinability Polishability
Cutting tools	High stresses at high temperature (high temperature strength) Shock loading (toughness) High temperature wear resistance

Tool wear results from a combination of several wear mechanisms, depending on the exact working conditions. It is machining tools that present the most demanding wear resistance requirements. In continuous high-speed cutting the region of the tool that comes in contact with the workpiece may reach temperatures in the range 600-900 °C [40]. The cutting edge temperature increases with increasing cutting speed and with increasing feed rate [41]. Wear appears both on the clearance face of the tool (flank wear) and on its rake face (crater wear). The main wear mechanisms involved are [40] adhesion, abrasion and chemical interaction. Adhesion between the tool and the workpiece is followed by shearing and fracture of the tool material, leading to the formation of debris. Chemical interaction and diffusion lead to the transfer of atoms from the tool to the workpiece and to the progressive softening of the tool surface. Abrasion is due to the action of hard particles embedded in the workpiece or of hard debris which slide over the tool surface. Diffusion-controlled wear is the major wear mechanism when cutting at high speed and high feed rates and causes rapid tool wear, whereas lower feed rates favor plasticity-controlled wear (adhesion and shearing) [42]. Adhesion is also the predominant wear mode in milling tools and drills because the cutting speed is lower than in lathe tools [41].

Hot rolling, hot forging and hot extrusion tools come in contact with the workpiece, at high pressure and temperature, for relatively long periods of time, resulting in rapid oxidation of the tool surface. At moderate pressure the thin oxide layer remains adherent and protects the surface from adhesive wear, but at high pressure the oxide film fractures producing hard particles which lead to intense abrasive wear. Chemical interaction, diffusion and strong surface deformation will occur if cooling is inefficient, the tool stays in contact with the workpiece too long, or the pressure is excessive. In cold forming tools, molds, casting dies and stamping tools, abrasion is the major wear mechanism. It is caused by the movement of a hard surface or a surface containing embedded hard particles against the tool surface (two-body abrasion), or when hard particles become entrapped between the workpiece and the tool (three-body abrasion).

Tool steels are able to resist wear due to the presence of large particles of hard primary carbides embedded in a relatively hard and tough matrix, formed by tempered martensite and a fine dispersion of secondary carbides. As the material must be wear resistant over long exposure to high temperature, all the microstructural elements must be heat resistant [43]. The surface roughness is also important [37]. If it is low, the contact between the two bodies tends to be elastic, and the predominant unlubricated sliding wear mechanisms are adhesion, surface fatigue or tribochemical wear, depending on the exact working conditions. Conversely, sliding contact between rough surfaces leads to extensive plastic deformation and favors abrasion. In tools the surface finish is often excellent, but mechanical or chemical damage to the surface may create conditions for abrasive wear. The unlubricated sliding wear resistance of steels depends on the hardness and the toughness of the material [37]. For similar toughness the wear resistance increases with hardness, but tough structures present better wear resistance than brittle structures with similar hardness. The influence of retained austenite and carbides depends on the working conditions. Under mild load (low deformation of the sliding surfaces) wear resistance increases with decreasing austenite proportion and with increasing carbide proportion, while under heavy load the opposite occurs [44]. This dependence is explained by the fact that austenite transforms to martensite under heavy loads, increasing the effective surface hardness, whereas carbides tend to fracture, leading to material losses. Conversely, if the pressure is low, austenite does not transform and its softness leads to low wear resistance whereas carbides harden the material and increase its wear resistance.

Resistance to abrasive wear depends on the material and on the abrasive [37]. Wear resistance increases as the work hardening coefficient of the matrix increases. The influence of second phase particles is complex. The wear loss caused by soft abrasive particles decreases with increasing carbide proportion. Conversely, large hard abrasive particles tend to deform the soft second phase particles or to detach them from the matrix, and the wear resistance will decrease with increasing proportion of second phase particles [37]. Fine dispersions of carbides may increase wear resistance because they strengthen the matrix while large carbide particles will decrease wear by resisting scratching by abrasives. As a consequence, second phase particles will enhance abrasion resistance mainly if they are larger and harder than the abrasive particles. If the second phase particles are larger but softer than the abrasive particles, the variation in abrasion resistance with its volume fraction will reach a maximum and then decrease, as microcracking and flaking of carbides becomes the main material loss mechanism [37].

Some tools are submitted to rolling wear, a situation where the prevailing wear mechanism is surface fatigue [45]. The rolling wear resistance of high-alloy steels increases with decreasing volume fraction of carbides [44] and with increasing hardness of the matrix [46]. Austenite seems to decrease the surface fatigue resistance of steel [47]. In martensitic steels with similar hardness surface fatigue resistance decreases with increasing average carbide size. Large carbide particles and inclusions increase wear since they fracture under repeated loading and promote pitting.

Crossed cylinder wear tests showed that the maximum wear resistance of high-speed steels is associated with the secondary hardening peak [48]. Furthermore, for similar hardness, the wear resistance of samples that were overtempered at temperatures beyond the hardness peak temperature is lower than that of samples tempered at temperatures below the hardness peak temperature. The low wear resistance of overtempered samples is due to the abrasive effect of coarse primary carbide particles. An interesting result of this work is that at the secondary hardening peak primary carbides have no major role in enhancing the wear resistance of high speed steels and so the wear resistance is independent of their proportion, if the hardness remains constant.

3. Experimental Procedure

The chemical composition of the steels used in the present study is indicated in Table 2. AISI 420 and 440C are martensitic stainless steels used in plastic mold manufacture and AISI T15 is a high-speed steel.

TABLE 2. Chemical composition of the steels (wt.%)

Steel	C	Cr	W	Mo	V	Co	Ni
420	0.47	12.8	0	0.26	0.19	0	0.19
440C	1.00	18.5	0	0.20	0.10	0	0.10
T15	1.55	4.41	12.8	0.24	4.27	4.24	0.033

The laser treatments were carried out with CW CO_2 lasers. Details of the laser treatment procedure have been reported elsewhere [32, 34, 49]. The roughness of the samples was evaluated using a profilometer. The microstructures were investigated using light microscopy, scanning and transmission electron microscopy, and X-ray diffraction. It was found that a very careful metallographic preparation was necessary to avoid transformation of austenite.

Transmission electron microscopy was performed using a 200 kV scanning transmission electron microscope. Details of the technique used to prepare the thin foils have been presented in previous papers [32, 49]. Compositional data were obtained using energy dispersive and wavelength dispersive spectrometers attached to the scanning and transmission electron microscopes.

The measurement of the amount of retained austenite in the material was carried out by X-ray diffraction, using the Averbach and Cohen method [50]. Since laser treated materials usually present a strong crystallographic texture, the volume fraction of retained austenite was calculated from the ratios of the intensities of γ_{200}, γ_{220}, α''_{200} and α''_{211} lines. The doublets of tetragonal martensite were only partially resolved and were treated as a single peak. The volume fraction of carbides in the laser melted material was estimated by quantitative metallography and their proportion taken into account in the calculation of the austenite volume fraction.

4. Experimental Results

4.1. SURFACE ROUGHNESS

If the laser treatment is carried out properly, laser treated surfaces are free from oxidation and other major defects. They present a characteristic topography, consisting of relatively flat zones alternating with rippled ones (Fig. 1a). Two kinds of ripples can be distinguished on the basis of their amplitude. Large amplitude waves do not exhibit a noticeable periodicity. The distance between them varies from 70 to 150 μm and their shape reflects the shape of the solidification front. By contrast, low amplitude ripples are periodic and present an average wavelength of about 11 μm. Low amplitude ripples are often curved and parallel to the large amplitude ones (Fig. 1a), but straight ripples, perpendicular to the scanning direction were also observed (Fig. 1b). The large amplitude waves are associated with surface tension effects [51] and with the periodic overflow of liquid metal [27]. Periodic ripples were attributed to the interference of the incident plane wave with surface electromagnetic waves [52].

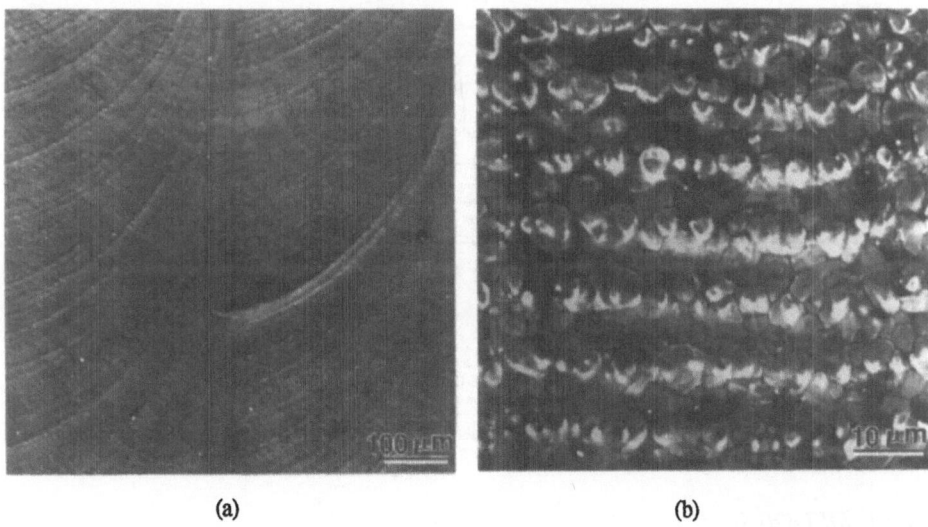

(a)   (b)

Figure 1. Surface topography of laser melted AISI 420 steel: (a) typical surface topography of a laser melted track; (b) straight surface ripples perpendicular to scanning direction.

The surface roughness of the laser melted samples depends on the finish prior to the laser treatment. Figure 2a shows the surface profile along a direction perpendicular to the laser beam pass of a sample of AISI 420 steel that was rough milled prior to the laser treatment. The deep valleys observed in this profile are milling traces that survived laser melting. The small amplitude undulations observed have an average wavelength of 0.37 mm, which corresponds to the lateral displacement of the table to overlap successive tracks. In polished surfaces only these fluctuations are detected (Fig. 2b). Ra values are within the range 0.3-2.50 μm for normal laser treatment conditions. The roughness decreases with increasing scanning speed and with decreasing laser power and does not depend on the initial structure of the material.

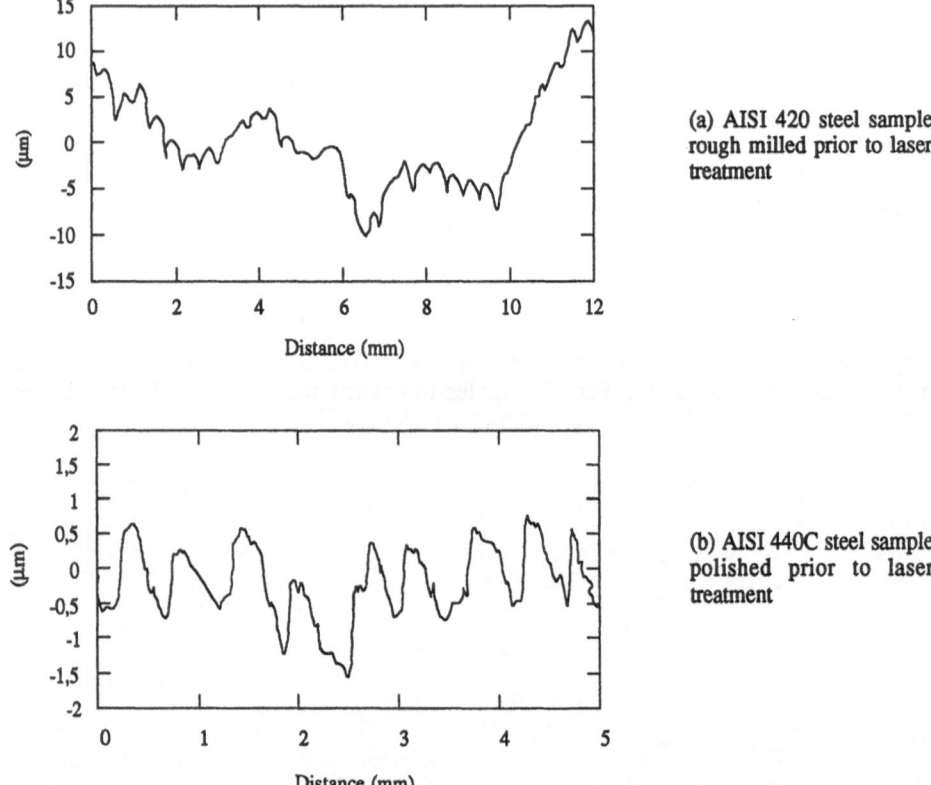

(a) AISI 420 steel sample rough milled prior to laser treatment

(b) AISI 440C steel sample polished prior to laser treatment

Figure 2. Roughness profiles of laser surface melted samples.

4.2. STRUCTURE AND PROPERTIES OF LASER SURFACE MELTED TOOL STEELS

4.2.1. *AISI 420 Steel*

This steel is used for plastic mold manufacturing. It is supplied in the annealed condition, with a microstructure consisting of $M_{23}C_6$ and M_7C_3 carbides dispersed in a matrix of α-ferrite. Its hardness is 250 HV0.2.

Topography. The dendritic structure is visible at the surface of the sample (Fig. 3a). Since the dendrites grow in the direction of the temperature gradient, they tend to be perpendicular to the successive positions of the solid-liquid interface, leading to a strong preferred orientation. In Fig. 3b the characteristic relief effect associated with martensitic transformations can be recognized, disclosing the presence of martensite.

The amount of this phase is larger in laser melted regions that were reheated during subsequent laser scans. Martensite needles are usually longer than the primary dendritic spacing and extend across several dendrites.

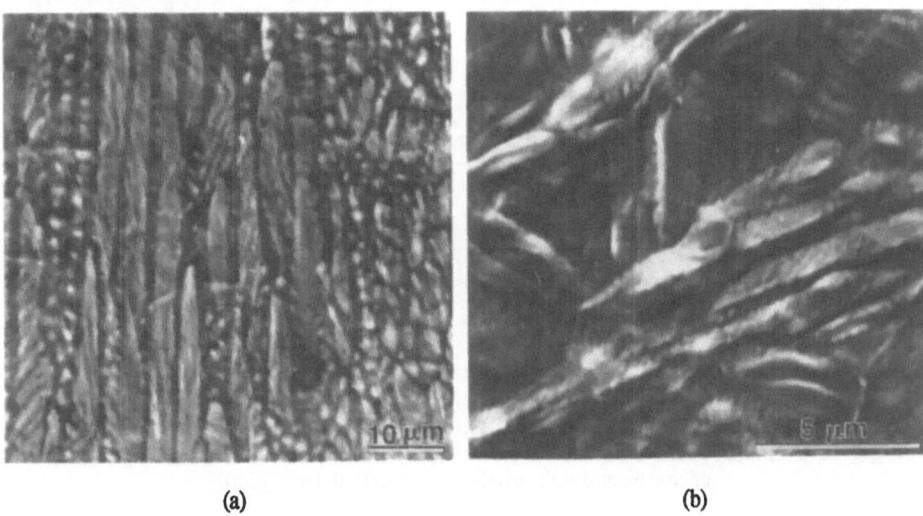

(a)   (b)

Figure 3. (a) Dendritic structure revealed by the retraction of the liquid at the surface; (b) martensitic needles.

Microstructure. A typical laser melted layer presents several different zones, depending on the maximum temperature attained during the laser treatment (Fig. 4a).

In the base material (A) the temperature stays under the critical temperature range and the structure is not modified. In the heat-affected zone (B) only solid-state phase transformations occur. Near the base material, α-ferrite transforms to austenite during heating and carbides dissolve partially in this phase. During cooling, austenite transforms, at least partially, to martensite. The microstructure of this region (Fig. 5a) consists of martensite, undissolved carbides and retained austenite. Closer to the fusion line δ-ferrite appears at the grain boundaries (Fig. 5b). Its proportion increases with increasing maximum temperature. δ-ferrite does not transform during cooling due to the high concentration of strong ferrite-forming alloying elements and the fast cooling rate [13]. In the transition region (C) the material is partially melted. δ−ferrite is the predominant phase here (Fig. 5c). The plane front solidification zone (D) and the cellular-dendritic zone (E) correspond to the layer that was completely melted. In zone D a plane solid/liquid interface was stabilized due to the large G/R ratio prevalent during the first stages of solidification [53]. In single tracks this zone consists of austenite and, presumably, some δ−ferrite (Fig. 5c). A significant amount of martensite (α'') resulting from $\delta \rightarrow \gamma \rightarrow \alpha''$ transformations forms in this region when the material is reheated during subsequent laser passes (Fig. 4b).

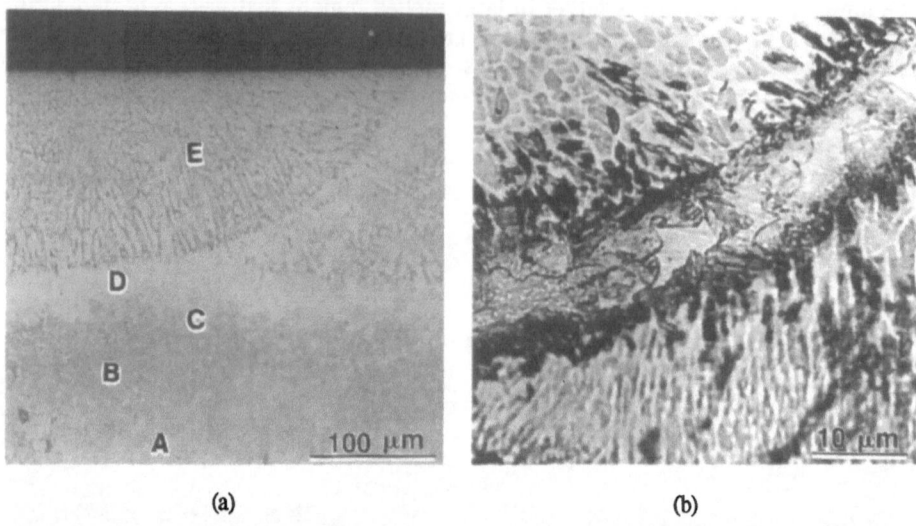

(a)   (b)

Figure 4. (a) Transverse cross section of the melted layer of AISI 420 (Murakami reagent); (b) microstructure at the overlapping zone (Vilella reagent).

The instability of the plane S/L interface leads to the development of a cellular-dendritic structure (E in Fig. 4a). The structure is very fine: the secondary arm dendritic spacing is within the range 1 to 2.5 μm for the laser treatment parameters used. A thin interdendritic film is observed for high scanning speeds. The dendrites contain 10 to 12% Cr and the Cr content in the interdendritic spaces is about 18%. This region contains δ–ferrite, austenite, martensite and $M_{23}C_6$ carbide in proportions which depend on the local solidification conditions. The volume fraction of retained austenite increases with increasing cooling rate. When austenite is destabilized by reheating it transforms to martensite. Since the proportion of austenite transformed increases with increasing reheating temperature, larger volume fractions of martensite are found in the heat-affected zone produced in previous melted tracks by subsequent laser scans (Fig. 4b). Also, the proportion of martensite increases with decreasing scanning speed, increasing laser beam power and increasing overlap between consecutive tracks. The volume fraction of retained austenite is within the range 30 to 50% for samples treated with scanning speeds between 3 and 13 mm/s. This value is much higher than those found when this steel is conventionally hardened (about 5%). The peaks of one or several BCC or BCT phases were observed by X-ray and electron diffraction, but it is difficult to discriminate between the different body-centered phases that may exist in the steel: BCC δ–ferrite, BCC martensite or tetragonal martensite with small c/a ratio (on the basis of diffraction results). Analysis of the (002) peak profile suggests that slightly tetragonal α" is predominant in the laser melted region, a result that was corroborated by TEM observations. The γ lattice parameter in a sample treated with 1320 W and at 6.7 mm/s was 3.606±0.001 Å, a value which is compatible with an austenite containing 0.5% C and 13% Cr [54]. The parameters of martensite in the same sample are a=2.87±0.01 Å and c=2.91±0.01 Å.

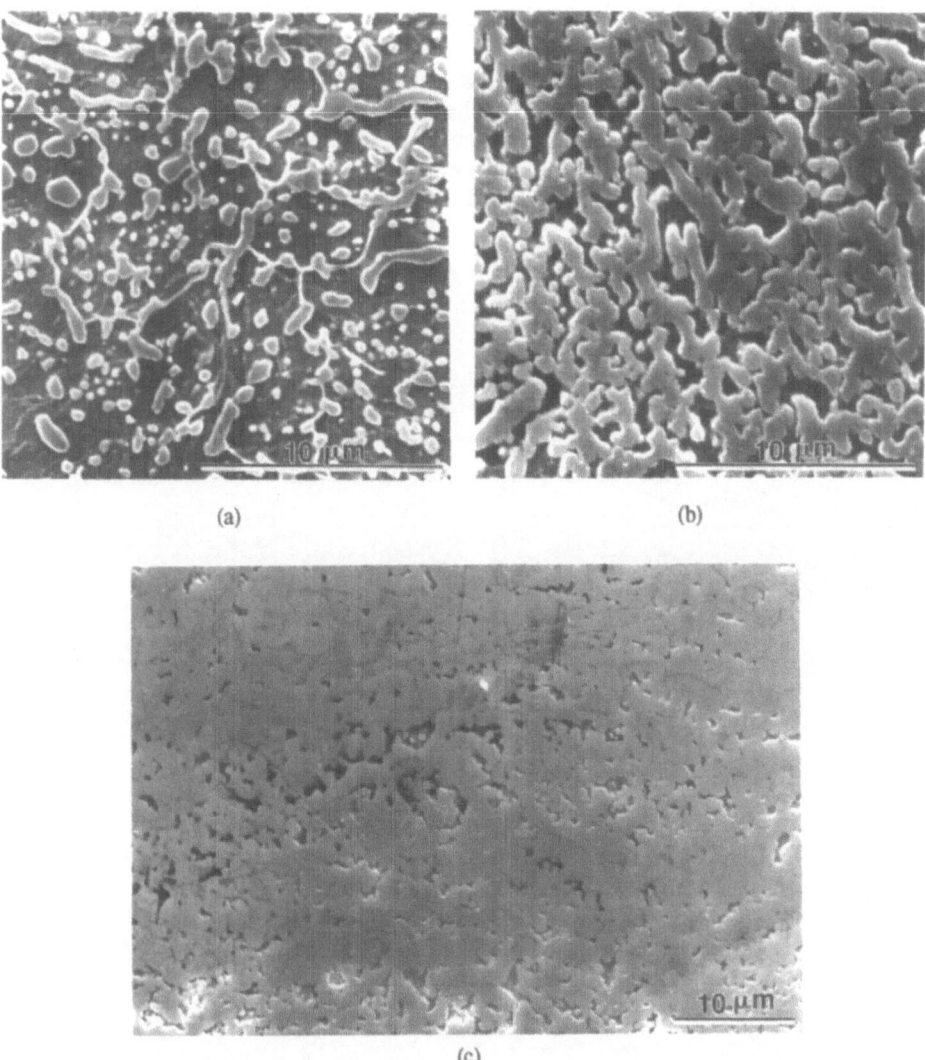

Figure 5. Microstructure of the heat-affected zone (a) near the base material and (b) near the fusion line, and (c) microstructure of the transition zone.

Representative examples of the microstructures observed by TEM are shown in Fig. 6. They correspond to the melted region of a sample treated with 700 W and 5 mm/s. The microstructure at a depth of 70 µm (Fig. 6a) contains mainly lath-type martensite with a distorted BCC or BCT lattice and a high dislocation density. Some twinned plate martensite was also detected (Fig. 6b). The retained austenite observed presents very high dislocation density. A few particles of $M_{23}C_6$ carbide were observed in the interdendritic regions. At a depth of about 160 µm (also in the cellular-dendritic region) the microstructure consists of elongated cells, separated by low angle boundaries (Fig.

6c). The predominant phases were a slightly distorted BCC or BCT phase and austenite as before, but the body-centered phase does not show a morphology typical of either lath- or plate-type martensite. It may be martensite with a different morphology or δ–ferrite. The dislocation density in both phases is lower than closer to the surface. Precipitates of $M_{23}C_6$ carbide are located at the cell boundaries. The carbide does not form a continuous interdendritic film in this sample but continuous films were observed in samples treated with high scanning speeds (>20 mm/s).

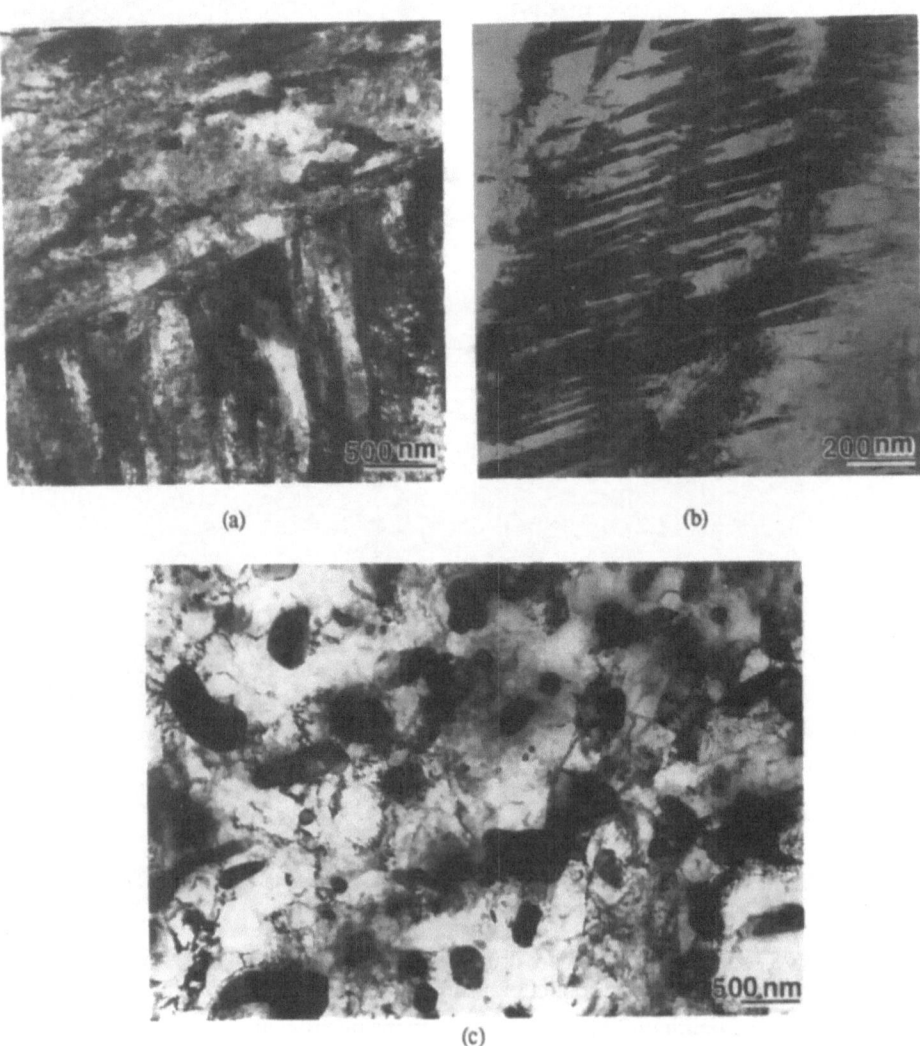

(a)

(b)

(c)

Figure 6. Microstructure of the melted zone at a depth of (a) and (b) 70 μm and (c) 160 μm.

According to the predictions of the generally accepted C-Cr-Fe phase diagram [55], the solidification of a 0.47 wt.% C and 13 wt.% Cr steel should start with the precipitation of δ–ferrite, followed by the monovariant peritectic reaction $L+\delta \rightarrow \gamma$ and, finally, by the quasi-peritectic reaction $L+\delta \rightarrow \gamma+M_7C_3$. However, the microstructures described show that austenite is the primary solidification phase, despite the presence of δ–ferrite in the heat-affected and transition zones, which should favor the epitaxial growth of δ–ferrite. Also, the carbide which precipitates is $M_{23}C_6$ instead of M_7C_3. It was found experimentally that an increase in the solidification rate causes a transition from δ–ferrite to austenite solidification modes in laser welding of austenitic stainless steels [56]. By contrast, when solidification occurs in liquid metal droplets, an increase in the solidification rate leads to the preferential growth of δ–ferrite [57]. These contradictory findings were explained by the kinetics of nucleation and growth of both phases. It was found [58] that for large undercoolings the nucleation barrier for homogeneous nucleation is always lower for δ–ferrite than for austenite. This explains why in small droplets, where the probability of homogeneous nucleation is high, solidification starts with ferrite. However, at high solidification rates, the tips of austenite dendrites grow at lower undercoolings than the tips of ferrite dendrites. As a consequence, when solidification occurs by epitaxial growth, as in welding and laser surface treatment, austenite tends to be the primary phase. A transition from ferrite to austenite primary solidification may occur as solidification rate increases. This argument explains why solidification in laser surface melted AISI 420 steels starts with the precipitation of austenite instead of ferrite. The solidification process in this steel is a non-equilibrium process, in spite of the relatively moderate solidification rates used.

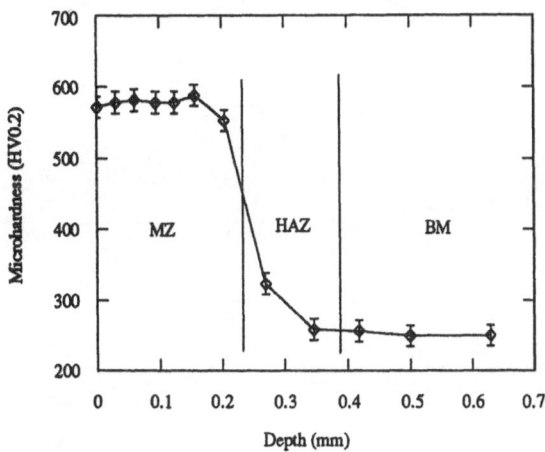

Figure 7. Depth microhardness profile for a sample of laser melted AISI 420 steel.

Hardness Measurements. The hardness profile of a laser melted sample of AISI 420 steel is shown in Fig. 7. The hardness of multiple scan laser treated samples of AISI 420 steel varies between 550 and 700 HV. It increases with increasing proportion of martensite, and the harder samples are those processed under conditions which favor the formation of martensite, i.e. with high laser power, low scanning speed and high overlap between consecutive tracks. Despite a proportion of retained austenite of about

45%, the hardness of the steel is remarkably high. This high hardness is explained by the small grain size, high solute content and high dislocation density in austenite.

4.2.2. AISI 440C Steel

AISI 440C martensitic stainless steel is used for bearings and guiding shafts, applications where surface fatigue is the prevalent mechanism of wear. The surface fatigue resistance of high-alloy steels increases with decreasing volume fraction of carbides and with increasing hardness of the matrix [37]. As LSM refines the microstructure and decreases the proportion of carbides, it may be expected to improve the properties of this steel. The tests were carried out on annealed or hardened and tempered samples. In the annealed condition, the steel consists of particles of $M_{23}C_6$ and M_7C_3 carbides, dispersed in a matrix of ferrite. After hardening and tempering, the matrix is formed of tempered martensite and undissolved $M_{23}C_6$ carbide particles [12].

Microstructure. Typical microstructures of laser melted AISI 440C steel are shown in Fig. 8. With scanning speeds higher than 4 mm/s the structure consists of dendrites containing austenite and a small proportion of martensite, surrounded by a fine lamellar eutectic of austenite, M_3C and probably M_7C_3 (Fig. 8a). Typical volume fractions of retained austenite, carbides and martensite are 90%, 6% and 4%, respectively. In samples treated at 4 mm/s the dendrites near the fusion line present a core of δ–ferrite (Fig. 8b), but far from the fusion line the structure is similar to the structure found in samples treated with higher scanning speeds.

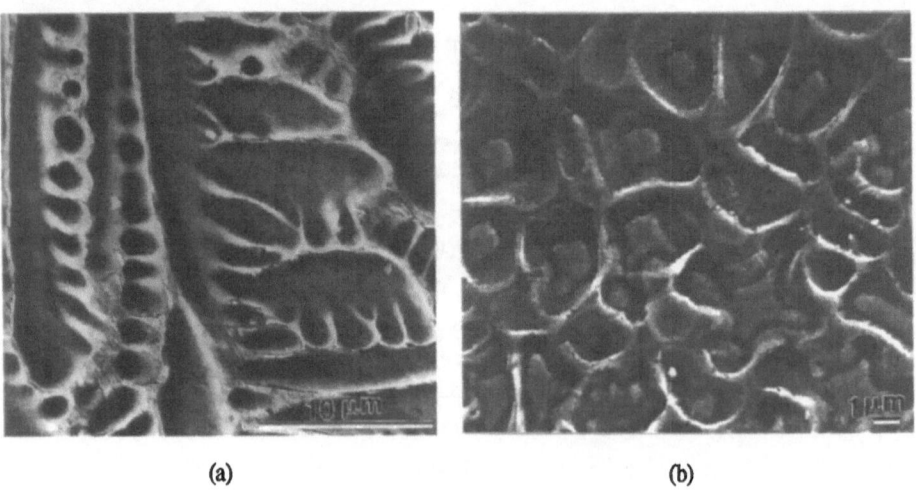

(a)	(b)

Figure 8. Microstructure of AISI 440C steel laser melted with scanning speeds (a) over 4 mm/s and (b) at 4 mm/s.

According to with the equilibrium Fe-C-Cr phase diagram [55], the sequence of transformations which can occur during solidification of laser melted 440C is the following: (1) precipitation of γ; (2) monovariant eutectic reaction $L \rightarrow \gamma + M_7C_3$; (3) quasi-peritectic reaction $L + M_7C_3 \rightarrow \gamma + M_3C$ and, finally, (4) if any liquid still exists, monovariant eutectic reaction $L \rightarrow \gamma + M_3C$, a solidification sequence which is compatible with the observed microstructure. During cooling a small proportion of austenite transforms to martensite. At low solidification rates δ is the leading phase and the solidification path becomes: (1) $L \rightarrow \delta$; $L + \delta \rightarrow \gamma$, thereafter as above. The possibility of a transition from δ to γ solidification modes as the solidification rate increases has already been discussed in relation to AISI 420 steel. It can be expected that ferrite is the leading phase deep in the molten pool, where solidification rates are low, while austenite becomes the leading phase near the surface, at higher solidification rates.

Hardness Measurements. The hardness profiles obtained on samples treated with a power density of 405 W/mm^2 and an interaction time of 0.05 s, either in the annealed or quenched and tempered condition, are shown in Fig. 9. The hardness of the melted zone (MZ) is lower than the hardness of the conventionally hardened steel but higher than the hardness of the annealed steel, and does not depend on the previous structure of the material. The hardness of the heat-affected zone (HAZ) varies between 600 and 650 HV. The relative softness of AISI 440C steel after laser melting is explained by the extremely high austenite content.

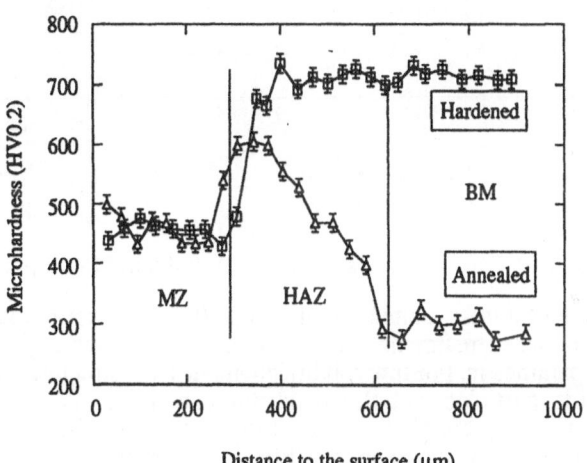

Figure 9. Depth microhardness profiles for samples of laser melted AISI 440C steel initially in the annealed and hardened conditions.

4.2.3. *Sintered AISI T15 Steel*

Sintering presents significant advantages as a technique to produce cutting tools, but the fatigue resistance [59], the fracture toughness [60] and the wear resistance [61] of sintered steels are usually inferior to the properties of similar materials produced by conventional routes, owing to residual sintering porosity. Laser surface melting was used to eliminate the residual porosity of sintered AISI T15 high-speed steel, a steel that is particularly difficult to produce by conventional routes due to its high content of alloying elements.

Microstructure. The techniques used for the production of the sintered AISI T15 steel samples and for their laser treatment has been described elsewhere [32]. After sintering, quenching and tempering this material contains 15% M_6C, $M_{23}C_6$ and MC carbides dispersed in a matrix formed of martensite and 15% retained austenite. Uniaxially compressed and sintered samples contain about 50 pores/cm^2, with an average diameter of 35 μm. Some pores with a diameter exceeding 100 μm were observed. After hot isostatic pressing the density of pores decreases to 35 pores/cm^2 and the average pore diameter to 10 μm. More importantly, no large pores remain.

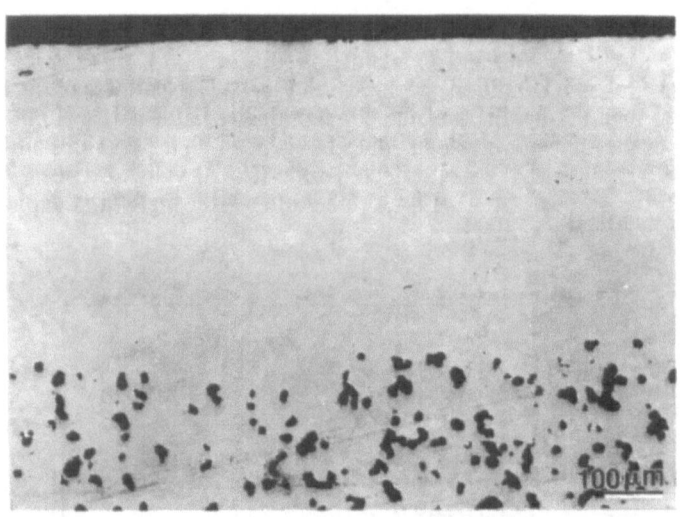

Figure 10. Cross-section of a laser surface melted AISI T15 steel sample.

The cross-section of a laser melted sample (Fig. 10) shows that porosity is completely eliminated by the laser treatment. The structure of the melted region depends on the laser treatment parameters. For low solidification rates, the structure consists of primary MC-type carbide particles, cellular dendrites containing martensite and retained austenite, and a poorly coupled eutectic formed by austenite, MC and M_6C-type carbides, which surrounds the dendrites (Fig. 11a). For higher solidification rates, no primary carbides are detected and a skeleton of δ–ferrite is observed in the core of the dendrites (Fig. 11b). Two types of eutectic were detected: a fine M_2C/γ feathery eutectic with an average interlamellar spacing in the range 0.1 to 0.2 μm (A) and a coarser, herring-bone type M_6C/γ eutectic, which degenerates to form a continuous interdendritic film in some regions (B).

Hardness Measurements. The hardness profile on the transverse cross-section of a sample treated with 700 W at 16 mm/s is illustrated in fig. 12. The hardness of the material after conventional hardening is 800 HV. After laser melting, the hardness is within the range 950 to 1100 HV. It increases with laser power and scanning speed, reflecting the influence of dendrite size and martensite and carbides volume fractions.

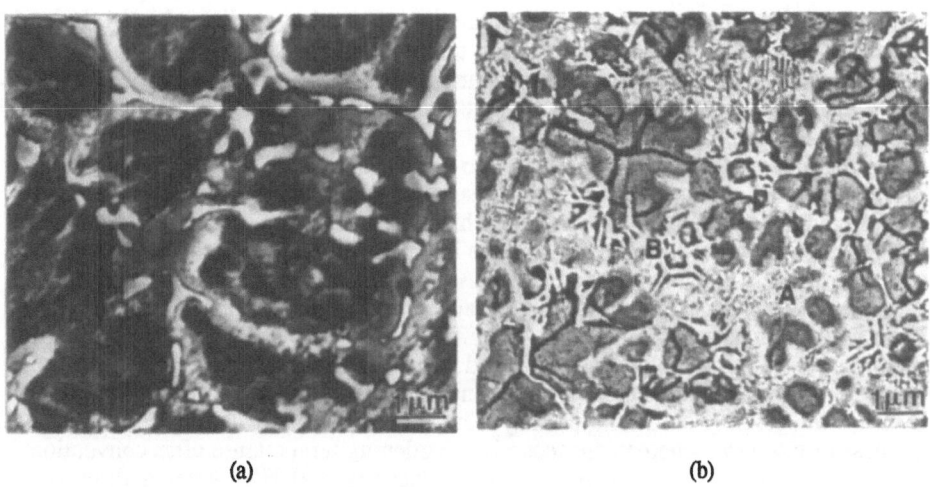

(a)   (b)

Figure 11. Microstructure of the melted region for (a) low solidification rate P = 700 W, v=4 mm/s and (b) high cooling rate P = 500 W, v=16 mm/s.

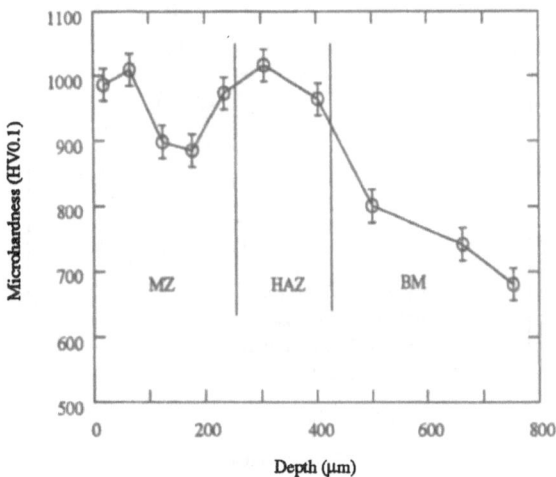

Figure 12. Microhardness profile of a laser melted AISI T15 steel sample.

5. Tempering of Laser Melted Tool Steels

Laser melted tool steels contain large proportions of retained austenite rich in alloying elements. The influence of retained austenite on tool steels is a matter of some controversy, but this phase is generally considered to have a detrimental influence on material properties and it must be eliminated. Depending on the composition of the steel retained austenite may be eliminated by subzero cooling, by isothermal transformation to bainite as a result of tempering at a suitable temperature, or transformed to martensite, after destabilization at high temperature. Subzero cooling and isothermal transformation are effective for carbon and low-alloy steels only. In high-alloy steels retained austenite is extremely stable and must be destabilized by tempering above 500 °C and transformed to martensite [62]. The tempering treatment has the additional advantage of further increasing the hardness of the steel by secondary hardening [12, 32, 63] and the formation of hard martensite.

The phase transformations and the variation in hardness resulting from tempering laser melted tool steels will be exemplified for AISI 440C steel, since this steel presents a particularly high proportion of retained austenite. The variation in hardness as a function of tempering temperature after two- and three-stage 1 hour tempering treatments at temperatures between 500 and 700 °C is depicted in Fig. 13. In the same figure we plotted the conventional tempering curve [64]. After a two-stage tempering treatment a secondary hardening peak appears at 600 °C, corresponding to a maximum hardness of 620 HV, whereas the secondary hardening temperature after conventional hardening is 510 °C and the maximum hardness is 570 HV. After a three-stage tempering the difference between secondary hardening peak temperature of laser melted and conventionally hardened steels decreases.

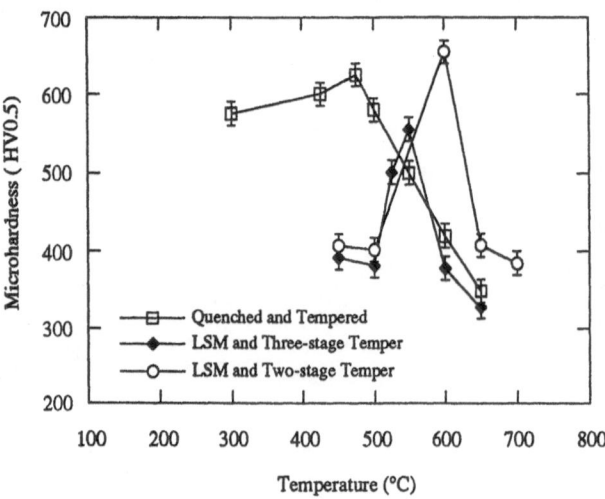

Figure 13. Tempering curve for AISI 440C steel.

X-ray diffraction and metallographic observations showed that tempering at 500 °C does not modify the structure of the material noticeably (Fig. 14a and 15a). By contrast, after tempering at 600 °C, the proportion of austenite decreases to 30% (as compared to 90% before tempering) and those of ferrite and M_3C increase (Fig. 14b). An intense precipitation of fine carbide particles is observed within the dendrites (Fig. 15b). Increasing the tempering temperature to 640 °C leads to the complete transformation of austenite to ferrite (Fig. 14c). The diffraction peaks of ferrite become narrow, showing that ferrite is recrystallized [65]. The proportion of M_3C decreases and this carbide is replaced by M_7C_3. The precipitation within the dendrites is now generalized (Fig. 15c). Tempering at 700 °C produces a further replacement of M_3C by M_7C_3 (Fig. 14d) and fragmentation, spheroidization and coarsening of the interdendritic carbide network (Fig. 15d). The lattice parameters of austenite and ferrite decrease with increasing tempering temperature up to 640 °C, denoting a progressive loss of carbon and alloying elements in both phases as carbides precipitate. Above 640 °C austenite disappears and the lattice parameter of ferrite does not vary noticeably, suggesting that the evolution of this phase is complete at that temperature. The precipitation of carbides within austenite is accompanied by the reduction of supersaturation in alloying elements leading to its transformation into martensite during cooling [62].

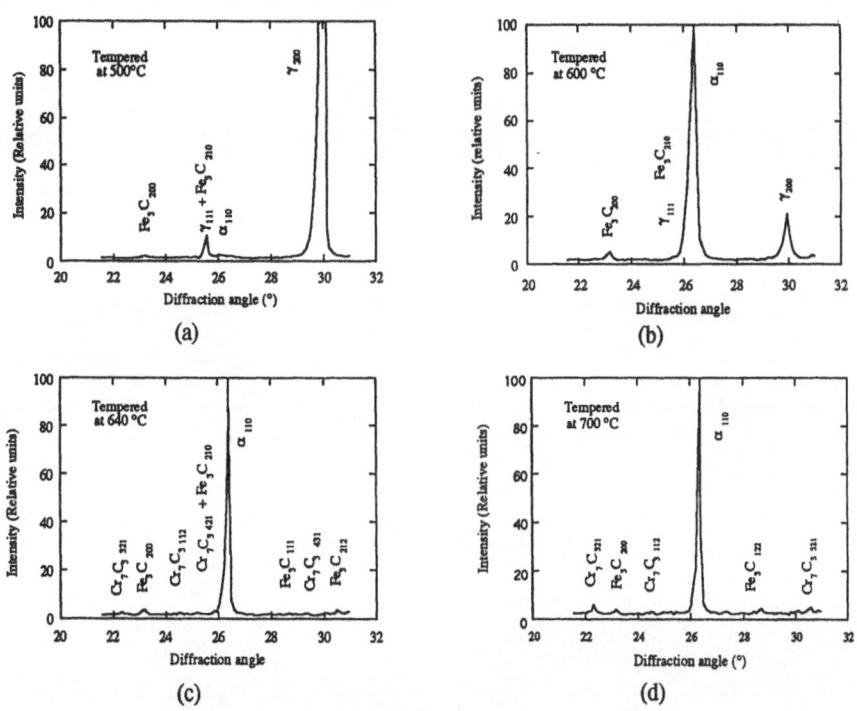

Figure 14. X-ray diffractograms for laser melted AISI 440C steel tempered at different temperatures.

The high secondary hardening temperature and peak hardness in rapidly solidified steels as compared to conventionally treated steels was previously reported by several authors [29, 33, 66, 67]. Rayment and Cantor [66] suggested that the shift in the secondary hardening peak is caused by a delay in onset of M_2C precipitation. Other authors [29, 33] explained this shift by the stabilizing effect of alloying element supersaturation in austenite.

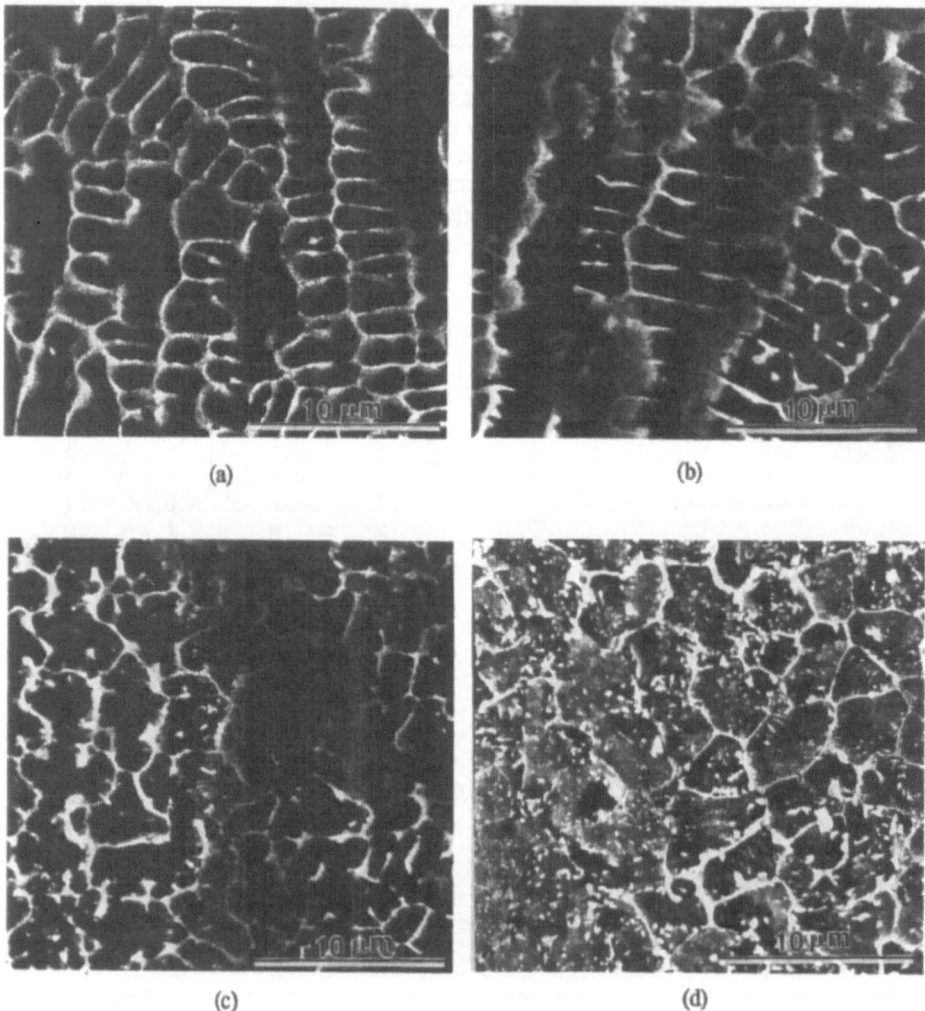

Figure 15. Microstructure of laser melted AISI 440C steel after tempering at (a) 500 °C, (b) 600 °C, (c) 640 °C and (d) 700 °C.

Tempering essentially involves age-hardening processes, but due to the complexity of the microstructure and of the chemical composition of the alloys, several aging processes occur simultaneously or sequentially. The precipitation sequences during tempering of conventionally hardened steels were described by Honeycombe [68]. In high-alloy medium to high carbon steels containing martensite and retained austenite tempering takes place in five distinct but partially overlapping stages:

- Stage 1 (T≤ 250 °C): precipitation of ε or η iron carbides at dislocations and formation of BCC martensite containing about 0.25% C. The kinetics of precipitation is controlled by carbon diffusion.

- Stage 2 (200≤T≤ 300 °C): decomposition of retained austenite into ferrite and Fe$_3$C. In high-alloy steels austenite is more stable and does not transform isothermally under 500 °C. It is destabilized by holding at higher temperatures and transforms to martensite during cooling.

- Stage 3 (200≤T≤ 350 °C): Fe$_3$C nucleates at ε/matrix interfaces, dislocations, twin or lath boundaries and original austenite grain boundaries, and gradually replaces ε carbide. The presence of carbide-forming alloying elements retards this transformation of ε carbide to Fe$_3$C up to 500 °C.

- Stage 4 (500≤T≤ 600 °C): secondary hardening. This stage appears only in steels which contain relatively large proportions of carbide-forming alloying elements and consists of the precipitation of a fine stable dispersion of alloy carbides that progressively replace Fe$_3$C. Alloy carbides only form above 500 °C because substitutional alloying elements cannot diffuse fast enough at lower temperatures. The carbides nucleate at the Fe$_3$C/matrix interfaces, dislocations, twin and lath boundaries and original austenite grain boundaries. The particles which nucleate at Fe$_3$C/matrix interfaces and at original austenite GBs do not contribute significantly to strength, because nucleation sites are too scattered, but those which nucleate at dislocations and twin and lath boundaries make a large contribution to strength.

- Stage 5 (T> 350 or T> 600 °C respectively in carbon and alloy steel): coalescence of carbides, recovery and recrystallization of the matrix. During this stage carbides coarsen by an Ostwald ripening process. The kinetics of coarsening of Fe$_3$C is controlled by vacancy diffusion whereas the coarsening of alloy carbides is controlled by substitutional atom diffusion. Above 500 to 600 °C dislocations and lath boundaries are eliminated progressively by recovery and, eventually ferrite recrystallizes. Recovery and recrystallization is facilitated by the coarsening of carbides that pin dislocations and lath boundaries. It is significantly delayed by the fine dispersion of alloy carbides formed during secondary hardening, which resist coalescence better than Fe$_3$C.

The microstructures of laser surface melted and quenched steels are different and this affects their evolution during tempering. The most noticeable differences are:

(a) the proportion of retained austenite is higher in LSM than in quenched steel. As a consequence, the properties of LSM steels depend critically on the transformations which occur within austenite during tempering, while the properties of conventionally treated steels are dominated by the evolution of martensite.

(b) the solute content in austenite and martensite is larger after LSM than after quenching. This is true, in particular, for vanadium, which exists mainly as primary V$_4$C$_3$ carbide in conventionally treated steel, but which goes into solid solution in LSM steels.

(c) austenite in LSM materials presents a high dislocation density, probably due to the plastic deformation induced by internal stresses. LSM steels also present a larger proportion of twinned martensite than similar quenched steels.

The larger proportion of strong carbide-forming elements in solid solution in the martensite found in LSM steel delays all the tempering stages, since alloying elements retard the precipitation of ε carbide, its transformation to Fe$_3$C and the coarsening of Fe$_3$C. Since larger amounts of alloying elements are available for precipitation, the density of secondary hardening carbides will be higher, leading to increased hardness. Moreover, vanadium is also in solution in the martensite, favoring the precipitation of V$_4$C$_3$. This carbide forms at a higher temperature than Cr, Mo or W carbides and is very resistant to coarsening, effectively retarding recovery and recrystallization of the matrix. All these effects tend to increase the secondary hardening temperature and the maximum hardness of the material and will, at least partially, explain the experimental results obtained.

The presence of large amounts of austenite also has a strong influence on tempering behavior. The austenite found in LSM steel presents a solute content higher than after quenching. As a result, austenite is very stable and will only transform at temperatures where precipitation of alloy carbides occurs. In conventionally hardened steels, the most likely nucleation sites for carbides are the original austenite grain boundaries, and precipitation from austenite will not contribute significantly to hardness, but in LSM steels austenite contains a high density of dislocations where carbides increase the hardness of the material. The results obtained on AISI 440C steel show an intense precipitation of M_7C_3 carbide within the austenite grains in the temperature range 500-640 °C. Since this precipitation probably occurs by nucleation and growth on dislocations and other crystallographic defects, the precipitates are finely dispersed and, as a result, they will contribute significantly to strengthening the austenite and to retarding its recovery. Furthermore, since the precipitation of carbides is accompanied by a decrease in supersaturation, austenite will be destabilized and will transform to martensite during cooling, leading to a further increase in the overall hardness of the material. Obviously, this last contribution occurs only if retained austenite is present before the tempering treatment. Another important fact that must be taken into account to explain the relative sluggishness of the precipitation reactions during tempering of LSM steels is that the diffusion coefficients of alloying elements in austenite are several orders of magnitude lower compared to ferrite or martensite (Table 3). As a consequence, secondary hardening will be delayed until higher temperatures.

TABLE 3. Diffusion coefficient of some alloying elements (cm^2/s) [69].

Element	$\alpha\,(\delta)$ - Iron			γ - Iron		
	Temperature °C			Temperature °C		
	20	400	800	20	400	800
C	10^{-17}	10^{-8}	10^{-5}	10^{-27}	10^{-13}	10^{-8}
Cr	10^{-42}	10^{-18}	10^{-11}	10^{-37}	10^{-18}	10^{-13}
Mo	10^{-46}	10^{-18}	10^{-11}	10^{-49}	10^{-21}	10^{-13}
V	-	-	10^{-12}	-	-	10^{-14}

6. Conclusions

1. To accomplish their tasks satisfactorily tool steels must be hard, tough and wear resistant throughout their operating temperature range. Since these properties are difficult to reconcile, surface engineering plays an important role in tool manufacturing.
2. Among the surface modification techniques available, laser hardening, laser alloying and laser cladding present significant advantages for the surface treatment of tools.
3. Successful laser hardening depends on the possibility of forming relatively homogeneous austenite during the short time of the laser induced thermal cycle. This treatment is only suited to carbon and low-alloy steels. For high-alloy steels, the

complete dissolution of carbides requires that the laser treatment is performed in the liquid state.

4. LSM improves the properties of tool steels by refining their microstructure, decreasing the proportion and average particle size of carbides, increasing the supersaturation of austenite and martensite, and eliminating inclusions. When applied to sintered steels, residual porosity can be completely eliminated.

5. If the ratio of austenite stabilizers to ferrite stabilizers is too high, austenite will be retained in large proportions and the steel will be too soft. A similar result is obtained if this ratio is too low, favoring the formation of metastable δ ferrite. The hardness will be equally low.

6. Since retained austenite has a deleterious influence on the properties of the material, it must be eliminated, either by multiple tempering treatments or cryogenic cooling. Tempering will be preferred for high-alloy steels, due to the beneficial effect of this treatment on the hardness of the material.

7. The secondary hardening peak temperature and maximum hardness after LSM are higher than after conventional heat treatment. The difference decreases as the number of tempering treatments increases. The delay in precipitation reactions during tempering of LSM steel can be explained by the high supersaturation of strong carbide-forming alloying elements in martensite and by the low diffusion coefficients of carbon and substitutional alloying elements in austenite.

References

1. Rickerby, D.S. and Matthews, A. (1991) Introduction, in D.S. Rickerby and A. Matthews (eds.), *Advanced Surface Coatings: A Handbook of Surface Engineering*, Blackie, Glasgow, pp. 1-13.
2. Bass, M., ed. (1983) *Laser Materials Processing. Materials Processing Theory and Practices*, F.F.Y. Wang (series ed.), Vol. 3, North-Holland Publishing Company.
3. Bell, T., Hancock, I.M. and Bloyce, A. (1987) Laser Surface Treatment of Tool Steels, in *Tool Materials for Molds and Dies*, Colorado School of Mines Press, St. Charles, Illinois, pp. 197-216.
4. Conde, O., Ferreira, M. L. G., Hochholdinger, P., Silvestre, A. J. and Vilar, R. (1992) CO_2 Laser Induced CVD of TiN, *Applied Surface Science* **54**, 130-134.
5. Silvestre, A. J., Conde, O., Vilar, R. and Jeandin, M. (1994) Structure and Morphology of Titanium Nitride Films Deposited by Laser Induced CVD, *Journal of Materials Science* **29**, 404-411.
6. Steen, W.M. (1988) Surface Treatment of Materials by Laser Beams - A Review, in *ECLAT '88 - 2nd European Conference on Laser Treatment of Materials*, DVS Verlag, Dusseldorf, pp. 60-64.
7. Barralis, J., Mathieu, G., Barreau, G. and Castex, L. (1990) Improvement of Mechanical Fatigue Strength by Optimization of Laser Surface Treatment, in A.P. Loureiro, O. Conde, L. Guerra-Rosa and R. Vilar (eds.), *Surface Engineering with High Energy Beams*, Trans Tech Publications Ltd., Zurich, pp. 111-120.
8. Åhman, L. (1984) Microstructure and its Effect on Toughness and Wear Resistance of Laser Surface Melted and Post Heat Treated High Speed Steel, *Metallurgical Transactions* **15A**, 1829-1935.
9. Amende, W. (1988) The Production of Wear Resistant Zones on Tools by Means of the CO_2 Laser, in *LIM5 - 5th International Conference on Lasers in Manufacturing*, IFS Publications, Bedford, UK, pp. 119-125.
10. De Beurs, H. and De Hosson, J.T. (1987) Wear Induced Hardening of Laser Processed Chromium-Carbon Steel, *Scripta Metallurgica et Materialia* **21**, 627-632.
11. Vilar, R., Miranda, R.M. and Oliveira, A.S. (1988) Laser Surface Treatment of AISI 420 Tool Steel, in O.D.D. Soares, S.P. Almeida and L.M. Bernardo (eds.), *Laser Technologies in Industry*, SPIE - Society of Photo-Optical Instrumentation Engineers, Bellingham, USA, pp. 713-718.

476

12. Colaço, R. and Vilar, R. (1992) Laser Surface Melting of Bearing Steels, in S. Martellucci and A. Scheggi (eds.), *Laser Applications for Mechanical Industry*, Kluwer Academic Publishers, Dordrecht, pp. 305-314.
13. Strutt, P.R., Nowotny, H., Tuli, M. and Kear, B.H. (1978) Laser Surface Melting of High Speed Tool Steels, *Materials Science and Engineering* **36**, 217-222.
14. Rabitsch, K., Ebner, R. and Major, B. (1994) Boride Laser Alloying of M2 High-Speed Steel, *Scripta Metallurgica et Materialia* **30**, 253-258.
15. Ebner, R. and Kriszt, B. (1992) Laser Alloyed High-Speed Steels, in B.L. Mordike (ed.), *Laser Treatment of Materials*, DGM Informationsgesellschaft mbH, Oberusel, pp. 187-192.
16. Colaço, R., Carvalho, T. and Vilar, R. M. (1994) Laser Cladding of Stellite 6 on Steel Substrates, *High Temperature Chemical Processes* **3**, 21-29.
17. Com-Nougué, J., Kerrand, E. and Hernandez, J. (1987) Chromium Steel Coating with Cobalt Base Alloy, in *LAMP '87 - International Conference on Laser Advanced Materials Processing*, High Temperature Society of Japan and Japan Laser Processing Society, Osaka, Japan, pp. 389-384.
18. Shen, J., Nowotny, S., Dausinger, F. and Hugel, H. (1992) Laser Surface Treatment of a Carbon Steel by Laser Cladding with Tungsten Carbide Composite Powders, in *LAMP '92 - International Conference on Laser Advanced Materials Processing - Science and Applications*, High Temperature Society of Japan, Nagaoka, Japan, pp. 755-760.
19. Conde, O., Mariano, J., Silvestre, A.J. and Vilar, R. (1990) Production of Hard Coatings by Laser CVD, in H.W. Bergmann and R. Kupfer (eds.), *ECLAT '90 - 3rd European Conference on Laser Treatment of Materials*, Sprechsaal Publishing Group, Coburg, pp. 145-154.
20. Cooper, K.P. and Ayers, J.D. (1984) Laser Melt-Particle Injection Processing, *Surface Engineering* **1**, 263-272.
21. Shafirstien, G., Bamberger, M., Langohr, M. and Maisenhalder, F. (1991) Laser Particle Alloying of Carbon Steel and α-Fe with CrB_2 Particles, *Surface & Coatings Technology* **45**, 417-423.
22. Löschau, W., Juch, K. and Reitzenstein, K., (1992) Alloying and Dispersion of TiC Hard Particles into Tool Steel Surfaces Using a High Power Laser, in B.L. Mordike (ed.), *Laser Treatment of Materials*, DGM Informationsgesellschaft mbH, Oberusel, pp. 281-286.
23. Escudero, M.L. and Belló, J.M. (1992) Laser Surface Treatment and Corrosion Behaviour of Martensitic Stainless AISI 420 Steel, *Materials Science and Engineering* **A158**, 227-233.
24. Ion, J.C., Moisio, T., Pedersen, T.F., Sorensen, B., and Hansson, C.M. (1991) Laser Surface Modification of a 13.5% Cr, 0.6% C Steel, *Journal of Materials Science* **26**, 43-48.
25. Meijer, J., Seegers, M., Vroegop, P.H., Wes, G.J.W (1985) Line Hardening by Low Power CO_2 Lasers, in C. Albright (ed.), *Laser Welding, Machining and Materials Processing, ICALEO '85 - International Conference on Applications of Lasers and Electro-Optics*, IFS Publications Ltd., San Francisco, pp. 229-238.
26. Kim, Y.W., Strutt, P.R. and Nowotny, H. (1979) Laser Melting and Heat Treatment of M2 Tool Steel: a Microstructural Characterization, *Metallurgical Transactions* **10A**, 881-886.
27. Strutt, P.R. (1980) A Comparative Study of Electron Beam and Laser Melting of M2 Tool Steel, *Materials Science and Engineering* **44**, 239-250.
28. Molian, P.A. and Rajasekhara, H.S. (1986) Analysis of Microstructures of Laser Surface Melted Tool Steels, *Journal of Materials Science Letters* **5**, 1292-1294.
29. Bloyce, A., Bell, T. and Hancock. I.M. (1988) Laser Surface Engineering of Tool Materials, in O.D.D. Soares, S.P. Almeida and L.M. Bernardo (eds.), *Laser Technologies in Industry*, SPIE - Society of Photo-Optical Instrumentation Engineers, Bellingham, Vol. 952 (Part 2), pp. 691-699.
30. Kusinski, J. (1988) Laser Melting of T1 High Speed Steel, *Metallurgical Transactions* **19A**, 377-382.
31. Vilar, R., Colaço, R. and Durão, L. (1992) Laser Surface Melting of Martensitic Stainless Tool Steels, in *LAMP '92 - International Conference on Laser Advanced Materials Processing - Science and Applications*, High Temperature Society of Japan, Nagaoka, Japan, pp. 779-782.
32. Vilar, R., Sabino, R. and Almeida, M.A. (1991) Laser Surface Melting of Sintered AISI T15 High-Speed Steel, in *ICALEO '91 - 10th International Congress on Applications of Lasers and Electro-Optics*, Laser Institute of America, Orlando, pp. 424-434.

33. Peng, Q.F., Shi, Z., Hancock, I.M. and Bloyce, A. (1989) Energy Beam Surface Treatment of Tool Steels and Their Wear, in A.P. Loureiro, O. Conde, L. Guerra-Rosa and R. Vilar (eds.), *Surface Engineering with High Energy Beams*, Trans Tech Publications Ltd., Zurich, pp. 229-244.
34. Vilar, R. and R. Colaço (1991) Laser Surface Melting of 440C Tool Steel, in *ICALEO '91 - 10th International Congress on Applications of Lasers and Electro-Optics*, Laser Institute of America, Orlando, pp. 435-444.
35. Costa, A.R., Anjos, M.A. and Vilar, R. (1992) Hardness Improvement in Laser Surface Melted Tool Steel by Cryogenic Quenching, in *ECLAT '92 - 4th European Conference on Laser Treatment of Materials*, DGM Metallurgy Information, New York, pp. 263-267.
36. Rayson, H.W. (1992) Tool Steels, in F. B. Pickering (ed.), *Constitution and Properties of Steels*, VCH, Weinheim, pp. 583-640.
37. Zum Gahr, K.H. (1987) *Microstructure and Wear of Materials*, Elsevier Scientific Publishing Company, Amsterdam.
38. Lim, S.C. and Ashby, M.F. (1987) Wear Maps, *Acta Metallurgica et Materialia* **35**, 1-24.
39. Ashby, M.F. and Lim, S. C. (1990) Wear Maps, *Surface & Coatings Technology* **24**, 805-810.
40. Trent, E.M. (1983) The Tribology of Metal Cutting, in M.H. Jones and D. Scott (eds.), *Industrial Tribology:The Practical Aspects of Friction, Lubrication and Wear*, Elsevier Scientific Publishing Company, Amsterdam, pp. 446-470.
41. Donaldson, C., LeCain, G. and Goold, V.C. (1973) *Tool Design*, McGraw-Hill, New York.
42. Lim, S.C., Liu, Y.B., Lee, S.H., and Seah, K.H.W (1993) Mapping the Wear of Some Cutting-Tool Materials, *Wear* **162-164**, 971-974.
43. Brandis, H., Haberling, E. and Weigand, H.H. (1980) Metallurgical Aspects of Carbides in High Speed Steel, in M.G.H. Wells and L.W. Lherbier (eds.), *Processing and Properties of High Speed Tool Steels*, The Metallurgical Society of AIME, Warrendale, pp. 1-18.
44. Zum Gahr, K.H. (1977) The Influence of Thermal Treatments on Abrasive Wear Resistance of Tool Steels, *Zeitung Metallkunde* **68**, 783-792.
45. Hertjen, D.J. and Jarvis, R.A. (1983) Rolling Element Bearings, in M.H. Jones and D. Scott (eds.), *Industrial Tribology:The Practical Aspects of Friction, Lubrication and Wear*, Elsevier Scientific Publishing Company, Amsterdam, pp. 132-183.
46. Kalousek, J., Fegredo, D.M. and Laufer, E.E. (1985) The Wear Resistance and Worn Metallography of Pearlite, Bainite and Tempered Martensite Rail Steel Microstructures of High Hardness, *Wear* **105**, 199-222.
47. Rice, S.L. (1987) Pitting Resistance of Some High Temperature Carburized Cases, Discussion Paper No. 780773, Society of Automotive Engineers.
48. El-Rakayby, A.M. and Mills, B. (1986) The Role of Primary Carbides in the Wear of High Speed Steels, *Wear* **112**, 327-340.
49. Vilar, R., Conde, O. and Colin, D. (1989) Laser Surface Melting of AISI 420 Stainless Tool Steel, in T.S. Sudarshan and D.G. Bhat (eds.) *Surface Modification Technologies III*, TMS, Warrendale, pp. 343-358.
50. Jatczak, J.F., Larson, J.A. and Shin, S.W. (1980) *Retained Austenite and its Measurement by X-ray Diffraction*, Society of Automotive Engineers, Warrendale, PA.
51. Anthony, T.R. and Cline, H.E. (1977) Surface Rippling Induced by Surface-Tension Gradients During Laser Surface Melting and Alloying, *Journal of Applied Physics* **48**, 3888-3894.
52. Keilmann, K. and Bai, Y.H. (1982) Periodic Surface Structures Frozen into CO_2 Laser-Melted Quartz, *Applied Physics* **29A**, 9-18.
53. Frenk, A. and Kurz, W. (1992) Microstructure Formation in Laser Materials Processing, *Lasers in Engineering* **1**, 193-212.
54. Dyson, D.J. and Holmes, B. (1970) Effect of Alloying Additions on the Lattice Parameter of Austenite, *Journal of Iron and Steel Institute* **5**, 469-474.
55. Forgeng, W.D. and Forgeng Jr., W.D. (1973) C-Cr-Fe (Carbon-Chromium-Iron), in T. Lyman (ed.), *Metals Handbook*, Vol. 8, American Society for Metals, Metals Park, OH, pp. 402-404.
56. Brooks, J.A., Baskes, M.I. and Greulich, F.A. (1991) Solidification Modeling and Solid-State Transformations in High-Energy Density Stainless Steel Welds, *Metallurgical Transactions* **22A**, 915-926.

57. Kelly, T.F., Cohen, M. and Sande, J.B.V. (1984) Rapid Solidification of a Droplet-Processed Stainless Steel, *Metallurgical Transactions* **15A**, 819-833.
58. Löser, W. and Herlach, D.M. (1982) Theoretical Treatment of the Solidification of Undercooled Fe-Cr-Ni Melts, *Metallurgical Transactions* **23A**, 1585-1591.
59. Franklin, P. and Davies, D.L. (1978) *Powder Metallurgy* **10**, 7.
60. Cotterell, B., He, S.Q. and Mai, Y.W. (1994) Fatigue of Sintered Steel, *Acta Metallurgica et Materialia* **42**, 99-104.
61. Vardavoulias, M., Jeandin, M. and Grillon, F. (1993) Effects of Tempering on the Sliding-Wear Behavior of a Sintered High Speed Steel: A Quantitative Analysis Approach, *Scripta Metallurgica et Materialia* **29**, 359-364.
62. Thelning, K.E. (1984) *Steel and its Heat Treatments*, Butterworths, London.
63. Peng, Q.F., Shi, Z., Hancock, I.M. and Bloyce, A. (1990) Energy Beam Surface Treatment of Tool Steels and Their Wear, *Key Engineering Materials* **46 & 47**, 229-244.
64. Roberts, G.A. and Cary, R.A. (1980) *Tool Steels*, American Society for Metals, Metals Park, OH.
65. Vilar, R.M., Pelletier, M. and Cizeron, G. (1981) Étude du Revenu de L'acier Z 10 C D NbV09-02 (type 9% Cr, 2% Mo + Nb, V) Envisagé pour la Réalisation de Certains Composants des Réacteurs à Neutrons Rapides, *Journal of Nuclear Materials* **105**, 237-247.
66. Rayment, J.J. and Cantor, B. (1981) The As-quenched Microstructure and Tempering Behavior of Rapidly Solidified Tungsten Steels, *Metallurgical Transactions* **12A**, 1557-1567.
67. Peng, Q.F., Shi, Z., Bloyce, A. and Bell, T. (1990) Surface Electron-Beam Melting and Alloying of Tool Steels, *Materials Science and Technology* **6**, 999-1004.
68. Honeycombe, R.W.K. (1981) *Steels: Microstructure and Properties, Metallurgy and Materials Science*, R.W.K. Honeycombe and P. Hancock (series eds.), Edward Arnold, London.
69. Folkhard, E. (1988) *Welding Metallurgy of Stainless Steels*, Springer-Verlag, Wien.

LASER TREATMENT FOR ADHESIVE BONDING IN COATED STEELS

M. OLFERT*, W. DULEY*, T. NORTH**

* Guelph-Waterloo Physics Program (GWP²) Waterloo Campus,
Waterloo, Ontario, N2L 3G1 Canada.
** Department of Metallurgy and Materials Science, University of
Toronto, Ontario.

ABSTRACT

Zinc coated steel sheet in the form of both temper rolled galvanized and galvanneal are used extensively in the automotive industry. Through a process of excimer laser surface treatment, we have succeeded in significantly enhancing the adhesion characteristics of the coated steel. Preliminary adhesion strength testing has been carried out in the form of simple T peel tests, using a hot melt nylon resin as the adhesive. In some cases, well over a two fold improvement over the untreated parent material has been observed. SEM micrographs indicate a wide range of new surface morphologies strongly dependent on both parent material and processing conditions. Adhesion strength vs. process parameters is discussed and potential trends for optimization of the treatment are indicated.

1. Introduction

Short pulse duration and high fluence which are characteristic of excimer lasers make them extremely useful for surface processing of materials without altering their bulk properties. The coupling of ultraviolet radiation with metal surfaces tends to be considerably greater than that of longer wave radiation, further strengthening the position of excimer lasers in the area of metallic surface processing.

Excimer laser surfacing of metals has been shown to significantly alter optical properties [1, 2]. Studies of adhesion properties of excimer laser treated aluminum revealed little if any significant change from untreated parent material [3].

In this paper we will present results of adhesion studies in excimer laser treated coated steels which indicate significant enhancement in adhesion properties over

J. Mazumder et al. (eds.), Laser Processing: Surface Treatment and Film Deposition, 479–490.
© 1996 Kluwer Academic Publishers.

untreated parent. The surfacing technique with its associated benefits and problems will be discussed as well as the parametric variables. The adhesion results will be correlated with surface topography measurements obtained with a confocal scanning laser microscope (SLM). In general, the adhesion characteristics of a given surface depend on both surface topography and surface chemistry. We will only consider effects associated with surface topography in this work, that is mechanical adhesion will be assumed to be the dominant mechanism.

2. Experimental

2.1. SAMPLE PRE-PREPARATION

Production line samples of 0.033 inch G90 temper rolled galvanized and galvanneal were obtained from Stelco Steel, and sectioned into strips measuring one half inch by four inches. The strips were then ultrasonically washed in acetone and stored in a dessicater until required for laser treatment or bonding.

2.2. LASER TREATMENT

Samples of galvanneal and galvanized sheet steel were surface processed with a Lumonics Hyper-Ex 460 laser fitted with a uniform beam electrode set. The laser was operated on a xenon chloride gas mixture which produces laser output at 308 nm. The output was filtered through a 1.9 cm square aperture and regulated to deliver 50 ± 10 mJ/pulse to the work. A two and a half inch plano-convex lens focused the beam to 0.40 mm^2 on the sample surface, resulting in energy deposition of roughly 13 J/cm^2/pulse. Pulse lengths were not measured, but in accordance with the laser operating manual, pulse lengths on the order of 40 to 60 ns would be expected.

The general surface treatment scheme utilized a bi-directional scan/step sample motion coupled with a constant repetition firing of the laser. A schematic diagram of the treatment scheme is shown in *Figure* 1. The parameters involving scan velocity and step size are treated as variables, thus allowing considerable range in the extent of the treatment. Treatments were performed in normal atmosphere with no assist gas. Treated samples were once again washed in an ultrasonic bath of acetone, and then placed in a dessicater until ready for bonding.

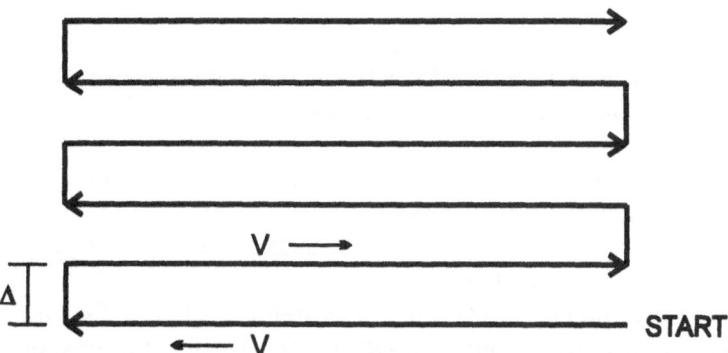

Figure 1. Schematic representation of treatment scheme. With this simple scan/step scheme, the variable parameters are scan velocity V, and step size Δ. The number of scans required to treat a given area will depend on the magnitude of the step chosen.

Unless otherwise noted, the scan direction was parallel to the short dimension of the metal strips. Three treatment regimes were investigated. The minimum treatment involved a linear pulse density of 79 pulses/cm and a step size of 0.051 cm resulting in an overall energy deposition of 78 J/cm^2 (based on laser output of 50 mJ/pulse). In the two remaining treatments the energy deposition was doubled to 155 J/cm^2 in one case by doubling the linear pulse density, and in the other case by doubling the scan density. From this point forth, the treatment regimes will be identified in terms of their relative pulse and scan densities. Low density treatment will indicate 79 pulses/cm and 20 scans/cm. High pulse density treatment refers to 157 pulses/cm and 20 scans/cm. High scan density treatment refers to 79 pulses/cm and 40 scans/cm.

2.3. BONDING

Pairs of like strips which had been subjected to identical treatments were bonded together over the treated regions in a geometry conducive to "T" peel testing of the bonded joint. Experimental limitations would not allow testing conditions to fulfill ASTM standards for "T" peel testing [4], so instead the following conditions were standardized throughout the experiments.

Like samples were bonded together over approximately one inch extending to one end with Elvamide 8061 nylon resin. The resin, in the form of 3mm×3mm×1.5mm pellets was sandwiched between pairs of strips at the treated end and held together with spring clips. The clamped samples were subsequently placed in an oven at 204°C for 10 minutes to flow the resin evenly. Upon removal from the oven, samples were quenched to room temperature in water, then dried with air. Extruded resin was trimmed from the joint, and samples were placed in a dessicater for no less than 20 hours before peel testing.

2.4. PEEL TESTING

Bonded samples were placed in a bending jig and bent into a "T" configuration with the bonded joint constituting the leg of the "T". To isolate and standardize peel force contributions associated with plastic-elastic deformation of the steel adherends [5], a peel jig was constructed which standardized the extent of deformation amongst the peel tests. Furthermore, the jig allows baseline deformation force measurements to be made by essentially performing a "peel" test on unbonded samples. *Figure* 3 details the geometry involved for the peel tests.

The loaded peel jig was mounted in an Instron configured for cross head speed of 2 cm/min and data was recorded to a strip chart recorder at a speed of 5 cm/min. Since the jig holds two half inch wide "T" joints back to back, the separation force recorded by the Instron corresponded directly with peel force in kg/inch and was scaled to give results in kg/cm.

2.5. SCANNING LASER MICROSCOPY

A confocal scanning laser microscope was used to measure the surface roughness of pre and post laser treatment specimens. The microscope was of the scanning beam type, utilizing a 633 nm Helium Neon Laser beam. The microscope was fitted with a 50× 1.0 NA objective, yielding field of view of roughly 400×400 µm (x-y plane). Each x-y image was stored into a 512×512 pixel array giving an x-y resolution of less than 1 µm. Surface profiling was accomplished by stepping a piezoelectric z stage through a series of 1.5 µm intervals which encompassed the full topography of the specimen. Images of the z slices were then post processed with software specifically designed to generate topographic information from families of confocal z slices. Resolution in the z direction is roughly twice the step size or 3 µm.

3. Results

3.1. INITIAL SEM SURVEY

Initial studies were carried out on non temper rolled galvanized steel sheet as well as galvanneal sheet investigating the morphological effects of scanning focused excimer laser pulses across the metal surfaces. Overall surface energy deposition density was varied by adjusting pulse overlap (by varying scan speed), scan overlap, and pulse energy. Topographical changes were induced in the metals by a wide range of processing conditions. These changes were examined with a scanning electron microscope to localize the parameter set for the survey of adhesion properties.

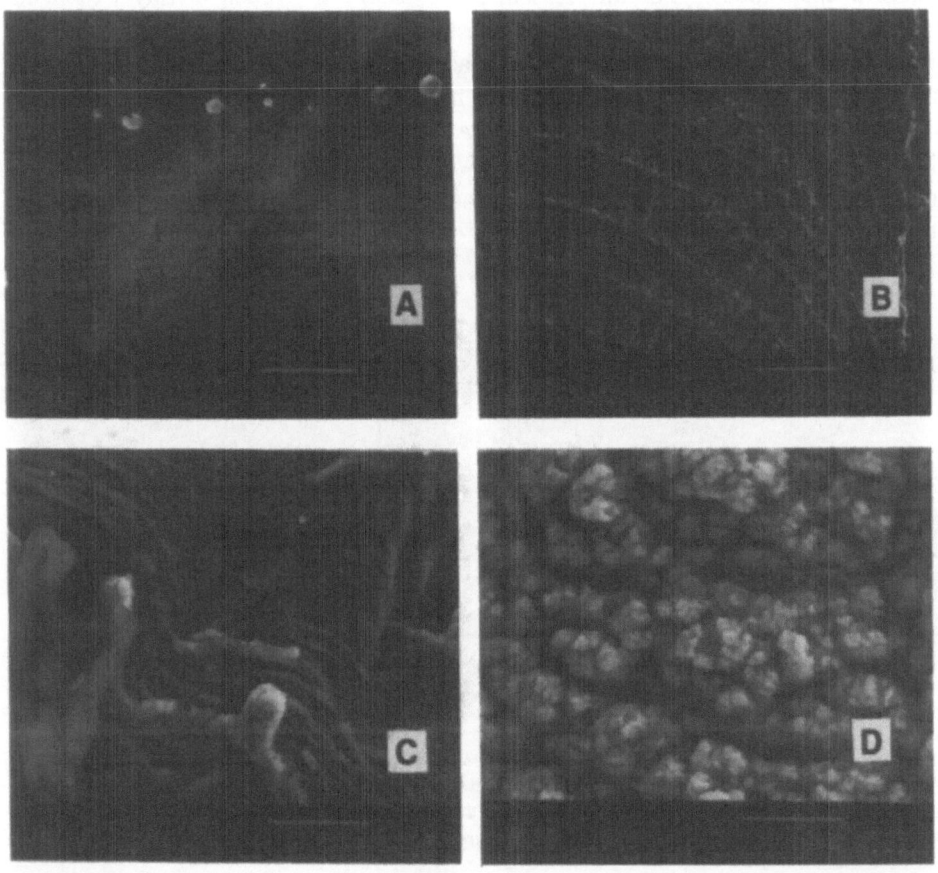

Figure 2. SEM micrographs of non temper rolled galvanized steel sheet. a) Untreated. b) surface treated with 190 J/cm² energy density. c) Higher magnification view of *Figure* 1b. d) Surface treated with 1200 J/cm² energy density.

Microscopic analysis revealed a range of surface topography dependent strongly on processing parameters. SEM images of non temper rolled galvanized sheet highlight the extent of surface alteration possible through variation of the processing parameters. *Figure* 2a shows the surface of non temper rolled galvanized steel before laser treatment. *Figure* 2b shows a region which has been processed with a surface energy deposition density of 190 J/cm². A higher magnification view of the same treatment (*Figure* 2c) shows that the alteration of the surface topography extends down to the micron scale. The micrograph shown in *Figure* 2d demonstrates the nature of the surface generated under much higher energy deposition density treatment (1200 J/cm²). In this case, the surface appears to be covered with a powdery deposition which would be undesirable for

484

adhesive applications. For this reason, lower energy deposition density treatments (≤ 155 J/cm^2), were chosen for the initial investigation of adhesion properties.

The material dependence of the surface topography is demonstrated in *Figure* 3 showing galvanneal which has undergone high energy deposition density treatment similar to the galvanized shown in *Figure* 2d. As well, dependence on the scan direction is highlighted in *Figure* 3a and 3b where the only difference in the treatment is a reversal of the scanning direction of the beam. This is attributed to asymmetries in the intensity distribution of the focal spot.

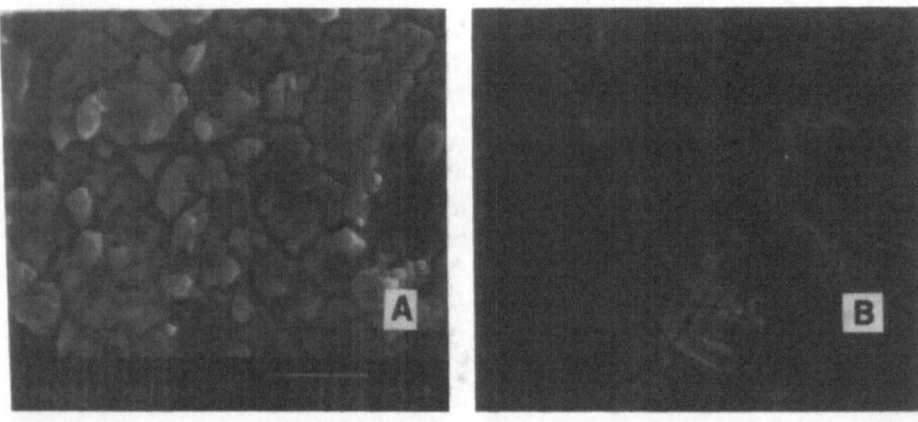

Figure 3. SEM micrographs of high energy density treated galvanneal. The treatment corresponds to that of the galvanized shown in *Figure* 1d. The only difference in treatment between *Figure* 2a and 2b is a reversal of the scanning direction of the laser. The magnification of *Figure* 2b is the same as that of *Figure* 2a.

3.2. PEEL TEST RESULTS

Peel tests carried out on treated samples were always accompanied by at least one "benchmark" test involving untreated material bonded at the same time and under the same conditions as the treated materials. Baseline measurements, performed by separating unbonded strips give an indication of the force involved solely in the deformation of the adherends. Peel force measurements are all calculated based on the difference between the peel test and the baseline for the particular material.

Figure 4 shows results of baseline, benchmark and low density treatment peel tests for galvanneal. The peel values are experimentally determined to be 5.63 kg/cm, 6.73 kg/cm, and 7.44 kg/cm respectively for the baseline, benchmark and treated specimens. This yields true adhesive peel values of 1.1 kg/cm and 1.8 kg/cm for the benchmark peel and the treated galvanneal respectively. Defining the margin of improvement (MOI) as the ratio of treated sample peel value to benchmark peel value we have for the low density treatment of galvanneal an MOI of 1.6.

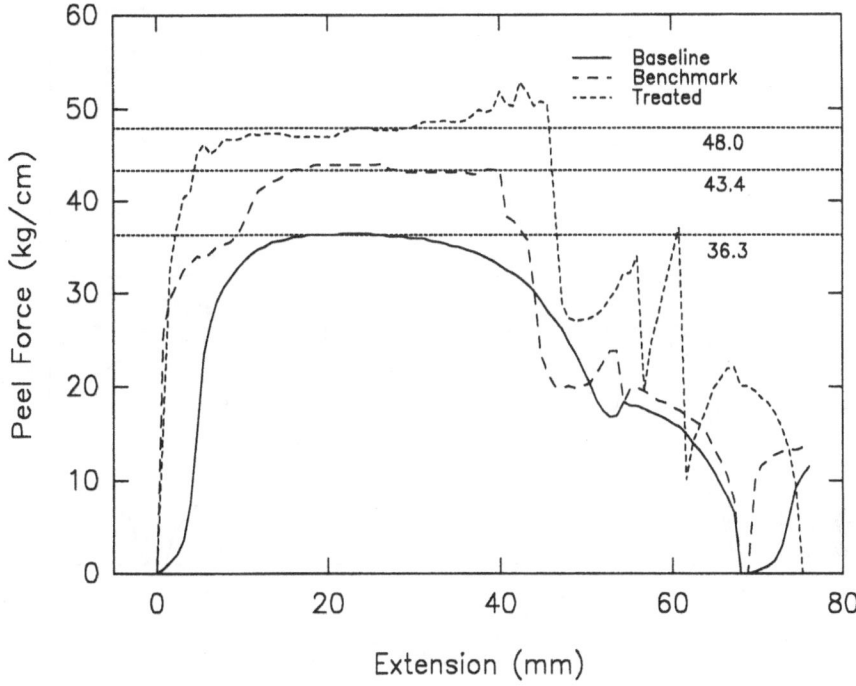

Figure 4. Comparative peel test results involving low density treatment in galvanneal including both baseline and benchmark peel results. The results indicate that the surface treatment improves the adhesion by a factor of 1.6.

Comparative peel tests performed on the temper rolled galvanized exhibited considerably different behaviour than the galvanneal tests. Peel tests of the treated samples tended not to plateau at all, having instead strongly increasing trends after the initial peak. This tendency can be explained only if the failure mode of the joint has changed from primarily interfacial failure to elastic and cohesive failure of the adhesive itself. In these situations, estimation of the margin of improvement associated with the treatment is obviously limited by the strength of the adhesive. *Figure* 5 shows an example of elastic/cohesive failure in a low density treatment on temper rolled galvanized.

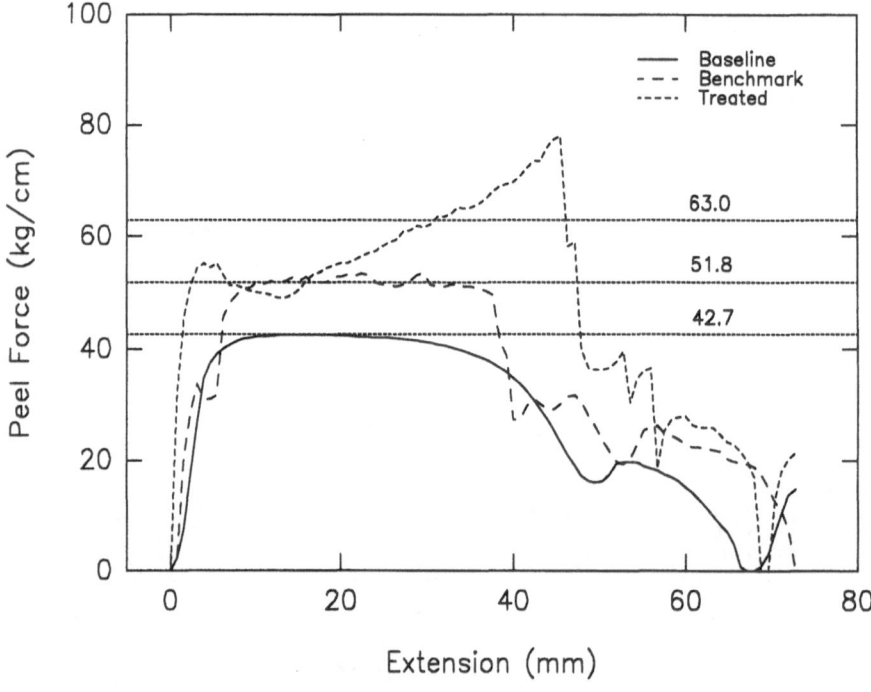

Figure 5. Peel force comparisons involving low density treated temper rolled galvanized. The increasing trend associated with the treated specimen is indicative of elastic/cohesive failure of the adhesive.

The results of the comparative peel tests are summarized in tables 1 and 2. Peel failures and margins of improvement which have been limited by elastic/cohesive failure of the adhesive are marked with asterisks.

Table 1. Summary of peel test results for laser surfaced galvanneal:

Treatment	Benchmark	Treated	MOI
low density	1.1 kg/cm	1.8 kg/cm	1.6
high pulse density	1.1 kg/cm	1.5 kg/cm	1.4
high scan density	1.1 kg/cm	1.8 kg/cm	1.6

Table 2. Summary of peel test results for laser surfaced temper rolled galvanized:

Treatment	Benchmark	Treated	MOI
low density	1.4 kg/cm	3.1 kg/cm*	2.2*
high pulse density	1.4 kg/cm	4.0 kg/cm*	2.8*
high scan density	1.4 kg/cm	3.2 kg/cm*	2.3*

* All of the treated temper rolled galvanized exhibited some degree of cohesive failure in peel tests.

3.3. SLM RESULTS

Surface topography profiles of untreated galvanneal are shown in *Figure* 6 indicating sharply varying surface structure extending beyond 30 microns in depth frequently. The horizontal lines traversing the accompanying SLM image indicate the locations of the profiles.

Surface profiles of low density treatment galvanneal demonstrate topography which is highly dependent on scan direction. *Figure* 7 shows an SLM image of low density treatment galvanneal showing segments of both scan directions (Separated by the dark vertical line). The surface profiles indicate that in one scan direction, the overall surface structure appears considerably smoother than that of untreated parent material, having the surface texture reduced to a depth of roughly 10 microns.

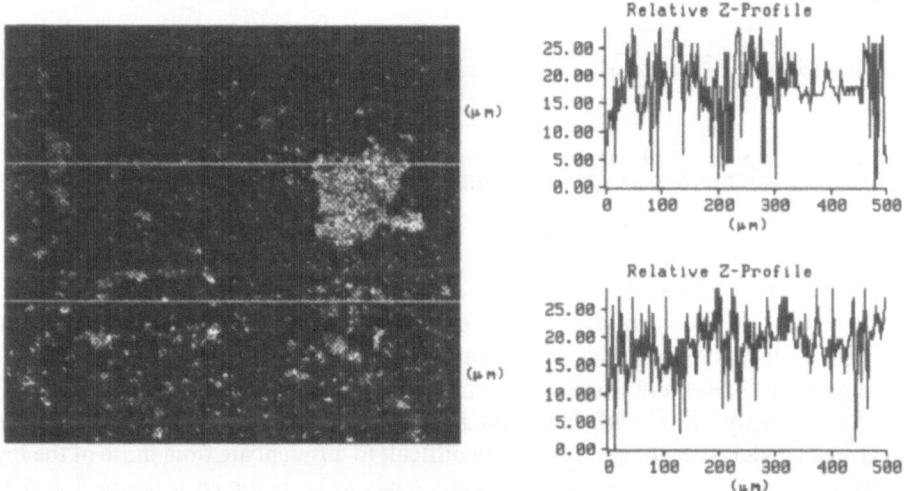

Figure 6. SLM image and z-profiles for untreated galvanneal. Locations of the surface profiles are marked with horizontal lines. The profiles reveal highly rough surface structure extending beyond 30 microns frequently.

Similar imaging of the high scan density and high pulse density treatment galvanneal indicate generally smoother surfaces than the parent material with large scale (≈30 micron) features occurring infrequently and average topography restricted to the order of 10 microns. It is worth noting that the largest topographical features are nearly always associated with either scan boundaries or trailing edge pulse boundaries.

SLM imaging of untreated temper rolled galvanized reveals surface structure which is seen to vary over at least a 30 micron range, but only in isolated regions. The remainder of the surface indicates roughness extending over less than 5 microns in general and as such, the temper rolled galvanized is considerably smoother than the untreated galvanneal.

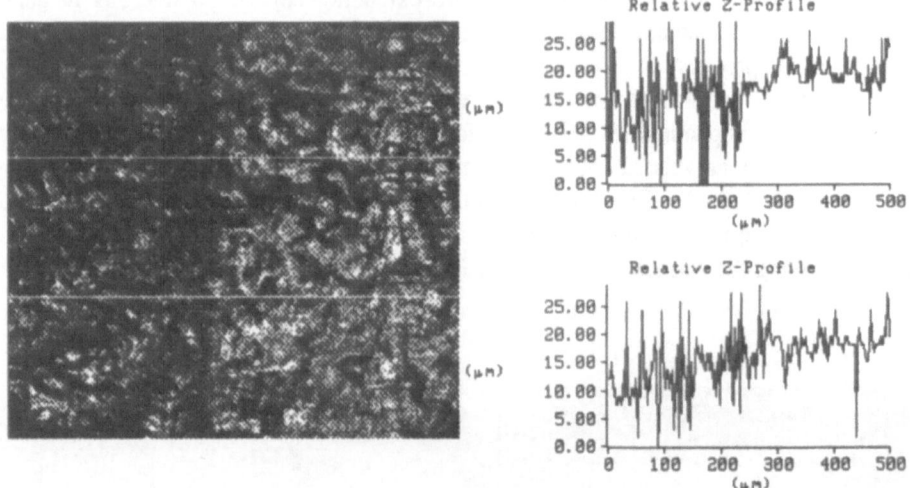

Figure 7. SLM image and z-profiles of low density treatment galvanneal. The dark vertical line just left of center is the scan direction boundary. Note the discontinuity in surface roughness associated with the scan direction.

Surface profiles associated with low density treatment galvanized reveal a surface with both consistent and greater overall roughness than the untreated parent material. Regions of the surface which correspond to either pulse crater boundaries, or scan boundaries generally have the largest scale topography extending beyond 30 microns frequently. SLM surface profiling for the remainder of the treatment regimes on galvanized yields results which are very difficult to differentiate from those of the low density treatment. In general, the roughness tends to be scattered between 5 and 15 microns with larger scale topography localised to boundary regions as discussed above.

4. Discussion

The results of the data set involving galvanneal indicate that at least for the elvamide nylon resin, adhesion qualities are not directly linked to large scale and high frequency surface roughness. Untreated galvanneal exhibits surface roughness which tends to be greater than any of the surfaces generated by laser treatment, yet in all cases (Table 1), adhesion quality is enhanced through laser surfacing. One possibility is that the resin does not fully wet the adherend and hence makes little or no use mechanically of the higher frequency features. Through laser surfacing, many of the high frequency features appear to be smoothed which may result in a more complete wetting for the given adhesive.

Results of peel tests on galvanneal showed no significant dependence on variation of processing parameters with margins of improvement between 1.4 and 1.6. Such an insensitivity to parameters is desirable from an industrial standpoint where production quality and volume are important.

Unlike galvanneal, temper rolled galvanized is characterized by a generally smoother surface with isolated regions of high roughness. SLM data of processed galvanized reveals increased roughness in all treatment regimes. The dependence of the peel strength on processing parameters was somewhat shrouded by the fact that all tests showed evidence of cohesive failure within the adhesive. However, with the exception of the high pulse density treatment with an MOI of 2.8, the margins of improvement were grouped between 2.2 and 2.3. If one assumes the major contribution to enhancement of adhesion comes from the large scale topographical features associated with pulse and scan boundaries, then it is no surprise that the high pulse density regime demonstrates the best adhesion qualities. The high scan density regime delivers the same surface energy density as the high pulse density regime, but because the scans overlap, much of the processing is simply destroying the large scale features from the previous scan.

In summary, through excimer laser processing of temper rolled galvanized and galvanneal sheet steel, we have enhanced the adhesion qualities of both materials significantly. Galvanneal experiences adhesion enhancements ranging from 140% to 160% and galvanneal experiences enhancements from 220% to 280% , being limited only by the strength of the chosen adhesive. The choice of a bi-directional scan step treatment technique was somewhat arbitrary and the results have indicated that the surface topography can be strongly dependent on scan direction. To maximize the desirable topography for adhesion, the treatment scheme should be restricted to a uni-directional type of scan step treatment.

Acknowledgments

The authors wish to gratefully acknowledge the financial support of the Ontario Centre For Materials Research. We extend our thanks to D. Sakai and J. Hood of Stelco Steel for many useful discussions, as well as Dr. S. Damaskinos for performing the Scanning Laser Microscope measurements.

References

1. Kinsman, G. and Duley, W. W. (1993) "Treatment of Metal Surfaces with Excimer Laser Radiation for Radiative Applications," *Applied Optics*, **32**, 7462-7470.

2. Kinsman, G. (1991) "Excimer Laser Irradiation of Metals," Ph.D. thesis, York University, North York Ontario, Canada.

3. Libertucci, M. Kinsman, G. North, T. and Duley, W.W. (1994) "Effect of Excimer Laser Treatment on the Strength of Aluminum/Polypropylene Joints," *Polymer Science and Engineering*.

4. ASTM document designation **D 1876-93,** (1993), "Standard Test Method for Peel Resistance of Adhesives (T-Peel test)".

5. Igarashi, T. (1982) "Peel Strength and Energy Dissipation," in Mittal, K. L. (ed.), *"Adhesive Joints,"* Plenum, pp 419-432.

STRUCTURE-PROPERTY RELATIONSHIP OF CERAMIC COATINGS ON METALS PRODUCED BY LASER PROCESSING

J.Th.M. DE HOSSON and M. VAN DEN BURG
Department of Applied Physics, University of Groningen,
Nijenborgh 4, 9747 AG Groningen, The Netherlands.

Abstract

This paper concentrates on the mechanical performance of various ceramic coatings of Cr_2O_3 on steel (SAF2205), as produced by CO_2 laser processing. The thickness of the coating that can be applied by laser coating is limited to about 200 µm setting a limit to the maximum strain energy release rate that can be measured in a 4 point flexure test before severe yielding occurs. In addition, a network of cracks with spacings of the order of 200 µm was always present in the laser applied coating preventing steady state crack growth along the interface.

It is concluded that a firmly bonded coating of Cr_2O_3 on steel could be produced by high power laser processing. The actual interface strength of a (Fe,Cr)-spinel applied to stainless steel by laser coating depends strongly on the composition of the substrate and coating materials. The the energy release rate was extremely high and delamination occurred by fracture through the coating and partially along the interface, indicating that the interface strength is similar to or higher than the fracture strength of (Fe,Cr)-spinel. The experimental observations are interpreted in terms of a microscopic description of emitting dislocations form an interfacial crack between dissimilar materials.

1.   Introduction

Performance of a laser applied coating is limited by its material property. However, it is quite rare that a material performance is determined by one single property. In particular the situations in which ceramic coatings are applied often demand a combination of corrosion and wear resistance, both of which are system properties rather than material properties as such. Unfortunately, standardised and specific experiments to test such combinations of system properties are not well developed. A specific problem with normalisation is the multitude of combinations that can be made from individual properties. As a result experimental efforts are directed to test one property at a time.

491

J. Mazumder et al. (eds.), Laser Processing: Surface Treatment and Film Deposition, 491–510.
© 1996 *Kluwer Academic Publishers.*

In practice the interface region between the coating and the substrate material often turns out to be the weakest link. Therefore our effort of evaluating the performance of the coating is aimed at testing these interface properties. There is also a more fundamental engineering interest in interface properties; as the properties of the individual substrate and coating are known or can be measured it is interesting to investigate possible correlations between the interface properties and the properties of the individual constituents.

The methods applied are based on the release of strain energy when creating a unit area of free surface. The advantage of this approach lies in the independence between the measured properties and the crack distribution in the ceramic coating.

2. Experimental

2.1. LASER COATING PROCESS

A CW-CO_2-laser (Spectra Physics 820) is used for the laser coating process with the following laser parameters: 1.0 kW laser power, a spot size of 1.27 mm, a laser scan velocity of 20 mm/s and an overlap of 75% between the tracks. As coating powder Cr_2O_3, Cr_2O_3 containing Mn and Cr_2O_3 containing extra Cr and Mo are used. The Mn is added to investigate the influence of Mn on the lattice parameters of the (Fe,Cr)-spinel coating and on the interface between the substrate and the coating. Cr and Mo are added to manipulate the crystal structure of the melt pool in the austenitic stainless steels from fcc to bcc in order to change the thermal expansion coefficient of the substrate. The compositions of the coating powders are listed in table 1.

TABLE 1. The composition of the coating powders.

composition (w%)	Cr_2O_3	Mn	Cr	Mo
1	100	-	-	-
2	92.5	7.5	-	-
3	75	25	-	-
4	66	-	20	14

As substrate the dual phase SAF2205 steel and the austenitic stainless steels SS304 and SS316 are used. Colour etching and X-ray diffraction methods indicate that phases in the laser melted substrate are the same before and after laser treatment when Cr_2O_3 powder or Cr_2O_3 powder containing Mn is used. After coating with Cr_2O_3 powder containing extra Cr and Mo the melt pool in the substrate is completely transformed to bcc in all substrate steels.

X-ray diffraction is used to identify the phases present in the laser coating. The peak positions of the $FeCr_2O_4$ and Cr_3O_4 phases [1] are used to calculate the lattice parameters of the (Fe,Cr)-spinel phases in the coating. Cross sectional TEM, SEM and SEM fractography as well as optical microscopy are applied to study the crack patterns and fracture surfaces in the laser applied coating. The composition of the different phases is analysed using semi-quantitative EDS (energy dispersive X-ray spectroscopy).

2.2. 3- AND 4-POINT FLEXURE TESTS: DETERMINATION OF $G_{INTERFACE}$ and $K_{C(INTERFACE)}$

Several test methods exist to determine the mechanical properties of bimaterial interfaces, most of them bear some disadvantages. Most tests require the attachment of a handle to the top of the coating in order to apply a force to the bimaterial interface [2]. When the interfacial strength of the bimaterial is very high it is impossible to attach a handle to the material in such a way that the handle-to-coating interface is stronger than the bimaterial interface.

The three- and four-point flexure tests show the best promise of circumventing the problem of attachment and at the same time of obtaining a significant quantity for the interfacial strength of a bimaterial [3, 4, 5, 6]. The specimen used in the flexure experiments consists of a bimaterial beam (fig. 1) with overall dimensions 40 x 3 x 3 mm. The sides of the specimen are polished to facilitate optical observations during and after testing but the surfaces of the coatings are as received to prevent the formation of surface cracks due to grinding or polishing. The actual experiments are performed on duplex SAF2205 coated with a 200 μm plasma sprayed Cr_2O_3 coating and with a 100 μm laser applied (Fe,Cr)-spinel coating. Specimen with laser tracks parallel as well as perpendicular to the direction of loading are tested.

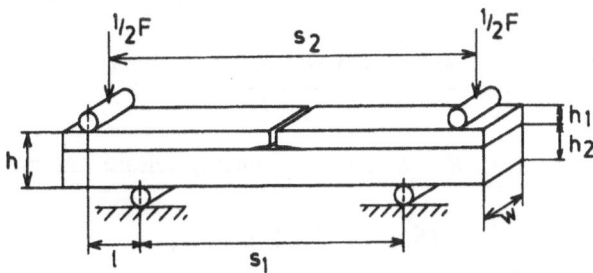

Figure 1. Four-point flexure set-up for the determination of the critical energy release rate of coating delamination. s_2 = 33 mm, s_1 = 19 mm, l = 7 mm.

The flexure test consists of three steps, the first of which is making one or more Knoop indentations with the long diagonal of the indentation along the width of the specimen. A precrack is then made through the thickness of the coating along the line of the Knoop indentations by three point flexure. The objective of three-point flexure is to induce a symmetrical precrack along the bimaterial interface. The specimen is then loaded in four-point flexure until delamination along the interface occurs. As the coating is only a small portion of the total system in the case of the plasma sprayed and laser applied ceramic coatings interfacial delamination is monitored by acoustic emission.

The four-point flexure test is based on the storage of a well known amount of elastic energy on bending and a release of this elastic energy on fracture. The specimen rests on two rollers with two other rollers exerting a force on the specimen (fig. 1) resulting in a constant bending moment in the region between the inner rollers:

$$M = \frac{Fl}{2w} \tag{1}$$

i.e. the same stress state is obtained throughout the region in between the inner rollers, making the experiment insensitive to the exact location of fracture. Consequently, the strain energy release rate upon delamination of the coating should exhibit steady state characteristics, at least when the interface crack length significantly exceeds the thickness of the coating. The steady state energy release rate G_{ss} can be deduced from the difference in the strain energy in the cracked and the uncracked beam. Since there is negligible strain energy in the beam above the crack, i.e. in the delaminated coating, the energy release rate is simply the difference in strain energy of the uncracked section and the section of the lower beam beneath the crack, i.e. of the substrate plus coating and the substrate respectively. From Euler-Bernoulli beam theory and plane strain conditions the strain energies can be expressed in terms of the applied moment:

$$U = \frac{(1 - v^2)M^2}{2EI} \quad , \tag{2}$$

where U is the strain energy per unit cross-section and I the second moment of area per unit width. The strain energy release rate upon coating delamination becomes [3, 7]:

$$G_{ss} = \frac{M^2(1 - v_2^2)}{2E_2}\left(\frac{1}{I_2} - \frac{\lambda}{I_c}\right) \tag{3}$$

For thin coatings, i.e. $h_1 << h_2 \approx h$, the strain energy release rate may be approximated by:

$$G_{ss} = \frac{18M^2(1 - v_2^2)^2 E_1}{E_2^2(1 - v_1^2)} \cdot \frac{h_1}{h^4}. \tag{4}$$

As can be seen from eq. (4) the strain energy release rate depends critically on the measured value of $h \approx h_2$. Interfacial crack propagation occurs when the strain energy release rate G_{ss} equals the critical energy release rate G_c of interfacial failure. The stress intensity factor K_c characterising the bimaterial interface is given by:

$$G_c = \frac{2\left(\dfrac{1-v_1}{\mu_1} + \dfrac{1-v_2}{\mu_2}\right)}{4\cos h^2(\pi\varepsilon)} K_c^2 \tag{5}$$

where ε represents the bimaterial constant [7].

3. Results

Plasma sprayed Cr_2O_3 coating on steel

The critical energy release rate and the stress intensity factor are found to be:
$G_c = 117$ [J/m^2],
$|K_c| = 4.6$ [MPa$\sqrt{m}$].

Laser applied (Fe,Cr)-spinel coating on steel

The thickness of the coating that can be applied by laser coating is limited to about 200 μm setting a limit to the maximum strain energy release rate that can be measured before severe yielding occurs.

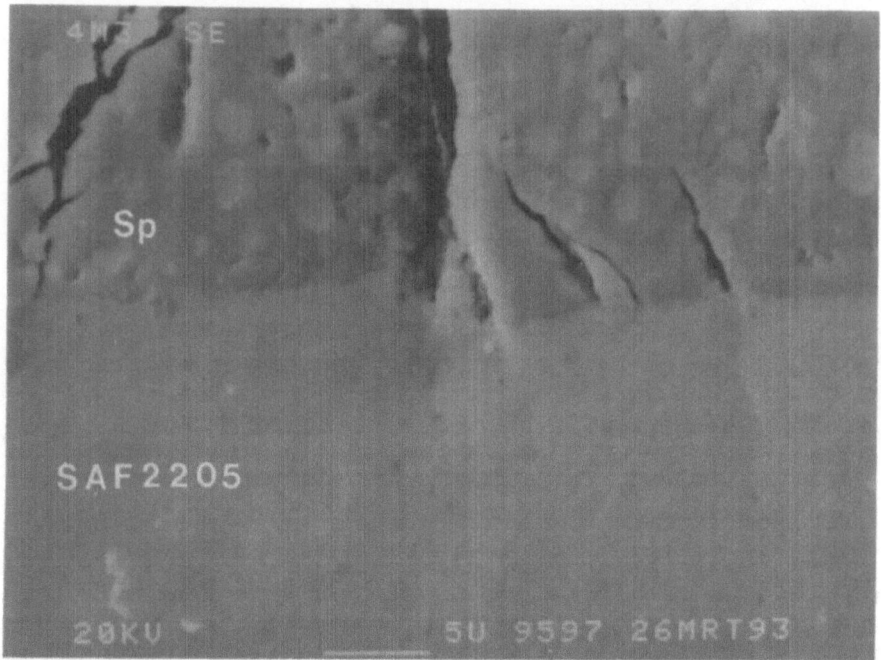

Figure 2. Side view of a four-point flexure specimen showing crack through the (Fe,Cr)-spinel coating blunting in the (Fe,Cr)-spinel SAF2205 matrix. Deformed volume is visualized by necking in the matrix.

Furthermore, a network of cracks with spacings of the order of 200 μm is always present in the laser applied coating preventing steady state crack growth along the interface. When the specimens having laser applied Cr_2O_3 coatings are set up in 4-point flexure two modes of failure may occur.First of all delamination along the interface is possible resulting in a value for the critical energy release rate and the stress intensity factor. The value obtained is a minimum value because of the large tensile stresses present in laser treated materials. In order to obtain a actual value of the critical energy release rate the strain energy present after laser treatment should be superimposed onto the strain energy stored during bending.The stress state after laser treatment is however inhomogeneous due to the localised character of the laser melting. When the interface is very strong the cracks through the coating will not extend along the interface but will blunt by plastic deformation of the substrate.

In the present case failure occurs by crack blunting in the substrate as can be seen from fig. 2. In fig. 2 the edge of the specimen is polished before 4-point flexure and the size of the deformed volume can be observed by the plane stress type of necking. Only in some cases delamination occurs after work hardening of the substrate (fig. 3). Assuming that the plastic deformation only occurs in a small volume near a crack through the coating the amount of energy stored, and thus a minimal value for the critical energy release rate, can still be obtained from the maximum load.

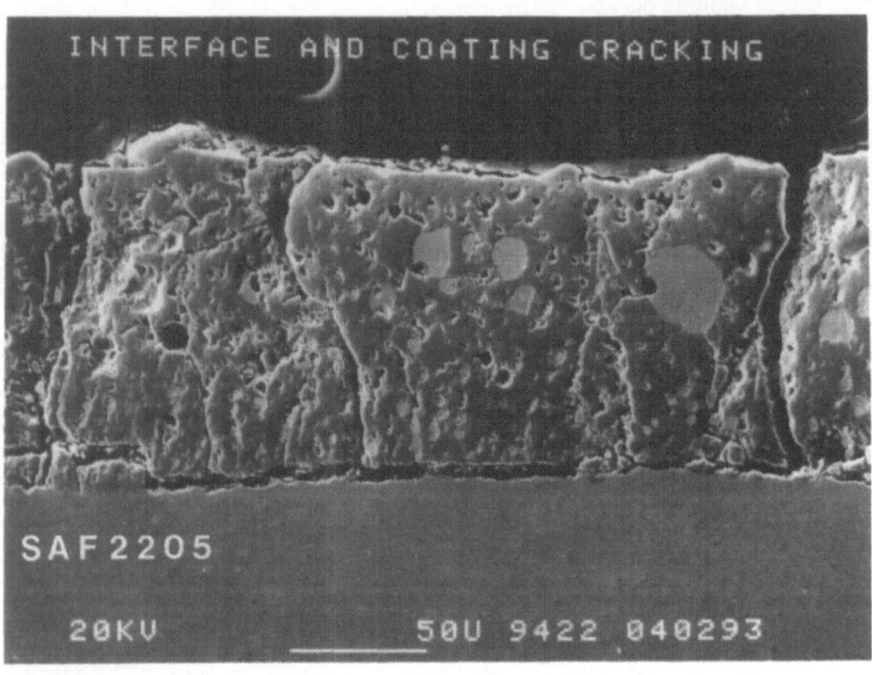

Figure 3. Side view of a four-point flexure specimen showing delamination of (Fe,Cr)- spinel coating after work hardening of the SAF2205 matrix.

Observations indicate that the critical strain energy release rate and stress intensity factor are larger than:

$$G_{ss}^{min} = 270 \quad [J/m^2],$$
$$\left| K_c^{min} \right| = 8.5 \quad [MPa\sqrt{m}].$$

4. Discussion

Results in the literature on the laser application of ceramic coatings are scarcely available [8, 9, 10] indicating that laser coating with ceramic materials is rather difficult. The problem is caused by the inhomogeneous heating and subsequent cooling during laser treatment which result in considerable cooling strains.

Temperature field during laser treatment

The thermal loading by the laser beam causes stresses being built up. It generates strain which will be multiplied by the appropriate Young's modulus to calculate the stress. Without melting or yielding all thermal strains will be accommodated elastically, i.e. disappear when returning to the starting temperature. Melting or yielding is necessary in order to attain a residual stress state. Because the cooling during laser treatment starts from the bottom to the surface of the specimen the residual stress state near the surface of the specimen, i.e. the ceramic coating, will be of tensile character. The residual stress adds up with the mechanically applied stresses causing failure at relatively low levels of applied stress. To study the formation of cracks during and after laser treatment we have to scrutinize the temperature field near the laser melt pool to obtain the stress during and after laser treatment.

Because the stresses being built up during and after laser treatment are caused by thermal strains our treatment of these stresses has to start from an inspection of the temperature field during laser treatment. In our calculation of the temperature field the method of Cline and Anthony [11, 12] is followed to calculate the field in steel during laser treatment. The results far from the centre of the laser beam are checked against the temperature field generated by a point source. The power P absorbed on the surface is calibrated by equating the calculated width to the experimental width of the laser track. The temperature field and thermal gradients in the melt pool itself have no influence on the stress so they are neglected in our treatment. The temperature field is given by:

$$T(\mathbf{r}, t) = \frac{P}{c_p} \cdot \int_0^t \frac{\exp\left[-\left\{ \left[(x + vt')^2 + y^2 \right] \left(2R^2 + 4\alpha_D t' \right)^{-1} + z^2 (4\alpha_D t')^{-1} \right\} \right]}{\sqrt{\pi^3 \alpha_D t'} \left(2R^2 + 4\alpha_D t' \right)^{-1}} dt'. \tag{6}$$

where α_D is the thermal diffusivity and c_p represents the specific heat. Eq. (6) is the general result for the temperature field of a Gaussian beam sweeping over the surface of a material in the x-direction at velocity v with total power P and radius R. The transformation method of Cline and Anthony is followed to obtain a form that can be integrated numerically. By taking derivatives with respect to x and y the thermal gradients $\frac{\partial T}{\partial x}, \frac{\partial T}{\partial y}$ and the absolute value of the total gradient is obtained. The temperature field and thermal gradients are also evaluated for a laser melt pool in material having a ten times lower thermal conduction coefficient, i.e. for a melt pool in (Fe,Cr)-spinel. Here the melt pool is very elongated in the x-direction and heat conduction is almost completely in the y- and z-direction.

Stress development in laser applied ceramic coatings

Although we did not solve this problem numerically for a bimaterial we may extrapolate the data from the two singular cases of steel and (Fe,Cr)-spinel using the diffusion equation together with the fact that the temperature field has to be continuous across the interface. This implies that the thermal gradients parallel to the interface, i.e. in the x- and y-direction, are equal on both sides of the interface ($z = 0$ on the interface):

$$\left(\frac{\partial T}{\partial x}\right)^{met}_{x,y,0} = \left(\frac{\partial T}{\partial x}\right)^{cer}_{x,y,0} \tag{7a}$$

$$\left(\frac{\partial T}{\partial y}\right)^{met}_{x,y,0} = \left(\frac{\partial T}{\partial y}\right)^{cer}_{x,y,0} \tag{7b}$$

We also use the fact that the heat flux is continuous over the interface:

$$k_{met}\left(\frac{\partial T}{\partial z}\right)^{met}_{x,y,0} = k_{cer}\left(\frac{\partial T}{\partial z}\right)^{cer}_{x,y,0} \Rightarrow \left(\frac{\partial T}{\partial z}\right)^{cer}_{x,y,0} \approx 10\left(\frac{\partial T}{\partial z}\right)^{met}_{x,y,0} \tag{8}$$

Observations of the equi-temperature lines on the interface perpendicular to the direction of the laser beam, i.e. in the y-direction, indicate that for the (Fe,Cr)-spinel coating on steel:

$$\left(\frac{\partial T}{\partial z}\right)^{met}_{x,y,0} \approx 0.26\left(\frac{\partial T}{\partial y}\right)^{met}_{x,y,0} \tag{9a}$$

$$\left(\frac{\partial T}{\partial z}\right)^{cer}_{x,y,0} \approx 2.6\left(\frac{\partial T}{\partial y}\right)^{cer}_{x,y,0} = 2.6\left(\frac{\partial T}{\partial y}\right)^{met}_{x,y,0} \tag{9b}$$

Eq. (9a) shows that, in the metal, the thermal gradient in the z-direction is only one quarter of the gradient in the y-direction validating the use of the temperature field calculated for a single phase steel having a zero thermal gradient in the z-direction on the surface. Eq. (9b) indicates that the large temperature gradient in the y-direction is accompanied by a more than twice as large thermal gradient in the z-direction in the ceramic coating. We further use the fact that the thermal gradients in the y-direction are much larger in single-phase (Fe,Cr)-spinel than in the uncoated steel causing the temperature gradient in the z-direction to invert further away from the melt pool, i.e. instead of heat transported from the coating to the steel substrate the coating is heated by the substrate.

In conclusion we can define the basic problem in the laser applied coating being the very large thermal gradients in the z-direction in the laser applied coating. From the thermal gradient in the y-direction of $1.8 \cdot 10^6$ °C/m in steel a thermal gradient in the z-direction of $4.7 \cdot 10^6$ °C/m is estimated in the (Fe,Cr)-spinel coating. For a 100 µm coating this results in a 470 °C temperature difference over the thickness of the coating. As these large thermal gradients occur at temperatures near the melting point they are accompanied by low stress levels. However, during cooling the temperature difference results in the build-up of residual stresses.

Because, in the region of interest, the coating cools from the interface towards the top of the coating the stress is of tensile character in the top of the coating, i.e. a bending moment is set up in the coating that may cause delamination of the ceramic coating. This is indeed observed for coating thicknesses larger than 200 µm when the coating curls up near the side of the laser track.

The residual stress state can be estimated from the solution of an bending plate on an elastic foundation of which the solution is known [11, 12]. When bending is restricted by the substrate, i.e. edges are clamped and delamination does not occur, the stress in the top of the coating is given by:

$$\sigma_{yy}(z) = \frac{\alpha_1 E_1 \frac{\partial T}{\partial z}}{(1 - \nu_1)} \cdot z, \tag{10}$$

where index 1 refers to the properties of the coating. The temperature field at a depth z upon cooling can be approximated by:

$$T(z, t) = T_o + (T_r - T_o) erfc(z/2 \sqrt{t \alpha_D}), \tag{11}$$

where α_D is the thermal diffusivity ($6.4 \cdot 10^7$ m²s⁻¹ in Cr_2O_3). Clearly there exists a stress distribution over the thickness of the coating which sets up a bending moment in the coating which can be approximately written as:

$$M_1 = \frac{\alpha_1 E_1 \frac{\partial T}{\partial z}}{12(1 - \nu_1)} \cdot h_1^3. \tag{12}$$

Using equations (10) and (11) the bending moment is found to be 14.2 Nm/m for a coating thickness of 200 µm. which is close to the outcome of eq.(12), namely 13.7 Nm/m.

When the coating does not delaminate edge moments have to be applied to prevent bending. These edge moments are effectively produced by stress components in the z-direction over the interface. The stress σ_{zz} at the edge of the bending plate is given by Hetenyi (fig. 4) and the maximum tensile stress is given by:

$$\sigma_{zz}^{max} = 2 \cdot \frac{\alpha_1 E_1 \frac{\partial T}{\partial z}}{12(1-v_1)} \cdot \sqrt{\frac{3k_2}{E_1}} \cdot h_1^{1\frac{1}{2}} , \qquad (13)$$

where k_2 is the modulus of deflection of the substrate ($E_1/k_2 = 6.6\cdot10^{-5}$). In the laser coating process several laser tracks have to be applied before the coating reaches its final thickness, i.e. its final stress state. Realising that an edge moment, and thus a stress σ_{zz}, is only generated at a free edge one crack through the thickness of the coating is always present.

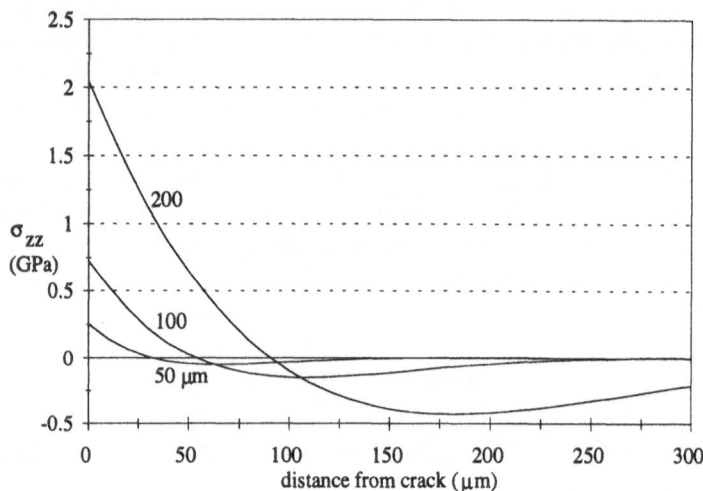

Figure 4. Stress component σ_z onto the interface as a function of distance y from the crack tip. Values for coating thickness of 50, 100 and 200 μ m are displayed.

After the first crack parallel to the laser track two things may happen. First of all delamination may start from the free edge and grow to the last laser track applied where the coating becomes thinner and crack propagation will stop. When delamination occurs the stress σ_{yy} is removed by bending of the coating so no further cracking through the thickness of the coating will occur. Another possibility is that the stress σ_{zz} at the free edge is not large enough to cause delamination. Then the stress σ_{yy} is locally reduced due to crack opening but is unchanged in the rest of the coating. As a result cracks more or less regularly spaced will appear through the thickness of the coating. The process of delamination versus perpendicular fracture is determined by the relative strength of the ceramic and the interface and by the thickness of the coating. As the stress σ_{yy}, causing fracture through the coating, scales with h_1 and σ_{zz}, causing delamination, scales with $h_1^{1.5}$ the coating can always be made thick enough so that delamination will occur. Ultimately, when the interface is very strong, delamination may occur by fracture through the ceramic parallel to the interface, instead of along the metal-ceramic interface, and both fracture processes are governed by the strength of the ceramic σ_f^c, being of the order of 1 GPa. If this is the case the fracture process, i.e. delamination vs. perpendicular fracture, is solely determined by the coating thickness and the critical coating thickness is determined by the elastic properties of the coating and the substrate.

Whether or not fracture will occur is of course related to the thermal gradient in the coating. Assuming that the thermal gradient is large enough for fracture to occur delamination is the governing process if:

$$h_1 > \frac{3E_1}{k_2}. \tag{14}$$

Using the experimental maximum coating thickness of 200 µm for the (Fe,Cr)-spinel coating on SAF2205, where delamination is mainly through the ceramic, the critical bending moment M_{1c} is calculated to be 14.2 Nm/m. From the calculated bending moment in the ceramic coating the critical energy release rate G_c of delamination can be calculated:

$$G_c = \frac{M_{1c}^2 (1 - v_1)^2}{2E_1} \cdot \frac{12}{h_1^3} = 320 \ \left[J/m^2 \right] \tag{15}$$

The question is whether the gradient of the residual stress distribution in a coating of a thickness of 200 µm would predict delamination. For this purpose, the energy release rate of eq.(14) is written in the usual form (eq. 5), i.e. in terms of the stress intensity factor K. The problem of delamination due to edge loads and moments was solved by Thouless et al. [14][12].

However, usually there is a significant amount of mode II component involved whereas we like to focus here on a critical thickness h_c corresponding to a pure mode I trajectory. The problem involves a non-linear algebraic equation, the solution of which we display in figure 5 as a function of $\kappa = h_c / \sqrt{\alpha_D t}$. The mode I stress intensity factor is found to be $K_{1c} = 0.189 \ \sigma \sqrt{h_c}$, forcing K_{IIc} equal to zero. From a critical energy release rate (eq. 5) expressed in terms of K_{1c} and the calculated value of 320 J/m² (eq. 15) follows for $h_c = 200$ µm a stress level of 1.3 GPa. Indeed the latter is about the fracture stress of a ceramic material.

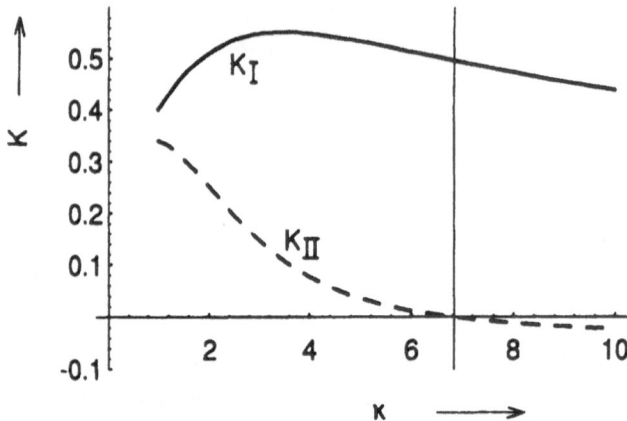

Figure 5. Stress intensity factors as a function of $\kappa = \dfrac{h_c}{\sqrt{\alpha_D t}}$ *in units of* $\sigma \sqrt{\kappa}$.
Solid line is K_I *, dashed line is* K_{II}

Therefore we may expect that a thick coating of 200 μm will delaminate by fracture through the ceramic parallel to the interface, instead of along the metal-ceramic interface which is in accordance to the experimental observations. It should be noted that the crack path will deviate from a planar delamination if K_{II} is not equal to zero. However, the coating thickness on SS304 and SS316 is limited to 50 μm indicating a maximum bending moment M_{1c} of only 0.21 Nm/m The maximum tensile stress σ_{zz}, i.e. the interface strength, is then 500 MPa. As the (Fe,Cr)-spinel coating material is equal to the (Fe,Cr)-spinel on SAF2205 the strength of the ceramic is of the order of 1 GPa and delamination will be mainly along the interface because the interface is by far the weakest part of the system. This is indeed observed in the experiments. In this case the critical energy release rate G_c = 4.6 [J/m²] which is equal to the energy release rate 2γ of cleavage fracture. If delamination does not occur the stress σ_{yy} is reduced due to crack opening. The wavelength of the stress reduction is related to the extent of the stress field across the interface (fig. 4) and scales with $h_1^{3/4}$. As the maximum value of the stress σ_{yy} scales with h_1, the distance over which the fracture stress σ_f^c is reached scales with $h_1^{1/4}$ indicating that the crack spacing is only weakly dependent on the thickness of the coating. For a coating of 100 μm the extension of the stress field over the interface is about 200 μm indicating a crack spacing of the same size (fig. 4). This is in agreement with experimentally observed crack spacings. However, these crack spacings also scale with the transversal displacement of the laser beam. This is partially because the thermal gradient peak in the metal causes an extended gradient peak in the ceramic. In this way a region of large stress parallel to the laser track is developed where fracture takes place preferentially. This effect is augmented by the fact that the ceramic coating is somewhat thicker at the edge of the metallic melt pool, i.e. at the same place where the gradients are largest. It can thus be concluded that the crack spacing is more strongly related to the stress peaks in the coating, i.e. to the transversal displacement of the laser beam, rather than to areas where the stress is reduced due to a prior crack.

A similar explanation can be given for the stress development in the x-direction, i.e. in the direction of movement of the laser beam. Although the thermal gradients in the z-direction in the coating can be very large due to the thermal gradients in the x-direction delamination in the x-direction is never observed because the coating thickness is always much smaller near the middle of the last laser track that is applied. However, the stress in the x-direction scales with the thermal gradient in the z-direction near the middle of the laser track so that fracture through the coating, more or less perpendicular to the direction of movement of the laser beam is always observed. For all other cases where the heat flow is not purely in the x- or y-direction the stress in the coating is determined by the total magnitude of the thermal gradient. Because of the low thermal conduction coefficient the flow of heat from the coating is however mainly in the y-direction making delamination in the y-direction the most significant process.

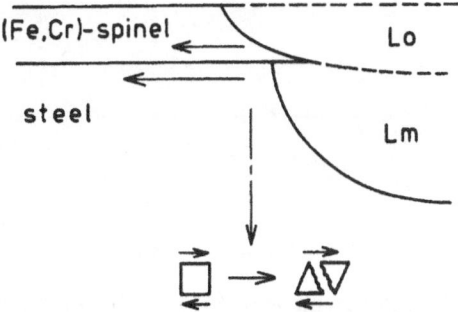

Figure 6. Schematic drawing of the cross-section of a laser track showing the shear stress, set-up during cooling. The stress is produced by a difference in thermal expansion coefficient between the (Fe,Cr)-spinel coating and the SS304 substrate. The small volume elements show a crack running from the upper left to the lower right due to the shear stress applied on the element.

Another point to consider is the stress development during laser treatment: influence of thermal expansion difference $\Delta\alpha$. A difference in thermal expansion coefficient between the coating and the substrate material can result in normal [15] and in shear stresses. In the present case of coating ceramic material on steel using a high power laser shear stresses will be set up due to the asymmetry of the process. As the laser track is restricted by the surrounding material a tensile stress will be present in both materials near the interface but, as no macroscopic shrinkage can occur the stress is not transmitted over the interface. However, because during laser treatment the material closer to the centre of the laser track is at a higher temperature, yielding will cause a net flow of material in the outward direction. The magnitude of the net displacement of material will scale with the thermal expansion coefficient α of the corresponding material. For the case of (Fe,Cr)-spinel on SS304 or SS316 the thermal expansion coefficient of the steel is larger than the coefficient of the (Fe,Cr)-spinel resulting in a shear stress on the interface, where the steel substrate exerts a force on the ceramic material in the outward direction (fig.6). Consequently, a crack perpendicular to the interface will bend towards the centre of the laser beam as is evident from the forces acting on a small volume of material.

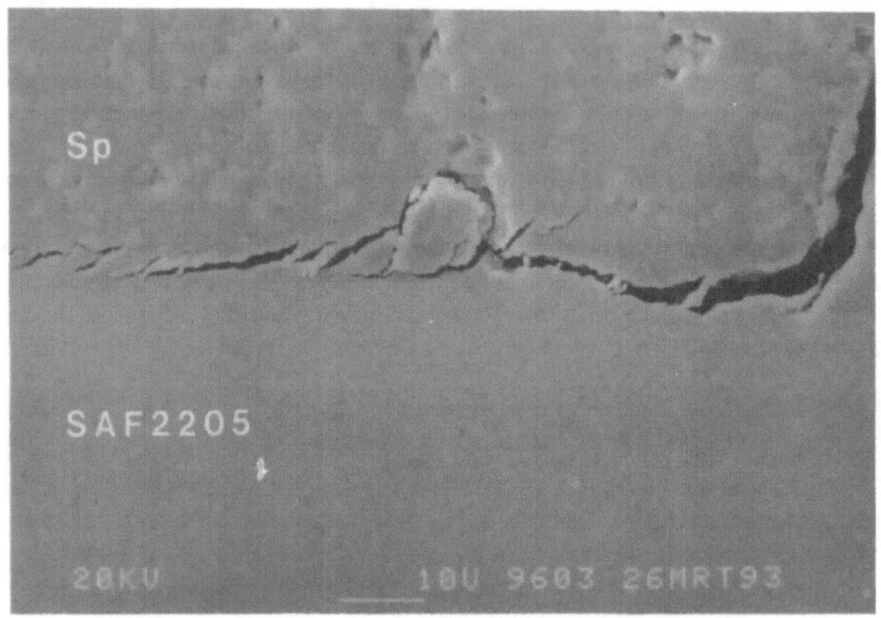

Figure 7. Fracture near the interface by connecting small mixed mode cracks.

Whether the actual crack path makes an angle of 45° with the interface or runs more or less parallel to the interface depends on the critical stress intensity factors of tensile (mode I) and shear (mode II) fracture. As the stress intensity factors will normally be higher for mode II than for mode I, due to the higher friction in mode II, a crack parallel to the interface often consists of small mixed mode (45°) cracks growing together (fig. 7).

Mechanical testing of ceramic coatings

Four-point flexure is an excellent method to test the interfacial properties of bimaterials. Because the stress state is constant in between the inner rollers steady state crack growth occurs and the sequential character of the crack propagation results in testing of a large interface area. However, there are a few limitation on the use of the 4-point flexure method, especially when testing thin coatings on ductile substrates. First of all a reliable value for the critical energy release rate requires steady state crack growth instead of dynamical crack growth depending critically on the onset of fracture. Secondly, this method has the same disadvantage as all bending methods; the values for the critical energy release rate scale with the power cube of the specimen thickness making it very sensitive to fluctuations in the thickness. However, as shown in eq. (4) the energy release rate scales linearly with the coating thickness, whereas during laser treatment the energy release rate scales with one over the power cube of the coating thickness (eq. (14)). Because the total specimen thickness for 4-point bending is more controllable than the coating thickness during laser treatment, more accurate values for the critical energy release rate can be obtained from 4-point bending than from the measurement of the maximum coating thickness. The third disadvantage is the limitation on the maximum energy release rate that can be measured when investigating thin coatings on ductile substrates. The prevention of fracture through the coating, perpendicular to the interface requires a substrate thickness being equal to the coating thickness. However, to prevent yielding of the substrate a thick substrate is required. The maximum stress in the substrate is given by:

$$\sigma_m = \frac{3F(s_1 - s_2)}{2w(h - h_1)^2} \tag{16}$$

and in the coating:

$$\sigma_c = \frac{3F(s_1 - s_2)}{2wh^2} \cdot \frac{E_1}{E_2}. \tag{17}$$

In order to get a reliable result σ_m should be smaller than the yield strength σ_y and σ_c smaller than the fracture stress σ_f. Given a certain coating thickness the maximum energy release rate that can be obtained reliably is found by equating the maximum energy release rate due to coating fracture to the maximum energy release rate due to substrate yielding. For a 100 μm (Fe,Cr)-spinel coating on duplex the maximum energy release rate is of the order of 160 J/m^2 for a specimen thickness of 3 mm. The value of 270 J/m^2 as obtained experimentally involves some substrate yielding making work hardening of the substrate and interface sliding possible. For a reliable measurement of the critical energy release rate this should be prevented but it can be used to obtain an indication of the quality of the interface. A further disadvantage of 4-point flexure for testing thin coatings is the phase angle of loading of the order of 45° as put forward by Charalambides [4], indicating that the stress at the crack tip has a large mode II component, i.e. a large shear component. As mode II failure is very sensitive to interface roughness the rough interface between the (Fe,Cr)-spinel and the steel after laser treatment makes the critical energy release rate much higher than it would be under mode I loading.

When using the 4-point bending to test the laser applied (Fe,Cr)-spinel coatings the problem is evident. The maximum coating thickness that can be applied is 200 μm but due to thickness variations the pratical coating thickness is limited to 100 μm. As calculated for the laser coating process, the critical energy release rate of delamination is of the order of 320 J/m^2. Because the bending moment after laser treatment is situated completely in the coating delamination occurs by mode I fracture along the interface which is expected to result in the lowest interface strength. The elastic energy stored in a 100 μm thick coating due to the laser treatment is only 37 J/m^2 so in order to get delamination during 4-point flexure about 280 J/m^2 should be applied to the coating by bending. However, yielding of the substrate limits the elastic energy stored in the coating to 160 J/m² making delamination of a 100 µm (Fe,Cr)-spinel coating by 4-point flexure impossible.

The question is whether we can connect the macroscopic observation of section 2 to a microscopic physical picture. The following might be considered as an attempt to make this connection. Our applied mechanics viewpoint is focused on two aspects:

- The dislocation nucleation and emission from a crack lying at the interface between spinel and fcc-Fe or bcc-Fe.
- The energy release rate of dislocation emission G_{disl} vs. that of cleavage fracture G_{cleav}. The basic idea is that if $G_{disl} < G_{cleav}$ the crack will be blunted by plasticity, whereas if $G_{cleav} < G_{disl}$ rupture of the interface takes place without a substantial dislocation production.

The analysis is based on work by Rice, Hutchinson and Thomson, Charalambides et al., Beltz et al. and Evans [16, 17, 3, 18, 19]. The present situation of interest is schematically drawn in fig. 8.

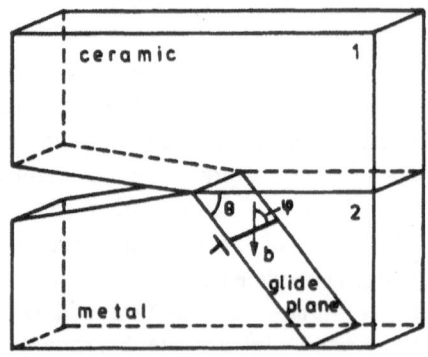

Figure 8. Schematical situation of a crack intersecting a glide plane.

$$G_{disl} = \frac{b^2\mu_2^2}{8\pi r_c(1-\nu_2)^2 E^* \cosh^2\pi\varepsilon} \cdot \left[\frac{\cos\varphi + (1-\nu_2)\sin\varphi\tan\varphi}{\sum_{r\theta}^I \cos\psi' + \sum_{r\theta}^{II} \sin\psi'} \right]^2 \tag{18}$$

where Ψ' is the so-called phase angle. Eq. (18) has been applied to the present situation of spinel(1) - steel(2) interfaces. The interface normal was taken in the [1 0 1] direction and the line of intersection between the crack and the interface along the [$\bar{1}$ 0 1] direction. G_{disl} as described in eq. (18) depends on ϕ, θ, ψ.

Fig. 9 displays the energy release rate G_{dis} fcc and bcc of dislocation emission into material 2 as a function of the angle between the Burgers vector and the dislocation line (= crack line). For fcc, emitting a Shockley partial dislocation the energy release rate is 1.49 J/m² whereas for a perfect dislocation 2.57 J/m² is calculated. In bcc the energy release rate turned out to be larger:

6.23 J/m² and 3.39 J/m² for 1/2[1 $\bar{1}$ 1] (1 2 1) and 1/2[1 $\bar{1}$ 1](1 3 2), respectively. It may be concluded that $G_{disl} < G_{cleavage}$, i.e. blunting of the tip is expected, which is in accordance with the experiments (Fig. 2).

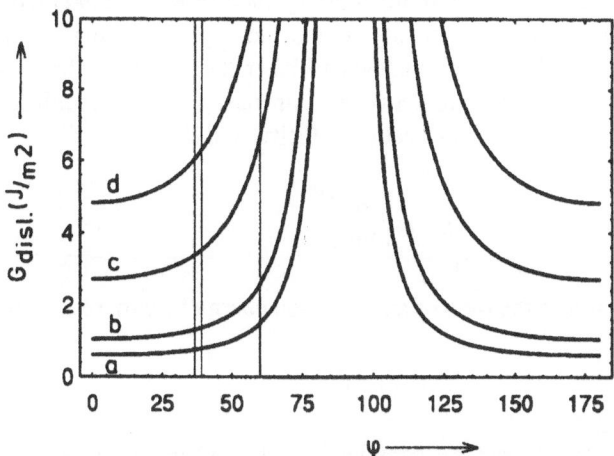

Figure 9. G_{disl} [θ, ψ, φ] : φ dependence.
a: G_{disl}[35.3, 55, 60] (1 $\bar{1}$ 1) partial dislocation in fcc.
b: G_{disl}[35.3, 55, 60] (1 $\bar{1}$ 1) perfect dislocation in fcc.
c: G_{disl}[56.3, 55, 36.8] (1 3 2) dislocation in bcc.
d: G_{disl}[63.4, 55, 39.2] (1 2 1) dislocation in bcc.

However, it should be realized that G_{disl} is significantly affected by the assumption concerning the critical distance r_c from which the total force exerted on the emitting dislocation $F_{tot} = F_{emit} - F_{image}$ should be larger than zero.

Further, one should bear in mind that our analysis is based on a description of a stationary crack that emits dislocations. Actually, when one considers cleavage fracture versus ductile fracture, dynamic elements should be incorporated as well. It is feasible that a stationary crack may emit dislocations, whereas a fast moving crack will only nucleate dislocations which are not moving at a high enough velocity to escape from the stress field of the moving crack. In such a situation, although dislocations are nucleated indeed they will not contribute to a relaxation of the cleavage fracture stress.

It was found that G_c for the plasma sprayed $Cr_2 O_3$ coating on steel was 117 J/m². Although for the laser processed coating on steel G_c could not be determined experimentally, it was concluded that the critical energy release rate should be much larger than (see 2) 270 J/m². Because plasticity is involved we reformulate G_c as consisting of a linear elastic contribution G_{elas} and a part that reflects dissipation of energy by plasticity G_{diss} [Kor93]. Further, it is assumed that the dissipation takes place starting at a critical value of the work of adhesion which is proportional to the energy release rate by dislocations, G_{disl}. G_c is rewritten as:

$$G_c = \alpha \, W_{adh} + (\alpha \, W_{adh} - G_{disl})^\lambda \, , \tag{19}$$

508

where W_{adh} represents the work of adhesion for brittle cleavage fracture, 2 γ. The constant α reflects the increase of the critical energy release rate due to the roughness and microplasticity that may still be involved in cleavage fracture. The experimentally determined value of α is at least of the order of 10. λ should depend on the dislocation activation energy of slip and velocity of the moving dislocation, for bcc with the multiple interacting slip systems somewhat higher than for fcc. The most frequently used law to describe the relation between strain rate $\dot{\gamma}$, which according to Orowan's equation is proportional to the dislocation velocity, and stress is:

$$\dot{\gamma} = A \left(\frac{\sigma}{\sigma_0} \right)^m \tag{20}$$

where m corresponds to the ratio of an activation energy F_0 with respect to kT:

$$\dot{\gamma} = \dot{\gamma}_0 \left(\frac{\sigma}{\tau} \right)^{F_0/kT} \tag{21}$$

Strictly speaking m can not be completely independent of the applied stress and therefore it is more appropriate to define m as:

$$m = \left(\frac{\delta \ln \dot{\gamma}}{\delta \ln \sigma} \right)_T \tag{22}$$

the so-called strain-rate sensitivity factor m, experimentally determined in the early stage of yielding at 78 K [21] for bcc iron was found to be 6.5; i.e. $F_0 = 0.042$ eV and m at 300 K is to a first approximation about 1.6. Now since the dissipation term in eq. (19) is proportional to velocity and the strain rate we take λequal to m = 1.6.

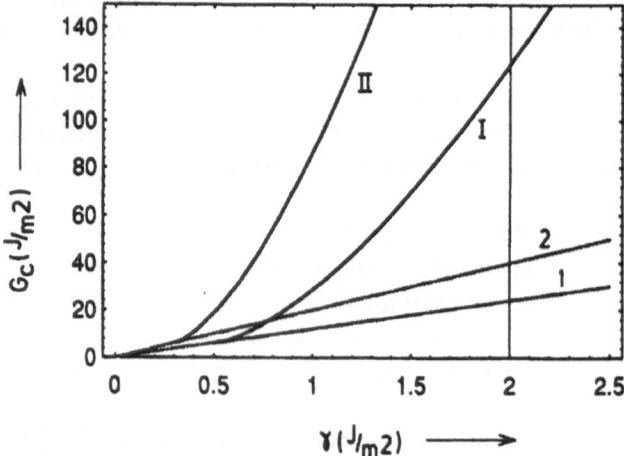

Figure 10 Critical energy release rate versus surface free energy as calculated using eq. (19). I and II refer to ductile fracture along the interface in plasma sprayed and laser processed coatings, respectively, whereas 1 and 2 denote brittle cleavage fracture.

Substituting the values for α (=10), m(= 1.6), $\gamma_{\{110\}} = 2$ J/m^2 and G_{disl} (6.23 J/m^2), the critical energy release rate is predicted to be 319 J/m^2, i.e. much larger than one is able to detect experimentally. G_c is depicted in fig. 10 as a function of γ.

The straight line displays the brittle cleavage fracture as a function of γ and the offset from linearity corresponds to G_{disl}. The critical energy release rate for the plasma sprayed coating has to be different. In particular it is expected that the contribution to the critical energy release rate of cleavage fracture due to roughness and microplasticity is less. The value of G_c equal to 117 J/m^2 can be reproduced by taking α equal to 6, i.e. smaller than the value of α for the laser processed coating.

5. Conclusion

The actual interface strength of a (Fe,Cr)-spinel applied to stainless steel by laser coating depends strongly on the composition of the substrate and coating materials. One reason is the solidification of the melt pool in the steel from the bottom to the interface pushing impurities preferentially to the interface. Another reason for the strong compositional dependence is the sensitivity of the martensitic transformation in (Fe,Cr)-spinel depending on the chemical composition. The actual energy release rate on coating delamination varies from 4.7 J/m^2 for the (Fe,Cr)-spinel coating on SS304 and SS316, corresponding to perfectly brittle cleavage fracture, to 320 J/m^2 for (Fe,Cr)-spinel on SAF2205, being of the same order of magnitude as the fracture energy of ductile fracture in steel. Since the energy release rate is extremely high and delamination occurs by fracture through the coating and partially along the interface, it indicates that the interface strength is similar to or higher than the fracture strength of (Fe,Cr)-spinel. The fracture energy of 320 J/m^2 for (Fe,Cr)-spinel on SAF2205 has to be compared with the value of 117 J/m^2 for plasma sprayed Cr_2O_3 on SAF2205 to appreciate the very high strength of the laser applied coating.

The strength of bimaterial interfaces can be tested by 4-point flexure. The advantage of the 4-point flexure method when testing bimaterial interfaces lies in the independence of the method on micro cracks present in the ceramic coating. The method results in a value for the fracture energy assuming a perfectly flat fracture surface. However, the fracture surface is hardly ever perfectly flat resulting in a larger fracture surface and a roughness related fracture resistance. These two factors partially contribute to an experimental fracture energy being much higher than twice the surface free energy. A large plasticity related term is, however, necessary to explain the extremely high values found for the fracture energies of ceramic coatings on metals. When testing the (Fe,Cr)-spinel coatings on SAF2205 failure occurs mostly by yielding in the steel substrate. This is related to the laser treatment being usually a more demanding test than the 4-point flexure. In order to obtain a reliable coating in the present systems the thickness is reduced to 100 μm and consequently the elastic energy that can be stored in the coating by bending is limited to 160 J/m^2, although higher values can be obtained when work hardening is allowed. However, the fact

510

remains that the loading angle during 4-point flexure is smaller than during laser treatment raising the fracture energy of coating delamination.

6. References

1. Hilty, D.C., Forgeng, W.D. and Folkman, R.L. (1955) Metall. Trans. **203,** 253.
2. Steinman, P.A. and Hintermann, H.E., (1989) J. Vac. Sci. Technol. **A7,** 2267.
3. Charalambides, P.G., Lund, J., Evans, A.G. and R.M. McMeeking (1989) J. Appl. Mech. **56,** 77.
4. Charalambides, P.G., Cao, H.C., Lund, J. and Evans, A.G. (1990) Mech. of Mtls. **8,** 269.
5. Evans, A.G., Rühle, M., Dalgleish, B.J. and Charalambides, P.G. (1990) Mat. Sci. & Eng. **A126,** 53.
6. Wang, J.S. and Suo, Z. (1990) Acta Metall. **38,** 1279.
7. Rice, J.R. (1988) J. Appl. Mech. **55,** 98.
8. Tsukamoto, K., Uchiyama, F., Okutomi, M., Shiratori, S. and Ohno, Y. (1987) Proceedings of LAMP'87, Osaka.
9. Bell, T., Sohi, M.H., Betz, J.R. and Bloyce, A. (1990) Key. Eng. Mater. **46-47,** 69.
10. Fellowes, F.C.J., Steen, W.M. and Coley, K.S. (1990) Key. Eng. Mater. **46-47,** 435.
11. Timoshenko, S.P. and Woinowsky-Krieger, S. (1959) Theory of Plates and Shells, McGraw-Hill, New York.
12. Hutchinson, J.W. (1993) Adv. Applied Mech. **29,** 63.
13. Hetenyi, M. (1946) Beams on Elastic Foundations, Oxford University Press, London.
14. Thouless, M.D., Evans, A.G., Ashby, M.F. and Hutchinson, J.W. (1987) Acta Metall., **35,** 1333.
15. Dreier, G. and Schmauder, S. (1993) Scripta Metall. **28,** 103.
16. Rice, J.R. and Thomson, R. (1974) Phil. Mag. **29,** 73.
17. Thomson, R.M. (1986) Solid State Physics **39,** 1.
18. Beltz, G.E. and Rice, J.R. (1992) Acta Metall. **40S,** 321.
19. Evans, A.G. and Hutchinson, J.W. (1989) Acta Metall. **37,** 909.
20. Korn, D. (1993) VDI Reihe **5,** Nr. 23, Düsseldorf.
21. Tomalin, D.S. and McMahon, C.J. (1975) Acta Metall. **21,** 1189.

MICROSTRUCTURE AND MECHANICAL PROPERTIES OF LASER TREATED ALUMINIUM ALLOYS

J.TH.M. DE HOSSON, L. DE MOL VAN OTTERLOO,
J. NOORDHUIS
Department of Applied Physics, University of Groningen,
Nijenborgh 4, 9747 AG Groningen, The Netherlands.

Abstract

Al-Cu alloys and an Al-Cu-Mg alloy, Al 2024-T3, were exposed to laser treatments at various scan velocities. In this paper the microstructural features and mechanical properties are reported. As far as the mechanical property of the Al-Cu-Mg alloy is concerned a striking observation is a minimum in the hardness value at a laser scan velocity of 1/2 cm/s. Usually an increasing hardness with increasing laser scan velocities is reported in the literature. This remarkable property could be explained based on the microstructural features observed by transmission electron microscopy.

After subsequent shot peening of the Al-Cu-Mg alloy, in all cases the formation of precipitates was observed, independent of the laser scan velocities originally applied. This phenomenon of precipitation, induced by shot peening afterwards is most striking at a high concentration of alloying elements in solid solution.

Further, it turned out that the hardness of the Al-Cu alloys could be increased substantially compared to the Al-Cu-Mg alloys under investigation upon increasing the Cu content. A Vickers hardness of around 500 could be attained in an Al-40wt% Cu alloy at a laser scan velocity of 0.125 m/s.

1. Introduction

Hardening by laser treatments is achieved primarily by the small grain sizes that are formed as a result of the rapid solidification process. The flow stress varies as $d^{-\alpha}$ where α depends on the cell size and cell wall-type. Since the cell size is inversely related to the solidification rate, commonly an increasing hardness is observed with increasing laser scan velocity.

J. Mazumder et al. (eds.), Laser Processing: Surface Treatment and Film Deposition, 511–527.
© 1996 *Kluwer Academic Publishers.*

The nucleation and growth of precipitates that may contribute even more considerably to the hardness of laser melted materials is usually suppressed by high cooling rates involved, typically in the order of 104-105 K/s. In an Fe-W-C tool steel however, we reported [1] the presence of precipitates in a surface that was "auto-tempered" by applying overlapping laser tracks. Nevertheless, in non-overlapping laser tracks precipitation processes are not likely to occur.

In this work the microstructural features and mechanical behaviour of laser treated Al-Cu-Mg and Al-Cu alloys were investigated. In Al-Cu alloys the precipitation of intermetallic compounds is known to proceed through a series, starting with fully coherent Guinier-Preston (GP) zones, that can initially be regarded as copper discs, one atomic layer in thickness. If the alloy also contains magnesium, the Al-Cu precipitates can partly be replaced by other ones called S' and S. These precipitates are modified with respect to composition, crystal structure and orientation relationship. With magnesium in excess to the amount of about 4 w% or more of an Al- 4.4 w% Cu alloy, freezing may follow the Al-S branch leading to the so-called T-phase (approximately $CuMgAl_6$) and Mg_2Al_3.

In this work the laser scan velocity was varied in order to obtain various stages in the precipitation sequence. Subsequently, shot peening was applied to study the work hardening behaviour of the laser treated material. In addition attention was paid also to the residual stress state in the samples before and after peening treatment.

2. Experimental

Samples of commercial Al 2024-T3 (4.4% Cu, 1.5% Mg, 0.6% Mn and Bal. Al.; solution treated, cold rolled and naturally aged) were sandblasted to obtain a rough well absorbing surface. After ultrasonically cleaning the samples were irradiated using a transverse flow Spectra Physics 820 CW-CO_2 laser under a protective argon atmosphere. At the surface the power of the beam was 1300 W. The focus point of the ZnSe lens with focal length of 127 mm lay 5 mm above the surface, resulting in a spot diameter of 0.75 mm. Single tracks were made at laser scan velocities ranging between 1/8 to 25 cm/s. On the samples for analysis by X-ray diffraction several adjacent tracks were laid down.

It should be emphasized that of all the elements in Al 2024, magnesium has by far the lowest boiling point (1390 K), and it is therefore to be expected that a substantial fraction of magnesium will actually evaporate.

Shot peening was carried out in a conventional blast cleaning apparatus, applying glass beads with an average diameter of 720 μm. The air pressure was 2.6 bar. Vickers hardness measurements were carried out just below the surface of a taper sectioned sample, with a 100 gram weight. TEM samples were prepared using a mechanical dimpler, followed by ion milling in the cold stage of a Gatan ion mill, in such a way that the electron transparent area is located in the centre of the laser track at a depth of approximately 30 μm. The hardness measurements provide information on this area as well.

X-ray analysis applying Cu-radiation, revealed the stress state after laser melting and subsequent shot peening, utilising the <422> reflection. Line profile analyses on the <111> reflection were carried out to obtain information about the dislocation density. For further experimental details we refer to [2].

3. Results

After laser melting a cellular structure develops during solidification. Cracks are occasionally observed in samples treated at low as well as at high laser scan velocities, indicating a reduced ductility at both ends of the scan velocity range. The measured cell sizes are depicted in Figure 1. The intermetallic compounds mainly consist of the $CuAl_2$ phase, and the $CuMgAl_2$ (S) phase. These phases were identified by X-ray diffraction.

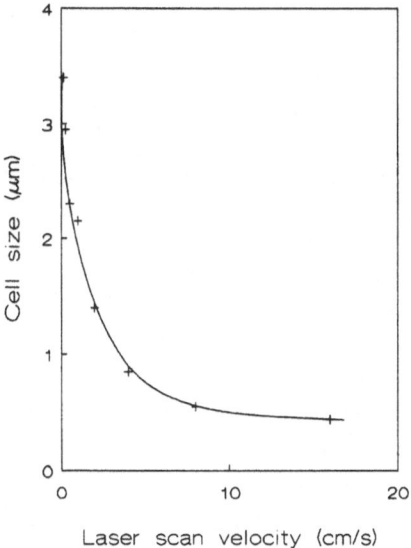

Figure 1. Cell size as a function of laser scan velocity.

Hardness measurements

Because natural ageing is known to play an important role in conventional heat treated aluminium alloys, attention should be paid to this aspect as well. After various ageing times, starting 5 hours after laser melting, a hardness profile is recorded. These measurements reveal that the hardness is significantly increasing, almost independent of the laser scan velocity. The amount of hardening is approximately 20% during the first week. After about one year another 10% increase is observed.

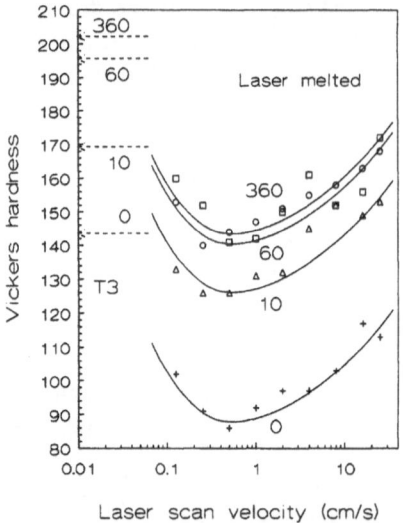

Figure 2. Hardness profile as a function of laser scan velocity and shot peening time(s). (Symbols refer to various shotpeening times: 0 s +, 10 s △, 60 s □ and 360 s o).

This increment is approximately the same as in conventional quenched alloys. Because of the rapid hardening right after the laser treatment, all measurements presented in the following are performed on samples aged for at least two weeks. The surface hardness as a function of laser scan velocity, is displayed in figure 2. This graph exhibits a parabolic shape, with a minimum value at a scan velocity of 1/2 cm/s. After subsequent shot peening the shape of the hardness curve remains approximately the same, although it shifts to higher values.

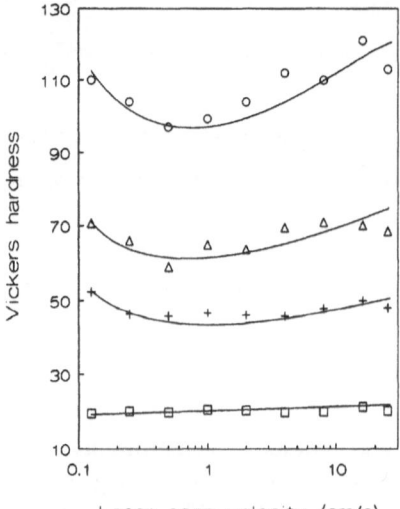

Figure 3. Hardness profiles as a funtion of laser scan velocity for pure Al □ , Al-2.32wt% Cu + , Al-4.5wt% Cu and Al-2024 o.

The results of hardness measurements of the laser treated pure Al and Al-Cu samples, are depicted in figure 3. The pure Al sample shows no minimum with respect to the laser scan velocity, but is slightly increasing. The other two Al-Cu samples, however, show just like the Al 2024 sample a minimum value at a laser scan velocity of 1/2 cm/s. The amplitude of the dip is larger in the 4.5 wt% sample compared to the 2.32 wt% sample. This indicates that presence of copper alone is enough to produce the dip in the hardness curve. The magnesium in Al 2024 thus causes primarily, as expected, a shift to higher hardness values with respect to the 4.5 wt% sample. The following research is focussed on the Al 2024 alloy.

X-ray measurements

X-ray stress measurements reveal, analogous to the hardness measurements, a parabolically shaped curve for the residual stress state, as is displayed in figure 4. Before shot peening the stress state has a tensile character with an average magnitude of 100 MPa. After shot peening the stress is inverted to a compressive one with a magnitude of the order of 250 MPa, while maintaining the parabolic shape. The magnitude of the stress after 360 s shot peening is lower than after 60 s. This is not due to an inaccuracy of the measurement. In fact it turned out to be reproducible by peening the 60 s sample for an extra 300 s period.

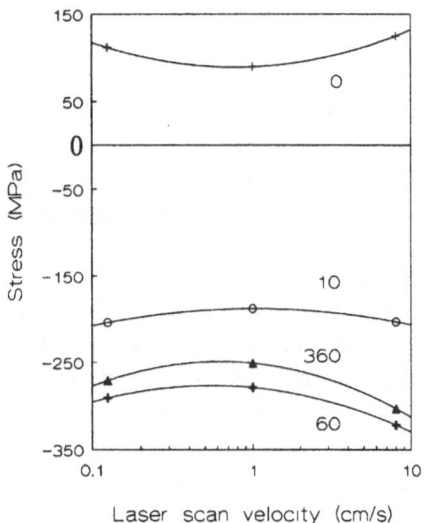

Figure 4. Residual stress as a function of laser scan velocity and shot peening time(s). (Symbols refer to various shotpeening times: 0 s +, 10 s o , 60 s + and 360 s Δ).

Line profile analysis was applied to obtain information on the dislocation density [3], [4]. This again shows a minimum as a function of laser scan velocity (see figure 5). The increase in dislocation density is approximately the same for all different samples except for the decrease after 360 s of shot peening of samples treated at a high scan velocity.

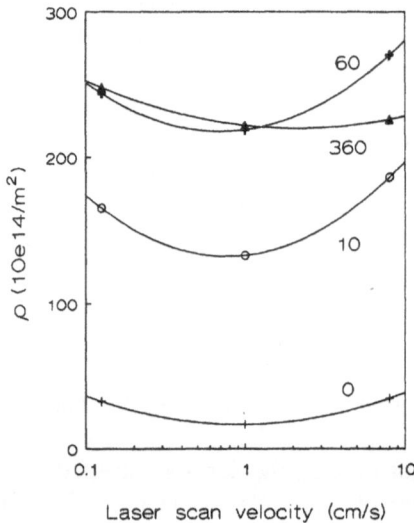

Figure 5. *Dislocation density as a function of laser scan velocity and shot peening time(s). (Symbols refer to various shotpeening times: 0 s +, 10 s o , 60 s + and 360 s Δ).*

Microscopic observations

TEM specimens taken from Al 2024 samples and exposed to different laser scan velocities, before shot peening, show marked differences in microstructural features. The samples treated at lower scan velocities, 1/8 cm/s to 1/2 cm/s, are mainly precipitate free although in some areas coarser precipitates could be observed. However, it is uncertain whether these are partly dissolved cell boundaries or real precipitates.

The most striking feature is the formation of helical dislocations, figure 6a and 6b, as being observed in all samples treated at low laser scan velocity. Most of these dislocations exhibit a uniform equilibrium shape with respect to pitch and radius. At the lowest scan velocity however, non uniform shapes are occasionally observed, figure 7.

In samples treated at laser scan velocities between 1 cm/s to 4 cm/s the precipitation of the tetragonal platelets, with an average size of 150 nm, is observed, as can be seen in a bright field TEM micrograph, figure 8. Since these platelets are oriented parallel to the {100} faces of the aluminium matrix, we may conclude that these precipitates are not the so-called S' precipitates (Al$_2$CuMg). The S' precipitates grow locally on {120} planes, giving {110}, {100} or {130} as an overall growth plane [5],[6]. Beside the platelets a more globular precipitate is also observed in these samples, mainly nucleating on the edges of the θ' plates and on dislocations. This could be the θ phase as well as the S or S' phase. Since in alloys with a copper-magnesium ratio of less than 8 to 1 Al₂CuMg precipitates can start to play a role [7], the existence of the S and S' phase cannot be excluded. Electron diffraction patterns have not yet revealed the nature of these precipitates.

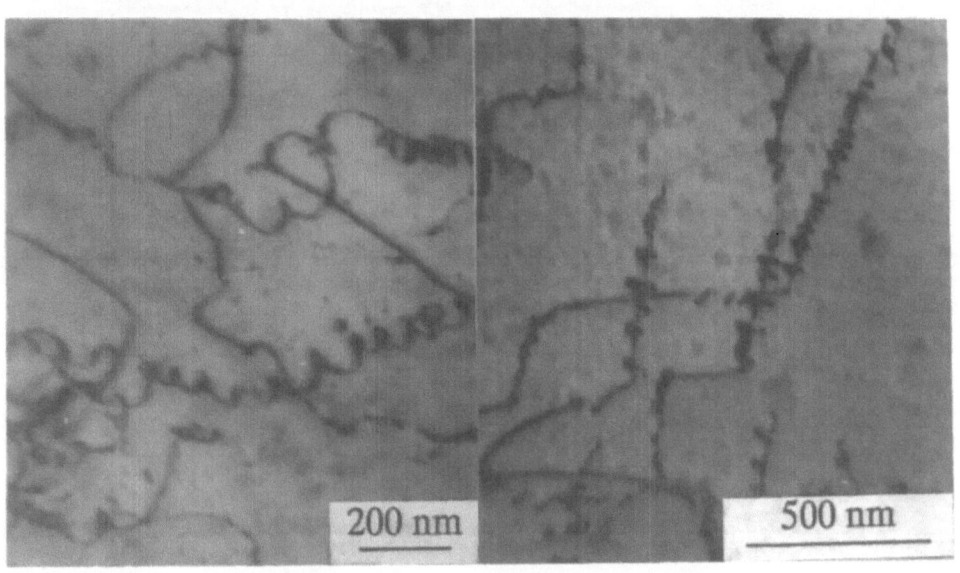

200 nm

500 nm

a

b

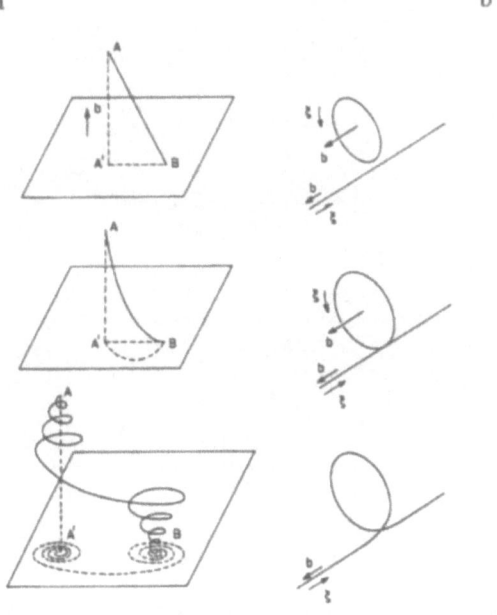

c

d

Figure 6. Helical dislocations in a sample treated at a laser scan velocity of 1/8 cm/s: grown by vacancy absorption (a), and by interaction of a screw dislocation with prismatic vacancy loops (b). Schematic representations of the operating mechanisms are depicted in (c) and (d) respectively [10].

518

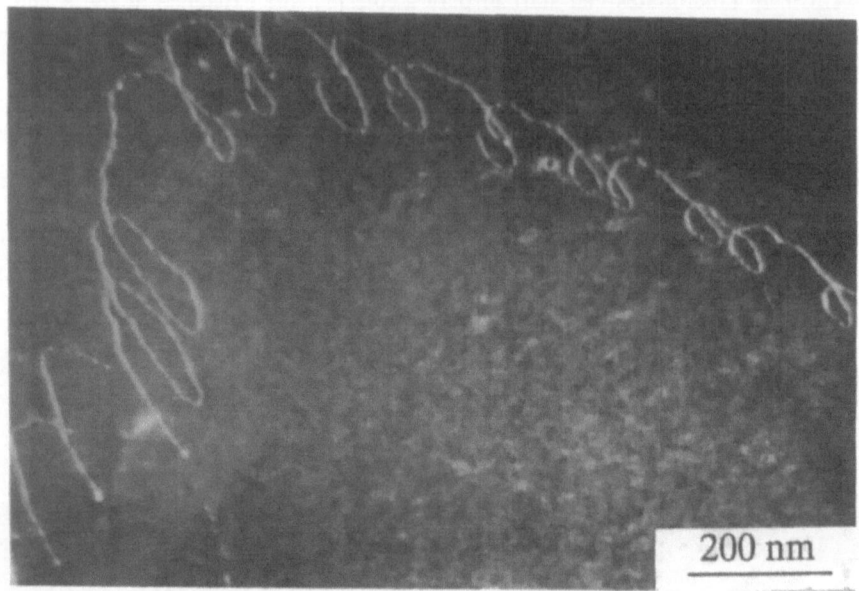

Figure 7. Dark-field TEM micrograph of a non equilibrium helical dislocation.

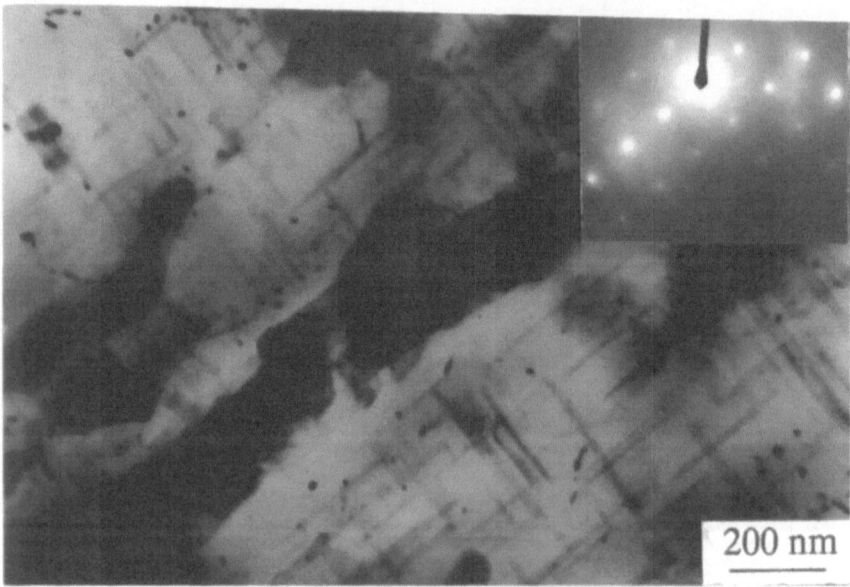

Figure 8. Bright field TEM micrograph showing platelets and some incoherent or S precipitates, in a sample treated at 2 cm/s.

Samples treated at laser scan velocities of 8 cm/s to 25 cm/s show only a cellular structure, in which it is seen that the cell boundaries act as obstacles for dislocation motion. GP-zones, that could precede the formation of the θ ' plates, are not observed.

After shot peening, in all samples a higher dislocation density was observed by TEM. Apart from the increased dislocation density however, now precipitates could be detected in samples that did not develop observable precipitates before shot peening. A typical example of this precipitation process induced by shot peening in a sample treated at a low laser scan velocity of 1/8 cm/s, is shown in Figure 9.

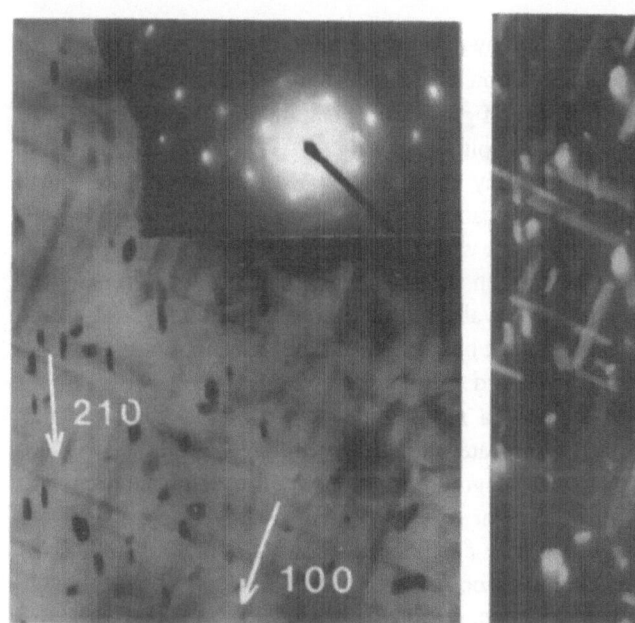 

Figure 9. Bright (a) and dark-field (b) TEM micrograph of shot peening induced and S' precipitates in a sample treated at a laser scan velocity of 1/8 cm/s, peened during 360 s. The chrystallographic planes on which the precipitates are observed to grow, are indexed.

The concentration of precipitates seems also to be higher than before peening in samples treated in the scan velocity range of 1 cm/s to 4 cm/s. This might indicate an increased solid solution in these samples, but the smaller average size, 100 nm, will also result in a higher concentration.

Again, most of these precipitates are θ '-platelets, aligned with the cubic axes. A second type of precipitate, smaller in size but also of the shape of a platelet, is identified as the S' phase, initially growing on [120] planes. These precipitates have a much smaller width-to-thickness ratio and will therefore contribute less to an increase of the flow stress than the θ '-plates (at the same volume fraction). The very broad size distribution, down to only 2 nm, makes this contribution more uncertain.

4. Discussion and analysis

To discuss the experimental results, the laser scan velocity range is divided into two regimes: the low scan velocity regime (1/8 cm/s to 1/2 cm/s) and the high scan velocity regime (1/2 cm/s to 25 cm/s).

Low scan velocity regime

Since the cooling rates are lower in the low laser scan velocity samples, compared to the 2 cm/s samples, in which precipitation of the θ ' plates is observed, one would expect to see an increasing nucleation and growth of precipitates in these samples. In contrast, below 1 cm/s no significant precipitation is observed. The explanation for this observation lies in the decreasing vacancy concentration with decreasing laser scan velocity. Apparently at the lower laser scan velocity the vacancies are annealed out before the temperature reaches values that would favour the nucleation and growth of observable precipitates, a.o. by absorption at dislocations causing the formation of helical dislocations. The rise in hardness values in the low velocity regime can not be caused by precipitation hardening but must be due to two other contributions:

First, the dissolution of the cell-walls and subsequent homogenization will play an important role. Since there exists only a small temperature interval during which homogenization takes place the cooling rate must be low enough to allow sufficient time for diffusion. The copper (and magnesium) brought into solid solution will contribute to an enhanced hardness but for reasons described above, copper will not form any precipitate. The dissolution may continue as long as not all cell walls have been fully dissolved, provided that the cooling rate is low enough. Obviously this situation has not been reached in practice, and it explains why the hardness is still increasing with decreasing laser scan velocity.

In the additional experiments performed on the alloy with 2.32 wt% Cu, it was observed that the cell walls were completely dissolved, only near the edges of the low scan velocity (1/8 cm/s) laser tracks. From this experiment we conclude that to a first approximation only half of the copper is in solid solution in Al 2024 treated with a scan velocity of 1/8 cm/s. A calculation of the diffusion distance in a 1/8 cm/s track, during cooling in the interval between 660 °C and 450 °C, applying the analytical model for the temperature profile [8], yields a distance of approximately 0.1 m. A calculation for magnesium yields a similar value. These values, which should only be regarded as being a first approximation, are too small for significant homogenization. Nevertheless, it shows that homogenization may play a role only at the lowest scan velocities.

An indication that supports homogenization as being the mechanism responsible for the hardness increase in the low scan velocity regime, lies in the observation that in a laser track of 1/8 cm/s scan velocity the hardness increases with depth, as displayed in figure 10. This is a result of the increased homogenization time with increasing depths.

In laser tracks where this mechanism does not operate, e.g. a 8 cm/s track, a hardness decrease is detected with increasing depth.

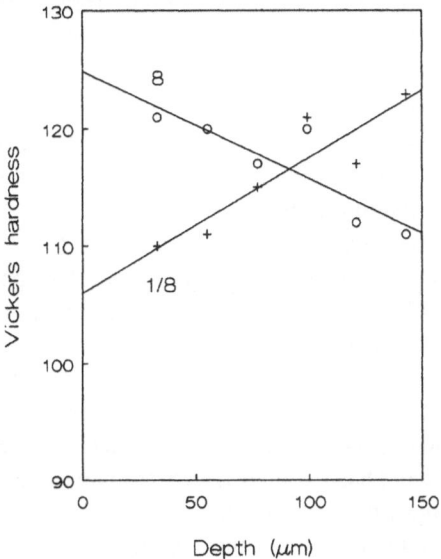

Figure 10. Hardness profile as a function of depth of two samples treated at a low (1/8 cm/s) and a high (8 cm/s) laser scan velocity.

The formation of helical dislocations is observed in all samples treated at low laser scan velocity. From the geometry of these dislocations it is concluded that they arise from both climb of the part of a mixed dislocation of edge character, figure 6a, and from the interaction with prismatic vacancy loops, figure 6b. Both mechanisms require a supersaturation of vacancies [9][10]. A large quantity of vacancies is present because of the substantial cooling rates. The reason that the formation is possible only in samples treated at the low laser scan velocity is that only here the cooling rate at intermediate temperatures is low enough to allow for adequate vacancy diffusion. Once the temperature is too low for significant vacancy diffusion, the dislocations become immobile over the distance that has been affected by climb or by the interaction with the vacancy loop. The equilibrium shape of a helical dislocation is one of a constant pitch and loop radius [10]. Since a non-equilibrium shape is sometimes observed, we conclude that in the sample treated at the lowest scan velocity the annealing time is still not long enough for all the dislocations to reach this equilibrium situation. This in turn can be caused by the continuous formation of new helical segments, which means that not all vacancies are used, or by an increased locked segment through relaxation to the equilibrium shape. These explanations are also in line with the fact that the hardness curve is still rising at the 1/8 cm/s point. Nevertheless, we do not expect that this mechanism will dominate over the role of homogenization as described previously.

High scan velocity regime

The increased hardness in the high velocity regime, reflects the generally observed trend, and is caused by two reasons. First, the cell size decreases upon increasing the laser scan velocity and causes a hardening that depends on K and in the following equation: $\tau = K/d^{\alpha}$. Unfortunately, the values of α and K are not easily derived. a may vary from 1/2 in the Hall-Petch relation, up to 2 as has been described in [11]. Furthermore, precipitates will contribute to the hardness, and this contribution may be even larger than the contribution arising from the small cell size. The presence of these precipitates is observed in samples with scan velocities of 1 cm/s to 4 cm/s.

The absence of GP zones in samples that have not yet developed precipitates, in the range between 8 cm/s and 25 cm/s, might be explained by the fact that the amount of Cu in solid solution is less compared to conventionally heat treated Al 2024. In Al with 2 wt% Cu, GP-I zones do not appear, and the precipitation sequence starts with GP-II [7]. Furthermore it has been reported that in rapidly quenched samples the early stages in the precipitation sequence are accelerated and that GP-II zone formation is suppressed in favour of the phase. If both effects are present in our samples the phase is indeed the first precipitate to be observed.

An additional point that might explain the absence of the GP zones, is that these precipitates are related to the stress condition of the matrix. Since GP zone formation is accompanied by a decrease in average atomic volume, the resultant hydrostatic tensile stresses will contribute to the formation energy. When a hydrostatic tensile stress is already present, the formation of GP zones will be hindered. This mechanism is assumed to explain the absence of GP zones in Al-Cu alloys containing Si or Al_2O_3 particles, which give rise to tensile stresses [12]. In our samples tensile stresses arise because of shrinkage during solidification. These stresses are in the order of 100 MPa, which is in the same order of magnitude as in Al-Cu containing Si or Al_2O_3 particles. As a consequence GP zone formation may also be hindered by the presence of the hydrostatic stresses in our case. As a matter of course this argument counts equally well at low laser scan velocity (figure 4).

When explaining the hardness curve vs laser scan velocity (figure 2) based on the above mentioned contributions, it will become clear that the contributions of cell boundary strengthening and solid solution strengthening do not cover the complete story of the decrease of the hardness with increasing laser scan velocity in the lower scan velocity regime. In fact the contribution of the cell boundary strengthening increases over the whole regime. Taking the proportionality constant K of the Hall-Petch relation for cell boundaries (upper limit) one half of that for grain boundaries (about 0.07 MN/ m3/2 [13]) the cell boundary strengthening may increase from 20 MPa to 35 MPa in the low velocity regime, whereas the contribution by solid solution strengthening decreases from a maximum of 19 MPa, based on 2.32 wt% Cu in solid solution in the samples treated at the lowest laser scan velocity of 1/8 cm/s. In total it results in a much too small increment of the hardness at the lowest scan velocity of 1/8 cm/s. So, in this regime the immobilizations of dislocations by the helical segments and by preferential nucleation of solutes have to be considered as well to explain the increase in the flow stress in a more quantitative way.

After shot peening an increase in hardness independent of the laser scan velocity is observed. This is not what one would expect since the microstructure is quite different at different scan velocities. The precipitation induced by shot peening, that was observed earlier in a laser melted Al-Si alloy, explains why the observed values are the same for the low and high scan velocities: After shot peening precipitates are present independent of the scan velocity regimes, whereas the amount of alloying elements in solid solution is expected to be much less than before shot peening.

The minimum value of the hardness of samples treated at a scan velocity of 1/2 cm/s is still present after shot peening, and is also visible in the curves of the residual stresses as well as of the dislocation density vs laser scan velocity. Apparently the amount of alloying elements brought into solid solution during laser melting has a minimum for this scan velocity. At higher scan velocities more copper and magnesium is retained during the rapid solidification process, and for the lower ones the homogenization time during cooling is longer.

Comparison of the absolute hardness values between the laser treated samples and the T3 samples, one might conclude that laser surface melting is not a suitable technique for obtaining hard surfaces. This conclusion is invalid since the amount of copper that is brought into solid solution is far less than what ultimately is possible. In fact, one should not use a hypo eutectic alloy when a high solid solution concentration is required. In the past it has been shown, by applying splat cooling of an eutectic Al-Cu alloy, that all the copper (33 %) can be retained in solid solution [14]. Furthermore, it was shown that the precipitation sequence was still the same as in the more dilute alloys. A higher concentration of copper obtained by laser melting of such an alloy will likely lead to a high density of platelets, resulting in substantially higher hardness values. For that reason we have concentrated our current work on Al-Cu alloys, in which the Cu content ranges between 0-40 wt.%, and is aimed at describing the mechanical and microstructural properties of these alloys upon variation of the laser scan velocity in the range of 0.0125 to 0.125 m/s. The Cu alloys containing 0, 5, 10, 15, 20, 30, 33 and 40 wt.% Cu are produced by melting the 99.999% pure components of the material with an arc-furnace. From figure 14 it is clear that the hardness increases in a logarithmic way with increasing laser scan velocity. The slopes, at which this increase takes place, rise with augmenting Cu contents. Each hardness value consists of at least five measurements for the laser tracks and at least eight measurements for the bulk. The hardness values are well above those found in Al-Cu-Mg alloys after corresponding laser treatment.

One of the important microstructural features is that the distance between the Cu rich material (bright contrast) decreases significantly with increasing the Cu content. In an Al 10 wt.% Cu alloy this distance is about 1 μm (figure 12) compared to about 0.1 μm in an Al 33 wt.% Cu alloy (figure 13). The appearance of the brighter areas change from a cellular to a lamellar network with increasing the Cu contents. This lamellar network is oriented perpen- dicular to the solidification front.

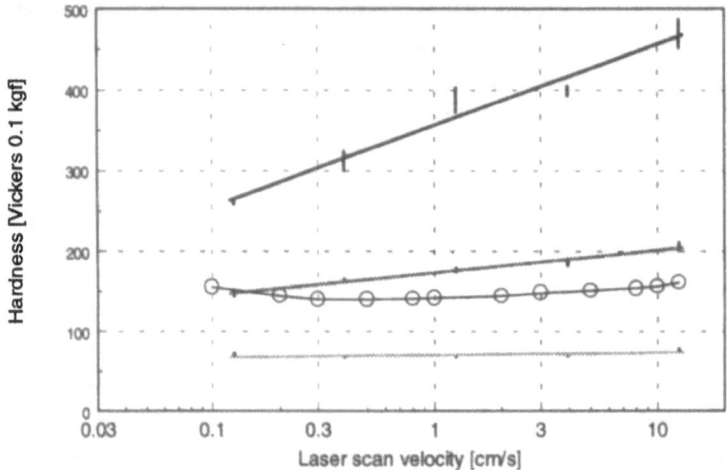

Figure 11. Hardness profiles as a function of the laser scan velocity and Cu contents. (Horizontal line: 5 wt% Cu; o Al 2024; dashed line 20 wt% Cu, solid line 40 wt% Cu).

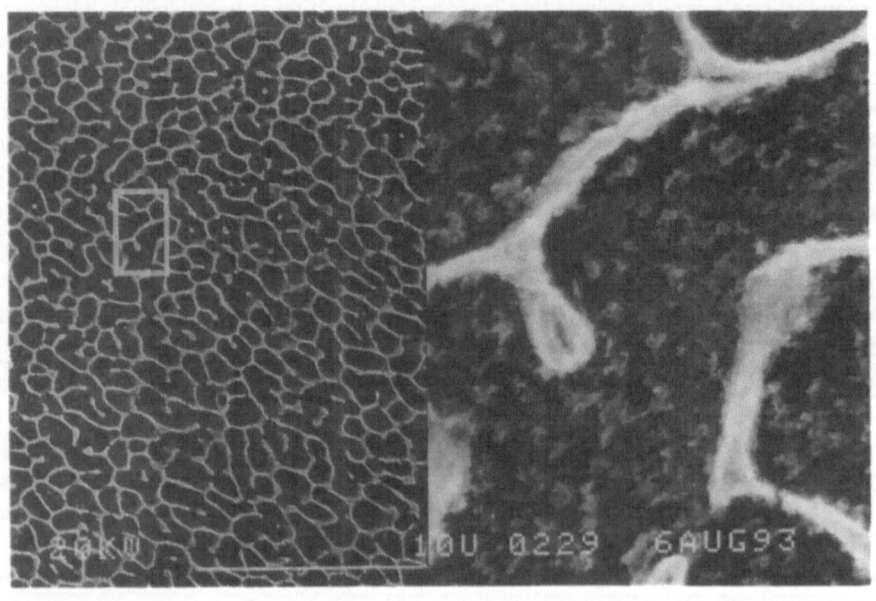

Figure 12. SEM micrograph of an etched Al-10 wt% Cu sample laser scanned at a velocity of 3.953 cm/s.

Secondly, the θ phase appears in an Al-Al$_2$Cu eutectic lamellar structure. This eutectic composition is formed in the Al matrix. However in the Al 40 wt.% Cu alloy, islands of more or less pure Al$_2$Cu are found, while the matrix displays a eutectic-like structure. As the volume fraction of the eutectic structure increases, the mobility of dislocations in the material decreases. In this respect one has to consider that the lamellar spacing is in the order of 1 μm. [15].

However, the increase in hardness due to laser treatment should be explained by a combination of the following phenomena. Firstly, a refinement in structure occurs, causing an increase in hardness [11]. This increase must also be explained by the change in the type of network which the Cu rich phase exhibits: cellular to lamellar (figures 12 and 13). As soon as this change in network is realized, the slope between Vickers hardness and the laser scan velocity in figure 11 rises considerably.

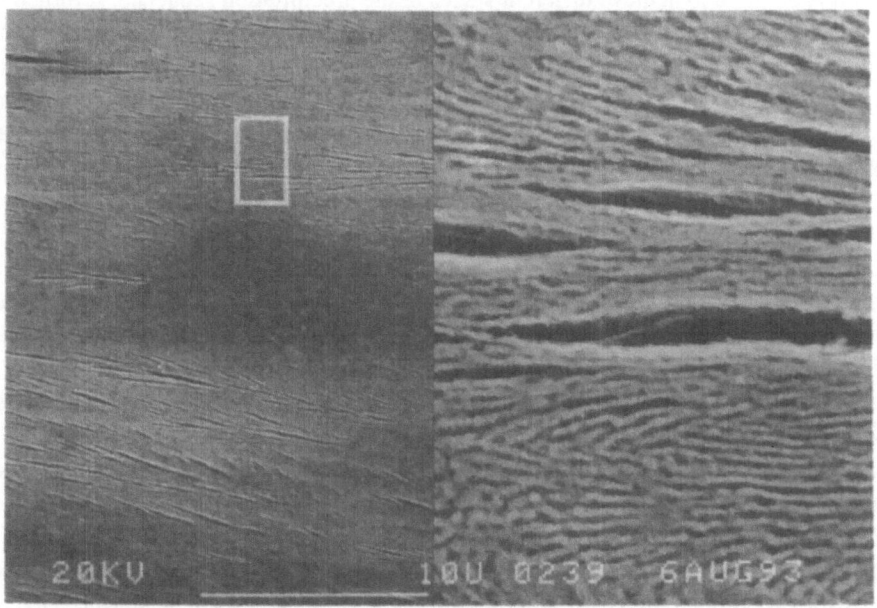

Figure 13. SEM micrograph of an etched Al-33 wt % Cu sample laser scanned at a velocity of 3.953 cm/s.

Secondly, the high quench rates, which can be attained by laser processing, will cause a larger fraction of the available Cu to be retained in solid solution [16]. Also, as mentioned earlier, precipitation starts with the formation of GP-zones followed by the θ'' and θ' phase. In this precipitation sequence, hardness will increase due to coherency loss between the precipitates and the Al matrix causing the matrix to be highly strained. On the other hand, hardness will decrease due to the larger spacing between the precipitates. The maximum hardness in conventional alloys will thus be associated with a combination of θ'' and θ' [14]. Nevertheless, during formation of GP-zones, as well as θ'' and θ' phases in the laser-treated alloys, the distance between these precipitates will be constrained by the refined cell size. Together with solid solution hardening, the decreasing cell size will contribute to an increase in hardness.

Finally, the increasing slopes in figure 11 are correlated with the Cu content in the alloys. As the amount of Cu increases, the thermal conductivity will rise, more or less proportionally, and as a consequence the quench rates. Both refinement in structure and solid solution hardening will become more easy, thereby causing a faster increase in hardness.

5. Conclusions

The hardness curve observed in this study on Al-Cu-Mg alloys, with a minimum at a laser scan velocity of 1/2 cm/s, is quite different from the monotonically increasing curves, usually detected in many other laser treated materials. The reason for this observation is the minimum in the solid solution concentration. By varying the quench rate it is possible to achieve different stages in the precipitation sequence: It turned out that depending on the laser scan velocity nucleation and growth of precipitates could either be enhanced or suppressed. The formation is suppressed by quenching too rapidly using a high laser scan velocity, whereas at too low a scan velocity the formation is suppressed due to a lack of sufficient vacancies. In the latter case, the vacancies are trapped by dislocations, leading to the formation of helical dislocations, i.e. before the temperature is low enough to develop a driving force for precipitation.

The mechanisms by which the hardening occurs are the same as the ones in conventionally hardened specimens, except for the contributions originating from the small cell sizes and from the formation of helical dislocations. The latter ones, however, do not seem to contribute significantly. Only at lower laser scan velocities dissolution of cell walls leads to an increased amount of copper and magnesium in solid solution and, therefore, to an increase of the hardness caused by solid solution hardening.

After subsequent shot peening, in all cases the formation of precipitates was observed, independent of the laser scan velocities originally applied. This phenomenon of precipitation induced by shot peening is most striking at a high concentration of alloying elements in solid solution. Since this is the case in samples treated at a low scan velocity (i.e. an increased homogenization time) as well as in samples treated at a high scan velocity (high quench rates), the parabolic shape is maintained even after shot peening. An increased dislocation density, due to shot peening, contributes to the increased hardness as well.

It turned out that the the hardness of Al-Cu alloys can be boosted even further by an increase in the Cu content. The hardness values found in this work are well beyond the values which have, been found in the Al-Cu-Mg system under investigation, i.e. Vickers hardness for a Al 2024 alloy treated with comparable laser scan velocities. The findings presented in the Al-Cu case on the other hand, testify that a Vickers hardness of around 470 can be reached, if an Al 40 wt.% Cu alloy is treated with a laser scan velocity of 0.125 m/s.

6. References

(1) Hegge, H.J., DeBeurs, H., Noordhuis, J., DeHosson, J.Th.M., (1990) Metall. Trans. **21A**, 987.
(2) Noordhuis, J., DeHosson, J. Th.M., (1992) Acta Met. et Metall. **40**, 3317.
(3) Williamson, G.K. and Smallman, R.E., (1956) Phil. Mag. **1**, 34.

(4) Mikkola, D.E., Cohen, J.B., (1965) in J.B. Cohen and J.E. Hilliard (eds.) *Local Arrangements studied by X-ray Diffraction,* Met. Soc. Conf. **36**, 271.

(5) Wilson, R.N. and Partridge, P.G. (1965) Acta Metallurgica et Materialia **13**, 1321.

(6) Stolz, R.E. and Pelloux, R.M. (1976) Met. Trans. **7A**, 1295.

(7) Mondolfo, R.M. (1976) *Aluminum alloys structure and properties*, 264.

(8) Ashby, M.F. and Easterling, K.E. (1984) Acta Metallurgica et Materialia **32**, 1935.

(9) Amelinckx, S., Bontinck, W., Dekeyser, W. and F. Seitz, (1957) Phil. Mag. 2, 355.

(10) Hirth, J.P. and Lothe, J. (1982) *Theory of dislocations*, 623.

(11) Hegge, H.J., DeBeurs, H., Noordhuis, J. and DeHosson, J.Th.M. (1990) Met. Trans. **21A**, 987.

(12) Starink, M.J., (1992) Thesis, University Delft.

(13) Embury, J.D. (1971) *Strengthening Methods in Crystals*, A. Kelly and R. B. Nicholson (eds.) Wiley, New York, and references therein.

(14) Scott, M.G. and Leake, J.A. (1975) Acta Metallurgica et Materialia **23**, 503.

(15) DeHosson, J.Th.M., Boom, G., Schlagowski, U., Kanert, O. (1986) Acta Metallurgica et Materialia **34**, 1571.

(16) Noordhuis, J. and DeHosson, J.Th.M. (1992) Acta Metallurgica et Materialia **41**, 2895.

REAL TIME PYROMETRY IN LASER SURFACE TREATMENT

I. SMUROV
Ecole Nationale d'Ingénieurs de Saint-Etienne,
58, rue Jean Parot, 42023 Saint-Etienne Cedex 2,
France

M. IGNATIEV
Institut de Science et de Génie des Matériaux et Procédés,
Centre National de la Recherche Scientifique,
B.P.5 Odeillo, 66125 Font-Romeu Cedex
France

ABSTRACT. The main principles of design of high speed high spatial resolution pyrometer are presented. The opportunities for temperature measurement by developed pyrometer system in a wide range of laser applications are illustrated by heat treatment, welding, cladding, alloying and laser assisted machining. Continuous CO_2 and pulsed Nd:YAG laser machining are considered. The influence of such factors as beam traverse speed, laser power (or pulse energy), slab thickness, melt droplets expulsion, distance from the seam center, etc on temperature evolution is analyzed. The advantages and the drawback of real time pyrometry in laser machining are discussed.

1. Introduction

The accurate measurement and control of temperature is of great importance in most materials manufacturing and processing applications. In many cases contact with subject is either not desirable, because it may significantly alter the temperature or other characteristics of the subject, or is not possible, because of the conditions of measurement. The precision technique of noncontact temperature measurements is required, for example, for the following processes: crystals growth, continuous steel casting, high temperature metallurgy and so on. This is of particular importance for High Power Energy Flows (HPEF) processes which are quite difficult to control in real time due to very fast rates of temperature variation, high temperature gradients and strong emission from the erosion plasma plume. The absence of on-line temperature control and monitoring is one of the main limitations of HPEF processes integration into modern industry.

529

J. Mazumder et al. (eds.), Laser Processing: Surface Treatment and Film Deposition, 529–564.
© *1996 Kluwer Academic Publishers.*

2. Theoretical and Experimental Background of Optical Pyrometry

Monochromatic Pyrometry. Any monochromatic optical pyrometer (brightness pyrometer) measures the spectral radiance (brightness) temperature of the source. The spectral radiance (radiance per unit wavelength) for a black body is given by the Planck radiation law

$$P_\lambda = \frac{C_1}{\lambda^5 [\exp(C_2/\lambda T)-1]} \tag{1}$$

where λ is the wavelength, P_λ is the hemispherical intensity of the spectral radiance of a black body, T is the absolute temperature and $C_1 = 3.7413 \times 10^{-16}$ Wm^2, $C_2 = 1.4388 \times 10^{-2}$ mK are the Planck constants. In accordance with Planck law, the maximum of the spectral radiance intensity P_λ^{max} is shifted to shorter wavelengths at temperature rise (for example, P_λ^{max} corresponds to $\lambda_{max} = 1.8$ μm at T=1600 K and to $\lambda_{max} = 6.0$ μm at T=500 K). The maximum spectral radiance intensity P_λ^{max} versus temperature can be obtained by the following formula

$$P_\lambda^{max} = b_1 T^5 \tag{2}$$

where $b_1 = 1.2864 \times 10^{-5}$ $Wm^{-3}K^{-5}$. As a result radiance measurements require the application of different types of detectors at low and high temperatures (low temperature range requires more sensitive detectors working in infra-red region).

A real body, however, emits only a fraction of what a black body emits at any given temperature. The total hemispherical emissivity is thus defined as

$$\varepsilon_\lambda = P_\lambda^* / P_\lambda \tag{3}$$

where P_λ^* is the hemispherical intensity of the spectral radiance of a real body.

The radiance (brightness) temperature of the real body T_r is defined as the temperature of the blackbody having the same radiance intensity at the same wavelength. The absolute temperature is higher than radiance temperature T_r by an amount dependent on the emissivity of the source

$$T^{-1} = T_r^{-1} + (\lambda/C_2)\ln \varepsilon(\lambda,T) \tag{5}$$

Lack of reliable emissivity data (especially for high temperatures) is the major drawback in brightness pyrometry. For example, take the case of iron, where the pure material exhibits an emissivity as low as 35%, whereas the oxidized surface can have an emissivity as high as 95% at high temperatures (λ=0.65 μm). Or, the case of aluminum, where the emissivity (λ=0.65 μm) can vary from 10% to 40% depending upon the degree of oxidation, surface treatment and so on. It is therefore not untypical to find relative emissivity uncertainties of 50% and even 100%. In our example, the associated uncorrected temperature errors would be between 50 K and 200 K at low temperatures (up to 2000 K) and could reach 800 K at high temperatures (more than 3000 K). As a result monochromatic pyrometer may be used for accurate temperature measurements only when the emissivity data are well known.

Ratio (two-color) Pyrometry. Ratio or two-color pyrometers can circumvent the emissivity measurement issue in certain specific cases. The two-color method uses an approximation of the Planck law called the Wien radiation relation

$$P_\lambda^* = \frac{\varepsilon_\lambda C_1}{\lambda^5 \exp(C_2/\lambda T)} \tag{6}$$

This approximation gives a deviation of less than 1% from Planck law if $\lambda T < 3000$ μmK [1]. The Wien relation can be solved for temperature at two different wavelengths to give

$$T = \frac{C_2(\lambda_2 \lambda_1)}{\lambda_1 \lambda_2 [5\ln(\lambda_2/\lambda_1) - \ln(P_{\lambda 1}^*/P_{\lambda 2}^*) + \ln(\varepsilon_1/\varepsilon_2)]} \tag{7}$$

If the emissivities at the both wavelengths are equal (this is so called "gray body" conditions), the emissivity dependence of the temperature measurement disappears; if they are not equal, their ratio must be known in order to derive the true temperature from measured color temperature T_c, that is defined as

$$T_c^{-1} = T^{-1} - \frac{\lambda_1 \lambda_2}{C_2(\lambda_2 - \lambda_1)} \ln \frac{\varepsilon_1(\lambda_1, T)}{\varepsilon_2(\lambda_2, T)} \tag{8}$$

Hence, some knowledge of emissivity is still required and two-color pyrometer is more accurate than monochromatic one if the wavelengths λ_1 and λ_2 are chosen such that "gray body" behavior can be assumed ($\varepsilon_1 = \varepsilon_2$ or $\varepsilon_1/\varepsilon_2$=const) then the emissivity term drops out and the temperature calculation is straight forward.

The two-color ratio pyrometry is quite sensitive to variation in emissivity if one emissivity varies more than the other [2,3]. In the case, for instance, where one emissivity varies, but the other one stays constant, differentiation of equation (8) with respect to ε_1 gives

$$\frac{\delta T_c}{T_c} = \frac{T_c \lambda_1 \lambda_2}{C_2(\lambda_2 - \lambda_1)} \frac{\delta \varepsilon_1}{\varepsilon_1} \qquad (9)$$

This equation allows a quantitative estimation of the emissivity error of the two-color ratio pyrometer. For example, in the two-color ratio pyrometer with $\lambda_1 = 0.65$ μm, $\lambda_2 = 1.3$ μm, $T_c = 1500$ K, a 10% variation in emissivity at one wavelength will give 1.5% variation in indicated temperature if the emissivity at the other wavelength does not change.

If the wavelengths of two-color ratio pyrometer are put closer together the situation becomes worse: for $\lambda_1 = 0.65$ μm, $\lambda_2 = 0.7$ μm, a 10% variation in emissivity at one wavelength uncompensated by a change at the other wavelength would give nearly 10% variation in indicated temperature. Of course, if two wavelengths are close together it is less likely that the emissivity at one wavelength will vary without some compensating variation in emissivity at the other wavelength. Unfortunately, as the wavelength are brought close together, the denominator of equation (8) grows small, and unavoidable small variations in the factors affecting the term in the numerator which should be constant produce unacceptable errors in the pyrometer readings.

That is the ratio pyrometer becomes very sensitive to measurement noise when wavelengths separation is decreased (particular for the measurements at longer wavelengths). According to the results reported in [1], even the addition of 0.2%-0.5% root-mean-square (rms) noise causes a significant increase in the uncertainty in the measured temperature (up to 50 K-100 K). This is of particularly important for HPEF processes when other sources of intense radiation (plasma, hot gases, etc...) can cause the high level of noise. The preliminary experiments are also required for accurate choice of wavelength separation to reach the acceptable accuracy of measurements.

Finally, for two-color ratio pyrometry the assumption of "gray body" behavior becomes more valid as $\Delta\lambda = (\lambda_1 - \lambda_2) \longrightarrow 0$, but as $\Delta\lambda \longrightarrow 0$ any errors in the radiance measurements become more significant. Increasing the separation of the wavelengths reduces the effects of radiance measurement errors but "gray body" assumption becomes less valid.

Multiwavelengths Pyrometry. To increase the accuracy of the pyrometer further, measurements of the spectral radiance at a large number of wavelengths are used (multiwavelengths pyrometry) [1,4-6] .

In this method, measurements of the spectral radiance emission of the object are taken at several wavelengths. The processing of these data is then performed by a variety of techniques [1,4-6], the most accurate of which have proven to be "least-squared-based" multiwavelengths technique [1]. Radiance from an object are detected by a spectrograph. These radiance values are input to a computer that curve-fits these signals to a theoretical emissivity function that has variable coefficients of the emissivity function and temperature.

The true temperature and emissivity can be simultaneously calculated on base of object radiance values detected by spectrograph at different wavelengths without any a priori information about the source. The error of temperature measurement is claimed to be less 0.5% in the temperature range 1200 K to 2600 K [1].

Usually, this method requires a lot of time for calculation procedure. It may limit the multiwavelengths pyrometry applications in real-time temperature monitoring and control (particularly for HPEF processes which are characterized by fast rates of temperature and emissivity variations).

In spite of well developed theoretical background of multiwavelengths pyrometry, a number of obvious, but sometimes neglected reasons can cause the large errors in absolute temperature measurements. These errors arise from: aberrations, scattering and diffraction in pyrometer optical components; gaseous or plasma absorbs along the optical path; influence of own radiation of hot gases or plasma (at wavelengths as used in pyrometer); extraneous radiation reflected into the pyrometer detector by target; distinctions between calibration and measurement conditions and so on.

Active Pyrometry. The active pyrometry that is a new method of determining the emissivity of a hot target from laser-based reflectance measurement which is conducted simultaneously with a measurement of heat target radiance [7,8]. By use of a laser beam directed to and reflected by the target surface a quantitative value of emissivity is obtained. The method of determining emissivity from reflectivity relates to Kirchhoff postulate that at thermal equilibrium all bodies in a closed environment emit as much radiation as they absorb. This postulate is described by the following equation

$$\varepsilon(\lambda,r)=1-\rho(\lambda,r) \qquad (10)$$

where $\varepsilon(\lambda,r)$ is the spectral emissivity for emission in the direction r; $\rho(\lambda,r)$ is the directional hemispherical reflectivity for radiation incident in the direction r.

The existing laser pyrometers can determine the emissivity from a reflectivity measurement which is normalized for distance to the target and laser power [7,8].

This method also allows to eliminate the influence of extraneous energy reflected by the target on accuracy of absolute temperature measurements. In many cases of interest the target is placed in an environment where other sources of intense radiation are present (for example, a target in a furnace, or plasma or hot gases are near the target surface). This radiation is reflected off the target adding itself to the target's self-emission and leading to false temperature readings. The active pyrometer have provision for the measurement of ambient radiance as well as the target emissivity and radiance. During the digital signal treatment each target measurement can be compensated first for ambient component and then for the target emissivity to yield the absolute target temperature [7].

The active laser pyrometer is good applied only for materials whose polar scattering patterns are uniformly diffuse (Lambertian). The usage of active pyrometer for mirror-like surfaces (particularly with variation of reflectivity direction during measurement cycle) requires the development of special methodology, approaches and equipment.

Conclusion. The application of optical pyrometry for accurate temperature measurements under actual conditions of materials manufacturing and processing requires the solution of the following methodological problems: (i) correct measurements of radiance temperature; (ii) recalculation of the true temperature.

3. Main Goals of Pyrometry Application for Laser Machining

Material machining by High Power Energy Flows, such as welding, cutting, cladding, drilling, transformation hardening, alloying, grain boundary modification and so on are becoming commercially more important as the advantages of the new technologies become more widely known.

The thermal state of material in the zone of laser action is one of the main integrated parameters to determine the evolution of physicochemical processes and structural phase transformations. In view of this, the information on the dynamics of surface temperature during the whole thermocycle is of major practical interest.

A number of methods can be used for real-time diagnostic of laser action: control of weld quality by acoustic emission [9,10], recording of stress pulses [11], and photopyroelectric signal detection [12]. Among the optical methods, the following can be mentioned: on-line photodiode monitoring of CO_2-laser welding [13], monitoring of pulsed laser melting and evaporation by means of measurements of deflection of a secondary diagnostic laser beam [14], measurement of UV-plasma radiation during laser cutting of ceramics [15], time-resolved reflectivity [16], pool shape data visualization [17], laser induced fluorescence (LIF) [18] and so on. The methods of process monitoring based on real physical variables (for example, surface temperature) and on indirect integrated parameters (signals connected with simultaneous variation of a number of physical variables, for example, total radiation intensity from the zone of laser action that is affected by both laser plume and surface temperature, which cannot be easily separated) must be distinguished. Generally speaking, methods of the optimization and on-line control based on indirect parameters monitoring are not universal and can only be used within a limited range of process parameters.

In practice, all laser machining processes require a comparatively narrow range of variation of main parameters (power density, beam traverse speed, shielding gas flow rate, focus spot dimension and so on). Several examples developed hereafter illustrate the necessity for knowledge of actual temperature values. For example, the critical temperature range of the solidified metal directly behind the melt pool must be within the interval 1650-1750 K to produce high-quality laser welding of flanges of concentric stainless steel cylinders [19]. A surface temperature deviation greater than 10% leads to significant variation of the hardened layer depth [20]. The spatial distribution of surface temperature determines the flow pattern in thermocapillary melt convection and, therefore, the concentration fields of alloying elements [21-23].The absence of experimental surface temperature thermocycles in real-time makes it difficult to understand the peculiarities of heat and mass transfer in a number of laser applications.

From the general point of view, the absence of on-line control and monitoring is one of the main factors limiting integration of laser-technology into modern industry. Data on surface temperature dynamics obtained by pyrometry could be widely used: (i)to understand and optimize the laser processes (namely, to define the links between surface-temperature history and the properties of produced materials); (ii)create process control including feedback from the pyrometer.

The development of a pyrometer system for recording the actual surface temperature during laser machining involves significant methodological difficulties that are defined by the following process parameters: wide temperature range (500 K-4000 K), high heating and cooling rates (10^3 -10^8 K/s), small size of the heated zone (0.2-5 mm), the influence of laser plume radiation, sharp variation of a material's optical and thermal properties and so on (a micromachining of materials by short laser pulses (1 ns - 100 μs) is not considered in the present paper).

The high speed high spatial resolution pyrometer (HSP) have been specially developed for applications in laser machining [24-25]. The combination of high speed, wide temperature range and spatial resolution makes the HSP a promising system for a number of applications. Small size and the opportunity of easy transportation and installation of HSP should be mentioned.

4. Design of High Speed High Spatial Resolution Pyrometer

The following basis for optical system and electronics design was selected [26] to satisfy the above mentioned requirements.

4.1. OPTICAL SYSTEM

The optical system scheme is shown in the Figure 1. The radiation from the heated surface (1), passing through the photo (2) and objective (3,4) lenses, strike a mirror (5) with pin-hole diaphragm. After the diaphragm the light beam is dispersed by a diffraction grating system (8,9). Two silicon photo diodes (10) are used to detect radiation from the heated-surface at selected wavelengths. The variations of the latter (from 0.5 to 1.1 μm) are obtained by moving the photo diodes along the grating image

536

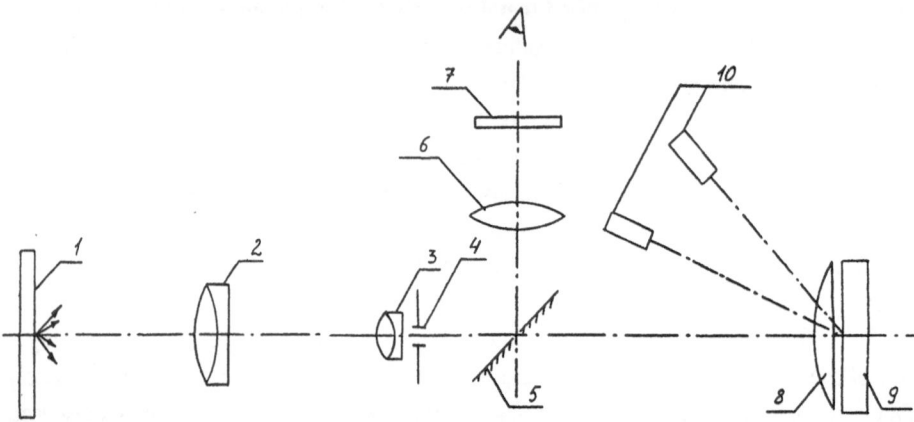

Figure 1. Optical scheme of the HSP. (1)-heated surface; (2)-photo lens; (3)-objective lens; (4)-diaphragm of the objective lens; (5)-mirror with pin hole diaphragm; (6)-lens of the eyepiece; (7)-shadow screen of the eyepiece; (8)-collimating lens; (9)-diffraction grate; (10)-photodiodes.

plane. This method of radiation filtering is chosen in order to change easily the operating wavelength of pyrometer, the later is important under actual conditions of laser machining. The pyrometer can be aimed at the required point with help of the object image observed through the eyepiece (6,7).

4.2. ELECTRONICS

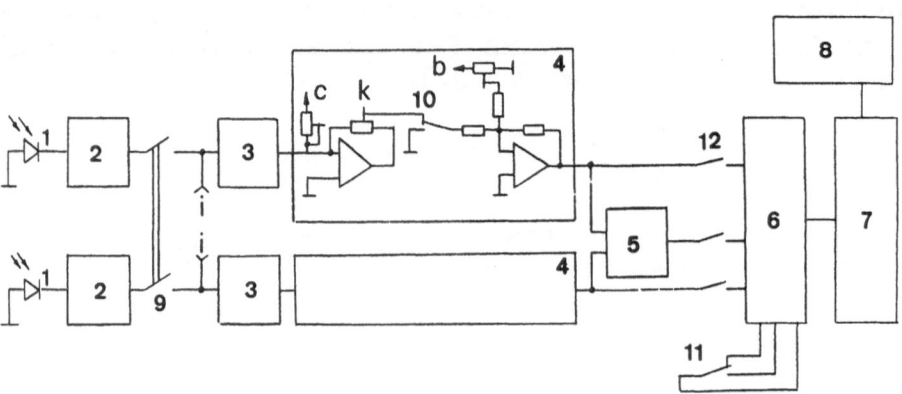

Figure 2. Block scheme of the HSP electronics: (1)-photodiodes; (2)-low noise preamplifier; (3)-logarithmic amplifiers; (4)-scaling amplifiers; (5)-analog processor; (6,7)-components of the data acquisition system; (8)-digital display; (9-12)-calibration switches.

The block scheme of the pyrometer electronics is represented in the Figure 2. The wide temperature range of the pyrometer (1150–3500 K) was obtained by calculating the logarithm (logarithmic amplifiers 3) of the preamplified (low noise preamplifiers 2) signals of photo diodes (1). The scaling amplifiers (4) and analogue processor (5) form the output signal which is proportional to radiance or color temperatures (1 mV per 1 K). The output signal is recorded by data acquisition system (6,7) directly into the computer memory. The digital display (8) and the switches (9-12) serve for pyrometer calibration.

5. Metrology of Temperature Measurements During Laser Machining

5.1. TEST OF THE PYROMETER PERFORMANCE

A pyrometric test set was used for trials (a product of Pyrometer Instrument Co. Inc., USA, model N104). It includes a tungsten strip lamp and a micro-optical pyrometer (MOP). The MOP operates at the wavelength $\lambda_1 = 0.656$ µm and its error is ± 5 K over the total temperature range [26].

To check the calibration linearity, the developed high-speed pyrometer (HSP) and the MOP were sighted at the same point of the tungsten strip lamp. The results of radiance temperature measurement were directly compared for the wavelength $\lambda_1 = 0.656$ µm. The threshold sensitivity ($\lambda_1 = 0.656$ µm) was found to be 1515 K.

The temperature dependencies of the tungsten emissivities [27] were used to calculate the radiance temperatures for the second wavelength $\lambda_2 = 0.956$ µm. The indications of the HSP were compared with the calculated values. The threshold sensitivity ($\lambda_2 = 0.956$ µm) for the second wavelength was found to be 1124 K.

The accuracy of the pyrometer near the sensitivity levels is ± 10 K (0.7%) and ± 5 K for wavelengths $\lambda_1 = 0.656$ µm and $\lambda_2 = 0.956$ µm respectively (Figure 3,4). The noise amplitude does not exceed 0.15% at higher radiance temperatures.

The calibration of the HSP has shown a good linearity (better than 5 K) at temperatures up to 2500 K. At higher temperatures, the calibration curve can be extrapolated by Wien approximation up to 3500 K.

To measure the response time, a rotating disc with slits was used to cut the light of the lamp and the radiation pulses with 1 kHz frequency and 10 µs leading edge were performed. The response time of the HSP was found to be 190 ms. The dynamic error (relative to the reference temperature) is less than 0.25% [26] (Figure 5).

The general procedure of the spatial resolution test performed is as follows. One end of fiber optic (400 µm diameter) is placed in front of the tungsten lamp. The indications of the HSP (from the other end of the fiber) depend on the distance between the center of its spot of vision and the center of the optical fiber cross section. The spatial resolution of the HSP was found to be 200 ± 10 µm (Figure 6).

538

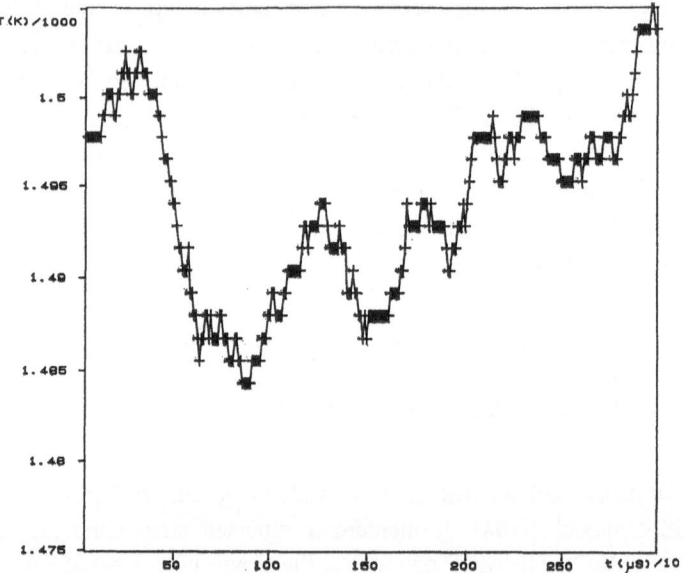

Figure 3. Noise amplitude near the sensitivity level. Wavelength $\lambda_1 = 0.656\ \mu m$.

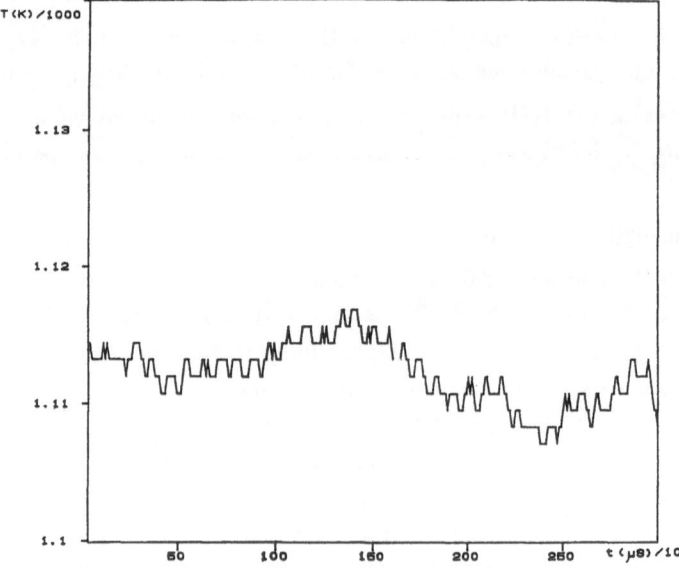

Figure 4. Noise amplitude near the sensitivity level. Wavelength $\lambda_2 = 0.956\ \mu m$.

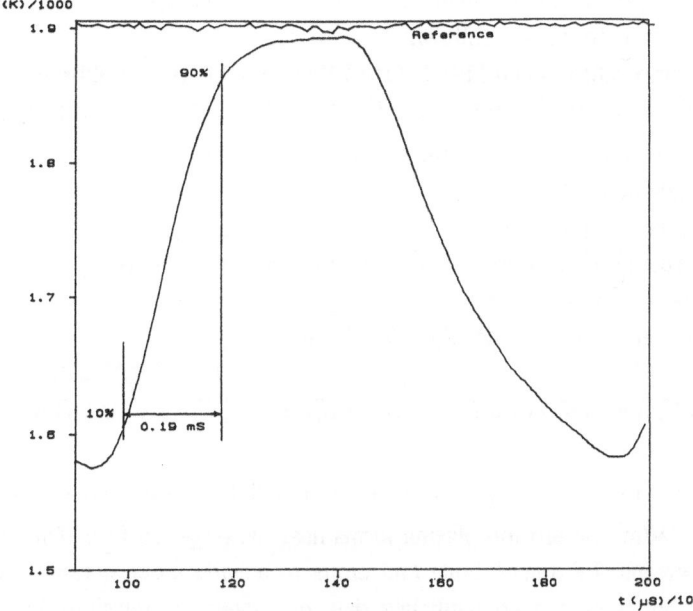

Figure 5. Response of the HSP to single radiation pulse. Reference radiance temperature $T_r(\lambda_1=0.656\ \mu m)=1900$ K. Frequency of the light flow modulation f=1 kHz.

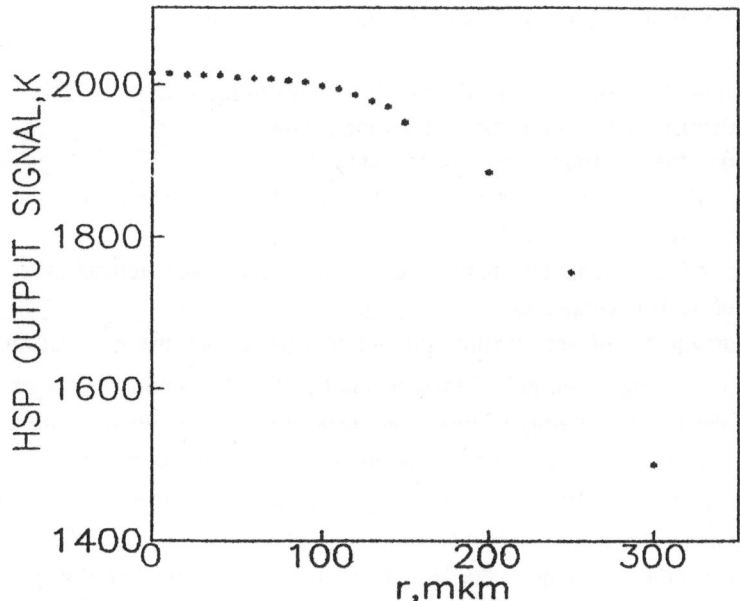

Figure 6. Dependence of the HSP indications on distance between the center of its spot of vision and the center of the fiber cross section.

As a result, the main technical parameters of the developed high speed high spatial resolution pyrometer are the following:

-radiance temperature range 1124 K (1515 K) to 3500 K (threshold sensitivity 1124 K corresponds to λ_2=0.956 μm, 1515 K corresponds to λ_1=0.656 μm, because the threshold sensitivity depends on the selected wavelength);

-measurement error 1%;

-response time 200 μs;

-200 μm spatial resolution at 350 mm distance between the pyrometer and the temperature measurement zone;

-range of the effective wavelengths 0.5-1.1 μm.

5.2. EXAMINING OF TEMPERATURE MEASUREMENT CONDITIONS

When the metals are exposed to the laser radiation at an average power density $q=10^9$-10^{13} W/m^2, an erosion plasma forms near the target surface. The plasma cloud partially screens the treated zone and emits in a wide spectral range. It should be remarked that there are no published data on materials emissivity behavior under conditions of laser treatment (high heating and cooling rates; high temperatures; variations of chemical compositions in the surface layers caused by intensive remixing, chemical reactions, evaporation and so on).

It is necessary to perform systematic research on the materials optical properties, as well as the effects of erosion plasma influence on the accuracy of temperature measurements.

The temperature-measurement method for laser machining includes [28,29]:

-recording the emission spectrum from the erosion plasma;

-determination of the spectral transparency windows;

-reflectometry with the auxiliary laser in the region of relative transparency;

-determination of the monochromatic emissivities $\varepsilon(\lambda$=const,T);

-measurement of the radiance temperatures and (or) color temperature and recalculation of the true temperature.

Spectral diagnostic of the erosion plasma is carried out using a diffraction spectroscope. For example, for pulsed laser action ($q\sim10^9$ W/m^2) on stainless steel in air, the spectrum consists mainly of lines from neutral excited atoms of elements Fe and Cr against a background of a weak continuum due to the target vapor. When the power density rises to 10^{10} W/m^2 or more, the plume shows lines from singly ionized Cr and Fe atoms. There are only lines from ions and atoms of the target whose ionization and excitation energies are less than 5.0 eV. The relative-transparency zones for the plasma plume from stainless steel in air lie around 0.67, 0.57, 0.55, and 0.49 μm [28, 29].

After preliminary determination of the spectral transmission zones of the erosion plasma, it is possible to select the effective pyrometer wavelengths and then to determine the temperature state of the treatment zone with an acceptable accuracy (note that signal to noise ratio should be examined).

The surface emissivity in the laser treatment region is determined by reflectometry [29] with laser probe and Ulbricht integrating sphere.

6. Test of the High Speed High Spatial Resolution Pyrometer under Actual Conditions of Laser Machining

The main aim of the present paragraph is to show the advantages and difficulties of application the high speed high spatial resolution pyrometer system in a wide range of laser machining processes.

6.1. EXPERIMENTAL CONDITIONS

5-kW carbon dioxide ($\lambda=10.6$ μm) and Nd:YAG pulse-periodic ($\lambda=1.06$ μm, pulse duration 1-4 ms, pulse energy 5-15 J) lasers were used in the experiments. The specimens were mounted on a numerical-control (CNC) X-Y table and were used in laser hardening, butt welding, cladding and laser assisted machining. During laser treatment helium was used to protect the metal surface from oxidation (the angle between gas flow and sample was 45°). The cladded material (Stellite 6) in the form of fine powder was injected by means of inert gas flow into the zone of laser action. Pure Fe with Mo coating (thickness 35 μm) and pure Ti samples were used in experiments on pulse-periodic laser alloying from pre-deposited coatings and from the gas phase. The experimental conditions of laser assisted machining are described in details in the paragraph 6.6. An infrared pyrometer was used, in addition, to investigate the peculiarities of cooling of the heat-affected zone (HAZ) in laser welding. This instrument has 150 μm spot resolution at 200 μm distance; temperature range 850-1170 K; response time 500 μs.

The opportunities for actual temperature measurements in various laser applications are illustrated by several examples in heat treatment, welding, cladding, alloying and laser assisted machining. The peculiarities of heat processes are shown and discussed.

6.2. LASER HEAT TREATMENT

The thermocycles during CO_2-laser hardening were investigated over a wide range of beam-traverse speed, namely 400-4000 mm/min.

The sharp change of thermocycles characteristics of mild steel was found at shielding gas flow rates less than 35 l/min. When the beam-traverse speed (BTS) was

542

more than 3100 mm/min (laser power 1.7 kW) the surface temperature was smaller than the threshold sensitivity (1150 K) for the selected wavelength. If the BTS value was 15-25 mm/min less, two situations were observed. In the first case, only the upper part of the heating curve from 1150 K to 1350 K could be seen (Figure 7, curve 3); in the second case, a small variation of beam traverse speed (nearly 10 mm/min) led to a sharp increase of heating (for example, curve 2 in Figure 7). The maximum temperature variations were unpredictable over the range 1400-1700 K. These abrupt changes of thermocycles were measured for exactly the same process parameters.

When BTS was decreased more than 100 mm/min intensive melting of the steel surface started (Figure 7, curve 1, "shelve" at the level of the steel melting point 1810 K). No smooth transition between these two situations (first T_{max} about 1350 K, second T_{max} about melting point) was not observed.

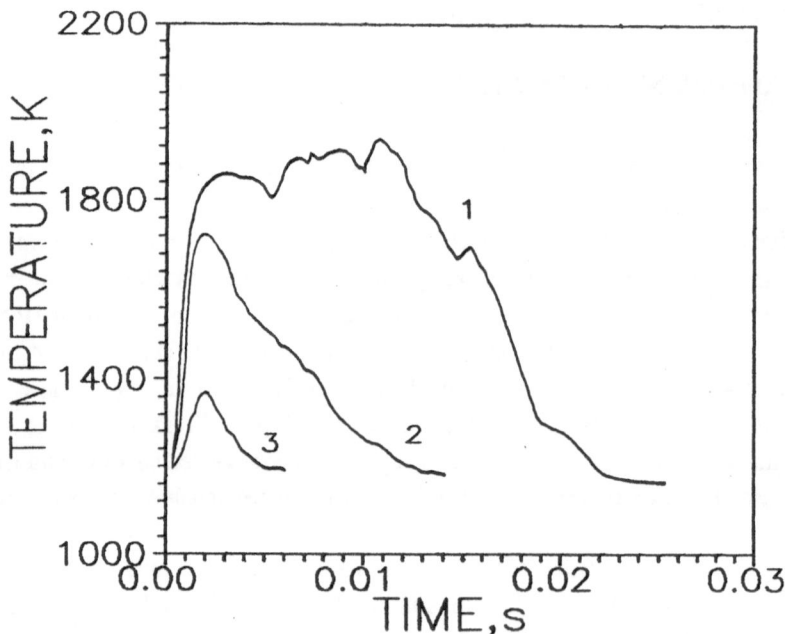

Figure 7. Surface thermocycles in the center of the laser beam scan. Mild steel. Laser power 1.7 kW. Beam traverse speed, mm/min: curve 1-2900; 2-3075; 3-3085. Shielding gas flow rate 28 l/min.

Such threshold type of thermocycle variation may be explained by: (i) temperature dependencies of absorptivity and thermal properties of metal and (ii) the influence of surface oxidation (due to remixing of shielding gas with ambient atmosphere) on the absorption coefficient.

For a number of metals, significant variations of absorptivity $A(T)$ and thermal conductivity $\lambda(T)$ with temperature can cause a sharp increase of surface temperature. As a rule, the strongest effect results from an increase of absorptivity with a simultaneous decrease in thermal conductivity. Also, a comparatively wide temperature interval for noticeable $A(T)$, $\lambda(T)$ variations is often required. Laser heating of Mo and W can be referred to as a typical example for the above mentioned requirements [30,31]. In this case, after a certain temperature value (or a certain period of heating) the surface temperature sharply increases with time, and the corresponding transient period can be relatively short. On the other hand, the analysis of laser heating of different steels has shown that the above mentioned nonlinear affects are relatively week [31]. Often, both absorption and thermal conductivity of steels increase with temperature, compensating one another. Nevertheless, for a certain group of steels $\lambda(T)$ decreases with temperature over a comparatively narrow interval, so the variations of $A(T)$ and $\lambda(T)$ are not large enough to explain the experimental results.

Heating of metal by continuous CO_2 laser irradiation in an oxidizing atmosphere stimulates the development of a layer of oxide film on the surface of the metal, whose rate of growth depends markedly on the value of laser power density. The absorptivity of the system "oxide film-metal substrate" increases markedly and, as a result, the rate of heating increases, which in turn, causes the growth of the oxide film thickness, and so on. This process may be considered as a typical example of nonlinear heating with positive feedback. The presence of an oxidation reaction on the surface leads to existence of characteristic heating stages, depending on the magnitude of incident radiation flux, interaction time, and so on. In the case of relatively small values of the laser power density absorptivity hardly increases at all and the transient temperature corresponds to the solution of the problem neglecting thermochemical reaction on the metal surface. With further increase in the power density, beginning at a certain time, there is a sharp growth in the temperature, which corresponds to a considerable increase in the absorptivity due to growth of the oxide film.

A decrease of BTS, that corresponds to increase of interaction time, results in growth of an oxide film and an increase in surface temperature. The low temperature thermocycle (Figure 7, curve 3) corresponds practically to the absence of oxidation. Melting of the surface (Figure 7, curve 1) is attained by increase of the interaction time (decrease of speed). In the intermediate case, the temperature starts to rise as a result of intensive oxidation, but the maximum value does not reach the melting point (Figure 7, curve 2). The results of temperature measurements in this BTS range differ from one another because of slight unpredicted variations of oxidation conditions (state of the surface, re-mixing of the shielding gas with ambient atmosphere, and so on) that shift the beginning of the intensive oxidation phase and, therefore, the position of the final sharp temperature rise.

It was possible to reduce the sharp influence of surface oxidation on thermocycle variation at high beam-scanning speed and shielding gas flow rate over 35 l/min. In this case, one type of thermocycle was observed with a fair increase of maximum temperature at BTS range from 3350 to 3100 mm/min (Figure 8, curve 3 and 2). When BTS was decreased to 1300 mm/min (with simultaneous decrease of laser power to 0.8 kW, but with the same shielding gas flow rate) fast heating and melting of the mild steel surface was observed (Figure 8, curve 1) as a result of surface oxidation that was re-initiated.

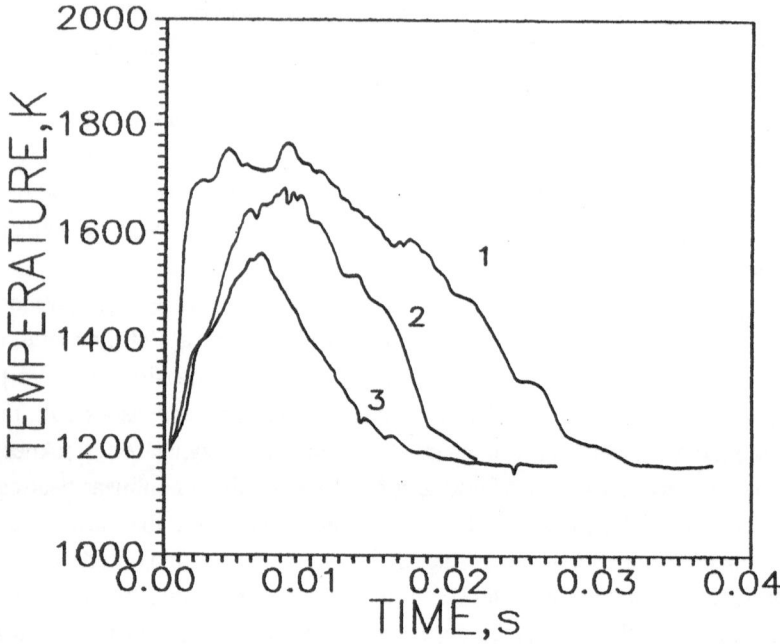

Figure 8. Surface thermocycles in the center of the laser beam scan. Mild steel. Laser power, kW: curve 1-0.8; 2,3-1.7. Beam traverse speed, mm/min: curve 1-1300; 2-3100; 3-3350. Shielding gas flow rate 35 l/min.

6.3. WELDING

In a number of cases, small variations of laser welding parameters (power, beam traverse speed, shielding gas flow rate, beam focus conditions, etc.) can lead to considerable degradation of the joint quality [30,31]. Therefore, the main parameters of this process should be controlled with high accuracy. A number of factors have great influence on final results, among them absence of porosity, rate of welded-seam cooling, size of heat-affected zone (HAZ), thermal and residual stresses and so on. Most of these factors are directly connected with surface temperature variation. A lot of experiments or complicated numerical simulation (for example, [32]) are usually required to find the optimum parameters. High-speed pyrometry could offer a real opportunity to clarify some features of laser welding processes. Measurements of real-time thermocycles during laser welding also allow to clarify the influence of different sources of noise (plasma radiation and its fluctuations, screening for a time of temperature measurement zone by plume as a result of gas-dynamic perturbations, metal droplets expulsion, and so on) on output signal of pyrometer.

In this chapter the attention specially is focused on the influence of different types of "physical noise" on pyrometer output signal.

Deep penetration CO_2-laser welding of mild steel plates (thickness 3, 6 and 9 mm) was investigated. The pyrometer was aimed close to the weld seam border. The pyrometer signal is characterized by intensive "noise" due to intensive radiation of laser plasma and metal-droplet expulsion from the melt (Figure 9, curve 1). A great number of solidified droplets have been found around the weld seam. As indicated by the measurements, the droplet temperature is mainly "colder" than the welded seam in the selected point. This is confirmed by several sharp downfalls of the pyrometer output signal. Under the conditions of this experiment, the droplets crossed the pyrometer field of view in the final part of their trajectories. That is why the droplet temperature is relatively "cold". Nevertheless, some "hot" droplets have been found, whose measured signal corresponds to their temperature in the initial part of the trajectory, just after they left the molten pool.

An increase of BTS led to a decrease of weld-seam size. In this case the pyrometer monitored a spot inside the HAZ where the surface temperature was a little higher than the threshold sensitivity (Figure 9, curve 2). Under the conditions of this experiment, only a few droplets have been found. The "noise" amplitude and frequencies differ rather from the previous case.

Simultaneous increase of laser power and BTS (keeping the ratio P/V=const) leads to further decrease of the weld-seam width [33,34]. Therefore, a sharp increase in the pyrometer signal due to increase of the melting pool temperature was observed (Figure 10, curve 1). The conditions of welding were characterized by intensive boiling and evaporation that could induce formation of porosity. The pyrometer output signal is strongly influenced by radiation of plasma and nonstability of its gas dynamic expansion. The higher traverse speed at the same value of power allowed minimization of weld seam size (keeping the conditions of full penetration welding). In this case, the temperature and the pyrometer output signal were not extremely high (Figure 10, curve 2). The measurements (Figure 10, curves 1,2) were carried out close to the border of welded seam. The nonstability of solid liquid interface and its influence on pyrometer output signal should be taken into account.

The welding of mild steel of greater thickness (6 mm) required smaller values of traverse speed. The weld-seam size reached 2 mm. The intensive expulsion of "hot" (with respect to the temperature of the base metal) particles and fluctuations of plume brightness were observed at these treatment conditions. This could be recognized by the sharp fluctuation in the pyrometer signal (Figure 11). This might cause some defects in the weld joint. The corresponding pyrometer signal was measured in HAZ region close to the border with the weld seam. The base metal temperature is lower than in Figure 9 and does not reach the melting point.

The "noise" of typical pyrometer signals can be eliminated by a signal-filtering technique without losing useful information. Note, that "noise" frequency and amplitude also depend on process parameters (scanning speed, power and plate thickness, see Figures 9-11). In the present experiments spikes were detected over the 125-500 Hz range. In case of droplets expulsion, the height and width of individual

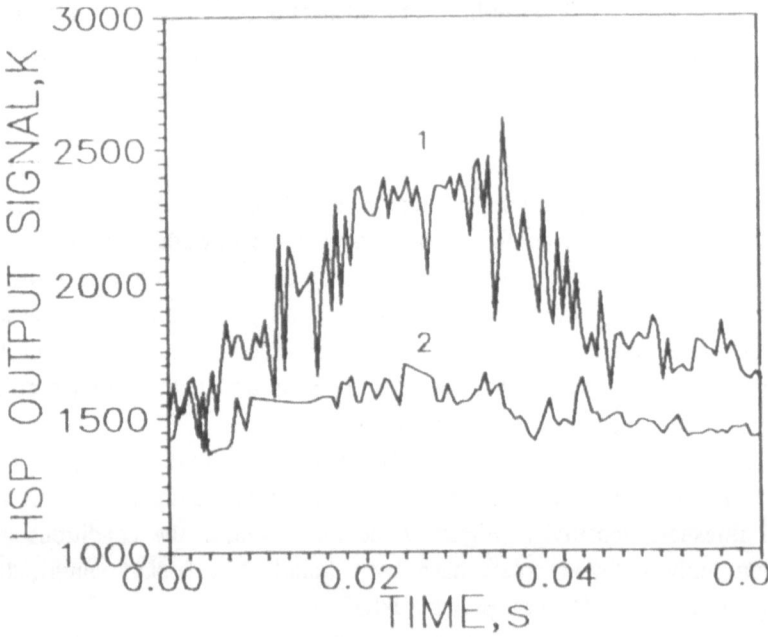

Figure 9. Pyrometer output signal for laser welding of mild steel plates. Thickness 3 mm. Laser power 2.5 kW. Beam traverse speed, mm/min: curve 1-1000; 2-2000. Distance from the weld-seam axis, mm: curve 1-1; 2-0.8.

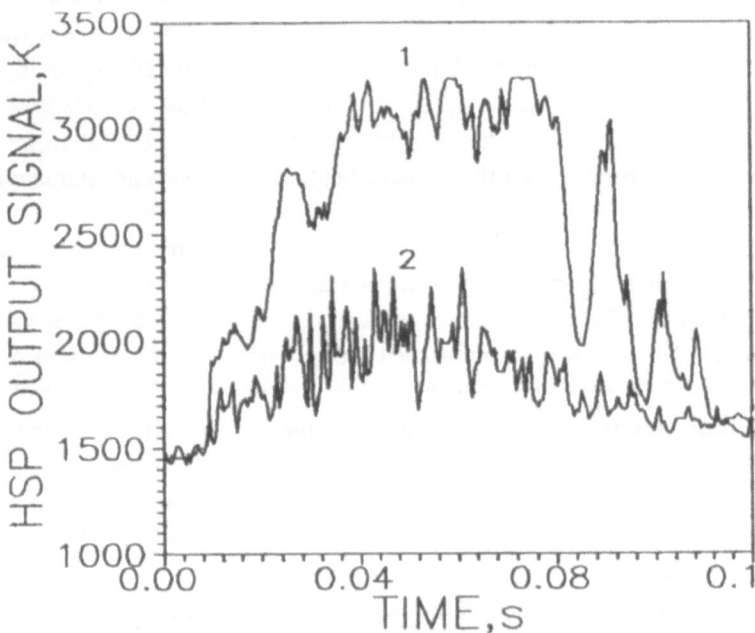

Figure 10. Pyrometer output signal for laser welding of mild steel plates. Thickness 3 mm. Laser power 5 kW. Beam traverse speed, mm/min: curve 1-2500; 2-4000. Distance from the weld-seam axis, mm: curve 1-0.5; 2-0.3.

peaks depend on particle size, temperature and velocity. The gas-dynamic perturbations of laser plasma also may be analyzed. The amplitude-frequency analysis of the "noise" signal can provide useful information. In the general case the sharp increase of the "noise" amplitude indicates the nonstability of the process. Therefore, the process control could be based on average temperature value together with amplitude-frequency parameters of the "noise".

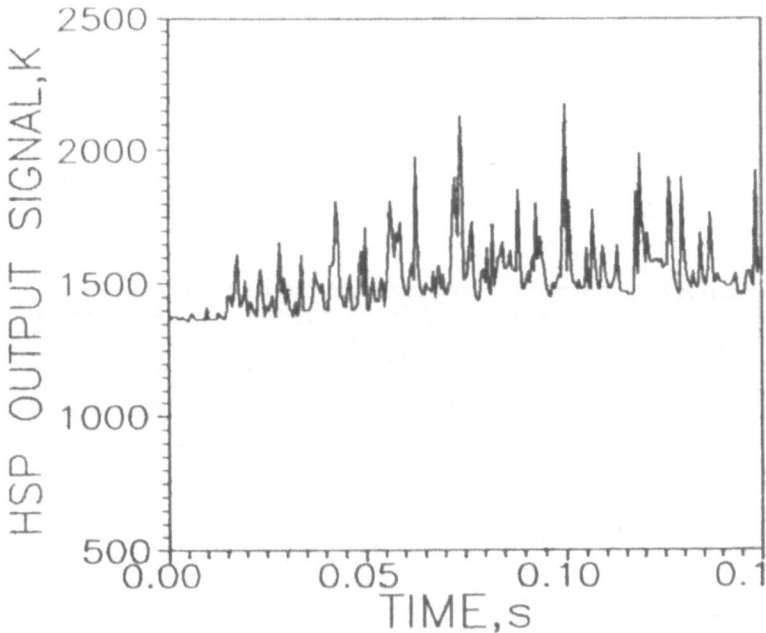

Figure 11. Pyrometer output noise during laser welding of mild steel plates. Thickness 6 mm. Beam traverse speed 500 mm/min. Distance from the weld-seam axis 2.1 mm.

The pyrometric measurements in full-penetration laser welding in the low-temperature band (900-1200 K) have demonstrated significantly different cooling rates as a function of position in the HAZ, for different sample thickness and BTS values. The experimental conditions reported in the Figures 12,13 were as follows: full penetration butt welding for slabs of 3, 6, and 9 mm thickness keeping the laser power constant (5 kW) and varying the BTS. The initial cooling rate decreases by half when the distance from the axis of the beam trace increases from 1 to 1.5 mm (Figure 12). The cooling rates increase from approximately 700 K/s up to 3300 K/s for material thickness variation from 9 to 3 mm and BTS from 500 to 2500 mm/min respectively (Figure 13). Note that, in this experiment, temperature was measured in HAZ close to the melt border, so the distances from the weld-seam axis were different for each temperature measurement, but the distance between the molten pool edge and the point of measurement was constant.

548

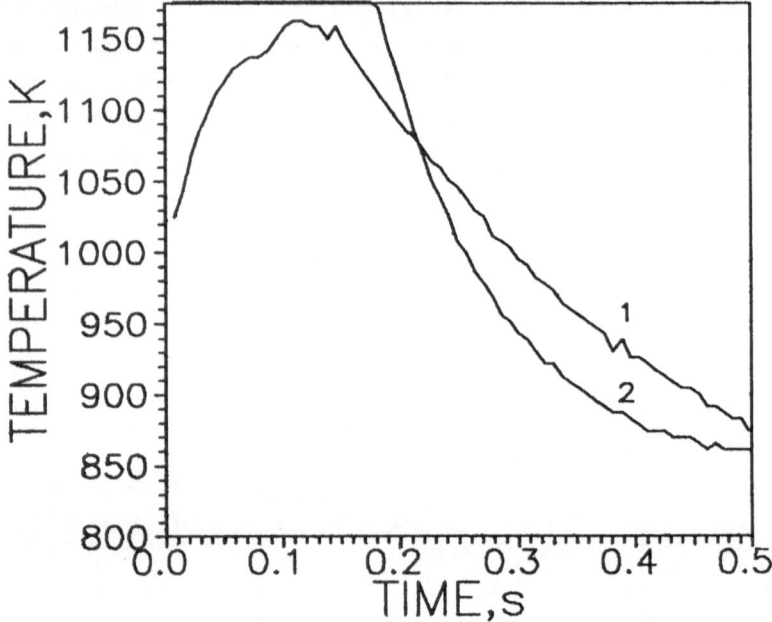

Figure 12. Cooling of heat affected zone after laser welding of 3 mm mild steel plates. Beam traverse speed 1000 mm/min. Distance from the weld-seam axis, mm: curve 1-1.5; 2-1. Laser power 5 kW.

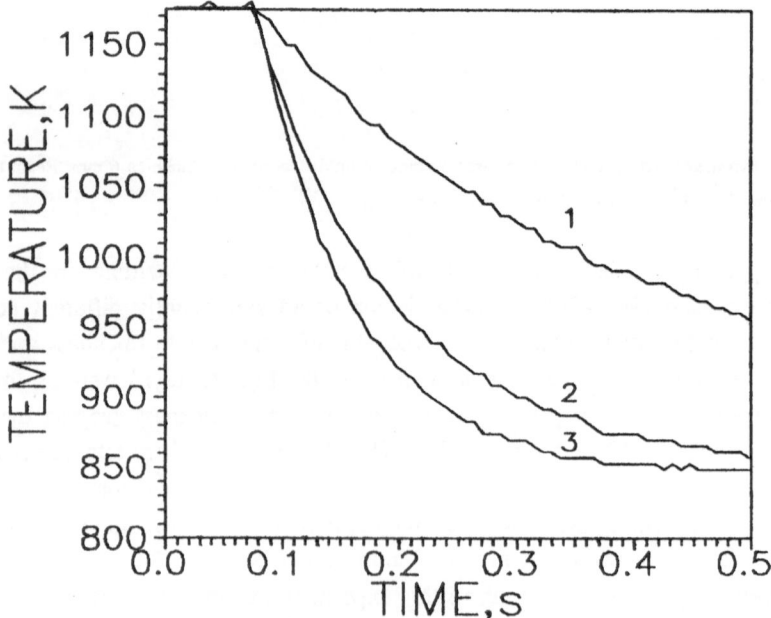

Figure 13. Cooling of heat affected zone after laser welding of mild steel plate with different thickness, mm: curve 1-9; 2-6; 3-3. Laser power 5 kW. Beam traverse speed, mm/min: curve 1-500; 2-1000; 3-2500.

For the simplicity of presentation the temperature curves in Figure 13 are superposed at the maximum temperature level detected by infrared pyrometer used in measurements.

The experimental results presented are in qualitative agreement with general rules of heat processes and evolution of surface temperature [33].

Deep penetration CO_2 laser welding may be considered as a typical example of strong influence of various "physical noises" on pyrometer output signal. The accurate measurement of the surface temperature here is the complicated procedure, that hardly could be realized in a general case. On the other hand, keeping in mind the opportunity to vary easily the pyrometer operating wavelengths, the output signal may provide an important information about the process dynamics: in particular, about its stability. In the general case the process control could be based on average temperature value together with amplitude-frequency parameters of the "noise".

6.4. CLADDING

In cladding, there exists low and high cladding speed limits (for a fixed energy input or energy input limits the fixed speed), depending on the laser power density and the powder mass-flow rate. The high speed limit corresponds to low cladded layer adhesion due to insufficient heat input into melting the base alloy. The low speed limit corresponds to intensive remixing of materials and substrate distortion to unacceptable levels [35]. In-process monitoring and control of cladded bead surface temperature is required to reach satisfactory quality (microhardness, thickness, low dilution level and so on).

Figure 14 shows the cross section of mild steel samples with cladded bead in optimum conditions. The microhardness of base material (measured near the treated surface) decreased from cladded border (220 Hv) to HAZ border (160 Hv). The microhardness of cladded bead (near the surface) is practically uniform and varies in the range 310-320 Hv.

Figure 15 illustrates typical thermocycles for laser cladding measured at different distances from the center of the laser beam scan. The influence of powders flow on temperature dynamics can be seen in temperature downfall in the central part of the curves 1 and 2. The sharp fluctuations of pyrometer signal (pyrometer "output noise") are not observed under the chosen cladding conditions. The maximum surface temperature of cladded layer reached Stellite 6 melting point with a small overheating, and the cooling rate was nearly 1.7 kK/s (Figure 15, curve 1). The cooling rates in the center of cladded zone (Figure 15, curve 1) and near its border (Figure 15, curve 2) are practically equal. In the HAZ, the cooling rate is sharply decreased from 1.2 kK/s to 0.6 kK/s when the distance from the center of the laser beam scan is increased from 1.6 mm to 1.8 mm (Figure 15, curves 3 and 4 correspondingly). This is the main reason for the decrease of microhardness from 220 to 160 Hv.

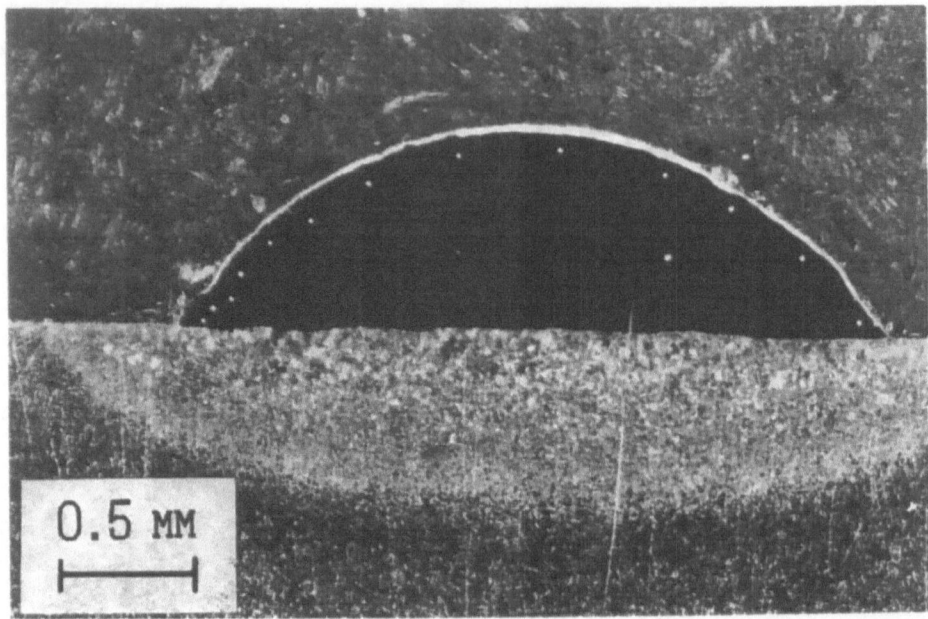

Figure 14. Cross section of cladded bead on the surface of mild steel. Laser power 1.9 kW. Beam traverse speed 300 mm min⁻¹. Scanning beam parameters: frequency 400 Hz; scanning width 3.0 mm.

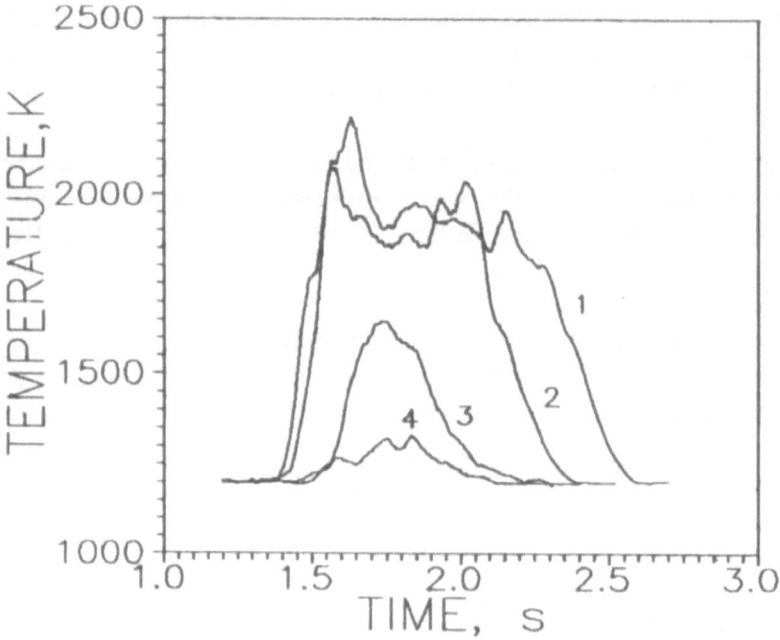

Figure 15. Thermocycles for laser cladding process. Laser power 1.9 kW. Beam traverse speed 300 mm/min; Scanning beam parameters: frequency 400 Hz; scanning width 3.0 mm. Distance from the center of laser beam scan, mm: curve 1-0; 2-1.4; 3-1.6; 4-1.8.

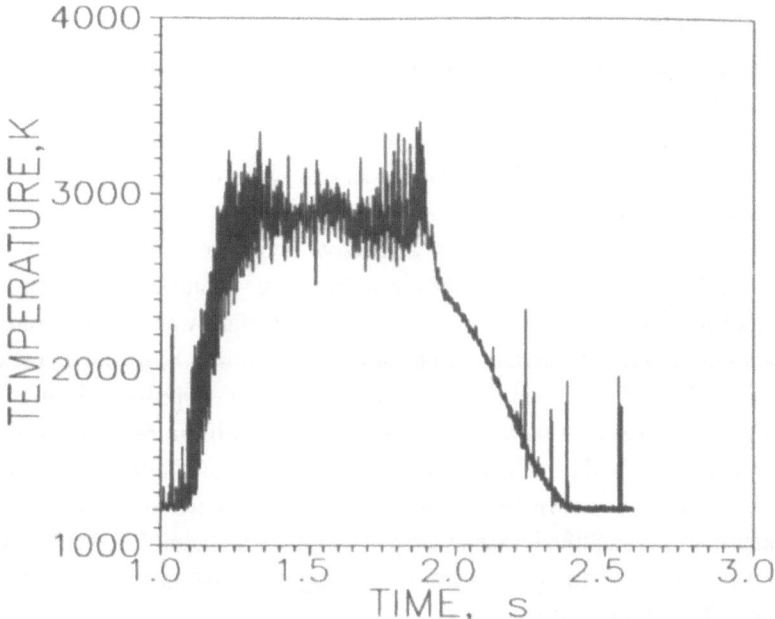

Figure 16. Thermocycle for laser cladding process at the center of the laser beam scan. Laser power 2.9 kW. Beam traverse speed 300 mm/min; Scanning beam parameters: frequency 400 Hz; scanning width 3.0 mm.

In the general case, cladding process could be enough "noisy" because of particles injection, scattering and laser beam - particles interaction (see Figure 16 as an example, but less "noisy" than deep-penetration laser welding because of the absence of laser plasma). The sharp individual spikes that are recorded before the temperature rise and after the temperature fall (Figure 16) correspond to the particles of cladded materials that crossed the pyrometer field of vision.

6.5. ALLOYING

High-repetition-rate laser alloying is one of the promising methods to increase operation parameters of materials [36-38]. Such processes are characterized by complex heat and mass transfer dynamics [39-41]. Comparatively small variations in treatment conditions can lead to significant variations in doping element distribution [42].

Samples of pure Fe with pre-deposited Mo coatings were treated by pulsed laser radiation. The pulse duration was 4 ms and the energy input range was 5-15 J. The pyrometer was aimed at the center of the laser focus spot.

A sudden temperature spike was observed during the action of a 5 J laser pulse (Figure 17-a). This spike corresponds to fast coating heating due to its imperfect thermal contact with substrate. In this case, the coating was exfoliated from the surface and subsequently melted, partially evaporated and destroyed. A bulge of solidified Mo was formed around the center of laser focus spot (Figure 18). The temperature of the exposed substrate did not reach the Fe melting point and alloying was absent. Increasing the laser energy up to 9.5 J led to substrate melting after the coating's destruction (Figure 17-b), and substrates were partially alloyed at the pulse end. The coating-substrate system was melted as one when the laser energy equaled 14.5 J. The temperature thermocycle looks like a classical temperature curve corresponding to pulsed laser action on a massive body except for the first stage (Figure 17-c) [36-42]. The first part of the thermocycle was characterized, as usual, by fast heating of the Mo coating. However, the substrate surface was melted before Mo layer exfoliation, and the temperature dropped due to improved heat contact between Mo layer and Fe base. The second part of the thermocycle corresponds to heating and melting of the Mo-Fe system as a whole. A practically uniform distribution of the doping element (Mo) in the molten pool was found after laser alloying at a pulse energy of 14.5 J. The pyrometric temperature measurements were confirmed by cross-section analysis of the modified zone and by numerical simulation [43-45] that have revealed: (i) coating exfoliation; (ii) substrate melting under the undestroyed coating; (iii) substrate alloying with the coating.

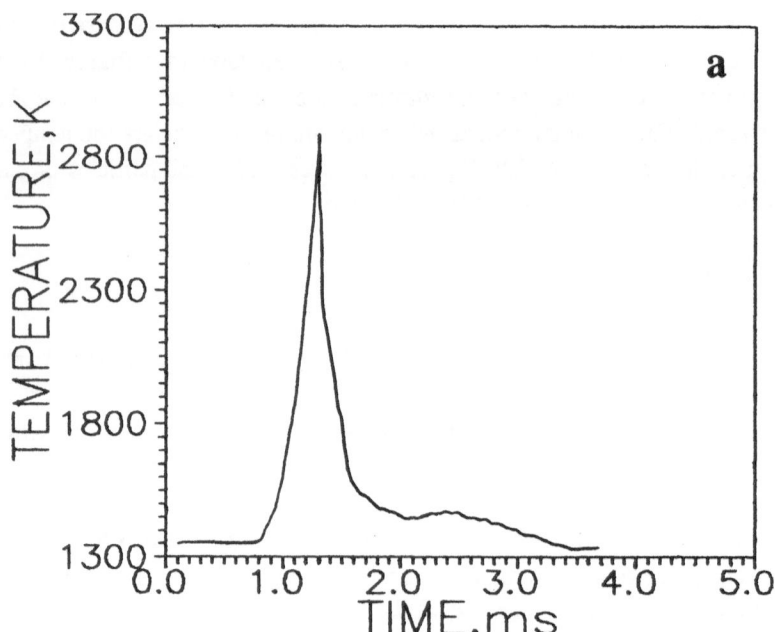

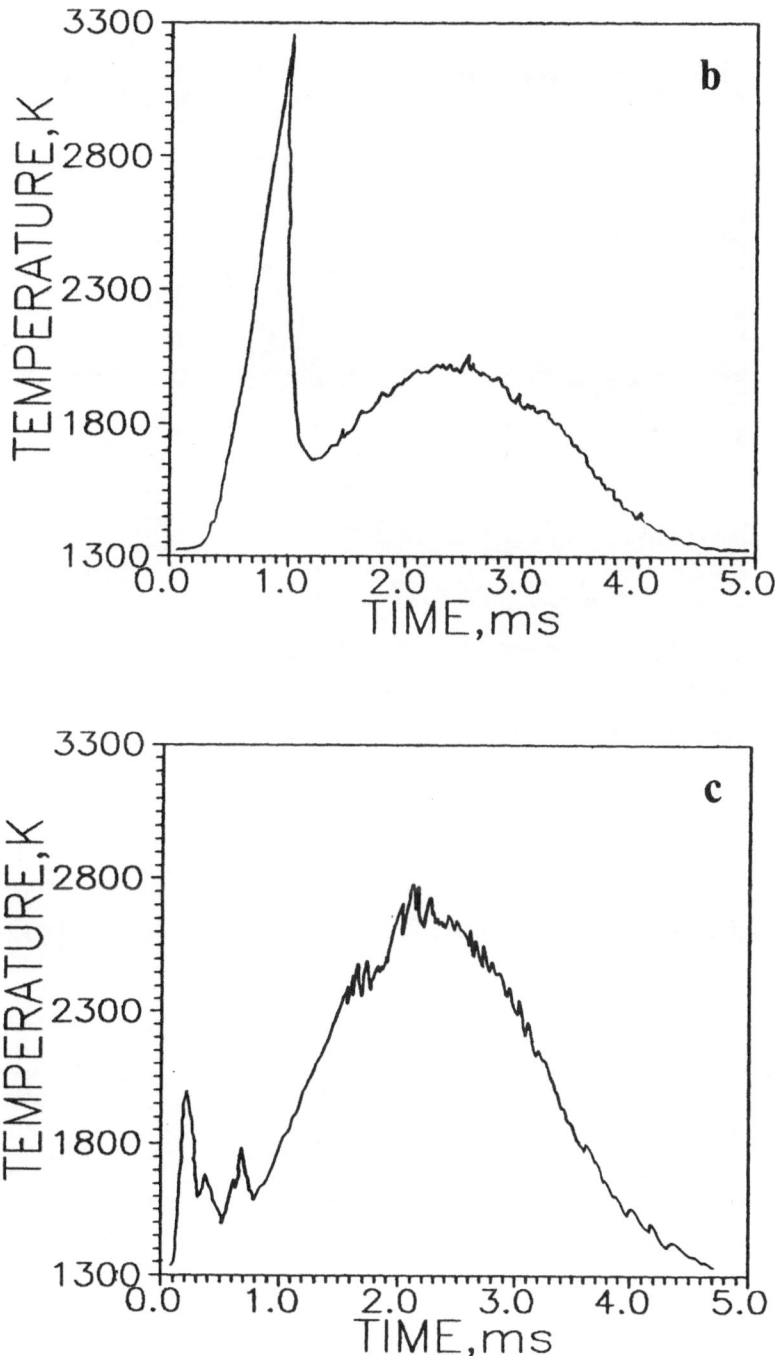

Figure 17. Surface thermocycles in the center of the laser spot for system "Mo coating on Fe substrate" during pulse laser alloying. Thickness of Mo coating 35 μm. Laser pulse energy, J: curve a-5; b-9.5; c-14.5.

554

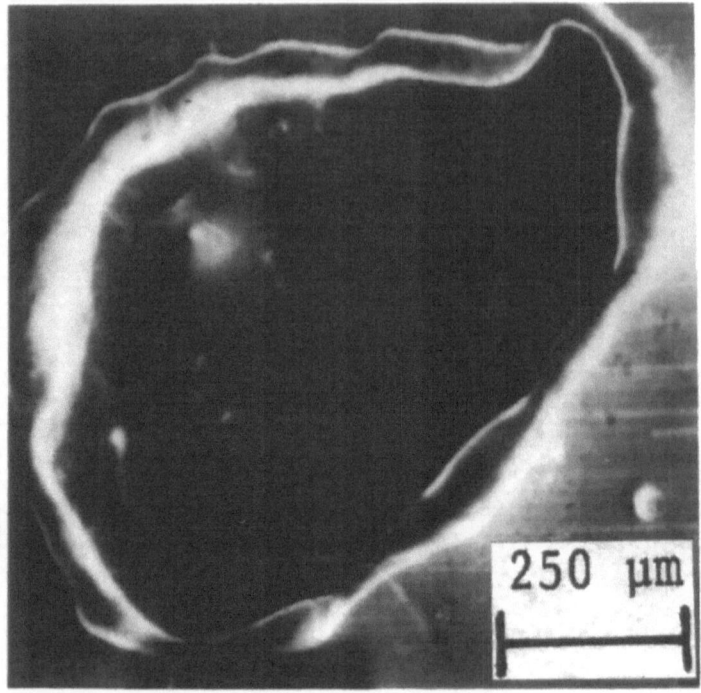

Figure 18. Top view of destroyed coating and solidified Mo bulge after action of a 5 J laser pulse.

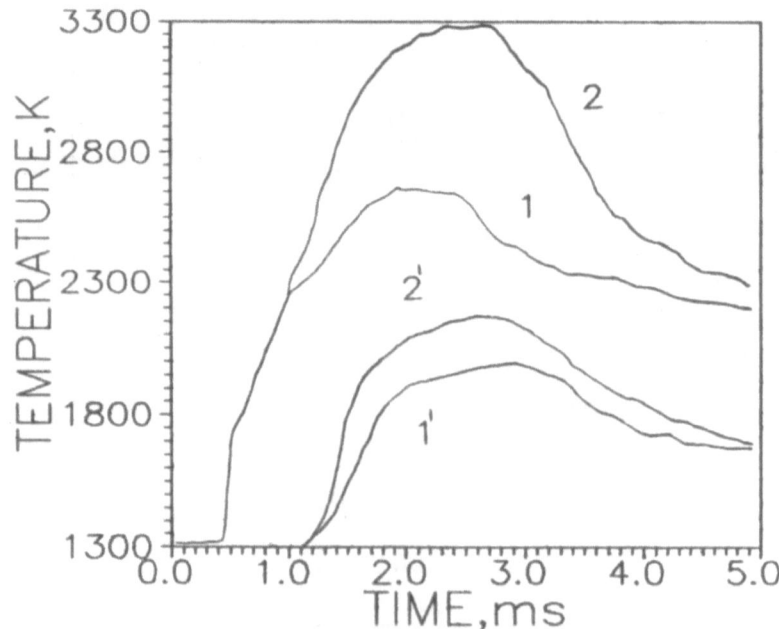

Figure 19. Thermocycles of Ti foil during laser action in nitrogen jet. Laser pulse energy 3 J. Curves 1,2-heated surface; 1',2'-back-side of foil; 1,1'-first pulse; 2,2'-second pulse.

Temperature measurements of Ti substrate during pulsed laser alloying with nitrogen were also carried out. The thin (200 μm) Ti foil was irradiated by a pulse-periodic laser. Co-axial nitrogen flow was used as source for the doping element. Typical surface thermocycles are presented in Figure 19. Curve 1 corresponds to laser-beam interaction with the Ti surface during the first laser pulse. A titanium nitride (TiN) layer was produced as a result of laser alloying [37]. The modified zone was irradiated by a second laser pulse. The surface temperature was significantly higher during the second pulse (Figure 19, curve 2) due to modification of the optical and thermophysical sample properties. The temperature of the foil back-side was also measured (Figure 19, curves 1',2'). The smaller difference between backside thermocycles for the first and second pulses is the result of the surface nature of laser radiation absorption and large temperature gradients across the slab.

6.6. LASER ASSISTED MACHINING

One of the methods for improvement the machinability of metals is to preheat the workpiece by a suitable source of energy. Resistance heating and plasma flow are the most widely used techniques to reduce the cutting force and increase the cutting speed [46-47].

Laser evident advantages for material preheating are the opportunity to focus the beam to a small spot. It is possible to note other promising features of laser, such as opportunity to vary and to control spatial and energy parameters on the irradiated surface, high stability of the process, general use for a wide class of materials, radiation supply to places where access is difficult, possibilities for process automation, etc.

Various physical phenomena which accompany laser assisted machining (LAM) of a wide range of materials (Mo, W, Ti, Ceramic materials, Special steels, etc.) are discussed in [48]: plasticization, structural and phase changes of materials, variation of friction conditions with temperature, laser plasma influence, etc. It is shown that at optimum conditions LAM significantly improves the quality of machining: roughness of the treated surface, shear and microcracks disappear, etc. LAM permits: interrupted cutting of materials with increased strength characteristics; turning of soft and/or thin-walled materials; finishing of materials for which structural and phase changes beneath the cut line are not desirable. Generally, the productivity of machining can be increased up to 4-6 times.

The cutting temperature (θ_c-temperature at the contact point of a tool with treated material) is the most important parameter for turning process. Friction conditions, tool wear, formation of a buildup at a cutter edge, surface quality (roughness, cracks, residual stresses, etc.) are depended on θ_c [48]. The cutting temperature θ_c comprises the additional surface heating temperature θ_h and the temperature θ_d caused by the processes of deformation and friction in zone of chip formation ($\theta_c \sim \theta_h + \theta_d$). It must be emphasized that θ_d, in general, is a function of θ_h. Thus, in-process monitoring and control of θ_h is the most relevant approach to guarantee the high quality and stability of LAM. Among the main aims of thermocycles measurements during LAM are the following:

-determination of dependencies of surface temperature versus main parameters of the process;

-revealing of the relationship between surface heating temperature θ_h and cutting temperature θ_c;

-choice of the most sensitive LAM parameters and corresponding laws for on-line control.

6.6.1. *Experimental Conditions*

LAM setup is shown in Figure 20. The installation is mounted on the base of a turning lathe. CO_2-laser (5 kW) was used in experiments.

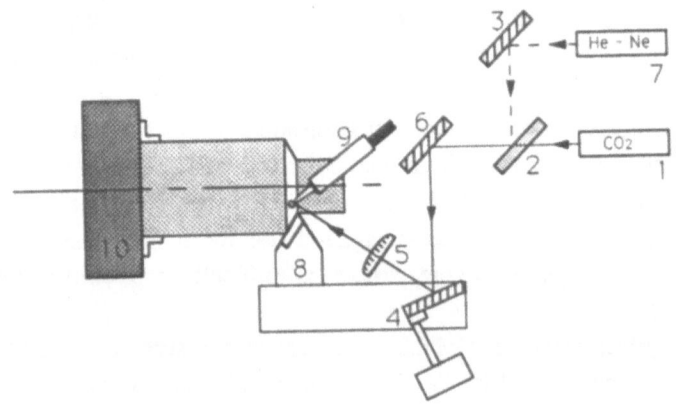

Figure 20. LAM setup diagram: (1)-CO_2- laser; (2)-(6)-optical system: mirrors, lens; (7)-He-Ne laser; (8)-cutter; (9)-pyrometer; (10)-lathe.

Design of the optical system makes possible the stabilization of the focal spot position relative to the tool and includes the piezodrive for the mirror (4) to scan the focal spot on the cutting surface. Alignment of the laser beam on the workpiece and adjustment of the optical system is carried out by a visible λ=0.63 μm He-Ne laser, the optical axis of which is made coincident with that of infrared laser by mirrors (2),(3) [48,49]. Focusing is realized by metal optics.

Cutters with welded plated from W-Co, W-Co+boron nitride were used in experiments. The following cutting edge geometry was used: tool clearance α=12°(8°), tool rake γ=15°(10°), tool cutting edge ϕ=45°, cutting edge radius ρ=0.2

mm (0.5 mm). For zirconium hydride cutting tools with W-Co tips was used: $\alpha=10°$, $\gamma=20°$, $\phi=90°$, $\phi_1=30°$, r=0.2 mm.

The workpieces were cylinders with length 50-150 mm and diameter 15-20 mm of various materials: Mo[111], W[111], Ti, zirconium hydride, ceramic materials, and special steels.

K_2SiO_2 coating was deposited on the treated surface to increase the absorption of metals at the wavelength $\lambda=10.6$ μm.

LAM main parameters were controlled during the process. Temperature in the heating spot was measured by high-speed high-spatial resolution pyrometer. The pyrometer was placed on the lathe support and was moved together with the tool. The additional surface heating temperature was measured at the different points of the heated spot. The cutting temperature was determined by the method of natural thermocouple [46]. The treated samples and cutting tool were isolated and the thermocurrent was registered at the contact point "cutter-treated surface".

The cutting force was measured by dynamometer, wear rate of the cutter - by acoustic emission (AE), and oscillation amplitude - by Balatron 2003 equipment.

After LAM, the roughness of the treated surface, depth and rate of cold hardening, presence of shears, microcracks, structural and phase transformation of the treated surface layers were analyzed.

6.6.2. Results and Discussion

The main results of the present work are illustrated by the examples of Mo and W. The dependencies of the average heating temperature on laser power for Mo and W are presented in Figure 21. Application of 5 kW laser makes possible to reach 2000 K (W) and 2900 K (Mo) at the central part of focus spot.

Figure 22 shows the average heating temperature at different distance from the center of the heated spot in the direction of motion. The temperature sharply slopes down at the front of the heating source; the cooling rate reaches about 10^6 K/s. Increase of the turning speed causes decrease of the heating temperature (Figure 23). Note that the surface temperature is slightly affected by turning speed variation (in the range 5-100 m/min). The fluctuations of the energy parameters of the process (caused by the changes of: laser power, diameter of the focus spot, absorption coefficient, etc.) lead to the important variations of heating and cutting temperatures. The oscillations of the surface temperature about the average value are evident in the ranges: 7-52 Hz, 100-300 Hz, 1000-2000 Hz. The laser power does not influence on the oscillation frequencies, but it is responsible for the increase of oscillation amplitude.

The measurements of cutting temperature θ_c were carried out both for "cold" turning and turning with laser heating. The optimum cutting ranges were determined from the tool wear resistance tests at conditions of high quality of machining. These ranges are shown by hatching in Figures 24, 25.

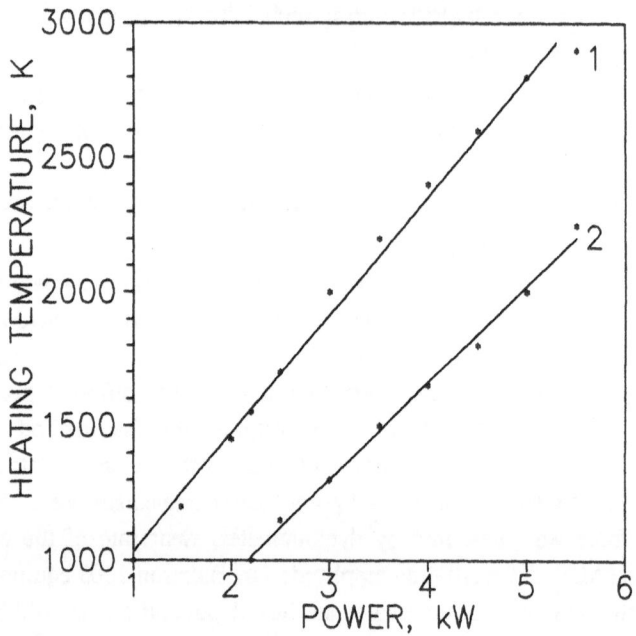

Figure 21. Average heating temperature (at the center of the focus spot) versus laser power. Cutting speed V=0.5 m/s. Diameter of the laser focus spot d=0.5 mm. (1)-Mo; (2)-W.

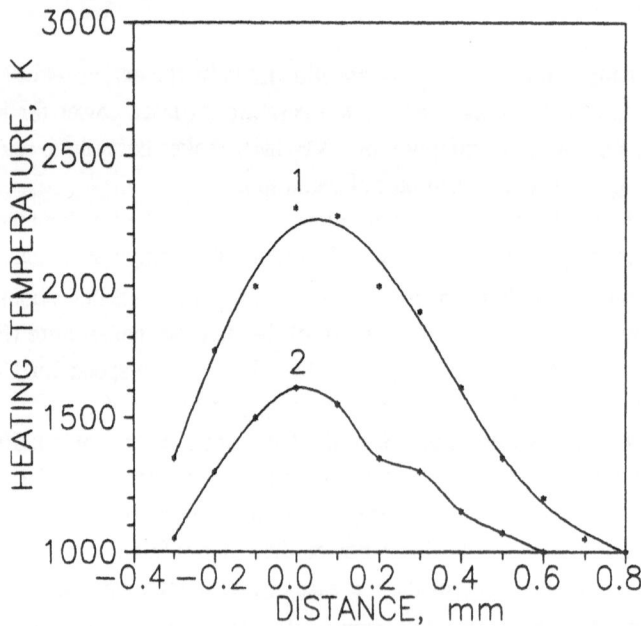

Figure 22. Average heating temperature versus distance to the laser focus spot. Laser power P=4 kW. V=0.5 m/s. d=0.5 mm. (1)-Mo; (2)-W.

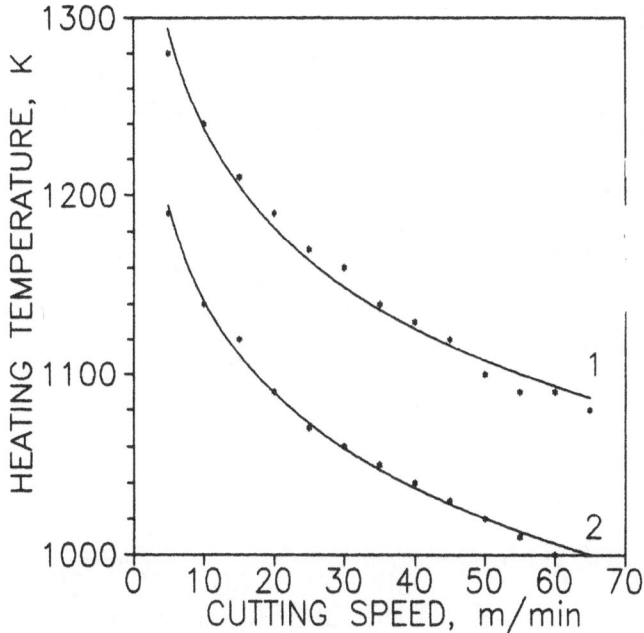

Figure 23. Average heating temperature (distance from the center of the focus spot L=0.5 mm) versus cutting speed. d=0.5 mm. (1)-Mo, P=3.4 kW; (2)-W, P=4.6 kW.

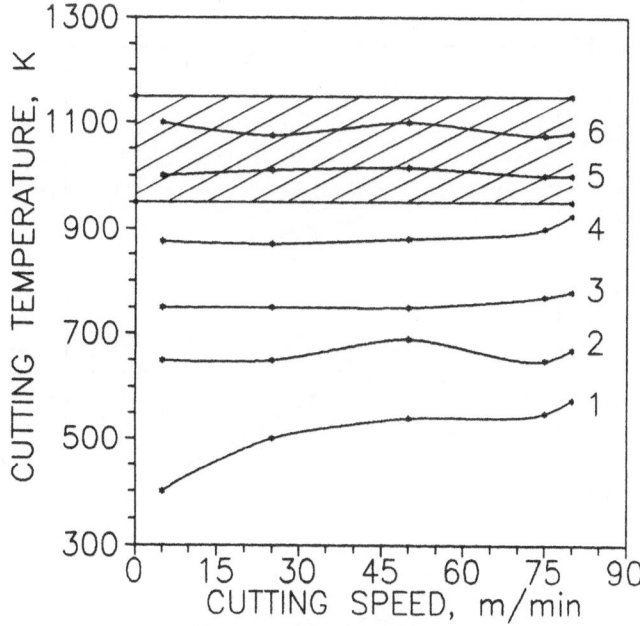

Figure 24. Cutting temperature versus cutting speed. Material-Mo. Cutting depth-z=0.2 mm. Cutting feed-S=0.05 mm/rev. L=0.5 mm. d=0.5 mm. (1)-θ_h=300 K; (2)-θ_h=1300 K; (3)-θ_h=1400 K; (4)-θ_h=1500 K; (5)-θ_h=1600 K; (6)-θ_h=1700 K;

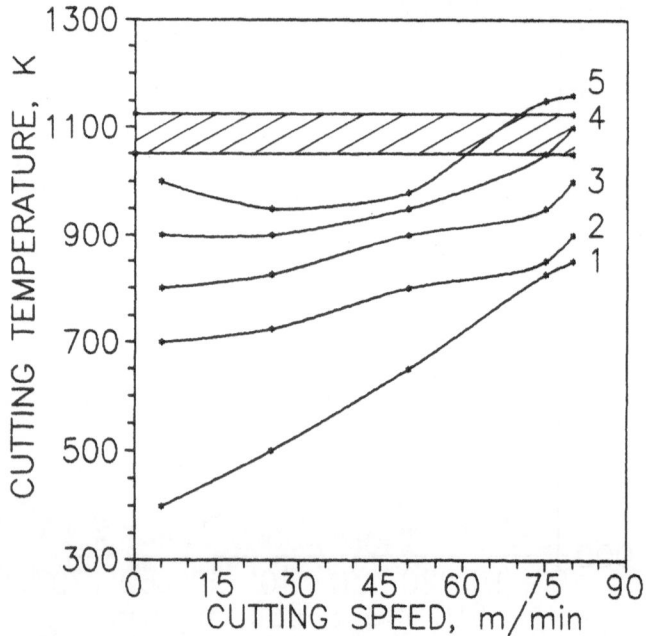

Figure 25. Cutting temperature versus cutting speed. Material-W. z=0.2 mm. S=0.05 mm/rev. L=0.5 mm. d=0.5 mm. (1)-θ_h=300 K; (2)-θ_h=1300 K; (3)-θ_h=1400 K; (4)-θ_h=1500 K; (5)-θ_h=1600 K.

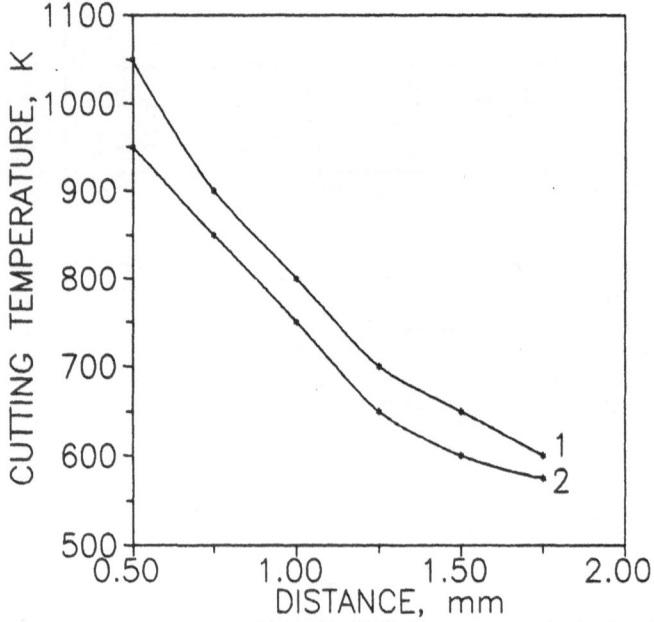

Figure 26. Cutting temperature versus distance between the center of the focus spot and the cutter edge. θ_h=1600 K. V=60 m/s. z=0.2 mm. S=0.05 mm/rev. (1)-W; (2)-Mo.

The optimum cutting temperature range can not be provided by "cold" turning at high quality of machining. LAM offers the solution of this problem. The required range for θ_c can be maintained by control of the heating temperature. The optimum heating temperature for Mo falls in the broad range (between 1600-1800 K) and it is independent on cutting speed. The optimum heating temperature for W should be kept at 1550 $\pm$25 K; in so doing the cutting speed can be varied in the narrow range 65-80 m/min.

Hence, the main principle of process control for LAM is to hold the energy parameters at the required level, and, consequently, to keep heating and cutting temperatures within optimum ranges. Any heating temperature deviation should be compensated by a laser power increase/decrease or variation of the distance L between center of the focus spot and the cutter edge. The second way has the minimum response time and the high sensitivity because of the sharp dependence of cutting temperature on the distance L (Figure 26). The variation of the distance L can be easy realized by rotation of the focusing mirror 4 (Figure 20).

Laser assisted machining seems to be the most convenient process to realize monitoring and control based on surface temperature measurements by pyrometer.

7. Conclusion

Opportunities for high speed high spatial resolution optical pyrometry have been examined in real-time temperature measurements in laser transformation hardening, welding, cladding, alloying and laser assisted machining. A test of this method with a wide range of parameters corresponding to different technological processes has shown that it can be used in a number of different laser applications and it is sensitive enough to reveal certain features of thermo physical phenomena in laser affected zone.

The developed pyrometer also can be used as a basic instrument for on-line process control in some laser applications. The successful implementation of optical pyrometry in laser machining requires the solution of the following metrological problems: knowledge of surface emissivity data under actual conditions of laser machining; elimination of strong influence of laser plasma; filtration of the "noise" induced by droplets expulsion, particles injection and so on. Based on the obtained results, it is possible to recommend the utilization of high speed high spatial resolution pyrometry in such processes of laser surface treatment as: transformation hardening, laser assisted machining, alloying and cladding.

8. References

1. Khan, M.A., Allemand, C.A. and Eagar, T.W.(1991) Noncontact temperature measurement, parts I,II, *Review off Scientific Instrument* **62** (2), 392-409.

562

2. Foley, G.M.(1978) Modification of emissivity response of two-color pyrometer, *High Temperatures-High Pressures* **10**, 391-398.

3. Bedford, R.E.(1972) High temperature thermometry (A review), *High Temperatures-High Pressures* **4**, 241-260.

4. Svet, D.Ya.(1972) Optimal utilization of redundant information in thermal radiation in thermophysical measurements, *High Temperatures-High Pressures* **4**, 715-722.

5. Azer, M.N., Mazumder, J.(1993) Substrate temperature measurement diagnostics during laser chemical vapor deposition (LCVD) of titanium on silicon, *in Proceedings of International Congress on Applications of Lasers and Electro-Optics ICALEO'93, Oct.24-28, 1993, Orlando, Florida USA*, Laser Institute of America, Orlando.

6. Gardner, J.L., Jones, T.P. and Davies, M.R.(1981) A six-wavelength radiation pyrometer, *High Temperatures-High Pressures* **13**, 459-466.

7. Stein, A.(1985) Laser Pyrometry, *Advertising documents of Quantum Logic Corporation, USA*.

8. Hernandez, D.(1994) Mesure de l'évolution de propriétés de surface par radiomÈtre rapide, *dans Récent Progrès en Génie des Procédés, Capteurs en Elaboration de Matériaux, volume 8-33, Une Collection dédire aux Congrès en France dans le domaine du Génie, Limoges-94*, Lavoisier Technique et Documentation, Paris, pp.5-18.

9. Manosa, L.I., Planes, A. and Cesari, E. (1982) *Journal of Physics D: Applied Physics* **22**, 977-981.

10. Li, L. and Steen, W.(1992) Non-contact acoustic emission monitoring during laser processing, *in Proceedings of International Congress on Applications of Lasers and Electro-Optics ICALEO'92, Oct.25-29, 1992, Orlando, Florida USA*, Laser Institute of America, Orlando, pp.719-728.

11. Krehl, P., Schwizke, F. and Cooper, A.W.(1975) Correlation of strength wave profiles and the dynamics of the plasma produced by laser irradiation of plane solid targets, *Journal of Applied Physics* **46**(10), 4400-4406.

12. Marinelli, M., Zammit, U., Mercuri, F. and Pizzoferato, R.(1992) High-resolution simultaneous photothermal measurements of thermal parameters at a phase transition with the photopyroelectric technique, *Journal of Applied Physics* **72**(3), 1096-1100.

13. Bagger, C.C. and Olsen, F.(1992), *in Proceedings of the 3rd Conference on Laser Material Processing in the Nordic Countries, August 1992, Lappeenranta, Finland*, Lappeenranta University of Technology, pp.189-199.

14. Shannon, M.A., Rostami, A.A. and Russo, R.E. (1992) Photothermal deflection measurements for monitoring heat transfer during modulated laser heating of solids, *Journal of Applied Physics* **71**(1), 53-63.

15. Tonshoff, H.K., Overmeyer, L. and Alvensleben, F.(1992) Computer controlled laser cutting of ceramics, in Mordike, B.L.(edt), *Laser Treatment of Materials. Proceedings of the European Conference on laser Treatment of Materials, ECLAT'92, October 1992, Gottingen, Germany*, DGM, Oberursel, pp.21-26.

16. Ediger, M.N. and Pettit, G.H.(1992) Time-resolved reflectivity of ArF laser-irradiated polyimide, *Journal of Applied Physics* **71**(7), 3510-3514.

17. Voelkel, D.D. and Mazumder, J.(1990), Visualization of a laser melt pool, *Applied Optics* **29**, 1718-1720.

18. Chen, X. and Mazumder, J. (1994), Laser chemical vapor deposition of titanium nitride and process diagnostic with laser induced fluorescence spectroscopy, to be published in Mazumder, J., Steen, W.M.

and Vilar, R.(eds.), *Laser Processing: Surface Treatment and Film Deposition, Nato ASI Series, Series E: Applied Science*, Kluwer Academic Publishers, Dortrecht.

19. Vaccari, J.(1988) Advances in laser welding, *American Machinist & Automated Manufacturing* **4**, 53-56.

20. Geissler, E. and Bergman, H.W.(1989), *in Surface Engineering, High Energy Beams Science and Technology: Proceedings of 2nd IFHT Seminar, Lisbon, September 25-26, 1989*, Lisbon, pp.121-130.

21. Smurov, I., Covelli, L., Tagirov, K. and Aksenov, L.(1992) Peculiarities of pulse laser alloying. Influence of beam spatial distribution, *Journal of Applied Physics* **71**(7), 3147-3158.

22. Uglov, A., Smurov, I., Tagirov, K. and Guskov, G.(1992) Modeling of thermocapillary mass transfer during pulse laser alloying, *International Journal of Heat and Mass Transfer* **35**(4), 783-793.

23. Postacioglu, N., Kapadia, P. and Dowdon, J.(1991), Journal of *Physics D: Applied Physics* **24**, 15-21. Hugel Ecaleo

24. Ignatiev, M., Ermolaev, A., Titov, V., Smurov, I. and Sturlese, S.(1992) The high speed pyrometer system for laser welding, cutting, heat treatment and alloying processes temperature control, in Mordike,B.L.(edt), *Laser Treatment of Materials. Proceedings of the European Conference on laser Treatment of Materials, ECLAT'92, October 1992, Gottingen, Germany*, DGM, Oberursel, pp.15-20.

25. Smurov, I., Martino, V., Ignatiev, M. and Flamant, G.(1994) On-line thermocycles measurements in laser applications, *Journal de Physique IY*, Colloque C4, supplément au Journal de Physique III, v.4, 147-150.

26. Ignatiev, M., Smurov, I., and Flamant, G.(1994) Real-time pyrometry in laser machining, *Measurement Science and Technology* **5**, 563-573.

27. Svet, D.Ya.(1968) *Objective Methods of High Temperature Pyrometry at Continuous Spectrum of Radiation*, Nauka, Moscow.

28. Uglov, A., Zavidei, V.I. and Ermolaev, A.(1990) Optical measurement of the surface temperature of metals under pulsed laser irradiation conditions, *Soviet Journal of Quantum Electronics* **20**(4), 453-455.

29. Uglov, A., Ermolaev, A., Zavidey, A. and Smurov, I. (1990) Apparatus for examining surface thermal and optical characteristics in exposure to concentrated energy fluxes, *Teplofizika Vysokikh Temperatur* **28**(4), 786-792.

30. Uglov, A., Smurov, I., Lashin, A. and Guskov, A.(1991) *Modeling of Thermal Processes under Pulse Laser Action on Metals*, Nauka Publishers, Moscow.

31. Biermann, B., Dierken, R. and Bergman, H.(1990), *Metallurgical Transaction* **45**(4), 328-334.

32. Kar, A. and Mazumder, J.(1989) Three-dimensional transient thermal analysis for laser chemical vapor deposition on uniformly moving finite slab, *Journal of Applied Physics* **65**, 2923-2934.

33. Rykalin, N.N.(1947) *Heat Basis of Welding*, AN SSSR Publishers, Moscow-Leningrad.

34. Rykalin, N.N.(1951) *Calculating of Heat Processes in Welding*, Mashgiz, Moscow.

35. Weerasinghe, V.W. and Steen, W.M.(1987) *Applied Laser Tooling*, Martinus Nijhoff, Dortrecht.

36. Ignatiev, M., Kovalev, E., Titov,V., Uglov, A., Smurov, I. and Sturlese, S.(1992) The application of laser technology for improvement high load friction joints tribotechnical parameters, in Mordike, B.L.(edt), *Laser Treatment of Materials. Proceedings of the European Conference on laser Treatment of Materials, ECLAT'92, October 1992, Gottingen, Germany*, DGM, Oberursel, pp.241-244.

37. Uglov, A., Ignatiev, M., Gnedovets, A. and Smurov, I. (1989) Nitride and carbide coatings synthesis on the surface of refractory metals by laser action, *Journal de Physique* **50**(5), 727-733.

564

38. Smurov, I. and Covelli, L.(1992) Synthesis of nitride and carbide compounds on titanium by means of a solid state laser source, in Mordike,B.L.(edt), *Laser Treatment of Materials. Proceedings of the European Conference on laser Treatment of Materials, ECLAT'92, October 1992, Gottingen, Germany*, DGM, Oberursel, pp.251-256.

39. Smurov, I., Uglov, A, Krivonogov, Yu., Sturlese, S. and Bartuli, C.(1992) Pulsed laser treatment of plasma sprayed thermal barrier coatings: effect of pulse duration and energy input, *Journal of Materials Science* 27(16), 4523-4530.

40. Smurov, I., Uglov, A., Lashin, A., Matteazi, P., Covelli, L. and Tagliaferri, V.(1991) Modeling of pulse-periodic energy flow action on metallic materials, *International Journal of Heat and Mass Transfer* 34(415), 961-971.

41. Smurov, I.(1993) Characteristics of laser heating of materials, in Martelluci, S., Chester, A.N. and Scheggi, A.M.(eds), *Laser Applications for Mechanical Industry, NATO ASI Series, Series E: Applied Sciences*, vol.238, Kluwer Academic Publishers, Dortrecht, pp.131-150.

42. Smurov, I. and Lashin, A.(1993) Heat processes in pulsed laser action, in Martelluci, S., Chester, A.N. and Scheggi, A.M.(eds), *Laser Applications for Mechanical Industry, NATO ASI Series, Series E: Applied Sciences*, vol.238, Kluwer Academic Publishers, Dortrecht, pp.165-206.

43. Smurov, I., Covelli, L., Flamant, G., Ignatiev, M. and Balat, M.(1993) Laser, plasma and concentrated solar energy modification of surface layers, *in Proceedings of the International Conference on Surface Engineering, March 9-11, 1993, Germany, Bremen*, pp.15-17.

44. Uglov, A., Ignatiev, M., Smurov, I., Konstantinov, S. and Titov, V.(1990) Peculiarities of heating of coating-substrate system in laser alloying of metal surface, *Teplofizika Vysokikh Temperatur* 29(3), 509-514.

45. Smurov, I., Flamant, G. and Konstantinov, S.(1992) Heat processes in laser treatment of coatings and joint materials, *in Proceedings of International Congress on Applications of Lasers and Electro-Optics ICALEO'92, Oct.25-29, 1992, Orland, Florida USA*, Laser Institute of America, Orlando, pp.111-120.

46. Reznikov, A.N.(1981) *Thermophysics of Machining of Metals*, Mashinostroenie, Moscow.

47. Stroshkov, A.N., Tesler, S.L. and Shabashov, S.P.(1977) *Cutting of Preheated Low-Machinability Materials*, Mashinostroenie, Moscow.

48. Smurov, I. and Okorokov, L.(1993) Laser assisted machining, in Martelluci, S., Chester, A.N. and Scheggi, A.M.(eds), *Laser Applications for Mechanical Industry, NATO ASI Series, Series E: Applied Sciences*, vol.238, Kluwer Academic Publishers, Dortrecht, pp.151-164.

49. Gavryushenko, B.S., Okorokov, L.V., Rykalin, N.N., Uglov, A.A., Halboshin, A.P. and Smurov, I.Yu.(1985), Laser assisted machining of metals, *Fizika i Khimiya Obrabotki Materialov* 19(2), 4-7.

MICROSTRUCTURES AND PROPERTIES OF AL-NB ALLOYS PRODUCED BY LASER SURFACE TREATMENT

P. PETROV[*], R. VILAR AND A. ALMEIDA

Instituto Superior Técnico

Departamento de Engenharia de Materiais

Av. Rovisco Pais 1096 Lisboa Codex, Portugal

Aluminium-niobium surface alloys were produced by the blown powder technique using a CW 3 kW CO_2 laser. The alloyed layers present porosity that was eliminated by remelting. Their niobium content varied between 48 and 56 wt% and an even distribution of niobium along the depth of the remelted layers was obtained. The material is formed by dendrites of Al_3Nb and a small volume fraction of interdendritic α–aluminium solid solution. The Vickers hardness of the niobium-rich layer varies in the range 550-650 HV, depending on the dendritic spacing and on the volume fraction of Al_3Nb. The large amount of this intermetallic compound explains the high hardness observed.

1. Introduction

High-strength aluminium alloys are widely used for structural applications in high-performance automobiles, railway cars, airplanes and spacecraft, light ships, etc., owing mainly to their excellent mechanical strength, low specific weight, good formability and

---

[*] on leave from:
Institute of Electronics
Bulgarian Academy of Sciences
72, Tzarigradsko Chausse, 1784 Sofia, Bulgaria

J. Mazumder et al. (eds.), Laser Processing: Surface Treatment and Film Deposition, 565–573.
© 1996 *Kluwer Academic Publishers.*

relatively low cost. However, the widespread use of these materials has been limited by their low resistance to corrosion and wear. This limitation accounts for the interest on surface modification of aluminium alloys using lasers and electron beams. Surface melting of aluminium alloys with high powered laser beams produces very fine microstructures with moderately improved hardness and good corrosion resistance [1]. However, the improvement of surface dependent functional properties is moderate and it will not always justify the increased cost due to the laser treatment. Laser surface alloying possesses all the features of surface melting, plus the additional capability of modifying the chemical composition by adding alloying elements which greatly enhance the wear and corrosion resistance of aluminium alloys [2-4]. Niobium is a very promising alloying element to improve the wear resistance of aluminium alloys, since it leads to the formation of the hard intermetallic compound Al_3Nb, even for small concentrations. Furthermore, several metastable niobium aluminides were referred to in the literature [5-9] that deserve attention. A detailed study of the microstructural characteristics of laser processed niobium aluminides was carried out by Sircar et al. [10, 11]. The authors reported the existence of metastable Al_3Nb (with base-centered tetragonal structure) and Nb_2Al (with base-centered orthorhombic structure) phases in laser cladded Al-25at.%Nb powder mixture. The formation of these metastable phases is attributed to the nonequilibrium solidification due to the rapid cooling rates involved in laser cladding [10]. In the present paper we report results of a study of laser alloying of industrial purity aluminium with niobium.

2. Experimental Methods

Commercial purity aluminium plates (100x50x5 mm) were used as substrates. Prior to the laser treatment, the plates were sand-blasted and cleaned with acetone in an ultrasonic bath. The laser treatment was carried out with a CO_2 laser, with a laser beam power of 2 kW. The laser beam was focused with water cooled copper mirrors. During the experiments, the focal plane was situated above the surface of the samples, leading to a laser beam diameter at the surface of 1.2 mm. Laser alloying was performed by the blown powder technique using a powder flow rate of 0.03 g/s. Due to the high melting point of niobium, long interaction times are needed to produce the alloys, therefore a scanning speed of 5 mm/s was used in all the experiments. To reduce oxidation of the melted pool, argon was blown over the surface of the samples during the laser treatment. Complete surface coverage was achieved by overlapping consecutive tracks 50% of the width of a single track. The powder used for laser alloying was a mixture of 25 wt% niobium and 75 wt% aluminium high purity powders. To reduce moisture the powder mixture was dried in an oven at 100 °C for 4 hours before the experiments.

In order to eliminate defects in the laser alloyed surface layers and to homogenize the structure most samples were remelted in a direction perpendicular to the direction used for alloying, with scanning speeds of 5, 10, 20 and 40 mm/s and laser processing parameters similar to those used previously.

The laser treated samples were cross-sectioned and cut longitudinally for metallographic study. The specimens were prepared using standard metallographic techniques and etched with Keller's reagent. The microstructure was observed by optical and scanning electron microscopy. The phases present were identified by X-ray diffraction. Chemical analysis of selected samples was carried out using an electron probe microanalyser (EPMA). Microhardness Vickers tests were made at the surface of the remelted layers, in a direction transverse to the direction of the tracks. Hardness depth profiles were established by measuring the microhardness on transverse cross-sections of the tracks, along their central line. All microhardness measurements were carried out under a load of 100 g. Each hardness value is the average of 5 measurements.

3. Results and Discussion

After laser alloying with niobium, surface layers about 1 mm thick are produced. The alloyed layers are heterogeneous and present pores and undissolved niobium particles (fig. 1). An examination of the cross-section of the alloyed layer reveals two different

zones: a top layer, about 700 μm thick, where the niobium concentrated forming an
aluminium-niobium alloy (A) and a bottom layer that is formed essentially by
resolidified aluminium solid solution (B).

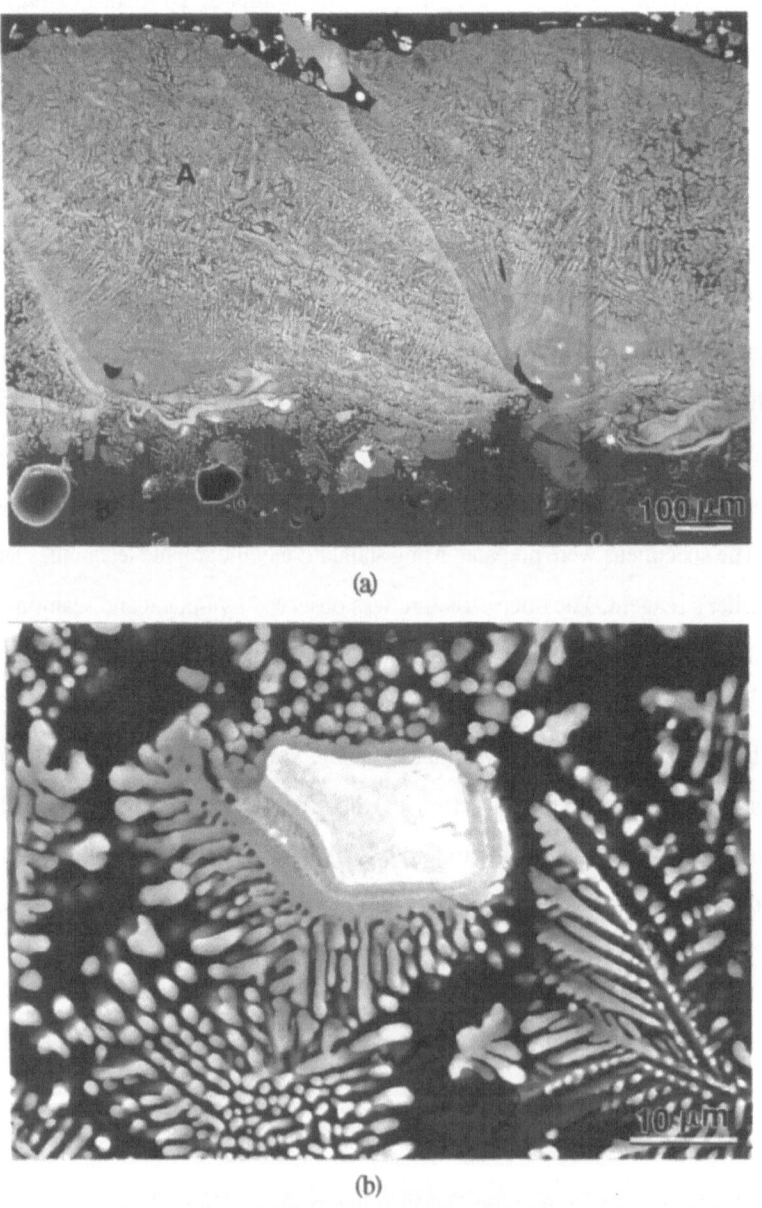

(a)

(b)

Figure 1 - Cross-section after (a) alloying (τ=0.24 s) and (b) partially dissolved Nb particle
surrounded by a layer of intermetallic compound.

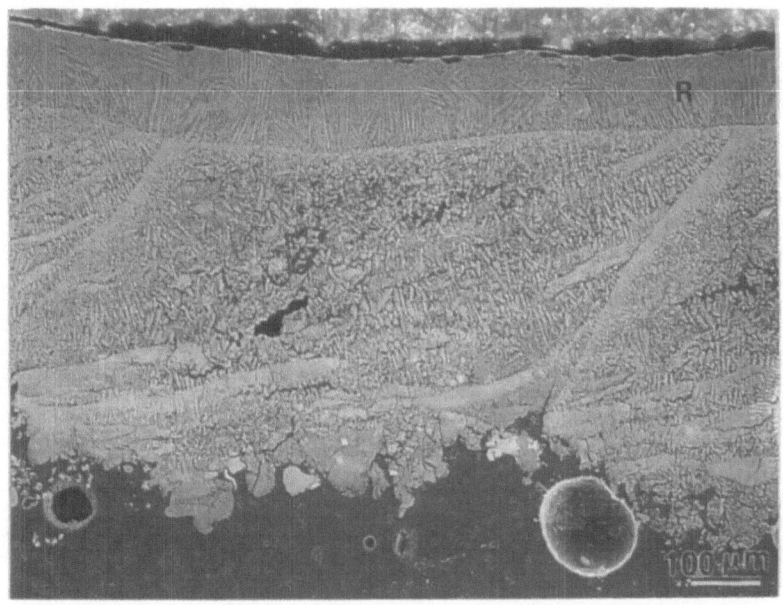

Figure 2 - Cross-section after alloying and remelting (τ= 0.03 s).

After remelting (R) the structure is homogenized and the pores and undissolved particles are eliminated (Fig. 2). The depth of the remelted layer varies between 320 μm and 550 μm and it increases with the interaction time (τ) during remelting.

In the top niobium-rich zone (A) the structure consists of dendrites of an intermetallic compound and α-aluminium solid solution in the interdendritic regions (fig. 3a). The intermetallic compound was identified as Al_3Nb by X-ray diffraction. Remelting does not modify the constitution of this niobium-rich layer. The structure becomes finer than in the alloyed material (fig. 3b), but the phases identified by X-ray diffraction are the same.

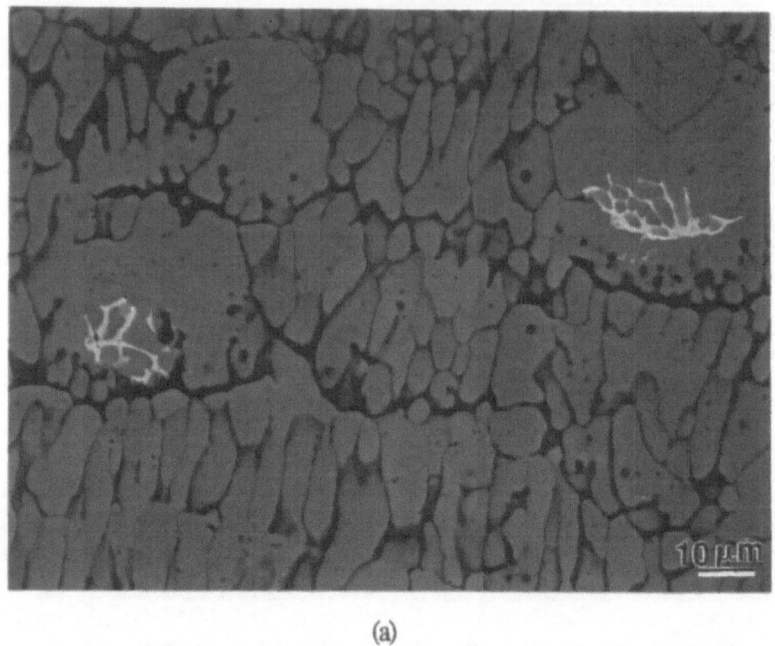

(a)

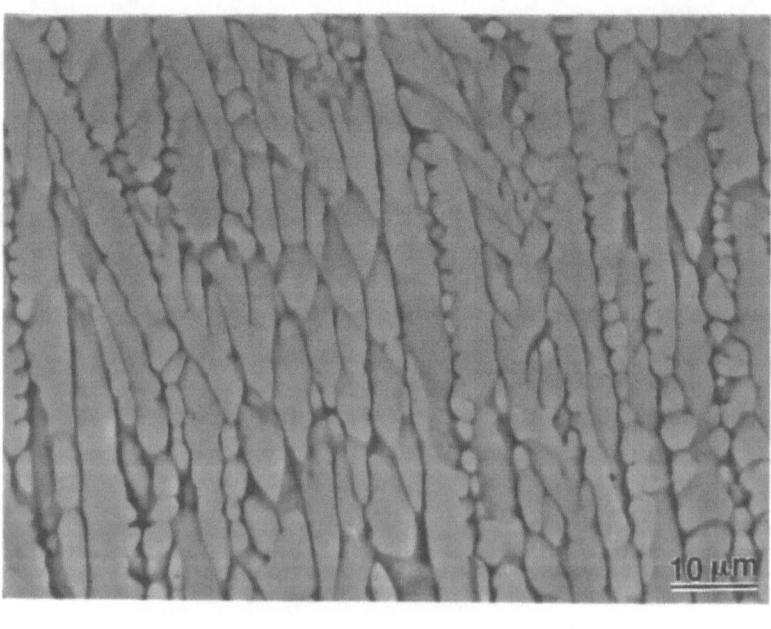

(b)

Figure 3 - Microstructure of Al-Nb layer after (a) alloying and (b) alloying and remelting

(τ=0.12 s), high magnification.

Fig. 4 shows concentration profiles of niobium along the depth of the alloyed layer before and after remelting. The results show that the distribution of niobium after laser alloying is a very heterogeneous whereas this element is evenly distributed after remelting. The results clearly demonstrate the efficiency of the remelting treatment in homogenizing the alloyed layer. The average concentration of niobium in the alloy ranges from 48 to 56 wt%. The concentrations of niobium and aluminium are different from the composition of the starting powder mixture due to the segregation of aluminium to the lower region of the alloyed layer.

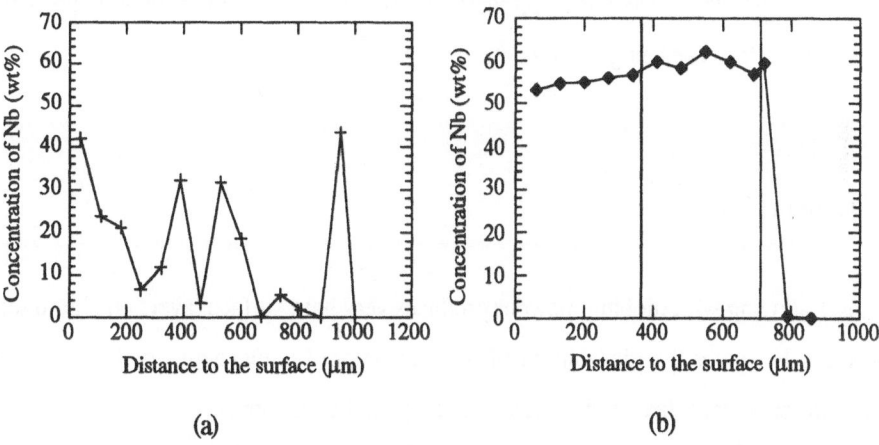

(a)   (b)

Figure 4 - Depth concentration profiles of niobium in the alloyed layer (a) before and (b) after remelting ($\tau = 0.06$ s).

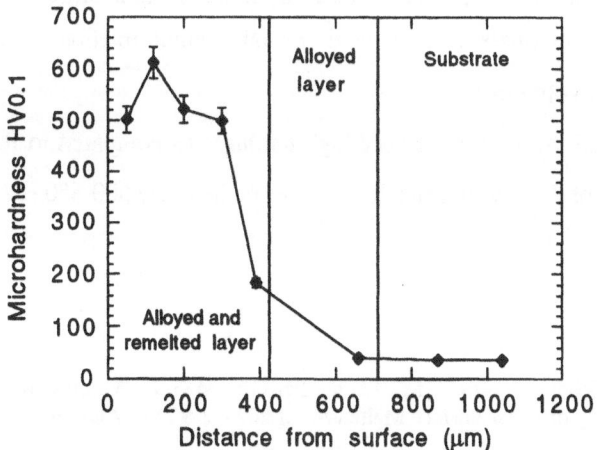

Figure 5 - Microhardness profile of laser alloyed aluminium with niobium ($\tau = 0.12$ s).

Figure 5 shows the microhardness profile measured along the depth of the cross-sections of an alloyed layer. The hardness of the alloyed layers is about 200-250 HV whereas it attains 500 to 650 HV after remelting. The hardness increases with increasing scanning speed during remelting. These high hardness values result from the large volume fraction of Al_3Nb present in the structure and also from the refinement of the microstructure resulting from the remelting treatment and are similar to those found in alloys produced by laser cladding [11].

The results obtained suggest that Al-Nb alloys produced by laser alloying may be interesting for wear resistance applications. To investigate this potential property abrasive wear tests are being carried out.

4. Conclusions

1. Laser alloying with niobium powder produces very hard surface layers on aluminium substrate. However the distribution of niobium is not homogeneous and the alloyed layers present defects like undissolved niobium particles and pores.

2. Laser remelting the alloyed layers eliminates most defects, the material is homogenized and produces a finer structure.

3. The homogenized surface layers contain 48-56 wt%. Their structure consists of dendrites of Al_3Nb intermetallic compound a small volume fraction of interdendritic α–aluminium solid solution.

4. The laser alloyed layers present very high hardness as compared to the untreated material. Hardness of the niobium-rich layer varies in the range 500-650 HV.

5. References

1. Almeida, A., Vilar, R., Anjos, M., Li, R., Ferreira, M.G.S., Watkins, K.G. and Steen, W. (1994) Study of Laser Surface Melted ANSI 2024 and 7175 Aluminium Alloys, (to be published).

2. Almeida, A., Vilar, R., Anjos, M., Li, R., Ferreira, M.G.S., Steen, W. and Watkins, K.G. (1994) Laser Alloying of Aluminium Alloys with Chromium, *Surface & Coatings Technology*, (in press).
3. A. Almeida, R. Vilar, R. Li, M.G.S. Ferreira, K.G. Watkins and W.M. Steen, Laser Alloying of Aluminium Alloys and 7175 Aluminium Alloy for Enhanced Corrosion Resistance, in P. Denney, I. Miyamoto and B.L. Mordike (eds.), Proceedings of the Laser Materials Processing Conference ICALEO'93, Laser Institute of America, Orlando, pp. 903-912.
4. Luft, U., Bergman, H. and Mordike, B. (1987) Laser Surface Melting of Aluminium Alloys, in B. Mordike (ed.), *Laser Treatment of Materials*, DGM Informationsgesellchaft mbH, Oberursel, Germany, pp. 147-162.
5. Kumar, K. (1990) Ternary Intermetallics in Aluminium-refractory Metal-X Systems (X=V,Cr,Mn,Fe,Co,Ni,Cu,Zn), *International Materials Reviews* **35**, 293-327.
6. Tunca, N. and Smith, R. (1989) Intermetallic Compound Layer Growth at the Interface of Solid Refractory Metals Molybdenum and Niobium with Molten Aluminum, *Metallurgical Transactions* **20A**, 825-836.
7. Murray, J. and German, R.(1992) Reactive Sintering and Reactive Hot Isostatic Compaction of Niobium Aluminide NbAl₃, *Metallurgical Transactions* **23A**, 2357-2364.
8. Bart, E. and Sanchez, J. (1993) Observation of a New Phase in the Niobium-Aluminium System, *Scripta Metallurgica et Materialia* **28**, 1347-1352.
9. Suryanarayna, C., Zhou, E., Peng, Z. and Froes, F. (1994) Synthesis of Ordered Al₃Nb Intermetallic by Mechanical Alloying, *Scripta Metallurgica et Materialia* **30**, 781-785.
10. Sircar, S., Chattopadhyay, K. and Mazumder, J. (1992) Nonequilibrium Synthesis of NbAl₃ and Nb-Al-V Alloys by Laser Cladding: Part I. Microstructure Evolution, *Metallurgical Transactions* **23A**, 2419-2429.
11. Haasch, R., Tewari, S., Sircar, S., Loxton, C. and Mazumder, J. (1992) Nonequilibrium Synthesis of NbAl₃ and Nb-Al-V Alloys by Laser Cladding: Part II. Oxidation Behavior, *Metallurgical Transactions* **23A**, 2631-2639.

Acknowledgments

Dr. P. Petrov gratefully acknowledges a fellowship of NATO Scientific Affairs Division. The authors acknowledge Cabot Performance Materials for kindly offering the niobium powder used in this work.

LASER SURFACE MELTING AND GAS-ALLOYING OF Ti-6Al-4V
Experimental Approach and Property Evaluation

J. M. ROBINSON, R. C. REED AND D. R. F. WEST
Department Of Materials
Imperial College Of Science,
Technology and Medicine
London SW7 2BP
United Kingdom

Abstract

The effect of CO_2 laser surface treatment of Ti-6Al-4V is discussed. Laser surface melting and gas alloying of the Ti-6Al-4V is undertaken in an atmosphere of argon or alternatively, nitrogen diluted with argon. The gas composition of the atmosphere is either 100%Ar, 90%Ar + 10% N_2 or 80%Ar + 20%N_2 by volume. The microstructure of the laser processed layer is martensitic when processed in pure Ar; for atmospheres containing nitrogen, a continuous TiN surface layer is present. All material melted in an atmosphere containing nitrogen exhibits some cracking in the processed layer. In spite of the cracking, results from performance evaluation illustrate that the laser treated material has both a substantially superior cavitation as well as water droplet erosion resistance.

1. Introduction

1.1 LASER SURFACE TREATMENT

Laser surface treatment has the potential to produce enhanced surface properties in an elegant single stage process. Laser gas alloying (LGA) is particularly appropriate for producing wear resistant surface layers on Ti alloys (eg [1]). Furthermore, the depth of modification attainable using LGA is substantially higher than conventional methods such as thermochemical, plasma vapour deposition or ion implantation treatments [2]. Precision control and process automation mean that consistent, repeatable results may be obtained. LGA relies on the rapid, localised melting produced by the intense optical energy of a laser beam. Convection currents in combination with the high diffusivity of solutes in the liquid state ensure that alloying occurs throughout the melt in the atmosphere of a suitably reactive gas. Typically, however, gradients of composition and microstructure exist in the alloyed regions (eg. [3, 4]).

In the case of Ti and Ti alloys, surface melting in an atmosphere of N_2 gas causes the formation of a nitrogen rich surface layer. The microstructure of the alloyed region, although affected by a range of process variables, consists primarily

575

J. Mazumder et al. (eds.), Laser Processing: Surface Treatment and Film Deposition, 575–585.
© *1996 Kluwer Academic Publishers.*

of TiN dendrites within a nitrogen enriched α-Ti matrix [4, 5]. Katayama et al originally surface melted various Ti alloys in a continuous stream of N_2 gas [6]. They noted the difficulty of producing a smooth, crack–free surface layer. Subsequently, mixtures of N_2 and inert gases have been considered (eg. [5, 7]). Bell et al have reported that surface cracking can be eliminated by working at high traverse velocities [3] and more recently, by preheating the substrate prior to laser processing [7]. The previously published work has indicated optimum values for specific variables such as scanning speed, approximate power density and intertrack overlap [7]. The work presented here demonstrates that in spite of the potentially improved properties, further fundamental research into the subtle effects of process parameters is required. To this end, the ongoing development of accurate process models against which experimental data may be compared is particularly important.

1.2. CAVITATION AND WATER DROPLET EROSION

Cavitation phenomena may occur in a broad range of hydrodynamic and turbomachine applications. The effect is almost universally deleterious resulting in a range of problems such as modification of hydrodynamic properties and the associated performance breakdown as well as the generation of vibration and noise. Three approaches are generally applied to minimise or avoid erosive damage [8]; the optimisation of hydrodynamic profiles, the development of cavitation erosion resistant materials and the surface modification of conventional materials. It has been recently suggested [9] that laser surface treatments may offer particularly encouraging possibilities for the surface modification of alloys for enhanced cavitation resistance. The cavitation erosion of Ti base alloys is of specific interest because of their application in corrosion-resistant pumps and valves as well as in marine and chemical industry piping. These applications subject the alloys to exposure to high flow rate fluids and therefore cavitation phenomena in some cases. It is clear that surface modification represents an appropriate approach because of the potential to produce materials with both optimised bulk and surface properties.

Water droplet erosion has long been identified as a significant problem in final and penultimate row blading in the low pressure end of steam turbines (eg. [10]). Conventionally, Stellite alloys have been utilised as protective erosion shields for 12Cr steel blades and this has provided a long standing satisfactory solution. However, concurrent with the more recent interest in Ti alloy last row blading, a need has been identified for the development of erosion protection for these alloys. Stellite no longer represents a satisfactory solution due to thermal expansion mismatch with Ti-based alloys and alternative protection methods are therfore necessary. In this regard, laser gas alloying represents a potentially viable solution. Indeed, enhanced water droplet erosion has been recently demonstrated and a full discussion of the results will be presented elsewhere [11].

For the purposes of this work, we report on the rationalisation of the starting parameters for the initial property evaluation of laser processed Ti-6Al-4V. In addition, we report briefly on the cavitation and water droplet erosion performance of Ti-6Al-4V processed using these parameters. Some of the potential problems associated with generalising the work between different research institutions are discussed. We propose methods for circumventing these practical problems.

2. Experimental Techniques

2.1. MATERIAL AND LASER PROCESSING

The commercial alloy, IMI 318 (Ti-6Al-4V) conforming to the composition limits outlined in table 1 was used in all experiments in the course of this work. Hot worked and annealed dual phase ($\alpha + \beta$) material was processed in the form of 10 mm thickness plate with sample dimensions 65 mm × 150 mm.

Table 1: Chemical compostion limits for commercial alloy Ti-6Al-4V.

Designation	Alloy	Al	V	N	C	H	Fe	O
IMI 318	Ti-6Al-4V	5.50	3.50	0.05	0.10	0.015	0.40	0.20
		6.75	4.50					

Laser surface alloying experiments were undertaken at 2kW using a transverse flow, continuous wave 5kW CO_2 laser. The TEM_{01}• beam was focussed using a 69° off-axis parabolic mirror with a working distance approximately 216 mm. The focussed beam was characterised using a hollow needle beam analyser (Prometec Laserscope UFF100) and the power measured at the workpiece surface using a calibrated calorimeter. Experiments were undertaken below the focal point using an average power density of $\sim 1 \times 10^5$ Wcm^{-2}. All samples were grit-blasted ($R_a \approx 7.5 \ \mu m$) prior to laser processing in order to increase the fraction of incident radiation absorbed. A traverse velocity of 20 mm/s was used throughout, resulting in conduction limited [12] melting to a depth of approximately 0.5 mm.

The gas delivery for oxidation protection and alloying was via a purpose designed shroud, illustrated in figure 1. The shrouding permitted accurate control of the composition of the local atmosphere both immediately prior to melting as well as during solidification. An integral gas jet was aimed directly at the point of interaction between the laser beam and the substrate. Total gas flow was controlled at 20 dm^3/min and divided evenly between the gas jet and the shroud system. Three discrete atmospheres were utilised i.e. pure Ar, Ar + 10%N_2 and Ar + 20%N_2 by volume. Surfaces were produced by overlapping adjacent melt tracks. The successive overlap between tracks was either 50% or 75% of the top surface width of the melted region as illustrated in figure 2. The top width was initially determined by producing a single melt track, sectioning the sample and measuring the dimensions of the melt pool. 50% overlap is the minimum requirement for producing a

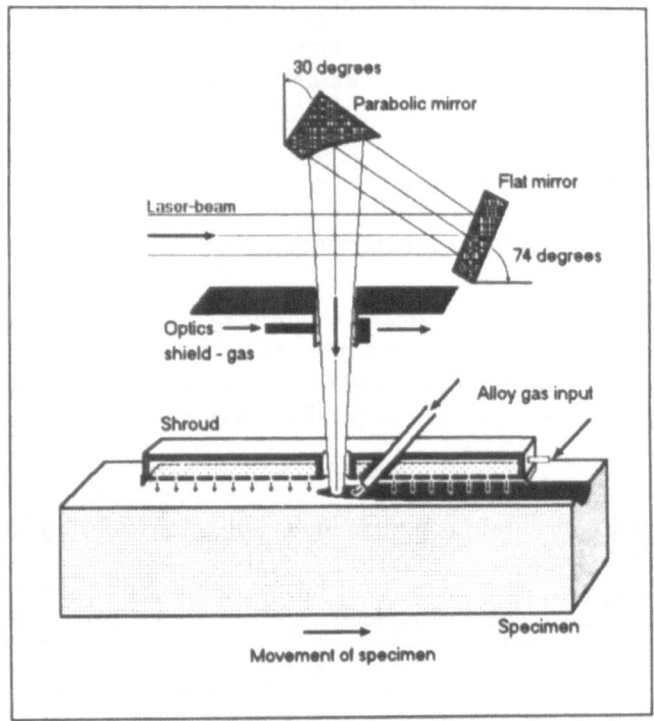

Figure 1: Schematic illustration of the experimental arrangement used to undertake laser surface treatments.

continuously treated surface layer and the literature [7] has indicated 75% overlap as an optimum for producing an alloyed layer with homogeneous microstructure.

2.2. EROSION EVALUATION

Vibratory cavitation erosion tests were undertaken using an ultrasonic laboratory cavitation rig (see eg. [13, 14]). The technique involves exposing the sample to controlled cavitation in deionised water at a stabilised temperature (25°C). The mass loss was determined at specific interuption points during the testing i.e. after 1, 2, 3, 4, 5, 6, 7.5, 10, 15 and 20 hours exposure to cavitation. Full details of the specific experimental conditions are published elsewhere [15].

For the evaluation of water droplet erosion, specimens were tested in an erosion rig consisting of two contra rotating shafts, each carrying a mild steel overhung disc. Four specimens are separately mounted on the perimeter of the larger diameter disc and the spray nozzles are located on the smaller. Primarily, the water

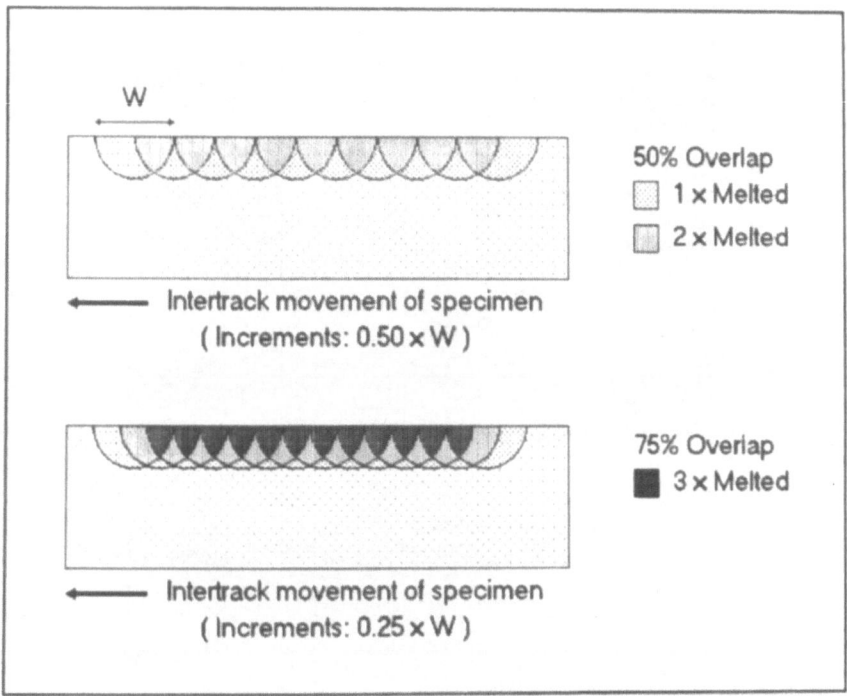

Figure 2: Schematic illustration of the production of a laser processed surface layer. The succesive movement of the sample is calculated as a percentage of the top width (W) of a single melt track.

droplet erosion apparatus is intended as a method for rapidly evaluating the relative erosion resistance of a range of materials. Conditions during testing are, however, designed to approximate the actual operating conditions in the low pressure stage of a steam turbine in terms of temperature, impact velocity ($\sim 500\ ms^{-1}$) and droplet size ($\sim 100\ \mu m$).

3. Results and Discussion

3.1. LASER PROCESSING

The surface of Ti-6Al-4V processed in pure Ar (melted) was optically highly reflective while the surfaces processed in a nitrogen containing atmosphere (alloyed) were a characteristic golden colour (eg. [16]). Scanning electron microscopy images of the surfaces produced with a 75% intertrack overlap are illustrated in figure 3. It is clear from the figure that the surface topography is substantially altered by increasing the nitrogen content in the alloy gas, even at the low dilutions employed.

580

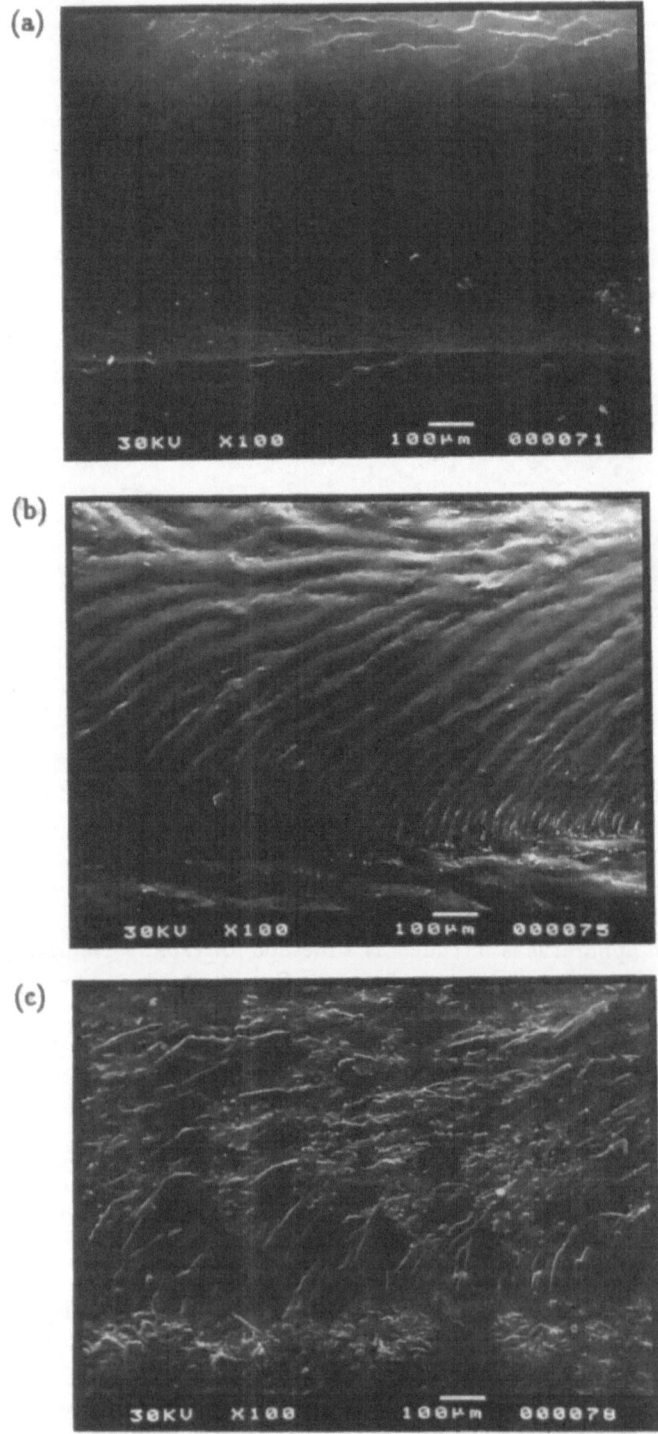

Figure 3: Scanning electron micrographs of Ti-6Al-4V surfaces, laser processed in atmospheres of (a) 100% Ar, (b) 10%N$_2$ and (c) 20% N$_2$.

As the nitrogen content is increased, the surface roughness is increased which can be attributed, at least in part to the higher viscosity TiN-containing melt pool [7]. Moreover it is clear from the micrograph shown in figure 4 that the surface layer of the material alloyed in a 20% N_2 atmosphere is cracked. SEM observations confirmed the results of dye penetrant NDT testing which had shown the surfaces of material processed in both 10 and 20% N_2 containing atmospheres to be cracked regardless of overlap. The density of cracking was highest in the material processed in 20% N_2 and the material processed in 100% Ar was totally crack free.

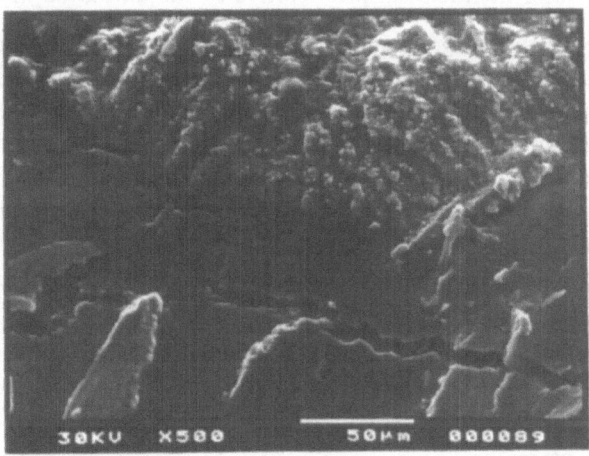

Figure 4: Scanning electron micrograph of Ti-6Al-4V showing surface cracking after laser processing in an atmosphere of 20% N_2.

The fact that the nitrided material was observed to be cracked at the low dilutions employed in this work requires rationalisation in order to resolve apparent contradictions with previously reported data. The nitrogen dilutions employed in the current work are substantially lower than [7] or at least equivalent to [17] any which have been previously discussed in the literature and reported to be crack free. It is possible that the shroud mechanism illustrated in figure 1 permitted a more efficient uptake of the alloy gas into the meltpool than in previously reported experiments. Previous laser nitriding work has utilised either a coaxial gas flow or an angled separate alloy gas nozzle (eg. [7]). A further possibility is that in the course of this work, the substrate temperature was measured and deliberately allowed to stabilise to less than 50°C prior to each successive laser track being laid. The purpose of temperature stabilisation is to ensure that each successive track is produced under directly comparable experimental conditions to the previous track. The net effect may however have been to minimise preheating of the substrate by the preceding track or tracks. Preheating has previously been demonstrated to remove the occurrence of surface cracking in laser nitrided Ti alloys [7]. Further experimentation is necessary to ascertain the relative importance of the above possibilities.

3.2. EROSION PERFORMANCE

In spite of the existence of cracking in the alloyed layers, an evaluation of the as-processed layers for both water droplet and cavitation erosion resistance was undertaken. This allowed for a clear initial assessment of the effectiveness of the laser surface treatment of Ti-6Al-4V for erosion resistance. Full discussions of both the cavitation [15] and water droplet erosion [11] performance of the laser treated Ti-6Al-4V are to be reported elsewhere. However, figures 5 and 6 respectively illustrate the cavitation and water droplet erosion performance of both the surface melted and the laser nitrided samples relative to the untreated material.

In can be seen from the erosion data that laser treatment substantially improves both the cavitation and the water droplet erosion performance of Ti-6Al-4V. The most significant improvement results from alloying in an atmosphere containing 10% nitrogen. SEM examinations of both the erosion surfaces and cross-sections have suggested that it is not the brittle continuous surface layer of TiN which enhances the erosion resistance of Ti-6Al-4V, but rather the production of a relatively deep, melted and nitrogen-alloyed layer.

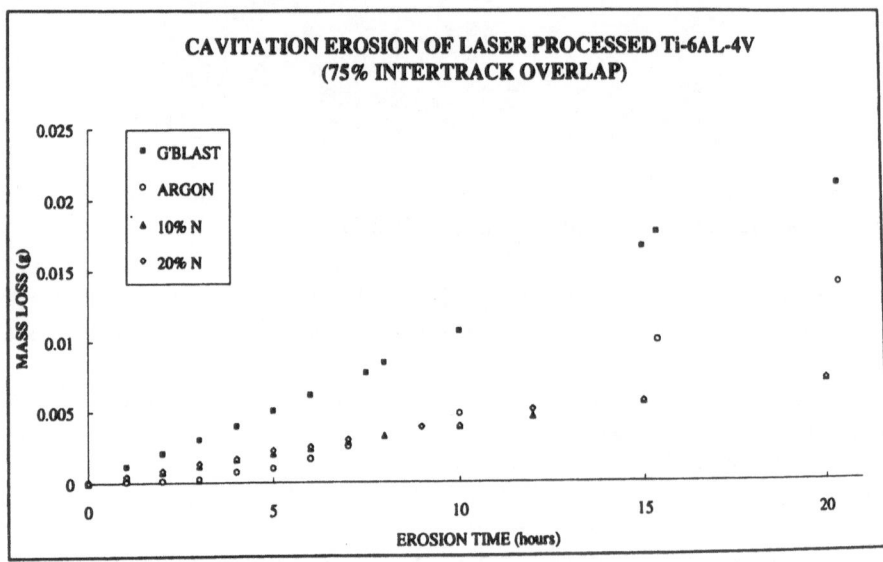

Figure 5: Cavitation erosion evaluation showing mass loss versus time over 20 hours for untreated, laser melted and laser nitrided Ti-6Al-4V.

Figure 5 and 6 illustrate the performance of laser treated surfaces produced using 75% intertrack overlap. An important result of the work is that there is no significant difference between the erosion performance of samples produced with either 50 or 75% overlap. This in turn has implications in terms of the economics of treating large surface areas.

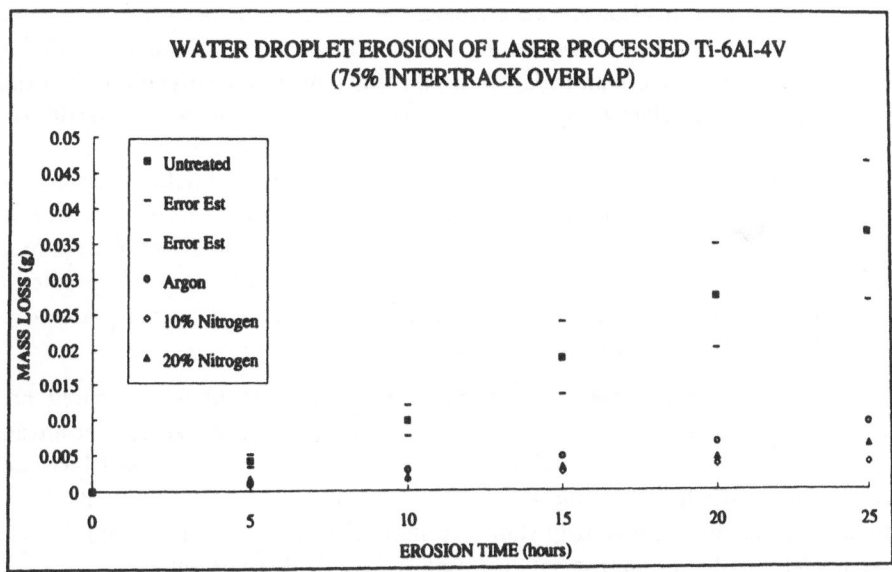

Figure 6: Water droplet erosion evaluation showing mass loss versus time over 25 hours for untreated, laser melted and laser nitrided Ti-6Al-4V.

There are strong similarities between the performance of laser treated alloy, Ti-6Al-4V under both cavitation and water droplet erosion conditions. This can be anticipated given the accepted similarities (eg. [13]) in performance ranking when materials are subjected to either mechanism of erosion. An anomaly, however, is that the steady state erosion rate (slope of mass loss vs time) of the material melted in argon was significantly improved under water droplet erosion but this was not the case under cavitation erosion.

4. Summary and Conclusions

It is evident from the results presented that laser surface treatment can substantially improve the resistance of Ti-6Al-4V alloy to both cavitation and water droplet erosion. Laser surface alloying of Ti-6Al-4V by melting in a dilute nitrogen atmosphere results in a substantial decrease in the steady state erosion rate but variation in the overlap between successive melt tracks does not have a significant effect. Similarly, varying the concentration of nitrogen in the alloy gas between 10% and 20% has a negligible effect on the erosion rate.

The production of laser melted and laser nitrided Ti-6Al-4V surfaces for the purposes of erosion evaluation has highlighted a number of points. Most importantly, it is evident that it is imperative that attempts are made to standardise the way in which experiments are carried out in order that comparable results are obtained. In terms of the production of multiple-track alloyed surfaces, allowing

584

a temperature stabilisation between successive tracks is one such factor relevant to producing consistent and repeatable results. In addition, in order to correlate the results of experiments undertaken in different laboratories it is important that more significance is placed on the measurement of power and power distribution of the laser beam as seen by the workpiece. In order to minimise the extensive and costly experimentation associated with research of this nature, efforts should possibly be made to utilise formal methods of experimentation which limit arbitary parameter variation. Finally, the use and concurrent development of process models particularly incorporating convective flow (eg. [18]) and ultimately prediction of solidification microstructures is especially important if the results are to be universal and applicable to variable component geometries.

In summary, there exists the potential to use the technique presented here to treat industrial components for enhanced cavitation and water droplet erosion resistance. However, some cracking has been observed in the processed layer and practicable methods to eliminate these are currently under development. While pre-heating has been previously shown to minimise or eliminate cracking [7], this may not always be an appropriate solution in practical situations

Acknowledgements - The authors are grateful to Parsons Steam Turbines Ltd., UK for the provision of the Ti-6Al-4V material and the water droplet erosion evaluation. Primary financial support for the work from the Science and Engineering Research Council, UK under grant ref: GR/H 36535 is also gratefully acknowledged.

5. References

1. Massoud J. P. and Coquerelle G. (1988) High power laser treatments on Ti-6Al-4V in order to improve its erosion resistance, in *Sixth World Conf. on Ti*, TMS, Pennsylvania, USA, 1847-1851.

2. Morton P. H. and Bell T. (1989) Surface engineering of titanium, *Memoires Etudes Scientifiques Revue de Metallurgie*, 639-645.

3. Bell T., Bergmann H. W., Lanagan J., Morton P. H. and Staines A. M. (1986) Surface engineering of titanium with nitrogen, *Surface Engineering* 2, 133-143.

4. Walker A., Folkes J., Steen W. M. and West D. R. F. (1985) Laser surface alloying of titanium substrates with carbon and nitrogen, *Surface Engineering* 1, 23-29.

5. Mordike B. L. (1985) Laser gas alloying, in C. W. Draper and P. Mazzoldi (eds), *Laser Surface Treatment of Metals*, San Miniato, Italy, 389-412.

6. Katayama S., Matsunawa A., Morimoto A., Ishimoto S. and Arata Y. (1983) Laser nitriding of titanium and its alloys, in *3rd Int. Colloquium on Welding and Melting by Electron and Laser Beam*, Lyon, France.

7. Morton P. H., Bell T., Weisheit A., Kroll J., Mordike B. L. and Sagoo K. (1992) Laser gas nitriding of titanium and titanium alloys, in T. S. Sudarshan and J. F. Braza (eds), *Surface Modification Technologies 5*, 593-609.

8. Karimi A. and Martin J. L. (1986), Cavitation erosion of materials, *International Materials Reviews* **31**, 1-26.

9. Hansson C. M. and Hansson I. L. H. (1992) Cavitation erosion, in *ASM Handbook Vol 18: Wear*, ASM, Ohio, USA, 214-220.

10. Pollard D., Lord M. J. and Stockton E. C. (1979) An evaluation of low-pressure steam turbine erosion, in *Steam Turbines for the 1980's*, Institute of Mechanical Engineers, London, UK, 413-419.

11. Robinson J. M., Reed R. C. and West D. R. F. (1994), Laser treatment of Ti-6Al-4V for enhanced water droplet erosion resistance, accepted for publication in *8th Int. Conf. on Erosion by Solid and Liquid Impact*, Cambridge, UK.

12. Paul A. and Debroy T. (1988) Free surface flow and heat transfer in condcution mode laser welding, *Metallurgical Transactions* **22B**, 851-858.

13. Preece C. M. and and Brunton J. H. (1980) A comparison of liquid impact erosion and cavitation erosion, *Wear* **60**, 269-284.

14. Zhou K. S. and Herman H. (1982) Cavitation erosion of titanium and Ti-6Al-4V: effects of nitriding, *Wear* **80**, 101-113.

15. Robinson J. M., Anderson S., Knutsen R. D. and Reed R. C. (1994) Cavitation erosion of laser melted and laser nitrided Ti-6Al-4V, accepted for publication in *Materials Science and Technology*.

16. Mridha S. and Baker T. N. (1991) Characteristic features of laser-nitrided surfaces of two titanium alloys, *Materials Science and Engineering* **A142**, 115-124.

17. Lee S. Z. and Bergmann H. W. (1988) Laser surface alloying of titanium and its alloys, in *Sixth World Conf. on Ti*, TMS, Pennsylvania, USA, 1811-1844.

18. Mazumder J. and Voelkel D. (1992) Challenges in modeling and measurement of laser materials processing, in *Laser Advanced Materials Processing (LAMP '92)*, Nagaoka, Japan, 373-380.

PROCESS DIAGRAMS FOR LASER TRANSFORMATION HARDENING

JOHN C. ION

Laser Processing Centre
Lappeenranta University of Technology
FIN-53851 Lappeenranta
Finland

ABSTRACT

Analytical mathematical models are applied to laser transformation hardening of engineering steels. The depth and maximum hardness of the transformed region are modelled in terms of the principal process variables. Dimensionless groups comprising material and processing variables are identified, and relationships with the properties of the transformed region are established. Good agreement between experimental results and model predictions is obtained. Data are presented in the form of various process diagrams. In their most practical form, these can be used to select the laser power and traverse rate required for hardening a particular material to a given depth with a fixed beam geometry. Alternatively, when constructed using dimensionless groups, the diagrams provide a means of displaying data for a range of materials, and evaluating the effects of variations in all the process variables on the properties of the transformed region. Thus process variables may be optimised with respect to various productivity criteria, and an estimate made of the sensitivity of the process to variations in process variables. The models can be used in a variety of ways: to expedite the trial and error testing phase of procedure development; as the core of an on-line adaptive control system; and as an educational tool.

J. Mazumder et al. (eds.), Laser Processing: Surface Treatment and Film Deposition, 587–612.

1. Introduction

The development of multikilowatt industrial lasers has provided new opportunities for surface engineering with a highly controllable heat source. Laser transformation hardening (LTH) is one such method of producing a hard, wear-resistant surface on components made from hardenable engineering alloys. The method involves scanning a shaped laser beam over the parts of the component to be hardened. In transformable ferrous alloys, surface regions are austenitised and hard martensite is subsequently formed through rapid self-quenching, whilst the structural properties of the bulk material are retained. The advantages of LTH over traditional surface hardening methods, such as gas carburising and induction hardening, are given in Appendix 1, and stem from the rapid, localised, non-contact nature of the treatment, the low distortion produced, and the process flexibility provided.

TABLE 1. Applications of laser transformation hardening

Component	Material	Reasons for Application (Appendix 1)	Reference
Conical Shaft	Steel	4d	[1]
Gear housing	Ferritic cast iron	1a 1f 1g 3c 3e	[2]
Gear wheel teeth	C steel	4e	[1]
Gear wheel teeth		4e	[3]
Valve guide	Grey cast iron	3e 3f	[4]
Valve seat	Steel	1a 1g 3c	[5]
Spacer	Malleable cast iron	2b	[6]
Camshaft lobes	Cast iron	1g	[7]
Motor shaft	Low C steels	1g 3c 2a	[8]
Shaft	Steel	1g 2a	[9]
Piston ring	Steel	3d 3e	[10]
Piston ring groove	Cast iron / steel	1g 2b 3d 3f	[11]
Piston ring groove	Cast iron / steel	1g 2b 3d 3f	[6]
Piston ring groove	Ductile cast iron	1g 2b 3d 3f	[12]
Spring		1a 2b	[9]
Hand brake ratchet	Low C steel	3e	[8]
Roll flute crowns		3c 3e	[10]
Rivet		3e	[8]
Cutting edge	Steel	3e	[8]
Cylinder bore	Cast Iron	3c	[13]
Cylinder Liner	Alloy cast iron	1a 3c 3e	[14]
Rivet	Steel	1g 3e	[9]
Turbine blade edge	Martensitic SS	1a	[15]
Tool bed	Cast iron	1g 3e	[9]

LTH was developed into a production line process in the 1970's [2]. Applications principally in the automotive, aerospace, and domestic goods industries have been reported, Table 1, which normally involve high volume production of discrete hardened patterns on machined components with a range of shapes and sizes. However, a comparison of Table 1 and Appendix 1 indicates that only limited advantage has been taken of the potential of the process. There are various reasons. Designers and production engineers may be more familiar with traditional hardening techniques, and little may be known of the properties, advantages and opportunities provided by LTH. There are few systematic methods by which process variables can be selected, and little quantitative data concerning design principles and cost estimation are available.

This report describes methods by which analytical mathematical models can be used to select and optimise process variables, and assist in a preliminary analysis of design and economic aspects of LTH.

2. Experimental Procedure

Experiments were designed in order to provide data for calibration and evaluation of the mathematical models developed.

2.1. MATERIAL

A hot rolled vanadium-microalloyed forging steel, designated V2906, was used. The composition is given in Table 2. The microstructure comprised pearlite with an average nodule diameter of 75 μm and average interlamellar spacing of 0.4 μm, surrounded by proeutectoid ferrite with an average colony width of 13 μm. 16 mm thick sections were cut from bar stock, and the surfaces ground and coated with graphite to increase absorption of the laser beam.

TABLE 2. Chemical composition of V2906 steel (wt%)

C	Mn	Si	P	S	Cr	Ni	Mo	V	Ti	Cu
0.47	0.72	0.31	0.01	0.056	0.11	0.09	0.02	0.082	0.002	0.16

2.2. LASER PROCESSING

A Rofin Sinar RS6000 CO_2 laser was used for processing. The design comprises fast axial gas flow, radio frequency excitation and a stable optical cavity. The parameters of the emitted beam were as follows: 10.6 μm wavelength, TEM_{20} mode, 40 mm nominal diameter, 500-6000 W power range, linearly-polarised beam. The beam was formed into a heating pattern of width 7 mm and length 5 mm and uniform power density using an optical kaleidoscope. Laser power levels between 1 and 3 kW were used for processing. The beam power delivered to the workpiece was calibrated using a Laser Craft P10K measurement device. The power stability of the laser during the course of a treatment was typically $\pm 2\%$.

Samples were handled in a CN-controlled work station. The beam was scanned across the stationary specimen surface with traverse rates in the range 0.05-10 m min^{-1}, in order to produce hardened tracks of length 100 mm. Velocities could be controlled with an accuracy better than $\pm 1\%$. The interaction zone was shrouded with argon to reduce oxide formation which may influence surface absorptivity in an unpredictable manner.

2.3. METALLURGICAL EXAMINATION

After treatment, transverse sections were cut, ground and polished for metallurgical examination. Sections were etched in a solution of 3% nital in order to reveal the microstructure of the hardened case by optical microscopy.

Hardness profiles were made perpendicular to the specimen surface on transverse sections, using a load of 300 g, and an indentation spacing of 0.05 mm. Three hardness indentations were also made on the specimen surface using a load of 1 kg.

3. Results

The extent of the transformed region was defined by a distinct boundary between transformed and untransformed pearlite, the A_{c1} isotherm, Fig. 1. The microstructure of the transformed region was found to depend strongly on the laser power and traverse rate. Low traverse rates produced a banded microstructure, Fig. 2, comprising: martensite in surface regions; proeutectoid ferrite and martensite (containing small undissolved cementite plates) close to the transformation boundary; and an intermediate region of ferrite and carbides in a martensitic matrix. The A_{c3} isotherm was located by the contour at which ferrite transformed completely. Only the latter features described above were present after treatments carried out with higher traverse rates, due to incomplete development of the microstructure. Martensite was observed in all cases except that produced with the lowest speed (0.05 m min^{-1}). Steel properties and experimental data are given in Tables 3 and 4.

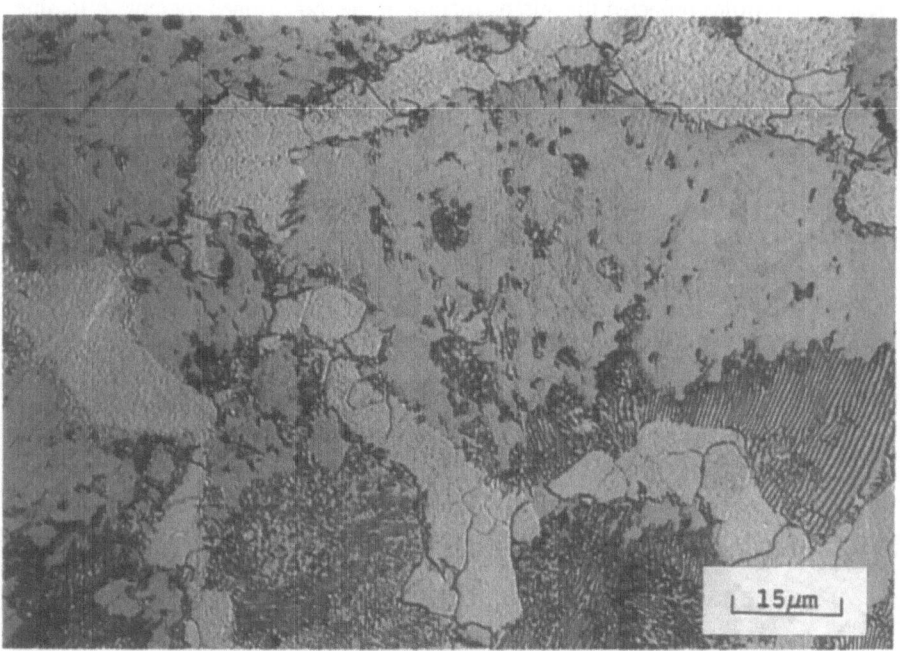

FIGURE 1. Boundary between pearlite and martensite (A_{c1} isotherm):
ferrite appears white and martensite grey (etched in 3% nital).

TABLE 3. Thermodynamic data for steel V2906

Property	Value	Reference
Thermal conductivity	30 J s^{-1} m^{-1} K^{-1}	[16]
Specific heat capacity	680 J kg^{-1}	[16]
Density	7860 kg m^{-3}	[17]
Melting Temperature	1768 K	[17]
Equilibrium A_{c1} Temperature	998 K	[18]
Equilibrium A_{c3} Temperature	1052 K	[18]
Activation energy for grain boundary diffusion of C in austenite	79.8 kJ mol^{-1}	[19], [20]
Activation energy for volume diffusion of C in austenite	142.4 kJ mol^{-1}	[19]
Diffusion frequency factor	1.5x10^{-5} m^2 s^{-1}	[19]

Hardness profiles reflected the microstructure of the transformed regions. A hardness profile from the case section of Fig. 2 is shown in Fig. 3, together with an indication of the development of the microstructure. Average hardness values obtained from samples having a homogeneous transformed region to a depth of at least 0.5 mm are given in Table 4.

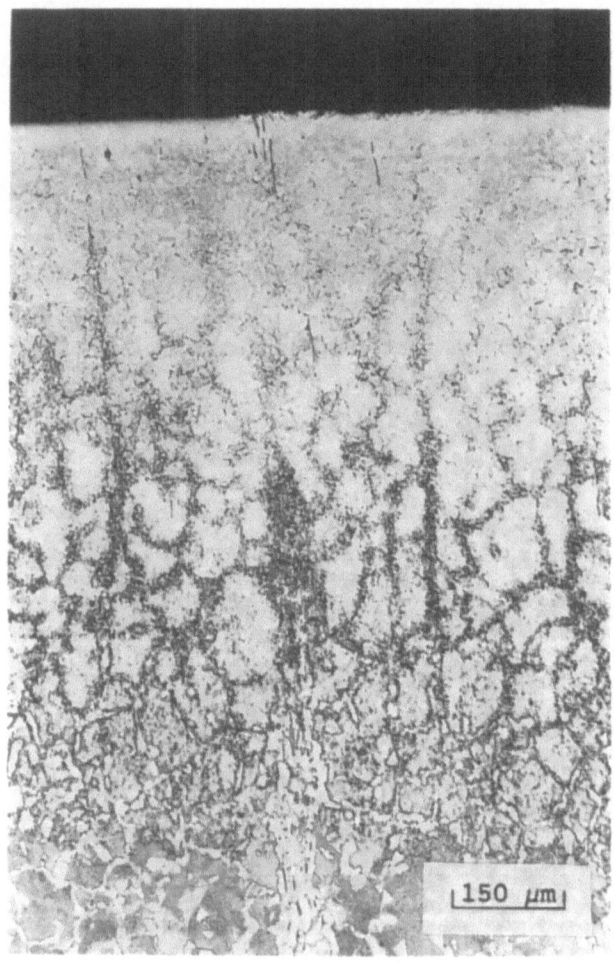

FIGURE 2. Transverse case section revealed by etching in 3% nital. 1 kW, 0.1 m min⁻¹, uniform 7 x 5 mm heating pattern.

TABLE 4. Processing parameters and case properties
produced by a uniform intensity beam of size 7mm x 5mm

Incident Power (kW)	Traverse Rate (mm min^{-1})	A_{c1} Depth (mm)	A_{c3} Depth (mm)	Homogeneous Hardness (HV1)
1	50	1.06	-	694
1	100	0.93	0.68	750
1	125	0.97	0.68	734
1	150	0.89	0.61	751
1	175	0.83	0.56	753
1	200	0.86	0.53	720
1	250	0.80	0.47	-
1	300	0.73	0.42	-
1	400	0.70	0.39	-
1	600	0.51	0.25	-
1	1000	0.27	-	-
1	1200	0.21	-	-
1	1500	0.10	-	-
1	2500	-	-	-
2	500	0.70	0.41	-
2	1000	0.46	0.16	-
2	1300	0.45	0.14	-
2	1500	0.40	0.12	-
2	2000	0.35	0.11	-
2	3000	0.24	-	-
2	4000	0.11	-	-
2	5000	0.04	-	-
2	6000	-	-	-
2	7000	-	-	-
3	3000	0.27	0.09	-
3	4000	0.21	0.04	-
3	6000	0.20	-	-
3	8000	0.10	-	-
3	10000	0.03	-	-

594

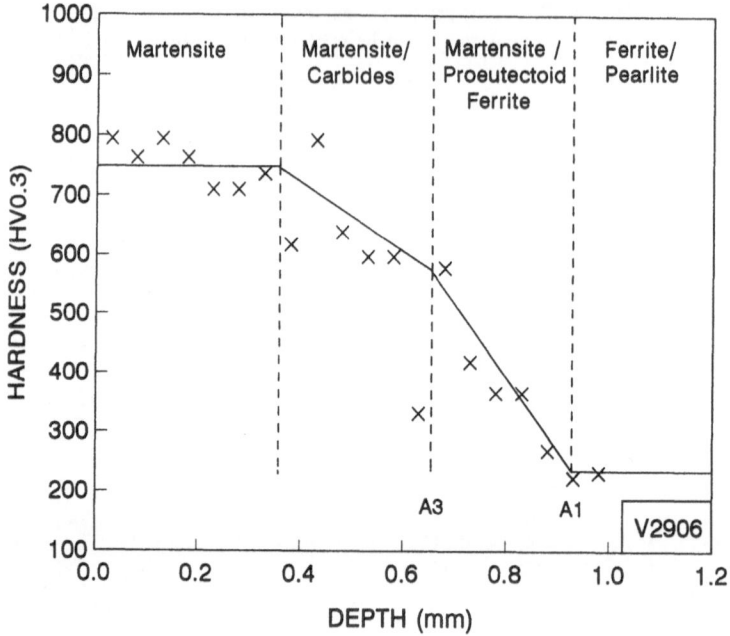

FIGURE 3. Hardness profile and microstructural development in typical case. Power 1 kW, traverse rate 0.1 m min⁻¹, uniform 7 x 5 mm heating pattern.

4. Mathematical Modelling

Mathematical models of processes have many uses: they provide an understanding of the underlying physical principles; treatments can be simulated, expediting procedure development; they may be used in process control systems; and they are a useful tool for design. Analytical and numerical mathematical models of LTH have been developed [21], although few provide a complete description of the process. Here analytical models of the temperature field, microstructural transformation and case properties are developed, the intention being to produce a system by which results may be obtained in time scales relevant to on-line control of the process.

4.1. THE TEMPERATURE FIELD

4.1.1 *Dimensional Equations*

The thermal cycle $T(t)$ induced at a depth z in a semi-infinite body, below the centre of a moving, continuous wave, circular Gaussian laser beam, can be written [22], adapted from [23]:

$$T - T_o = Aq \; / \; (2\pi\lambda v[t(t+t_o)]^{1/2}) \; . \; \exp\text{-}[(z+z_o)^2/4at] \tag{1}$$

where: T_o is initial temperature (K), A is the fraction of incident beam power absorbed, q is beam power (J s^{-1}), λ is thermal conductivity (J s^{-1} m^{-1} K^{-1}), v is beam traverse rate (m s^{-1}), t is time (s), z is depth (m), and a is thermal diffusivity (m^2 s^{-1}). Notation is given in Appendix 2.

Equation (1) contains two reference parameters. The first, t_o, is a characteristic heat transfer time, related to the beam width, defined by $t_o = r_B^2/4a$, where r_B is the radius at which the beam intensity has fallen to $1/e$ of the peak value. The second, z_o, is a characteristic length; its function is to limit the surface temperature to a finite value due to the finite energy input time - the true surface is taken to be a distance z_o below the "model" surface. The method of determining z_o is described below. Heat flow in the direction of travel is neglected - this is a reasonable assumption for practical rates of LTH for which $v >> ar_B$. The power distribution in the beam then has only a small effect on the thermal cycle. Heat reflected from the underside is also neglected, since the case depth is normally small in comparison with the sample thickness.

The heating and cooling rates are found by differentiating equation (1) with respect to time:

$$dT/dt = (T-T_o)/t \; . \; \{[(z+z_o)^2/(4at)-[(2t+t_o)/(2t+2t_o)]]\}. \tag{2}$$

An expression for the time taken to attain the peak temperature, t_p, at a depth, z, is obtained by setting equation (2) equal to zero, and solving the resulting quadratic equation

$$t_p = t_o \; / \; 4\{(z+z_o)^2/r_B^2 - 1 + [4(z+z_o)^4/r_B^4 + 12(z+z_o)^2 + 1]^{1/2}\}. \tag{3}$$

The peak temperature, T_p, may then be obtained by substituting t_p into equation (1). z_o is found by iteration, by equating the peak temperature at the surface obtained from equation (1) with that given by a standard solution for a stationary beam acting for a time, t_p: [24]

$$T\text{-}T_o = Aq \; / \; (r_B\lambda\pi^{3/2}).\tan^{-1}(4at_p/r_B^2)^{1/2}. \tag{4}$$

The average thermal properties corresponding to a temperature of 60% of the melting temperature are used, Table 3, since these have been shown to reproduce thermal cycles well in such steels and treatments [25]. A constant average value has been used for the fraction of incident energy absorbed, as described in Section (5.1).

4.1.2 Dimensionless Equations
Equations (1 and 3) may be expressed in dimensionless terms using the following groups which are derived from consideration of the temperature field:

$$q^* = Aq/[r_B\lambda(T_m-T_o)] \quad \text{(dimensionless beam power)} \quad (5)$$

$$v^* = vr_B/a \quad \text{(dimensionless beam traverse rate)} \quad (6)$$

$$T^* = (T-T_o)/(T_m-T_o) \quad \text{(dimensionless temperature rise)} \quad (7)$$

$$t^* = t/t_o \quad \text{(dimensionless time)} \quad (8)$$

$$z^* = z/r_B \quad \text{(dimensionless depth).} \quad (9)$$

Equations (1 and 3) then become:

$$T^* = (2/\pi).(q^*/v^*)/[t^*(t^*+1)]^{1/2} . \exp[-[(z^*+z_o^*)^2/t^*] \quad (10)$$

$$t_p^* = 1/4\{(z^*+z_o^*)^2 - 1 + [4(z^*+z_o^*)^4 + 12(z^*+z_o^*)^2 + 1]^{1/2}\}. \quad (11)$$

4.2. MICROSTRUCTURE AND HARDNESS OF THE TRANSFORMED ZONE

Microstructural transformations, and the case hardness developed, depend on the degree of austenitisation of the initial microstructure, the cooling rate in the transformed region, and the composition of the material. The cooling rate is expressed in terms of the time taken to cool from 800 to 500 °C, Δt; this represents the temperature interval for C-Mn steels in which austenite transforms to phases stable at room temperature (ferrite, pearlite, bainite and martensite). The influence of material composition on phase transformations during cooling is expressed in terms of a carbon equivalent, C_{eq}, which sums the contributions of the various alloying elements towards increasing the hardenability of the steel. Empirical methods are then used to calculate the hardness of the individual phases, in terms of composition and cooling rate. The average hardness is then obtained by using a weighted average.

4.2.1 Austenitisation of the Initial Microstructure
Due to the rapid nature of the thermal cycles, austenitisation takes place under conditions far from equilibrium. Pearlite transforms to austenite by dissolution of the cementite lamellae, and growth of the transformation front into regions of high carbon

concentration. The rate-controlling step is carbon migration, assumed to occur predominantly by volume diffusion in austenite, over a characteristic diffusion distance, d, taken to be half the interlamellar spacing.

Transformation of ferrite to austenite is assumed to occur by two mechanisms which act simultaneously: shrinkage of the ferrite grains, controlled by volume diffusion of carbon in austenite over a characteristic distance of half the ferrite colony width; and growth of austenite nucleated at internal ferrite grain boundaries, controlled by grain boundary diffusion of carbon. An average activation energy for grain boundary and volume diffusion has therefore been assumed.

The equilibrium A_{c1} and A_{c3} temperatures for the steel given in Table 3 have been calculated from empirical relationships, based on the nominal steel composition. Under isothermal conditions, the time taken for transformation, t_t, is given by $t_t = d^2/2D$, where D is the diffusion coefficient ($m^2 s^{-1}$) for carbon migration at the dissolution temperature. The variation of D with temperature is given by $D = D_o \exp{-(Q/RT)}$, where D_o is the frequency factor ($m^2 s^{-1}$), Q is the activation energy (J mol^{-1}) and R is the gas constant (J mol^{-1} K^{-1}). For the case of a thermal cycle $T(t)$, transformation is just complete when:

$$d^2 = 2D_o {}_{t2}\int^{t1} \exp{-(Q/(RT(t)))}dt. \tag{12}$$

The limits of the integral denote the time spent above the nominal transformation temperature. The peak temperature of a thermal cycle which satisfies equation (12) then defines the superheated transformation temperature. A_{c1} and A_{c3} temperatures are thus calculated using appropriate diffusion distances, and the thermodynamic properties given in Table 3, in order to establish transformation temperatures relevant to LTH.

4.2.2. Volume Fractions of Microstructural Phases

The products of austenite transformation on cooling are determined by the intersection of the cooling curve with the phase transformation curves of the relevant CCT (Continuous Cooling Transformation) diagram. The cooling time between 800 and 500 °C, Δt, is obtained by evaluating equation (1) for the corresponding temperatures during the cooling period. Critical cooling times can be defined which result in HAZ microstructures containing: 100% martensite ($\Delta t_m{}^{100}$); 50% martensite ($\Delta t_m{}^{50}$); 0% martensite ($\Delta t_m{}^0$); 0% ferrite ($\Delta t_f{}^0$); 0% pearlite ($\Delta t_p{}^0$); 50% bainite ($\Delta t_b{}^{50}$) and 0% bainite ($\Delta t_b{}^0$). $\Delta t_m{}^{50}$ and $\Delta t_b{}^{50}$ have a significant effect on the microstructure developed, and depend on the material composition. Relationships can be established empirically from CCT diagrams [26] constructed for welding:

$$\Delta t_m{}^{50} = \exp(11.685\, C_{eq} - 2.506) \tag{13}$$

$$\Delta t_b{}^{50} = \exp[\ln(\Delta t_b{}^0.\Delta t_f{}^0)/2] \tag{14}$$

where
$$\Delta t_b^0 = \exp(12.209 \, C_{eq} + 1.510) \tag{15}$$

and
$$\Delta t_f^0 = \exp(13.390 \, C_{eq} - 3.548). \tag{16}$$

C_{eq} is defined as [27]:

$$C_{eq} = C + Mn/6 + (Cr+Mo+V)/5 + (Cu+Ni)/15. \tag{17}$$

Element symbols refer to nominal steel composition in wt%.

Phase transformations are modelled in the manner described in [28], to give the following equations for the volume fractions of martensite, V_m, bainite, V_b, and ferrite/pearlite, V_{fp}, present in the homogeneous transformed region, in terms of the critical cooling times defined above:

$$V_m = \exp\{\ln(0.5).(\Delta t/\Delta t_m^{50})^2\} \tag{18}$$

$$V_b = \exp\{\ln(0.5).(\Delta t/\Delta t_b^{50})^2\} - V_m \tag{19}$$

$$V_{fp} = 1 - (V_m + V_b). \tag{20}$$

4.2.3. Hardness of Microstructural Phases

It is unrealistic to calculate the hardness of individual phases from first principles. Empirical relationships derived for a range of alloy steels [29] are therefore adopted in order to calculate H_m, H_b and H_{fp}, the hardness of martensite, bainite and a ferrite/pearlite mixture, respectively:

$$H_m = 127 + 949C + 27Si + 11Mn + 8Ni + 16Cr + 21 \log V' \tag{21}$$

$$H_b = -323 + 185C + 330Si + 153Mn + 65Ni + 144Cr + 191Mo$$
$$+ (89 + 53C - 55Si - 22Mn - 10Ni - 20Cr - 33Mo) \log V' \tag{22}$$

$$H_{fp} = 42 + 223C + 53Si + 30Mn + 12.6Ni + 7Cr + 19Mo$$
$$+ (10 - 19Si + 4Ni + 8Cr + 130V) \log V'. \tag{23}$$

Element symbols refer to nominal steel composition in wt%, and V' is the cooling rate at 700 °C (K/hr), related to Δt (s) by:

$$V' = 1.08 \times 10^6 / \Delta t. \tag{24}$$

Equations (21-23) were determined using steels with compositions within the following limits (wt%) [29]: $0.1 < C < 0.5$; $Si < 1$; $Mn < 2$; $Cr < 3$; $Mo < 1$; $Ni < 4$; $Cu < 0.5$; $0.01 < Al < 0.05$; $V < 0.2$; $Mn+Ni+Cr+Mo < 5$.

4.2.4. *Homogeneous Case Hardness*

The homogeneous case hardness, H_{hom}, is determined by summing the contributions from the individual phases using a simple rule of mixtures:

$$H_{hom} = H_m V_m + H_b V_b + H_{fp} V_{fp}. \tag{25}$$

5. Process Diagrams

The models described above are now used to construct diagrams which show the influence of process variables on the properties of the transformed case.

5.1. SELECTION OF PRACTICAL PROCESS PARAMETERS

Practical LTH normally involves the production of a hardened track of a specified width on a given component. The principal process variables are therefore the beam power and the beam traverse rate. Figure 4 shows a practical parameter selection diagram, based on the requirement of a given case depth, constructed for a given beam geometry and material.

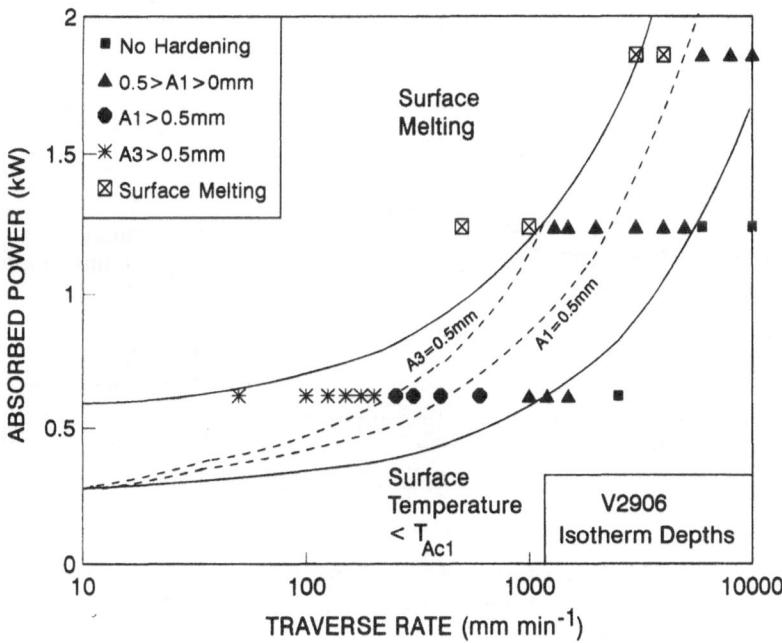

FIGURE 4. Process diagram illustrating conditions for hardening and case geometry for transformation hardening of hypoeutectoid steel. Uniform 7 x 5 mm heating pattern. Data points are experimental results; A_1 and A_3 are A_{c1} and A_{c3} isotherm depths.

The surface melting limit is found by solving equations (1) and (3) for a surface peak temperature corresponding to the material melting temperature, Table 3. The limit of hardening is also obtained by solving equations (1) and (3), using the elevated A_{c1} isotherm temperature (Section 4.2.1). These two limits define a theoretical operating region for transformation hardening, based on the surface peak temperature.

In practice a hardened case with a specified depth is required. This is obtained by solving equations (1) and (3) for the required case depth value (z). In Figure 4, the boundary corresponding to a superheated A_{c1} isotherm depth of 0.5 mm is shown, which defines a practical operating region. Within this region, the contour representing the superheated A_{c3} isotherm depth of 0.5 mm bounds the region of uniform hardening.

Experimental data of hardening depth (Table 4) are plotted in Fig. 4 by assuming a value for A of 0.62. This is obtained from a representative data point at which the case depth and process variables are known; it calibrates the model to the particular processing condition. Data are seen to fall in appropriate regions of the plot.

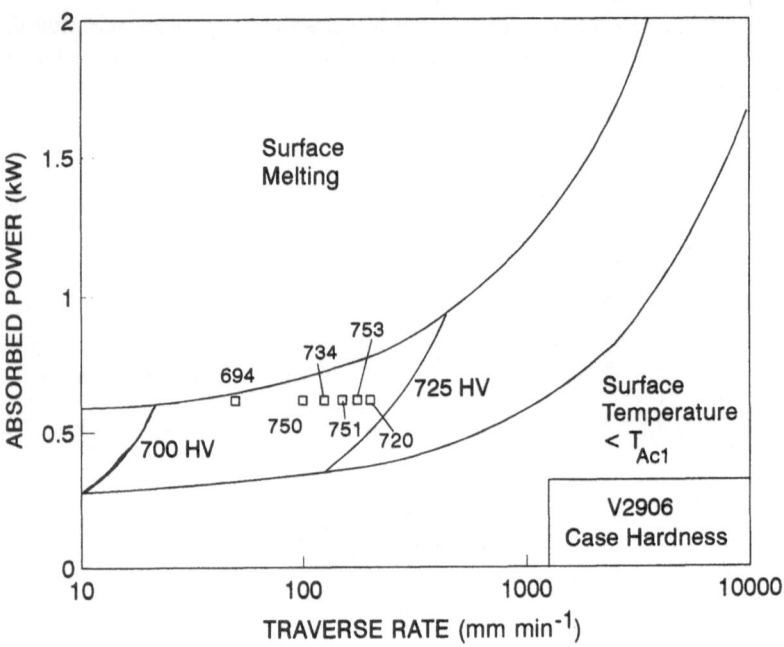

FIGURE 5. Process diagram illustrating homogeneous case hardness for transformation hardening of hypoeutectoid steel. Uniform 7 x 5 mm heating pattern.

Figure 5 shows a similar plot to Fig. 4, in which contours of constant hardness are shown, constructed from the models described. Experimental data, plotted by assuming a value for A of 0.62, are seen to correspond well with the model predictions.

5.2. DIMENSIONLESS PROCESS VARIABLES

It may be necessary to assess the influence of changes in a large number of process variables on the properties of the transformed case. Figure 6 shows a plot in which the dimensionless power and traverse rate are used as axes, and presents the conditions for hardening with a given case geometry. The theoretical limits on process variables for successful hardening (solid lines), described in Section 5.1, are constructed by solving equations (10) and (11) for an appropriate surface peak temperature. For simplicity, a constant average value of $T_p^* = 0.6$ has been used for the normalised superheated A_{c1} isotherm temperature. $T_p^* = 1$ denotes the normalised melting temperature. By solving equations (10) and (11) for the normalised case depth, l^*, at which $T_p^* = 0.6$, normalised case depth contours can be plotted for practical ranges of q^* and v^*.

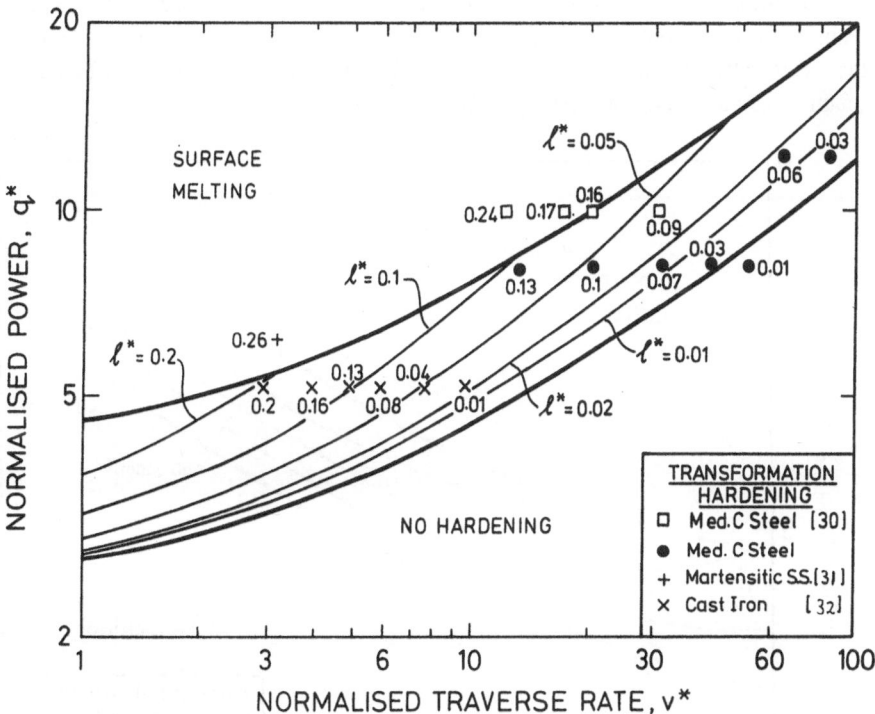

FIGURE 6. Dimensionless process diagram showing model predictions of the onset of surface hardening and melting, with contours and experimental data for normalised case depth.

A selection of experimental data determined here is plotted in Fig. 6 (●), using the value of $A = 0.62$. Additional published data obtained from a range of ferrous alloys are also included, plotted using an average value of $A = 0.5$ in the absence of a more exact value. Data are seen to fall in appropriate regions of the chart.

5.3. OPTIMISING THE USE OF LASER ENERGY

In a previous empirical processing diagram [33], contours of constant energy density ($J\,m^{-2}$) were plotted, there being a simple relationship between this quantity and the axes of the plot - power density ($W\,cm^{-2}$) and beam interaction time (s). In practice the amount of laser energy used per unit *volume* of transformed case may be an important production criterion. The models allow this to be determined in terms of the process variables.

For a case depth, l, the volume of material raised above the A_{c1} temperature during transformation hardening is approximately $2lr_B$ per unit length of track. The energy used per unit volume is then $q/(2vlr_B)$, which is proportional to $q^*/(v^*l^*)$. Hence contours of $q^*/(v^*l^*)$ give equally energy-efficient treatments. These contours may be constructed from the model using the intercepts of contours of q^*/v^* and l^*, Fig. 7. The figure shows that for optimum use of energy, hardening should be carried out with the highest combination of power and speed which will give the required case depth. However, other restraints, such as the need to homogenise the microstructure, must be taken into account when selecting process variables.

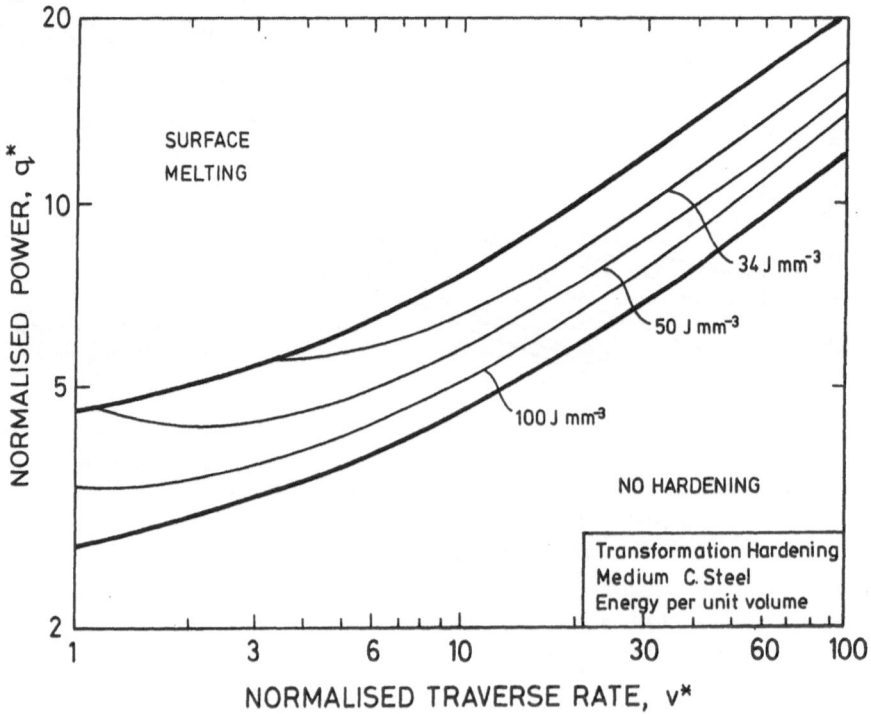

FIGURE 7. Process diagram showing hardening limits and contours of constant energy per unit volume of hardened case.

5.4. MAXIMISING THE COVERAGE RATE

From a production point of view, it may be necessary to know which combination of q and r_B maximises the area coverage rate, for a particular case depth, l. For a specified l, the contours of l^* become contours of r_B, and contours of constant q may be constructed for a given material, using the r_B contours and the q^* axis. An operating window may then be defined by the maximum power available, and the region of the diagram for hardening - see Fig. 8. To maximise the coverage rate, the maximum value of v^* is sought - the diagram then indicates the influence of beam size on this criterion at any power up to the maximum available.

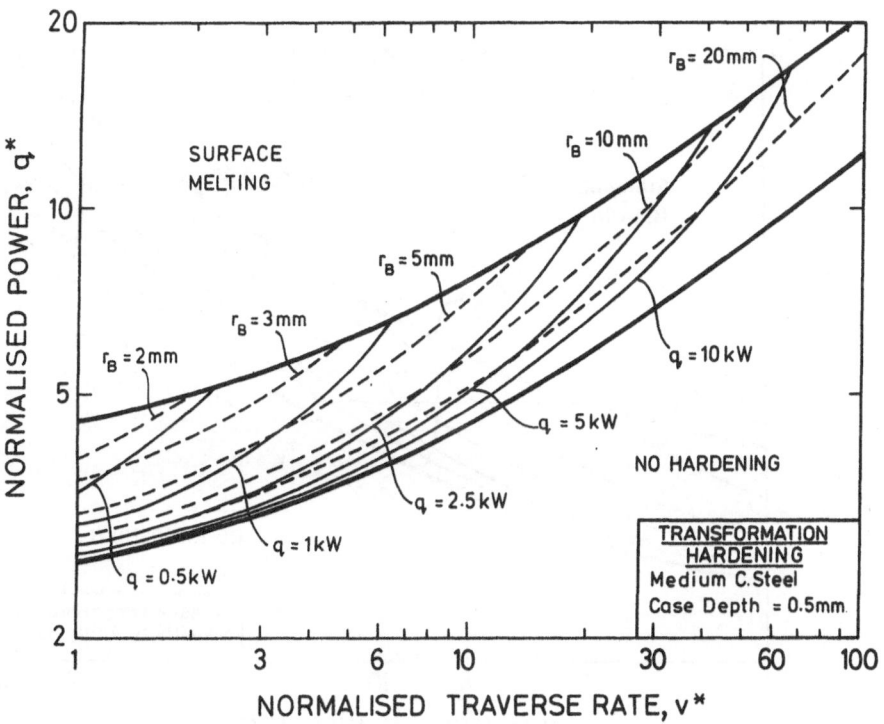

FIGURE 8. Process diagram showing limits of hardening, with contours of power (solid lines) and beam size (broken lines) required to produce a hardened case 0.5 mm deep in a medium C steel.

5.5. SENSITIVITY ANALYSIS

It is likely that much LTH is carried out under conditions far from optimum. One reason is that at the limit of efficiency the process may become sensitive to fluctuations in process variables, e.g. a 10% reduction in power. The model allows a sensitivity analysis of the influence of changes in process variables to be conducted. Figure 9 shows a processing diagram, with contours of normalised case depth, with a horizontal band representing a nominal beam power and a value 10% lower. Note that on log scales this band is of constant width at all values of q^*. By comparing values of l^* at a given v^* at these two power levels, it can be seen that the influence on l^* is stronger at higher values of v^*. For example, at $v^*=7$, a 10% reduction in power causes a reduction in case depth of 30%, whereas at $v^*=15$, the effect is a reduction of 75%. In order to maintain a case depth of e.g. 0.5 ± 0.1mm, an operator may therefore have to select a lower traverse rate than is desirable, solely on the grounds of sensitivity.

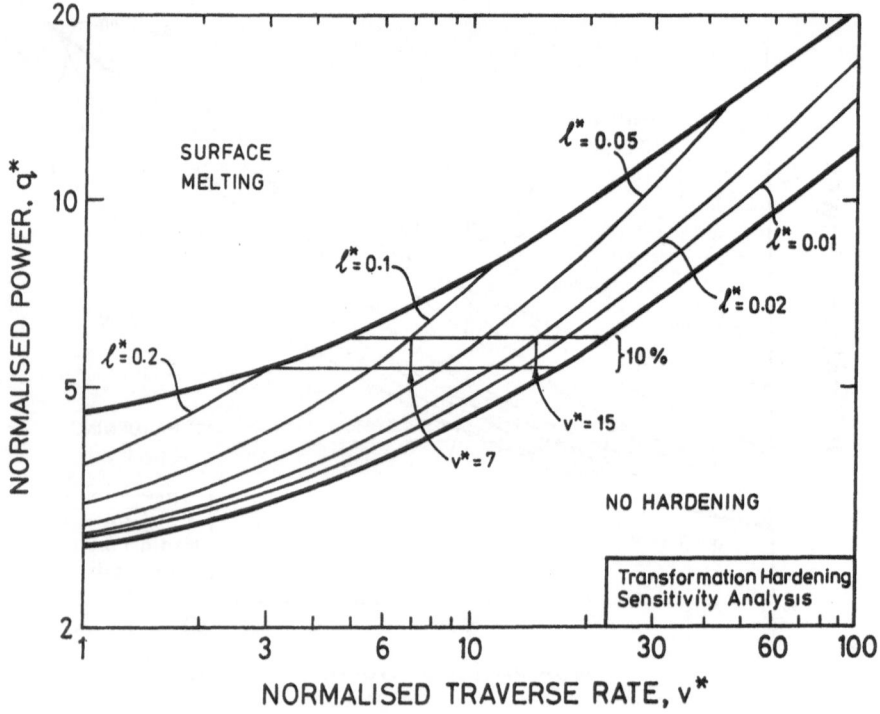

FIGURE 9. Process diagram showing the limits of hardening, with contours of normalised case depth, and a band representing the effect of a 10% reduction in power.

5.6. DESIGN AND PROCESS ECONOMICS

An obstacle to the uptake of any new process is the lack of information concerning processing costs, and the implications of the process on component design. Here a simple comparison is made between the costs of a conventional bulk heat treatment and LTH.

The cost of a bulk heat treatment is proportional to the weight of the component, since the cost is based on the time taken to heat the entire component. In contrast, the cost of LTH depends on the surface area to be treated. By equating these costs for a particular component, we obtain:

$$C_b W_c = C_l f A_c$$

$$\therefore \quad C_b \rho V_c = C_l f A_c$$

where: C_b = bulk treatment cost ($ kg^{-1}); C_l = LTH cost ($ m^{-2}); W_c = weight of component (kg); ρ = density (kg m^{-3}); V_c = component volume (m^3); A_c = component surface area (m^2); f = fraction of surface area to be hardened. Rearrangement gives:

$$f = (C_b/C_l).\rho.(V_c/A_c). \tag{26}$$

Consider a cylindrical shaft with a size characterised by its radius, r. Let the length of the shaft be $10r$. Suppose that we wish to know what fraction of the curved surface of the shaft can be hardened by LTH before a bulk heat treatment becomes more economical. Let the cost of the bulk treatment be $1 kg^{-1}, and the cost of LTH be $500 m^{-2}. Equation (26) is plotted in Fig. 10. The region above the line indicates the conditions for LTH to be more economical than the bulk treatment. It can be seen that LTH is more economical for hardening small areas of large components.

The above is a simplified model of processing economics, but it contains all the relevant variables, and can therefore be applied to a variety of component geometries, and allows variations in processing costs to be included. In practice, additional factors such as material selection and post-treatment finishing influence the overall economics.

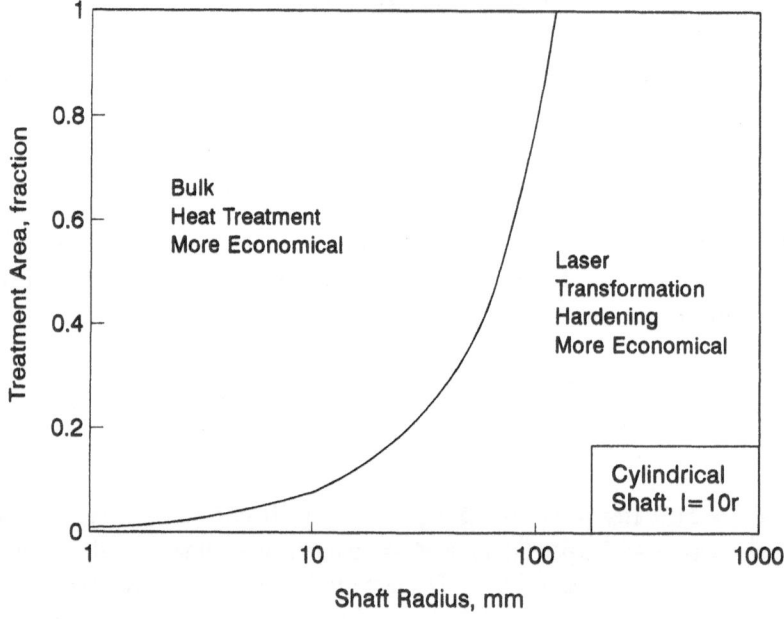

FIGURE 10. Diagram indicating relative costs of LTH and a bulk heat treatment for hardening a given fraction of the surface area of a cylindrical shaft.

6. Future Development

The models described are currently being incorporated into a PC-based computer package, which runs under the Microsoft Windows™ (trademark of Microsoft Corporation) graphical environment. The package contains a standardised record of process parameters, and allows the results of LTH to be simulated, providing information on the case depth and hardness obtained with a given material and a given set of laser parameters. Procedures can thus be developed more rapidly, by eliminating unsatisfactory material and parameter combinations at an early stage of the development process. Routines for process economics will also be included. The program contributes to increased quality assurance, and can be used by engineers to optimise process parameters, educationalists designers with a limited knowledge of the possibilities of LTH.

7. Summary

1.   Simple analytical models can be used to predict the case depth and hardness produced by laser transformation hardening, provided a suitable calibration is made to determine the principal unknown in the process - the fraction of incident energy absorbed. Since the models have a physical basis, they can be used to predict the effects of variations in both material properties and laser beam parameters over a relatively wide range.

2.   The models described provide a framework in which both predictions and existing experimental data can be displayed. In their most practical form, the diagrams can be used in a processing environment to select candidate materials and the process variables most easily controlled. When constructed in terms of dimensionless groups, the effect of variations in a larger number of process variables can be considered.

3.   Since the methods used are analytical, results can be obtained quickly, using a desktop PC. Use at the design stage for preliminary material and parameter selection, expedition of qualification testing procedures, inclusion in on-line adaptive control systems, and as an educational tool are all potential application areas.

Acknowledgements

The models were developed in collaboration with Prof. M.F. Ashby and Dr. H.R. Shercliff, Department of Engineering, University of Cambridge. Financial support from the Nordic Industrial Fund is gratefully acknowledged.

References

1.   Sandven, O. (1981) Laser surface transformation hardening, in *Metals Handbook*, 9th edn., vol. 4, American Society for Metals, Ohio, pp. 507-517.

2.   Miller, J.E. and Wineman, J.A. (1977) Laser hardening at Saginaw steering gear, *Metal Progress* **111**, 38-43.

3.   Creal, R. (1980) Research institute cuts risks in shopping for lasers, *Heat Treating*, **12(12)**, 20-22.

4.   Yessik, M. and Schmatz, D.J. (1975) Laser processing at Ford, *Metal Progress* **107(5)**, 61-66.

5.   Ready, J.F. (1978) *Industrial Applications of Lasers*, Academic Press, New York.

6.   Seaman, F.D. and Gnanamuthu, D.S. (1975) Using the industrial laser to surface harden and alloy, *Metal Progress* **108(8)**, 67-74.

7.   Obrzut J.J. (1978) *Iron Age* July 10 39-42.

8.   Bellis, J. (1980) Laser processing of metals, in *Lasers: Operation, Equipment, Application and Design*, McGraw-Hill, New York, pp. 124-135.

9.   Stanford, K. (1980) Lasers in metal surface modification, *Metallurgia* **47(3)**, 109-116.

10.   Creal, R. (1980) Laser specialist discovers boom market in heat treating, *Heat Treating*, **12(12)** 24-27.

11.   Bransden, A.S., Gazzard, S.T., Inwood, B.C., and Megaw, J.H.P.C. (1986) Laser hardening of ring grooves in medium speed diesel engine pistons, *Surface Engineering* **2** 107-113.

12.   Asaka, Y., Kobayashi, H. and Arita, S. (1987) Laser heat treatment of piston ring groove, in *Proceedings of Laser Advanced Materials Processing (LAMP '87)*, pp. 555-560.

13.   Amende, W. (1989) Oberflächenbehandlung von Werkzeugen und Motorkomponenten mit dem Laser, *VDI-Z*, **131(6)**, 80-83.

14.   Anon (1978) Laser treating of cylinder liners for diesel engines to increase wear resistance, *Industrial Heating* **45(9)**, 38-39.

15.   Camoletto, A., Molino, G., and Talentino, S. (1991) *Materials and Manufacturing Processes*, **6(1)**, 53-65.

16.   Edwards, A.L. (1969) For computer heat conduction equations - a compilation of thermal properties data", Report UCRL-50589, University of California Research Laboratory, February 1969.

17.   Smithells, C.M. (1976) *Metals Reference Book*, 5th edn., Butterworths, London.

18.   Andrews, K.W. (1965) Empirical formulae for the calculation of some transformation temperatures, *Journal of the Iron and Steel Institute* **203**, 721-727.

19.   Mrowec, S. (1980) *Defects and diffusion in solids - an introduction*, Elsevier, Amsterdam.

20.   Brown, A.M. and Ashby, M.F. (1980) Correlations for Diffusion Constants, *Acta Metallurgica* **28**, 1085-1101.

21.   Ion, J.C. (1992) Modeling of laser material processing, in D. Belforte and M. Levitt (eds.), *The Industrial Laser Handbook*, Springer-Verlag, New York, pp. 39-47.

22.   Ashby, M.F. and Easterling, K.E. (1984) The transformation hardening of steel surfaces by laser beams - I. hypoeutectoid steels, *Acta Metallurgica* **32**, 1935-1948.

23.   Rykalin, N., Uglov, A., and Kokora, A. (1978) *Laser Machining and Welding*, Chapter 3, Pergamon Press, Oxford.

24.   Bass, M. (1983) Laser heating of solids, in M. Bertolotti (ed.), *Physical Processes in Laser-Materials Interactions*, Plenum Press, New York, pp. 77-115.

25.   Kou, S., Sun, D.K., and Le, Y.P (1983) A fundamental study of laser transformation hardening, *Metallurgical Transactions* **14A**, 643-653.

26.   Inagaki, M. and Sekiguchi, H. (1960) Continuous cooling transformation diagrams of steels for welding and their applications, *Transactions of National Research Institute for Metals* **2**, 102-125.

27.   IIW Document IX-535-67, IXG-13067, Technical Report, International Institute of Welding, 1967.

28.   Ion, J.C., Easterling, K.E., and Ashby, M.F. (1984) A second report on diagrams of microstructure and hardness for heat-affected zones in welds, *Acta Metallurgica* **32**, 1949-1962.

29.   Blondeau, R., Maynier, Ph., Dollet, J. and Viellard-Baron, B. (1976) Mathematical model for the calculation of mechanical properties of low-alloy steel products: a few examples of its applications", in *Heat Treatment '76*, The Metals Society, London, pp. 189-200.

30.   Bradley, J.R. and Kim, S. (1988) Laser transformation hardening of iron-carbon and iron-carbon-chromium steels, *Metallurgical Transactions* **19A**, 2013-2025.

31.   Ion, J.C., Moisio, T.J.I., Pedersen, T.F, Sørensen, B., and Hansson, C.M. (1991) Laser surface modification of a 13.5%Cr, 0.6%C steel, *Journal of Materials Science* **26**, 43-48.

32.   Trafford, D.N.H., Bell, T., Megaw, J.H.P.C., and Bransden, A.S. (1983) Laser treatment of grey iron, *Metals Technology* **10**, 69-77.

33.   Breinan, E.M., Kear, B.H., and Banas, C.M. (1976) Processing materials with lasers, *Physics Today* **29**, 44-50.

APPENDIX 1

Advantages of laser transformation hardening over traditional hardening methods

---

1   Higher productivity through:
  a) reduction in process cycle time
  b) reduction in use of materials
  c) use of cheaper details
  d) reduction in labour costs
  e) reduction in equipment maintenance costs
  f) reduction in scrap
  g) reduction in post fabrication treatments
  h) reduction in response time to orders
  i) high equipment availability time
  j) reduced energy costs

2.   Higher product quality through:
  a) improved service properties
  b) improved tolerances
  c) accurate control of process parameters
  d) selection of new materials
  e) product redesign

3.   Opportunity to manufacture due to:
  a) nature of laser processing (non-contact, clean etc)
  b) ability to process conventionally non-hardenable materials
  c) ability to treat large components
  d) precise treatment of small components
  e) treatment of selected areas
  f) accessibility of laser beam to concealed locations

4.   Increased flexibility through:
  a) adaptability to automation and flexible manufacturing
  b) wide range of processing options
  c) rapid switching between different treatment geometries
  d) processing of a wide range of shapes
  e) ease of beam shaping

APPENDIX 2

Notation

A	fraction of incident beam power absorbed (-)
A_c	component area (m^2)
A_1	A_{c1} temperature of a ferrous alloy (K)
A_3	A_{c3} temperature of a ferrous alloy (K)
C_b	bulk treatment cost ($\$$ kg^{-1})
C_{eq}	carbon equivalent (wt%)
C_l	cost of laser transformation hardening ($\$$ m^{-2})
D_o	diffusion frequency factor (m^2 s^{-1})
H_b	hardness of bainite (HV)
H_{fp}	hardness of ferrite/pearlite (HV)
H_m	hardness of martensite (HV)
H_{hom}	hardness of homogeneous transformed region (HV)
Q	activation energy (J mol^{-1})
R	gas constant (J mol^{-1} K^{-1})
T	temperature (K)
T^*	normalised temperature (-)
T_o	initial temperature (K)
T_p	peak temperature (K)
T_p^*	normalised peak temperature (-)
T_m	melting temperature (K)
V'	cooling rate at 700 °C (K s^{-1})
V_b	volume fraction of bainite (-)
V_c	component volume (m^3)
V_{fp}	volume fraction of ferrite/pearlite (-)
V_m	volume fraction of martensite (-)
a	thermal diffusivity (m^2 s^{-1})
c	specific heat capacity (J kg^{-1} K^{-1})
d	characteristic diffusion distance (m)
e	base of natural logarithms
f	fraction of surface area to be treated (-)
l	depth of treatment (m)
l^*	normalised depth of treatment (-)
$q^{\cdot}$	beam power (J s^{-1})
q^*	normalised beam power (-)
r_B	beam radius or half-width (m)
t	time (s)
t^*	normalised time (-)
t_o	characteristic heat transfer time (s)
t_p	time to attain peak temperature (s)
$t_p^{\cdot *}$	normalised time to attain peak temperature (-)
t_l	time for completion of a diffusion-controlled transformation (s)
Δt	cooling time from 800 to 500 °C (s)
$\Delta t_b^{\,0}$	Δt for 0% bainite formation (s)
$\Delta t_b^{\,50}$	Δt for 50% bainite formation (s)
$\Delta t_f^{\,0}$	Δt for 0% ferrite formation (s)
$\Delta t_m^{\,0}$	Δt for 0% martensite formation (s)

Δt_m^{50}	Δt for 50% martensite formation (s)
Δt_m^{100}	Δt for 100% martensite formation (s)
Δt_p^{0}	Δt for 0% pearlite formation (s)
v	beam traverse rate (m s^{-1})
v^*	normalised beam traverse rate (-)
z	depth (m)
z_o	characteristic heat flow length (m)
z^*	normalised depth (-)
z_o^*	normalised characteristic heat flow length (-)
λ	thermal conductivity (J s^{-1} m^{-1} K^{-1})
ρ	density (kg m^{-3})

THE USE OF LASERS IN THE FORMATION OF VITREOUS LAYERS ON SURFACES.

K.J. BLAIR[1], J.T.SPENCER[2], W.M. STEEN[1].

[1]*The Laser Group, Department of Mechanical Engineering, University of Liverpool, P.O. Box 147, Liverpool, L69 3BX. UK.*
[2]*Company Research Laboratory, British Nuclear Fuels plc, Springfield Works, Salwick, Preston, PR4 0XJ. UK.*

Summary.

The action of a laser beam on a cement-based material, such as concrete, can provide the necessary energy to convert the cement into a glass. The concrete surface is non-uniform, owing to the presence of aggregate pieces. Consequently, a non-uniform glass results. If an additional, uniform cement layer is applied to the surface prior to laser processing, a uniform vitreous layer can be formed on the concrete surface.

This technique can be applied to the "fixing" of hazardous materials, such as radioactive dust, to stable substrates. The material is immobilised, reducing the danger of inhalation or ingestion. A similar cement coating can be applied to other building materials, such as steel and brick, resulting again in a uniform, vitreous sealing layer.

The heating effect of the laser in forming this glass can have a detrimental effect on the mechanical integrity of the coating and substrate. Direct glazing of concrete or the application of a thin coating layer shows that dehydration can occur, which destroys bonds and thus reduces mechanical strength. To avoid this, an insulating layer can be applied beneath the vitrifiable layer, in order to thermally insulate the substrate from excessive temperature rises. Several interfaces are therefore formed within the system, dependent both upon the temperature reached, and the specific material. The heat-affected zone (HAZ) provides the "weak link" in the coated material, and generally mechanical failure occurs within it.

The substrate/coating system can be characterised in terms of these interfaces and the temperatures attained. The subsequent performance of the system is dictated by the nature of the materials and processing applied.

J. Mazumder et al. (eds.), Laser Processing: Surface Treatment and Film Deposition, 613–628.
© 1996 *Kluwer Academic Publishers.*

The effect of various processing parameters (e.g. laser type, power density, sample traverse speed) has been investigated. The effect of incorrect processing is mechanical failure of the coating or substrate, resulting in cracking and spalling of the material. Microscopic inspection and mechanical pull-off tests have been used to determine the optimum processing conditions.

1. Introduction & Background.

The decommissioning of a nuclear reprocessing plant is a carefully planned exercise which involves several stages and much careful assessment of risk. After all the pressure vessels are emptied and have been removed, the problem remains of what to do with the building structure. This stage of the decommissioning involves the safe long-term management and ultimate decontamination of any radioactive areas.

It is possible that the buildings themselves are contaminated by a large variety of radionuclides, e.g. ^{60}Co, ^{90}Sr, ^{137}Cs, PuO_2 and UO_2. The nature of the species present depends on the history of the particular area. These materials could be on the surface in the form of dust, which can easily be removed, but it is possible they could have penetrated up to 120mm into the surface [1]. This cannot be easily removed.

This paper presents work done to investigate the possibility of laser glazing the surface of the concrete to form an impermeable layer. This layer will "tie-down" the particulate contamination, and prevent water leaching of such material to other sites. The coating does not act as a biological shield to prevent the escape of radiation, but only to limit the movement of the radioactive particles. The effect of a laser is to provide a heat source which will melt the surface of the material and thus achieve this "tie-down".

Concrete is the most widely used building material within the nuclear power industry. It is therefore important to have an understanding of the structure and behaviour of this material under the conditions it will suffer during laser processing. The nature of the material poses several problems relating to its behaviour at high temperatures. It is a composite material consisting of an array of some type of aggregate pieces embedded within a cementitious matrix. The particular substrate material used for these experiments is an Ordinary Portland Cement (OPC) concrete. It consists of limestone aggregates ranging in size from 5 to 20 mm. These aggregate pieces are embedded in a Portland cement matrix. The cement is not a homogeneous array, but contains pores, air bubbles and unhydrated cement particles. A

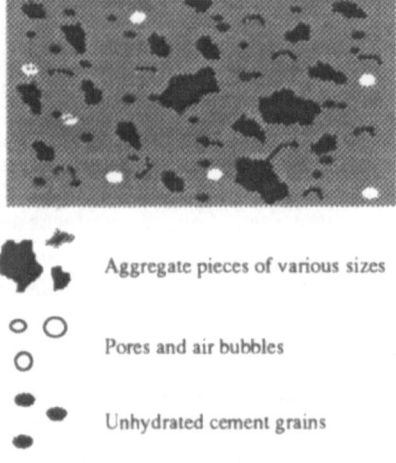

Aggregate pieces of various sizes

Pores and air bubbles

Unhydrated cement grains

Figure 1. **Schematic Representation of Concrete Structure.**

schematic diagram of a section of concrete is shown in *Figure 1*.

Concrete contains silicates which, on heating, will form an amorphous, glassy layer. This will prevent the diffusion of particles out of the concrete and into the surrounding atmosphere.

Concrete gains its strength by the setting and hardening of the cement matrix. The chemistry of this process is the same for cements, grouts or composites made from OPC. The reactions which cause cement to harden and set are *exothermic hydration* reactions. The compound composition of a cement and the manner in which each of its phases reacts with water determine its physical and chemical properties. Since the reactions which take place are hydration reactions, the reverse reaction, i.e. *dehydration* will occur with the application of heat. This can be shown when a relatively low power laser beam is incident upon the concrete surface.

Section 2 describes the effect of laser interaction with bare concrete. The problems which occur can be overcome by using a coating material. Section 3 describes the development of such a coating, and the interaction of it with the laser beam.

2. Laser Interaction with Bare Concrete.[2]

2.1. EXPERIMENTAL WORK.

Initial experiments were performed to assess the effect of direct laser action upon the bare surface of the concrete substrate. *Figure 2* shows schematically the experimental set up. Several important process parameters were varied, including laser power, traverse speed of the specimen, etc. Both CO_2 and Nd:YAG lasers were used to process the material. The quality of the glaze was judged by surface appearance and the strength of the bond between the glazed layer and the unaffected concrete. This bond strength was measured using a proprietry adhesion tester as shown in *Figure 3*.

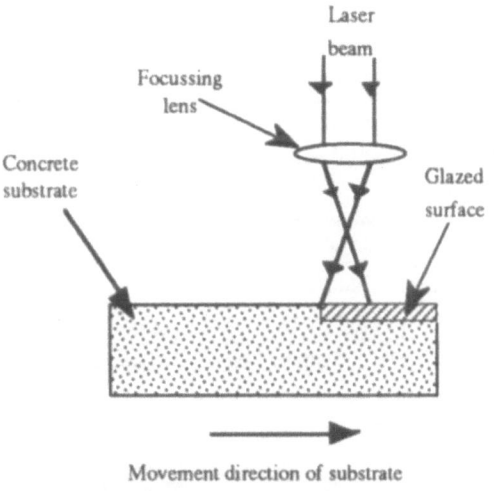

Figure 2. Schematic Representation of Laser Glazing of Bare Concrete.

2.2. RESULTS.

2.2.1. Surface Quality

616

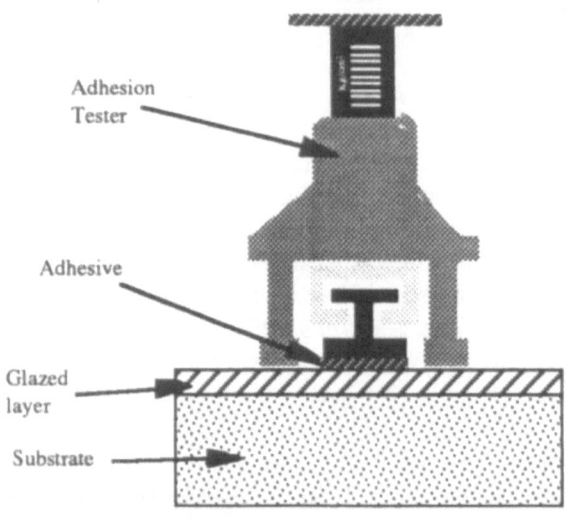

Figure 3. Testing Pull-off Strength using *Elcometer* Adhesion Tester.

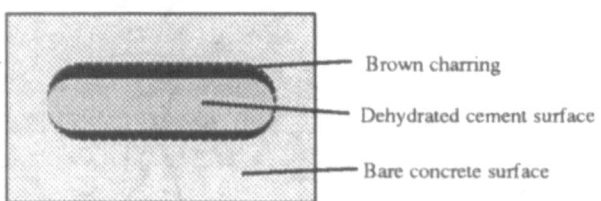

Brown charring

Dehydrated cement surface

Bare concrete surface

Figure 4a. Schematic Diagram to Show Heat Affected Zone at Low Powers.

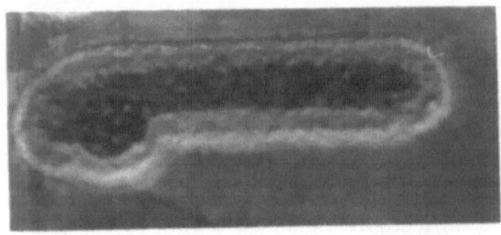

Figure 4b. Photograph to Show Dehydration in HAZ.

With relatively low power density(<200 Wcm^{-2}) and low traverse speed (<3 mms^{-1}) a continuous vitreous (glassy) surface coating was achieved. XRD experiments show this layer to have no crystalline character. This is due to the presence of silicon dioxide (SiO_2) in the concrete matrix which, on heating produces the amorphous layer observed. This glassy layer, however, was cracked with poor adhesion to the surface. For the same power density, there is no principle difference between the quality of processing by CO_2 and Nd:YAG. However, the shorter wavelength YAG beam does have lower absorbance than the CO_2.

The use of higher powers or lower traverse speeds resulted in fracturing of the surface, with ejection of pieces of material. Conversely lower powers or higher traverse speeds resulted in no glazing, but solely dehydration of the surface. This can be observed as a colour change from dark to light grey, with some brown charring around the edge of the affected area. *Figure 4a* shows this effect schematically, for the sample shown in *Figure 4b*. The depth to which the concrete is heat affected is proportional to the laser power, and inversely proportional to the traverse speed. *Figures 5a and 5b* show the effect of various process parameters on the concrete surface. The effect of such variation is discussed in more detail in Section 2.3.

Figure5a. Surface of Concrete after Processing at 1mm/s and 200W.

Figure 5b. Surface of Concrete after Processing at 1mm/s and 500W.

Figure 6a. Surface of As-cast Concrete after Processing at 2mm/s and 150W.

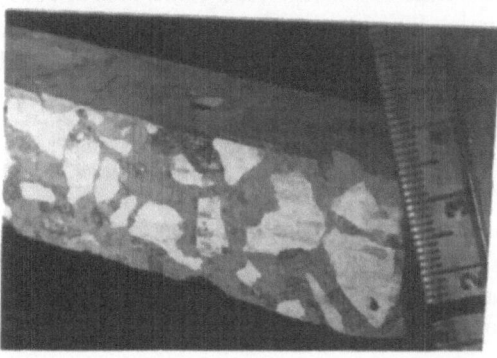

Figure 6b. Surface of Cut Concrete after Processing at 2mm/s and 150W.

The specimens used in this work were sectioned from a larger, cast concrete blocks. The cut surfaces had exposed pieces of limestone aggregate, whereas the as-cast faces were continuous cement. The aggregate in the latter was therefore covered by a thin layer (a few mm) of cement. The behaviour of the two types of surface differed in that the cement forms the glassy layer, and the aggregate pieces heat up, fracture and pieces are ejected from the concrete at considerable speeds. *Figure 6a and b* show the contrast between the different substrate surfaces using the same process parameters. The surfaces with exposed aggregates react more violently than do the as-cast, since the aggregates are not protected by any cement. If higher powers (or lower traverse speeds) are applied to as-cast surfaces, their behaviour becomes like that of the cut surface, i.e. violent ejection of pieces of aggregate from just below the cement protection.

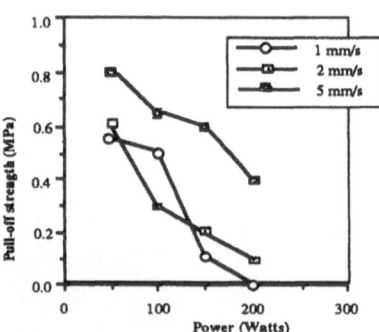

Figure 7 . **Pull-off Strengths for Bare Concrete.**

2.2.2. Pull-off Testing

Pull-off test data could only be obtained for some of the more successful samples, since on the remainder, the surface was so poorly attached, that it proved impossible to achieve a measurement of its pull-off strength. The most successful samples were those processed at low speeds and powers (<200 Wcm^{-2}, <3 mms^{-1}). Not surprisingly, these samples also had the best surface condition. *Figure 7* shows the pull-off strengths for various process parameters.

2.3. HEAT AFFECTED ZONE.

2.3.1. Cement heating

The heat affected zone (HAZ) includes both the glazed area, and the *dehydrated* material below it. *Table 1* shows the effect of increased temperature on the cement matrix. Temperatures near the surface of the concrete are in excess of 1000°C, decreasing, with depth, through the material. Since the traverse speed required to glaze the material is slow, the temperature within the bulk material can rise enough for it to be dehydrated as in equation 1.

$$Ca(OH)_2 \quad \longrightarrow \quad CaO + H_2O \tag{1}$$

This dehydration of free lime {Ca(OH)2} in the cement matrix occurs at about 500°C, and therefore a significant depth within the material is affected. Thus the HAZ extends below the glassy layer into the bulk of the concrete. The increase in temperature in this area results in dehydration, and consequently, a loose interface within the substrate.

The mass of material (i.e. H2O and CO2lost during this process has been monitored and is found to increase with laser power, and decrease with traverse speed. *Figures 8a and b* show

Table 1. **Effect of Increasing Temperature on Cement Matrix.**

Constituent	Effect	Temperature
Ettringite	Water ↑	100 -150 ° c
C-S-H	Water ↑	150 - 450 ° c
Ca(OH)2	Water ↑	450 - 550 ° c
CaCO3	CO2 ↑	700 ° c

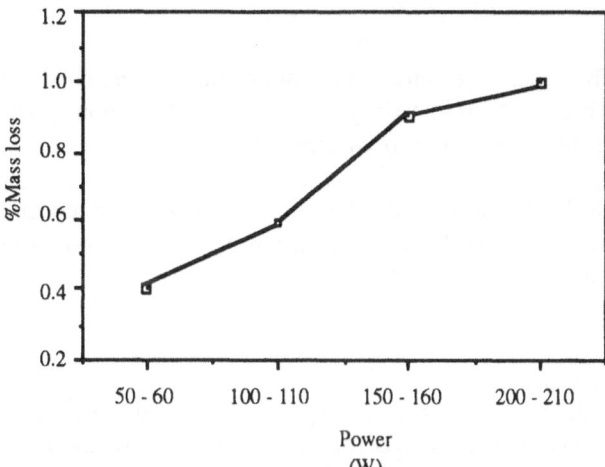

Figure 8b - Graph to Show Increase in Water Loss with Increasing Power.

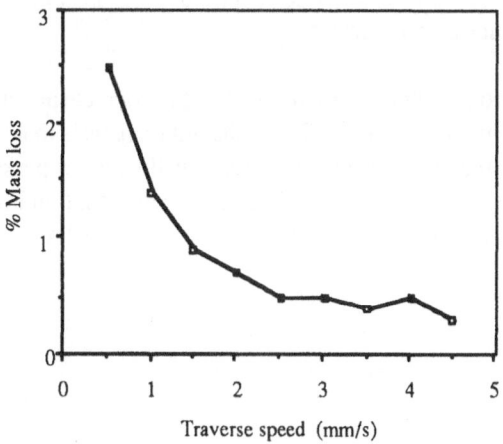

Figure 8 **a. Graph to Show Decrease in Water Loss With Increasing Traverse Speed.**

the change in mass with power and traverse speed. The samples are immersed in water after glazing and being allowed to cool. It is found that water is absorbed by the samples until they almost regain their preprocessing mass. This implies that the material lost during glazing is primarily water.

The CaO formed, exists as a loose powder and capillaries open up in the structure due to the dehydration reaction. This results in a mechanically weak area. Thermal gradients are set up in the surface of the material, which also contribute to this loss of structural integrity. Mechanical failure appears to occur at the limit of the HAZ, wherever this dehydration has occurred. The limit of dehydration is at the *thermal interface*.

The dehydration reaction which $Ca(OH)_2$ undergoes results in a change in pH. This change is an indication of the extent of the HAZ. Phenolphthalein is an acid-base indicator which is colourless in acid, and bright pink in alkali. Thus, when applied to the surface of the material, the position of the HAZ can be estimated, by the extent of the pink colour. Such experiments have shown that the pH change can occur up to 10 mm below the surface of the concrete. The accuracy of this method of determining the HAZ depth is yet to be confirmed, but it does give an indication of the extent of dehydration, and hence the amount of protection the substrate requires from excessive heating.

The extent of the HAZ can be controlled to some degree, by reducing the power or increasing the processing speed. However, the limiting factor is the time taken to melt and vitrify the surface layer. A balance between excessive heating and poor glazing has to be reached.

2.3.2. Aggregate behaviour

The behaviour of the aggregate pieces is less predictable than that of the cement matrix. The aggregate may be one of many types of rock, but in this case it is limestone. The behaviour of this rock under high temperatures is such that any moisture held in the structure rapidly vaporises and expands, resulting in ejection of rock fragments. Limestone does not contain any vitrifiable materials and therefore will not glaze under the laser action. Thus, direct glazing of the concrete surface is ineffective because the substrate is inhomogeneous, which results in an inhomogeneous surface.

As discussed previously, the as-cast surfaces of the concrete are able to withstand greater power densities than the cut faces. This behaviour suggests that the thin cement layer is protecting the substrate to some degree. Thus, vitrification is possible if the substrate can be protected. The next section describes the development and behaviour of a suitable coating to both protect the substrate and formed a vitreous layer.

3. Laser Interaction with Coated Materials.[3]

3.1. MATERIALS SELECTION.

Glazing of bare concrete has shown that cement contains certain constituents that form a glass when heated. Therefore, cement-based materials were used to coat the concrete surface. Such a material is similar in nature to the substrate and has comparable thermal and mechanical properties. It will also bond well to concrete, forming a bond that is stronger than the bonding within the bulk coating material. Thus, the purpose of applying a coating can be summarised as:

(i)   The coating must act as a thermal insulator and protect the concrete from excessive temperature rises which would result in dehydration and fracture.
(ii)   The coating must provide a source of vitrifiable material which can be glazed effectively during laser processing.

The material selected to coat the concrete substrate must meet the following criteria:
(i)   A good bond must be formed between the coating and the substrate with no cracking, spalling etc.
(ii)   The coating must be long-lived and not deteriorate with time.
(iii)   The coating must be chemically resistant to any agents likely to be present in its proposed environment.
(iv)   The melting point of the coating constituents must be less than that of any of the contaminants.

Additional factors which must be borne in mind concerning compatibility with substrate are; melting point, thermal expansion coefficient, degree of thermal conductivity.

Initially, individual materials were tested both for their bonding ability to concrete, and the effect of laser action. The results of these tests are shown in *Table 2* The results of the pozzolan and chamotte experiments had strong factors in their favour: Chamotte produces a coating with very few cracks and porosities, but it is crystalline and thus susceptible to chemical or moisture attack. The pozzolan is amorphous, but it is brittle and full of cracks. A combination of these two materials in a suitable ratio should optimise the beneficial properties and reduce the detrimental properties of the two. This was the next step in the development of the coating.

Various ratios of pozzolan and chamotte were mixed in order to find the optimum combination. The quality was judged on adherence to the substrate and ability to form a continuous glassy coating. It was found that the ratio which gave the best glass surface differed from the ratio to provide the best insulation of the substrate. Therefore, a two layer coating was applied, with differing ratios of PPC to chamotte. The lower layer has a higher

Table 2 - Results of Individual Materials Used as Coatings.

MATERIAL	COMMENTS	COATING RESULTS	LASER ACTION
POZZOLANA	Volcanic dust consisting mainly of Al_2O_3 & SiO_2	Water-glass necessary as a setting agent	Produces a Dark glassy surface. Vary brittle. Full of cracks. Expands a lot on heating. Amorphous structure
PORTLAND CEMENT		Dependent on curing methods	Needs to be mixed 1:3 with sand to prevent cracking
POZZOLAN	aka TRASS CEMENT or POZZOLANIC PORTLAND CEMENT (PPC)	Needs wet curing to set. Will not set with water-glass	as for pozzolana
CHAMOTTE	Consists mainly of Al_2O_3, TiO_2 & SiO_2	Water-glass necessary as a setting agent	Produces a needle-like ceramic network. No porosities. Very few cracks. Crystalline, hence susceptible to moisture and chemical attack.
ALUMINOUS CEMENTS	Various types with varying amounts of Al_2O_3	Large number of cracks	Laser remelts material to eliminate cracks, but too many bubbles and porosities
SANDS	i) BLACK VOLCANIC ii) RED iii) WHITE	Adheres well, no cracks visible before melting	Coating adheres well but too many pores, bubbles and fine cracks
GLASS POWDERS	i) LEAD GLASS ii) PYREX GLASS	Adheres well, no cracks	Pb glass produces poisionous fumes, and does not adhere well. Both contain many bubbles and cracks
ALUMINA	Al_2O_3	Needs waterglass to set. Some cracks present (maybe owing to CaO contamination)	Crystalline structure
MAGNESIA	MgO	Needs waterglass to set	Crystalline structure
ZIRCONIA	ZrO_2	Needs waterglass to set	Crystalline structure. Degree of crystallinity dependent upon purity

proportion of chamotte which acts as an insulator, whilst the top layer contains more sand and PPC.

3.2. Experimental Work.

The composition of the two coating layers are presented in *Table 3* The dry powders of the insulating undercoat were mixed thoroughly before the water and waterglass were added.

Table 3 . **Composition of Coating.**

	Chamotte	Pozzolana	Sand	Waterglass	Water
Undercoat	10%	5%	35%	45%	5%
Topcoat	5%	30%	30%	30%	5%

This ensured the even distribution of the various materials throughout the coating. The liquids were then added to the dry materials, and mixed thoroughly.

The heat generated during the hydration of cement causes the water present to evaporate from the cement surface. This can lead to shrinkage cracking and crazing of the material.

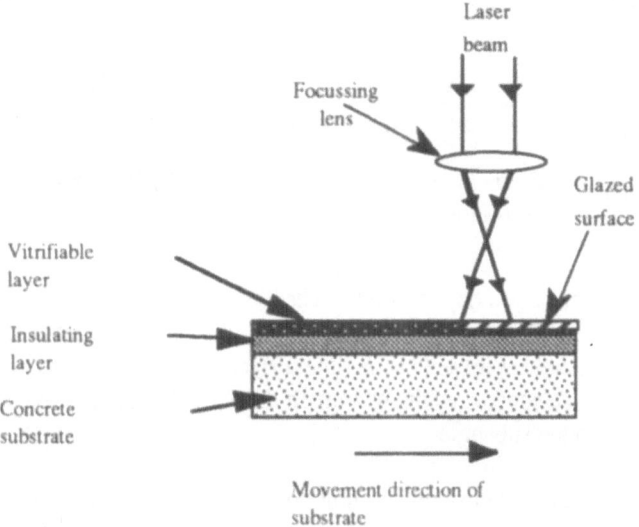

Figure 9. **Experimental Set-up to Glaze Coated Surfaces.**

624

A mechanically weak matrix is thus established which is poorly bonded to both the concrete substrate and itself. To avoid such cracking and to achieve a strong cement layer, the coating must be kept damp throughout the curing period. The samples were kept in airtight containers, and misted regularly with water. After 7 - 10 days, the top coat is applied upon the cured under coat. The same mixing and curing regime as described above is used for this layer.

Once cured, the material must be allowed to dry out as much as possible since any excess water will vaporise during glazing resulting in holes and bubbles in the glass.

Figure 9 shows the experimental set up for glazing the coating surface. Several process parameters were varied, including laser power, traverse speed of the specimen, etc. Again, both CO_2 and Nd:YAG lasers were used for processing. The quality of the results was judged on surface appearance and the adhesion of the glass to the unaffected coating material. This adhesion was tested using an adhesion tester as with the bare concrete specimens.

3.3. RESULTS.

Figure 10 shows the operating window to produce a smooth well adhered glass on the coated concrete. XRD experiments show this layer to have no crystalline character. The light grey area shows the optimum parameters to achieve this. The dark grey region shows samples which initially appear successful, but the

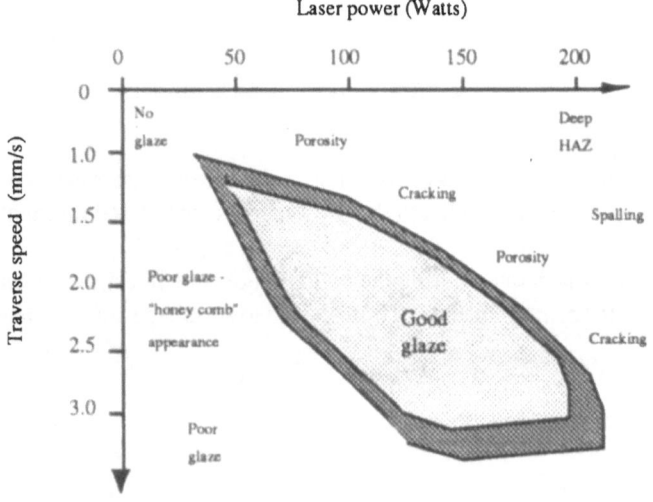

Figure 10. Schematic Diagram of Operating Window for Successful Glazing.

coating spalls off after only a few days or weeks. The laser used for this was the CO_2. There was no principle difference between this and the YAG results, but time limitations made the use of the CO_2 more practical. *Figure 11* shows the surface of such a glazed sample. Higher powers or slower traverse speeds results in a much deeper HAZ and consequently, poor adhesion of the glass to the substrate.

The extent of the HAZ can be controlled, by choice of the experimental parameters. A high power density and low traverse speed would result in a deep HAZ. Conversely, lower powers or higher traverse speeds result in poor glazing, which is discontinuous and the surface appears "honeycombed". There is a large number of pores in the surface, which

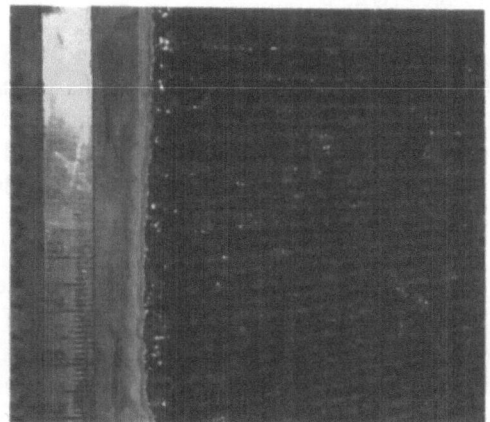

Figure 11. Surface of Glazed, Coated Specimen. (2mm/s and 150 W).

would allow the penetration of water through to the substrate and thus permit the possibility of chemical leaching of the radionuclides.

The pull-off strengths of the glass to cement bond are shown in *Figure 12*. It can be seen that the highest bond strengths correspond to the well-glazed area of *Figure 10*.

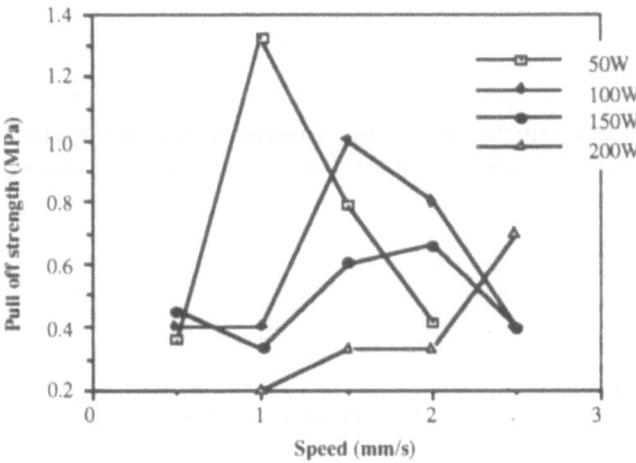

Figure 12 . Pull-off Strengths of Glazed, Coated
Samples.

3.4. HEAT AFFECTED ZONE (HAZ).

The laser action on concrete causes a temperature rise in the substrate as discussed previously. This heating effect of the laser can still affect the substrate, in that the material will still be heated. The limiting factor in the choice of material and thickness for the insulator is the extent of heating which the substrate can withstand without damage to its mechanical integrity.

The application of a two layer coating introduces two additional interfaces into the system; the *chemical* interface and the *cement-cement* interface. The position of these is fixed, but the position of the *thermal* interface varies with processing parameters. For the insulator to be sufficient in its role, the *thermal* interface must be above the *chemical* interface. *Figure 13* shows the relative positions of these interfaces. If conditions are such that the thermal interface is within the substrate, then the problems described in 2.2 & 2.3 occur, i.e. the substrate can dehydrate and a mechanically weak zone is introduced.

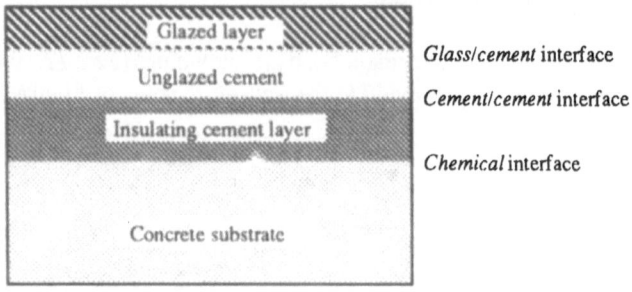

Figure 13. **Schematic to Show Various Interfaces Within Coated, Glazed Specimen.** The *thermal* interface can be anywhere below the glazed layer, depending upon power and traverse speed.

3.5. DISCUSSION.

Application of an additional cement layer improves the glazing results, whilst reducing the heat damage to the substrate. The quality of the glaze is dependent on the composition of the coating and also its condition, i.e. the amount of pore water present. Excess water vapourises and leaves holes in the glazed surface. Thus, careful preparation of the specimens is vital to the success of the glazing. The coating must be thoroughly cured, then dried to remove as much excess water as possible. Temperatures for this drying must be kept below 100°C, to prevent water being lost from the ettringite phase (see *Table 1*).

3.5.1. Processing Speed.

The processing is slow since a balance must be reached for glazing without excess heating. It would seem logical to increase power and speed to process more quickly. However this is not effective. There is a significant time lag for melting of the material and its vitrification. Thus, the traverse speed must be kept fairly low ($\approx$ 2mm/s) since if the power were to be increased, then dehydration of the substrate would occur.

3.5.2. Laser Power.

The power must be limited or the HAZ is too deep and mechanical failure results. The strength of the bond between the glass and unglazed cement is the "weak link" in the mechanical integrity. Optimum results are obtained with a thin glass layer. Less shrinkage occurs as it cools and consequently there are less cracks and spalling is reduced. Relatively low powers are required for this ($\approx$200 W for a laser spot diameter of 5mm).

3.5.3. Large Area Processing.

The use of a larger spot size to increase processing rates fail because the surface tension of the softened glass is too high. The glass does not flow over the surface to form a thin layer, rather it clumps into "beads". This leaves much of the cement surface unglazed. The use of a line beam created with a rapidly rotating polygon was more successful. The area melted at any time is large, but the line beam is long and thin so the clumping does not occur. Such a beam shape would also facilitate the glazing of ceilings where melting a large area could result in droplets falling from the surface.

3.5.4. Materials.

If the fluidity of the glass could be increased (and hence surface tension reduced) then large area processing could be developed. Some material addition to the coating may increase processing rates in this way. Alternatively, finer powders, with a greater surface area could be used. Such particles will melt more quickly allowing more rapid traverse speeds.

A decrease in the melting temperature of the coating would reduce the time needed to melt and vitrify a particular area. Thus, the same power but a higher traverse speed could be used. This would reduce the laser interaction time and hence decrease the HAZ, whilst increasing the processing rate. This decrease in melting temperature could again be achieved by some material addition or variation of the coating. Further development is required to achieve a lower melting temperature and higher fluidity of the coating.

3.5.5. Heat Affected Zone.

The HAZ extends well below the glazed layer and can result in mechanical weakness. The use of phenolphthalein gives an indication of the extent of the HAZ, but further characterisation of this important area is necessary. Quantitative data on the temperature rise at different depths within the substrate along with specific information on the temperature of dehydration is necessary. Experiments are in progress to produce this data with the use of embedded thermocouples. Subsequently, modelling of this heat flow and consequent dehydration can then be performed to fully characterise the HAZ.

4. Conclusions.

1 - Laser action on cementitious materials results in vitrification.

2 - Direct glazing of concrete surfaces results in a glazed layer, but it is poorly adhered. The presence of aggregate peices near the concrete surface results in a non-uniform coating being produced.

3 - A heat affected zone is produced which includes the glazed layer and the dehydrated material below it.

4 - Application of a cement coating a few mm thick prior to glazing can protect the substrate and provide vitrifiable material.

5 - Glazing of coated samples results in a smooth, well adhered, glassy layer when the processing parameters are within the operating window.

5. References.

1. Li, L. (1993) Nuclear Decontamination with Laser Beams, *Proceedings of ISLOE*, 25 - 30.
2. *U.K Patent application* 9200107.2, 4th Jan 1992 by BNFL plc.
3. *U.K Patent application* 9209473.9, May 1992 by BNFL plc.

DIRECT SYNTHESIS OF METAL NITRIDE BY LASER

C.BOULMER-LEBORGNE, A.L.THOMANN AND J.HERMANN
GREMI, Université d'Orléans
BP6759, 45067 - ORLEANS CEDEX2, FRANCE

In this work titanium nitride synthesis is carried out by direct laser irradiation at Ti sample surface in presence of ambient nitrogen. The experimental procedure is performed in a chamber allowing plasma study by emission spectroscopy. Two pulsed laser types are used, a TEA-CO$_2$ (10 600nm) and a XeCl excimer (308nm) in order to compare the laser-material coupling influence on the layer synthesis process depending on the laser wavelength. In both cases the titanium nitride synthesis is evidenced.

1. Introduction

Metal nitrides are interesting compounds for technology applications especially as hard coatings. At present time these coatings are produced by chemical vapour deposition, ion implantation, plasma or thermal nitriding. Nevertheless the piece size is limited by the chamber volume. It would be interesting to demonstrate that the direct synthesis of metal nitride can be carried out outside using a laser beam focused on the piece covered by N$_2$ gas jet. This the goal of this work.

Layer synthesis induced by laser-plasma depends on the material characteristics (absorptivity, thermal conductivity, melting temperature...) and also on the wavelength and power delivered by the laser. In this work, the experiment is successively performed with a (pulsed) Transverse Excited Atmospheric (TEA) CO$_2$ laser (λ=10.6μm) and a XeCl excimer laser (λ=308nm). In both cases, TiN synthesis is obtained. The plasma created by the laser-material interaction has in different composition and behaviour depending on the large difference between the laser wavelength values.

After processing the synthesised films are analysed : the surface states are controlled by scanning electron microscopy (SEM) and roughness measurements, chemical composition is investigated on the surface (50Å depth) by X-ray photoelectron spectroscopy (XPS), and in the material bulk by nuclear analysis, Rutherford Back-Scattering (RBS) with α particles and Back-Scattering (BS) with protons.

2. Experimental-set-up

The experiments are respectively performed with a TEA-CO2 (Gen-Tec DD-250-A Head) laser source emitting at 10.6μm wavelength and an excimer laser source XeCl (Questek 2520Vβ) emitting at λ=308nm. Both lasers are pulsed working, the laser pulse shape are different for each one. The CO$_2$ laser presents a 200ns width peak followed by a 4μs tail, 70% of the energy is in the peak. The XeCl laser pulse shape is a 20ns width rectangle shape.

J. Mazumder et al. (eds.), Laser Processing: Surface Treatment and Film Deposition, 629–636.
© *1996 Kluwer Academic Publishers.*

The Ti targets are placed in a chamber containing N_2 gas with a pressure varying from 100Torr to 760Torr. The best N_2 pressure condition to obtain the thickest TiN films is 760Torr with the CO_2 laser, whereas a 380Torr or 760Torr N_2 pressure yields to similar characteristics for TiN films synthesised with the XeCl laser. The laser beam is focused perpendicularly to the Ti target (10mm x 10mm x 1mm). The targets are placed over the focal point to obtain the largest treated zone with the necessary surface laser power density to obtain the nitride. Thus the laser beam spot area is ≈1mm^2 with the CO_2 laser and ≈6mm^2 with the XeCl laser. The laser power density for TiN synthesis is in the respective ranges [30MW/cm^2 - 70MW/cm^2] for the CO_2 laser and [40MW/cm^2 - 100MW/cm^2] with the XeCl laser. The targets are step by step displaced after irradiation of a surface spot to nitride all the sample area. The following scheme presents the set-up.

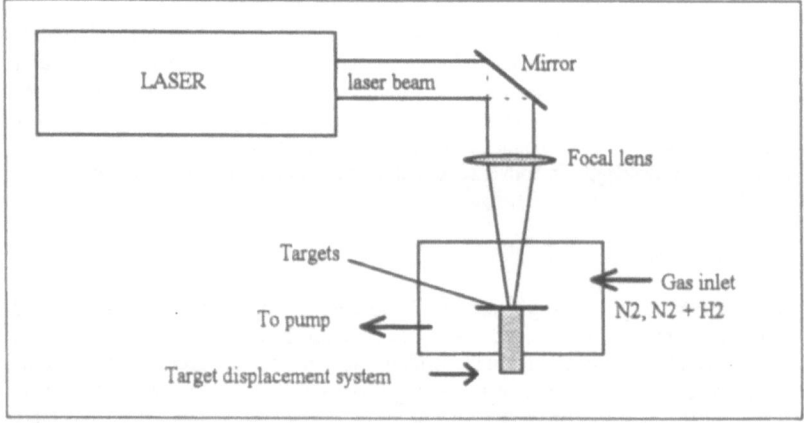

Figure 1. Experimental set-up : the laser is either a CO_2 or a XeCl laser ; optical lens material is adapted to each laser wavelength (ZnSe for λ=10.6µm and quartz for λ=308nm).

3. Plasma formation

3.1. VAPORISATION STAGE

The plasma formation requires the vaporisation of the material surface as a first step. When the laser radiation is absorbed at the surface of the target, the light energy is transformed into heat and the surface temperature increases. Simultaneously heat conduction takes place into the metal bulk, thus increasing the thickness of the heated layer.

The quantity of vaporised mass depends on the laser wavelength and on the laser energy absorbed by the target. It is clear from [1] that short laser wavelengths favour the ablation process. The CO_2 laser power density threshold for titanium target vaporisation has been experimentally determined to be 50MW/cm^2. With XeCl laser, the titanium vaporisation has been observed for power densities over 25MW/cm^2.

3.2. PLASMA STAGE

The vapour is ionised and electrons gain energy in the laser field through inverse Bremsstrahlung absorption. The electron temperatures in plasmas created in vacuum in front of a solid target have been calculated, for different laser wavelengths [2]. It is shown that the electron temperature is larger for the CO_2 laser (10.6μm) than for excimer laser (308nm), because inverse Bremsstrahlung process heating the electrons is more efficient for IR. Indeed the laser radiation absorption coefficient varies in λ^{-2}.

Electrons will lose energy by elastic and inelastic collisions with neutral particles. Some electrons will be lost by attachment. Lateral gas dynamic expansion of the vapour out of the laser beam and diffusion of electrons out of the breakdown volume are also a source of electron losses. The frequencies characteristic of these electron loss processes are generally negligible compared to the energy loss frequency if the ambient gas pressure is larger than 100Torr [3].

If the laser irradiance is high enough, primary electrons in spite of the energy loss processes will gain an energy larger than the ionisation energy. These electrons will generate new electrons by impact ionisation of atoms of the vapour, thereby leading to cascade growth. The condition for initiating the electron avalanche is that the rate of electron energy increase exceeds the rate of energy loss. The threshold intensity to produce a gas breakdown is deduced from this criterion. Assuming that the electron mean energy is roughly equal to $\Delta/3$ where Δ stands for the ionisation potential of neutral gas atoms, the criterion becomes

$$Ii\left[MW/cm^2\right] > 2.10^3 * \frac{\Delta}{\lambda^2 M} \qquad (1)$$

with Δ[eV], λ[μm], M[amu] specie mass, for ambient gas pressures larger than 100Torr.

Therefore, Ii is three orders of magnitude larger for λ=308nm than for λ =10.6μm. Thus only 20MW/cm^2 laser power density is necessary to obtain nitrogen plasma with the CO_2 laser whereas a two times larger laser energy is needed to vaporise the target. The IR laser beam interacts with the target, the primary electrons are produced from the vaporisation of the surface defects, then the gas plasma breakdown occurs and ionisation avalanche holds. The surface defect vaporisation stops as the plasma becomes denser, the laser beam is completely absorbed in the plasma and does not reach anymore the target surface ; the nitrogen plasma is self-sustained.

With the XeCl laser, the power density necessary to obtain the gas breakdown is higher than the used intensity, thus only a vapour plasma is observed. The plasma formation is due to the emission of primary electrons during the first ns of laser-target interaction, then followed by electron-Ti atom collisions at the sample surface as soon as the vaporisation process begins yielding to vapour ionisation (during the laser pulse). The plasma is dominated by titanium. No nitrogen line can be observed for gas pressure over 0.1Torr.

Laser absorption waves have been observed in nitrogen plasma created by CO_2 laser on the titanium target [4]. Pressure pulses related to these laser absorption waves produce stress waves in the target material which can lead to deformation or fracture.

4. Surface state

In order to understand the mechanism of TiN synthesis, we have performed chemical analysis of the targets submitted to various laser pulse numbers (corresponding to various irradiation times).

The irradiated zone have been examined by scanning electron microscopy. The sample surfaces irradiated by the CO_2 laser show a variation of colours from grey to gold yellow as a function of the laser pulse number. This colour change is not observed when the Ti targets are irradiated by the XeCl laser. It can be deduced that TiN film crystal phase is different according to each laser treatment.

With the TEA-CO2 laser treatment, the surface roughness (measurements with a Talysurf 10, Taylor-Hobsen) varies from ±0.2μm for 500 pulses to ±2μm for 1000-3000 pulses. For the weakest laser pulse numbers, pores, splits and cracks appear and for the largest, the surface is composed of compact piles (5 to 10μm width) separated by cracks.

With the excimer laser treatment, the roughness is limited to ±0.7μ for the largest laser pulse numbers. The sample surface begins to be covered by cracks (500 laser pulses) then becomes homgeneously spongy and keeps this appearence for larger laser irradiation times.

5. Chemical analysis

5.1 XPS RESULTS

XPS measurements have been performed in a XPS VG ESCALAB UHV chamber (10^{-8}Pa) with a non momochromatised Mg Kα X-ray source (1253.6eV). The XPS spectrometer was calibrated to give the Au($4f_{7/2}$) line located at 84.0eV.

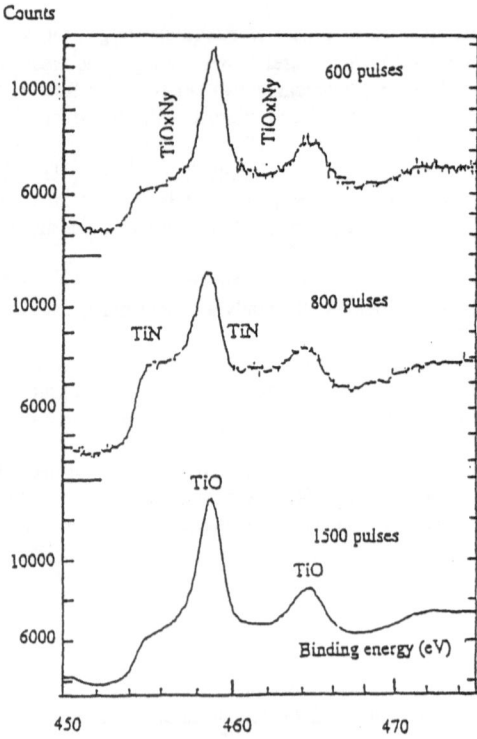

Figure 2. XPS spectra in the Ti2p region for different pulse numbers CO_2 laser irradiation, 45MW/cm^2

The program used to fit XPS spectra uses an algorithm based on the convolution of lorentzian and gaussian curves and a Shirley background subtraction of experimental signals. In the titanium spectrum region, there are 2 doublets (Ti2p$_{1/2}$ and Ti2p$_{3/2}$ levels) relative to the Ti-O bounds in TiO$_2$ oxide (458.5–464.4eV) and to the metallic bounds (454.3–460.4eV). The doublet with main peak (Ti2p$_{3/2}$) located at 455eV stands for titanium nitride (maybe no-stoichiometric) and the peaks relative to titanium oxinitrides with various oxigen quantities are located between Ti-O and Ti-N bond peaks.

5.1.1 CO_2 laser irradiation

Evolution with the laser pulse number   Whatever the laser power density during the target irradiation, the same species are synthesised. That means that the doublets are always present and located at the same binding energies. With increasing the laser pulse number, the TiN quantity seems to decrease in comparison with the oxinitride amount as seen in Fig.2.

Laser power density influence   We have performed the experiments for several power densities 35 to 70W/cm^2 with the short laser pulse shape and the targets have been submitted to different pulse numbers in each case. The same kind of spectrum that the one presented in Fig.2 is obtained. No significant difference in peak intensities is observed as far as the samples are treated with the same plasma time duration.

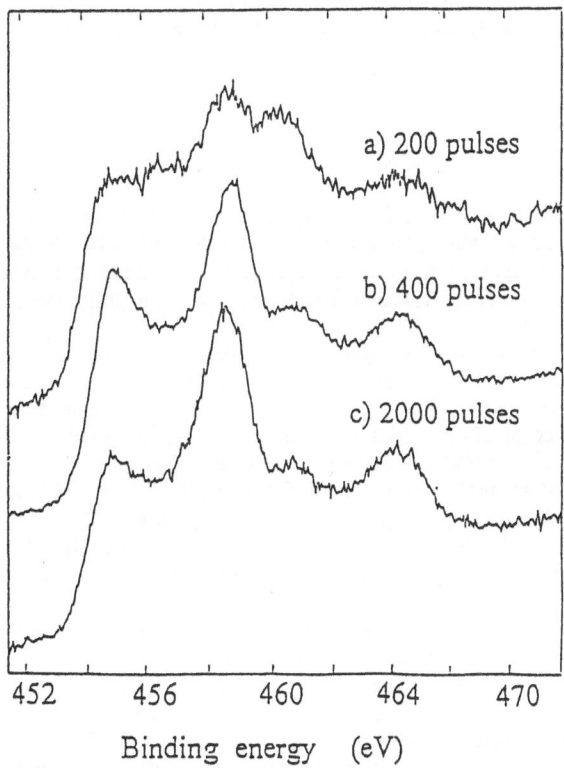

Figure 3. XPS spectra for different pulse numbersXeCl irradiation (60MW/cm^2) 760Torr N$_2$

Whatever the laser power, there is always a pulse number for which TiN and TiNxOy quantity is maximum. When this irradiation time is exceeded, destruction of synthesised TiN and TiOxNy species occurs. This seems to be due to the plasma recoil action on the surface. Thus we conclude that the most important parameter for an optimum synthesis with the CO_2 laser, is the N_2 plasma time duration on the surface. Best results are obtained when the plasma acts several seconds on titanium target. Working at high laser power densities allows a faster plasma ignition and a weaker treatment time duration.

5.1.2 *XeCl laser irradiation*

Evolution with the laser pulse number   Figure 3 shows the XPS result for Ti targets submitted to XeCl laser irradiation ($60MW/cm^2$ in 760Torr N_2) for various pulse numbers. The three species, TiN, TiOxNy and TiO_2 are present. For the weaker laser irradiation time (200 pulses), the TiN and TiOxNy peak intensities of the spectrum are the largest in comparison with TiO_2 peaks. When the laser pulse number increases (400), the oxinitride peak intensities decrease, then when the pulse number reaches 1000 the nitride peak intensities decrease too : the intensities of the nitride and oxinitride peaks become equivalent. This holds for larger laser pulse numbers, excepted that the whole peak intensity of the spectrum decreases. Spectra obtained in the same experimental conditions but for 380Torr N_2 pressure show similar results.
Laser power density influence   Different laser power densities have been used for experiment ($40-100$ W/cm^2). The XPS spectra show that the same species are synthesised whatever the laser power density, it appears that the largest nitride and oxinitride peak intensity is obtained for the weakest laser power.

6. Nuclear analysis

6.1 CO2 LASER IRRADIATION

The samples obtained with the CO_2 laser irradiation at $45MW/cm^2$, 760Torr N_2 for different laser pulse numbers, have been analysed by Rutherford BackScattering (RBS) method with 1.3 Mev α particles. The nitrogen concentration profile in the synthesised layer is deduced from the calculation with rump program fitting the experimental signals.

Thus for the lowest laser pulse number (100-200) corresponding to the surface heating, the RBS signal is similar to those of pure Ti sample. At intermediate laser pulse number (300-500), the nitrogen concentration can be represented by a decreasing exponential function of the z depth from the surface $C=0.25\exp(-10^{-5}.z)$. That means that there is 3 Ti for 1 N at the surface (z=0). For larger laser pulse number (600-800), this concentration profile varies to $C=0.4\exp(-0.75 \ 10^{-4}.z)$. The surface TiN stoichiometry becomes $Ti_1N_{0.7}$. This holds for larger pulse numbers.

These results show that N amount in the layer increases with the laser pulse number from 100-200 to 600-800, then saturates for larger number. In addition to XPS results, it can be deduced that the plasma created on the surface yields the nitrogen diffusion into the bulk, this process is limited to a $\approx 3\mu m$ thickness layer, for larger plasma time, the surface is damaged and the nitrogen diffusion process stops.

BackScatering (BS) analysis have been also performed on the targets irradiated with 2MeV protons. Fig 4 presents a typical result for targets submitted to 600 and 800 laser pulses ($45MW/cm^2$).

The nitrogen peak increases with the pulse number. The presence of an oxigen peak (on the right of the nitrogen peak) must be noted, it intensity increases too with the laser peak number.

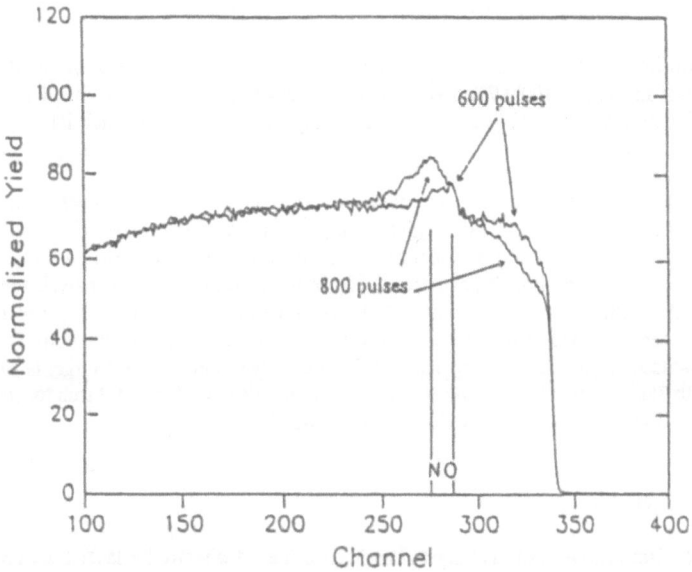

Figure 4. Backscattering spectra for TiN films for 600 and 800 pulses - CO$_2$ laser irradiation, 45MW/cm^2

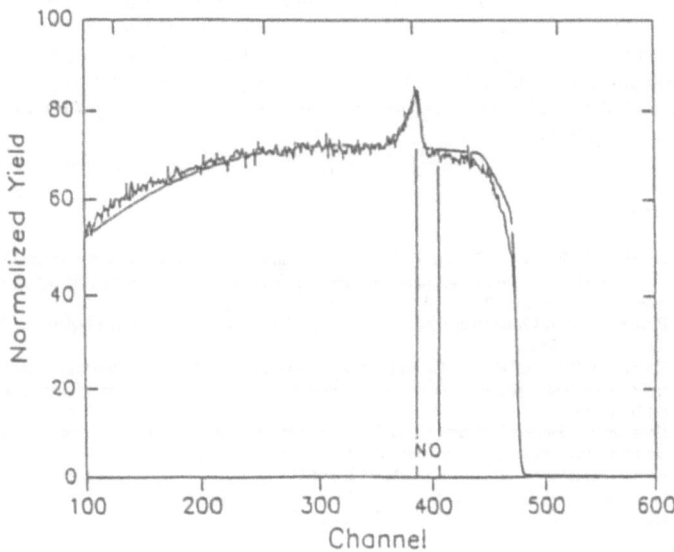

Figure 5. BS spectrum for TiN films 1000 pulses XeCl laser 760Torr N$_2$, 60MW/cm^2

6.2 XeCl LASER IRRADIATION

With XeCl laser, the TiN film thickness ($\approx$2-3μm) is maximum for 2000 to 1000 laser pulses in the [40-100MW/cm^2] laser power density range scanned. In BS spectra no oxide peak is observed as seen in Fig 5. Rump fit of the experimental data has shown that the titanium nitride becomes stoichiometric for a special laser pulse number depending on the laser power
density. Then this stoichiometry does not vary when the pulse number is increased. This number is smaller and smaller as the laser power is increased. That means that working at higher laser power density allows to obtain the TiN layer in a shorter time.

The experiments have been achieved with 380Torr N_2 and analysis results are similar than those obtained with 760Torr. Consequently, it can be deduced that the nitrogen adsorbed at the target surface before laser treatment reacts with Ti vapour produced by laser irradiation. The nitrogen quantity absorbed on the target is sufficient with 380Torr N_2 to yield stoichiometric titanium nitride. This is not true for lower N_2 pressures as a substoichiometric titanium nitride is then obtained.

7. Conclusion

TiN film synthesis on Ti targets has been achieved directly by laser treatment. The layer is 2-3μm thick, but its composition and crystal phase depend on the laser wavelength used for nitriding operation.
With CO_2 laser irradiation the main parameter to control the TiN synthesis is the time duration N_2 plasma interacting with the target. The best result for film synthesis is obtained at least for 760Torr N_2. The synthesised layer is composed of no-stoichiometric TiN and TiNxOy. Synthesised layer surface (100Å thickness) exhibits a large oxigen contamination due to TiO_2 native oxide on the metal sample.

With XeCl laser irradiation, no N_2 plasma is observed but Ti vapour plasma. It seems that TiN synthesis is due to this Ti plasma interaction with the previously adsorbed N_2 on the target surface. The synthesised TiN film is stoichiometric for 2000 to 1000 pulses depending on the laser power density in the [40-100 MW/cm^2] range either for 760Torr or 380Torr N_2 pressure. No TiNxOy appears excepted in the surface contamination layer.

8. References

1. Geerstsen C., Briand A., Chartier F., Lacour J.L., Mauchien P., Sjöström S. and Mermet J.M. (1994) Comparison between infrared and ultraviolet laser ablation at atmospheric pressure, *Journal of analytical atomic spectrometry* 9, 17-22

2. Richter A (1190) Characteristic features of laser-produced plasmas for thin film deposition, *Thin Solid Films* 188, 275-292

3. Hermann J, Boulmer-Leborgne C., Mihailescu I.N. and Dubreuil B. (1993) Multistage plasma initiation process by pulsed CO2 laser irradiation of a Ti sample in an ambient gas, *Journal of Applied Physics* 73 (1) 1091-1099

4. Hermann J., Boulmer-Leborgne C., Dubreuil B. and Mihailescu I.N.(1993) Influence of irradiation conditions on plasma evolution in laser-surface interaction, *Journal of Applied Physics* 74(5), 3071-3079

5. Hermann J., Boulmer-Leborgne C., Dubreuil B. and Mihailescu I.N. (1992) Investigation by laser-induced fluorescence of surface vaporization during the pulsed CO2 laser irradiation of a titanium sample in an ambient gas, *Journal of Applied Physics* 71(11), 5629-5634

LASER PROCESSING OF HIGH T_C SUPERCONDUCTORS

K. MUKHERJEE, C.W. CHEN, J. YOO, I. OH, AND S. KUDAPA
Department of Materials Science and Mechanics
Michigan State University
East Lansing, MI 48864, USA

1. INTRODUCTION

Three most important high-T_C (high critical temperature) oxide superconductor systems are Y-Ba-Cu-O (yttrium-barium cuprates or rare earth-barium cuprates), Bi-(Pb)- Sr-Ca-Cu-O (bismuth-lead-strontium-calcium cuprates), and Tl-Ba-Ca-Cu-O (thallium- barium-calcium cuprates). All of these three systems have complicated crystal structures with copper-oxygen planes, which are believed to be responsible for the superconducting properties of these compounds. Among these cuprates Tl-Ba-Ca-Cu-O system has the highest T_C, about 125 K[1,2], but the potential for industrial application is limited due to severe health problem associated with the toxicity of thallium compounds.

Material processing utilizing high-power lasers, especially high-energy pulsed lasers, have been widely adopted to produce high T_C superconductors. These techniques include:

1. Pulsed laser vapor deposition (PLD) or ablation[3-23]

2. Thin film patterning

      (a) by pulsed laser ablation[24-26]
      (b) by direct laser writing of precursor films[27]
      (c) by laser assisted etching[28]

3. Laser calcining of high T_C superconducting powder[29,30]

4. Laser re-melting/texturing[31].

5. Laser annealing of superconducting thin films[32].

Of the above mentioned techniques, we will emphasize laser assisted vapor deposition of thin and thick films. An objective is to develop a low cost high speed process for

637

J. Mazumder et al. (eds.), Laser Processing: Surface Treatment and Film Deposition, 637–664.
© 1996 *Kluwer Academic Publishers.*

fabrication of these oxide superconductors. Before proceeding any further, we will briefly introduce the candidate superconducting materials suitable for thin film applications.

1.1 HIGH TEMPERATURE OXIDE SUPERCONDUCTORS

The discovery of oxide high temperature superconductors opens numerous opportunities for applications that were previously impossible. At the present time, potential applications of bulk oxide superconductors are very restricted because of their poor mechanical properties and low critical current densities. Enormous effort has been made to produce thin films with correct stoichiometry, high transition temperature and high critical current density.

Three most important high-T_C (high critical temperature) oxide superconductor systems are Y-Ba-Cu-O (yttrium-barium cuprates or rare earth-barium cuprates), Bi-(Pb)- Sr-Ca-Cu-O (bismuth-lead-strontium-calcium cuprates), and Tl-Ba-Ca-Cu-O (thallium- barium-calcium cuprates). These systems are often referred as YBCO, BSCCO and TBCCO systems. All these three systems have complicated crystal structures with copper-oxygen planes, which are believed to be responsible for the superconducting properties of these compounds. Among these cuprates Tl-Ba-Ca-Cu-O system has the highest T_C, about 125 K[1,2]. Recently a Hg bearing oxide superconductor has been reported, which has an even higher critical temperature; nearly 133 K. However, because of fabrication difficulty, and toxicity of Hg, this system is considered not to be so viable for commercial applications.

$YBa_2Cu_3O_x$ (1-2-3 compound), with T_C around 95 K, is considered to be the most favorable oxide superconductor for thin film applications because, in comparison with other systems, it has a less complicated chemistry and structure, good phase stability and a high critical current density[3]. On the other hand, $Bi_{2-x}Pb_xSr_2Ca_2Cu_3O_Y$ (2-2-2-3 compound) of the Bi-(Pb)-Sr-Ca-Cu-O system, with T_C about 110 K, is a good candidate for tape application due to its thin plate shaped grains that facilitate mechanical alignment of the grains to obtain higher transport critical current density[33-34].

$Bi_{2-x}Pb_xSr_2Ca_2Cu_3O_Y$ (2223 phase) of the Bi-(Pb)-Sr-Ca-Cu-O (BSCCO) system, with T_C about 110 K, is a good candidate for tape application due to its thin plate shaped grains that facilitate mechanical alignment of the grains to obtain higher critical current density[33-34]. However, the BSCCO superconductor system has a number of phases with different critical temperatures. The three major phases are referred to by their cation ratios as 2201 (T_C below 77 K), 2212 ($T_C \approx 85$ K), and 2223 ($T_C \approx 110$ K)[35]. Since the discovery of superconductivity in this system[36], many efforts have been made to maximize the amount of high T_C-2223 (110 K) phase. A major enhancement in the formation of high T_C-2223 phase was made by doping Bi-Sr-Ca-Cu-O compound with Pb[37-41]. However, With Pb doping, about 200 hours of total processing time is required to obtain a single or nearly a single high T_C phase. In this study, a new laser processing technique was developed with an objective to enhance the kinetics of formation of the high T_C phase of a Pb-doped BSCCO superconductor. With laser calcination process, near single-phase high T_C phase samples were obtained in about 100 hours. In our study, the onset critical temperature of the laser calcined sample was found to be about 110 K, although the zero resistance temperature was about 98 K.

1.1.1. Crystal Structures of High T_C Superconductors

1.1.1. Crystal Structures of High T_C Superconductors All three important high T_C systems can be visualized as stacking of various perovskite cells with some defects in the cells or in stacking. Figure 1 shows two different ways of drawing a perovskite unit cell that has a ABO_3 formula. Many perovskite metal oxides, e.g. $BaTiO_3$, are well known for their ferroelectricity. Two distinct cation sites exist in the structure. A cation occupies the larger site with coordination number of twelve, on the other hand, the smaller B cation sits in the oxygen octahedron. In high T_C superconducting cuprates, copper ions occupy the B cation sites while other larger cations occupy the A cation sites. Due to oxygen deficiency or other variation from the standard perovskite structure, the nearest oxygen ions of the copper cation is often less than six. In fact, copper-oxygen planes with four oxygen ions per copper cation are the key feature of high T_C oxide superconductors.

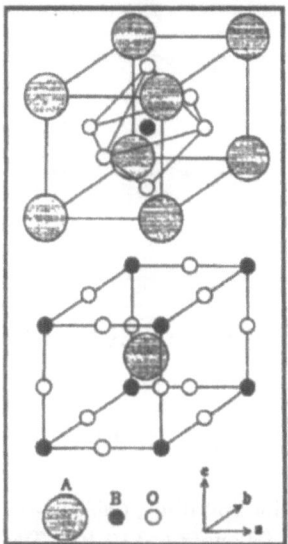

Figure 1 Perovskite unit cell.

There is an important relation between T_C and structure of these superconducting cuprates, that T_C increases as the number of adjacent Cu-O planes (n) increases, at least up to n = 3. These adjacent planes , that sandwich sparsely occupied Y-atom or Ca-atom plane, are separated by a distance of about 3.2 Å. This distance is approximately the same for all high T_C cuprates and is smaller than the standard perovskite cell size or lattice parameters a or b of the high T_C cuprates. T_C of superconducting cuprates is less than 77 K for n = 1, about 85 to 95 K for n = 2, and more than 100 K for n = 3.

Crystal Structure of $YBa_2Cu_3O_{7-d}$ (YBCO). An YBCO unit cell consists of three oxygen deficient perovskite cells. Figure 2 shows the unit cell of an orthorhombic YBCO with seven oxygen ions per unit cell, i.e. d = 0, where the top and bottom Cu-O plans have no oxygen ions in "a" direction, and the middle cell has no oxygen ions on the side edges. Two types of copper ions exist. Copper ions in the middle of the unit cell form Cu-O plans, with four planar nearest oxygen ions, while the copper ions at the top and bottom of the unit cell (also having four planar nearest oxygen ions) form one-dimensional chains in the "b" direction. The two adjacent Cu-O planes are separated by a sparsely occupied Y-atom plane. The distance between these adjacent Cu-O planes is about 3.2 Å while the separation between two sets of these adjacent planes is > 8.2 Å.

At high temperatures, $YBa_2Cu_3O_{7-d}$ loses additional oxygen accompanied by a decrease of the oxygen ordering. The structure of YBCO becomes tetragonal. The orthorhombic-tetragonal phase transition is reversible with a transition temperature around 700 °C. The tetragonal phase is semiconducting.

Crystal Structures in BSCCO System. The high T_C Bi-Sr-Ca-Cu-O superconducting system has a number of phases with different critical temperatures (T_C). The three major phases are referred to by their cation ratios as 2201 (T_C below 77 K), 2212 ($T_C \sim 85$ K), and 2223 ($T_C \sim 110$ K) phases[35]. The structures of these phases are very similar to the structure of $(La_{2-x}Sr_x)CuO_4$. For simplicity the following discussion is based on the un-doped La_2CuO_4 (La-Cu-O). Conventional unit cell of La-Cu-O contains two chemical formula units, i.e. $2(La_2CuO_4)$. Similar to the YBCO unit cell, this cell is also composed of three perovskite cells; however the top and bottom cells are displaced as shown in Figure 3.

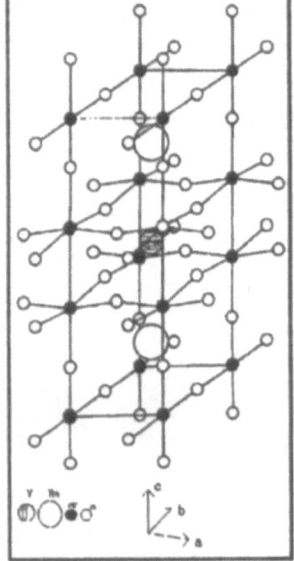

Figure 2. YBCO unit cell.

2. PULSE LASER DEPOSITION (PLD) OF SUPERCONDUCTING OXIDES

Among the techniques adapted to grow high T_C superconductor thin films, pulsed laser vapor deposition is the most widely studied and a most successful one[3-6,9,10,19]. The key reasons for the success of the PLD process are its stoichiometry conservation property, and the production of excited high energy vapor species[42-46].

It is well known that high quality YBCO thin films can be obtained by using an excimer laser with a 193, 248, or 308 nm wavelength, nanosecond pulse width and a few J/cm² pulse energy density. However, the deposition rate is very low, typically a few tenth of nm per second. Moreover, the deposited area coverage is very small, only a few millimeters in width, due to the narrow angular distributions of the vapor, as sharp as $cos^8 \theta$ to $cos^{12} \theta$[15,16,42]. On the other hand, high deposition rates up to 100 nm/s[15] and a large uniform deposition coverage, due to a broader (cos q) angular distribution of the evaporated species can be obtained by using millisecond pulsed lasers. These enhanced deposition rates make millisecond-PLD (ms-PLD) a practical process to produce continuous coating of large substrates.

Almost all high T_C superconductor thin film deposition techniques can be categorized as physical or chemical vapor depositions. One example of physical vapor depositions is laser evaporation or ablation[3-23] (plasma or ion beam sputtering, electron beam or olecular beam epitaxy, and thermal evaporation are other examples). Chemical vapor deposition involves decomposition of organometallic chemical vapors. All vapor deposition processes involve three major steps which are:

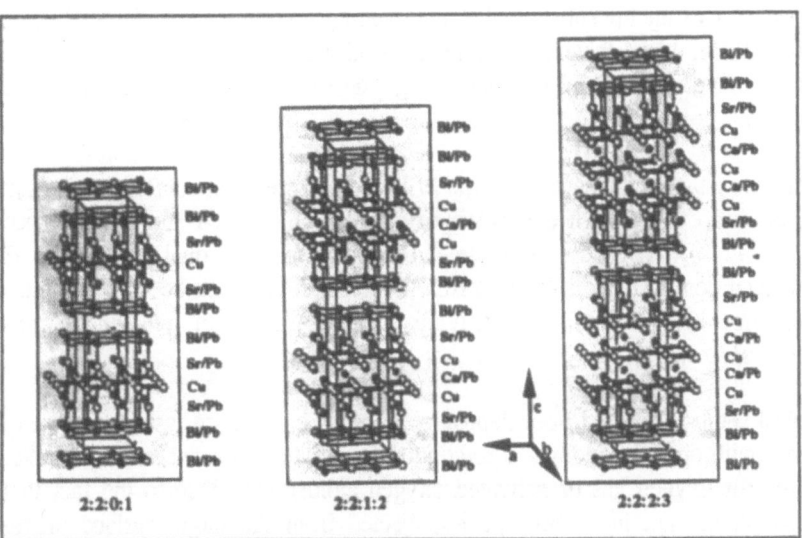

Figure 3. Unit cells of various BSCCO compounds.

1. Creation of vapor phase species by thermal energy, photon energy or kinetic energy.

2. Transport of the vapor species from source to substrate: Dependent on the deposition technique used, this transport can occur without much collisions between atoms and molecules, i.e. under molecular-flow condition, or with many collisions in the vapor phase species, i.e. higher partial pressure of the vapor and/or gas species or some ionized species.

3. Film growth on the substrate: Dependent on deposition rate and mobility of atoms especially in the surface and near-surface, the as deposited film can be amorphous or crystalline. Increasing energy of the vapor species (e.g. producing species in excited state or increasing velocity of vapor species), increasing substrate temperature and/or bombardment of the growing film by ions in the vapor specieswill increase the mobility of atoms on the growing film; thus, promote nucleation and growth of crystalline films.

Deposition rate, stoichiometry and crystalline quality are the most important parameters in depositing high quality superconductor thin films. The first parameter mainly depends on the first two steps of the vapor deposition process; while the crystalline quality depends primarily on the film growth step of the above model. Therefore, it will be much easier to optimize these three parameters if three major steps in the vapor deposition process can be independently controlled. For all physical vapor deposition techniques mentioned above, these three deposition steps can, more or less, be independently controlled. On the other hand, in the chemical vapor deposition technique, it is difficult to separate these three steps because the deposition species are created right on the surface of the hot substrate, i.e. all three major steps of vapor deposition occurred simultaneously. Nevertheless, if complicated substrate geometry is necessary, chemical vapor deposition is the only non line-of-sight deposition method among all techniques mentioned.

The key reasons for the success of the PLD process are its stoichiometry conservation property, and excited high energy vapor species, i.e. excited and ionized species, it produced[42-46]. Because of these two properties, high quality epitaxial film of complicated composition, such as YBCO superconductor, can be made in situ at low temperature (400 - 650 °C) with a relatively large degree of freedom in processing parameters[4,12,14].

A typical setup for pulsed laser deposition consists of a pulsed laser source, a vacuum chamber with optical windows, a rotating target holder, a heated substrate holder, and a source for oxygen gas or activated oxygen atoms/ions. Due to the fact that laser induced plume, i.e. the vapor species, ejected from the target surface moves in a direction normal to the target surface, the substrate holder is positioned parallel to the target and centered to the laser-target interaction spot.

Substrate distance from the target, typically a few centimeters, is critical because of the non-uniform spatial distribution of the vapor species and declining energy of vapor species as they move away from target. For a given deposition condition, the substrate distance also controls the area coverage of deposited film on the substrate. Because stoichiometry is usually preserved during pulsed laser ablation process, a single stoichiometry bulk target is normally used. The target is usually rotating at a low speed to expose different parts of the target to the laser pulse in order to prevent local overheating and erosion of the target. This is because overheating will result in compositional variation of the target by segregation and evaporation, and non-planar target surface will direct the plume away from the substrate.

2.1. FUNDAMENTAL ASPECTS OF PLD PROCESS

PLD is a flexible manufacturing tool. This technique allows processing of thin or thick films of supeconductor under a controlled set of conditions. Various process parameters can be monitored and analyzed. Most importantly, this is a noncontacting, and thus ultra clean processing and in-situ fabrication of devices is possible.

2.1.1. *The Mechanism.*
Evaporation and/or ablation of presintered bulk high T (target) provides the vapor.

Vaporized material is deposited on a suitable substrate placed parallel to the target Substrate is heated a temperature at which in-situ crystal growth can occur. Atomic mobility plays an important role in during the deposition process. The deposition temperature must be carefully chosen for the appropriate phase stability.

Plasma Formation. A plasma is formed by laser interaction with the target (in the present case a presintered superconductor bulk sample of correct stoichiometry). The plasma consists of neutral and/or ionized atoms and/or molecules of the constituents of the target material. The evolving plasma interacts with the laser beam. This plasma-laser interaction produces heat transfer to the plasma plume, and as a result the plasma expands rapidly.

The Deposition Rate. The film deposition rate, film thickness, and film quality are determined by the following laser and processing parameters:

* The focal spot size of the laser beam incident on the target
* The laser energy density
* Pulse repetition rate of the laser beam
* The pulse width (FWHM)
* Laser wavelength (or the frequency)
* Distance of substrate from the target
* Oxygen partial pressure
* Substrate temperature

Type of Lasers used in PLD processes. Typically Nd:YAG and Eximer lasers are used for PLD process. The importance of the type of the laser is in the wavelength (or frequency) of the radiation. Following are the common lasers, used in PLD, and their inherent wavelengths:

Laser	Typical Pulse Duration	Wavelength
Eximer Laser	10-50 ns	193-308 nm
Nd: YAG	5 ns- 2ms	1064 nm
CO_2	25 ns- CW	10.6 µm

Effect of Pulse Width and Wavelength. In general long pulse and long wavelength (high frequency) laser radiation gives rise to inhomogeneous and rough surface of the deposited film. When the wavelength is short, the photon energy is dissipated near the surface of target. Thus, evaporation or ablation occurs from a *very thin surface layer*. Hence, short wavelength pulses $\Rightarrow$ near-surface reaction $\Rightarrow$ homogeneous chemistry of the deposited layer. However, at high energy density and short pulse length target gouging occurs. Also at high energy density a poor lateral dispersion of the evaporated species is observed. The angular dispersion of vapor species in PlD is given by:

$$Cos \; \Theta^n \; ,$$

where $n \approx 3\text{-}12$ compared with $n=1$ for thermal vapor deposition.

Steps in Laser Interaction Process. The Laser Interaction Process Steps. The entire laser interaction mechanism associated with PLD can be summarized by the following 3-stages:

     1. Interaction between solid surface-laser $\Rightarrow$ evaporation/ablation

     2. Laser-vapor interaction $\Rightarrow$ isothermal plasma expansion

     3. At the trailing edge of the laser pulse $\Rightarrow$ adiabatic plasma expansion

Reflection Loss and Energy Absorption. Initially electrons acquire all of the energy of the incident photons of the laser beam. Relatively speaking, this energy is gradually transferred to the lattice atoms whereby the thermal energy of the lattice increases. The depth of penetration of laser beam depends on the wavelength of the laser and the dielectric constant of the target. Photon energy which are not coupled with the lattice atoms may be considered as lost by reflection.

The thermal effects associated with laser-materials interactions can be summarized as follows:

* Heat is carried away from the surface by ablation and evaporation, thus the surface temperature of the target is less than the subsurface temperature. That is, heat energy is accumulated below the surface.

* High sub-surface temp $\Rightarrow$ micro-explosions $\Rightarrow$ fragmented particles are erupted from the target. These fragments are then embedded on he film surface, giving rise to an undesirable roughness.

* Typically these fragments are on the order of 0.5-5.0 μm in diameter.

* Figure 4 shows a schematic diagram of the temperature distribution as a function of depth at various energy densities.

* Vapor in plasma is heated by laser.

* Energy absorption by plasma is determined by the absorption coefficient of the plasma.

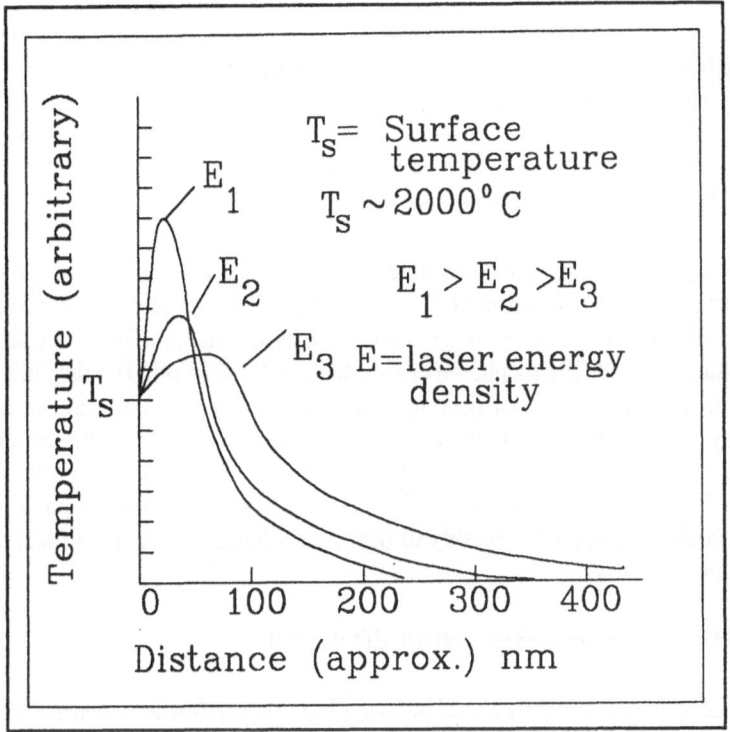

Figure 4. Variation of surface temperature with depth for high energy laser pulse heating (arbitrary scale).

Absorption coefficient, γ, of a plasma is given by[47]:

$$\gamma \propto n_i^2 / T^m \nu^p ,$$

where T is the plasma temperature , n_i is the ion concentration in he plasma plume, ν is laser frequency (c/λ, where c is the velocity of light, and λ is the laser wavelength).

m and p are 1.5 and 2 respectively for kT >> hν, and

m and p are 0.5 and 3 respectively for kT << hν

thus absorption can be enhanced by increasing the number of ionized species in the plasma or by low frequency laser , ie., long wavelength radiation. Note that hotter the plasma better is the vapor transport [47].

2.2. EFFECT OF PULSE WIDTH AND WAVELENGTH

Laser pulses used for PLD process typically have a pulse width of about ten to a few tens of nanoseconds, focused to a few millimeters spot size on the target, and a pulse-energy density of a few J/cm² that produces a peak power density in the order of tens to hundreds of million Watts per cm². Due to extremely high peak power density and extremely short duration, the laser pulses only affect a very thin surface layer of the target and spontaneously convert it into a vapor or fragments without compositional variation or segregation. Such variation might be expected from excessive melting of the High T_C superconducting materials[4,48].

Although superconducting thin films have been successfully deposited with a wide range of wavelengths, i.e. from infrared CO_2 and Nd:YAG laser to ultraviolet excimer laser, highest quality in situ epitaxial films are produced by UV lasers. This is because shorter wavelength lasers have higher photon energy and greater absorption by the plasma plume resulting in a higher degree of decomposition, excitation and ionization of the vapor species by photo-ionization and other related processes[16,43]; thus, the quality of deposited film is enhanced by increased mobility of atoms near the surface of the growing film. Deposition using infrared lasers will result in higher absorption depth and higher deposition rate, but the quality of deposited films are lower than films deposited with UV lasers[9,16,84].

2.3. NATURE OF THE LASER INDUCED PLUME

Once vapor species formed by initial part of a laser pulse, they are further decomposed, ionized and heated by interacting with the remaining laser pulse. Due to rapid heating and expanding of the laser induced plasma, an intense plume consists of high energy vapor species ejected from the target with very high velocities toward the substrate[42-46].

Important information about laser induced plume, such as vapor species, principle luminescent species, expansion velocities, and extent of ionization, have been determined by many groups using emission spectroscopy[43,45,46], mass spectrometry[44], and ion probe[43]. These parameters are very important in terms of understanding the PLD process and improving properties of laser deposited films.

Following characteristics of laser induced plume from high T_c superconductor target have been reported by several groups under typical deposition conditions, i.e. several Joule/cm² energy density and a few tens of nanoseconds pulse-width.

1. Angular distribution of vapor species consisted of two distinct components. One of the components is non-stoichiometric and has a cos q angular distribution which is characteristic of evaporation; the other forward direction component is stoichiometric and has a $cos^8 \theta$ to $cos^{12}\theta$ distribution[15,16,42]. The forward component increases with respect to the evaporation component as laser energy density increases[42].

2. The major luminescent species in the plume are neutral and positively ionized atoms

and dimers[43].

3. By observing the change in the emission intensity of laser induced plume, the plume initially undergoes a one-dimensional expansion for about one spot diameter and then a three-dimensional expansion, in which the vapor density rapidly falling off and the excitation diminishing[3,43].

4. The expansion velocities of the vapor species are about 10^4 m/s, and the corresponding particle energies are in the 25-50 eV range. Relative velocities for Cu and Y are found to confirm the $M^{-1/2}$ dependence predicted by vacuum free expansion model[13,43,45,46].

5. Degree of ionization of the plume is estimated to be 1.4 - 4 % of ionized species from integrated ion signal and mass removal per pulse, assuming atomic species only and a cos θ angular distribution of the species respected to the target normal[43]. This 1.4 - 4 % ionization estimation is likely to be much underestimated because the sharp $cos^8θ$ to $cos^{12}θ$ distribution of the forward vapor component.

6. Plasma temperature is estimated, from expansion velocity and local thermodynamic equilibrium, to be in the range of 6400-13000 K[43].

2.4. PROBLEMS WITH HIGH TEMPERATURE POST ANNEALING

Correct crystallinity and oxygen content is essential for good superconducting properties of the high T_C oxide superconductors. Low deposition rate in the order of angstrom per laser pulse, adequate substrate temperature about 600 to 750 °C, and sufficient oxygen partial pressure of a few hundred millitorrs are important criteria for obtaining in situ high T_C superconducting films. If these criteria are not satisfied, the as-deposited film will not be superconducting or will not have high T_C and high J_C (critical current density). Therefore, a high temperature, above 850 °C, post-annealing is necessary to restore the superconductivity. Care has to be taken to minimize the following undesirable properties associated with high temperature post-annealing process.

1. Chemical reaction with the substrate resulting in the formation of a non-superconducting interface layer[5,49-51].

2. Loss of film stoichiometry after annealing due to surface evaporation and film-substrate interaction[4,51,52]. Barium and/or copper deficiency has been observed after post-annealing process in YBCO films.

3. Crack and/or roughing during thermal cycling resulting in low critical current densities[6].

3. METHODS TO IMPROVE QUALITY OF LOW TEMPERATURE GROWN FILMS

Lowering substrate temperature, yet preserving the superconducting properties, is the key in growing multi-layer component with different materials or to integrate high T_C superconducting thin film process into existing semiconductor processes. Several variations of the basic PLD process, i.e. plasma assisted PLD[10,46], dc-biased PLD[14], and atomic oxygen assisted PLD[13], have been successfully able to lower the in situ deposition temperature to as low as 400 °C[10]. Reduction of the deposition temperature is made possible by creating reactive oxygen species such as atomic oxygen, oxygen ion beam, or oxygen plasma and/or by enhancing mobility of surface atoms with dc-bias or ion bombardment.

3.1. PROBLEMS WITH SURFACE PARTICLES

One unique feature of PLD deposited film is the presence of surface particles on otherwise smooth film surface. The formation of these surface particles has been attributed to sub-surface micro explosion caused by overheating of the sub-surface layer during laser-target interaction[3,53,54]. Koren et al. found that particles are fewer and smaller as the wavelength of the laser decrease due to higher absorption of laser energy by the plasma plume and higher degree of fragmentation in the vapor phase[16].

For a given laser wavelength, the surface particle problem can be minimized by increasing the absorption coefficient and thermal conductivity of the target, which reduce the sub-surface overheating, and by improving the strength of the target, which increases the resistance to explosion[55]. Increased density of the bulk target by silver addition, laser surface melting, melt process, or hot isostatic pressing (HIP) is expected to reduce the surface particle formation. Recently, Eidelloth et al. reported a method to remove surface particles by chemical polishing with diluted HF solution without degrading the superconducting properties of the films[56].

4. PULSED LASER ABLATION (PLA) USING A MILLISECOND Nd:YAG LASER

Although high quality YBCO thin films can be obtained by using nanosecond UV lasers, the deposition rate is very low, typically a few angstroms per laser pulse, and the deposited coverage is very small, a few mm in width, due to angular distributions as sharp as $\cos^8\theta$ to $\cos^{12}\theta$[42,15,16]. On the other hand, high deposition rate up to 100 nm/s[15] and large uniform deposition coverage due to a broader angular distribution, i.e. $\cos\theta$, of the evaporated species can be obtained by using milli-second pulsed lasers. These unique properties make millisecond PLA a potential process to be used to produce continuous coating of tapes. An ion beam assisted millisecond pulsed laser ablation process has been developed to fabricate $YBa_2Cu_3O_x$ high T_c superconductor films with polycrystalline pure silver as primary substrate material.

In this study, a millisecond pulsed neodymium:yttrium aluminum garnet (Nd:YAG) laser with 1064nm wavelength was used. The advantages of using a Nd:YAG laser over an excimer laser are as follows: Firstly, high power Nd:YAG lasers can be operated at a high repetition rate; thus a larger deposition rate can be expected. Secondarily, it has a wide range of possible output wavelengths by harmonic generation, namely near-infrared 1064nm fundamental, green visible 532nm 2nd harmonic, and ultraviolet 355nm 3rd harmonic. Thirdly, there are choices of a wide range of operation modes such as high repetition rate mode, long pulse (ms) mode and Q-switched (ns) mode. Finally, due to its solid state design, it does not have complicated moving no need for corrosive gases.

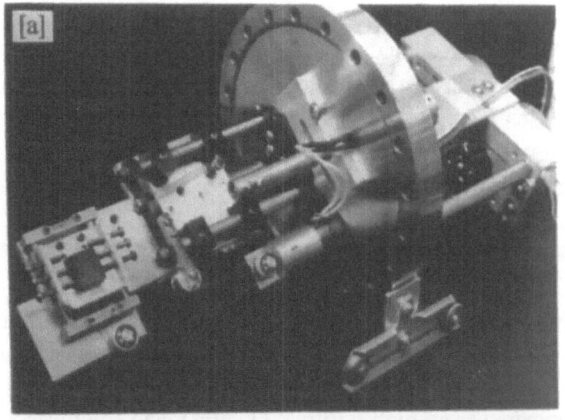

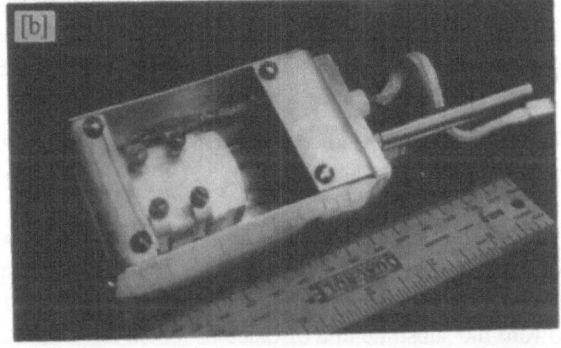

Figure 5. Figure 5(a) shows the target holder with x-y motion; (b) shows the substrate holder with heat shield.

Due to orders of magnitude longer pulse-width, hundreds of $Joule/Cm^2$ pulse energy density is required to achieve peak power density close to mega $Watt/cm^2$, which is still much lower than peak power density produced by nanosecond laser pulses at typically a few $Joule/cm^2$ energy density. With a rotating target, overheating and severe local melting are major problems which are greatly reduced by using a zigzag scanning pattern of the target with an x-y target manipulator designed by us. Figure 5 shows a photograph of this sample stage. A penning ion gun is used to introduce high energy oxygen ions to the laser induced plume and the growing film. Oxygen ion beam is found not only to help replenish oxygen but also help to blow vapor species onto the substrate.

An uniform coating of YBCO with large area coverage (more than 1" diameter) has been successfully deposited on a fused silica substrate. As- deposited films are composed of various compounds of Y_2O_3, BaO and CuO (such as $xBaO \cdot yY_2O_3$) as well as Y_2BaCuO_5. Due to microscopic chemical homogeneity of films, films with majority 1-2-3 phase are formed after just a half hour of post annealing at 850°C. Further annealing results in a decrease of the non- superconducting phases and a higher degree of c-axis preferential orientation. Silver buffer layer is found to effectively minimize degradation caused by film/substrate reaction of (001)YSZ substrate during post annealing.

4.1. EXPERIMENTAL PROCEDURE

A vacuum deposition chamber,Figure 6, with a penning ion gun and a x-y target manipulator was specially designed for the ion beam assisted laser vapor deposition process. YBCO targets with stoichiometric composition (1-2-3 or $YBa_2Cu_3O_x$) were prepared from high purity Y_2O_3, BaO, and CuO using conventional solid state sintering process, i.e. double calcined at 930°C for 10h, then pressed into rectangular pellets and sintered at 950°C for 36h. Only $YBa_2Cu_3O_x$ phase was detected in final sintered YBCO target by using an x-ray diffraction method. Substrates used were metallographically polished polycrystalline silver, optically polished (001) yttrium stabilized zirconia (YSZ) and (001) YSZ with 0.2mm silver buffer layer. Silver buffer layers were prepared by vacuum thermal evaporation.

A pulsed Nd:YAG laser with 50J pulse energy was used in this study. An 8" quartz lens focuses the laser beam to a 1mm x 1.4mm elliptical spot on the YBCO target at 45° to the laser beam. Laser induced plasma plume ejected perpendicularly from the target toward the substrate at a distance of 3.5cm from the target. Substrates were held at 650 to 750°C during deposition by a resistance heater. During deposition, the chamber was kept at 100mtorr of oxygen after flushing with oxygen several times initially. A stream of oxygen ion, generated by the penning ion gun operated at about 1kV, was used to supply activated oxygen to the plume and the growing film. An optical multichannel analyzer was used to study the plasma plume. Simultaneous photographic recording of the plasma also supplied some additional information. Figure 7 shows a schematic diagram of this setup.

The nature of plasma plume, and oxygen ion beam (plasma) are shown in Figure 8. Figures 9(a) and (b) show the variation of the plasma plume at different oxygen partial pressures.

Since the as-deposited films were not superconducting, post annealing was made at 850°C for 0.5 to 2 hours in air. Loss of film, especially from the corners and edges of samples, was occasionally observed. This can be prevented by covering the thin film with a piece of YBCO sintered bulk sample during annealing. Figure 10 shows a scanning electron micrograph (SEM) of typical film of YBCO superconductor produced by ion-beam assisted pulse laser deposition (IBPLD) on Ag substrate. Figure 11 shows

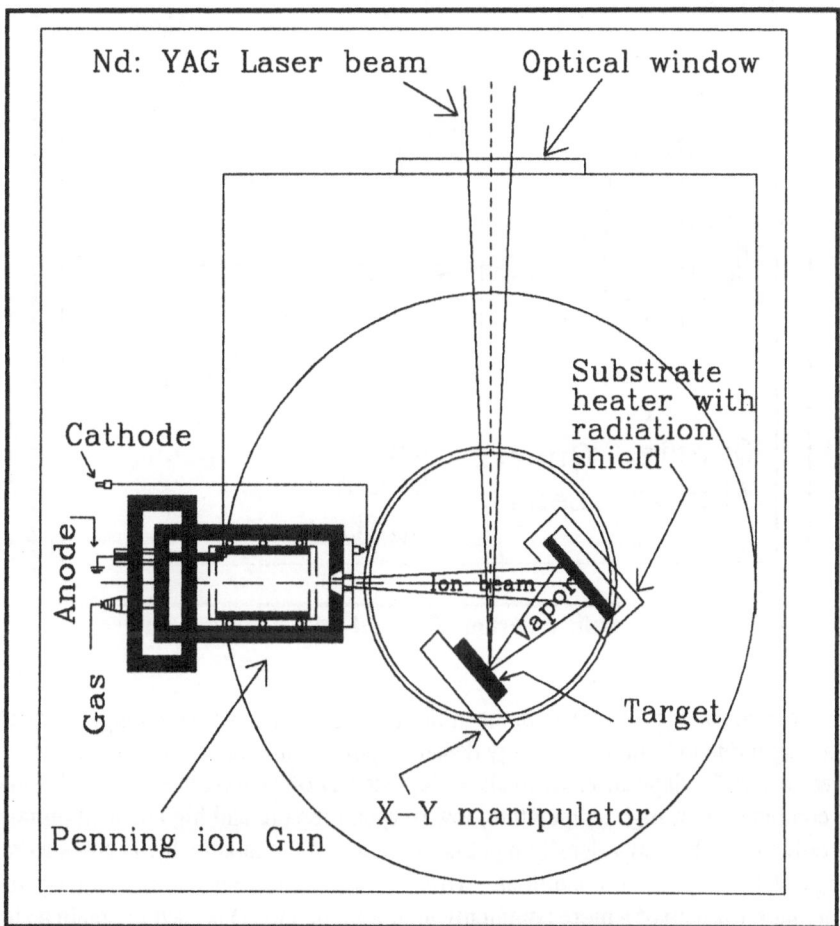

Figure 6. A schematic diagram of the IBPLD apparatus.

a similar thin film IBPLD deposited on (001) surface of YSZ crystal , and Figure 12 is that for a film deposited on (001) YSZ surface with a buffer layer of Ag.

Phases and crystal structures of the films were analyzed by x-ray diffraction (XRD) with $Cu_{K\alpha}$ radiation. Microstructure and chemical composition were characterized by using a scanning electron microscope (SEM) equipped with an energy dispersive spectrum analyzer (EDS). Quantitative analysis with the EDS was made by using a YBCO sintered target as standard and assuming the target to be of stoichiometric composition.

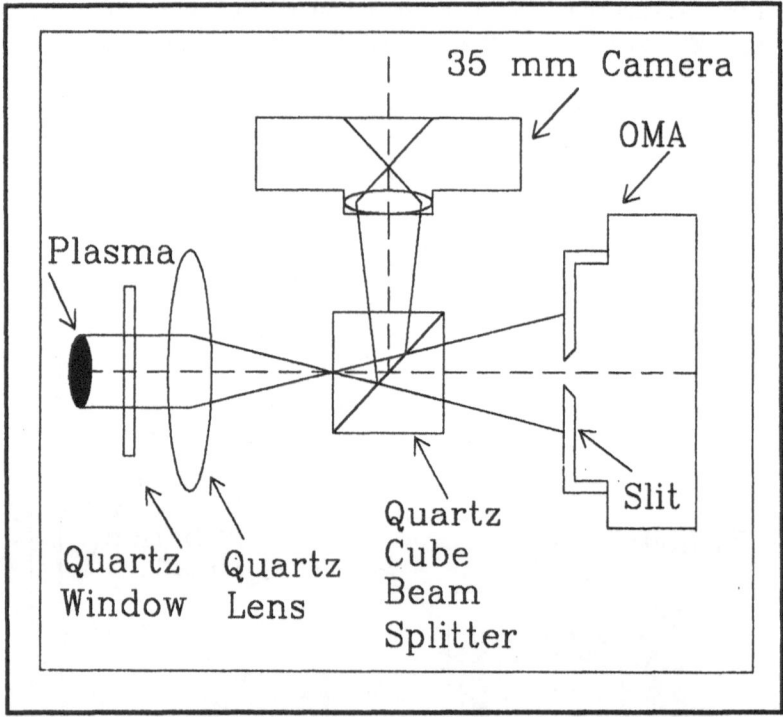

Figure 7. A schematic diagram of the set-up for plasma studies.

There are two major problems with the millisecond pulse deposition. Firstly, overheating of the target due to high pulse energy density input required to create a reasonable peak power density for deposition. Secondary, depletion of oxygen on the deposited film due to a combination of high temperature, low oxygen pressure and high deposition rate. A side effect of high energy density input is cavitation of the target which causes a drastic change of the deposition conditions. To ease the overheating problem, an x-y target manipulator (instead of a more commonly used rotating target) is used to obtain a zigzag scanning pattern on the target to allow more cooling time on each spot of the target surface and to maintain a relative flat target surface. A penning ion gun is used to introduce high energy oxygen ions to the laser induced plume and the growing film.

Very large difference is found between films deposited under identical conditions except with or without oxygen ion beam. When deposited without oxygen ion beam, only a very light coating of light brown color is deposited on the substrate which is about 0.5cm recessed in the heater. This color indicates oxygen deficiency. The light coating may be explained by the location of the substrate which is outside the plume because plume produced by millisecond pulses extended only about 2cm from the target surface. Under same conditions but with the oxygen ion beam turned on, a uniform black coating is obtained on the substrate. This observation indicates that oxygen ion beam not only helps to replenish oxygen but it also helps to blow the vapor species onto the substrate.

Uniform coating of YBCO with a large area coverage, more than 1" diameter, has been successfully deposited on fused silica substrate by ion beam assisted laser ablation. Composition of the film is rather uniform throughout, Figure 13. Results of quantitative composition variation across the film, obtained by using EDS, is shown in Figure 14.

As-deposited film is slightly high in barium and low in yttrium and copper. However, increase in yttrium and further decrease in copper are observed after post annealing. Further investigation is continuing to control the compositional variation during deposition and post annealing.

Data on x-ray diffraction (XRD) of IBPLD as-deposited, and annealed films on silver substrate for YBCO, and Ag-doped YBCO thin films are shown in Figures 15 and 16 respectively. Due to high deposition rate of the process, x-ray diffraction patterns of as-deposited films only show peaks from the substrate plus a few broad peaks around 30° of 2θ angle which are attributed to various compounds of Y_2O_3, BaO and CuO as well as Y_2BaCuO_5, also known as 2-1-1 phase[54]. Just half hour of post annealing at 850°C converts a majority of the film to 1-2-3 phase which is believed to be assisted by microscopic chemical homogeneity of the film. Further annealing results in decreasing of the non-superconducting phases and a higher degree of c-axis preferential orientation which is clearly indicated by the increasing ratio between 002 peak near 15° and 103/110 peak near 33°. As deposited films on (001)YSZ and (001)YSZ with 0.2mm silver buffer layer show similar result as the film on silver substrate. However, after post annealing, strong film/substrate reaction is observed in films deposited on (001)YSZ which is indicated by an additional peak of $xBaO \cdot yZrO_2$ near 43°. It is found that 0.2mm of silver buffer layer effectively minimizes degradation caused by film/substrate reaction of (001)YSZ substrate.

4.1.1. *Summary of PLA Using Millisecond Nd: YAG Laser*

Target overheating problem is greatly reduced by using a zigzag scanning pattern of the target with an x-y target manipulator. A penning ion gun is used to introduce high energy oxygen ions to the laser induced plume and the growing film. Oxygen ion beam is found not only to help to replenish oxygen but also to help blow vapor species onto the substrate. Uniform coating of YBCO with a large area coverage, more than 1" diameter, has been successfully deposited on fused silica substrate by ion beam assisted laser vapor deposition. Compositional variation is observed during deposition and post annealing. Further investigation is needed to control the compositional variation during deposition and post annealing.

As-deposited films are composed of various compounds of Y_2O_3, BaO and CuO as well as Y_2BaCuO_5. Due to microscopic chemical homogeneity of films, nearly pure 1-2-3 phase can be induced by post annealing at 850°C for only about half an hour. Further annealing results in decreasing of the amount of non-superconducting phases and a higher degree of c-axis preferential orientation. Silver buffer layer is found to effectively minimizes degradation caused by film/substrate reaction of (001)YSZ substrate during post annealing. In addition to our PLD, and IBPLD work, we conducted some further investigation on laser calcination of BSCCO superconductor compound. These results are briefly discussed next.

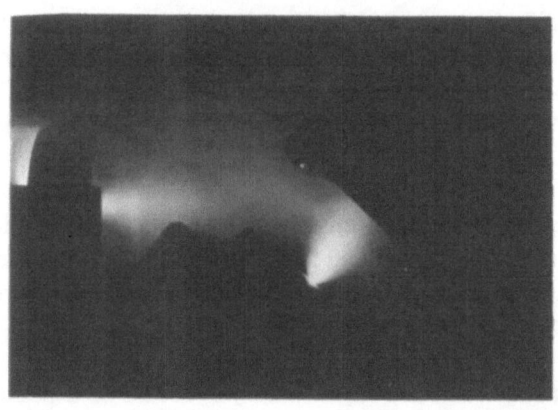

Figure 8. A photograph of the plasma plume during IBPLD proces.

5. LASER CALCINATION OF PB DOPED BI-SR-CA-CU-O SUPERCONDUCTOR

The high T_c Bi-Sr-Ca-Cu-O superconducting compound has a number of phases with different critical temperatures (T_c). The three major phases are referred to by their cation ratios as 2201 (T_c below 77K), 2212 (T_c~85K), and 2223 (T_c~110K) phases[35]. Since the discovery of superconductivity in this compound[36], many efforts have been made to maximize the amount of high T_c-2223 (110K) phase. A major enhancement in the formation of high T_c- 2223 phase was made by doping Bi-Sr-Ca-Cu-O compound with Pb[37-41]. However, With Pb doping, about 200 hours of total processing time is required to obtain a single or nearly a single high T_c phase. It has been shown that addition of Pb enhances the formation of 110K (2223) phase in Bi- Sr-Ca-Cu-O system because of kinetic reasons[37]. With the premise that a local temperature spike, exceeding the melting temperature of the compound, might enhance the nucleation and growth of the 110K phase, a new laser processing technique was developed with an objective to enhance the kinetics of formation of the high T_c phase of a Pb doped Bi-Sr-Ca-Cu-O superconductor. With laser calcination process, we were able to fabricate a near single-phase high T_c Bi-Pb-Sr-Ca-Cu-O superconductor in about 100 hours of sintering, following the laser calcination. The onset critical temperature of the laser calcined sample was found to be about 110K. However, the zero resistance temperature was about 98K.

5.1. EXPERIMENTAL PROCEDURE

Starting materials of Bi_2O_3, PbO, $SrCO_3$, $CaCO_3$, and CuO were weighed and mixed to have Bi:Pb:Sr:Ca:Cu cation ratio of 1.5:0.5:2:2:3. This powder was thoroughly ground with a pestle and mortar to ensure through mixing. The mixed fine powder was then placed in a specially designed slow rotating pan and irradiated with a de-focused pulsed Nd:YAG laser beam. A laser with 2.2ms pulse width, 100Hz pulse rate, $0.5cm^2$ spot size, and power level of 50 or 90W was used. During laser irradiation, the powder was constantly stirred by a stationary vane attached to the rotating pan. Periodically, calcined powder was re-ground to ensure uniform laser-powder interaction. The average linear speed of

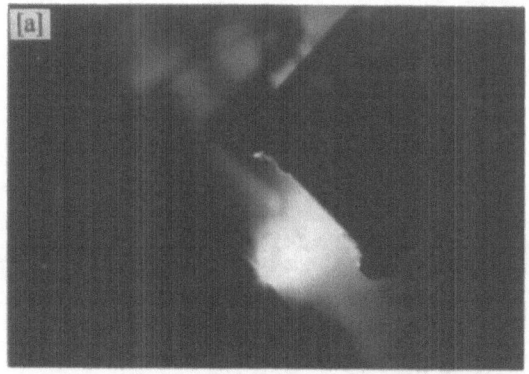

Figure 9. The effect of partial pressure of oxygen on the nature of the plusma. (a) 200 mtorr, (b)300 mtorr.

the rotating pan with respect to laser beam was about 2.5 cm/s. For comparison, an identical batch of powder mixture was also calcined conventionally in a furnace set at 830°C for 7 hours.

The laser calcined and conventionally calcined powders were pelletized and these pellets were sintered at 865°C in air for different durations up to 240 hours, then cooled at a rate of 100°C/s. A Scintac-XDS-2000 x-ray diffractometer (XRD) withaCu target, a Hitachi S-2500C scanning electron microscope with Link energy dispersive x-ray spectroscopy (EDS), were used to characterize these superconductors. Resistivity was measured by using an auto-balancing ac bridge with a lock-in amplifier (Linear Research) using a standard four-probe technique. The mutualinductance, which is proportional to the ac magnetic susceptibility of the in-phasecomponent (χ'), was measured directly using an auto-balancing Hartshorn bridge.

Figure 10. SEM picture of IBPLD YBCO on Ag substrate.

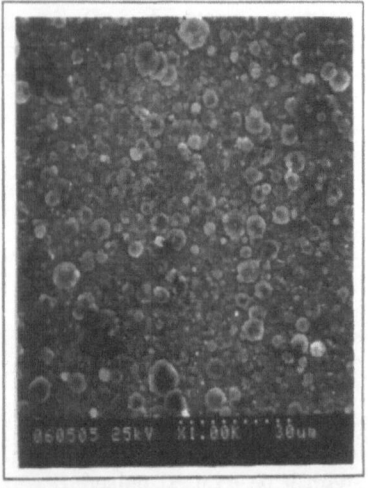

Figure 11. Same as in Figure 10, but on (001) YSZ

5.1.1. *Results and Discussions.*

Melt-quench effect was obtained when Bi-Pb-Sr-Ca-Cu-O powder mixture was exposed to high energy and short duration laser pulses. Coarse particles were formed by localized melting. Due to a localized rapid heating and quenching, chemical homogeneity of the well mixed starting powder was preserved. Therefore, we were able to greatly increase the rate of formation of the high-T_c phase ($T_c \sim 110$K).

The mechanism of formation of high T_c phase by laser and conventional processes appeared to be quite different by observation of the XRD data. *ure 7*, The conventionally processed sample initially showed a mixture of low T_c-2212 phase and Ca_2PbO_4. As the duration of sintering increased, the amount of Ca_2PbO_4 decreased, and

Figure 12. SEM of IBPLD YBCO film on (001) YSZ with a buffer layer of Ag.

a phase cycling behavior was observed[1]. The high T_c-2223 phase was then formed by a reaction between the low T_c phase and Ca_2PbO_4 and Ca_2CuO_3 *[58,59]*.

Different results were observed for two laser processed samples. At an early stage of sintering, predominately a low T_c-2212 phase was formed. After prolonged sintering, the high T_c phase, $Bi_2Sr_2Ca_2Cu_3O_x$, was formed via an unbalanced reaction of the following form[60]:

$$Bi_{22}Sr_2CaCu_2O_x \Rightarrow Bi_2Sr_2Ca_2Cu_3O_y + Bi_2Sr_2CuO_z$$

The initial formation of 2212 phase leaves some excess Ca and Cu ions of the starting stoichiometric 2223 composition. The 2201 ($Bi_2Sr_2CuO_z$) phase formed, then combines with the unused Ca and Cu ions to form the low T_c-2212 phase. Therefore, compared to the high T_c-2223 phase, the amount of retained 2201 phase in laser calcined samples was not large. Due to the high energy density of the laser beam, some weight loss in the form of particulate ejection of the starting powder is unavoidable during laser calcination process. Thus, the final compositions of laser calcined and conventionally calcined samples were analyzed by EDS rather than by a weight loss measurement. No significant compositional difference was found through the chemical analysis.

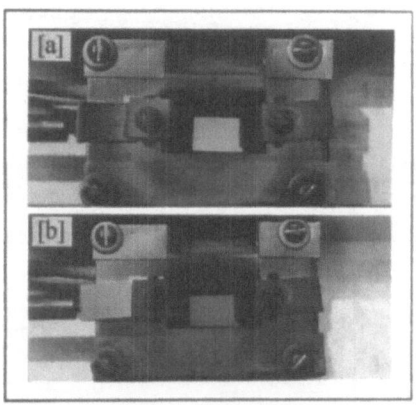

Figure13. Two thin films.(a) Without ion beam, (b) with ion beam. In (b) an uniform grey color film is clearly visible.

Characteristic plate-like grain structures ere observed for both laser and conventionally processed samples; the plate size of laser processed sample, 10-30mm wide and about 1mm thick, was about 2-3 times as large as the conventionally processed one. All these samples show superconducting onset temperature of about 110K. The zero resistance temperature of laser processed sample, after 100 hours sintering, is about 98K.

5.1.2.Summary of Laser Calcination of Pb Doped Bi-Sr-Ca-Cu-O superconductor.
Superconductors of Bi-Pb-Sr-Ca-Cu-O system of near single high T_c phase were made by using a new laser calcining process. The total processing time was reduced to about 100 hours. Both resistance and magnetic susceptibility data showed an onset of superconducting transition at about 110K. A sharp magnetic susceptibility drop was observed above 106K. The zero resistance temperature was about 98K. It is concluded that the high T_c phase was formed via a different kinetic path in laser calcined sample compare with the conventionally processed sample.

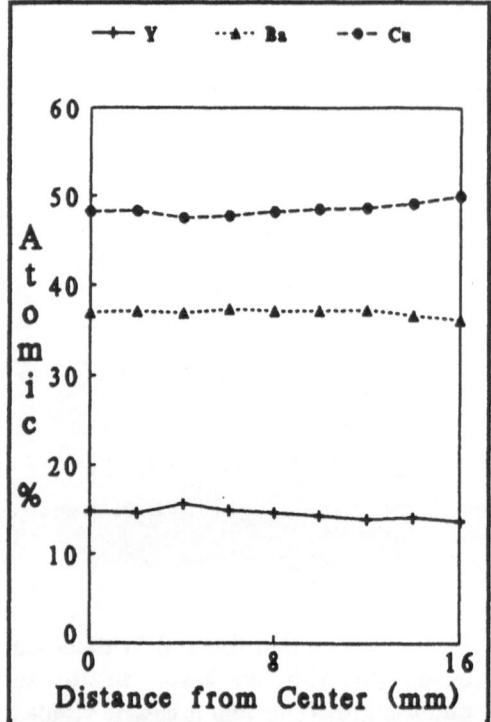

Figure 14. Composition profile across the surface of IBPLD film of YBCO. Data from EDS.

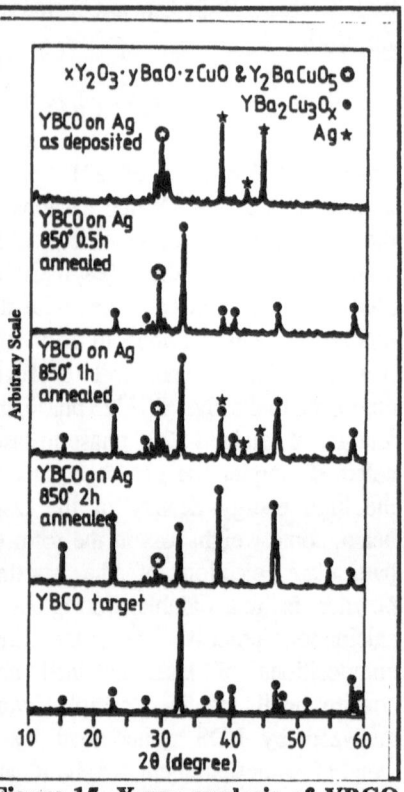

Figure 15. X-ray analysis of YBCO on Ag substrate.

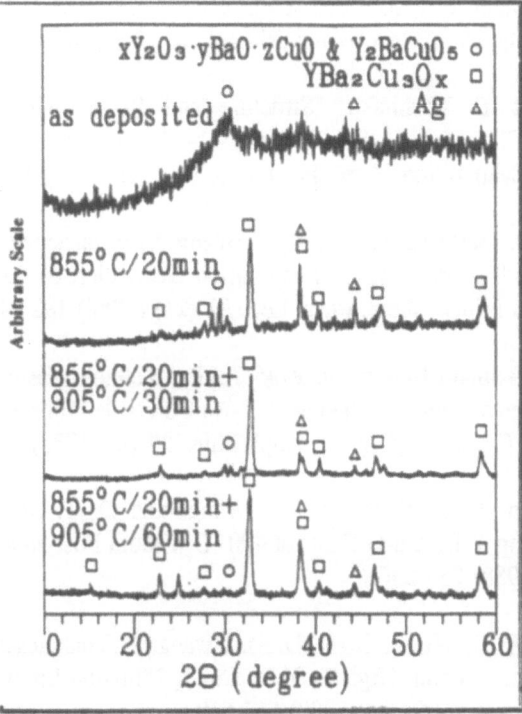

Figure 16. X-ray diffraction data from as-deposited and annealed YBCO with 15% Ag doping.

REFERENCES

1. R. Beyers et al., "Crystallography and Microstructure of Tl-Ca-Ba-Cu-O Superconducting Oxides," Appl. Phys. Lett., **53 (5)** (1988) 432-434.

2. W. Y. Lee et al., "Low-temperature Formation of Epitaxial $Tl_2Ca_2Ba_2Cu_3O_{10}$ Thin Films in Reduced O_2 Pressure," Appl. Phys. Lett., **60 (6)** (1992) 774-774.

3. R. K. Singh and J. Narayan, "The Pulsed-Laser Deposition of Superconducting Thin Films," JOM, **no. 3** (1991) 13-20.

4. D. Dijkkamp and T. Venkatesan, "Preparation of Y-Ba-Cu Oxide Superconductor Thin Films Using Pulsed Laser Evaporation from High T_c Bulk Materials," Appl. Phys. Lett. **51 (8)** (1987) 619-621.

5. X. D. Wu et al, "Epitaxial Ordering of Oxide Superconductor Thin Films on $(100)SrTiO_3$ Prepared by Pulsed Laser Evaporation, "Appl. Phys. Lett. **51 (11)** (1987) 861-863.

6. X. D. Wu, A. Inam, T. Venkatesan et al., "Low-temperature Preparation of High T_C Superconducting Thin Films," Appl. Phys. Lett. **52 (9)** (1988) 754-756.

7. S. Miura and T. Yoshitake, "Structure and Superconducting Properties of $Y_1Ba_2Cu_3O_{7-d}$ Films Prepared by Transversely Excited Atmospheric Pressure CO_2 Pulsed Laser Evaporation,"Appl. Phys. Lett. **52 (12)** (1988) 1008-1010.

8. H. S. Kwok, P. Mattocks, L. Shi, X. W. Wang, S. Witanachchi, Q. Y. Ying, J. P. Zheng, and D. T. Shaw, "Laser Evaporation Deposition of Superconducting and Dielectric Thin Films," Appl. Phys. Lett. **52 (21)** (1988) 1825-1827.

9. Osamu Eryu, Kouichi Murakami, Koki Takita, Kohzoh Masuda, Hiromoto Uwe, Hiroshi Kudo, and Tsunetaro Sakudo, " Y-Ba-Cu Oxide Films Formed with Pulsed-Laser Induced Fragments," Jpn. J. Appl. Phys. **27 (4)** (1988) L628-L631.

10. S. Witanachchi, H. S. Kwok, X. W. Wang, and D. T. Shaw, "Deposition of Superconducting Y-Ba-Cu-O Films at 400 °C without Post-annealing" Appl. Phys. Lett. **53 (3)** (1988) 234-236.

11. C. C. Chang, X. D. Wu, A. Inam, D. M. Hwang, T. Venkatesan, P. Barboux, and J. M. Tarascon, "Smooth High T_C $Y_1Ba_2Cu_3O_x$ Films by Laser Deposition at 650 °C," Appl. Phys. Lett. **53 (6)** (1988) 517-519.

12. B. Roas, L. Schultz, and G. Endres, "Epitaxial Growth of $YBa_2Cu_3O_{7-x}$ Thin Films by a Laser Evaporation Process," Appl. Phys. Lett. **53 (16)** (1988) 1557-1559.

13. G. Koren, A. Gupta, and R. J. Baseman, "Role of Atomic Oxygen in Low-temperature Growth of $YBa_2Cu_3O_{x-d}$ Thin Films by Laser Ablation Deposition," Appl. Phys. Lett. **54 (19)** (1989) 1920-1922.

14. R. K. Singh, J. Narayan, A. K. Singh, and J. Krishnaswamy, "In situ Processing of Epitaxial Y-Ba-Cu-O High T_C Superconducting Films on $(100)SrTiO_3$ and $(100)YS$-ZrO_2 Substrates at 500-650 °C," Appl. Phys. Lett. **54 (22)** (1989) 2271-2273.

15. M. Balooch, D. R. Olander, and R. E. Russo, "Y-Ba-Cu-O Superconducting Films Produced by Long-pulsed Laser Vaporization," Appl. Phys. Lett. **55 (2)** (1989) 197-199.

16. G. Koren, A. Gupta, R. J. Baseman, M. I. Lutwyche, and R. B. Laibowitz, "Laser Wavelength Dependent Properties of $YBa_2Cu_3O_{x-d}$ Thin Films Deposited by Laser Ablation," Appl. Phys. Lett. **55 (23)** (1989) 2450-2452.

17. B. H. Moeckly, S. E. Russek, D. K. Lathrop, R. A. Buhrman, Jian Li, and J. W. Mayer, "Growth of $YBa_2Cu_3O_7$ Thin Films on MgO: The Effect of Substrate Preparation," Appl. Phys. Lett. **57 (16)** (1990) 1687-1689.

18. J.P. Zheng, S.Y. Dong, and H.S.Kwok,"Texturing of Epitaxial in situ Y-Ba-Cu-O Thin Films on Crystalline Substrates," Appl. Phys. Lett. **58 (5)** (1991) 540-542.

19. R. K. Singh and D. Bhattacharya, "Improvement in the Properties of High T_c Films Fabricated in situ by Laser Ablation of $YBa_2Cu_3O_7$-Ag Targets," Appl. Phys. Lett. **60 (2)** (1992) 255-257.

20. Y. Iijima, N. Tanabe, O. Kohno, and Y. Ikeno, "In-plane Aligned $YBa_2Cu_3O_{7-x}$ Thin Films Deposited On Polycrystalline Metallic Substrates," Appl. Phys. Lett. **60 (6)** (1992) 769-771.

21. D. K. Fork and K. Nashimoto, "Epitaxial $YBa_2Cu_3O_{7-d}$ on GaAs(001) Using Buffer Layers," Appl. Phys. Lett. **60 (13)** (1992) 1621-1623.

22. J. A. Alarco,G. Brorsson, Z. G. Ivanov, P.-A, Nilsson, and E. Olsson, "Effect of Substrate Temperature on The Microstructure of $YBa_2Cu_3O_{7-d}$ Films Grown on (001)Y-ZrO_2 Substrates," Appl. Phys. Lett. **61 (6)** (1992) 723-725.

23. S. Komuro, Y. Aoyagi, T. Morikawa and S. Namba, "Preparation of High-T_c Superconducting Films by Q-switched YAG Laser Sputtering," Jpn. J. Appl. Phys. **27 (1)** (1988) L34-L36.

24. Arun Inam and X. D. Wu, "Pulsed Laser Etching of High T_c Superconducting Films," Appl. Phys. Lett. **51 (14)** (1987) 1112-1114.

25. J. Mannhart, M. Scheuermann, C. C. Tsuei, M. M. Oprysko, C. C. Chi, C. P. Umbach, R. H. Koch and C. Miller, "Micropatterning of High T_c Films with an Excimer Laser," Appl. Phys. Lett. **52 (15)** (1988) 1271-1273.

26. R. G. Humphreys, J. S. Satchell, N. G. Chew, and J. A. Edwards, "Narrow Tracks in $YBa_2Cu_3O_7$ Thin Films Defined by Laser Ablation," Appl. Phys. Lett. **54 (1)** (1989) 75-77.

27. Arunava Gupta and Gad Koren, "Direct Laser Writing of Superconducting Patterns of $Y_1Ba_2Cu_3O_{7-d}$," Appl. Phys. Lett. **52 (8)** (1988) 665-666.

28. B. W. Hussey and A. Gupta,"Laser-assisted Etch of $YBa_2Cu_3O_{7-d}$," Appl. Phys. Lett. **54 (13)** (1989) 1272-1274.

662

29. G. E. Jang, K. Mukherjee, P. A. A. Khan and M. Tayal, "Manufacturing of High Temperature Superconductor by Using A Nd:Yag Laser," Laser Materials Processing III (Warrendale, PA: The Minerals, Metals & Materials Society, 1988), 159-169.

30. C. W. Chen, P. A. A. Khan and K. Mukherjee, "Laser Fabrication of Pb Doped Bi-Sr-Ca-Cu-O Superconductor," J. Mat'ls Sci. **27** (1992) 3221-3224.

31. M. Levinson, S. S. P. Shah, and D. Y. Wang, "Laser Zone-melted Bi-Sr-Ca-Cu-O Thick Films," Appl. Phys. Lett. **55 (16)** (1989) 1683-1685.

32. Naoaki Aizaki, Koichi Terashima, Jun-ichi Fujita and Sinji Matsui, "YBa$_2$Cu$_3$O$_y$ Superconducting Thin Film Obtained by Laser Annealing," Jpn. J. Appl. Phys. **27 (2)** (1988) L231-233.

33. P. Haldar, J. G. Hoehn, Jr., and J. A. Rice, "Enhancement in Critical Current Density of Bi-Pb-Sr-Ca-Cu-O Tapes by Thermomechanical Processing: Cold Rolling versus Uniaxial Pressing," Appl. Phys. Lett., **60 (4)** (1992) 495-497.

34. S. Jin, G. W. Kammlott, T. H. Tiefel and S. K. Chen, "Formation of Layered Microstructure in the Y-Ba-Cu-O and Bi-Sr-Ca-Cu-O Superconductors," Physica C **198** (1992) 333-340.

35. J. M. Tarascon, W. R. McKinnon, P. Barboux, D. M. Hwang, B. G. Bagley, L. H. Greene, G. Hull, Y. LePage, N. Stoffel, and M. Giroud, "Preparation, Structure, and Properties of the Superconducting Compound Series Bi$_2$Sr$_2$Ca$_{n-1}$Cu$_n$O$_y$ with n = 1, 2, and 3," Phys. Rev. **38 (13)** (1988) 8885-8892.

36. H. Maeda, Y. Tanaka, M. Fukutomi, and T. Asano, "A New High-T$_C$ Oxide Superconductor without a Rare Earth Element," J. Appl. Phys. **27 (2)** (1988) L209-L210.

37. D. Shi, M. S. Boley, J. G. Chen, M. Xu, K. Vandervoort, Y. X. Liao, A. Zangvil, J. Akujieze, and C. Segre, "Origin of Enhanced Growth of the 110 K Superconducting Phase by Pb Doping in the Bi-Sr-Ca-Cu-O System," Appl. Phys. Lett. **55 (7)** (1989) 699-701.

38. Y. Yamada and S. Murase, "Pb Introduction to the High-T$_C$ Superconductor Bi-Sr-Ca-Cu-O," Jpn. J. Appl. Phys. **27 (6)** (1988) L996-L998.

39. M. Takano, J. Takada, K. Oda, H. Kitaguchi, Y. Miura, Y.Ikeda, Y. Tomii, and H. mazaki, "High-T$_c$ Phase Promoted and Stabilized in the Bi, Pb-Sr-Ca-Cu-O System," Jpn. J. Appl. Phys. **27 (6)** (1988) L1041-L1043.

40. M. Mizuno, H. Endo, J. Tsuchiya, N. Kijima, A. Sumiyama, and Y. Oguri, "Superconductivity of $Bi_2Sr_2Ca_2Cu_3Pb_xO_Y$ (x=0.2, 0.4, 0.6)," Jpn. J. Appl. Phys. **27** **(7)** (1988) L1225-L1227.

41. S. A. Sunshine, T. Siegrist, L. F. Schneemeyer, D. W. Murphy, R. J. Cava, B. Batlogg, R. B. van Dover, R. M. Fleming, S. H. Glarum, S. Nakahara, R. Farrow, J. J. Krajewski, S. M. Zahurak, J. V. Waszczak, J. H. Marshall, P. March, L. W. Rupp, Jr. ,and W. F. Feck, "Structure and Physical properties of Single Crystals of the 84-K Superconductor $Bi_{2.2}Sr_2Ca_{0.8}Cu_2O_{8+d}$," Phys Rev. B**38** (1988) 893.

42. T. Venkatesan, X. D. Wu, A. Inam, and J. B. Wachtman, "Observation of Two Distinct Components During Pulsed Laser Deposition of High T_C Superconducting Films," Appl. Phys. Lett. **52 (14)** (1988) 1193-1195.

43. P. E. Dyer, R. D. Greenough, A. Issa, and P. H. Key, "Spectroscopic and Ion Probe Measurements of KrF Laser Ablated Y-Ba-Cu-O Bulk Samples," Appl. Phys. Lett. **53 (6)** (1988) 534-536.

44. T. Venkatesan et al., "Nature of The Pulsed Laser Process for The Deposition of High T_C Superconducting Thin Films," Appl. Phys. Lett. **53 (15)** (1988) 1431-1433.

45. J. P. Zheng, Z. Q. Huang, D. T. Shaw, and S. Kwok, "Generation of High-energy Atomic Beam in Laser-Superconducting Target Interactions," Appl. Phys. Lett. **54 (3)** (1989) 280-282.

46. J. P. Zheng, Q. Y. Ying, S. Witanachchi, Z. Q. Huang, D. T. Shaw, and S. Kwok, "Role of The Oxygen Atomic Beam in Low-temperature Growth of Superconducting Films by Laser Deposition," Appl. Phys. Lett. **54 (10)** (1989) 954-956.

47. John F. Ready, "Effects of High-Power Laser Radition", Academic Press, p. 188, (1971).

48. O. Auciello, A. R. Krauss, J. Santiago-Aviles, A. F. Schreiner, and D. M. Gruen, "Surface Compositional and Topographical Changes resulting from Excimer Laser Impacting on $YBa_2Cu_3O_7$ Single Phase Superconductors," Appl. Phys. Lett. **52 (3)** (1988) 239-241.

49. M. Gurvitch and A. T. Fiory, "Preparation and Substrate Reaction of Superconducting Y-Ba-Cu-O Films," Appl. Phys. Lett. **51 (13)** (1987) 1027-1029.

50. M. Aslam, R. E. Soltis, E. M. Logothetis, R. Ager, M. Mikkor, W. Win, J. T. Chen and L. E. Wenger, "Rapid Thermal Annealing Of YBaCuO Films on Si and SiO_2 Substrates," Appl. Phys. Lett. **53 (2)** (1988) 153-155.

664

51. M. J. Cima, J. S. Schneider, and S. C. Peterson, "Reaction of $Ba_2YCu_3O_{6.9}$ Films with Yttria-stabilized Zirconia Substrates," Appl. Phys. Lett. **53** (8) (1988) 710-712.

52. S. I. Shah, "Annealing Studies of $YBa_2Cu_3O_{7-x}$ Thin Films," Appl. Phys. Lett. **53** (7) (1988) 612-614.

53. A. Zherikhin, V. Bagratashvili, V. Burimov, E. Sobol, G. Shubnii and A. Sviridov, "The Action of Powerful Laser Radiation on 1-2-3 Superconducting Thin Films and Bulk Materials," Physica C **198** (1992) 341-348.

54. Shigeru Otsubo, T. Minamikawa, Y. Yonezawa, A. Morimoto, and T. Shimizu, "Thermal Analysis of Target Surface in the Ba-Y-Cu-O Film Preparation by Laser Ablation Method," Jpn. J. Appl. Phys. **29** (1) (1990) L73-L76.

55. Rajiv K. Singh, D. Bhattacharya, and J. Narayan, "Control of Surface Particle Density in Pulsed Laser Deposition of Superconducting $YBa_2Cu_3O_7$ and Diamondlike Carbon Thin Films," Appl. Phys. Lett. **61** (4) (1992) 483-485.

56. W. Eidelloth, R. L. Sandstrom and M. M. Plechaty, "Polishing of Highly Oriented $YBa_2Cu_3O_{7-d}$ Thin Films," Physica C **197** (1992) 389-393.

57. Powder Diffraction File (Philadelphia, PA: American Society for Testing and Materials, 1988).

58. T. Uzumaki, K. Yamanaka, N. Kamehara, and K. Niwa, "The Effect of Ca_2PbO_4 Addition on Superconductivity in a Bi-Sr- Cu-O System," Jpn. J. Appl. Phys. **28** (1) (1989) L75-L77.

59. H. Endo, J. Tsuchiya, N. Kijima, A. Sumiyama, M. Mizuno, and Y. Oguri, "Thermal Stability of the High-T_c Superconductor in the Bi-Sr-Ca-Cu-O System," Jpn. J. Appl. Phys. **27** (10) (1988) L1906-L1909.

60. N. Kijima, H. Endo, J. Tsuchiya, A. Sumiyama, M. Mizuno, and Y. Oguri, "Reaction Mechanism of Forming the High-T_c Superconductor in the Pb-Bi-Sr-Ca-Cu-O System," Jpn. J. Appl. Phys. **27** (10) (1988) L1852-L1855.

LASER CHEMICAL VAPOUR DEPOSITION
OF TITANIUM-BASED HARD COATINGS

O. CONDE, A.J. SILVESTRE, M.L. PARAMÊS
Department of Physics, University of Lisbon
Campo Grande, Ed. C1, 1700 Lisboa, Portugal

Abstract. This article is primarily concerned with the laser-assisted chemical vapour deposition (LCVD) of titanium nitride and titanium carbide coatings. The principles of the LCVD technique are briefly outlined and the literature on the LCVD of TiN and TiC films is reviewed. Emphasis is given to the growth mechanisms and to the morphological and structural aspects of the films.

1. Introduction

One of the most widespread technologies for fabricating thin films and coatings is the chemical vapour deposition (CVD) technique. It has long been used in the manufacture of very different types of films with applications in different industrial fields, such as cutting tools, microelectronics, optoelectronics, optics, refractory fibers, etc. However, the two major areas of application of CVD have been in the semiconductor industry and in the so-called metallurgical hard and wear resistant coatings industry. For a recent general review of the CVD process covering its different aspects from background to applications, the reader is referred to the textbook by Pierson [1]. Other review papers by various authors [2-5] give a comprehensive summary of the main features of the CVD technique.

In a CVD process, a solid product is formed as the result of a chemical reaction occurring in a gas or vapour phase near or on the surface of a substrate. In conventional CVD the reaction is thermally activated, i.e., the substrate is directly and uniformly

665

J. Mazumder et al. (eds.), Laser Processing: Surface Treatment and Film Deposition, 665–691.
© 1996 *Kluwer Academic Publishers.*

heated, and an extended uniform film of the deposited material is obtained. The advantages of CVD over other deposition methods (physical vapour deposition - PVD) are numerous. For instance, uniform film thickness, conformal step coverage, radiation-damage-free deposition, control of the stoichiometry, morphology and crystal orientation of the deposited film, are some of the features commonly achieved [6]. Further, the CVD technique is characterised by high throwing power, i.e., it is not restricted to a line of sight deposition [1], and by low cost processing [1, 6]. Perhaps the most important reason for the success of this technique is its ability to coat pieces of very different shapes and sizes.

However, the increasing complexity and miniaturisation of systems needed in such areas as microelectronics and integrated optics require controlled area coating techniques which avoid excessive application of chemicals. This is achieved in CVD, and other conventional deposition techniques, by employing a sequential application of mechanical masking and/or lithographic methods. Nevertheless, these procedures are time-consuming and not always possible, especially if the substrate is non-planar. Further, for many of the chemical reactions used in the deposition of materials with interesting properties - silicon or carbon, as well as compounds such as carbides, nitrides and oxides - the chemical reaction threshold temperature is relatively high, about 1000 °C [2, 4], which limits the ability to coat heat-sensitive substrates.

Laser-assisted chemical vapour deposition (LCVD) aims at minimizing the drawbacks of CVD when selective area growth is desirable, and presents a number of additional advantages. Although the main field of application of LCVD has been microelectronics, the characteristics of this technique are particularly convenient for the localised deposition of hard coatings for micromechanics and metallurgical applications.

2. The LCVD process

Laser-induced chemical processing (LCP) of materials has been developing rapidly because of its new and unique processing characteristics which are of fundamental importance in many areas of technology [7]. In chemical processing the laser light activates a chemical reaction or enhances the reaction rate. Such reactions can take place in adsorbed layers, at gas/solid or liquid/solid interfaces, or within the substrate itself. As a result, material deposition, etching, doping and material synthesis may be carried out.

Among the applications of LCP, laser-assisted chemical vapour deposition has been one of the most extensively studied, both experimentally and theoretically. The extremely high deposition rates achieved in LCVD [8] together with strongly localised heat and chemical treatments, make LCVD an attractive technique for the deposition of a wide variety of materials useful in microelectronics, micromechanics, etc.

Laser induced deposition from the gas phase is a complex process because different microscopic mechanisms may be involved simultaneously. Laser chemical vapour deposition can be based on reactions which are initiated by photolysis (photochemical), by pyrolysis (photothermal) or both.

2.1. PHOTOLYTIC LCVD

Photo-induced chemical vapor deposition has attracted much attention during the last few years as a low temperature technique for growing thin films. Both lamps and lasers have been used as photon sources. Because laser beams provide a much higher power density, film growth rates can be increased compared to those obtained by employing conventional light sources. Photolytic LCVD is thus a powerful technique for the growth of extended coatings on temperature sensitive materials, such as GaAs, InP and polymers. Direct writing of microstructures may also be achieved by this method [9].

In photolytic LCVD, film deposition occurs by direct bond breaking in the precursor molecules due to resonant absorption of the laser radiation. Because most molecular bond energies are of several eV, ultraviolet laser light (e.g. frequency doubled Ar^+ or Kr^+ and excimer lasers) is generally required for dissociative-electronic excitation. But photolytic deposition can also be accomplished by multiphoton vibrational excitation with IR laser light, by using an appropriate laser type/chemical precursor combination, for example a CO_2 laser with BCl_3, SiH_4 or NH_3 gases.

The usual experimental set-up for photolytic laser deposition is based on a parallel configuration, i.e., the laser beam passes parallel to the substrate surface at a certain distance, and the chemical reaction starts after the laser photons have been absorbed by the reactive gas phase. This geometry allows for independent control of the substrate temperature during the deposition process. Therefore, growth rates and properties of the deposited films can be optimised within a broader range.

The reader will find in the review article by Jackson et al. [10] and in the textbook by Eden [11] detailed and valuable information on the subject of photochemical LCVD.

2.2. PYROLYTIC LCVD

In pyrolytic LCVD the gas phase is transparent to the laser radiation. The laser beam is directed perpendicular to the substrate, and acts as the driving force of the chemical reaction by raising the temperature of the substrate surface locally. With a stationary laser beam three-dimensional material structures can be grown [12]. By moving the substrate relative to the laser beam single-step deposited patterns may be generated. For example, in the case of a single translation, one obtains stripes. The width and height of the stripes depend on a large number of variables including the optical and thermal properties of the substrate and film materials, the energy distribution profile of the beam, the laser power density, the scanning velocity, the chemical reaction threshold temperature and the reaction kinetics.

The microscopic mechanism for deposition is expected to be essentially the same as in conventional large area CVD, namely the thermal activation of the chemical reaction. As a consequence, all reactions observed in large area deposition can, in principle, be used in pyrolytic LCVD. However, a number of fundamental new kinetic aspects arise in laser thermal deposition *versus* traditional CVD owing to the 3D character of the diffusion of chemical species into the microreaction zone generated by the localised heat source. A major consequence of this is the observed enhancement of the deposition rate by two to four orders of magnitude [13].

Pyrolytic LCVD work was firstly carried out with an IR laser and was reported in 1972 [14]. Since that time almost all commercially available lasers have been utilised as radiation sources. When high spatial resolution is required, in particular submicron size patterns for microelectronic devices, Ar^+ and Kr^+ are preferentially utilised. Otherwise, Nd:YAG, Nd:glass and CO_2 lasers can be successfully employed [15]. Among these, CO_2 lasers are preferred because they offer greater efficiency and possess better beam qualities than other continuous mode lasers [16].

This paper is restricted to the pyrolytic LCVD technique, i.e., thermally activated laser CVD.

3. LCVD of titanium based hard coatings

3.1. GENERAL PROPERTIES

Titanium is a very unstable metal with respect to higher oxidation states and, therefore, it reacts readily with, e.g., boron, carbon, nitrogen, oxigen and silicon to form refractory compounds. Three of these compounds - titanium nitride (TiN), titanium carbide (TiC) and titanium diboride (TiB$_2$) - have similar properties, although they exhibit some differences which turn out to be very useful when these materials are combined in multilayer coatings [17]. They are extremely refractory, with a melting point around 3000 ^{o}C, very hard and wear resistant. They present low diffusivity, a low coefficient of friction, high thermal and electrical conductivities, and are generally chemically inert. Table 1 shows the general properties of these compounds and some property tendencies are presented in Table 2.

TABLE 1. General properties [17, 18]

	TiN	TiC	TiB$_2$
Crystal structure	cubic fcc	cubic fcc	hexagonal
Density (g/cm^3)	5.39	4.91	4.49
Melting point (^{o}C)	2950	3250	2900
Colour	gold	silver	grey
Vickers hardness at 25 ^{o}C (Kg/mm^2)	2800	3200	3400
Electrical resistivity ($\mu\Omega$.cm)	25	60-250	15
Thermal conductivity at 20 ^{o}C (W/mK)	23	17	26
Thermal expansion 25-300 ^{o}C (x 10^{-6}/K)	9.5	7.6	6.6

Beyond their more traditional field of application as protective coatings against wear and corrosion for high performance cutting tools, they have also been widely used in semiconductor devices (TiN [6, 19], TiC [6, 20], TiB$_2$ [21]), as low-resistivity contacts (TiN [22]), as optical coatings (TiN [18]), as thermal barrier coatings in fusion reactors (TiC [23]) and as thermal resistant materials in aerospace heat shields (TiB$_2$ [21]). In addition, because of its attractive gold colour, titanium nitride is very often used as a decorative finish coating in the jewellery industry.

TABLE 2. Property tendencies [17]

	High	Medium	Low
Melting point	TiC	TiN	TiB_2
Thermal expansion	TiN	TiC	TiB_2
Hardness	TiB_2	TiC	TiN
Wear resistance	TiC	TiB_2	TiN
Toughness	TiN	TiC	TiB_2
Lubricity	TiN	TiC	TiB_2
Reactivity	TiB_2	TiC	TiN

Elders and Voorst [21] recently reported on their work on the deposition of TiB_2 using the LCVD technique. It is the purpose of this work to present an overview of the work done by several authors on the laser-assisted deposition of TiN and TiC.

3.2. TITANIUM NITRIDE

A literature review on the deposition of TiN films by LCVD has shown that a few research groups are involved with the subject, and that the first consistent results were only published in 1990 [24, 25]. Therefore, this is a very recent investigation with many fundamental aspects not yet fully understood. We have summarized in Table 3 the characteristic features of each deposition process regarding the type of laser used to induce deposition, substrate materials, gaseous or vapour precursors and their flow dynamics, e.g., static atmosphere in a batch mode reactor or continuous flow in an open reactor, and the dominant deposition process, i.e., thermal *versus* photochemical deposition. Most often, a stationary laser beam has been employed yielding spot-shaped deposits [24-29, 31] and in only one study [30] stripes of TiN have been produced by scanning the substrate under the laser beam, i.e., by laser direct writing.

TABLE 3. Deposition of TiN by Laser CVD

Laser	Atmosphere	Substrate	Process	Ref.
CO_2	$TiCl_4 + N_2 + H_2$ static	Incoloy 800H	pyrolytic	[24, 26]
CO_2	$TiCl4 + NH3$ dynamic	Mo, 304 SS	pyrolytic	[25]
CO_2	$TiCl_4 + N_2 + H_2$ static	Carbon fibres	pyrolytic	[27]
Nd-YAG	$TiCl_4 + N_2 + H_2$ dynamic	Mo, Din 1.2601 and Din 1.7405 tool steel	pyrolytic	[28]
CO_2	$TiCl_4 + N_2 + H_2$ static	AISI M2 tool steel	pyrolytic	[29]
CO_2	$TiCl_4 + N_2 + H_2$ static	Mild steel	pyrolytic	[30]
Argon / Excimer	$TDMATi + N_2 + H_2$ $TiCl_4 + N_2 + H_2$ $TiCl_4 + NH_3$ dynamic	Din 1.3343 tool steel	pyrolytic/photolytic	[31]

3.2.1. *TiN spots*

Surface morphology and microstructure. The surface profiles of the films are strongly influenced by the deposition temperature, which in pyrolytic LCVD depends on the laser/materials interaction parameters, notably laser power density and interaction time. Figure 1 shows the thickness profiles of spots deposited on AISI M2 tool steel substrates [29] with a cw CO_2 laser operating dominantly in the TEM_{00} mode (approximately Gaussian distribution of energy in the beam), for two different values of laser irradiance and different interaction times. As can be seen from the figure, when low power densities are used ($I \le 2.7 \times 10^4$ W.cm^{-2}), the spots show a broad Gaussian type profile, even for the longest interaction time considered (Fig. 1a, b). Otherwise, the thickness profiles show a central depression (Fig. 1c) due to the development of the well-known volcano-like behaviour [32 - 34]. In the limiting case of high power densities and long interaction times, melting of the film/substrate as a whole can occur at the centre of the deposits (Fig. 1d).

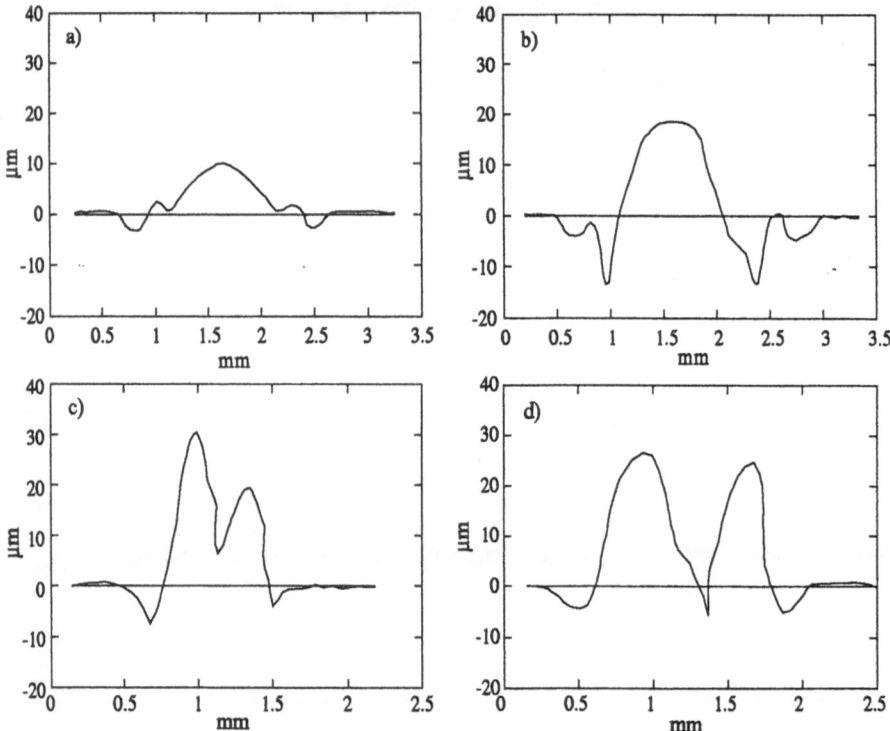

Figure 1 - Thickness profiles of LCVD TiN spots deposited onto AISI M2 steel at various laser power densities and interaction times. a) $I = 2.7 \times 10^4$ W.cm^{-2}, $t_{int} = 2$ s; b) $I = 2.7 \times 10^4$ W.cm^{-2}, $t_{int} = 8$ s; c) $I = 6.7 \times 10^4$ W.cm^{-2}, $t_{int} = 6$ s; d) $I = 6.7 \times 10^4$ W.cm^{-2}, $t_{int} = 10$ s [29].

Similar surface morphologies have been found in TiN films deposited on different metallic substrates [26, 28], even with different wavelength radiation [28]. The laser power density value associated with the transition from a Gaussian to a volcano-like profile is different in each of these studies because of differences in the coupling between the laser beam and the materials. However, its order of magnitude remains the same. For instance, Croonen and Verspui [28] reported a threshold laser power of 35 W for this transition observed on TiN films deposited on molybdenum substrates with a Nd/YAG laser. The corresponding laser power density can be estimated as 1.07×10^4 W.cm^{-2}.

Explanations given to account for the development of double-humped profiles have been based on high chemical reaction rates, convection from the surface, low sticking coefficients, melting, and even evaporation, at the centre of the heated spots.

Among the several theoretical models proposed in order to understand the basic mechanisms of LCVD processes, some [33, 35-37] yield deposits with volcanic morphology. Conde *et al.* [38] used the model developed by Kar *et al.* [37], with additional assumptions, to study the spatial variation of the thickness of TiN spots deposited onto Incoloy 800H substrates by pyrolytic LCVD with a CO_2 laser, from a reactive atmosphere consisting of $TiCl_4$, N_2 and H_2. Assuming that the rate of formation of TiN is first order with respect to the $TiCl_4$ concentration, and that the sticking coefficient of TiN at the substrate surface is temperature dependent, the volcanic profiles observed at high deposition temperatures [26] were reproduced in [38]. Figure 2 illustrates two examples of modeling the experimental thickness profiles obtained at two different interaction times. A description of the model and other results can be found in the chapter by Kar and Mazumder in this proceedings volume.

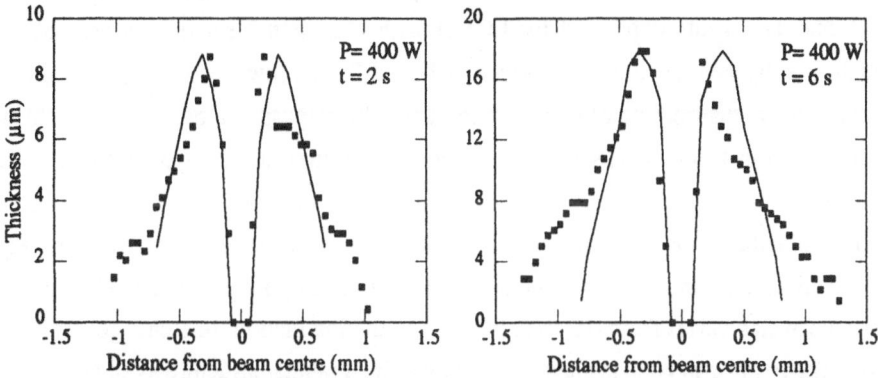

Figure 2 - Thickness profiles of LCVD TiN spots deposited onto Incoloy 800H substrate at constant laser power density, $I = 1.7 \times 10^4$ W.cm^{-2}, and different interaction times. ■ experimental; — calculated [38].

Deposition of films by laser heating of a substrate is an extension of traditional thermal CVD processes [39]. Therefore, we should expect that the chemical reactions leading to deposit formation are essentially the same, if the same precursors as of CVD are employed in LCVD. Also, in laser deposited films many of the characteristic features of CVD films should be observed. Carlsson [3] investigated the CVD of tungsten on various metal substrates and showed that an iron substrate in a gaseous atmosphere of hydrogen and a halide vapour (WF_6), at low temperatures, competes with the hydrogen as a reducing agent, leading to consumption of substrate material, which means corrosion of the substrate. A similar phenomenon appears in the profiles of Fig. 1 where the lack of material at the periphery of the spots suggests that etching of the substrate has occurred. Michalski and Wierzchon [40] showed experimentally that the presence of an iron source in a CVD process from a gas mixture of $TiCl_4$, nitrogen and hydrogen is a prerequisite for the synthesis of TiN at temperatures below 1200 K. Therefore, the lack of material observed in the outer region of the laser-deposited titanium nitride spots can be explained by the predominance of the reduction reaction of $TiCl_4$ by the iron contained in the HSS substrate at low temperatures (less than 1200 K) [29], yielding the formation of iron dichloride which evaporates when the temperature increases. X-ray image maps of these spots performed with an electron probe microanalyser (EPMA) showed the presence of iron and oxygen in the etched region, confirming the ease of oxidation of the etched portion of the substrate after removal of the samples from the deposition chamber.

The TiN spots deposited by LCVD with Gaussian laser beams are usually characterised by several concentric regions where different microstructures grow due to the non-uniform distribution of temperatures across the deposits. In order to illustrate the typical structures found in TiN films, the scanning electron micrographs in Fig. 3 show the microstructures developed in the different regions of the TiN spots. The central zone is formed of cube-shaped crystals with a mean edge of a few tenths of a micrometre (Fig. 3a). The microstructure of the intermediate zone consists of juxtaposed platelets (Fig. 3b) whose mean size varies from 0.2 to 2.2 µm when the interaction time changes from 0.5 to 10 s. The peripheral zone is characterised by a fine granular structure as shown in Fig. 3c. However, in a few cases, a columnar microstructure could also be observed in this zone (Fig. 3d). Analogous results on the radial microstructure change induced by an approximately Gaussian distribution of temperature at the substrate surface were found in the films deposited on Incoloy 800H [24, 26] and on molybdenum [28]. The main results on the microstructures observed in TiN laser-assisted deposited films are summarised in Table 4.

We have investigated the influence of laser power density, interaction time and partial pressure of the titanium tetrachloride precursor on TiN crystallite size [29]. Mesured values of the mean grain size are plotted in Fig. 4 as a function of the interaction time for two different values of laser irradiance (Fig. 4a) and of $TiCl_4$ partial pressure (Fig. 4b). For a given laser power density, the mean size of TiN crystallites increases

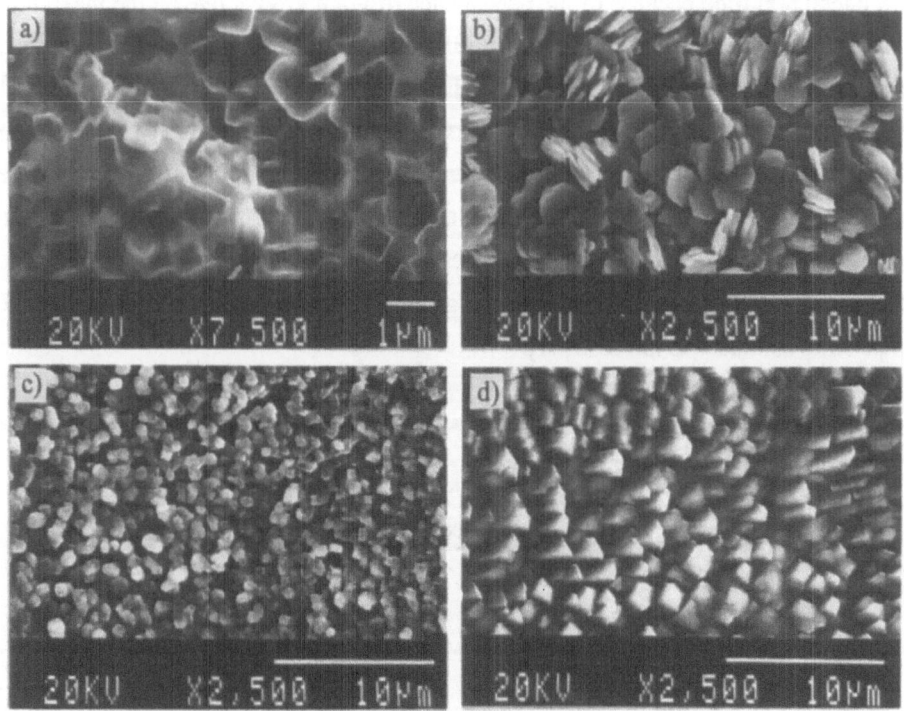

Figure 3 - Scanning electron micrographs of the surface of TiN spots grown on AISI M2 steel at
$I = 6.7 \times 10^4$ W.cm^{-2}. a) central zone, t_{int} = 10s; b) intermediate zone, t_{int} = 10s; c) peripheral
zone, t_{int} = 4s; d) peripheral zone, t_{int} = 10s [29].

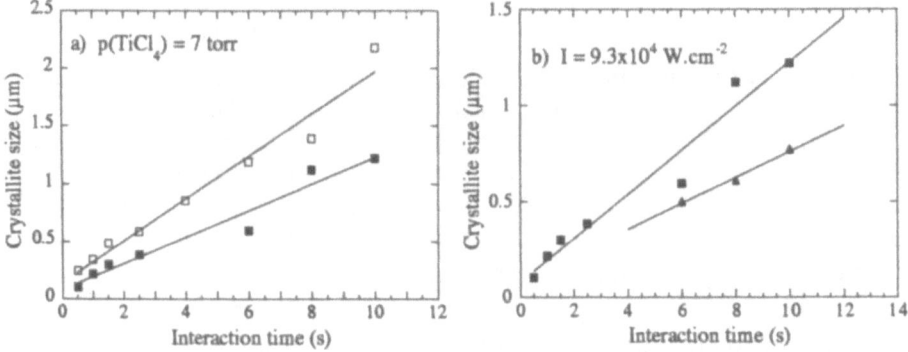

Figure 4 - Dependence of crystallite size on interaction time, laser irradiance and precursor partial
pressure. a) □ $I = 6.7 \times 10^4$ W.cm^{-2}, ■ $I = 9.3 \times 10^4$ W.cm^{-2}; b) ■ p(TiCl$_4$) = 7 torr, ▲ p(TiCl$_4$) =
10 torr [29].

with interaction time because of the dependence of grain growth rate on the maximum temperature attained, which increases with irradiation time. By contrast, it can be seen from Fig. 4a that when the laser irradiance increases from 6.7×10^4 $W.cm^{-2}$ to 9.3×10^4 $W.cm^{-2}$, the crystal size decreases by a factor of almost two. In the range of interaction times studied, the temperatures attained under a laser irradiance of 6.7×10^4 $W.cm^{-2}$ were estimated to be lower than those attained at 9.3×10^4 $W.cm^{-2}$. Therefore, a possible explanation for this result could be the existence of a maximum on the grain growth rate as a function of temperature at which the growth rate would start to decrease as the temperature continues to increase [41].

TABLE 4. Typical microstructures observed on TiN films deposited by LCVD

Laser	Atmosphere	Substrate	Microstructure	Ref.
CO_2	$TiCl_4 + N_2 + H_2$	Incoloy 800H	platelets/octahedral crystallites (central zone) columnar structure (middle zone) equiaxed fine grain (peripheral zone)	[24, 26]
CO_2	$TiCl_4 + NH_3$	304 SS	equiaxed fine grain average size 200Å	[25]
CO_2	$TiCl_4 + N_2 + H_2$	Carbon fibres	columnar structure	[27]
Nd-YAG	$TiCl_4 + N_2 + H_2$	Molybdenum	platelets (central zone) nodular structure (middle zone) granular structure (peripheral zone)	[28]
CO_2	$TiCl_4 + N_2 + H_2$	AISI M2 tool steel	cube-shaped crystals (central zone) platelets (middle zone) granular/columnar structure (peripheral zone)	[29]
CO_2	$TiCl_4 + N_2 + H_2$	Mild steel	equiaxed fine grain (~ 100 nm)	[30]
Excimer	$TDMATi + N_2 + H_2$	Din 1.3343 tool steel	ball-like structure	[31]

Regarding the TiN crystal size dependence on the partial pressure of $TiCl_4$, when partial pressure increases, crystal size decreases. This result brings out the strong dependence of the nucleation rate of the titanium nitride on the partial pressure of the $TiCl_4$ precursor. As supersaturation increases, the nucleation rate increases, decreasing the crystal size. This behaviour was also observed by Kim and Chun [42] in the case of TiN films deposited by conventional CVD.

3.2.2. Laser writing of TiN lines

Surface morphology and microstructure. Deposition of TiN on mild steel was performed by scanning the substrates under a CO_2 laser beam. The laser-written TiN lines, 15 mm in length, were grown from a static gas mixture of titanium tetrachloride, nitrogen and hydrogen (see Table 3). For a given composition of the gas phase and scanning velocities within the range 0.5-5.0 mm.s^{-1}, Silvestre et al. [30] determined experimentally the threshold irradiance value (about 1.5×10^4 W.cm^{-2}) for TiN formation, and the upper limit (about 2.3×10^4 W.cm^{-2}) above which melting of the film/substrate was observed.

All the lines deposited within the selected parameters exhibit the characteristic TiN golden colour, good adherence and a broad Gaussian-type profile. These films also show a very fine and uniform grain structure. A typical microstructure of the films as observed by SEM, at high magnification, is illustrated in Fig. 5. As can be seen, the mean grain size is about 100 nm, and the grains seem to be equiaxed and to have originated from a high density of nucleation centres. In contrast to the behaviour found on the TiN spots [24, 26, 28, 29], the mean grain size of the deposited lines does not show any dependence on either laser irradiance or scanning velocity owing, probably, to the short interaction times used in this study.

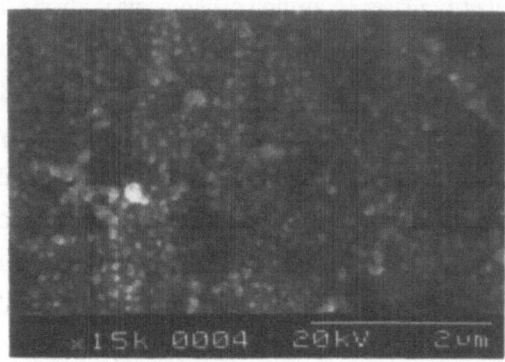

Figure 5 - SEM image of the surface of a TiN line.
$I = 2.3 \times 10^4$ W.cm^{-2}, v = 4 mm.s^{-1} [30].

Chemical characterisation. Chemical characterisation by electron probe microanalysis (EPMA) revealed that the deposited material is titanium nitride slightly over stoichiometric. The atomic ratio N:Ti was measured at the centre of lines deposited at varying laser power density and scanning velocity. The chemical composition is scattered around an average value of [N]/[Ti] = 1.1, and does not depend on the deposition temperature range implied by the laser processing parameters considered in the experimental study. Therefore, the important parameter in determining the stoichiometry of the films should be the gas phase composition. Incorporation of oxygen in the films was also detected, which is likely to be present mostly as surface contamination, as in the case of the deposition of spots [28].

Rate limiting step during laser chemical vapour deposition. Although a laser thermally-driven CVD reaction may present a more complex structure than conventional CVD, because the intermediate species that are frequently formed may absorb the laser radiation and contribute photochemically to the final product, mass transport and surface kinetics are still the most important rate-limiting mechanisms. Arrhenius diagrams, i.e., the logarithm of the deposition rate *versus* the reciprocal temperature curves, are a convenient way to test these reaction control mechanisms. Surface chemical reactions are thermally activated, and the activation energy, E_a, deduced from the slope of the Arrhenius plots, is usually a few hundreds of kJ.mole^{-1}. By contrast, when the deposition process is controlled by the transport of gaseous species into or from the chemical reaction zone, the growth rate is independent of or weakly dependent on temperature, yielding a low value for E_a.

Drawing an Arrhenius plot using data from laser deposition experiments is always a difficult task because normally the temperature cannot be measured due to the reduced size of the deposits and, even in those cases where such an experimental measurement could be envisaged, the temperature is not homogeneous across the deposit since nearly Gaussian beams are usually employed. Only recently, Azer and Mazumder [45] reported an experimental measurement of the temperature distribution across the top surface of a titanium film deposited by thermal LCVD with a Nd-YAG laser. Therefore, the current practice in LCVD is to predict the deposition temperature by using analytical or numerical models [46-49]. Also, because apparent growth rates are determined from film thickness measurements, and because the thickness profiles change with the heat input to the film/substrate composite, from nearly flattened or Gaussian to volcanic shape, a criterion has to be set for the best measure of thickness. Most frequently, the film thickness in the centre is used. We have followed this criterion for the TiN lines since, as mentioned above, broad Gaussian-type profiles were always observed. It will be seen below that for asymmetric profiles a 3D representation of the deposited material should be considered.

By using the steady-state region of the laser-written TiN lines, apparent growth rates at the laser spot centre (maximum thickness) were determined. The temperature at the same position was calculated via the three-dimensional transient heat transfer model developed by Kar and Mazumder [48] for uniformly moving finite slabs. These results are shown in figure 6, together with the best fit to the data by the Arrhenius equation. The fact that only one straight line is needed to fit the data indicates that only one mechanism is responsible for the TiN film growth, within the deposition temperature range investiga-

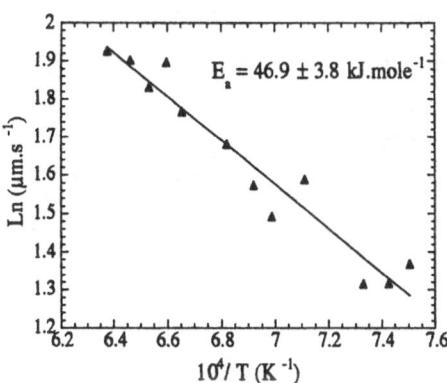

Figure 6 - Logarithm of the growth rate of LCVD TiN lines as a function of reciprocal temperature (Arrhenius plot). Gas phase parameters: p(total) = 207 torr, $p(TiCl_4)$ = 7 torr, $p(N_2)$ = $p(H_2)$.

ted and for the chemical composition of the reactive atmosphere considered. From the slope of the straight line the apparent activation energy was calculated to be 46.9 kJ.mole^{-1}. This value is in very good agreement with reported activation energy values for conventional CVD using the same gas phase precursors, for conditions under which transport of the gaseous species is the rate limiting step for titanium nitride film growth [42, 44].

3.3. TITANIUM CARBIDE

Research work on laser-assisted chemical vapour deposition of titanium carbide thin films dates back to 1979 [50]. Since that time, other LCVD studies on TiC have been conducted [32, 43, 51-56], employing CO_2 but also Nd-YAG and Ar$^+$ lasers and, thus, different laser processing parameters - irradiance, spot size, deposition dwell time. While $TiCl_4$ has been used in all these works as the precursor for titanium, several hydrocarbons such as CH_4, C_2H_2, C_2H_4, etc., have been employed to provide the carbon content in the films. In their experiments, Hopfe et al. [27, 55] took as carbon source the substrate material itself, consisting of carbon fibres. However, the results regarding the tensile strength of the TiC coated fibres were poorer than those obtained by adding benzene to the gas mixture because of carbon diffusion into the films in the former case, leading to a too high adhesion of the layers and, probably, to fibre surface defects [55]. Table 5 summarises the experimental conditions reported by the different research groups.

TABLE 5. Deposition of TiC by Laser CVD

Laser	Ti and C precursors	Substrate	Process	Ref.
CO_2	$TiCl_4$, CH_4 /C_2H_2 /C_4H_{10}	Carbon steel, Stainless steel, Quartz	Pyrolytic (photolytic)	[32, 50-52]
Nd-YAG	$TiCl_4$, CH_4	Stainless steel	Pyrolytic	[53]
Argon	$TiCl_4$, C_2H_4	Cemented carbide inserts (precoated with TiC)	Pyrolytic	[43, 54]
CO_2	$TiCl_4$, C from substrate /C_6H_6 /C_2H_2	Carbon fibres	Pyrolytic	[27, 55]
CO_2	$TiCl_4$, CH_4	Silica	Pyrolytic	[56]

3.3.1. Phase and chemical characterisation

Phase analysis of the deposits has been performed by X-ray diffractometry (XRD) and the results presented in the above cited publications. Westberg et al. [54] found that their Ar^+ laser-deposited stripes consisted of single-phase TiC close to the stoichiometric composition. We obtained a similar result for our films deposited with a stationary CO_2 laser beam [56] and we showed that, under the gas phase composition used in that work (the $TiCl_4$:CH_4:H_2 molar ratio was 7:21:130), an enhancement of the (200) diffraction line was always observed revealing a (200) preferentially oriented crystal growth. Umezawa et al. [53], who employed a Nd-YAG laser to induce film deposition, but the same reactants as in [56], also found a (200) line intensity enhancement for a CH_4:H_2 molar ratio of 1:6, close to our own ratio value. However, when they increased the amount of hydrogen in relation to methane to 10 to 1, a preferred growth along the [111] direction was observed (see figure 9 in ref. [53]). Besides the titanium carbide phase, the XRD spectra included low intensity peaks which were assigned [56] to various titanium oxides, predominantly TiO_2, also reported in ref. [53], and less Ti_3O_5. A colour analysis of the deposited material showed that the oxides are formed at the outer rim of the metallic-grey central film of TiC, where the deposition temperature is lower.

In order to investigate the influence of hydrogen concentration on the growth features of TiC deposits, e.g., growth rate, composition and morphology, we have recently carried out deposition experiments where the laser working parameters (power density and irradiation time) were chosen as in previous work, but the hydrogen concentration in the gaseous atmosphere was either varied at the expense of the methane content [57] or totally replaced by argon or helium. The total pressure of the reactants always had the same value, in order that any variation in the deposition rate which might occur could be considered to be entirely due to gas phase composition rather than to total pressure changes. The results regarding phase analysis are compared in Fig. 7 which shows two X-ray diffraction patterns obtained at 1° glancing incidence (GIXRD) of films grown from gas mixtures of $TiCl_4$, methane, and argon (a) or hydrogen (b). It is readily seen that the low intensity Ti_xO_y peaks present in spectrum b) are no longer visible in spectrum a), i.e., when argon is used. A similar result has been obtained for helium gas.

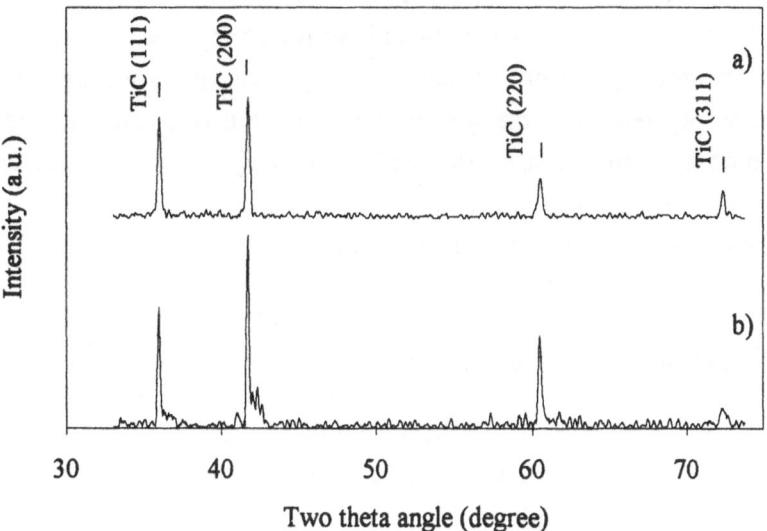

Figure 7 - Typical GIXRD patterns of as-deposited coatings from the following gas mixtures: a) $TiCl_4$, CH_4 and argon; b) $TiCl_4$, CH_4 and hydrogen.

In CVD of Ti-based materials from a $TiCl_4$ and H_2 gas mixture, it is generally accepted [40, 58-60] that above a threshold temperature of about 1170 K, $TiCl_4$ is reduced by hydrogen to titanium subchlorides, which are not stable compounds and can easily give rise to titanium oxides. The formation of these oxides was discussed by Alexandrescu et al. [61] who considered that the reactant oxygen is provided by the heated silica substrate. Conversely, when argon or helium is used as the buffer gas, the reduction reaction of $TiCl_4$ is retarded and the decomposition of methane is favoured, yielding preferentially the formation of titanium carbide. This explains why the films deposited with H_2 assistance exhibit oxide contamination in contrast to those deposited without hydrogen.

The presence of free carbon in the films was demonstrated by Westberg *et al.* [54] using Auger electron spectroscopy (AES). The authors concluded that the carbon content in the films decreases when the laser power density or the reactor temperature are raised, and explained this result by considering that the carbon is preferentially deposited in the grain boundaries. They also concluded that the free carbon is deposited at the later stages of the deposition process, after the formation of the TiC film, due to the decomposition of the C_2H_4 present in excess in the gas phase once it has been depleted of $TiCl_4$. This conclusion was drawn based on the observation that at longer deposition times (lower scanning velocities) more carbon was detected. Because XRD has a low sensitivity for carbon, the XRD patterns of our films did not reveal any free carbon deposition. However, for some experimental conditions, notably when argon was used as the buffer gas, a visual observation of the specimens revealed a darker colour and the SEM investigations of the microstructure showed that a powdery film was deposited on top of the TiC layer (see figure 11a) below), which may be ascribed to carbon film formation. AES studies of these films are currently under way.

3.3.2. *Surface morphology and microstructure*

As well as the dependence of the deposit profiles on the laser generated surface temperature distribution, as we have seen for the TiN deposits, the composition of the gas phase may also strongly influence the morphology of the films.

In general, the TiC films deposited on silica substrates at lower laser power densities or irradiation times show a flat thickness distribution. Increasing each of these parameters leads to the development of Gaussian films, followed by the appearance of double-humped profiles [56]. Figure 8 illustrates this behaviour for the deposition of TiC at constant laser power. In the case of this particular system, one can explain the volcano formation by considering that, for each two molecules consumed in the reaction, four of hydrogen chloride are produced, yielding an accumulation of reaction products in the chemical reaction zone, which will increase the local pressure and will thus decrease the diffusive flow of fresh reactants to the hot zone, favouring deposition in cooler areas. This explanation does not exclude other phenomena taking place at the same time, for instance decrease of the sticking coefficient for the TiC molecules as temperature increases. This argument was successfully used to model the LCVD of TiN films at high temperatures, as mentioned in § 3.2.1.

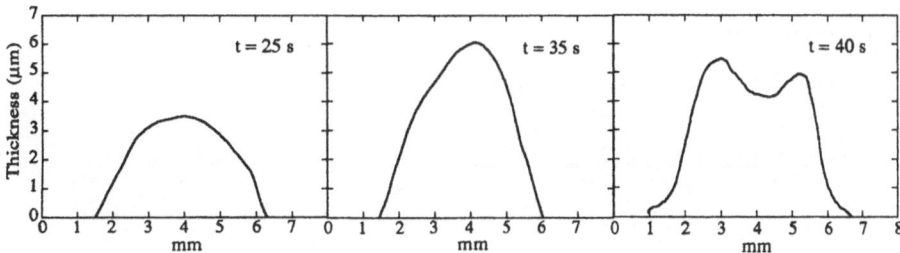

Figure 8 - Spatial variation of the thickness of TiC films deposited at a constant laser power of 250 W for different irradiation times. Gas phase parameters: $p(TiCl_4) = 7$ torr, $p(CH_4) = 21$ torr, $p(H_2) = 130$ torr [56].

This type of profile evolution is much more drastic when argon is used as the buffer gas, instead of hydrogen or helium. The double-humped shape starts to appear at much lower values of incident laser power or irradiation time. This is due to the different behaviour of these gases in regions of large thermal gradients. Because of thermal diffusion, heavier molecules diffuse towards cooler regions, while lighter molecules accumulate near the hotter reaction zone [62]. This will affect the local partial pressure of the reactive gases at the substrate surface and, as a consequence, the deposition rate will be higher for a gas phase containing argon rather than hydrogen or helium. Therefore, the influence of the local pressure of by-products will be felt sooner, with the subsequent formation of volcano-like deposits.

SEM analyses of the surface of the films revealed a radial microstructure differentiation related to the local induced temperature distribution. In the films deposited with hydrogen, three types of microstructure can usually be found [56]: faceted crystallites in the central region (the most extended), a nodular region surrounding this, and a platelet habit at the edge of the TiC spots (Fig. 9). For low laser power and interaction time values, only the nodular and the platelet microstructures develop. Growth of the same kind of faceted crystallites in the highest temperature zone of TiC spots has also been observed by Umezawa et al. [53] (see Table 5 for experimental conditions), as well as on the TiC strings deposited with an Ar^+ laser from a C_2H_4 precursor [54]. The grain size at the edge of the strings is smaller than in the centre, and increases by increasing laser power density. On the other hand, the development of platelets outside the centre of the films is in agreement with low temperature conventional CVD results for titanium carbide deposition [5].

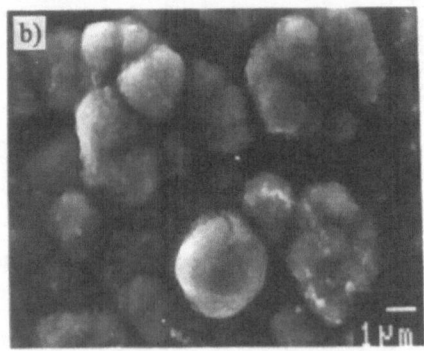
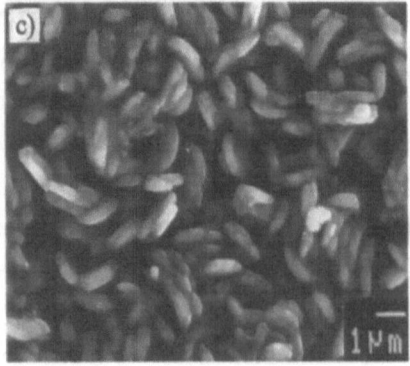

Figure 9 - Scanning electron micrographs of the surface of a TiC film grown at 175 W for 40 s with H_2 assistance. a) faceted crystallites, b) nodules, c) platelets.

In the cases of argon and helium assisted deposition, only a granular structure may be observed all over the films (Fig. 10) which is more compact at the centre than at the edges. This different type of microstructure is also related to the different behaviour of the three buffer gases. The fact that argon tends to flow away from the heated region will facilitate access of the reactive gases to the reaction zone and will increase the number of adsorption sites for these gases at the substrate surface.

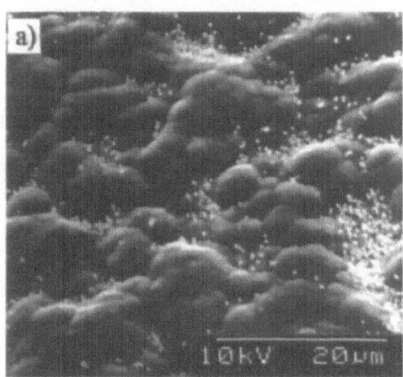 

Figure 10 - Scanning electron micrographs of the surface of a TiC film grown at 250 W for 30 s with a) argon and b) helium atmospheres.

3.3.3. Growth kinetics

The kinetics of the TiC film growth was investigated by means of the thickness profiles, which are very useful in providing a source not only for the evaluation of the total amount of deposited material, but also for estimating surface temperature across each spot diameter [56].

For Gaussian laser beams, the deposition rate can be a strong function of radial position. Because of the exponential dependence of deposition rate on surface temperature, the deposit profile is narrower than the incident laser beam and always remains narrower than the temperature profile [47]. Fig. 11-a) shows the influence of laser power on the diameter (taken from the thickness profiles) of the TiC spots deposited with constant irradiation time and different gas mixtures (TiCl4, CH4, and H2 or He or Ar). Independently of the buffer gas utilised, the diameters always increase with laser power up to a maximum value that is much smaller than the spot size of the laser beam at the substrate surface, ~ 13 mm. This means that the threshold temperature for the chemical reaction is reached inside the Gaussian energy distribution in the laser beam.

Then, by assuming that the threshold temperature is 1173 K [63] and is attained at the edge of the TiC deposits, the surface temperature at the spot centre can be predicted for each pair of laser power/interaction time values. This temperature was thus considered as the deposition temperature for the LCVD process.

Because the shape of the profiles changes with temperature, it becomes difficult to set a criterion for the best measure of thickness (e.g., maximum thickness above substrate surface *versus* thickness at film centre). Therefore, it is more accurate to evaluate the apparent deposition rate for the TiC films in terms of the total mass deposited per unit time. To obtain this value we integrated the 2D experimental cross-sections of the thickness profiles over an angle ϕ which describes the rotation of the profile around an axis perpendicular to the substrate. The results of the calculations are shown in Fig. 11-b) for the three different gas phases, where we used the bulk density of TiC, $\rho = 4.93$ g.cm^{-3}.

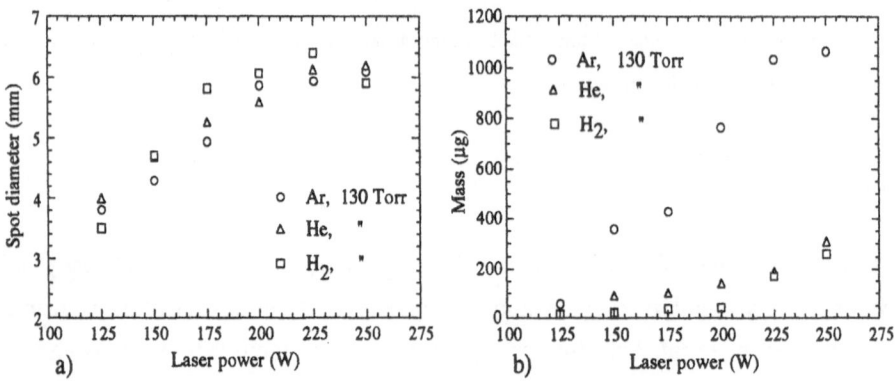

Figure 11 - Influence of incident laser power on a) lateral extension and b) mass of TiC films deposited on silica by CO_2 laser induced CVD. Processing parameters: $t = 40$ s, $p(TiCl_4) = 7$ torr, $p(CH_4) = 21$ torr.

As can be seen, when all the processing parameters are maintained, the total amount of material produced under argon assisted deposition is significantly larger than for hydrogen or helium atmospheres, owing to the strong convection in the gas phase above the hot spot, as discussed previously.

The Arrhenius plots for these three cases are shown in Fig. 12, where the deposition rates and the temperatures are those evaluated as above. Each set of data was fitted by a single straight line using a least-squares method, yielding apparent activation energies of 100 kJ.mole^{-1} when hydrogen is used and 130 kJ. mole^{-1} for the He and Ar containing atmospheres. This difference in activation energy indicates that the choice of buffer gas does affect the chemistry of the process, as observed in the XRD patterns, i.e. the overall chemical reaction leading to TiC formation proceeds according to different sequences of intermediate chemical reactions. Nevertheless, in every case the deposition process is controlled by the diffusion of the gaseous species into the chemical reaction zone, as can be deduced from the comparison of our results for E_a with the values reported for CVD of TiC at T ≥ 950°C [64,65].

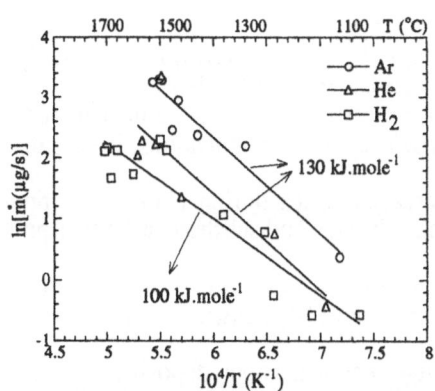

Figure 12 - Effect of temperature at beam centre on TiC deposition rate (Arrhenius plot). Gas phase parameters as in Fig. 11.

While these results demonstrate that mass transport is the unique mechanism responsible for TiC film growth from methane under CO_2 laser irradiation, the data obtained by Westberg et al. [54], from a C_2H_4 precursor with an Ar$^+$ laser, clearly show a transition from a kinetically controlled deposition mechanism at low laser power densities (temperature) to a mass transport limiting process at higher temperature values. This different behaviour may be expected if the differences in gas phase composition and total pressure, 158 torr in this work and 37.5 torr in [54], are taken into account.

4. Summary

We have described recent work in the field of laser-assisted CVD applied to the growth of titanium nitride and carbide thin films. The first studies have emphasised the microstructural and chemical composition aspects of the films as well as the deposition mechanism(s) in relation to the processing parameters. Additional research devoted to the physical and mechanical properties of the films produced is essential for future applications.

5. References

1. Pierson, H.O. (1992) *Handbook of Chemical Vapor Deposition - Principles, Technology and Applications*, Noyes Publications, Park Ridge, N.J..
2. Blocher Jr., J.M. (1982) Chemical Vapor Deposition, in R.F. Bunshah (ed.), *Deposition Technologies for Films and Coatings*, Noyes Publications, Park Ridge, N.J.
3. Carlsson, J.-O. (1985) Processes in interfacial zones during chemical vapour deposition: aspects of kinetics, mechanisms, adhesion and substrate atom transport, *Thin Solid Films* **130**, 261-282.
4. Kern, W. (1989) Chemical Vapor Deposition, in R.A. Levy (ed.), *Microelectronics Materials and Processes*, NATO ASI Series, vol.164, Kluwer Academic Publ., Dordrecht.
5. Carlsson, J.-O. (1991) Thermally activated chemical vapor deposition, in D.S. Rickerby and A. Matthews (eds.), *Advanced Surface Coatings - A Handbook of Surface Engineering*, Blackie and Son Ltd., Glasgow.
6. Pauleau, Y. (1989) Interconnect materials, in R.A. Levy (ed.), *Microelectronics Materials and Processes*, NATO ASI Series, vol.164, Kluwer Academic Publ., Dordrecht.
7. Bauerle, D. (1986) *Chemical Processing with Lasers*, Springer Series in Materials Science., vol. 1, Springer, Berlin.
8. Allen, S.D., Trigubo, A.B., and Jan, R.Y. (1983) Direct writing using laser chemical vapor deposition, *Mater. Res. Soc. Symp. Proc.* **17**, 207-214.
9. González, P., Fernández, D., Pou, J., García, E., Serra, J., Leon, B., and Pérez-Amor, M. (1992) Photo-induced chemical vapour deposition of silicon oxide thin films, *Thin Solid Films* **218**, 170-181.
10. Jackson, R.L., Baum, T.H., Kodas, T.T., Ehrlich, D.J., Tyndall, G.W., and Comita, P.B. (1989) Thermally activated chemical vapor deposition, in D.J. Ehrlich and J.Y. Tsao (eds.), *Laser Microfabrication: Thin Film Processes and Lithography*, Academic Press, Boston.
11. Eden, J.G. (1992) *Photchemical Vapor Deposition*, John Wiley & Sons, Inc., NY.
12. Bauerle, D. and Leyendecker, G. (1982) Laser induced chemical vapor deposition of C and Si, *Appl. Phys. B* **28**, 267-268.
13. Bauerle, D. (1984) Laser-induced chemical vapor deposition, in D. Bauerle (ed.), *Laser Processing and Diagnostics*, Springer Series in Chemical Physics, vol. 39, Springer Verlag, Berlin.
14. Lydtin, D. (1972) in F. A. Glaski (ed.), *Proc. of the 3rd Intern. Conf. on CVD*, Am. Nucl. Soc., Hinsdale, p. 127.
15. See ref. 7, p. 64.
16. Molian, P.A. (1988) Principles and applications of lasers for wear-resistant coatings, in T.S. Sudarshan and D.G. Bhat (eds.), *Proc. of the 1st Intern. Conf. on Surface Modification Technologies*, The Metallurgical Society, Inc., Warrendale, pp. 237-265.

17. Pierson, H.O. (1993) A review of the chemical vapor deposition (CVD) of the refractory compounds of titanium - a unique family of coatings, *Materials & Manufacturing Processes* **8**, 519-534.
18. Sundgren, J.-E. (1985) Structure and properties of TiN coatings, *Thin Solid Films* **128**, 21-44.
19. Dapor, M., Elena, M., Girardi, S., Ginuta, G., Guzman, L. and Narsale, M. (1987) Stoichiometry in Ti-N barrier layers studied by X-ray emission spectroscopy, *Thin Solid Films* **153**, 303-311.
20. Rist, O. and Murray, P.T. (1991) Growth of TiC thin films by pulsed laser evaporation, *Mater. Lett.* **10**, 323-328.
21. Elders, J. and Voorst, J.D.W. (1994) Laser-induced chemical vapor deposition of titanium diboride, *J. Appl. Phys.* **75**, 553-562.
22. Ostling, M., Nygren, S., Petersson, C.S., Norstrom, H., Buchta, R., Blom, H.-O. and Berg, S. (1986) A comparative study of the diffusion barrier properties of Tin and ZrN, *Thin Solid Films* **145**, 81-88.
23. Abe, T., Murakami, Y., Obara, K., Hiroki, S., Nakamura, K., Mizoguchi, T., Doi, A. and Inagawa, K. (1985) Development of TiC coated wall materials for JT-60, *J. Nucl. Mater.* **133**, 754-759.
24. Conde, O., Mariano, J., Silvestre, A.J., and Vilar, R. (1990) Production of hard coatings by laser CVD, in H.W. Bergman and R. Kupfer (eds.), *Proc. of the 3rd European Conf. on Laser Treatment of Materials*, Sprechsaal Publ. Group, Coburg, vol. 1, pp. 145-154.
25. Chen, B., Biunno, N., Singh, R.K., and Narayan, J. (1990) Laser chemical vapor deposition of TiN films, in T.M. Besman and B.M. Gallois (eds.), *Mater. Res. Soc. Symp. Proc.* **168**, 287-292.
26. Conde, O., Ferreira, M.L.G., Hochholdinger, P., Silvestre, A.J., and Vilar, R. (1992) CO_2 laser induced CVD of TiN, *Appl. Surf. Sci.* **54**, 130-134.
27. Hopfe, V., Tehel, A., Baier, A., and Scharsig, J. (1992) IR-laser CVD of TiB_2, TiC_x and TiC_xN_y coatings on carbon fibres, *Appl. Surf. Sci.* **54**, 78-83.
28. Croonen, Y.H., and Verspui, G. (1993) Laser induced chemical vapour deposition of TiN coatings at atmospheric pressure, *J. Physique IV* **3**, 209-215.
29. Silvestre, A.J., Conde, O., Vilar, R., and Jeandin, M. (1994) Structure and morphology of titanium nitride films deposited by laser-induced chemical vapour deposition, *J. Mater. Sci.* **29**, 404-411.
30. Silvestre, A.J., Paramês, M.L.F., and Conde, O. (1994) Investigation of the microstructure, chemical composition and lateral growth kinetics of TiN films deposited by laser-induced chemical vapour deposition, *Thin Solid Films* **241**, 57-60.
31. Illmann, U., Ebert, R., Reiss, G., Freller, H., and Lorentz, P. (1994) Laser based chemical vapour deposition of titanium nitride coatings, *Thin Solid Films* **241**, 71-75.
32. Allen, S.D. (1981) Laser chemical vapor deposition: A technique for selective area deposition, *J. Appl. Phys.* **52**, 6501-6505.
33. Herman, I.P., Hyde, R.A., McWilliams, B.M., Weisberg, A.H., and Wood, L.L. (1983) Wafer-scale laser litography: I. Pyrolytic deposition of metal microstructures, in R.M. Osgood, S.R. Brueck and H.R. Schlossberg (eds.), *Mater. Res. Soc. Symp. Proc.* **17**, 9-18.
34. Moylan, C.R., Baum, T.H., and Jones, C.R. (1986) LCVD of copper: deposition rates and deposit shapes, *Appl. Phys.* **A40**, 1-5.

35. Allen, S.D., Jan, R.Y., Edwards, R.H., Mazuk, S.M., and Vernon, S.D. (1984) Optical and thermal effects in laser chemical vapor deposition, *SPIE Laser assisted deposition, Etching and doping* **459**, 42-48.

36. Skouby, D.C., and Jensen, K.F. (1987) Modeling of energy and mass transport in laser-assisted CVD. I: surface morphology, *J. Appl. Phys.* **63**, 140-148.

37. Kar, A., Azer, M.N., and Mazumder, J. (1991) Three-dimensional transient mass transfer model for laser chemical vapor deposition of titanium on stationary finite slabs, *J. Appl. Phys.* **69**, 757-766.

38. Conde, O., Kar, A., and Mazumder, J. (1992) Laser chemical vapor deposition of TiN dots: A comparison of theoretical and experimental results, *J. Appl. Phys.* **72**, 754-761.

39. Jackson, R.L., Baum, T.H., Kodas, T.T., Ehrlich, D.J., Tyndall, G.W., and Comita, P.B. (1989) Laser deposition: energetics and chemical kinetics, in D.J. Ehrlich and J.Y. Tsao (eds.), *Laser Microfabrication - Thin Film Processes and Lithography*, Academic Press Inc., London, p. 385.

40. Michalski, J., and Wierzchon, T. (1989) Formation of titanium nitride from a gaseous phase with the participation of a chemical reaction, *J. Mater. Sci. Lett.* **8**, 779-780.

41. Machlin, E.S. (1991) *An introduction to aspects of thermodynamics and kinetics relevant to materials science*, Gyro Press, New York, p. 237.

42. Kim, M.S., and Chun, J.S. (1983) Effects of the experimental conditions of chemical vapour deposition on a TiC/TiN double-layer coating, *Thin Solid Films* **107**, 129-139.

43. Boman, M., and Carlsson, J.-O. (1991) Laser-assisted chemical vapor deposition of hard and refractory binary compounds, *Surf. Coat. Technol.* **49**, 221-227.

44. Bhat, D.G. (1990) A thermodynamic and kinetic study of CVD TiN coating on cemented carbide, in K.E. Spear and G.W. Cullen (eds.), *Proceedings of the 11th International Conference on Chemical Vapor Deposition*, The Electrochemical Society Inc., Pennington, p. 648, and references therein.

45. Azer, M.N., and Mazumder, J. (1993) Substrate temperature measurement diagnostics during laser chemical vapor deposition (LCVD) of titanium on silicon, in P. Denney, I. Miyamoto and B.L. Mordike (eds.), *Proc. of the Laser Materials Processing Conference - ICALEO'93*, vol. 77, Laser Institute of America, Orlando, pp. 584-593.

46. Burgener, M.L., and Reedy, R.E. (1982) Temperature distribution produced in a two-layer structure by a scanning cw laser or electron beam, *J. Appl. Phys.* **53**, 4357-4363.

47. Allen, S.D., Goldstone, J.A., Stone, J.P., and Jan, R.Y. (1986) Transient nonlinear laser heating and deposition: a comparison of theory and experiment, *J. Appl. Phys.* **59**, 1653-1657.

48. Kar, A., and Mazumder, J. (1989) Three-dimensional transient thermal analysis for laser chemical vapor deposition on uniformly moving finite slabs, *J. Appl. Phys.* **65**, 2923-2934.

49. Garrido, C., Leon, B., and Perez-Amor, M. (1991) A model to calculate the temperature induced by a laser, *J. Appl. Phys.* **69**, 1133-1140.

50. Mazumder, J., and Allen, S. D. (1979) Laser chemical vapor deposition of titanium carbide, *SPIE* **198**, 73-80.

51. Allen, S. D., Trigubo, A. B., and Teisinger, M. L. (1982) Properties of several types of films deposited by laser CVD (Summary abstract), *J. Vac. Sci. Technol.* **20**, 469-470.

52. Allen, S. D. (1983) Laser chemical vapor deposition, in M. Bertolotti (ed.), *Physical Processes in Laser - Materials Interaction*, Plenum Press, N.Y., p.455.
53. Umezawa, A., Kikuchi, K., and Shikata, N. (1991) Formation of titanium carbide films by YAG laser chemical vapor deposition, *J. Mech. Eng. Laboratory* **45**, 257-264.
54. Westberg, H., Boman, M., and Carlsson, J.-O. (1992) Kinetics in thermal laser-assisted chemical vapour deposition of titanium carbide, *Thin Solid Films* **218**, 8-14.
55. Hopfe, V., Bohm, S., Wieghardt, G., and Schulze, A. (1993) Laser CVD on carbon fibres: structure of layers and tensile strength of fibres, *Appl. Surf. Sci.* **69**, 380-387.
56. Parames, M. L. F., and Conde, O.(1993) Structure and morphology of laser assisted chemical vapour deposited TiC coatings, *J. de Physique IV* **3**, 217-224.
57. Conde, O., Paramês, M.L.F., and Silvestre, A.J. (1994) Laser assisted deposition of TiN and TiC coatings, in *Proc. of the 4th Intern. Conf. on Optics - ROMOPTO'94,* SPIE- The International Society for Optical Engineering (in press).
58. Jang, H., and Paik, Y.H. (1983) Kinetics of Ti-deposition on stainless steel at elevated temperatures, *J. Kor.Inst. Met.* **21**, 38-43.
59. Kato, A., and Tamari, N. (1975) Crystal growth of titanium nitride by chemical vapor deposition, *J. Cryst. Growth* **29**, 55-60.
60. Sadahiro, T., Cho, T., and Yamaya, S. (1977) Chemical vapor deposition of titanium nitride on cemented carbides, *J. Jpn. Inst. Met.* **41**, 542-545.
61. Alexandrescu, R., Cireasa, R., Dragnea, B., Morjan, I., Voicu, I., and Andrei, A. (1993) Deposition of titanium-based films by laser-assisted thermal CVD of titanium tetrachloride, *J. de Physique IV* **3**, 265-272.
62. Chapman, S. and Cowling, T. G. (1970) *The Mathematical Theory of Non-Uniform Gases* , Cambridge University Press, London, p. 272.
63. Boving, H.J. and Hintermann, H.E. (1990) Wear-resistant hard titanium carbide coatings for space applications, *Tribology International* **23**, 129-133.
64. Hara, A., Yamamoto, T., and Tobioka, M.(1978) Effect of some properties of cemented carbide substrates on the chemical vapour deposition of titanium carbide, *High Temp.-High Pressures* **10**, 309-314.
65. Rossignol, J. Y, Langlais, F., and Naslain, R.(1984) A tentative modelization of titanium carbide C.V.I. within the pore network of two-dimensional carbon-carbon composite preforms, in *Proc. of the 9th Intern. Conf. on CVD* , Electrochem. Soc., pp. 596-614.

LASER CHEMICAL VAPOR DEPOSITION OF TITANIUM NITRIDE AND PROCESS DIAGNOSTICS WITH LASER INDUCED FLUORESCENCE SPECTROSCOPY

X. CHEN, M. N. AZER, K. M. EGLAND, AND J. MAZUMDER
Center for Laser-Aided Materials Processing
Department of Mechanical and Industrial Engineering
University of Illinois at Urbana-Champaign
Urbana, Illinois 61801
USA

Abstract

With gaseous precursors of $TiCl_4$, H_2, and N_2, titanium nitride (TiN) films have been deposited on Ti-6Al-4V alloy substrates by a cw CO_2 laser chemical vapor deposition (LCVD) process. Transient Ti atomic concentration above the substrate is measured by laser induced fluorescence (LIF) spectroscopy. Surface temperature during deposition is obtained by multi-wavelength pyrometry. Stylus profilometry and Auger electron spectroscopy (AES) yield film growth rates and compositions. Relationships between the growth rate and $TiCl_4$, H_2, and N_2 partial pressures are established, from which the rate-controlling reactions and activation energy are obtained.

1. Introduction

Chemical vapor deposition (CVD) of TiN has been carried out with $TiCl_4$ vapor and either $H_2 + N_2$ gases or NH_3 gas [1-10]. Growth rates of these processes are in the order of 10 Å/sec. Only a few studies have been reported so far on the laser chemical vapor deposition (LCVD) of TiN [11-14]. The major characteristic of this technique is the greatly enhanced deposition rate (2 to 4 orders of magnitude higher) due largely to the localization of laser heating. There are still some fundamental aspects of the process that are not well understood. Most notably, the reaction pathways, chemical kinetics, and film growth mechanisms are worthy of further study.

The goal of this research is to develop *in situ* nondestructive diagnostic techniques to study the kinetics of LCVD processes. TiN films are grown on Ti-6Al-4V substrates with a cw CO_2 laser (TEM_{00} mode, 2 mm beam diameter) and precursor gas mixtures of $TiCl_4$, H_2, and N_2. This substrate has many aerospace applications. Pulsed dye laser induced fluorescence (LIF) spectroscopy is employed to obtain gas phase Ti species concentrations at different instances relative to the processing laser. Post-processing materials analyses such as stylus profilometry and Auger electron spectroscopy yield film growth rate and composition that are to be linked to the diagnostic results. Multi-wavelength pyrometry is also applied simultaneously to measure the substrate surface temperature.

693

J. Mazumder et al. (eds.), Laser Processing: Surface Treatment and Film Deposition, 693–701.
© 1996 *Kluwer Academic Publishers.*

2. Theory and Experiment

Figure 1 is a schematic of the LCVD chamber. The overall chemical reaction for the LCVD of TiN is: $TiCl_4(g) + 2H_2(g) + 1/2 \ N_2(g) = TiN(s) + 4 \ HCl(g)$. With the CO_2 laser as the energy source, the LCVD of TiN is believed to be pyrolytic. From studies of conventional CVD with the same precursors, it is believed that deposition of TiN occurs at substrate temperatures above 900 °C. Our literature survey finds that, among all the possible species during LCVD of TiN, only the Ti atomic species have the energy levels and transition probabilities thoroughly documented [15-18]. Therefore, quantitative concentration measurements are possible only for Ti atomic species, together with measurements of substrate surface temperature by a multi-wavelength pyrometry technique [19-20]. The lack of data on the quenching rate is circumvented by the use of the saturated fluorescence technique [21-22]. Since LCVD of TiN is a surface reaction and TiN dissociates to its elements upon vaporization [23], the Ti atomic species in the gas phase above the substrate is believed to be a result of the TiN desorption from the deposited film. The concept of TiN desorption is also supported by the model in Ref.[13] that incorporated TiN desorption to make excellent predictions on the film thickness profiles. The Ti concentrations detected by LIF are in the order of $10^{12} \ cm^{-3}$. An equilibrium computation [5] of the Ti-H-N-Cl system at comparable experimental conditions shows that the Ti concentration produced by the gas-phase reaction is 6 orders of magnitude smaller than that detected by LIF in our experiment. Hence, we can ignore the gas-phase production of Ti.

With a dye laser pumping from level 1 to level 2, the steady state solution of the rate equations for a three-level model yields the saturated fluorescence intensity for the transition from level 2 to level 3 as [24]:

$$F_{23} = A_{23}N_1^o \ \frac{1}{\left(1 + \dfrac{g_1}{g_2}\right)} \tag{1}$$

in which N_1^o is the Ti ground state population, A_{23} is the fluorescence transition rate, and g_1 and g_2 are the level degeneracies. Therefore, in the saturated regime, the fluorescence intensity is independent of pump laser intensity, and quenching rates are not necessary to calculate the lower level population N_1^o. The total number density of Ti is then obtained assuming a Boltzmann distribution. In our experiments, the dye laser beam is positioned at 0.5 mm above the substrate. At that location, the gas temperature T_g is expected to be only ≤ 15 °C smaller than the surface temperature T_S [25]. Therefore, the measured surface temperature T_S is used in place of the gas temperature T_g. From Boltzmann distribution, the population fraction of the Ti atoms that are in the ground state (level 1) is calculated to be $f_1 = (29.25 \pm 0.89)\%$ for the temperature range of 1473 K ± 200 K. All the surface temperatures measured fall into this range. Hence, the total Ti concentration C(Ti) is obtained by dividing N_1^o by the fraction f_1.

The diagnostic systems are schematically shown in Figure 2. The dye laser beam ($\lambda = 3962.85$ Å, 0.5 mm in diameter) is delivered to the LCVD chamber to excite the gas phase Ti species at 0.5 mm above the substrate surface. The fluorescence from the

(a) Side View:

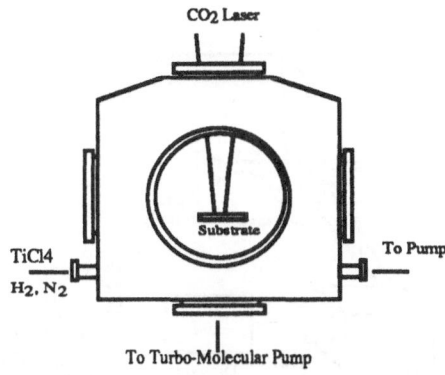

(b) Top View:

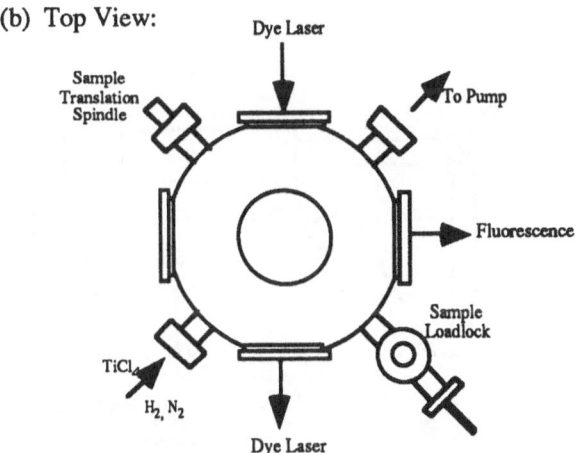

Figure 1. Schematics of the laser chemical deposition chamber.

Ti: a ^{3}F - y ^{3}F^O transition at 3989.76 Å is directed into a 0.33 m spectrograph. The detector is an intensified CCD camera (Princeton Instruments ICCD-576). A light calibration source with a known intensity (Optronics Laboratories 453-1) is used for measuring the overall collection efficiency that is necessary for calculating the absolute fluorescence intensity. Synchronization between the lasers and the detector is provided by a digital delay pulse generator. Both the probe laser and the ICCD detector are operated at 60 Hz. Fluorescence signals from 30 consecutive dye laser pulses are accumulated that gives an average LIF intensity within 0.5 sec. The signal accumulation is necessary due to the weakness of the LIF signal generated by a single dye laser pulse. A full description of the multi-wavelength pyrometry technique can be found in Ref. [19-20].

3. Results and Discussion

It is found that the film growth rate, R_{TiN}, and the rate of Ti concentration increase, dC/dt, increase linearly with P_{TiCl_4}, when P_{TiCl_4} < 27 Torr. Further increase in P_{TiCl_4}

(a)

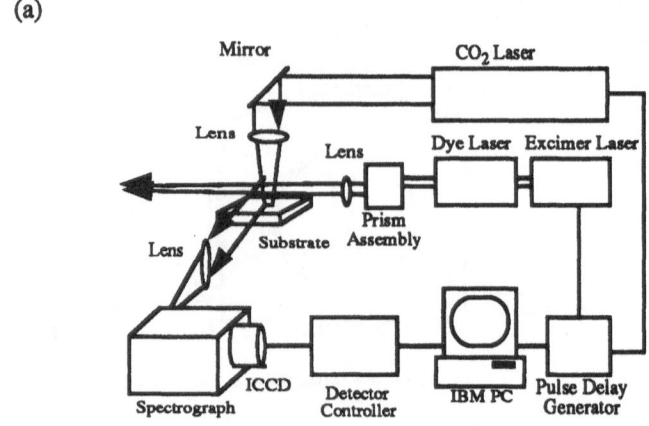

(b)

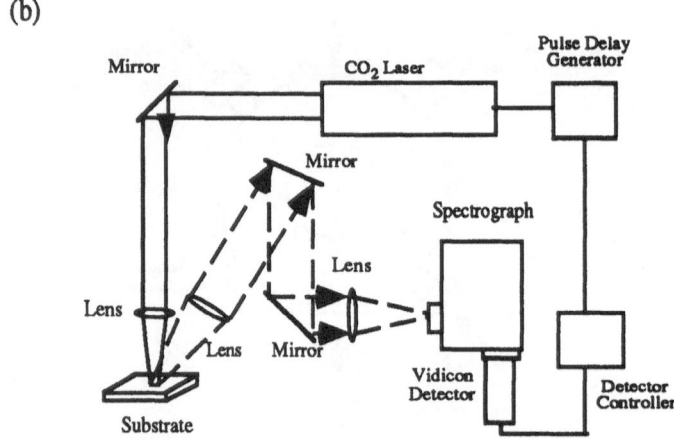

Figure 2. Schematics of (a), the laser-induced fluorescence (LIF) diagnostic system, and (b), the multi-wavelength pyrometry system.

will result in a decrease in R_{TiN} and dC/dt. Similar observations for R_{TiN} have been reported for conventional CVD of TiN [1-2, 4, 9-10]. The increase of deposition as P_{TiCl_4} increases is easy to understand, because $TiCl_4$ is the source of Ti. The decrease of deposition at higher P_{TiCl_4} is believed to be a result of competitive adsorption between reactant molecules $TiCl_4$, N_2, and H_2. After the initial growth, further film deposition will be onto the TiN covered substrate. You, Nakanishi and Kato [6] showed that the N_2 and H_2 adsorption equilibrium constant is about 10^4 times smaller than that of $TiCl_4$. Hence, surface coverages of N_2 and H_2 are much smaller than that of $TiCl_4$ at similar partial pressures. When P_{TiCl_4} is raised above a certain value, $TiCl_4$ takes up so much of the available surface sites that it is possible that there will be not enough surface adsorbed N and H to carry out the reactions, leading to a decrease in R_{TiN}.

To study the dependencies of R_{TiN} and dC/dt on the N_2 and H_2 partial pressures, P_{N_2} and P_{H_2}, experiments are performed in which P_{N_2} (or P_{H_2}) is fixed while changing P_{H_2} (or P_{N_2}) and adding inert Ar gas to maintain the total pressure P_{Total}. All other parameters are kept constant: a TiCl$_4$ partial pressure of P_{TiCl_4} = 27 Torr, a total pressure of P_{Total} = 600 Torr, and a CO$_2$ laser power of 400 W. The results are shown in Figures 3 and 4. TiN films of a few micron thickness can be deposited in less than 10 sec. Evidently the growth rate of LCVD is much greater than that of conventional CVD processes. The behaviors of the Ti concentration and the film thickness are identical, indicating that the mechanism for producing the Ti species above the substrate may be the same as that for film growth. All films are characterized by AES as TiN$_x$ (x = 0.8 ± 0.1), with Cl contamination as low as 0.5 at.% and O and C 1 at.%.

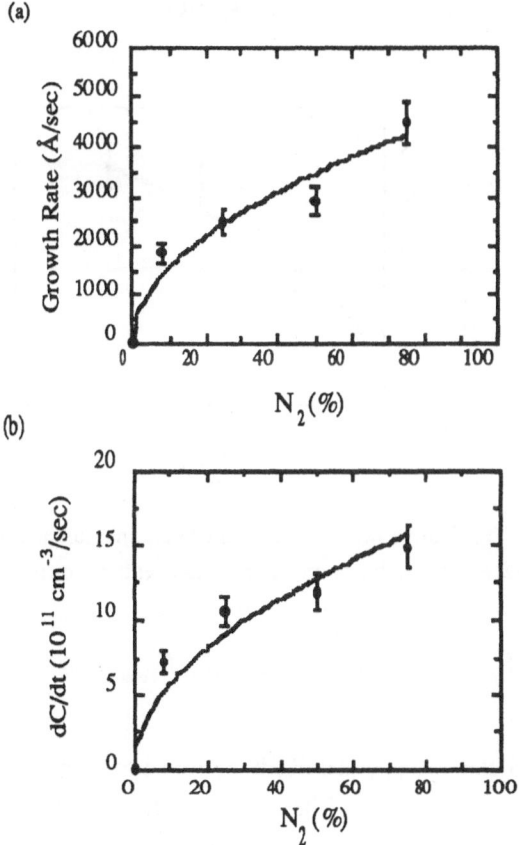

Figure 3. (a) Growth rates based on thickness measurements, and (b) rates of Ti concentration increase based on LIF measurements for different P_{N_2} when P_{H_2} is fixed at 25% of $(P_{N_2} + P_{H_2} + P_{Ar})$. P_{N_2} is varied as a percentage of $(P_{N_2} + P_{H_2} + P_{Ar})$. P_{TiCl_4} = 27 Torr, P_{Total} = 600 Torr, and Laser Power = 400 W.

(a)

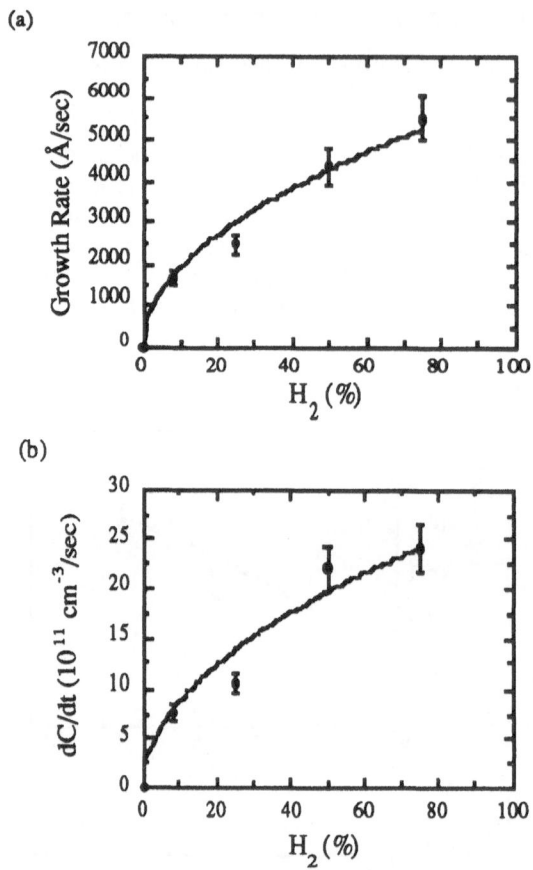

(b)

Figure 4. (a) Growth rates based on thickness measurements, and (b) rates of Ti concentration increase based on LIF measurements for different P_{H_2} when P_{N_2} is fixed at 25% of $(P_{N_2} + P_{H_2} + P_{Ar})$. P_{H_2} is varied as a percentage of $(P_{N_2} + P_{H_2} + P_{Ar})$. P_{TiCl_4} = 27 Torr, P_{Total} = 600 Torr, and Laser Power = 400 W.

From Figures 3 and 4, R_{TiN} and dC/dt are found to be proportional to the square root of P_{N_2} and P_{H_2}, respectively. The square root dependencies on P_{N_2} and P_{H_2} are also observed by other researchers in conventional CVD processes [3, 5, 7-8]. As to the detailed reaction pathways, there have been no studies before this research on the laser initiated CVD process. However, there have been different suggestions in the literature on conventional hot wall CVD [2-4, 5, 7], each explaining its own experimental observations. After careful examinations, it is found that the possible surface reaction pathways listed by Nakanishi et al [5] are the most suitable in interpreting our experimental results. Since the surface coverages of N and H atoms (when dissociatively adsorbed) are proportional to the square roots of P_{N_2} and P_{H_2} at low surface coverages, the results in Figures 3 and 4 indicate that the rate limiting step in the reaction process is the dissociative adsorption of N_2 and H_2 molecules on the substrate surface.

The effect of changing CO_2 laser power on the LCVD process is also studied. The laser power is set at 250, 400, and 550 W. All other conditions are constant: $P_{TiCl_4} = 27$ Torr, $P_{N_2}:P_{H_2} = 3:1$, and $P_{Total} = 600$ Torr. Because the reactant gas composition remains unchanged for this set of experiments, and the only effect of the changing laser power is on the substrate surface temperature, the reaction kinetics can be studied using these results. R_{TiN} and dC/dt are plotted with respect to the substrate temperature in Figure 5. The horizontal axis is chosen to be $1/T_{peak}$, with T_{peak} being the peak substrate temperature measured by the simultaneous multi-wavelength pyrometry for the different laser powers. Here, T_{peak} is chosen because it is found to represent the substrate temperature at which the majority of the film growth occurs. It follows from the Arrhenius equation that,

(a)

(b)

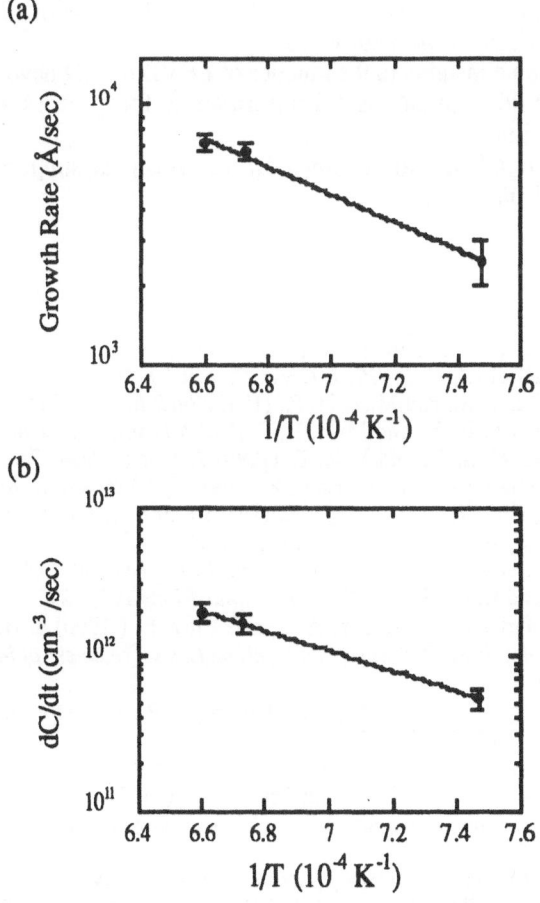

Figure 5. (a) Growth rates based on thickness measurements, and (b) rates of Ti concentration increase based on LIF measurements for different peak substrate temperatures. $P_{N_2}:P_{H_2} = 3:1$, $P_{TiCl_4} = 27$ Torr, and $P_{Total} = 600$ Torr.

$$R_{TiN} = K_1 \exp\left(-\frac{E}{RT}\right) \quad \text{and} \quad \frac{dC}{dt} = K_2 \exp\left(-\frac{E}{RT}\right) \tag{2}$$

where K_1 and K_2 are constants related to the Arrhenius constant and reactant pressures, E is the apparent activation energy, and R the molar gas constant. From the slopes of the straight lines in Figure 5, the apparent activation energy is calculated to be 104.3 kJ/mol from part (a), the film thickness data, and 125.7 kJ/mol from part (b), the LIF data, under the following conditions: P_{TiCl_4} = 27 Torr, $P_{N_2}:P_{H_2}$ = 3:1, P_{Total} = 600 Torr, and T = 1330 - 1520 K. The similarity between the two values seems to confirm that the Ti species probed by the LIF technique are directly resulting from the film growth on the surface. There are large discrepancies in the reported activation energy values for conventional hot wall CVD. Under different experimental conditions, they range from 39.3 kJ/mol to 308.9 kJ/mol [1-5, 7-10]. The values depend greatly on the ranges of substrate temperature at which the experiments are performed. At temperatures close to that of our experiment, Cao *et al* [3,4] reported a value of 120.4 ± 24.2 kJ/mol which is in good agreement with our value.

In summary, some insights to the kinetics of LCVD of TiN have been gained. The new findings would provide useful information for process optimization and mathematical modeling.

The authors would like to acknowledge the financial support provided by the National Science Foundation.

4. References

1. Peterson, J. R. (1974) *J. Vac. Sci. Technol.* **11**, 715
2. Kato, A. and Tamari, N. (1975) *J. Cryst. Growth* **29**, 55
3. Cao, Z. R., Du, Y. S., and Miao, H. F., (1989) *Surf. Engr.* **5**, 315
4. Tsao, C. J., Chen, E. B., and Miao, H. F. (1986) *Trans. Inst. Mining Met.* **95**, 63
5. Nakanishi, N., Mori, S., and Kato, E. (1990) *J. Electrochem. Soc.* **137**, 322
6. You, M. S., Nakanishi, N., and Kato, E. (1991) *J. Electrochem. Soc.* **138**, 1394
7. Sadahiro, T., Cho, T., and Yamaya, S. (1977) *J. Jpn. Inst. Met.* **41**, 542
8. Yoshikawa, N., Aikawa, H., and Kikuchi, A. (1992) *J. Jpn. Inst. Met.* **56**, 1132
9. Takahashi, T. and Suzuki, Y. (1974) *J. Jpn. Inst. Chem.* **30**, 1043
10. Kim, M. S. and Chun, J. S. (1983) *Thin Solid Films* **107**, 129
11. Conde, O., Mariano, J., Silvestre, A. J., and Vilar, R. (1990) in H. W. Bergmann and R. Kupfer (eds.), *Proc. 3rd European Conf. on Laser Treatment of Materials*, Sprechsal, Coburg, 145
12. Conde, O., Ferreira, M. L. G., Hochholdinger, P., Silvestre, A. J., and Vilar, R. (1992) *Appl. Surf. Sci.* **54**, 130
13. Conde, O., Kar, A., and Mazumder, J. (1992) *J. Appl. Phys.* **72**, 754
14. Hopfe, V., Tehel, A., Baier, A., and Scharsig, J. (1992) *Appl. Surf. Sci.* **54**, 78
15. Martin, G. A., Fuhr, J. R., and Wiese, W. L. (1988) *J. Phys. Chem. Ref. Data* **17**, **Suppl. 3**, 85
16. Salih S. and Lawler, J. E. (1990) *Astron. Astrophys.* **239**, 407
17. Lowe, R. M. and Hannaford, P. (1991) *Z. Phys. D - Atoms, Molecules and Clusters* **21**, 205
18. Forsberg, P. (1991) *Physics Scripta* **44**, 446
19. Azer, M. N. (1994) *Ph. D. Thesis*, University of Illinois at Urbana-Champaign, Urbana, Illinois, USA

20. Chen, X., Azer, M. N., and Mazumder, J. (1993) in J. Mazumder and K. Mukherjee (eds.) *Laser and Materials Processing IV*, TMS, Warrendale, Pennsylvania
21. Piepmeier, E. H. (1972) *Spectrochimica Acta* **27B**, 431
22. Daily, J. W. (1978) *Appl. Opt.* **17**, 225
23. Davis, K. A., Brezinsky, K., and Glassman, I. (1991) *Combustion Sci. Tech.* **77**, 171
24. Eckbreth, A. C. (1988) *Laser Diagnostics for Combustion Temperature and Species, Energy and Engr. Sci. Ser. 7*, Abacus, Tunbridge Wells, UK.
25. Breiland, W. G., Coltrin, M. E., and Ho, P. J. (1986) *Appl. Phys.* **59**, 3267

PROCESS CONTROL OF LASER CHEMICAL VAPOUR DEPOSITION OF TiC ON TOOL STEEL

A. Hatziapostolou[1], I. Zergioti[1], E. Hontzopoulos[1], A. Zervaki[2]
and G. Haidemenopoulos[3]

[1] Foundation for Research and Technology-Hellas, Institute of
Electronic Structure and Laser, P.O.Box 1527, Heraklion 71110, Greece
[2] Metallurgical Industrial Research and Technological Development
Centre, A Industrial area of Volos, Volos 38500, Greece
[3] Department of Mechanical Engineering, University of Thessaly, Volos
38334, Greece

1. Introduction

Titanium carbide is a refractory material [1] which combines ceramic properties such as high melting point (3067°C), high hardness (2800HV), thermal and chemical stability, wear and corrosion resistance, and certain metallic properties such as low friction coefficient, high electrical and thermal conductivity. Consequently, it can be used as protective coating against thermal, chemical, and mechanical wear. Typical applications include high performance cutting and forming tools, high corrosion resistance coatings for molten metal containers, thermal barrier in fusion reactors and chemical reactors and diffusion barrier in semiconductor technology.

Conventionally the formation of TiC [2,3] is based on the reaction between precursor compounds under appropriate temperature and pressure. $TiCl_4/C_2H_6/H_2$, $TiCl_4/CCl_4/H_2$ and $TiCl_4/CH_4/H_2$ are used as gas mixture at reaction temperatures in the range of 850° to 1250°C. These temperatures can cause property changes and/or thermal damages in certain substrates.

The LCVD techniques can be divided into two main categories depending on whether the laser beam interacts with the reactant gases or the substrate material. In the photolytic LCVD the laser beam is absorbed by the reactant gases which undergo photodissociation, whereas in the pyrolytic process the substrate is heated locally by the laser beam and the chemical reaction is thermally induced.

In the present application the demand was the TiC deposition on D2 tool steel. This has undertaken a certain heat treatment, namely austenitizing, quenching and double tempering, in order to improve the mechanical and microstructure properties. The tempering temperature of this steel is 500°C which is the maximum temperature which the bulk material can withstand without considerable loss in hardness. The main motivation behind the present work is to develop an LCVD technique which will enable the local deposition of uniform TiC layers with minimum thermal loading of the

703

J. Mazumder et al. (eds.), Laser Processing: Surface Treatment and Film Deposition, 703–710.
© 1996 Kluwer Academic Publishers.

substrate. In addition to the inherently localized thermal effect of any LCVD technique, particular emphasis was put on the selection of the precursor molecules in order to minimize the appropriate temperature needed for TiC formation

LCVD of TiC on stainless steel has been demonstrated for the first time by Mazumder and Allen [4] where a $TiCl_4$ / CH_4 reaction gas mixture and a 1.4KW CO_2 laser were used. Allen [5] compared various combinations of substrate/coating material and concluded that TiC on stainless steel is the most difficult due to the steel's high reflectivity at the 10.6 μm wavelength and high thermal conductivity. Umezada et al [6] have also deposited TiC films on stainless steel by laser chemical vapour deposition using a Nd:YAG laser. Other investigators deposited TiC on various substrate materials such as silica [7] and carbon fibers [8] using 200-400W CO_2 lasers. Westberg et al [9] have investigated the growth kinetics of TiC strings using a combination of a hot wall CVD reactor and a Ar^+ laser.

In this paper, we have obtained TiC films on D2 steel specimens and miniature tools using a 100W CO_2 laser and a gas mixture of $TiCl_4/CCl_4/H_2/Ar$. The effect of irradiation with a combination of ArF excimer laser and CO_2 laser, defined as twin beam processing technique, is also examined in order to enhance the deposition rate. Emphasis was placed on the application of the technique on real industrial tools in order to demonstrate its future potential.

2. Experimental procedure

A thermodynamic analysis of the TiC formation from gaseous precursors $TiCl_4$, CH_4 or CCl_4 and H_2 was performed in order to calculate the heterogeneous equilibrium reaction pressures and temperatures. The calculations were carried out with the aid of the computer program "THERMOCALC" edited by the Royal Institute of Sweden. Considering that the total pressure was 5 mbar, it was found that the temperature at which the reaction is thermodynamically feasible is lower in the case of $TiCl_4/CCl_4/H_2$ (400°K) than in the case of $TiCl_4/CH_4/H_2$ (800°K) as precursor gases.

$$TiCl_4 + CCl_4 + H_2 \rightarrow TiC + 8HCl$$

For the twin beam processing technique and considering that the $TiCl_4$ and CCl_4 absorb at 193nm radiation of excimer laser [10,11,12], the one photon fragmentation is described by the photochemical reactions:

$$TiCl_4 \xrightarrow[\text{hv (6.4eV)}]{} TiCl_3 + Cl$$

$$CCl_4 \xrightarrow[\text{hv (6.4eV)}]{} \begin{array}{l} CCl_3 + Cl \\ CCl_2 + Cl_2 \end{array}$$

According to the thermodynamic analysis there is a strong possibility that formation of TiC can be achieved efficiently at lower temperatures by the reaction of the photofragments:

$$TiCl_3 + CCl_3 + 3H_2 \rightarrow TiC + 6HCl$$

A schematic diagram of the LCVD reactor is shown in Fig. 1. The RF excited CO_2 laser with TEM_{oo} mode and maximum power 100W at a wavelength of 10.6 μm was incident at 45° angle to the substrate. The $1/e^2$ spot size at the laser exit was 4mm, and was focused using a ZnSe lens. Power density has been controlled in the region of 1000 - 17000 W/cm^2. In the case of the twin beam processing setup an ArF excimer beam was incident through a quartz window perpendicular to the substrate. Both laser entrance windows were flushed with Ar during the deposition process.

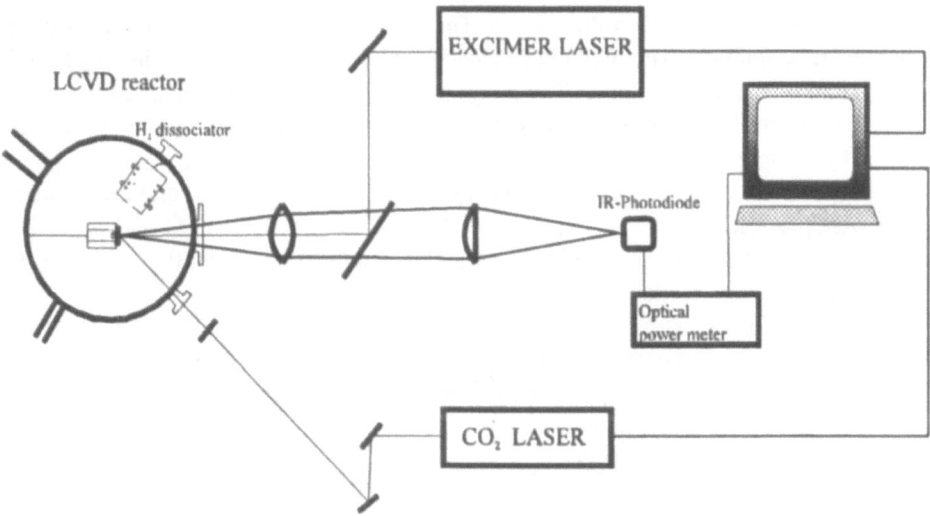

Figure 1: Experimental setup for the twin-beam processing method with the feed-back loop system

The $TiCl_4$, CCl_4 in liquid phase were heated inside glass containers / bubblers at 40°-50°C. Carrier gas Ar was passing through the $TiCl_4$ and CCl_4 bubblers and the resulting gas-vapor mixtures were passed through reflux coolers maintained at steady temperatures (20-30°C); this system ensured that the degree of saturation of the carrier gas is 100%. The reactant gas supplies were thus controlled by the cooler temperature and the Ar mass flow rates adjusted accurately by means of mass flow controllers.

A simple atomic hydrogen source has been built in order to produce reactive atomic hydrogen by thermal dissociation of H_2 [13]. The source consisted of a small water cooled cylindrical stainless-steel chamber which contained a W filament shaped in a coil form. Molecular H_2 presents a significant degree of dissociation at 1400-2300°C under the catalytic action of W at pressures below 1 Torr, despite the high binding energy of the H_2 molecule. This device was used in most of the experiments as it was proved effective in eliminating Cl contamination of the deposited layers.

Before the introduction of the reaction gas mixture, the reactor was evacuated to a pressure of 10^{-5} mbar. The substrate samples were ground with 600grit SiC metallographic paper and cleaned with acetone/ethanol in an ultrasonic bath.

In the pyrolytic LCVD the substrate temperature is the dominant factor of the deposition growth rate. The temperature depends on the heat source, i.e. the CO_2 laser power, the reflectivity of the substrate at the laser wavelength, the reflectivity of the coating, the thermal conductivity of the substrate and the coating and finally the power losses due to convection and radiation of the substrate. Additionally, at the early stages of the deposition the absorption coefficient of the substrate changes considerably due to multiple reflection of the laser beam in the microcavities formed by microstructure grown during the LCVD. The critical role of the temperature created the need to control it during the deposition. An infrared pyrometry system was used as the sensor of a PC-based feed back loop system that controls the CO_2 laser power (Fig. 1). An estimation of the temperature was performed by measuring the IR emission from the hot spot created by the laser beam on the sample. The IR emission was collected and focused onto an IR photodiode. The photodiode analog signal was fed to an A/D interface card installed in a PC microcomputer. The program compared the digital signal to a desired value and controlled accordingly the timer-counter card which triggered the CO_2 laser power by comparing the measured temperature with the desired value. The computer also triggered the excimer laser, while the synchronization of the CO_2 and the excimer laser were possible.

Table 1 shows the range of experimental conditions used in the experiments discussed in the next section

TABLE 1. Experimental conditions for the CO_2 laser and the gas supply system

CO_2 laser:	power density	1000 - 17000 W/cm^2
	Repetition rate	5 KHz
	% duty cycle	20 - 90 %
Precursors:	flow of carrier gas Ar for the $TiCL_4$	100 sccm
	flow of carrier gas Ar for the CCl_4	10 sccm
Partial pressure of H_2 / H		5 mbar
Total pressure		8 mbar

3. Results and discussion

The substrates used in the experiments were rectangular D2 steel samples (10x10x5mm), cylindrical D2 steel samples (Ø5x10-20mm length) representing miniature forming tools and textile needles as real industrial tools. The morphology, thickness, structure, and properties of the layers have been analyzed by SEM/EDAX, TEM, XRD and microhardness measurements.

Initials experiments performed with the rectangular samples and without the use of the feed-back loop system and the atomic hydrogen source, resulted in films grown both inside and outside of the irradiated area with variable thickness and morphology. A typical result is depicted in the SEM photograph of Fig.2. There is a melted central region where EDAX analysis showed low Ti concentration, surrounded by an annular area, where a 3μm-thick film was deposited, exhibiting high Ti presence. Further away from the central area the Ti presence is again minimized and a Cl-rich phase appears.

Figure 2: SEM photograph of the film deposited on the rectangular sample

Transmission electron microscopy was used for determining the phases present in the annular region film. Samples were taken by extraction replica and were examined after suitable preparation. The dominating phase detected had an fcc structure with a lattice constant of a_o=3.87Å, while the theoretical value for TiC is 4.32Å; it can be deduced that the phase detected was non-stoichiometric TiC (TiC$_x$). Other weak phases present in the film had very large values of a_o, characteristic of compounds such as TiO$_2$ or TiCl$_x$ which were also detected by XRD analysis

All subsequent experiments were performed with the feed-back loop system and the atomic hydrogen source. The former eliminated the local melting of the sample while the latter minimized the presence of Cl in the films. Cylindrical samples with a diameter of 5mm and 10-20mm length were irradiated on the flat face. The modified temperature profile on the surface as well as the small size of the irradiated area led to the development of uniform crystalline films with particle-like morphology (Fig.3). The film thickness was within the range of 1 to 3μm and covered all the flat surface without any sign of the laser spot which was less than 1mm. The deposition rate was estimated around 1μm/min. EDAX analysis showed significant presence of Ti all over the film and virtually no Cl. TEM analysis revealed the presence of TiC with a lattice constant a_o=4.20Å, very close to the theoretical value of 4.32Å. Bright and dark field images showed that the film had a fine dense structure and their crystalline size were of the order of nanocrystalls (10-250nm).

KNOOP microhardness measurements were performed using a load of 25gr. For higher accuracy the diagonal measurements were carried out in the SEM and values of around 2000HK were found. The TiC hardness value is 2800HV [1] and the reduced measured values can be attributed to the substrate influence which, as thermally treated D2 tool steel exhibits a value of 500HV. Another important factor for the film characterization is the adhesion strength. In the present case no spallation was observed around a Rockwell C indentation, which led us to the conclusion that the adhesion was moderate.

Figure 3: SEM photograph of the film deposited on the flat surface of the cylindrical sample

The bulk substrate material was carefully examined in terms of both microstructure and hardness; these tests showed that no change had been caused during the deposition process.

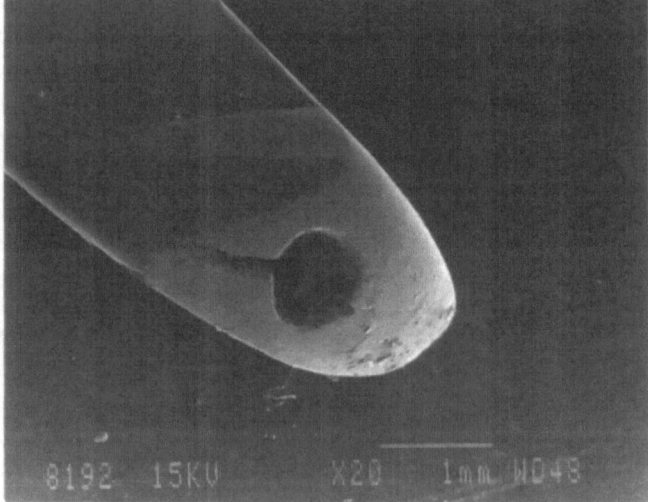

Figure 4: SEM photograph of the textile needle after deposition

The next series of experiments was performed with real industrial tools and more specifically needles used in the textile industry. These tools suffer from wear due to the cotton thread passing at high speeds through their holes. The laser spot was focused on the needle hole and the optimized result is depicted in the SEM photograph of Fig. 4. The film was found to be near-stoichiometric TiC with microhardness values of around

2500HK. The deposition rate was estimated around 7μm/min. The deposited layer covered not only the area around the hole but all the curved surface of the needle; moreover, the specific sample selected for deposition was heavily worn and the damaged area was partly repaired. This shows the potential of this technique for repairing miniature tools.

The twin beam processing technique was applied in the case of the cylindrical D2 tool specimens as well as the textile needles. The optimum experimental conditions for the CO_2 laser, defined in the previous experiments, were applied, while the energy density of the ArF was adjusted up to 240mJ/cm^2; higher energy densities led to the destruction of the deposited layer. The results of these experiments indicated that the ArF excimer laser enhanced the deposition rate by around 10%. The present LCVD reactor did not allow parallel irradiation of the samples by the excimer laser, which would enable higher excimer laser energy densities to be used, possibly leading to further enhancement of the deposition rate. Elders and Voorst [14] also observed enhancement of the deposition rate by the ArF laser action in the case of the LCVD of TiB_2.

4. Summary and conclusions

TiC layers have been successfully deposited locally on D2 tool steel specimens with the aim to improve their mechanical and chemical surface properties. The deposition process was based on pyrolytically-induced chemical reaction of $TiCl_4$ and CCl_4 compounds combined with an atomic hydrogen source. Local heating of the substrate was achieved by means of a 100W CO_2 laser controlled by a microcomputer-based feedback loop system, involving on-line substrate temperature monitoring. SEM/EDAX, XRD and TEM techniques showed that the films consisted of stoichiometric TiC with an fcc structure and lattice constant of 4.2Å, exhibited structural uniformity and moderate adhesion and had a thickness of 1-3 μm. Microhardness values were found in the range of 2000-2500HK, compared to the 500HV hardness of the tempered D2 tool steel. The deposition rate was enhanced by 10% when an ArF excimer laser was employed in combination with the CO_2 laser.

The application of the present method for the local deposition of TiC on miniature industrial tools resulted in uniformly covered curved surfaces around the irradiated area. This suggests that the LCVD technique can produce protective films with good properties on sensitive points of small three-dimensional tools and improve their work life and reliability.

5. Acknowledgments

The authors would like to acknowledge the valuable assistance of Mr. A. Petrakis for the implementation of the feed-back loop system and Ms. A. Patentalaki for the SEM work. This work was supported by the Commission of the European Communities within the BRITE-EURAM project BE-3327 under contract BREU-0049-C(GDF).

710

6. References

1. Holleck, H. (1986) Material selection for hard coatings, *J.Vac. Sci. Technol.* **A 4** (6), 2661-2669
2. Goto, T., Guo, C.-Y., Takeya H. and Hirai T. (1992) Coating of titanium carbide films on stainless steel by chemical vapor deposition and their corrosion behaviour in a Br_2-O_2-Ar atmosphere, *Journal of Materials Science*, vol. 27, no.1, 233- 239.
3. Delblanc Bauer, A., Carlsson, J.-O. (1991) Corrosion of chemically vapor deposited titanium carbide on an inert substrate, *Journal de physique IV*, vol.1, no.C2, 641-648.
4 Mazumder, J. and Allen, S.D. (1979) Laser chemical vapor deposition of titanium carbide, *Proceedings of the Society of Photo-Optical Instrumentation Engineers*, vol.198, 73-80.
5 Allen, S.D., Laser chemical vapor deposition: A technique for selective area deposition, *J.Appl.Phys.* **52**(11), (1981), 6501-6505
6. Umezada, A., Kikuchi, K. and Shikata K. (1991) Formation of titanium carbide films by Yag laser chemical laser deposition, *Journal of Mechanical Engineering Laboratory*, vol 45, no.6, 257-264.
7. Parames M.L.F.and Conde O. (1993) Structure and morphology of laser assisted chemical vapor deposited TiC coatings, *Journal de physigue IV*, vol.3, no C3, 217-224.
8. Hopfe, V., Tehel, A.,Baier, A. and Scharsig (1992) IR-laser CVD of TiB_2, TiCx and TiCxNy coatings on carbon fibres, *Applied Surface Science* **54**, 78-83.
9. Westeberg, H., Boman, M. and Carlsson J.O. (1992) Kinetics of thermal assisted chemical vapor deposition of titanium carbide, *Thin Solid Films*, vol. 218, no.1-2, 8-14
10. Tsao J.Y., Becker R.A., Ehrlich D.J. and F.J.Leonberger (1983) Photodeposition of Ti and application to direct writing of Ti:LiNbO₃ waveguides, *Appl. Phys. Lett.* **42** (7), 559-561.
11. Okabe, H. (1978) *Photochemistry of small molecules*, J. Wiley & Sons, New York
12. Gupta, A., West, G.A. and Beeson, K.W. (1985) Excimer laser-induced chemical vapor deposition of titanium silicide, *J. Appl. Phys.* **58** (9), 3573-3582
13. Franghiadakis, Y. and Tzanetakis, P. (1989), Hydrogen dissociation and incorporation into amorphous silicon produced by electron beam evaporation, *J. Vac. Sci. Technology*, **A7**(2), 136-143.
14. Elders, J. and v. Voorst, J.,D.,W. (1993) Laser-induced chemical vapor deposition of titanium diboride, *J. Appl. Phys.*, **75** (1), 553-562.

ULTRAFINE METAL PARTICLES IN POLYMERS AND IN VAPOR PHASE NUCLEATION

M. SAMY EL-SHALL
Department of Chemistry
Virginia Commonwealth University
Richmond, VA 23284-2006

Abstract

A novel technique for the synthesis of cationic polymers containing ultrafine metal particles is being developed. In the experiments, laser vaporization of metal targets is used to generate ultrafine metal particles and cations which are capable of catalyzing the cationic polymerization of isobutylene. High molecular weight polymers containing submicron metal particles have been obtained. This method can lead to the generation of new polymeric materials with unique properties.

In another study, a novel method which combines laser vaporization of metal targets with controlled condensation in a diffusion cloud chamber is used to synthesize nanoscale metal oxide, carbide and nitride particles (5 - 10 nm) of homogeneous size and well - defined composition. The following oxides have been synthesized: ZnO, SiO_2, Fe_2O_3, Bi_2O_3, PdO, NiO, AgO, TeO, Sb_2O_3, TiO_2, ZrO_2, Al_2O_3, CuO, In_2O_3, SnO, V_2O_5, MgO. The microscale structures of the SiO_2 and Al_2O_3 particles exhibit interesting web-like matrices with a significant volume of voids. Raman, IR, XPS, Mass Spectrometric and electron microscopic studies of these particles will be presented. These materials may have special applications in catalysis and as reinforcing agents for liquid polymers.

1. Introduction

The last decade has witnessed explosive growth in the research on the synthesis of new materials with unique properties. The research on atomic and molecular clusters has raised hopes for an understanding of how the properties of matter evolve as the size of a material system ranges from molecular to macroscopic dimensions[1-4]. This includes clusters of a few atoms to nanostructures of a few thousand atoms. The reduced size and dimensionality of these atomically engineered materials are responsible for their unique electronic, magnetic, chemical and mechanical properties. The applications of laser ablation, laser processing and laser surface treatment have been widely used for the synthesis and characterization of nanomaterials and film deposition. Of particular interest are the combinations of laser ablation techniques with other chemical and physical processes to synthesize new materials with novel properties. In this paper, we specifically address two areas of Materials Chemistry which promise far-reaching progress for the development of novel materials with unusual properties. The first area deals with the synthesis of new polymeric materials containing ultrafine metal particles. This has been achieved by combining recent advances in laser vaporization/ionization of metals with the very fast propagation rates characteristic of ionic polymerization.

711

J. Mazumder et al. (eds.), Laser Processing: Surface Treatment and Film Deposition, 711–726.
© 1996 *Kluwer Academic Publishers.*

In the second area, combination of laser vaporization of metals with gas phase chemical reactions followed by controlled condensation from the vapor phase is used to synthesize nanoscale metal oxide, carbide and nitride particles with controlled sizes and compositions. These two areas are presented and discussed under two different headings.

2. Ultrafine Metal particles in Polymers

In the present work, we present the results from a new approach based on polymerization of liquid isobutylene using energetic gas phase ions[5,6]. In this approach, energetic metal ions are generated in the gas phase by pulsed laser vaporization/ionization of a metal target using the second harmonic of a Nd:YAG laser (532 nm). The laser vaporization method typically releases more than 10^5 metal ions per pulse (20 ns pulse width). The ions are pulled toward a monomer liquid (isobutylene) by applying appropriate electric fields across the reaction chamber as shown in Figure 1. Under different experimental conditions (temperature, pressure of a carrier gas, laser energy, electric field strength, reaction time) different polymeric products are obtained with average molecular weights up to 10^6. It is expected that, with the proper development and initial application of this technique to the ionic polymerization of isobutylene, this method could lead to a powerful approach to the synthesis of a wide variety of polymeric materials with unique properties such as stability, strength and photo and electric conductivity.

Isobutylene was chosen as an appropriate monomer for a prototype study since it is known to be polymerized in the bulk phase only by cationic mechanisms[7,8]. Using our new technique, polyisobutylene polymers were synthesized by laser desorption of any of the following metals: Zn, Ti, Fe, Ni, Cu, Zr, Pd, Pt, Ag, In, Sn and Al. Copolymerization of an isoprene (10%) - isobutylene mixture was also achieved using the same technique. Among the metal targets used, Zn showed an apparent tendency for higher yields and high molecular weight polymers. This is due to efficient charge transfer from Zn^+ to isobutylene to generate the $C_4H_8^+$ ions in the gas phase. The $C_4H_8^+$ ions then undergo the known ion-molecule reaction with isobutylene to form the t-butyl carbocations[9,10] which initiate the polymerization in the liquid phase.

The polymeric materials also contain micron- and submicron- sized metal particles. An examples of an electron micrograph of the polymer film (SEM) is shown in Figure 2. The surface morphology of the polymer film obtained is dependent on experimental conditions such as laser power, temperature, pressure and electric field strength. This is a significant result since the incorporation of ultrafine metal particles and clusters into the polymer matrices would greatly extend the scope of these polymeric materials in, for example, electrical, magnentic and optical applications. Under certain experimental conditions (application of an electric field and the deposition of polymers on copper surfaces) we were able to deposit polymer wires or lines as shown in Figure 3. The wires are 0.4 μm wide and spaced 2 μm apart. This morphology can be obtained by applying a second electric field across the liquid monomer during the laser vaporization experiment.

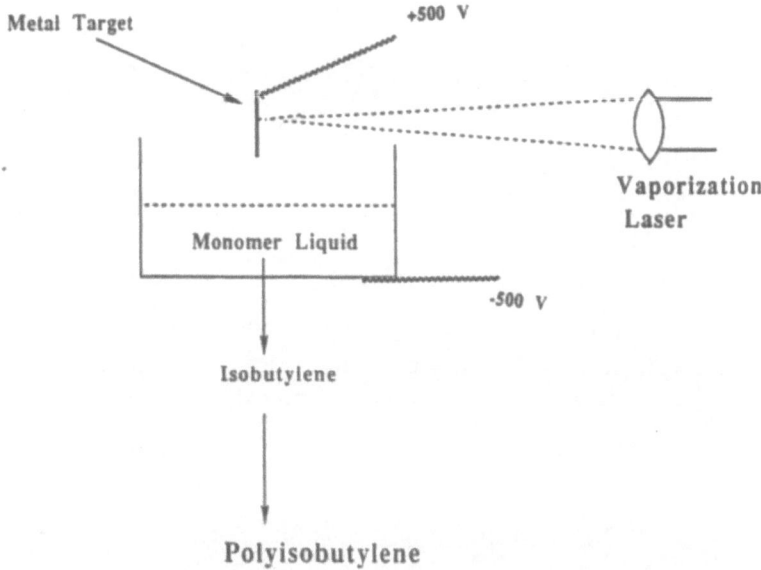

Figure 1. Experimental Set-up for Laser Vaporization Cationic Polymerization.

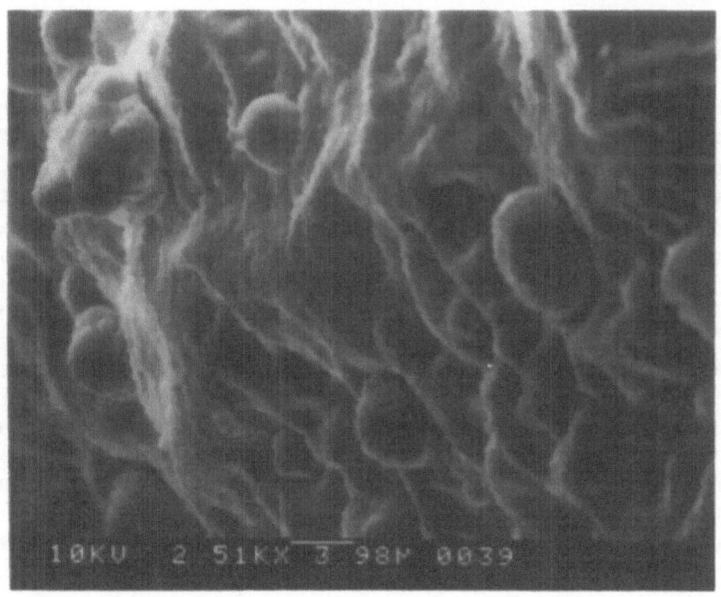

Figure 2. Electron Micrograph of Polyisobutylene Film Produced by Laser
Vaporization of Zn metal.

714

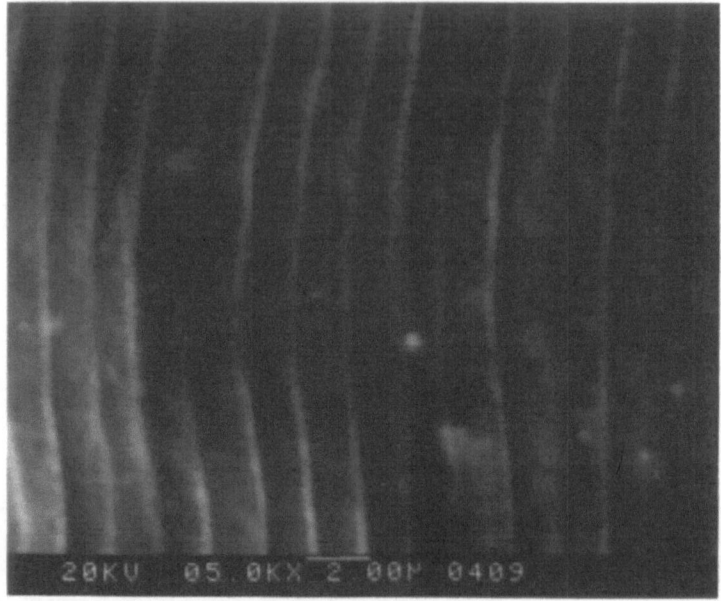

Figure 3. Electron Micrograph of Polyisobutylene Film Produced by Laser Vaporization in presence of an electric field within the monomer liquid.

2.1. GAS PHASE REACTIONS

In order to understand the mechanism of polymerization we investigated the gas phase reactions between isobutylene and the atomic metal cations produced by Laser Vaporization in a High Pressure Mass Spectrometric (LVHPMS) source[11]. A schematic diagram of the LVHPMS source is given in Figure 4. The source consists of a 4 x 4 cm^2 aluminum block mounted on a copper base wrapped in coolant coils for ion source temperature control. The source is fitted with a quartz window through which the laser beam enters. Opposite to the window is a metal rod from which the metal cations are generated. The output of the second harmonic of a Nd:YAG laser (532 nm, energy < 10 mJ/pulse) pulsed at 20 Hz is focused by a 30 cm lens, traverses the ion source, then vaporizes the titanium rod surface to create a metal atomic vapor, including a fraction of atomic ions. Gas samples are admitted to the ion source at selected pressures via an adjustable needle valve and pass through a 5 cm long 0.3 cm diameter tube bored inside the copper cooling block in order to ensure efficient thermal contact between the gas sample and the cooling block. The pressure is measured with a Baratron capacitance manometer coupled with the gas inlet tube. Ions exit the source through a 0.02 cm diameter hole. The quadrupole mass filter (Extrel C-50) is mounted coaxially to the ion exit hole.

The operating pressure in the mass spectrometer region is typically $1 - 8 \times 10^{-6}$ torr. The titanium rod and isobutylene were obtained from Aldrich with stated purities of 99.99% and 99.9%, respectively.

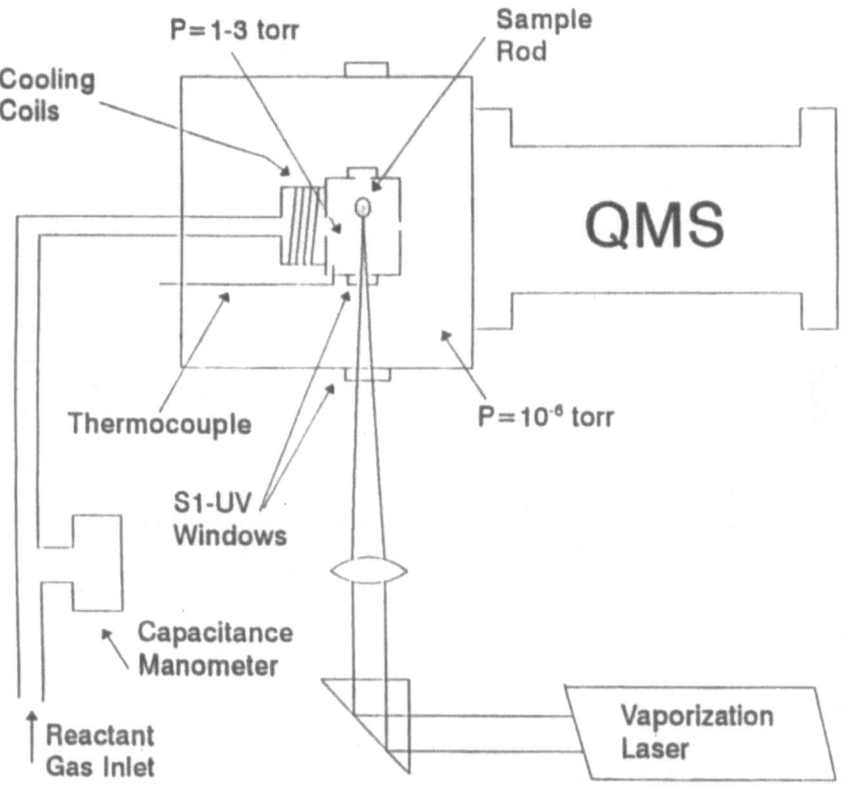

Figure 4. Schematic for the high pressure laser vaporization source.

Figure 5-a displays a typical mass spectrum observed for pure isobutylene at a total source pressure of 0.05 torr and a temperature of 288 K. Similar mass spectra are observed for mixtures of 4% isobutylene in different carrier gases such as Ar and N_2. The first product observed corresponds to a dehydrogenation reaction thus generating the $Ti^+C_4H_6$ ion. The intact product $Ti^+C_4H_8$ is not observed even at a source pressure as high as 1.5 torr and a temperature as low as -50°C. The second reaction step yields the product ions $Ti^+(C_4H_6)(C_4H_8)$, $Ti^+(C_4H_6)_2$ and $Ti^+(C_4H_4)(C_4H_6)$ while the third step generates mainly $Ti^+(C_4H_6)(C_4H_8)_2$ and $Ti^+(C_4H_6)_2(C_4H_8)$.

Other peaks in the mass spectrum correspond to the ions $TiC_{16}H_{26}^+$, $TiC_{15}H_{24}^+$, $TiC_{14}H_{22}^+$, $TiC_{13}H_{24}^+$, $TiC_9H_{16}^+$, $TiC_7H_{10}^+$ and $TiC_5H_8^+$. By increasing the total pressure in the source and keeping the temperature and the laser power at constant values, the ion intensity of the Ti^+ containing ions decreases and a new series corresponding to ions with the molecular formula $C_4H_9(C_4H_8)_n^+$ starts to appear as shown in Figure 5-b. The time profiles of the observed ions contain information on the relative rates of the sequential reactions. Figure 6 shows the temporal profiles of the $C_4H_9(C_4H_8)_n^+$ series measured at total source pressures of 0.3 torr.

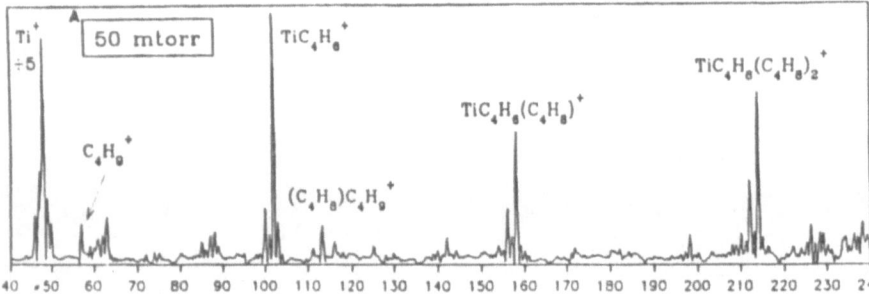

Figure 5-a. Mass spectrum of Ti^+ produced at a total source pressure of 5.0×10^{-2} torr (pure isobutylene) and temperature of 288 K.

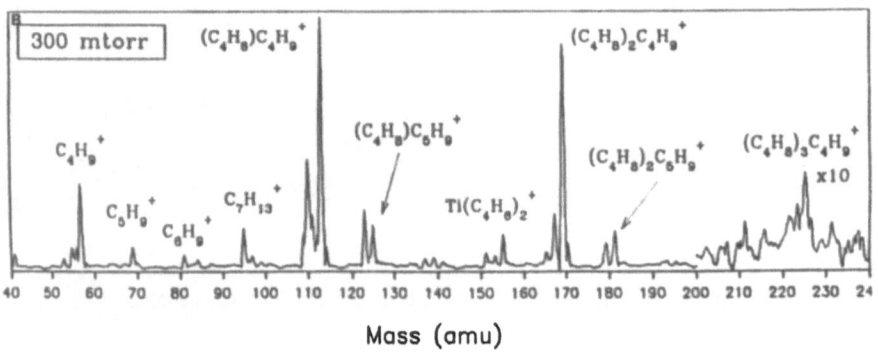

Mass (amu)

Figure 5-b Mass spectrum obtained at identical conditions as in (a) except the total pressure is 0.3 torr.

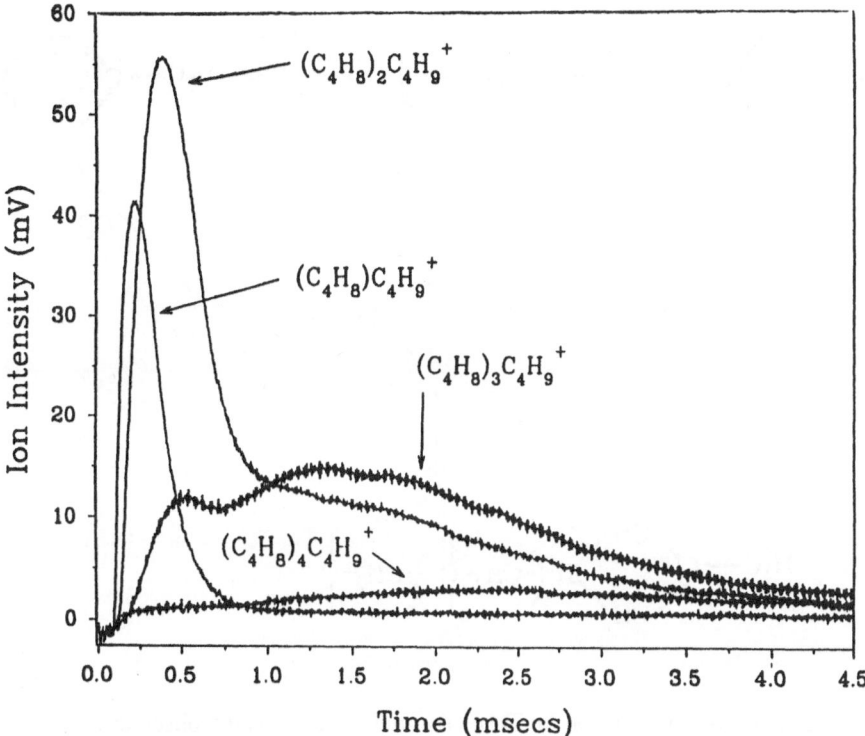

Figure 6. Temporal profiles of $C_4H_9^+(C_4H_8)_n$ at a source pressure of 0.3 torr isobutylene.

At lower pressure Ti^+ reacts with isobutylene by H_2 elimination to give $TiC_4H_6^+$ according to:

$$Ti^+ + C_4H_8 \longrightarrow TiC_4H_6^+ + H_2 \qquad (1)$$

The reactions of Ti^+ with several other olefins have been observed and H_2 eliminations constitute the most common reactions[12]. Similar reactions have been reported for "cation-like" active centers such as $CP_2ZrCH_3^+$ with a number of unsaturated hydrocarbons[13]. A preliminary reaction scheme consistent with our data is given below.

$$\text{Ti} \;+\; \underset{\text{CH}_2}{\overset{\displaystyle \text{C}}{\|}} \;\rightleftharpoons\; \overset{+}{\text{Ti}} \cdots \cdots \underset{\text{CH}_2}{\overset{\displaystyle \text{C}}{\|}} \;\longrightarrow\; H_3C-\underset{\text{CH}_2}{\overset{\text{CH}}{C}}\!\diagdown\!Ti$$

$$\searrow C_4H_8$$

$$(I) \qquad H_3C-\underset{\text{CH}_2}{\overset{\text{CH}}{C}}\!\diagdown\!Ti-CH_2-\overset{+}{\underset{|}{C}}$$

$$\downarrow\; n\,C_4H_8$$

$$H_3C-\underset{\text{CH}_2}{\overset{\text{CH}}{C}}\!\diagdown\!Ti\!\left(CH_2-\underset{|}{\overset{|}{C}}\right)_{\!n}\!CH_2-\overset{+}{\underset{|}{C}}$$

$$(2)$$

Another intriguing feature of the Ti^+/isobutylene system is the observation of the $C_4H_9^+ (C_4H_8)_n$ ions only under high pressure conditions. The $C_4H_9^+$ ions which could be generated by a chain transfer reaction[11] sequentially add isobutylene molecules to generate a polymeric ion of the form:

$$H_3C-\underset{|}{\overset{|}{C}}\!\left(CH_2-\underset{|}{\overset{|}{C}}\right)_{\!n-2}\!CH_2-\overset{+}{\underset{|}{C}}$$

This is consistent with the generally acceptable conclusion that cationic polymerization of isobutylene in condensed phases proceeds via propagation of the n-butyl carbocation $(C_4H_9^+)$[7,10]. We now turn to a discussion of the kinetics of the observed sequential reactions. The variation of the ion intensity with time for the $C_4H_9^+(C_4H_8)_n$ series suggests that the rate constants for cationic propagation vary significantly among the first few reaction steps. The time profile of the decay of $C_4H_9^+$ is essentially the same as that of $C_4H_9^+(C_4H_8)$ suggesting that the first two steps are very fast. However, the decay rates for the ions $C_4H_9^+(C_4H_8)_n$ with n = 2 - 5 are relatively slow. The strong dependence of the rate of cationic propagation on size can be explained in terms of steric or structural factors which may result in a situation where the charge becomes less accessible to the incoming monomers, thus slowing down further propagation steps. It will be interesting to design new experimental techniques to allow the measurements of the slow reaction steps.

We have also investigated the gas phase reactions of isobutylene with other metal cations. Figure 7 displays the mass spectra obtained from the reactions of isobutylene with Zn^+. It is clear that direct charge transfer takes place between Zn^+ and isobutylene to generate $C_4H_8^+$ which reacts with neutral isobutylene to produce $C_4H_9^+$. The reactions with Pt^+ involve several dehydrogenation and association channels. These reactions are currently under investigation in our laboratory.

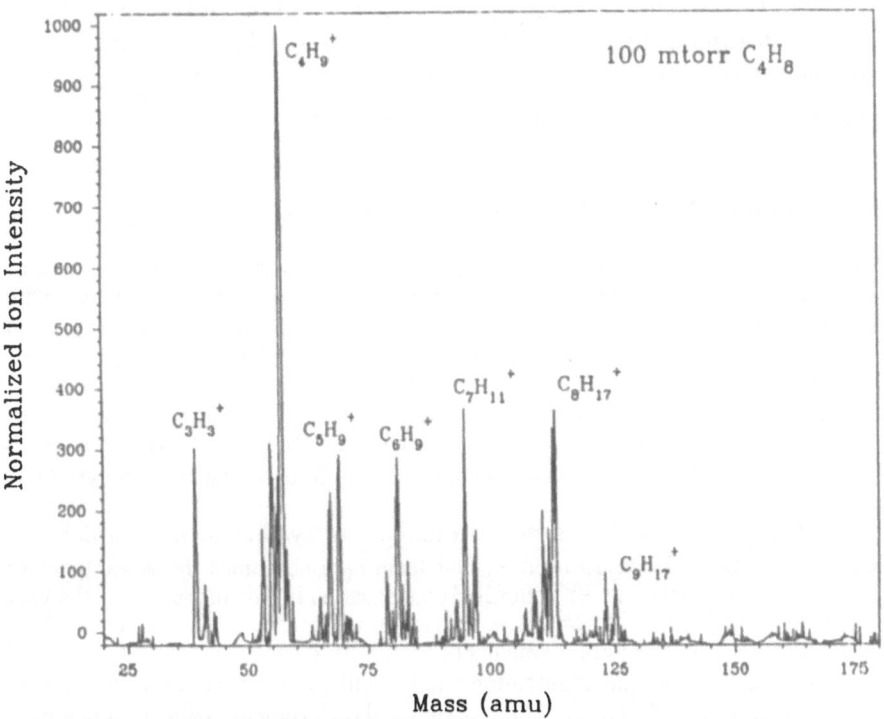

Figure 7. Mass spectrum of the Zn^+ gas phas reactions with isobutylene.

2.2. CONCLUSIONS AND OUTLOOK OF THE POLYMERS WORK

In conclusion, we have discovered that high molecular weight polyisobutylene can be synthesized using a laser vaporization technique. We are currently using this method for polymerization initiated by ions generated from other involatile materials such as fullerenes (C_{60}, C_{70}, C_{84}, etc), refractive, semiconductor and superconductor materials. In this case, the laser desorbs the ions of interest from their solid precursors which are coated on a glass or a metal substrate. These new directions, which are being explored for the first time, could lead to a powerful approach for the synthesis of a wide variety of polymeric materials with unusual properties. The method could also evolve into a valuable technique for the in-situ doping of ultrafine metal particles into the polymer matrices during their synthesis and also for the preparation of polymeric thin films. Systematic experimentation on a range of important monomers that can be polymerized by a cationic mechanism and matched doped metals would make available a base of results upon which the properties of future polymeric materials could be reliably assessed.

Detailed information regarding the mechanism of polymerization associated with laser vaporization can be obtained from gas phase and clusters studies. Sequential reactions are observed between atomic metal cations and isobutylene suggesting a polymerization behavior induced by the metal cation. For Ti^+, the reactions appear to be initiated by the ionic species $Ti^+C_4H_6$. The formation of stable ionic species (probably cyclic structures) interrupts the general pattern of successive addition reactions. For Zn^+, direct charge transfer to generate $C_4H_8^+$ followed by an ion-molecule reaction to produce $C_4H_9^+$ has been observed. For other metal cations such as Al^+ and Ag^+, generation of $C_4H_9^+$ has been observed as a result of the reaction of $C_4H_8^+$ most likely produced by laser plasma ionization of isobutylene.

3. Ultrafine Metal particles in Vapor Phase Nucleation

The synthesis of nanoscale particles has received considerable attention in view of the potential for new materials with novel properties and the design of catalysts with specific dimensions and compositions[17-20]. Nanoscale particles possess several unique properties such as large surface areas, unusual adsorptive properties, surface defects and fast diffusivities. The characterization of these properties can ultimately lead to identifying many potential uses, particularly in the field of catalysis. For example, particles of metal oxides and mixed oxides such as SiO_2/TiO_2 exhibit unusual acidic properties and can be used as acidic catalysts[21].

Recently we described a novel technique to synthesize nanoparticles of controlled size and composition[22]. Our technique combines the advantages of pulsed laser vaporization with controlled condensation in a diffusion cloud chamber under well defined conditions of temperature and pressure. Using this method, monodisperse particles with diameters of ≈ 5 nm have been synthesized. The SiO_2 particles exhibit a unique aggregation pattern with well defined pores and a very high surface area of 500 m^2/g. These particles show efficient visible luminescence at room temperature which could arise from the photophysics of certain surface states. Another class of nanoparticles that has been prepared is metal carbides. For example, molybdenum carbide particles with a stoichiometry MoC_4, different from the known bulk form Mo_2C, have been produced. These particles show special catalytic activity for the production of larger fullerenes and carbon nano-tubes.

In the experiments, a modified upward thermal diffusion cloud chamber is used for the synthesis of the nanoscale particles. This chamber has been commonly used for the production of steady state supersaturated vapors for the measurements of homogeneous and photo-induced nucleation rates of a variety of substances[23,24]. A sketch of the diffusion cloud chamber with its relevant components is shown in Figure 8-a.

The chamber consists of two, horizontal, circular stainless steel plates, separated by a circular glass ring. A metal target of interest sets on the lower plate, and the chamber is filled with a carrier gas such as He or Ar containing a known composition of the reactant gas (e.g. O_2 in case of oxides, N_2 or NH_3 for nitrides, CH_4 or C_2H_4 for carbides, etc.). The metal target and the lower plate are maintained at a temperature higher than that of the upper one (temperatures are controlled by circulating fluids). The top plate can be cooled to less than 120 K by circulating liquid nitrogen. Nichrome heater wires are wrapped around the glass ring and provide sufficient heat to prevent condensation and fogging. The large temperature gradient between the bottom and top plates results in a steady convection current which can be enhanced by using a heavy carrier gas such as Ar or Kr under high pressure conditions ($1-10^3$ torr). The metal vapor is generated by pulsed laser vaporization using the second harmonic (532 nm) of a Nd-YAG laser (15-30 mJ/pulse). Laser vaporization typically releases more than 10^{14} metal atoms per pulse (2×10^{-8} s). The hot metal atoms react with O_2 (in the case of oxides syntheses) to form vapor phase metal oxide molecules which undergo several collisions with the carrier gas thus resulting in efficient cooling via collisional energy loss. Under the total pressure employed in these experiments (100 - 800 torr), it is expected that the metal atoms and the oxide molecules approach the thermal energy of the ambient gas within several hundred microns from the vaporization target. The unreacted metal atoms and the less volatile metal oxide molecules are carried by convection to the nucleation zone near the top plate of the chamber. The temperature T, partial pressure P of diffusant, saturation pressure P_e, and supersaturation, $S = P/P_e$ vary with elevation in the chamber in the manner demonstrated in Figure 8-b. By controlling the temperature gradient, the total pressure and the laser power (which determines the number density of the metal atoms released in the vapor phase), it is possible to control the size of the condensing particles. This is the basic principle of our method.

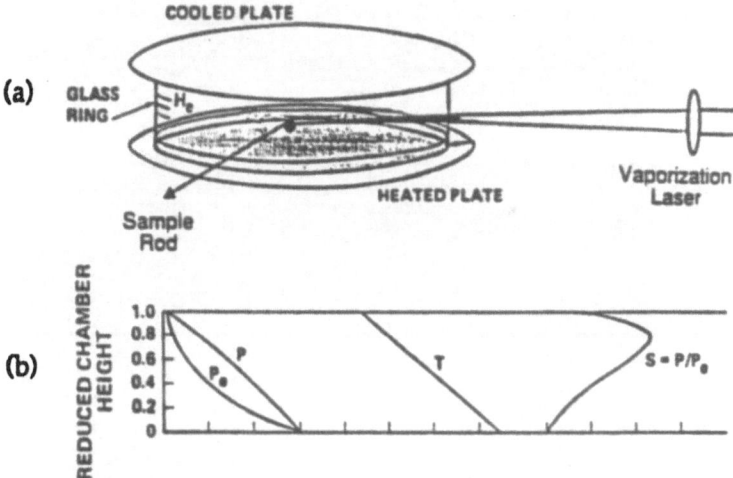

Figure 8-a. Sketch of diffusion cloud chamber.
Figure 8-b. Typical courses of diffusant partial pressure P, saturation pressure P_e, temperature T, and supersaturation S in the chamber.

Figures 9-a and 9-b display SEM micrographs obtained for ZnO and Al_2O_3 particles synthesized using 20% O_2 in He at a total pressure of 800 torr and top and bottom plate temperatures of -100°C and 20°C, respectively. It is clear that the particles exhibit a unique agglomerate pattern which appears as a web-like matrix. Based on the TEM analyses of these, individual particle's sizes are estimated between 5-10 nm.

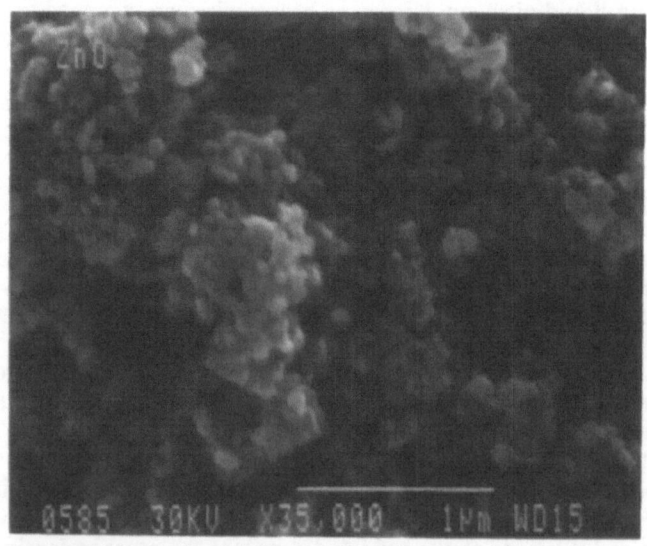

Figure 9-a. SEM micrograph of ZnO particles.

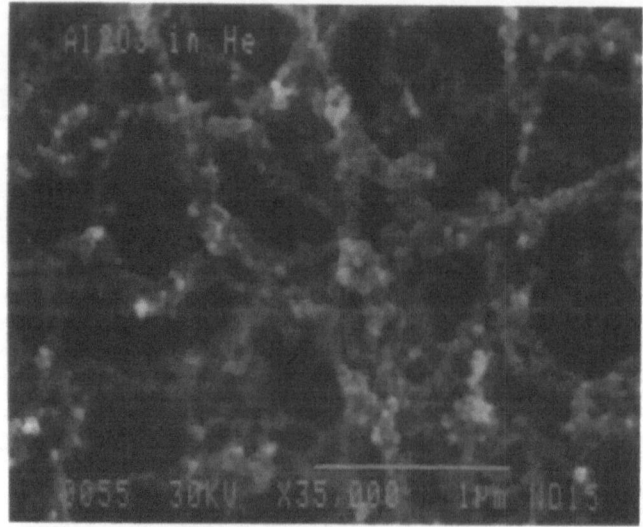

Figure 9-b. SEM micrograph of Al_2O_3 particles.

The Raman spectrum of the TiO₂ nanoparticles synthesized in Ar (shown in Figure 10) exhibits the characteristic peaks of the nanoscale anatase phase of TiO₂ (the low temperature phase)[25,26]. By varying the experimental conditions such as the partial pressure of O_2 and the carrier gas we were able to synthesize both the rutile and anatase phases of TiO_2 [22].

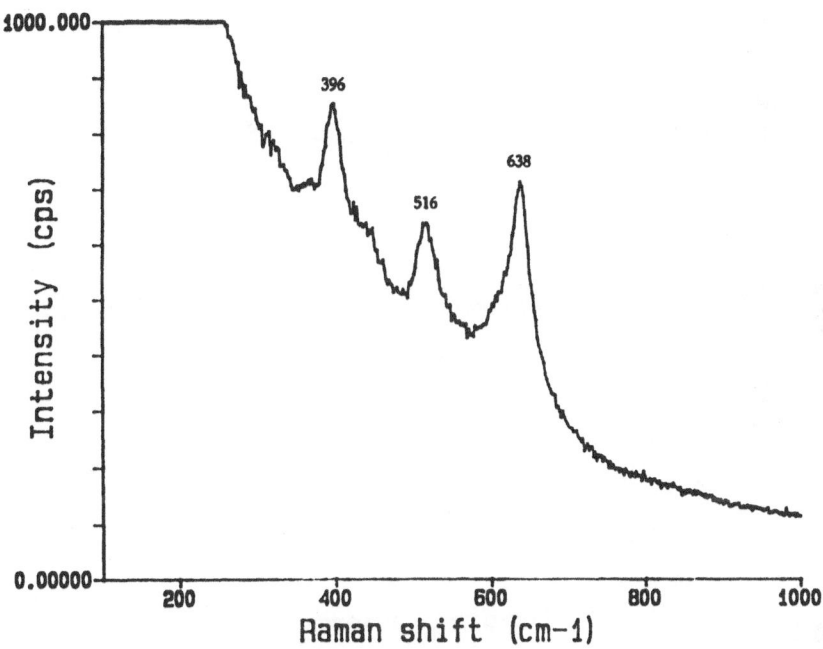

Figure 10. Raman spectra of TiO₂ particles synthesized in O₂/Ar mixture.

To illustrate the effect of the O_2 / carrier gas mixture on the ratio of the oxidized-to-nonoxidized metal particles, we measured the x-ray photoelectron spectra (XPS) of the particles prepared under different experimental conditions. Figure 11 illustrates the results for SiO₂ particles synthesized using different concentrations of O_2 in He. By decreasing the O_2 concentration in the carrier gas the Si 2p peak ($\approx$ 99.5 eV) is enhanced due to the unoxidized Si particles. From the data presented in Figure 14 (a, b, c), the SiO₂ / Si ratios are estimated as 0.99, 0.76 and 0.60, respectively. Therefore, the data demonstrates the control of the compositions of the nanoparticles. The oxide / metal ratio in the nanoparticles also depends on the type of the carrier gas as well as on the O_2 concentration. By using different carrier gases containing different concentrations of O_2, it was possible to affect the efficiency of the oxidation process. In this way, nanoparticles containing more than 99% oxide molecules or more than 60% unoxidized metal have been synthesized.

724

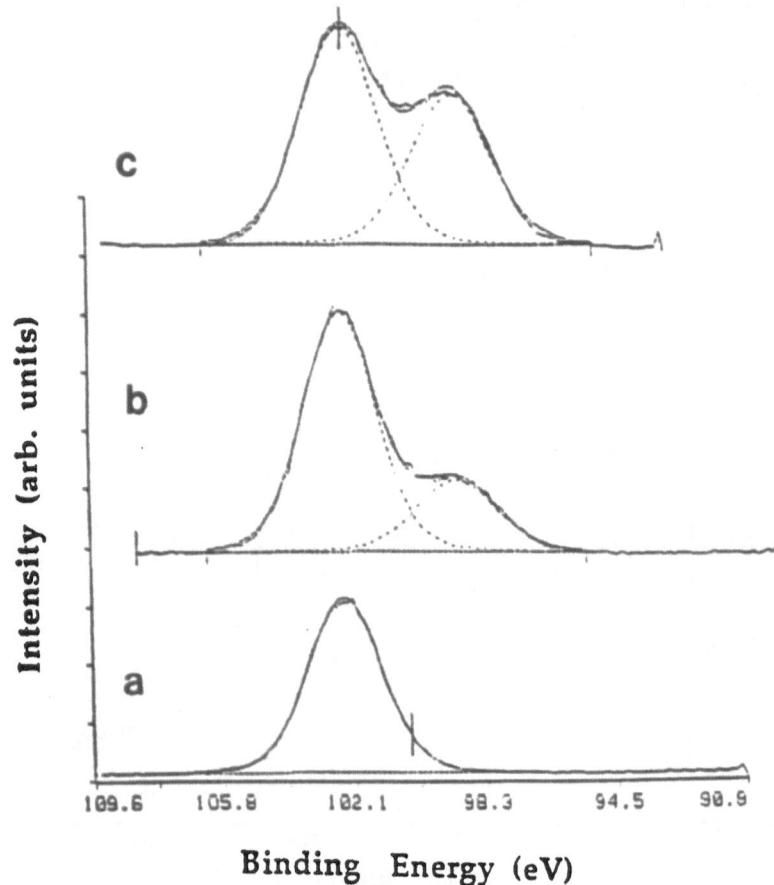

Figure 11. XPS spectra of SiO_2 particles synthesized using : (a) 30% O_2 in He, (b) 7% O_2 in He, (c) 2% O_2 in He.

3.1. CONCLUSIONS OF THE NANOPARTICLES WORK

The results of this study appear to demonstrate the feasibility and the promise of the laser vaporization / condensation method in the synthesis of nanoparticles with well defined properties. The combination of the laser synthetic approach and the spectroscopic characterizations of the bulk and surface properties of the nanoparticles will lead to significant results particularly in two areas: (i) the synthesis of specific catalysts (e.g. metal/metal oxide systems) of controlled sizes and compositions and (ii) the design of new nanomaterials such as mixed oxides/nitrides or carbides/nitrides and other refractive, semiconductor and superconductor materials.

4. Acknowledgement

The author gratefully acknowledges financial support from NSF Grants No. DMR 9207433 and and CHE 9311643. Acknowledgement is also made to the donors of the Petroleum Research Fund, administered by the American Chemical Society, the Thomas F. and Kate Miller Jeffress Memorial Trust, the Dow Corning Corporation and the Exxon Education Foundation for the partial support of this research.

5. References

1. Sugano, S., Nishina, Y. and Ohnishi, S. (eds.) (1987) *Microclusters*, Springer, New York.
2. Jena, P., Rao B. K. and Khana, S. N. (eds) (1987) Physics and Chemistry of Small Clusters, *NATO ASI Series No 158*. Plenum, New York.
3. See for example: Moskovits, M. (ed.) (1986) *Metal Clusters*, Wiley, New York; Johnson, B. F. G. (ed.) (1980) *Transition Metal Clusters*, Wiley, New York.
4. Jena, P., Rao, B. K. and Khanna, S. (1992) *The Physics and Chemistry of Finite Systems: From Clusters to Crystals*, Kluwer Academic Publishers.
5. Vann, W. and El-Shall, M. S. (1993) *J. Am. Chem. Soc.* **115**, 4385; see also *Chemical& Engineering News* May 24, 1993, p. 31.
6. Vann, W., Daly, G. M. and El-Shall, M. S.(1993) Laser Ablation in Materials Processing: Fundamentals and Applications, in B. Braren, J. Dubowski and D. Norton (eds.), Vol. **285**, *Materials Research Society Symposium Proceedings Series,* pp. 593-598 .
7. Kennedy, J. P. and Marechal, E. (1982) *Carbocationic Polymerization*, John Wiley & Sons, New York.
8. Goethals, E. J. (ed.) (1984) *Cationic Polymerization and Related Processes*, Academic Press, New York.
9. Mitchell, A. L. and Tedder, J. M. (1986) *J. Chem. Soc. Perkin Trans.II*, 1197; Henis, J. M. S. (1970) *J. Chem. Phys.* **52**, 282; Sieck, L. W. and Ausloos, P. (1972) *J. Chem. Phys.* **56**, 1010; Koyano, I. (1966) *J. Chem. Phys.* **45**, 706; Herman, J., Herman, K. and Ausloos, P. J. (1970) *J. Chem. Phys.* **52**, 28; Abramson, F. P. and Futrell, J. H. (1968) *J. Phys. Chem.* **72**, 1994.
10. Sparapany, J. J. (1966) *J. Am. Chem. Soc.* **88**, 1357; Aquilant,V., Galli, A., Giordini-Guidoni, A., Volpi, G. G. (1967)*Trans. Faraday Soc.* **63**, 926.
11. Daly, G. and El-Shall, M. S. (1994) *J. Phys. Chem.* **98**, 696 .
12. Allison, J. and Ridge, D. P. (1977) *J. Am. Chem. Soc.* **99**, 35.
13. Christ, C. S., Eyler, J. R. and Richardson, D. E. (1988) *J. Am. Chem. Soc.* **110**, 4038; (1990) **112**, 597.
14. El-Shall, M. S. and Marks, C. (1991) *J. Phys. Chem.* **95**, 4932.
15. El-Shall, M. S. and Schriver, K. E. (1991) *J. Chem. Phys.* **95**, 3001.
16. Daly, G. M. and El-Shall, M. S. (1993) Z. Phys. D. **26S**, 186.
17. See for example: (1992) Frontiers in Materials Science, *Science* **255**, 1049.
18. Siegel, R. W. (1991) *Annu. Rev. Mater. Sci.* **21**, 559,.
19. Pool, R. (1990) *Science* **248**, 1186.
20. Che, M. and Bennett, C. O. (1989) The Influence of Particle Size on the Catalytic Properties of Supported Metals, *Adv. in Cataylsis* **36**, 55.
21. Stakheev, A., Yu., Shapiro, E. S., Apijok, J. (1993) *J. Phys. Chem.* **97**, 5668.

22. El-Shall, M .S., Slack, W., Vann, W., Kane, D., Hanley, D. (1994) *J. Phys. Chem.* **98**, 3067.
23. Wright, D., Caldwell, R., Moxely, C., El-Shall, M. S. (1993) *J. Chem. Phys.* **98**, 3356.
24. El-Shall, M. S., Rabeony, H. M., Reiss, H. (1989) *J. Chem. Phys.* **91**, 7925.
25. Parker, J. C., Siegel, R. W., (1990) *J. Mater. Res.* **5**, 1246.
26. Capwell, R. J., Spagnolo, F., DeSesa, M. A. (1972) *Appl. Spectrosc.* **26**, 537.

PULSED LASER DEPOSITION OF Nb5Te4 THIN FILMS

Ablation Plume Analyses

F. GRANGEON, H. SASSOLI[†], L. LAMBERT,

Y. MATHEY[†], M. AUTRIC, W. MARINE[†]

Laboratoire Interdisciplinaire Ablation Laser et applications

U.M. 34 C.N.R.S. Institut de Mécanique des Fluides (case 918)

[†]URA C.N.R.S. 783 Département de Physique (case 901)

Parc Scientifique et Technologique de Luminy

163, Avenue de Luminy 13009 Marseille

Abstract

Niobium telluride thin films were produced by excimer laser vaporisation of $NbTe_2$ bulk targets and deposition of the ablation plume on heated silicon substrates. Precise control of the laser energy density and of the substrate temperature leads to good quality, well-crystallized thin films. For a specific couple of fluence and temperature, films were crystallized in the Nb_5Te_4 phase. In order to characterize the influence of laser parameters, analyse of the ablation plume by optical spectroscopy was performed using an Optical Multichanel Analyzer.

1. Introduction

Chalcogenides of transition metals showing low dimensional properties represent significant areas of applications such as micro-electronics (1D and 2D conductors), solid state batteries (cathodic materials) and micro-mechanics (solid lubricants) [1, 2, 3, 4]. The growth of high quality thin films [5] showing, for example, both conductivity and

727

J. Mazumder et al. (eds.), Laser Processing: Surface Treatment and Film Deposition, 727–737.
© *1996 Kluwer Academic Publishers.*

lubrication behaviour appears interesting for many high precision applications,especially in micro-electromechanical systems. Chalcogenides of transition metals are the compounds of general formula MX_n where M is a transition metal of the groups IV, V or VI, and X a chalcogen atom (S, Se, Te). In particular, several niobium tellurides are known with either bidimensional ($NbTe_2$) or quasi-unidimensional (Nb_5Te_4, Nb_3Te_4, $NbTe_4$) structure.

In the Pulsed Laser Deposition (P.L.D) process, the kinetic properties of the ablation plume play a major role. Knowing the species in the plume, the energies of neutral and ionized particles and molecules and, eventually, chemical reactions is of first importance in the understanding of the film formation. In order to characterize the ablation process, the emission spectroscopy is a preferential mean of investigation.

2. Thin films deposition

P.L.D was performed in a stainless steel vacuum chamber (pressure 2.10^{-4} pascal). The irradiation of a bulk, sintered $NbTe_2$ target was performed by a KrF laser delivering 500 mJ in 20 ns pulse duration at a repetition rate of 1 Hz. The beam was focused on the rotating target at an angle of 45° relatively to the target surface. The energy density ranged from 2 J/cm^2 to 8 J/cm^2. The substrates were silicon (100) and (111) monocrystals and were heated in the range from ambient temperature up to 450°C. The target - substrate distance was 50 mm. X-Ray diffraction and X-Ray fluorescence were used to determine the films composition and crystallinity as functions of substrate temperature and laser energy density.

X-Ray diffraction analyse of the films deposited at a substrate temperature below 290°C shows an amorphous or very minimally crystallized structure. In order to produce crystallized films, the substrate must be heated to a higher temperature, above the melting-evaporation temperature of Te. In this case we could deposit only substoichiometric ($x < 2$) films relative to the target composition ($NbTe_2$).

X-Ray diffractograms of crystallized films show four large peaks representative of a microcrystalline structure. Only one interreticular distance d can be calculated from these peaks, the few large angle lines being the harmonics of the unique and strong low angle ones. Figure 1 shows the variation of d with the substrate temperature. The unique interreticular distance decreases when the substrate temperature increases. Despite the large lattice mismatch between the lattice parameters of $NbTe_x$ compounds and Si substrates, we can note an influence of the orientation of the substrate on the crystallization process. The interreticular distance changed when the films were deposited on the (111) or the (100) face of the silicon. The laser fluence increase leads to a better crystallisation, as can be seen from figure 2, where we present the d variation versus laser fluence. At low substrate temperature (320°C), the slope of the d variation is much higher than at 410°C substrate temperature. This shows the influence of kinetics parameters of the laser induced plume on the crystallization of $NbTe_x$ compounds.

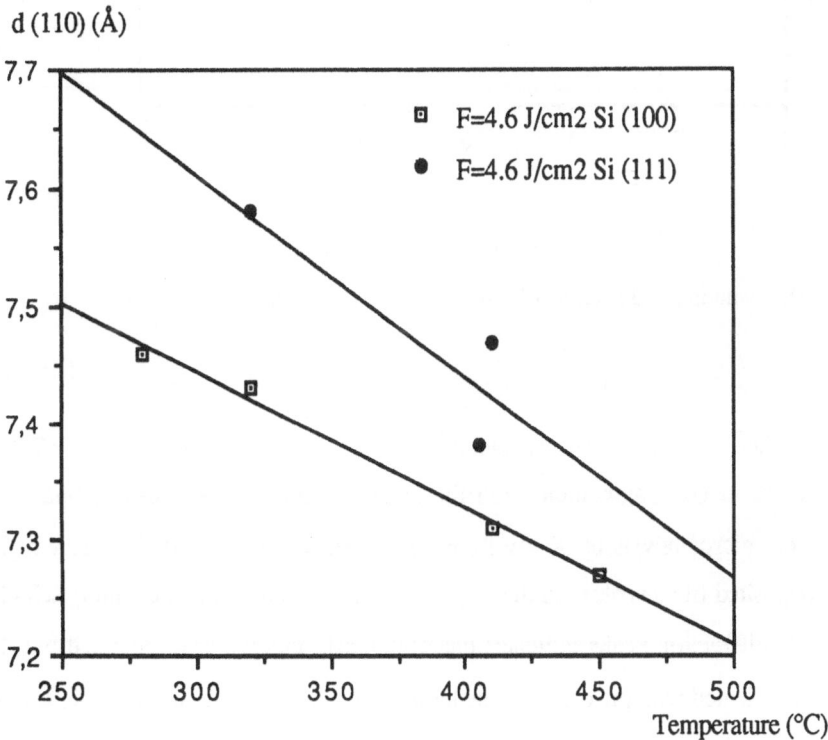

d (110) (Å)

□ F=4.6 J/cm2 Si (100)

● F=4.6 J/cm2 Si (111)

Temperature (°C)

Fig. 1 : Interreticular distance d versus substrate temperature

d (110) (Å)

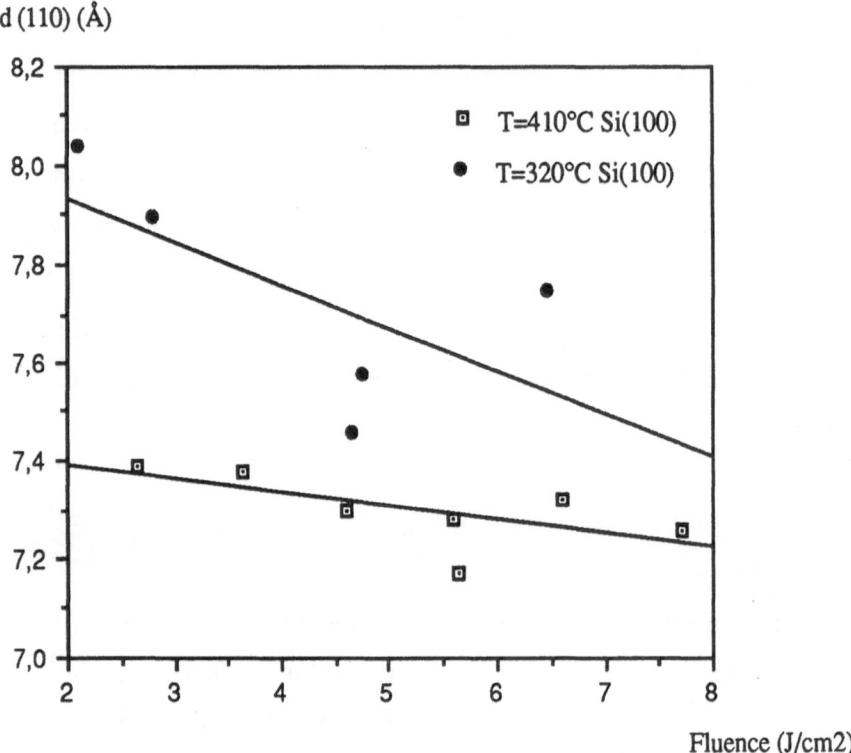

Fig. 2 : Interreticular distance versus fluence.

The evolution of d is plotted for two substrate temperatures (320°C and 410°C)

The interreticular distances, calculated from X rays diffractograms in the 7.2-8.0 Å range are in good agreement with the (110) distance of the Nb_5Te_4 phase [6]. Moreover, microanalysis by X ray fluorescence show that the stoichiometry of our P.L.D deposited films is close to the Nb_5Te_4 phase. The observation of (hh0), with h = 1, 2, 3, 4, diffraction peaks indicates unambiguously that we have obtained Nb_5Te_4 thin films oriented with the c axis of the tetragonal cell parallel to the substrate surface. A sketch of a constituting chain lying in the substrate plane is given in figure 3. White dots are niobium atoms and black dots are tellurium atoms. Tetragonal cells are bounded along their c axis.

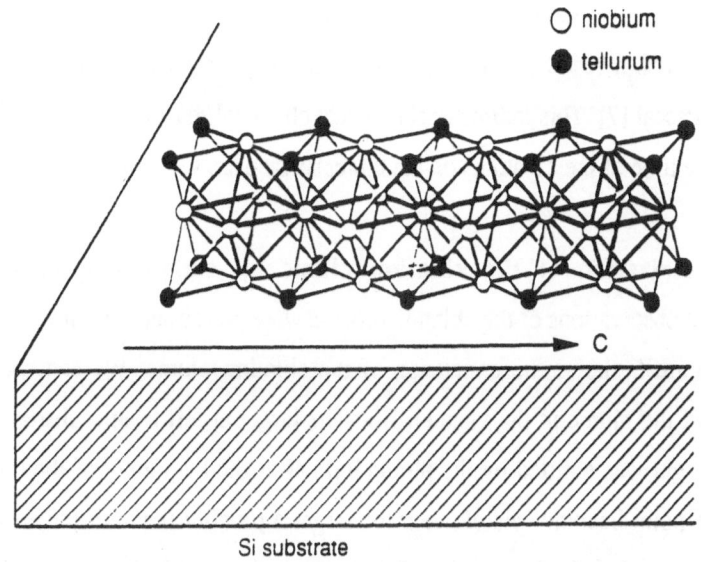

Fig. 3 : Nb$_5$Te$_4$ chain on Si substrate sketched from [6]

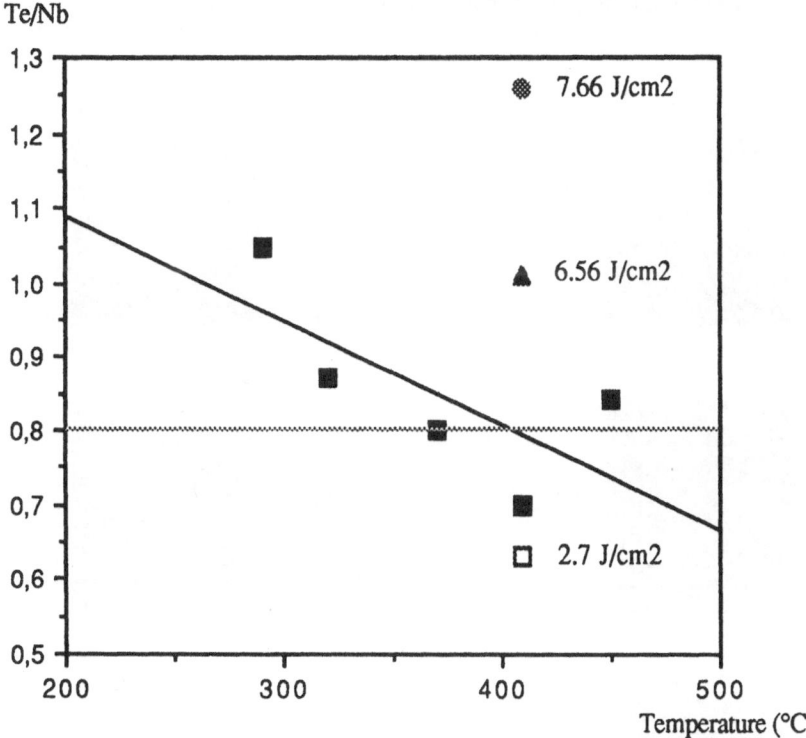

Fig. 4 : Te/Nb ratio versus substrate temperature at constant fluence (4.6 J/cm^2). The horizontal line represents the stoichiometry of the Nb$_5$Te$_4$ compound. The square, the triangle and the open circle show the ratio for other fluences (2.7 J/cm^2, 6.56 J/cm^2, 7.66 J/cm^2 respectively) at T = 410°C.

As shown in figure 4, increasing substrate temperature induces a decrease in the Te content. Such a result is not surprising as the vapour pressure of Te at 300°C is about 10^{-2} Pascal [7]. This indirectly shows that chemical reaction of Nb and Te and/or the residence time of the Te atoms on the target surface are very low.

3. Characterization of the ablation plume by emission spectroscopy

Global observations of the ablation process were performed by using a rapid CCD camera. Figure 5 shows the expansion of the ablation plume. We can see the free expansion of the plume in the vacuum from the target at left to the substrate at right. No visible light was emitted by the plume 9 µs after the beginning of the interaction. A luminous point becomes visible near the target 4 µs after the laser pulse and is still luminous 30 µs after the laser pulse. This point moves slowly in the direction of the substrate (few 100 m/s). It may be constituted of heavy clusters [8].

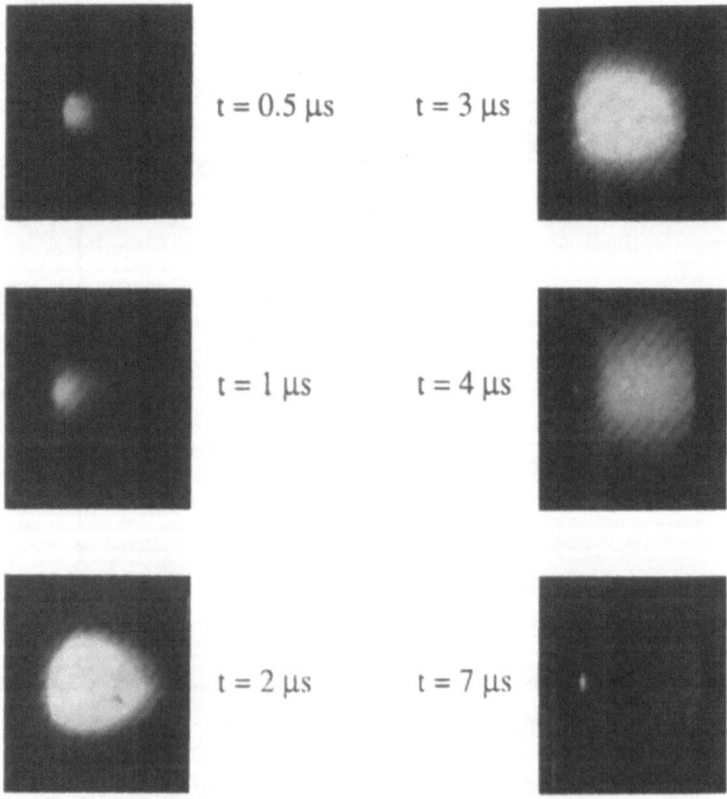

$t = 0.5 \ \mu s$ $t = 3 \ \mu s$

$t = 1 \ \mu s$ $t = 4 \ \mu s$

$t = 2 \ \mu s$ $t = 7 \ \mu s$

Fig. 5 : CCD images of the expansion of the ablation plume

Characterization of the ablation plume by emission spectroscopy was performed using a 1 m focal length spectrograph (Jobin-Yvon THR 1000 MSL) with an Optical Multichanel Analyzer (EG&G O.M.A III). This set-up allows the accumulation of the emitted light by intensification of the detector during exposure times ranging from 100 ns to 10 ms. A delay generator allows time-resolved observations by delaying the beginning of the acquisition. The image of the plasma was formed on the slit of the spectrograph. The observed area was determined by the slit width (300 μm) and the image magnification (0.9).

The first studies using the O.M.A were concerning the determination of spectral lines emitted by the plume. The selected exposure time was about 100 μs in order to obtain a time-integrated spectrum showing all emitted lines. We can see on figures 6a and 6b emission spectra in the range 4860 - 4980 Å and 5610 - 5730 Å showing spectral lines of NbI (4967.78 Å, 4973.14 Å, 5628.26 Å,...), TeII (4866.24 Å, 5708.12 Å), and Te$_2$ (4909.29 Å, 4957.75 Å) species.

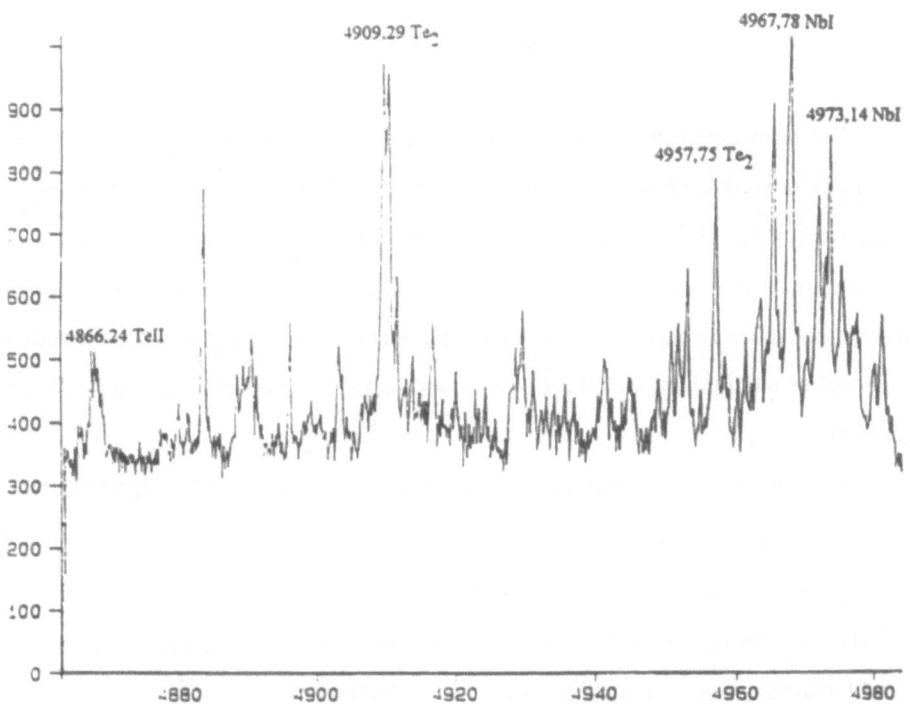

Fig. 6a : Emitted spectrum in the range 4860-4980 Å

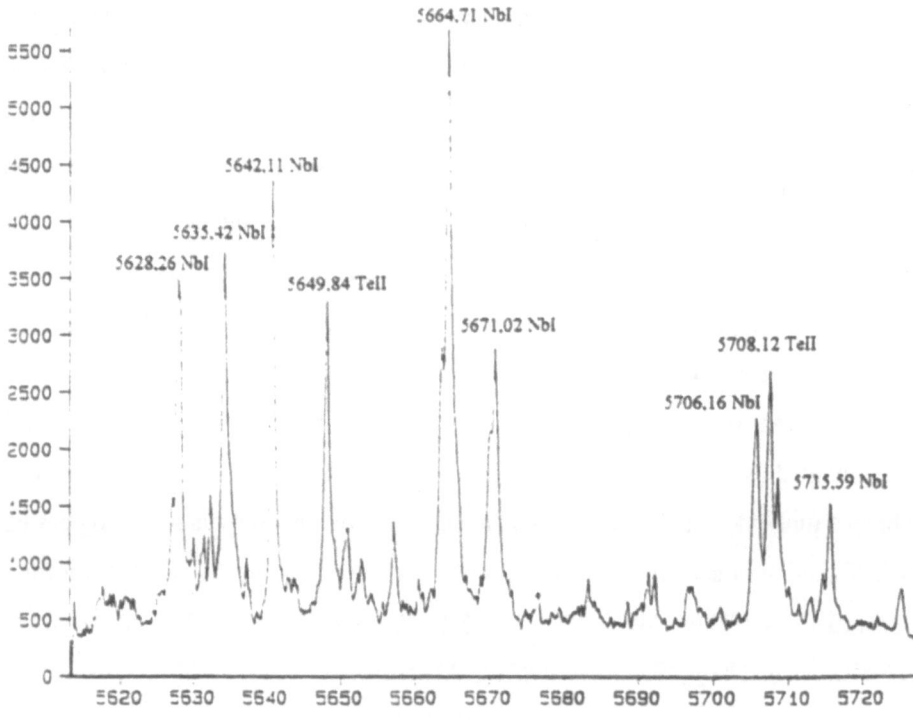

Fig. 6b : Emitted spectrum in the range 5610-5730 Å

We have noted that only few spectral lines of TeI species (not shown here : 6456.7 Å, 6790.0 Å, 6854.7 Å) could be observed due to the relatively low intensity of these ones in the visible range [9]. In the same way, no NbII spectral lines could be observed with this set-up, because there is no NbII spectral lines in the visible range, but we can suppose that the plume contains ionized niobium because the ionization potential of niobium atoms (6.88 eV) is lower than the one of tellurium atoms (9.009 eV) [10].

Several unidentified lines were observed in this experiment. These lines (4904 Å, 4978 Å,...) are not corresponding to known lines of neutral or ionized niobium and tellurium or molecules. These lines disappeared when we used a short integration time (<1 μs). By observation of the slow luminous point, we have observed few lines (see fig. 7), corresponding to the unidentified lines at 4904 and 4978 Å.

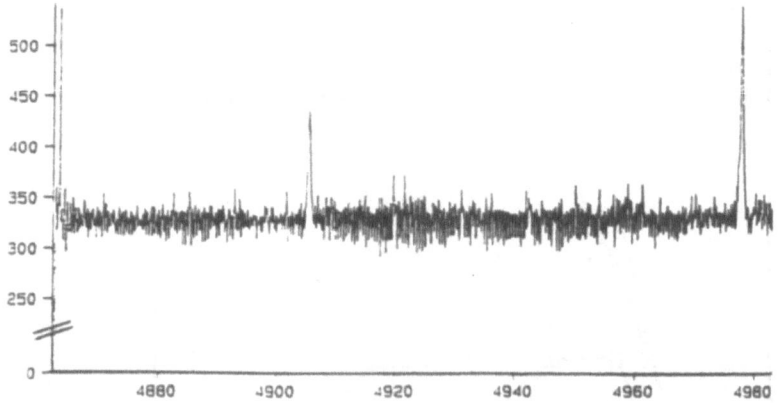

Fig. 7 : Unidentified lines observed by delaying the acquisition

By accumulation of the emitted light, the O.M.A allows us to observe the total emission of several spectral lines in a same acquisition. By using a long integration time (100 μs) and beginning the acquisition immediately after the laser pulse, we have observed the variation of the time-integrated intensity of three lines of different species (TeII 4866.24 Å, NbI 4967.78 Å, Te$_2$ 4909.29 Å) with the distance to the target. Figures 8a and 8b show the time-integrated intensity of these lines as a function of the distance from the target for two fluences (4.1 and 6.2 J/cm^2).

Intensity (a.u.)

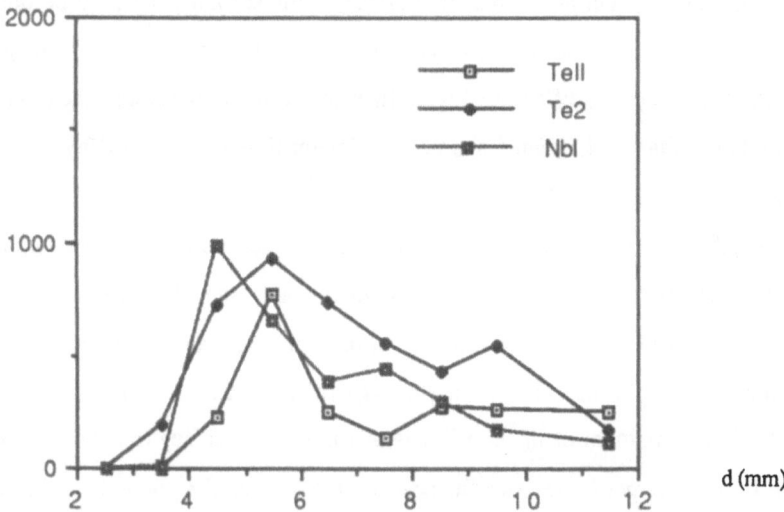

Fig. 8a : Time - integrated intensity of spectral lines of NbI, TeII, Te$_2$ species

for a fluence of 4.1 J/cm^2

736

Intensity (a.u.)

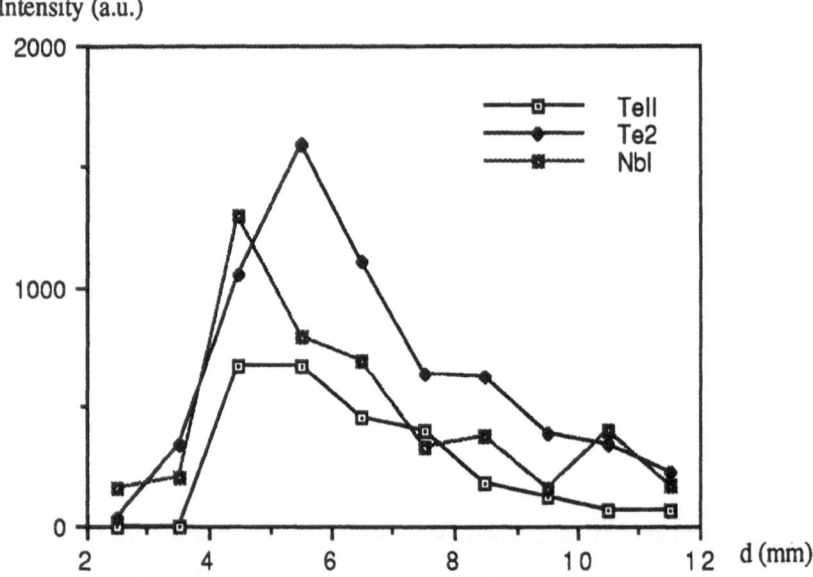

Fig. 8b : Time - integrated intensity of spectral lines of NbI, TeII, Te$_2$ species for a fluence of 6.2 J/cm^2

We can see that the maximum of intensity for the NbI line is located at 4.5 mm from the target for the two fluences, at 5 mm (4.1 J/cm^2) and at 5.5 mm (6.2 J/cm^2)for the TeII 4866 Å line, and at 5.5 mm for the two fluences for the Te$_2$ 4909.29 Å line. We can note the existence of a second peak of intensity of the NbI 4967.78 Å line at 7.5 mm (4.1 J/cm^2) and at 8.5 mm (6.2 J/cm^2). This second peak may be due to neutral niobium atoms reexcited by collisions and/or chemical reaction after cooling the plume, the position being different with the fluence because at 6.2 J/cm^2, the plume is probably hotter than at 4.1 J/cm^2 and needs a longer time to be cooled by radiative cooling.

The difference in the position of the maximum of the TeII line is probably due to the higher temperature of the plasma for high fluences, the plasma is cooled and recombine farther [11]. Moreover, the peak of intensity is broader for a high fluence, indicating that the plasma exists farther at high fluence. A first explanation for the position of the maximum of the Te$_2$ line is that molecules of Te$_2$ may be created at 5.5 mm from the target for the two fluences, probably when the plume is sufficiently cooled by radiative cooling. Second, Te$_2$ could be excited by numerous gas-phase collisions at this position.

4. Conclusion

Crystallized Nb_5Te_4 thin films were produced for the first time by using Pulsed Laser Deposition method. At the present state of this study, the best P.L.D-deposited films show good crystallization in the Nb_5Te_4 phase with preferential orientation along the c-axis. The experimental conditions were a fluence of 5.5 J/cm^2 and a substrate temperature of 410°C. Relatively to its applications, this compound appears interesting in thin film form. Analyses of the ablation plume by emission spectroscopy show the formation of Te_2 molecules and heavy clusters in the plume. The presence of these kind species must strongly influence the film growth. Moreover, chemical reactions in the gas phase lead to the plume cooling, decrease the energies of species arriving on the substrate.

References

1. Brown, B.E. (1966) Acta Cryst. **20** 264

2. Selte, K. and Kjekshus, A. (1963) Acta Chem. Scand. **17** 2560

3. Balchin, A.A. (1976) Crystallography and chemistry of materials with layered structure Ed. F. Levy, Reidel, Dordrecht

4. Meerschaut, A. and Rouxel, J. (1986) Crystal structure and properties of materials with a quasi-onedimensional structure Ed. J. Rouxel, Reidel, Dordrecht

5. Langlade, C., Fayeule, S., Mathey, Y., Pailharey, D., and Kassem, M. (1993) Surface and Coating Technology **62** 417

6. Hughbanks, T. (1989) Prog. Solid St. Chem. **19** 329

7. Honig, R.E. and Kramer, D.A. (1969) Vapor pressure data for the solid and liquid elements, RCA Review 285

8. Geohegan, D.B. (1994) AIP Conference proceeding, 288, COLA 93, Ed. J.C. Miller and D.B. Geohegan, AIP Press, N.Y.

9. Handbook of chemistry and physics, 62nd edition, Ed. R.C. Weast and M.J. Astle, CRC Press, Boca Raton, Florida, E-300

10. Handbook of chemistry and physics, 62nd edition, Ed. R.C. Weast and M.J. Astle, CRC Press, Boca Raton, Florida, E-65

11. Singh, R.K. and Narayan, J. (1990) Physical Review B **41** 8843

THIN FILM DEPOSITION BY LASER ABLATION

W.W. DULEY
GWP[2]
University of Waterloo
Waterloo, Ontario
Canada   N2L 3G1

ABSTRACT

Most refractory materials can be deposited in thin film form by laser sputtering or ablation of precursor solids. Ablation of solids using UV excimer laser radiation can yield high deposition rates, deposition of monolayers and stoichiometric deposits. This paper summarizes available data on the deposition of a variety of non metallic solids using laser ablation. Mechanisms involved in the deposition of diamond-like carbon films are discussed as a special case.

1.   INTRODUCTION

Deposition of solid films by laser evaporation of a solid precursor has been investigated since 1965 (Smith and Turner 1965). Ruby, Nd:YAG and CO_2 lasers have all been shown to be effective in laser deposition by vaporization (see Duley 1983 for a review). The CO_2 laser is particularly useful in vaporizing insulators and some semiconductors (Duley 1976) while ruby and Nd:YAG laser radiation is more efficient in vaporizing metals. CW laser radiation with powers as low as ~ 25 watts can be used for controlled deposition of highly absorbing materials (Groh 1968, Hess and Milkosky 1972), but optimized results are usually obtained using pulsed laser sources. However, these sources can lead to the sputtering of large particles which can contaminate the deposited film. In addition, the structure and composition of thin films deposited by laser vaporization may differ significantly from those of the parent material.

The high intensity, short wavelength and narrow pulse width of excimer laser radiation has made these devices ideal as vaporization sources for metals, insulators, semiconductors and superconductors (Paine and Bravman 1990, Ashby et al. 1992, Braren et al. 1993). A comparison of typical parameters for deposition with free-running Nd:YAG (1.064 μm) and KrF (248 nm) laser sources is given in Table 1 (Lynds and Weinberger 1990). The useful

J. Mazumder et al. (eds.), Laser Processing: Surface Treatment and Film Deposition, 739–750.

	Nd:YAG (1.064 μm)	KrF (248 nm)
Ablation rate (atoms/sec)	10^{20}	10^{22}
Gas phase density (atom/cm^3)	$10^{15} - 10^{16}$	$10^{15} - 10^{16}$
Ablation rate per pulse (atoms/pulse)	10^{17}	10^{15}
Kinetic energy of ejecta (eV)	$2 - 20$	$40 - 850$
Composition of ejecta	Neutral Clusters	Single atoms /ions

Table 1. Comparison of Nd:YAG and KrF ablation sources for the deposition of $YBa_2 Cu_3 O_{7-x}$ films by laser ablation (Lynds and Weinberger 1990).

characteristics of excimer laser deposition sources are then: -

1) High ablation rates

2) Stoichiometric deposition

3) Monolayer deposition

4) Ability to vaporize all materials

A comparative study of the effect of wavelength and pulse duration on the properties of laser deposited $BaTiO_3$ films (Gibson et al. 1990) has shown that films deposited with psec pulses contain many spattered particles. These particles were observed when ablation occurred at either 532 or 266 nm, although the morphology of the deposited film was roughest at 532 nm. Deposition using nsec pulses at 532 nm resulted in films with a number of $0.5 - 3$ μm inclusions. Clean, clear films were deposited with 266 nm, nsec, pulses. Shorter wavelength excitation for ablative deposition sources then optimizes film quality and minimizes particulate contamination when the pulse length lies in the nsec range.

2. EXPERIMENTAL DATA

A summary of materials that have been deposited in thin film form by direct laser ablation of a solid precursor is given in Table 2. In many instances, optimum results are obtained using either 193 or 248 nm radiation. This is particularly true when ablating materials with a wide bandgap (eg. SiO_2).

Table 2. Representative data on deposition of solids by laser ablation.

Material	Substrate	Wavelength (nm)	Fluence (J/cm²)	Rate (nm/pulse)	Reference
AℓN	Sapphire (500-670 C)	248	2	0.025	Norton et al. (1991)
B_4C	(100)Si (300 C)	248	~2		Donley et al. (1992)
BN	(100)Si	248	3.9	0.018	Doll et al. (1990)
	(100)Si	266		6×10^{-3}	Knapp (1992)
	(100)Si (400 C)	248	2.5	7×10^{-3}	Friedmann et al. (1993)
$BaTiO_3$	MgO (700 C)	266		4×10^{-3}	Gibson et al. (1990)
	SiO_2 (620 C)	308	3	0.03	Davis and Gower (1989)
$BaFe_{12}O_{19}$	Sapphire (900 C)	248	2		Horwitz et al. (1993)
$CaAℓ_2Si_2O_8$	$Aℓ_2O_3$ (530-750 C)	248		0.01	Mallamaci et al. (1993)
CdS	$Aℓ_2O_3$	532	1.1		Shi et al. (1991)
aCN	Si	193	15	0.1-0.2	Xiong and Chang (1993)
$CoSi_2$	(100)Si (200-600 C)	248			Tiwari et al. (1993)
CuO/Cu_2O	(100 Si)	193	~10		Ortiz et al. (1992)
GeO	(100)Si	193	~1		Wolf et al. (1993)
InSnO (ITO)	Glass	193	~1	0.01	Zheng and Kwok (1993)

Material	Substrate	Wavelength (nm)	Fluence (J/cm^2)	Rate (nm/pulse)	Reference
KTaNbO$_3$ (KTN)	SrTiO$_3$ (700-750 C)	248	~1	0.02	Yilmaz et al. (1991)
	(100)MgO (300-700 C)	248	1.5	0.08	Cotell and Leuchtner (1993)
LiNbO$_3$	GaAs (680 C)	308	0.8-1.3		Fork and Anderson (1993)
	Aℓ_2O$_3$ (500-800 C)	193		0.1	Shibata et al. (1993)
LiTaO$_3$	Sapphire	248	1-2	0.1	Agostinelli et al. (1993)
MgO	(100)Si	248			Kanetar et al. (1991)
MoS$_2$	440 C stainless steel	248/193	0.5-1		John et al. (1991)
NbSe$_2$	440 C stainless steel	248	1.3-2.8		Day et al. (1993)
Pb$_5$Ge$_3$O$_{14}$	Sapphire	248	5	2.2	Peng and Krupanidhi (1992)
PbNbMgO$_x$ (PMN)	Pt/SiO$_2$	248	1.5	0.05	Saenger et al. (1993)
PbO	440 C stainless steel (20-300 C)	248	0.53		Zabinski et al. (1992)
PbTiO$_3$/ Pb(MgW)O$_3$	Pt/Si (300-650 C)	193	16	0.3-0.4	Lee et al. (1993)
PbZrTiO$_3$ (PZT)	Pt/Si	248	1-3		Roy et al. (1991)
	MgO (550 C)	248	0.5-1.4		Leuchtner et al. (1992)

Material	Substrate	Wavelength (nm)	Fluence (J/cm^2)	Rate (nm/pulse)	Reference
	$SrTiO_3$ (200-550 C)	248	2		Chrisey et al. (1990)
	Sapphire (400-750 C)	193	3-8.5		Morimoto et al. (1990)
SiC	Si (800 C)	308	8.7	0.2	Balooch et al. (1990)
	Si (800 C)	308	8.7	0.5	Tench et al. (1990)
	SiO_2 Sapphire (800 C)	351	1.5		Rimai et al. (1993)
Si_3N_4	(100)Si (20-450 C)	193	0.5-10	0.01	Fogarassy et al. (1993)
	(100)Si (210 C)	248	1-6	0.02	Xu et al. (1993)
SiO	(100)Si	193	0.5-10	0.04-0.05	Fogarassy et al. (1993)
SiO_2	(100)Si	193	1-10	$5-10 \times 10^{-3}$	Slaoui et al. (1992)
$SrFeO_{2.5}$	Quartz	248	0.7-3.9	0.25-0.4	Sanders and Post (1993)
$SrTiO_3$	(001)MgO (600 C)	248		3	Hiratani et al. (1993)
TiC	440 C Stainless steel	248	~2		Donley et al. (1992)
TiN	(100)Si (25-550 C)	308	4-5	0.05	Biunno et al. (1989)
	(100)GaAs (350 C)	248	10		Zheleva et al. (1993)
TiO_xN_y	(100)Si (400-800 C)	248/532	4-5		Craciun et al. (1993)
TiO_2	(100)Si (500 C)	532			Chen and Murray (1990)

Material	Substrate	Wavelength (nm)	Fluence (J/cm^2)	Rate (nm/pulse)	Reference
ZnO	(100)Si (250-600 C)	248	2.5-3		Amirhaghi et al. (1993)
ZnS/ ZnSe	(001)GaAs (300-325 C)	248	0.35	0.2	McCamy et al. (1993)
	GaAs (420 C)	308	2-8	~0.3	Rajakarun-anàyake et al. (1993)
ZrO_2	$A\ell_2O_3$ (20-500 C)	248	1-4		Smith et al. (1992)

Despite the large number of reports on film deposition by laser ablation little quantitative data has been published on the physical properties of such deposits, other than their structure and stoichiometry. Where comparisons are possible, laser ablation seems to yield deposits that have similar or somewhat less favorable properties to those of samples prepared by conventional plasma or photo-deposition. Under most conditions, near stoichiometric deposition is possible, even for deposition of compounds such as $BaTiO_3$ and PZT. Substrate-oriented growth has been demonstrated in many systems.

3.   DEPOSITION OF CARBON FILMS

Deposition of carbon films by UV laser sputtering yields diamond like carbon (DLC) with a composition intermediate between pure diamond and crystalline graphite. This variability can be expressed in terms of the ratio of diamond like bonding (sp^3 hybridized bonding) to graphitic bonding (sp^2 hybridized bonding). This variability is reflected in such parameters as the value of the optical bandgap energy Eg, hardness, resistivity and IR absorption. These parameters depend on laser fluence, the composition of the ambient atmosphere and the temperature of the deposition substrate (Duley 1984, Ogmen and Duley 1988, Collins et al. 1989, Krishnaswamy et al. 1989, Davanloo et al. 1990, Richter and Klose 1992, Pappas et al. 1992, Charyshkin and Sakipov 1992). Some properties of DLC films deposited from graphite using excimer laser ablation are summarized in Table 3. It is evident that the imaginary component, k, of the complex refractive index is particularly sensitive to deposition conditions. In the visible part of the spectrum, k, which is proportional to the optical absorption coefficient, is a strong function of the average size of aromatic clusters (Robertson 1986). An increase in k then accompanies an increase in the sp^2 content of the film and a decrease in Eg.

Table 3. Optical and electrical properties of DLC films prepared
         by laser sputtering of graphite (data from Duley 1984,
         Ogmen and Duley 1988, Pappas et al. 1992) n, k at 632 nm.

Laser	Intensity (w/cm^2)	n	k	% sp^2	Eg (eV)	Resistivity (Ω - cm)
NdYAG (1.064 μm)	5 x 10^{11}	2.35	0.32	25	1.0	>3.3 x 10^7
XeCℓ (308 nm)	1.25 x 10^8	2.40	0.13		1.27	
	8 x 10^8	2.40	0.60		0.30	
	3 x 10^8	2.2	0.042		1.4	
KrF (248 nm)	1.4 x 10^8	2.55	0.035	15.27	1.70	>10^6
	1.4 x 10^8	2.53	0.14	32	1.50	>10^6
C$_2$H$_2$ (plasma)		1.74	0.0065		2.2	10^{10}

Table 4. Properties of a-C:N films prepared by ArF laser
         sputtering of pyrolytic graphite in N$_2$ gas (from Xiong et
         al. 1993).

Deposition Rate	0.5 - 2 nm/sec
N/C Ratio	0.3 - 0.66
Microhardness	8 - 18 GPa
Density (N/C = 0.45)	2.5 ± 0.2 gm/cm^3
Young's Modulus (N/C = 4.5)	153 GPa
Bandgap	0.25 eV
Refractive Index (600 nm)	n = 2.7 k = 0.6

While direct laser sputtering is effective in the production
of DLC films, there have been several studies of the effect of an
electrical bias on the properties of the deposited film
(Krishnaswamy et al. 1989, Collins et al. 1989, Davanloo et al.
1990, Collins et al. 1993). Such systems use an external voltage
to extract ions from the laser plume over the sputter target. This
voltage can either be continuous or in the form of a pulse. There
has been some discussion on the effects of this treatment on film

hardness and uniformity (Davanloo et al. 1990, Pappas et al. 1992).
Laser sputtering of graphite in an atmosphere of N_2 can yield
carbon films with up to 66% N content (Xiong et al. 1993). The a-
C:N films have high mechanical hardness, good wear resistance and
chemical inertness. Nitrogen content increases linearly with N_2
pressure over the range 30 - 100 m Torr. These films are
essentially hydrogen-free and contain less sp^3 bonded carbon than
pure DLC films prepared under the same conditions. Some properties
of these deposited films are summarized in Table 4. Surprisingly,
a study of the preparation of DLC films by KrF laser ablation of
graphite in the presence of H atoms (Thebert-Peeler et al. 1992)
has shown that the characteristics of the deposited film is
independent of the presence of H atoms.

CONCLUSIONS

Laser ablation is rapidly developing into a valuable technique
for thin film deposition. Even complex stoichiometries can be
maintained during deposition of solids using ablation with excimer
laser radiation. This would appear to occur via cluster ejection
from the target during ablation followed by explosive decomposition
in the gas phase. Similar processes occur during deposition of a
variety of non-metallic solids.

ACKNOWLEDGEMENTS

This research was supported by the NSERCC and OCMR.

REFERENCES

Agostinelli, J.A., G.H. Brannstein and Blanton, T.N. 1993. Appl.
Phys. Lett. 63, 123.

Amirhaghi, S., Craciun, V., Beech, F., Vichers, M., Tarling, S.,
Barnes, P., and Boyd, I.W. 1993. Mat. Res. Soc. Symp. Proc.
285, 489.

Ashby, C.I.H., Brannon, J.H. and Pang, S.W. 1992. "Photons and
Low Energy Particles in Surface Processing" Mat. Res. Soc.
Symp. 236.

Balooch, M., Tench, R.J., Siekhaus, W.J., Allen, M.J., Connor, A.L.
and Olander, D.R. 1990. Appl. Phys. Lett. 57, 1540.

Biunno, N., Narayan, J., Hofmeister, S.K., Srivatsa, A.R. and
Singh, R.K. 1989. Appl. Phys. Lett. 54, 1519.

Braren, B., Dubowski, J.J. and Norton, D.P. 1993. "Laser Ablation
in Materials Processing: Fundamentals and Applications" Mat.
Res. Soc. Symp. 285, 1.

Buhay, H., Sinharoy, S., Kasner, W.H., Francombe, M.H., Lampe, D.R.
and Stephe, E. 1991. Appl. Phys. Lett. 58, 1470.

Charyshkin, Y.V. and Sakipov, N.Z. 1992. J. Appl. Phys. 72, 2508.

Chen, M.Y. and Murray, P.T. 1990. Mat. Res. Soc. Symp. Proc. 191, 43.

Chrisey, D.B., Horwitz, J.S. and Grabowski, K.S. 1990. Mat. Res. Soc. Symp. Proc. 191, 25.

Collins, C.B., Davanloo, F., Juengerman, E.M., Osborn, W.R. and Jander, D.R. 1989. Appl. Phys. Lett. 54, 216.

Collins, C.B., Davanloo, F., Lee, T.J., Yon, J.H. and Park, H. 1993. Mat. Res. Soc. Symp. Proc. 285, 547.

Cotell, C.M. and Leuchtner, R.E. 1993. Mat. Res. Soc. Symp. Proc. 285, 367.

Craiciun, V., Craciun, D., Amirhaghi, S., Vickers, M., Tarling, S., Barnes, P. and Boyd, I.W. 1993. Mat. Res. Soc. Symp. Proc. 285, 337.

Davanloo, F., Juengermann, E.M., Jander, D.R., Lee, T.J. and Collins, C.B. 1990. J. Appl. Phys. 67, 2081.

Davis, G.M. and Gower, M.C. 1989. Appl. Phys. Lett. 55, 112.

Day, A.E., Lanbe, S.J.P., Donley, M.S. and Zabinski, J.S. 1993. Mat. Res. Soc. Symp. Proc. 285, 539.
Doll, G.L., Sell, J.A., Salamanca-Riba, L. and Ballal, A.K. 1990. Mat. Res. Soc. Symp. Proc. 191, 55.

Donley, M.S., Zabinski, J.S., Sersler, W.J., Dyhouse, V.J., Walck, S.D. and McDevitt, N.T. 1992. Mat. Res. Soc. Symp. Proc. 236, 461.

Duley, W.W. 1976. "CO_2 Lasers: Effects and Applications" Academic Press, N.Y.

Duley, W.W. 1983. "Laser Processing and Analysis of Materials" Plenum Press, N.Y.

Duley, W.W. 1984. Astrophys. J. 287, 694.

Fogarassy, E., Fuchs, C., Slaoni, A., de Unamuno, S., Stoquert, J.P. and Marine, W. 1993. Mat. Res. Soc. Symp. 285, 319.

Fork, D.K. and Anderson, G.B. 1993. Mat. Res. Soc. Symp. Proc. 285, 355.

Friedmann, T.A., McCarty, K.F., Klans, E.J., Johnsen, H.A., Medlin, D.L., Mills, M.J., Ottesen, D.K. and Stulen, R.H. 1993. Mat. Res. Soc. Symp. Proc. 285, 513.

748

Gibson, U.J., Ruffner, J.A., McNally, J.J. and Peterson, G. 1990. Mat. Res. Soc. Symp. Proc. 191, 19.

Groh, G. 1968. J. Appl. Phys. 39, 5804.

Hess, M.S. and Milkosky, J.F. 1972. J. Appl. Phys. 43, 4680.

Hiratani, M., Tarutani, Y., Fukazawa, T., Okamoto, M. and Takagi, K. 1993. Thin Solid Films 227, 100.

Horwitz, J.S., Chrisey, D.B., Grabowski, K.S., Carosella, C.A., Lubitz, P. and Edmondson, C. 1993. Mat. Res. Soc. Symp. Proc. 285, 391.

John, P.J., Dyhouse, V.J., McDevitt, N.T., Safriet, A., Zabinski, J.S. and Donley, M.S. 1991. Mat. Res. Soc. Symp. Proc. 201, 117.

Kanetkar, S.M., Sharan, S., Tiwari, P., Matera, J. and Narayan, J. 1991. Mat. Res. Soc. Symp. Proc. 201, 189.

Knapp, J.A. 1992. Mat. Res. Soc. Symp. Proc. 236, 473.

Lee, B.W., Lee, H.M., Cook, L.P., Schench, P.K., Paul, A., Wong-Ng, W., Chiang, C.K., Brody, P.S., Rod, B.J. and Bennett, K.W. 1993. Mat. Res. Soc. Symp. Proc. 285, 403.

Leuchtner, R.E., Grabowski, K.S., Chrisey, D.B. and Horowitz, J.S. 1992. Appl. Phys. Lett. 60, 1193.

Lynds, L. and Weinberger, B.R. 1990. Mat. Res. Soc. Symp. Proc. 191, 3.

Maffei, N. and Krupandhi, S.B. 1992. Appl. Phys. Lett. 60, 781.

Mallamaci, M.P., Bentley, J. and Carter, C.B. 1993. Mat. Res. Soc. Symp. Proc. 285, 433.

McCamy, J.W., Lowndes, D.H., Budai, J.D., Jellison, G.E., Herman, I.P. and Kim, S. 1993. Mat. Res. Soc. Symp. Proc. 285, 471.

Morimoto, A., Otsubo, S., Shimizu, T., Minamikawa, T., Yonezawa, Y., Kidoh, H. and Ogawa, T. 1990. Mat. Res. Soc. Symp. Proc. 191, 31.

Ortiz, C., Vega, F. and Solis, J. 1992. Thin Solid Films 218, 182.

Paine, D.C. and Bravman, J.C. 1990. "Laser Ablation in Materials Synthesis" Mat Res. Soc. Symp. 191.

Pappas, D.L., Saenger, K.L., Bruley, J., Krakow, W., Cuomo, J.J., Gu, T. and Collins, R.W. 1992. J. Appl. Phys. 71, 5675.

Peng, C.J. and Krupanidhi, S.B. 1992. Thin Solid Films 219, 162.

Rajakarunanayaki, Y., Luo, Y., Adkins, B.T. and Compaan, A. 1993. Mat. Res. Soc. Symp. Proc. 285, 477.

Ramesh, R., Luther, K., Wilkens, B., Hart, D.L., Wang, E., Tarascom, J.M., Inam, A., Wu, X.D. and Venkatesan, T. 1990. Appl. Phys. Lett. 57, 1505.

Rimai, L., Ager, R., Hangas, J., Logothetis, E.M., Abu-Agell, N. and Aslam, M. 1993. Mat. Res. Soc. Symp. Proc. 285, 695.

Robertson, J. 1986. Adv. Phys. 35, 317.

Roy, D., Krupanidhi, S.B. and Dougherty, J.P. 1991. J. Appl. Phys. 69, 7930.

Saenger, K.L., Roy, R.A., Beach, D.B. and Etzold, K.F. 1993. Mat. Res. Soc. Symp. Proc. 285, 421.

Sanders, B.W. and Post, M.L. 1993. Mat. Res. Soc. Symp. Proc. 285, 427.

Shi, L., Hashishin, Y., Dong, S.Y. and Kwok, H.S. 1991. Mat. Res. Soc. Symp. Proc. 201, 171.

Shibata, Y., Kaya, K., Akashi, K., Kanai, M., Kawai, T. and Kawai, S. 1993. Mat. Res. Soc. Symp. Proc. 285, 361.

Slaoui, A., Fogarassy, E., Fuchs, C. and Siffert, P. 1992. J. Appl. Phys. 71, 590.

Smith, G.A., Chen, Li-C and Chuang, M.C. 1992. Mat. Res. Soc. Symp. Proc. 236, 429.

Smith, H.M. and Turner, A.F. 1965. Appl. Optics 4, 147.

Tench, R.J., Balooch, M., Connor, A.L., Bernandez, L., Olson, B., Allen, M.J., Siekhaus, W.J. and Olander, D.R. 1990. Mat. Res. Soc. Symp. Proc. 191, 61.

Thebert-Peeler, D., Murray, P.T., Petry, L. and Haas, T.W. 1992. Mat. Res. Symp. Proc. 236, 467.

Tiwari, P., Chowdhury, R. and Narayan, J. 1993. Mat. Res. Soc. Symp. Proc. 285, 533.

Wolf, P.J., Christensen, T.M., Coit, N.G. and Swinford, R.W. 1993. Mat. Res. Soc. Symp. Proc. 285, 439.

Xiong, F. and Chang, R.P.H. 1993. Mat. Res. Soc. Symp. Proc. 285, 587.

Xiong, F., Chang, R.P.H. and White, C.W. 1993. Mat. Res. Soc. Symp. Proc. <u>285</u>, 587.

Xu, X., Seki, K., Chen, N., Okabe, H., Frye, J.M. and Halpern, J.B. 1993. Mat. Res. Soc. Symp. Proc. <u>285</u>, 331.

Yilmaz, S., Venkatesan, T. and Gerhard-Multhaupt, R. 1991. Appl. Phys. Lett. <u>58</u>, 2479.

Zabinski, J.S., Donley, M.S., Dyhouse, V.J., Moore, R. and McDevin, N.T. 1992. Mat. Res. Soc. Symp. Proc. <u>236</u>, 437.

Zheleva, T., Jagannadham, K., Kumar, A. and Narayan, J. 1993. Mat. Res. Soc. Symp. Proc. <u>285</u>, 343.

J.P. Zheng and H.S. Kwok 1993. Appl. Phys. Lett. <u>63</u>, 1.

LASER ABLATION AND APPLICATION TO THIN FILM DEPOSITION

A.CATHERINOT, B.ANGLERAUD, J.AUBRETON, C.CHAMPEAUX, C.GERMAIN, C.GIRAULT.

*Equipe Plasma, Laser, Matériaux, LMCTS URA n° 320
Faculté des Sciences, 123 Ave A. Thomas
87060 Limoges Cedex, (France).*

I-. INTRODUCTION

Since the beginning of works concerning lasers, numerous theoretical and experimental works have been undertaken in order to understand the phenomena that govern the interaction of a laser radiation with a material and excellent review books have been published on this subject as those proposed in the references [1-11].

Generally, the interaction and the effects that result of the interaction, depend on numerous parameters such as the nature, the physical and chemical characteristics of the material and of its surface, the fluence, the wavelength and the irradiation duration of the laser radiation and finally the nature and the pressure of the surrounding atmosphere. Schematically, three regions may be distinguished :

- the target bulk,
- the surface of the target ,
- the volume located above the target that can contain chemically active elementary species and where a "plasma plume" can be observed under convenient interaction conditions.

One of the main problems encountered in understanding of the involved phenomena is that the physical and chemical characteristics of these three regions are strongly interdependent and vary rapidly during and after the interaction.

However, we can distinguish three regimes of interaction between laser radiation with materials, according to the laser fluence value at the material surface :

Low fluence regime :

In this case, the vaporisation rate remains negligible and the deposited laser energy is diffused within the material and/or used to activate chemical processes at the material surface. The problem may be treated by solving the heat transfer equation and/or the equations describing the photo induced chemical kinetics at the material surface.

J. Mazumder et al. (eds.), Laser Processing: Surface Treatment and Film Deposition, 751–796.
© *1996 Kluwer Academic Publishers.*

Intermediate fluence regime :

The vaporisation becomes important but the vapour remains transparent to the laser radiation and the expansion of the ejected material is determined by its state near the surface (temperature, pressure and density). The energy brought by the laser beam is used essentially to equilibrate losses by heat diffusion within the material and fusion-vaporisation processes. The expansion of the ejected material may be described by solving the flux conservation equations of mass, momentum and energy.

High fluence regime :

The problem becomes much more complex, according to the apparition of a "plasma plume" above the target surface. This plume is partially absorbent to the laser radiation and the phenomena within the material, at the target surface and in the plasma plume are all interdependent. Moreover, the ambient atmosphere may react with the plasma plume.

Finally, we observe experimentally the existence of threshold fluence values corresponding to the beginning of vaporisation or to the apparition of a plasma. These values are strongly dependent on the nature and the state of the involved surface and on the nature and pressure of the ambient atmosphere.

In the second and the third interaction regimes, a certain amount of material is ejected from the target. This material can be collected on a convenient substrate placed in front of the considered target and we can take advantage of these phenomena to deposit thin films.

This method of film deposition by laser ablation (**FDLA**) has been largely developed, these last past years, according to the spectacular success encountered in deposition of high Tc superconducting thin films [12]. Despite these results and the performances reach now in deposition of various materials, the involved phenomena are far to be well understood. Particularly, we observe that the characteristics and the properties of the as-deposited films are rather different of those deposited using "classical" methods. For instance, FDLA allows the possibility to transfer a given chemical composition (even complex) from a target to a substrate without noticeable modification. The measured thickness distributions are generally strongly peaked around the axis normal to the target surface (in $\cos^n \theta$, with n from $\cong 8$ to $\cong 12$). Finally it is possible to modify the chemical composition of the ejected material during its transport [13], by a judicious choice of nature and pressure of the ambient atmosphere.

A typical experiment of FDLA is schematically presented in Figure 1. The radiation beam delivered by a pulsed laser (generally the pulse duration is of some 10 ns ; Nd-YAG, excimers, etc.) is focused on a target located within a vacuum vessel. This target is moved in order to insure a surface renewal between shots and a good reproducibility of the interaction. The substrate is placed on a heating substrate holder and the target-substrate distance varies between 3 and 12 cm. According to the considered experiment, the laser fluence at the surface changes from some

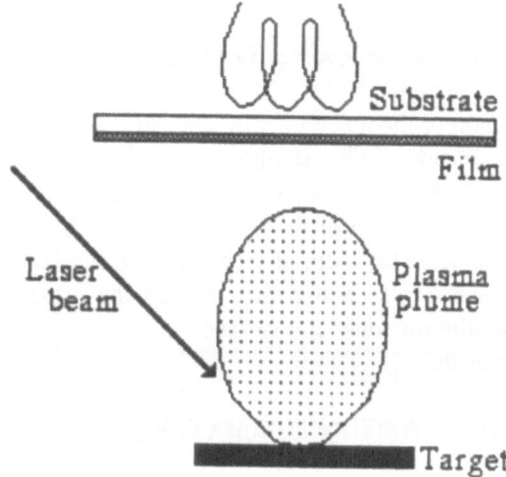

Figure 1 : Typical experiment of film deposition by laser ablation.

100MW/cm^2 to some 100GW/cm^2, values easily attainable with existing commercial lasers.

The different phenomena involved in FDLA may be decomposed in four successive steps :

1/ The interaction of the radiation with the material leading to the ablation.

2/ The first moments of the ejection of material and the eventual interaction of the vapour cloud with the laser radiation so far the laser pulse is finished.

3/ The adiabatic expansion of the ejected material cloud and the transport of species up to the substrate. This step, during which ejected particles may react with the ambient atmosphere, play a deciding role on the nature and fluxes of species that impinge on the film (or substrate) surface.

4/ The film growth.

II-. PHENOMENA WITHIN THE MATERIAL

Basically, we can distinguish between two types of interaction :

- *interaction with a metal.*

The target contains a large concentration of free carriers, inducing a large value of the absorption coefficient "a". The radiation is then absorbed within a very thin skin layer (of order 1/a) and in a first approximation, we can consider that energy is deposited at the target surface and that heat diffuses classically within the material. The thermally affected thickness is of the order of $(2\kappa t_L)^{1/2}$ ($>>1/a$), where κ is the thermal diffusivity of the material and t_L is the laser pulse duration. If R is the reflexion coefficient of the material at the considered wavelength, the energy balance may be written :

$$(1-R)E-(Eloss) = \Delta x_c [\rho \ C \Delta T + \Delta H] \qquad (1)$$

Where Δx_c is the vaporised thickness by the laser pulse and ΔH the heat of vaporisation of the target material. Eloss represents the lost energy, corresponding to the amount supplied to the ejected vapour, conduction within the material and radiation emission. We observe in equation (1) that the ablation phenomenon takes place above a threshold Eth of incident energy. At low fluence values, Eloss is almost independent on E (Eloss $\cong$ Eth) and in a first approximation, Δx_c is proportional to E. On the contrary, at high fluence values, Eloss depends strongly and non-linearly on E and we observe saturation phenomena that result especially from incident laser radiation shielding by the plasma created above the surface.

- *interaction with an insulator or a semi-conductor.*

In this case, the thermally affected thickness during the laser pulse duration is small compared to 1/a (the medium presents a certain transparency to the incident radiation and the thermal diffusivity is relatively low). The vaporised material thickness depends essentially on the absorption coefficient "a" and the ablated depth per laser pulse varies with the incident fluence logarithm, at least when this depth remains small. At high fluence values, the creation of free carriers by interband transitions, followed by avalanche phenomena within the material tends to transform this type of interaction into the metallic case previously discussed.

III-. PHENOMENA ABOVE THE TARGET

III-.1 Preliminary approach

In the following, we present a simple approach [14] that allows an estimation of some characteristic parameters of the interaction and to point out important phenomena such as the creation of a plasma medium. We consider a target irradiated by a laser beam and in the following the index g, 0 and s indicate respectively the gas phase, the surface of the target and the solid phase.

Now we write the conservation laws of mass, momentum and energy at the target surface :

- Mass :
$$\frac{dM}{dt} = \overset{\bullet}{M} = \rho_{g,0} \, u_{g,0} \tag{2}$$

- Momentum :
$$P_{s,0} = \rho_{g,0} + \overset{\bullet}{M} u_{g,0} \tag{3}$$

- Energy :
$$\overset{\bullet}{M}(\, Lv + \frac{1}{2} u_{g,0}^2) = \alpha \, F_0 \tag{4}$$

Where Lv is the "effective" energy of material ablation, F_0 the laser fluence at the surface, $u_{g,0}$ the vapour velocity and $\alpha = (1-R)$ is the proportion of absorbed laser energy.

We assume the vapour is an ideal gas :
$$P_{g,0} = \rho_{g,0} RT_{g,0} \tag{5}$$

R is the ideal gas constant and vapour pressure connected to the temperature $T_{g,0}$ by the Clausius-Clapeyron relation :
$$T_{g,0} = T_v^{ref.} (1 - b \, Ln(\frac{P_{g,0}}{P_{ref.}}))^{-1} \tag{6}$$

Where $T_v^{ref.}$ is the vaporisation temperature of the material at the reference pressure $P_{ref.}$, b constant is given by :
$$b = R \, T_v^{ref.} / Lv .$$

The obtained results correspond to characteristics that may be expected at the beginning of the interaction and can be used as initial conditions for a description of the creation of a plasma medium. As an example, for an absorbed power density (α F_0) of 1 GW/cm^2 on an aluminium target, calculation gives : $P_{g,0} \cong 6000$ bar, $u_{g,0} \cong 2. \, 10^3$ m/s, M $\cong$ 60 kg/s.cm^2 and $T_{g,0} \cong 6000$ K.

At the initial time (t=0), a thermodynamic equilibrium is assumed in the vapour (electronic temperature T_e = gas temperature T_g = T) and the electronic density n_e is calculated using the Saha law :
$$\frac{n_e(0)n_i(0)}{n_a(0)} = 2 \frac{g_i}{g_a} (\frac{2\pi m_e \, kT}{h^2})^{3/2} . \, \exp(-\frac{Ei}{kT}) \tag{7}$$

Where Ei is the ionisation energy of the neutral of concentration n_a. The ion concentration is n_i and we assume neutrality of the plasma : $n_e = n_i$

Temporal evolutions of T_e and n_e may be described by the following equations :

- The balance equation of electron concentration :

$$\frac{dn_e}{dt} = k_i \, n_e n_a \, - \, k_r n_e^2 n_i \tag{8}$$

Where k_i and k_r are respectively the ionisation coefficient of neutral by electronic collisions and recombination coefficient of ions stabilised by electrons.

- The energy balance of electrons :

$$\frac{dT_e}{dt} = \frac{2 \, K_L F}{3 \, k \, n_e} \, - (T_e - T_g) \nu_{tr.} \, - (\frac{1}{n_e} \frac{dn_e}{dt})(T_e + \frac{2}{3} E_i) \tag{9}$$

The first term of the right member of equation (9) represents heating of electrons by inverse bremsstrahlung induced by laser radiation in field of ions (coef. k_{e-i} (n_e, T_e, λ_L)) and in field of neutrals (coef. k_{e-a} (n_e, T_e, n_a, λ_L)), the two contributions are grouped in the global coefficient K_L. The second term represents the effect of energy loss by electron-neutral elastic collisions and the last term the loss of energy by ionising collisions.

The problem has been solved for various situations [15, 16], examples of the obtained results are given in Figures 2 (CO_2 laser ; wavelength $\lambda_L = 10.6\mu m$) and 3 (YAG laser working at 2ω ; $\lambda_L = 0.53\mu m$) for a titanium target and for a same laser fluence F of 0.5 GW/cm². Examination of these results points out three important facts. Firstly, we observe a similarity of behaviour in both cases : a rapid growth of T_e up to a "plateau", the value of which depends on λ_L (and on F),

. $\lambda_L = 10.6\mu m$, F_0=0.5 GW/cm², α=0.08 ; T_e=13 800K,
. $\lambda_L = 0.53\mu m$, F_0=0.5 GW/cm², α=0.42 ; T_e=6 800K,

then T_e starts to growth again. In parallel, the electronic density begins to growth significantly at the end of the electronic temperature plateau. The rise time of n_e is very short, according to "avalanche" ionisation phenomena in the vapour. If we consider the "breakdown" is achieved for a ionisation rate of some percent, we observe that, for a given fluence, a plasma medium appears with a characteristic delay t_c strongly depending on λ_L. In the cases under study we found $t_c \cong 2.10^{-9}$ s and 6.10^{-11} s respectively for $0.53\mu m$ and $10.6\mu m$.

For irradiation using U.V. lasers (excimer lasers or YAG at 3ω or 4ω) the problem of "breakdown" phenomenon becomes more complex according to the possibility of large contribution of direct photoionisation processes from low lying energy levels of atoms in the vapour.

III-.2 First step of the material ejection : creation of the cloud

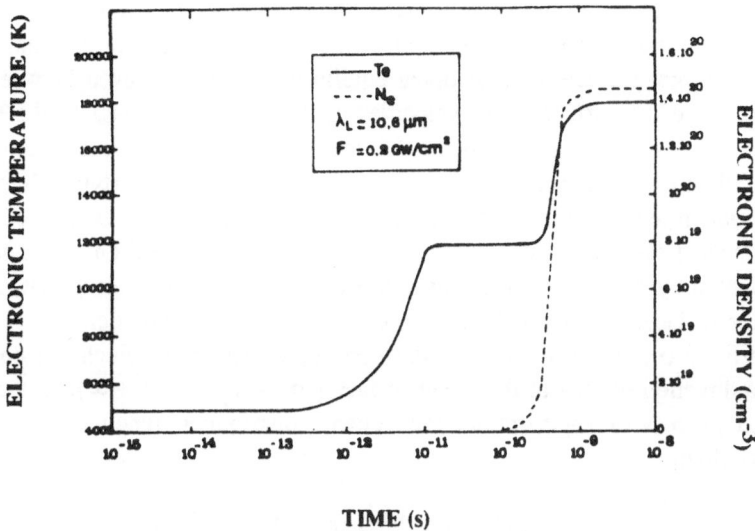

Figure 2 : Temporal evolutions of T_e and n_e above a titanium target irradiated by a CO_2 laser radiation beam ($\lambda_L = 10.6\mu m$) at a fluence of 0.5 GW/cm^2.

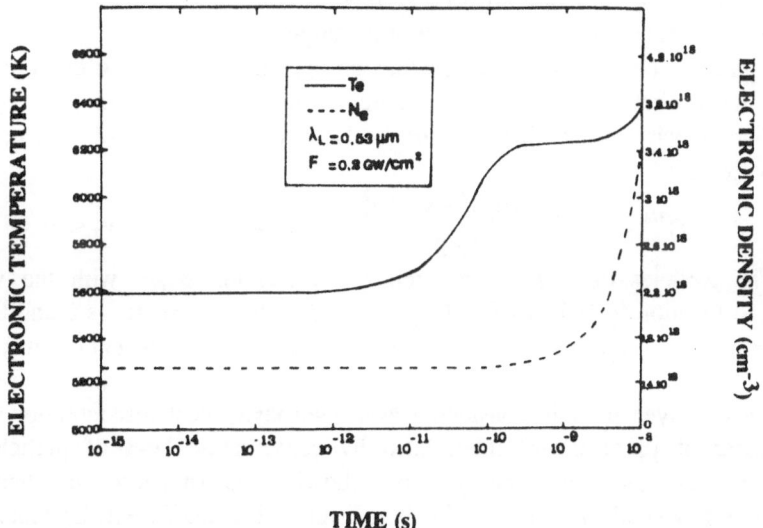

Figure 3 : Temporal evolutions of T_e and n_e above a titanium target irradiated by a Nd-YAG laser radiation beam working at 2ω ($\lambda_L = 0.53\mu m$) and at a fluence of 0.5 GW/cm^2.

In the following, the vaporisation rate is always considered as relatively large and the material thickness vaporised is assumed of at least some monolayers. In the immediate vicinity of the target surface a translational equilibrium is not achieved between the particles that leave the surface, their velocities in the direction normal to the target surface must be positive. These particles can reach a translational equilibrium only after some collisions, that is to say beyond some mean free paths (typically three) [17]. The problem may be treated in terms of the formation of a Knudsen layer (KL) and has stimulated very interesting works [19-24], essentially based on the approach initially proposed by Anisimov [18]. This approach consists in considering that, in the vicinity of the surface, we can distinguish three classes of vapour particles :

- 1/ The particles coming directly from the surface, velocities of which are positive ($v_x > 0$) in a direction normal to the target surface ($-\infty < v_y$, $v_z < +\infty$). In the case of a thermal process of vaporisation, the velocity distribution function is a "half-range" Maxwellian :

$$f_1 = \rho_s (2\pi RT_s)^{-3/2} \exp(-\frac{v_x^2 + v_y^2 + v_z^2}{2RT_s}) \text{ , with } v_x > 0 \text{ et } -\infty < v_y, v_z < +\infty \quad (10)$$

and the averaged velocity along the direction perpendicular to the target surface is given by:

$$<v_x> = (2kT_s/\pi m)^{1/2} \quad (11)$$

The index "s" indicates parameter values at the target surface.

- 2/ The particles coming from the previous group, that have crossed the limit of the Knudsen layer after some collisions. These particles have relaxed to a complete Maxwellian velocity distribution function in the center of mass (CM) frame, with a CM velocity u_K :

$$f_2 = \rho (2\pi RT)^{3/2} \exp(-\frac{(v_x - u_K)^2 + v_y^2 + v_z^2}{2RT}) \quad ; \quad -\infty < v_x, v_y, v_z < +\infty \quad (12)$$

- 3/ The particles of class 2 that return towards the target with the velocity distribution function f_2 and $v_x < 0$. The proportion of these particles is β and they are often supposed to recondense onto the target with a sticking coefficient of unity.

The Knudsen layer limit is considered as a hydrodynamical discontinuity and the three classes of particles are interrelated by conservation laws of particle flux, momentum flux and total energy flux, allowing determination of density ρ, temperature T beyond the limit of Knudsen layer and β coefficient, as functions of density ρ_s, temperature T_s at the target surface and of the heat capacity ratio γ $= C_p/C_v$ [18,22]:

$$\frac{T}{T_s} = (\sqrt{1 + \pi(\frac{\gamma-1}{\gamma+1}\frac{m}{2})^2} - \sqrt{\pi}\frac{\gamma+1}{\gamma-1}\frac{m}{2})^2$$

$$\frac{\rho}{\rho_s} = \sqrt{\frac{T_s}{T}}\left[(m^2 + 1/2) e^{m^2} \text{erfc}(m) - m/\sqrt{\pi}\right] + \frac{T_s}{2T}(1 - \sqrt{\pi}m e^{m^2} \text{erfc}(m)) \quad (13)$$

$$\beta = \left[(2m^2 + 1) - m\sqrt{(\pi T_s / T)}\right] e^{m^2} \frac{\rho_s}{\rho} \sqrt{\frac{T_s}{T}}$$

with : $m = u_K / 2RT = M\sqrt{\gamma/2}$ and $\mathrm{erfc}(m) = \frac{2}{\sqrt{\pi}} \int_m^\infty \exp(-x^2)\, dx$

In (13), M is the Mach number at the KL exit. In the cases considered here, $M = 1$ and the CM velocity u_K at this position is [19, 20]:

. $u_K \approx c_K = (\gamma\, k\, T / m)^{1/2}$ where c_K is the sound velocity at the Knudsen layer exit,

. $\gamma = C_p / C_v = (j + 5) / (j + 3)$, the heat capacity ratio, with j the number of activated degrees of freedom of the considered particles.

This approach of "vaporisation" in terms of Knudsen layer formation leads to important conclusions on the ejected particles characteristics. As previously quoted, the translational equilibrium is achieved on very short distances ($\cong$ typically three mean free paths). Moreover, the ratio between particle flux that really crosses the limit of Knudsen layer and particle flux ejected from the target surface is 0.816 (for a neutral monoatomic gas ($\gamma = 5/3$) and a Mach number $M = 1$) [22, 23]. The proportion of particles ($\beta.f_2$ velocity distribution function) returning to the target (where they should be recondensed) depends on the number j of degrees of freedom (from 18% for an atom to 26% for a large molecule [19]) and on their mass. For a target of complex chemical composition, the lighter components should recondense preferentially. As quoted in Ref. [20], this last remark implies a converse change in the chemical composition of the material that finally leaves the Knudsen layer. This point must be taken into account when we are interested on deposition of films by laser photoablation of targets of complex chemical composition.

III-.3 First step of the transport phenomena of ejected particles.

As previously quoted, the particles that finally leave the Knudsen layer have a Maxwellian velocity distribution function in the CM frame moving at the velocity u_K. Two cases may now be considered. Indeed, we have seen, in paragraph III-.1, that irradiation of the vapour by the laser beam can lead to the creation of a plasma medium with characteristic times depending on laser fluence and wavelength. At high laser fluences the plasma may be highly ionised and consequently highly absorbent to the incident radiation. For lower fluence values the ionisation "avalanche" should not have enough time to be fully developed during the laser pulse duration, and in this case the plume can be considered as transparent to the laser radiation.

- *High fluences.*

We consider the ionisation "avalanche" characteristic time t_C as very short compared to the laser pulse duration t_L [25-27]. This leads to the rapid formation of a dense and highly ionised plasma close to the target surface. According to the high expansion velocities (10^3 to 10^4 m/s) of this medium, continuously supplied by vaporisation, the particle concentrations decrease strongly beyond the central region and a dynamical equilibrium is quickly achieved between radiation absorption by the plasma medium and absorbed energy is transferred into kinetic energy of particles. These phenomena lead to the formation of an expanding isothermal plasma medium [25-27]. This step will stop at the end of the laser pulse. This approach, based on the initial works of Dawson [25], has been recently developed for FDLA by Singh and Narayan [26-27]. They show that evolution of the characteristic dimensions $(X(t), Y(t), Z(t))$ of the expanding isothermal plasma cloud is described by the equations :

$$X(t)(\frac{1}{t}\frac{dX}{dt} + \frac{d^2X}{dt^2}) = Y(t)(\frac{1}{t}\frac{dY}{dt} + \frac{d^2X}{dt^2}) = Z(t)(\frac{1}{t}\frac{dZ}{dt} + \frac{d^2Z}{dt^2}) = kT_0/m \qquad (14)$$

Where T_0 is the temperature of the "isothermal plasma", m the mass of the considered material and $t < t_L$. Concentration distribution of particles in the cloud may be written :

$$n(x,y,z,t) = \frac{N_T\, t}{\sqrt{2}\ \pi^{3/2}\ X(t)Y(t)Z(t)\ t_L}\ \exp(-\frac{x^2}{2X(t)^2} - \frac{y^2}{2Y(t)^2} - \frac{z^2}{2Z(t)^2}) \qquad (15)$$

Where N_T is the total number of ejected particles during laser pulse duration t_L. Velocities and accelerations of the plasma in the three directions are determined by the initial dimensions of the cloud, that is to say the millimeter in transverse directions and the micrometer in the direction normal to the target surface.

- Relatively low fluences
Let us just recall that ionisation avalanche has not enough time to be completely developed during the laser pulse and that the plume of ejected material remains transparent to the laser radiation during the interaction duration. The problem has been treated by several authors in term of an unsteady adiabatic expansion (UAE) considered, during this step, as monodimensional (1D), according to the very strong gradients in the x direction normal to the target surface compared to the transverse gradients. Kelly [20, 21, 24] points out the analogy of the encountered phenomena with the gas expansion in a gun and suggests analytical approximations of the solutions, whereas Knight [22,23] solves the problem numerically after an analytical development, taking into account the eventuality of presence of an ambient atmosphere. Finally, Kools [28,29] uses a similar approach to describe the beginning of the expansion but stops this 1D description when the cloud dimensions are of the same order in the three directions.

The equations describing this 1D unsteady adiabatic expansion may be written :

$$\frac{\partial \rho}{\partial t} + \frac{\partial \rho u}{\partial x} = 0$$

$$\frac{\partial u}{\partial t} + u\frac{\partial u}{\partial x} + \frac{c^2}{\rho}\frac{\partial \rho}{\partial x} = 0 \qquad (16)$$

$$\frac{\rho}{\rho_K} = (\frac{c}{c_K})^{2/(\gamma-1)}$$

u and c (local sound velocity) may be calculated as functions of c_K [20], assuming that the characteristic time t_K required for KL crossing is negligibly small compared to t_L :

$$c = c_K (1 - \frac{\gamma - 1}{\gamma + 1}\frac{x}{c_K t})$$

$$u = c_K (1 + \frac{2}{\gamma + 1}\frac{x}{c_K t}) \qquad (17)$$

The space- and time- dependent Mach number $M = u(x,t) / c(x,t)$ and the maximum velocity (expansion under vacuum) corresponding to the maximum spatial extent (where $c = 0$) may be calculated :

$$\hat{u} = \frac{\gamma + 1}{\gamma - 1} c_K \qquad (18)$$

giving the position of the expansion front : $x_f = \hat{u}\, t$.

As in Ref. [30], velocity of the gas cloud center of mass may be estimated by calculating the average value of the Mach number :

$$<M> = \frac{\int M(x)\, \rho(x)\, dx}{\int \rho(x)\, dx} = \frac{\int \frac{u}{c}(\frac{c}{c_K})^{2/(\gamma-1)}\, dx}{\int (\frac{c}{c_K})^{2/(\gamma-1)}\, dx} \qquad (19)$$

The integrals run up to x_f.

III-.4 Three dimensional expansion of ejected material cloud

The following stage of the ejected material transport consists in a three dimensional adiabatic expansion [25-30]. This last step is particularly important in the determination of particle fluxes impinging a substrate located in front of the target and consequently in the film growth. In this regime, the following solutions have been proposed to the flow equations [25-28] for the particle concentration and velocity distributions :

$$n(x,y,z,t) = \frac{N_T}{\sqrt{2}\, \pi^{3/2}\, X(t)Y(t)Z(t)}\, \exp(-\frac{x^2}{2\, X(t)^2} - \frac{y^2}{2\, Y(t)^2} - \frac{z^2}{2\, Z(t)^2}) \qquad (20)$$

$$v(x,y,z,t) = \frac{x}{X(t)}\frac{dX(t)}{dt}\, i + \frac{y}{Y(t)}\frac{dY(t)}{dt}\, j + \frac{z}{Z(t)}\frac{dZ(t)}{dt}\, k \qquad (21)$$

n(x,y,z,t) et v(x,y,z,t) are respectively the concentration and velocity of particles in the center of mass frame .

Taking into account the adiabatic expansion relation :

$$\frac{c}{c_K} = (\frac{n}{n_K})^{(\gamma-1)/2} \quad , \tag{22}$$

we found [26-29] that cloud dimensions X(t), Y(t) et Z(t) are solutions of the system :

$$X(t)\frac{d^2X}{dt^2} = Y(t)\frac{d^2Y}{dt^2} = Z(t)\frac{d^2Z}{dt^2} = \frac{kT_0}{m}(\frac{X_0 Y_0 Z_0}{X(t)Y(t)Z(t)})^{(\gamma-1)} \tag{23}$$

Where $X(t=0)=X_0$, $Y(t=0)=Y_0$, $Z(t=0)=Z_0$, T_0, and n_0 are respectively dimensions, temperature and particle density of the cloud at the end of the previous stage (§ III-.3). These relations just express conservation of the ellipsoidal shape of the ejected material cloud. Equations (23) have been solved numerically by Singh and Narayan [26, 27] and analytically by Kools [28] in asymptotic regime. For $(t\to\infty)$, Kools finds that the characteristic dimensions of the gas cloud may be written :

$$X(t) = t(\frac{2kT_x}{m})^{1/2} \quad ; \quad Y(t) = t(\frac{2kT_y}{m})^{1/2} \quad ; \quad Z(t) = t(\frac{2kT_z}{m})^{1/2} \tag{24}$$

Where the "temperatures" T_x, T_y et T_z are characteristic of the expansion velocities of the ejected material in the three directions. These "temperatures" verify the asymptotic relation:

$$T_x + T_y + T_z = \frac{2}{\gamma-1}T_0 \tag{25}$$

In these conditions $(t \to \infty)$, The concentration distribution of particles in the gas cloud may be written (in the laboratory frame :

$$n(x,y,z,t) = \frac{N_T m^{3/2}}{(2\pi k)^{3/2}(T_x T_y T_z)^{1/2} t^3} \exp((-\frac{m}{2k})(\frac{1}{T_x}(\frac{x}{t} - u)^2 + \frac{1}{T_y}(\frac{y}{t})^2 + \frac{1}{T_z}(\frac{z}{t})^2)) \tag{26}$$

These relations allow calculation of the angular distribution of ejected material intensity under an easily tractable form [28]:

$$I(\theta) \cong \cos^p(\theta) \tag{27}$$

Assuming an axisymetric geometry of the system $(T_y=T_z)$, an approximation of the p exponent is given in [28] by the relation :

$$p = 1.3 * M^{3/2} * (1 + 2(\frac{X_0}{Y_0})^{1.2})^{3/4} * (\frac{Y_0}{X_0})^{1.2} \tag{28}$$

This interesting analytical approximation is in good agreement with angle resolved time of flight measurements [29, 30] but it is difficult to give definitive conclusions relative to thickness distribution of deposited films. Indeed, calculations assume a low ionisation rate in the gas cloud whereas plasma effects must be taken into account to describe involved phenomena when laser fluence is rather high [31,32]. In this latter case more or less complex semi analytical [26,27] or full numerical calculations [33,-,36] are necessary but, at our knowledge, no available simulation is able to describe completely the involved phenomena.

Moreover, the unsteady interaction of the gas cloud with the substrate or with the growing film are far to be well known. Particularly, the sticking coefficient of a particle impinging the substrate depends on one hand on nature (charged or neutral, atom or molecule) and on kinetic energy of the incident particle and on the other hand on nature and on temperature of the considered substrate.

IV-. EXPERIMENTS

The results presented in the following are essentially taken out of works developed in our laboratory, as well for diagnostics of the plasma plume as for film deposition.

IV-.1 Experimental devices

- Experimental set-up
The experimental set-up is schematically shown in Figure 4 and has been described in previous papers. The experiments are performed in an ultra-high-vacuum stainless-steel vessel equipped with fused silica windows for spectroscopic investigations and with an introduction chamber for target and substrates manipulation. Vacuum is ensured by turbomolecular pumping systems and pressure controlled using baratron, pirani and penning gauge. Targets are irradiated at an angle of 35° by the laser beam and are translated or rotated between each laser shot using motors. The laser source is an excimer laser (Lambda Physik EMG 101or EMG 150) operating at 248nm (KrF), at a repetition rate of 10 Hz, with a pulse duration of about 15 ns and an energy per pulse in the range 20-300 mJ. After spatial filtering, the laser beam is focused onto the target using a suprasil quartz lens (focal length 250mm). According to the 45° angle of incidence, the laser spot is elliptical and fluence can be changed by varying the lens-to-target distance.

For thin film deposition, the material ejected in a perpendicular direction to the target surface is collected on a substrate located at a distance varying from 30mm to 70mm from the target. The substrates can be heated up to 900°C by an halogen lamp and gases can be added during the deposition and the cooling process.

- Spectroscopic diagnostic of the plume
For temporally and spatially resolved spectroscopy experiments, the laser-induced plume is imaged onto the entrance slit of a high resolution spectrometer (THR 1000 Jobin-Yvon, resolving power = 100000) and observed slice by slice with a spatial resolution better than 0.1mm (Figure 5). This arrangement allows emissive particle detection as well as spectroscopic time-of-flight (STOF) measurements. The spectrometer is equipped with a RCA 7265 photomultiplier tube (rise time < 2 ns) connected to a boxcar averager (E.G&G PAR 4400 temporal resolution = 2 ns) for spectrum recording and to a Tektronix 2440 digitizer (500 MHz) for temporal evolution studies.

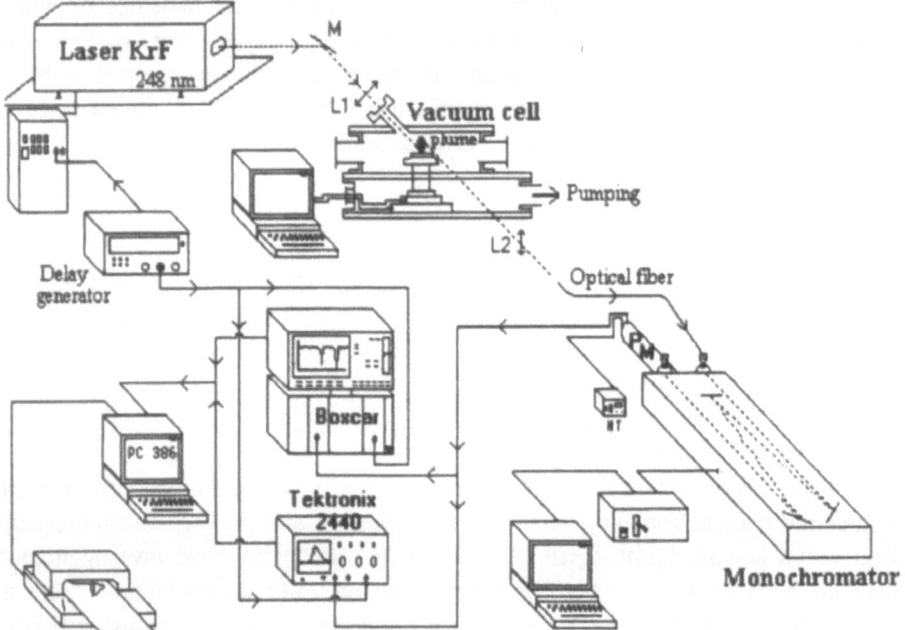

Figure 4 : Typical experimental set up for film deposition by laser ablation and diagnostic of the plasma plume.

Another possibility is the use of a fast intensified CCD camera (Princeton) in place of the photomutiplier, allowing the simultaneous recording of a spectrum range of about 100Å and spatial (along the plume radius) resolution with a temporal resolution down to 5ns. This system permits temporally and spatially resolved calculation of the local emission coefficient by Abel inversion measured intensities giving rise to the determination of radial distribution of emitting particles or Stark broadening of emitted lines [42, 43].

The electronic devices are connected to desktop computers for data acquisition and treatment and are synchronised with the laser pulse using a delay generator (Stanford DG 535).

- *Mass spectrometry*

For mass spectrometric investigations, we use a quadrupolar mass analyser (QMG 420 Balzers) equipped with a differential turbomolecular pumping system (sampling hole of 50μm in diameter), which is substituted to the substrate holder. Measurements of the angular distributions of ionic species concentrations have been carried out by angular tilting of the target.

- *Time of flight measurements*

Numerous time of flight have been developed by different authors using more or less sophisticated detection systems. Particularly, nice angle resolved experiments have been done on various materials [see for instance 19, 29, 30, 40, 41]. Interesting results on characteristics of the plasma plume have been obtain simply, by introducing a detector in front of a carbon target under high vacuum conditions, and by time resolved analysis of the detected electric signal [42, 43].

- *Fast photography*

Fast Photography of the plasma plume may be obtained using the Intensified CCD camera equipped with telephoto lens. Propagation of the cloud of ejected material and influence of background atmosphere can be investigated with temporal resolution down to 5ns. Analysis of two-dimensional digitised images of the visible plume emission leads to informations on the different ablation regime [38, 39, 44].

- *Film characterisations*

Deposited films are characterised using various methods depending of the material nature and of the researched properties. As examples :

. For YBaCuO superconducting film deposition, EDX microprobe and RBS have been used for composition determination, X-ray characterisations have been achieved using a two circles diffractometer (Siemens D 5000) for crystallographic structure, and electric measurements have been done by the classical four points method in a temperature regulated cryogenerator (Leybold).

. For carbon films, composition and structure have been analysed by XPS, hardness has been studied using nano-indentation and optical properties have been investigated by ellipsometry and IR absorption.

V-. DEPOSITION OF OXIDE THIN FILMS

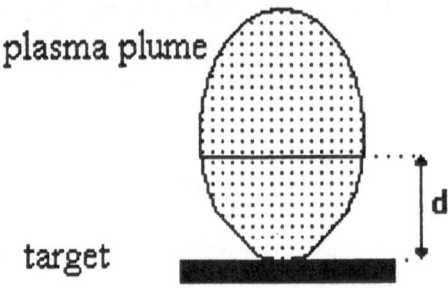

Figure 5 : Scheme of the temporally resolved spectroscopic observation of a plasma slice (thickness of about 0.1mm) located at a distance d from the target surface.

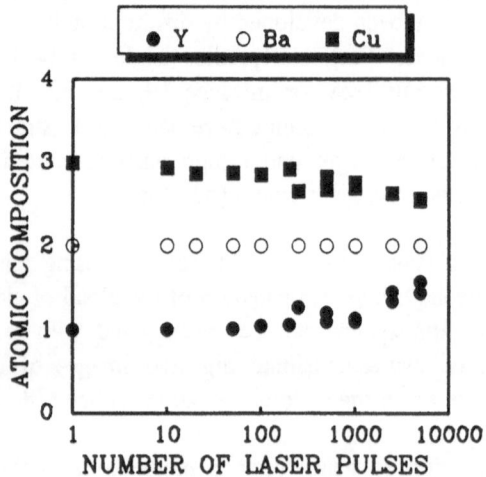

Figure 6 : Composition evolution (EDX microprobe) of a YBaCuO superconducting target surface irradiated by successive laser shot (248nm) at the same location.

Film deposition by laser ablation of materials of complex chemical composition and particularly oxide ceramics has been largely stimulated by the spectacular success obtained for high Tc superconducting materials. The obtained results have stimulated tentative film deposition of numerous other oxide material such as PZT or PLZT.

V-.1 Deposition of YBaCuO superconducting films

Reliable deposition of high quality superconducting YBaCuO films strongly depends on the chosen parameters and on the process control. This implies a detail investigation and control of each step involved in film deposition, that is :
- the interaction of the radiation with the material leading to the ablation,
- the transport of species from target to substrate and eventual reactive phenomena with the ambient atmosphere,
- the film growth.

These points will be examined successively in the following.

V-.1-.1 Interaction of the KrF laser beam with the target [45]

Though the target is displaced between each laser shot, at the end of the deposition process, each point of the target surface has been irradiated several times. Starting from a "home made" sintering superconducting YBaCuO target, we have analysed the surface composition after successive laser shots at the same position.

The measured composition evolution of cationic species is shown in Figure 6. We observe a rapid change of composition during the first 20 shots that may be ascribed both to Knudsen layer formation (cf. § III-.2) and to segregation phenomena within the target material. During the 1000 next shots the composition remains constant (permanent working regime) and change beyond (decrease of Cu and increase of Y relative to Ba). In conclusion, an adequate rotation speed of the target and a determined repetition rate of the laser have to be chosen to insure a permanent ablation regime.

We have also studied the influence of KrF laser fluence on the characteristic of the deposited material. Measurements indicate that best results are obtained for a laser fluence of about 3 J/cm^2. These observations may be correlated to results of investigations of the plume by mass spectrometry.

V-.1-.2 Spectroscopic investigations of the plume [37, 46, 47]

As previously quoted, temporally and spatially resolved high resolution spectroscopy has been used to provide informations on the nature and transport characteristics of ablated species from the YBaCuO target. These parameters have been particularly studied as a function of the oxygen ambient pressure.

First, investigations have been carried out to determine the nature of species using a boxcar gate width of 40ns located at a given delay after the beginning of the interaction and at a given distance d from the target surface. The thickness of the observed plasma plume slice is of about 0.1mm.

From a qualitative point of view, we observe that an increase of the oxygen pressure value leads to a colour change of the plasma plume which from green white colour becomes reddish.

Typical spectra obtained at a pressure of 10^{-5} mbar, a distance d= 2 mm from the YBaCuO target surface, a laser fluence of 4 J/cm^2 and recorded with a spectral resolution of 0.3 A, are shown in Figure 7. The observed spectra are quite similar to those previously shown by Dyer et al. [48]. In the region 350 nm - 550 nm, the most prominent lines are due to transitions involving neutral and ionised atomic species, namely, Y* (404.76, 408.37, 410.24, 412.83, and 414.39 nm), Y^{+*} (377.43 nm), Ba* (553.55 and 577.77 nm), Ba^{+*} (455.4 nm), Cu* (578.21 nm), and Cu^{+*} (490.97 and 493.16 nm).

Between 550 nm and 650 nm, emission of diatomic molecules are observed. In particular, transitions in the A$^2\Pi$ - X$^2\Sigma$ system of YO at 597.2, 598.77, 600.36, and 616.51 nm, of the A$^2\Sigma$ - X$^2\Pi$ system of CuO at 639.25 and 640.15 nm, and of the A$^1\Sigma$ - X$^1\Sigma$ of BaO at 603.96 nm, are the most readily recognisable. Moreover, examination of the spectrum has also pointed out emissions originating from atomic oxygen at the wavelengths 777.2, 777.4 and 777.54 nm. As reported by Wu et al. [49], the emission intensities of almost all the detected atomic lines and molecular bands are significantly enhanced under oxygen atmosphere compared to vacuum conditions. Emissions of these species are observed at various distances from the target, however, detection becomes difficult close to the target surface (d < 0.3 mm), due to the emission of a strong continuum background radiation that covers all the wavelength range under study.

Second, our investigations have been concerned by spectroscopic time-of-flight (STOF) measurements.

As shown schematically in Figure 5, for each measurement, we can study the temporal evolution of the intensity emitted at a given wavelength, by the considered atoms or molecules contained in a slice of about 0.1mm in thickness, located at a distance d from the target surface. Time-of-flight studies are carried out by varying the distance d. Indeed, the temporal location of the maximum emission intensity depends on the distance d of the observed slice from the target surface. This dependency, linear in a first approximation at least for the monoatomic species, leads to the possibility of an estimation of the expansion velocity "u" of the considered species from the target surface to the observed slice in the plume [37, 46, 47].

As an example, the temporal evolution of the intensity of the BaI spectral line at 553.55 nm is shown in Figure 8 for different values of the distance d. The emission intensity of this line decreases slowly up to a distance d of about 1mm and then decreases rapidly for d > 1 mm. Beyond a distance d = 5mm, no significant emission is detected in our experimental conditions.

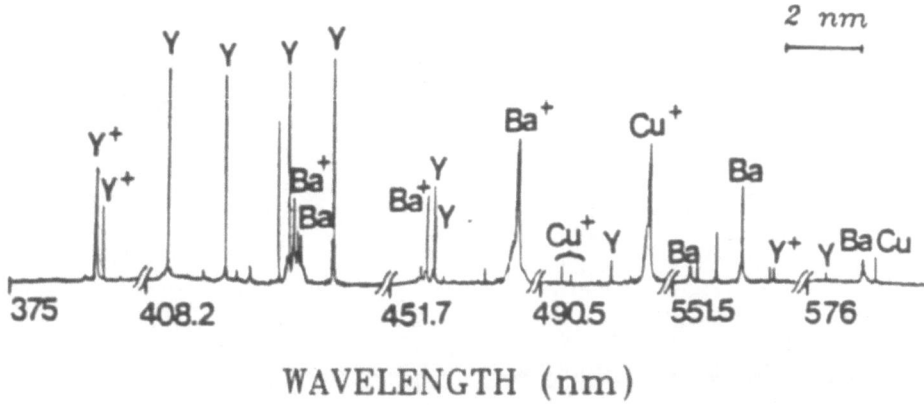

WAVELENGTH (nm)

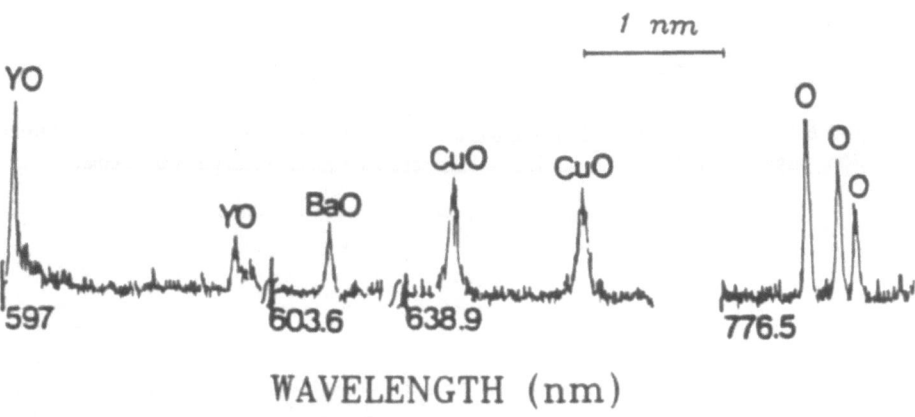

WAVELENGTH (nm)

Figure 7 : Spectra recorded with a spatial resolution of 0.3Å, at a distance d=2mm from the YBaCuO target surface, at an ambient pressure of 10^{-5} mbar, for a laser fluence of 4 J/cm^2.

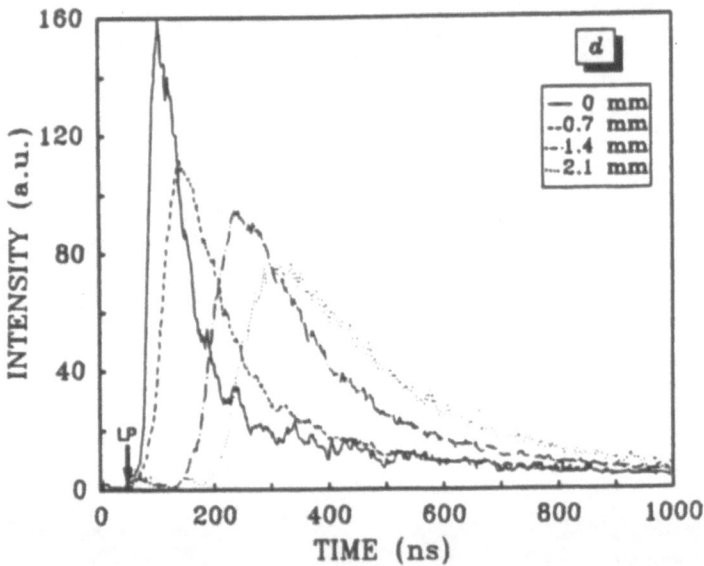

Figure 8 : Temporal evolution of the BaI line intensity at 553.55 nm, for several distances d from the YBaCuO target. The KrF laser fluence is of 4 J/cm^2 and the ambient pressure of 0.05 mbar.

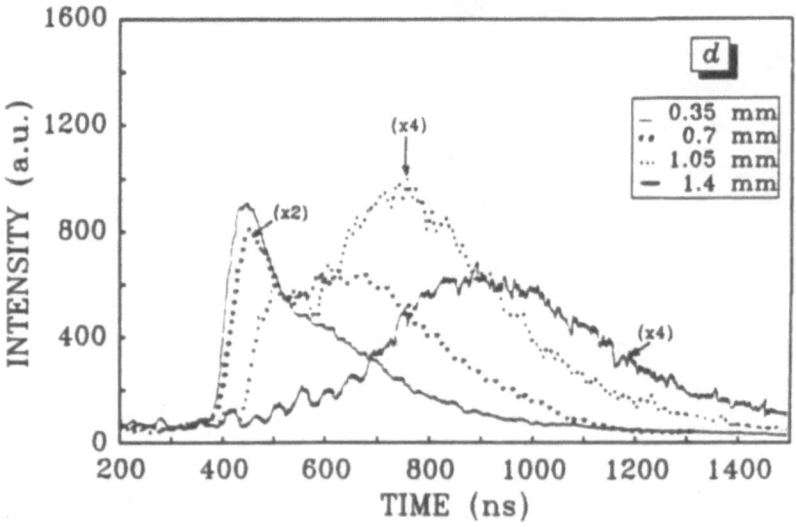

Figure 9 : Temporal evolution of the 613.2 nm YO band head intensity, for different distances d from the target surface. Same experimental conditions as in Fig. 8.

Similar results are obtained for the other detected species excepted for YO. Indeed, as shown in Figure 9, the relaxation of the 613.2 nm ($A^2\Pi$ - $X^2\Sigma$) band head of YO exhibits clearly two components when the observed plasma plume slice corresponds to d < 1mm and for d > 1mm, these two components are gradually mixed in only one broad component when d increases. We have ascribed this result [47] to the presence of two YO molecule populations within the plasma plume. The first and sharp component may be attributed to YO molecules ejected from the YBaCuO target with a highly directional velocity at a right angle to the surface, whereas the delayed and broad emission component may be attributed to YO molecules created by collisional reactive processes within the high-pressure near-surface region of the plasma plume and having a near isotropic velocity distribution. These phenomena exist probably also for BaO and CuO molecules but have never been clearly observed under our experimental conditions, despite numerous investigations.

Measurements of the temporal location of the maximum emission intensity, as a function of distance d, give rise to an estimation of the STOF of the considered particles and consequently to an estimation of the velocity components "u" of these particles in a direction normal to the YBaCuO target surface. Such measurements have been carried out for all the spectroscopically detected species under vacuum and as functions of oxygen ambient pressure .

The obtained results are summarised in Figure 10. We observe immediately that, at measurements accuracy, all the elemental particles have the same velocity "u", whatever the oxygen pressure is. This velocity "u" is quite constant (about $1.2*10^4$ m/s) from vacuum conditions up to an oxygen pressure of about 0.1 mbar and decreases beyond. The observed evolution of measured velocities "u" above 0.1 mbar, may be ascribed to a change in the gas-dynamic regime of expansion of the ejected material.

For diatomic molecules the measured velocity component "u" decreases regularly from about $1.2*10^4$ m/s under vacuum. As previously discussed for YO, for which the phenomenon is pronounced and well temporally resolved, we have observed that collisional reactive processes can lead to a broadening of the measured intensity temporal evolutions and a delaying of the maximum emission intensity location for diatomic molecules. Consequently, for these particles, the measured velocity components "u" depend on the characteristic frequency of collisional reactive processes. Since this characteristic frequency obviously increases with the ambient oxygen pressure, these phenomena may explain the evolutions of the as-measured velocity components "u" for diatomic molecules observed in Figure 10(c) and indicate an increasing amount of collisional reactive processes within the plasma plume, when the oxygen pressure increases.

Temporally and spatially resolved investigations of the plasma plume have given rise to the determination of the nature of emitting ejected species, on their velocities and on the influence of an oxygen ambient pressure on these parameters.

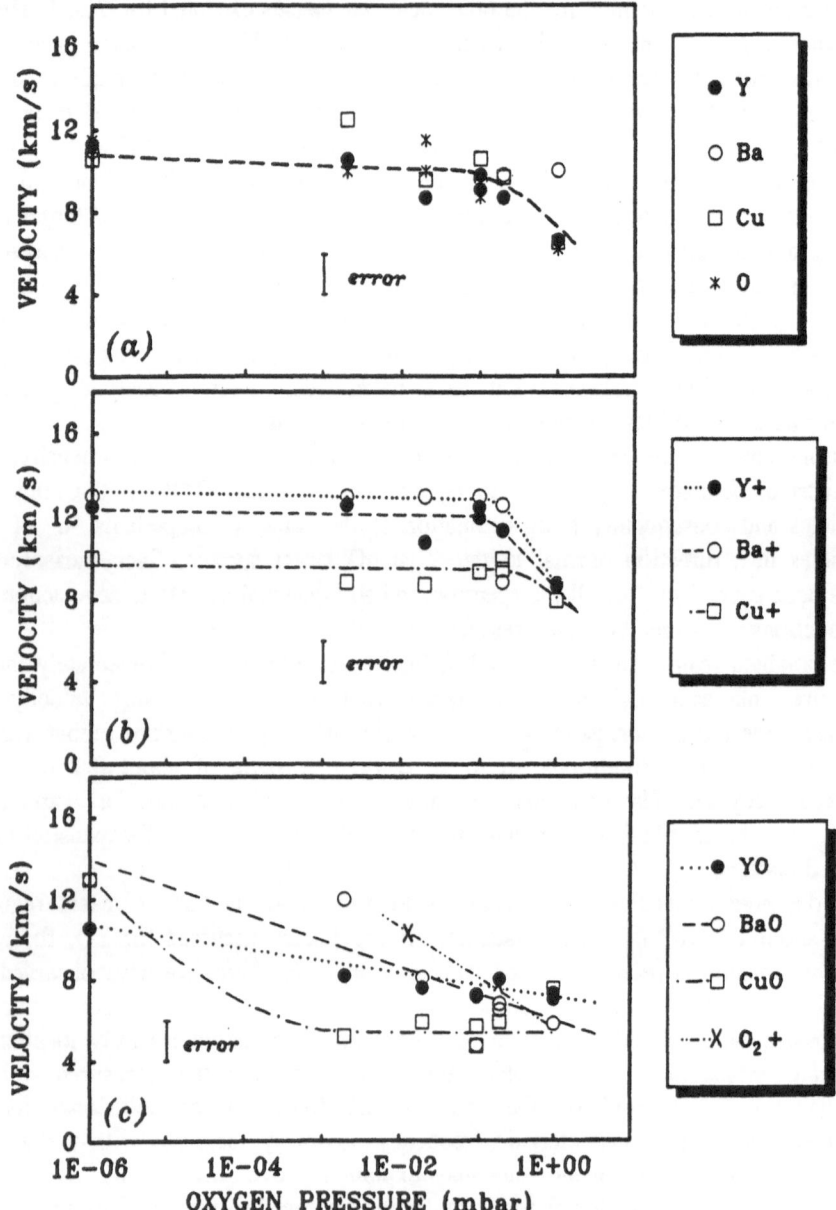

Figure 10 : Velocities of particles in the direction normal to the target surface ; (a) atomic neutrals, (b) atomic ions, (c) diatomic molecules.

Moreover, from these measurements, informations may be deduced on the chemical kinetic processes within the plasma plume and consequently during the transport of ejected species from the target to the substrate. Particularly, we can deduced that a large amount of oxide molecules is formed within the plasma plume by collisional reactive elementary processes for YBaCuO targets, providing a substantial aid to obtain the "good" oxygen stoichiometry in the deposited films.

V-.1-.3 Mass spectrometric investigations

Mass spectrometric investigations have been undertaken in order to complete optical spectroscopic studies. Indeed if optical emission spectra from the plasma plume have allowed to monitor monoatomic and diatomic species in excited states, a mass spectrometric analysis of the plume is required to detect the presence of clusters and ground state particles.

These measurements have been carried out under high-vacuum conditions (residual pressure of about 10^{-7} mbar in the chamber). According to the poor signal to noise ratio observed for the detection of neutral particles (ioniser on), the present work is limited to the study of ions (ioniser off).

When the YBaCuO target is irradiated by the KrF laser beam the positively charged ions listed in table 1 are detected. Each measurement corresponds to an average over, at least, 500 laser shots. For mass values higher than 308, the bad signal to noise ratio does not allow sufficiently reproducible measurements.

TABLE 1 :

Positively charged species detected by mass spectrometric measurements in the KrF laser-induced plasma plume on a YBaCuO target; background pressure is of about 10^{-7} mbar and the target is in constant rotation.

Detected Species	Molecular Weight
O^+	16
Cu^+	63
Cu^+	65
CuO^+	79
CuO^+	81
Y	89
YO	105
YO_2	121
Ba^+ (isotopes)	130...137
Ba^+	138
BaO^+ (isotopes)	146...153
BaO^+	154
BaO_2^+	170
Y_2O^+	194
$Y_2O_2^+$	210
Ba_2O^+	292

$Ba_2O_2^+$	308

Our results are compared in Figure 11 with those obtained by Dietze et al. [50] for YAG laser-induced plasma plume on YBaCuO targets. Detection of O^+, Cu^+, and CuO^+ ions is not reported by Dietze et al., and significant differences are observed between the measured relative concentrations for several oxide molecular ions such as YO, Y_2O, and BaO_2. This comparison indicates that the interactions are probably quite different at the target level and within the plasma plume.

Moreover relative concentration evolutions of the detected ionic species have been studied as functions of the KrF laser fluence at the target level.

Monoatomic species
For Ba^+, O^+, Y^+, and Cu^+, the detected signal intensities I, as a function of the laser fluence, are well described by relations of the type:
$$I = I_0 * \exp (F/F_0) \tag{29}$$
where F_0 is the threshold fluence for the considered ion detection. The obtained results are summarised in Figure 12.

Diatomic oxide species
For the diatomic oxide ionised species, the evolutions of detected signal intensities, as a function of laser fluence, are found very similar to those of monoatomic species, namely a near exponential evolution is observed, as shown in Figure 13. These results seem to indicate that both monoatomic and diatomic ionised species have been submitted to similar phenomena.

"Aggregates"
Some results are given in Figure 14 and excepted for $Y_2O_2^+$, the concentration evolution of which is found similar to those observed in Figures 12 and 13, the detected mass spectrometric signal intensities corresponding to ionic aggregates increase up to a maximum reached for a laser fluence of about 3 J/cm^2. This value just corresponds (cf. § V-.1-.1) to the laser fluence that we currently use to deposit thin films with "good" superconducting characteristics.

As previously reported, the experimental device allows measurements of angular concentration distributions by tilting the YBaCuO target. Some results are shown in Figure 15, exhibiting a very strong forward peaking. Indeed, measurements have been fitted by $\cos^n\theta$ curves, where θ is the angle relative to the direction normal to the target surface. Values of the cosine exponent n as large as 90 (for Ba^+, O^+, and Y^+) and 170 (for BaO^+, Cu^+, and Ba_2O^+) are found for the best approximations, as shown in Figure 15.

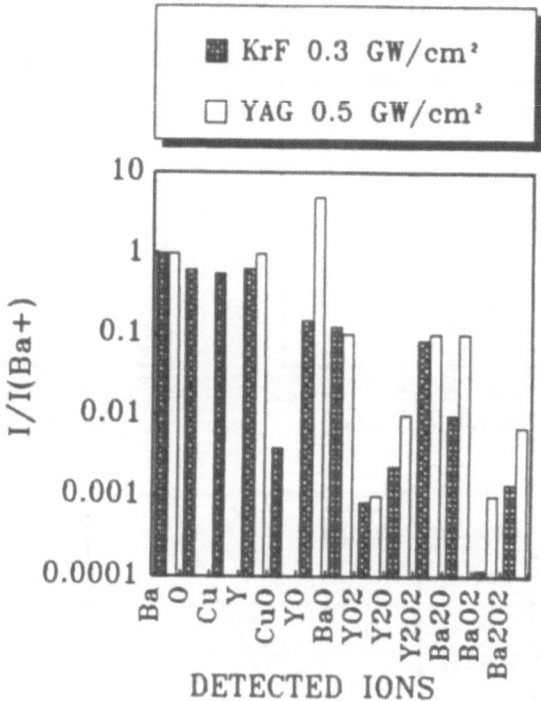

Figure 11: Comparison between relative concentrations of detected positive ionic species using mass spectrometry in the laser induced plasma plume above a YBaCuO target, respectively with a YAG laser from Dietze et al. [50] and with a KrF laser [37]. Results are normalised to measured concentration of Ba^+.

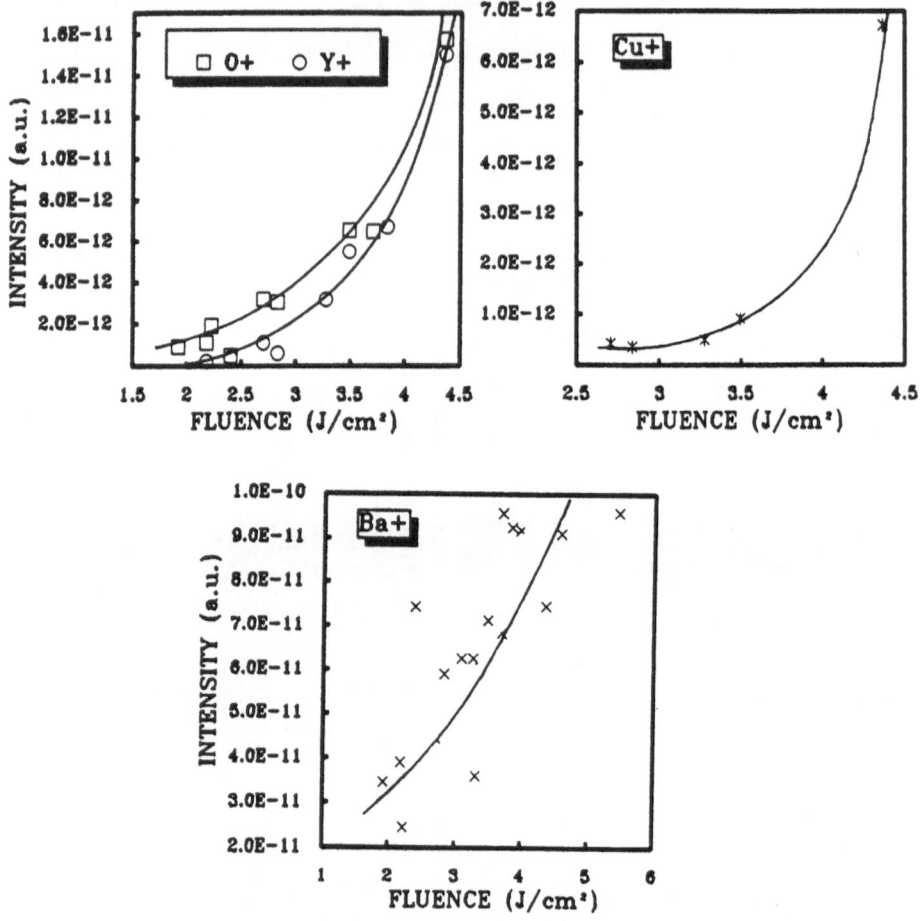

Figure 12: Evolution of mass spectrometric signal intensities corresponding to monoatomic ionised species, as functions of KrF laser fluence. Background pressure is of 10^{-7} mbar and measurement accuracy estimated at 20%.

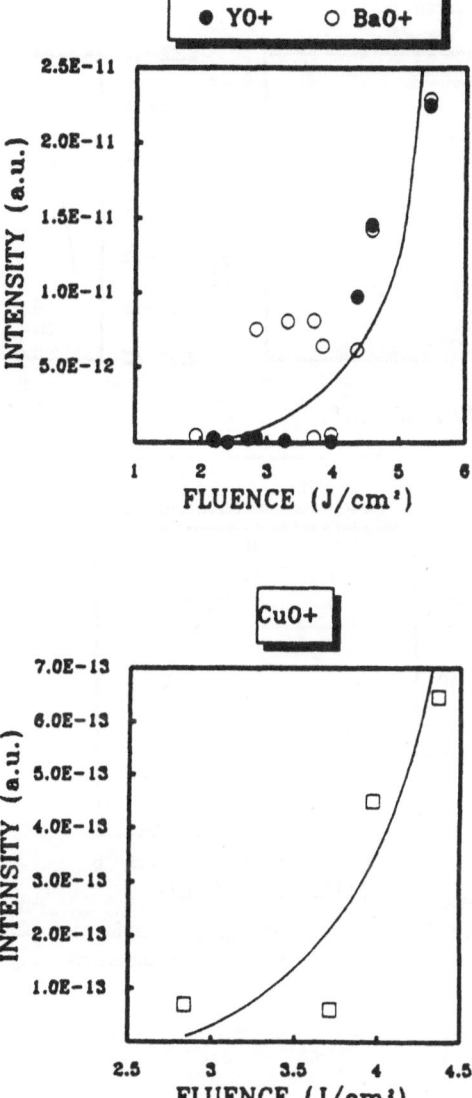

Figure 13 : Evolution of mass spectrometric signal intensities corresponding to diatomic ionised species, as functions of KrF laser fluence. Same conditions as in Fig. 12.

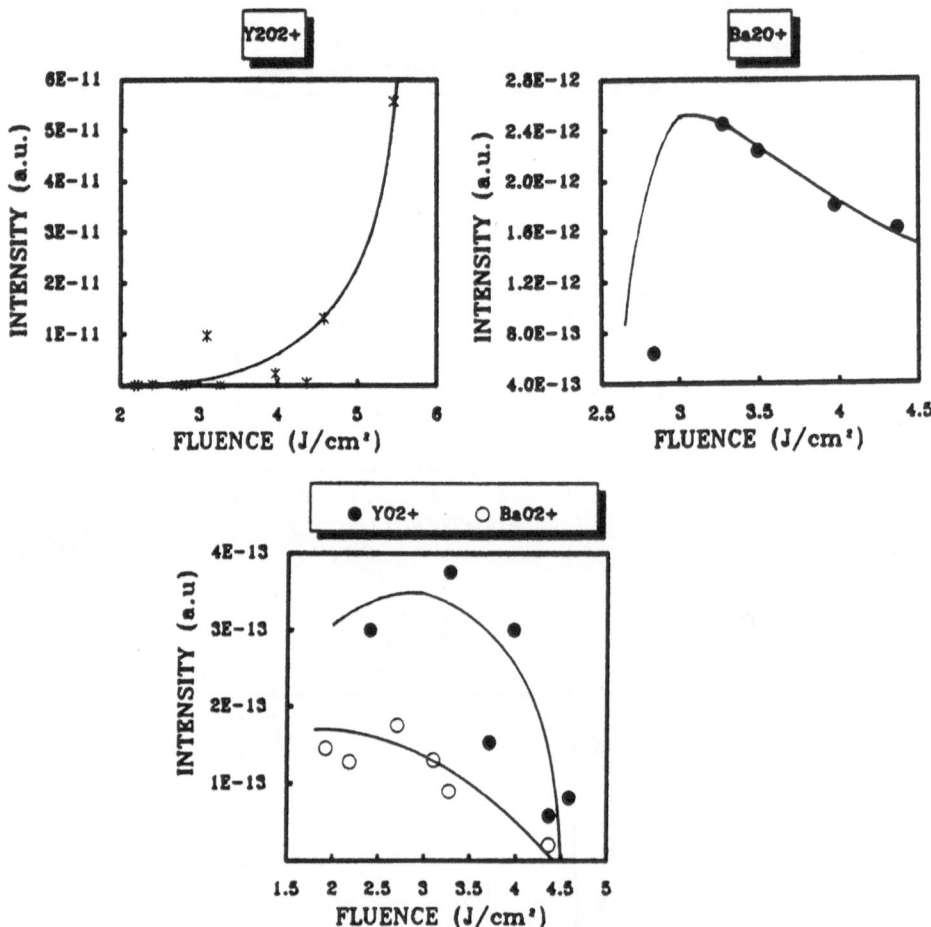

Figure 14 : Evolution of mass spectrometric signal intensities corresponding to detected ionised "aggregates", as functions of KrF laser fluence. Same conditions as in Fig. 12.

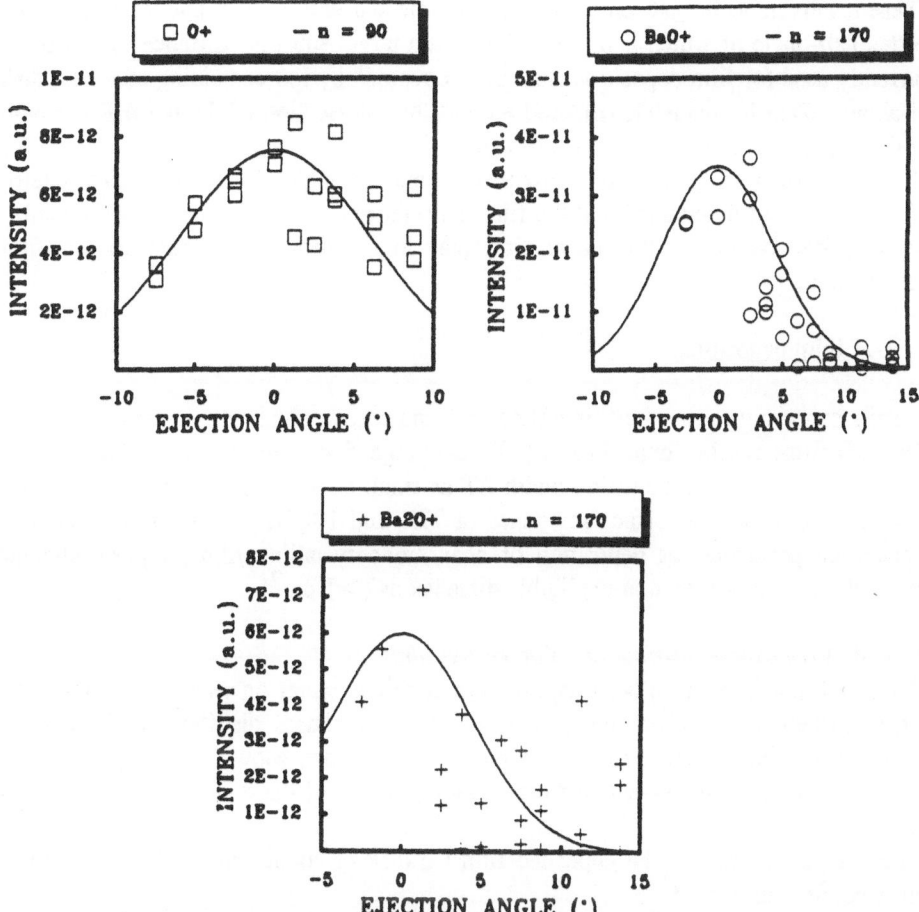

Figure 15 : Angle resolved concentration distributions of some ejected ionised species (accuracy 20%), for a laser fluence of 4 J/cm^2 and a background pressure of 10^{-7}mbar.
Continuous curve : best fitting by a curve in $\cos^n\theta$.

These results indicate that ionised species are ejected within the plume in a cone, the vertex half-angle of which is of about 10°. At a target-substrate distance $d = 3.6$cm, currently used for film deposition, the area "covered" by ion-containing cone is a disk of about 1.2cm in diameter, centered around the intersection of the normal direction to the target with the substrate. These results may be correlated with the observations of Venkatesan et al. [51] which show that beyond an angle of 10°, the stoichiometry of the film is not maintained and the film is no more superconducting. Consequently, we may deduced from this correlation that ionic species contribute to the film composition.

V-.1-.4 Film deposition

Details concerning procedure, investigations and results obtained on superconducting YBaCuO films can be found in Ref.[53]. Our objective is to obtain superconductive films with a high Tc, a transition width ΔT as small as possible, a metallic behaviour in normal state, and a surface resistance at 77K and 10GHz as low as possible. To reach such performances deposition of near monocrystalline films, c-perpendicular oriented, is necessary on non-negligible dimensions (> 1 cm^2).

V-.1-.4-.1 Thickness distribution of deposited films

YBaCuO thin films have been deposited on a silicon wafer on which was positioned a grid shaped mask. Then, measurements of the thickness distribution have been carried out using a DEKTAK profilometer. Results are shown in Figure 16, and analysis of these measurements indicates that a good fit of the thickness distribution is found:

- for the central part of the deposited film (a disk of about 1cm in diameter), by a curve $\cos^p\theta$ with $p = 16$.
- for the external part (beyond the central disk of about 1cm in diameter), by a curve $\cos^p\theta$ with $p = 12$.

As suggested by Sajjadi et al. [52], we may assume a similarity of the film deposition problem with a photometric one. The ejected material flux originating from the target element δA_T and received in the x position on the substrate (Figure 17), is given by :

$$\frac{dF}{dA_T} = I * \cos\phi * \cos^3\beta * \frac{x}{a^2} * d\alpha * dx \tag{30}$$

where I is the source "intensity".

If we assume the intensity given by equation (27) :

$$I = I_0 * \cos^n\theta$$

The deposited thickness d(x) is given by:

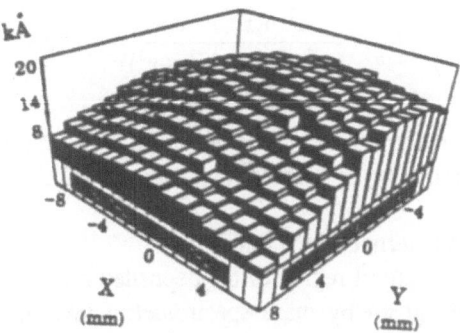

Figure 16 : Measured thickness distribution of a YBaCuO film deposited through a grid shaped mask on a silicon wafer.

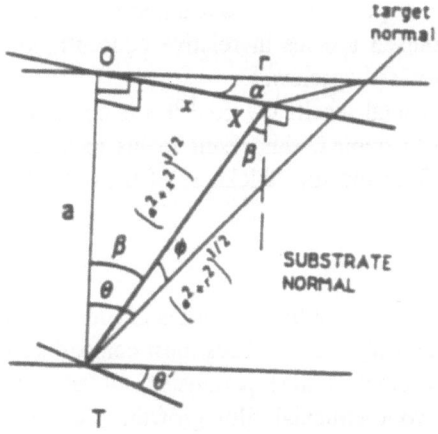

Figure 17 : Geometrical representation of the laser ablation system from Sajjadi et al. [52].

$$d(x) = \frac{I_0 * a}{2\pi(a^2 + x^2)^{3/2}} \int_0^{2\pi} \left(\frac{a^2 + x*r*\cos\alpha}{((a^2 + x^2)*(a^2 + r^2))^{1/2}}\right)^{n+1} d\alpha \tag{31}$$

Comparison between measurements given in Figure 16 and thickness distribution calculated using relation (31) indicates that experimental results correspond to "intensities" :

- $I = I_0 * \cos^9\theta$, in the external region and,
- $I = I_0 * \cos^{13}\theta$, in the central region of the deposited film.

According to the results given by mass spectrometric investigations, let us assume now that the film thickness in the external part is a consequence of deposition of neutral particles only and that the central region results from contributions of both neutrals and ionised species. The "intensity" I corresponding to this central region may be roughly approximated by:

$$I = I_0 * \{ (1-X) * \cos^9\theta + X * \cos^{90}\theta \} , \tag{32}$$

In (32), the first term corresponds to the contribution of neutrals and the second term to the contribution of ionised species in relative concentration X (see §III-.2-.2 for angular distributions of ionised species).

The best fit with experimental results ($\cong \cos^{13}\theta$) is obtained for $X \cong 10\%$. Under the assumptions previously quoted, this result seems to indicate that ionised species may contribute up to 10% to the film thickness of the central region of the deposited film.

V-.1-.4-.2 *Substrates*

Substrate choice is determinant on the properties of the deposited film. The substrate material physical and chemical characteristics must compatible with those of YBaCuO (Thermal dilatation coefficient, lattice parameter,...) and monocrystalline material have to be chosen to favour epitaxial film growth. Moreover the chosen substrate must convenient for considered applications in microwave domain. It appears that monocrystalline MgO :

. a-lattice parameter = 0.421 nm (compared to 0.382 for YBaCuO),
. dielectric constant = 9,
. $\tan\delta = 4.\ 10^{-5}$,

at 80K and 10GHz, is a good candidate and our efforts are particularly focused on deposition on this material.

V-.1-.4-.3 *Substrate temperature and influence of nature of the ambient atmosphere.*

Numerous experiments have been done to determine the more convenient "deposition" substrate temperature Ts. We found that for Ts < 620K crystallisation of films is bad, that between 620K and 700K crystallisation is random (semi-conductor films) and that for Ts > 725K films are superconductor and c-oriented for Ts $\cong$ 790K.

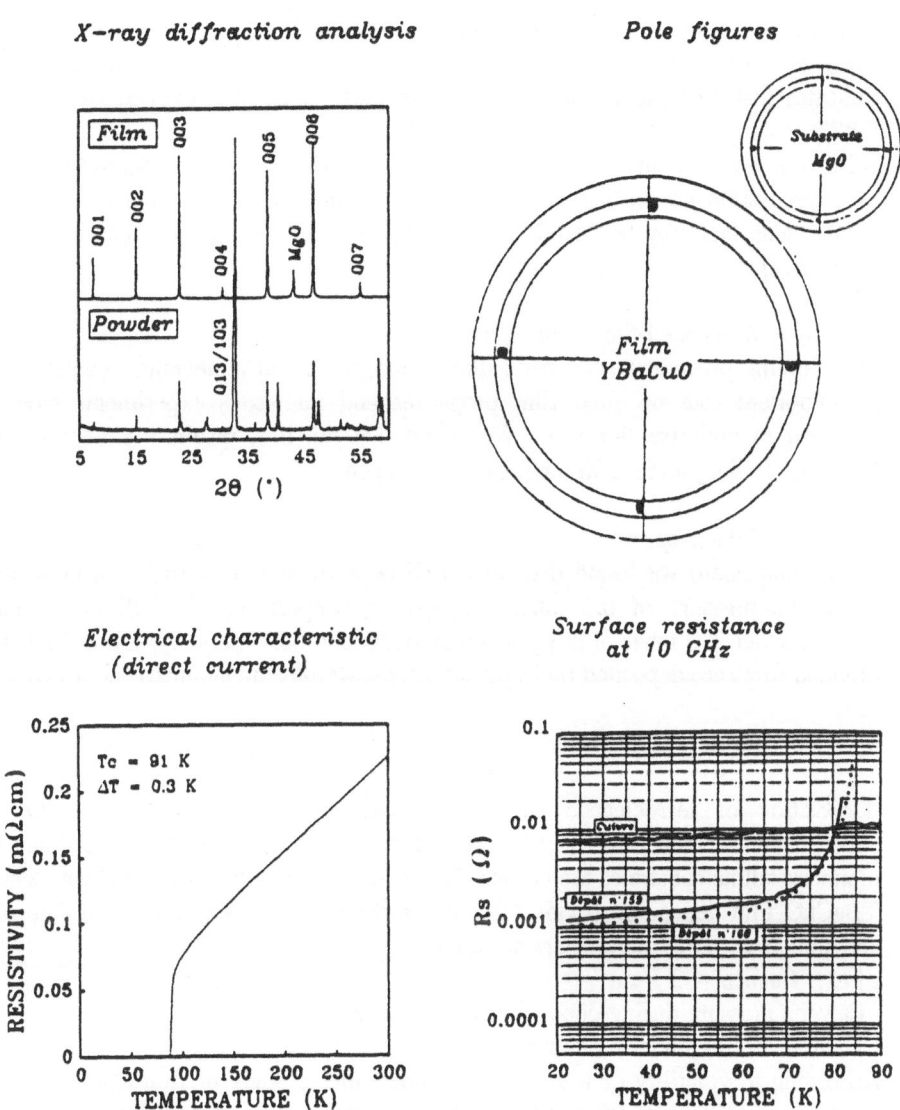

Figure 18 : Typical crystallographic and electric (DC and microwave) characteristics of superconducting YBaCuO films deposited on MgO substrates by photoablation at 248nm.

The influence of the nature of the oxidant atmosphere have been studied in details using two oxidant vectors : pure oxygen and a oxygen-ozone mixture. Results indicates that c-oriented films are only obtained under pure oxygen atmosphere (at 790K).

Moreover experiments have been done to determine the optimal deposition pressure of oxygen atmosphere. The best results are obtained for an ambient pressure of 0.3 mbar, value that may be correlated to transport and kinetic phenomena in the plume (§ V-.1-.2).

V-.1-.4-.4 *Influence of the cooling step*

The cooling procedure after deposition, atmosphere and temperature evolution, play an important role on final film properties and numerous experiments have been achieved to optimise this step. We found that the best results are obtained for a "natural" cooling under 200 mbar of pure oxygen.

V-.1-.4-.5 *Film properties*

As a conclusion, we found that for a KrF laser fluence of 3 J/cm^2, a pure oxygen deposition pressure of 0.3 mbar, a substrate temperature of 790K and a natural cooling under 200 mbar of oxygen, superconducting near monocrystalline YBaCuO c-oriented films are deposited on MgO substrates. Results are summarised in Figure 18.

V-.2 Deposition of PZT films

Deposition of ferroelectric PZT films have been undertaken using the same experimental method. $PbZr_{0.52}Ti_{0.48}O_3$ ferroelectric bulk targets are used and MgO monocrystalline substrates are chosen. Difficulties are similar to those encountered for YBaCuO film deposition (oxide of complex chemical composition, perovskite phase) and similar working parameters are used :
. Laser fluence = 3 J/cm^2,
. Oxygen pressure during deposition = 0.3 mbar,
. Natural cooling under 330mbar of oxygen.
According to volatility of lead at high temperature, a study of film characteristics as function of substrate temperature was necessary. A systematic study has been undertaken in the range 400-800K and results on composition and crystallographic structure of the deposited films are given respectively in Figure 19 and Figure 20. These results indicate that in situ obtaining of PZT films with perovkite structure is possible in the substrate temperature range 600-650K.

VI-. DEPOSITION OF CARBON FILMS

It is now well known that laser ablation is a very powerful tool for carbon film deposition. Very promising results have been recently obtained [43, 54-59] and

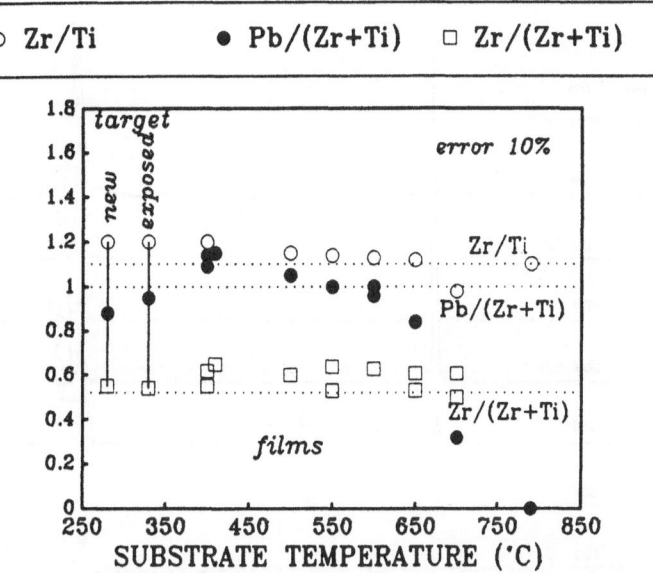

Figure 19 : Influence of substrate temperature on PZT film composition.

786

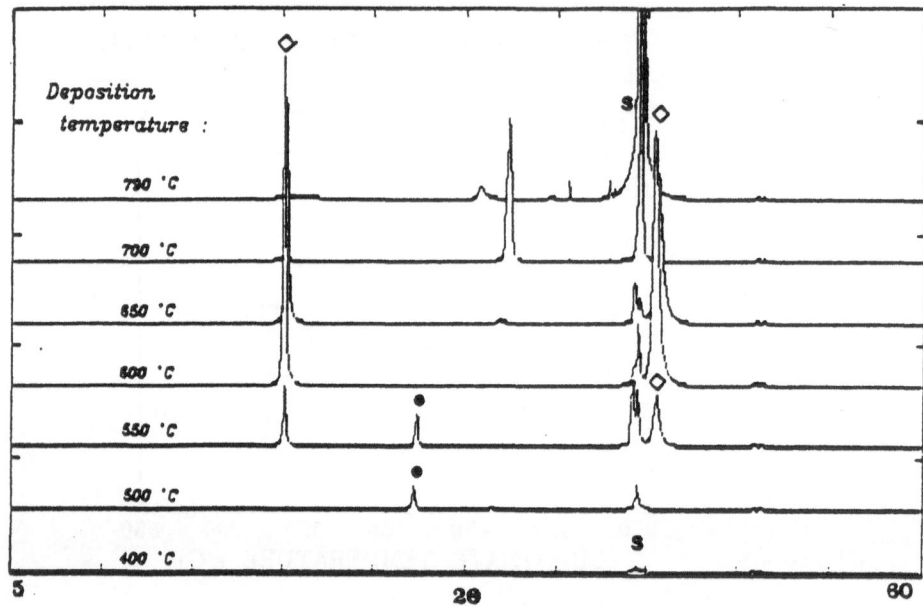

Figure 20 : X-ray diffraction spectra of PZT films as function of substrate temperature
(PZT pyrochlore phase (●), PZT perovkite (◊), MgO substrate (S)).

tentative modellings of the laser induced plasma plume created above a carbon target have been proposed [60].

The experimental device [43,59] is similar to those previously described excepted that an EMG 150 MSC (Lambda Physik) excimer laser is used, the beam qualities of which allow very high focusing on the carbon target. Consequently, energy densities up to 10^4 J/cm^2 are attainable on the target. Film deposition is achieved under high vacuum conditions ($< 10^{-7}$mbar) and the high-purity carbon target is displaced between laser shots.

VI-.1 Spectroscopic investigations of the plasma plume

The plasma plume has been studied by spatially and temporally resolved spectroscopy in order to obtain informations on the nature and transport phenomena of ejected species [43, 61].

For low laser fluence (typically < 7J/cm^2) the recorded spectrum is dominated by emission bands (A$^3\Pi$-X$^3\Pi$ Swan system) of C$_2$ molecules that may be directly ejected from the target or created within the plasma plume. For fluences larger than 7J/cm^2 the emission of C$_2$ gradually disappears and we observe successively emissions originating from C$^+$ and C^{++} when the fluence increases. Surprisingly, no emissions originating from C have been detected in the wavelength range under study, whatever the laser fluence was.

At low fluences, the detection of emission bands of C$_2$ molecules is of particular interest since it provides the possibility of estimates the plasma plume temperature. Indeed, comparison between recorded and synthetic spectra allows estimations of "rotational temperature", known to be close to the kinetic temperature of heavy particles, which is found larger than 6000K [61].

At high fluences, recording of atomic ion spectral lines profiles indicates large line broadenings that have been ascribed to Stark effect. From measurements of these line broadenings we have deduced estimations of charged particle densities, at least in the regions close to the target surface. These results have been completed for larger distances by Langmuir probe measurements [59] and an example of the electronic concentration evolution, along the axis normal to target surface, is presented in Figure 21.

Moreover, measurements of radial distribution of emission intensities give rise by Abel inversion to the local emission coefficient and consequently to the population distribution of the emitting excited state. Results concerning excited states of C$^+$ are presented, for two fluences values, are shown in Figure 22. We observe an important depletion in the central region that increases with the laser fluence, indicating a probable increase of the ionisation degree (C$^+$→C^{++} or more) in the central region of the plasma cloud.

As in paragraph V, velocities "u" of ejected particles perpendicularly to target surface have been studied by spectroscopic time-of-flight (STOF). However,

788

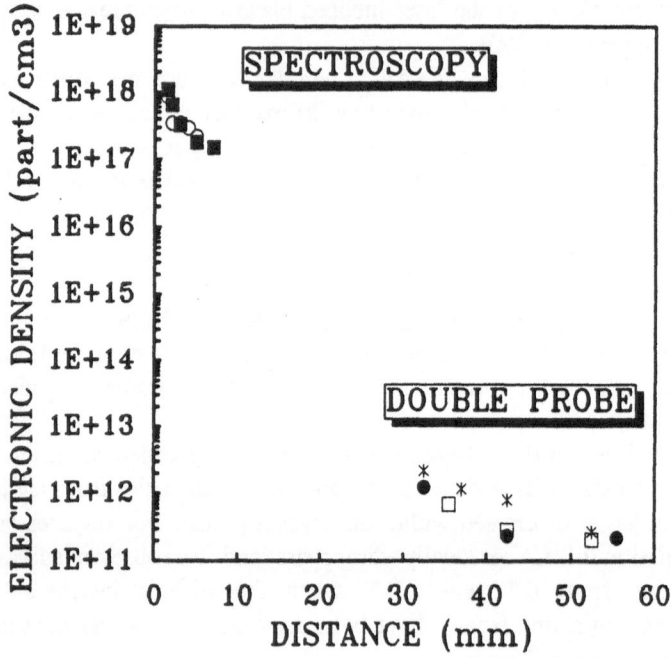

Figure 21 : Evolution of charged particle concentration within the plasma plume determined either by emission spectroscopy (near the target) or by double Langmuir probe, as function of distance to the target surface. Fluence values vary between 30 and 250 J/cm^2.

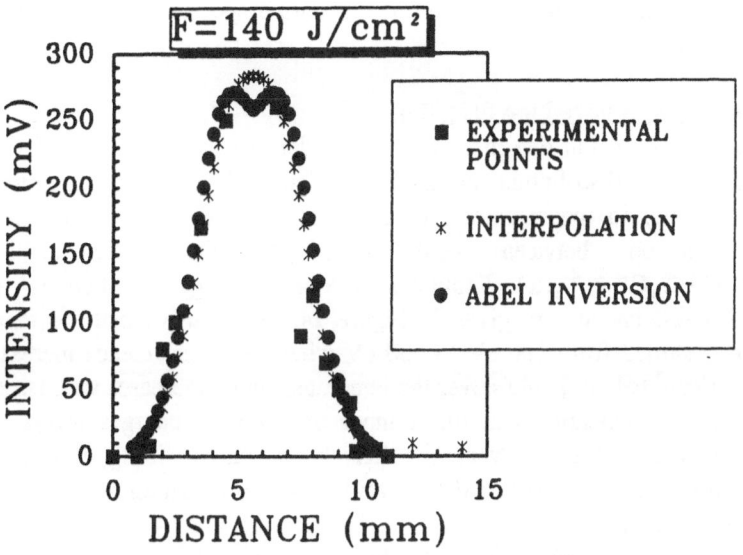

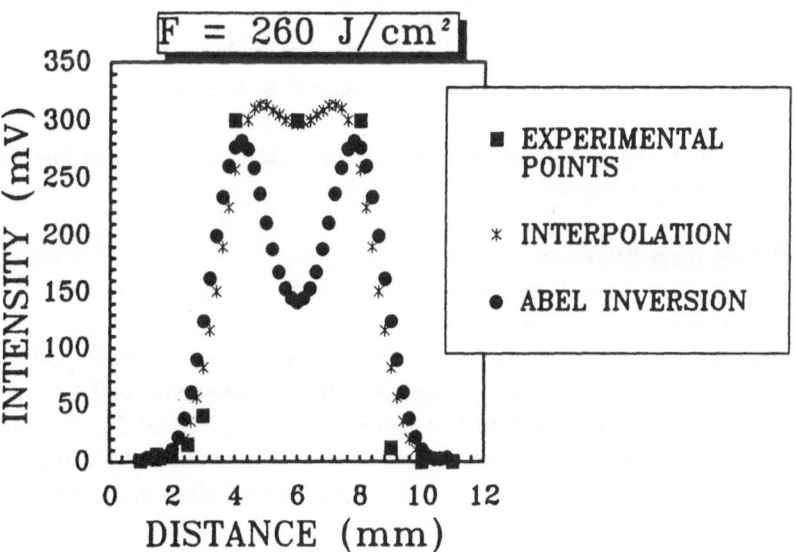

Figure 22 : Radial intensity distribution of the 426.7 spectral line of C$^+$, local emission coefficient '
obtained after Abel inversion.

measurements have been carried out for ionised species only C^+ and C^{++}. Results are shown in Figure 23.

VI-.2 Time resolved ion detection

Time resolved ion detection has been undertaken to determine TOF velocities and kinetic energy distributions of charged particles. The detector is a simple Faraday cup leading only to measurements of the total collected current intensity without any discrimination between contributions of the different ionic species $(C^+, C^{++}, C^{+++},...)$. Examples of kinetic energy distributions deduced from TOF measurements are given in Figure 24. We observe that the maximum energy peak is shifted from 100 eV to 600 eV when the laser fluence increases from 10 to 2000 J/cm^2 [61, 62]. Moreover the integrated signal increases with laser fluence up to 200 J/cm^2, indicating that the number of collected charges grows and for larger fluence values does not vary significantly. Tentative fitting of measurements have been done assuming a shifted Maxwellian energy distribution function at the plasma temperature T:

$$f(v)dv = N(\frac{m}{2 \pi kT})^{3/2} * \exp(-\frac{m}{2kT}((v_x - u)^2 + v_y^2 + v_z^2))dv_x\, dv_y\, dv_z \quad (33)$$

According to Kelly and Dreyfus [19], the detected signal can be fitted by a function :

$$S(x,t) = A\frac{x^2}{t^5}\exp(-\frac{m}{2kT}(x - ut)^2) \quad (34)$$

Figure 24 indicates that, if a good fit is found at high fluence for a temperature of approximately 37000K, in the low fluence regime it is impossible to correctly fit the experimental results. This suggests that at high fluence an equilibration of translational energy is achieved within the initial plasma, whereas at low fluence this equilibration is not attained.

VI-.3 Film deposition

Film deposition have been extensively studied as function of laser fluence and substrate temperature [43, 59]. Various substrates have been used and corresponding film properties have been studied in details. Let us just recall some important results.

The Raman spectra of the films deposited at high temperature (600°C) are typical of microcrystalline graphite but at low temperature (below 100°C), two broad peaks appear, centered around 1570 cm^{-1} and 1350 cm^{-1}, and a small and sharp peak at 1332 cm^{-1} characteristic of diamond. As proposed by Collins et al. [58], these results may be explained by the presence of very small (nanometer-sized) particles of diamond within the film.

The influence of deposition temperature has been also studied by XPS analysis of deposited films. In Figure 25, XPS spectra of films deposited with a fluence of 5000 J/cm^2, at different substrate temperatures, has been reported. Let us recall that the C1s peak of graphite (sp$_2$) is located at 284.4 eV, and that the peak corresponding to sp$_3$ hybridation of carbon atoms (diamond) is at 285.2 eV.

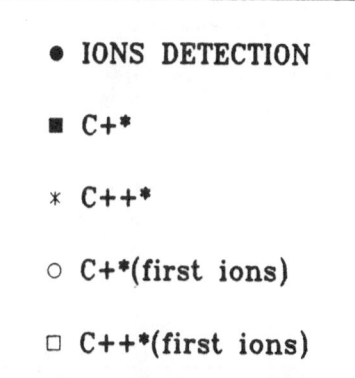

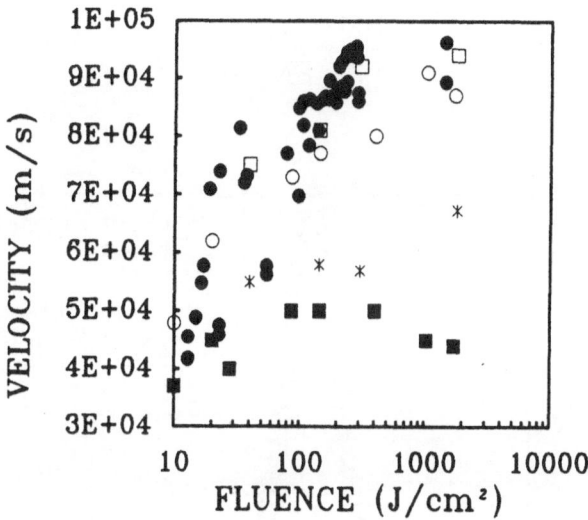

Figure 23 : Time-of-flight velocities of ions determined by ion detection (global ionic signal), and by .
spatially and temporally resolved spectroscopy for C^+ and C^{++}, considering the maximum of emitted
intensity and considering the beginning of emission.

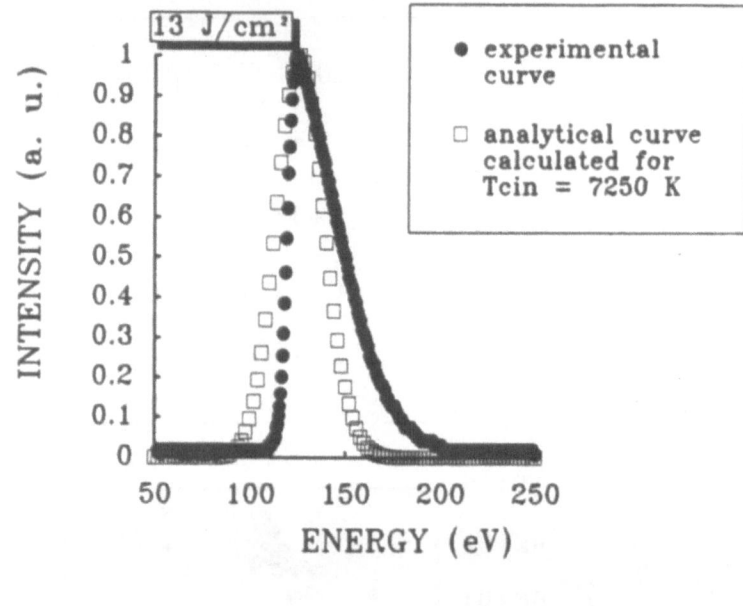

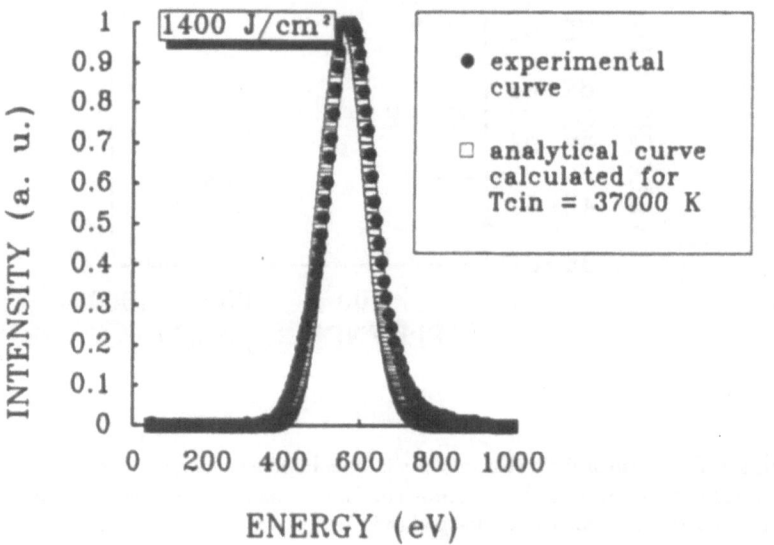

Figure 24 : Kinetic energy distributions of ionic species deduced from TOF measurements for two fluence values and comparison with calculated results obtained when a displaced Maxwellian velocity distribution is assumed.

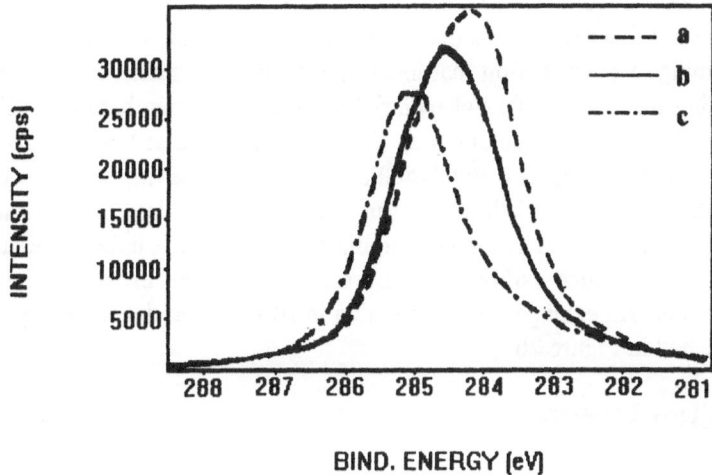

Figure 25 : XPS spectra of C1s region of films deposited on Si wafer with a laser fluence of 5000J/cm^2 at various substrate temperature : (a) T=600°C, (b) T=100°C, (c) T=22°C.

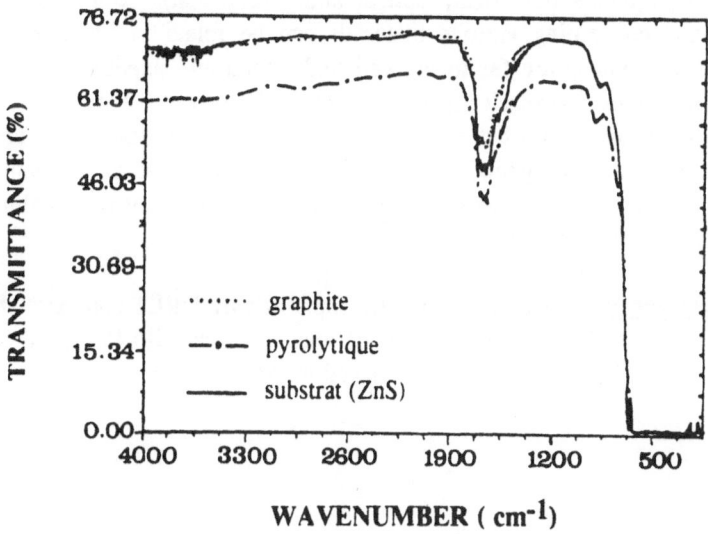

Figure 26 : IR transmission spectra of two films deposited in identical conditions (5000J/cm^2 ,22°C) with two different targets : classical graphite and pyrolytic graphite.

Examination of XPS spectra indicates clearly that the C1s peak is shifted from the graphite position towards the diamond one when substrate temperature is decreased.

Concerning deposited film hardness, preliminary estimations have been done by scratching them with different material powders of known hardness. Obtained results indicate, in a rough approximation, that films deposited at low temperatures and high fluence are harder than c-BN (45GPa). Recently, measurements performed using a nanoindentation system allow more precise results on films hardness which is found about 80 GPa [63]. As a comparison, diamond hardness is between 80 and 100 GPa.

Finally the deposited carbon films are found transparent in the IR and visible region of spectrum. An example of transmission in IR of a film deposited on a ZnS substrate is presented in Figure 26.

VII-. CONCLUSION

The results presented in this lecture show that laser ablation (particularly using excimer lasers) is a very powerful tool for thin film deposition. These techniques have been succeeded in deposition of a large variety of materials, even of complex chemical composition, and lead to deposition of films with very promising properties in various domains (electrical, optical and mechanical). Numerous works on this subject and concerning various materials can be found in literature. However the involved phenomena are far to be well understood and experimental and theoretical works are required to reach a good control of the process. Particularly, great efforts must be done on diagnostics and modelling of the expansion plume and phenomena that governs transport phenomena of ejected species, and on elementary processes involved during the interaction of ejected particles and substrate (/film) surface that determine the film growth.

Acknowledgements : These works were supported by DRET contracts 88-084 and 90-118, and by CNRS in the frame of the interface SPI/Sciences Chimiques on "Photonique and spatial resolution" created in 1993.

REFERENCES

1/ J.F. READY, "Effects of high power laser radiation", Academic Press, New York, (1971).

2/ W.W.DULEY, "CO_2 effects and applications", Academic Press, New York, (1976).

3/ W.W.DULEY, "Laser processing and analysis of materials", Plenum Press, (1983).

4/ M.VON ALLMEN, "Laser beam interactions with materials", Springer Series in material Science, (1987).

5/ I.W.BOYD, "Laser processing of thin films and microstructures", Springer Series in material Science, (1987).

6/ D.BAUERLE, "Chemical processing with lasers", Springer Series in material Science, (1986).

7/ A.B.VANNES ed. "Laser de puissance et traitement des matériaux", Presses Polytechniques et Universitaires Romandes, (1991).

8/ A.M.PROKHOROV, V.I.KONOV, I.URSU, I.N.MIHAILESCU, "Laser Heating of metals", Adam Hilger Series on Optics and Optoelectronics, (1990).

9/ I.W.BOYD, E.FOGARASSY, M.STUKE Eds, "Surface processing and laser assisted chemistry", Proc. of Symposium E, E-MRS spring conference, North-Holland, (1990).

10/ M.STUKE, E.E.MARINERO,I.NISHIYAMA eds, "Material surface processing", Proc. of Symposium B, E-MRS spring conference, (1992).

11/ E.FOGARASSY and S.LAZARE Eds, " Laser ablation of electronic materials", North Holland Pub., (1992).

12/ N.G. DHERE in "Physics of Thin Films", Vol. 16, M.H. Francombe an J.L.Vossen Eds, Academic Press, N.Y., p 2-143, (1992) ; J.J.POUCH, S.A.ALTEROVITZ, R.R.ROMANOFSKY, A.F.HEPP, Eds. " Synthesis and characterisation of high temperature superconductors", Materials Science Forum, Vols 130-132, Trans. Tech.Publications, Aedermannsdorf Suisse, (1993).

13/ A.GUPTA, J. Appl. Phys., 73, 7811, (1993).

14/ R.J.HARRACH, "Theory of laser-induced breakdown over a vaporising target surface"
     Lawrence Livermore Lab. Rep. UCRL-S2 389, Calif. Univ. Livermore, (1977)

15/ C.GIRAULT, PHD Thesis, Limoges Univ., (1990).

16/ A.CATHERINOT, D.DAMIANI, C.CHAMPEAUX, C.GIRAULT, "Photoablation par laser", Chap. 2 in Ref. 7.

17/ I.NOORBATCHA, R.LUCCHESE, and Y.ZEIRI, J.Chem.Phys. 86, 5816, (1987).
     Phys.Rev.B36, 4978, (1987) ; J.Chem.Phys., 89, 5251, (1988).

18/ S.I.ANISIMOV, Sov. Phys. JETP, 27, 182, (1968).

19/ R.KELLY and R.W.DREYFUS, Surf. Scien., 198, 263, (1988) ; Nucl. Instr. and Meth. in Phys. Res., B32, 341, (1988).

20/ R.KELLY, J.Chem.Phys., 92, 5047, (1990).

21/ R.KELLY, A.MIOTELLO, B.BRAREN, A.GUPTA, K.CASEY, Nucl. Instr. and Meth. in Phys. Res., B65, 187, (1992).

22/ Ch.J.KNIGHT, A.I.A.A. Jour., 17, 519, (1979).

23/ Ch.J.KNIGHT, A.I.A.A. Jour., 20, 950, (1982).

24/ R.KELLY, Nucl. Instr. and Meth. in Phys. Res., B46, 441, (1990).

25/ J.DAWSON, P.KAW, B.GREEN, The Phys. of Fluids, 12, 875, (1969).

26/ R.K.SINGH and J.NARAYAN, Phys. Rev. B, 41, 8843, (1990).

27/ R.K.SINGH, O.W.HOLLAND, J.NARAYAN, J. Appl. Phys., 68, 233, (1990).

28/ J.C.S.KOOLS, T.S.BALLER, S.T.DEZWART, J.DIELEMAN, J. Appl. Phys., 71, 4547, (1992).

29/ J.C.S.KOOLS, E.VAN DE RIET, J.DIELEMAN, Appl. Surf. Sci., 69, 133, (1993).

30/ J.C.S.KOOLS and J.DIELEMAN, J. Appl. Phys., 74, 4163, (1993).

31/ C.PHIPPS and R.W.DREYFUS, Bull. Amer. Phys. Soc., 37, 83, (1992)

32/ R.W.DREYFUS, C.PHIPPS, A.VERTES, in "Laser ablation : mechanisms and applications - II", p285, AIP conf. Procee. 288, J.C.MILLER and D.B.GEOHEGAN eds., Knoxville, TN, (1993).

796

33/ A.VERTES, in "Laser ablation : mechanisms and applications - II", p285, AIP conf. Procee. 275, J.C.MILLER and D.B.GEOHEGAN eds., Knoxville, TN, (1993).

34/ E.Y.LO, W.T.LAUGLIN, E.R.PUGH, in "Laser ablation : mechanisms and applications - II", p 291, AIP conf. Procee. 275, J.C.MILLER and D.B.GEOHEGAN eds., Knoxville, TN, (1993).

35/ M.ADEN, E.BEYER, G.HERZIGER, H.KUNZE, J. Phys. D, **25**, 57, (1992).

36/ M.ADEN, E.W.KREUTZ, A.VOSS, J. Phys. D, **26**, 1545, (1993).

37/ C.CHAMPEAUX, D.DAMIANI, C.GIRAULT, P.MARCHET, J.AUBRETON, J.P.MERCURIO and A.CATHERINOT, p 141-165 in Ref. 11.

38/ D.B.GEOHEGAN, p 73-88 in Ref. 11.

39/ D.B.GEOHEGAN, Appl. Phys. Lett., **60**, 2732, (1992).

40/ A.RUPP and K.ROHR, J. Phys. D, **24**, 2229, (1991).

41/ D.SIBOLD and H.M.URBASSEK, J. Appl. Phys., **73**, 8544, (1993).

42/ C.GERMAIN, C.GIRAULT, R.GISBERT, J.AUBRETON, A.CATHERINOT, Diam. and Rel. Mat., **3**, 598, (1994).

43/ C.GERMAIN, PHD Thesis, Univ. of Limoges, (1993).

44/ B.ANGLERAUD, C.GERMAIN, C.GIRAULT, C.CHAMPEAUX, J.AUBRETON, A.CATHERINOT, Procee. of UVX 94 conference, Nouan le Fuzellier, june 13-16, J.M. Pouvesle ed., (1994).

45/ C.CHAMPEAUX, P.MARCHET, A.CATHERINOT, in "Laser ablation : mechanisms and applications - II", p 433, AIP conf. Procee. 275, J.C.MILLER and D.B.GEOHEGAN eds., Knoxville, TN, (1993).

46/ C.GIRAULT, D.DAMIANI, J.AUBRETON, A.CATHERINOT, Appl. Phys. Lett., **54**, 2035, (1989).

47/ C.GIRAULT, D.DAMIANI, J.AUBRETON, A.CATHERINOT, Appl. Phys. Lett., **55**, 182, (1989).

48/ P.E. DYER, R.D.GREENOUGH, A.ISSA, P.H.KEY, Appl. Phys. Lett., **53**, 534, (1988).

49/ X.D.WU, B.DUTTA, M.S.HEDGE, A.INAM, T.VENKATESAN, E.W.CHASE, C.C.CHANG, R.HOWARD, Appl. Phys. Lett., **54**, 179, (1989).

50/ H.J.DIETZE, S.BECKER, Inter. J. of Mass Spectro. and Ion Proc., **82**, R1, (1988).

51/ T.VENKATESAN, X.D.Wu, A.INAM, J.B.WATCHMAN, Appl. Phys. lett., **52**, 14, (1988).

52/ A.SAJJADI, K.KUEN-LAU, F.SABA, F.BEECH, I.W.BOYD, Appl. Surf. Sci., **46**, 84, (1990).

53/ C.CHAMPEAUX, PHD thesis, Univ. of Limoges, (1992).

54/ T.SATO, S.FURUNO, S.IGUSHI, M.HANABUSA, Appl. Phys. A , **45**, 355, (1988).

55/ A.A.GORBUNOV, V.I.KONOV, Surf. Coat. Technol., **47**, 503, (1991).

56/ J.J.CUOMO, J.BRULEY, J.P.DOYLE, D.L.PAPPAS, K.L.SAENGER, J.C.LIU, R.E.BATSON, Mat. Res. Soc. Symp. Procee., **202**, 247, (1991).

57/ F.DAVANLOO, E.M.JUENGUERMANN, D.R.JANDER, T.J.LEE, and C.B.COLLINS, J. Mater. Res. **5**, 1398, (1990).

58/ C.B.COLLINS, F.DAVANLOO, D.R.JANDER, T.J.LEE, H.PARK, J.H.YOU, J. Appl. Phys., **69**, 7862, (1991).

59/ C.GERMAIN, C.GIRAULT, R.GISBERT, J.AUBRETON, A.CATHERINOT, Diam. and Rel. Mater., **3**, 598, (1994).

60/ J.STEVEFELT, C.B.COLLINS, J. Phys. D, **24**, 2149, (1991).

61/ C.GERMAIN, C.GIRAULT, J.AUBRETON, A.CATHERINOT, Appl. Surf. Sci., **69**, 359, (1993).

62/ C.GERMAIN, C.GIRAULT, R.GISBERT, J.AUBRETON, A.CATHERINOT, in "Laser ablation: mechanisms and applications - II", p 183, AIP conf. Procee. 275, J.C.MILLER and D.B.GEOHEGAN eds., Knoxville, TN, (1993).

63/ Private communication , Ecole Centrale de Lyon.

THEORETICAL DESCRIPTION OF LASER ABLATION

C. KÖRNER, H.W. BERGMANN
Institut für Werkstoffwissenschaften II, Martensstr. 5
Universität Erlangen-Nürnberg, D-91058 Erlangen, Germany

Abstract

The expansion and the relevant parameters determining the heat affected zone (HAZ) for the laser ablation processes is discussed. It is shown that the essential parameter to minimize the HAZ is the power density. The limitation of the ablation depth for non-metals is explained with the help of the hyperbolic heat conduction equation which takes into account the finite velocity of heat transport. Ablation velocities faster than this velocity are not possible.

1. Introduction

The application of laser radiation for microstructuring is discussed worldwide. Serveral different techniques are investigated. The laser radiation can be used either for ablation or constructive processes. The laser beam can be applied for ablation directly or be combined with chemical methods. In addition, focussing and imaging techniques can be employed. In spite of the multitude of different methods the aims are, however, identical: retaining geometrical form, miniaturizing of structures, no degeneration of properties. Besides these requirements a high efficiency must be maintained. The alteration of properties of the worked material can be either of thermal, mechanical or chemical nature and is in general undesired. A high efficiency combined with a small HAZ is often accompanied by high ablation pressures inducing elastic shock waves moving into the material and producing twins and other defects. Hence, a better understanding of the ablation process is necessary in view of a theoretical description of mechanical phenomena.

There are many different laser types, each characterized by the wavelength of the laser light, the pulse energy, the pulse length, the frequency, and last but not least, the power density. Temperature profiles for long pulse durations are based on classical heat conduction, whereas optical penetration dominates in the domain of very short pulse durations. That is, the development of mathematical models for heating and ablation processes depend critically on the respective laser parameters. This paper gives theoretical explanations for phenomena observed for the ablating processes with pulsed lasers with pulse durations of 1-100ns and power densities in the range of 10^8-$10^{11}\mathrm{W/cm^2}$ without chemical support.

J. Mazumder et al. (eds.), Laser Processing: Surface Treatment and Film Deposition, 797–808.
© 1996 *Kluwer Academic Publishers.*

2. The Heat Affected Zone

For processes where the ablation depth is much smaller than the thermal influence the HAZ can be estimated with the help of the thermal penetration depth δ:

$$\delta = \sqrt{4k\tau}$$ (1)

(τ: pulse duration, k: thermal diffusivity)

50ns-pulses lead to values for δ in the micrometer range. The HAZ is even larger if the influence of preceding pulses are not neglected. To minimize the thermal input, the pulse duration can be reduced until δ reaches the optical penetration length of the laser radiation. This results in the requirement of very high power densities and high frequencies in order to keep efficiency.

The thermal penetration depth as defined above is not the relevant quantity for processes combined with high ablation rates. In this case material ablation during the pulse has to be taken into account. With the help of a very simple model the HAZ can be estimated. The underlying assumption is that the time necessary to reach vaporization temperature is small compared with the pulse duration. For this reason a "quasi-stationary" state with continuous removal of material during the laser pulse is assumed. Furthermore, it is assumed that a one-dimensional solution of the heat conduction equation is justified. This stationary problem is solved in a moving frame where the material surface and the origin of the new coordinate system coincide (coordinate transformation: $x \rightarrow x' = x-ut$, $t \rightarrow t' = t$, u: ablation velocity). The stationary parabolic heat conduction equation (surface source) reads [3]:

$$- k \frac{\partial^2}{\partial x'^2} T(x') - u \frac{\partial}{\partial x'} T(x') = \frac{k}{\lambda} S \delta(x')$$ (2)

(λ: thermal conductivity, u: ablation velocity, S: absorbed power density, δ: Delta function)

The stationary state is characterized by the fact that the absorbed energy and the energy necessary for vaporization are identical, i.e. energy conservation determines u:

$$u = \frac{S}{(L + c_p T_0) \rho}$$ (3)

(L: latent heat, ρ: density, c_p: specific heat capacity, T_0: ablation temperature)

T_0 is an additional parameter to determine. The solution of the heat conduction equation on condition that $T(x<0)=T_0$ is depicted in figure 1. The HAZ is proportional to 1/u and does not depend on the pulse duration. The pulse duration is irrelevant as long as a stationary description is allowed. The HAZ and the energy deposited into the workpiece can be reduced by increasing of the power density, i.e. increasing of the ablation rate. Starting from this idea the reduction of the pulse length is not the essential point in order to minimize the thermal input. Important are high power densities.

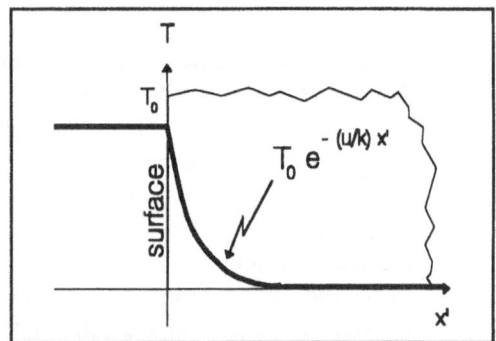

Fig.1 Solution of the stationary parabolic heat conduction equation in a moving frame.

3. The Processing Velocity

3.1. EXPERIMENTAL RESULTS

The increase in the processing velocity by increasing the power density will eventually reach a limitation. At high power densities a laser induced plasma tends to screen the laser light by reflection, absorption and scattering. The developement of a screening plasma depends critically on the power density and the wavelength of the respective laser light. A background gas favours the plasma formation and leads to laser supported absorption waves (LSA-wave). In this way the efficiency is reduced dramatically and the ablation is limited or impossible (e.g. TEA-CO_2-Laser in air). The formation of a screening plasma in vacuum due to the ablated material requires much higher power densities.

Ablation experiments with three different lasers were performed:

	TEA-CO_2-Laser	CVL-Laser	XeCl-Excimer-Laser
wavelength	$10,6\mu$	511/578nm	308nm
pulse length	70ns	50ns	55ns
pulse energy	6J	6mJ	2J
frequency	10Hz	6,5kHz	20Hz
beam area	$1,3mm \times 1,3mm$	$\sim 1000\mu m^2$	$1.6mm \times 1,6mm$

In order to avoid the formation of LSA-waves the experiments with the TEA-CO_2-Laser were carried out under vacuum conditions which are not necessary for the Excimer- or CVL-Laser.

800

For all the experiments similar results were found:

- for all materials (metals, glasses, ceramics, plastic) there exists a material dependent critical power density S_{crit} below which ablation is not possible
- S_{crit} is a function of the pulse length and the material, i.e. absorption coefficient, thermal conductivity, vaporization energy and temperature
- at low power densities substances with a small thermal conductivity are ablated best since the heat lost due to thermal conduction into the bulk material is small
- increasing the power density shows that materials with high ablation thresholds also have high ablation rates and vice versa
- the ablation rate of all non-metals reaches a limiting value with increasing power density; this behaviour is not observed for metals up to power densities of $10^{10} W/cm^2$

Fig. 2 shows schematically the ablation depth as a function of the power density for different materials. The higher S_{crit} is, the higher the ablation rate and the later the limiting value is reached. Ablation rates as a function of the power density for different glasses obtained with the excimer laser clearly demonstrate this behaviour (fig. 3). Figure 4 shows the ablation depths (TEA-CO_2-Laser) for several metals and non-metals as a function of power density.

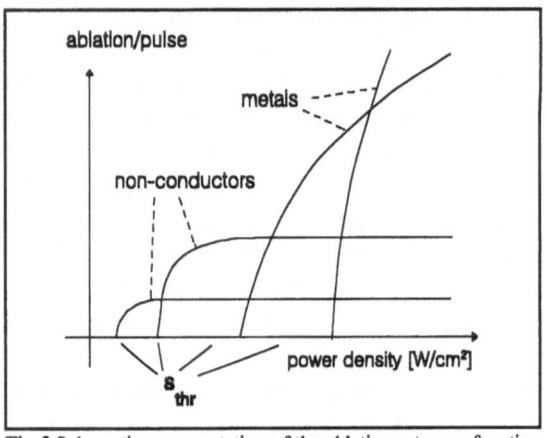

Fig.2 Schematic representation of the ablation rate as a function of the power density

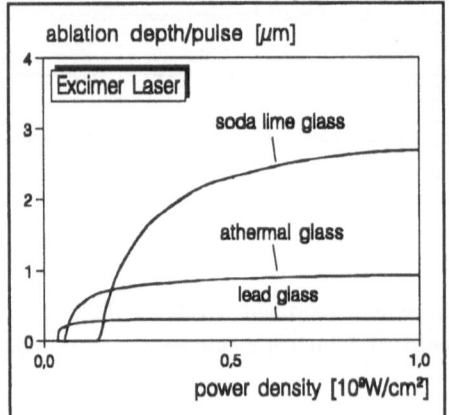

Fig.3: Ablation rates for glass treatment with the Excimer Laser [2]

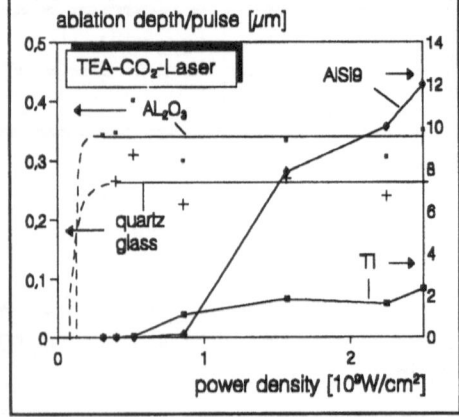

Fig.4: Ablation rates for different materials with the TEA-CO_2-Laser

From a purely thermal point of view the existence of a limiting value for all materials with low ablation thresholds can not be explained. Whether the ablation depth for metals also approaches a limiting value is not clear. Up to power densities of $10^{10}W/cm^2$ (CVL-Laser) the values increase continuously with S (AL: $80\mu m$/pulse i.e. $u = 1600m/s$).

Usually, the restriction of the ablation rate is ascribed to a screening material plasma. Unfortunately, this explanation raises some open questions:

- Why should the screening effect for non-metals be higher than for metals although the ablation rate for metals is much higher?
- Why should screening lead to a *constant* ablation rate?
- Why are there no large differences between the different lasers observed although plasma absorption is strongly dependent on the wavelength of the laser light?

3.2. THEORETICAL CONSIDERATIONS

Ablation requires energy to be deposited into the material. There are two processes to distinguish. On the one hand energy is transported by heat conduction, which takes time. On the other hand the laser radiation has a finite, temperature dependent optical penetration depth. In principle, this transport process requires no time. The optical penetration depth depends strongly on temperature and is reduced to a few nanometers at the vaporization temperature. For this reason the main part of the energy transport takes place by diffusion processes. If this process is not fast enough, energy accumulates at the surface and can not be used for further ablation, i.e. there is much more energy present than can penetrate into the material. As a consequence the ablation temperature and pressure rise and a further increase of the power density will not increase the ablation rate. That means that the limiting value for the ablation rate could result from a limitation on energy transport.

A fundamental difference between metals and non-metals exists in their heat transport mechanisms. In non-metals energy is transported by phonons, in metals heat is mainly transferred by electrons which are about **three** orders of magnitude faster than phonons. Usually, this is not important but for very high heating rates and temperature gradients this distinction becomes important because the velocity of heat transport is limited in this way. A fundamental derivation of the heat conduction equation with the help of the Boltzmann-Equation (transport equation) shows that the usually assumed Fourier-Law must be extended by an additional term in order to take into account the finite electron/phonon velocity.

This **modified Fourier-Law** [5][7] reads:

$$\vec{q} + \tau \frac{\partial}{\partial t}\vec{q} = -\lambda \nabla T \tag{4}$$

(q: heat flux, τ: relaxation time)

The relaxation time τ is identical with that describing the collision term in the Boltzmann equation. The **hyperbolic heat conduction equation** results using energy conservation is:

$$\frac{1}{c^2}\frac{\partial^2}{\partial t^2}T + \frac{1}{k}\frac{\partial}{\partial t}T - \Delta T = \frac{1}{\lambda}\left(S + \tau\frac{\partial}{\partial t}S\right) \tag{5}$$

($c^2 = k/\tau$, S: energy source)

The hyperbolic heat conduction equation is a damped wave equation. It takes into account the wave-like propagation of heat, i.e. a thermal disturbance spreads with velocity c. For large times the solution of the hyperbolic heat conduction equation tends towards the parabolic one. Fig.5 shows the one-dimensional temperature fields for different times after the beginning of a temporal constant laser pulse (for details concerning the solution see appendix). Due to the finite velocity c of heat propagation the wave front can only penetrate into the material the distance ct. This is in contrast to

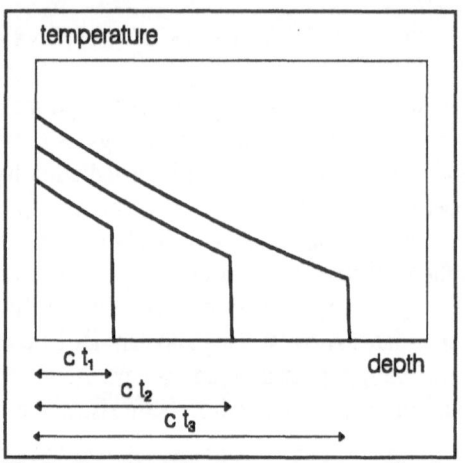

Fig.5: One-dimensional temperature profile for different times (hyperbolic heat conduction equation).

the parabolic formulation of heat conduction which implies infinite propagation velocities of thermal disturbances [1]. As a consequence the hyperbolic equation predicts a faster and higher heating of the material surface compared with the parabolic description.

The wave-like propagation appears when the heat conduction problem is solved for a slab. The superpositon of the incoming and reflected temperature wave is possible leading to the strange situation that the temperature field oscillates. This phenomenon is depicted in figure 6.

In order to answer the question whether hyperbolic effects are actually observable for pulse durations about 50ns and power densities in the range of 10^8-10^{11}W/cm^2 the

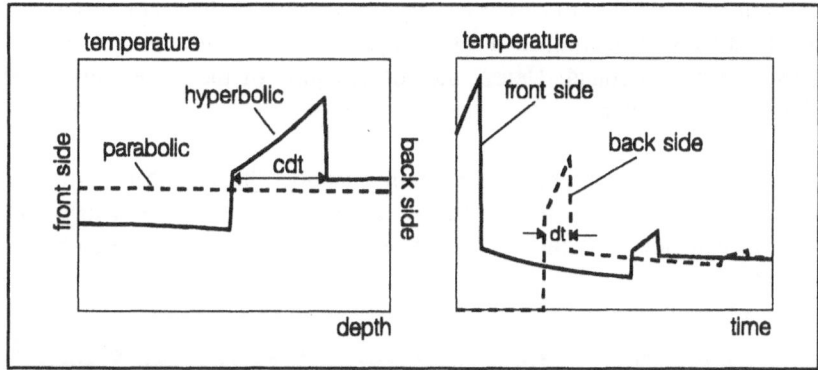

Fig.6: Solution of the hyperbolic heat conduction equation in a slab and an infinite extented surface heat source.

introduction of dimensionless space and time coordinates is useful. This shows that hyperbolic effects are restricted to times and distances smaller than the relaxation time τ and the mean free electron/phonon path Λ, respectively. How large τ and Λ are depends on the heat transport mechanism, i.e. conduction by electrons and phonons has to be distinguished. The determination of τ and Λ for electrons is rather easy due to two reasons (e.g. [4]):

- only electrons at the fermi surface are able to contribute to the thermal current; the fermi velocity can be calculated on the assumption of free electrons with a uniform effective mass
- the thermal and electric current are connected via the Wiedemann-Franz-Law

This yields: $\tau \approx 10^{-14}s$, $c \approx 10^6 m/s$, $\Lambda \approx 100\text{Å}$ (at room temperature) Since the relaxation time τ is **five** orders smaller than the relevant time scale (nanoseconds) hyperbolic effects can be excluded if electrons are the carriers of the thermal current.

For phonons the situation is much more complicated. On the one hand all phonons contribute to the thermal current. These have different velocities and relaxation times, i.e. the assumption of mean values for the velocities and relaxation times is questionable. On the other hand this velocity is hard to determine. The velocity of sound can serve as upper limit but the actual value of the heat transport velocity is probably much lower.

Due to the lack of the Wiedemann-Franz-Law the relaxation time must be estimated with the help of the thermal diffusivity and the velocity of sound:

$$\tau \geq \frac{k}{c^2} \approx 10^{-12}s \tag{6}$$

In fact, this time is likely to be much larger. For example, the heat transport velocity in liquid helium was determined to 19m/s [6] - more than one order of magnitude less than the velocity of sound! Hence, the consideration of the finite phonon velocity might be quite important.

Also, for the hyperbolic heat conduction equation the stationary problem with continuous removal of material can be solved. The stationary hyperbolic heat conduction equation in a moving frame and a pointlike moving source S is (see appendix):

$$- k \left(1 - u^2/c^2\right)\frac{\partial^2}{\partial x'^2}T(x') - u \frac{\partial}{\partial x'}T(x') = \frac{k}{\lambda} S \left(\delta(x') - \tau u\frac{\partial}{\partial x'}\delta(x')\right) \qquad (7)$$

This differential equation is equivalent to the parabolic one where $\partial^2 T/\partial x^2$ has been multiplied by $(1-u^2/c^2)$ and an additional source term has been added. For $u<c$ the following holds:

$$T(x') = T_0 \cdot \begin{cases} \dfrac{1}{1 - u^2/c^2} \; e^{-\frac{ux'}{k(1-u^2/c^2)}} & x' \geq 0 \\ \\ 1 & x' < 0 \end{cases} \quad for \qquad (8)$$

The finite velocity of heat transport results in an increased surface temperature and a faster exponential drop of the temperature field (fig. 7). For u approaching c the energy is no more able to penetrate into the material and is accumulated at the surface. Similar results are found when taking into account the finite optical penetration depth. In this case the relation between the thermal penetration k/u and the optical penetration determines the maximal temperature. For $u>c$

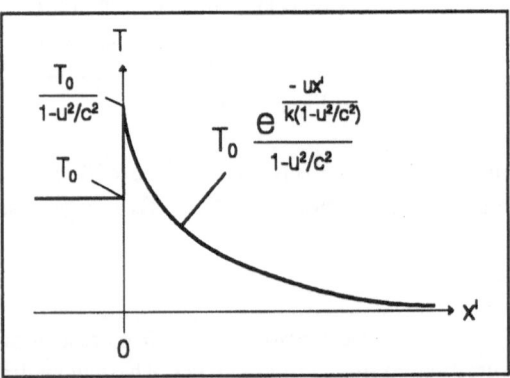

Fig.7 : Stationary solution for the hyperbolic heat conduction equation

there exists no solution of the upper differential equation with the condition that $T(x'<0)=T_0$. This is also true if a finite optical penetration depth is taken into account. A further rise of the power density must lead to an increased ablation temperature; ablation velocities higher than the speed of heat propagation are not possible.

Following these arguments the diverse ablation behaviours can be explained. Materials with high thermal conductivities show high ablation thresholds due to the large amount of heat lost into the bulk material. If electrons are the main carriers of the thermal current, large ablation rates are possible since the electron velocity is always much higher than the ablation velocity. For all materials where phonons carry the thermal current the ablation threshold is determined by the thermal diffusivity and absorption length. A limiting value of the ablation rate is reached with increasing power density since phonons which transport energy are comparatively slow.

4. Mechanical damage

The mechanical damage caused by laser induced shock-waves in the bulk material is inseparable from the thermal influence. Two generating mechanism have to be distinguished: the ablation pressure and the radition pressure. Up to power densities of 10^{10}W/cm^2 the radiation pressure can be neglected. The ablation pressure increases with u and T_0. The resulting shock-wave can be observed experimentally. Several reflections of elastic waves in thin slabes have been detected. The wave dissipates by the creation of twins or other defects or by heat generation (fig.8). In terms of shock-

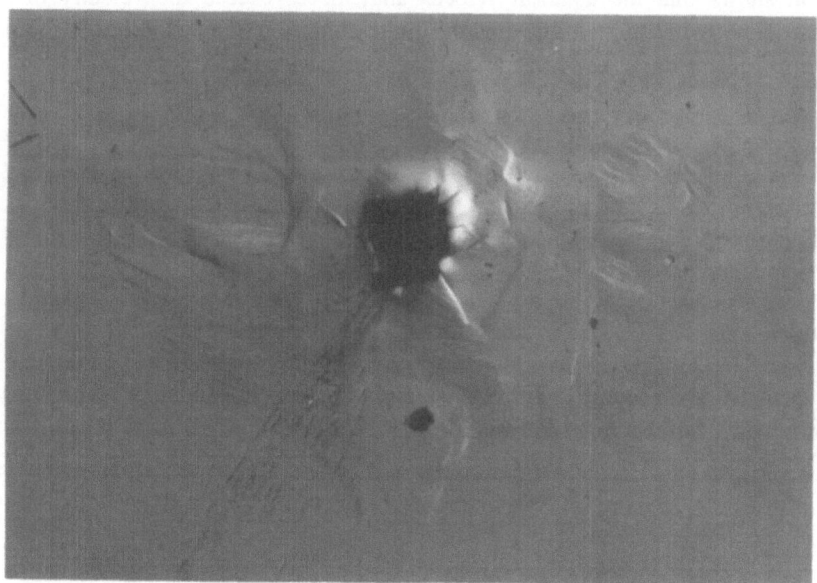

Fig.8: Photograph of a hole drilled in Ti (backside, thickness 220μm, diameter $\approx$ 100μm)

hardening effects this might be desirable, but for avoiding mechanical damage this must be prevented. Above all for materials where there exists a limiting value of the ablation rate the actual value of the power density determines the mechanical input although the ablation rate is fixed. For this reason it is absolutely necessary to understand the connection between the thermal and mechanical effects in order to prohibit mechanical damage.

5. Conclusion

The minimum expansion of the HAZ is given by the optical penetration depth if energy conduction by diffusion processes can be neglected. In order to reach this situation very short pulses combined with high power densities are necessary. The description of the ablation process as a quasi-stationary process has shown that the HAZ decreases with increasing power density, i.e. ablation velocity, demonstrating that not the pulse length but the power density is the essential parameter.

The existence of a limited ablation depth with increasing power density for all non-metals can be explained with the help of the hyperbolic heat conduction equation. This extension of the classical heat conduction equation takes into account the finite velocity of heat transport which is given for metals by the electron velocity and for non-metals by the phonon velocity. The solution of the stationary hyperbolic heat conduction equation shows that the ablation velocity may never exceed the velocity of heat transport. Following these arguments the finite ablation rate of non-metals is a heat transport problem rather than a problem of plasma screening.

Acknowledgements

Part of this work was funded by the "Deutsche Forschungsgemeinschaft (DFG)". This support is gratefully appreciated.

Appendix

This appendix is concerned with the solution of the hyperbolic heat conduction equation which follows from the **modified fourier law**

$$\vec{q} + \tau \frac{\partial}{\partial t} \vec{q} = -\lambda \nabla T \tag{9}$$

and the equation of **energy conservation:**

$$c_p \rho \frac{\partial}{\partial t} T = -\nabla \vec{q} + S(\vec{x}, t) \tag{10}$$

(ρ: density)

Hyperbolic heat conduction equation

$$\frac{1}{c^2}\frac{\partial^2}{\partial t^2}T + \frac{1}{k}\frac{\partial}{\partial t}T - \Delta T = \frac{1}{\lambda}\left(S(\vec{x},t) + \tau\frac{\partial}{\partial t}S(\vec{x},t)\right) \tag{11}$$

With regard to laser material working with a temporal and spatial arbitrary source S solutions using Green's functions are best suited. he Green's function G is the solution of a differential equation for a pointlike source. That is:

$$\frac{1}{c^2}\frac{\partial^2}{\partial t^2}G + \frac{1}{k}\frac{\partial}{\partial t}G - \Delta G = \delta(\vec{x})\,\delta(t) \tag{12}$$

G can be found by a transformation into momentum space and the use of methods of complex analysis. The one-dimensional Green's function is given by:

$$G(x,x',t,t') = \tfrac{1}{2}c\,\exp\left(-\frac{c^2(t-t')}{2k}\right)I_0\left(\frac{c}{2k}\sqrt{c^2(t-t')^2 - |x-x'|^2}\right)\Theta(c(t-t')-|x-x'|) \tag{13}$$

(I_0: modified Bessel function, Θ: Heavyside step function)
For simple geometries (semi infinite solid, slab, edge, rod) solutions fulfilling adiabatic boundary conditions follow by superposition of several mirror sources. For example, the temperature field due to an arbitrary source S for a semi infinite solid ($z \geq 0$) is given by integration:

$$T(\vec{x},t) = T_0 + \frac{1}{\lambda}\int_0^t dt'\int d\vec{x}' \left(G(\vec{x}-\vec{x}',t-t') + G(x-x',y-y',z+z',t-t') \right)\left(S(\vec{x}',t') + \tau\frac{\partial}{\partial t'}S(\vec{x}',t')\right) \tag{14}$$

(T_0: initial temperature, i.e. $T_0 = T(t=0)$)
The transformation of the hyperbolic heat conduction equation to a moving frame of velocity u gives:

$$\left[\frac{1}{c^2}\frac{\partial^2}{\partial t'^2} + \left(\frac{u^2}{c^2} - 1\right)\frac{\partial^2}{\partial x'^2} - \frac{2u}{c^2}\frac{\partial}{\partial t'}\frac{\partial}{\partial x'} + \frac{1}{k}\frac{\partial}{\partial t'} - \frac{u}{k}\frac{\partial}{\partial x'}\right]T = \frac{S}{\lambda}\left(\delta(x') - \tau u\frac{\partial}{\partial x'}\delta(x')\right) \tag{15}$$

where a pointlike moving source is assumed. The presence of the spatial derivative in the source term is essential.

808

For the stationary state the time derivatives vanish:

$$\left[\left(\frac{u^2}{c^2}-1\right)\frac{\partial^2}{\partial x'^2} - \frac{u}{k}\frac{\partial}{\partial x'}\right] T = \frac{S}{\lambda}\left(\delta(x') - \tau u\frac{\partial}{\partial x'}\delta(x')\right) \tag{16}$$

The solution of this differential equation can be found by integration with the help of the solution for a pointlike source. There are three cases to distinguish:

v < c:

$$T(x') = T_0 \cdot \begin{cases} \dfrac{1}{1 - u^2/c^2}\, e^{-\frac{ux'}{k(1 - u^2/c^2)}} & x' \geq 0 \\ \qquad\qquad\qquad\qquad for \\ 1 & x' < 0 \end{cases} \tag{17}$$

v = c:

$$T(x') = T_0\left[\Theta(-x') + \frac{k}{c}\delta(x')\right] \tag{18}$$

v > c: There exists no solution on condition that T=const. for x' < 0.

References

1. Boley B.A. (1964) *The Analysis of Problems of Heat Conduction and Melting, High Temperature Structures and Materials*, Pergamon Press, New York, pp. 260-315
2. Buerhop C. (1994), *Glasbearbeitung mit Hochleistungslasern* Dissertation, Erlangen
3. Carslaw H.S., Jaeger J.C.(1959) *Conduction of heat in solids,* Oxford Science Publications, Oxford
4. Ibach H., Lüth H. (1990) *Festkörperphysik* Springer Verlag
5. Morse P.M., Feshbach H. (1953) *Methods of Theoretical Physics,* Part I McGraw-Hill, New York, pp. 865-869
6. Peshov V. (1944) Second Sound in He II *Journal of Physics*, USSR, Vol. 8, p.381
7. Taitel Y., On the Parabolic, Hyperbolic and Discrete Formulation of the Heat Conduction Equation, *Int. J. Heat Mass Transfer*, Vol. 15, pp. 269-371

LASER REACTIVE ABLATION: ONE STEP PROCEDURE FOR THE SYNTHESIS AND DEPOSITION OF COMPOUND THIN FILMS

Maria Dinescu, N. Chitica, V.S. Teodorescu, Adriana Lita,

Institute of Atomic Physics, P.O. Box MG-6 Bucharest-76900, ROMANIA

A. Luches, M. Martino, A. Perrone,

Physics Department, University of Lecce, Lecce, ITALY

Maria Gartner,

Institute of Physical Chemistry, Bucharest, ROMANIA

Abstract

Laser Reactive Ablation (LRA) was proved to be an efficient one-step procedure for the deposition of thin films of compounds like TiN and TiC. We extended this method to the deposition of silicon nitride films on Si wafers by laser ablation of a Si target in ambient NH_3 reactive gas. Amorphous Si_3N_4 films were deposited on Si wafers at a pressure of 1 mbar and a temperature of the collector of 200° C. The mechanisms involved in this case seem to be essentially different to those involved in the deposition of TiN or TiC films. A possible explanation is proposed.

1. Introduction:

Lasers have been applied to the synthesis and deposition of compound thin films. A special mention should be made to Pulsed Laser Deposition (PLD), Laser Chemical Vapor Deposition (LCVD) and Laser Direct Synthesis (LDS). PLD is successfully used to deposit stoichiometric thin films of alloys or compounds by ablating already-prepared solid targets. LCVD is an excellent way to deposit thin films, combining the advantages of the conventional CVD (very good control of the composition, good quality of the surface, uniformity, etc.) with a very high deposition rate. LDS gives the possibility to produce local modifications to the composition of the surface of a solid sample.

J. Mazumder et al. (eds.), Laser Processing: Surface Treatment and Film Deposition, 809–821.
© *1996 Kluwer Academic Publishers.*

Recently, a new method, which combines (in one step) the synthesis of the compound and the deposition, was derived from the PLD method. A solid target consisting of the material to be reacted is ablated in a reactive gas atmosphere. The resulting material is collected on a substrate placed at a given distance from the target. This method is referred as Laser Reactive Ablation (LRA). Several advantages of this method should be mentioned:

1. High purity. Being a "cold wall" reactor type method, a very low level of contamination is achievable.

2. Low thermal budget. The substrate is not subjected to high temperatures.

3. Versatility. The synthesized compound can be easily changed by changing the target and / or the reactive gas. This way, multilayer structures and superlattices can be deposited.

4. No "memory" effect. The probability of contamination from a previous deposition is very low.

5. Ensures sharp interfaces.

6. Does not require already-prepared compound targets, or expensive or hazardous gases.

The successful application of this method to the deposition of titanium nitride [1, 2] and carbide [3], germanium oxide [4], silicon dioxide [5], or silicon nitride films [6] was already reported.

For a better understanding of the synthesis and deposition processes involved in the LRA, we performed a parametric study of the deposition of silicon nitride layers on Si wafers by LRA of a Si target in ambient NH_3.

2. Experimental:

The experimental setup used for LRA is schematically presented in Fig. 1. The irradiations were performed with a Lambda Physik LPX 315i excimer laser (denoted by 1 in Fig. 1), operated with $XeCl^*$ active mixture ($\lambda = 308$ nm, $t_{FWHM} = 30$ ns). All irradiation were performed at a repetition rate of 10 Hz.

The targets (2), with dimensions of 15 x 15 mm^2, were cut from Si wafers. During irradiation, the targets mounted on the holder (12) were rotated with 3 Hz. The laser radiation was incident on the polished surface of the targets under an angle of 45°, on eccentric locations. The laser fluence on the surface of the targets E_s was in the range of 5 - 6 J/cm^2. The radiation was focused with the Suprasil lens (3). The ablated material was collected on Si wafers (4) mounted on the holder (5), which could be heated with the electrical heater (13). During deposition, the Si wafer collector was

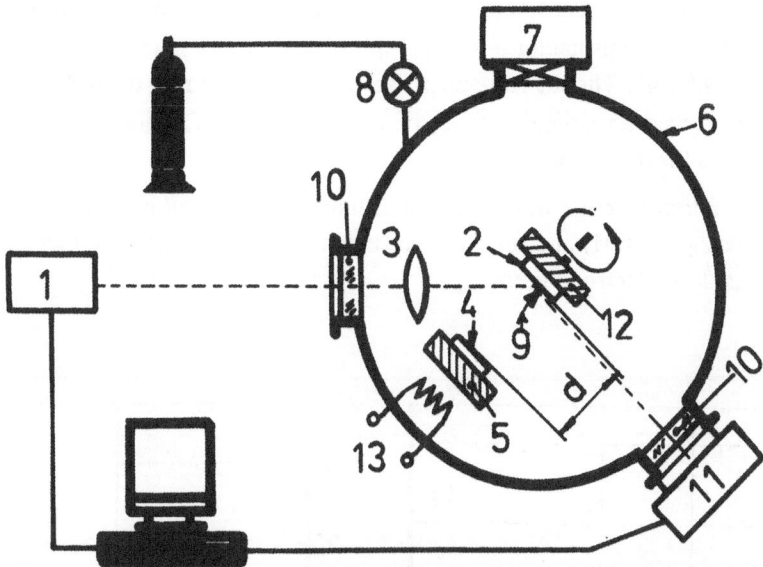

Figure 1.   The experimental setup: (1) XeCl* excimer laser, (2) Si target, (3) Focusing lens, (4) Si wafer collector, (5) Heated holder, (6) High vacuum enclosure, (7) Pumping system, (8) Gas flow controlling valve, (9) Plasma plume, (10) Suprasil window, (11) Spectrograph, (12) Target's holder, (13) Electrical heater.

heated at 200° C. The distance d between the target (2) and the collector (4) was set at 14, 21 and 30 mm, respectively. Each sample was prepared by collecting the material ablated by 10000 laser pulses.

After loading a new substrate, the high vacuum enclosure (6) was heated to 100° C and submitted to several evacuation - refilling cycles, in order to reduce the residual pressure of the contaminants (oxygen and water vapors). Each time the enclosure was pumped down to 10^{-6} mbar with the aid of an oil-free pumping system and refilled to 1 bar with the high purity (99.98 %) NH_3 reactive gas used for the deposition. During irradiation, the pressure p of the NH_3 gas was set at different values in the range 10^{-3} mbar - 1 mbar. The pressure was maintained in a dynamic equilibrium by partially closing the entrance valve of the pumping system and by controlling the gas flow with the valve (8). This procedure was used in order to ensure a low partial pressure of the residual impurity gases.

The high vacuum enclosure has two Suprasil windows (10), one for the laser radiation, the other for the spectrograph (11).

The profiles of the films deposited on Si wafers by LRA of Si in NH_3 were mapped with an Alphastep 200 (Tencor Instruments) profilometer.

812

Spectroscopic ellipsometry (SE) investigations of the films were performed with a multiangle ellipsometer in UV-Vis spectral range (foton energy $E = h\nu$ in the range 1.5 - 4.5 eV). The measurements were made on a 2 mm diameter zone. The apparatus allowed a reading accuracy of 1 min for the two azimuths (polarizer and analyzer) and for the incidence angle.

The microstructure of the deposited films was analyzed by transmission electron microscopy (TEM) and selected area electron diffraction (SAED) with a JEOL TemScan 200CX electron microscope.

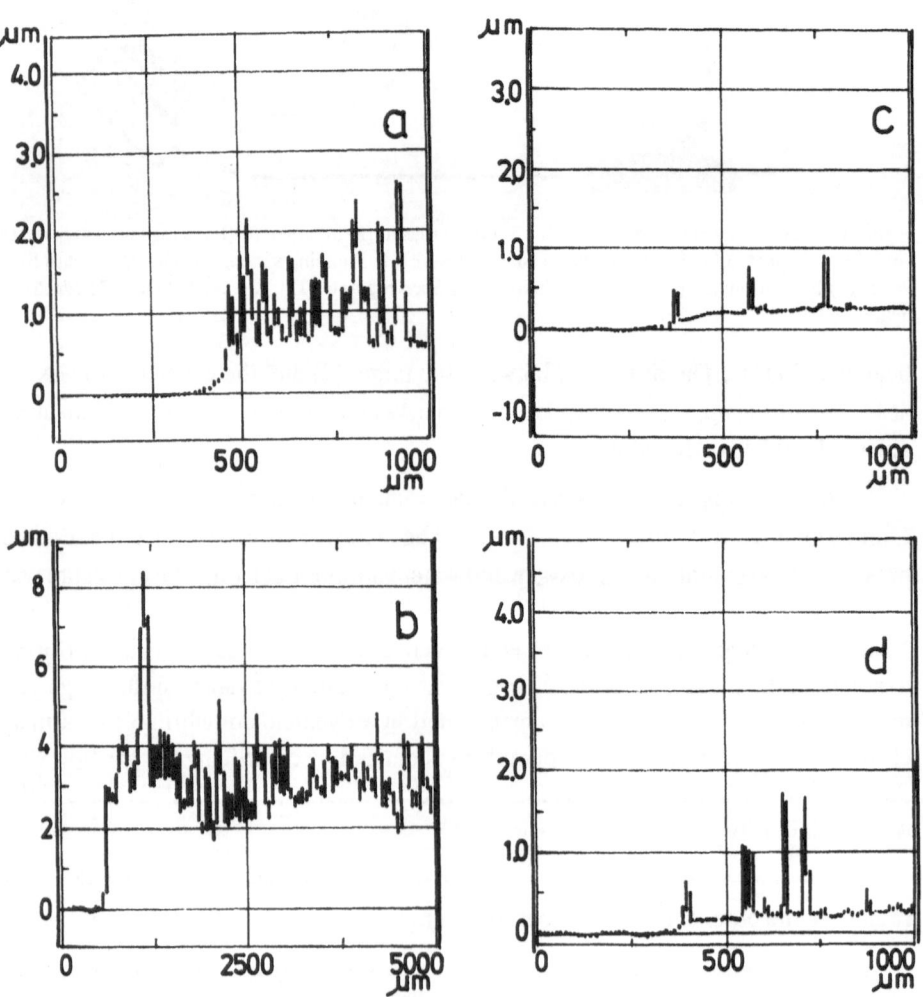

Figure 2. Profilometer recordings for the films deposited at: a) d = 14 mm, p = 1 mbar; b) d = 14 mm, p = 10^{-3} mbar; c) d = 30 mm, p = 1 mbar; d) d = 30 mm, p = 10^{-3} mbar.

3. Results:

3.1. DEPOSITION PROFILE:

Typical profilometer recordings of the deposited films are presented in Figs. 2a-d. The profilometer recordings evidenced the following characteristics of the deposited films:

- At a given distance, no variation of the film thickness with the gas pressure p could be evidenced (compare Fig. 2a with Fig. 2b and Fig. 2c with Fig. 2d).

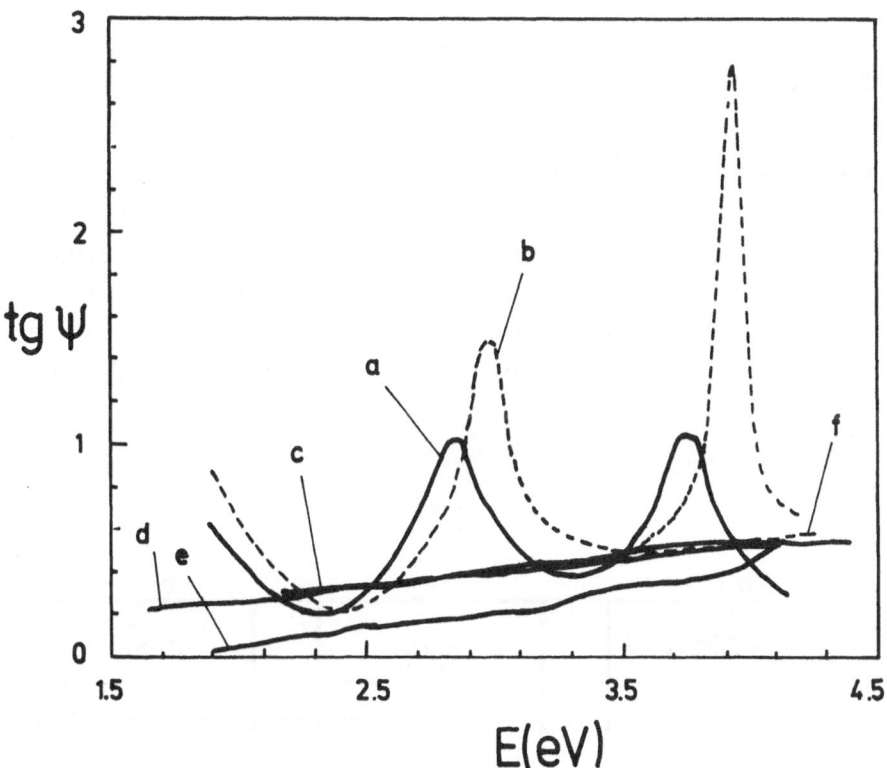

Figure 3.   Spectroscopic ellipsometry recordings: (a) Experimental data corresponding to the sample deposited at $p = 1$ mbar and $d = 30$ mm. This curve coincides with the best fit, obtained with the structure in Fig. 4a. (b) The curve computed for the structure in Fig. 4b, consisting of a Si wafer substrate coated with a 3 nm SiO_2 layer and a 280 nm $a\text{-}Si_3N_4$ layer. (c) Experimental data corresponding to the sample deposited at $p = 10^{-3}$ mbar and $d = 21$ mm. (d) Experimental data corresponding to the sample deposited at $p = 10^{-3}$ mbar and $d = 30$ mm. (e) Experimental data corresponding to the sample deposited at $p = 10^{-6}$ mbar and $d = 21$ mm. (f) The curve computed for the structure in Fig. 4c, consisting of a Si wafer substrate coated with a 3 nm SiO_2 layer and a 280 nm a-Si layer.

- The thickness of the films strongly varies with the target - collector distance. Its value is in the range 2 - 3 μm for $d = 14$ mm, 1 - 2 μm for $d = 21$ mm and in the range 0.2 - 0.3 μm for $d = 30$ mm.

- Peaks with various dimensions are observed on the surface of the films. The density of these peaks is larger at low target - collector distances (Figs, 2a, b), decreasing significantly with increasing distance (Figs. 2c, d). No variation with the gas pressure could be evidenced.

3.2. SPECTROSCOPIC ELLIPSOMETRY STUDIES:

The ellipsometric measurements were made with a 2 mm diameter sampling spot. The sampling spot was placed on different homogeneous zones on the surface of each sample.

In Fig. 3 we present the SE curves (a, c, d, e) recorded for samples deposited at different values of p and d.

The experimental data were processed with the "Ellipsometric Simulation Program" version 4.01 of Allan R. Heyd (1988).

The best fit of the data corresponding to the sample deposited at $d = 30$ mm and $p = 1$ mbar (curve a) was obtained with the structure presented schematically in Fig. 4a. This structure consists of a crystalline Si wafer substrate coated with a 3 nm SiO_2 layer and a 280 nm Si_3N_4 layer containing 2.3 % a-Si. The 3 nm SiO_2 layer is the natural oxide covering a Si wafer. We note that the thickness of the deposited layer

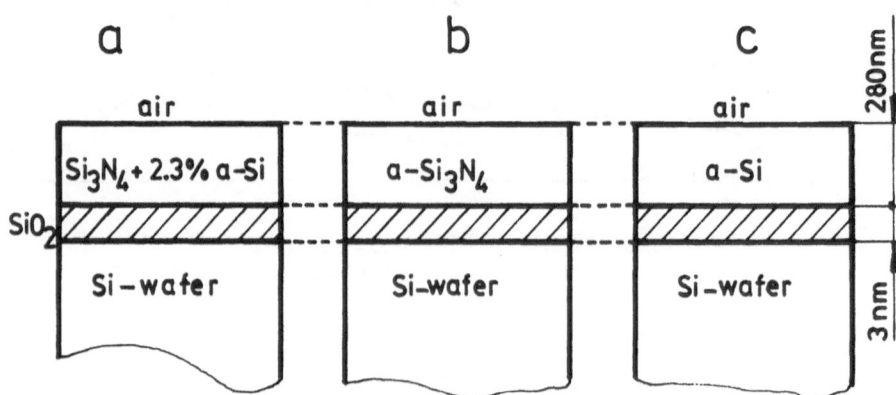

Figure 4.   Schematic representation of: a) the structure corresponding to the best fit of the experimental SE data recorded for the sample deposited at p = 1 mbar and d = 30 mm, b) the structure for which was computed curve (b) in Fig. 3, c) the structure for which was computed curve (f) in Fig. 3.

matches very well the value obtained by profilometry (see Fig. 2d). The value of the refractive index of the deposited layer is $n_{ex} = 2.6$ at $\lambda = 547$ nm.

For comparison, we have simulated (with the above-mentioned program) two other structures schematically presented in Figs. 4b and c. The second layer of the structure in Fig. 4b consists of pure Si_3N_4. For this structure we have obtained the curve (b) in Fig. 3. Curve (f) in Fig. 3 was obtained for the structure in Fig. 4c having a second layer of pure a-Si. We note that the experimental curve (a) (Fig. 3) is similar to the curve (b). The experimental SE curves corresponding to the samples deposited at $p = 10^{-3}$ mbar and $d = 21$ mm (c) or $d = 30$ mm (d) are almost identical to the curve (f) computed for the structure in Fig. 4c. We note that the SE curve (e), recorded for the sample deposited at $p = 10^{-6}$ mbar and $d = 21$ mm, is significantly different from the curves (a) and (b).

3.3. ELECTRONIC MICROSCOPY STUDIES:

The TEM and SAED investigations evidenced that the deposited films are mainly amorphous. Three amorphous phases are present in close vicinity, practically mixed, on distances of the order of $10\,\mu$m. The relative amount of each phase varies with the deposition conditions. Droplets with dimensions up to $1\,\mu$m are present in the films.

The SAED patterns corresponding to films deposited in various conditions were analyzed in comparison with the SAED patterns of 3 standards: a-Si (prepared by thermal evaporation in high vacuum) and a-Si_3N_4 (prepared by LCVD). The positions of the diffraction maxima in the SAED pattern of each amorphous phase were compared.

A first amorphous phase was identified to be present in all deposited films. A typical SAED pattern of this phase is presented in Fig. 5a. It has two maxima centred at 3.22 nm^{-1} and 7.14 nm^{-1}, very close to the maxima of the SAED pattern of the a-Si_3N_4 standard centred at 3.12 nm^{-1} and 7.41 nm^{-1}, respectively (Fig. 6a). The relative amount of this amorphous compound increases with the pressure p of NH_3. At $p = 1$ mbar, only the pattern of this phase could be evidenced in the SAED image.

Amorphous Si is also present in the deposited films. A typical SAED pattern of a zone containing this phase is presented in Fig. 5b. It is identical to the pattern in Fig. 6b corresponding to the a-Si standard. The two maxima are centred at 3.17 nm^{-1} and 5.618 nm^{-1}, respectively. We note that in the film deposited at $p = 10^{-6}$ mbar and $d = 21$ mm, only this phase was identified. The relative amount of a-Si phase is decreasing when increasing the pressure p.

The pattern corresponding to a third amorphous phase was identified in the films deposited at low and intermediate pressure. It has two maxima centred at   3.17

816

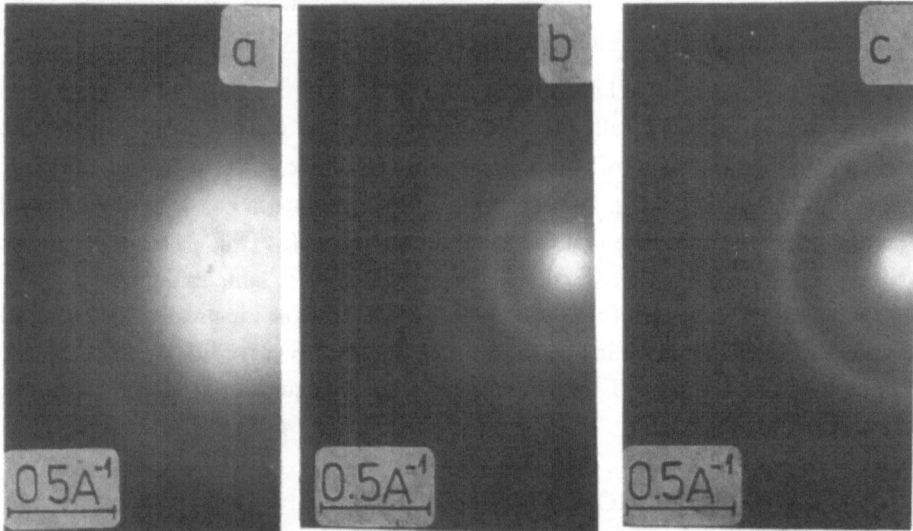

Figure 5. Typical SAED patterns of the phases present, in close vicinity, in the deposited films: a) a-Si₃N₄ (in the film deposited at $p = 1$ mbar, d = 30 mm). b) a-Si (in the film deposited at $p = 10^{-3}$ mbar, d = 21 mm). c) a-SiN$_x$, x < 4/3 (in the film deposited at $p = 10^{-2}$ mbar, d = 21 mm).

nm^{-1} and 4.87 nm^{-1}, respectively (Fig. 5c), being significantly different from the SAED patterns of the a-Si standard (Fig. 6b). We note that the SAED pattern of this phase is similar to that of a-Si₃N₄ standard (Fig. 6a). This could be an amorphous non-stoichiometric silicon nitride phase (SiN$_x$, x < 4/3).

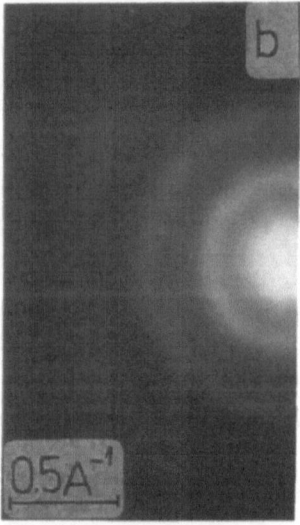

Figure 6. SAED patterns of the standards: a) a-Si₃N₄ compound deposited by LCVD b) a-Si deposited by thermal evaporation.

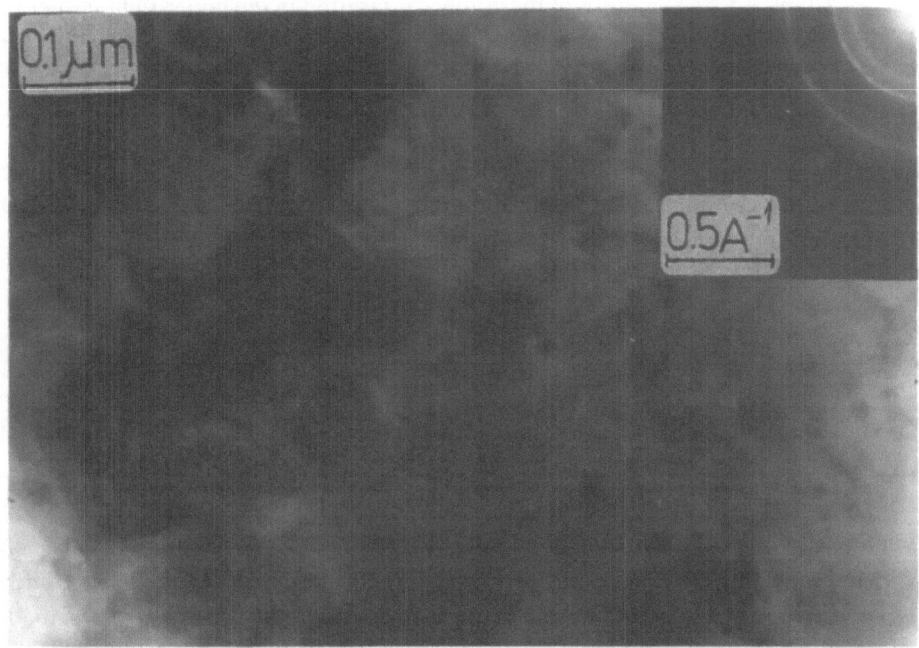

Figure 7.　Typical TEM image of a small polycrystalline island embedded in an amorphous film deposited at $p = 10^{-2}$ mbar and d = 21 mm. The corresponding SAED pattern is inserted in the upper right corner.

Small polycrystalline zones are present in the amorphous films deposited at small pressures (from 10^{-3} mbar to 0.1 mbar). As visible from Fig. 7, the crystallites within these areas have dimensions not exceeding 10 nm. The crystallinity is poor and varies within different areas on the sample.

TABLE 1.　Experimental data from Fig. 7 and ASTM data for α-Si, β-Si, β-Si$_3$N$_4$

Experimental	β - Si (Si:N) (hkl)		β - Si$_3$N$_4$ (hkl)		a - Si (hkl)	
3.29 ± 0.02						(111)
3.27 ± 0.02	3.29	(200)	3.31	(200)		
3.14					3.14	(220)
2.72						
2.68	2.69	(211)	2.668	(101)		
1.92					1.92	(311)
1.78						
1.77	1.768	(312)	1.753	(301)		
1.67					1.67	
1.59			1.59	(221)		
1.64	1.649	(400)				

A typical SAED pattern of these zones is inserted in the upper right corner of Fig. 7. From Table I, one observes that these zones have a structure very close to that of β-Si, the high pressure phase of Si. The lattice is distorted with 1 % as compared to the β-Si structure. This structure could be related to a nitrogen-stabilized β-Si phase.

Droplets are present on the deposited films (Fig. 8). We note that these droplets consist of polycrystalline α-Si only, with no traces of other phases. A typical SAED pattern is presented in the upper right corner of Fig. 8.

4. Discussion:

LRA of Si in ambient NH_3 at $p = 1$ mbar and $d = 30$ mm results in the deposition of amorphous Si_3N_4 films, as proved by spectroscopic ellipsometry and electron micros-copy investigations. The best fit of the SE data indicates a content of 2.3 % a-Si as impurity. At $p \leq 0.1$ mbar or $d \leq 21$ mm, the deposited films mainly consist of a-Si, with low amounts of a-Si_3N_4 and a-SiN_x ($x < 4/3$). Small polycrystalline islands with a structure very close to β-Si (attributed to a nitrogen-stabilized β-Si phase) are embed-ded in the films. In all cases, droplets of polycrystalline Si only are present on the deposited films.

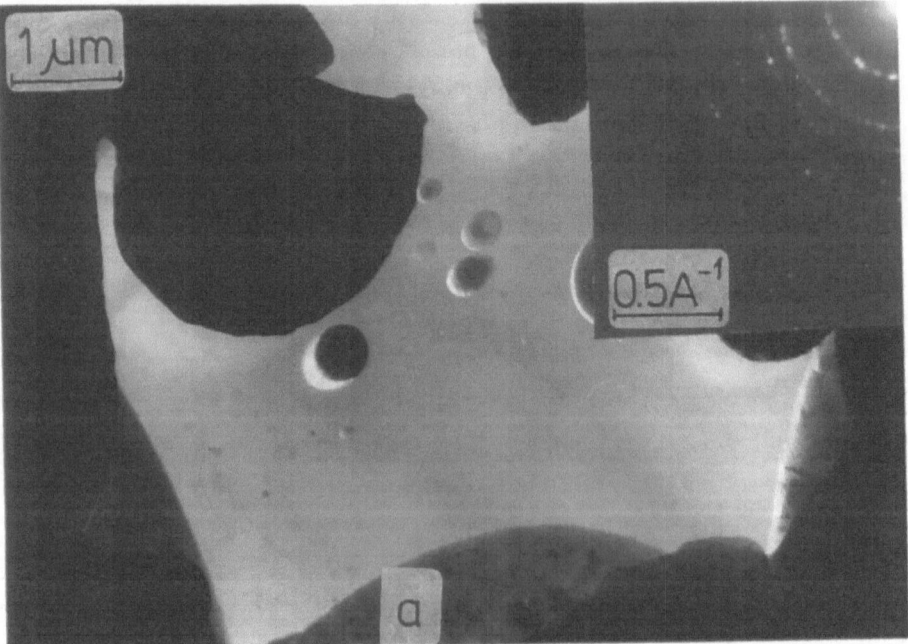

Figure 8. Typical TEM image of a droplet present in a film deposited at p = 0.1 mbar and d = 21 mm. The corresponding SAED pattern is presented in the upper right corner.

On the other hand, it was proved [1, 2] that LRA of a Ti target in N_2 ambient gas results in the deposition of a polycrystalline film of stoichiometric TiN (fcc phase). No trace of unreacted Ti or non-stoichioetric nitrides (like Ti_2N) could be evidenced. The compound presents no oxygen contamination. The droplets present on the deposited films entirely consist of the stoichiometric TiN phase. It was demonstrated [1] that the stoichiometric TiN compound forms on the Ti target.

The droplets produced during LRA of Si in NH_3 consist of unreacted polycrystalline Si only (as proved by electron diffraction). This observation leads to the conclusion that only Si is present in the molten layer from which the ablation is promoted and the droplets are insignificantly nitrided after their formation. We underline that this situation is completely different from that observed in the case of LRA of Ti in N_2 or CH_4 [1, 2, 3]

We shall present a possible explanation of these different behaviors.

Ti is a very good getter for gases. In particular, nitrogen can be dissolved in Ti bulk up to a 1:1 ratio [7]. Nitrogen atoms have a large diffusivity in Ti bulk [8]. Thus, between two laser pulses, N_2 molecules stick on the Ti target surface. The nitrogen is gettered in the bulk, leaving the surface free for new molecules to stick. At the working pressure of $p = 30$ mTorr, for $t = 100$ ms between two subsequent laser pulses, the corresponding exposure E of the Ti target surface is equal to $E = 3 \times 10^3$ L. If we assume a N:Ti ratio of 1:1 (which is also the stoichiometric ratio in the TiN compound), this exposure should provide the nitrogen to "complete" a number of atomic layers in the Ti bulk, of the order of 10^3. The corresponding thickness of the Ti layer "completed" with N atoms will be of the order of 10000 A. We note that the thickness of the layer ablated by one laser pulse is less than 1000 A.

During heating, Ti and N atoms which are in very close vicinity (at distances of the order of Ti lattice constant) can react with the formation of TiN molecules. TiN molecule has a large binding energy and the probability to be subsequently dissociated by the laser irradiation is quite low.

In conclusion, stoichiometric TiN could result during heating and ablation of a Ti layer which contains N atoms up to a ratio N:Ti = 1:1.

Oppositely, only a very low concentration of nitrogen can be present in the layer ablated from the Si target. When the target surface is in solid phase, the gas molecules can only be adsorbed. Nitrogen will not enter in the Si bulk as it does in Ti bulk. During the presence of a liquid Si layer, the exposure of the surface is of the order of 1 L. This means that only one layer of molecules could be adsorbed and eventually dissolved in the liquid.

We note that nitrogen has a very low solubility limit in silicon [8]. If the nitrogen concentration is increased over this limit by a mechanism related to a liquid layer present on the target's surface (plasma recoil action, convective fluxes in the molten layer, etc), the excess nitrogen will segregate and will be eliminated at the surface, or will precipitate during solidification. The very large cooling rate subsequent to the laser irradiation will increase the solubility limit, but with not more than 1-2 orders of magnitude.

In conclusion, during the heating and ablation of Si from the surface of the target, the nitridation reaction is very limited by the available nitrogen.

An important observation is that Si_3N_4 compound dissociates at 1100 $^\circ$C and does not exist in liquid or vapour phase [8]. We note that one cannot refer to a "Si_3N_4 molecule" which could form and travel through the gas as an individual, like in the case of TiN molecule. One can refer to Si_3N_4 only in condensed phase.

In this case, the main contribution to nitridation is expected from the reaction on the collector.

The ablated material bombards the collector's surface. The mean kinetic energy of the incident particles of 20-30 eV is sufficient to initiate a reaction with the gas molecules adsorbed on the collector's surface. We note that at a typical deposition rate of 1 A / pulse, 1 monolayer of adsorbed gas molecules is sufficient for the formation of a stoichiometric compound.

The nitridation reaction could continue during the time between two laser pulses, sustained by the thermal energy supplied by the heated collector and by the continuous exposure of the surface to the reactive gas. In this case, an improvement of the stoichiometry should be expected at high temperatures of the collector. We expect that the reaction on the collector should be more efficient in NH_3 than in N_2 due to the lower dissociation energy of the NH_3 molecule of $E_{NH_3 \to NH_2+H} = 4.78$ eV as compared to $E_{N_2 \to N+N} = 9.82$ eV.

Another argument for the nitridation on collector surface is given by the improvement of the stoichiometry (increase of the nitrogen content) at larger distances d. From the profilometer recordings, we deduce a mean deposition rate of 0.2-0.3 nm/pulse at $d = 14$ mm, 0.1-0.2 nm/pulse at $d = 21$ mm and of only 0.02-0.03 nm/pulse at $d = 30$ mm. Thus, the exposure of the suface to gas molecules before being covered with a new monolayer of ablated material is much larger at $d = 30$ mm than at $d = 14$ or 21 mm.

We should note that the nitridation reaction on the collector could be limited by the fact that the ablated substance falling on the collector's surface consists not only of ions and atoms but also of clusters and nanodroplets. There exists the possibility that a cluster or nanodroplet of Si cannot be completely nitrided. This could be a possible explanation for the presence on the collector of unreacted Si and of SiN_x ($x < 4/3$) phases.

5. Conclusions:

We have shown that Laser Reactive Ablation can be used for the one-step synthesis and deposition of silicon nitride thin films on Si wafers.

The LRA of Si in NH_3 results in the deposition of films containing several phases: Si_3N_4, SiN_x ($x < 4/3$), Si. The deposited films are mainly amorphous. The composition improves with increasing the gas pressure p and the target - collector distance d.

At $p = 1$ mbar and $d = 30$ mm, a-Si_3N_4 films with only 2.3 % a-Si (as impurity) were obtained.

Most probably, in this case, the main contribution to compound formation is given by the reactions taking place on the collector substrate.

References:

[1] I.N. Mihailescu, N. Chitica, L.C. Nistor, M. Popescu, V.S. Teodorescu, I. Ursu, A. Andrei, A. Barborica, A. Luches, M.L. De Giorgi, A Perrone, B Dubreuil, J. Hermann (1993) J. Appl. Phys, 74, pp. 5781-5789,

[2] I.N. Mihailescu, N. Chitica, V.S. Teodorescu, M. Luiza De Giorgi, G. Legieri, A. Luches, M. Martino, A. Perrone, B. Dubreuil (1993) J. Vac. Sci. Technol. A, 11 (5), pp. 2577-2582

[3] G.Leggieri, A.Luches, M.Martino, A.Perrone, G.Majni, P.Mengucci and I.N. Mihailescu (1994) Thin Solid Films (in press)

[4] F. Vega, C. N. Afonso, C. Ortega, and J. Siejka (1993) J. Appl. Phys. 74, 963.

[5] E. Fogarassy, A. Slaouy, C. Fuchs, and J. P. Stoquert (1990) Appl. Surf. Sci. 46, 195.

[6] A.Luches, G.Leggieri, M.Martino, A.Perrone, G.Majni, P.Mengucci and I.N. Mihailescu (1994) Appl. Surf.Sci. 79/80, 244

[7] Gmelins Handbook of Anorganic Chemistry, 8th ed. (1951) edited by R.J. Meyer and E.H. Erich Pietsch, Verlag Chemie, Weinheim (in German)

[8] C.V. Samsonov and I.M. Vinitskii (1980) Hanbook of Refractory Compounds, IFI/Plenum New York-Washington-London,

PULSED LASER ABLATION OF COPPER

R. Jordan, D. Cole and J. G. Lunney,
Department of Physics, Trinity College, Dublin 2, Ireland.

K. Mackay and D. Givord,
Laboratoire de Magnetisme Louis Neel, C.N.R.S., Grenoble, France.

Abstract

The laser ablation of copper with a 532 nm, 6 ns laser has been investigated in the regime normally used for pulsed laser deposition. The ablation depth per pulse and the flux and energy distribution of the ions in the plume were measured and compared to the deposition rate as measured by a quartz microbalance. These measurements were compared with an analytic model of ablation via a laser sustained plasma. It is shown that self-sputtering of the growing film is significant.

Introduction

There is growing interest in the use of pulsed laser evaporation for the deposition of thin solid films. The technique has been applied to a wide range of materials[1-3] including metals, semiconductors, insulators and superconductors. The lasers most commonly used have pulse durations of 10 -30 ns, but for metals, the value of irradiance required for significant evaporation is about the same as the threshold value for plasma formation. Thus, for the pulsed laser deposition (PLD) of metals, the plume of vapour produced is significantly ionised, and it has been shown that the ions present have energies ranging up to several hundred eV.

In order to properly control the growth conditions, it is essential to know how the ablation rate depends on the laser fluence, how the composition of the vapour is partitioned between neutral and ionised species, and what their energy distribution is, arriving on the growing film.

There is an extensive literature on the study of pulsed laser deposition [1-3]. However, more recently PLD of metals has been less extensively investigated than of high-temperature superconductor materials. Nevertheless it is an important technique for the preparation of many metallic systems e.g. intermetallic compounds or reactive or refractory metals.

The ablation of metals using excimer lasers has been studied using ion probes [4,5], emission spectroscopy [6] and laser induced fluorescence [7]. All these studies show that for values of laser fluence greater than about 2 Jcm^{-2} a plasma, with an electron temperature of several eV, is formed.

The physical description of laser ablation has been a subject of study since the invention of the laser. The basic mechanism of ablation via a laser sustained plasma provides theoretical framework for discussing laser ablation as it occurs in PLD. The physical description of laser ablation has recently been reviewed by Phipps et al. [8] and Phipps

823

J. Mazumder et al. (eds.), Laser Processing: Surface Treatment and Film Deposition, 823–829.
© 1996 Kluwer Academic Publishers.

and Dreyfus [9], and analytic expressions were derived for the electron temperature and density and the ablation depth in terms of experimental parameters.

In this paper we describe a range of measurements on the ablation of solid copper using a pulsed laser at 532 nm. The ablation depth per pulse and the energy distribution of the ions were measured, and compared with the predictions of an analytic model of laser ablation via a laser sustained plasma. At the substrate, the ion dose was compared with the deposition rate, and the extent of re-sputtering was estimated.

Experiment

The ablation measurements were made in a vacuum chamber at a pressure of 10^{-4} mbar using a frequency-doubled Nd-YAG laser operating at 532 nm and a pulse length of 6 ns. Due to the amount of material removed per pulse, oxidising effects between pulses is considered negligible. A 35cm focal length lens was used to focus the beam, at 45°, onto a target of Cu metal. The fluence (F) was varied using calibrated filters in the beam, or by moving the target relative to the focus of the lens, thus changing the size of the beam on the target. The spot size was measured by noting the dimensions of an ablated region after 100 laser pulses. Spot areas of 0.01 and 0.003 cm^2 were used, and knowing the energy on target the average fluence was calculated. The laser beam was rastered over a 5 x 5 mm^2 area using a vibrating mirror. The ablation depth per pulse was determined by measuring the mass loss from the target. A Faraday cup was used to measure the ion current and the time-of-flight of ions in the plasma plume. This Faraday cup is similar to that reported by Raven et al. [10] designed to overcome secondary electron emission by using a honeycombed collector, and has been described previously [11]. The Faraday cup was placed at 12.8 cm from the target and could be rotated about the target normal. A retarding potential analyser[12], placed at 25 cm from the target to minimise space charge effects, was used to determine the ion energy, and thus the charge-to-mass ratio. A fixed quartz crystal thickness monitor was used to measure the mass deposition rate in the centre of the ablation plume.

Results

Figure 1 shows the variation of the ablation depth per pulse with fluence, F; above 1 Jcm^{-2} this depth varies as $F^{0.66}$. Figure 2 (insert) shows the temporal variation of the ion signal for three values of fluence using a 0.01 cm^2 spot. The arrival time of the plasma for the 2.1 Jcm^{-2} case corresponds to an energy of 500 eV if the ions are assumed to have unit atomic mass. Using the retarding potential analyser it was shown that, at 1.9 Jcm^{-2} most of the ions have a charge-to-mass ratio corresponding to single ionisation and unit atomic mass. The time of flight spectra can then be converted to corresponding energy spectra and are shown in Figure 2. From such curves the average ion energy can be computed, and the variation of this with laser fluence is plotted in

Fig. 3. This tends to a power law $\bar{E} \approx F^{0.73}$. In Fig. 4 the mass deposited per pulse, as measured by the quartz microbalance, is compared with the integrated ion signal, for two different spot sizes. As expected, the deposition rate and ion dose, for a given fluence, are much lower for the smaller spot size due to the smaller area being ablated, in agreement with experiments performed by Kools et al.[14].

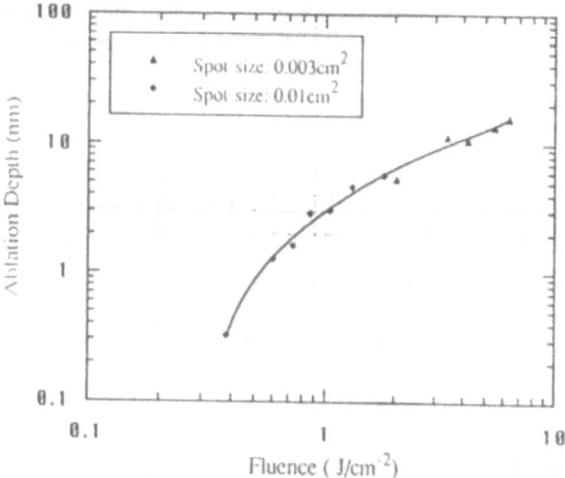

Figure 1 :   Measured ablation rate versus fluence using two different spot sizes. (Line is a guide to the eye)

At the lower values of fluence the ion dose is less than or comparable to the net deposition, while at the higher end of the measured range, the ion dose is higher than the mass deposited by about a factor 2. It was found that a negatively biased grid placed at 11 cm from the target could be used to completely suppress the ion signal on the Faraday cup. For a fluence of 1.9 J cm^{-2} a voltage of -250 V on this grid reduced the ion signal to almost zero while the deposition rate was reduced to 30% of the zero-voltage value.

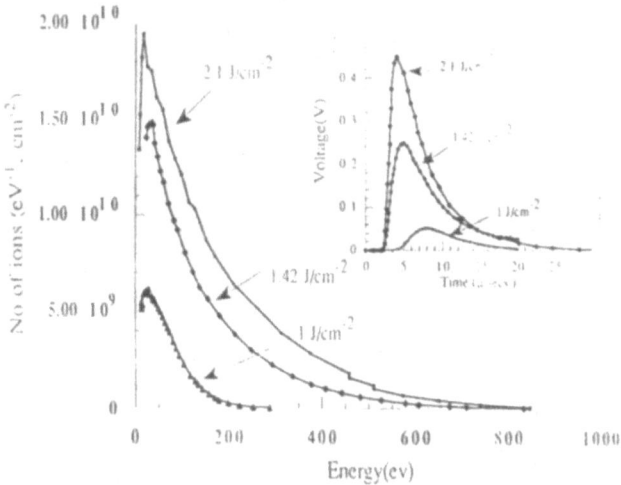

Figure 2 :   Kinetic energy spectra of ions for different fluences calculated assuming a mass/charge ratio of 1. Inset: As measured time of flight signals from Faraday cup.

Discussion

It is expected that ablation will have a rapid onset when the temperature of the heated surface reaches the boiling point, since the vapour pressure is a strong function of temperature. This temperature is achieved for a critical value of fluence and for a copper target and the laser conditions used here, this is calculated to be around 1.7 Jcm^{-2}. Fig. 1 shows detectable vaporisation at a much lower value of about 0.3 Jcm^{-2}. As the fluence is increased above this value, the density of the vapour increases and experiment shows that there is a transition to ablation via a laser produced plasma. An analytic description of such ablation [8] yields the following expressions for the electron temperature (T_e), electron density (n_e) and mass ablation rate

$$T_e = 2.98 \times 10^4 \, A^{1/8} \, (Z+1)^{-5/8} \, Z^{3/4} \, (I\lambda)^{1/2} \, \tau^{1/4} \quad K \tag{1}$$

$$n_e = 3.5 \times 10^{11} \, A^{5/16} \, Z^{-1/8} \, (Z+1)^{-9/16} \, I^{1/4} \, \tau^{-3/8} \, \lambda^{-3/4} \quad cm^{-3} \tag{2}$$

$$m = 2.66 \times 10^{-6} \, A^{-1/4} \, \psi^{9/8} \, (I/\lambda)^{1/2} \, \tau^{-1/4} \quad g \, cm^{-2} \, s^{-1} \tag{3}$$

where A is atomic weight, Z is the average ion charge, $\psi = 0.5A[Z^2(Z+1)]^{-1/3}$, I is the laser intensity in W/cm^2, λ(cm) is the wavelength and τ (s) is the laser pulse duration. Eqs. (1) and (2), together with a Local Thermal Equilibrium (LTE) description of the ionisation balance, predicted that for 2 J/cm^{-2} the plasma near the target has an electron temperature of 3.8 eV, electron density of 1.9×10^{20} cm^{-3} and an average ion charge of 1.5.

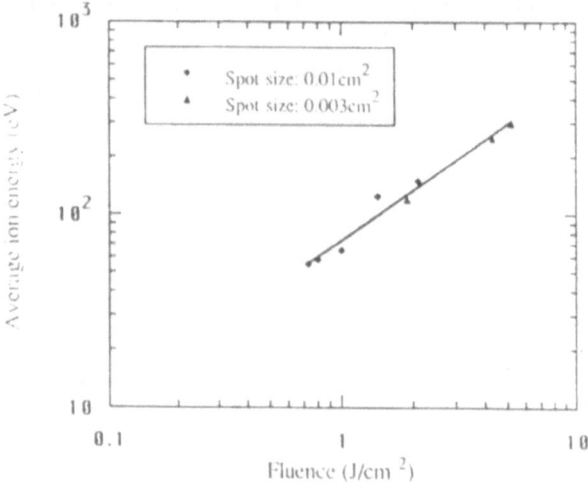

Figure 3 :   Average kinetic ion energy versus fluence.

When the plasma expands adiabatically most of the energy will be carried by the more massive ions, and the average energy $(\bar{E})$ of these ions will given by

$$\bar{E} = 3.33\ (Z+1)kT_e = 32\ eV$$

for $Z=1.5$ and $T_e = 3.8$ eV. This is substantially lower than the value from Fig. 3 of 150 eV for 2 Jcm^{-2}. The ablation depth per pulse can be estimated as $m\tau/\rho$; and for 2 Jcm^{-2} the expected value is 45 nm which is much larger than the value of 6 nm shown in Fig. 1. However above this fluence, the ablation depth varies as $I^{0.8}$ and tends to $I^{0.5}$ at higher fluences, as expected from the model. Thus it seems that the plasma temperature is higher, and the ablation depth lower, than predicted by the model. Two likely causes of this discrepancy can be identified. Firstly it was noted that, because of longitudinal mode beating, the 6 ns laser pulse actually comprised three 1 ns pulses separated by about 1 ns. A higher temperature and lower ablation depth is expected for a succession of such pulses as compared to a single longer pulse. A second cause may be non-uniformity of the intensity distribution in the laser beam giving a higher effective intensity over a smaller effective area. Laser burn marks did show the beam to be non uniform, though the level of non-uniformity was not quantified.

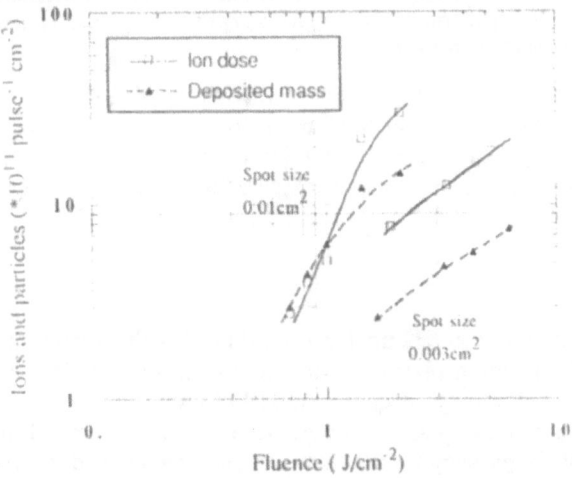

Figure 4 :   Ion dose and deposited particles (as measured by a quartz microbalance) versus fluence for two different spot sizes. (Lines are a guide to the eye)

In Fig. 2 it can be seen that even at fluences of 2 Jcm^{-2}, the ion energies range up to 700 eV. It is thus necessary to consider the effects of re-sputtering of material from the substrate by the incoming energetic ions. Using the sputtering yield formula of Zalm [12] it is estimated that the fraction of this ion distribution actually deposited is only 0.38 at this fluence. Fig. 5 is a replot of Fig.4, with all points normalised to the 0.01 cm^2 spot size, and shows the **deposited** ion dose as a function of fluence calculated by this method, compared to the mass deposited. These two curves agree very

828

well. Van de Reit et al. [13] have also found reasonable agreement between the above formula and experiment.

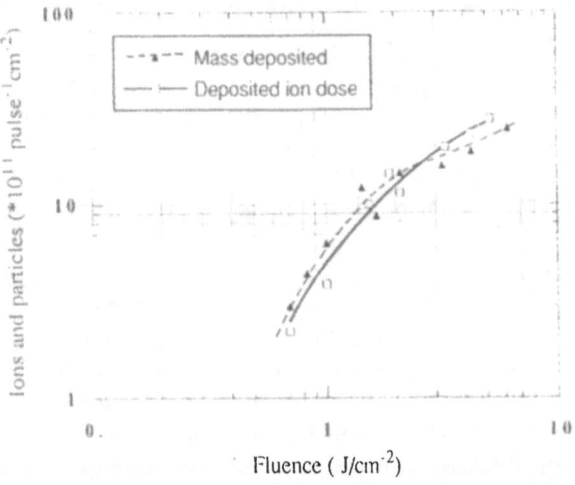

Figure 5 : Replot of fig. 4 with normalisation to account for different spot sizes (see text). Here the ion dose is also corrected by (1- the sputtering yield) to give a *deposited* ion dose. (Lines are a guide to the eye)

At low fluence, the deposited ion dose is less than the total deposit (0.67 at 1 J/cm^2) which is consistent with the neutral fraction of the total deposit being 0.33 as found from the ion repulsion experiments. There are however, some discrepancies still at higher fluences perhaps due to multiply charged species at these higher temperatures.

Conclusion

The laser ablation of copper at 532 nm has been studied with particular emphasis on the ionic species in the ablation plasma. The ablation depth is lower, and average ion energy higher than predicted by an analytic model of laser ablation, which we ascribe to the temporal structure and spatial inhomogeneity of the laser beam. It has been shown that re-sputtering of the growing film is significant and accounts for the large differences between the measured deposition rate and the ion dose seen at higher fluences.

Acknowledgements

The authors would like to acknowledge the following contributions to the work: Universite de Grenoble for supporting the visit of J.G.Lunney to Laboratoire Louis Neel, P.David for technical assistance, the European Union who supported the work as part of the BRITE-EURAM ALADIN research contract and R.Dreyfus for some helpful discussion.

References

1   J. C. Miller and R. F. Haglund, Jr. (Eds.), Laser Ablation, (Springer-Verlag, Berlin, 1991).
2   E. Fogarassy and S. Lazare (Eds.), Laser Ablation of Electronic Materials, (North-Holland, Amsterdam, 1992).
3   Chrisey and Hubler (Eds.), Pulsed Laser Deposition of Thin Films, (J. Wiley, in press).
4   R. J. Gutfield and R. W. Dreyfus, Appl. Phys. Lett. 54 (1989) 1212.
5   P. E. Dyer, Appl. Phys. Lett. 55 (1988) 1630.
6   G. Mehlman, P. G. Burkhalter, J. S. Horwitz and D. A. Newman, J. Appl. Phys. 74 (1993) 53.
7   R. W. Dreyfus, J. Appl. Phys. 69 (1991) 1721.
8   C. R. Phipps et al., J. Appl. Phys. 64 (1988) 1083.
9   C. Phipps and R. W. Dreyfus, "Laser Ablation and Plasma Formation", in Laser Microprobe Mass Analysis (Wiley, ?)
10   A. Raven, P. T. Rumbsy and J. Watson, Rev. Sci. Instrum. 51 (1980) 351.
11   K. Mann and K. Rohr, Laser and Particle Beams 10 (1992) 435.
12   P. C. Zalm, J. Vac. Sci. Technol. B2 (1984) 151.
13   E. van de Riet, J.C.S. Kools, and J. Dieleman, J. Appl. Phys. 73 (1993) 8290
14   J.C.S. Kools, T.S. Baller, S.T. De Zwart, and J. Dieleman, J.Appl. Phys. 71 (1992) 4547

ADAPTIVE BEAM DELIVERY
FOR CO_2 LASER MATERIAL PROCESSING

M. GEIGER[1], N. NEUBAUER[2], P. HOFFMANN[2] AND
J. HUTFLESS[1]
[1] *Chair of Manufacturing Technology*
Egerlandstr. 11, 91058 Erlangen,
University Erlangen-Nuremberg, Germany

AND

[2] *Bayerisches Laserzentrum*
Haberstr. 2, 91058 Erlangen, Germany

Abstract. Adaptive optics are known as a powerful method of optimizing
the working quality in CO_2 laser material processing. The integration of two
adaptive mirrors with variable focal length in the beam delivery system of
a laser machine allows the independent adjustment of focus radius and
position within the physical limits defined by the propagation properties
of laser beams. During the layout of a beam delivery system with adaptive
optics these limits are calculated. The determination of the range of focus
radius and position which can be realized by the adaptive beam delivery
system is possible. The process-optimized adjustment of the characteristic
quantities of the focussed laser beam is explained for the example of laser
ablation.

1. Introduction

An important factor for the successful use of CO_2 laser technology in industrial applications is the exact adjustment of the characteristic quantities of
the focussed beam and an optimum adaption to the specific process. Using
conventional beam guiding systems with mirrors of fixed focal length, the
accuracy of the adjustment of the focus parameters is limited by the propagation properties of laser beams. A well-known problem is the variation

J. Mazumder et al. (eds.), Laser Processing: Surface Treatment and Film Deposition, 831–843.

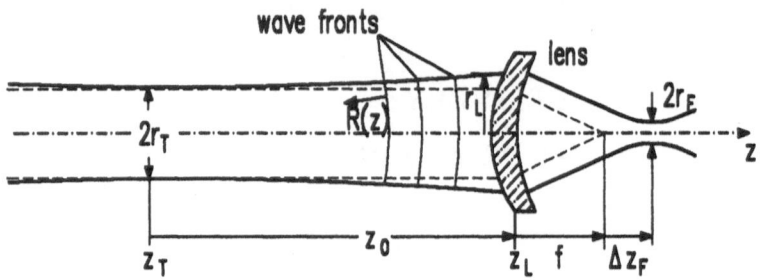

z_T : position of the beam waist

r_T : radius at the beam waist

$R(z)$: radius of curvature of the wavefront

r_L : beam radius on the lens

z_L : position of the lens

z_0 : distance between beam waist and lens

f : focal length of the lens

Δz_F : focus shift

r_F : focus radius

Figure 1. Propagation and focussing of a Gaussian beam.

of the focus radius over the working area in systems with varying beam path lengths due to the so-called flying optics [1]. Especially the properties flexibility and accuracy of a laser machine can be improved by the use of new components for beam delivery and beam forming.

A versatile system component is the adaptive optics, a mirror with variable focal length. The on-line control of the focus geometry by a beam delivery system with coupled adaptive mirrors is described in [2, 3]. The resulting focus geometry is dicussed for different focal lengths of the adaptive mirrors in a specific beam delivery system. In this paper the different integration strategies for the adjustment of the focus radius and the focus position in CO_2 laser material processing are discussed. The presentation of methods for the calculation of the required focal lengths of the adaptive optics and the estimation of constraints given by physical limits allow the design of an optimized beam delivery system.

Examples for new processing technologies using adaptive optics in laser beam welding and cutting have been previously discussed. In [4] an intelligent processing head with adaptive optics for the flexible laser beam welding of spatially formed workpieces is proposed. Experimental results for the improved performance of laser cutting machines are given in [5]. Here an example for the laser ablation process is given. By a process and workpiece optimized focussing using adaptive optics new processing strategies are possible.

2. Fundamentals on laser beam propagation

The propagation laws for laser beams are described in [6]. A short review of the equations which are important in the context of adaptive optics is given. The characteristic quantities of a Gaussian beam are depicted in fig. 1. The evolution of the beam radius r(z) and the radius of curvature of the wavefront R(z) are given by:

$$
\begin{aligned}
r(z) &= r_T \cdot \sqrt{1 + (z - z_T)^2 / z_R^2}, \\
R(z) &= (z - z_T) \cdot \left(1 + \frac{z_R^2}{(z - z_T)^2}\right), \\
z_R &= \pi \cdot r_T^2 / \lambda
\end{aligned}
\tag{1}
$$

where z is the axis of propagation, r_T the radius at the beam waist, z_T the waist position, z_R the Rayleigh length and λ the wave length. The focussing of the Gaussian beam with a lens of the focal length f_L at the position z_L leads to a focus radius r_F at the distance Δz_F from the back focal plane of the lens. Focus radius r_F and focus shift Δz_F are found by:

$$
\begin{aligned}
r_F &= \frac{r_T \cdot f_L}{\sqrt{(z_0 - f_L)^2 + z_R^2}} \approx \frac{\lambda \cdot f_L}{\pi \cdot r_L}, \\
\Delta z_F &= \frac{(z_0 - f_L) \cdot f_L^2}{(z_0 - f_L)^2 + z_R^2} = -\frac{f_L^2}{R(z_L - f_L)}
\end{aligned}
\tag{2}
$$

The focus radius r_F is in a good approximation inversely proportional to the beam radius r_L on the lens, if the focal length f_L is small compared to the distance z_0 between beam waist and lens: $f_L \ll z_0$.

The given formulas remain valid for real laser beams with transversal resonator modes of higher order. The laser beam propagation can be described by an embedded Gaussian beam with the known propagation behaviour. For a link with the beam radius r_{real} of the real laser beam the radius of the embedded Gaussian beam simply has to be multiplied by the well known beam quality factor M^2. The following discussion is limited to Gaussian beams but can easiliy be transfered to real laser beams.

3. Aspects for the integration of adaptive optics in beam delivery systems

Adaptive optics as a one-channel-system are known from several applications in laser material processing [7, 8]. Fig. 2 shows an adaptive mirror which has been developed in a cooperation of Diehl GmbH with the application laboratory for laser material processing at the University Erlangen-Nuremberg and is manufactured and distributed by Diehl. The shape of the mirror plate with a clear aperture of 38 mm can be adjusted by a single

Figure 2. Adaptive optics for CO_2 laser beam delivery.

piezoactuator which deformes centrically the back side. This nearly spherical deformation by a maximum actuator stroke of 40 μm results in focal
lengths in a range from infinity to -1.5 m. By the integration of adaptive
optical elements in the beam guiding system of a laser machine, the focus
parameters can be adjusted in a wide range.

A model for the functional connection between the actuator stroke s_{AO}
and the resulting focal length f_{AO} has been given in [9]:

$$f_{AO} = -\frac{d_{eff}^2 + 4s_{AO}^2}{16s_{AO}} \approx -\frac{d^2_{eff}}{16s_{AO}} \tag{3}$$

d_{eff} is the effective diameter of the mirror plate of the adaptive optics. In
this model the surface of the adaptive optics is assumed as a section of a
sphere with the height of the actuator stroke. In consideration of the forces
from the clamping at the edge of the mirror plate an effective diameter is
introduced. From experiments with a high power laser this diameter has
been determined as $d_{eff} = 29mm$.

Though the effects of adaptive optics in the beam delivery system are
more complex, two important cases can be distinguished (see fig. 3), which
will be discussed in the following sections.

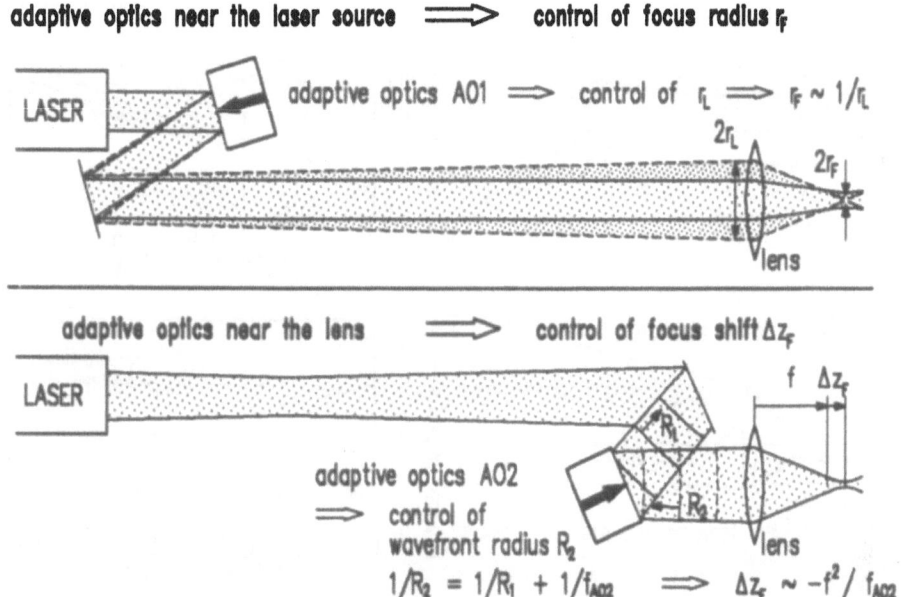

Figure 3. Control of the focus parameters by the use of adaptive optics. The integration near the laser source allows the adjustment of the focus radius. The focus position can be controlled by an adaptive optics near the lens.

Focus radius and position are regarded as independent parameters in laser material processing. For a given laser beam the focus radius is adjusted by the focussing optics focal length f_L. The focus position can be varied by a relative movement of lens and nozzle. For an optimum ease of use of adaptive optics in a beam delivery system, two different setups are aimed at, corresponding to the variation of focus radius and position.

3.1. CONTROL OF FOCUS RADIUS

In the upper part of fig. 4 the backpropagation through the lens of focussed laser beams with different focus radii and identical focus position is depicted. The distance between adaptive optics and lens for an optimum adjustment of the focus radius is given by the position of similar beam radii of the backpropagated beams. The equal beam radius at the location of the adaptive optics is demanded because after the adaptive optics, the backpropagated beams must fit the same beam, i.e. the beam of the laser source.

In fig. 4 there is no position where all of the back propagated beams have the same radius. Therefore the adjustment of the focus radius by the integration of a single adaptive optics in the beam delivery system is

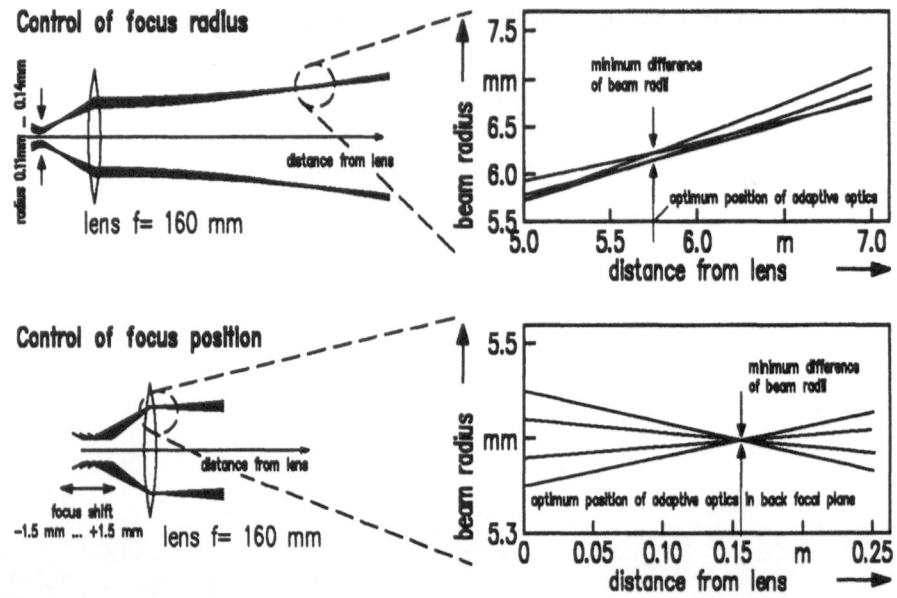

Figure 4. Backpropagation of the focussed laser beam for varying focus radius and focus position. The location with minimum difference of the backpropagated beams indicates the optimum integration position of the adaptive optics into the beam delivery system.

connected with a slight defocussing. Nevertheless it can be seen that the optimum integration position for the adaptive optics in order to adjust the focus radius is far from the focussing lens.

In fig. 3 the mode of action of an adaptive optics AO1, integrated near the laser source at a large distance from the focussing lens, is depicted. By changing the focal length f_{AO1} of the adaptive optics AO1 the laser beam is expanded or focussed causing a larger or smaller beam radius on the lens depending on the sign of f_{AO1}. From equation 2 can be concluded that the main effect of the adaptive optics AO1 in this case is the variation of the focus radius.

3.2. CONTROL OF FOCUS POSITION

It could be seen that the variation of the focus radius by an adaptive optics integrated far from the focussing optics is also connected with a focal shift. The compensation by a second adaptive optics AO2 integrated in the processing head at a short distance in front of the focussing lens is possible. The lower part of fig. 4 shows the backpropagation of focussed laser beams with identical focus radius and varying focus positions. All of the backpropagated beams have the same radius in the back focal plane of the lens, which is therefore the optimum location of an adaptive optics for the adjustment of the focus position.

In the lower part of fig. 3 another approach to the problem is depicted. The adaptive optics AO2 with focal length f_{AO2} changes the radius of curvature R_1 of the incoming wavefront to R_2:

$$1/R_2 = 1/R_1 + 1/f_{AO2}$$

If the adaptive optics is located in the front focal plane of the focussing optics, the focus radius remains constant. The focus shift Δz_F from the back focal plane of the focussing lens to the focus position is given by:

$$\Delta z_F = -f_L^2/R_1 - f_L^2/f_{AO2} \tag{4}$$

The first term is independent of the parameters of the adaptive optics AO2 and gives the focus shift of eq. 2. The second term gives a focus shift which is proportional to the refractive power $1/f_{AO2}$ or with eq. 3 to the actuator stroke s_{AO} of the second adaptive optics AO2.

The focus shift in eq. 4 is proportional to the square of the focal length f_L of the focussing optics. Therefore the focal length should be chosen as large as possible for applications where large focus shifts are required. But it should be kept in mind that according to eq. 2 for large focal lengths also an increase of the focus radius is obtained.

The effects of one adaptive optics integrated in the beam delivery system of a laser machine on the focus parameters have been discussed seperately for two different situations. In systems where focus radius and position have to be controlled, the combination of the two cases allows the independent adjustment of the focus parameters. As this system is the most common one it will be disscussed in the following section.

4. Calculation of the layout of a beam delivery system with adaptive optics

A typical setup of a beam delivery system with adaptive optics is depicted in fig. 5. The laser beam has the waist radius r_0 at the position z_0. The first adaptive optics AO1, which is installed at the fixed position z_{AO1}, transformes the beam to a new waist given by its position z_1 and radius r_1. The beam is transformed again by the second adaptive optics which is located at position z_{AO2} to the waist with radius r_2 and position z_2. Finally the focussing optics at position z_L with focal length f_L generates a focus with the radius r_F at the distance Δz_F from its back focal plane. The adaptive optics AO2 and the focussing optics are integrated into the processing head of the laser machine at a fixed distance from each other. The distance between the first adaptive optics AO1 and the processing head varies in machines with flying optics.

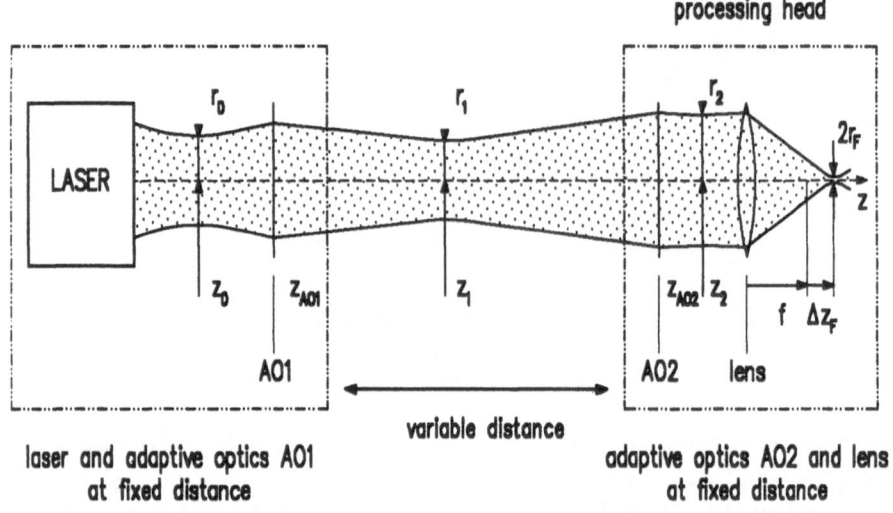

Figure 5. Typical setup of a beam delivery system with adaptive optics.

A list of constant quantities for the calculation of the appropriate focal lengths f_{AO1} and f_{AO2} for the independent adjustment of the focus parameters r_F and Δz_F is given:

z_0 position of the beam waist of the laser beam

r_0 radius of the beam waist of the laser beam

z_{AO1} position of adaptive optics AO1

f_L focal length of the focussing optics

Δz_F focus shift: distance from the back focal plane of the lens to the focus

r_F focus radius

With $z_{R,F} := \pi \cdot r_F^2 / \lambda$ (Rayleigh length of the focussed beam) position z_2 and radius r_2 of the beam between adaptive optics AO2 and lens can be calculated:

$$r_2 = \frac{r_F \cdot f_L}{\sqrt{\Delta z_F^2 + z_{R,F}^2}}, \qquad z_2 = z_L - f_L - \frac{\Delta z_F \cdot f_L^2}{\Delta z_F^2 + z_{R,F}^2}$$

With the Rayleigh length $z_{R,2} := \pi \cdot r_2^2 / \lambda$ the beam radius r_{AO2} on the adaptive optics AO2 is given by:

$$r_{AO2} = r_2 \sqrt{1 + \frac{(z_{AO2} - z_2)^2}{z_{R,2}^2}}$$

The beam radius r_{AO1} on the adaptive optics AO1 is found with $z_{R,0} :=$ $\pi \cdot r_0^2 / \lambda$ by:

$$r_{AO1} = r_0 \sqrt{1 + \frac{(z_{AO1} - z_0)^2}{z_{R,0}^2}}$$

With the knowledge of the beam radius r_{AO1} on the adaptive optics only a set of combinations of waist radius r_1 and waist positions z_1 is possible. This set is given by

$$z_1 = z_{AO1} \pm \frac{\pi \cdot r_1}{\lambda} \sqrt{r_{AO1}^2 - r_1^2} \tag{5}$$

Another constraint for the possible values of r_1 and z_1 is given by the determined beam radius on the adaptive optics AO2. The set is described by:

$$z_1 = z_{AO2} \pm \frac{\pi \cdot r_1}{\lambda} \sqrt{r_{AO2}^2 - r_1^2} \tag{6}$$

An example for the sets defined by equations 5 and 6 is depicted in fig. 6. The intersection points of the two sets give waist radius r_1 and positon z_1 of the Gaussian beams which propagate from AO1 to AO2 and have the radius r_{AO1} on the adaptive optics AO1 and r_{AO2} respectively. If the sets do not intersect the desired values r_F and Δz_F cannot be realised by the beam delivery system.

Therefore the evaluation of the sets defined by eqs. 5 and 6 allows the estimation if the desired focus parameters r_F and Δz_F can be achieved by the beam delivery system. If the sets have no intersection, the beam delivery system has to be revised to achieve the previously defined focus parameters.

After some calculations an equivalent condition for the existence of an intersection of the two sets can be found:

$$(z_{AO2} - z_{AO1}) - \frac{\pi}{\lambda} \cdot r_{AO1} r_{AO2} < 0$$

The second term is similar to a Rayleigh length. Thus the inequation compares the distance of the adaptive optics with a Rayleigh length, which is defined by the beam radii on the adaptive optics.

After the calculation of r_1 and z_1 the focal lengths f_{AO1} and f_{AO2} of the adaptive optics are found by:

$$f_{AO1} = \frac{1}{r_1^2 - r_0^2} \cdot \left(r_1^2 (z_{AO1} - z_0) - c_1 \cdot r_1 \sqrt{r_0^2 (z_{AO1} - z_0)^2 + z_{R,0}^2 (r_1^2 - r_0^2)} \right)$$
$$f_{AO2} = \frac{1}{r_2^2 - r_1^2} \cdot \left(r_2^2 (z_{AO2} - z_1) - c_2 \cdot r_2 \sqrt{r_1^2 (z_{AO2} - z_1)^2 + z_{R,1}^2 (r_2^2 - r_1^2)} \right)$$

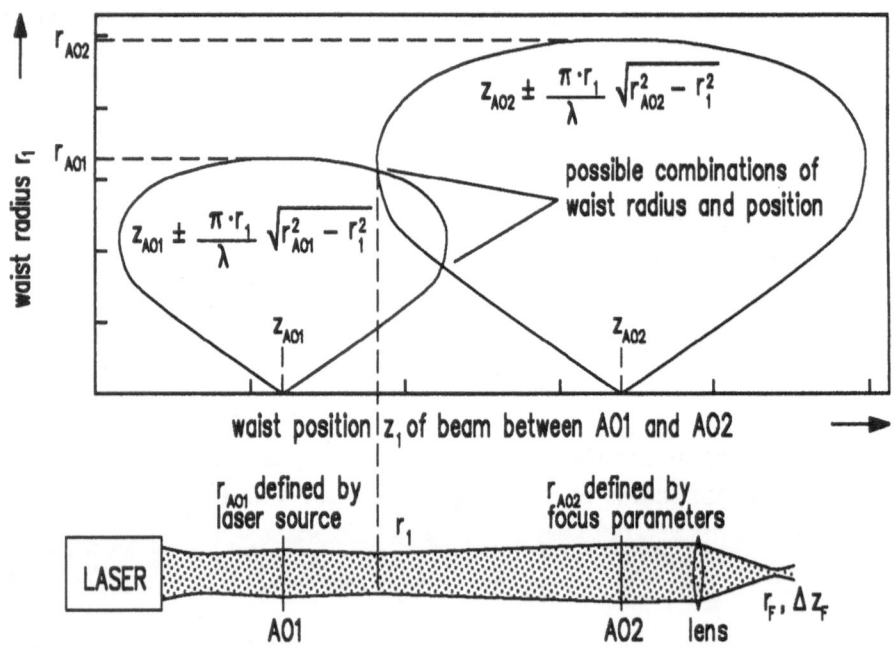

Figure 6. Possible combinations of waist radius r_1 and waist position z_1, when the beam radius r_{AO1} on the adaptive optics AO1 at location z_{AO1} and the beam radius r_{AO2} on the adaptive optics AO2 at location z_{AO2} are given.

where the constants c_1 and c_2 are defined as:

$$c_1 := \text{sign}\left(r_0 \cdot \sqrt{1 + (z_{AO2} - z_0)^2 / z_{R,0}^2} - r_{AO2}\right)$$
$$c_2 := \text{sign}\left(r_1 \cdot \sqrt{1 + (z_L - z_1)^2 / z_{R,1}^2} - r_L\right)$$

Generally the sets defined by eqs. 5, 6 have two points of intersection of which the solution with the larger beam waist radius r_1 is preferred. The use of a small beam radius is not desirable because a flying mirror between the adaptive optics could be thermally deformed thus affecting the further propagation of the laser beam or even be damaged by the higher intensity. Furthermore the surfaces of the adaptive optics have to be stronger deformed to yield the required focal lengths for the solution with the smaller value of r_1. With a stronger deformation more aberrations are introduced limiting the systems performance.

Laser machines often have varying beam path lengths between the laser source and the focussing head. Due to the propagation properties of laser beams the focus parameters are then nonconstant over the working area. The effect can be compensated by the use of adaptive optics [5]. The calculation of the adaptive optics' focal lengths has simply to be repeated for different distances accordingly. Then besides the process-optimized adjustment of the focus parameters, a compensation of the varying beam path lengths can be achieved by the implementation of the position-dependent characteristic function for the actuator strokes of the adaptive optics into the machine control.

5. Process optimized focussing in laser material processing

New machining strategies in laser materials processing due to the process-optimized focussing using adaptive optics have been presented in various applications. In [1] the improvement of the processing results in laser beam welding with a machine with flying optics could be achieved using one adaptive optics for the adjustment of the focus radius. The variation of the focus position was less than 0.5 mm and had no effects on the process. The mass-free workpiece tracking of the focus in laser beam welding at complex workpiece structures has been demonstrated [8] in a setup with one adaptive mirror integrated in the focussing head of a laser machine.

In [4] the adjustment of focus radius and position according to the technological requirements, which stem from the type of material and its thickness is, described. In combination with a height sensor the optimized processing of three-dimensional workpieces in a setup with automatic massfree workpiece tracking in laser beam cutting and welding becomes possible. By the integration of two adaptive mirrors in the beam delivery system of an industrial laser cutting machine with flying optics, focus position and radius can be kept constant [5].

Finally for the laser ablation technology it shall be shown exemplarily, that the use of adaptive optics is a real and effective enrichment for processing possibilities and leads to enhanced efficiency and working quality. State of the art is the control of the process by the parameters pulse width and energy, the feed rate and sort and pressure of assisting gases [10]. The ablation rate is proportional to the area of the focal spot and the intensity in the interaction zone. The roughness of the processed surface depends on the medium laser power. With low laser powers and small focus areas optimum roughness values are obtained at low ablation rates.

In conventional processing, the desired processing quality determines the choice of focus radius and ablation rate. The laser power, which can be used for the process, is limited by the specified value of roughness. The quantity and quality of the process is determined by the feed rate and the overlap of the processing paths.

By the integration of adaptive optics in the beam delivery system, a new processing strategy becomes possible (see fig. 7). The ablation is performed in two steps with different ablation rates and quality requirements. The

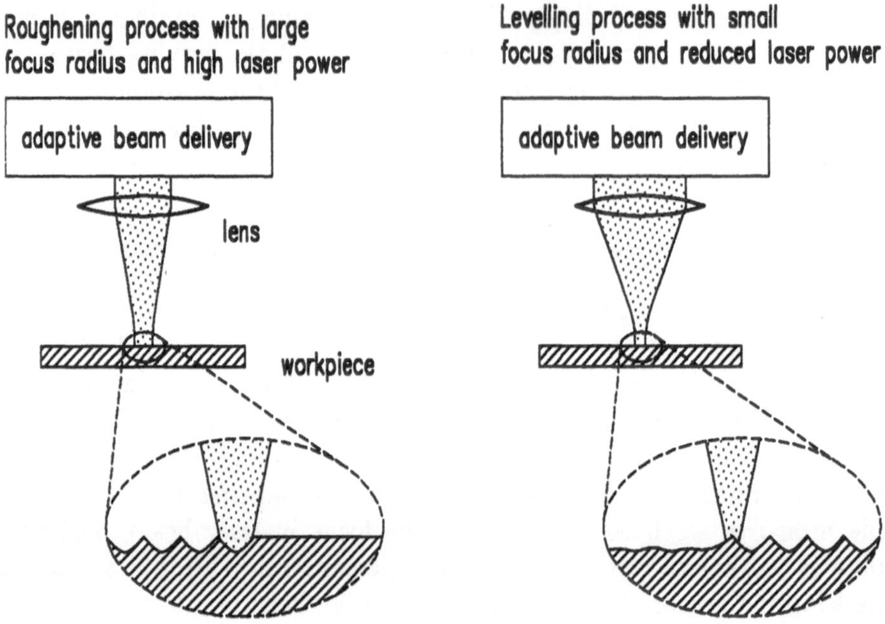

Roughening process with large
focus radius and high laser power

Levelling process with small
focus radius and reduced laser power

Figure 7. New laser ablation strategy by the use of adaptive optics.

first step is a type of roughening with high laser powers and a large focus diameter and therefore a high ablation rate. The levelling is performed with a reduced laser power and a small focus radius and minimum surface roughness. With this strategy a high processing speed in combination with a good processing quality can be obtained.

6. Acknowledgement

The investigations are supported by Diehl GmbH & Co., Nuremberg and the "Bayerische Forschungsstiftung" within the FORLAS program.

References

1. Schottelius, H.U., Geiger, M. and Hoffmann, P. (1992) Adaptive Optik für die prozeßoptimierte Fokussierung in Laserbearbeitungsmaschinen, in Waidelich, W. (ed.) *Laser in der Technik, Proceedings Laser 1991*, Springer, Berlin.
2. (1993) Bearbeitungskopf mit variabler Fokusgeometrie, in Hügel, H. (ed.) *Hochdynamische Strahlführungs- und Strahlformungseinrichtungen für die räumliche Bearbeitung mit Laserstrahlen, Sonderforschungsbereich 349*, Universität Stuttgart, Report 1990-1993, Stuttgart.
3. Bea, M., Giesen, A. and Hügel, H. (1994) On-line control of the focus geometry by coupled adaptive systems, *Laser und Optoelektronik* 26, Nr. 4, 43-49.
4. Geiger, M, Neubauer, N. and Hoffmann, P. (1994) Intelligent processing head for CO_2 laser material processing, in *Production Engineering (Annals of the German*

Academic Society for Production Engineering), Vol. I/2, Hanser, München, 93-98.

5. Neubauer, N. and Hoffmann, P. (1994) Enlarged productivity in laser material processing by the use of adaptive optics, in *Optics for productivity in manufacturing*, Proc. SPIE Vol. 2246.

6. Herziger, G. and Ripper, G. (1984) Werkstoffbearbeitung mit Laserstrahlung, Teil 5: Fokussierung von Laserstrahlung am Beispiel des CO_2-Lasers, in *Feinwerktechnik und Meßtechnik* **92**, 297-302.

7. Bea, M., Borik, S. and Giesen, A. (1990) Adaptive Optik für CO_2-Hochleistungslaser, in Waidelich, W. (ed.) *Laser und Optoelektronik in der Technik, Proceedings Laser '91*, Springer, Berlin, 446-451.

8. Hoffmann, P., Schuberth, S., Geiger, M. and Kozlik, C. (1992) Process optimizing optics for the beam delivery of high power CO_2 lasers, in *Laser energy distribution profiles: Measurements and Applications*, Proc. SPIE Vol. 1834.

9. Hoffmann, P. (1992) *Verfahrensfolge Laserstrahlschneiden und -schweißen - Prozeßführung und Systemtechnik in der 3D-Laserstrahlbearbeitung von Blechformteilen*, Hanser (Reihe Fertigungstechnik Erlangen Bd. 29), München.

10. Hügel, H. (1992) *Strahlwerkzeug Laser*, Teubner, Stuttgart.

LASER BEAM WELDING

FUME PROPAGATION AND SEAM PROPERTIES

M. DAHMEN, E.W. KREUTZ

Lehrstuhl für Lasertechnik
der Rheinisch-Westfälischen Technischen Hochschule Aachen
Steinbachstraße 15, 52074 Aachen, Germany

W. GILLNER

Institut für Neuroinformatik der Ruhr-Universität Bochum
Universitätsstraße 150, 44780 Bochum, Germany

Abstract. Flow visualization experiments are performed to examine the processing gas flow and the propagation of the fumes during laser beam welding of metals as a function of various processing variables. The influence of the processing gas flow, given by the arrangement of the gas supply, on the propagation of the fume plume is investigated. To ensure best welding performance the seam properties are evaluated. The results serve as a tool for the conceptual design of exhaust devices and as a data acquisition for computational modeling of contamination of the working area with hazardous aerosols.

1. Introduction

The most important features in materials processing with laser radiation is, on one hand, the disposal of hazardous emissions resulting from the process, and, on the other hand, to get a maximum efficiency of the process. Both aims are influenced by the process itself, the material to be processed, the workpiece, and the processing plant. Figure 1 [1] shows in a simplified way a Plat for laser beam processing and its components. The process, for example surface treatment, cutting or welding, determines the formation and the kind of emissions. Furthermore the type of handling system in conjunction with the design of the process has a significant influence on the formation, the propagation , and the decay of the plume.

845

J. Mazumder et al. (eds.), Laser Processing: Surface Treatment and Film Deposition, 845–854.
© 1996 *Kluwer Academic Publishers.*

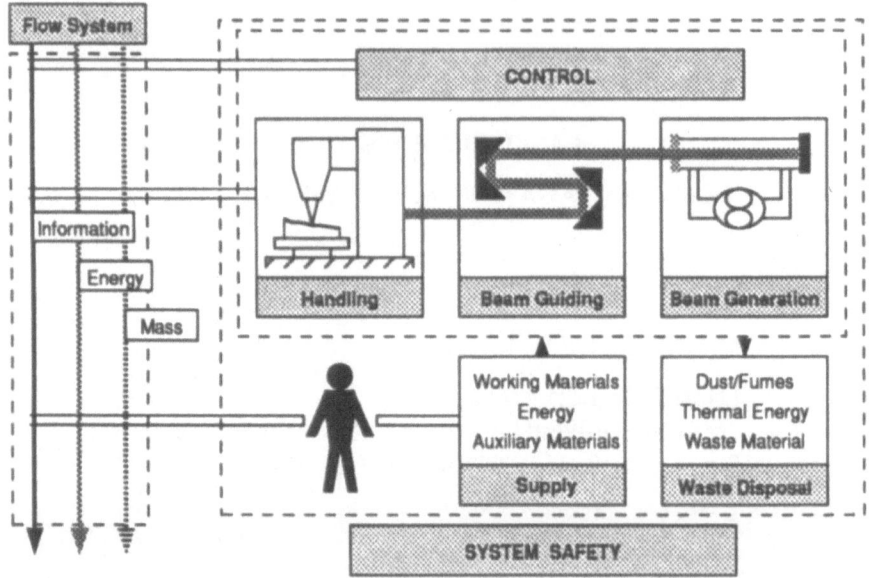

Figure 1. Schematical drawing of a laser processing plant and its compounds

From this the conclusion can be drawn that it is necessary to get first the process with a maximum efficiency to reduce the emissions as far as possible. Secondly there is a need to investigate the propagation of the emitted fumes to design an effective exhaust device [2]. In this work examinations of laser beam welding regarding the fume flow and the related seam quality in order to design the best process are presented.

2. Experimental Set-up

The experiments were carried out for partial penetration laser-beam welding at 6 mm thick non-alloyed steel St 52-3 (1.0570) in order to detect the influence of the flow of vapour and plasma out of the keyhole. A CO_2-laser with a nominal output power of 5 kW was used. The laser beam was directed by a beam guiding system using plane mirrors to a three axis handling system in cantilever construction, which allows to operate the workpiece in the X- and Y-direction as well as to move the focusing optic in the Z-direction (Fig. 2).

The laser beam is focused by an off-axis parabolic mirror with a focal length of 200 mm at an aperture F = 7.1, resulting in an focal radius $r_F = 175$ μm. Processing gas for the control of the laser-induced plasma was supplied by an external nozzle with an inside diameter of 4 mm in stitching arrangement. Helium was chosen because of its properties giving the most stable process in welding steel materials.

For flow visualization the laser light-sheet method was used parallel and perpendicular to the direction of welding. The beam of an Argon-ion laser was dispersed in one plane by a set of cylindrical lenses. The maximum power amounts to 10 W, reduced to 7.3 W in the case of separation of the line at $\lambda = 514$ nm.

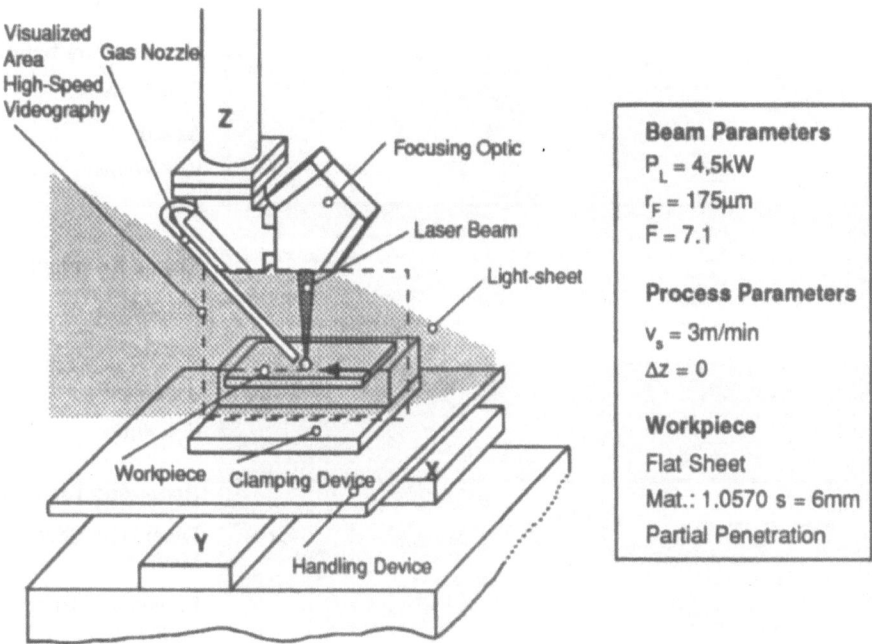

Figure 2. Experimental setup for flow visualization

The fumes are detected by recording the patterns generated by Mie-scattering at the aerosol particles. Recording was undertaken by high-speed videography with a rate of 200 pps. Furthermore photographic recording on high speed film (3200 ASA) was taken to get an insight in the mean direction and distribution of the emitted plume.

Evaluations of the welding results were done by visual inspection, metallographical examination, and measuring the seam parameters penetration depth and cross-sectional area in order to get informations about the efficiency of processing.

3. Results

3.1 FUME PROPAGATION

Figure 3 shows the propagation of the plume under conditions of different flow rates of the processing gas. The mean direction of the flow depends on the flow rate. For a helium flow of 10 l/min (Fig. 3a) the plume flows nearly parallel to the beam axis. A slight curvature in the upper part occurs since the bottom of the focusing optic acts as an obstacle to the flow. If the flow rate of the processing gas exceeds 20 l/min (Fig. 3b) a significant inclination of the plume can be observed, which increases with respect to the beam axis if the flow rate increases (Fig 3 c,d).

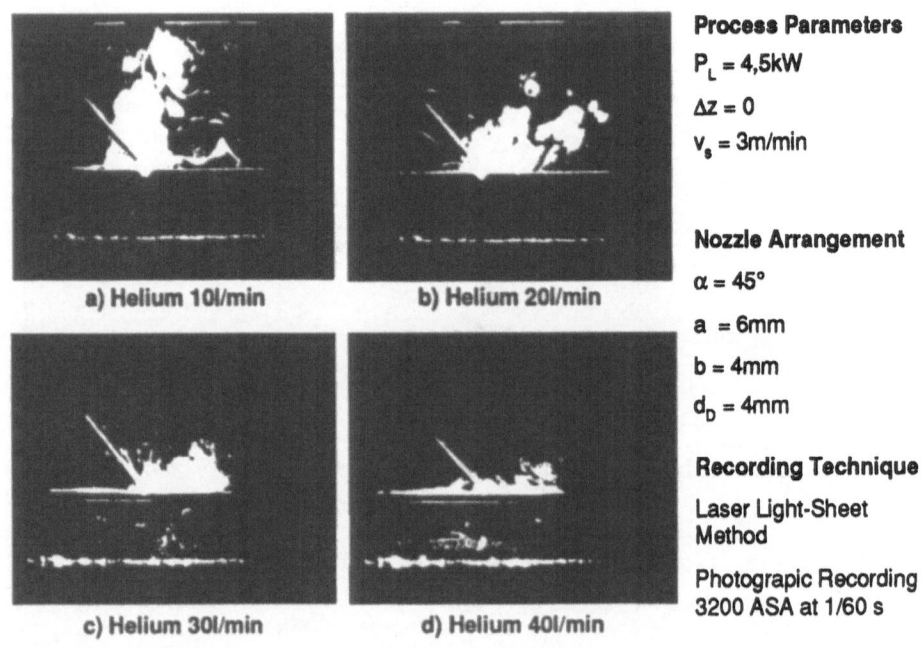

Process Parameters

$P_L = 4,5kW$

$\Delta z = 0$

$v_s = 3m/min$

Nozzle Arrangement

$\alpha = 45°$

$a = 6mm$

$b = 4mm$

$d_D = 4mm$

Recording Technique

Laser Light-Sheet Method

Photograpic Recording 3200 ASA at 1/60 s

a) Helium 10l/min   b) Helium 20l/min

c) Helium 30l/min   d) Helium 40l/min

Figure 3. Fume distribution at different flow rates at an inclination angle $\alpha = 45°$

The change of the inclination angle α from forty-five to thirty degrees causes a flattening of the plume at processing gas flow rates of 20 and 30 l/min due to the higher amount of the momentum component parallel to the sheet surface. The main effect of the inclination of the processing gas flow is that the separation of the plume from the material surface is advanced at a shorter distance L to the point of impingement with increasing inclination angle. No influence on the mean direction of the plume flow occurs at flow rates above 40 l/min (Fig. 4a).

After a closer glance at the light-sheet recordings it is seen that the flow is divided into two parts. The first one is a free jet, driven by the pressure inside the capillary, which is deflected by the momentum of the processing gas flow. The second part, parallel to the sheet surface, consists of a horseshoe vortex [3] which is generated because of the free jet acting as an obstacle on the helium flow.

The flow driven by the keyhole is decelerated by the pressure of the processing gas. Its velocity decreases from 175 m/s in the orifice plane to 12 m/s at a distance of 12 mm along the mean streamline [4]. At a distance of L = 30 mm the velocity is in the order of magnitude of 1 m/s. A vortex structure develops similar to the Kármán's vortex-street. The single vortices move with a velocity of 0.1 to 0.3 m/s. The mean deflection Θ_K decreases nearly linear with increasing inclination of the nozzle (Fig. 4b). With decreasing α the component of momentum in X-direction (parallel to the light-sheet) is reduced. Furthermore the volume in which the interaction of the jets takes place decreases for increasing α. Both are leading to a reduction of the momentum exchange and hence, to a steep inclination of the vortex street.

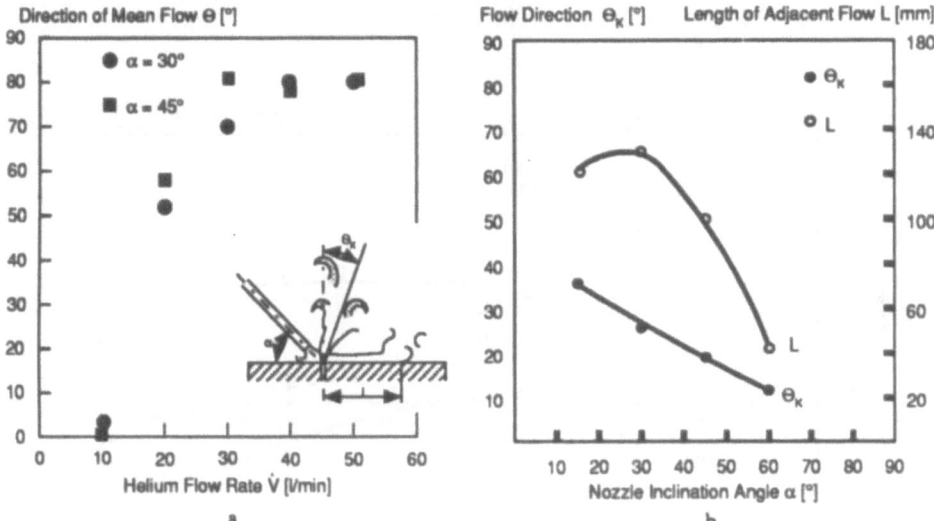

Figure 4. Flow of the fume plume

The flow parallel to the workpiece surface is characterized by the length of adjacent flow until separation occurs. The separation is influenced by two different mechanisms depending on the nozzle inclination α. If α is less or equal to 30 ° buoyancy forces are important. The processing gas is heated as it passes the interaction zone and mixed with hot aerosols emitted from the capillary. At a distance of about 130 mm buoyant flow will be dominant and the flow separates.

If α is greater than 30° a boundary-layer separation [5] occurs. After separation a vortex is formed, which swims off in a direction dependent on α. With increasing α the point of separation moves towards the interaction zone (Fig. 4b). At α = 60° both partial flows combine to one plume. This flows under an angle of 11° with respect to the axis of the CO_2-laser-beam and separates at approximately 30 mm behind the focal point. This can be explained by the energy-loss of the forward directed flow due to a high lateral flow at steep inclinations.

3.2 SEAM QUALITY

The seam quality mainly is judged by the seam geometry, i.e. penetration depth t_s and cross-sectional area A_s of the seam as a measure of the process efficiency. The results described are based on a horizontal displacement of the nozzle b with respect to the focal point of the CO_2 laser beam. Elevating the nozzle normal to the workpiece surface will lead to an increasing gas consumption, so the nozzle is fixed at a distance a = 5 mm normal to the workpiece surface. Furthermore at this distance collisions between nozzle and workpiece are avoided.

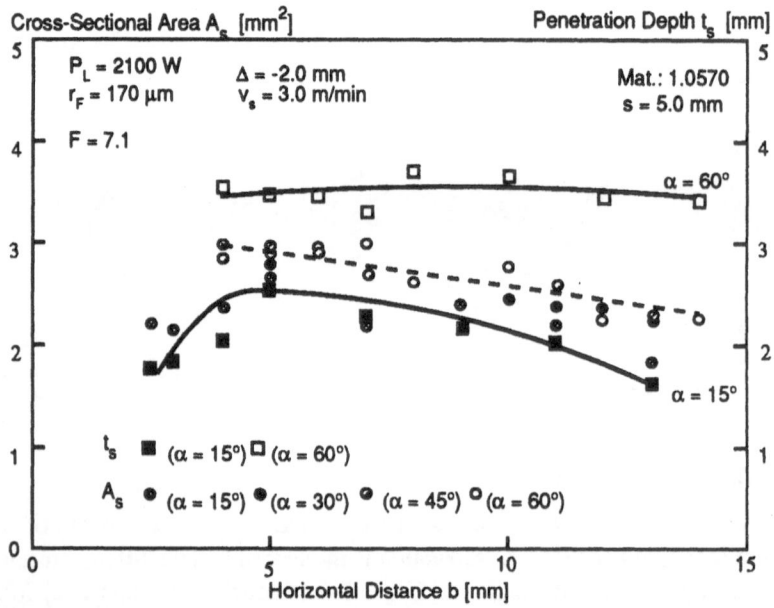

Figure 5. Seam geometry dependent on the nozzle distance

Figure 5 depicts the dependency of the penetration depth on the horizontal distance b for the extreme inclination angles 15 and 60° (full lines). The energy coupled into the workpiece is much lower at $\alpha = 15°$ than at $\alpha = 60°$, especially if the displacements are small (b $\leq$ 5 mm). The reason for this is an enhancement of the optical density of the laser induced plasma due to ambient nitrogen sucked into the interaction zone by the gas flow. These results can be confirmed by comparing the cross-sectional areas for the different angles of incidence. In the case of $\alpha = 15°$ a strong correlation between the penetration depth and the cross-sectional area was observed. For $\alpha \geq 30°$ the values follow the dotted line with an error of $\pm$ 10%. Hence the cross-sectional area is independent on α, and weakly dependent on b.

For $\alpha = 15°$ the penetration depth is strongly dependent on b. The seam is characterized by a wide and shallow shape with an aspect ration of $t_s/b_s = 0.92$. This aspect ratio is increased to 1.75 by increasing the inclination angle of the nozzle to 30°.

The penetration depth amounts to 2.25 ±0.3 mm. A further stabilizing of the process connected to an enhancement of the penetration depth to 2.9 mm is observed at $\alpha = 45°$. Maximum penetrtrion of $t_s = 3.5$ mm at an aspect ratio of 2.95 is yielded for $\alpha = 60°$.(Fig. 6) Probably the effects at steep inclination ($\alpha \geq 45°$) are evoked by the action of the vertical component of the dynamic pressure of the assist gas flow $\rho_G u^2 \sin\alpha/2$. It is assumed that the capillary is stabilized if this pressure is sufficient high. However at $\alpha = 60°$ the formation of spatters and pores was observed. This can be avoided by reducing the volume flow of the assist gas to 20 l/min at most.

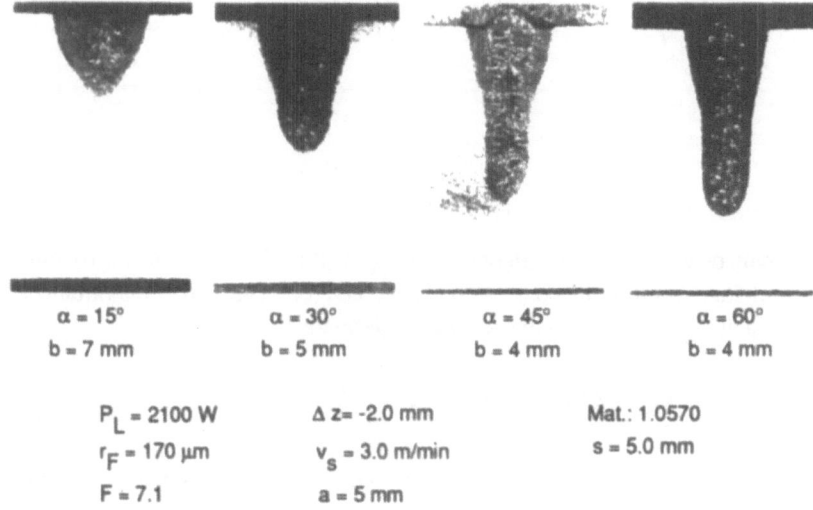

| $\alpha = 15°$ | $\alpha = 30°$ | $\alpha = 45°$ | $\alpha = 60°$ |
| b = 7 mm | b = 5 mm | b = 4 mm | b = 4 mm |

P_L = 2100 W	Δz = -2.0 mm	Mat.: 1.0570
r_F = 170 μm	v_s = 3.0 m/min	s = 5.0 mm
F = 7.1	a = 5 mm	

Figure 6. Seam geometries in cross-sectional view

4. Method of Evaluation

As mentioned before from the photographs only mean properties of the plume flow can be derived. This drawback can be mastered by recording the flow patterns by high-speed videography. A sufficient high temporal resolution is reached at a frequency of 200 pps. Fig. 7 shows the development of the flow within a time step of 0.005 s. This short exposure time (5 ms) in connection with an higher speed of the recording device (valve) allows to visualize the vortex street occurring in the partial flow driven by the keyhole and the separation structure of the horseshoe vortex (Fig. 8).

To evaluate these sequences with respect to direction and local velocities a correlation-based optical flow algorithm was utilized. This algorithm has been successfully tested for object tracking tasks on an autonomous mobile robot system. The algorithm measures the visible movement of the grey values in two-dimensional image sequences.

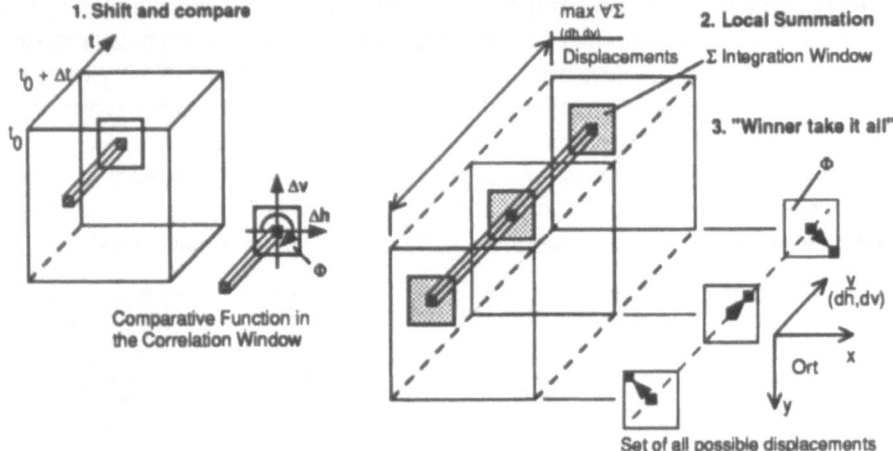

Figure 7. Scheme of the tracking algorithm

The scheme of working of the algorithm is depicted in Figure 7 and shortly outlined in the following. Within the first step "shift and compare" the maximum area of displacement D_m as a rectangular window is defined:

$$D_m = \left\{ (dh, dv) \mid (dh. dv) \in [-\Delta h, +\Delta h] \times [-\Delta v, \Delta v] \right\}$$

The correlation is defined as the absolute value of the difference in the grey-scale values:

$$m(s, y, dh, dv) = \Phi \left((I_t(x, y), I_{t+\Delta t}(x + dh, y + dv)) \right)$$
$$= |I_t(x, y) - I_{t+\Delta t}(x + dh, y + dv)|$$

The second step contains the local integration of the correlation values. A vicinity for the correlation of a certain displacement a rectangular window

$$D_M = \left\{ (ch, cv) \mid (ch. cv) \in [-\Delta ch, +\Delta ch] \times [-\Delta cv, \Delta cv] \right\}$$

is chosen. Result of the second step is a weighted correlation over the vicinity of each pixel:

$$M(x, y, , dh, dv) = \sum_{(x,y) \in D_M} m(x, y, dh, dv)$$

In the last step the maximum excitation of the resulting histogram is defined as the real displacement

$$V(x,y) = \max_{(dh,dv) \in D_M} M$$

by voting, the winner take it all:

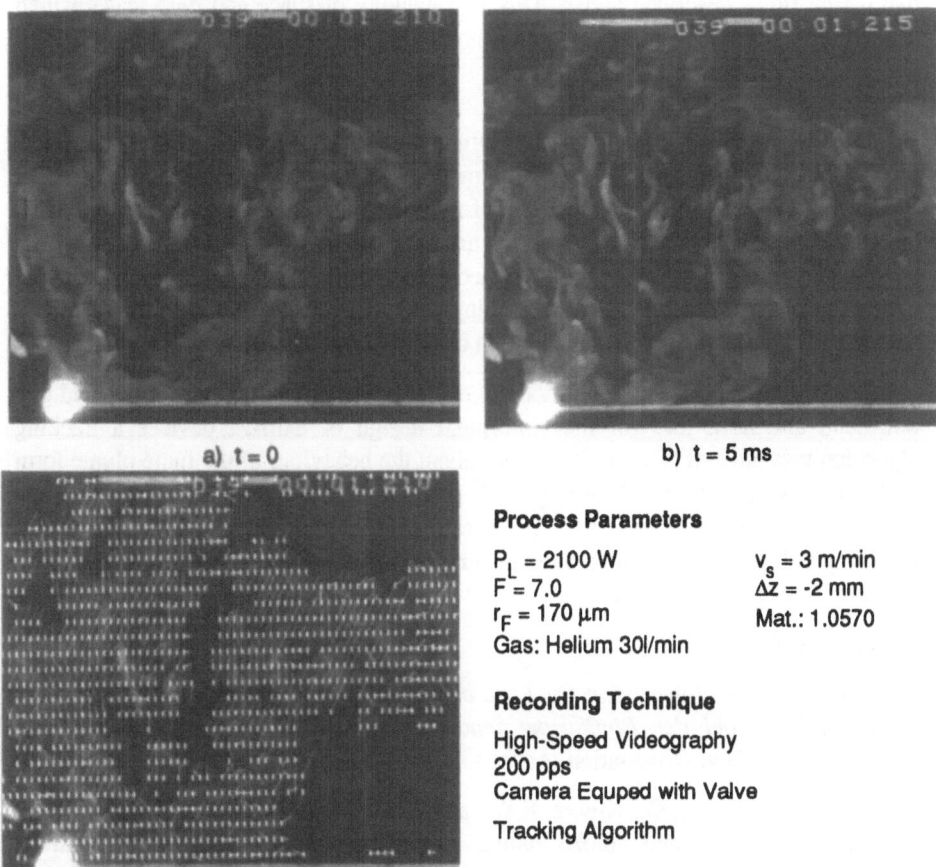

a) t = 0

b) t = 5 ms

c) Computed Displacements

Process Parameters

P_L = 2100 W v_s = 3 m/min
F = 7.0 Δz = -2 mm
r_F = 170 μm   Mat.: 1.0570
Gas: Helium 30l/min

Recording Technique

High-Speed Videography
200 pps
Camera Equped with Valve

Tracking Algorithm

Figure 8. Light-sheet videographs and result of the tracking computation

In Figure 8 one result of the application of the algorithm on light-sheet recordings is represented. The upper two videographs show the development within an time step of 5 ms duration (Fig. 8 a,b). After processing this two images a qualitative representation is generated which contains informations about the direction of the single parts of the the plume in the flow area. Based on Fig. 8a the calculated "vectors" are displayed. In the figure the nadir is represented as a white dot and the lines contains informations of the flow direction. In Fig. 8c the computed displacemtents after calculation of mean vectors, in an 8×8 vicinity for clearness, are shown. Maximum displacement window is restricted to $D \in$ [-6;6] × [-6;6] in pixel. The computation time is about 300 s. One way to reduce the calculation-time for optical computing is to transform the images into a gaussian pyramid. This preprocessing is done by lowpass-filtering and subsampling the image sequence, which results in a representation of flow fields in different coarse scales [6]. Another way to speed up the processing is a certain partitioning of the area of visible flow which restricts the area of computation to the two partial flows described above. This, and gauging of space and time scales which will lead to a real vector-representation is now work in progress.

5. Conclusions

Experiments were performed to get informations about the most safe and efficient proscess of laser beam welding. Examinations of the fume plume revealed that optimum conditions for the application of an effficient exhaust system are yielded if the inclination of the external gas nozzle amounts to 60°. The fume flow separates in one plume from the surface of the workpiece. Furthermore at this steep inclination the highest penetration depth and, correspondingly, the highes aspect ratio of the seam was observed. Besides of this, the consumption of processing gas is reduced.

For acquisition of boundary conditions in numerical simulations of the flow field one one hand and as a tool for the conceptual design of exhaust devices a tracking algorithm was utilized to get informations about the behaviour of the fume plume form high-speed videographic recordings. In primary tests informations of the flow directions was yielded. Steps of further investigations are the partitioning of the area of recording before computation and the gauging of the algorithm in order to get the hole information of the flow field.

6. References

1. Wolff, U. (1994) *Integration der Laserbearbeitung in ein Produktionssystem - Zur Vorgehensweise bei der Einführung innovativer Produktionstechnik - Fügen mit CO_2-Laserstrahlung*, Dissertation, Aachen

2. Dahmen, M., Funke, G., Kreutz, E.W., and Maischner, D. (1993) Investigations of prosessing gas flow and smoke plume propagation during laser beam welding, *Proceedings of the 2nd EUREKA Industrial Laser Safety Forum '93*, Coventry, U.K.

3. Baker, C.J. (1980) The turbulent horseshoe vortex, *Journal of Wind Engineering and Industrial Aerondynamics* 6, 9 - 13

4. Bungartz, K. (1992) *Bestimmung der Strömungsgeschwindigkeit des Metalldampfes im Bereich des Keyholes beim Schweißen mit CO_2-Lasern*, Diplomarbeit, Aachen

5. Schlichting, H. (1982) *Grenzschicht-Theorie*, Verlag G. Braun, Karlsruhe

6. Gillner,W. (1992) pyramidal optical flow computation and objekt tracking, in C. v.d.Malsburg and W. v.Seelen, *Institut für Neuroinformatik - Annual Report 1992*, Bochum

REPRODUCTION OF 3D OBJECT DISPLAYS WITH LASER SYSTEMS

S.AKKURT and M.AKKURT
Istanbul Technical University
Mechanical Engineering Faculty
Gümüşsuyu Istanbul, Turkey

ABSTRACT : In order to obtain an accurate production of mold for complex surfaces, a laser non-contacting measuring machine (CMM), integrated in a CAD/CAM/CNC system was used. In this paper the realization principles of such a system is studied.

1. INTRODUCTION

In this paper an application of laser technology in a Flexible Manufacturing System is presented. The classical manufacture of a mold product involves the following steps :

1. Design of product-CAD;
2. NC programming-CAM;
3. Mold machinning-CNC machines;
4. Control of mold.

Due to the complex surface characteristics of products (Figure 1), especially in plactics, forging, jewelry, eye wear, aerospace, toy industries, glassware, producing an accurate surface presents a serious problem. Final configurations can be produced only after a series of paintstaking, test and manual modification. For this reason producing accurate molds with this method is a time consumming trial and error process.

855

J. Mazumder et al. (eds.), Laser Processing: Surface Treatment and Film Deposition, 855–868.
© 1996 *Kluwer Academic Publishers.*

856

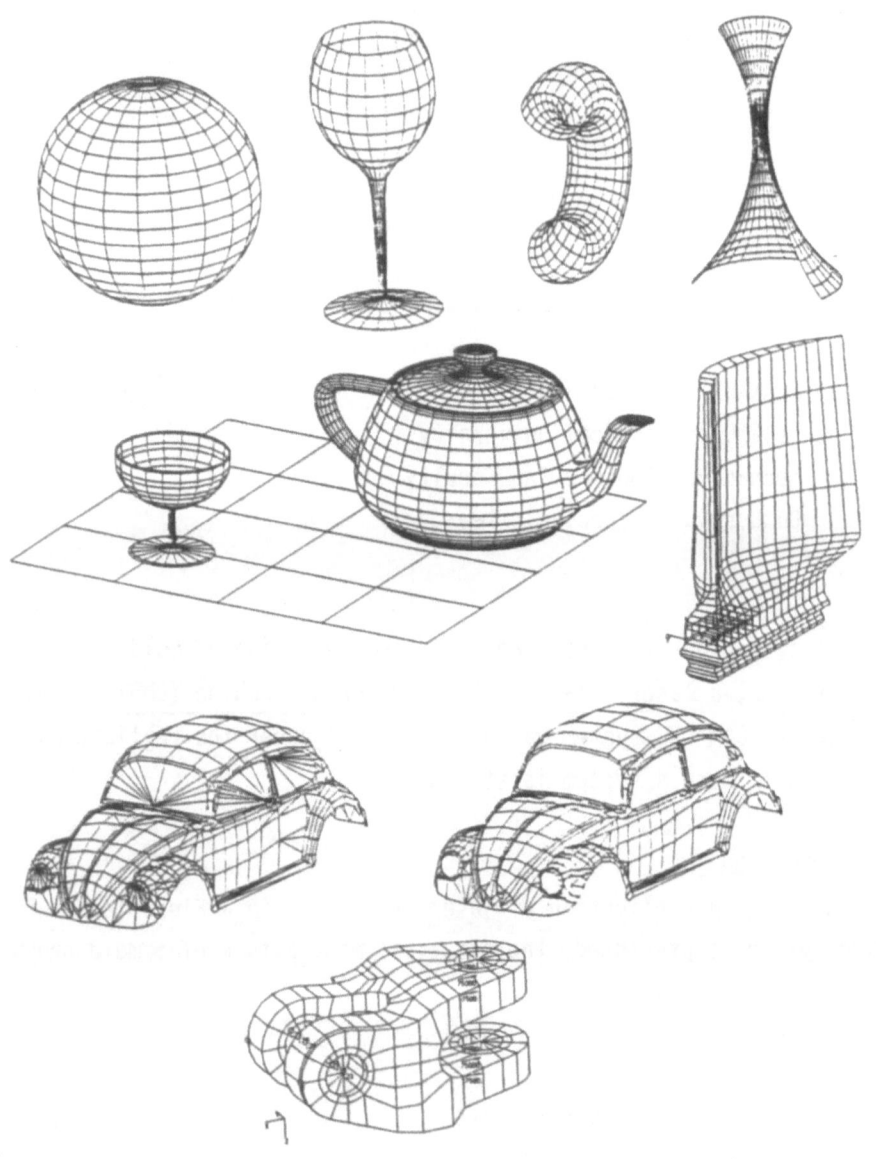

Figure 1. Complex surfaces of some products.

Another example is given in Figure 2. for the forging process. In this process a workpiece is deformed between dies by the application of compresive forces. The process itself is quite simple: starting from

very simple shape of billets (Figure 2a) we can obtain complex shape
(Figure 2b). But the production of dies are very complicate. Here in
closed-die forging, the metal flow is greatly influenced by the die
geometry. If the final component shape is complex and intricate, the

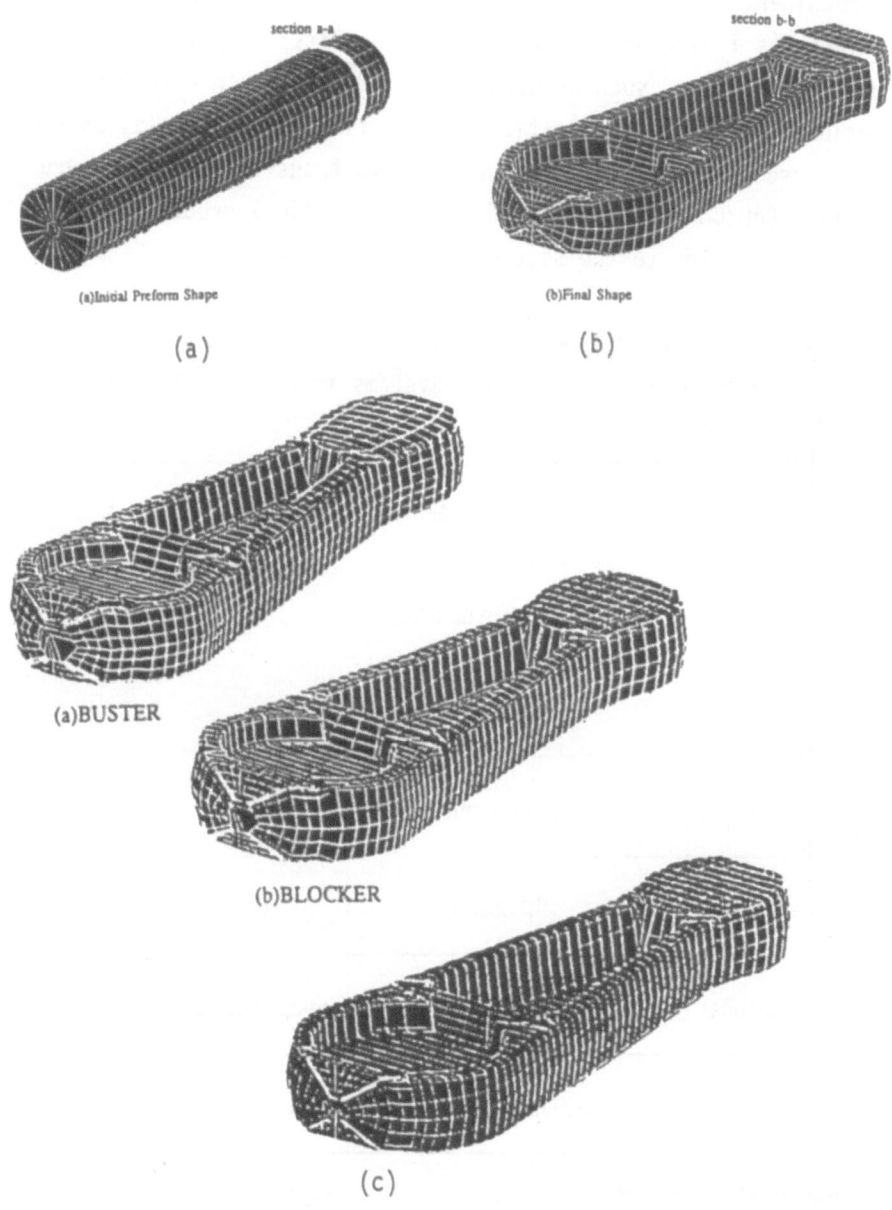

section a-a

(a)Initial Preform Shape

(a)

section b-b

(b)Final Shape

(b)

(a)BUSTER

(b)BLOCKER

(c)

Figure 2. The forging process.

billet cannot be deformed to the final shape in a single operation. The workpiece is deformed through several intermediate shapes generally referred to as the buster, blocker and finisher shapes (Figure 2c).

An intermediate solution is to utilise mechanical digitizing and/or copy machines. Mechanical systems with contact stylus probes have many disadvantages including: low accuracy, inability to digitize sharp and hard-to-reach areas, such as corners and $90°$ angles in a full three dimensions, inability to digitize soft or flexible materials, as epoxy which is used as model, and incapability to transfer data to modern production CAD/CAM systems. In addition, mechanical probes limit cutting offset capabilities to the size of the probe itselfs.

2. PRINCIPLE OF LASER MEASURING MACHINE

A classification of Laser Measuring Systems is given in Figure 3.

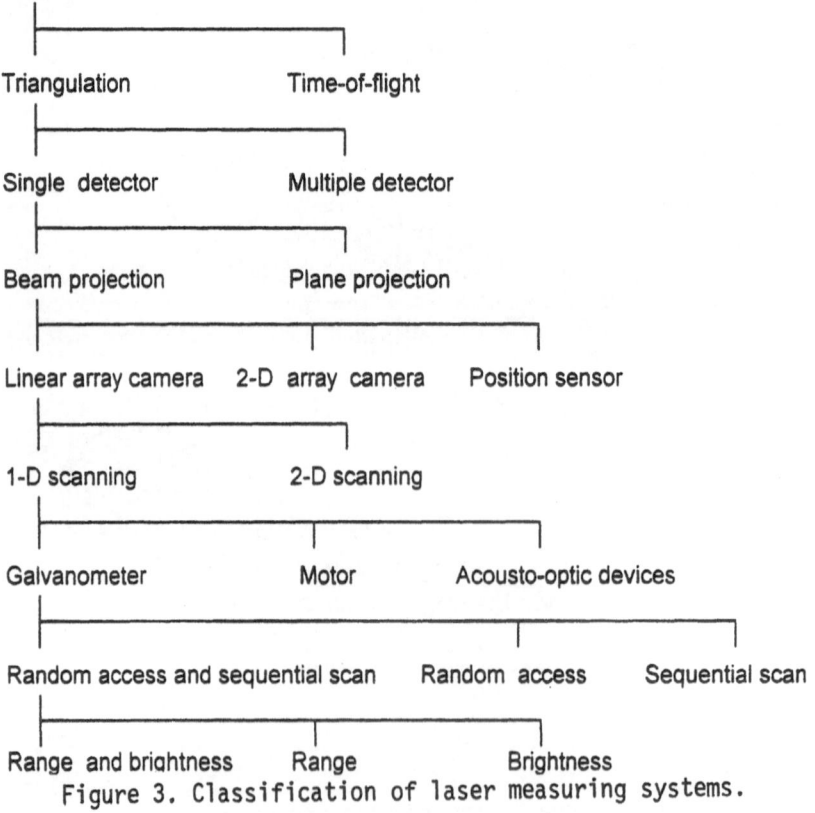

Figure 3. Classification of laser measuring systems.

In a laser non-contacting measuring machine, the laser probe measure
the distance to the part being measured by using a triangulation method
(Figure 4).

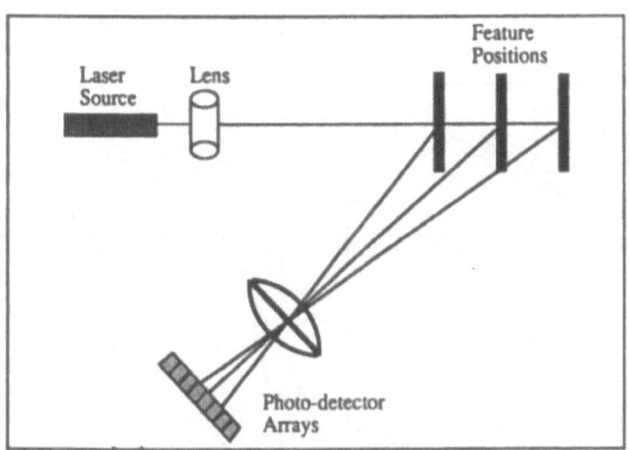

Figure 4. Principle of laser measuring machine.

This distance, is the basis for dimensional measurement that are
interpolated within control system to define the shape and form of
the part to be machined III. The laser beam emmited from the center of
the probe creates a spot on the part. This spot is reflected, through
two "eyes", onto two separate sensors. The sensors are graduated in a
way similar to that used in glass scales for readouts. Each graduation
represents 0.01 mm. (0.0004") difference in the Z dimension. As the
distance between the probe and the part changes, the beam reflects
through the eyes at a different angle, thus hitting a different point
on the sensor (Figure 5).

Since the eye is mounted at an angle, the reflected beam does not
always have a clear path to the eye. The path may be blocked by another
areaof the part. For this reason, two eyes are used, and usually, at
least one of the eyes can see the part. However, conditions may exist
where the view of both eyes is obstructed.

The laser probe is a solid state diode type for semi-continous scanning a high density of points which can be measured quickly, enhancing
both machining time and permitting very accurate duplicate.

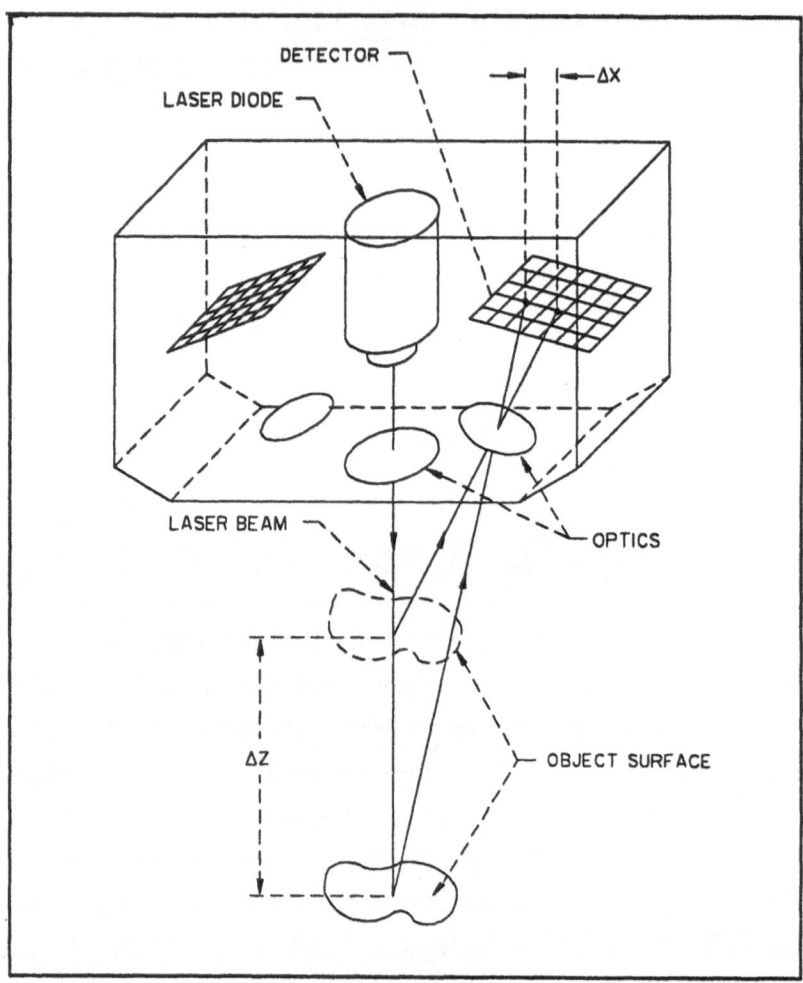

Figure 5. Laser measuring machine.

The Laser Digitizing System enables the digitizing of three-dimensional parts and subsequent reproduction of these parts to greatly
decrease the time from conception to finished part. The hardware configuration of digitizing system includes (Figure 6) :

- Laser probe with a laser safety device (touch switch).
- A 386/25 MHz PC controller with VGA screen and at least 40 MB hard
  disk.
- 3 axis switching boxes, to switch between laser probe control and CNC
  machine control.

- Laser interface board (to be inserted into the computer), laser con-
nector (parallel port), I/O board and interfaces.

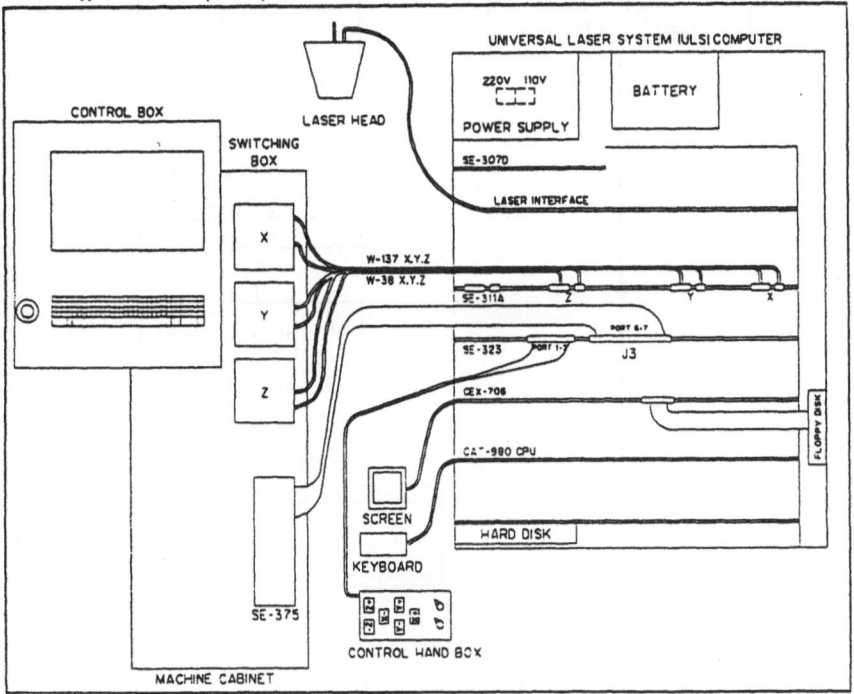

Figure 6. Laser digitizing system.

To this system has been included a software which is capable of mas-
saging the digitized data to include verifying, smoothing the surface,
and point editing from within the graphic file plotting utility. Laser
digitized files can then be enlarged or reduced with a scale factor,
and male/female converted. The system will then process the data for
direct cutter path generation for a variety of operations to include
roughing, semi-finish, and finish cutter path for flat bottom, ball
nose, or tapered tools.

3. THE MEASUREMENT PRINCIPLE

When a part has been manufactured it must be checked for defects. Idealy
such checking should be done against the orijinal design of part and
any out-of-tolerance difference between the two shoud be reported. A
diagram of the measurement principle is shown down :

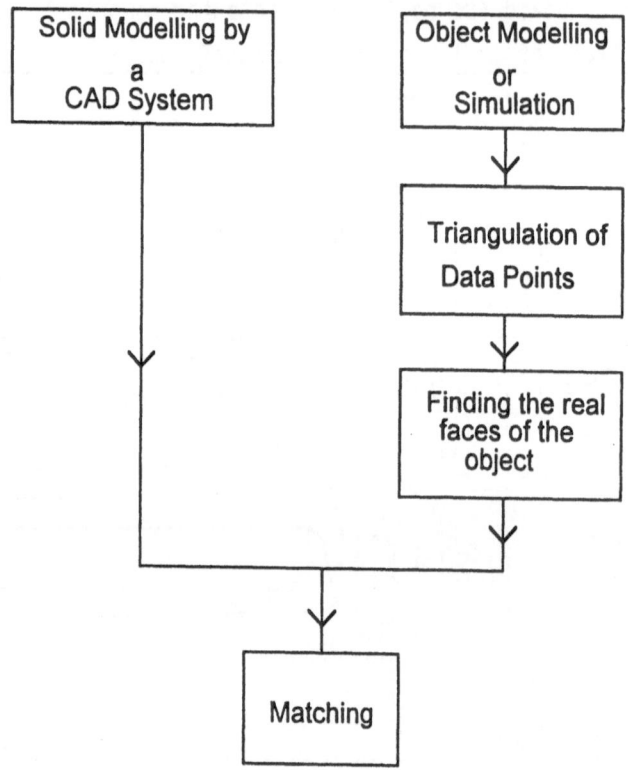

In operation, the part to be measured is first clamped to the CNC machine table. The tool in the machine spindle is replaced by the laser probe. The machine moves the probe in a scanning orbit, taking dimensional readings which are collected and stored on the hard disk of the computer. Through the digitizing program, the system "maps" the surface of a three dimensional object by measuring Z value for predetermined grid of X, Y coordinates.

This information can be used for different purpose, including cutting a similar or modified part, reverse engineering etc. Data can be converted to different formats and transmitted to CAD/CAM system for redesigns.

For cutting, the computer automatically creats a tool path from dimensional data collected and the software controls and offers tool path
for both roughing and finishing operations. The process is divided into
3 main stages:

  a) The digitizing stage, in which the information data is aquired.
  b) The processing stage, in which the data is manipulated and prepa-
     red for milling.
  c) The milling stage, in which the desired part is machined.

4. PROCESING THE DATA
4.1 Shape Reconstruction.

For inspection application, measured data need to be compared with reference data produced from a model. Initially the parts are described by
a set of coordinates (x,y,z) of surface scattered all over the part.
Such data are in general irregularly located in 3D space, generated by
laser coordinate measuring machine. These data need to be processed
further to form a shape which will then be compared with the master
solid model of the same part to find any defect. This process is named
shape reconstruction [2,3,4].

The problem of shape reconstruction is that of finding an interpolation over a set of point lying on its boundary. A lot of methods have
been given about this problem, but one very useful is the Delaunay
Triangulation. To obtain topological information about measured part
the Voroni diagram (or Dirichlet tessellation) of the measured points
is constructed. The geometrical dual of the Voroni diagram, obtained by
linking the points whose Voroni polyhedra are adjacent across a common
face, is called the Delaunay Triangulation (Figure 7). As a result of
this process the surface triangles and the surface points obviously are
clustered toghether in different clusters representing a face. Principle
components analysis is then used on the measured points making up triangle vertices in each collection to obtain a best-fit plane through
them all.

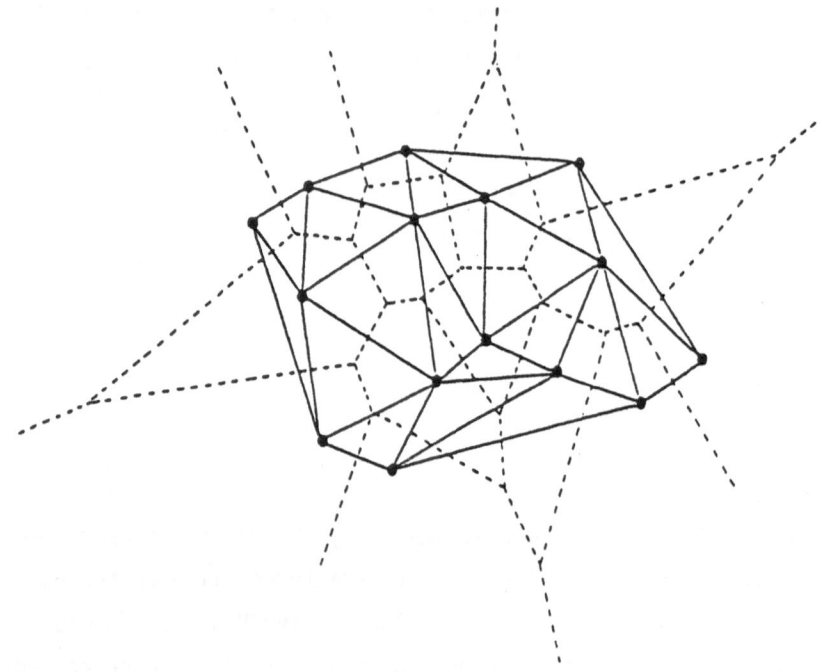

Figure 7. Voroni diagram (dottet lines) and Delaunay triangulation (bold lines).

4.2. Digitizing

Then process of <u>shape reconstruction</u> by <u>laser</u> system is named <u>digitizing</u> process. Here are some definitions related with this process.

The area to be digitized is called the master (Figure 8). The master is enclosed within a polygon, which defines the boundary of the digitized region (Figure 9).

If, within the master boundary, there are some sections that need to be digitized at a high point density, a window may be used to define such an area. The boundaries of windows are also defined by closed polygons. The windows must be completly contained within the boundaries of the master polygon.

A polygon is defined by its set of vertices, either by two points (rectangle) or by multiple points. The <u>digitizing</u> process begins with the <u>digitizing</u> of these vertices. Searching for the Z coordinate is only in the upward direction from the point defined.

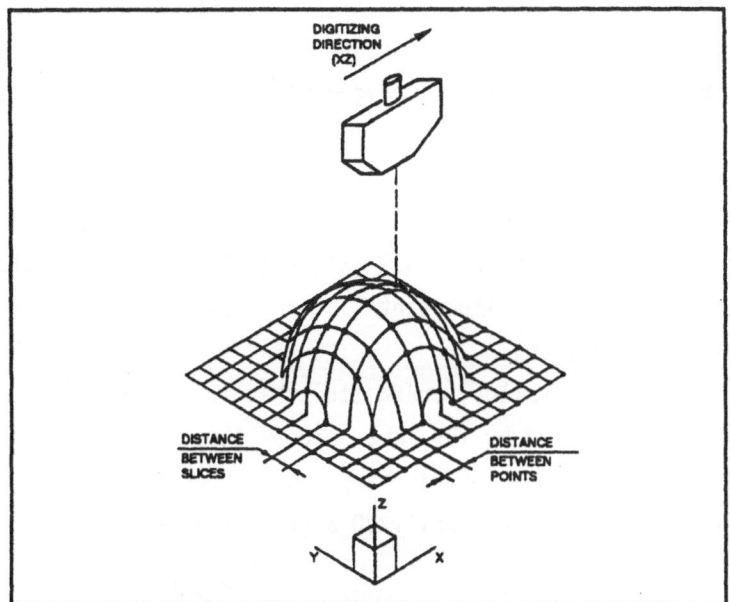

Figure 8. The digitized area.

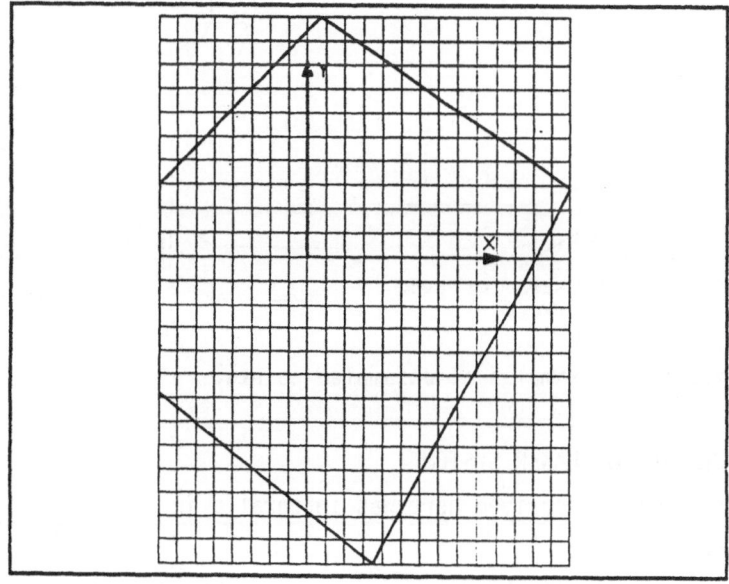

Figure 9. A polygon bound master/window

A valid (legal) polygone must fulfill the following conditions :

a) Edges shall not cross each other.

b) Maximum points defining a polygon shall not exceed 100.

c) The polygon must be defined so that each of its slices is "one piece" and no broken.

- A rectangular polygon can be easly defined by specifying only two points along the diagonal (X1, Y1; X2, Y2). The Z value of these points is not important (in contrast to the case of a multipoint polygon).

- A hole, slot, valley, crack etc. that is too deep and/or narrow can not be sampled by the laser probe. These points are named unsampled points. To enable sampling of these points the <u>laser probe</u> must be rotate usually 90 degrees (Figure 10a, b).

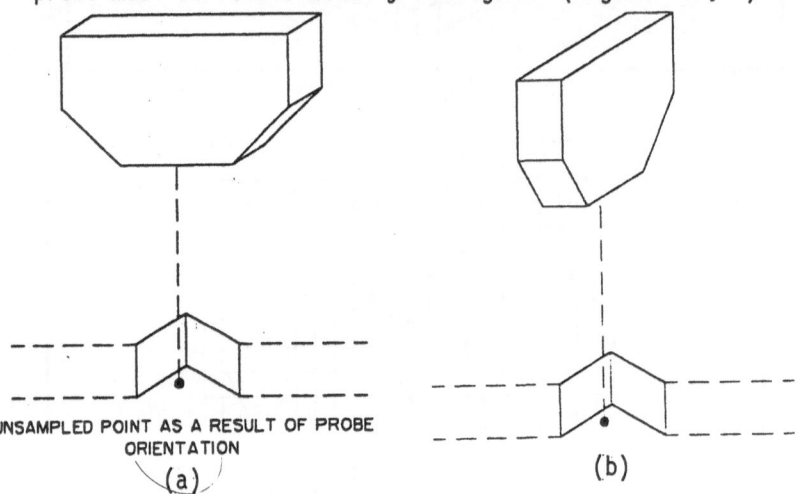

UNSAMPLED POINT AS A RESULT OF PROBE ORIENTATION

(a) (b)

Figure 10. Examples of unsampled points.

5. INTEGRATION OF LASER SYSTEM

The Laser System described here was integrated in a <u>CAD/CAM/CNC</u> system which includes :

- A PC-486/222 with 16 RAM and running in MS-DOS and Windows environment.
- AutoCAD 12/AutoSURF/AutoMILL/ANSYS CAD/CAM/FEA package software.
- A CNC Machining Center (Figure 11).

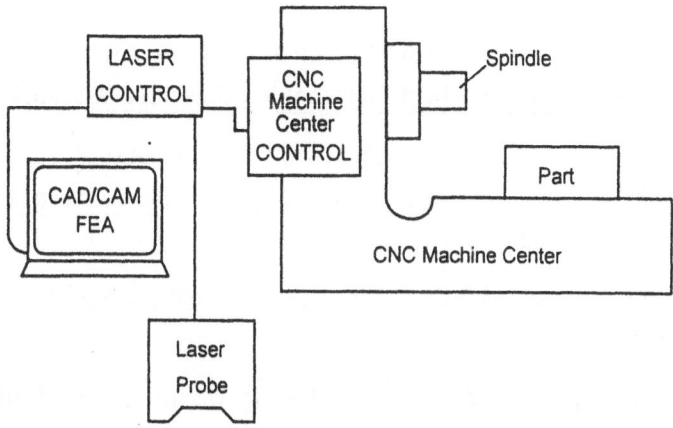

Figure 11. CAD/CAM/CNC/Laser CMM system.

The algorithms which allow a group of point on the surface of a part to be matched automatically with a real solid model or created by CAD system have been generated and coded; also all the necessary connec- tions software have been written. The programming language used was C++. One important point is that, in the connection systems most of the electronic boards used up until now have been eliminated and instead software packages are incorporated. This increase PC-CNC capability, which enables easy and flexible programming, fast performance and reac- tion, and large memory capacity.

Information that was gathered by the probe and processed by the com- puter may be transmitted to the machine cotroller or CAD/CAM/FEA system. If it is trasmitted to the machine controller, the machine operator enters cutting parameters and then directs the milling operations to begin. Once measurements are taken and stored, the part can be machined in many variations. A direct copy of the part can be machined, or a scaled copy produced. Other possibilities from the digitized dimensions are a male-to-female conversion or vice versa, a left or right part, or even a part or mold that is streched to allow for shrinkage. If the digitized data is transmitted to the CAD/CAM/FEA system more analysis and shape manipulation may be performed for an even finer, more precise mold or prototype part .

868

6. REFERENCES

1. Woodwark, J.R.and Bowyer,A.(1986) Better and faster pictures from solid models, IEE Computer Aided Engineering Journal, V3, No.2

2. Boissonnat,J.D.(1985) Shape reconstruction from planar cross-sections, Proceedings of the IEEE conf. on Computer Vision and Pattern Recognition, San Fransisco, 393-397.

3. Bowyer, A.(1981) Computing Dirichlet tessellations, Computer Journal V24, No.2, 162-166.

4. Bowyer, A., Graham, D., and Henry, G. (1985) The measurement of 3-D features using laser triangulation, Proc. 7th International Conference on Automated Inspection and Product Control, Birmingham, IFS Publications.

LASER PROCESSING OF INTEGRATED OPTIC WAVEGUIDES FOR PHOTONIC APPLICATIONS

ANADI MUKHERJEE, BEN JOY EAPEN, SWAPAN K. BARAL

Laser-Matter Interaction Labs Inc.
1423 Madison NE, Albuquerque, NM 87110, USA
and
Center for High Technology Materials
University of New Mexico, Albuquerque, NM 87131, USA

1. Introduction

Very low loss integrated waveguides of optically transparent materials are desirable for many passive and active photonic waveguide devices. In the past, there have been numerous reports on several waveguide fabrication techniques applied to organic, inorganic and semiconducting materials. This article focuses on laser fabrication of integrated optic waveguides of organic and inorganic materials. Choice of material and laser processing technique are application specific. While inorganic glass waveguides are appropriate for high temperature environments such as space applications, waveguides of organic polymers are not. The flexibility of film deposition and laser processing of organic polymers are however very attractive for photonic applications at moderate temperatures. Waveguide fabrication may be broadly classified under two categories, namely, lithographic patterning and resist-free patterning. Lithographic patterning is usually followed by ion diffusion or some form of etching (wet chemical or dry). Although lithographically patterned waveguides is an established technology, it involves time consuming mask transfer process and use of chemicals that are not environmentally safe. Laser fabrication of integrated waveguide structures is the dominant resist-free patterning technique, which offers flexibility, reproducibility and rapid manufacturing conditions, besides being environmentally safe. Also, as described in a later section, laser fabrication allows variation of film thickness and refractive index across the wafer which may be needed in improving coupling and bending losses. This article will describe pulsed and cw UV laser processing techniques and the inorganic and organic materials commonly used for the fabrication of channel waveguides for photonic applications. Our research effort on a unique laser processing of very low loss (0.08 dB/cm) channel waveguides on polymethylmethacrylate (PMMA) will be given in detail.

869

J. Mazumder et al. (eds.), Laser Processing: Surface Treatment and Film Deposition, 869–880.
© 1996 *Kluwer Academic Publishers.*

2. Pulsed and cw Laser Processing

Pulsed UV lasers, particularly excimer lasers have been used during the last decade for ablation of materials for patterning. R. Srinivasan and B. Braren[1] have reviewed UV laser ablative decomposition of polymers. Although the detail of the energetics involved is complicated, a fairly good understanding exists on the basic processes involved in the interaction. When an UV pulse is focused on a material, e.g. a thin film on a substrate in integrated optics, the pulse is absorbed and "bond breaking" takes place depending on the photon energy and the bond strength of the material. Typically, the sample reaches threshold of ablation after being exposed to a few hundred pulses. At this time, the fragments absorb the residual UV photons after bond breaking, and excites the rotational-vibrational manifolds which thermalize rapidly causing localized heating. This local heating results in a rapid propulsion of the fragments from the etched region. Since this entire process takes place in a time comparable to the laser pulsewidth (ps to ns), the thermal damage to the substrate is usually minimum, unless it is highly heat sensitive. This highly intensity dependent etching process allow patterning with high spatial resolution. However, due to the "brute force" patterning, the side walls of the etched regions usually have roughness more than the wavelength of light. For very low loss channel waveguide applications in integrated optics, the large scattering from the side walls usually increases the linear propagation loss. Direct UV laser patterning of PMMA films by ablation using a cw Ar-laser has been reported[2]. The ablation process of the cw laser beam too has resulted in the undesirable rough side walls. As indicated below, a more "gentle patterning" with cw UV laser beams are preferred for the fabrication of low loss waveguides needed in integrated optics.

Two types of "gentle patterning" of integrated optic waveguides using lasers have been reported in the last few years. First type involves a direct patterning of micron lines of a metal or its oxide on a substrate. This is followed by an indiffusion process into the substrate to define the buried waveguide. This process is usually applied to inorganic materials like Ti-indiffused $LiNbO_3$[3,4]. The second type involves either direct or indirect UV laser crosslinking (or polymerization) of large molecules to define the waveguide pattern. This technique is obviously applicable to organic materials like transparent optical epoxies[5,6], PMMA[7] etc. Some reports on these two types of laser processing technologies are summarized below.

A UV laser assisted chemical vapor deposition technique for directly writing micron lines of Ti on $LiNbO_3$ substrates has been developed by Tsao et. al.[3]. Surface adsorbed metal bearing molecules like $TiCl_4$ were photodecomposed to deposit the metal along the lines scanned by a frequency doubled Ar-laser. Later, the Ti was indiffused into the substrate following standard procedures. This process allows thickness variation of the Ti film along the length of the waveguides which may be applied towards mode matched coupling or graded index structures. This process however involves corrosive vapors not required in the laser written TiO_2 lines reported by Haruna et al.[4]. In the experiment reported by Haruna et al., thin films (0.1 μm thick) of Ti was RF sputtered onto $LiNbO_3$ substrates. An Ar-laser beam at 514.5 nm was focused onto the sample placed on a computer controlled translation stage in air ambient. At the right fluence, the Ti film along the laser scanned lines was thermally oxidized to TiO_2. We have used a similar laser induced oxidation process to reversibly pattern micron lines of high temperature superconducting (HTSC) material of $YBa_2Cu_3O_x$ films on $SrTiO_3$ and $LaAlO_3$ substrates[8]. The surrounding Ti film was removed by wet chemical etching before indiffusing the TiO_2 lines into the $LiNbO_3$ substrates to form waveguides. Propagation loss measurements[4] on waveguides (0.6 dB/cm) of such laser fabricated and conventionally patterned (lithographic) Ti:LiNbO3 waveguides showed no marked difference.

Commercially available spin-on polymers such as PMMA, polyimides, optical epoxies etc. offer well known flexibility for processing, fabrication of photonic integrated circuits on board and low production costs. Laser fabrication of channel waveguides on spin-on polymers have been reported by many authors[5,6,7]. The Bellcore group[5,6] used UV laser polymerization to pattern waveguides in commercially available transparent optical epoxies (Norland 61). Typical linear propagation loss achieved by them has been reported[6] to be 0.4 - 0.6 dB/cm. Such laser write process however, relies on the polymerization (i.e. large molecular size) of organic materials increasing bulk scattering in the medium. A novel technique of UV laser processing of extremely low loss (<0.08 dB/cm) channel waveguides on thin films of PMMA has been developed in our laboratories. This technique allows the advantages of smooth edges of a direct laser write technique while drastically reducing the bulk scattering loss by avoiding polymerization.

3. Polymeric Waveguides: Our work on PMMA

Planar and channel waveguides on polymers[9,10], optical epoxies[5,6,11], polyimides[12,13] and PMMA[2,14] have been reported. Several authors have reported[6,9,11,15] propagation loss of 0.3 - 1.2 dB/cm at 630 to 1550 nms over crossections in the range of 14 - 4000 $μm^2$ outside the infrared resonances of polymeric materials. The major factors preventing further reduction of the loss floor of polymeric channel waveguides besides the bulk absorption loss, are surface and bulk scattering from the waveguides. Although several authors have indicated that careful processing of the surface ought to be made to avoid surface scattering loss[9,13], no report was found, which addressed the bulk scattering loss of polymeric waveguides. As described below, our unique laser processing technique avoids polymerization in the waveguide region and address the role of bulk scattering in linear propagation loss through integrated polymeric waveguides and its reduction for the first time.

Efficient operation of some polymeric waveguide devices such as Mu-Lasers[17] depend critically on the linear propagation loss than other conventional active and passive devices. Besides other well known applications, the development of Mu-Lasers is the main motivation for this work. Mu-Laser principle involves infrared two-photon pumped[17,18] visible upconversion (particularly blue) in dye-doped polymer waveguides. The gain from a two-photon pumping process being inherently weak, requires even lesser linear propagation loss through the waveguides for efficient upconversion.

The desired target of propagation loss[6,9] < 0.1 dB/cm, reported for many systems applications has been achieved in this work[7]. Reduction of both surface and bulk scattering losses have yielded a linear propagation loss of 0.08 dB/cm at 632.8 nm for waveguides of crossection 20 μm^2. Experimental and preliminary theoretical results show the possibility of further reduction of linear propagation loss in integrated PMMA waveguides.

The substrate chosen for this study was SiO_2 on <100> silicon. Standard technique of wet thermal oxidation was used to grow 1.5 μm films of SiO_2 on 2" and 4" wafers of silicon. PMMA has been chosen as the polymer material due to its well known optical transparency, mechanical strength, high optical damage threshold[19] and integrability with optoelectronic circuit boards. PMMA dissolved in a solvent allow homogeneous doping with active species like dyes. Thin films (0.1 - 10 microns thick) of commercially available (OCG Microelectronics Inc.) PMMA dissolved in chlorobenzene (9% solids, 496 K) were spin coated on SiO_2 on Si substrates. A spin rate of 600 rpm resulted in films of 3.0 μm thickness. The films were baked at 90°C for 24 hours to evaporate the solvent. Films that were baked over a shorter time showed incomplete removal of solvent, causing the films to be soft and of lower refractive index. The surface roughness of the SiO_2 and PMMA films were measured with an atomic force microscope (Digital Instruments, Nanoscope II). The RMS roughness of the SiO_2 and PMMA film surfaces were 0.31nm and 0.38 nm respectively over an area of 5 μm x 5 μm. The surface roughness measurements show good optical qualities of (1) standard wet thermally grown oxide on silicon and (2) spin coated PMMA films. The spin coated polymer films were also seen to be of good optical quality (i.e. no color change) over the entire wafer as observed under a Nomarski optical microscope (Nikon Optiphot II).

The schematic used for the laser fabrication of the waveguides is shown in Fig. 1. The UV laser at 257 nm used to pattern the waveguides was obtained from an intracavity double Ar-laser. The UV beam was focused to a ~ 10 microns spot size on the film that was mounted on a computer controlled translation stage. Two parallel lines were scanned on the thin films by the UV beam. The space between the scanned lines defined the waveguide. The crosslinked PMMA, due to UV exposure, becomes highly soluble in a developer solution (IPA:MIBK, 1.8:1) compared to unexposed material. On development, the exposed areas were completely removed, leaving behind clean channel waveguides as shown in Fig. 1. Typical development time was 1 - 2 minutes. Longer development times showed cracking of the PMMA film, with the cracks starting from the fabricated side walls. The clean removal of exposed material was seen to be strongly dependent on the fluence of the UV beam. Optimized crosslinking was obtained at a fluence of 700 J/cm^2. A reduction in fluence of 20 % or more caused underexposure and incomplete crosslinking which resulted in partial removal of the material on development. A 20% or more increase in fluence caused overexposure and the well known "orange peel" structure was seen. The underexposure and overexposure were undesirable to achieve waveguides with clean side walls. Typical scan speed for optimal fluence was 16 mm/min. Substantially higher scan speeds with higher average power of the UV beam, maintaining same fluence, showed undesirable film ablation characteristics. The three essential steps of spin-expose-develop required for the fabrication of these waveguides is shown in Fig. 1. The two advantages derived out of this processing method are: (1) minimum top surface scattering of the film, since it was left unchanged and as deposited and (2) minimum scattering from the side walls of the waveguides, which were smooth due to the "gentle processing" of the laser beam with a Gaussian spatial profile. A scanning electron microscope picture showing smooth edges and clean facet of a typical waveguide is given in Fig. 2. Complete removal of PMMA on development at the laser written lines and the channel waveguide between them can be seen clearly. Integrated optic devices of such PMMA waveguides of any length and shape can be fabricated by an appropriate computer control of the translation stage[20].

Multimodal (TE$_0$, TE$_1$ and TE$_2$ modes) linear propagation loss in the fabricated PMMA channel waveguides (3.0 μm thick) at 632.8 nm was done by endfire coupling a He-Ne laser beam and recording the streak of the scattered light perpendicular to the waveguide. The streak was recorded by a CCD camera and stored in the computer. Linear propagation loss was estimated from a straight line fit to the ln of the scattered signal versus the length of the waveguide. Linear loss of waveguides fabricated from unfiltered PMMA solution was typically 2.1 dB/cm. It is notable that the laser fabrication process used (Fig.1) in this study does not use the conventional scheme of laser written waveguides[5], where the UV laser exposed and polymerized regions form the waveguide.

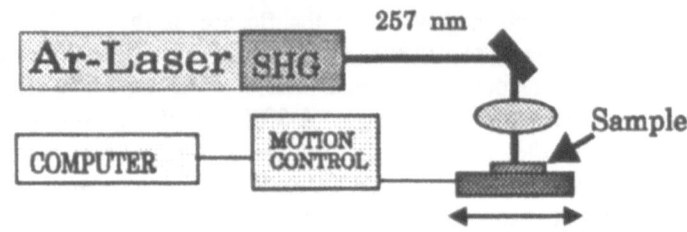

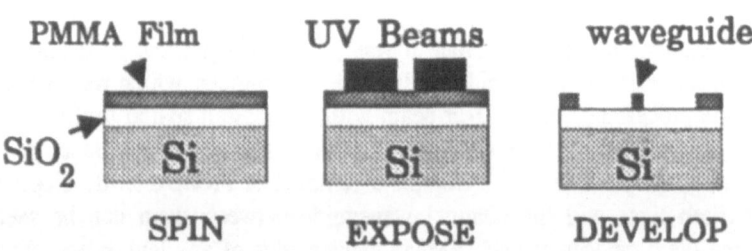

Figure 1. The schematic used for the laser processing of the waveguides. The optimized fluence at 257 nm used for the fabrication of waveguides shown in Figure 2 was 700 J/cm^2.

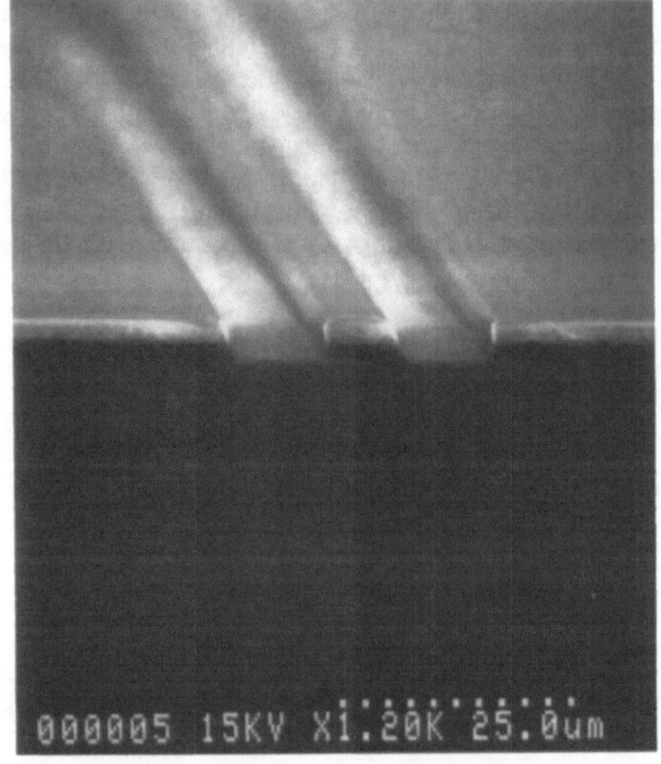

Figure 2. A scanning electron microscope picture of the waveguide. Laser processed areas of the film was completely removed on development. Clean facet and smooth walls of the channel waveguide can be seen.

Laser polymerization increases the molecular size, which enhances bulk scattering of the propagating beam. Commercially available PMMA consists of a distribution of different size polymers of different molecular weights, the minimum size being ~ 7 nm of the monomer MMA. To investigate the role of bulk scattering in polymeric materials and its effect on the linear optical propagation loss in the waveguides, we filtered the PMMA solution through filters of different pore size. The pore dimension effectively limited the maximum size of the polymers before it was spin coated. Polymers of PMMA are long chain molecules which coil up in roughly spherical shapes which percolate through circular pores of the filter. Hence it is assumed that the maximum radius of the spherical polymer molecules which passes through the filter is approximately same as the radius of the filter pore. The strong effect of polymer size on linear propagation loss of fabricated PMMA waveguides is shown in Fig. 3. These loss measurements were taken in waveguides that were fabricated on thin films of PMMA filtered through 0.5, 0.2 and 0.1 micron filters, which determined the maximum radius of the polymers in the film. All other fabrication parameters were kept the same. One order of magnitude reduction of propagation loss through the waveguides was attributed to the reduction of bulk scattering from polymers of reduced dimension. The linear propagation loss data for PMMA waveguides filtered through 0.1 micron pore, is shown in Fig. 4. The loss data has been taken over the section of the waveguide where the leaky modes at the coupling end had decayed. The loss of 0.08 dB/cm obtained in these waveguides meet the target[6] of several polymeric waveguide devices. The loss obtained here is also comparable to that of integrated glass waveguides[16] and plastic optical fibers[6].

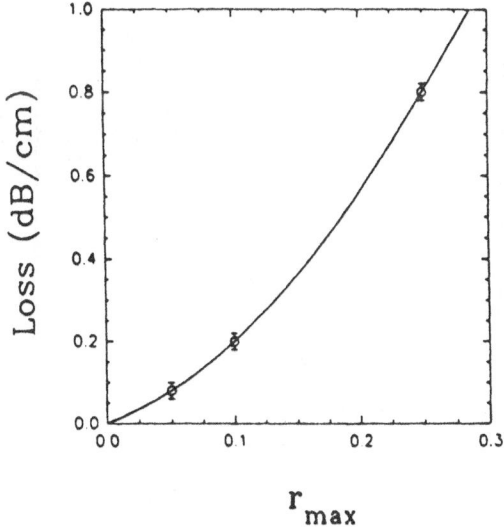

Figure 3. Linear propagation loss data for waveguides processed from filtered PMMA. The pore size of the filter was taken to be the diameter of the largest polymers used. r_{max} is the radius in microns of the largest polymer.

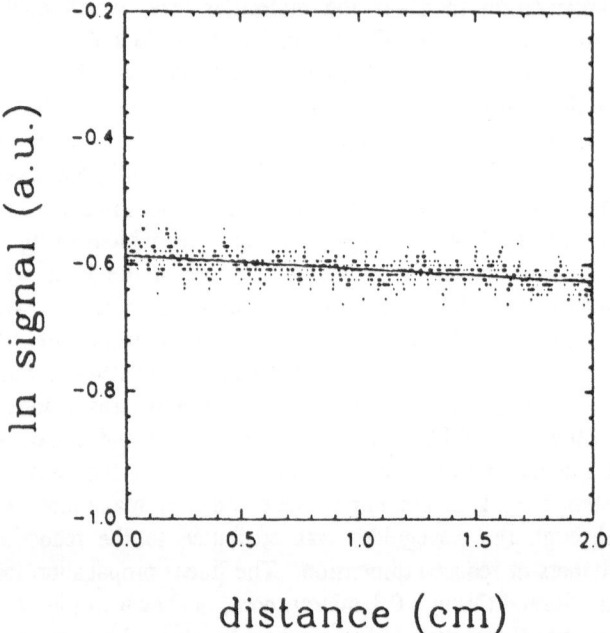

Figure 4. Linear propagation loss data of channel waveguides of PMMA at 632.8 nm filtered through 0.1 micron filter. The loss was measured to be 0.08 dB/cm.

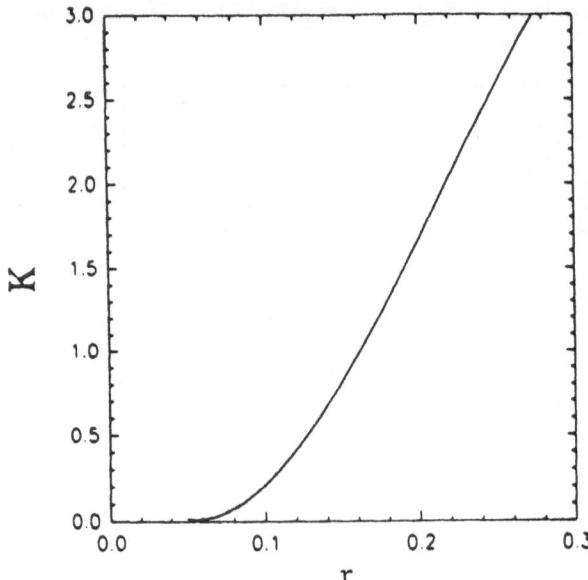

Figure 5. Mie scattering coefficient K as a function of r, the radius in microns of scattering polymer particles. Here it is assumed that all the particles are of the same size and any interference of scattering due to close proximity of the particles is neglected.

The fact that the loss reduction in the waveguides was due to a reduced scattering from smaller size polymers is seen from Mie theory of scattering from dielectric spheres[21,22]. The total scattering coefficient K is defined as the total energy scattered per second per unit crossectional area of a particle illuminated at unit intensity . The functions are defined as:

$$K = 2 / \alpha \sum_{1}^{\infty} (2n + 1)(|a_n|^2 + |b_n|^2)$$

where

$$a_n = (-1)^{n+\frac{1}{2}} \frac{S_n(\alpha)S_n{}'(\beta) - mS_n(\beta)S_n{}'(\alpha)}{\phi_n(\alpha)S_n{}'(\beta) - mS_n(\alpha)\phi_n{}'(\beta)}$$

$$b_n = (-1)^{n+\frac{1}{2}} \frac{mS_n(\alpha)S_n{}'(\beta) - S_n(\beta)S_n{}'(\alpha)}{m\phi_n(\alpha)S_n{}'(\beta) - S_n(\alpha)\phi_n{}'(\beta)}$$

$$S_n = (\frac{\pi z}{2})^{\frac{1}{2}} J_{n+\frac{1}{2}}(z)$$

$$C_n = (\frac{\pi z}{2})^{\frac{1}{2}} (-1)^n J_{-n-\frac{1}{2}}(z)$$

$$\phi_n = S_n(z) + iC_n(z)$$

$$\alpha = 2\pi r/\lambda$$

$$\beta = m\alpha$$

m = 1.485 is the refractive index of PMMA at 632.8 nm, r is the radius of the polymers, z represents α or β and primes denote the first derivatives. Assuming same size of all polymer particles in the medium i.e. neglecting the realistic distribution of polymer size in PMMA, the total scattering coefficient K was computed for media of different polymer size. Fig. 5 shows K versus r, the polymer size in bulk PMMA. This simplified calculation has also neglected any interference of the scattered beams from particles being close to each other. Qualitatively we see from Fig. 5 that as the size of the constituent particles in the polymer medium is made smaller, the bulk scattering reduces roughly at the same rate as has been observed in the experimental data of Fig.3. This shows that the loss in the PMMA waveguides is dominated by bulk scattering in the medium and it may be further reduced by using smaller size polymers. This may be done by using an appropriate oligomer of MMA where most of the polymers are smaller than that reported here. The mechanical strength and the bulk properties of polymers however reduces drastically with the reduction of degree of polymerization. A more rigorous calculation needs to account for the distribution of molecular size and any interference of scattering beams from the polymer particles being too close to each other.

We have measured linear propagation loss in laser processed PMMA waveguides at different wavelengths ranging from 488 nm to 740 nm. This data shown in Fig. 6, suggests almost no change in the linear propagation loss from ~0.1 dB/cm in these PMMA waveguides over a wavelength span of 250 nm. The explanation of such observation is given below. The waveguide used in these measurements was fabricated from PMMA passed through 0.1 micron pore filter. The radius of the largest polymers expected in this waveguide is ~ 50 nm. which is much smaller than the wavelength of light in the medium. In this case, we would have expected the bulk scattering (and therefore the linear loss) to decrease as the fourth power of wavelength, i.e. dominated by Rayleigh scattering ($\sim 1/\lambda^4$). This is true only if the transmission curve of PMMA is flat over the entire wavelength region. Commercially available PMMA fibers show an increase in absorption in the near infrared wavelengths. This absorption causes an increase in the linear loss which counteracts the expected lowering of linear loss due to reduced Rayleigh scattering at longer wavelengths. The result of this opposing factors of linear loss resulted the almost straight line behaviour of linear loss shown in Fig. 6.

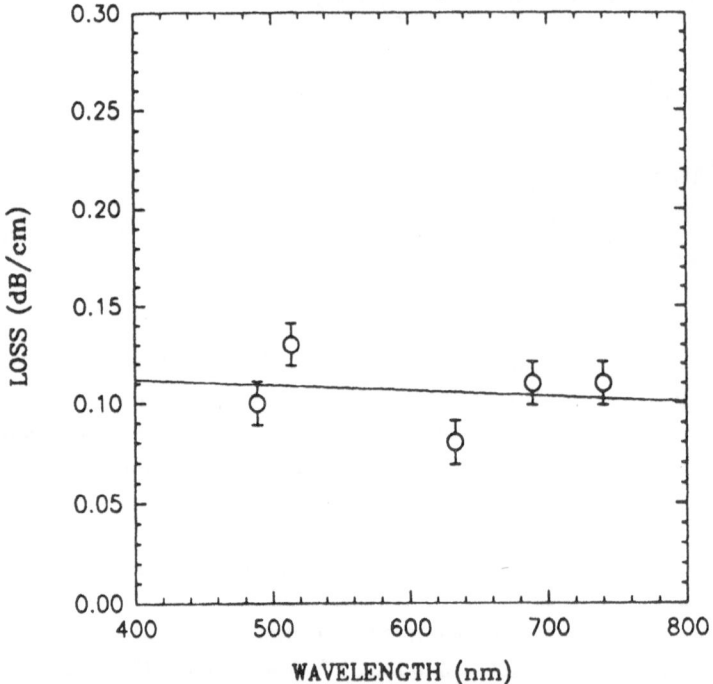

Figure 6. The measured linear propagation loss through laser processed PMMA waveguides as a function of wavelength. Note that the reduction of linear loss with wavelength expected from Rayleigh scattering is opposed by the increase of absorption of PMMA at longer wavelengths.

4. Conclusion

In conclusion, we have briefly outlined pulsed and cw UV laser processing techniques used for the fabrication of integrated optic waveguides in inorganic and organic materials. Also, our research efforts on fabrication of very low loss integrated channel waveguides on PMMA was reported. Simple high volume and low cost spin coating technique has been used for the fabrication of the high optical quality thin films. A unique laser processing technique has been used to pattern the waveguides, which does not polymerize the waveguide regions. This process allows minimum surface and side wall scattering of the propagating beam. Spatial filtering of the UV beam (not done in this experiment) may result in smoother side walls and encapsulation of the surfaces (e.g. an overlayer of spin on glass) may reduce the surface scattering further. Most importantly we have shown for the first time, the role of bulk scattering in limiting the propagation loss in polymeric waveguides. The experimental data and preliminary Mie scattering theory show that further reduction of linear propagation loss in polymeric waveguides may still be possible using oligomers, provided the mechanical strength of the polymer does not deteriorate substantially. It is expected that bulk scattering may play a similar role in other polymeric waveguides (e.g. polyimides, optical epoxies etc.) and further reduction of linear loss in waveguides of those materials may be possible. Typical application areas of this fast growing technology will be in the fabrication of integrated modulators, mode converters, optical fiber couplers & splitters as well as switches and waveguide lasers.

The authors would like to acknowledge financial support of SBIR grant no. F33615-93-C-5340 of Wright Laboratory, Wright Patterson AFB, OH and partial support of U.S. Air Force Office of Scientific Research for this project. Technical assistance from A. Frauenglass and helpful discussions with Dr. Bruce A. Reinhardt and Dr. Paras N. Prasad are highly appreciated. Also, the authors are indebted to Dr. Steven R. J. Brueck for permitting use of laboratory equipments.

5. References

1. Srinivasan R. and Braren B. (1990) *Lasers in Polymer Science and Technology: Applications, Vol III*, CRC Press, Boca Raton, Florida.
2. Bozhevolnyi, S.I., Potemkin, I.V., and Svetovoy, V.B. (1992) Direct writing in polymethyl methacrylate films using near-ultraviolet light of Ar^+ laser, *J. Appl. Phys.* 71, 2030 -2032.
3. Tsao, J.Y., Becker, R.A., Ehrlich, D.J., and Leonberger, F.J. (1983) Photodeposition of Ti and application to direct writing of Ti:LiNbO$_3$ waveguides, *Appl. Phys. Lett.* 42 (7), 559 -561.
4. Haruna, M., Murata, Y., and Nishihara, H. (1992) Laser -beam direct writing of TiO$_2$ Channels for Fabrication of Ti:LiNbO$_3$ waveguides, *Jpn. J. Appl. Phys.* 31. 1593 - 1596.
5. Krchnavek, R.R. Lalk, G.R., and Hartman, D.H., (1989) Laser direct writing of channel waveguides using spin-on polymers, *J. Appl. Phys.* 66, 5156 -5160.
6. Hartman, D.H., Lalk, G.R. Howse, W.,and Krchnavek, R.R. (1989) Radiant cured polymer optical waveguides on printed circuit boards for photonic interconnection use, *Appl. Opt.* 28, 40 - 47.
7. Baral, S.K., Eapen, B.J.,and Mukherjee, A., (1994) Laser fabrication of very low loss channel waveguides of polymethylmethacrylate for photonic applications , *Conference on Lasers and Electrooptics (CLEO)*, Paper No. CThI24.

8. Dye, R.C., Muenchausen, R.E., Nogar, N.S. Mukherjee, A and Brueck, S.R.J. (1990) Laser writing of superconducting patterns on $YBa_2Cu_3O_x$ films, Appl. Phys. Lett. 57 (11), 1149

9. Sullivan, C.T., Booth, B.L., and Husain, A. (1992) Polymeric waveguides, *IEEE Circuits and Devices*, 8, 27-31.

10. Booth, B.L. (1989) Low loss channel waveguides in polymers, *IEEE J. Lightwave Tech.* 7, 1445 - 1453.

11. Olsen, C.M., Trewhella, J.M., Fan, B., and Oprysko, M.M. (1992) Propagation properties in short lengths of Rectangular epoxy waveguides, *IEEE Photon. Tech. Lett.* 4, 145 - 148.

12. Reuter, R., Franke, H., and Feger, C. (1988) Evaluating polyimides as lightguide materials, *Appl. Opt.* 27, 4565 - 4570.

13. Selvaraj, R., Lin, H.T., and McDonald, J.F. (1988) Integrated optical waveguides in polymide for wafer scale integration, *IEEE J. Lightwave Tech.* 6, 1034 -1044.

14. Kawamura, Y., Toyoda, K., and Namba, S. (1982) Effective deep ultraviolet photoetching of polymethyl methacrylate by an excimer laser, *Appl. Phys. Lett.* 40, 374 - 375.

15. Thakara, J.I., Lipscomb, G.F., Stiller, M.A., Ticknor, A.J.,and Lytel, R. (1988) Poled electrooptic waveguide formation in thin-film organic media, *Appl. Phys. Lett.* 52, 1031- 1033.

16. Zhenguang, H., Srivastava, R., and Ramaswamy, R.V. (1989) Low-loss small-mode passive waveguides and near-adiabatic tapers in BK7 glass, *IEEE J. Lightwave Tech.* 7, 1590 - 1596.

17. Mukherjee, A. (1993) Two-photon pumped upconverted lasing in dye doped polymer waveguides, *Appl. Phys. Lett.* 62, 3423 - 3425.

18. Mukherjee, A. (1993) , *Quantum Electronics and Laser Science Conference (QELS)*, Paper No. QTuA7.

19. O'Connel, R.M. (1990) *Lasers in Polymer Science and Technology: Applications, Vol III*, CRC Press, Boca Raton, Florida.

20. Haruna, M., Yoshida, S., Toda, H., and Nishihara, H. (1987) Laser-beam writing system for optical integrated circuits, *Appl. Opt. 26*, 4587 - 4592.

21. Born, M. and Wolf, E (1987) *Principles of Optics*, Pergamon Press.

22. Goldberg, B. (1953) New computation of the Mie scattering functions for spherical particles, *J. Opt. Soc. America*, 43, 1221-1222.

OPTICAL AND ACOUSTICAL TECHNIQUES FOR THE CHARACTERIZATION OF MATERIAL ABLATION BY Q-SWITCHED ND:YAG AND EXCIMER LASER RADIATION.

C. STAUTER*, J. FONTAINE*, Th. ENGEL* and A. BIERNAUX**

* *Laser Application Laboratory, Ensais*
24 Bld de la victoire 67000 Strasbourg, France
** *Irepa-Laser, Pôle d'Innovation- 67400 Illkirch, France*

ABSTRACT. Characterization of the effects of the interaction between short laser pulses and a material has been made in the conditions of surface cleaning and surface micro-machining. Fast heating of the target surface results in a pressure wave that propagates inside the sample and a shock wave in the surrounding gas medium. By monitoring the energy of the shock wave using a microphone and a piezoelectric transducer, it is possible to relate the ablation rate to the observed signals. Experiments have been made using Nd:YAG and KrF lasers. At high fluences, in particular at 1,06 µm, gas breakdown drastically alters the process conditions and disperses the recorded data. By using an additionnal optical beam deflection technique, we have determined the range of validity for monitoring of the process using simple detection devices. Experimental observations and theoritical modellings show that the ablation mechanism is of photolytic type during irradiation by excimer laser and is based on explosive boiling with Q-switched Nd:YAG laser.

1. Introduction

The use of high power continuous CO2 and Nd:YAG lasers for treatment and machining of materials has grown rapidly during the past decade. More recently, another class of lasers, those capable of generating short pulses, like Q-switched solid-state and excimer lasers, has found applications on the production line. Of particular importance is the fact that, with short pulses (around 10 ns), matter can be removed through ablation with little thermal effects, compared to processing with continuous (or long pulses) lasers. In many cases this ablative mechanism results in a higher degree of accuracy in terms of depth of removed matter, which makes laser processing a promising tool in microelectronics, microoptics, or micromechanics. Current applications include : surface cleaning, surface structuring, blind hole drilling, thin film patterning, diffractive elements machining. There are however several drawbacks associated with the use of short pulse lasers for micromachining of materials.

- The pulse to pulse stability is not always sufficient to obtain submicron accuracy in the removal process. The instability is particularly important with laser pulses at wavelength obtained via doubling or tripling main wavelength of a solid state medium.

- With some lasers, in particular excimer lasers, the active medium evolves with time and the process parameters are not constant.

881

J. Mazumder et al. (eds.), Laser Processing: Surface Treatment and Film Deposition, 881–894.
© 1996 *Kluwer Academic Publishers.*

- The laser irradiation causes a variation in the physico-chemical properties of the sample : absorption may change during process due to change in surface roughness or surface chemical composition.

Although these phenomena have been widely studied and are well understood, they cannot be overcome in the commercial devices; delicate and frequent adjustments are necessary to reach repeatable results in term of ablation rate of the machined material. In this situation, for purpose of automatic process control, real time diagnostic devices are a useful addition to the laser tool. Several techniques have been developped. A piezo-electric transducer can be fixed on the sample to record the pressure wave [1-2]. Microphone and laser beam deflection have been used to detect the variation of pressure above the irradiated sample [3-4].

In this paper we describe the use of simple sensing systems for monitoring the two following processes. The first one is the selective removal of one or several layers from a surface for cleaning purpose of industrial parts ; here a process control is particularly important, in order to avoid substrate damage. The second one is the drilling and structuring of a ceramic surface.

2. Interaction between short laser pulses and materials

The interaction between medium energy short pulse, high fluence laser beams and various materials has been widely described in the literature [5-6].The effects which occur are related mainly to the intense evaporation of the irradiated sample. Transport phenomena and mechanical effects are then predominant.

Figure 1 summarizes the conditions of material irradiation by short pulse lasers . A thin surface layer of material is heated by the laser energy. Depending on the nature of the sample, the thickness of this layer is of the order of a few tens of nanometers. Due to the high irradiance (of the order of 1 GW/cm^2), the evaporation temperature is reached within a few nanoseconds. The vapor pressure increases significantly during the pulse duration. A supersonic gas flow eventually arises close to the surface. At a short distance away from the surface, the flow becomes subsonic and a shock front appears and can be detected by a microphone for diagnostic purpose.

The ejection of material in the vapor phase into the surrounding gas medium, causes a recoil force on the solid sample. The pressure wave that propagates inside the sample can be related to the interaction conditions. Another consequence of the recoil forces is the violent agitation of the molten material. The result is a surface topography altered from one pulse to the other ; some matter is also blown sideways.

At highest irradiance level, 10 GW/cm^2 or more depending on the material, ionization phenomena in the vapor mixture brings more complexity and instability in the interaction. The plasma medium thus formed may become highly absorbing to the incoming laser radiation and may shield the target surface. As a result, a higher input of laser energy does not necessarily lead to a higher amount of ablated matter. Our experimental results show that there is a range of pulse fluence which gives optimum removal rates.

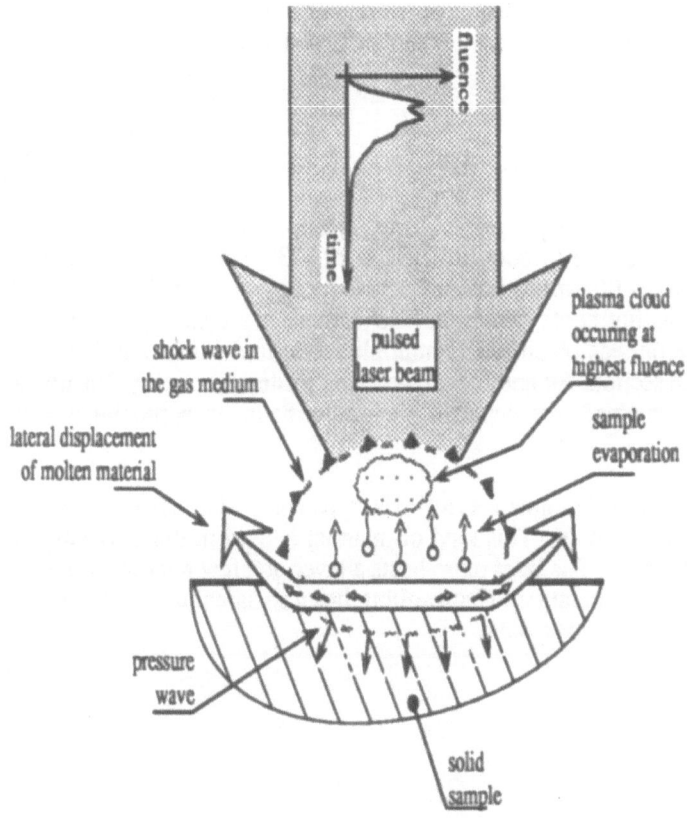

Figure 1 : Interaction phenomena occuring during irradiation of a material
with laser pulses in the nanosecond range.

3. Laser surface cleaning

3.1 INTRODUCTION

A short pulse laser beam is an efficient tool to remove selectively or completely one or
several layers of organic material on top of another organic, composite or metallic
substrate. The pulse duration for this process should be in the nanosecond range.
Depending of the nature of the coating materials, effect leading to the cleaning process
are based on vaporization, chemical decomposition or separation by subsurface heating.
Applications are found :
- in the artistic domain : cleaning of historical monuments and paintings [7].
- in the industrial domain : painting removal on automotive and aeronautic parts,
  decontamination of nuclear installation parts; cleaning of electronic circuits
  solder joints, cleaning of critical mechanical parts, descaling of automobile
  motor parts, cleaning of casting moulds, cleaning of storage tanks, ...

Advantages in using lasers for cleaning purposes are :
- Absence of mechanical contact and hence of deformation ;
- Negligible thermal effects on substrate material, particularly appreciated with
  sensitive parts like electronic circuits or paintings ;

- Versatility : a wide range of materials can be processed ;
- Safety and environment protection : no solvent or chemical products are needed.

3.2 EXPERIMENTAL SET-UP

The schematic of our experimental arrangement is shown on Figure 2. We used a Q-switched Nd:Yag laser (Model BMI 502-DNS 7730) delivering pulses of 15 ns FWHM duration, at the 1,06 μm wavelength ; pulse energy can be adjusted up to 750 mJ. The beam contains hot spots which are detrimental to a cleaning or stripping process ; homogenization was obtained via multiple reflections inside a hollow reflecting tube with a square section. An homogenized beam profile with energy density variations less than 15 % is obtained ; efficiency of the transformation is approximately 75 %.

The irradiated sample was a Kevlar type composite material covered with a two-layer polyurethane paint. This kind of material is used in the aeronautic industry as a part of the airframe. The laser pulse heats a layer of a few microns of the paint up to its decomposition temperature. The resulting hot gas flares up in a bright yellow plume generating a strong shock wave.

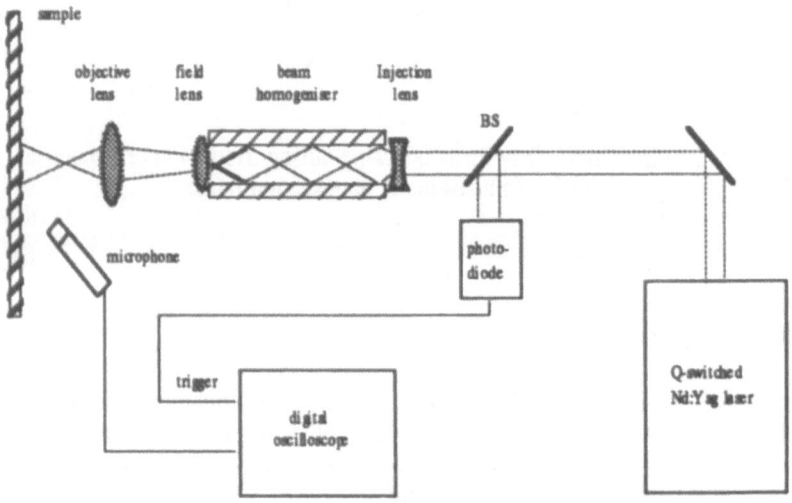

Figure 2 : Experimental set-up

3.3 PROCESS MONITORING

The shock wave was detected by a 10 kHz bandwidth microphone placed 40 millimeters away from the irradiated surface. The output of a photodiode receiving a part of the beam triggered the oscilloscope. As it was demonstratred and measured by other authors [8], the spectrum of the real waveform lies in the range of 30 to 300 kHz; the response of the microphone is pulsed and the signal waveform does not depend on the applied laser pulse energy. However, the amplitude of the signal reflects the strength of the shock wave, which itself, depends strongly of the laser-matter coupling coefficient, a parameter defined by the nature of the irradiated target. Figure 3 shows an example of the recorded signal.

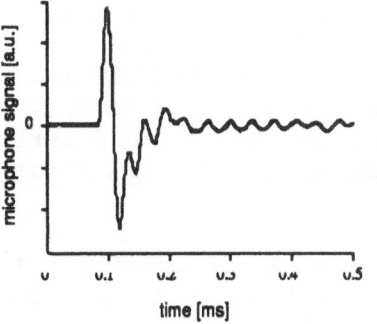 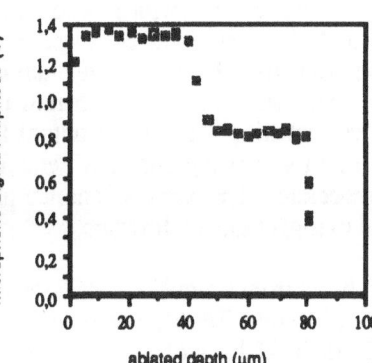

Figure 3 : Signal obtained from a micro-phone placed at 40 mm of the irradiated sample surface

Figure 4 : Monitoring of Nd:YAG laser removal depth from a painted surface with a microphone. The 1rst drop of the signal corresponds to the transition from the 1rst to the 2nd layer. The second drop is the substrate surface.

Figure 4 shows the variation of the amplitude of the recorded signal versus the ablated depth . The signal slightly increases over the first few microns due to an increase of absorption because of surface microstructure change. After that, the signal does not change over a depth of 45 microns ; this part of the curve corresponds to the removal of the first layer of paint. The sharp transition observed further, is recorded at the separation surface between the first and the second layer. The difference in microphone signal indicates a difference in absorption coefficient and hence, a difference in composition of the two layers. The second vertical drop in the signal indicates the position of the substrate surface. This result shows that a direct monitoring of the ablated thickness can be done during a laser surface cleaning process.

4. LASER DRILLING AND STRUCTURING OF CERAMICS

4.1 INTRODUCTION

Laser ceramics processing has recently received considerable attention due to the fact that ceramics cannot be easily machined by conventional tools. In particular, short pulses duration lasers processing has the advantage of avoiding thermal loading which may lead to the formation of microcracks. Experiments were performed with the experimental setup shown on Figure 5. The laser used in this work was either the previously described Q-switched Nd:Yag laser or an excimer laser (Model Lumonics EX748). The excimer laser generates 0,5 Joule per pulse of 15 ns duration at the wavelength of 248 nm (KrF). The maximum pulse repetition rate is 200 Hz. In both cases a uniform intensity portion of the beam was selected using an aperture and then imaged onto the sample by a projection lens. The diameter of the irradiated spot was about 0.8 mm. The laser fluence was determined by a GENTEC joulemeter.

The strong acoustical waves which propagate from the interaction spot into the solid sample on one side and into the surrounding gas on the other side, were characterized respectively with a piezo-electric transducer and a probe HeNe laser beam. The piezo-electric transducer consisted of a 9 µm thick film of PVDF having a 50 nm Ni-Al electrode film, bonded to the alumina sample to be machined. The other side of the transducer was bonded to a 5 mm thick piece of unpoled PVDF for acoustic impedance matching. The signal output from the transducers was fed into a digital oscilloscope for viewing and to a Personal Computer through a GPIB interface for further processing. The output of another photodiode, receiving a part of the laser beam was used to trigger the oscilloscope.

A beam from a 5 mW He-Ne laser was passed parallel to the substrate surface, at a distance of 15 mm. The blast wave generated by the interaction causes a gradient index that deflects the HeNe probe beam. The deflection was detected by a knife edge and a photodiode.

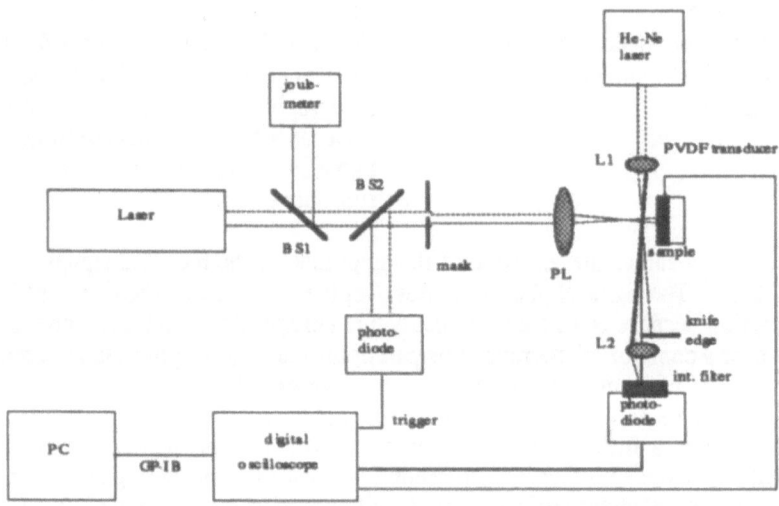

Figure 5 : Experimental set-up used to characterize the machining process of a ceramic sample using short pulses from a Q-switched Nd:YAG or an excimer laser. The two diagnostic tools shown are a PVDF transducer fixed on the back side of the sample and a probe HeNe beam deflected by the shock wave above the irradiated sample surface.

4.2 BLAST WAVE MODELLING

The idealized problem of a strong explosion in a homogenous atmosphere has already been solved by Sedov [9] who has considered a perfect gas with constant specific heats and density in which a large amount of energy is liberated in a small volume during a short time interval. In the conditions of irradiation of a material with focused short laser pulses, the energy release can be assumed to be both instantaneous and occuring at a point ; the point explosion model seems then to be valid to describe the occuring phenomena. This model is helpful to evaluate the amount of energy transfered from the laser beam into the shock wave. Several solutions exist depending on the pressure gradient. With the laser pulse energies we are considering (between 5 and 50 mJ), and in the region accessible to the observation, atmospheric pressure cannot be negected [10]. In this domain of weak shock wave, the solution to the problem of the propagation of the wave has been obtained by numerically integrating the appropriate partial differential equations. Numerical results, tables, and graphs of the distributions of the flow at different times can be found in References [9], [11] and [12]. These data have enabled us to draw the curve shown on Figure 6, that gives the relation between the shock wave energy and the transit time over the fixed distance of 15 mm.

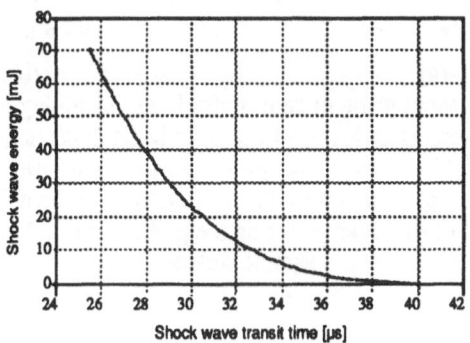

Figure 6 : Shock wave energy vs transit time over a fixed distance

4.3 EXPERIMENTAL RESULTS WITH EXCIMER LASER

Figures 7a shows the response of the piezo-electric transducer in terms of laser fluence ; Figure 7b displays the variation of shock wave energy versus the fluence. From these data we can determine a threshold for the the ablation which is about 1,5 J/cm^2. From a process control point of view, the important relation is that between the ablation rate and the recorded signals. For this reason, we have established the amount of matter removed at various laser fluences ; the result is displayed on Figure 8. By combining the data on Figure 7 and 8, it is possible to relate the monitoring signals to the depth of removed matter (Fig. 9a and 9b).

We see on Figure 9a that the piezo transducer signal displays a threshold below which no significant amount of matter is removed. The signal obtained in this zone where no ablation is observed can be attributed to thermoelastic stresses in the sample. Above threshold, the ablation thickness is linearly related to the recorded amplitude of piezo electric transducer. The measurement of the shock wave transit time (Fig. 9b) displays a similar behaviour. This type of response conveniently lends itself to process automatization, since by signal integration, it is possible at any time to know the thickness of ablated matter with better than 10 % accuracy and this, independantly of the fluctuations of process parameters. Figure 8 shows the relation between the ablation rate of the alumina sample and the laser fluence applied on its surface.

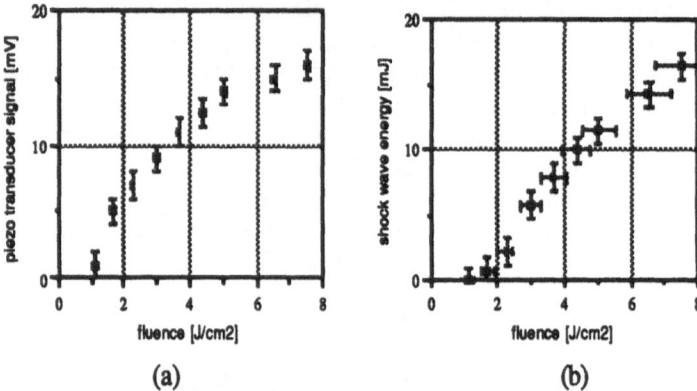

(a)   (b)

Figure 7: Monitoring of short pulse laser machining of ceramic using
a piezo-electric transducer and a beam deflection technique.
(a) Response of the piezo- electric transducer ;
(b) Shock wave energy evaluated from the measurement of the delay between the laser
pulse and the arrival of the shock wave on the probe beam at 15 mm above the solid
sample surface.

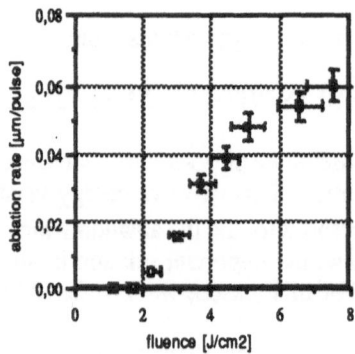

Figure 8 : Ablation curve of an alumina sample using KrF laser pulses

We see on Figure 9a that the piezo transducer signal displays a threshold below
which no significant amount of matter is removed. The signal obtained in this zone
where no ablation is observed can be attributed to thermoelastic stresses in the sample.
Above threshold, the ablation thickness is linearly related to the recorded amplitude of
piezo electric transducer. The measurement of the shock wave transit time (Fig. 9b)
displays a similar behaviour. This type of response conveniently lends itself to process
automatization, since by signal integration, it is possible at any time to know the
thickness of ablated matter with better than 10 % accuracy and this, independantly of the
fluctuations of process parameters. Figure 8 shows the relation between the ablation rate
of the alumina sample and the laser fluence applied on its surface.

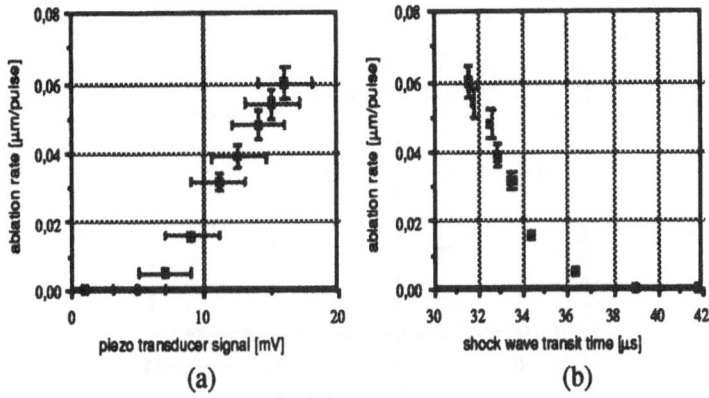

Figure 9 : Relation between ablation rate and
(a) Shock wave energy   (b) Shock wave transit time

4.4 EXPERIMENTAL RESULTS WITH ND:YAG LASER

The figures 10a and 10b show the responses of the piezo transducer and the laser beam deflection signal interpreted in term of acoustic energy. The behaviours of these two transducers are different. With the piezo-electric transducer, a good correlation with the ablation depth is obtained, as it was the case with excimer laser. The beam deflection technique however, can be used only in a short range of fluence (0 to 4 J/cm2). The deviation occuring at higher energy density can be attributed to the formation of an absorption wave. Actually, the projection setup used is a favorable configuration for the formation of an absorption wave since the energy density increases as the wave is moving away from the sample surface to the focal point. Since the coefficient of absorption varies as the square of the wavelength, this effect is stronger at 1,06 μm than at 0,248 μm. The formation of this absorption wave causes the screening of the sample from further irradiation ; as a consequence the ablation rate reaches saturation. One can see this phenomenon on Figure 11.

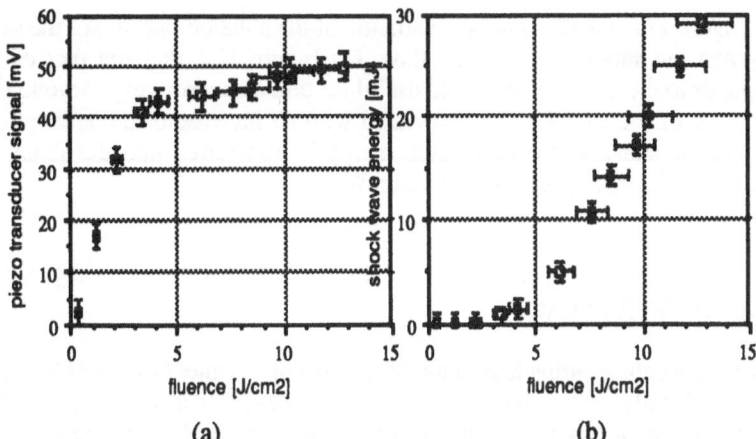

Figure 10 : Piezo-electric transducer response (a) and shock wave energy evaluated from transit time between irradiation zone and HeNe probe beam (b).

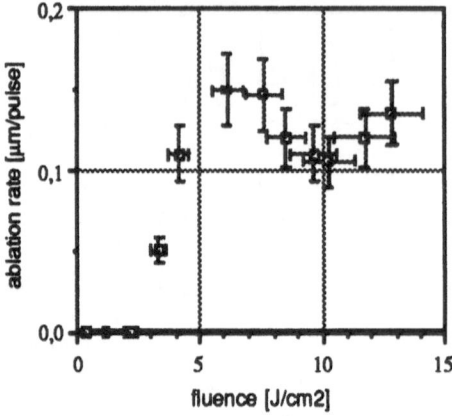

Figure 11 : Ablation curve of an alumina sample with a Q-switched Nd:YAG laser. The saturation observed above 6 J/cm2 is due to screening of laser energy by the plasma.

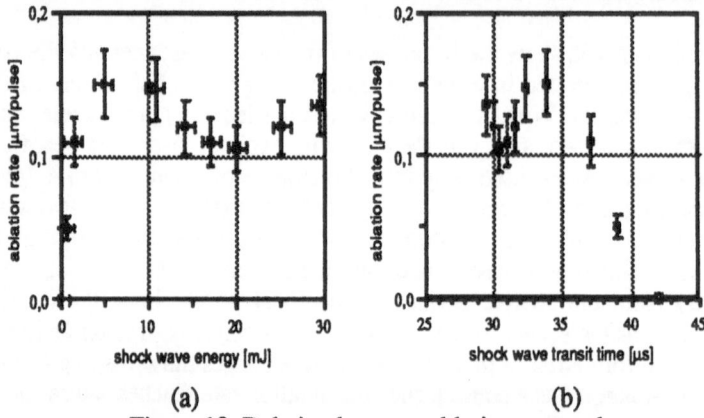

(a)   (b)

Figure 12: Relation between ablation rate and
12.a : Shock wave energy   12.b : Shock wave transit time

Figure 12a and 12b show the variation of the ablation rate versus the shock wave energy and the shock wave transit time. On Figure 12a, the first part of the curve corresponds to the range of the signal useful for process monitoring. Beyond this region the increase of the signal is not associated with an increase of the amount of ablated matter. In this zone, additionnal laser energy is transferred into the shock wave via formation of a Laser Supported Detonation Wave.

4.5 COMMENTARY ON THE INTERACTION MECHANISMS VS LASER RADIATION WAVELENGTH

Phenomena occuring during irradiation of alumina by excimer laser and Q-switched laser differ significantly. In the first case, absorption takes place at the sample surface [13] , inside a layer of a few tenths of nanometers thickness; ablation mechanism is then essentialy photolytic [14, 15].

In the second case, the material is partially transparent to the laser wavelenth (1,06 μm). Absorption takes place on the impurities located at the grains boundaries; as a result, absorption depth is of the order of 1 to 2 μm, roughly the grain size. Ablation in this case is of thermal origin. According to theoretical modelling published in the literature [16, 17], maximum temperature is reached inside the sample, the surface being cooled via evaporation. Overheating of the material, in the solid and liquid phase, can reach several hundreds of degrees [18] and leads to a explosive boiling. Observation of the morphology of the irradiated surfaces and measurements of the shock wave energy corroborate these theoretical descriptions.

Figures 13 and 14 are SEM photographs of an alumina sample irradiated respectively with a Q-switched Nd:YAG and an excimer laser at a fluence of 4 J.cm^{-2} . Each zone has received 200 pulses. In the of case of 1,06 μm irradiation (Fig.13), the bottom of the ablated area shows a high roughness. On the edges, long radial corrugations can be seen, that are due to explosion of the molten zones ; many droplets with an average size of 3 to 5 μm are also observed. In the case excimer irradiation, the ablated surface shows much smaller roughness and the redeposited material covers a smaller area.

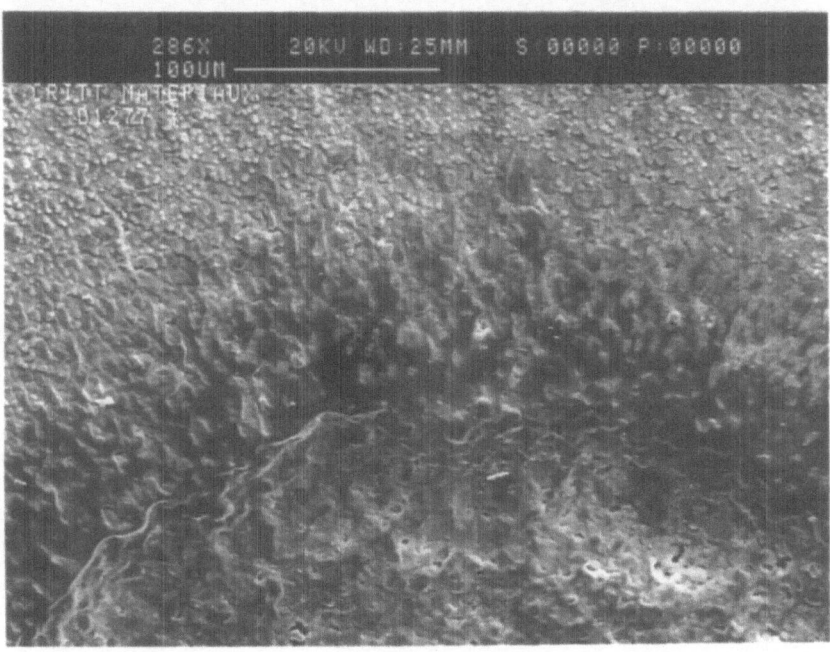

Figure 13: Surface morphology of alumina sample irradiated with 200 Q-switched Nd:YAG laser pulses at 4 J/cm2.

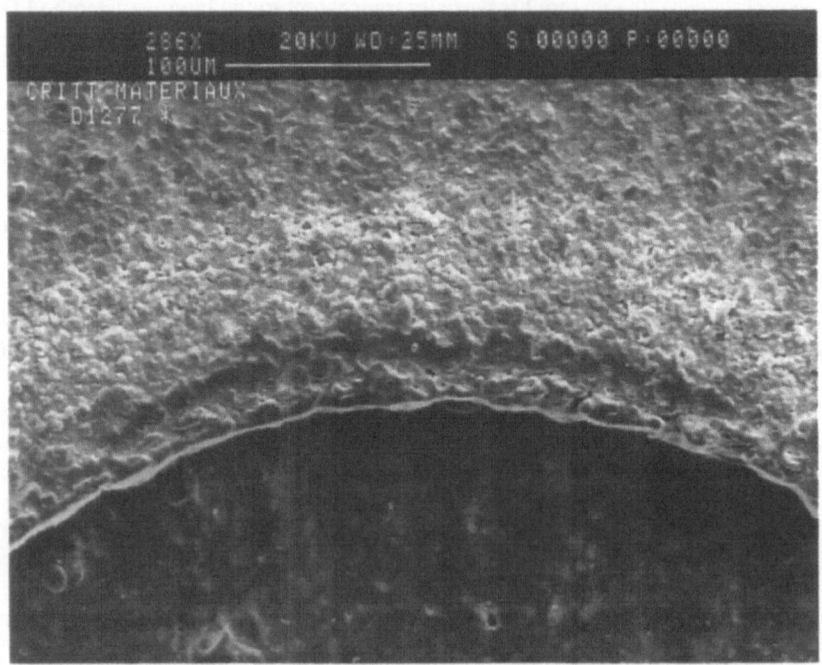

Figure 14: Surface morphology of alumina sample irradiated with 200 KrF excimer laser pulses at 4 J/cm2.

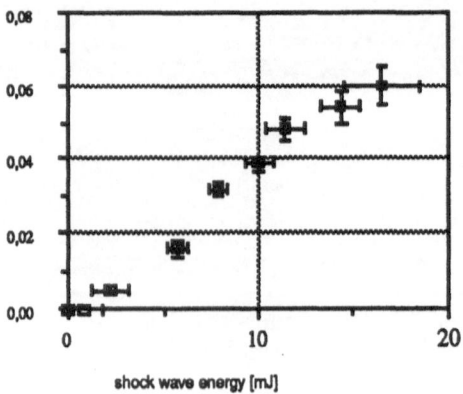

shock wave energy [mJ]

Figure 15: Relation between ablation rate and shock wave energy

Figure 15 shows the variation of the shock wave energy versus the ablated mass in the case of excimer laser irradiation. The relation is quasi-linear. By assuming that the shock wave energy corresponds to the cinetic energy of the debris as in [19-20] and taking a Maxwellian distribution of the debris speeds, we find an average speed of the debris of 14 km.s^{-1}, in agreement with reference [15]. In the case of Nd:YAG irradiation in the regime of low fluence, the shock wave is not influenced by the vapor breakdown

same hypothesis on the shock wave energy applies. Figure 12 shows that for a given ablated mass, the shock wave energy is an order of magnitude lower that for excimer laser irradiation. This is due to a higher size for the debris and an ejection speed, 3 to 4 times lower.

5. CONCLUSION

Short laser pulses from an excimer and a Q-switched Nd: YAG laser have been used for two applications: surface cleaning and micro-machining. The main effect of the laser pulses on the material is mechanical : a pressure wave appears inside the target and a shock wave is generated in the surrounding gas. These phenomena have been observed using three different diagnostic techniques : microphone, probe laser beam deflection and piezo-electric transducer. In the case of surface layers removal, we have shown that it is possible to discriminate the different layers from the substrate. In the case of micro-machining with excimer laser, the reponses of two sensors allow to evaluate on-line the amount of ablated matter with an accuracy between 2 and 10 % depending on the number of pulses used. During machining with Nd:YAG laser beam, the range of predictbility is reduced due to absorption of laser energy in the breakdown zone in the gas medium. As a consequence, it appears more interesting to increase the pulse repetition rate instead of pulse energy in order to improve ablation efficiency. The effect of wavelength on the interaction mechanism has been shown. In contrary to excimer laser processing Q-switched Nd:YAG laser irradiation leads to overheating, explosive boiling and ejection of larger size debris.

ACKNOWLEDGMENTS

This work was done at the Irepa-Laser Institute with a grant from the European Community.

6. REFERENCES

1. Dyer, P. E., Farrar, S.R., and Key, P.H. (1992) Fast time response photoacoustic studies and modelling of KrF laser ablated YBa2Cu3O7, *Applied Surface Science* **54**, 255-263.
2. Ghost, A.P., and Hurst, J.E. (1988) Photoacoustic studies of excimer laser induced ablation of polymethylmethacrylate, *J. Appl. Phys.* **64(1)**, 287-290.
3. Sell, J.A., Heffelginger, D., Ventzek, P, and Gilgenbach, R. (1991) Photoacoustic and photothermal beam deflection as a probe of laser ablation of materials, *J. Appl. Phys.* **69(3)**, 1330-1335.
4. Ventzek, P., Gilgenbach, R., Heffelginger, D., and Sell, J. (1991) Laser beam deflection measurements and modeling of pulsed laser ablation rate and near surface plume densities in vacuum, *J. Appl. Phys.* **70(2)**, 587-593.
5. Ready, J.F. (1971) *Effects of High-Power Laser Radiation*, Academic Pess, New York.
6. Von Allmen, M.(1986) *Laser-Beam Interactions with Materials*, Springer-Verlag.
7. Asmus, J. F., Seracini, M., and Zetler, M. J. (1976) Surface morphology of laser-cleaned stone, *Lithoclastia* **1**, 23-46.

894

8. Diaci, J. and Mozina, J. (1992) A study of blast wave waveforms detected simultaneously by a microphone and a laser probe during laser ablation, *Appl. Phys.* A55, 352-358.

9. Sedov, L. I. (1959) *Similarity and Dimensional Methods in Mechanics*, Academic Press, New York.

10. Zel'dovich, Y.B. and Raizer, Y.P. (1966) *Physics of Shock Waves and High-Temperature Hydrodynamic Phenomena*, Academic Press, New York.

11. Goldstine, H.H. and Von Neumann, J. (1955) Blast wave calculation, *Commun.in Pure Appl. Math.*8,327-353.

12. Okhotsimskii, I.L., Kondrasheva, Z.P., Vlasova, R.K., and Kasakova, R.K.(1957) Calculation of a point explosion with counterpressure, *Tr. Mat. Inst. Akad. Nauk. SSSR* 50.

13. Rothenberg, J.E. and Koren, G. (1984) Laser produced plasma in crystalline «--Al$_2$O$_3$ and aluminium metal, *Appl. Phys. Lett* 44(7), 664-666.

14. Dreyfus, R.W., Mc Donald, F. A., and Von Gutfeld, R.J. (1987) Laser energy deposition at sapphire surfaces studied by pulsed photothermal deformation, *Appl. Phys. Lett.* 50(21), 1491-1493.

15. Dreyfus, R. W., Kelly, R., and Walkup, R.E. (1986) Laser-induced fluorescence studies of excimer laser ablation of Al$_2$O$_3$, *Appl. Phys. Lett* 49(21), 1478-1480.

16. Dabby, F.W., and Un-Chul Paek (1972) High-intensity laser induced vaporization and explosion of solid material, *IEEE J. Quantum Electronics* QE-8(2),106-111.

17. Singh, R.K., Bhattacharya, D., and Narayan, J. (1990) Subsurface heating effects during pulsed laser evaporation of materials, *Appl. Phys. Lett* 57(19),.2022-2024.

18. Smurov, C., Surry, C., Mazhukin, V.I., and Flamant, G. (1994) Overheated metastable states in pulsed laser deposition versus laser radiation wavelength, *J. de Physique* C4 vol 4, 151-154.

19. Zyung, T., Kim, H., Postlewaite, J.C., and Dlott, D.D. (1989) Ultrafast imaging of 0.532 µm laser ablation of polymers: Time evolution of surface damage and blast wave generation, *J. Appl. Phys.* 65(12), 4548-4562.

20. Grad, L., and Mozina, J. (1992) Acoustic in situ monitoring of excimer laser ablation of different ceramics, *Appl. Surf. Sci.* 69,370-375

ADVANCES IN LASER DIRECT WRITING

T. M. BLOOMSTEIN, S. T. PALMACCI, R. H. MATHEWS,
N. NASSUPHIS and D. J. EHRLICH
Lincoln Laboratory, Massachusetts Institute of Technology
Lexington, Massachusetts 02173-9108

1. Abstract

Recent improvements in the laser direct write technology are reviewed with particular emphasis on new applications for micromechanics and advanced electronic packaging. The laser stereo etching method has been scaled so as to permit the accurate three-dimensional etching of silicon in a chlorine ambient at a rate of one-half million cubic micrometers per second. Laser deposition processes have been found for the direct writing of metal interconnects on copper/polyimide substrates and similar materials for multichip modules. Apparatus to implement these processes on highly three-dimensional substrates has been developed.

2. Introduction

The laser direct writing technology has been developed for basic microelectronic fabrication steps such as deposition, etching, passivation, and doping.[1] These steps are accomplished in a maskless fashion directly under software control. They therefore provide powerful methods for prototyping and for direct alteration of microsystems. For integrated circuit (IC) applications the laser techniques supplement the capability of focused ion beam processing. The microchemical writing speed of the laser techniques is 2 to 6 orders of magnitude greater than the ion beam analogs. Additionally, the electronic materials qualities of the laser-deposited thin films are much higher, e.g., resistivity is typically 2 orders of magnitude greater than the best ion beam deposited films. In this paper we review the extensions of the laser technology to microlectromechanical systems (MEMS) and to multichip modules (MCMs).

J. Mazumder et al. (eds.), Laser Processing: Surface Treatment and Film Deposition, 895–906.
Kluwer Academic Publishers.

Laser direct write processes offer an attractive option as primary patterning methods for the emerging field of MEMS.[2] Conventional lithographic techniques for fabricating these inherently three-dimensional mechanical devices are limited to unidirectional extensions of two-dimensional patterns, for example, by x-ray lithography in thick photoresist[3] or by crystallographic selective etching. Only simple mechanical structures are possible with such extended shapes. More truly three-dimensional structures are required to optimize structural characteristics such as mechanical function, strength, fracture resistance, flow properties, and out-of-plane coupling. As mentioned above the laser direct writing has been found to have important applications in IC debug and testing. The optimization of the technology for MCMs requires two main areas of refinement, namely, (1) development of processes and instrumentation to permit writing on highly three-dimensional substrates with topography greatly exceeding that of ICs, and (2) development of materials compatibility with MCMs, specifically compatibility with copper/polymer interconnect metallurgy.

3. Apparatus

The applications in MEMS and MCMs both require laser writing instrumentation with significant three-dimensional patterning capability. Despite many similarities, the requirements for the two applications have some differences.

For the MEMS application scanning speed is a dominant consideration. One system is illustrated in Fig. 1. An acoustooptic scanner is used to digitally access a 256 ¥ 256-pixel scanning field with 0.1-pixel accuracy. Random access rates up to 50,000 pixels/s are achieved and are limited by acoustic wave stabilization in the crystals. The address grid is spaced at 1-μm increments in the x-y plane. Intensity variations due to the nonlinear transmission of the deflection crystals over the scanning field are compensated using closed-loop feedback by an additional acoustooptic crystal and a photodiode that senses a portion of the output beam. Focusing of the 900-mW 488-nm laser light is through a 0.4 NA, 20¥ flat-field-corrected objective to ~1.0-μm beam size, although the effective reaction zone may be slightly larger because of heat transfer in the highly conducting bulk.[4] The beam is introduced through a quartz glass cover into a stainless steel vapor cell containing the sample. Circularly polarized 488-nm light from a 15-W argon-ion laser is used as the source, and a helium-neon laser beam is also introduced into the system for autofocusing on the surface using a spot minimization algorithm. The reaction is observed through the focusing optic with a CCD camera.

A second scanning system for MEMS utilizes a high-performance x-y galvanometer scanner to deflect the output of a 15-W argon-ion laser running multiline (488- to 514-nm wavelength). The laser beam is expanded to 16 mm and focused to a 10-µm spot through a lens having 250-mm focal length. The beam is introduced through a quartz cover glass into a stainless steel sample cell that is backfilled with chlorine gas. The sample is biased to 300 °C using a halogen lamp incident on the back side of the wafer.

The MEMS scanning systems are driven by computer-aided-design/computer-aided-manufacturing (CAD/CAM) software in a manner similar to recent demonstrations of rapid prototyping of macroscopic plastic parts by solution polymerization. Three-dimensional structures are first constructed using a commercial solid-modeling software package. Additional software has been developed to digitize the structures into a stack of planar masks, each comprising an array of pixel addresses. The stored arrays are read out from an 80386-based computer in real time during fabrication and converted into analog voltages for activating the appropriate driver circuitry for each scanner. Drivers for controlling shuttering, intensity leveling, and focus position have also been integrated. The vapor cell is located on an x-y stage, allowing multifield structures to be written.

For the MCM applications a scanning system with accurate large-excursion travel but slower scan speed is needed. This instrumentation, as illustrated in Fig. 2, is constructed around a precision x,y,z servo-motor-controlled stage system. All three axes are position-encoded to 0.15-µm resolution and are governed along with the lasers by a dedicated multiaxis microprocessor that is embedded in a personal computer. Two lasers are used: a diode-pumped frequency-doubled Nd:YLF laser at 525-nm wavelength and an argon-ion laser operating multiline (main output 488 and 514 nm). The 525-nm output is a Q-switched pulse train at up to 50,000 pulses per second and nominal pulse width 10 ns. Typical average power at the substrate is 1-50 mW. The argon-laser output is continuous wave and is typically at a power of 10-500 mW at the substrate. The substrate is enclosed in a flowing vapor cell connected to a roughing pump (base pressure 25 mTorr).

4. Stereo Laser Etching

The flexibility of laser microchemical processing is achieved by utilizing highly localized laser microchemical reactions and accurate scanning of the laser beam. Depth contouring is accomplished by dynamically refocusing the laser beam. Initial investigation has concentrated on laser microchemical etching of silicon because of its excellent mechanical properties and its predominance in MEMS fields. Laser-induced metallization of resulting structures as well as replication through compression molding have been demonstrated.

The process of the greatest initial interest has been the thermal microreaction of silicon in a chlorine ambient. Silicon, when heated to its melting point, reacts nearly at the gas-transport-limited rate with chlorine. Volatile silicon chlorides are swept away in the slowly flowing gas (~1 sccm). Figure 3 shows an etched structure in silicon. Nominally 1-μm-thick planes are removed by acoustooptically scanning the laser beam at 7500 μm/s (133-μs dwell per 1-μm pixel) over the areas to be etched in a 100-Torr chlorine ambient. After a plane is etched, the focusing objective is lowered 1 μm, and a new pattern is etched. Depths are accurately controlled through precise timing of the dwells of each scan plane. Further smoothing is possible if desired by reducing the pressure or laser intensity during the final laser scans. Atomically smooth (faceted) surfaces are possible in the low-laser-power photochemical etching regime. Additionally, structures can be replicated using simple compression molding techniques. The duplicated version of an array is shown in Fig. 4. This was transferred into poly-vinyl-chloride through thermocompression at 185 °F under 1000 psi of loading.

Structures larger than the 256 ¥ 256-pixel field are fabricated using field stitching with x-y stage motion. The latter capability is essential for many applications. For example, a class of microfluidic flow-plate devices has been developed around the ability of the laser microchemical technique to contour three-dimensionally. A library of structural cells similar to IC standard cells has been designed. The CAD system allows fast layout of a complex flow system through assembly of these predefined shapes.

The rate of the chlorine reaction evolves in time and scales with pressure, power density, and laser spot size. For example, with an ~1.0-μm reaction zone (laser power of 900 mW), the measured removal rate in bulk silicon is $2.0 ¥ 10^4$ μm^3/s at 400 Torr and is only weakly dependent on scanning speed and laser power. Changing the laser dwell time at a pixel alters the etched depth in a nearly linear fashion. This is because thermal and mass transport processes have achieved steady state, although the etch eventually terminates because of beam divergence past the depth of focus and sidewall scattering in the progressing etch. A larger dynamic range in etch rate is obtained by changing the chlorine pressure.

The galvanometer system is scanned up to angular velocities of 40 rad/s. This translates to linear velocities on the surface of 10 m/s with the current focusing optic. These optical and scan parameters were chosen to allow changes in thermal and mass transport dynamics to be evident. The mass diffusion coefficient of chlorine and the thermal diffusion coefficient of silicon near melt are both ~ 0.1 cm^2/s. For a reaction zone on the order of 10-μm diameter, the characteristic transport time, $t_{transport} \approx w_o^2/D$ where w_o is the spot size and D is the diffusion coefficient, is approximately 10 μs. The resident time of the laser beam is $t_{residence} \approx w_o/v_s$ and is a factor of 10 below the transport time for scanning velocity v_s, equal to 10 m/s. A volumetric removal rate of 5 ¥ 10^5 μm^3/s can now be achieved by optimized scan speed and pressure conditions. Similarly, a surface roughness average of less than 0.1 μm can be achieved on laser-etched surfaces.

5. Processing for Multichip Modules

Important metallurgical and materials problems specific to laser processing of MCMs are laser deposition on the thermally fragile polymer dielectrics, particularly on polyimide, and electrical performance. Adhesion and stress testing is necessary to qualify the laser processes for characteristic MCM substrates. The laser formation of vias in polyimide is a well-studied problem and is not treated here.

Laser deposition processes were characterized on copper/polyimide, SiO$_2$, and ceramic MCM substrates from different sources. Polyimide layers from various suppliers were typically 25-75 μm thick and were solvent degassed after spin-on by various mild- to hard-bake conditions as specified by the manufacturers. Qualitatively similar results were found for all materials. Laser deposition experiments were performed dominantly with the argon laser, since the continuous-wave source provides the greatest writing speed. Deposited lines were characterized by mechanical stylus and electrical probe. Table 1 collects results for five deposition processes on polyimide. The precursors used were platinum fluorophosphine, gold acetylacetonate, trimethylvinyl silyl hexofluoro acetylacetate copper (I), molybdenum fluorophosphine, and tungsten hexafluoride.[5] The reaction mechanisms and detailed chemistry are under further characterization and will be reported separately. Figure 5 illustrates a typical result for platinum deposition with the argon laser. Because of its high writing speed and excellent metallurgical stability, platinum deposition was selected as the favored process for most MCM applications. Gold and copper have superior resistivity and are available when this is the dominant consideration. Laser-written interconnect is suitable for applications in design iteration, trimming, and repair. In all of these areas the laser-written length is a small fraction of the total MCM interconnect.

An important application of laser direct writing will be the fast-turnaround (or real-time) modification of MCM interconnections. Ion beams do not have a sufficient writing speed or deposited-material conductivity for MCMs, and therefore the laser techniques will be the clear approach of choice. From the IC analogy there will be large benefits from maskless iteration of interconnect designs to perform system testing and to optimize parametric performance of system prototypes. The laser techniques are entirely reversible since both cutting and patching of interconnect are available and in many instances they may be performed in an active test environment. A full interconnect rework process was developed including deposition, cutting, via formation, and contact formation. Figure 6 illustrates a typical example of conductor reconfiguration.

For high-speed applications the circuit terminations at individual ICs (traditionally bond pads) become an important design consideration. Frequently, process control is not adequate to reproducibly achieve the needed accuracy for these terminations. Laser trimming of terminations can be implemented as a post-test or active-test (in situ) operation. We have developed a method to make such adjustment on terminating TaN thin-film resistors. Radiation from a Nd:YLF laser is used to convert the TaN resistor material to an oxidized insulator by solid-state reaction. The process can be implemented on fully passivated MCM substrates without compromising the passivation layer.

Figure 7 shows a serpentine resistor converted to insulator at isolated circular regions along its centerline. Each spot is produced by either a single laser pulse or a gated series of pulses. Other patterns for converting the TaN are possible, but this method of controlling dose offers the advantage that each converted spot raises the resistance by a fixed amount. The trim accuracy is ~ 1 W for the current apparatus.

6. Conclusions

Three-dimensional patterning and replication of microstructures have been demonstrated by precision laser machining. An etching rate of up to $5 ¥ 10^5$ μm^3/s has been achieved. Processing of partially fabricated silicon parts, including those with prefabricated damage-prone electronic devices should be possible. Replication of silicon master shapes by thermocompression of polymers has been demonstrated. Software for conversion of three-dimensional-shape files to laser scan routines and for rapid design of complex microfluidic devices from standard cells has been developed. A complete set of laser direct write interconnection processes has also been developed for emerging MCM applications. The greater topography of the substrate has required the construction of

METAL	TYPICAL LINE DIMENSIONS	RESISTIVITY (μΩ • cm)	WRITING RATE (μm/s)
PLATINUM	14-55 μm WIDE 0.5-8 μm THICK	17-20 (Bulk = 10.6)	200
GOLD	20-45 μm WIDE 2.5-20 μm THICK	3-5 (Bulk = 2.4)	10
COPPER	25-80 μm WIDE 1.2-4 μm THICK	7-10 (Bulk = 1.7)	10
MOLYBDENUM	45-150 μm WIDE 0.2-3 μm THICK	25-80 (Bulk = 5.2)	100
TUNGSTEN	5-40 μm WIDE 1-4 μm THICK	12-22 (Bulk = 5.6)	200

Table 1.

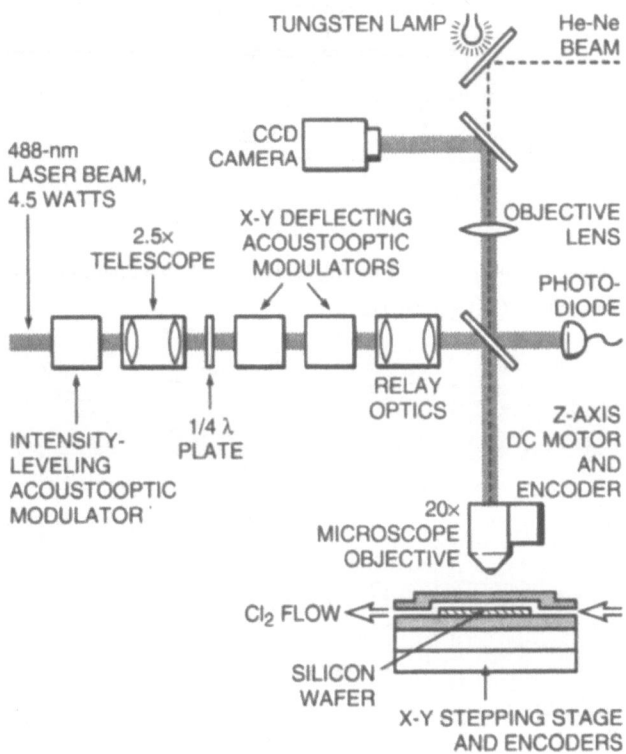

Figure 1. Schematic diagram of the acoustooptic scanning apparatus. The x-y deflecting modulators cover a digitally controlled 256 x 256-element scan field that is addressable in a vector scan at up to 50,000 elements/s.

902

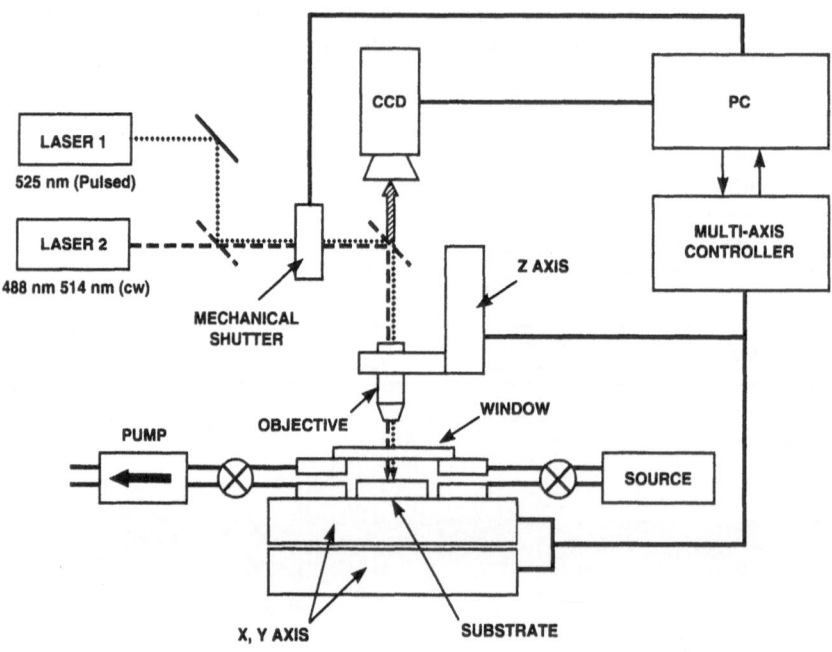

Figure 2. Schematic diagram of laser direct writing instrumentation with three-dimensional servo motion control for multichip module (MCM) applications.

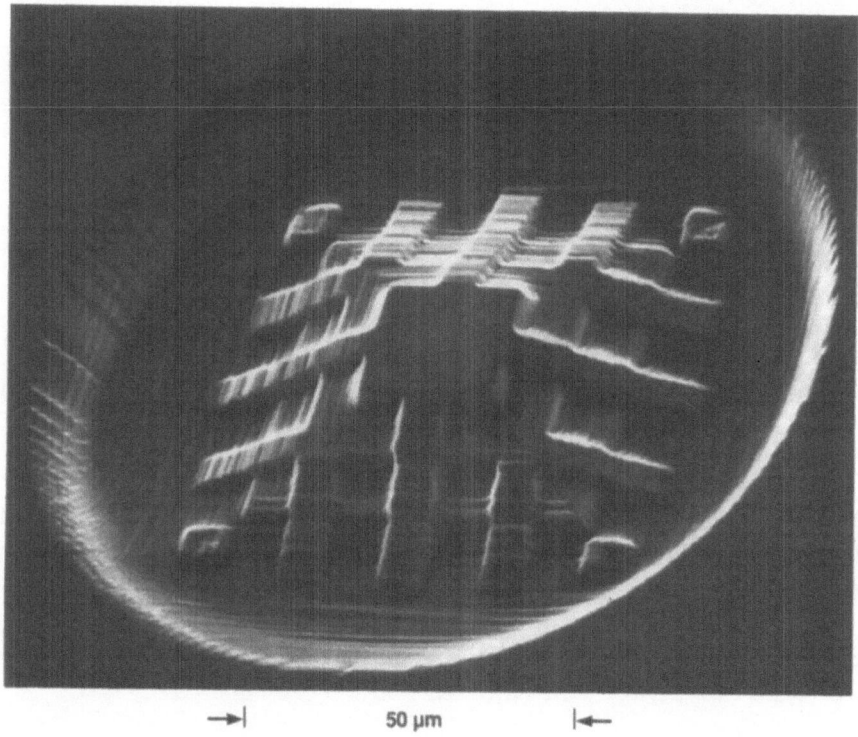

Figure 3. Scanning electron micrograph of a typical structure produced in silicon by the laser stereo etching technique.

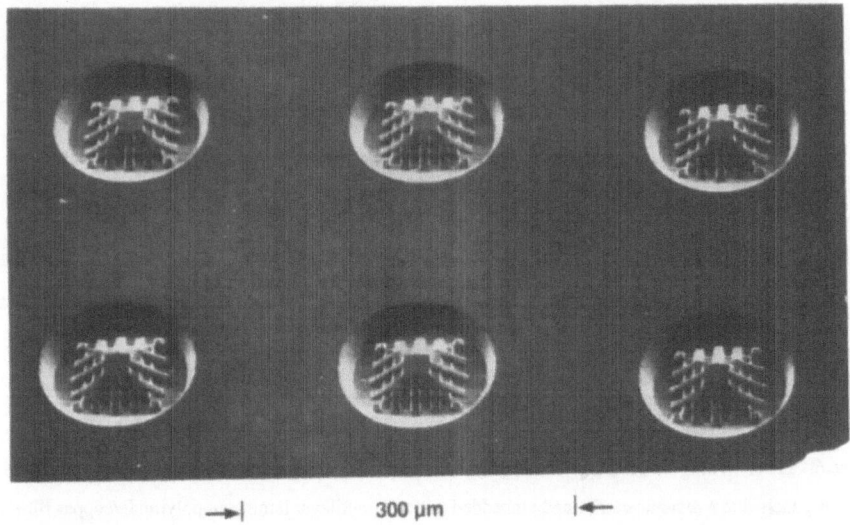

Figure 4. Replicated array in poly-vinyl-chloride from a silicon mold using thermocompression. Replication is at 185 °F under 1000 psi of pressure for a few seconds.

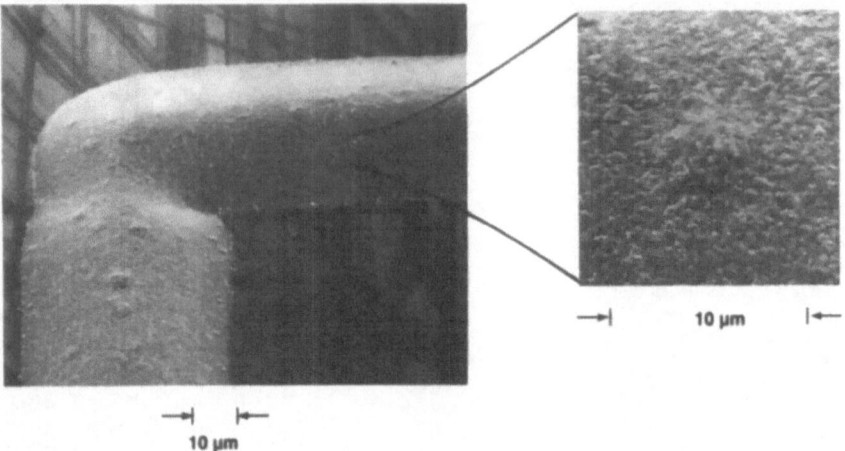

Figure 5. Platinum line laser deposited over 60-μm-thick polyimide (Hitachi L110) substrate. An argon laser (488 nm) was used at 125-mW power. The deposited interconnect is 6.9 μm thick and has a measured resistivity of 18.4 μΩ cm or a resistance per unit length of 0.84 Ω/mm.

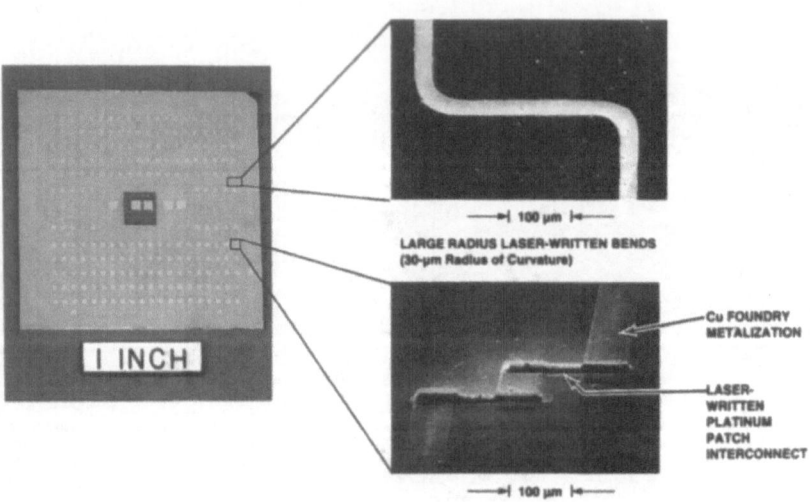

Figure 6. Laser writing of platinum conductors on a polyimide/copper MCM test substrate. A silicon chip center is attached to a ceramic carrier and embedded under a multilayer laminated polyimide/copper film. Laser writing is used to modify the copper interconnect pattern.

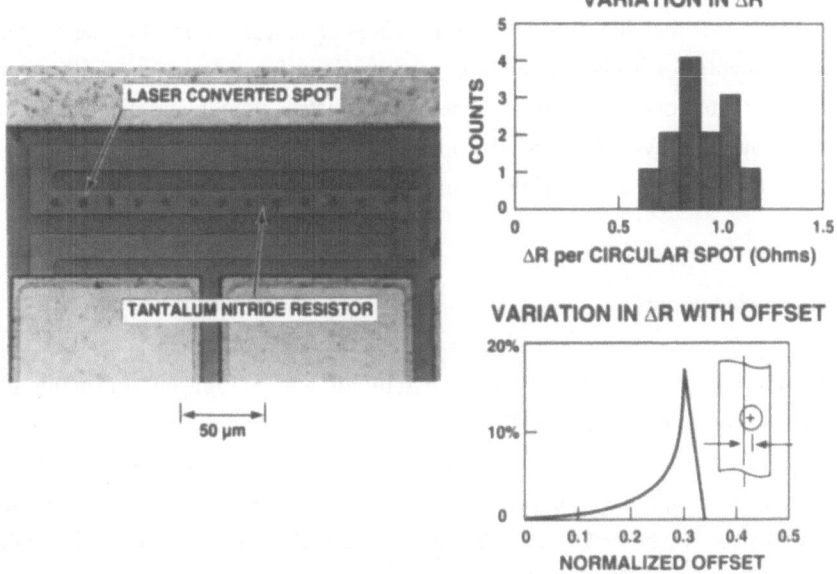

Figure 7. Serpentine TaN terminating resistor converted by spot irradiations in order to trim resistance value. Trimming is by solid-state reaction without disturbing passivation. The histogram at the upper right shows the increase in resistance ΔR for 13 individual conversions of approximately circular regions near the center line of a single planar resistor using a gated series of laser pulses. The measured variance in ΔR is 0.85 Ω. The plot in the lower right shows the predicted variation in ΔR for changes of the offset distance from resistor center line (normalized to resistor width).

laser apparatus for writing three-dimensional scanning trajectories. Laser deposition processes produce metallic conductors of excellent electrical and mechanical properties, and are available for MCM system applications requiring up to many millimeters of discretionary interconnect. All other necessary dielectric etching and conductor cutting operations are also available by laser operations. These methods provide the leading alternative for real-time engineering, trimming, and substrate recovery.

7. Acknowledgments

We would like to thank Professors M. Schmidt and K. Jensen for many useful suggestions, and S. Palmacci and W. DiNatale for expert technical assistance. We also gratefully acknowledge help from G. Loney in setting up the steering mirror and J. Bartley and W. Delaney for performing the compression molding. This work was supported by the Air Force Office of Scientific Research. T. M. Bloomstein is supported under a fellowship from Rockwell International Corporation.

References

1.   See, for example, *Laser Microfabrication: Thin Film Processes and Lithography*, edited by Ehrlich, D.J. and Tsao, J Y. (Academic, Boston, 1989).

2.   Bloomstein, T.M. and Ehrlich, D.J. in *Proceedings of MEMS '91* (IEEE, New York, 1991), pp. 202-203; Bloomstein, T.M. and Ehrlich, D.J. in *Technical Digest of Transducers '91* (IEEE, New York, 1992), pp. 507-511; Bloomstein, T.M. and Ehrlich, D.J., Appl. Phys. Lett. **61**, 708 (1992).

3.   Ehrfeld, W., Götze, F., Münchmeyer, D., Schelb, W. and Schmidt, D., *in Technical Digest of the Solid-State Sensor and Acutator Workshop* (IEEE, New York, 1988), pp. 1-4; Guckel, H., Christenson, T.R., Skrobis, K.J., Denton, D.D., Choi, B., Lovell, E.G., Lee, J.W., Bajikar, S.S. and Chapman, T.W., in *Technical Digest of the Solid-State Sensor and Actuator Workshop* (IEEE, New York, 1990), pp. 118-122.

4.   See Ashby, C.I.H. and Tsao, J.Y., Ref. 1, pp. 268-271.

5.   Black, J.G., Doran, S.D., Rothschild, M. and Ehrlich, D.J., *Appl. Phys. Lett.* **56**, 1072 (1990).

LASER CLEANING IN ART RESTORATION : A REVIEW

K. G. Watkins[†], J. H. Larson [††], D. C. Emmony [†††] and W. M. Steen[†]

[†]Department of Mechanical Engineering, University of Liverpool, UK
[††]Conservation Division, National Museums and Galleries on Merseyside,
Liverpool, UK
[†††]Department of Physics, Loughborough University of Technology, UK

Abstract

This paper presents a review of previous work on laser cleaning from the perspective of conservation and restoration in the art world. Particular attention is given to laser cleaning mechanisms, including selective vaporisation, ablation, selective removal of small particles, steam cleaning and special features for metals.

Results of recent work in our laboratories on the cleaning of one of the world's first aluminium statues, the Liverpool Gilbert *Eros* statue, using a Q-switched Nd:YAG laser and the investigation of the effect of process control features (laser power, working distance, number of pulses per position) on the morphology and composition of the cleaned surface (as determined by EDX, EPMA, XRD and SEM) are also presented.

1. Introduction

1.1 BACKGROUND ON THE USE OF LASERS IN ART RESTORATION AND CONSERVATION

The application of lasers in conservation and restoration is increasing as the advantages of the use of optical energy over conventional techniques becomes clearer with the development of more portable, more effective and more flexible laser systems. Lasers have been successfully applied in cleaning [1], consolidation [2] and imaging applications [3]. Laser cleaning is the most widely applied technique and has been used in the successful removal of black sulphation from marble and limestone, tarnish from silver threads in textiles, overpaints from monochromatic upholstery, fungi from leather and vellum, calcareous deposits from pottery and lead, corrosion from bronze, encrustations from stained glass, slaked lime from frescos and graffiti from concrete [4]. In all these cases, successful action was defined as the removal of the surface encrustation with minimal damage to the underlying material. Since this is a physical process which

J. Mazumder et al. (eds.), Laser Processing: Surface Treatment and Film Deposition, 907–923.
© 1996 *Kluwer Academic Publishers.*

is applied under conditions in which there is a strictly limited interaction with the substrate material, laser cleaning is often more effective than traditional cleaning methods based on chemical or mechanical action [5-9].

1.2 ADVANTAGES OF LASER CLEANING

Conventional techniques involve particle/air abrasion techniques or chemical techniques. Particle/air abrasion techniques pose environmental hazards to the operative and introduce changes in surface profile. Chemical techniques involve the above difficulties and also pose problems over the continuing action of the chemicals involved, perhaps over long periods after restoration has ended. Laser cleaning has the following advantages over these processes[1,7] since it is:

1) a physical process which ceases shortly after the laser pulse has ended.

2) a selective process which can be tuned for the removal of specific substances,

3) a non-contact process - can be automated and produces no contact wear,

4) a process that preserves surface relief,

5) versatile - most materials can be removed by correct selection of operating conditions,

6) controllable - a specific thickness of material can be removed.

2. Mechanisms of Laser Cleaning

The following mechanisms have been proposed:

1) Selective evaporation of encrustations,

2) Scouring of the surface by the action of rapidly expanding vapours (especially localised steam cleaning that takes place when a thin layer of water is added to the surface prior to laser irradiation),

3) Thermal and photo decomposition of superficial layers that are subsequently easily removed by washing,

4) Removal of small adherent particles as a result of selective excitation of the particles or the substrate.

5) Delamination of superficial deposits as a result of thermal expansion mismatch with the substrate,

6) Spallation induced by a laser induced shock wave (ablation mechanism).

Laser cleaning can be tuned to emphasise one or more of these interactions by selecting laser parameters (wavelength, pulse length, power density), or the cover fluid (air, argon, water, nitrogen).

2.1 BACKGROUND AND EARLY WORK BY ASMUS

The origins of this work can be traced to 1972 when the Italian Petroleum Institute funded John Asmus from University of California, San Diego, to visit Venice to study laser holography for the recording of the city's decaying treasures. During this work Asmus was asked to observe the effect of the interaction of a focussed ruby laser (which had until then been used for holographic recording) with an encrusted stone statue. It was found that the darker encrustations were selectively removed from the surface, resulting in no apparent damage to the underlying, white stone.

Asmus returned to the United States and began researching laser cleaning of art works, laying the basis for a powerful series of techniques, particularly in the use of pulsed ruby and pulsed Nd:YAG lasers. Interestingly, the connection between the use of lasers in cleaning and in holographic recording of art works remains, with the two technologies developing side by side in the conservation world.

From work on the laser cleaning of stone and marble using a ruby laser [1-4,6,7,9], Asmus concluded that there are two principal cleaning mechanisms [4,9]. In normal pulse mode (pulse duration approximately 1 μs - 1 msec), cleaning occurred as a result of the selective vaporisation of the surface contaminants compared with the underlying material which remained almost wholly unaffected. This in turn occurred when the absorption coefficient of the darker encrustation was sufficiently large to lead to a temperature rise favouring vaporisation while the absorption coefficient of the underlying material was sufficiently small to limit temperature rises to moderate values that did not allow the occurrence of cracking (as a result of differential thermal expansion), melting or vaporisation - conditions that are frequently obtained with dark encrustations on marble or stone. However, in Q-switched mode (pulse duration approximately 5-30 nsec) an ablation mechanism was responsible for the cleaning effect. These two mechanisms as developed by Asmus are discussed in 2.2 and 2.3

2.2 SELECTIVE VAPORISATION MECHANISM IN THE LASER CLEANING OF STONE

For laser cleaning by selective vaporisation and considering a one dimensional model, as schematically illustrated in Figure 1, the following assumptions are made: 1) The laser beam is uniform with no transverse variation in intensity, 2) The encrustation is uniform and planar, 3) The beam diameter is much larger than the encrustation thickness, 4) The beam diameter is much larger than the thermal diffusion distance.

For practical cleaning using a Nd:YAG laser, the beam diameter is typically 0.5 cm.

The thermal diffusivity is given by

$$\alpha = k/\rho.C_v \tag{1}$$

where α = thermal diffusivity
             k = thermal conductivity
             ρ = density
             C_v = specific heat

For typical minerals $\alpha = 10^{-2}$ cm^2.s^{-1}

The distance z that a thermal wave will travel into the material is given by

$$z = (\alpha \, t)^{1/2} \tag{2}$$

where t = pulse length = 10^{-3} s

$$z = (10^{-3}. \, 10^{-2})^{1/2} = 3 \times 10^{-3} \text{ cm}$$

This shows that assumption 4 is valid and also that the thermal effect is strongly localised in the surface.

For heat flow in one dimension:

$$\frac{\partial^2 T_{(z,t)}}{\partial z^2} - \frac{1}{\alpha}\frac{\partial T_{(z,t)}}{\partial t} = -\frac{A_{(z,t)}}{k} \tag{3}$$

where $T_{(z,t)}$ = temperature at distance z after time t,
        $A_{(z,t)}$ = heat produced per unit volume and per unit time as a function of position and time.

If a constant flux Fo is absorbed at the surface (z = 0) and there is no phase change in the material, the solution of the above equation is:

$$T_{(z,t)} = 2 \frac{Fo(\alpha \, t)^{1/2}}{k} \, \text{ierfc} \left[\frac{z}{2(\alpha \, t)^{1/2}} \right] \tag{4}$$

at the surface (z = 0)

$$T_{(0,t)} = 2 \frac{Fo}{k} \left[\frac{\alpha \, t}{\pi} \right]^{1/2}$$

But
$$F_0 = (1-R)I_0 = \beta I_0$$

where
$$R = \text{reflectance}$$
$$\beta = \text{absorptance}$$
$$I_0 = \text{incident flux}$$

and
$$T_{(0,t)} = \frac{2 \beta I_0}{k} \left[\frac{\alpha t}{\pi}\right]^{1/2} \tag{5}$$

For a typical encrusted statue,

$$\beta_{encrustation} = 0.6$$
$$\beta_{stone} = 0.2$$

For the case where $t = 10^{-3}$ s, $k = 2 \times 10^{-2}$ J/K.cm.s and $\alpha = 10^{-2}$ cm^2.s^{-1} and plotting contours of constant temperature for Equation 5 gives the profiles shown in Figure 2-4. Hence if the encrustation and the stone have a similar boiling temperature of 1000 °C, incident power Io greater than min and less than max will result in selective removal by evaporation of the encrustation (Figure 2). If the encrustation and the stone have different boiling temperatures, the respective isotherms should be used. The operating window is larger if the underlying stone has a higher vaporisation temperature (e.g. 1500°C) than the encrustation (e.g 1000° C), as shown in Figure 3. Poorly absorbing encrustations (e.g. $\beta = 0.3$) can still be selectively removed from substrates of higher absorptance (e.g. $\beta = 0.6$) if the vaporisation temperature of the encrustation is sufficiently low (e.g. 500 °C) compared with that of the substrate (e.g 1500° C), as shown in Figure 4. Regimes of encrustation removal are thus summarised in Figure 5.

2.3 LASER CLEANING BY ABLATION

It has been observed that the rate of cleaning is improved by approximately an order of magnitude if a Q-switched laser providing short pulses (approx. 5 - 20 ns) is used. An ablation mechanism has been proposed. At this high flux level (10^7 - 10^{10} W/ cm^2), even relatively reflective surfaces absorb sufficient energy to reach the vaporisation temperature.

High temperatures (typically 10^4 - 10^5 K) are produced in the vaporised material produced from the surface and at these temperatures this vapour becomes partially ionised and absorbs the laser energy strongly. The initial surface vaporisation stops as the target is shielded from the laser by the partially ionised ("plasma") vapour. As the pulse continues, the vapour is further heated and high pressures (1 - 100 Kbar) can be produced, resulting in a shock wave which produces microscopic compression of the surface of the target material. When the laser pulse ends, the plasma expands away from the surface, the material surface relaxes and a thin surface layer (1- 100 μm) is removed resulting in spallation. Whilst cleaning is more rapid in this case, there is a greater propensity for damage to the material underlying the surface encrustation that requires removal. A

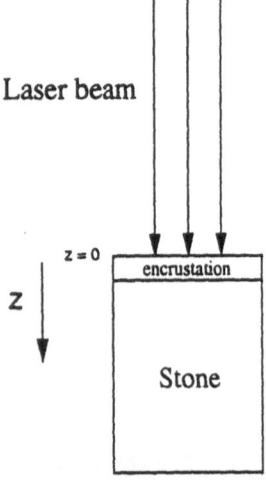

Figure 1. Schematic of one dimensional surface heating model

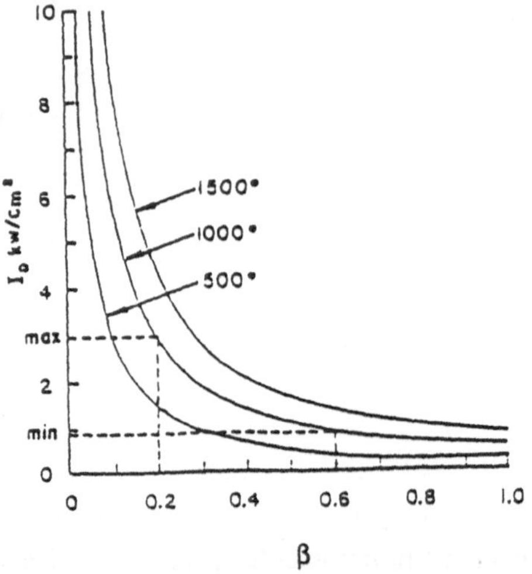

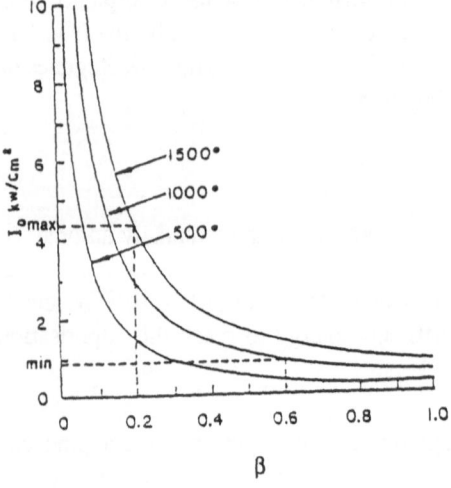

Figure 3. Contours of constant temperature for equation 5 showing effect of difference in evaporation temperature for deposit and substrate (adapted from Asmus [1])

Figure 2. Contours of constant temperature for equation 5 (after Asmus [1])

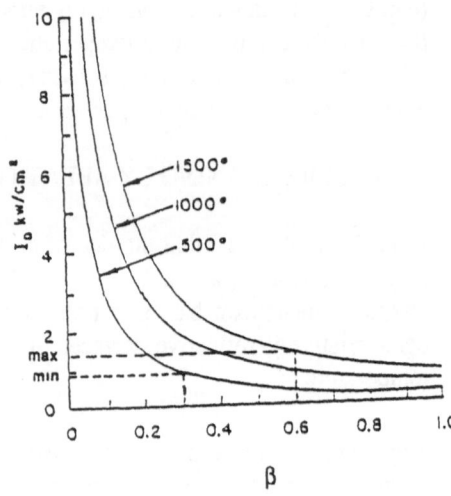

Figure 4. Contours of constant temperature for equation 5 showing removal of poorly absorbing deposit (adapted from Asmus [1])

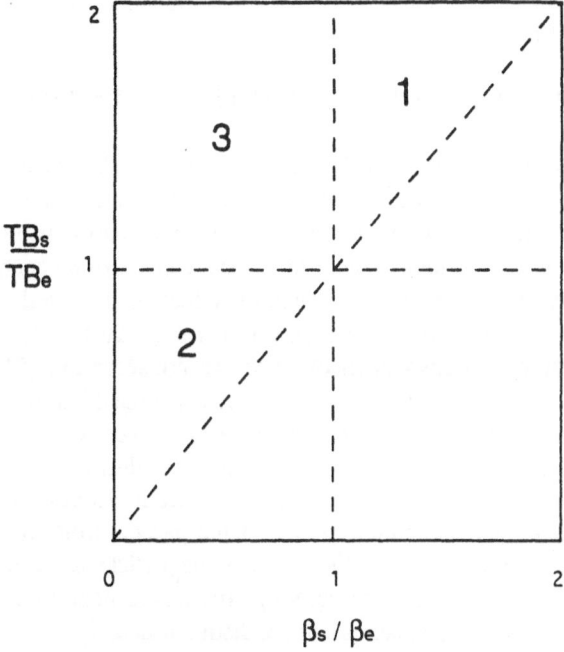

Figure 5. Regimes of selective cleaning (from Equation 5)
(1 - cleaning by boiling point discrimination, 2 - cleaning by
absorptance discrimination,3 - contribution of both 1 and 2).

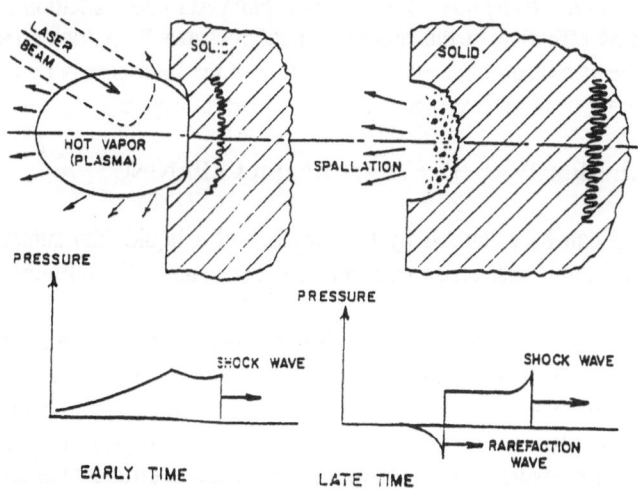

Figure 6. Schematic of laser cleaning by abaltion [4, 9]

schematic diagram showing interaction of giant pulse laser radiation with a solid surface is shown in Figure 6 [4,9].

2.4. SELECTIVE REMOVAL OF SMALL PARTICLES - DRY LASER CLEANING

Work in the electronics industry on the effect of pulsed laser radiation on submicron surface particulates [10, 11] has shown that the major part of the incident energy can be absorbed on a particular surface (the particulates or the substrate) depending on the wavelength of the laser used, offering potential for selective cleaning by control of laser wavelength. Absorptivity of laser energy was found to be increased when the target was covered with a thin film of liquid, usually water, offering potential for water enhanced laser cleaning. This effect has also been observed in conservation [5] when the application of a thin layer of water by brush prior to laser treatment led to an enhanced rate of cleaning. A simple analysis shows that the adhesion forces between a substrate and a small particle (arising from Van der Vaal's forces, the electrostatic double layer force and capillary attraction in the presence of atmospheric moisture) are very large compared to gravitational forces. For the removal of the particle from the surface an acceleration that is inversely proportional to the square of the particle diameter is required As a result the removal of very small (sub micron) particles is difficult by conventional techniques. For dry laser cleaning, two conditions are distinguished.

2.4.1. Strong Substrate absorption. Rapid pulsed heating of the dry substrate can lead to the ejection of micron and submicron particles as a result of the sudden expansion of the substrate surface. For example, 20 ns excimer laser pulses at a fluence of 350 mJ/cm^2 removed 0.3 μm alumina particles from Si surfaces. Although the expansion amplitude is small, the time is also very short, resulting in a strong acceleration.

2.4.2 Strong particle absorption. 20 ns pulsed Nd:YAG laser radiation at a fluence of 650 mJ/cm^2 was effective in the removal of micron-size W particles from a lithium niobate substrate as a result of selective absorption of the laser energy by the particles. A schematic diagram of these two mechanisms is shown in Figure 7.

2.5 PARTICLE REMOVAL BY STEAM LASER CLEANING

Particle removal can be enhanced by the presence of a liquid film subject to different types of laser heating. These types are summarised schematically in Figure 8 [10,11].

2.5.1. Strong substrate absorption. For a short wavelength laser such as an excimer and using short pulses (about 16 ns) the thermal diffusion distance in a material such as Si is about 1 μm and in water is about 0.1 μm. Hence, irradiation of Si covered with a thin film of water results in very efficient heating of the liquid substrate interface; superheating and explosive evaporation of the water leads to efficient particle removal.

2.5.2. Strong liquid film absorption. Removal of 0.2 μm Au particles from Si in the presence of a water film by use of a Q switched Er:YAG (2.94 μm wavelength) laser

915

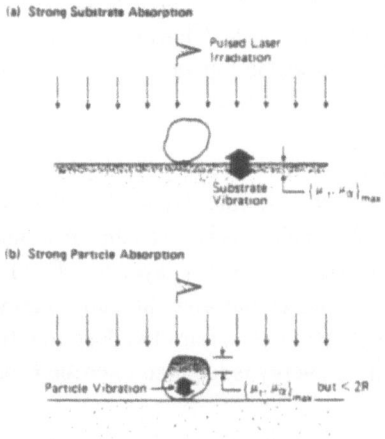

Figure 7. Mechanisms of particle removal by dry laser cleaning [10, 11]

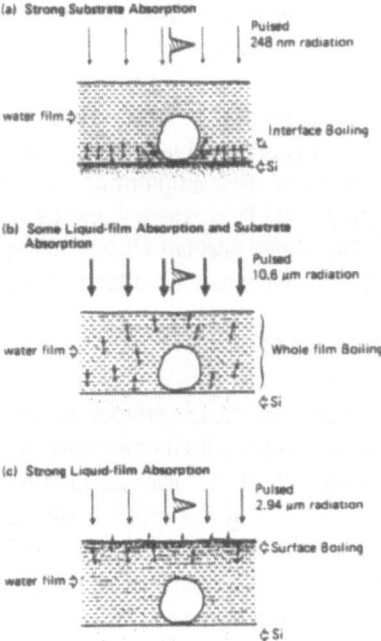

Figure 8. Mechanisms of particle removal by steam laser cleaning [10, 11]

with 10 ns pulse length resulted in strong absorption in the water with a penetration depth of 0.8 μm and weak absorption in the Si. Cleaning was less effective than in 2.5.1 since peak temperature was achieved at the top surface of the liquid rather than at the substrate / liquid interface.

2.5.3. Joint liquid / substrate absorption. Use of a pulsed 10.6 μm CO_2 laser to remove alumina particles on Si in the presence of a water layer resulted in particle removal but at a lower energy efficiency than 2.5.1. This was because the absorption depth of this laser in water is about 20 μm and hence, if the water layer is a few microns thick, only a fraction of the laser energy is absorbed by the water and this energy is distributed in the bulk of the water. The remaining energy penetrates too deep into the Si to be useful for interface heating. Hence, more laser energy is required to create explosive evaporation of the liquid.

Tam et al conclude that the selection of laser process conditions to favour strong substrate absorption produces the most efficient steam cleaning technique. The ejection velocity of the exploding water droplets is estimated at 10^4 cm/s, causing the formation of a jet of water droplets and ejected particles that is visible up to 1 cm from the irradiated surface and which causes a shock pulse in the air which is audible as a characteristic snapping sound. The peak temperature is estimated as about 370 °C together with a high transient pressure (up to a few hundred atmospheres). However, since these conditions exist for only a very short time in pulsed laser operation, the extent of substrate damage is strictly limited.

While we know of no detailed account of laser steam cleaning applied to art restoration, it is known that this technique is highly effective in the removal of certain encrustations [5,12], for example the removal of dark sulphurous encrustations from limestone sculpture using a Q switched Nd:YAG laser where the water is applied as a thin surface layer by periodic brushing. The above account closely mirrors the experience of the conservator in this respect and it is likely that a similar mechanism is operative.

2.6 LASER CLEANING OF METALS

In the case of metallic materials with clean surfaces, it has been reported by many researchers that while most metals show initially low absorption to laser light (an effect that is dependent on the wavelength of the laser in question and on the metal type, among other factors), a strong increase in absorption is subsequently observed. Studies [13-19] have linked this increase in absorption to the formation of a partially ionised cloud of metal vapour ("plasma") above the metal target when irradiated. For pulsed laser radiation, two relatively distinct regimes have been observed - a laser surface combustion (LSC) regime which tends to be favoured at low intensity and a laser surface detonation (LSD) regime which is favoured at higher intensity. Hence, in the laser cleaning of metallic materials with pulsed laser radiation it would be anticipated that "plasma" formation would play an important role and that at high flux density a spallation mechanism could be operative.

It has been found that the absorption of laser energy is increased for aluminium, for example, by the modification of the oxide layer [18] and hence various oxide corrosion layers on aluminium alloys may be expected to enhance energy absorption during laser cleaning of the metal. In addition, absorptivity is affected by surface irregularities and by inhomogeneity of the laser beam spatial profile and hence irregularities in the corroded surface of metals during cleaning may be of particular importance in determining the extent of effect on the substrate. If surface melting takes place, removal of material by ejection of molten metal could take place [19]. In summary, laser-materials interaction may be expected to be complex in the case of the cleaning of corroded metals.

Bergmann [20] has considered the excimer laser interaction with a 'technical' aluminium allloy surface (i.e. one that contains a grease layer and deformation layers beneath the surface oxide layer) and has found that different processing results are obtained with variation in laser energy density since the energy necessary to produce plasma formation is different for organic absorbates, oxide layers and the unbderlying substrate. At low intensities, cleaning is possible below the plasma threshold of the substrate by the removal of organic contaminants.

2.7 MATERIAL REMOVAL BY THERMOELASTIC STRESS PRODUCTION

A number of workers have observed the effect of short pulse lasers in generating thermoelastic stresses in the surface of solids [21-26]. The removal of small particles as a result of this effect has already been discussed in Section 2.4. As Ready has pointed out [27], if the magnitude of these induced stresses in the substrate become sufficiently large so that the fracture stress of the material is exceeded, removal of material may take place by physical fracture. Although this could be a useful mechanism for the removal of surface material in situations where a relativley large amount of surface removal is required (for example in the treatment of surfaces for decontamination), it is unlikely that this mechanism will be useful in cleaning in art restoration and conservation because of the extent of material removal and the unpredictability of the effect on surface texture. Indeed, these effects may be regarded as a mechanism of surface damage taking place under conditions which are to be avoided in conservation.

3. Work at Liverpool on the Laser Cleaning of the Gilbert *Eros* Statue

An opportunity to investigate laser cleaning of material from corroded aluminium statuary was presented when in 1992 the National Galleries and Museums on Merseyside were commissioned to restore the Liverpool Gilbert *Eros* statue. This statue, which was cast in 1929 from the same mould as the original (1893) London statue of *Eros (The Shaftesbury Memorial)* by Albert Gilbert [28], had been on outdoor display in Liverpool's Sefton Park continuously since 1932 and hence had suffered intensive internal and external corrosion. The investigations reported here [29] were carried out on a number of small pieces showing general external corrosion which had been temporarily removed from the statue.

The cleaning system used was based on a pulsed Nd:YAG SL400/SL800 laser manufactured by Spectron Laser Systems, operated in Q-switched mode. The laser operates in the infra red region of the spectrum at a wavelength of 1.06 μm with a pulse duration of 6 ns, maximum pulse frequency of 10 Hz and a maximum energy output of approximately 300 mJ. The beam delivery system had been specially developed to make it as flexible and controllable as possible in a conservation environment [12] and consisted of mirrors mounted in a movable arm unit with several flexible joints and a final focussing lens with an effective focal length of 15cm. Use of the movable arm unit allowed pulses of laser energy to be directed onto the target object to be cleaned in a way which allowed manual control over laser position, duration at a given position (number of pulses per position) and distance of the object from the focussing lens.

The samples to be cleaned had become detached from the left-side wing of the Liverpool Gilbert *Eros* statue and represented external surfaces which had suffered mainly general corrosion which took the form of a relatively uniform, dark grey layer. As Asmus [1] has noted (and as was found in this work), effective cleaning in Q-switched mode is often accompanied by an audible report from each laser pulse and in practical cleaning by an experienced operator the establishment of conditions for the creation of such a report (by changing the incident flux by altering the target-laser working distance or the laser power) is often used as an indicator of effective cleaning.

In the work reported here, a more systematic variation of the parameters influencing the removal of corrosion deposits was carried out while retaining procedures characteristic of practical cleaning with this system. Each sample was supported firmly and an area of approximately 0.5 cm^2 was cleaned during a period of operation of some 30s in which the laser spot was traversed over this area by means of manual movement of the flexible jointed beam focussing system. By varying the pulse frequency, the total number of pulses applied to this area could be varied and by varying the laser rod voltage the laser fluence at a given working distance could be varied. In addition, variation of the working distance (the distance between the target and the final focussing mirror at the tip of the movable arm) allowed variation of the laser spot size and hence, at a given rod voltage, a further means of varying the laser fluence. Under these conditions, the following variables were investigated:

i) *Variation of laser pulse energy.* With the working distance maintained constant at 15 cm (and hence with a laser spot area of 0.16 cm^2) and the number of pulses maintained at 90, the laser voltage applied in separate tests was 75, 80, 90, 120 or 150 mJ,

ii) *Variation of number of pulses.* With the working distance maintained constant 15 cm (and hence with a laser spot area of 0.16 cm^2) and the laser voltage maintained at 75 mJ, the number of pulses applied in separate tests was 60, 120, 180, 240, 300, 360 or 720.

iii) *Variation of working distance* . With the laser voltage maintained constant at 75 mJ and the number of pulses maintained at 90, the working distance in separate tests was maintained at 3, 5, 10, 15 or 20 cm (producing a laser spot area of approximately 0.10, 0.12 0.13, 0.16 or 2.0 cm², respectively).

The relationship between working distance and spot size was determined in a series of initial experiments in which photographic paper was placed on the surface of the target and irradiated at different working distances. The relationship between laser rod voltage and laser fluence was determined from data supplied by the manufacturer. Because of the method of measurement in the former case and because of possible long term instability in the laser in the latter case, both of these parameters should be regarded as approximations. A summary of the various laser parameters investigated in these tests is given in Table 1.

The surface of the as-received statuary material was analysed by XRD and EPMA techniques in order to characterise the corrosion products. XRD, EPMA, light microscopy and SEM was used to characterise the cleaned surfaces.

3.1. CHARACTERISATION OF THE CORRODED LAYER AND THE BASE ALLOY

Initial investigation of the corrosion layer by SEM and EDX and XRD revealed that corrosion products made up from S, Cl, Ca, P and K in addition to Al, Cu, Si, Fe and Zn arising from the base material. Lighter elements such as O, N and H which were assumed to be present were undetected by these techniques. The corroded surface took the form of relatively uniform, dark grey layer which had formed on atmospheric exposure of the outer surface of the statue which, when observed at high magnification by SEM, revealed micro-cracking in a crazed formation. Other corrosion products composed of more voluminous white and brown coloured deposits were not considered. The overall composition of the uncorroded alloy was found by EPMA to be an aluminium alloy with approximately 2.5 wt % Zn and 2.5 wt % Cu. Si and Fe were detected in much smaller amounts. The second phase precipitates present were identified by EDX as a copper-aluminium compound of probable composition $CuAl_3$.

3.2. VARIATION OF WORKING DISTANCE AND NUMBER OF PULSES

With the laser voltage maintained constant at 75 mJ and the number of pulses maintained at 90, tests carried out at different working distance showed that cleaning was most effective when the distance was close to 15 cm, the focal length of the laser focussing system.

With the working distance maintained constant at 15 cm (and hence with a laser spot area of 0.16 cm²) and the laser voltage maintained at 75 mJ, the number of pulses applied in separate tests was 60, 120, 180, 240, 300, 360 or 720. It was found by XRD that after the application of 1-2 laser pulses at each position in the cleaned area, a

Working Distance (cm)	Voltage (15cm,3Hz) (V)	Pulse Energy (mJ)	Spot Area (cm²)	Energy Density (J/cm²)	Power Density W/cm² x 10⁶
15	950	75	0.16	0.5	78
"	1000	75	"	0.5	78
"	1050	80	"	0.5	83
"	1100	90	"	0.6	93
"	1200	120	"	0.8	125
"	1300	150	"	0.9	157
20	950	75	0.20	0.4	63
15	"	"	0.16	0.5	78
10	"	"	0.13	0.6	97
5	"	"	0.10	0.8	125
3	"	"	0.10	0.8	125

Table 1. Laser cleaning conditions for Liverpool *Eros* statue

majority of the corrosion products was removed. Application of further pulses resulted in a more complete removal of the corrosion products.

3.3. VARIATION OF PULSE ENERGY

A laser voltage of 75 mJ was found to be the effective threshold for any cleaning action to take place, the corroded surface remaining unaffected at voltage levels below this value. At 75 mJ and 80 mJ the cleaned surface was relatively uniform and showed no evidence of preferential attack of particular regions by the laser beam. It was clear that no surface melting has taken place in this case with corrosion product removal occurring solely as a result of the ablative effect of the laser induced partially ionised vapour formed above the sample. The irregularities in the surface represented the effect of the unequal extent of corrosion at certain regions of the surface in roughening the surface to this extent - that is, the natural state of the surface following removal of the corrosion product has been revealed by laser cleaning with little or no subsequent damage to the surface being caused by the action of the laser. The laser flux density in this voltage range was approximately 10^7 W/cm^2.

In contrast, for laser voltages above 120mJ where laser flux density in this voltage range is approximately 10^8 W/cm^2 there was localised damage of the surface as a result of laser cleaning. The surface was roughened to a greater extent than is implied by the roughening action of corrosion prior to laser cleaning. Examination of the surface cleaned at 150 mJ at higher magnification revealed selective removal of material in the grain boundaries of the alloy. In a typical cast microstructure of an aluminium alloy containing Cu and Si, it would be expected that a grain boundary precipitated second phase based on intermetallic compounds of aluminium combined with these elements would be formed. A network of grain boundary precipitates of this type was found in this case. Selective removal of the grain boundary precipitates has occurred as a result of laser cleaning at laser voltages greater than 120 mJ. The reason for enhanced laser absorption at these sites may be connected with the craze cracking that existed in the overlying corrosion product prior to cleaning. The dimensions of the craze cracks in the corrosion layer are of the same order as the regions of selective removal during laser cleaning. This raises the possibility that at high laser fluence, enhanced absorption has taken place at the cracks in the corrosion layer and that this has resulted in localised melting of the substrate beneath the cracks. The connection between this observation and the earlier observation of removal of second phase material at grain boundaries could be accounted for if cracking in the corrosion layer had itself taken place selectively over the second phase as a result of a less favourable mismatch in thermal expansion coefficient between the substrate and the corrosion layer in this case. In this case, this work provides a further example of the influence of inhomogeneity in the corrosion layer on aluminium on laser absorption. Since surface damage is to be avoided in conservation, it can be concluded that laser cleaning at the lower voltage range (75 - 90 mJ) is preferable.

The extent of laser cleaning was confirmed by comparing XRD spectra before and after cleaning for material treated at 75 mJ with the XRD spectrum for the uncorroded base alloy. In both cases the spectrum of the cleaned material is essentially similar to that of the uncorroded base material and the change in spectrum compared with the corroded material can be attributed to the removal of the corrosion products.

3.4. CONCLUSIONS ON LASER CLEANING OF *EROS*

For the laser cleaning of generally corroded sections of Liverpool's Gilbert *Eros* statue:

1) The polluted surface investigated contained corrosion products made up from S, Cl, Ca, P and K (in addition to elements arising from the substrate alloy). These elements were removed on treatment with a Q-switched Nd:YAG laser,

2) The mechanism of cleaning is thought to be based on ablation engendered by shock waves produced in a partially ionised vapour shielding the target from the laser,

3) Corrosion products were removed with minimum damage to the substrate at a laser voltage of 75 - 90 mJ, a working distance of 15 cm and 90 pulses per test area. SEM observations of the cleaned area revealed a surface profile consistent with that the natural surface underlying corrosion products with little or no additional damage being caused by the removal of the corrosion products by the laser,

4) Removal of corrosion products by laser cleaning at higher voltages (120 - 150 mJ) resulted in a roughened surface. SEM observations implied selective removal of second phase material at grain boundaries and that this resulted from enhanced absorption of laser energy at cracks in the corrosion layer formed selectively above these precipitates prior to laser cleaning.

4. References

1. Asmus, J. F. (1986) *More Light for Art Conservation* IEEE Circuits and Devices Magazine, March , 6-15

2. Asmus, J. F. (1974) *Laser Consolidation Tests* International Fund for Monuments (Final Report), May

3. Asmus, J. F. (1973) *Holography in the Conservation of Statuary* Studies in Conservation **18** pp

4. Asmus, J. F. (1978) *Light Cleaning - Laser Technology for Surface Preparation in the Arts* Technology and Conservation **13**, 14

5. Larson, J. H. (1985) *The Conservation of Stone Sculpture in Museums* The Conservation of Building and Decorative Stone (Butterworths) **2** 197

6. Asmus, J. F., Seracini, M. and Zetler, M. J. (1976) *Surface Morphology of Laser-Cleaned Stone* Lithoclastia No 1, 23

7. Asmus, J. F. (1987) *Light for Art Conservation* Interdisiplinary Science Reviews 12 171

8. Cooper, M. I., Emmony, D. C.and Larson, J. H. (1992) *The Use of Laser Energy to Clean Polluted Stone Sculpture* J Photographic Science **40** , 55

9. Asmus, J. F., Munk, W. H. and Murphy, C. G. (1973) *Studies on the Interaction of Laser Radiation with Art Artefacts* Proc Soc Photo-Optical Instrumentation Engineers **41**

10. Zapka,W., Zeimlich, W. and Tam, A. C.(1991) *Efficient Pulsed Laser removal of 0.2 mm Sized Particles from a Solid Surface* Appl Phys Lett. **58** 2217

11.Tam, A. C., Leung, W. P., Zapka,W. and Zeimlich, W. (1992) *Laser Cleaning Techniques for the Removal of Surface Particulates* J Appl Phys **71** 3515

12. Cooper, M.I. (1994) *Laser Cleaning of Limestone Sculpture* PhD Thesis, University of Loughborough

13. Allmen, M. Von, Blaser, P., Affolter, K.and Sturmer, E. (1978) *Absorption Phenomena in Metal Drilling with Nd-Lasers* J of Quantum Electronics **QE-14** 85

14. Bonch-Breuvich, A.M., Imas, Ya A.,Romanov, G. S., Libenson, M. N. and Mal'tser, L. N. (1968)*Effect of Laser Pulse on the Reflecting Power of a Metal* Sov Phys -Tech Phys **13** 640

15. Bergel'son,V. I.,Golub, A. P., Loseva,T. V., Newchinov, I. V., Orlova, T. I., Popov, S .P. and Svettsov,V. V. (1974) *Appearance of a Layer Absorbing Laser Radiation Near the surface of a Metal Target* Sov J Quant Electron **4** 704

16. Metz, S. A., Hettche, L. R., Stegman, R. L. and Shreimpf, J. T. (1975) *Effect of Beam Intensity on Target response to High Intensity Pulsed CO_2 Laser Radiation* J Appl Phys **46** 1634.

17. Hettche, L. R.,Tucker, T.R., Shreimpf, J. T. , Stegman, R. L. and Metz, S. A (1976) *Mechanical Response and Thermal Coupling of metallic Targets to High Intensity 1.06 µm Laser Radiation* J Appl Phys **47** 1415

18. Patel, R. S. and Brewster, M. Q. (1990) *Effect of Oxidation and Plume Formation on Low Power Nd:YAG Laser Metal Interaction* Trans ASME, J of Heat Transfer **112** 170.

19. Bass, M. Nasser, M.A. and Swimm, R. T. (1987) *Impulse Coupling to Aluminium Resulting from Nd:glass Laser Irradiation Induced Metal Removal* J Appl Phys **61** 1137.

20. Bergmann, H. W. and Schubert, E. (1990) *Review on Materials Processing with Excimer Lasers* ECLAT '90, 813.

21. White, R. M. (1963) *Elastic Wave Generation by Electron Bombardment or Electromagnetic Wave Absorption* J Applied Physics **34** 2123

22. Percival, C.M. (1967) *Laser-Generated Stress Waves in a Dispersive Elastic Rod* J Applied Physics **38** 5315

23. Lee, R.E. and White, R. M. (1968) *Excitation of Surface Elastic Waves by Transient Surface Heating* Applied Physics Letters **12** 12

24. Bushnell, J.C. and McCloskey, D.J. 1968) *Thermoelastic Stress Production in Solids* J Applied Physics **39** 5541

25. Brienza, M.J. and DeMaria, A. .J. (1967) *Laser-Induced Microwave Sound by Surface Heating* Applied Physics Letters **11** 44

26. White, R. M. (1963) *Generation of Elastic Waves by Transient Surface Heating* J Applied Physics **34** 3559

27. Ready, J. F. (1971) *Effects of High-Power Laser Radiation* Academic Press, New York, 116 ff.

28. Dorment, R.(1985) *Alfred Gilbert* (Yale University Press)

29 Dalton. M. E., Watkins, K. G., Larson, J. H. , Green, A. and Emmony, D.C. (1994, in press) *Q-Switched Laser Cleaning of Aluminium Statuary Material* Studies in Conservation.

The experience of laser heat treatment application in industry of Ukraine

Volodymyr S.Kovalenko

Laser Technology Research Institute, Kiev, Ukraine

Abstract

Laser Heat treatment is used now at different plants of Ukraine to increase wear resistance of cutting tools and machine parts. Laser hardening of end mills, broaches, gear cutters as well as different dies and punches gives the durability increase up to 2-6 times. Highly efficient is the technology of laser hardening and cladding of long slides (up to 3 m long) for polymer film producing machines, crank shafts and sleeves of diesel cylinders. Laser heat treatment is usually performed using YAG or CO_2 laser industrial systems. For hardening and cladding gear shafts and coupling threads of boring columns the fully automated robotized CO_2 laser equipment has been developed.

Laser heat treatment is very critical to surface absorptivity and surface temperature control in the interaction zone of the material. Special measuring devices have been developed to deal with this problem with time constant $10-5s$.

To increase hardening efficiency through surface absorptivity increase different types of coatings have been developed. This factor was used as well in development of special powder compounds for laser cladding and different types of powder feeders.

To control the value and sign of residual stresses formation different techniques were proposed. One of them -- laser irradiation combined with plastic deformation, is shown in presentation. Some problems of further introduction of the developed laser heat treatment processes into industry are discussed as well.

I. INTRODUCTION

In the Former Soviet Union (FSU) Ukraine was considered as the most developed and highly industrialized republic. According to UN statistics for 1992 the intellectual potential of Ukraine (the total amount of engineers and scientists in the country) was around 6.8% of the world intellectual power. New developments in science, technology and engineering usually were and still now are of top priority in national economic policy in spite of the great economic problems in the last few years. From the very beginning of laser era the substantial efforts were directed to the R & D programs in this field. Up to now many institutions are working in the field of lasers and laser technology. More so it is easier to list the institutions which are not connected with laser developments problems. Among

925

J. Mazumder et al. (eds.), Laser Processing: Surface Treatment and Film Deposition, 925–932.
© 1996 Kluwer Academic Publishers.

industrial laser applications laser surface treatment is considered to be the most developed technology and quite wide spread in the industry. Because of some specific reasons great attention is paid to different research programs in this field. More and more companies especially which were born on the wave of economic restructuring in free market principles development are eager to use laser surface treatment technology to get the advantages with competitors. To help this high technology speed up in their way to production floor special attention should be paid to the initiation of different teaching programs at the universities in the field of laser technology.

II. LASER & LASER TECHNOLOGY R & D IN UKRAINE.

Among the main Institutions working in the field of laser and laser technology developments are the following:
(1) Institutes of the National Academy of Science of Ukraine
   -- Paton Welding Institute
   -- Institute of Material Science Problems
   -- Institute of Super hard Materials
   -- Institute of Metal Physics
   -- Institute of Reliability Problems
   -- Institute of Physics
   -- Institute of Electrodynamics
   -- Institute of Semiconductors Engineering, etc. (more than 30)
(2) Universities and Institutes
   -- Kiev Polytechnic
   -- Lwiv Polytechnic
   -- Kharkov Polytechnic
   -- Kiev University
   -- Uzhgorod University
   -- Kharkov University
   -- Odessa University, etc., (more than 20)
(3) Plants
   -- "Rotor Co.", Cherkassy
   -- "Malyshev Tank Co.", Kharkov
   -- "Arsenal Co.", Kiev
   -- "Kirov geology Co.", Brovary"
   -- "KRAZ automotive Co.", Kremenchuk
   -- "Yuzhmash Co.", Dnepropetrovsk
   -- "Nikolaev Shipyard Co.", Nikolaev
   -- "Izumrud" Diamond Co., Kiev
   -- "Bolshevik" Machine building Co., Kiev, etc. (more than 150).
Many research and developments programs in the field of laser technology have been initiated in the framework of State Committee of Science and High Technology, National Academy of Science, Ministry of Industry and others. The most substantial program at the

level of State Program have been sponsored by the Ministry of Machine building and Conversion of Ukraine.

This Program was called "MACH LASER-UKRAINE" and initially had 288 different projects.

117 of them were devoted to the laser industrial systems and accessories development, other 113 were directed to the laser technology processes introduction into industry. The main topic for other 26 projects was the new laser technology processes development. For 29 projects that topic was measurement and metrology and the rest was directed to laser safety, ecology and standardization problems.

Among R & D projects in laser industrial applications quite big accent is made on laser surface treatment study. And there are specific reasons for this. First, laser surface treatment is a very versatile technology. Second, because of quite wide spread deviations from standard quality of steels produced in Ukraine and FSU there is a necessity to compensate the quality drawbacks of such steels with additional treatment, and laser treatment is very good instrument in this respect. Third, in Ukraine and in FSU there was always great demand for the techniques to restore the working properties of the worn off components instead of manufacturing the new one. Fifth, because of lack of tungsten and other refractory metals in Ukraine and other FSU countries high speed cutting steels have poor quality and thus cutting tools made of such steels need additional hardening treatment. The last, fifth, the opportunity to get the unique combination of alloys at the surface layers of the components. All these features of laser surface treatment attract the attention both of researchers and industrialists.

III. R & D IN LASER SURFACE TREATMENT

The main problems of interest to solve in laser surface treatment are as follows:
-- modeling
-- heat exchange control from laser irradiated zone
-- temperature and absorption control at laser interaction with material surface
-- development of new composition for laser cladding and alloying
-- development of new techniques to introduce alloying (cladding) composition into
    laser irradiated zone
-- development of new combined processes of surface treatment.

Modeling. Besides different theoretical models based on heat transfer theory much work is done on development of experimental models convenient to use at industrial floor. Created nomograms helps to choose the optimal working conditions more adequate to reality but need quite broad data base. For practical use such nomograms are preferable but more deep and precise experimental research must be performed to improve further these models.

Heat exchange control from laser irradiated zone. The main problem at laser surface heat treatment is to have enough base material for fast heat transfer. In case of large massive components to be laser hardened this problem doesn't arise. But when the component has limited dimensions or very complicated shape then heat transfer from the irradiated part deep into the material is becoming more difficult. For high quality laser hardening in this case such limitations must to be taken into account. In Laser Technology Research Institute (LTRI) the study was performed on finding the proper conditions for irradiation of the component of limited thickness and of the angular components with different angles. Theoretically and experimentally the regions of optimal working conditions were found to get the guaranteed hardening effect in given volume of the component.

Temperature and absorption control at laser interaction with material surface. Laser surface treatment is multyfactor process which depends upon more than 30 factors. It is understandable that to measure and to control all this factors practically very difficult or even impossible. So the way out of this situation is to determine the main factors which influence the other factors and the process in general. Among such factors are the following: surface temperature, duration and speed of heating, cooling speed. Together with surface temperature one of the most significant factors is the surface absorptivity. To control and to increase the hardening efficiency one has to find the way how to control the surface absorptivity and temperature simultaneously with the irradiation process. In LTRI (former Laser Technology Lab)substantial efforts were devoted to the developments different devices and systems to measure and to control the mentioned parameters. One of the first version of such devices was based on the calorimeter type system. The device has double wall hemisphere with central opening to pass laser beam. Internal surface of hemisphere has absorbing coating. Special gas working media is pumped between wall space. Reflected from the irradiated surface laser beam gas mixture causes gas mixture heating and thus mixture pressure changes between the walls of hemisphere. These data serve as initial information to measure absorptance changes during the irradiation.

Next version of the device uses as a sensor the pyroelectric detector, fixed at the focusing system over the treated surface and electrically connected with the information developing block (IDB) and than with the register device. IDB is connected with laser power meter as well. As a resulted of comparison of two signals proportional to the power of incident and reflected radiation the surface absorptivity is measured and than is shown through interface at the computer monitor. The program developed enables to analyze the absorptivity changes in dynamic. The time constant of the developed device is 10-5 s. This device helped to evaluate the efficiency of different developed coatings to increase the surface absorptivity. The results obtained show the wide opportunities to use such devices installed in the systems for the processes adaptive control.

Further improvement of the device brought to the development the system for measurement not only the absorptivity but instant temperature in the irradiated zone as well. In this device the problem of separation of reflected laser radiation and radiation from heated surface was solved with help of their modulation and by use of special filters. Tests of the device have shown that it may be used in automated control system for the process of laser surface treatment.

New composition for laser cladding and alloying. For laser cladding and alloying in many practical cases researchers use the compositions originally developed for plasma cladding. The most wide spread powders are produced on nickel base. The main drawbacks of such powders are high price, gripping of cladded layer at dry friction and formation of significant tensile stresses.

New composition based on Fe-B-C-alloy have been developed in LTRI. Due to iron base the tensile stresses, formed in cladded layer are lower. Boron coating of carbon steels give high wear resistance but has very low plasticity. So, to increase the boron plasticity extra elements (Si and Cr) have been added to the powder. Tests has shown the following advantages of the new developed compositions for laser cladding:

-- 4-5 times cheaper than nickel based alloys and 8-9 times cheaper than cobalt based alloys,

-- because of phosphorous presence in nickel based alloys at cladding some toxic oxides are

present. New alloys doesn't have such drawback,

-- cladded layer hardness is not less than that for nickel based alloys (62-63 HRC),

-- adhesion is better because of cladded and matrix materials similarity.

New techniques to introduce alloying (cladding) composition into laser irradiated zone. Except the conventional techniques to introduce alloying (cladding) elements under the laser beam at the treated surface, like powder distribution on the surface, painting, etc. new techniques have been proposed and investigated:

-- laser irradiation of the surface under the liquid alloying media,

-- alloying in gas media,

-- precoating by electrospark alloying,

-- precoating with plasma spraying,

-- detonation precoating,

-- electrolytic (electrochemical) precoating,

-- powder injection,

-- laser surface powder alloying in magnetic (electromagnetic) fields, etc.

For example, the last technique is very attractive in the cases when there are problems with preserving the powder layer at vertical or ceiling surfaces. Changing the electromagnetic field intensity it possible to control the absorptivity and thus the final parameters of alloyed (cladded) layer.

New combined processes of treatment. To intensify the hardening effect at laser surface treatment different combination of laser technology with other techniques of treatment may be proposed. In LTRI research was done on the following combined processes of treatment: Laser-plastic deformation hardening, Laser ultra-sonic hardening, Laser hardening in the liquid nitrogen media.

(1) Laser plastic deformation(LPD) hardening combines two processes in one -- laser hardening and thermo-plastic deformation hardening. The expected benefits of such combination are: first, possibilities to get hardened layer with specific structure for laser quenching; second, possibilities to get guaranteed compressed stresses favorable for increasing fatigue strength and wear resistance. In proposed process during laser irradiation of material plastic deformation was caused with roller. The main problem was to determine the magnitude of deforming force which depends upon deformation temperature, resistance to deformation and value of material surface cold working. For Steel 45 (1045) the value of deforming force was accepted at the level 500-600 N. Using calculated

temperature distribution in irradiated zone the distance from the center of irradiated spot and roller position was found. These data were used to design the device for surface laser plastic deformation. The study of the process and analysis of the obtained results have brought to following conclusions:

-- due to simplicity and high efficiency the laser plastic deformation hardening is a prospective method to increase fatigue strength and wear resistance of machine components,

-- in comparison with only laser hardening the LPD hardening gives 1500 MPa higher the micro hardness and 100-200 m higher the hardened layer depth,

-- to get the good combination of mechanical properties after LPD treatment the austanite cold working must be done before quenching in the conditions which allow to get not only high dislocations density but their more uniform distribution as well.

(2)   Laser ultra-sonic (LUS)hardening combines laser irradiation with vibration hardening at ultra-sonic frequency. Such combination may cause the increase in micro hardness on 3000-4000 MPa in comparison with pure laser hardening. Surface topography is improving as well due to such treatment.

(3)   Laser hardening in liquid nitrogen is quite efficient technique because at this process the heat removal from irradiated zone is improved and conditions for better nitrogen impregnation into matrix material are created. This gives the substantial increase of treated material micro hardness.

IV. LASER SURFACE TREATMENT INTRODUCTION INTO INDUSTRY.

One of the first big scale application of laser heat treatment was initiated in "Bolshevik" machine building Company in Kiev. First it was used for hardening of different types of cutting tools--broaches of large size, end mills, cutters, knives, etc. Usually after laser hardening wear resistance of treated instruments was 3-5 times higher than that of untreated one. Such results encouraged the plant experts to initiate more broad R & D program in laser technology for this plant. Finally the modern laboratory - laser shop was organized with different types of industrial laser systems. Step by step the variety of components and tools increased and now laser technology is considered as reliable and flexible technology for this plant..

One of the most interesting hardening application at this plant is laser treatment of long slides for polymer film producing equipment. Volume quenching of 3 m long slides with conventional hardening technique is quite a difficult problem. Because of thermal distortion additional grinding operation must to be performed. Taking into account that only inner part of the slides needs hardening laser treatment becomes the only one most effective means to increase surface micro hardness and thus the life time of the component.

At "Kirov geology" Co. laser technology was developed to increase wear resistance of boring tube couplings. The external and internal threads of such couplings are the most weak part of the component and breakage and damage occurs just in this part. Due to laser hardening wear resistance has increased twice and even three times. Special robotized industrial laser system was developed which performs hardening treatment fully automatically.

Another fully automated laser industrial system was designed and manufactured at LTRI for machine building company "Wishard" in Poland. Laser hardening was used for wear resistance increase of gear blocks for airplane gearbox. The developed technology and equipment appeared to be very efficient and few systems have been sold in Europe.

At "KRAZ" automotive Co. laser hardening also has found use for life time increase of different components of heavy trucks. Specialised shop is functioning now at this plant.

Serious R & D program in laser technology is accepted at "Malyshev tank Co." The specialized shop equipped with few laser industrial systems is using laser treatment for hardening large components of diesel engines for diesel locomotives. Wear resistance of large crankshafts and cylinder sleeves increase twice.

Many other examples of laser surface treatment application in industry and different hardening processes and equipment developed by LTRI in cooperation with different enterprises of Ukraine may be described as well.

V. EDUCATIONAL PROGRAMS IN LASER TECHNOLOGY.

To introduce into industry high technology and modern equipment based on lasers would be impossible without development of special educational program in this field in the country. In 1984 new specialty "Equipment & Technology for Laser Machining" was opened in Kiev Polytechnic Institute. Later The Department of Laser Technology and Material Science was organized. And now every year we have 50 newcomers on this specialty The curricula for 5.5 years study includes different scientific and engineering subjects among which the following considered to be the most specific:
-- conventional manufacturing systems & technology,
-- electrophysical & electrochemical machining methods,
-- physics of high power lasers,
-- physics of laser beam interaction with matter
-- laser optical systems,
-- laser power suppliers,
-- Laser dimensional machining,
-- laser surface treatment,
-- laser & electron beam welding,
-- laser industrial system design,
-- advanced laser applications,
-- research on laser technology.

There are Bachelor, Master and Ph.D. programs at the Department of Laser Technology and Material Science.

For foreign student there is special preparatory course for getting knowledge of Russian or Ukrainian language and to refresh knowledge in math, physics or other general subjects. There are about 300 Ukrainian and 70 foreign students from 12 different countries now studying laser technology at the Laser Technology and Material Science Department of Kiev Polytechnic Institute

VI. CONCLUSIONS

-- Ukraine has high scientific and engineering potential in laser technology field and cheep working power,

-- Ukraine has high industrial potential of converted military plants and good access to the huge market of FSU,

-- laser surface treatment is of significant interest for Ukrainian industrialists and has good prospects for further development,

-- there is readiness of some institutions in Ukraine to assemble or to manufacture the foreign laser industrial systems of good design and high quality on the base of joint ventures,

-- there is readiness and willingness of some institutions for common scientific programs in the field of laser technology (like "COPERNICUS", "INTAS", "NATO","SOROS" and others).

APPLICATIONS OF THE Nd:YAG LASER IN ELECTRONIC ASSEMBLY
PROCESSES.

A. Flanagan and T.J. Glynn
Physics Dept., University College, Galway, Ireland.

Abstract

Laser soldering has the potential to solve many of the problems which arise in bonding
fine pitch components onto various substrates. A 65 W Nd:YAG laser under computer
control has been used to solder a quad flat pack device and a TAB device onto circuit
boards. Designed experiments were used to identify optimum laser parameters, and
finite element thermal models of the test devices were used to investigate the heat flow
paths in the soldering process. The results were used to assess the feasibility of laser
repair and de-soldering of components.

1. Introduction

The use of lasers for soldering in electronic assembly was first demonstrated in 1977
using a CO_2 laser and in 1982 with a Nd:YAG laser [1]. The advantages of laser
soldering are:
(i) Localized heating makes it possible to solder heat sensitive devices in densely
populated boards while also reducing bridging between the solder joints [2].
(ii) The rapid rise and fall in temperature of the solder creates a fine microstructure and
virtually no intermetallic in the solder joint making the joints more reliable [2,3].
(iii) It is a non-contact process avoiding tool wear and corrosion.
(iv) The technique can be easily automated.
   The major disadvantage of laser soldering is speed; conventional mass soldering
technologies raise the temperature of the whole PWB and components making thousands
of solder joints simultaneously, whereas laser soldering is sequential. The greatest use
of laser soldering has been in aerospace and military applications for high reliability
circuit boards [4]; mass soldering technology has dominated in more general electronic
assembly where a high through-put is essential for competitiveness in the marketplace.
   Due to the continued reduction in size of integrated circuits, however, devices with
a high lead count (> 300) and new packaging methods, such as tape automated bonding
(TAB), are available which cause problems when they are used with the mass soldering

933

J. Mazumder et al. (eds.), Laser Processing: Surface Treatment and Film Deposition, 933–944.
© 1996 *Kluwer Academic Publishers.*

technologies. In this paper, the laser soldering of a quad flat pack component with 224 leads and a 16 leaded TAB component is investigated. Finite element thermal models were developed for the process, and pull strength and temperature measurements were made during the laser soldering experiments in order to assess the quality of the joints.

2. Experimental Details

The laser soldering experiments were undertaken with a Nd:YAG laser (Spectron Model SL902TQ) rated at 65 W multimode and 15 W TEM_∞ mode. The laser delivery and control schematic is shown in Figure 1; an intracavity acousto-optic modulator was used to switch the laser on for time periods ranging from milliseconds to seconds, and X-Y galvanometer mirrors were used to move the focused laser beam across the workpiece. The (A-O) modulator and the (X-Y) mirrors were computer controlled, making it possible to laser solder multiple components to a circuit board.

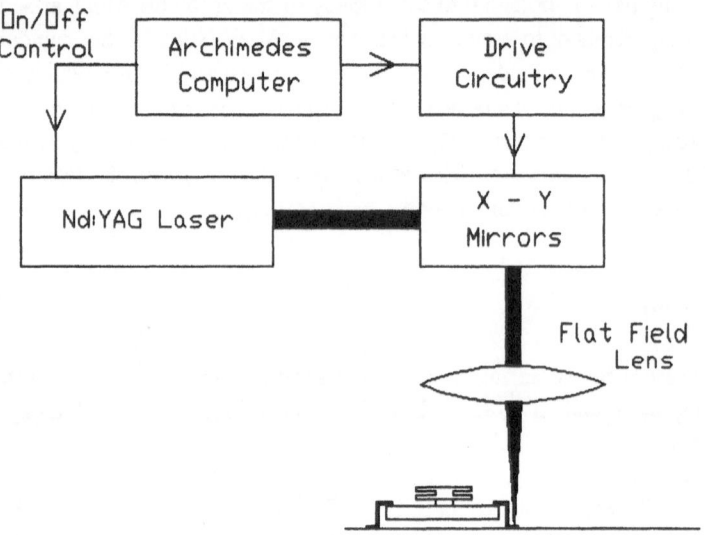

Figure 1. Schematic of the arrangement used in the laser soldering experiments.

The components used for the experiments were a 224 leaded quad flat pack (QFP) with 0.025" pitch leads, and a two sided tape automated bonding (TAB) component with 16 leads at a pitch of 0.02". The circuit boards used with both types of component were FR-4 based multi-layer boards. The solder plating present on the copper pads was used in both cases for making the solder joints. Before soldering a fixed quantity of low residue flux was applied to the solder pads with a syringe. After laser soldering the solder joints were tested by using pull tests where the average force needed to pull the solder joint apart was measured.

Temperature measurements were made of the solder joints during laser soldering by using tiny thermocouples (bead diameter 35 μm) attached to a solder pad before placement of the component; fast computer acquisition of the thermocouple readings enabled temperature measurements to be made every 0.5 ms.

The experimental units were expensive and in short supply so all the experiments were planned using experimental design theory; this allowed a minimum number of experimental trials to be undertaken using certain combinations of the laser parameters. A model of the response factor can then be fitted as a function of the laser parameters. A centre cubic faced (CCF) experimental design for three factors was used in all the experiments as shown in Figure 2; the filled circles indicate the combinations

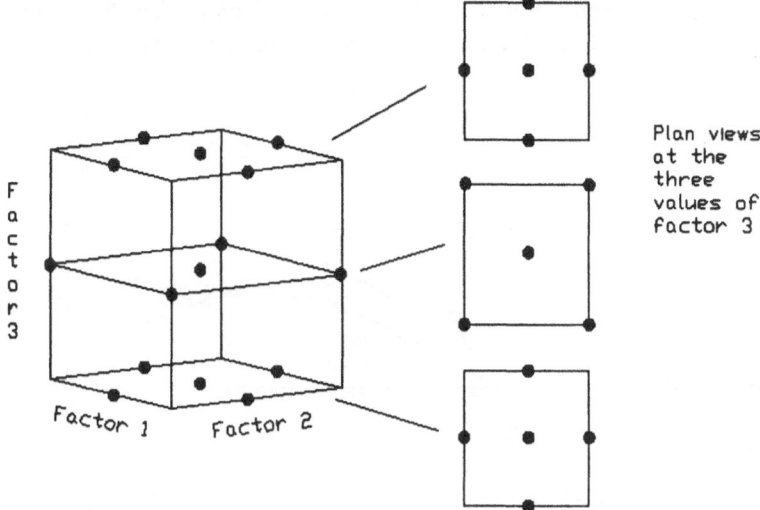

Figure 2. The parameter space of a CCF design for three factors; the experimental trials were performed at the parameter combinations shown by the dots.

at which the experimental trials were undertaken. With this design a quadratic model could be fitted to the response values using multiple regression techniques which was then used to predict the response values throughout the whole of the parameter space, assuming a quadratic interpolation. The experimental designs and multiple regression analysis of the results were facilitated by using the RS/DISCOVER software package [5].

A rigorous heat transfer analysis was also undertaken using the finite element method with the software package ABAQUS [6].

3. Experimental Results

In the laser soldering experiments, the three independent parameters were laser power, beam diameter, and scanspeed (or pulse-time) of the laser beam. The temperature rise during soldering and the pull strength of the solder joints were the two response factors measured.

3.1 QUAD FLAT PACK COMPONENT

A designed experiment was performed with the ranges of laser power from 10 to 40 W, beam diameter from 1 to 2.5 mm, and a scanspeed from 2 to 14 mm/s. The average pull strength of the solder joints was measured and it was found that good quality solder joints (pullstrength ≥ 1.2 N) were achieved below the line shown in Figure 3 for all the beam diameters. At high powers delamination of the FR-4 circuit board occurred because of overheating; this is also represented in Figure 3 by a line to the right of which delamination occurs for a beam diameter of 1.75 mm. Thus the optimum laser soldering window was located in the lower left-hand corner of the parameter space as shown.

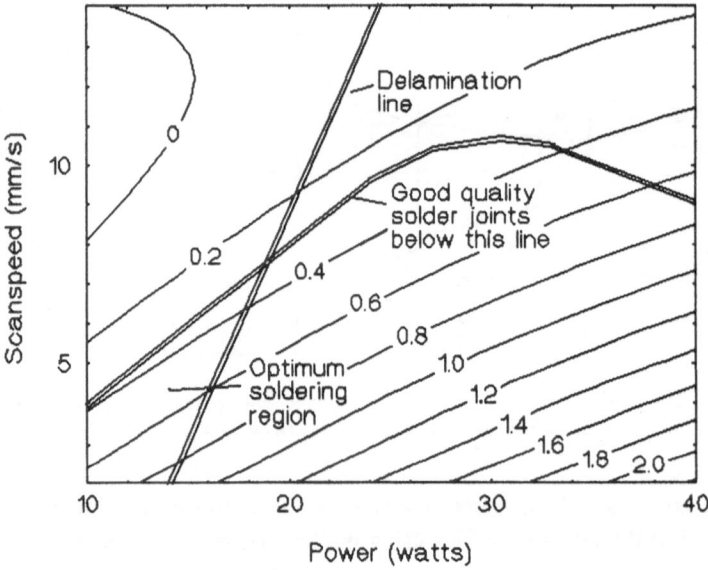

Figure 3. Diagram showing contours of the solder molten time for the beam diameter of 1.75 mm. The optimum soldering region, where no delamination occurs and the boundary for good quality joints (pull strength ≥ 1.2 N) are also shown.

The temperature rise of the solder joint was also measured for the same laser parameter ranges. The solder molten time was calculated from these temperature measurements and is compared to the pull strength results in Figure 3. It was deduced that for this component good quality solder joints were made when the solder molten time was ≥ 0.4 seconds. Thus the maximum soldering speed is limited for this component and similar SMT components. Due to the much greater throughput of mass soldering processes the niche for laser soldering with this type of packaging technology may be restricted to repair and de-soldering processes, or to the soldering of heat sensitive devices after the primary soldering operation.

3.2 TAB COMPONENT

With this component, the laser beam was aimed at the lead to be soldered and was pulsed rather than scanned as with the QFP component. The beam diameter ranges used in laser soldering the TAB component were from 0.2 to 2 mm, with laser powers from 5 to 20 W, and with pulse-times from 20 to 400 ms. By comparing the resultant models of temperature rise and pull strength it was concluded that the best quality solder joints were produced when the temperature rise was from 300 to 400 °C. The production of good quality solder joints was not found to have a minimum solder molten time within the parameter ranges investigated; the smallest pulse-time that produced good quality solder joints was 30 ms, with a beam diameter of 0.2 mm, and a laser power of 20 W. This short soldering time was possible because TAB leads have small solder joints and the solder has only a short distance to move to form acceptable solder joints. Laser soldering of TAB leads can thus achieve greater speeds than that of conventional SMT components such as the QFP. It is also much faster than other soldering methods for TAB such as hot bar or single point bonding which require a soldering time of up to 2 seconds for each solder joint. Laser soldering therefore has the potential to become the primary soldering method in TAB assembly.

The quality of solder joint produced by laser soldering depends on the temperature rise in the solder joint which is in turn dependent on the absorption of the laser radiation, and on the thermal properties of the solder pad, circuit board, and the component leads. In the laser soldering experiments performed a large variation was found in the quality and the temperature rise of the solder joints; for example, in the QFP component the RMS error of the pull strength of the joints was measured to be 0.4 N, and the RMS error of the temperature rise was 30 °C. In order to investigate the temperature rise and heat flow paths in the laser soldering process more thoroughly a thermal analysis was performed and is described in the next section.

4. Thermal Analysis

Electronic circuit boards contain many materials with widely different thermal properties; when Nd:YAG laser radiation is incident on a circuit board the heat generation includes both surface heating of the solder pads (surface absorption) and distributed body heating of the FR-4 material (FR-4 is semi-transparent to 1.064 μm); the finite element method allows the inclusion of complex structures and boundary conditions and is thus an ideal tool for modelling laser heating of electronic circuit boards.

4.1 TAB COMPONENT

The model is shown in Figure 4; a small part of the silicon chip is included to the right of the model and a lead is connected to this on top of the chip (inner lead bond) which bends down to touch the solder pad. A copper track is connected from the right of the

938

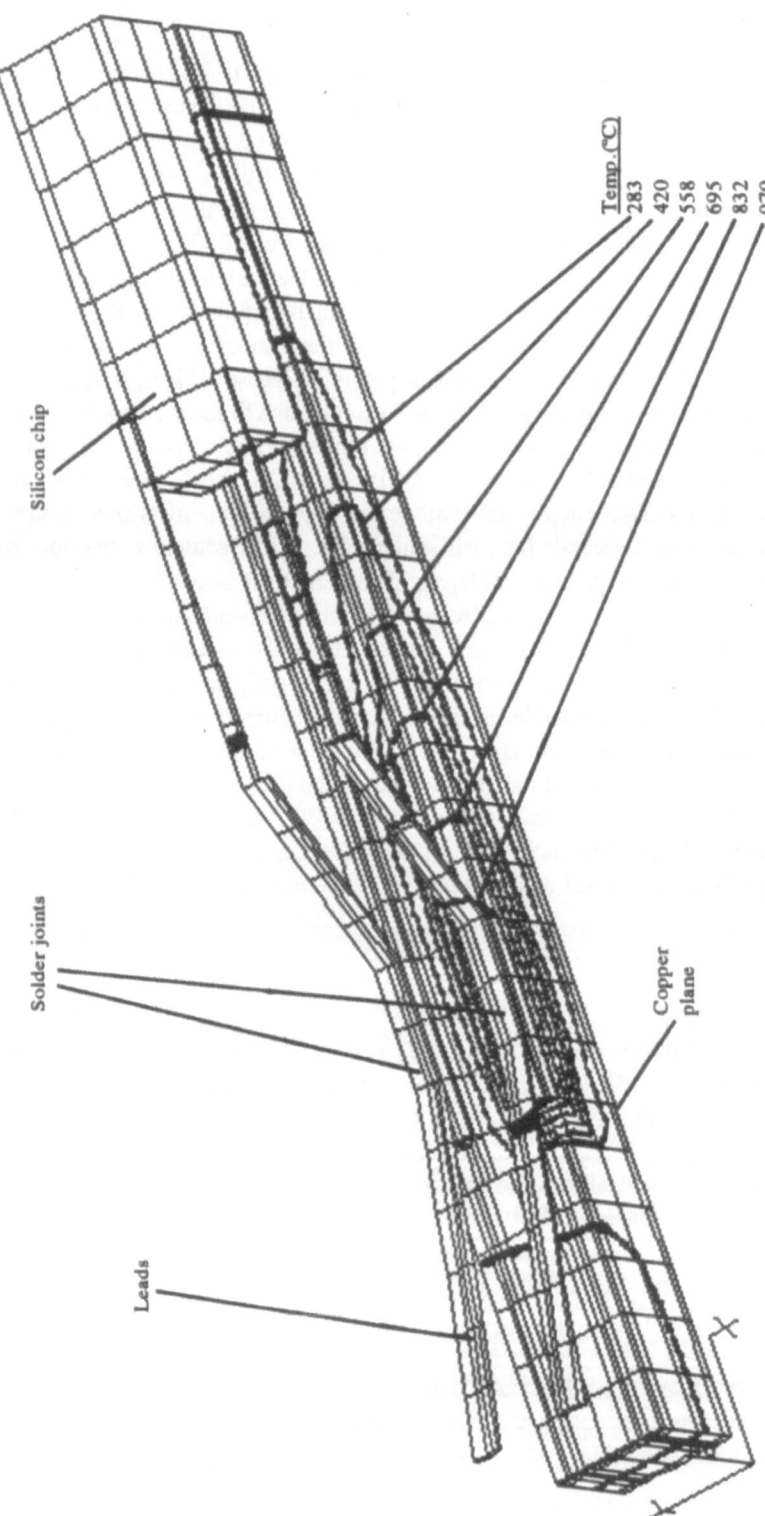

Figure 4. The finite element model of the TAB component with two solder joints. Contour lines are isotherms predicted for a 0.22 mm laser beam incident on the lower joint.

pad underneath the chip to a via, which is imbedded in the PWB and connects to a copper plane at the bottom of the model. The FR-4 material is between this plane and the top of the model.

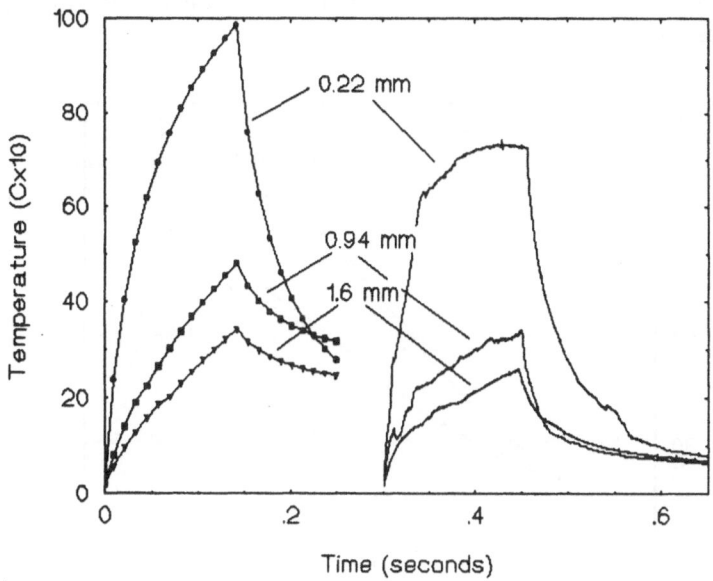

Figure 5. The temperature rise vs. time of the solder joint at three beam diameters as predicted by the model with a laser power of 12.5 W and a pulse-time of 150 ms. The set of curves to the right are the corresponding temperature rises measured by thermocouples.

The boundary conditions used were radiation on outside surfaces and a heat flux simulating the laser beam. The heat flux was applied in the model as a body flux to the FR-4 elements and as a surface flux to the opaque solder, lead, and copper plane elements. The beam diameters chosen were of 0.22, 0.94, and 1.6 mm with an incident laser power of 12.5 watts and a pulse-time of 150 ms (note that these were not the optimum soldering conditions). The smallest beam diameter used was a rectangular area of 0.215 × 0.235 mm because this was the smallest size of one element on the lead. For the other two beam diameters a circle was drawn on the top of the model and a heat flux was applied to those elements falling inside the circumference. The isotherms predicted by the model for a 0.22 mm incident laser beam are included in Figure 4. The close contours extending towards the copper plane and between the two solder pads indicate a large temperature gradient due to the thermal insulating properties of the FR-4 material. The main heat flow path is the copper track and via, which spreads the heat through the PWB to the copper plane below. Note also that there is very little heat transfer along the copper lead into the silicon chip due to the small conduction path of the lead (i.e. small cross-section of the lead). Thus the major factor influencing the rise and fall of the solder joint temperature in laser soldering is the copper track and via.

The temperature rise versus time of a node in the solder is shown in Figure 5 for the three different beam diameters. The real-time temperature rise of the solder joint during soldering was measured by small thermocouples and is superimposed on the figure for comparison. The temperature rise predicted by the model is greater than that measured by the thermocouples. The major reasons for this difference are that the model was limited in size, convection was not included, and the vaporization of the flux applied before soldering was not taken into account.

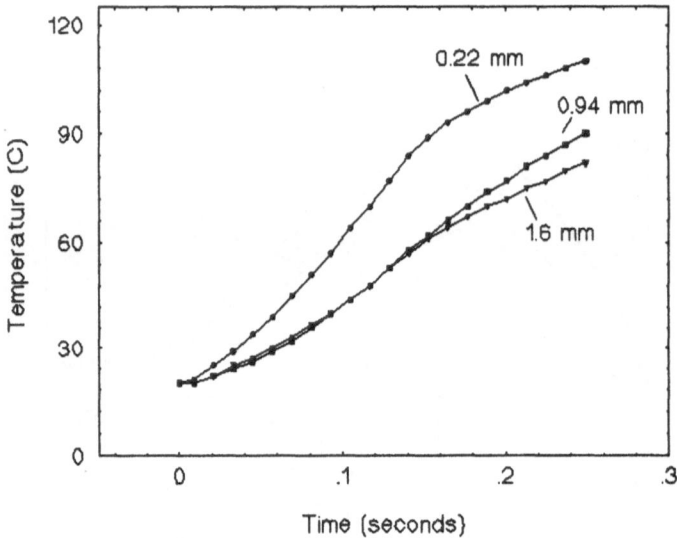

Figure 6. The temperature rise predicted by the model for the lead/chip contact.

The temperature rise in the chip during soldering is of vital importance for heat sensitive devices such as this TAB component. It is known that using a laser for soldering creates a very localized temperature rise, but no data has been previously available as to the resulting temperature rise in the chip during laser soldering. For the memory device, a prediction of the model for this temperature rise is given in Figure 6. The temperature recorded at a node just under the lead/chip contact is below that for damage to occur in all cases; this is because the conduction path along the lead is small allowing only a small heat transfer to the chip. The temperature rise in the chip was also measured in real-time using thermocouples and this also showed a small temperature rise of ≤ 40 °C.

4.2 QUAD FLAT PACK COMPONENT

A 3D model of the Quad flat-pack was constructed as shown in Figure 7. It consists of four leads/solder pads, underlying FR-4, full length via's, and one copper plane. The model was constructed primarily to examine the differences in temperature rise

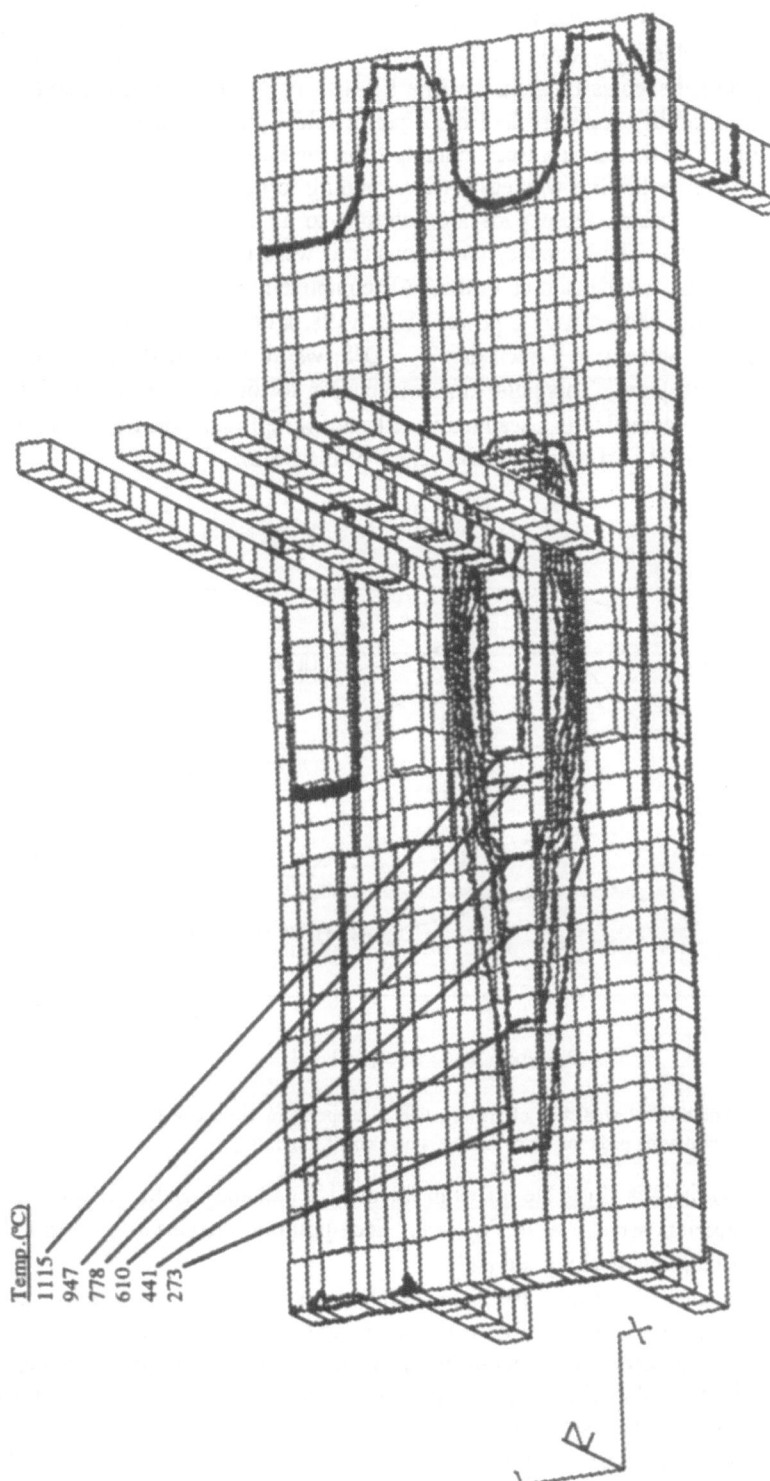

Temp.(°C)
1115
947
778
610
441
273

Figure 7. The finite element model of the QFP with isotherms predicted for a 1 mm incident laser beam.

between adjacent solder pads and the temperature distribution patterns in the circuit board during laser soldering. Only the most influential elements to the heat transfer process were included such as those in the first FR-4 layer, the copper layer, and the vias. The direction of the copper tracks from adjacent solder pads alternate in direction to the front and back of the pads; a track running from the back of a pad is underneath the component during soldering whereas a track running to the front of a pad can be exposed to the laser beam; these pads are referred to as inside and outside pads respectively (Figure 7). The boundary conditions used were radiation and an applied heat flux simulating the laser beam as in the TAB component model.

The parameter values chosen were a power of 25 watts, a beam diameter of 1, 1.75, and 2.5 mm, and the pulse-time T such that T = D/S, where D is the diameter of the beam and S is its scanspeed. This pulse-time was chosen to roughly compare with the experimental results obtained with a scanned laser beam; it represents the time taken for the whole beam to travel past a point at the centre of the beam.

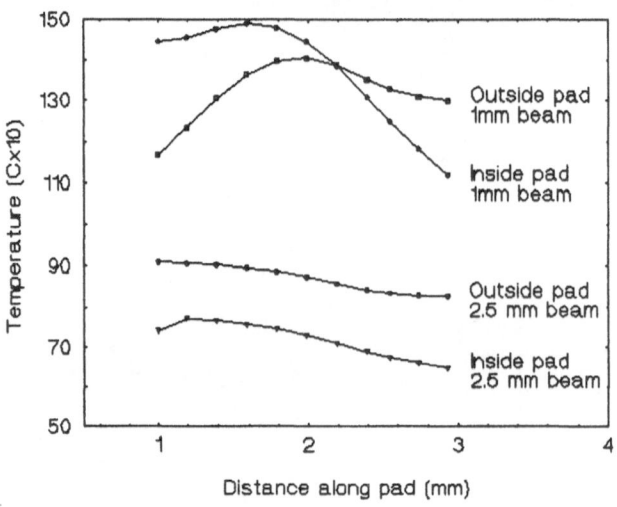

Figure 8. The thermal prediction of the variation in temperature rise across the inside and outside pads for beam diameters of 1 and 2.5 mm.

The temperature variations along the nodes in the outside pad and inside pad for 1 and 2.5 mm beam diameters are shown in Figure 8. Each point on the curves corresponds to a node located lengthwise along the pad.

For the 1 mm beam both types of pad demonstrate the heat sink effect of the track in reducing the temperature substantially on the track side of the pad. The inside pad shows the greatest variation in temperature due to the effects of (i) the laser being incident nearer to the blind end of the pad, and (ii) the heat sinking effects of the upright lead section and the track at the back of the pad.

With a 2.5 mm beam diameter the temperature variation across the pad is reduced.

The temperature rise in the outside pad, however, is greater for all nodes than with the inside pad. This is because the 2.5 mm beam is incident on part of the track of the outside pad, heating the track and thereby reducing its heat sinking effect on the pad. When the laser beam is incident on the inside pad it overlaps onto the FR-4 in front of the pad and has little effect on the pads temperature rise.

The variation in temperature rise between adjacent solder joints shown by this analysis would result in a variation in solder joint quality as measured in the experimental trials. It is thus important to carefully design a circuit board, with similar solder pads, for laser soldering to produce repeatable quality solder joints.

5. Repair Work and De-soldering of Components

On expensive circuit boards the repair of faulty joints is more economical than rejecting the whole board. Using a laser for repair simply requires a small tool the lead in contact with the solder pad while the laser radiation melts the solder. From the thermal analysis it was found that small beam diameters should be used to keep the temperature rise highly localized to the target solder joint making use of the insulating properties of the FR-4 between the pads. The optimum laser parameters are found by using process models developed for each component as shown above. When using a small beam diameter the large temperature variation across the solder pad predicted by the thermal model could be reduced by dithering the beam up and down the solder pad as reported by Whitehead et al [7].

De-soldering of a component is required if the integrated circuit is faulty, or if the component leads are misaligned with their solder pads during the soldering process. All the solder joints must be simultaneously above the melting temperature of solder before the component can be lifted off the PWB. This requires the laser beam to be scanned across all the solder joints fast enough so that upon returning to the initial joint the temperature has not decreased to below the melting temperature. The minimum scan-speed required can be achieved by heating each joint to below the temperature at which delamination of the board occurs. In the case of the QFP component this temperature is approximately in the range 250-300 °C for each beam diameter. If the RMS error in the temperature rise of the solder joints (± 30 °C) is taken into account then a suitable maximum mean temperature rise to be achieved for de-soldering is ~ 250 °C. After the laser beam has passed over a solder joint its temperature can be allowed to decrease by ~ 40 °C to 210 °C again, taking into account the RMS error. An estimate of the time taken for the temperature to fall from 250°C to 210 °C was made from a temperature measurement; a value of 60 ms was obtained. Thus the scan-speed required to de-solder the Quad flat-pack is that required for the laser to scan along all four sides in 60 ms; this was calculated as $S \approx 2900$ mm/s. Laser beam scan-speeds of 2900 mm/s could only be attained by using galvanometer controlled mirrors or rotating mirrors to scan the laser beam.

6. Conclusion

This paper described the applications of laser soldering in electronic assembly. Experimental results using temperature and pull strength measurements on QFP and TAB components showed that for typical surface mount devices laser soldering has a niche in the repair and de-soldering processes, but for the smaller TAB assembly it has the potential to become the primary soldering method. The use of the laser in repair and de-soldering processes was also discussed. Thermal models of laser soldering were constructed using the finite element method; these explored the heat flow paths governing the temperature rise of the solder joint during laser soldering, and the reasons for variations in temperature between the solder joints for several beam diameters. The thermal models also showed that the temperature rise in the chip during laser soldering was small and that the method was thus suitable for soldering heat sensitive components.

7. References

1. Lish, E.F. (1985) Application of microsoldering to printed wiring assemblies, IPC Technical review, **26**, part 7, 10-20.
2. Lea, C. (1989) Laser soldering production and microstructural benefits for SMT, Soldering and Surface mount technology, No.2, June, 13-21.
3. Yamada, T., Barrett, J., Doyle, R., and Boetti, A. 1994 Quality optimization of fine pitch surface mount solder joints using Taguchi experimental design techniques, Soldering and Surface mount technology, No.16, February, 15-20.
4. Keeler, R. (1987) Lasers for high reliability soldering, Electronic Packaging and Production, **27**, part 10, 1987, 29-31.
5. RS/Discover, Version 2.0, BBN Software Products Corporation, Cambridge Mass.
6. ABAQUS, Version 5.2, Hibbitt, Karlsson and Sorenson, Inc., Providence R.I., 1989.
7. Whitehead, D.G., Polijanczuk, A.V., and Beckett, P.M. (1990) Reflow soldering by laser, ASME Heat transfer conference, **143**, 47-56.